LIFE

The Science of Biology

EIGHTH EDITION

Sinauer Associates, Inc.

W. H. Freeman and Company

EIGHTH EDITION

LIFE

The Science of Biology

DAVID SADAVA
The Claremont Colleges
Claremont, California

H. CRAIG HELLER
Stanford University
Stanford, California

GORDON H. ORIANS
Emeritus, University of Washington
Seattle, Washington

WILLIAM K. PURVES
Emeritus, Harvey Mudd College
Claremont, California

DAVID M. HILLIS
University of Texas
Austin, Texas

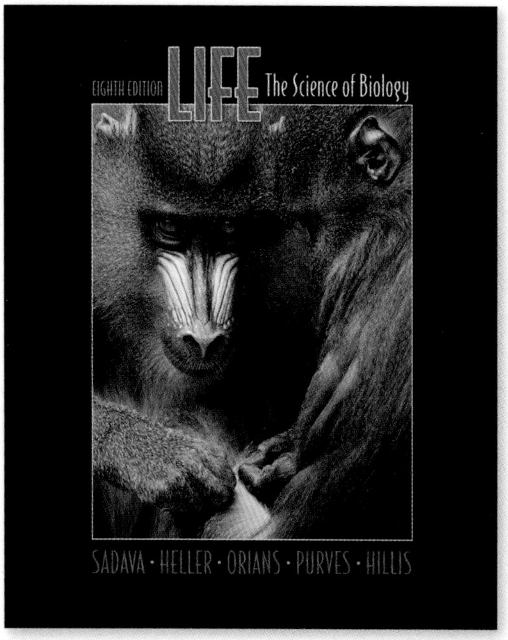

Cover Photograph
"Amor de Madre," photograph of mandrill (*Mandrillus sphinx*) mother and young.

Frontispiece
Grizzly bears (*Ursus arctos horribilis*) hunt salmon in an Alaskan river.

LIFE: The Science of Biology, Eighth Edition

Address editorial correspondence to:
Sinauer Associates Inc., 23 Plumtree Road, Sunderland, MA 01375 U.S.A.
www.sinauer.com
publish@sinauer.com

Address orders to:
VHPS/W.H. Freeman & Co., Order Dpt., 16365 James Madison Highway,
U.S. Route 15, Gordonsville, VA 22942 U.S.A.
www.whfreeman.com

Examination copy information: 1-800-446-8923
Orders: 1-888-330-8477

Library of Congress Cataloging-in-Publication Data
Life: the science of biology / David Sadava ... [et al.]. — 8th ed.
p. cm.
Includes index.
ISBN-13: 978-0-7167-7671-0 (hardcover) – ISBN 978-0-7167-7673-4 (Volume I) –
ISBN 978-0-7167-7674-1 (Volume 2) – ISBN 978-0-7167-7675-8 (Volume 3)
1. Biology. I. Sadava, David E.
QH308.2.L565 2007
570—dc22 2006031320

Printed in China
Second Printing October 2007

To our students, especially the more than 30,000
we have collectively instructed in introductory biology over the years.

The Authors

Craig Heller Gordon Orians Bill Purves David Sadava David Hillis

David Sadava is the Pritzker Family Foundation Professor of Biology at the Keck Science Center of Claremont McKenna, Pitzer, and Scripps, three of The Claremont Colleges. Twice winner of the Huntoon Award for superior teaching, Dr. Sadava has taught courses on introductory biology, biotechnology, biochemistry, cell biology, molecular biology, plant biology, and cancer biology. He is a visiting scientist in medical oncology at the City of Hope Medical Center. He is the author or coauthor of five books on cell biology and on plants, genes, and crop biotechnology. His research has resulted in over 50 papers, many coauthored with undergraduates, on topics ranging from plant biochemistry to pharmacology of narcotic analgesics to human genetic diseases. For the past 15 years, he and his collaborators have investigated multi-drug resistance in human small-cell lung carcinoma cells with a view to understanding and overcoming this clinical challenge. Their current work focuses on new anti-cancer agents from plants.

Craig Heller is the Lorry I. Lokey/Business Wire Professor in Biological Sciences and Human Biology at Stanford University. He earned his Ph.D. from the Department of Biology at Yale University in 1970. Dr. Heller has taught in the core biology courses at Stanford since 1972 and served as Director of the Program in Human Biology, Chairman of the Biological Sciences Department, and Associate Dean of Research. Dr. Heller is a fellow of the American Association for the Advancement of Science and a recipient of the Walter J. Gores Award for excellence in teaching. His research is on the neurobiology of sleep and circadian rhythms, mammalian hibernation, the regulation of body temperature, and the physiology of human performance. Dr. Heller has done research on sleeping kangaroo rats, diving seals, hibernating bears, and exercising athletes. Some of his recent work on the effects of temperature on human performance is featured in the opener to Chapter 40.

Gordon Orians is Professor Emeritus of Biology at the University of Washington. He received his Ph.D. from the University of California, Berkeley in 1960 under Frank Pitelka. Dr. Orians has been elected to the National Academy of Sciences and the American Academy of Arts and Sciences, and is a Foreign Fellow of the Royal Netherlands Academy of Arts and Sciences. He was President of the Organization for Tropical Studies, 1988–1994, and President of the Ecological Society of America, 1995–1996. He is a recipient of the Distinguished Service Award of the American Institute of Biological Sciences. Dr. Orians is a leading authority in ecology, conservation biology, and evolution. His research on behavioral ecology, plant–herbivore interactions, community structure, and environmental policy has taken him to six continents. He now devotes full time to writing and to helping apply scientific information to environmental decision-making.

Bill Purves is Professor Emeritus of Biology as well as founder and former Chair of the Department of Biology at Harvey Mudd College in Claremont, California. He received his Ph.D. from Yale University in 1959 under Arthur Galston. A fellow of the American Association for the Advancement of Science, Dr. Purves has served as head of the Life Sciences Group at the University of Connecticut, Storrs, and as Chair of the Department of Biological Sciences, University of California, Santa Barbara, where he won the Harold J. Plous Award for teaching excellence. His research interests focused on the hormonal regulation of plant growth. Dr. Purves elected early retirement in 1995, after teaching introductory biology for 34 consecutive years, in order to concentrate entirely on research directed at learning and science education. He is currently participating in the development of a virtual technical high school, with responsibility for curriculum design in scientific reasoning and health science.

David Hillis is the Alfred W. Roark Centennial Professor in Integrative Biology and the Director of the Center for Computational Biology and Bioinformatics at the University of Texas at Austin, where he also has directed the School of Biological Sciences. Dr. Hillis has taught courses in introductory biology, genetics, evolution, systematics, and biodiversity. He has been elected into the membership of the American Academy of Arts and Sciences, awarded a John D. and Catherine T. MacArthur Fellowship, and has served as President of the Society for the Study of Evolution and of the Society of Systematic Biologists. His research interests span much of evolutionary biology, including experimental studies of evolving viruses, empirical studies of natural molecular evolution, applications of phylogenetics, analyses of biodiversity, and evolutionary modeling. He is particularly interested in teaching and research about the practical applications of evolutionary biology.

Preface

As active scientists working in a wide variety of both basic and applied biology, we are fortunate to be part of a field that is not only fascinating but also changes rapidly. It is apparent not just in the time span since we started our careers—we see it every day when we open a newspaper or a scientific journal. As educators of both introductory and advanced-level students, we desire to convey our excitement about biology's dynamic nature.

This new edition of *Life* looks, and is, quite different from its predecessors. In planning the Eighth Edition, we focused on three fundamental goals. The first was to maintain and enhance what has worked well in the past—an emphasis on not just what we know but how we came to know it; the incorporation of exciting new discoveries; an art program distinguished by its beauty and clarity; plus a unifying theme. As should be the case in any biology textbook, that theme is evolution by natural selection, a 150-year-old idea that more than ever ties together the living world. We have been greatly helped in this endeavor by the addition of a new author, David Hillis. His knowledge and insights have been invaluable in developing our chapters on evolution, phylogeny, and diversity, and they permeate the rest of the book as well.

Our second goal has been to make *Life* more pedagogically accessible. From the bold new design to the inclusion of numerous learning aids throughout each chapter (see New Pedagogical Features), we have worked to make our writing consistently easy to follow as well as engaging.

Third, between editions we asked seven distinguished ecologists—all of whom teach introductory biology—to provide detailed critiques of the Ecology unit. As a result of their extensive suggestions, Part Nine, Ecology, has a fresh organization (see The Nine Parts). And one of the seven, May Berenbaum, has agreed to join the *Life* author team for the Ninth Edition. The other six stalwarts are thanked in the "Reviewers of the Eighth Edition" section.

Enduring Features

As stated above, we are committed to a blending of a presentation of the core ideas of biology with an emphasis on introducing our readers to the process of scientific inquiry. Having pioneered the idea of depicting seminal experiments in specially designed figures, we continue to develop this here, with 96 EXPERIMENT figures (28 percent more than in the Seventh Edition). Each follows the structure: Hypothesis, Method, Result, and Conclusion. Many now include "Further Research," which asks students to conceive an experiment that explores a related question.

A related feature is the RESEARCH METHOD figures, depicting many laboratory and field methods used to do this research. All the Experiment and Research Method figures are listed in the endpapers at the back of the book.

Another much-praised feature—which we pioneered ten years ago in *Life*'s Fifth Edition—is the BALLOON CAPTIONS used in our figures. We know that many students are visual learners. The balloon captions bring explanations of intricate, complex processes directly into the illustration, allowing the reader to integrate the information without repeatedly going back and forth between the figure and its legend.

Life is the only introductory biology book for science majors that begins each chapter with a story. These OPENING STORIES, most of which are new to this edition, are meant to intrigue students while helping them see how the chapter's biological subject relates to the world around them.

New Pedagogical Features

There are several new elements in the Eighth Edition chapters. Each has been designed as a study tool to aid the student in mastering the material. In the opening page spread, IN THIS CHAPTER previews the chapter's content, and the CHAPTER OUTLINE gives the major section headings, all numbered and all framed as questions to emphasize the inquiry basis of science.

Each main section of a chapter now ends with a RECAP. This key element briefly summarizes the important concepts in the section, then provides two or three questions to stimulate immediate review. Each question includes reference to pertinent text or a figure or both.

The CHAPTER SUMMARY boldfaces key terms introduced and defined in the chapter. We have kept the highlighted references to key figures and to the Web tutorials and activities that support a topic in the chapter.

Another new element, BIOBITS, is not strictly a learning aid, but offers intriguing (occasionally amusing) supplemental information. BioBits, like the opening stories, are intended to help students appreciate the interface between biology and other aspects of life.

Redesigned WEB ICONS alert the reader to the tutorials and activities on *Life*'s companion website (www.thelifewire.com). Each of these study and review resources, many of which are new for the Eighth Edition, has been created specifically for *Life*. A full list, by chapter, is found in the front endpapers of the book.

The Nine Parts

We have reorganized the book into nine parts. Part One sets the stage for the entire book, with the opening chapter on biology as an exciting science, starting with a student project, and how evolution unites the living world. This is followed by chapters on the basic chemical building blocks that underlie life. We have tried to tie this material together by relating it to theories on the origin of life, with new discoveries of water in our solar system as an impetus.

In Part Two, Cells and Energy, we present an integrated view of the structure and biochemical functions of cells. The discussions of biochemistry are often challenging for students; thus we have reworked both the text and illustrations for greater clarity. These discussions are presented in the context of the latest discoveries on the origin of life and evolution of cells.

Part Three, Heredity and the Genome, begins with continuity at the cellular level, and then outlines the principles of genetics and the identification of DNA as the genetic material. New examples, such as the genetics of coat color in dogs, enliven these chapters. This is followed by chapters on gene expression and on the prokaryotic and eukaryotic genomes. Many new discoveries have been made in this new field of genomics, ranging from tracking down the bird flu virus to the genomes of wild cats, such as the cheetah.

Part Four, Molecular Biology: The Genome in Action, reinforces the basic principles of classical and molecular genetics by applying them to such diverse topics as cell signaling, biotechnology, and medicine. We use many new experiments and examples from applied biology to illustrate these concepts. These include the latest information on the human genome and the emerging field of systems biology. The chapter on natural defenses now includes a discussion of allergy.

Part Five, The Patterns and Processes of Evolution, has been updated in several important ways. We have emphasized the importance of evolutionary biology as a basis for comparing and understanding all aspects of biology, and have described numerous practical applications of evolutionary biology that will be familiar and relevant to the everyday lives of most students.

Recent experimental studies of evolution are described and explained, to help students understand that evolution is an ongoing, observable process. The chapters on phylogenetics and molecular evolution have been completely rewritten to reflect recent advances in those fields. Other changes reflect our growing knowledge of the history of life on Earth and the mechanisms of evolution that have given rise to all of biodiversity.

Part Six, The Evolution of Diversity, reflects the latest views on phylogeny. It continues to emphasize groups united by evolutionary history over classically defined taxa. This emphasis is now supported by an appendix on the Tree of Life that clearly maps out and describes all groups discussed in the text, so that students can quickly look up unfamiliar names and see how they fit into the larger context of life. We now discuss aspects of phylogeny that are still under study or debate (among major groups of eukaryotes, plants, and animals, for instance).

In Part Seven, Flowering Plants: Form and Function, we report on several exciting new discoveries. These include the receptors for auxin, gibberellins, and brassinosteroids as well as great progress on the florigen problem. We have updated our treatment of signal transduction pathways and of circadian rhythms in plants. The already strong treatment of environmental challenges to plants has been augmented by new Experiment figures on plant defenses against herbivores, one confirming that nicotine does help tobacco plants resist certain insects.

Part Eight, Animals: Form and Function, is about how animals work. Although we give major attention to human physiology, we embed it in a background of comparative animal physiology. Our focus is systems physiology but we also introduce the underlying cellular and molecular mechanisms. For example, our explanations of nervous system phenomena—whether they be action potentials, sensation, learning, or sleep—are discussed in terms of the properties of ion channels. The actions of hormones are explained in terms of the molecular mechanisms. Maximum athletic performance is explained in terms of the underlying cellular energy systems. Throughout Part Eight we try to help the student make the connections across all levels of biology, from molecular to behavioral, and to see the relevance of physiology to issues of health and disease. Of central importance in each chapter is mechanisms of control and regulation.

Part Nine, Ecology, begins with a new chapter that describes the scope of ecological research and discusses recent advances in our understanding of the broad patterns in the distribution of life on Earth. The next chapter, also new, combines Behavior and Behavioral Ecology. It shows how the decisions that organisms make during their lives influence both their survival and reproductive success, and also the dynamics of populations and the structure of ecological communities. The chapter on Population Ecology has new material that explains how ecologists are able to mark and follow individual organisms in the wild to determine their survival and reproductive success. Following a chapter on Community Ecology, another new chapter, Ecosystems and Global Ecology, shows how ecologists are expanding the scope of their studies to encompass the functioning of the global ecosystem. This discussion leads naturally to the final chapter in the book, Conservation Biology, which describes how ecologists and conservation biologists work to reduce the rate at which species are becoming extinct as a result of human activities.

Full Books, Paperbacks, or Loose-Leaf

We again provide *Life* both as the full book and as a cluster of paperbacks. Thus, instructors who want to use less than the whole book, or who want their students to have more portable units, can choose from these split volumes:

Volume I, The Cell and Heredity, includes: Part One, The Science and Building Blocks of Life (Chapters 1–3); Part Two, Cells and Energy (Chapters 4–8); Part Three, Heredity and the Genome (Chapters 9–14); and Part Four, Molecular Biology: The Genome in Action (Chapters 15–20).

Volume II, Evolution, Diversity, and Ecology, includes: Chapter 1, Studying Life; Part Five, The Patterns and Processes of Evolution (Chapters 21–25); Part Six, The Evolution of Diversity (Chapters 26–33); and Part Nine, Ecology (Chapters 52–57).

Volume III, Plants and Animals, includes: Chapter 1, Studying Life; Part Seven, Flowering Plants: Form and Function (Chapters 34–39); and Part Eight, Animals: Form and Function (Chapters 40–51).

Note that each volume also includes the book's front matter, Appendixes, Glossary, and Index.

Life is also available in a loose-leaf version. This shrink-wrapped, unbound, 3-hole punched version is designed to fit into a 3-ring binder. Students take only what they need to class and can easily integrate any instructor handouts or other resources.

Media and Supplements for the Eighth Edition

The media and supplements for *Life,* Eighth Edition have been assembled with two main goals in mind: (1) to provide students with a collection of tools that helps them effectively master the vast amount of new information that is being presented to them in the introductory biology course; and (2) to provide instructors with the richest possible collection of resources to aid in teaching the course—preparing, presenting the lecture, providing course materials online, and assessing student comprehension.

All of the *Life* media and supplemental resources have been developed specifically for this textbook. This gives the student the greatest degree of consistency when studying across different media. For example, the animated tutorials and activities found on the Companion Website were built using textbook art, so that the manner in which structures are illustrated, the colors used to identify objects, and the terms and abbreviations used are all consistent.

The rich collection of visual resources in the Instructor's Media Library provides instructors with a wide range of options for enhancing lectures, course websites, and assignments. Highlights include: layered art PowerPoint® presentations that break down complex figures into detailed, step-by-step presentations; a collection of approximately 200 video segments that can help capture the attention and imagination of students; and the new set of PowerPoint® slides of textbook art with editable labels and leaders that allow easy customization of the figures.

For a detailed description of all the media and supplements available to accompany the Eighth Edition, please turn to "Life's Media and Supplements package" on page xiii.

Many People to Thank

One of the wisest pieces of advice ever given to a textbook author is to "be passionate about your subject, but don't put your ego on the page." Considering all the people who looked over our shoulders throughout the process of creating this book, this advice could not be more apt. We are indebted to many people, who gave invaluable help to make this book what it is. First and foremost are our colleagues, biologists from over 100 institutions. Some were users of the previous edition, who suggested many improvements. Others reviewed our chapter drafts in detail, including advice on how to improve the illustrations. Still others acted as accuracy reviewers when the book was almost completed. Our publishers created an advisory group of introductory course coordinators. They advised us on a variety of issues, ranging from book content and design to elements of the print and media supplements. All of these biologists are listed in the Reviewer credits.

We needed a fresh editorial eye for this edition, and we were fortunate to work with Carol Pritchard-Martinez as development editor. With a level head that comes from years of experience, she was a major presence as we wrote and revised. Elizabeth Morales, our artist, was on her second edition with us. This time, she extensively revised almost all of the prior art and translated our crude sketches into beautiful new art. We hope you agree that our art program remains superbly clear and elegant. Once again, we were lucky to have Norma Roche as the copy editor. Her firm hand and encyclopedic recall of our book's many chapters made our prose sharper and more accurate. For this edition, Norma was joined by the capable and affable Maggie Brown. Susan McGlew coordinated the hundreds of reviews that we described above. David McIntyre was a truly proactive photo editor. Not only did he find over 500 new photographs, including many new ones of his own, that enrich the book's content and visual statement, but he set up, performed, and photographed the experiment shown in Figure 36.1. The elegant new interior design is the creation of Jeff Johnson. He also coordinated the book's layout and designed the cover. Carol Wigg, for the eighth time in eight editions, oversaw the editorial process. Her influence pervades the entire book—she created many BioBits, shaped and improved the chapter-opening stories, interacted with David McIntyre in conceiving many photo subjects, and kept an eagle eye on every detail of text, art, and photographs.

W. H. Freeman continues to bring *Life* to a wider audience. Associate Director of Marketing Debbie Clare, the Regional Specialists, Regional Managers, and experienced sales force are effective ambassadors and skillful transmitters of the features and unique strengths of our book. We depend on their expertise and energy to keep us in touch with how *Life* is perceived by its users.

Finally, we are indebted to Andy Sinauer. Like ours, his name is on the cover of the book, and he truly cares deeply about what goes into it.

DAVID SADAVA

CRAIG HELLER

GORDON ORIANS

BILL PURVES

DAVID HILLIS

Reviewers for the Eighth Edition

Between-Edition Reviewers (Ecology and Animal Parts)

May Berenbaum, University of Illinois, Urbana-Champaign
Carol Boggs, Stanford University
Judie Bronstein, University of Arizona
F. Lynn Carpenter, University of California, Irvine
Dan Doak, University of California, Santa Cruz
Jessica Gurevitch, SUNY, Stony Brook
Margaret Palmer, University of Maryland
Marty Shankland, University of Texas, Austin

Advisory Board Members

Heather Addy, University of Calgary
Art Buikema, Virginia Polytechnic Institute and State University
Jung Choi, Georgia Technical University
Rolf Christoffersen, University of California, Santa Barbara
Alison Cleveland, Florida Southern University
Mark Decker, University of Minnesota
Ernie Dubrul, University of Toledo
Richard Hallick, University of Arizona
John Merrill, Michigan State University
Melissa Michael, University of Illinois
Deb Pires, University of California, Los Angeles
Sharon Rogers, University of Nevada, Las Vegas
Marty Shankland, University of Texas, Austin

Manuscript Reviewers

John Alcock, Arizona State University
Charles Baer, University of Florida
Amy Baird, University of Texas, Austin
Patrice Boily, University of New Orleans
Thomas Boyle, University of Massachusetts, Amherst
Mirjana Brockett, Georgia Institute of Technology
Arthur Buikema, Virginia Polytechnic Institute and State University
Hilary Callahan, Barnard College
David Champlin, University of Southern Maine
Chris Chanway, University of British Columbia
Mike Chao, California State University, San Bernardino
Rhonda Clark, University of Calgary
Elizabeth Connor, University of Massachusetts, Amherst
Deborah A. Cook, Clark Atlanta University
Elizabeth A. Cowles, Eastern Connecticut State University
Joseph R. Cowles, Virginia Polytechnic Institute and State University
William L. Crepet, Cornell University
Martin Crozier, Wayne State University
Donald Dearborn, Bucknell University
Mark Decker, University of Minnesota
Michael Denbow, Virginia Polytechnic Institute and State University
Jean DeSaix, University of North Carolina, Chapel Hill
William Eldred, Boston University
Andy Ellington, University of Texas, Austin
Gordon L. Fain, University of California, Los Angeles
Kevin M. Folta, University of Florida
Miriam Goldbert, College of the Canyons
Kenneth M. Halanych, Auburn University
Susan Han, University of Massachusetts, Amherst
Tracy Heath, University of Texas, Austin
Shannon Hedtke, University of Texas, Austin
Mark Hens, University of North Carolina, Greensboro
Albert Herrera, University of Southern California
Barbara Hetrich, University of Northern Iowa
Erec Hillis, University of California, Berkeley
Jonathan Hillis, Austin, Texas
Hopi Hoekstra, University of California, San Diego
Kelly Hogan, University of North Carolina, Chapel Hill
Carl Hopkins, Cornell University
Andrew Jarosz, Michigan State University
Norman Johnson, University of Massachusetts, Amherst
Walter Judd, University of Florida
David Julian, University of Florida
Laura Katz, Smith College
Melissa Kosinski-Collins, Massachusetts Institute of Technology
William Kroll, Loyola University of Chicago
Marc Kubasak, University of California, Los Angeles
Josephine Kurdziel, University of Michigan
John Latto, University of California, Berkeley
Brian Leander, University of British Columbia
Jennifer Leavey, Georgia Institute of Technology
Arne Lekven, Texas A&M University
Don Levin, University of Texas, Austin
Rachel Levin, Amherst College
Thomas Lonergan, University of New Orleans
Blase Maffia, University of Miami
Meredith Mahoney, University of Texas, Austin
Charles Mallery, University of Miami
Ron Markle, Northern Arizona University
Mike Meighan, University of California, Berkeley
Melissa Michael, University of Illinois, Urbana-Champaign
Jill Miller, Amherst College
Subhash Minocha, University of New Hampshire
Thomas W. Moon, University of Ottawa
Richard Moore, Miami University of Ohio
John Morrissey, Hofstra University
Leonie Moyle, University of Indiana
Mary Anne Nelson, University of New Mexico
Dennis O'Connor, University of Maryland, College Park
Robert Osuna, SUNY, Albany
Cynthia Paszkowski, University of Alberta
Diane Pataki, University of California, Irvine
Ron Patterson, Michigan State University
Craig Peebles, University of Pittsburgh
Debra Pires, University of California, Los Angeles
Greg Podgorski, Utah State University
Chuck Polson, Florida Institute of Technology

Donald Potts, University of California, Santa Cruz
Jill Raymond, Rock Valley College
Ken Robinson, Purdue University
Sharon L. Rogers, University of Nevada, Las Vegas
Laura Romano, Denison University
Pete Ruben, Utah State University, Logan
Albert Ruesink, Indiana University
Walter Sakai, Santa Monica College
Mary Alice Schaeffer, Virginia Polytechnic Institute and State University
Daniel Scheirer, Northeastern University
Stylianos Scordilis, Smith College
Kevin Scott, University of Calgary
Jim Shinkle, Trinity University
Denise Signorelli, Community College of Southern Nevada
Thomas Silva, Cornell University
Jeffrey Tamplin, University of Northern Iowa
Steve Theg, University of California, Davis
Sharon Thoma, University of Wisconsin, Madison
Jeff Thomas, University of California, Los Angeles
Christopher Todd, University of Saskatchewan
John True, SUNY, Stony Brook
Mary Tyler, University of Maine
Fred Wasserman, Boston University
John Weishampel, University of Central Florida
Elizabeth Willott, University of Arizona
David Wilson, University of Miami
Heather Wilson-Ashworth, Utah Valley State College

Accuracy Reviewers

John Alcock, Arizona State University
John Anderson, University of Minnesota
Brian Bagatto, University of Akron
Lisa Baird, University of San Diego
May Berenbaum, University of Illinois, Urbana-Champaign
Gerald Bergtrom, University of Wisconsin, Milwaukee
Stewart Berlocher, University of Illinois, Urbana-Champaign
Mary Bisson, SUNY, Buffalo
Arnold Bloom, University of California, Davis
Judie Bronstein, University of Arizona
Jorge Busciglio, University of California, Irvine
Steve Carr, Memorial University of Newfoundland
Thomas Chen, Santa Monica College
Randy Cohen, California State University, Northridge
Reid Compton, University of Maryland, College Park
James Courtright, Marquette University
Jerry Coyne, University of Chicago
Joel Cracraft, American Museum of Natural History
Joseph Crivello, University of Connecticut, Storrs
Gerrit De Boer, University of Kansas, Lawrence
Arturo DeLozanne, University of Texas, Austin
Stephen Devoto, Wesleyan University
Laura DiCaprio, Ohio University
John Dighton, Rutgers Pinelands Field Station
Jocelyne DiRuggiero, University of Maryland, College Park
W. Ford Doolittle, Dalhousie University
Emanuel Epstein, University of California, Davis
Gordon L. Fain, University of California, Los Angeles
Lewis J. Feldman, University of California, Berkeley
James Ferraro, Southern Illinois University
Cole Gilbert, Cornell University
Elizabeth Godrick, Boston University
Martha Groom, University of Washington
Kenneth M. Halanych, Auburn University
Mike Hasegawa, Purdue University
Mark Hens, University of North Carolina, Greensboro
Richard Hill, Michigan State University
Franz Hoffman, University of California, Irvine
Sara Hoot, University of Wisconsin, Milwaukee
Carl Hopkins, Cornell University
Alfredo Huerta, Miami University
Michael Ibba, The Ohio State University
Walter Judd, University of Florida
Laura Katz, Smith College
Manfred D. Laubichler, Arizona State University
Brian Leander, University of British Columbia
Mark V. Lomolino, SUNY College of Environmental Science and Forestry
Jim Lorenzen, University of Idaho
Denis Maxwell, University of Western Ontario
Brad Mehrtens, University of Illinois, Urbana-Champaign
John Merrill, Michigan State University
Allison Miller, Saint Louis University
Clara Moore, Franklin and Marshall College
Julie Noor, Duke University
Mohamed Noor, Duke University
Theresa O'Halloran, University of Texas, Austin
Norman R. Pace, University of Colorado
Randall Packer, George Washington University
Walt Ream, Oregon State University
Eric Richards, Washington University
Steve Rissing, The Ohio State University
R. Michael Roberts, University of Missouri, Columbia
Pete Ruben, Simon Fraser University
David A. Sanders, Purdue University
Mike Sanderson, University of California, Davis
Marty Shankland, University of Texas, Austin
Jeff Silberman, University of Arkansas
Margaret Silliker, DePaul University
Dee Silverthorn, University of Texas, Austin
M. Suzanne Simon-Westendorf, Ohio University
Alastair G.B. Simpson, Dalhousie University
John Skillman, California State University, San Bernardino
Frederick W. Spiegel, University of Arkansas
John J. Stachowicz, University of California, Davis
Heven Sze, University of Maryland
E.G. Robert Turgeon, Cornell University
Mary Tyler, University of Maine
Mike Wade, Indiana University
Leslie Winemiller, Texas A&M University
Mimi Zolan, Indiana University

Tree of Life Appendix Reviewers

John Abbott, University of Texas, Austin
Joseph Bischoff, National Center for Biotechnology Information
Ruth Buskirk, University of Texas, Austin
David Cannatella, University of Texas
Joel Cracraft, American Museum of Natural History
Scott Federhen, National Center for Biotechnology Information
Carol Hotton, National Center for Biotechnology Information
Robert Jansen, University of Texas, Austin
Brian Leander, University of British Columbia
Detlef Leipe, National Center for Biotechnology Information
Beryl Simpson, University of Texas, Austin
Richard Sternberg, National Center for Biotechnology Information
Edward Theriot, University of Texas
Sean Turner, National Center for Biotechnology Information

Supplements Authors

Dany Adams, The Forsyth Institute
Erica Bergquist, Holyoke Community College
Ian Craine, University of Toronto
Ernest Dubrul, University of Toledo
Edward Dzialowski, University of North Texas
Donna Francis, University of Massachusetts, Amherst
Jon Glase, Cornell University
Lindsay Goodloe, Cornell University
Celine Muis Griffin, Queen's University
Nancy Guild, University of Colorado at Boulder
Norman Johnson, University of Massachusetts, Amherst
James Knapp, Holyoke Community College
Jennifer Knight, University of Colorado, Boulder
David Kurjiaka, University of Arizona
Richard McCarty, Johns Hopkins University
Betty McGuire, Cornell University
Nancy Murray, Evergreen State College
Deb Pires, University of California, Los Angeles
Catherine Ueckert, Northern Arizona University
Jerry Waldvogel, Clemson University

LIFE's Media and Supplements Package

For the Student

Companion Website www.thelifewire.com

(Also available as a CD, optionally packaged with the book)

The *Life*, Eighth Edition Companion Website is available free of charge to all students (no access code required). The site features a variety of study and review resources designed to help students master the wide range of material presented in the introductory biology course. Features of the site include:

- *Interactive Summaries.* These summaries combine a review of important concepts with links to all the key figures from the chapter as well as all of the relevant animated tutorials and activities.
- *Animated Tutorials.* Over 100 in-depth animated tutorials present complex topics in a clear, easy-to-follow format that combines a detailed animation with an introduction, conclusion, and quiz.
- *Activities.* Over 120 interactive activities help the student learn important facts and concepts through a wide range of activities, such as labeling steps in processes or parts of structures, building diagrams, and identifying different types of organisms.
- *Flashcards.* For each chapter of the book, there is a set of flashcards that allows the student to review all the key terminology from the chapter. Students can review the terms in the study mode, and then quiz themselves on a list of terms.
- *New! Experiment Links.* New for the Eighth Edition, each experiment featured in the textbook has a corresponding treatment on the companion website that links to further information about the experiment, further research that followed, and applications derived from the research.
- *Interactive Quizzes.* Every question includes an image taken from the textbook, thorough feedback on both right and wrong answer choices, references to textbook pages, and links to electronic versions of book pages, where the related material is highlighted.
- *Online Quizzes.* These quizzes test the student's comprehension of the chapter material, and the results are stored in the online gradebook.
- *Key Terms.* The key terminology introduced in each chapter is listed, with definitions and audio pronunciations from the Glossary.
- *Suggested Readings.* For each chapter of the book, a list of suggested readings is provided as a resource for further study.
- *Glossary.* The language of biology is often difficult for students taking introductory biology, so we have created a full glossary with audio pronunciations.
- *Math for Life* (Dany Adams, *The Forsyth Institute*). *Math for Life* is a collection of mathematical shortcuts and references to help students with the quantitative skills they need in the laboratory.
- *Survival Skills* (Jerry Waldvogel, *Clemson University*). *Survival Skills* is a guide to more effective study habits. Topics include time management, note-taking, effective highlighting, and exam preparation.

Study Guide (ISBN 978-0-7167-7893-6)

Edward M. Dzialowski, *University of North Texas;* Jon Glase, *Cornell University;* Lindsay Goodloe, *Cornell University;* Nancy Guild, *University of Colorado;* and Betty McGuire, *Smith College*

For each chapter of the textbook, the *Life* Study Guide offers a variety of study and review tools. The contents of each chapter are broken down into both a detailed review of the Important Concepts covered and a boiled-down Big Picture snapshot. In addition, Common Problem Areas and Study Strategies are highlighted. A set of study questions (both multiple-choice and short-answer) allows students to test their comprehension. All questions include answers and explanations.

Lecture Notebook (ISBN 978-0-7167-7894-3)

This invaluable printed resource consists of all the artwork from the textbook (more than 1,000 images with labels) presented in the order in which they appear in the text, with ample space for note-taking. Because the Notebook has already done the drawing, students can focus more of their attention on the concepts. They will absorb the material more efficiently during class, and their notes will be clearer, more accurate, and more useful when they study from them later.

MCAT® Practice Test (ISBN 0-7167-5907-1)

A complete printed MCAT exam, with answers, allows students to test their knowledge of the full range of introductory biology content as they prepare for the medical school entrance exams.

CatchUp Math & Stats

Michael Harris, Gordon Taylor, and Jacquelyn Taylor

This primer will help your students quickly brush up on the quantitative skills they need to succeed in biology. Presented in brief, accessible units, the book covers topics such as working with powers, logarithms, using and understanding graphs, calculating standard deviation, preparing a dilution series, choosing the right statistical test, analyzing enzyme kinetics, and many more.

***Student Handbook for Writing in Biology*, Second Edition**

Karen Knisely (ISBN 0-7167-6709-0)

This book provides practical advice to students who are learning to write according to the conventions in biology. Using the standards of journal publication as a model, the author provides, in a user-friendly format, specific instructions on: using biology databases to locate references; paraphrasing for improved comprehension; preparing lab reports, scientific papers, posters; preparing oral presentations in PowerPoint®, and more.

Bioethics and the New Embryology: Springboards for Debate

Scott F. Gilbert, Anna Tyler, and Emily Zackin (ISBN 0-7167-7345-7)

Our ability to alter the course of human development ranks among the most significant changes in modern science and has brought embryology into the public domain. The question that must be asked is: Even if we *can* do such things, *should* we do such things?

BioStats Basics: A Student Handbook

James L. Gould and Grant F. Gould (ISBN 0-7167-3416-8)

BioStats Basics provides introductory-level biology students with a practical, accessible introduction to statistical research. Engaging and informal, the book avoids excessive theoretical and mathematical detail to focus on how core statistical methods are put to work in biology.

Laboratory Manuals

W. H. Freeman publishes a range of high-quality biology lab texts, all of which are available for bundling with *Life,* Eighth Edition. Our laboratory texts are available as complete paperback texts, or as Freeman Laboratory Separates.

- *Biology in the Laboratory,* Third Edition
  Doris R. Helms, Carl W. Helms, Robert J. Kosinski, and John C. Cummings (ISBN 0-7167-3146-0)
- *Laboratory Outlines in Biology-VI*
  Peter Abramoff and Robert G. Thomson
  (ISBN 0-7167-2633-5)
- *Anatomy and Dissection of the Frog,* Second Edition
  Warren F. Walker, Jr. (ISBN 0-7167-2636-X)
- *Anatomy and Dissection of the Rat,* Third Edition
  Warren F. Walker, Jr. and Dominique Homberger
  (ISBN 0-7167-2635-1)
- *Anatomy and Dissection of the Fetal Pig,* Fifth Edition
  Warren F. Walker, Jr. and Dominique Homberger
  (ISBN 0-7167-2637-8)
- *Atlas and Dissection Guide for Comparative Anatomy,* Sixth Edition
  Saul Wischnitzer and Edith Wischnitzer
  (ISBN 0-7167-6959-X)

Custom Publishing for Laboratory Manuals

http://custompub.whfreeman.com

Instructors can build and order customized lab manuals in just minutes, choosing material from Freeman's biology laboratory manuals, as well as their own material.

For the Instructor

Instructor's Media Library

In order to give you the widest possible range of resources to help engage students and better communicate the material, we have assembled an unparalleled collection of media resources. The Eighth Edition of *Life* features an expanded Instructor's Media Library (available on a set of CDs and DVDs) that includes:

- *Textbook Figures and Tables.* Every image from the textbook is provided in both JPEG (high- and low-resolution) and PDF formats.
- *Unlabeled Figures.* Every figure in the textbook is provided in an unlabeled format. These are useful for student quizzing and custom presentation development.
- *Supplemental Photos.* The supplemental photograph collection contains over 1,500 photographs (all in addition to those in the text), forming a rich resource of visual imagery.
- *Animations.* A collection of over 100 in-depth animations, all of which were created from the textbook's art program, and which can be viewed in either narrated or step-through mode.
- *Videos.* This collection of approximately 200 video segments covering topics across the entire textbook helps demonstrate the complexity and beauty of life.
- *PowerPoint® Resources.* For each chapter of the textbook, we have created several different types of PowerPoint® presentations. These give instructors the flexibility to build a presentation in the manner that best suits their needs. Included are:
  - Figures and Tables
  - Lecture Presentation
  - New! Editable Labels
  - Layered Art
  - Supplemental Photos
  - Videos, Animations
- *Clicker Questions.* A set of questions written specifically to be used with classroom personal response systems ("clickers") is provided for each chapter. These questions are designed to reinforce concepts, gauge student comprehension, and provide an outlet for active participation.
- *Chapter Outlines, Lecture Notes,* and the complete *Test File* are all available in Microsoft Word® format for easy use in lecture and exam preparation.
- An intuitive *Browser Interface* provides a quick and easy way to preview all of the content on the Instructor's Media Library.
- *Computerized Test Bank.* The entire printed Test File, plus the textbook end-of-chapter Self-Quizzes, the Companion Website Online Quizzes, and the Study Guide questions are all included in Brownstone's easy-to-use Diploma® software.
- *Instructor's Website.* A wealth of instructor's media, as well as electronic versions of other instructor supplements, are available online for instant access anytime.
- *Online Quizzes.* The Companion Website includes an Online Quiz for each chapter of the textbook. Instructors can choose to use these quizzes as assignments, and can view the results in the online gradebook.

- *Course Management System Support.* As a service for adopters using WebCT, Blackboard, or ANGEL for their courses, full electronic course packs are available.

Instructor's Resource Kit

The *Life,* Eighth Edition Instructor's Resource Kit includes a wealth of information to help instructors in the planning and teaching of their course. The Kit includes:

- *Instructor's Manual,* featuring:
  - A "What's New" guide to the Eighth Edition
  - A brief chapter overview
  - A key terms section with all the boldface terms from the text
  - Chapter outlines
- *Lecture Notes*—detailed notes for each chapter that can serve as the basis for lectures, including references to figures and media resources.
- *Media Guide*—A visual guide to the extensive media resources available with the Eighth Edition of *Life.* The guide includes thumbnails and descriptions of every video, animation, PowerPoint®, and supplemental photo in the Media Library, all organized by chapter.
- *Lab manual* and custom lab manual information.

Overhead Transparencies

This set includes over 1,000 transparencies—all the four-color line art and all the tables from the text—along with convenient binders. Balloon captions have been removed and colors have been enhanced for clear projection in a wide range of conditions. Labels and images have been resized for improved readability.

Test File

Ernest Dubrul, *University of Toledo;* Jon Glase, *Cornell University;* Norman Johnson, *University of Massachusetts;* Catherine Ueckert, *Northern Arizona University*

The test file offers more than 5,000 questions, including fill-in-the-blank and multiple-choice test questions. The electronic version of the Test File also includes all of the textbook end-of-chapter Self-Quiz questions, all of the Student Website Online Quiz questions, and all of the Study Guide questions.

iclicker

Developed for educators by educators, iclicker is a hassle-free radio-frequency classroom response system that makes it easy for instructors to ask questions, record responses, take attendance, and direct students through lectures as active participants. For more information, visit www.iclicker.com.

The *Life* eBook

The *Life,* Eighth Edition eBook is a complete online version of the textbook that can be purchased online, or packaged with the printed textbook. This online version of *Life* is a substantially less expensive alternative that gives your students an efficient and rich learning experience by integrating all of the resources from the companion website directly into the eBook text. In addition, the eBook offers instructors unique opportunities to customize the text with their own content. Key features of the eBook include:

- *For Students:*
  - Integration of all website **activities** and **animated tutorials**
  - In-text **self quiz questions**
  - Interactive **summary exercises**
  - Custom text **highlighting**
  - A **notes** feature that allows students to annotate the text
  - Complete **glossary**, **index**, and **full-text search** features
- *For Instructors:*
  - A powerful notes feature that can incorporate text, Web links, and documents directly into the text
  - Easy integration of instructor media resources, including videos and supplemental photographs
  - Ability to link directly to any eBook page from other sites

BIOPORTAL

New for the Eighth Edition, BioPortal is the digital gateway to all of the teaching and learning resources that are available with *Life.* BioPortal integrates the *Life* textbook, all of the student and instructor media, extensive assessment resources, and course planning resources all into a powerful and easy-to-use learning management system. All of this means that your students get easy access to learning resources, presented in the proper context and at the proper time, and you get a complete learning management system, ready to use, without hours of prep work. Features of BioPortal include:

- *eBook*
  - Completely integrated with all media resources
  - Customizable with notes, sections, images, and more
- *Student Resources*
  - Animated Tutorials and Activities
  - Assessment: Online Quizzes and Interactive Quizzes
- *Instructor Resources*
  - Complete Test Bank
  - All quizzes
  - Media resources: Videos, PowerPoints, Supplemental Photos, and more
- *Assignments*
  - Quizzes and exams
  - Assignable textbook sections
  - Assignable animations and activities
  - Custom assignments
- *Easy-to-Use Course Management*
  - Complete course customization
  - Custom resources/Document posting
  - Announcements/Calendar/Course Email/Discussions
  - Robust Gradebook

Contents in Brief

Part One ■ The Science and Building Blocks of Life

1 Studying Life 2
2 The Chemistry of Life 20
3 Macromolecules and the Origin of Life 38

Part Two ■ Cells and Energy

4 Cells: The Working Units of Life 68
5 The Dynamic Cell Membrane 96
6 Energy, Enzymes, and Metabolism 118
7 Pathways That Harvest Chemical Energy 138
8 Photosynthesis: Energy from Sunlight 160

Part Three ■ Heredity and the Genome

9 Chromosomes, The Cell Cycle, and Cell Division 180
10 Genetics: Mendel and Beyond 206
11 DNA and Its Role in Heredity 232
12 From DNA to Protein: Genotype to Phenotype 256
13 The Genetics of Viruses and Prokaryotes 282
14 The Eukaryotic Genome and Its Expression 306

Part Four ■ Molecular Biology: The Genome in Action

15 Cell Signaling and Communication 332
16 Recombinant DNA and Biotechnology 352
17 Genome Sequencing, Molecular Biology, and Medicine 374
18 Immunology: Gene Expression and Natural Defense Systems 400
19 Differential Gene Expression in Development 426
20 Development and Evolutionary Change 448

Part Five ■ The Patterns and Processes of Evolution

21 The History of Life on Earth 464
22 The Mechanisms of Evolution 486
23 Species and Their Formation 508
24 The Evolution of Genes and Genomes 524
25 Reconstructing and Using Phylogenies 542

Part Six ■ The Evolution of Diversity

26 Bacteria and Archaea: The Prokaryotic Domains 560
27 The Origin and Diversification of the Eukaryotes 582
28 Plants without Seeds: From Sea to Land 610
29 The Evolution of Seed Plants 630
30 Fungi: Recyclers, Pathogens, Parasites, and Plant Partners 650
31 Animal Origins and the Evolution of Body Plans 670
32 Protostome Animals 690
33 Deuterostome Animals 716

Part Seven ■ Flowering Plants: Form and Function

34 The Plant Body 744
35 Transport in Plants 764
36 Plant Nutrition 780
37 Regulation of Plant Growth 796
38 Reproduction in Flowering Plants 818
39 Plant Responses to Environmental Challenges 836

Part Eight ■ Animals: Form and Function

40 Physiology, Homeostasis, and Temperature Regulation 854
41 Animal Hormones 874
42 Animal Reproduction 896
43 Animal Development: From Genes to Organisms 920
44 Neurons and Nervous Systems 942
45 Sensory Systems 964
46 The Mammalian Nervous System: Structure and Higher Function 984
47 Effectors: How Animals Get Things Done 1004
48 Gas Exchange in Animals 1024
49 Circulatory Systems 1044
50 Nutrition, Digestion, and Absorption 1068
51 Salt and Water Balance and Nitrogen Excretion 1092

Part Nine ■ Ecology

52 Ecology and The Distribution of Life 1112
53 Behavior and Behavioral Ecology 1140
54 Population Ecology 1166
55 Community Ecology 1184
56 Ecosystems and Global Ecology 1204
57 Conservation Biology 1226

Contents

Part One ■ The Science and Building Blocks of Life

1 Studying Life 2

1.1 What Is Biology? 3
- Living organisms consist of cells 4
- The diversity of life is due to evolution by natural selection 5
- Biological information is contained in a genetic language common to all organisms 7
- Cells use nutrients to supply energy and to build new structures 7
- Living organisms control their internal environment 7
- Living organisms interact with one another 9
- Discoveries in biology can be generalized 9

1.2 How Is All Life on Earth Related? 10
- Life arose from nonlife via chemical evolution 10
- Biological evolution began when cells formed 11
- Photosynthesis changed the course of evolution 11
- Eukaryotic cells evolved from prokaryotes 11
- Multicellularity arose and cells became specialized 12
- Biologists can trace the evolutionary Tree of Life 12

1.3 How Do Biologists Investigate Life? 13
- Observation is an important skill 13
- The scientific method combines observation and logic 13
- Good experiments have the potential of falsifying hypotheses 14
- Statistical methods are essential scientific tools 15
- Not all forms of inquiry are scientific 16

1.4 How Does Biology Influence Public Policy? 16

2 The Chemistry of Life 20

2.1 What Are the Chemical Elements That Make Up Living Organisms? 21
- An element consists of only one kind of atom 21
- Protons: Their number identifies an element 22
- Neutrons: Their number differs among isotopes 23
- Electrons: Their behavior determines chemical bonding 23

2.2 How Do Atoms Bond to Form Molecules? 25
- Covalent bonds consist of shared pairs of electrons 25
- Multiple covalent bonds 27
- Ionic bonds form by electrical attraction 27
- Hydrogen bonds may form within or between molecules with polar covalent bonds 29
- Polar and nonpolar substances: Each interacts best with its own kind 29

2.3 How Do Atoms Change Partners in Chemical Reactions? 30

2.4 What Properties of Water Make It So Important in Biology? 31
- Water has a unique structure and special properties 31
- Water is the solvent of life 32
- Aqueous solutions may be acidic or basic 33
- pH is the measure of hydrogen ion concentration 34
- Buffers minimize pH change 34
- Life's chemistry began in water 34

An Overview and a Preview 36

3 Macromolecules and the Origin of Life 38

3.1 What Kinds of Molecules Characterize Living Things? 39
- Functional groups give specific properties to molecules 39
- Isomers have different arrangements of the same atoms 40
- The structures of macromolecules reflect their functions 40
- Most macromolecules are formed by condensation and broken down by hydrolysis 41

3.2 What Are the Chemical Structures and Functions of Proteins? 42
- Amino acids are the building blocks of proteins 42
- Peptide bonds form the backbone of a protein 43
- The primary structure of a protein is its amino acid sequence 44
- The secondary structure of a protein requires hydrogen bonding 45
- The tertiary structure of a protein is formed by bending and folding 46

The quaternary structure of a protein consists of subunits 46
Both shape and surface chemistry contribute to protein specificity 46
Environmental conditions affect protein structure 48
Chaperonins help shape proteins 48

3.3 What Are the Chemical Structures and Functions of Carbohydrates? 49
Monosaccharides are simple sugars 49
Glycosidic linkages bond monosaccharides 50
Polysaccharides store energy and provide structural materials 51
Chemically modified carbohydrates contain additional functional groups 53

3.4 What Are the Chemical Structures and Functions of Lipids? 54
Fats and oils store energy 55
Phospholipids form biological membranes 55
Not all lipids are triglycerides 56

3.5 What Are the Chemical Structures and Functions of Nucleic Acids? 57
Nucleotides are the building blocks of nucleic acids 57
The uniqueness of a nucleic acid resides in its nucleotide sequence 59
DNA reveals evolutionary relationships 60
Nucleotides have other important roles 60

3.6 How Did Life on Earth Begin? 61
Could life have come from outside Earth? 61
Did life originate on Earth? 61
Chemical evolution may have led to polymerization 62
RNA may have been the first biological catalyst 62
Experiments disproved spontaneous generation of life 63

Part Two ■ Cells and Energy

4 Cells: The Working Units of Life 68

4.1 What Features of Cells Make Them the Fundamental Unit of Life? 69
Cell size is limited by the surface area-to-volume ratio 69
Microscopes are needed to visualize cells 70
Cells are surrounded by a plasma membrane 72
Cells are prokaryotic or eukaryotic 72

4.2 What Are the Characteristics of Prokaryotic Cells? 72
Prokaryotic cells share certain features 72
Some prokaryotic cells have specialized features 73

4.3 What Are the Characteristics of Eukaryotic Cells? 74
Compartmentalization is the key to eukaryotic cell function 74
Organelles can be studied by microscopy or isolated for chemical analysis 75
Some organelles process information 75
The endomembrane system is a group of interrelated organelles 79
Some organelles transform energy 82
Several other organelles are surrounded by a membrane 83
The cytoskeleton is important in cell structure 86

4.4 What Are the Roles of Extracellular Structures? 90

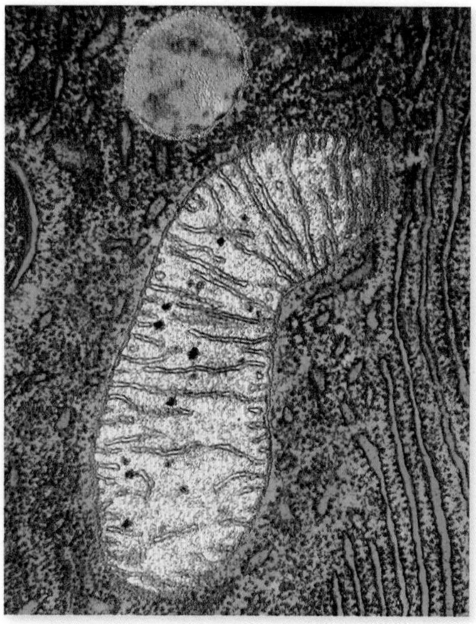

The plant cell wall is an extracellular structure 90
The extracellular matrix supports tissue functions in animals 91

4.5 How Did Eukaryotic Cells Originate? 92
The endosymbiosis theory suggests how eukaryotes evolved 92
Both prokaryotes and eukaryotes continue to evolve 92

5 The Dynamic Cell Membrane 96

5.1 What Is the Structure of a Biological Membrane? 97
Lipids constitute the bulk of a membrane 97
Membrane proteins are asymmetrically distributed 99
Membranes are dynamic 100
Membrane carbohydrates are recognition sites 101

5.2 How Is the Plasma Membrane Involved in Cell Adhesion and Recognition? 102
Cell recognition and cell adhesion involve proteins at the cell surface 102
Three types of cell junctions connect adjacent cells 102

5.3 What Are the Passive Processes of Membrane Transport? 105
Diffusion is the process of random movement toward a state of equilibrium 105
Simple diffusion takes place through the phospholipid bilayer 106
Osmosis is the diffusion of water across membranes 106
Diffusion may be aided by channel proteins 108
Carrier proteins aid diffusion by binding substances 110

5.4 How Do Substances Cross Membranes against a Concentration Gradient? 111
Active transport is directional 111
Primary and secondary active transport rely on different energy sources 111

5.5 How Do Large Molecules Enter and Leave a Cell? 113
Macromolecules and particles enter the cell by endocytosis 113
Receptor-mediated endocytosis is highly specific 113

Exocytosis moves materials out of the cell 114

5.6 What Are Some Other Functions of Membranes? 114

6 Energy, Enzymes, and Metabolism 118

6.1 What Physical Principles Underlie Biological Energy Transformations? 119

There are two basic types of energy and of metabolism 119

The first law of thermodynamics: Energy is neither created nor destroyed 120

The second law of thermodynamics: Disorder tends to increase 120

Chemical reactions release or consume energy 122

Chemical equilibrium and free energy are related 123

6.2 What Is the Role of ATP in Biochemical Energetics? 123

ATP hydrolysis releases energy 124

ATP couples exergonic and endergonic reactions 124

6.3 What Are Enzymes? 125

For a reaction to proceed, an energy barrier must be overcome 126

Enzymes bind specific reactant molecules 126

Enzymes lower the energy barrier but do not affect equilibrium 127

6.4 How Do Enzymes Work? 128

Molecular structure determines enzyme function 129

Some enzymes require other molecules in order to function 129

Substrate concentration affects reaction rate 130

6.5 How Are Enzyme Activities Regulated? 131

Enzymes can be regulated by inhibitors 131

Allosteric enzymes control their activity by changing their shape 132

Allosteric effects regulate metabolism 133

Enzymes are affected by their environment 133

7 Pathways That Harvest Chemical Energy 138

7.1 How Does Glucose Oxidation Release Chemical Energy? 139

Cells trap free energy while metabolizing glucose 139

An overview: Harvesting energy from glucose 140

Redox reactions transfer electrons and energy 140

The coenzyme NAD is a key electron carrier in redox reactions 141

7.2 What Are the Aerobic Pathways of Glucose Metabolism? 142

The energy-investing reactions of glycolysis require ATP 142

The energy-harvesting reactions of glycolysis yield NADH + H^+ and ATP 144

Pyruvate oxidation links glycolysis and the citric acid cycle 144

The citric acid cycle completes the oxidation of glucose to CO_2 145

The citric acid cycle is regulated by concentrations of starting materials 147

7.3 How Is Energy Harvested from Glucose in the Absence of Oxygen? 147

7.4 How Does the Oxidation of Glucose Form ATP? 148

The electron transport chain shuttles electrons and releases energy 149

Proton diffusion is coupled to ATP synthesis 150

7.5 Why Does Cellular Respiration Yield So Much More Energy Than Fermentation? 153

7.6 How Are Metabolic Pathways Interrelated and Controlled? 154

Catabolism and anabolism involve interconversions of biological monomers 154

Catabolism and anabolism are integrated 155

Metabolic pathways are regulated systems 155

8 Photosynthesis: Energy from Sunlight 160

8.1 What Is Photosynthesis? 161

Photosynthesis involves two pathways 162

8.2 How Does Photosynthesis Convert Light Energy into Chemical Energy? 163

Light behaves as both a particle and a wave 163

Absorbing a photon excites a pigment molecule 163

Absorbed wavelengths correlate with biological activity 163

Photosynthesis uses energy absorbed by several pigments 164

Light absorption results in photochemical change 165

Excited chlorophyll in the reaction center acts as a reducing agent 166

Reduction leads to electron transport 166

Noncyclic electron transport produces ATP and NADPH 166

Cyclic electron transport produces ATP but no NADPH 168

Chemiosmosis is the source of the ATP produced in photophosphorylation 168

8.3 How Is Chemical Energy Used to Synthesize Carbohydrates? 169

Radioisotope labeling experiments revealed the steps of the Calvin cycle 169

The Calvin cycle is made up of three processes 170

Light stimulates the Calvin cycle 172

8.4 How Do Plants Adapt to the Inefficiencies of Photosynthesis? 172

Rubisco catalyzes RuBP reaction with O_2 as well as with CO_2 172

C_4 plants can bypass photorespiration 173

CAM plants also use PEP carboxylase 175

8.5 How Is Photosynthesis Connected to Other Metabolic Pathways in Plants? 175

Part Three ■ Heredity and the Genome

9 Chromosomes, The Cell Cycle, and Cell Division 180

9.1 How Do Prokaryotic and Eukaryotic Cells Divide? 181
- Prokaryotes divide by binary fission 181
- Eukaryotic cells divide by mitosis or meiosis 182

9.2 How Is Eukaryotic Cell Division Controlled? 184
- Cyclins and other proteins trigger events in the cell cycle 184
- Growth factors can stimulate cells to divide 186

9.3 What Happens during Mitosis? 187
- Eukaryotic DNA is packed into very compact chromosomes 187
- Overview: Mitosis segregates exact copies of genetic information 188
- The centrosomes determine the plane of cell division 188
- Chromatids become visible and the spindle forms during prophase 189
- Chromosome movements are highly organized 189
- Nuclei re-form during telophase 191
- Cytokinesis is the division of the cytoplasm 192

9.4 What Is the Role of Cell Division in Sexual Life Cycles? 193
- Reproduction by mitosis results in genetic constancy 193
- Reproduction by meiosis results in genetic diversity 193
- The number, shapes, and sizes of the metaphase chromosomes constitute the karyotype 195

9.5 What Happens When a Cell Undergoes Meiosis? 195
- The first meiotic division reduces the chromosome number 197
- The second meiotic division separates the chromatids 199
- The activities and movements of chromosomes during meiosis result in genetic diversity 199
- Meiotic errors lead to abnormal chromosome structures and numbers 199
- Polyploids can have difficulty in cell division 200

9.6 How Do Cells Die? 202

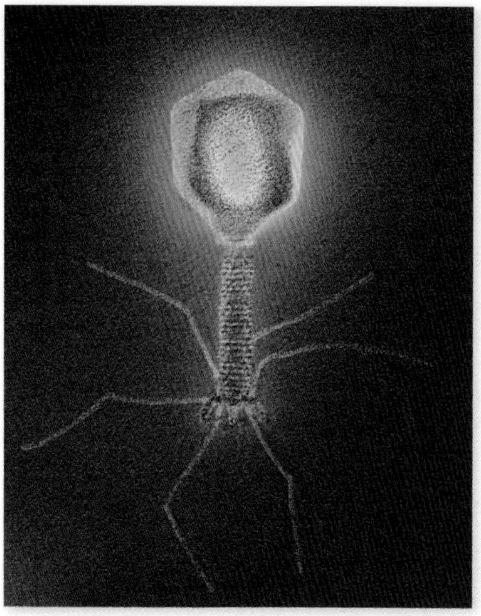

10 Genetics: Mendel and Beyond 206

10.1 What Are the Mendelian Laws of Inheritance? 207
- Mendel brought new methods to experiments on inheritance 208
- Mendel devised a careful research plan 208
- Mendel's first experiments involved monohybrid crosses 209
- Alleles are different forms of a gene 211
- Mendel's first law says that the two copies of a gene segregate 211
- Mendel verified his hypothesis by performing a test cross 213
- Mendel's second law says that copies of different genes assort independently 213
- Punnett squares or probability calculations: A choice of methods 214
- Mendel's laws can be observed in human pedigrees 216

10.2 How Do Alleles Interact? 217
- New alleles arise by mutation 217
- Many genes have multiple alleles 218
- Dominance is not always complete 218
- In codominance, both alleles at a locus are expressed 219
- Some alleles have multiple phenotypic effects 219

10.3 How Do Genes Interact? 219
- Hybrid vigor results from new gene combinations and interactions 220
- The environment affects gene action 220
- Most complex phenotypes are determined by multiple genes and the environment 221

10.4 What Is the Relationship between Genes and Chromosomes 222
- Genes on the same chromosome are linked 223
- Genes can be exchanged between chromatids 223
- Geneticists can make maps of chromosomes 224
- Linkage is revealed by studies of the sex chromosomes 225
- Genes on sex chromosomes are inherited in special ways 226
- Humans display many sex-linked characters 227

10.5 What Are the Effects of Genes Outside the Nucleus? 228

11 DNA and Its Role in Heredity 232

11.1 What Is the Evidence that the Gene is DNA? 233
- DNA from one type of bacterium genetically transforms another type 233
- The transforming principle is DNA 234
- Viral replication experiments confirmed that DNA is the genetic material 235
- Eukaryotic cells can also be genetically transformed by DNA 237

11.2 What Is the Structure of DNA? 238
- The chemical composition of DNA was known 238
- Watson and Crick described the double helix 238
- Four key features define DNA structure 239
- The double-helical structure of DNA is essential to its function 241

11.3 How Is DNA Replicated? 241
- Three modes of DNA replication appeared possible 241
- Meselson and Stahl demonstrated that DNA replication is semiconservative 241
- There are two steps in DNA replication 242
- DNA is threaded through a replication complex 243

Part Four ▪ Molecular Biology: The Genome in Action

15 Cell Signaling and Communication 332

15.1 What Are Signals, and How Do Cells Respond to Them? 333
- Cells receive signals from the physical environment and from other cells 333
- A signal transduction pathway involves a signal, a receptor, transduction, and effects 334

15.2 How Do Signal Receptors Initiate a Cellular Response? 336
- Receptors have specific binding sites for their signals 336
- Receptors can be classified by location 336

15.3 How Is a Response to a Signal Transduced through the Cell? 339
- Protein kinase cascades amplify a response to ligand binding 339
- Second messengers can stimulate protein kinase cascades 340
- Second messengers can be derived from lipids 341
- Calcium ions are involved in many signal transduction pathways 343
- Nitric oxide can act as a second messenger 344
- Signal transduction is highly regulated 345

15.4 How Do Cells Change in Response to Signals? 345
- Ion channels open in response to signals 345
- Enzyme activities change in response to signals 346
- Signals can initiate gene transcription 347

15.5 How Do Cells Communicate Directly? 348
- Animal cells communicate by gap junctions 348
- Plant cells communicate by plasmodesmata 348

16 Recombinant DNA and Biotechnology 352

16.1 How Are Large DNA Molecules Analyzed? 353
- Restriction enzymes cleave DNA at specific sequences 353
- Gel electrophoresis separates DNA fragments 354

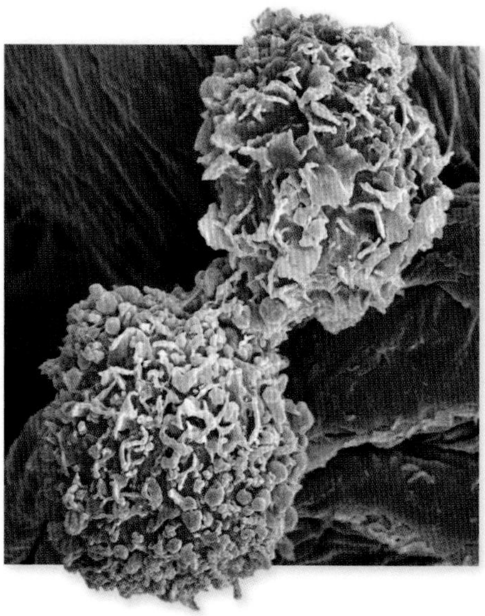

- DNA fingerprinting uses restriction analysis and electrophoresis 355
- The DNA barcode project aims to identify all organisms on Earth 356

16.2 What Is Recombinant DNA? 358

16.3 How Are New Genes Inserted into Cells? 359
- Genes can be inserted into prokaryotic or eukaryotic cells 359
- Vectors carry new DNA into host cells 359
- Reporter genes identify host cells containing recombinant DNA 361

16.4 What Are the Sources of DNA Used in Cloning? 362
- Gene libraries provide collections of DNA fragments 362
- cDNA libraries are constructed from mRNA transcripts 363
- DNA can be synthesized chemically in the laboratory 363
- DNA mutations can be created in the laboratory 363

16.5 What Other Tools Are Used to Manipulate DNA? 364
- Genes can be inactivated by homologous recombination 364
- Antisense RNA and interference RNA can prevent the expression of specific genes 365
- DNA chips can reveal DNA mutations and RNA expression 365

16.6 What Is Biotechnology? 367
- Expression vectors can turn cells into protein factories 367
- Medically useful proteins can be made by biotechnology 368
- DNA manipulation is changing agriculture 369
- There is public concern about biotechnology 371

17 Genome Sequencing, Molecular Biology, and Medicine 374

17.1 How Do Defective Proteins Lead to Diseases? 375
- Genetic mutations may make proteins dysfunctional 375
- Prion diseases are disorders of protein conformation 378
- Most diseases are caused by both genes and environment 379
- Human genetic diseases have several patterns of inheritance 379

17.2 What Kinds of DNA Changes Lead to Diseases? 380
- One way to identify a gene is to start with its protein 380
- Chromosome deletions can lead to gene and then protein isolation 381
- Genetic markers can point the way to important genes 381
- Disease-causing mutations may involve any number of base pairs 382
- Expanding triplet repeats demonstrate the fragility of some human genes 382
- DNA changes in males and females can have different consequences 383

17.3 How Does Genetic Screening Detect Diseases? 384
- Screening for disease phenotypes can make use of protein expression 384
- DNA testing is the most accurate way to detect abnormal genes 384

17.4 What Is Cancer? 386
- Cancer cells differ from their normal counterparts 386
- Some cancers are caused by viruses 387
- Most cancers are caused by genetic mutations 387
- Two kinds of genes are changed in many cancers 388
- Several events must occur to turn a normal cell into a malignant cell 389

17.5 How Are Genetic Diseases Treated? 391
- Genetic diseases can be treated by modifying the phenotype 391
- Gene therapy offers the hope of specific treatments 391

17.6 What Have We Learned from the Human Genome Project? 393
There are two approaches to genome sequencing 393
The sequence of the human genome contained many surprises 393
The human genome sequence has many applications 395
The use of genetic information poses ethical questions 395
The proteome is more complex than the genome 395
Systems biology integrates data from genomics and proteomics 396

18 Immunology: Gene Expression and Natural Defense Systems 400

18.1 What Are the Major Defense Systems of Animals? 401
Blood and lymph tissues play important roles in defense systems 402
White blood cells play many defensive roles 402
Immune system proteins bind pathogens or signal other cells 403

18.2 What Are the Characteristics of the Nonspecific Defenses? 404
Barriers and local agents defend the body against invaders 404
Other nonspecific defenses include specialized proteins and cellular processes 405
Inflammation is a coordinated response to infection or injury 406
A cell signaling pathway stimulates the body's defenses 407

18.3 How Does Specific Immunity Develop? 407
The specific immune system has four key traits 407
Two types of specific immune responses interact 408
Genetic changes and clonal selection generate the specific immune response 408
Immunity and immunological memory result from clonal selection 409
Vaccines are an application of immunological memory 409
Animals distinguish self from nonself and tolerate their own antigens 410

18.4 What Is the Humoral Immune Response? 411
Some B cells develop into plasma cells 411
Different antibodies share a common structure 411
There are five classes of immunoglobulins 412
Monoclonal antibodies have many uses 413

18.5 What Is the Cellular Immune Response? 414
T cell receptors are found on two types of T cells 414
The MHC encodes proteins that present antigens to the immune system 415
Helper T cells and MHC II proteins contribute to the humoral immune response 417
Cytotoxic T cells and MHC I proteins contribute to the cellular immune response 417
MHC proteins underlie the tolerance of self 417

18.6 How Do Animals Make So Many Different Antibodies? 418
Antibody diversity results from DNA rearrangement and other mutations 418
The constant region is involved in class switching 419

18.7 What Happens When the Immune System Malfunctions? 420
Allergic reactions result from hypersensitivity 420
Autoimmune diseases are caused by reactions against self antigens 421
AIDS is an immune deficiency disorder 421

19 Differential Gene Expression in Development 426

19.1 What Are the Processes of Development? 427
Development proceeds via determination, differentiation, morphogenesis, and growth 427
Cell fates become more and more restricted 428

19.2 Is Cell Differentiation Irreversible? 429
Plant cells are usually totipotent 429
Among animals, the cells of early embryos are totipotent 430
The somatic cells of adult animals retain the complete genome 431
Pluripotent stem cells can be induced to differentiate by environmental signals 432
Embryonic stem cells are potentially powerful therapeutic agents 433

19.3 What Is the Role of Gene Expression in Cell Differentiation? 435
Differential gene transcription is a hallmark of cell differentiation 435
Tools of molecular biology are used to investigate development 435

19.4 How Is Cell Fate Determined? 436
Cytoplasmic segregation can determine polarity and cell fate 436
Inducers passing from one cell to another can determine cell fates 437

19.5 How Does Gene Expression Determine Pattern Formation? 439
Some genes determine programmed cell death during development 439
Plants have organ identity genes 440
Morphogen gradients provide positional information 441
In the fruit fly, a cascade of transcription factors establishes body segmentation 442
Homeobox-containing genes encode transcription factors 445

20 Development and Evolutionary Change 448

20.1 How Does a Molecular Tool Kit Govern Development? 449
Developmental genes in diverse organisms are similar, but have different results 449

20.2 How Can Mutations with Large Effects Change Only One Part of the Body? 451
Genetic switches govern how the molecular tool kit is used 451
Modularity allows differences in the timing and spatial pattern of gene expression 451

20.3 How Can Differences among Species Evolve? 453

20.4 How Does the Environment Modulate Development? 454
Organisms respond to signals that accurately predict the future 454
Some signals that accurately predict the future may not always occur 456
Organisms do not respond to signals that are poorly correlated with future conditions 456
Organisms may lack appropriate responses to new environmental signals 457

20.5 How Do Developmental Genes Constrain Evolution? 457
Evolution proceeds by changing what's already there 457
Conserved developmental genes can lead to parallel evolution 458

Part Five ■ The Patterns and Processes of Evolution

21 The History of Life on Earth 464

21.1 How Do Scientists Date Ancient Events? 465
Radioisotopes provide a way to date rocks 466
Radioisotope dating methods have been expanded and refined 467

21.2 How Have Earth's Continents and Climates Changed over Time? 468
Oxygen has steadily increased in Earth's atmosphere 468
Earth's climate has shifted between hot/humid and cold/dry conditions 470
Volcanoes have occasionally changed the history of life 471
Extraterrestrial events have triggered changes on Earth 471

21.3 What Are the Major Events in Life's History? 471
Several processes contribute to the paucity of fossils 472
Precambrian life was small and aquatic 472
Life expanded rapidly during the Cambrian period 472
Many groups of organisms diversified 473
Geographic differentiation increased during the Mesozoic era 476
The modern biota evolved during the Cenozoic era 477
Three major faunas have dominated life on Earth 479

21.4 Why Do Evolutionary Rates Differ among Groups of Organisms? 479
"Living fossils" exist today 479
Evolutionary changes have been gradual in most groups 479
Rates of evolutionary change are sometimes rapid 480
Rates of extinction have also varied greatly 480

22 The Mechanisms of Evolution 486

22.1 What Facts Form the Base of Our Understanding of Evolution? 487
Adaptation has two meanings 489
Population genetics provides an underpinning for Darwin's theory 489
Most populations are genetically variable 490

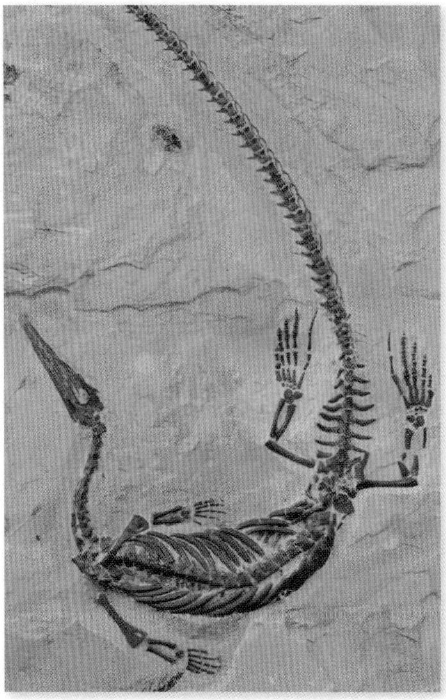

Evolutionary change can be measured by allele and genotype frequencies 491
The genetic structure of a population does not change over time if certain conditions exist 492
Deviations from Hardy–Weinberg equilibrium show that evolution is occurring 493

22.2 What Are the Mechanisms of Evolutionary Change? 494
Mutations generate genetic variation 494
Gene flow may change allele frequencies 494
Genetic drift may cause large changes in small populations 494
Nonrandom mating changes genotype frequencies 495

22.3 What Evolutionary Mechanisms Result in Adaptation? 497
Natural selection produces variable results 497
Sexual selection influences reproductive success 498

22.4 How Is Genetic Variation Maintained within Populations? 501
Neutral mutations may accumulate within populations 501
Sexual recombination amplifies the number of possible genotypes 501
Frequency-dependent selection maintains genetic variation within populations 501
Environmental variation favors genetic variation 502
Much genetic variation is maintained in geographically distinct subpopulations 503

22.5 What Are the Constraints on Evolution? 503
Developmental processes constrain evolution 504
Trade-offs constrain evolution 504
Short-term and long-term evolutionary outcomes sometimes differ 504

22.6 How Have Humans Influenced Evolution? 505

23 Species and Their Formation 508

23.1 What Are Species? 509
We can recognize and identify many species by their appearance 509
Species form over time 510

23.2 How Do New Species Arise? 511
Allopatric speciation requires almost complete genetic isolation 511
Sympatric speciation occurs without physical barriers 513

23.3 What Happens when Newly Formed Species Come Together? 515
Prezygotic barriers operate before fertilization 515
Postzygotic barriers operate after fertilization 516
Hybrid zones may form if reproductive isolation is incomplete 517

23.4 Why Do Rates of Speciation Vary? 518

23.5 Why Do Adaptive Radiations Occur? 520

24 The Evolution of Genes and Genomes 524

24.1 What Can Genomes Reveal about Evolution? 525
Evolution of genomes results in biological diversity 525
Genes and proteins are compared through sequence alignment 526
Models of sequence evolution are used to calculate evolutionary divergence 527
Experimental studies examine molecular evolution directly 527

24.2 What Are the Mechanisms of Molecular Evolution? 530

Much of evolution is neutral 531
Positive and stabilizing selection can be detected in the genome 532
Genome size and organization also evolve 533
New functions can arise by gene duplication 534
Some gene families evolve through concerted evolution 535

24.3 What Are Some Applications of Molecular Evolution? 536

Molecular sequence data are used to determine the evolutionary history of genes 537
Gene evolution is used to study protein function 538
In vitro evolution produces new molecules 538
Molecular evolution is used to study and combat diseases 538

25 Reconstructing and Using Phylogenies 542

25.1 What Is Phylogeny? 543

All of life is connected through evolutionary history 544
Comparisons among species require an evolutionary perspective 544

25.2 How Are Phylogenetic Trees Constructed? 545

Parsimony provides the simplest explanation for phylogenetic data 546
Phylogenies are reconstructed from many sources of data 547
Mathematical models expand the power of phylogenetic reconstruction 548
The accuracy of phylogenetic methods can be tested 549
Ancestral states can be reconstructed 550
Molecular clocks add a dimension of time 551

25.3 How Do Biologists Use Phylogenetic Trees? 551

Phylogenies help us reconstruct the past 551
Phylogenies allow us to compare and contrast living organisms 552
Biologists use phylogenies to predict the future 553

25.4 How Does Phylogeny Relate to Classification? 554

Phylogeny is the basis for modern biological classification 555
Several codes of biological nomenclature govern the use of scientific names 555

Part Six ▪ The Evolution of Diversity

26 Bacteria and Archaea: The Prokaryotic Domains 560

26.1 How Did the Living World Begin to Diversify? 561

The three domains differ in significant ways 561

26.2 Where Are Prokaryotes Found? 563

Prokaryotes generally form complex communities 563

26.3 What Are Some Keys to the Success of Prokaryotes? 565

Prokaryotes have distinctive cell walls 565
Prokaryotes have distinctive modes of locomotion 566
Prokaryotes reproduce asexually, but genetic recombination can occur 566
Some prokaryotes communicate 566
Prokaryotes have amazingly diverse metabolic pathways 567

26.4 How Can We Determine Prokaryote Phylogeny? 569

Size complicates the study of prokaryote phylogeny 569
The nucleotide sequences of prokaryotes reveal their evolutionary relationships 569
Lateral gene transfer may complicate phylogenetic studies 570

The great majority of prokaryote species have never been studied 570
Mutations are a major source of prokaryotic variation 571

26.5 What Are the Major Known Groups of Prokaryotes? 571

Spirochetes move by means of axial filaments 571
Chlamydias are extremely small parasites 572
Some high-GC Gram-positives are valuable sources of antibiotics 572
Cyanobacteria are important photoautotrophs 573
Not all low-GC Gram-positives are Gram-positive 573
The Proteobacteria are a large and diverse group 574
Archaea differ in several important ways from bacteria 575
Many Crenarchaeota live in hot, acidic places 576
The Euryarchaeota live in many surprising places 577
Korarchaeota and Nanoarchaeota are less well known 577

26.6 How Do Prokaryotes Affect Their Environments? 578

Prokaryotes are important players in element cycling 578
Prokaryotes live on and in other organisms 579
A small minority of bacteria are pathogens 579

27 The Origin and Diversification of the Eukaryotes 582

27.1 How Do Microbial Eukaryotes Affect the World Around Them? 583

The phylogeny and morphology of the microbial eukaryotes both illustrate their diversity 583
Phytoplankton are the primary producers of the marine food web 584

Some microbial eukaryotes are endosymbionts 585
Some microbial eukaryotes are deadly 585
We continue to rely on the products of ancient marine microbial eukaryotes 586

27.2 How Did the Eukaryotic Cell Arise? 588
The modern eukaryotic cell arose in several steps 588
Chloroplasts are a study in endosymbiosis 589
We cannot yet account for the presence of some prokaryotic genes in eukaryotes 590

27.3 How Did the Microbial Eukaryotes Diversify? 591
Microbial eukaryotes have different lifestyles 591
Microbial eukaryotes have diverse means of locomotion 591
Microbial eukaryotes employ vacuoles in several ways 591
The cell surfaces of microbial eukaryotes are diverse 592

27.4 How Do Microbial Eukaryotes Reproduce? 593
Some microbial eukaryotes have reproduction without sex, and sex without reproduction 593
Many microbial eukaryote life cycles feature alternation of generations 593
Chlorophytes provide examples of several life cycles 594
The life cycles of some microbial eukaryotes require more than one host species 595

27.5 What Are the Major Groups of Eukaryotes? 596
Alveolates have sacs under their plasma membrane 596
Stramenopiles have two unequal flagella, one with hairs 598
Red algae have a distinctive accessory photosynthetic pigment 601
Chlorophytes, charophytes, and land plants contain chlorophylls *a* and *b* 602
Diplomonads and parabasalids are excavates that lack mitochondria 603
Heteroloboseans alternate between amoeboid forms and forms with flagella 603
Euglenids and kinetoplastids have distinctive mitochondria and flagella 603
Foraminiferans have created vast limestone deposits 604
Radiolarians have thin, stiff pseudopods 605
Amoebozoans use lobe-shaped pseudopods for locomotion 605

28 Plants without Seeds: From Sea to Land 610

28.1 How Did the Land Plants Arise? 611
There are ten major groups of land plants 611
The land plants arose from a green algal clade 612

28.2 How Did Plants Colonize and Thrive on Land? 613
Adaptations to life on land distinguish land plants from green algae 613
The nonvascular plants usually live where water is available 614
Life cycles of land plants feature alternation of generations 614
The sporophytes of nonvascular plants are dependent on gametophytes 616

28.3 What Features Distinguish the Vascular Plants? 616
Vascular tissues transport water and dissolved materials 617
Vascular plants have been evolving for almost half a billion years 618
The earliest vascular plants lacked roots and leaves 619
The vascular plants branched out 619
Roots may have evolved from branches 619
Pteridophytes and seed plants have true leaves 620
Heterospory appeared among the vascular plants 621

28.4 What Are the Major Clades of Seedless Plants? 622
Liverworts may be the most ancient surviving plant clade 622
Hornworts have stomata, distinctive chloroplasts, and sporophytes without stalks 622
Water and sugar transport mechanisms emerged in the mosses 623
Some vascular plants have vascular tissue but not seeds 624
The club mosses are sister to the other vascular plants 624
Horsetails, whisk ferns, and ferns constitute a clade 625

29 The Evolution of Seed Plants 630

29.1 How Did Seed Plants Become Today's Dominant Vegetation? 631
Features of the seed plant life cycle protect gametes and embryos 631
The seed is a complex, well-protected package 633
A change in anatomy enabled seed plants to grow to great heights 634

29.2 What Are the Major Groups of Gymnosperms? 634
The relationship between gnetophytes and conifers is a subject of continuing research 635
Conifers have cones but no motile gametes 636

29.3 What Features Distinguish the Angiosperms? 638
The sexual structures of angiosperms are flowers 638
Flower structure has evolved over time 640
Angiosperms have coevolved with animals 641
The angiosperm life cycle features double fertilization 642
Angiosperms produce fruits 643

29.4 How Did the Angiosperms Originate and Diversify? 645
The basal angiosperm clade is a matter of controversy 645
The origin of the angiosperms remains a mystery 646

29.5 How Do Plants Support Our World? 647
Seed plants are our primary food source 647
Seed plants have been sources of medicines since ancient times 647

30 Fungi: Recyclers, Pathogens, Parasites, and Plant Partners 650

30.1 How Do Fungi Thrive in Virtually Every Environment? 651
The body of a multicellular fungus is composed of hyphae 651
Fungi are in intimate contact with their environment 652
Fungi exploit many nutrient sources 653
Fungi balance nutrition and reproduction 654

30.2 How Are Fungi Beneficial to Other Organisms? 655
Saprobic fungi dispose of Earth's garbage and contribute to the planetary carbon cycle 655
Mutualistic relationships are beneficial to both partners 655
Lichens can grow where plants cannot 655
Mycorrhizae are essential to most plants 657
Endophytic fungi protect some plants from pathogens, herbivores, and stress 658
Some fungi are food for the ants that farm them 658

30.3 How Do Fungal Life Cycles Differ from One Another? 659
Fungi reproduce both sexually and asexually 659
The dikaryotic condition is unique to the fungi 662
The life cycles of some parasitic fungi require two hosts 662
"Imperfect fungi" lack a sexual stage 663

30.4 How Do We Tell the Fungal Groups Apart? 663
Chytrids are the only fungi with flagella 663
Zygomycetes reproduce sexually by fusion of two gametangia 664
Glomeromycetes form arbuscular mycorrhizae 665
The sexual reproductive structure of ascomycetes is the ascus 665
The sexual reproductive structure of basidiomycetes is a basidium 667

31 Animal Origins and the Evolution of Body Plans 670

31.1 What Evidence Indicates the Animals Are Monophyletic? 671
Animal monophyly is supported by gene sequences and morphology 671
Developmental patterns show evolutionary relationships among animals 672

31.2 What Are the Features of Animal Body Plans? 674
Most animals are symmetrical 674
The structure of the body cavity influences movement 674
Segmentation improves control of movement 675
Appendages enhance locomotion 676

31.3 How Do Animals Get Their Food? 676
Filter feeders capture small prey 676
Herbivores eat plants 677
Predators capture and subdue large prey 678
Parasites live in or on other organisms 679

31.4 How Do Animal Life Cycles Differ? 679
All life cycles have at least one dispersal stage 680
No life cycle can maximize all benefits 680
Parasite life cycles evolve to facilitate dispersal and overcome host defenses 681

31.5 What Are the Major Groups of Animals? 682
Sponges are loosely organized animals 683
Ctenophores are radially symmetrical and diploblastic 684
Cnidarians are specialized carnivores 685

32 Protostome Animals 690

32.1 What Is a Protostome? 691
Trochophores, lophophores, and spiral cleavage evolved among the lophotrochozoans 692
Ecdysozoans must shed their exoskeletons 693
Arrow worms retain some ancestral developmental features 694

32.2 What Are the Major Groups of Lophotrochozoans? 695
Ectoprocts live in colonies 695
Flatworms, rotifers, and ribbon worms are structurally diverse relatives 695
Phoronids and brachiopods use lophophores to extract food from the water 697
The annelids and the mollusks are sister groups 698
Annelids have segmented bodies 698
Mollusks have undergone a dramatic evolutionary radiation 700

32.3 What Are the Major Groups of Ecdysozoans? 702
Several marine groups have relatively few species 702
Nematodes and their relatives are abundant and diverse 703

32.4 Why Do Arthropods Dominate Earth's Fauna? 705
Arthropod relatives have fleshy, unjointed appendages 705
Jointed legs first appeared in the trilobites 706
Crustaceans are diverse and abundant 706
Insects are the dominant terrestrial arthropods 708
Myriapods have many legs 712
Most chelicerates have four pairs of legs 712

An Overview of Protostome Evolution 714

33 Deuterostome Animals 716

33.1 What is a Deuterostome? 717

33.2 What Are the Major Groups of Echinoderms and Hemichordates? 718
Echinoderms have a water vascular system 719
Hemichordates have a three-part body plan 721

33.3 What New Features Evolved in the Chordates? 722
Adults of most urochordates and cephalochordates are sessile 722
A new dorsal supporting structure replaces the notochord in vertebrates 723
The vertebrate body plan can support large animals 724
Fins and swim bladders improved stability and control over locomotion 725

33.4 How Did Vertebrates Colonize the Land? 728
Jointed fins enhanced support for fishes 728
Amphibians adapted to life on land 728
Amniotes colonized dry environments 730
Reptiles adapted to life in many habitats 731
Crocodilians and birds share their ancestry with the dinosaurs 731
The evolution of feathers allowed birds to fly 733
Mammals radiated after the extinction of dinosaurs 734
Most mammals are therians 735

33.5 What Traits Characterize the Primates? 737
Human ancestors evolved bipedal locomotion 738
Human brains became larger as jaws became smaller 740
Humans developed complex language and culture 741

Part Seven ■ Flowering Plants: Form and Function

34 The Plant Body 744

34.1 How Is the Plant Body Organized? 745
Roots anchor the plant and take up water and minerals 746
Stems bear buds, leaves, and flowers 746
Leaves are the primary sites of photosynthesis 747
The tissue systems support the plant's activities 748

34.2 How Are Plant Cells Unique? 749
Cell walls may be complex in structure 749
Parenchyma cells are alive when they perform their functions 749
Collenchyma cells provide flexible support while alive 750
Sclerenchyma cells provide rigid support 750
Cells of the xylem transport water and minerals from roots to stems and leaves 750
Cells of the phloem translocate carbohydrates and other nutrients 752

34.3 How Do Meristems Build the Plant Body? 752
Plants and animals grow differently 753
A hierarchy of meristems generates a plant's body 754
The root apical meristem gives rise to the root cap and the root primary meristems 755
The products of the root's primary meristems become root tissues 756
The products of the stem's primary meristems become stem tissues 757
Many eudicot stems and roots undergo secondary growth 758

34.4 How Does Leaf Anatomy Support Photosynthesis? 760

35 Transport in Plants 764

35.1 How Do Plants Take Up Water and Solutes? 765
Water moves through a membrane by osmosis 765
Aquaporins facilitate the movement of water across membranes 767

Uptake of mineral ions requires membrane transport proteins 767
Water and ions pass to the xylem by way of the apoplast and symplast 768

35.2 How Are Water and Minerals Transported in the Xylem? 769
Experiments ruled out xylem transport by the pumping action of living cells 769
Root pressure does not account for xylem transport 770
The transpiration–cohesion–tension mechanism accounts for xylem transport 770
A pressure chamber measures tension in the xylem sap 771

35.3 How Do Stomata Control the Loss of Water and the Uptake of CO_2? 773
The guard cells control the size of the stomatal opening 773
Transpiration from crops can be decreased 774

35.4 How Are Substances Translocated in the Phloem? 775
The pressure flow model appears to account for translocation in the phloem 776
The pressure flow model has been experimentally tested 776
Plasmodesmata allow the transfer of material between cells 777

36 Plant Nutrition 780

36.1 How Do Plants Acquire Nutrients? 781
Autotrophs make their own organic compounds 781
How does a stationary organism find nutrients? 782

36.2 What Mineral Nutrients Do Plants Require? 782
Deficiency symptoms reveal inadequate nutrition 783
Several essential elements fulfill multiple roles 784
Experiments were designed to identify essential elements 784

36.3 What Are the Roles of Soil? 785
Soils are complex in structure 785
Soils form through the weathering of rock 786
Soils are the source of plant nutrition 786
Fertilizers and lime are used in agriculture 787
Plants affect soil fertility and pH 787

36.4 How Does Nitrogen Get from Air to Plant Cells? 788
Nitrogen fixers make all other life possible 788
Nitrogenase catalyzes nitrogen fixation 789
Some plants and bacteria work together to fix nitrogen 789
Biological nitrogen fixation does not always meet agricultural needs 789
Plants and bacteria participate in the global nitrogen cycle 790

36.5 Do Soil, Air, and Sunlight Meet the Needs of All Plants? 792
Carnivorous plants supplement their mineral nutrition 792
Parasitic plants take advantage of other plants 792

37 Regulation of Plant Growth 796

37.1 How Does Plant Development Proceed? 797
Several hormones and photoreceptors play roles in plant growth regulation 797
Signal transduction pathways are involved in all stages of plant development 798

The seed germinates and forms a growing seedling 798
The plant flowers and sets fruit 798
The plant senesces and dies 798
Not all seeds germinate without cues 799
Seed dormancy affords adaptive advantages 799
Seed germination begins with the uptake of water 800
The embryo must mobilize its reserves 800

37.2 What Do Gibberellins Do? 801
"Foolish seedling" disease led to the discovery of the gibberellins 801
The gibberellins have many effects on plant growth and development 802

37.3 What Does Auxin Do? 803
Phototropism led to the discovery of auxin 803
Auxin transport is polar and requires carrier proteins 803
Light and gravity affect the direction of plant growth 805
Auxin affects plant growth in several ways 805
Auxin analogs as herbicides 807
Auxin promotes growth by acting on cell walls 807
Auxin and gibberellins are recognized by similar mechanisms 808

37.4 What Do Cytokinins, Ethylene, Abscisic Acid, and Brassinosteroids Do? 809
Cytokinins are active from seed to senescence 809
Ethylene is a gaseous hormone that hastens leaf senescence and fruit ripening 810

Abscisic acid is the "stress hormone" 811
Brassinosteroids are hormones that mediate effects of light 811

37.5 How Do Photoreceptors Participate in Plant Growth Regulation? 812
Phytochromes mediate the effects of red and far-red light 812
Phytochromes have many effects on plant growth and development 813
Multiple phytochromes have different developmental roles 813
Cryptochromes, phototropins, and zeaxanthin are blue-light receptors 814

38 Reproduction in Flowering Plants 818

38.1 How Do Angiosperms Reproduce Sexually? 819
The flower is an angiosperm's structure for sexual reproduction 820
Flowering plants have microscopic gametophytes 821
Pollination enables fertilization in the absence of water 821
Some flowering plants practice "mate selection" 822
A pollen tube delivers sperm cells to the embryo sac 822
Angiosperms perform double fertilization 822
Embryos develop within seeds 823
Some fruits assist in seed dispersal 824

38.2 What Determines the Transition from the Vegetative to the Flowering State? 825
Apical meristems can become inflorescence meristems 825
A cascade of gene expression leads to flowering 825
Photoperiodic cues can initiate flowering 826
Plants vary in their responses to different photoperiodic cues 826
The length of the night is the key photoperiodic cue determining flowering 827
Circadian rhythms are maintained by a biological clock 828
Photoreceptors set the biological clock 829
The flowering stimulus originates in a leaf 829
In some plants flowering requires a period of low temperature 831

38.3 How Do Angiosperms Reproduce Asexually? 832
Many forms of asexual reproduction exist 832
Vegetative reproduction has a disadvantage 833
Vegetative reproduction is important in agriculture 833

39 Plant Responses to Environmental Challenges 836

39.1 How Do Plants Deal with Pathogens? 837
Plants seal off infected parts to limit damage 837
Some plants have potent chemical defenses against pathogens 838
The hypersensitive response is a localized containment strategy 838
Systemic acquired resistance is a form of long-term "immunity" 839
Some plant genes match up with pathogen genes 839
Plants develop specific immunity to RNA viruses 840

39.2 How Do Plants Deal with Herbivores? 840
Grazing increases the productivity of some plants 840
Some plants produce chemical defenses against herbivores 841
Some secondary metabolites play multiple roles 841
Some plants call for help 842
Many defenses depend on extensive signaling 842
Recombinant DNA technology may confer resistance to insects 842
Why don't plants poison themselves? 844
The plant doesn't always win 844

39.3 How Do Plants Deal with Climate Extremes? 845
Some leaves have special adaptations to dry environments 845
Plants have other adaptations to a limited water supply 846
In water-saturated soils, oxygen is scarce 846
Plants have ways of coping with temperature extremes 847

39.4 How Do Plants Deal with Salt and Heavy Metals? 848
Most halophytes accumulate salt 848
Halophytes and xerophytes have some similar adaptations 848
Some habitats are laden with heavy metals 849

Part Eight ■ Animals: Form and Function

40 Physiology, Homeostasis, and Temperature Regulation 854

40.1 Why Must Animals Regulate Their Internal Environments? 855
An internal environment makes complex multicellular animals possible 855
Homeostasis requires physiological regulation 856
Physiological systems are made up of cells, tissues, and organs 857
Organs consist of multiple tissues 859

40.2 How Does Temperature Affect Living Systems? 860
Q_{10} is a measure of temperature sensitivity 860
Animals can acclimatize to a seasonal temperature change 861

40.3 How Do Animals Alter Their Heat Exchange with the Environment? 861
How do endotherms produce so much heat? 861
Ectotherms and endotherms respond differently to changes in temperature 862
Energy budgets reflect adaptations for regulating body temperature 863
Both ectotherms and endotherms control blood flow to the skin 864
Some fish elevate body temperature by conserving metabolic heat 864
Some ectotherms regulate heat production 865

40.4 How Do Mammals Regulate Their Body Temperatures? 866
Basal metabolic rates are correlated with body size and environmental temperature 866
Endotherms respond to cold by producing heat and reducing heat loss 867
Evaporation of water can dissipate heat, but at a cost 868
The vertebrate thermostat uses feedback information 868
Fever helps the body fight infections 869
Turning down the thermostat 870

41 Animal hormones 874

41.1 What Are Hormones and How Do They Work? 875
Hormones can act locally or at a distance 875
Hormonal communication arose early in evolution 876
Hormones from the head control molting in insects 876
Juvenile hormone controls development in insects 877
Hormones can be divided into three chemical groups 878
Hormone receptors are found on the cell surface or in the cell interior 878
Hormone action depends on the nature of the target cell and its receptors 879

41.2 How Do the Nervous and Endocrine Systems Interact? 880
The pituitary connects nervous and endocrine functions 880
The anterior pituitary is controlled by hypothalamic hormones 882
Negative feedback loops control hormone secretion 883

41.3 What Are the Major Mammalian Endocrine Glands and Hormones? 883
Thyroxine controls cell metabolism 883
Thyroid dysfunction causes goiter 885
Calcitonin reduces blood calcium 885
Parathyroid hormone elevates blood calcium 885
Vitamin D is really a hormone 886
PTH lowers blood phosphate levels 887
Insulin and glucagon regulate blood glucose levels 887
Somatostatin is a hormone of the brain and the gut 887
The adrenal gland is two glands in one 887
The sex steroids are produced by the gonads 889
Changes in control of sex steroid production initiate puberty 889
Melatonin is involved in biological rhythms and photoperiodicity 890
The list of hormones is long 890

41.4 How Do We Study Mechanisms of Hormone Action? 890
Hormones can be detected and measured with immunoassays 891
A hormone can act through many receptors 892
A hormone can act through different signal transduction pathways 893

42 Animal Reproduction 896

42.1 How Do Animals Reproduce Without Sex? 897
Budding and regeneration produce new individuals by mitosis 897
Parthenogenesis is the development of unfertilized eggs 898

42.2 How Do Animals Reproduce Sexually? 899
Gametogenesis produces eggs and sperm 899
Fertilization is the union of sperm and egg 901
Mating bring eggs and sperm together 903
A single body can function as both male and female 904
The evolution of vertebrate reproductive systems parallels the move to land 904
Reproductive systems are distinguished by where the embryo develops 905

42.3 How Do the Human Male and Female Reproductive Systems Work? 906
Male sex organs produce and deliver semen 906
Male sexual function is controlled by hormones 909

Female sex organs produce eggs, receive sperm, and nurture the embryo 909
The ovarian cycle produces a mature egg 910
The uterine cycle prepares an environment for the fertilized egg 911
Hormones control and coordinate the ovarian and uterine cycles 911
In pregnancy, hormones from the extraembryonic membranes take over 912
Childbirth is triggered by hormonal and mechanical stimuli 912

42.4 How Can Fertility Be Controlled and Sexual Health Maintained? 913

Human sexual responses have four phases 913
Humans use a variety of methods to control fertility 914
Reproductive technologies help solve problems of infertility 916
Sexual behavior transmits many disease organisms 917

43 Animal Development: From Genes to Organisms 920

43.1 How Does Fertilization Activate Development? 921

The sperm and the egg make different contributions to the zygote 921
Rearrangements of egg cytoplasm set the stage for determination 922
Cleavage repackages the cytoplasm 922
Cleavage in mammals is unique 923
Specific blastomeres generate specific tissues and organs 925

43.2 How Does Gastrulation Generate Multiple Tissue Layers? 926

Invagination of the vegetal pole characterizes gastrulation in the sea urchin 927
Gastrulation in the frog begins at the gray crescent 928
The dorsal lip of the blastopore organizes embryo formation 928
The molecular mechanisms of the organizer involve multiple transcription factors 929
The organizer changes its activity as it migrates from the dorsal lip 931
Reptilian and avian gastrulation is an adaptation to yolky eggs 931
Placental mammals have no yolk but retain the avian–reptilian gastrulation pattern 932

43.3 How Do Organs and Organ Systems Develop? 932

The stage is set by the dorsal lip of the blastopore 932
Body segmentation develops during neurulation 933
Hox genes control development along the anterior–posterior axis 933

43.4 What Is the Origin of the Placenta? 934

Extraembryonic membranes form with contributions from all germ layers 934
Extraembryonic membranes in mammals form the placenta 935
The extraembryonic membranes provide means of detecting genetic diseases 936

43.5 What Are the Stages of Human Development? 936

The embryo becomes a fetus in the first trimester 936
The fetus grows and matures during the second and third trimesters 937
Developmental changes continue throughout life 938

44 Neurons and Nervous Systems 942

44.1 What Cells Are Unique to the Nervous System? 943

Neuronal networks range in complexity 943
Neurons are the functional units of nervous systems 944
Glial cells are also important components of nervous systems 946

44.2 How Do Neurons Generate and Conduct Signals? 946

Simple electrical concepts underlie neuronal function 947
Membrane potentials can be measured with electrodes 947
Ion pumps and channels generate membrane potentials 947
Ion channels and their properties can now be studied directly 950
Gated ion channels alter membrane potential 950
Sudden changes in Na^+ and K^+ channels generate action potentials 951
Action potentials are conducted along axons without loss of signal 953
Action potentials can jump along axons 954

44.3 How Do Neurons Communicate with Other Cells? 955

The neuromuscular junction is a model chemical synapse 955
The arrival of an action potential causes the release of neurotransmitter 955
The postsynaptic membrane responds to neurotransmitter 956
Synapses between neurons can be excitatory or inhibitory 957
The postsynaptic cell sums excitatory and inhibitory input 957
Synapses can be fast or slow 958
Electrical synapses are fast but do not integrate information well 958
The action of a neurotransmitter depends on the receptor to which it binds 958
Glutamate receptors may be involved in learning and memory 959
To turn off responses, synapses must be cleared of neurotransmitter 960

45 Sensory Systems 964

45.1 How Do Sensory Cells Convert Stimuli into Action Potentials? 965

Sensory receptor proteins act on ion channels 965
Sensory transduction involves changes in membrane potentials 966
Sensation depends on which neurons receive action potentials from sensory cells 967
Many receptors adapt to repeated stimulation 967

45.2 How Do Sensory Systems Detect Chemical Stimuli? 967

Arthropods provide good examples for studying chemoreception 968
Olfaction is the sense of smell 968
The vomeronasal organ senses pheromones 969
Gustation is the sense of taste 969

45.3 How Do Sensory Systems Detect Mechanical Forces? 970

Many different cells respond to touch and pressure 970
Mechanoreceptors are found in muscles, tendons, and ligaments 971
Auditory systems use hair cells to sense sound waves 972
Hair cells provide information about displacement 974

45.4 How Do Sensory Systems Detect Light? 976

Rhodopsins are responsible for photosensitivity 976
Invertebrates have a variety of visual systems 977
Image-forming eyes evolved independently in vertebrates and cephalopods 978
The vertebrate retina receives and processes visual information 980

46 The Mammalian Nervous System: Structure and Higher Function 984

46.1 How Is the Mammalian Nervous System Organized? 985
A functional organization of the nervous system is based on flow and type of information 985
The vertebrate CNS develops from the embryonic neural tube 986
The spinal cord transmits and processes information 987
The reticular system alerts the forebrain 988
The core of the forebrain controls physiological drives, instincts, and emotions 988
Regions of the telencephalon interact to produce consciousness and control behavior 989

46.2 How Is Information Processed by Neuronal Networks? 992
The autonomic nervous system controls involuntary physiological functions 992
Patterns of light falling on the retina are integrated by the visual cortex 992
Cortical cells receive input from both eyes 996

46.3 Can Higher Functions Be Understood in Cellular Terms? 997
Sleep and dreaming are reflected in electrical patterns in the cerebral cortex 997
Some learning and memory can be localized to specific brain areas 999
Language abilities are localized in the left cerebral hemisphere 1000
What is consciousness? 1001

47 Effectors: How Animals Get Things Done 1004

47.1 How Do Muscles Contract? 1005
Sliding filaments cause skeletal muscle to contract 1005
Actin–myosin interactions cause filaments to slide 1007
Actin–myosin interactions are controlled by calcium ions 1008
Cardiac muscle causes the heart to beat 1010
Smooth muscle causes slow contractions of many internal organs 1010
Single skeletal muscle twitches are summed into graded contractions 1012

47.2 What Determines Muscle Strength and Endurance? 1013
Muscle fiber types determine endurance and strength 1013
A muscle has an optimal length for generating maximum tension 1014
Exercise increases muscle strength and endurance 1014
Muscle ATP supply limits performance 1014

47.3 What Roles Do Skeletal Systems Play in Movement? 1016
A hydrostatic skeleton consists of fluid in a muscular cavity 1016
Exoskeletons are rigid outer structures 1016
Vertebrate endoskeletons provide supports for muscles 1017
Bones develop from connective tissues 1018
Bones that have a common joint can work as a lever 1019

47.4 What Are Some Other Kinds of Effectors? 1020
Chromatophores allow an animal to change its color or pattern 1020
Glands secrete chemicals for defense, communication, or predation 1021
Electric organs generate electricity used for sensing, communication, defense, or attack 1021
Light-emitting organs use enzymes to produce light 1021

48 Gas Exchange in Animals 1024

48.1 What Physical Factors Govern Respiratory Gas Exchange? 1025
Diffusion is driven by concentration differences 1025
Fick's law applies to all systems of gas exchange 1026
Air is a better respiratory medium than water 1026
High temperatures create respiratory problems for aquatic animals 1026
O_2 availability decreases with altitude 1026
CO_2 is lost by diffusion 1027

48.2 What Adaptations Maximize Respiratory Gas Exchange? 1028
Respiratory organs have large surface areas 1028
Transporting gases to and from the exchange surfaces optimizes partial pressure gradients 1028
Insects have airways throughout their bodies 1028
Fish gills use countercurrent flow to maximize gas exchange 1029
Birds use unidirectional ventilation to maximize gas exchange 1030
Tidal ventilation produces dead space that limits gas exchange efficiency 1031

48.3 How Do Human Lungs Work? 1032
Respiratory tract secretions aid ventilation 1034
Lungs are ventilated by pressure changes in the thoracic cavity 1034

48.4 How Does Blood Transport Respiratory Gases? 1036
Hemoglobin combines reversibly with oxygen 1036
Myoglobin holds an oxygen reserve 1037
The affinity of hemoglobin for oxygen is variable 1037
Carbon dioxide is transported as bicarbonate ions in the blood 1038

48.5 How is Breathing Regulated? 1039
Breathing is controlled in the brain stem 1039
Regulating breathing requires feedback information 1039

49 Circulatory Systems 1044

49.1 Why Do Animals Need a Circulatory System? 1045
Some animals do not have circulatory systems 1045
Open circulatory systems move extracellular fluid 1046
Closed circulatory systems circulate blood through a system of blood vessels 1046

49.2 How Have Vertebrate Circulatory Systems Evolved? 1047
Fish have two-chambered hearts 1047
Amphibians have three-chambered hearts 1048
Reptiles have exquisite control of pulmonary and systemic circulation 1048
Birds and mammals have fully separated pulmonary and systemic circuits 1049

49.3 How Does the Mammalian Heart Function? 1050

Blood flows from right heart to lungs to left heart to body 1050
The heartbeat originates in the cardiac muscle 1052
A conduction system coordinates the contraction of heart muscle 1053
Electrical properties of ventricular muscles sustain heart contraction 1054
The ECG records the electrical activity of the heart 1055

49.4 What Are the Properties of Blood and Blood Vessels? 1055

Red blood cells transport respiratory gases 1055
Platelets are essential for blood clotting 1056
Plasma is a complex solution 1056
Blood circulates throughout the body in a system of blood vessels 1057
Materials are exchanged in capillary beds by filtration, osmosis, and diffusion 1057
Blood flows back to the heart through veins 1059
Lymphatic vessels return interstitial fluid to the blood 1060
Vascular disease is a killer 1060

49.5 How Is the Circulatory System Controlled and Regulated? 1061

Autoregulation matches local blood flow to local need 1061
Arterial pressure is controlled and regulated by hormonal and neuronal mechanisms 1061
Cardiovascular control in diving mammals conserves oxygen 1063

50 Nutrition, Digestion, and Absorption 1068

50.1 What Do Animals Require from Food? 1069

Energy can be measured in calories 1069
Energy budgets reveal how animals use their resources 1070
Sources of energy can be stored in the body 1071
Food provides carbon skeletons for biosynthesis 1072
Animals need mineral elements for a variety of functions 1073
Animals must obtain vitamins from food 1073
Nutrient deficiencies result in diseases 1074

50.2 How Do Animals Ingest and Digest Food? 1075

The food of herbivores is often low in energy and hard to digest 1075
Carnivores must detect, capture, and kill prey 1075
Vertebrate species have distinctive teeth 1076
Animals digest their food extracellularly 1076
Tubular guts have an opening at each end 1077
Digestive enzymes break down complex food molecules 1077

50.3 How Does the Vertebrate Gastrointestinal System Function? 1078

The vertebrate gut consists of concentric tissue layers 1078
Mechanical activity moves food through the gut and aids digestion 1079
Chemical digestion begins in the mouth and the stomach 1080
What causes stomach ulcers? 1081
The stomach gradually releases its contents to the small intestine 1082
Most chemical digestion occurs in the small intestine 1082
Nutrients are absorbed in the small intestine 1083
Absorbed nutrients go to the liver 1084
Water and ions are absorbed in the large intestine 1084
The problem with cellulose 1084

50.4 How Is the Flow of Nutrients Controlled and Regulated? 1085

Hormones control many digestive functions 1086
The liver directs the traffic of the molecules that fuel metabolism 1086
Regulating food intake is important 1087

50.5 How Do Animals Deal with Ingested Toxins? 1089

The body cannot metabolize many synthetic toxins 1089
Some toxins are retained and concentrated 1089

51 Salt and Water Balance and Nitrogen Excretion 1092

51.1 What Roles Do Excretory Organs Play in Maintaining Homeostasis? 1093

Water enters or leaves cells by osmosis 1093
Excretory organs control extracellular fluid osmolarity by filtration, secretion, and reabsorption 1094
Animals can be osmoconformers or osmoregulators 1094
Animals can be ionic conformers or ionic regulators 1095

51.2 How Do Animals Excrete Toxic Wastes from Nitrogen Metabolism? 1095

Animals excrete nitrogen in a number of forms 1095
Most species produce more than one nitrogenous waste 1096

51.3 How Do Invertebrate Excretory Systems Work? 1097

The protonephridia of flatworms excrete water and conserve salts 1097
The metanephridia of annelids process coelomic fluid 1097
The Malpighian tubules of insects depend on active transport 1098

51.4 How Do Vertebrates Maintain Salt and Water Balance? 1098

Marine fishes must conserve water 1099
Terrestrial amphibians and reptiles must avoid desiccation 1099
Birds and mammals can produce highly concentrated urine 1099
The nephron is the functional unit of the vertebrate kidney 1099
Blood is filtered in the glomerulus 1100
The renal tubules convert glomerular filtrate to urine 1101

51.5 How Does the Mammalian Kidney Produce Concentrated Urine? 1101

Kidneys produce urine and the bladder stores it 1101
Nephrons have a regular arrangement in the kidney 1101
Most of the glomerular filtrate is reabsorbed by the proximal convoluted tubule 1103
The loop of Henle creates a concentration gradient in the surrounding tissue 1103
Water permeability of kidney tubules depends on water channels 1104
Water reabsorption begins in the distal convoluted tubule 1104
Urine is concentrated in the collecting duct 1104
The kidneys help regulate acid–base balance 1105
Kidney failure is treated with dialysis 1105

51.6 What Mechanisms Regulate Kidney Function? 1107

The kidneys maintain the glomerular filtration rate 1107
Blood osmolarity and blood pressure are regulated by ADH 1107
The heart produces a hormone that influences kidney function 1108

Part Nine ■ Ecology

52 Ecology and The Distribution of Life 1112

52.1 What Is Ecology? 1113

52.2 How Are Climates Distributed on Earth? 1114

Solar energy drives global climates 1114
Global oceanic circulation is driven by wind patterns 1115
Organisms must adapt to changes in their environment 1115

52.3 What Is a Biome? 1116

Tundra is found at high latitudes and in high mountains 1118
Evergreen trees dominate most boreal forests 1119
Temperate deciduous forests change with the seasons 1120
Temperate grasslands are widespread 1121
Cold deserts are high and dry 1122
Hot deserts form around 30° latitude 1123
The chaparral climate is dry and pleasant 1124
Thorn forests and tropical savannas have similar climates 1125
Tropical deciduous forests occur in hot lowlands 1126
Tropical evergreen forests are species-rich 1127
The distribution of biomes is not determined solely by climate 1128

52.4 What Is a Biogeographic Region? 1128

Three scientific advances changed the field of biogeography 1129
A single barrier may split the ranges of many species 1133
Biotic interchange follows fusion of land masses 1134
Both vicariance and dispersal influence most biogeographic patterns 1135

52.5 How Is Life Distributed in Aquatic Environments? 1135

Currents create biogeographic regions in the oceans 1135
Freshwater environments may be rich in species 1136

53 Behavior and Behavioral Ecology 1140

53.1 What Questions Do Biologists Ask about Behavior? 1141

53.2 How Do Genes and Environment Interact to Shape Behavior? 1142

Experiments can distinguish between genetic and environmental influences on behavior 1142
Genetic control of behavior is adaptive under many conditions 1144
Imprinting takes place at a specific point in development 1145
Some behaviors result from intricate interactions between inheritance and learning 1145
Hormones influence behavior at genetically determined times 1146

53.3 How Do Behavioral Responses to the Environment Influence Fitness? 1147

Choosing where to live influences survival and reproductive success 1147
Defending a territory has benefits and costs 1148
Animals choose what foods to eat 1149
An animal's choice of associates influences its fitness 1152
Responses to the environment must be timed appropriately 1152
Animals must find their way around their environment 1154

53.4 How Do Animals Communicate with One Another? 1156

Visual signals are rapid and versatile 1157
Chemical signals are durable 1157
Auditory signals communicate well over a distance 1157
Tactile signals can communicate complex messages 1157
Electric signals can convey messages in murky water 1157

53.5 Why Do Animal Societies Evolve? 1158

Group living confers benefits but also imposes costs 1158
Parental care can evolve into more complex social systems 1159
Altruism can evolve by contributing to an animal's inclusive fitness 1159

53.6 How Does Behavior Influence Populations and Communities? 1161

Habitat and food selection influence the distributions of organisms 1161
Territoriality influences community structure 1161
Social animals may achieve great population densities 1161

54 Population Ecology 1166

54.1 How Do Ecologists Study Populations? 1167

Ecologists use a variety of devices to track individuals 1167
Population densities can be estimated from samples 1168
Birth and death rates can be estimated from population density data 1169

54.2 How Do Ecological Conditions Affect Life Histories? 1171

54.3 What Factors Influence Population Densities? 1173

All populations have the potential for exponential growth 1173
Population growth is limited by resources and biotic interactions 1173
Population densities influence birth and death rates 1174
Several factors explain why some species are more common than others 1175

54.4 How Do Spatially Variable Environments Influence Population Dynamics? 1178
Many populations live in separated habitat patches 1178
Distant events may influence local population densities 1178

54.5 How Can We Manage Populations? 1180
Demographic traits determine sustainable harvest levels 1180
Demographic information is used to control populations 1180
Can we manage our own population? 1181

55 Community Ecology 1184

55.1 What Are Ecological Communities? 1185
Communities are loose assemblages of species 1185
The organisms in a community use diverse sources of energy 1186

55.2 What Processes Influence Community Structure? 1188
Predation and parasitism are universal 1189
Competition is widespread because all species share resources 1192
Commensal and amensal interactions are widespread 1193
Most organisms participate in mutualistic interactions 1194

55.3 How Do Species Interactions Cause Trophic Cascades? 1195
One predator can affect many different species 1195
Keystone species have wide-ranging effects 1196

55.4 How Do Disturbances Affect Ecological Communities? 1197
Succession is a change in a community after a disturbance 1197
Species richness is greatest at intermediate levels of disturbance 1199
Both facilitation and inhibition influence succession 1199

55.5 What Determines Species Richness in Ecological Communities? 1200
Species richness is influenced by productivity 1200
Species richness and productivity influence ecosystem stability 1200

56 Ecosystems and Global Ecology 1204

56.1 What Are the Compartments of the Global Ecosystem? 1205
Oceans receive materials from the land and atmosphere 1206
Water moves rapidly through lakes and rivers 1206
The atmosphere regulates temperatures close to Earth's surface 1208
Land covers about one-fourth of Earth's surface 1208

56.2 How Does Energy Flow Through the Global Ecosystem? 1209
Solar energy drives ecosystem processes 1209
Human activities modify flows of energy 1210

56.3 How Do Materials Cycle Through the Global Ecosystem? 1211
Water transfers materials from one compartment to another 1211
Fire is a major mover of elements 1212
The carbon cycle has been altered by industrial activities 1213
Recent perturbations of the nitrogen cycle have had adverse effects on ecosystems 1216
The burning of fossil fuels affects the sulfur cycle 1217
The global phosphorus cycle lacks an atmospheric component 1218
Other biogeochemical cycles are also important 1219
Biogeochemical cycles interact 1220

56.4 What Services Do Ecosystems Provide? 1221

56.5 What Options Exist to Manage Ecosystems Sustainably? 1223

57 Conservation Biology 1226

57.1 What Is Conservation Biology? 1227
Conservation biology is a normative scientific discipline 1227
Conservation biology aims to prevent species extinctions 1228

57.2 How Do Biologists Predict Changes in Biodiversity? 1229

57.3 What Factors Threaten Species Survival? 1231
Species are endangered by habitat loss, degradation, and fragmentation 1231
Overexploitation has driven many species to extinction 1232
Invasive predators, competitors, and pathogens threaten many species 1233
Rapid climate change can cause species extinctions 1234

57.4 What Strategies Do Conservation Biologists Use? 1235
Protected areas preserve habitat and prevent overexploitation 1235
Degraded ecosystems can be restored 1236
Disturbance patterns sometimes need to be restored 1237
New habitats can be created 1238
We use markets to influence exploitation of species 1238
Ending trade is crucial to saving some species 1238
Controling invasions of exotic species is important 1239
Biodiversity can be profitable 1240
Living lightly helps preserve biodiversity 1241
Captive breeding programs can maintain a few species 1243
The legacy of Samuel Plimsoll 1243

Appendix A: The Tree of Life 1246

Appendix B: Some Measurements Used in Biology 1252

Glossary G-1

Answers to Self-Quizzes A-1

Illustration Credits C-1

Index I-1

PART ONE
The Science and Building Blocks of Life

CHAPTER 1 Studying Life

Why are frogs croaking?

In August of 1995, a group of Minnesota middle school students on a field trip were hiking through some local wetlands when they discovered a horde of young frogs, most of them with deformed, missing, or extra legs. The students' find made the national news and focused public attention on amphibian population declines, an issue already being studied by many scientists.

There are a number of possible reasons for the problems amphibians are facing. Water pollution is an obvious possibility, since these animals breed and spend their early lives in ponds and streams. Acidic rain resulting from air pollution could also affect their watery homes. Could ultraviolet radiation be causing "mutant" frogs? Is global warming adversely affecting amphibians? Is some disease attacking them? Evidence exists to support each of these possibilities, and there is no single answer. In one case, a college undergraduate came up with an answer, and in the process gave scientists a whole new perspective on the question.

In 1996, Stanford University sophomore Pieter Johnson was shown a collection of Pacific tree frogs (also known as chorus frogs) with extra legs growing out of their bodies. He decided to focus his research project on finding out what caused these deformities. The frogs came from a pond in an agricultural region near abandoned mercury mines; thus two possible causes of the deformities were agricultural chemicals, and heavy metals from the old mines.

Pieter applied the *scientific method*. Based on what he knew and on his library research, he proposed a logical explanation for the monster frogs—environmental water pollution—and designed an experiment to test his idea. His experiment compared ponds where there were deformed frogs with ponds where the frogs were normal and tested for the presence or absence of pollutants. As frequently happens in science, his proposed explanation, or *hypothesis*, was disproved by his experiment. But his field work led to a new hypothesis: that the deformities are caused by a parasite. Pieter conducted laboratory experiments, the results of which supported the conclusion that a certain type of parasite is present in some ponds. These parasites burrow into newly hatched tadpoles and disrupt the development of the adult

Frogs Are Having Serious Problems These preserved Pacific tree frogs (*Hyla regilla*) exhibit multiple deformities of the hind legs. Similar deformities have been found in frogs from different regions of the world.

CHAPTER OUTLINE

A Biologist at Work As a college sophomore, Pieter Johnson studied numerous ponds that were home to Pacific tree frogs, trying to discover why some ponds had so many deformed frogs.

frog's hind legs. Pieter's research did not explain the global decline in amphibians, but it did illuminate one type of problem amphibians encounter. Science usually progresses in such small but solid steps.

Biologists use the scientific method to investigate the processes of life at all levels, from molecules to ecosystems. Some of these processes happen in millionths of seconds and others cover millions of years. The goals of biologists are to understand how organisms and groups of organisms function, and sometimes they use that knowledge in practical and beneficial ways.

IN THIS CHAPTER we examine the most common features of living organisms and put those features into the context of the major principles that underlie all biology. Next we offer a brief outline of how life evolved and how the different organisms on Earth are related. We then turn to the subjects of biological inquiry and the scientific method.

1.1 What Is Biology?

Biology is the scientific study of living things. Biologists define "living things" as all the diverse organisms descended from a single-celled ancestor that appeared almost 4 billion years ago. Because of their common ancestry, living organisms share many characteristics that are not found in the nonliving world. Most living organisms:

- consist of one or more cells
- contain genetic information
- use genetic information to reproduce themselves
- are genetically related and have evolved
- can convert molecules obtained from their environment into new biological molecules
- can extract energy from the environment and use it to do biological work
- can regulate their internal environment

This list can serve as a rough guide to the major themes and unifying principles of biology that you will encounter in this book. A simple list, however, belies the incredible complexity and diversity of life. Some forms of life may not display all of these characteristics all of the time. For example, the seed of a desert plant may go for many years without extracting energy from the environment, converting molecules, regulating its internal environment, or reproducing; yet the seed is alive.

And what about viruses? Although they do not consist of cells, viruses probably evolved from cellular organisms, and many biologists consider them to be living organisms. Viruses cannot carry out physiological functions on their own, but must parasitize the machinery of host cells to do those jobs for them—including reproduction. Yet viruses contain genetic information, and they certainly evolve (as we know because evolving flu viruses require annual changes in the vaccines we create to combat them). Are viruses alive? What do you think?

This book explores the characteristics of life, how these characteristics vary among organisms, how they evolved, and how they work together to enable organisms to survive and reproduce. *Evolution* is a central theme in biology, and therefore in this book. Through differential survival and reproduction, living systems evolve and become adapted to Earth's many environments. The processes of evolution have generated the enormous diversity that we see today as life on Earth (**Figure 1.1**).

Living organisms consist of cells

Cells and the chemical processes within them are the topic of Part Two of this book. Some organisms are *unicellular,* consisting of a single cell that carries out all the functions of life, while others are *multicellular,* made up of a number of cells that are specialized for different functions.

The discovery of cells was made possible by the invention of the microscope in the 1590s by father and son Dutch spectacle makers Zaccharias and Hans Janssen. The first biologists to improve on their technology and use it to study living organisms were Antony van Leeuwenhoek (Dutch) and Robert Hooke (English) in the middle to late 1600s. It was van Leeuwenhoek who discovered that drops of pond water teemed with single-celled organisms, and he made many other discoveries as he progressively improved his microscopes over a long lifetime of research. Hooke carried out similar studies. From observations of plant tissues (cork, to be specific) he concluded that the tissues were made up of repeated units he called *cells* (**Figure 1.2**).

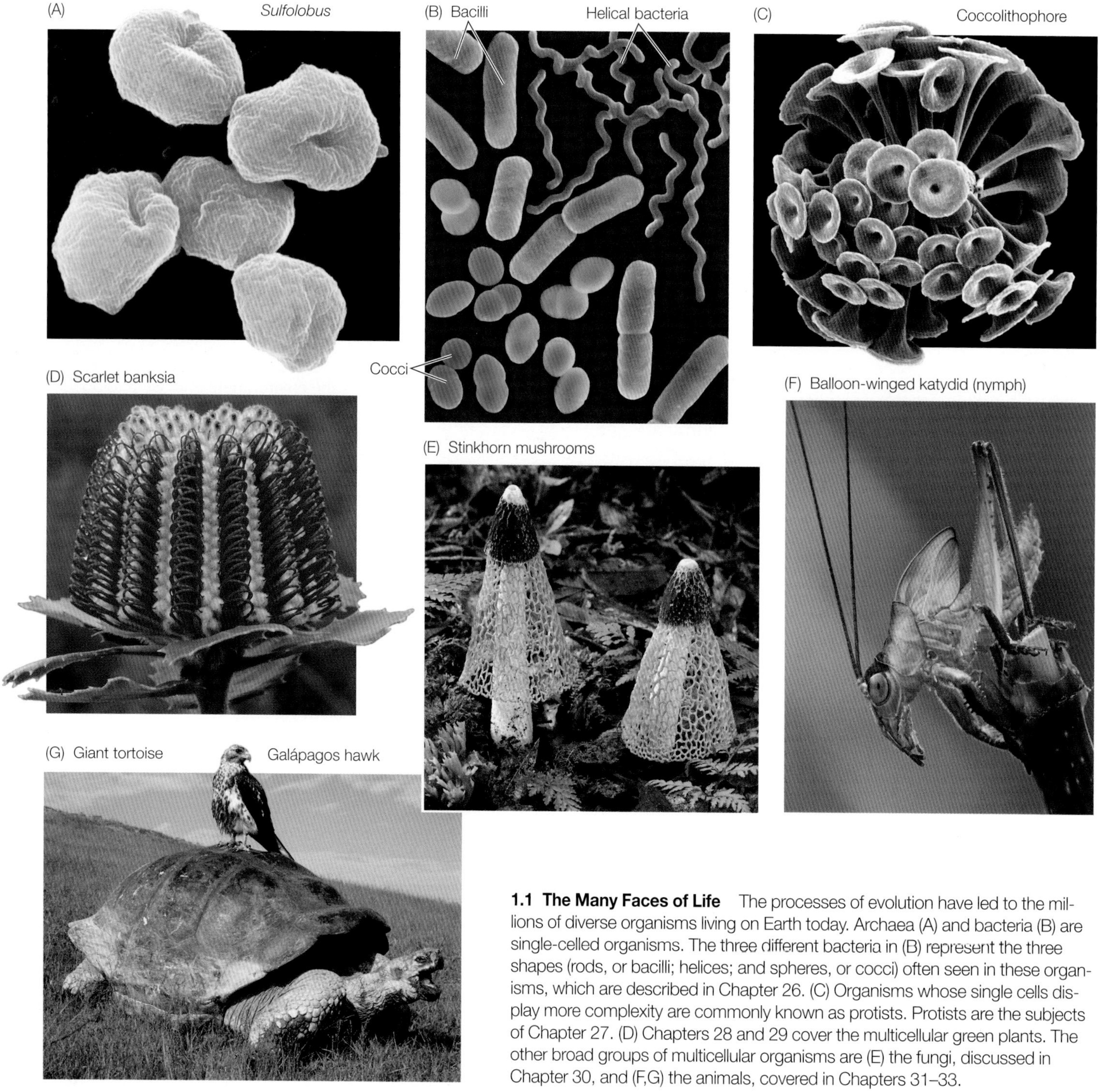

1.1 The Many Faces of Life The processes of evolution have led to the millions of diverse organisms living on Earth today. Archaea (A) and bacteria (B) are single-celled organisms. The three different bacteria in (B) represent the three shapes (rods, or bacilli; helices; and spheres, or cocci) often seen in these organisms, which are described in Chapter 26. (C) Organisms whose single cells display more complexity are commonly known as protists. Protists are the subjects of Chapter 27. (D) Chapters 28 and 29 cover the multicellular green plants. The other broad groups of multicellular organisms are (E) the fungi, discussed in Chapter 30, and (F,G) the animals, covered in Chapters 31–33.

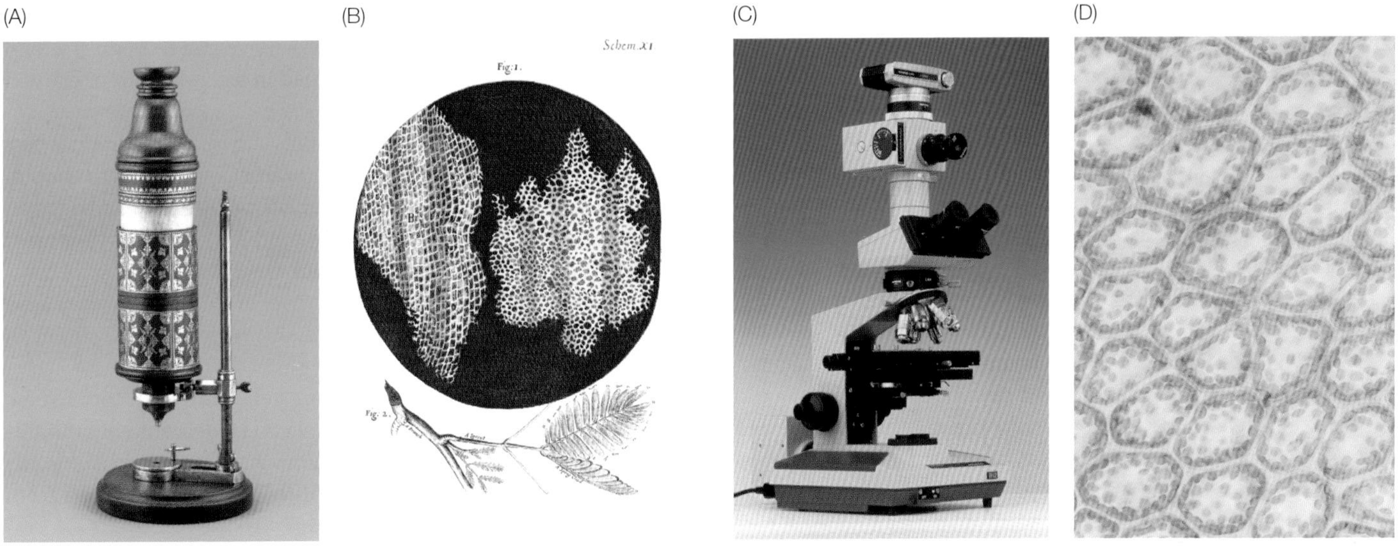

1.2 All Life Consists of Cells (A) The development of microscopes, such as this instrument of Robert Hooke's, revealed the microbial world to seventeenth-century scientists. (B) Hooke was the first to propose the concept of cells, based on his observations of thin slices of plant tissue (cork) under his microscope. (C) A modern version of the optical, or "light" microscope. (D) A modern light micrograph reveals the intricacies of cells in a leaf.

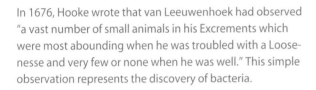

In 1676, Hooke wrote that van Leeuwenhoek had observed "a vast number of small animals in his Excrements which were most abounding when he was troubled with a Loosenesse and very few or none when he was well." This simple observation represents the discovery of bacteria.

More than a hundred years passed before studies of cells advanced significantly. In 1838, Matthias Schleiden, a German biologist, and Theodor Schwann, from Belgium, were having dinner together and discussing their work on plant and animal tissues, respectively. They were struck by the similarities in their observations and came to the conclusion that the structural elements of plants and animals were essentially the same. They formulated their conclusion as the **cell theory**, which states that:

- Cells are the basic structural and physiological units of all living organisms.
- Cells are both distinct entities and building blocks of more complex organisms.

But Schleiden and Schwann did not understand the origin of cells. They thought cells emerged by the self-assembly of nonliving materials, much as crystals form in a solution of salt. This conclusion was in accordance with the prevailing view of the day that life arises from nonlife by spontaneous generation—mice from dirty clothes, maggots from dead meat, insects from mixtures of straw and pond water. The debate over whether or not life could arise from nonlife continued until 1859, when the French Academy of Sciences sponsored a contest for the best experiment to prove or disprove spontaneous generation. The prize was won by the great French scientist Louis Pasteur; his experiment proving that life must be present in order for life to be generated is described in Figure 3.30. Pasteur's vision of microorganisms led him to propose the germ theory of disease and explain the role of single-celled organisms in the fermentation of beer and wine. He also designed a method of preserving milk by heating it to kill microorganisms, a process we now know as pasteurization.

Today we readily accept the fact that all cells come from preexisting cells. In addition, we understand that the functional properties of organisms derive from the properties of their cells. We also understand that cells of all kinds share many essential mechanisms because they share a common ancestry that goes back billions of years. We therefore add a few more elements to the cell theory:

- All cells come from preexisting cells.
- All cells are similar in chemical composition.
- Most of the chemical reactions of life occur within cells.
- Complete sets of genetic information are replicated and passed on during cell division.

At the same time Schleiden and Schwann were building the foundation for the cell theory, Charles Darwin was beginning to understand how organisms undergo evolutionary change.

The diversity of life is due to evolution by natural selection

Evolution by **natural selection**, as proposed by Charles Darwin, is perhaps the major unifying principle of biology and is the topic of Part Five of this book.

Darwin proposed that living organisms are descended from common ancestors and are therefore related to one another. He did not have the advantage of understanding the mechanisms of genetic inheritance that you will learn about in Part Three, but even so he surmised that such mechanisms existed because offspring resembled their parents in so many different ways. That simple fact is the basis for the concept of a **species**. Although the precise definition of a species is complicated, in its most widespread usage it refers to a group of organisms that look alike ("are morphologically similar") and can breed successfully with one another.

But offspring also differ from their parents. Any population of a plant or animal species displays variation, and if you select breeding pairs on the basis of some particular trait, that trait is more likely to be present in their offspring than in the general population. Darwin himself bred pigeons, and was well aware of how pigeon fanciers selected for unusual feather patterns, beak shapes, and body sizes. He realized that if humans could select for specific traits, the same process could operate in nature; hence the term *natural selection*.

How would selection function in nature? Darwin postulated that different probabilities of survival and reproductive success would do the job. He reasoned that the reproductive capacity of plants and animals, if unchecked, would result in unlimited growth of populations, but we do not observe such unlimited population growth in nature; therefore only a small percentage of offspring must survive to reproduce. So, any trait that confers even a small increase in the probability that its possessor will survive and reproduce would be strongly favored and would spread in the population. Darwin called this phenomenon *natural selection*.

Because organisms with certain traits survive and reproduce best under specific sets of conditions, natural selection leads to **adaptations**: structural, physiological, or behavioral traits that enhance an organism's chances of survival and reproduction in its environment (**Figure 1.3**). The many different environments and ecological communities organisms have adapted to over evolutionary history have led to a remarkable amount of diversity, which we will survey in Part Six of this book.

If all cells come from preexisting cells, and if all the diverse species of organisms on Earth are related by descent with modification from a common ancestor, then what is the source of information that is passed from parent to daughter cells and from parental organisms to their offspring?

Many leaves are wide and flat, a configuration that presents a maximum of photosynthetic surface to the sun. Some trees, such as this sugar maple, lose their leaves in response to cold or dry weather.
The leaves of many evergreen conifers, such as spruce trees, are waxy-coated needles that resist water loss and are not shed on a yearly basis.
These water lilies are rooted in the pond bottom; their large leaves are flat "pads" that float on the surface.
The leaves of pitcher plants form a vessel that holds water. The plant receives extra nutrients from the decomposing bodies of insects that drown in the pitcher.
The ability to climb can be advantageous to a plant, enabling it to reach above other plants to obtain more sunlight. Some of the leaves of this climbing cucumber are tightly furled tendrils that wrap around a stake.

1.3 Adaptations to the Environment The leaves of all plants are specialized for photosynthesis—the sunlight-powered transformation of water and carbon dioxide into larger structural molecules called carbohydrates. The leaves of different plants, however, display many different adaptations to their individual environments.

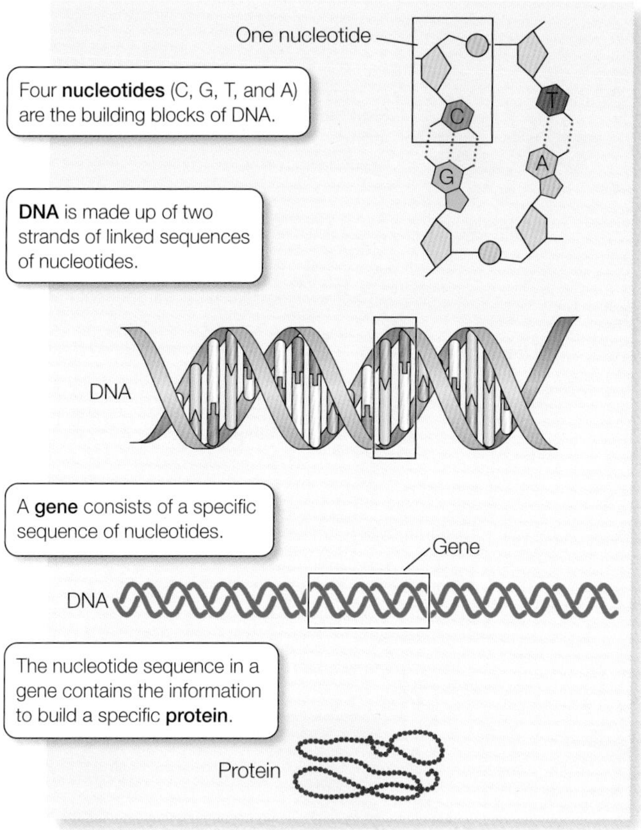

1.4 The Genetic Code Is Life's Blueprint The instructions for life are contained in the sequences of nucleotides in DNA molecules. Specific DNA sequences comprise genes, and the information in each gene provides the cell with the information it needs to manufacture a specific protein. The average length of a single human gene is 16,000 nucleotides.

Biological information is contained in a genetic language common to all organisms

A cell's instructions—or "blueprints" for existence—are contained in its **genome**, which is the sum total of all the DNA molecules in the cell. **DNA** (deoxyribonucleic acid) molecules are long sequences of four different subunits called **nucleotides**. The sequence of the nucleotides contains genetic information. Specific segments of DNA called **genes** contain the information the cell uses to make proteins (**Figure 1.4**). **Proteins** make up much of an organism's structure and are the molecules that govern the chemical reactions within cells. By analogy with a book, the nucleotides of DNA are like the letters of an alphabet. Protein molecules are the sentences they spell. Combinations of proteins that constitute structures and control biochemical processes are the paragraphs. The structures and processes that are organized into different systems with specific tasks (such as digestion or transport) are the chapters of the book, and the complete book is the organism. Natural selection is the author and editor of all the books in the library of life.

If you were to write out your own genome using four letters to represent the four nucleotides, you would have to write a total of more than 3 billion letters. If you used the same size type as you are reading in this book, your genome would fill about a thousand books the size of this one.

All the cells of a multicellular organism contain the same genome, yet different cells have different functions and form different structures. Therefore, different types of cells in an organism must express different parts of their genome. How the control of gene expression enables a complex organism to develop and function is a major focus of current biological research.

The genome of an organism consists of thousands of genes. If the nucleotide sequence of a gene is altered, it is likely that the protein that gene encodes will be altered. Alterations of genes are called *mutations*. Mutations occur spontaneously; they can also be induced by many outside factors, including chemicals and radiation. Most mutations are deleterious, but occasionally a change in the properties of a protein alters its function in a way that improves the functioning of the organism under the environmental conditions it encounters. Such beneficial mutations are the raw material of evolution.

Cells use nutrients to supply energy and to build new structures

Living organisms acquire substances called *nutrients* from the environment. Nutrients supply the organism with energy and raw materials for building biological structures. Cells take in nutrient molecules and break them down into smaller chemical units. In doing so, they can capture the energy contained in the chemical bonds of the nutrient molecules and use that energy to do different kinds of work. One kind of cellular work is the building, or *synthesis*, of new complex molecules and structures from smaller chemical units. For example, we are all familiar with the fact that carbohydrates eaten today may be deposited in the body as fat tomorrow (**Figure 1.5A**). Another kind of work cells do is mechanical work—for example, moving molecules from one cellular location to another, or even moving whole cells or tissues, as in the case of muscles (**Figure 1.5B**).

Living organisms control their internal environment

Life depends on thousands of biochemical reactions that occur inside cells. These reactions require materials to be moved into and out of cells in a controlled manner. Within the cell, those reactions are linked in that the products of one are the raw materials of the next. If this complex network of reactions is to be properly integrated, reaction rates within a cell must be precisely controlled. A large proportion of the activities of cells are directed toward the regulation of the multiple chemical reactions continuously in progress inside the cell.

Organisms that are made up of more than one cell have an *internal environment* that is not cellular. That is, their individual cells are bathed in extracellular fluids, from which they receive nutrients and into which they excrete wastes. The cells of multicellular organisms are specialized to contribute in some way to the maintenance of that internal environment. However, with the evolution of specialized

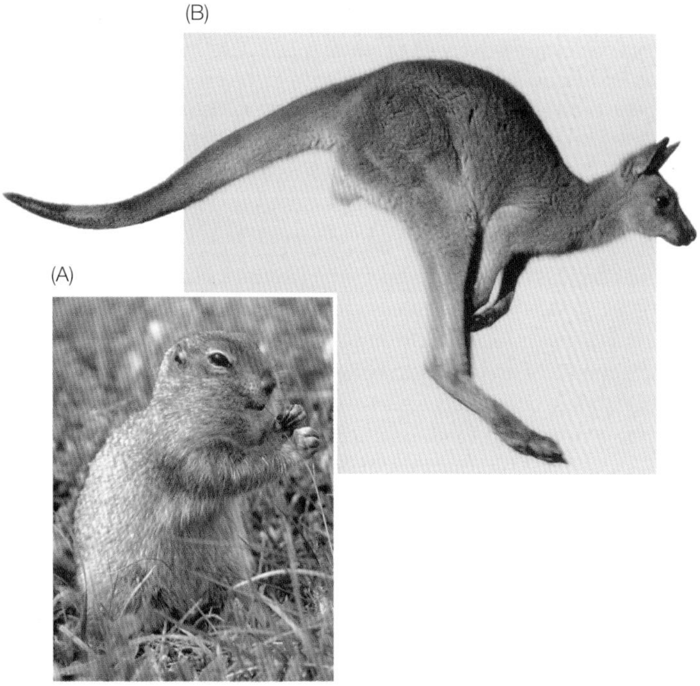

1.5 Energy from Nutrients Can Be Stored or Used Immediately (A) The cells of this Arctic ground squirrel have broken down the complex carbohydrates in plants and converted their molecules into fats, which are stored in the animal's body to provide an energy supply for the cold months. (B) The cells of this kangaroo are breaking down food molecules and using the energy in their chemical bonds to do mechanical work—in this case, to jump.

functions, these cells lost many of the functions carried out by single-celled organisms, and therefore depend on the internal environment for essential services. The interdependence of the different kinds of cells in a multicellular organism can by expressed in the famous motto of the Three Musketeers: "One for all, and all for one!"

To accomplish their specialized tasks, assemblages of similar cells are organized into *tissues.* For example, a single muscle cell cannot generate much force, but when many of these cells combine to form the tissue of a working muscle, considerable force and movement can be generated (see Figure 1.5B). Different tissue types are organized to form *organs* that accomplish specific functions. Familiar organs include the heart, brain, and stomach. Organs whose functions are interrelated can be grouped into *organ systems.* The functions of cells, tissues, organs, and organ systems are all integral to the multicellular *organism* (**Figure 1.6**). The biology of organisms is the subject of Parts Seven and Eight of this book.

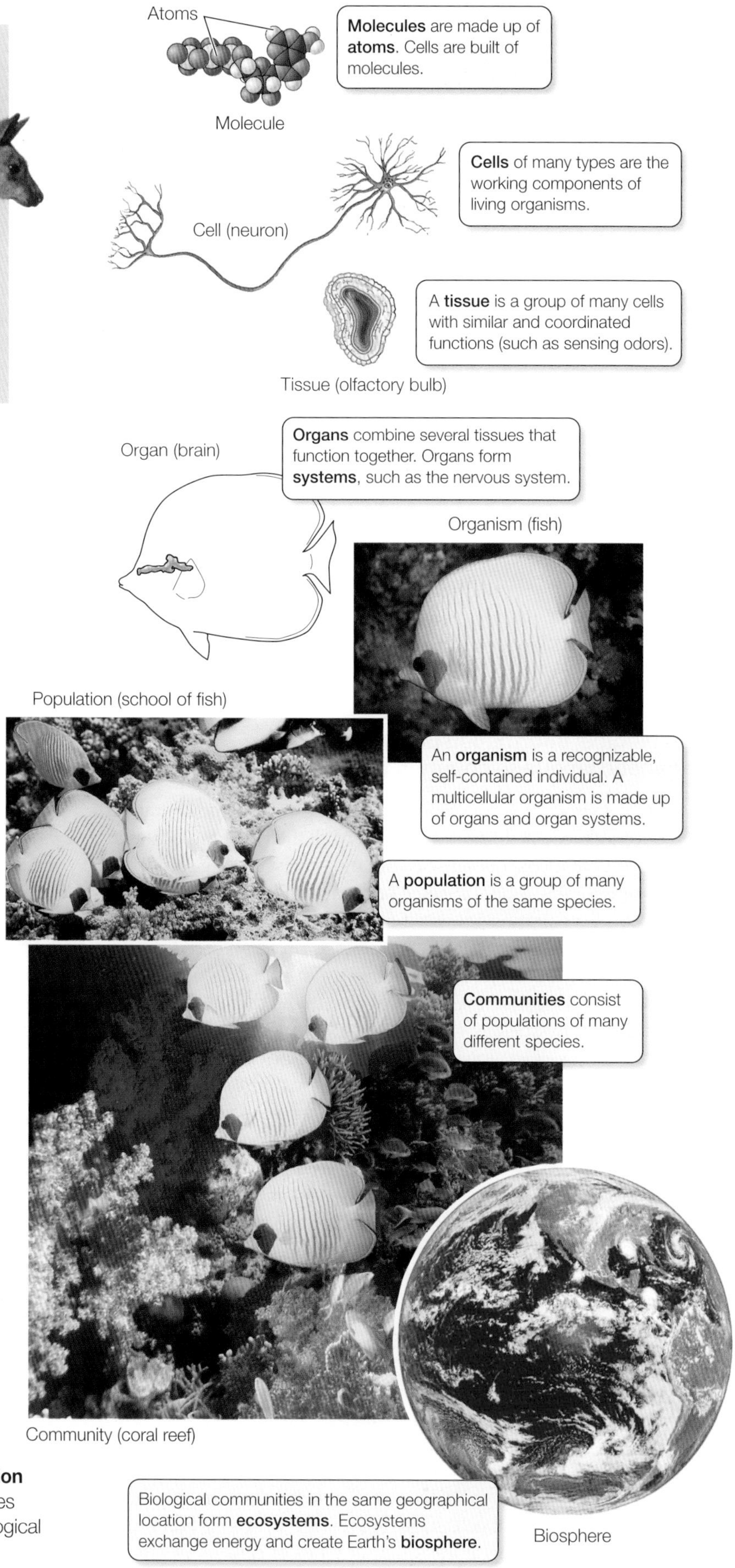

1.6 Biology Is Studied at Many Levels of Organization Life's properties emerge when DNA and other molecules are organized in cells. Energy flows through all the biological levels shown here.

(A)

(B)

1.7 Conflict and Cooperation Organisms of the same species interact with one another in various ways. (A) Territorial elephant seal bulls defend stretches of beach from other males. The single male who controls a stretch of beach is able to mate with the many females (seen in the background) that live there. (B) Members of a meerkat colony are usually related to one another. Meerkats cooperate in many ways, such as watching out for predators and giving a warning bark if one appears.

As Figure 1.6 shows, individual organisms do not live in isolation, and the hierarchy of biology continues well beyond that level.

Living organisms interact with one another

Organisms interact with their external environments as well as the internal environment. Individual organisms are part of *populations* that interact among themselves and with populations of different organisms.

Organisms interact in many different ways. For example, some animals are *territorial* and will try to prevent other individuals of their species from exploiting the resource they are defending, whether it be food, nesting sites, or mates (**Figure 1.7A**). Animals may also *cooperate* with members of their species, forming social units such as a termite colony, a school of fish, or a meerkat colony (**Figure 1.7B**). Such interactions among individuals have resulted in the evolution of social behaviors such as communication.

The interaction of populations of many different species forms a *community*, and interactions between different species are a major evolutionary force. Adaptations that give an individual of one species an advantage in obtaining members of another species as food (and the converse, adaptations that lessen an individual's chances of becoming food) are paramount in evolutionary history. Organisms of different species may compete for the same resources, resulting in natural selection for specialized adaptations that allow some individuals to exploit those resources more efficiently than others can.

In any given geographic locality, the interacting communities form *ecosystems*. Organisms in the ecosystem can modify the environment in ways that affect other organisms. For example, in most terrestrial (land) environments, the dominant plants greatly modify the environmental conditions in which animals and other plants must live. The ways in which species interact with one another and with their environment is the subject of *ecology*, the topic of Part Nine of this book.

Discoveries in biology can be generalized

Because all life is related by descent from a common ancestor, shares a genetic code, and consists of similar building blocks—cells—knowledge gained from investigations of one type of organism can, with care, be generalized to other organisms. Therefore, biologists can use **model systems** for research, knowing that they can extend their findings to other organisms and to humans. For example, our basic understanding of the chemical reactions in cells came from research on bacteria, but is applicable to all cells, including those of humans. Similarly, the biochemistry of photosynthesis—the process by which plants use sunlight to produce biological molecules—was largely worked out from experiments on *Chlorella* (a type of pond scum; see Figure 8.12). We learned much of what we know about the genes that control plant development from work on a single species of plant (see Chapter 19). Knowledge about how animals develop has come from work on sea urchins, frogs, chickens, roundworms, and fruit flies. And recently, the discovery of a major gene controlling human skin color came from work on zebrafish. Being able to generalize from model systems is a powerful tool.

1.1 RECAP

Living organisms are made of cells, evolve by natural selection, contain genetic information, extract energy from their environment and use it to do biological work, control their internal environment, and interact with one another.

- Can you describe the relationship between evolution by natural selection and the genetic code? See pp. 5–7
- Do you understand why results of biological research on one species can be generalized to very different species? See p. 9

Now that we have an overview of the major features of life that will be explored in depth in this book, we can ask how and when life first emerged. In the next section we will describe the history of life from the earliest simple life forms to the complex and diverse organisms that inhabit our planet today.

1.2 How Is All Life on Earth Related?

What do biologists mean when they say that all organisms are *genetically related*? They mean that species on Earth share a *common ancestor*. If two species are similar, as dogs and wolves are, then they probably have a common ancestor in the fairly recent past. The common ancestor of two species that are more different—say, a dog and a deer—probably lived in the more distant past. And if two organisms are very different—such as a dog and a clam—then we must go back to the *very* distant past to find their common ancestor. How can we tell how far back in time the common ancestor of any two organisms lived? In other words, how do we discover the evolutionary relationships among organisms?

For many years, biologists investigated the history of life by studying the *fossil record*—the preserved remains of organisms that lived in the distant past (**Figure 1.8**). Geologists supplied knowledge about the ages of fossils and the nature of the environments in which they lived. Biologists then inferred the evolutionary relationships among living and fossil organisms by comparing their anatomical similarities and differences. The development of modern molecular methods for comparing genomes, described in Chapter 24, has enabled biologists to more accurately establish the degrees of relationship between living organisms and use that information to help us interpret the fossil record.

In general, the greater the differences between the genomes of two species, the more distant was their common ancestor. Using molecular techniques, biologists are exploring fundamental questions about life on Earth. What were the earliest forms of life? How did those simple organisms give rise to the great diversity of organisms alive today? Can we reconstruct the family tree of all life?

Life arose from nonlife via chemical evolution

Geologists estimate that Earth is about 5 billion years old. For the first billion years or so, the planet was not a very hospitable place. Life probably arose toward the end of this time frame, a little over 4 billion years ago. If we picture the history of Earth as a 30-day calendar, this event would occur toward the end of the first week (**Figure 1.9**).

When we consider how life might have arisen from nonliving matter, we must take into account the properties of the young Earth's atmosphere, oceans, and climate, all of which were very differ-

1.8 Fossils Give Us a View of Past Life The most prominent of the many fossilized organisms in this rock sample are ammonoids, an extinct mollusk group whose living relatives include squids and octopus. Ammonoids flourished between 200 million and 60 million years ago; this particular group of fossils is about 185 million years old.

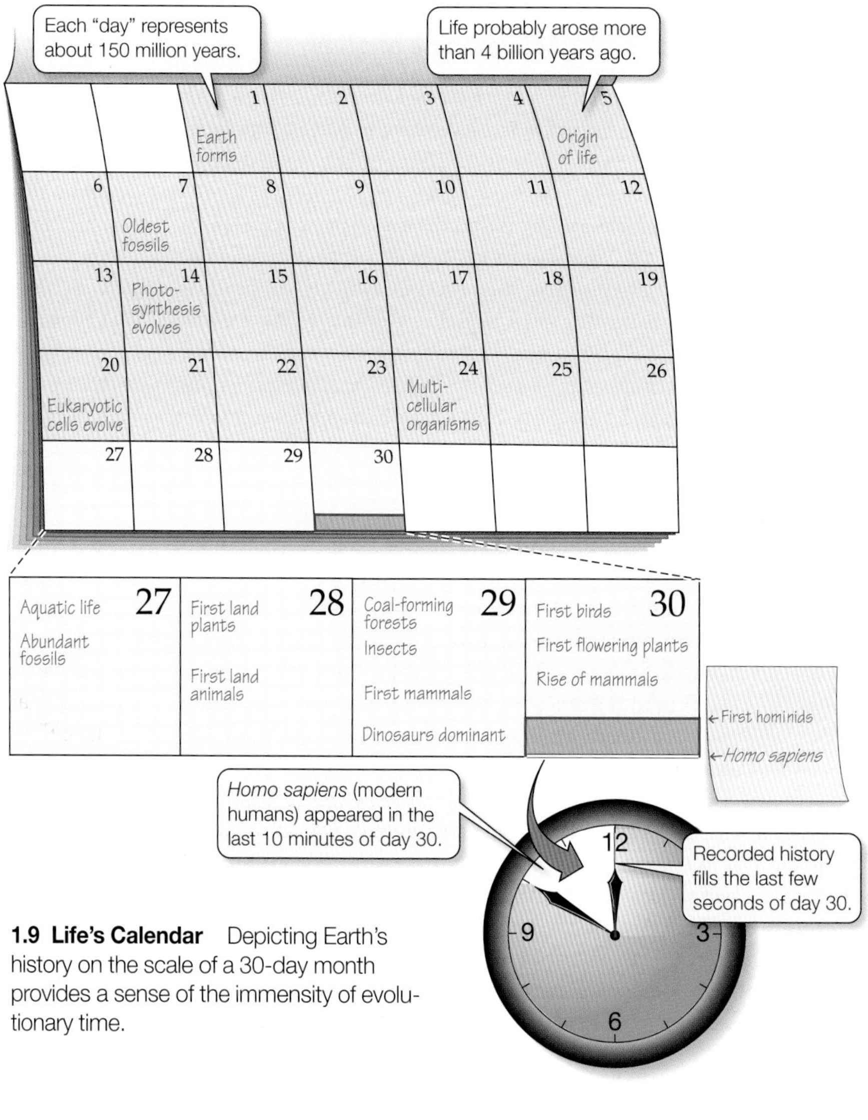

1.9 Life's Calendar Depicting Earth's history on the scale of a 30-day month provides a sense of the immensity of evolutionary time.

ent from today's. Biologists postulate that complex biological molecules first arose through the random physical association of chemicals in that environment. Experiments that simulate the conditions on early Earth have confirmed that the generation of complex molecules under such conditions is possible, even probable. The critical step, however, for the evolution of life had to be the appearance of molecules that could reproduce themselves and also serve as templates for the synthesis of large molecules with complex but stable shapes. The variation of the shapes of these large, stable molecules (described in Chapter 3) enabled them to participate in increasing numbers and kinds of interactions with other molecules: chemical reactions.

Biological evolution began when cells formed

The second critical step in the origin of life was the enclosure of complex biological molecules in *membranes*, which kept them close together and increased the frequency with which they interacted. Fat-like molecules were the critical ingredient: because these molecules are not soluble in water, they form membrane-like films. These films tend to form spherical *vesicles*, which could have enveloped assemblages of other biological molecules. Scientists postulate that about 3.8 billion years ago, this natural process of membrane formation resulted in the first cells with the ability to replicate themselves—an event that marked the beginning of biological evolution.

For 2 billion years after cells originated, all organisms consisted of only one cell. These first unicellular organisms were (and are, as multitudes of their descendants exist in similar form today) **prokaryotes**. Prokaryotic cell structure consists of DNA and other biochemicals enclosed in a membrane.

These early prokaryotes were confined to the oceans, where there was an abundance of complex molecules they could use as raw materials and sources of energy. The ocean shielded them from the damaging effects of ultraviolet light, which was intense at that time because there was no oxygen in the atmosphere, and hence no protective ozone layer.

Photosynthesis changed the course of evolution

The sum total of all the chemical reactions that go on inside a cell constitutes the cell's **metabolism**. To fuel their metabolism, the earliest prokaryotes took in molecules directly from their environment, breaking these small molecules down to release the energy contained in their chemical bonds. Many modern species of prokaryotes still function this way, and very successfully.

An extremely important step that would change the nature of life on Earth occurred about 2.5 billion years ago with the evolution of **photosynthesis**. The chemical reactions of photosynthesis (which are explained in Chapter 8) transform the energy of sunlight into a form of energy that can power the synthesis of large biological molecules. These large molecules become the building blocks of cells; they can also be broken down to provide metabolic energy. Because its energy-capturing processes provide food for other organisms, photosynthesis is the basis of much of life on Earth today.

Early photosynthetic cells were probably similar to present-day prokaryotes called *cyanobacteria* (**Figure 1.10**). Over time, photosynthetic prokaryotes became so abundant that vast quantities of oxygen gas—O_2, which is a by-product of photosynthesis—began slowly to accumulate in the atmosphere. O_2 was poisonous to many of the prokaryotes that lived at that time. However, those organisms that tolerated O_2 were able to proliferate as the presence of oxygen opened up vast new avenues of evolution. Metabolism based on the use of O_2, called *aerobic metabolism*, is more efficient than the *anaerobic* (non-oxygen-using) *metabolism* that characterized earlier organisms. Aerobic metabolism allowed cells to grow larger, and today it is used by the majority of Earth's organisms.

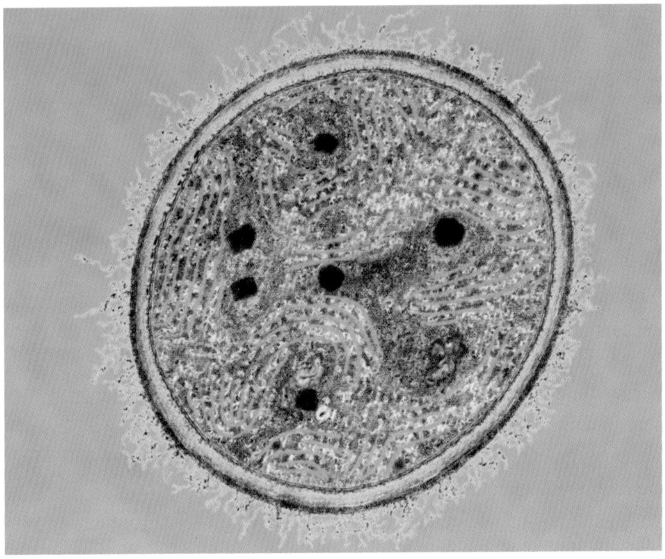

1.10 Photosynthetic Organisms Changed Earth's Atmosphere This modern cyanobacterium may be very similar to the early photosynthetic prokaryotes that introduced oxygen into Earth's atmosphere.

Over millions of years, the vast quantities of oxygen released by photosynthesis formed a layer of ozone (O_3) in the upper atmosphere. As the ozone layer thickened, it intercepted more and more of the sun's deadly ultraviolet radiation. Only in the last 800 million years has the presence of a dense ozone layer allowed organisms to leave the protection of the ocean and live on land.

Eukaryotic cells evolved from prokaryotes

Another important step in the history of life was the evolution of cells with discrete intracellular compartments, called **organelles**, that were capable of taking on specialized cellular functions. This event happened about 3 weeks into our calendar of Earth's history (see Figure 1.9). One of these organelles, the *nucleus*, came to contain the cell's genetic information. The nucleus has the appearance of a dense kernel, giving these cells their name: **eukaryotes** (from the Greek *eu*, "true," and *karyon*, "kernel"), as distinguished from the cells of prokaryotes, which lack internal compartments (*pro*, "before").

Some organelles are hypothesized to have originated when cells ingested smaller cells (see Figure 4.26). For example, the organelle specialized to conduct photosynthesis, the *chloroplast*, could have originated as a photosynthetic prokaryote that was ingested by a larger eukaryote. If the larger cell failed to break down this intended food object, a partnership could have evolved in which the ingested prokaryote provided the products of photosynthesis and the host cell provided a good environment for its smaller partner.

Multicellularity arose and cells became specialized

Until slightly more than 1 billion years ago, all the organisms that existed—whether prokaryotic or eukaryotic—were unicellular. Yet another important evolutionary step occurred when some eukaryotes failed to separate after cell division, remaining attached to each other. The permanent association of cells made it possible for some cells to specialize in certain functions, such as reproduction, while other cells specialized in other functions, such as absorbing nutrients and distributing them to neighboring cells. This **cellular specialization** enabled multicellular eukaryotes to increase in size and become more efficient at gathering resources and adapting to specific environments.

Biologists can trace the evolutionary Tree of Life

If all the species of organisms on Earth today are the descendants of a single kind of unicellular organism that lived almost 4 billion years ago, how have they become so different? And why are there so many species?

As long as individuals within a population mate at random, structural and functional changes may evolve within that population, but the population will remain one species. However, if some event isolates some members of a population from the others, structural and functional differences between them may accumulate over time. In short, the evolutionary paths of the two groups may diverge to the point where their members can no longer reproduce with each other. They have evolved into different species. This evolutionary process, called *speciation*, is detailed in Chapters 22 and 23.

Biologists give each species a distinct scientific name formed from two Latinized names (a *binomial*). The first name identifies the species' *genus*—a group of species that share a recent common ancestor. The second is the name of the species. For example, the scientific name of the human species is *Homo sapiens*: *Homo* is our genus and *sapiens* is our species. Scientific names usually refer to some characteristic of the species. *Homo* is derived from the Latin word for "man," and *sapiens* is derived from the Latin word for "wise" or "rational."

As many as 30 million species of organisms may exist on Earth today. Many times that number lived in the past but are now extinct. Many millions of speciation events created this vast diversity, and the unfolding of these events can be diagrammed as an evolutionary "tree" showing the order in which populations split and eventually evolved into new species. An evolutionary tree traces the descendants of ancestors that lived at different times in the past. The organisms on any one branch share a common ancestor at the base of that branch. The most closely related groups are placed together on the same branch; more distantly related organisms are on different branches. In this book, we adopt the convention that time flows from left to right, so the tree in Figure 1.11 (and other trees in this book) lies on its side, with its root—the ancestor of all life—at the left. Although many details remain to be clarified, the broad outlines of the Tree of Life have been determined. Its branching patterns are based on a rich array of evidence from fossils, structures, metabolic processes, behavior, and molecular analyses of genomes.

No fossils exist to help us determine the earliest divisions in the lineage of life because those unicellular organisms had no parts that could be preserved as fossils. However, molecular evidence has been used to separate all living organisms into three major **domains**: Archaea, Bacteria, and Eukarya (**Figure 1.11**). The organisms of each domain have been evolving separately from organisms in the other domains for more than a billion years.

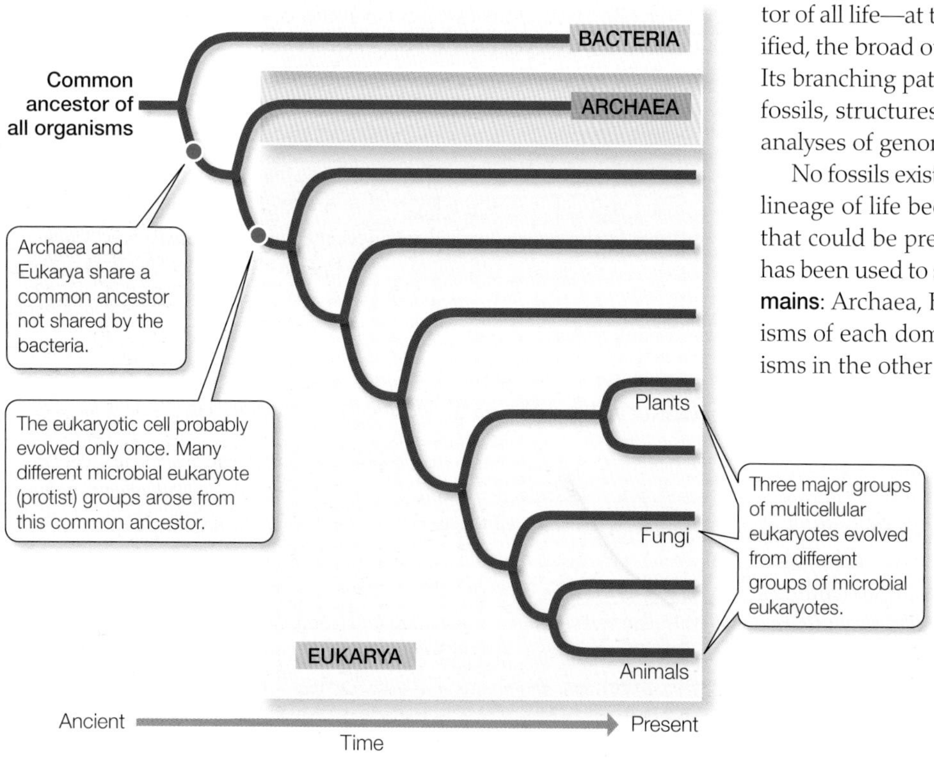

1.11 The Tree of Life The classification system used in this book divides Earth's organisms into three domains: Bacteria, Archaea, and Eukarya. The unlabeled blue branches within the Eukarya represent various groups of microbial eukaryotes, more commonly known as "protists." (See Appendix A for a more detailed version of the tree.)

Organisms in the domains **Archaea** and **Bacteria** are all prokaryotes. Archaea and Bacteria differ so fundamentally from each other in their metabolic processes that they are believed to have separated into distinct evolutionary lineages very early.

Members of the third domain—**Eukarya**—have eukaryotic cells. Three major groups of multicellular eukaryotes—plants, fungi, and animals—all evolved from unicellular *microbial eukaryotes*, more generally referred to as *protists*. The photosynthetic protist that gave rise to plants was completely distinct from the protist that was ancestral to both animals and fungi, as can be seen from the branching pattern of Figure 1.11.

Some bacteria, some archaea, some protists, and most plants are capable of photosynthesis. These organisms are called *autotrophs* ("self-feeders"). The biological molecules they produce are the primary food for nearly all other living organisms.

Fungi include molds, mushrooms, yeasts, and other similar organisms, all of which are *heterotrophs* ("other-feeders")—that is, they require a source of molecules synthesized by other organisms, which they then break down to obtain energy for their own metabolic processes. Fungi break down energy-rich food molecules in their environment and then absorb the breakdown products into their cells. Some fungi are important as decomposers of the waste products and dead bodies of other organisms.

Like fungi, animals are heterotrophs, but unlike fungi they ingest their food source, then break down the food in a digestive tract. Animals eat other forms of life, including plants, fungi, and other animals. Their cells absorb the breakdown products and obtain energy from them.

1.2 RECAP

The first cellular life on Earth was prokaryotic and arose about 4 billion years ago. The complexity of the organisms that exist today is the result of several important evolutionary events, including the evolution of photosynthesis, eukaryotic cells, and multicellularity. The genetic relationships of all organisms can be shown as a branching Tree of Life.

- Can you explain the evolutionary significance of photosynthesis? See p. 11
- What do the domains of life represent? What are the major groups of eukaryotes? See p. 12 and Figure 1.11

In February of 1676, Robert Hooke received a letter from the physicist Sir Isaac Newton. In this letter Newton famously remarked to Hooke, "If I have seen a little further, it is by standing on the shoulders of giants." We all stand on the shoulders of giants, building on the research of earlier scientists. By the end of this course, you will know more about evolution than Darwin ever could have, and you will know infinitely more about cells than Schleiden and Schwann did. Let's look at the methods biologists use to expand our knowledge of life.

1.3 How Do Biologists Investigate Life?

Biologists use many tools and methods in their research, but regardless of the methods they use, biologists take two basic approaches to their investigations of life: they observe and they conduct experiments.

Observation is an important skill

Biologists have always observed the world around them, but today their abilities to observe are greatly enhanced by many sophisticated technologies, such as electron microscopes, DNA chips, magnetic resonance imaging, and global positioning satellites. Advances in technology have been responsible for most major advances in biology. For example, not too long ago it was extremely difficult and time-consuming to decipher the nucleotide sequence that makes up a single gene. New technologies enabled biologists to sequence the entire human genome in only 13 years (1990–2003). Scientists now use these methods routinely, sequencing the genomes of organisms (including organisms that cause serious diseases) in only days. We will explore some of these technologies and what we have learned from them in Part Four of this book.

Our ability to observe the distributions of organisms, such as fish in the world's oceans, has also improved dramatically. A short time ago, researchers could put physical tags on fish and then only hope that someday a fisherman would catch one and send back the tag, which would at least reveal where the fish ended up. Today, electronic recording devices attached to fish can continuously record not only where the fish is, but also how deep it swims at different times of day and the temperature and salinity of the water around it (**Figure 1.12**). At set intervals, these tags download their information to a satellite, which relays it back to researchers. Suddenly we are acquiring a great deal of new knowledge about the distribution of life in the oceans.

The scientific method combines observation and logic

Observations lead to questions, and scientists make additional observations and do experiments to answer those questions. The conceptual approach that underlies the design and conduct of most modern scientific investigations is called the **scientific method**. This

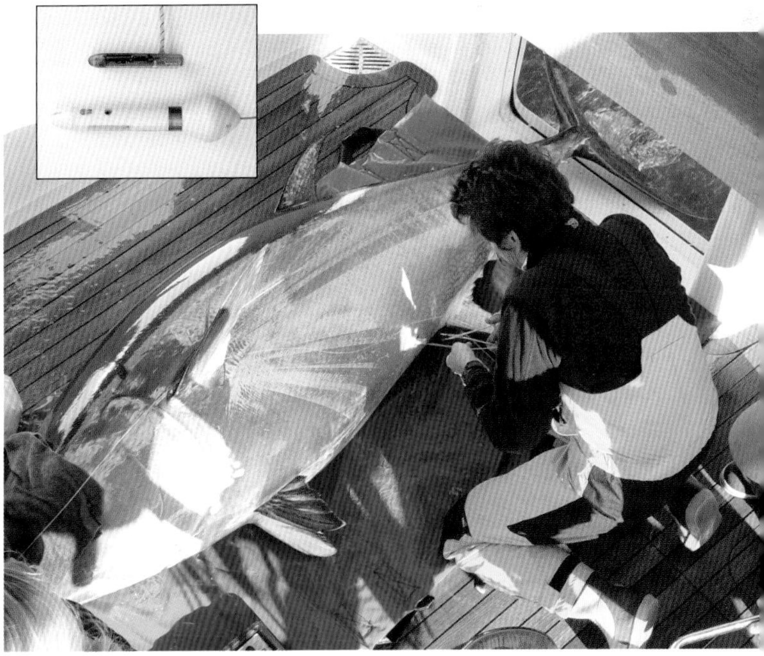

1.12 Tuna Tracking Marine biologist Barbara Block attaches a computerized data recording tag (inset) to a bluefin tuna. The use of such tags makes it possible to track individual tuna wherever they travel in the world's oceans.

powerful tool, also called the *hypothesis–prediction (H–P) method,* is a conceptual approach that provides a strong foundation for making advances in biological knowledge. The scientific method has five steps:

- Making *observations*
- Asking *questions*
- Forming *hypotheses,* or tentative answers to the questions
- Making *predictions* based on the hypotheses
- *Testing* the predictions by making additional observations or conducting experiments

Once a question has been posed, a scientist uses *inductive logic* to propose a tentative answer to the question. That tentative answer is called a **hypothesis**. For example, at the opening of this chapter, you learned that Pieter Johnson was shown abnormal frogs gathered in certain ponds. The first question stimulated by this observation was, is there something in these ponds that caused frogs to develop such extreme anatomical abnormalities?

In formulating a hypothesis, scientists put together the facts they already know to formulate one or more possible answers to the question. Pieter knew that there were likely to be contaminants in the ponds where deformed frogs were found because agricultural pesticides were used heavily in the region. In addition, mercury had once been mined nearby, and the abandoned mines could be a source of heavy metals in the water. He also knew that there were nearby ponds in which the frogs were normal. His first hypothesis, therefore, was that contaminants in the water caused mutations in the frog eggs.

The next step in the scientific method is to apply a different form of logic—*deductive logic*—to make predictions based on the hypothesis. Based on his hypothesis, Pieter predicted (1) that he would find contaminants in the ponds with the abnormal frogs, and (2) that eggs from those ponds would produce abnormal frogs when they were hatched in the laboratory.

EXPERIMENT

HYPOTHESIS: Something in the environment is causing developmental limb abnormalities in Pacific tree frogs (*Hyla regilla*).

METHOD

1. Identify a test area of small ponds in an area where abnormal tree frogs have been found (agricultural land in Santa Clara County, California).
2. Collect and analyze water samples from the ponds.
3. Census the organisms in the ponds.
4. Look for correlations between the presence of frog abnormalities and the characteristics of the ponds.

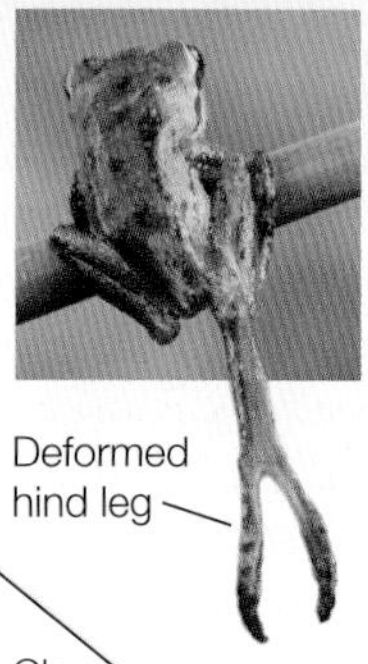

RESULTS

Pacific tree frogs were found in 13 of 35 ponds. Frogs with limb abnormalities were found in 4 of these 13 ponds. Water and census analyses of the 13 ponds containing frogs revealed no difference in water pollution, but did reveal the presence of snails infested with parasitic flatworms of the genus ***Ribeiroia*** in the 4 ponds with abnormal frogs.

	Pesticide residues in water?	Heavy metals in water?	Industrial chemicals in water?	Snails in water?	*Ribeiroia* in water?	*Ribeiroia* larvae in frogs?
Ponds with normal frogs	No	No	No	No	No	No
Ponds with abnormal frogs	No	No	No	**Yes**	**Yes**	**Yes**

CONCLUSION: Infection by parasitic *Ribeiroia* may cause abnormalities in the limb development of Pacific tree frogs.

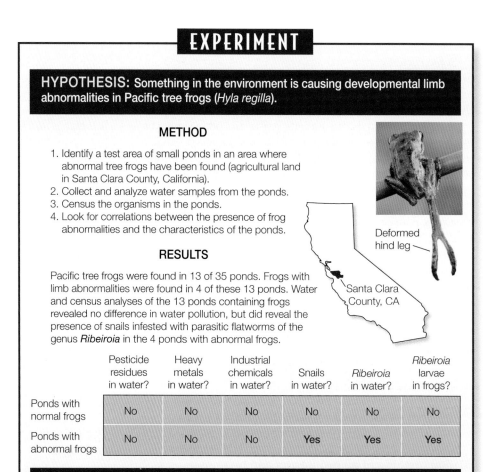

1.13 Comparative Experiments Look for Differences between Groups Pieter Johnson analyzed the differences between ponds in which deformed frogs were present versus nearby ponds in which there were no deformed frogs. Such comparisons can result in valuable insights.

Good experiments have the potential of falsifying hypotheses

Once predictions are made from a hypothesis, **experiments** can be designed to test those predictions. The most informative experiments are those that have the ability to show that the prediction is wrong. If the prediction is wrong, the hypothesis must be questioned, modified, or rejected.

Both of Pieter Johnson's initial predictions proved to be wrong. He counted frogs and other organisms in 35 ponds in the region where the deformed frogs had been found and measured chemicals in the water. Thirteen of the ponds were home to Pacific tree frogs, but he found deformed frogs in only four ponds. To Pieter's surprise, analysis of the water samples failed to reveal higher amounts of pesticides, industrial chemicals, or heavy metals in the ponds with deformed frogs. Also surprisingly, when he collected eggs from those ponds and hatched them in the laboratory, he always got normal frogs. The original hypothesis that contaminants caused mutations in the frog eggs had to be rejected. A new hypothesis had to be formulated and new experiments had to be conducted.

There are two general types of experiments and Pieter used both:

- In a **comparative experiment,** we predict that there will be a difference between samples or groups based on our hypothesis. We then test whether or not the predicted difference exists.
- In a **controlled experiment,** we also compare samples or groups, but in this case we start the experiment with groups that are as similar as possible. We predict on the basis of our hypothesis that some factor, or *variable,* plays a role in the phenomenon we are investigating. We then use some method to manipulate that variable in an "experimental" group while leaving the "control" group unaltered. We then

test to see if the manipulation created the predicted difference between the experimental and control groups.

COMPARATIVE EXPERIMENTS Comparative experiments are valuable when we do not know or cannot control the critical variables. Pieter Johnson performed a comparative experiment when he tested the water in the ponds (**Figure 1.13**). His challenge was to find some variable that differed between the ponds with normal and abnormal frogs. Finding no differences in the water chemistry of the two types of ponds, he had to reject his hypothesis that environmental contaminants were causing mutations in the frogs. So he compared the two types of ponds to see what variables *were* different between them.

Pieter found that a species of freshwater snail was present in the ponds with abnormal frogs, but absent from the ponds with normal ones. Freshwater snails are hosts for many parasites. His new hypothesis was that a parasite infecting the snail was in some way responsible for the frogs' deformities. To test that hypothesis, he performed controlled experiments.

CONTROLLED EXPERIMENTS In controlled experiments, one variable is manipulated while others are held constant. The variable that is manipulated is called the *independent variable* and the response that is measured is the *dependent variable*. A good controlled experiment is not easy to design because biological variables are so interrelated that it is difficult to alter just one.

Many parasites go through complex life cycles with several stages, each of which requires a specific host animal. Pieter focused on the possibility that some parasite that used freshwater snails as one of its hosts was infecting the frogs and causing their deformities. Pieter found a candidate parasite with this type of life cycle: a small flatworm called *Ribeiroia*, which was present in the ponds where the deformed frogs were found.

In Pieter's controlled experiment, the independent variable was the presence or absence of the snail and the parasite (**Figure 1.14**). He controlled all other variables by collecting frog eggs from ponds where there were no snails or flatworm parasites and hatching them in the laboratory. He divided the resulting tadpoles into two groups and placed them in separate tanks. He introduced snails and parasites into half of the tanks (the experimental group) and left the other tanks (the control group) free of snails and parasites. His dependent variable was the frequency of abnormalities in the frogs that developed under the different sets of conditions. He found that 85 percent of the frogs in the experimental tanks with *Ribeiroia* alone, but none of the frogs in the other tanks, developed abnormalities. Thus Pieter's results supported his hypothesis, and he could go on to investigate how the parasites caused abnormalities in the developing frogs.

Ribeiroia uses three hosts in California ponds: snails, frogs, and predatory birds such as herons. For the parasite to complete its life cycle and reproduce, it must be able to move from a frog to a bird. The limb deformities *Ribeiroia* causes may actually make infected frogs easier for predatory birds to capture and eat.

EXPERIMENT

HYPOTHESIS: Infection of Pacific tree frog tadpoles by the parasite *Ribeiroia* causes developmental limb abnormalities.

METHOD

1. Collect *Hyla regilla* eggs from a site with no record of abnormal frogs.
2. Allow eggs to hatch in laboratory aquaria. Randomly divide equal numbers of the resulting tadpoles into control and experimental groups.
3. Allow the control group to develop normally. Subject the experimental groups to infection with *Ribeiroia*, a different parasite (*Alaria*), and a combination of both parasites.
4. Follow tadpole development. Count and assess the resulting adult frogs.

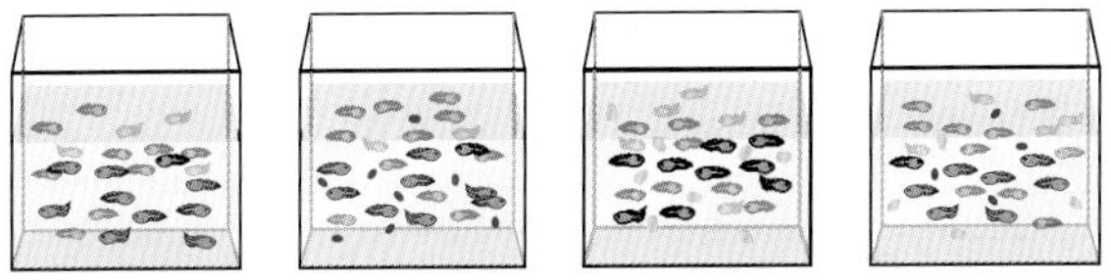

RESULTS

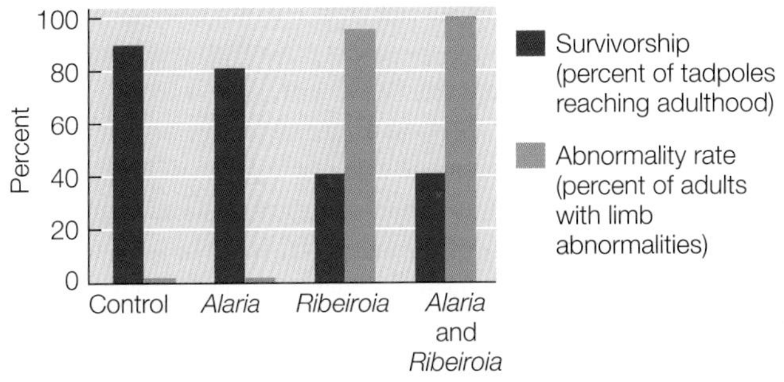

CONCLUSION: *Ribeiroia* causes developmental limb abnormalities in Pacific tree frogs.

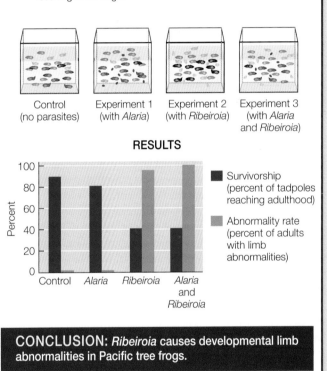

1.14 Controlled Experiments Manipulate a Variable The variable Johnson manipulated was the presence or absence of two species of parasitic flatworm. Other conditions of the experiment remained constant.

Statistical methods are essential scientific tools

Whether we are doing comparative or controlled experiments, at the end we have to decide whether there is a difference between the samples, individuals, groups, or populations in the study. How do we decide whether a measured difference is enough to support or falsify a hypothesis? In other words, how do we decide in an unbiased, objective way that the measured difference is significant?

Significance can be measured with statistical methods. Scientists use statistics because they recognize that variation is ubiquitous. Statistical tests analyze that variation and calculate the probability that the differences observed could be due to random variation. The results of statistical tests are therefore probabilities. A statistical test starts with a **null hypothesis**—the premise that no difference exists. When quantified observations, or **data**, are collected, statistical methods are applied to those data to calculate the likelihood that the null hypothesis is correct.

More specifically, statistical methods tell us the probability of obtaining the same results by chance even if the null hypothesis were true. Put another way, we need to eliminate insofar as possible the chance that any differences showing up in the data are merely the result of random variation in the samples tested. Scientists generally conclude that the differences they measure are significant if the statistical tests show that the *probability of error* (the probability that the results can be explained by chance) is 5 percent or lower. In particularly critical experiments, such as tests of the safety of a new drug, scientists require much lower probabilities of error, such as 1 percent or even 0.1 percent.

Not all forms of inquiry are scientific

Science is a unique human endeavor that is bounded by certain standards of practice. Other areas of scholarship share with science the practice of making observations and asking questions, but scientists are distinguished by what they do with their observations and how they answer their questions. Data, subjected to appropriate statistical analysis, are critical in the testing of hypotheses. The scientific method is the most powerful way humans have devised for learning about the world and how it works. Scientific explanations for natural processes are objective and reliable because the hypotheses proposed *must be testable* and *must have the potential of being rejected* by direct observations and experiments. Scientists clearly describe the methods they have used to test hypotheses so that other scientists can repeat their observations or experiments. Not all experiments are repeated, but surprising or controversial results are always subjected to independent verification. All scientists worldwide share this built-in process of testing and rejecting hypotheses, so they all contribute to a common body of scientific knowledge.

If you understand the methods of science, you can distinguish science from non-science. Art, music, and literature are activities that contribute to the quality of human life, but they are not science. They do not use the scientific method to establish what is fact. Religion is not science, although religions have historically purported to explain natural events ranging from unusual weather patterns to crop failures to human diseases and mental afflictions. Many such phenomena that at one time were mysterious are now explicable in terms of scientific principles.

The power of science derives from the uncompromising objectivity and absolute dependence on evidence that comes from *reproducible and quantifiable observations*. A religious or spiritual explanation of a natural phenomenon may be coherent and satisfying for the person or group holding that view, but it is not testable, and therefore it is not science. To invoke a supernatural explanation (such as an "intelligent designer" with no known bounds) is to depart from the world of science.

Science describes the facts about how the world works, not how it "ought to be." Many of the recent scientific advances that have contributed so much to human welfare also raise major ethical issues. Developments in genetics and developmental biology, for example, now enable us to select the sex of our children, to use stem cells to repair our bodies, and to modify the human genome. Although scientific knowledge allows us to do these things, science cannot tell us whether or not we should do them, or if we choose to do so, how we should regulate them.

Making wise decisions about such issues requires a clear understanding of the implications of available scientific information. Success in surgery depends on an accurate diagnosis. So does success in environmental management. However, to make wise decisions about public policy, we also need to employ the best possible ethical reasoning in deciding which outcomes we should strive for. For a bright future, society needs both good science and good ethics, as well as an educated public that understands the importance of both and the critical differences between them.

1.3 RECAP

The scientific method of inquiry starts with the formulation of hypotheses based on observations and data. Comparative and controlled experiments are carried out to test hypotheses.

- Can you explain the relationship between a hypothesis and an experiment? See p. 14
- What features characterize questions that can be answered only by using a comparative approach? See p. 14 and Figure 1.13
- What is controlled in a controlled experiment? See p. 15 and Figure 1.14
- Do you understand why arguments must be supported by quantifiable and reproducible data in order to be considered scientific? See p. 16

The vast amount of scientific knowledge accumulated over centuries of human civilization allows us to understand and manipulate aspects of the natural world in ways that no other species can. These abilities present us with challenges, opportunities, and, above all, responsibilities. Let's look at how knowledge of biology can affect the formulation of public policy.

1.4 How Does Biology Influence Public Policy?

The study of biology has long had major implications for human life. Agriculture and medicine are two important human activities that depend on biological knowledge. Our ancestors unknowingly applied the principles of evolutionary biology when they domesticated plants and animals. People have also been speculating about the causes of diseases and searching for methods of combating them since ancient times. Long before the causes of diseases were known, people recognized that diseases could be passed from one person to another. Isolation of infected persons has been practiced as long as written records have been available, but most so-called cures were not effective until scientists found out what caused diseases.

Today, thanks to the deciphering of genomes and the ability to manipulate them, vast new possibilities exist for improvements in the control of human diseases and agricultural productivity. At the same time, these capabilities have raised important ethical and policy issues. How much and in what ways should we tinker with the genetics of humans and other species? Does it matter whether our crops and domesticated animals are changed by traditional breeding experiments or by gene transfers? What rules should govern the release of genetically modified organisms into the environment? Science alone cannot provide answers to those questions, but wise policy decisions must be based on accurate scientific information.

Another reason for studying biology is to understand the effects of the vastly increased human population on its environment. Our use of natural resources is putting stress on the ability of Earth's ecosystems to continue to produce the goods and services on which our society depends. Human activities are changing global climates, causing the extinctions of a large number of species, and spreading new diseases while facilitating the resurgence of old ones. The rapid spread of the SARS and West Nile viruses, for example, was facilitated by modern modes of transportation, and the recent resurgence of tuberculosis is the result of the evolution of bacteria that are resistant to antibiotics. Biological knowledge is vital for determining the causes of these changes and for devising wise policies to deal with them. An understanding of biology also helps people appreciate the marvelous diversity of living organisms that provides goods and services for humankind and also enriches our lives aesthetically and spiritually.

Biologists are increasingly called on to advise government agencies concerning the laws, rules, and regulations by which society deals with the increasing number of problems and challenges that have at least a partial biological basis. As an example of the value of scientific knowledge for the assessment and formulation of public policy, let's return to the tracking study of bluefin tuna introduced in Section 1.3. Prior to this study, both scientists and fishermen knew that bluefins had a western Atlantic breeding ground in the Gulf of Mexico and an eastern Atlantic breeding ground in the Mediterranean Sea. Overfishing was endangering the western breeding population. Everyone assumed that the fish from the two breeding populations had geographically separate feeding grounds as well as separate breeding grounds, so an international commission drew a line down the middle of the Atlantic Ocean and established stricter fishing quotas on the western side of the line. The intent was to allow the western population to recover. However, new data revealed that in fact the eastern and western bluefin populations mix freely on the feeding (and hence fishing) grounds across the entire North Atlantic (**Figure 1.15**). Thus a fish caught on the eastern side of the line could be from the western breeding population, so the established policy was not appropriate for achieving its intended goal.

Throughout this book we will share with you the excitement of studying living things and illustrate the rich array of methods that biologists use to determine why the world of living things looks and functions as it does. The most important motivator of most biologists is curiosity. People are fascinated by the richness and diversity of life and want to learn more about organisms and how they interact with one another. The trait of human curiosity

1.15 Bluefin Tuna Do Not Recognize the Lines Drawn on Maps by International Commissions Because it was assumed that western (red dots) and eastern (gold dots) breeding populations of bluefin tuna also fed on their respective sides of the Atlantic Ocean, separate fishing quotas were established to either side of 45°W longitude (dashed line). It was believed this would allow the endangered western population to recover. However, tracking data showed that the two populations mix freely, especially in the heavily fished waters of the northernmost Atlantic (blue circle); so in fact the established policy does not protect the western population.

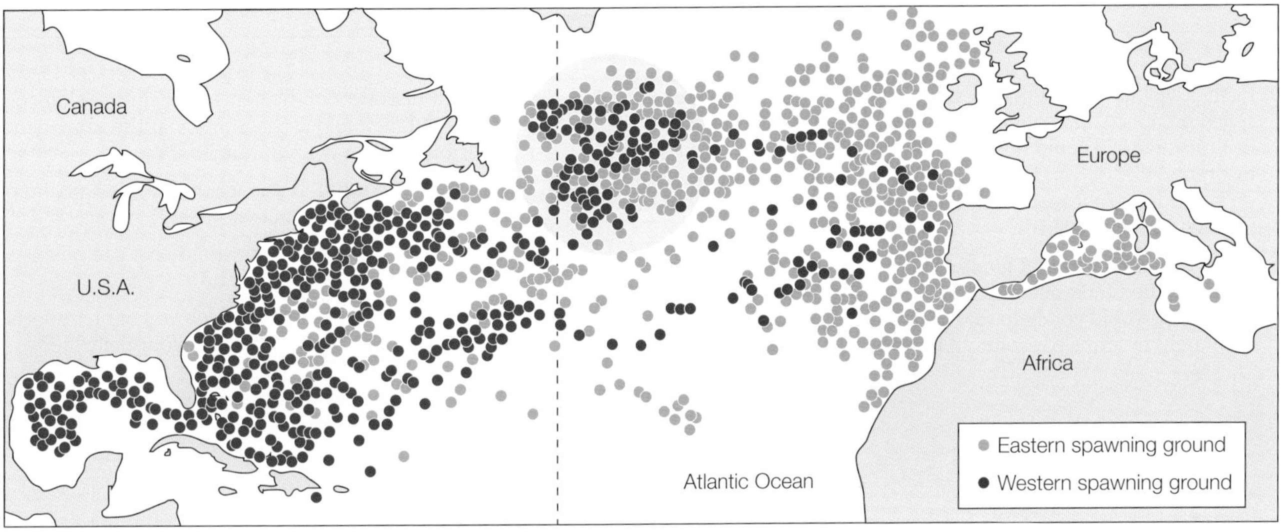

might even be seen as adaptive, and could have been selected for if individuals who were motivated to learn about their surroundings were likely to have survived and reproduced better, on average, than their less curious relatives!

There are vast numbers of questions for which we do not yet have answers, and new discoveries usually engender questions no one thought to ask before. Perhaps you will eventually pose and answer one or more of those questions.

CHAPTER SUMMARY

1.1 What is biology?

Biology is the study of life at all levels of organization, ranging from molecules to the biosphere.

The **cell theory** states that all life consists of cells, and all cells come from preexisting cells.

All living organisms are related to one another through descent with modification. **Evolution** by **natural selection** is responsible for the diversity of **adaptations** found in living organisms.

The instructions for a cell are contained in its **genome**, which consists of DNA molecules made up of sequences of **nucleotides**. Specific segments of DNA called **genes** contain the information the cell uses to make **proteins**. Review Figure 1.4

Cells are the basic structural and physiological units of life. Most of the chemical reactions of life take place in cells. Living organisms control their internal environment. They also interact with other organisms of the same and different species. Biologists study life at all these levels of organization. Review Figure 1.6, Web/CD Activity 1.1

Biological knowledge obtained from a **model system** may be generalized to other species.

1.2 How is all life on Earth related?

Biologists use fossils, anatomical similarities and differences, and molecular comparisons of genomes to reconstruct the history of life. Review Figure 1.9

Life first arose by chemical evolution. Biological evolution began with the formation of cells.

Photosynthesis was an important evolutionary step because it changed Earth's atmosphere and provided a means of capturing energy from sunlight.

The earliest organisms were **prokaryotes**; organisms with more complex cells, called **eukaryotes**, arose later. Eukaryotic cells have discrete intracellular compartments, called **organelles**, including a **nucleus** that contains the cell's genetic material.

The genetic relationships of **species** can be represented as an evolutionary tree. Species are grouped into three **domains**: **Archaea**, **Bacteria**, and **Eukarya**. The domains Archaea and Bacteria consist of unicellular prokaryotes. The domain Eukarya contains the microbial eukaryotes (protists), plants, fungi, and animals. Review Figure 1.11, Web/CD Activity 1.2

1.3 How do biologists investigate life?

The **scientific method** used in most biological investigations involves five steps: making observations, asking questions, forming hypotheses, making predictions, and testing those predictions.

Hypotheses are tentative answers to questions. Predictions made on the basis of a hypothesis are tested with additional observations and two kinds of **experiments**: **comparative** and **controlled experiments**. Review Figures 1.13 and 1.14

Statistical methods are applied to **data** to establish whether or not the differences observed are significant or whether they could be expected by chance. These methods start with the **null hypothesis** that there are no differences.

Science can tell us how the world works, but it cannot tell us what we should or should not do.

1.4 How does biology influence public policy?

Wise public policy decisions must be based on accurate scientific information. Biologists are often called on to advise governmental agencies on the solution of important problems that have a biological component.

FOR DISCUSSION

1. Even if we knew the sequences of all of the genes of a single-celled organism and could cause those genes to be expressed in a test tube, we still could not create one of those organisms in the test tube. Why do you think this is so? In light of this fact, what do you think of the statement that the genome contains all of the information for a species?
2. If someone told you that giraffes developed long necks because they stretched their necks to reach leaves higher and higher on trees, how would you help that person think about giraffes more accurately, in terms of evolution by natural selection?
3. In a recent discovery of the genes that control skin color in zebrafish, why did the biologists assume that the same genes might be responsible for skin color in humans?
4. Why is it so important in science that we design and perform tests capable of falsifying a hypothesis?
5. What features characterize questions that can be answered only by using a comparative approach?
6. Can you think of any ways in which you apply aspects of the scientific method to solve problems in your daily life?

FOR INVESTIGATION

1. The abnormalities of frogs in Pieter Johnsons's study were associated with the presence of a parasite. How would you investigate how the parasites induced the formation of monster frogs? Hint: When tadpoles were exposed to the parasites *after* they began to develop legs, they did not show abnormalities.

2. Just as all cells come from preexisting cells, mitochondria—the cell organelles that convert energy in food to a form of energy that can do biological work—all come from preexisting mitochondria. Cells do not synthesize mitochondria from the genetic information in their nuclei. What investigations would you carry out to understand the nature of mitochondria?

PART FIVE

The Patterns and Processes of Evolution

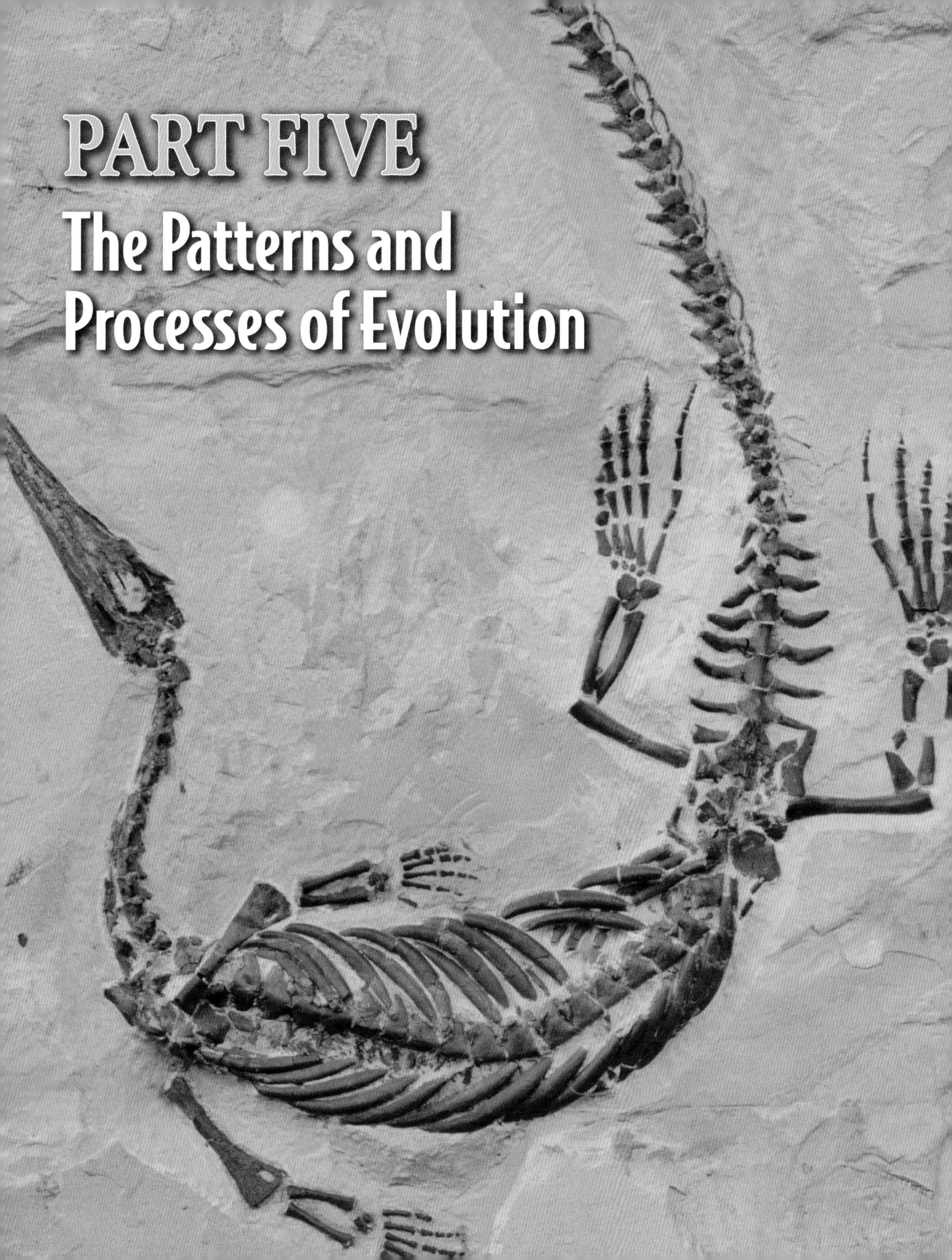

CHAPTER 21 The History of Life on Earth

Giant rodents

There are more species of rodents than any other group of mammals, and most of them are very small. House mice, for example, weigh about 30 grams, Norway rats weigh about 300 grams, and most squirrels weigh between 300 and 600 grams. South America, however, is home to a diverse group of rodents, including the familiar guinea pigs and chinchillas, that are, on average, significantly larger than rodents elsewhere. The largest living rodent, the capybara, weighs in at about 50 *kilograms* and lives in the marshes of South America.

Large rodents have been a distinctive feature of South America's evolutionary history. During the Miocene epoch (about 10 million years ago), the continent was home to a rodent the size of a buffalo. *Phoberomys pattersoni*, as it has been named, weighed about 700 kilograms—at least 10 times more than a capybara. The shapes of its fossilized teeth told paleontologists that *Phoberomys* was indeed a rodent, and also indicated that it fed on the grass of marshes and swamps, as the capybara does today. Paleontologists determined the weight of this giant rodent by applying a known, mathematically constant relationship between body mass and limb-bone diameter in living rodents to the fossilized bones.

How do biologists know *Phoberomys* lived 10 million years ago? The *Phoberomys* fossils were found in association with fossils of other mammals that no longer live today, in a specific layer of rock that was laid down a long time ago. But how long ago? The stratigraphic layering of different rocks allows us to tell their ages relative to each other, but does not indicate a given layer's absolute age.

One of the remarkable achievements of twentieth-century science was the development of sophisticated techniques that use the decay rates of various radioisotopes, changes in Earth's magnetic field, and the presence or absence of certain molecules to infer conditions and events in the remote past, and to date them accurately. It is those methods that provided the ages of the rocks in which the fossils of *Phoberomys* were found.

We are so accustomed to having time-measuring devices all around us that we forget how recently those devices were invented. When Galileo studied the motion of a ball rolling down an inclined plane about 400 years ago, he used his pulse to mark off equal intervals of time. The development of the science of biology is inti-

The World's Largest Rodents The South American capybara (*Hydrochaeris hydrochaeris*) is the largest extant rodent species. Adult capybaras, such as the one shown here with two young offspring, can weigh as much as an adult human.

Younger Rocks Lie on Top of Older Rocks In the Grand Canyon, the Colorado River has cut through and exposed many strata of ancient rocks. The oldest rocks visible here formed about 540 million years ago. The youngest rocks, at the top, are about 500 million years old. Fossilized remains of organisms that existed during the same evolutionary timeframe are found together in the same stratum.

mately linked to changing concepts of time, especially of the age of Earth. Biology as we know it could not and did not develop until about 150 years ago, when geologists first provided solid evidence that Earth was ancient. Before 1850, most people believed that Earth was no more than a few thousand years old. Charles Darwin could not have developed his theory of evolution by natural selection if he had not known that Earth was very old and that millions of years were available for life's evolution.

IN THIS CHAPTER we first describe how scientists assign dates to events in the distant evolutionary past. We then review the major changes in physical conditions on Earth during the past 4 billion years and look at how those changes affected life. We will describe the major patterns in the evolution of life and explain why rates of evolution change over time within and among groups of organisms. Finally we discuss the importance of evolutionary processes that are operating today.

CHAPTER OUTLINE

21.1 How Do Scientists Date Ancient Events?

Many evolutionary changes happen rapidly enough to be studied directly and manipulated experimentally. Plant and animal breeding by agriculturalists and the evolution of resistance to pesticides are good examples of rapid, short-term evolution. Other changes, such as the appearance of new species and evolutionary lineages, usually take place over much longer time frames.

To understand the long-term patterns of evolutionary change that we will trace throughout Parts Five and Six of this book, we must think in time frames spanning many millions of years, and we must imagine events and conditions very different from those we observe today. Earth of the distant past is, to us, a foreign planet inhabited by strange organisms. The continents were not where they are today, and climates were sometimes dramatically different from those of today.

As the opening of this chapter shows, **fossils**—the preserved remains of ancient organisms—can tell us a great deal about the body form, or *morphology*, of organisms that lived long ago, as well as how and where they lived. But to understand patterns of evolutionary change, we must also understand how life changed over time.

Earth's history is largely recorded in its rocks. We cannot tell the ages of rocks just by looking at them, but we can determine the ages of rocks relative to one another. The first person to formally recognize that this could be done was the seventeenth-century Danish physician Nicolaus Steno. Steno realized that in undisturbed *sedimentary* rock (rocks formed by the accumulation of grains on the bottoms of bodies of water), the oldest layers, or **strata** (singular *stratum*), lie at the bottom, and successively higher strata are progressively younger.

Geologists, particularly the eighteenth-century English scientist William Smith, subsequently combined Steno's insight with their observations of fossils contained within sedimentary rocks. They concluded that:

- Fossils of similar organisms were found in widely separated places on Earth.
- Certain organisms were always found in younger rocks than certain other organisms.

- Organisms found in higher, more recent strata were more similar to modern organisms than were those found in lower, more ancient strata.

These patterns revealed much about the relative ages of sedimentary rocks as well as patterns in the evolution of life. But the geologists still could not tell how old the rocks actually were. A method of dating rocks did not become available until after radioactivity was discovered at the beginning of the twentieth century.

Radioisotopes provide a way to date rocks

Radioactive isotopes of atoms (see Section 2.1) decay in a predictable pattern over long time periods. During each successive time interval, or **half-life**, half of the remaining radioactive material of the radioisotope decays, either changing into another element or becoming the stable isotope of the same element (**Figure 21.1**).

Each radioisotope has a characteristic half-life (**Table 21.1**). To use a radioisotope to date a past event, we must know or estimate the concentration of the isotope at the time of that event. In the case of carbon, the production of new ^{14}C in the upper atmosphere (by the reaction of neutrons with ^{14}N) just balances the natural radioactive decay of ^{14}C. Therefore, the ratio of ^{14}C to its stable isotope, ^{12}C, is relatively constant in living organisms and their environment. However, as soon as an organism dies, it ceases to exchange carbon compounds with its environment. Its decaying ^{14}C is no longer replenished, and the ratio of ^{14}C to ^{12}C in its remains decreases through time. *Paleontologists* (scientists who study fossils) can use the ratio of ^{14}C to ^{12}C in fossil organisms to date fossils that are less than 50,000 years old (and thus the sedimentary rocks that contain those fossils) with a fair degree of certainty. After that time so little ^{14}C remains that the limits of detection are reached.

TABLE 21.1

Half-Lives of Some Radioisotopes

RADIOISOTOPE	HALF-LIFE
Phosphorus-32 (^{32}P)	14.3 days
Tritium (^{3}H)	12.3 years
Carbon-14 (^{14}C)	5,700 years
Potassium-40 (^{40}K)	1.3 billion years
Uranium-238 (^{238}U)	4.5 billion years

TABLE 21.2

Earth's Geological History

RELATIVE TIME SPAN	ERA	PERIOD	ONSET	MAJOR PHYSICAL CHANGES ON EARTH
	Cenozoic	Quaternary	1.8 mya	Cold/dry climate; repeated glaciations
		Tertiary	65 mya	Continents near current positions; climate cools
	Mesozoic	Cretaceous	145 mya	Northern continents attached; Gondwana begins to drift apart; meteorite strikes Yucatán Peninsula
		Jurassic	200 mya	Two large continents form: Laurasia (north) and Gondwana (south); climate warm
		Triassic	251 mya	Pangaea begins to slowly drift apart; hot/humid climate
	Paleozoic	Permian	297 mya	Continents aggregate into Pangaea; large glaciers form; dry climates form in interior of Pangaea
		Carboniferous	359 mya	Climate cools; marked latitudinal climate gradients
		Devonian	416 mya	Continents collide at end of period; meteorite probably strikes Earth
		Silurian	444 mya	Sea levels rise; two large continents form; hot/humid climate
		Ordovician	488 mya	Massive glaciation, sea level drops 50 meters
		Cambrian	542 mya	O_2 levels approach current levels
Precambrian	Precambrian		600 mya	O_2 level at >5% of current level
			1.5 bya	O_2 level at >1% of current level
			3.8 bya	O_2 first appears in atmosphere
			4.5 bya	

Note: mya, million years ago; bya, billion years ago.

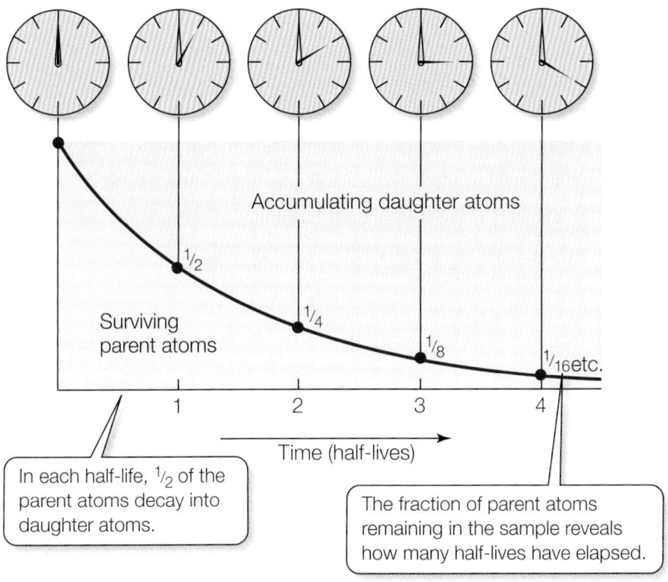

21.1 Radioactive Isotopes Allow Us to Date Ancient Rocks The decay of radioactive "parent" atoms into stable "daughter" isotopes happens at a steady rate known as the half-life. Different radioisotopes have different but characteristic half-lives that allow us to measure how much time has elapsed since the rocks containing such isotopes were laid down.

MAJOR EVENTS IN THE HISTORY OF LIFE
Humans evolve; many large mammals become extinct
Diversification of birds, mammals, flowering plants, and insects
Dinosaurs continue to diversify; flowering plants and mammals diversify; mass extinction at end of period (≈76% of species disappear)
Diverse dinosaurs; radiation of ray-finned fishes
Early dinosaurs; first mammals; marine invertebrates diversify; first flowering plants; mass extinction at end of period (≈65% of species disappear)
Reptiles diversify; amphibians decline; mass extinction at end of period (≈96% of species disappear)
Extensive "fern" forests; first reptiles; insects diversify
Fishes diversify; first insects and amphibians; mass extinction at end of period (≈75% of species disappear)
Jawless fishes diversify; first ray-finned fishes; plants and animals colonize land
Mass extinction at end of period (≈75% of species disappear)
Most animal phyla present; diverse photosynthetic protists
Ediacaran fauna
Eukaryotes evolve; several animal phyla appear
Origin of life; prokaryotes flourish

Radioisotope dating methods have been expanded and refined

Sedimentary rocks are formed from materials that existed for varying lengths of time before being transported, sometimes over long distances, to the site of their deposition. Therefore, the isotopes in a sedimentary rock do not contain reliable information about the date of its formation. Dating rocks more ancient than 50,000 years requires estimating isotope concentrations in *igneous* rocks (rocks formed when molten material cools). To date sedimentary rocks, geologists search for places where volcanic ash or lava flows have intruded into beds of those rocks.

A preliminary estimate of the age of the igneous rock determines which isotope is used to date it. The decay of potassium-40 to argon-40 has been used to date most of the ancient events in the evolution of life. Fossils in the adjacent sedimentary rock that are similar to those in other rocks of known ages provide additional clues.

Radioisotope dating of rocks, combined with fossil analysis, is the most powerful method of determining geological age. But in places where sedimentary rocks do not contain suitable igneous intrusions and few fossils are present, paleontologists turn to other dating methods. One method, known as *paleomagnetic dating*, relates the ages of rocks to patterns in Earth's magnetism, which change over time. Earth's magnetic poles move and occasionally reverse themselves. Because both sedimentary and igneous rocks preserve a record of Earth's magnetic field at the time they were formed, paleomagnetism helps determine the ages of those rocks. Other dating methods, which we will describe in later chapters, use continental drift, sea level changes, and molecular clocks.

Using these methods, geologists divided the history of life into geological *eras*, which in turn are subdivided into *periods* (**Table 21.2**). The boundaries between these divisions are based on the striking differences scientists observed in the assemblages of fossil organisms contained in successive layers of rocks; hence the suffix *zoic* ("of life") for the eras. Geologists established and named these divisions before they knew the ages of the eras and periods, and we are constantly refining the dates used for these boundaries as new discoveries are made.

21.1 RECAP

Fossils in sedimentary rocks enabled geologists to determine the relative ages of organisms, but absolute dating was not possible until the discovery of radioactivity. Geologists divide the history of life into eras and periods, based on assemblages of fossil organisms found in successive layers of rocks.

- What observations about fossils suggested to geologists that they could be used to determine the relative ages of rocks? See p. 465
- How is the rate of decay of radioisotopes used to estimate the absolute ages of rocks? See pp. 466–467 and Figure 21.1

The scale at the left of Table 21.2 gives a relative sense of geological time, especially the vast expanse of the Precambrian era, during which early life evolved amid stupendous physical changes. Earth continued to undergo physical changes that influenced the evolution of life, and these physical events and important milestones are listed in the table. Now we will describe the most important of these changes in more detail.

21.2 How Have Earth's Continents and Climates Changed over Time?

The maps and globes that adorn our walls, shelves, and books give an impression of a static Earth. It would be easy for us to assume that the continents have always been where they are. But we would be wrong. The idea that Earth's land masses have changed position over the millennia, and that they continue to do so, was first put forth in 1912 by the German meteorologist and geophysicist Alfred Wegener. His book, *The Origin of Continents and Oceans*, was initially met with skepticism and resistance. By the 1960s, however, physical evidence and increased understanding of the geophysics of *plate tectonics* had convinced virtually all geologists of the reality of Wegener's vision.

Earth's crust consists of a number of solid *plates* approximately 40 kilometers thick, which collectively make up the *lithosphere*. The lithospheric plates float on a fluid layer of molten rock, or *magma* (**Figure 21.2**). The magma circulates because heat produced by radioactive decay deep in Earth's core sets up convection currents in the fluid. The plates move because magma rises and exerts tremendous pressure. Where plates are pushed together, either they move sideways past each other, or one plate slides under the other, pushing up mountain ranges and carving deep *rift valleys*. (When they occur under water, such valleys are known as *trenches*.) Where plates are pushed apart, ocean basins may form between them. The movement of the lithospheric plates and the continents they contain is known as **continental drift**.

The idea of continental drift captured Wegener as he studied the close fit between the coastlines of western Africa and eastern South America. Geological evidence linking rock formations in Appalachia to similar formations in the Scottish highlands, and geological phenomena found in both South Africa and Brazil, further spurred his vision of continents that were once joined together.

We now know that at times, the drifting of the plates has brought continents together and that at other times, continents have drifted apart. The positions and sizes of the continents influence oceanic circulation patterns, sea levels, and global climates. Mass extinctions of species, particularly marine organisms, have usually accompanied major drops in sea level, which exposed vast areas of the continental shelves, killing the marine organisms that lived in the shallow seas that had covered them (**Figure 21.3**).

Oxygen has steadily increased in Earth's atmosphere

The continents have moved irregularly over Earth's surface, but some physical changes, such as the increase in atmospheric oxygen, have been largely unidirectional. The atmosphere of early Earth probably contained little or no free oxygen gas (O_2). The increase in atmospheric oxygen came in two big steps more than a billion years apart. The first step occurred about 2.4 billion years ago, when certain bacteria evolved the ability to use water as the source of hydrogen ions for photosynthesis. By chemically splitting H_2O , these bacteria generated atmospheric O_2 as a waste product. They also made electrons available for reducing CO_2 to form organic compounds (see Section 8.3).

One group of oxygen-generating bacteria, the *cyanobacteria*, formed rocklike structures called *stromatolites*, which are abun-

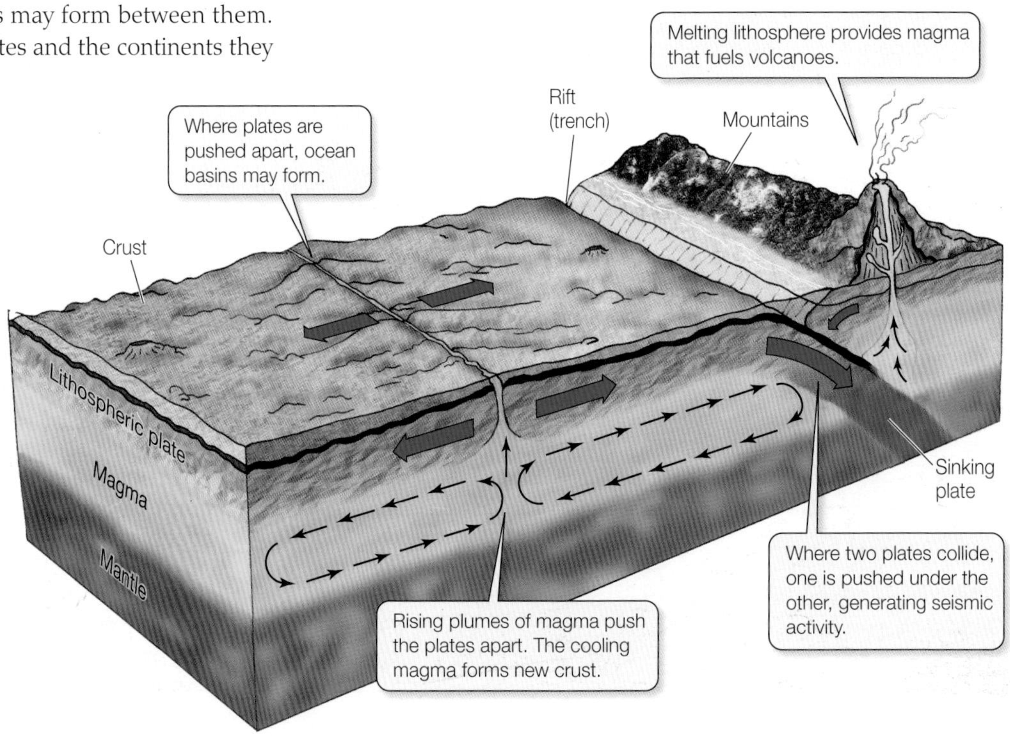

21.2 Plate Tectonics and Continental Drift The heat of Earth's core generates convection currents (arrows) that push the lithospheric plates, along with the land masses lying on them, together or apart. When lithospheric plates collide, one slides under the other. The resulting seismic activity can create mountains and deep rift valleys (oceanic trenches).

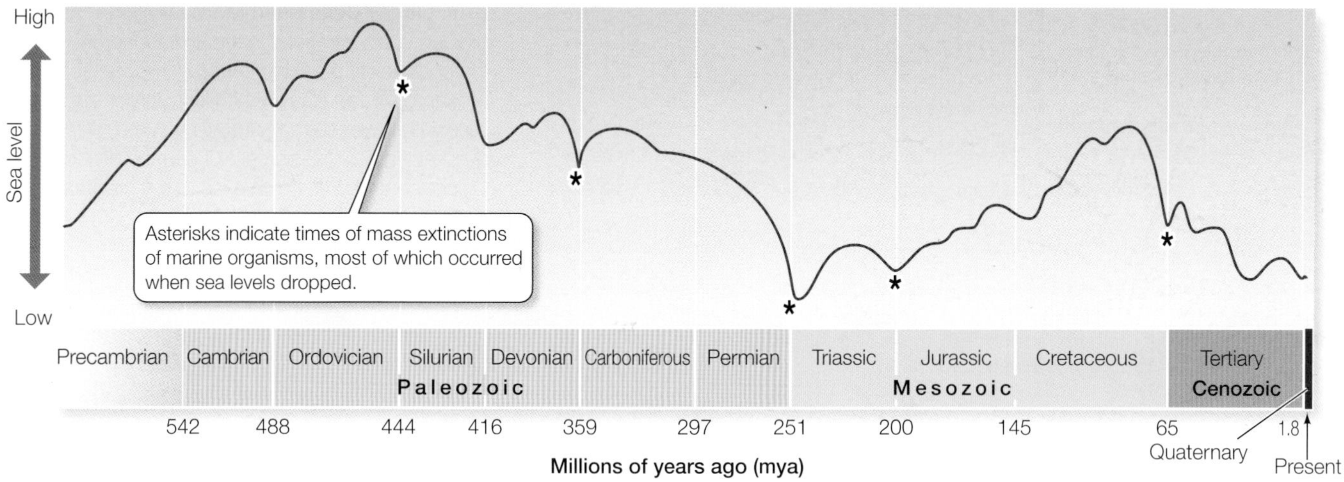

21.3 Sea Levels Have Changed Repeatedly Most mass extinctions (indicated by asterisks) have coincided with periods of low sea levels.

dantly preserved in the fossil record. Cyanobacteria are still forming stromatolites today in a few very salty places on Earth (**Figure 21.4**). Cyanobacteria liberated enough O_2 to open the way for the evolution of oxidation reactions as the energy source for the synthesis of ATP (see Section 7.1). Their ability to split water doubtless contributed to their extraordinary success.

Thus the evolution of life irrevocably changed the physical nature of Earth. Those physical changes, in turn, influenced the evolution of life. When it first appeared in the atmosphere, oxygen was poisonous to the anaerobic prokaryotes that inhabited Earth at the time. Those prokaryotes that evolved the ability to metabolize O_2 not only survived, but also gained a number of advantages. Aerobic metabolism proceeds at more rapid rates and harvests energy more efficiently than anaerobic metabolism (see Section 7.5). Consequently, organisms with aerobic metabolism replaced anaerobes in most of Earth's environments.

An atmosphere rich in O_2 also made possible larger cells and more complex organisms. Small unicellular aquatic organisms can obtain enough O_2 by simple diffusion even when O_2 concentrations are very low. Larger unicellular organisms have lower surface area-to-volume ratios (see Figure 4.2). In order to obtain enough O_2 by simple diffusion, they must live in an environment with a relatively high concentration of O_2. Bacteria can thrive on 1 percent of the current atmospheric O_2 levels; eukaryotic cells require oxygen levels that are at least 2 to 3 percent of current atmospheric concentrations. (For concentrations of dissolved O_2 in the oceans to reach these levels, much higher atmospheric concentrations were needed.)

Probably because it took many millions of years for Earth to develop an oxygenated atmosphere, only unicellular prokaryotes lived on Earth for more than 2 billion years.

(A)

(B)

21.4 Stromatolites (A) A vertical section through a fossil stromatolite. (B) These rocklike structures are living stromatolites that thrive in the very salty waters of Shark Bay, Western Australia. Layers of cyanobacteria are found in the uppermost parts of the structures.

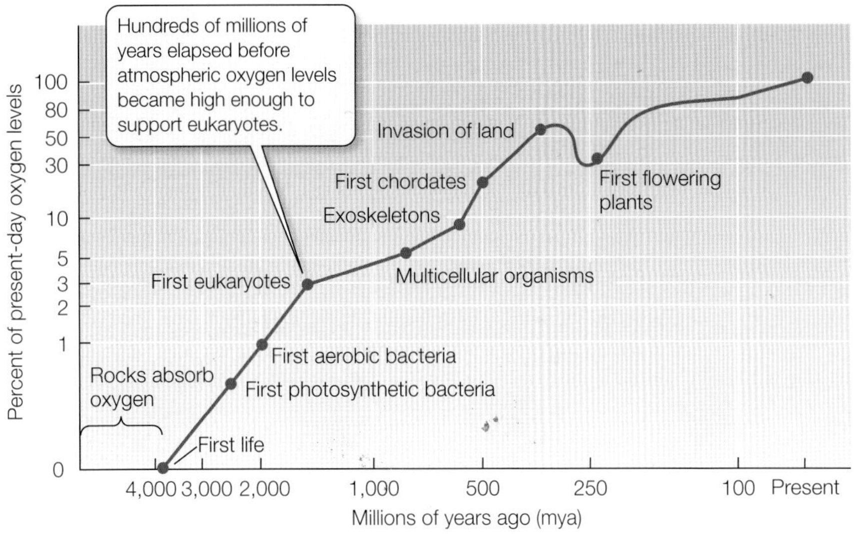

21.5 Larger Cells Need More Oxygen As oxygen concentrations in the atmosphere rose, the complexity of life increased. Although aerobic prokaryotes can flourish with less, larger eukaryotic cells with lower surface area-to-volume ratios require at least 2 to 3 percent of current atmospheric O_2 concentrations. (Both axes of the graph are logarithmic scales.)

About a billion years ago (1 bya), atmospheric O_2 concentrations became high enough for large eukaryotic cells to flourish (**Figure 21.5**). Further increases in atmospheric O_2 levels 700 to 570 million years ago (mya) enabled multicellular organisms to evolve.

In contrast to this largely unidirectional change in atmospheric O_2 concentration, most physical conditions on Earth have oscillated in response to the planet's internal processes, such as volcanic activity and continental drift. Extraterrestrial events, such as collisions with meteorites, have also left their mark. In some cases, as we will see later in this chapter, these events caused **mass extinctions**, during which a large proportion of the species living at the time disappeared. After each mass extinction, the diversity of life rebounded, but recovery took millions of years.

Earth's climate has shifted between hot/humid and cold/dry conditions

Through much of its history, Earth's climate was considerably warmer than it is today, and temperatures decreased more gradually toward the poles. At other times, however, Earth was colder than it is today. Large areas were covered with glaciers during the end of the Precambrian and during parts of the Carboniferous and Permian periods. These cold periods were separated by long periods of milder climates (**Figure 21.6**). For Earth to be in a cold, dry state, atmospheric CO_2 levels had to have been unusually low, but scientists do not know what caused such low concentrations. Because we are living in one of the colder periods in the history of Earth, it is difficult for us to imagine the mild climates that were found at high latitudes during much of the history of life. During the Quaternary period there have been a series of glacial advances, interspersed with warmer interglacial intervals during which the glaciers retreated.

Weather often changes rapidly; climates usually change slowly. Major climatic shifts have taken place over periods as short as 5,000 to 10,000 years, however, primarily as a result of changes in Earth's orbit around the sun. A few climatic shifts have been even more rapid. For example, during one Quaternary interglacial period, the Antarctic Ocean changed from being ice-covered to being nearly ice-free in less than a hundred years. Such rapid changes are usually caused by sudden shifts in ocean currents. Some climate changes have been so rapid that the extinctions caused by them appear "instantaneous" in the fossil record.

21.6 Hot/Humid and Cold/Dry Conditions Have Alternated over Earth's History Throughout Earth's history, periods of cold climates and glaciations (white depressions) have been separated by long periods of milder climates.

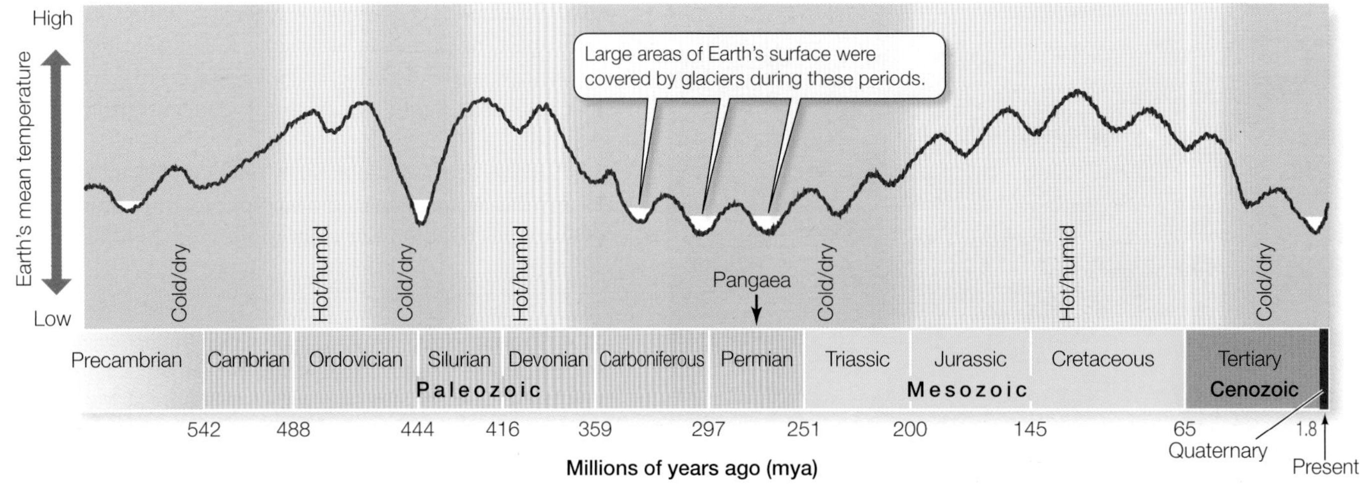

We are living today in a time of rapid climate change caused by a buildup of atmospheric CO_2, primarily from the burning of fossil fuels. The current concentration of atmospheric CO_2 is greater than at any time in many thousands of years, except for one warm interval 5,000 years ago, when the level may have been slightly higher than it is today. A doubling of the atmospheric CO_2 concentration, which may happen during the current century unless major efforts are made to reduce human consumption of fossil fuels, would probably increase the average temperature of Earth, cause droughts in the central regions of continents, increase precipitation in coastal areas, melt glaciers and ice caps, and result in sea level rises that would flood coastal cities and agricultural land. The possible consequences of such climate changes will be discussed in Chapters 56 and 57.

Volcanoes have occasionally changed the history of life

Most volcanic eruptions produce only local or short-lived effects, but a few very large volcanic eruptions have had major consequences for life. The collision of continents during the Permian period (about 275 mya) to form a single, gigantic land mass, called **Pangaea**, caused massive volcanic eruptions. The ash ejected by volcanoes into the atmosphere reduced the penetration of sunlight to Earth's surface, lowering temperatures, reducing photosynthesis, and triggering massive glaciation. Massive volcanic eruptions also occurred as the continents drifted apart during the late Triassic period and at the end of the Cretaceous.

Extraterrestrial events have triggered changes on Earth

At least 30 meteorites between the sizes of baseballs and soccer balls hit Earth each year. Collisions with large meteorites are rare, but large meteorites have probably been responsible for several mass extinctions. Several types of evidence tell us about these collisions. Their craters, and the dramatically disfigured rocks that resulted from their impact, are found in many places. Geologists have also discovered giant molecules that contain trapped helium and argon with isotope ratios characteristic of meteorites, which are very different from the ratios found on Earth.

A meteorite caused or contributed to a mass extinction at the end of the Cretaceous period (about 65 mya). The first clue that a meteorite was responsible came from the abnormally high concentrations of the element iridium in a thin layer separating rocks deposited during the Cretaceous from those deposited during the Tertiary (**Figure 21.7**). Iridium is abundant in some meteorites, but is exceedingly rare on Earth's surface. Subsequently, scientists discovered a circular crater 180 kilometers in diameter buried beneath the northern coast of the Yucatán Peninsula of Mexico (see p. 161). When it collided with Earth, the meteorite released energy equivalent to that of 100 million megatons of high explosives, creating great tsunamis. A massive plume of debris swelled to a diameter of up to 200 kilometers, spread around Earth, and descended. The descending debris heated the atmosphere to several hundred degrees, ignited massive fires, and blocked the sun, preventing plants from photosynthesizing. The settling debris formed the iridium-rich layer. About a billion tons of soot, which has a composition that matches smoke from forest fires, was also deposited. Many fossil species, particularly dinosaurs, found in Cretaceous rocks are not found in the Tertiary rocks of the next layer.

A thin band rich in iridium marks the boundary between rocks deposited in the Cretaceous and Tertiary periods.

21.7 Evidence of a Meteorite Impact Iridium is a metal common in some meteorites, but rare on Earth. Its high concentration in sediments deposited about 65 million years ago suggests the impact of a large meteorite.

21.2 RECAP

Conditions on Earth have changed dramatically over time. Some changes, such as increases in atmospheric concentrations of oxygen, have been primarily unidirectional, but other factors, such as climate, have repeatedly oscillated.

- Can you describe how increases in atmospheric concentrations of oxygen affected the evolution of multicellular organisms? See pp. 469–470 and Figure 21.5
- How have volcanic eruptions and collisions with meteorites influenced the course of life's evolution? See p. 471

The many dramatic physical events of Earth's history have influenced the nature and timing of evolutionary changes among Earth's living organisms. We now will look more closely at some of the major events that characterize the history of life on Earth.

21.3 What Are the Major Events in Life's History?

Life first evolved on Earth about 3.8 billion years ago. By about 1.5 billion years ago, eukaryotic organisms had evolved (see Figure 21.5). The fossil record of organisms that lived prior to 550 million years ago is fragmentary, but it is good enough to show that the total number of species and individuals increased dramatically in late Precambrian times. As we saw above, pre-Darwinian geologists divided geological history into eras and periods based on their distinct fossil assemblages. Biologists refer to the assemblage of all organisms of all kinds living at a particular time or place as the **biota** of that time or place. All of the plants living at a particular time or place are its **flora**; all of the animals are its **fauna**. Table 21.2 describes some of the physical and biological changes, such as mass extinctions and dramatic increases in the diversity of major groups of organisms, associated with each unit of time.

About 300,000 species of fossil organisms have been described, and the number is growing steadily. However, this number is only a tiny fraction of the species that have ever lived. We do not know how many species lived in the past, but we have ways of making reasonable estimates. Of the present-day biota, approximately 1.7 million species have been named. The actual number of living species is probably at least 10 million, because most species of insects and mites (the animal groups with the largest numbers of species; see Chapter 32) have not yet been described. So the number of described fossil species is less than 2 percent of the probable minimum number of living species. Life has existed on Earth for about 3.8 billion years. Species last, on average, less than 10 million years; therefore, Earth's biota must have turned over many times during geological history. So the total number of species that have lived over evolutionary time must vastly exceed the number living today. Why have so few species been described from fossils?

Several processes contribute to the paucity of fossils

Only a tiny fraction of organisms ever become fossils, and only a tiny fraction of those fossils are ever studied by paleontologists. Most organisms live and die in oxygen-rich environments in which they quickly decompose. They are not likely to become fossils unless they are transported by wind or water to sites that lack oxygen, where decomposition proceeds slowly or not at all. Furthermore, geological processes often transform rocks, destroying the fossils they contain, and many fossil-bearing rocks are deeply buried and inaccessible. Paleontologists have studied only a tiny fraction of the sites that contain fossils, but they find and describe many new fossils every year.

The number of known fossils is especially large for marine animals that had hard skeletons (which resist decomposition). Among the nine major animal groups with hard-shelled members, approximately 200,000 species have been described from fossils—roughly twice the number of living marine species in these same groups. Paleontologists lean heavily on these groups in their interpretations of the evolution of life. Insects and spiders are also relatively well represented in the fossil record (**Figure 21.8**). The fossil record, though incomplete, is good enough to demonstrate clearly that organisms of particular types are found in rocks of specific ages and that newer organisms appear sequentially in younger rocks.

By combining information about geological changes during Earth's history with evidence from the fossil record, scientists have composed portraits of what Earth and its inhabitants may have looked like at different times. We know in general where the continents were and how life changed over time, but many of the details are poorly known, especially for events in the more remote past. In this section we will provide an overview of how life has changed during its history on Earth. In Part Six of this book we will discuss the evolutionary history of particular groups of organisms in more detail.

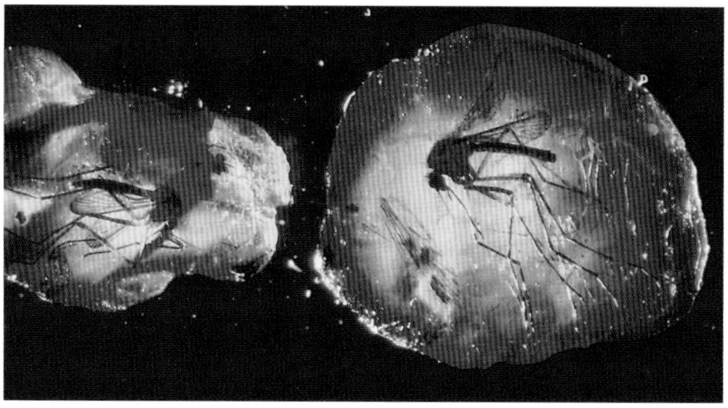

21.8 Insect Fossils These chunks of amber—fossilized tree resin—contain insects that were preserved when they were trapped in the sticky resin some 50 million years ago.

Precambrian life was small and aquatic

For most of its history, life was confined to the oceans, and all organisms were small. Over the long ages of the Precambrian—more than three billion years—the shallow seas slowly began to teem with life. For most of the **Precambrian**, life consisted of microscopic prokaryotes; eukaryotes probably evolved about two-thirds of the way through the era. Unicellular eukaryotes and small multicellular animals fed on floating photosynthetic microorganisms. Small floating organisms, known collectively as *plankton*, were eaten by slightly larger animals that filtered them from the water. Other animals ingested sediments on the seafloor and digested the remains of organisms within them. By the late Precambrian (about 650 mya), many kinds of multicellular soft-bodied animals had evolved. Some of them were very different from any animals living today, and may be members of groups that have no living descendants (**Figure 21.9**).

Life expanded rapidly during the Cambrian period

The **Cambrian** period (542–488 mya) marks the beginning of the **Paleozoic** era. The O_2 concentration in the Cambrian atmosphere was approaching its current level, and the continents had come together to form several large land masses. The largest of these land masses was called *Gondwana* (**Figure 21.10A**). A rapid diversification of life took place that we now refer to as the **Cambrian explosion** (although in fact it began before the Cambrian). Most of the major groups of animals that have species living today appeared during the Cambrian.

For the most part, fossils tell us only about the hard parts of organisms, but in three Cambrian fossil beds—the Burgess Shale in British Columbia, Sirius Passet in northern Greenland, and the Chengjiang site in southern China—the soft parts of many ani-

Spriggina floundersi

Mawsonites

21.9 Ediacaran Animals These fossils of soft-bodied invertebrates, excavated at Ediacara in southern Australia, were formed 600 million years ago. They illustrate the diversity of life that evolved in the Precambrian era.

mals were preserved (**Figure 21.10B**). Arthropods (crabs, shrimps, and their relatives) are the most diverse group in the Chinese fauna; some of them were large carnivores. Trilobites, members of an arthropod group that was abundant and diverse during the Cambrian (see Figure 32.21), suffered a major reduction at the end of the Cambrian, but they recovered and continued to be abundant until the end of the Permian, when they became extinct.

Many groups of organisms diversified

Geologists divide the remainder of the Paleozoic era into the Ordovician, Silurian, Devonian, Carboniferous, and Permian periods (see Table 21.2). Each period is characterized by the diversification of specific groups of organisms. Mass extinctions marked the ends of the Ordovician, Devonian, and Permian.

THE ORDOVICIAN (488–444 MYA) During the **Ordovician** period, the continents, which were located primarily in the Southern Hemisphere, still lacked multicellular plants. Evolutionary radiation of marine organisms was spectacular during the early Ordovician, especially among animals, such as brachiopods and mollusks, that lived on the seafloor and filtered small prey from the water. At the end of the Ordovician, as massive glaciers formed over Gondwana, sea levels were lowered about 50 meters, and ocean temperatures dropped. About 75 percent of the animal species became extinct, probably because of these major environmental changes.

THE SILURIAN (444–416 MYA) During the **Silurian** period, the northernmost continents coalesced, but the general positions of the continents did not change much. Marine life rebounded from the mass extinction at the end of the Ordovician. Animals able to swim and feed above the ocean bottom appeared for the first time, but no new major groups of marine organisms evolved. The tropical sea was uninterrupted by land barriers, and most marine organisms were widely distributed. On land, the first vascular plants appeared late in the Sil-

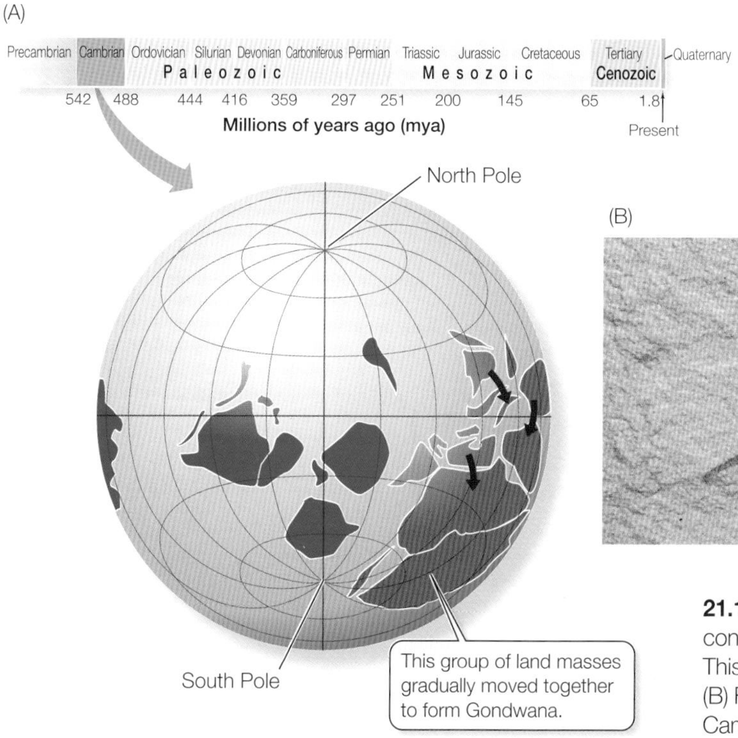

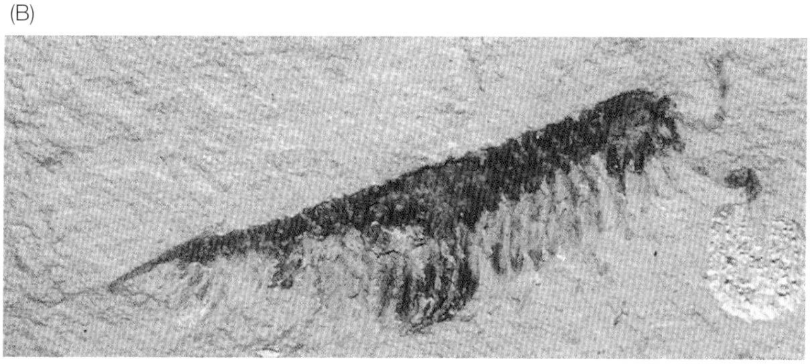

21.10 Cambrian Continents and Fauna (A) Positions of the continents during the middle of the Cambrian period (542–488 mya). This view of Earth has been distorted so that you can see both poles. (B) Fossil beds in China have yielded well-preserved remains of Cambrian animals such as this one, called *Jianfangia*.

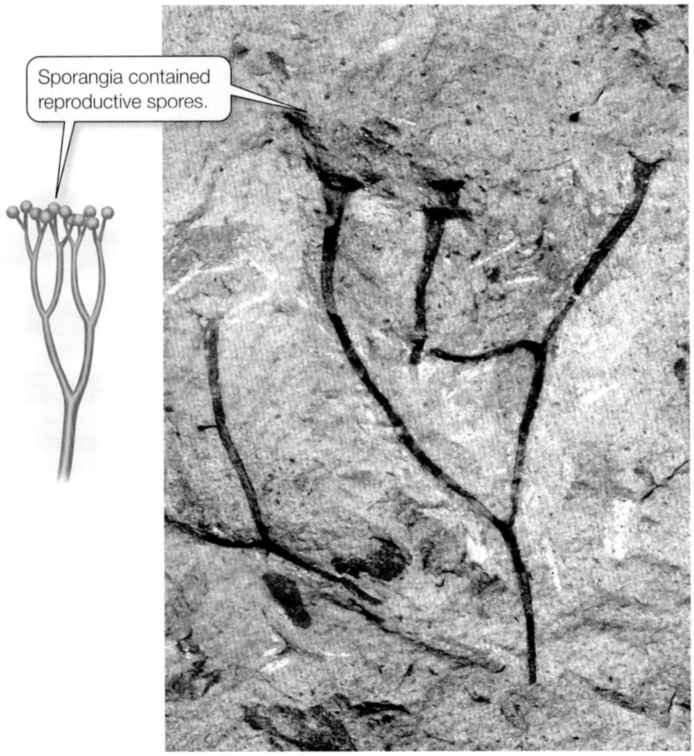

21.11 *Cooksonia*, the Earliest Known Vascular Plant These early plants were small and very simple in structure. However, they were true vascular plants (tracheophytes) with internal water-conducting cells (tracheids), well equipped to make the move from the aquatic to the terrestrial environment. This fossil of *Cooksonia pertoni* is from the Silurian period (about 420 mya).

urian period (about 420 mya). These plants were less than 50 cm tall and lacked roots and leaves (**Figure 21.11**). The first terrestrial arthropods—scorpions and millipedes—appeared at about the same time.

THE DEVONIAN (416–359 MYA) Rates of evolutionary change accelerated in many groups of organisms during the **Devonian** period. The northern land mass (called *Laurasia*) and the southern land mass (Gondwana) moved slowly toward each other (**Figure 21.12A**). There were great evolutionary radiations of corals and shelled squidlike cephalopods (**Figure 21.12B**). Fishes diversified as jawed forms replaced jawless ones, and heavy armor gave way to the less rigid outer coverings of modern fishes. All current major groups of fishes were present by the end of the period.

Terrestrial communities also changed dramatically during the Devonian. Club mosses, horsetails, and tree ferns became common toward the end of the Devonian; some attained the size of trees. Their deep roots accelerated the weathering of rocks, resulting in the development of the first forest soils. Distinct floras evolved on Laurasia and Gondwana toward the end of the period. A wind-pollinated precursor of seed-bearing plants, *Runcaria*, was found in sediments in Belgium laid down 385 million years ago. The ancestors of gymnosperms, the first plants to produce seeds, appeared later in the Devonian. The first known fossils of centipedes, spiders, mites, and insects date to this period. Fishlike amphibians began to occupy the land.

An extinction of about 75 percent of all marine species marked the end of the Devonian. Paleontologists are uncertain about the cause of this mass extinction, but two large meteorites that collided with Earth at that time, one in present-day Nevada and the other in Western Australia, may have been responsible, or at least a contributing factor.

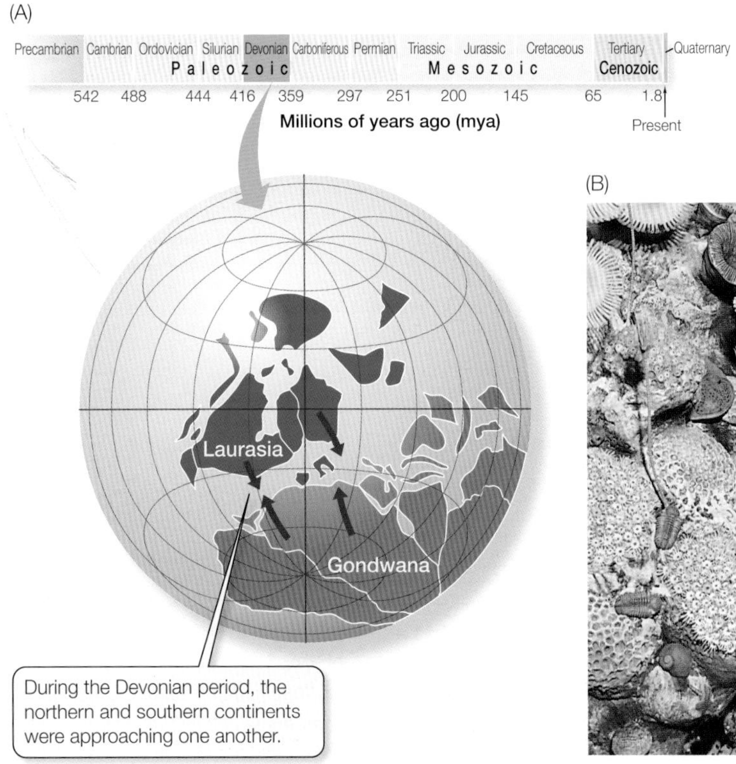

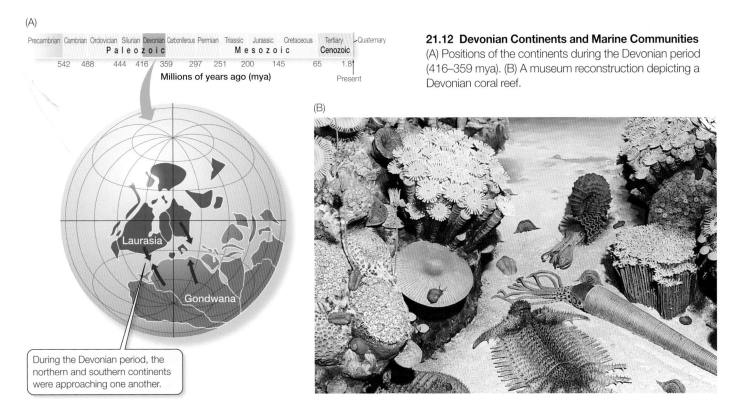

21.12 Devonian Continents and Marine Communities (A) Positions of the continents during the Devonian period (416–359 mya). (B) A museum reconstruction depicting a Devonian coral reef.

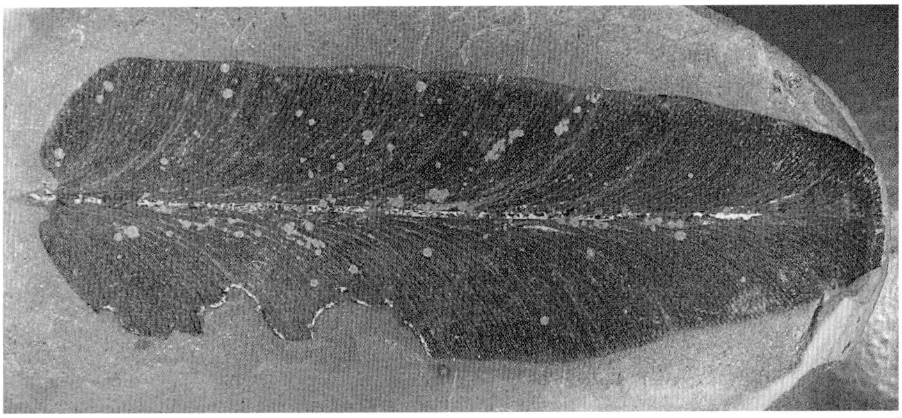

21.13 Evidence of Insect Diversification The margins of this fossil fern leaf from the Carboniferous period have been chewed by insects.

THE CARBONIFEROUS (359–297 MYA) Large glaciers formed over high-latitude Gondwana during the **Carboniferous** period, but extensive swamp forests grew on the tropical continents. These forests were not made up of the kinds of trees we know today, but were dominated by giant tree ferns and horsetails with small leaves (see Figure 28.8). Fossilized remains of those trees formed the coal we now mine for energy.

The diversity of terrestrial animals increased greatly during the Carboniferous. Snails, scorpions, centipedes, and insects were abundant and diverse. Insects evolved wings, becoming the first animals to fly. Flight gave them access to tall plants; plant fossils from this period show evidence of chewing by insects (**Figure 21.13**). Amphibians became larger and better adapted to terrestrial existence after splitting from the lineage leading to the *amniotes*, vertebrates with well-protected eggs that can be laid in dry places. In the seas, crinoids (sea lilies and feather stars) reached their greatest diversity, forming "meadows" on the seafloor (**Figure 21.14**).

THE PERMIAN (297–251 MYA) During the **Permian** period, the continents coalesced into the supercontinent *Pangaea* (**Figure 21.15**). Permian rocks contain representatives of most modern groups of insects. By the end of the period, one amniote group, the reptiles, greatly outnumbered the amphibians. Late in the period, the lineage leading to mammals diverged from one reptilian group. In fresh waters, the Permian period was a time of extensive diversification of ray-finned fishes.

Conditions for life deteriorated toward the end of the Permian. Massive volcanic eruptions resulted in outpourings of lava that covered large areas of Earth. The ash the volcanoes produced blocked sunlight and cooled the climate, resulting in the largest glaciers in Earth's history. Atmospheric oxygen concentrations gradually dropped from about 30 percent to about 12 percent. At such low oxygen concentrations, most animals would have been unable to survive at elevations above 500 meters; thus about half of the Permian land area would have been uninhabitable. The com-

21.14 A Carboniferous "Crinoid Meadow" Crinoids—the flowerlike organisms—were dominant marine animals during the Carboniferous period and may have formed communities similar to this one.

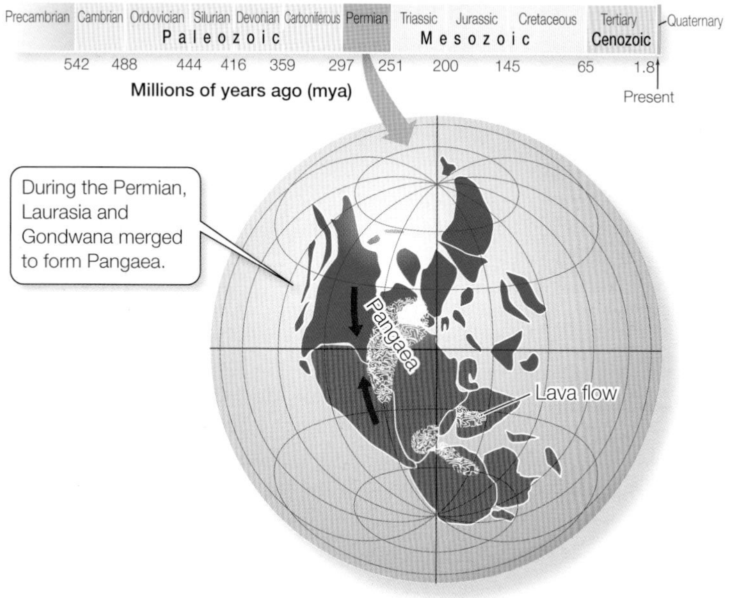

21.15 Pangaea Formed in the Permian Period At the end of the Permian period, massive lava flows spread over Earth, and the largest glaciers in Earth's history formed.

bination of these changes resulted in the most drastic mass extinction event in Earth's history. Many species became extinct simultaneously at the end of the period; others disappeared gradually over a period of several million years.

The Permian extinction may have come perilously close to wiping out life on Earth. Scientists estimate that about 96 percent of all species became extinct at that time.

Geographic differentiation increased during the Mesozoic era

The few organisms that survived the Permian mass extinction found themselves in a relatively empty world at the start of the **Mesozoic** era (251 mya). As Pangaea slowly separated into individual continents, the oceans rose and reflooded the continental shelves, forming huge, shallow inland seas. Atmospheric oxygen concentrations gradually rose to their former levels. Life again proliferated and diversified, but different groups of organisms came to dominate Earth. The three groups of *phytoplankton* (floating photosynthetic organisms) that dominate today's oceans—dinoflagellates, coccolithophores, and diatoms—became ecologically important at this time. New seed-bearing plants replaced the trees that had dominated the Permian forests.

During the Mesozoic, Earth's biota, which until that time had been relatively homogeneous, became increasingly **provincialized**; that is, distinct terrestrial biotas evolved on each continent. The biotas of the shallow waters bordering the continents also diverged from one another. The provincialization that began during the Mesozoic continues to influence the geography of life today. By the end of the era, the continents were close to their present positions, and many organisms looked similar to those living today.

The Mesozoic era is divided into three periods: the Triassic, Jurassic, and Cretaceous. The Triassic and Cretaceous were terminated by mass extinctions, probably caused by meteorite impacts.

THE TRIASSIC (251–200 MYA) Pangaea began to break apart during the **Triassic** period. Many invertebrate groups became more species-rich, and many burrowing animals evolved from groups living on the surfaces of seafloor sediments. On land, conifers and pteridosperms became the dominant trees. The first frogs and turtles appeared. A great radiation of reptiles began, which eventually gave rise to crocodilians, dinosaurs, and birds. The end of the Triassic was marked by a mass extinction that eliminated about 65 percent of the species on Earth. A large meteorite that crashed into what is now Quebec may have been responsible.

THE JURASSIC (200–145 MYA) During the **Jurassic** period, the land was once again divided in two large continents—Laurasia in the north, and Gondwana in the south. Ray-finned fishes began the great radiation that led to their dominance of the oceans. The first salamanders and lizards appeared, and flying reptiles (pterosaurs)

21.16 Jurassic Parkland The dinosaurs of the Mesozoic have captured human imaginations ever since their fossils were first discovered. This illustrations depicts dinosaurs that lived some 160 million years ago (the Jurassic period) on what are now the western plains of North America. In the foreground, a *Ceratosaurus* and two small *Coelurus* feed on the carcass of an *Apatosaurus.* In the background are (left to right) two *Camptosaurus*, *Stegosaurus*, *Brachiosaurus*, and another *Apatosaurus.*

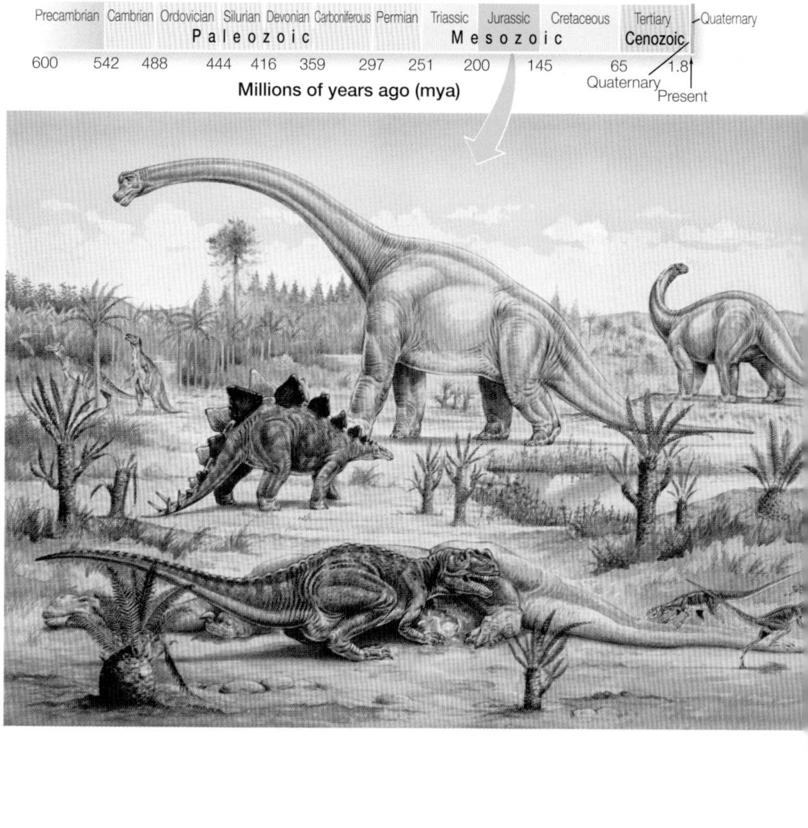

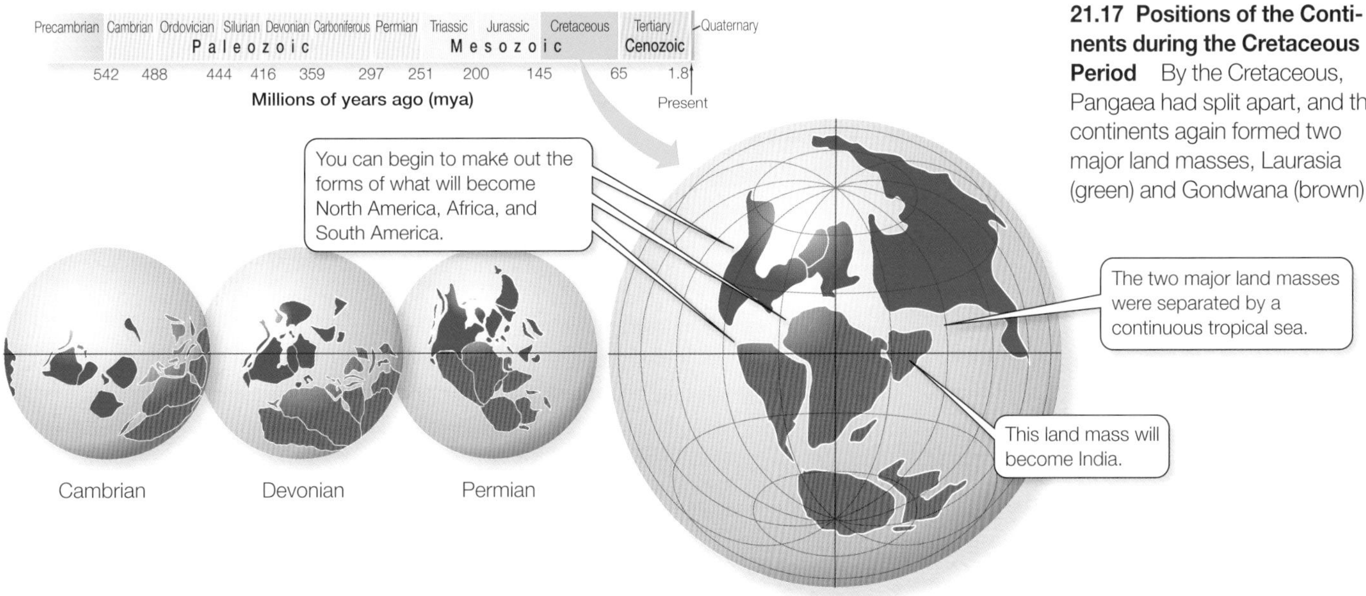

21.17 Positions of the Continents during the Cretaceous Period By the Cretaceous, Pangaea had split apart, and the continents again formed two major land masses, Laurasia (green) and Gondwana (brown).

evolved. Dinosaur lineages evolved into predators that walked on two legs and large herbivores that walked on four legs (**Figure 21.16**). Several groups of mammals first appeared during this time. Plant evolution continued with the emergence of the flowering plants that dominate Earth's vegetation today.

THE CRETACEOUS (145–65 MYA) By the early **Cretaceous** period, Laurasia was completely separate from Gondwana, which was beginning to break apart. A continuous sea encircled the tropics (**Figure 21.17**). Sea levels were high, and Earth was warm and humid. Life proliferated both on land and in the oceans. Marine invertebrates increased in diversity and in number of species. On land, dinosaurs continued to diversify. The first snakes appeared during the Cretaceous, but the modern groups with the most species resulted from a later radiation. Early in the Cretaceous, flowering plants began the radiation that led to their current dominance on land. Fossils of the earliest known flowering plants, dated at 124 million years ago, were recently discovered in Liaoning Province in northeastern China (**Figure 21.18**). By the end of the period, many groups of mammals had evolved. Most of them were small, but one species recently discovered in China, *Repenomamus giganticus*, was large enough to capture and eat young dinosaurs.

As described earlier in this chapter, another meteorite-caused mass extinction took place at the end of the Cretaceous period. In the seas, many planktonic organisms and bottom-dwelling invertebrates became extinct. On land, all animals larger than about 25 kilograms in body weight apparently became extinct. Many species of insects died out, perhaps because the growth of their food plants was greatly reduced following the impact. Some species survived in the northern parts of North America and Eurasia, areas that were not subjected to the devastating fires that engulfed most low-latitude regions.

The modern biota evolved during the Cenozoic era

By the early **Cenozoic** era (65 mya), the positions of the continents resembled those of today, but Australia was still attached to Antarctica, and the Atlantic Ocean was much narrower. The Cenozoic era was characterized by an extensive radiation of mammals, but other groups were also undergoing important changes.

Flowering plants diversified extensively and came to dominate world forests, except in cool regions. Mutations of two genes in one group of plants allowed them to use atmospheric N_2 directly by forming symbioses with a few species of nitrogen-fixing bacteria (see Section 36.4). The evolution of this symbiosis between certain early Cenozoic plants and these specialized bacteria was

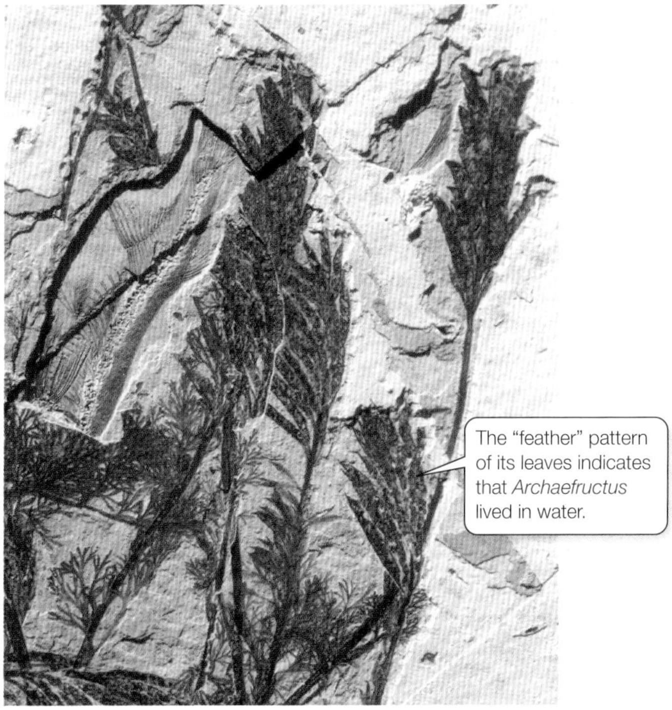

21.18 Flowering Plants of the Cretaceous These fossils of *Archaefructus* are the earliest known examples of flowering plants, the type of plants most prevalent on Earth today.

TABLE 21.3

Subdivisions of the Cenozoic Era

PERIOD	EPOCH ONSET (MYA)	EPOCH ONSET (MYA)
Quaternary	Holocene[a]	0.01 (~10,000 years ago)
	Pleistocene	1.8
Tertiary	Pliocene	5.3
	Miocene	23
	Oligocene	34
	Eocene	55.8
	Paleocene	65

[a] The Holocene is also known as the Recent.

the first "green revolution" and dramatically increased the amount of nitrogen available for terrestrial plant growth.

The Cenozoic era is divided into two periods, the Tertiary and the Quaternary. Because both the fossil record and our subsequent knowledge of evolutionary history become more extensive closer to our own time, paleontologists have subdivided these periods into *epochs* (**Table 21.3**).

THE TERTIARY (65–1.8 MYA) During the **Tertiary** period, Australia began its northward drift. By 20 million years ago it had nearly reached its current position. The early Tertiary was a hot and humid time, during which the ranges of many plants shifted latitudinally. The tropics were probably too hot for rainforests, and were clothed in low-lying vegetation instead. In the middle of the Tertiary, however, Earth's climate became considerably drier and cooler. Many lineages of flowering plants evolved herbaceous (nonwoody) forms; grasslands spread over much of Earth.

By the beginning of the Cenozoic era, invertebrate faunas resembled those of today. It is among the vertebrates that evolutionary changes during the Tertiary period were most rapid. Snakes and lizards underwent extensive radiations during this period, as did birds and mammals. Three waves of mammals dispersed from Asia to North America across the land bridge that has intermittently connected the two continents during the past 55 million years ago. Rodents, marsupials, primates, and hoofed mammals appeared in North America for the first time.

THE QUATERNARY (1.8 MYA TO PRESENT) The current geological period, the **Quaternary**, is subdivided into two epochs, the *Pleistocene* and the *Holocene* (also known as the *Recent*). The Pleistocene was a time of drastic cooling and climate fluctuations. During four major and about 20 minor "ice ages," massive glaciers spread across the continents, and the ranges of animal and plant populations shifted toward the equator. The last of these glaciers retreated from temperate latitudes less than 15,000 years ago. Organisms are still adjusting to these changes. Many high-latitude ecological communities have occupied their current locations for no more than a few thousand years. Interestingly, relatively few species became extinct during these climate fluctuations.

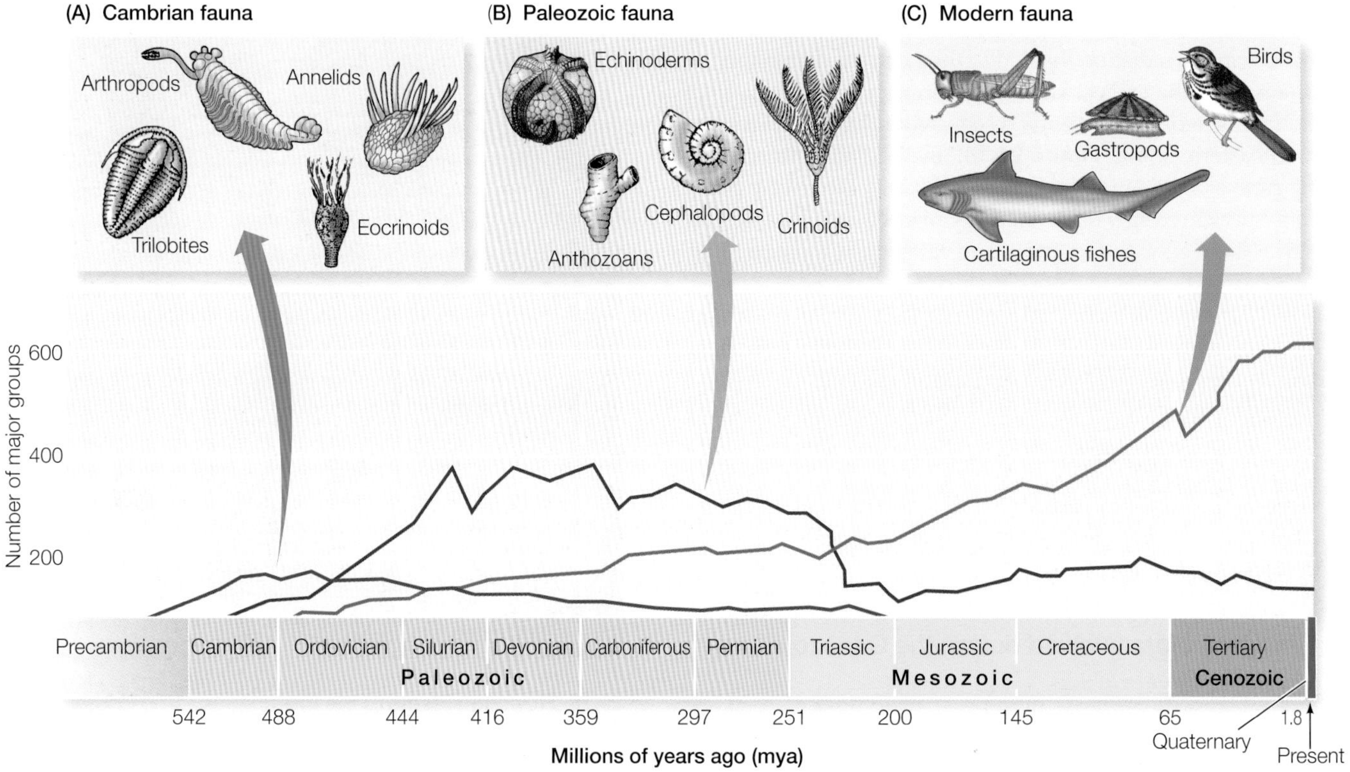

21.19 Evolutionary Faunas Representatives of the three great evolutionary faunas are shown, together with a graph illustrating the number of major groups in each fauna over time.

The Pleistocene was the time of hominoid evolution and radiation, resulting in the species *Homo sapiens*—modern humans (see Section 33.5). Many large bird and mammal species became extinct in Australia and in the Americas when *H. sapiens* arrived on those continents about 40,000 and 15,000 years ago, respectively. These extinctions were probably the result of hunting by humans, although the existing evidence does not convince all paleontologists.

Woolly mammoths survived on Siberia's Wrangel Island as recently as 4,000 years ago, long after hunters exterminated them from the mainland. The mammoths may have maintained their dry steppe environment on the island by trampling tundra mosses and recycling nitrogen, whereas the mainland environment became a moist tundra.

Three major faunas have dominated life on Earth

The fossil record reveals three great evolutionary radiations, each of which resulted in the evolution of major new faunas (**Figure 21.19**). The first one, the Cambrian explosion, actually began before the beginning of the Cambrian period. The second, about 60 million years later, resulted in the Paleozoic fauna. The great Permian extinctions 300 million years later were followed by the third great radiation, called the Triassic explosion, which led to our modern fauna.

During the Cambrian explosion, organisms ancestral to most of the present-day animal groups appeared, along with a number of groups that subsequently became extinct. The Paleozoic and Triassic explosions resulted in considerable diversification of the existing major groups of animals, but all of them were modifications of body plans that were already present when these great biological diversifications began (see Chapters 31–33).

21.3 RECAP

Life evolved in the Precambrian oceans. Life diversified as atmospheric oxygen approached its current level and the continents came together to form several large land masses. Numerous climate changes and rearrangements of the continents, as well as meteorite impacts, contributed to five mass extinctions.

- Why have so few of the multitudes of organisms that have existed over millenia become fossilized? See p. 472
- What do we mean when we refer to the "Cambrian explosion"? See pp. 472–473
- Can you identify the five mass extinctions and their possible causes? See pp. 473–477 and Table 21.2

The fossil record reveals broad patterns in life's evolution. It shows that many species changed very little over many millions of years, others changed only gradually, and still others underwent rapid changes that were followed by long periods of slow change. In short, the rate of evolutionary change has differed greatly at different times and among different lineages. Let's look at some examples of evolutionary patterns to determine why rates of evolutionary change are so variable.

21.4 Why Do Evolutionary Rates Differ among Groups of Organisms?

Fossils can reveal information about rates of change within particular lineages of organisms. Evolutionary change in a lineage may incorrectly appear to be rapid if its fossil record is very incomplete, but some rapid changes are well documented by excellent series of fossils.

Changes in the physical and biological environment are likely to stimulate evolutionary change. Organisms that live in environments that are changing are likely to evolve more rapidly than organisms living in relatively constant environments. When climates change, the ranges of some organisms may shift, and other organisms may find themselves with different predators or competitors. Similarly, predators may change in response to changes in their prey. In contrast, the morphology of organisms that live in relatively unchanging environments often changes slowly, if at all.

"Living fossils" exist today

Species whose morphology has changed little over millions of years are known as "living fossils." For example, the horseshoe crabs living today are almost identical in appearance to those that lived 300 million years ago (see Figure 32.30B). The sandy coastlines where horseshoe crabs spawn feature extremes in temperatures and salt concentrations that are lethal to many organisms. These harsh environments have changed relatively little over millennia, and the horseshoe crabs likewise remain relatively unchanged as they maintain the very specific adaptations that allow them to survive.

Similarly, the chambered nautiluses of the late Cretaceous are nearly indistinguishable from living species (see Figure 32.15F). Chambered nautiluses spend their days in deep, dark ocean waters, ascending to feed in food-rich surface waters only under the protective cover of darkness. Their intricate shells provide little protection against today's visually hunting fish; they survive, however, because they are adapted to a stringent, relatively unchanging environment in which recently evolved potential predators cannot survive.

The leaf shapes of many plants have changed little over time; fossilized leaves of *Ginkgo* trees from the Triassic, for example, are very similar to those of living trees (**Figure 21.20**). This may be because the physical nature of sunlight is unchanging, and the intricate photosynthetic mechanisms that harvest sunlight (see Chapter 8), once evolved, have remained relatively constant over millions of years.

Evolutionary changes have been gradual in most groups

The most striking feature of life's evolution is that rates of change are, on average, very slow. The fossil record contains many series of fossils that demonstrate gradual change in lineages over time. A good example is the series of fossils showing changes in the

21.20 "Living Fossils" Fossilized *Ginkgo* leaves from the Triassic appear very similar to the leaves of living ginkgo trees.

number of ribs on the exoskeleton in eight lineages of trilobites during the Ordovician (**Figure 21.21**). Rates of change differed among the lineages, and they did not all change at the same time, but all of the changes were gradual.

Why does slow, gradual change appear to dominate the fossil record? A likely reason is that climates have usually changed slowly. In addition, as climates changed, the ranges of most organisms shifted accordingly, so that the environments in which individuals lived actually changed very little. For example, as the climate warmed and glaciers retreated about 10,000 years ago, many species expanded their ranges northward (**Figure 21.22**).

Rates of evolutionary change are sometimes rapid

If the physical or biological environment changes rapidly, some lineages may also change rapidly. Good examples of rapid evolutionary change are provided by species that have been introduced into new regions that differ strikingly from the environment from which they came. For example, as recently as 1939, the house finch was confined to the arid and semiarid parts of western North America. That year, some captive finches were released in New York City. Many of them survived to form a small breeding population in the immediate vicinity of the city. During the early 1960s, that population began to grow and increase its range. By the 1990s, the house finch had spread across all of the eastern United States and southern Canada (**Figure 21.23**), colonizing regions with climates that differ dramatically from those of its original western range. Remarkably, by 2000, birds in different eastern finch populations that had been separated for only a few decades were as different in size as birds in western finch populations that had been separated for thousands of years.

Rates of extinction have also varied greatly

More than 99 percent of the species that have ever lived are extinct. Species have become extinct throughout the history of life, but extinction rates have fluctuated dramatically. Section 21.3 described at least five major extinction events that severely reduced the planet's biota. These events were often followed by high rates of evolution, as surviving organisms responded to a new environment with a different composition of predators, prey, and competitors.

Some groups have had high extinction rates while others were proliferating. For example, among the mollusks of the Atlantic

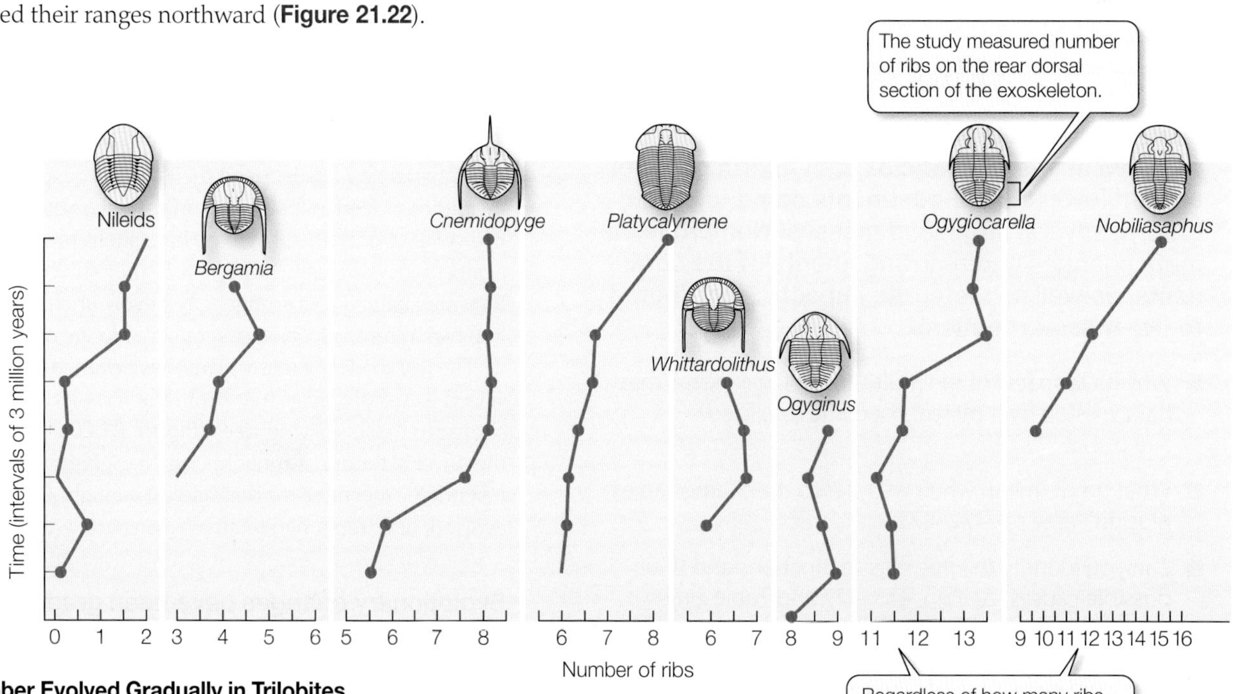

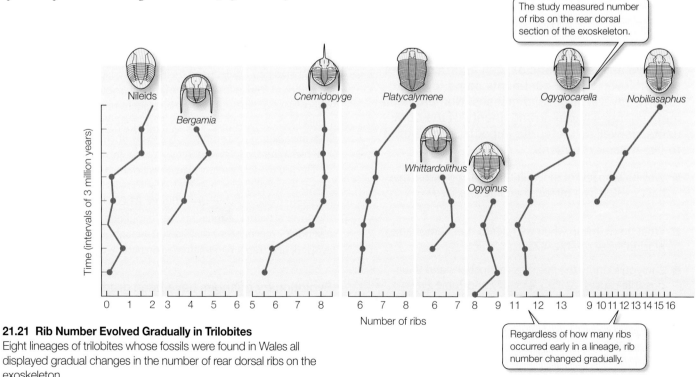

21.21 Rib Number Evolved Gradually in Trilobites
Eight lineages of trilobites whose fossils were found in Wales all displayed gradual changes in the number of rear dorsal ribs on the exoskeleton.

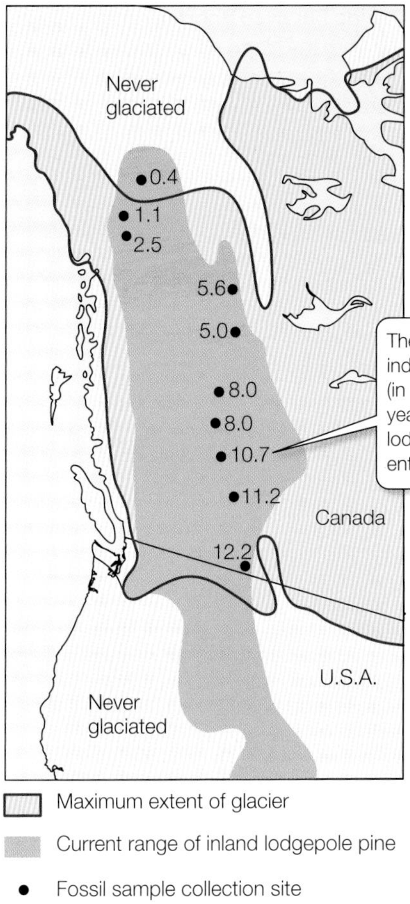

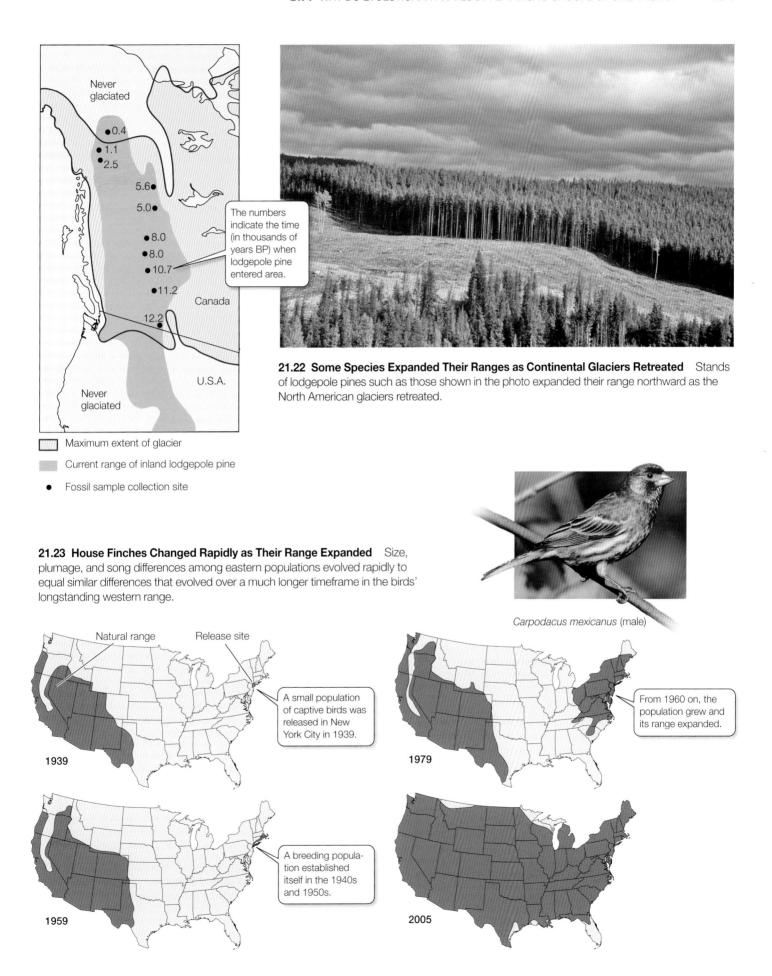

21.22 Some Species Expanded Their Ranges as Continental Glaciers Retreated Stands of lodgepole pines such as those shown in the photo expanded their range northward as the North American glaciers retreated.

21.23 House Finches Changed Rapidly as Their Range Expanded Size, plumage, and song differences among eastern populations evolved rapidly to equal similar differences that evolved over a much longer timeframe in the birds' longstanding western range.

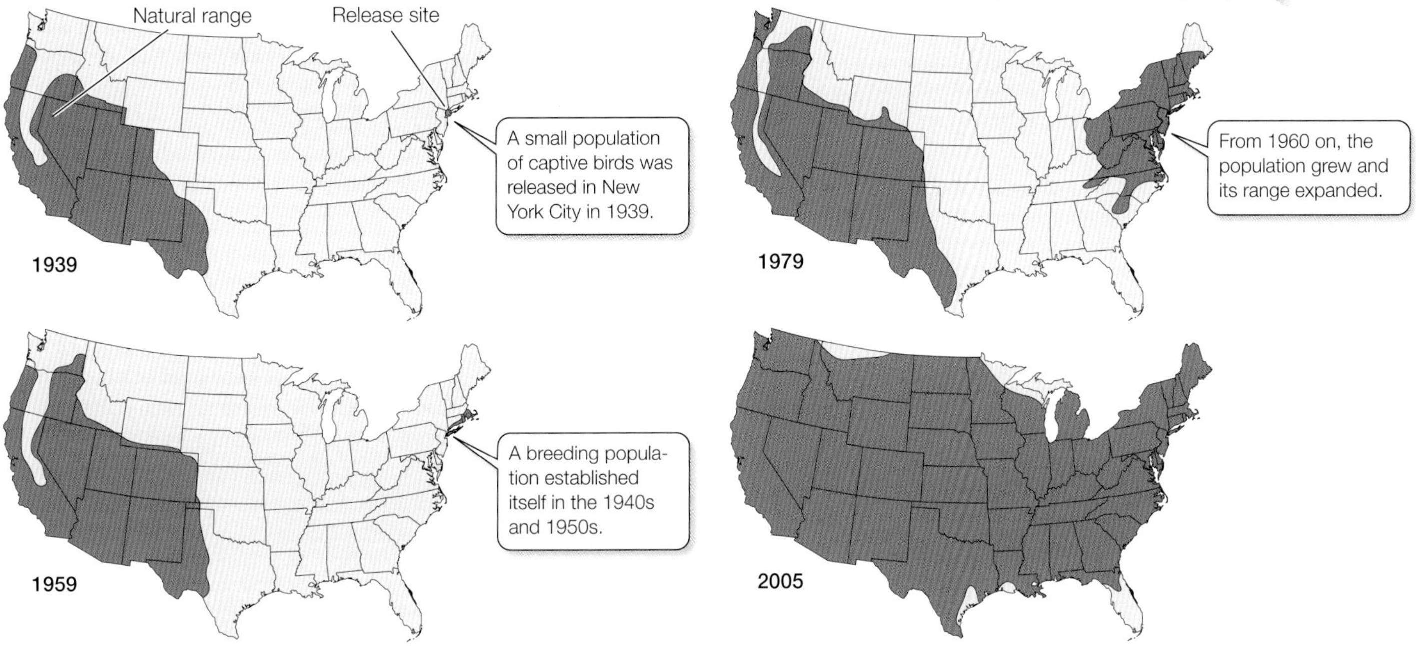

EXPERIMENT

HYPOTHESIS: Large size and correlated dietary specialization result in more rapid extinction.

METHOD

1. Use features of fossil jaws (jaw depth, length of shearing teeth) to infer the size and diet of individuals of different species.
2. Determine the length of time each species was present in the fossil record.

RESULTS

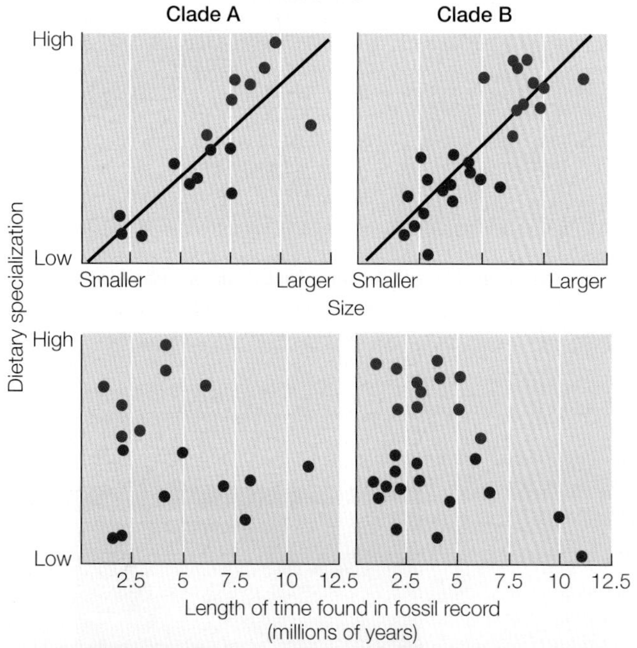

CONCLUSION: Large, specialized canid species survived for shorter times than smaller, less specialized ones.

21.24 Large, Specialized Canids Survived Shorter Times Each dot represents a canid species from the North American fossil records of two clades. In these two successive groups of canids, larger species are linked to more specialized diets (top graphs) and were more likely to become extinct (bottom graphs).

coastal plain of North America, species with broad geographic ranges were less likely to become extinct during normal times (when no mass extinctions were taking place) than were species with small geographic ranges. During the mass extinction at the end of the Cretaceous, however, groups of closely related mollusk species with large geographic ranges survived better than groups with small ranges, even if the individual species within the group had small ranges.

The diets of organisms may also influence their extinction rates. The fossil record shows that natural selection often favors larger size in carnivores (flesh-eating animals). However, larger carnivores typically have more specialized diets than smaller carnivores, and animals with specialized diets may be more vulnerable to loss of their food supply than are those with more generalized diets. A possible consequence is that large, specialized carnivores would be more likely to become extinct than smaller species with more generalized diets. An impressive fossil record of successive lineages of canids (dogs, wolves, and their relatives) enabled Blaire Van Valkenburgh and her colleagues to use the comparative method (see Section 1.3) to test and confirm this hypothesis (**Figure 21.24**).

At the end of the Cretaceous period, extinction rates on land were much higher among large vertebrates than among small ones. The same was true during the Pleistocene epoch, when extinction rates were high only among large mammals and large birds. During some mass extinctions, marine organisms were heavily hit, but terrestrial organisms survived well. Other mass extinctions affected organisms living in both environments. These differences are not surprising, given that major changes on land and in the oceans did not always coincide.

21.4 RECAP

Rates of evolutionary change vary greatly among lineages of organisms. Changes in the physical and biological environments typically induce evolutionary changes. Extinction rates have also fluctuated dramatically over evolutionary history.

- Why do slow, gradual changes dominate the fossil record? See pp. 479–480
- How can the diets of organisms influence their extinction rates? See p. 482 and Figure 21.24

Although the agents of evolutionary change are operating today as they have been since life first appeared on Earth, major changes are being driven by the dramatic increase of Earth's human population. Predation (hunting) by humans has caused the extinction of many large mammals in the recent past, and continues doing so today. Humans are also changing the physical and biological environment by dramatically altering Earth's vegetation, converting forests and grasslands to crops and pastures. Deliberately or inadvertently, we are moving thousands of species around the globe, reducing the provincialization of Earth's biota that began during the Mesozoic era. Humans have also taken charge of the evolution of certain species by means of artificial selection and biotechnology. As we saw in Chapter 16, modern molecular methods enable us to modify organisms by moving genes among even distantly related species. In short, humans have become a dominant agent of evolutionary change. How we wield our massive influence will powerfully affect the future of life on Earth.

CHAPTER SUMMARY

21.1 How do scientists date ancient events?

The relative ages of organisms can be determined by the dating of **fossils** and the **strata** of sedimentary rocks in which they are found.

Paleontologists use a variety of radioisotopes with different **half-lives** to date events at different times in the remote past. Review Figure 21.1

Geologists divide the history of life into eras and periods, based on major differences in the assemblages of fossils found in successive layers of rocks. Review Table 21.2

21.2 How have Earth's continents and climates changed over time?

See Web/CD Tutorial 21.1

Earth's crust consists of solid lithospheric plates that float on fluid magma. **Continental drift** caused by convection currents in the magma moves these plates and the continents they contain. Review Figure 21.2

Conditions on Earth have changed dramatically over time. Some changes, such as increases in atmospheric concentrations of oxygen, have been primarily unidirectional. Other changes, such as changes in climate, have undergone repeated oscillations. Review Figures 21.5 and 21.6

Oxygen-generating cyanobacteria liberated enough O_2 to open the door to oxidation reactions in metabolic pathways. The aerobic prokaryotes were able to harvest more energy than anaerobic organisms and began to proliferate. Increases in atmospheric O_2 levels supported the evolution of large eukaryotic cells.

Major physical events on Earth, such as the collision of continents that formed the supercontinent **Pangaea**, have affected Earth's surface, climate, and atmosphere. In addition, extraterrestrial events such as meteorite strikes created sudden and dramatic environmental shifts. All these changes have affected the history of life.

21.3 What are the major events in life's history?

Paleontologists use fossils and evidence of geological changes to determine what Earth and its **biota** may have looked like at different times.

During most of its history, life was confined to the oceans. Multicellular life diversified during the **Cambrian explosion**. Review Figure 21.10

The periods of the **Paleozoic** era were each characterized by the diversification of specific groups of organisms. Amniotes—vertebrates whose eggs can be laid in dry places—first appeared during the Carboniferous period.

During the **Mesozoic** era, Earth's biota became increasingly **provincialized**, as distinct terrestrial **biotas** evolved on each continent.

Five episodes of **mass extinction** punctuated the history of life in the Paleozoic and Mesozoic eras.

Three major **faunas** have dominated life on Earth, stemming from evolutionary radiations occurring in the Cambrian, Paleozoic, and Mesozoic eras. Review Figure 21.19

Earth's **flora** has been dominated by flowering plants since the **Cenozoic** era.

21.4 Why do evolutionary rates differ among groups of organisms?

Most lineages evolve slowly and gradually over time; the morphology of many species has changed little over many millions of years. Review Figure 21.21

Changes in the physical and biological environments typically induce evolutionary changes.

Extinction rates have fluctuated dramatically during the history of life, but some groups have had high extinction rates while others were proliferating.

See Web/CD Activity 21.1 for a concept review of this chapter.

SELF-QUIZ

1. The number of species of fossil organisms that have been described is about
   a. 50,000.
   b. 100,000.
   c. 200,000.
   d. 300,000.
   e. 500,000.
2. In undisturbed strata of sedimentary rock,
   a. the oldest rocks lie at the top.
   b. the oldest rocks lie at the bottom.
   c. the oldest rocks are in the middle.
   d. the oldest rocks are distributed among the strata of younger rocks.
   e. None of the above
3. Carbon-14 can be used to determine the ages of fossil organisms because
   a. all organisms contain many carbon compounds.
   b. carbon-14 has a regular rate of decay to carbon-12.
   c. the ratio of ^{14}C to ^{12}C in living organisms is always the same as that in the atmosphere.
   d. the production of new ^{14}C in the atmosphere just balances the natural radioactive decay of ^{14}C.
   e. All of the above
4. An important, generally unidirectional change in Earth during its history is a
   a. steady increase in volcanic activity.
   b. gradual coming together of the continents.
   c. steady increase in the oxygen content of the atmosphere.
   d. gradual warming of the climate.
   e. steady increase in Earth's precipitation.
5. The total of all species of organisms in a given region is known as the region's
   a. biota.
   b. flora.
   c. fauna.
   d. flora and fauna.
   e. diversity.
6. The coal beds we now mine for energy are the remains of
   a. trees that grew in swamps during the Carboniferous period.
   b. trees that grew in swamps during the Devonian period.
   c. trees that grew in swamps during the Permian period.
   d. small plants that grew in swamps during the Carboniferous period.
   e. None of the above

7. The cause of the mass extinction at the end of the Ordovician period was probably
   a. the collision of Earth with a large meteorite.
   b. massive volcanic eruptions.
   c. massive glaciation in Gondwana.
   d. the uniting of all continents to form Pangaea.
   e. changes in Earth's orbit.
8. The cause of the mass extinction at the end of the Mesozoic era probably was
   a. continental drift.
   b. the collision of Earth with a large meteorite.
   c. changes in Earth's orbit.
   d. massive glaciation.
   e. changes in the salt concentration of the oceans.
9. The times during the history of life when many new evolutionary lineages appeared were the
   a. Precambrian, Cambrian, and Triassic.
   b. Precambrian, Cambrian, and Tertiary.
   c. Cambrian, Paleozoic, and Triassic.
   d. Cambrian, Triassic, and Devonian.
   e. Paleozoic, Triassic, and Tertiary.
10. At which of the following times was there *no* mass extinction?
   a. The end of the Cretaceous period
   b. The end of the Devonian period
   c. The end of the Permian period
   d. The end of the Triassic period
   e. The end of the Silurian period

FOR DISCUSSION

1. Some groups of organisms have evolved to contain large numbers of species; other groups have produced only a few species. Is it meaningful to consider the former groups more successful than the latter? What does the word "success" mean in evolution? How does your answer influence your thinking about *Homo sapiens*, the only surviving representative of a clade that never had many species in it?
2. Scientists date ancient events using a variety of methods, but nobody was present to witness or record those events. Accepting those dates requires us to believe in the accuracy and appropriateness of indirect measurement techniques. What other basic scientific concepts are also based on the results of indirect measurement techniques?
3. Why is it useful to be able to date past events absolutely as well as relatively?
4. If we are living during one of the cooler periods in Earth's history, why should we be concerned about human activities that are contributing to global climate warming?
5. Large meteorites colliding with Earth have caused massive climate and evolutionary changes in the past. Should we attempt to take steps to prevent future impacts? What actions might we undertake? What adverse effects might such actions trigger?

FOR INVESTIGATION

The experiment in Figure 21.24 showed that, over evolutionary history, large canid species with specialized diets survived for shorter times than did smaller but less specialized species. Large herbivores also survived for shorter times that smaller ones. What hypotheses would you propose to explain these facts? How would you test your hypotheses?

CHAPTER 22 The Mechanisms of Evolution

Snake eats poisonous newt—and lives!

Newts and other salamanders move slowly, so they are easy prey for garter snakes and other predators. But some of these amphibians have evolved chemical defenses against predation—in a word, they are poisonous. The rough-skinned newt, *Taricha granulosa*, is a salamander that lives on the Pacific Coast of North America. The newt sequesters in its skin a potent neurotoxin called tetrodotoxin (TTX). TTX paralyzes nerves and muscles by blocking sodium channels (see Section 5.3). Most vertebrates, including predatory garter snakes, will die if they eat a rough-skinned newt.

But some snakes can eat rough-skinned newts and survive. In certain populations of the garter snake *Thamnophis sirtalis*, most individuals have TTX-resistant sodium channels in their nerves and muscles—but they pay a price for this attribute. TTX-resistant snakes can move only slowly for several hours after eating a newt, and they never move as fast as nonresistant snakes. Thus TTX-resistant snakes are more vulnerable to their own predators than are TTX-sensitive snakes that simply don't encounter poisonous newts.

Pufferfish, octopuses, tunicates, and some species of frogs also use TTX as a defensive chemical. Other animals and many plants use a variety of chemicals to defend themselves against predators, and many predators have evolved resistance to those chemicals. Both the production of and resistance to defensive chemicals are evolutionary *adaptations*. But adaptations that benefit an individual in one way may reduce the ability of the organism to carry out other activities, as happens with the decreased ability of TTX-resistant garter snakes to move quickly. And adaptations such as the ability to produce neurotoxins in the skin may be energetically costly for an organism to develop and maintain. In other words, to improve its performance in one area, an organism typically experiences reduced performance in some other area—there is a *trade-off* between the benefits of an adaptation and what it costs the organism that expresses it.

One tool of evolutionary biologists is to try to identify and measure the trade-offs that different adaptations impose, because the nature and strength of those trade-offs influence the extent to which different adaptations succeed, and thus how populations of organisms evolve. If TTX resistance had no cost, we might expect to find resistant garter snakes everywhere, even in environments where there are no toxic salamanders. This is not the case. If toxic defensive chemicals had no cost, individuals

Evolutionary War Rough-skinned newts (below) evolved the ability to secrete a paralytic poison in their skin, which deters most predators. Some common garter snakes (above) have evolved resistance to the poison.

CHAPTER OUTLINE

A Workable Plan Shows Up Elsewhere The slow-swimming map pufferfish (*Arothron mappa*), shown here feeding in the seas off Thailand, produces TTX in several of its organs, notably the liver. The level of toxicity appears to vary seasonally, and does not stop people from regarding the flesh of these fish as a dangerous delicacy.

of many prey species would probably be poisonous. They are not.

Preeminent among Charles Darwin's many contributions to biology was putting forth a plausible and testable hypothesis for a mechanism that could result in the adaptation of organisms to their environments. In effect, Darwin offered a mechanistic explanation for the evolution of all forms of life, including humans. Some people find it difficult to accept that the same mechanistic processes that determined the evolutionary pathways of plants, insects, and bacteria also guided human evolution. But as Darwin noted, "there is grandeur in this view of life."

IN THIS CHAPTER we will see how Charles Darwin developed his ideas, and then turn to the advances in our understanding of evolutionary mechanisms since Darwin's time. We will discuss the genetic basis of evolution and show how genetic variation within populations is measured. We will describe the mechanisms of evolution and show how biologists design studies to investigate them. Finally, we will discuss constraints on the pathways evolution can take.

22.1 What Facts Form the Base of Our Understanding of Evolution?

Today, a rich array of geological, morphological, and molecular data support and enhance the factual basis of evolution; but when Charles Darwin was a youth, it was not evident to him (or to almost anyone else) that life had evolved. Darwin was passionately interested in both geology and natural history—the scientific study of how different organisms function and carry out their lives in nature. Despite having these interests, he had planned to become a doctor, but he was nauseated by surgery conducted without anesthesia. He gave up medicine to study at Cambridge University for a career as a clergyman of the Church of England. Darwin was more interested in science than in theology, however, and he became a companion of scientists on the faculty, especially the botanist John Henslow. In 1831, Henslow recommended Darwin for a position on the H.M.S. *Beagle*, which was preparing for a survey voyage around the world (**Figure 22.1**).

Whenever possible during the 5-year-long voyage, Darwin (who was often seasick) went ashore to study rocks and to observe and collect specimens of plants and animals. He noticed how strikingly the species he saw in South America differed from those of Europe. He observed that the species of the temperate regions of South America (Argentina and Chile) were more similar to those of tropical South America (Brazil) than they were to temperate European species. When he explored the Galápagos Islands, west of Ecuador, he noted that most of its animal species were found nowhere else, but were similar to those of mainland South America. Darwin also recognized that the animals of the archipelago differed from island to island. He postulated that some animals had come to the archipelago from mainland South America and then undergone different changes on each of the islands. But what mechanism could account for these changes?

When he returned to England in 1836, Darwin continued to ponder his observations. Within a decade he had developed the

(A)

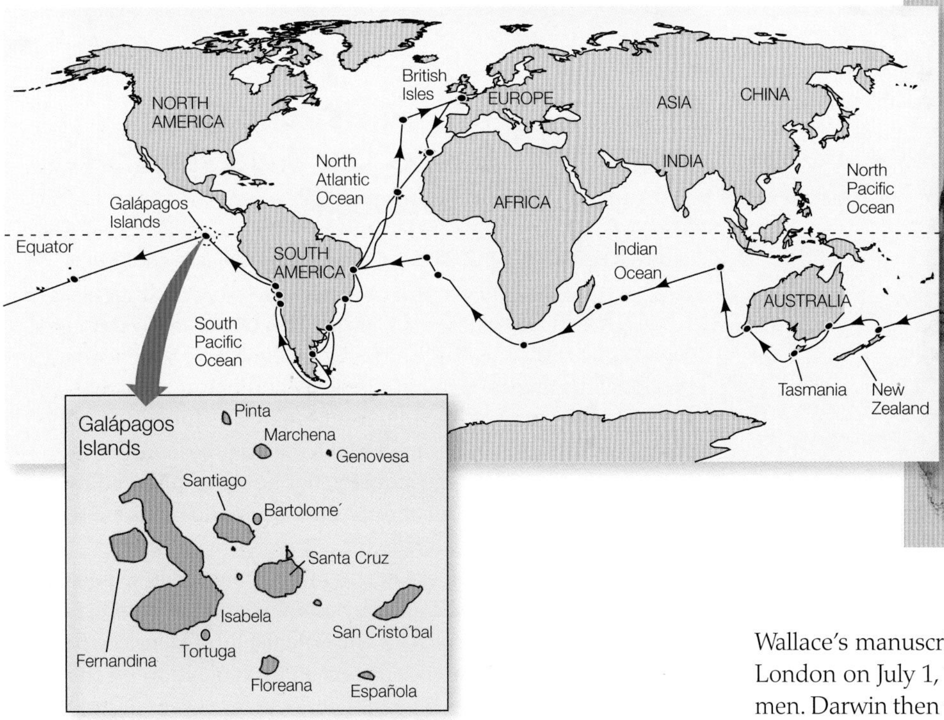

22.1 Darwin and the Voyage of the *Beagle* (A) The mission of H.M.S. *Beagle* was to chart the oceans and collect oceanographic and biological information from around the world. The map indicates the ship's path, and the inset shows the Galápagos Islands, whose organisms were an important source of Darwin's ideas on natural selection. (B) Charles Darwin at age 24, shortly after the *Beagle* returned to England.

(B)

major features of an explanatory theory for *evolutionary change* based on two major propositions:

- Species are not immutable; they change over time.
- The process that produces these changes is *natural selection*.

Darwin asserted that evolution is a historical fact that can be demonstrated to have taken place (his first proposition). In 1844, he wrote a long essay on natural selection, the process he described as the cause of evolution (his second proposition), but, despite urging from his wife and colleagues, he was reluctant to publish it, preferring to assemble more evidence first.

Darwin's hand was forced in 1858 when he received a letter and manuscript from another traveling naturalist, Alfred Russel Wallace, who was studying the biota of the Malay Archipelago. Wallace asked Darwin to evaluate the manuscript, in which Wallace proposed a theory of natural selection almost identical to Darwin's. At first Darwin was dismayed, believing that Wallace had preempted his idea. But parts of Darwin's 1844 essay, together with Wallace's manuscript, were presented to the Linnaean Society of London on July 1, 1858, thereby giving credit for the idea to both men. Darwin then worked quickly to finish his own book, *The Origin of Species*, which was published the next year.

Although both Darwin and Wallace independently articulated the concept of natural selection, Darwin developed his ideas first. Furthermore, *The Origin of Species* provided exhaustive evidence from many fields to support both natural selection and evolution itself; thus both concepts are more closely associated with Darwin than with Wallace.

The facts that Darwin used to conceive and develop his theory of evolution by natural selection were familiar to most contemporary biologists. His insight was to perceive the significance of relationships among them. Both Darwin and Wallace were influenced by the ideas of the economist Thomas Malthus, who in 1838 published *An Essay on the Principle of Population*. Malthus argued that because the rate of human population growth is greater than the rate of increase in food production, unchecked growth inevitably leads to famine. Darwin saw parallels throughout nature.

He recognized that populations of all species have the potential for rapid increases in numbers. To illustrate this point, he used the following example:

> Suppose ... there are eight pairs of birds, and that only four pairs of them annually ... rear only four young, and that these go on rearing their young at the same rate, then at the end of seven years ... there will be 2048 birds instead of the original sixteen.

Yet such rates of increase are rarely seen in nature. Therefore, Darwin reasoned that death rates in nature must also be high. Without high death rates, even the most slowly reproducing species would quickly reach enormous population sizes.

Darwin also observed that, although offspring tend to resemble their parents, the offspring of most organisms are not identical to one another or to their parents. He suggested that slight variations among individuals affect the chance that a given individual will survive and reproduce. Darwin called this differential survival and reproduction of individuals **natural selection**. Natural selection is formally defined as *the differential contribution of offspring to the next generation by various genetic types belonging to the same population*.

Darwin may have used the words "natural selection" because he was familiar with the **artificial selection** of individuals with certain desirable traits by animal and plant breeders. Many of Darwin's observations on the nature of variation came from domesticated plants and animals. Darwin was a pigeon breeder, and he knew firsthand the astonishing diversity in color, size, form, and behavior that pigeon breeders could achieve (**Figure 22.2**). He recognized close parallels between selection by breeders and selection in nature. As he argued in *The Origin of Species*,

> How can it be doubted, from the struggle each individual has to obtain subsistence, that any minute variation in structure, habits or instincts, adapting that individual better to the new conditions, would tell upon its vigour and health? In the struggle it would have a better chance of surviving; and those of its offspring which inherited the variation, be it ever so slight, would have a better chance.

That statement, written almost 150 years ago, still stands as a good expression of the process of evolution by natural selection.

It is important to remember, as Darwin clearly understood, that *individuals do not evolve; populations do*. A **population** is a group of individuals of a single species that live and interbreed in a particular geographic area at the same time. A major consequence of the evolution of populations is that their members become adapted to the environments in which they live. But what do biologists mean when they say that an organism is adapted to its environment?

Adaptation has two meanings

In evolutionary biology, the term **adaptation** refers both to the *processes* by which characteristics that appear to be useful to their bearers evolve—that is, the evolutionary mechanisms that produce them—and to the *characteristics* themselves. With respect to characteristics, an adaptation is a phenotypic characteristic that has helped an organism adjust to conditions in its environment. We will discuss the processes that result in adaptation in great detail in this chapter.

Biologists regard an organism as being adapted to a particular environment when they can demonstrate that a slightly different organism reproduces and survives less well in that environment. To understand adaptation, biologists compare the performance of individuals that differ in their traits. For example, biologists could assess the adaptive role of the chemical defenses of the rough-skinned newts described at the opening of this chapter by comparing the survival and reproductive rates of newts that had different concentrations of TTX in their skins.

When Darwin proposed his theory of evolution by natural selection, he could point to many examples of evolutionary mechanisms operating in nature, but none were supported by experiments. Since then, many observational and experimental studies of evolutionary mechanisms have been conducted. Biologists have also documented changes over time in the genetic composition of many populations, and our understanding of the mechanisms of inheritance has improved enormously.

Population genetics provides an underpinning for Darwin's theory

For a population to evolve, its members must possess heritable genetic variation, which is the raw material on which mechanisms of evolution act. In everyday life, we cannot directly observe the genetic compositions of organisms. What we do see in nature are *phenotypes*, the physical expressions of organisms' genes. The features

22.2 Many Types of Pigeons Have Been Produced by Artificial Selection Charles Darwin raised pigeons as a hobby, and he noted similar forces at work in artificial and natural selection. The pigeons shown here are just some of the more than 300 varieties that have been artificially selected by breeders to display different forms of characters such as color, size, and feather distribution.

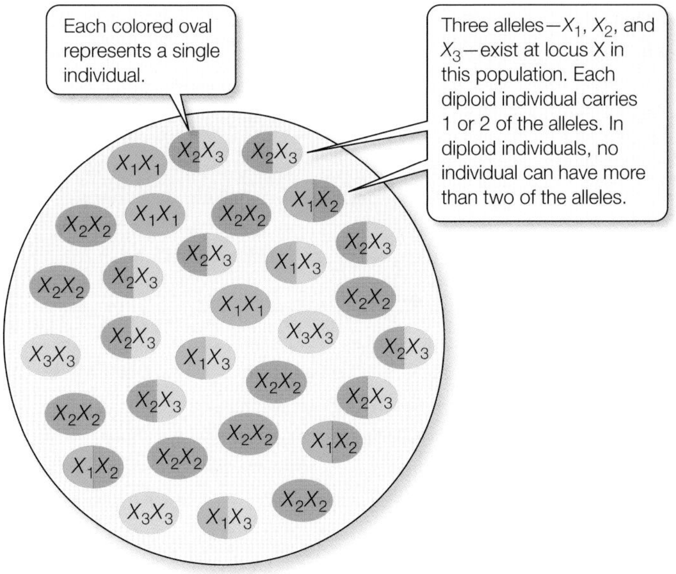

22.3 A Gene Pool A gene pool is the sum of all the alleles found in a population. The gene pool for only one locus, *X*, is shown in this figure. The allele frequencies in this gene pool are 0.20 for X_1, 0.50 for X_2, and 0.30 for X_3.

of a phenotype are its *characters*—eye color, for example. The specific form of a character, such as brown eyes, is a *trait*. A **heritable** trait is a characteristic of an organism that is at least partly determined by its genes. The genetic constitution that governs a character is called its *genotype*. *A population evolves when individuals with different genotypes survive or reproduce at different rates.*

The rediscovery of Gregor Mendel's publications in the early 1900s (see Section 10.1) paved the way for the development of the field of **population genetics**. Population genetics has three main goals:

- To explain the origin and maintenance of genetic variation
- To explain the patterns and organization of genetic variation
- To understand the mechanisms that cause changes in allele frequencies in populations

The perspective of population genetics complements the insights into evolutionary processes provided by developmental biology, which we discussed in Chapter 20.

As described in Section 10.1, different forms of a gene, called *alleles*, may exist at a particular locus. At any particular locus, a single individual has only some of the alleles found in the population to which it belongs (**Figure 22.3**). The sum of all copies of all alleles at all loci found in a population constitutes its **gene pool**. (We can also refer to the "gene pool" for a particular locus or loci.) The gene pool contains the genetic variation that produces the phenotypic traits on which natural selection acts. To understand evolution and the role of natural selection, we need to know how much genetic variation populations have, the sources of that genetic variation, and how genetic variation changes in populations over space and time.

Most populations are genetically variable

Nearly all populations have genetic variation for many characters. Artificial selection on different characters in a single European species of wild mustard produced many important crop plants (**Figure 22.4**). Agriculturalists could achieve these results because the original mustard population had genetic variation for the characters of interest.

Laboratory experiments also demonstrate the existence of considerable genetic variation in populations. In one such experiment, investigators attempted to breed populations of fruit flies (*Drosophila melanogaster*) with high or low numbers of bristles on their abdomens from an initial population with intermediate numbers of bristles. After 35 generations, all

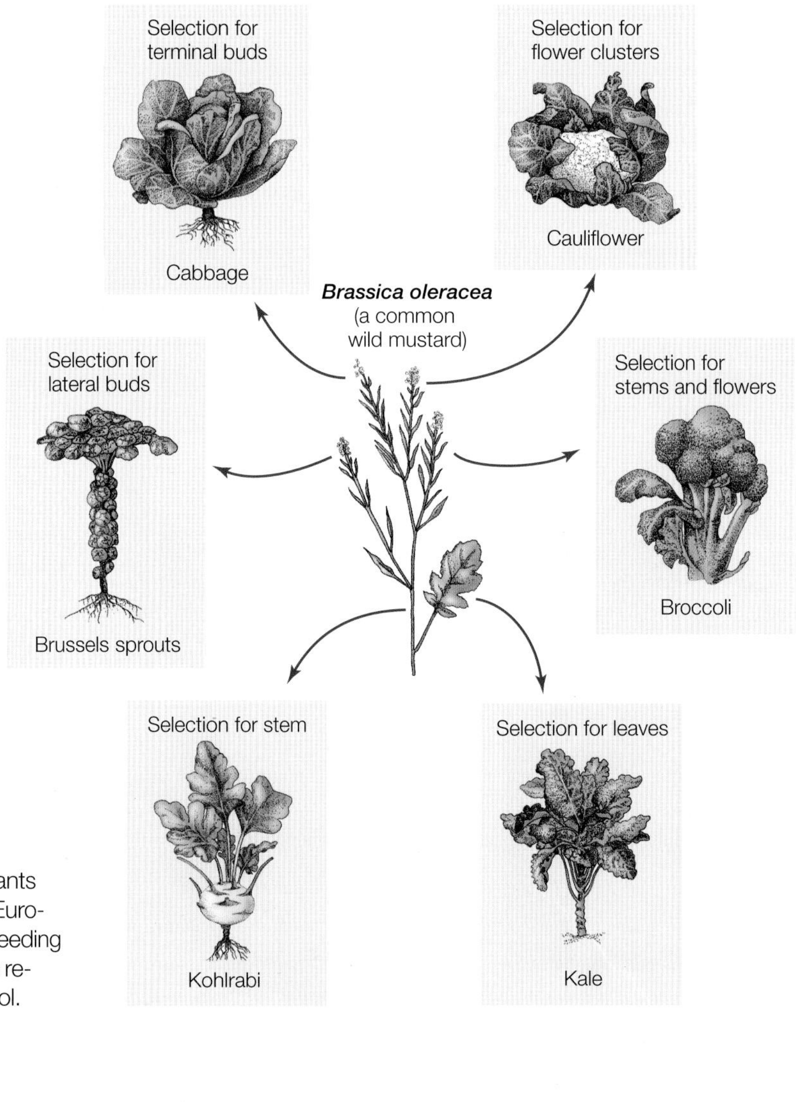

22.4 Many Vegetables from One Species All of the crop plants shown have been derived from a single wild mustard species. European agriculturalists produced these crops by choosing and breeding plants with unusually large buds, stems, leaves, or flowers. The results illustrate the vast amount of variation present in a gene pool.

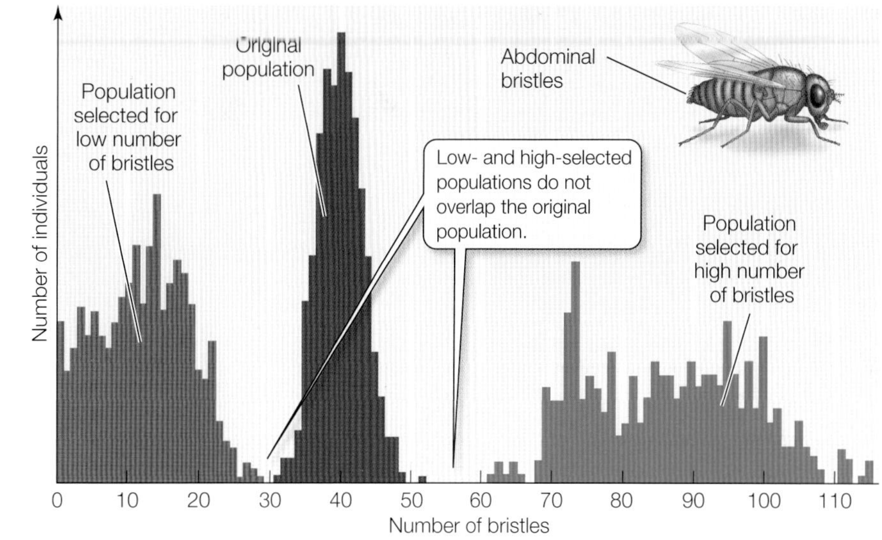

22.5 Artificial Selection Reveals Genetic Variation In artificial selection experiments with *Drosophila melanogaster*, bristle numbers evolved rapidly. The graphs show the number of flies with different numbers of bristles after 35 generations of artificial selection.

flies in both the high-bristle and low-bristle lineages had bristle numbers that fell well outside the range found in the original population (**Figure 22.5**). Thus there must have been considerable genetic variation in the original fruit fly population for selection to act on.

The study of the genetic basis of natural selection is difficult because genotypes alone do not uniquely determine all phenotypes. With dominance, for example, a particular phenotype can be produced by more than one genotype (e.g., *AA* and *Aa* individuals may be phenotypically identical). Similarly, as we described in detail in Section 20.4, different phenotypes can be produced by a given genotype, depending on the environment encountered during development. For example, the cells of all the leaves on a tree or shrub are usually genetically identical, yet leaves of the same plant often differ in shape and size.

Evolutionary change can be measured by allele and genotype frequencies

Allele frequencies are usually estimated in locally interbreeding groups, called **Mendelian populations**, within a geographic population of a species. To measure allele frequencies in a Mendelian population precisely, we would need to count every allele at every locus in every individual in it. By doing so, we could determine the frequencies of all alleles in the population. The word **frequency** in population genetics means *proportion*, so the frequency of a given allele or genotype is simply the proportion of the gene pool at that locus.

Fortunately, we do not need to make complete measurements, because we can reliably estimate allele frequencies for a given locus by counting alleles in a sample of individuals from the population. The sum of all allele frequencies at a locus is equal to 1, so measures of allele frequency range from 0 to 1.

An allele's frequency is calculated using the following formula:

$$p = \frac{\text{number of copies of the allele in the population}}{\text{sum of alleles in the population}}$$

If only two alleles (we'll call them *A* and *a*) for a given locus are found among the members of a diploid population, they may combine to form three different genotypes: *AA*, *Aa*, and *aa*. Such a population is said to be *polymorphic* at that locus, since there is more than one allele. Using the formula above, we can calculate the relative frequencies of alleles *A* and *a* in a population of *N* individuals as follows:

- Let N_{AA} be the number of individuals that are homozygous for the *A* allele (*AA*).
- Let N_{Aa} be the number that are heterozygous (*Aa*).
- Let N_{aa} be the number that are homozygous for the *a* allele (*aa*).

Note that $N_{AA} + N_{Aa} + N_{aa} = N$, the total number of individuals in the population, and that the total number of copies of both alleles present in the population is $2N$ because each individual is diploid. Each *AA* individual has two copies of the *A* allele, and each *Aa* individual has one copy of the *A* allele. Therefore, the total number of *A* alleles in the population is $2N_{AA} + N_{Aa}$. Similarly, the total number of *a* alleles in the population is $2N_{aa} + N_{Aa}$.

If p represents the frequency of *A*, and q represents the frequency of *a*, then

$$p = \frac{2N_{AA} + N_{Aa}}{2N}$$

and

$$q = \frac{2N_{aa} + N_{Aa}}{2N}$$

To show how this formula works, **Figure 22.6** calculates allele frequencies in two hypothetical populations, each containing 200 diploid individuals. Population 1 has mostly homozygotes (90 *AA*, 40 *Aa*, and 70 *aa*), whereas population 2 has mostly heterozygotes (45 *AA*, 130 *Aa*, and 25 *aa*).

The calculations in Figure 22.6 demonstrate two important points. First, notice that for each population, $p + q = 1$. If $p + q = 1$, then $q = 1 - p$. So when there are only two alleles at a given locus in a population, we can calculate the frequency of one allele and then easily obtain the second allele's frequency by subtraction. If there is only one allele at a given locus in a population, its frequency is 1: the population is then *monomorphic* at that locus, and the allele is said to be *fixed*.

The second thing to notice is that population 1 (consisting mostly of homozygotes) and population 2 (consisting mostly of heterozygotes) have the same allele frequencies for *A* and *a*. Thus they have the same gene pool for this locus. However, because the alleles in the gene pool are distributed differently among individuals, the *genotype frequencies* of the two populations differ. Geno-

RESEARCH METHOD

1 Determine the allele frequencies in the population.

In any population:

$$\text{Frequency of allele } A = p = \frac{2N_{AA} + N_{Aa}}{2N} \qquad \text{Frequency of allele } a = q = \frac{2N_{aa} + N_{Aa}}{2N}$$

where N is the total number of individuals in the population.

2 Compute allele frequencies for different populations.

For population 1 (mostly homozygotes):

$N_{AA} = 90$, $N_{Aa} = 40$, and $N_{aa} = 70$

so

$$p = \frac{180 + 40}{400} = 0.55$$

$$q = \frac{140 + 40}{400} = 0.45$$

For population 2 (mostly heterozygotes):

$N_{AA} = 45$, $N_{Aa} = 130$, and $N_{aa} = 25$

so

$$p = \frac{90 + 130}{400} = 0.55$$

$$q = \frac{50 + 130}{400} = 0.45$$

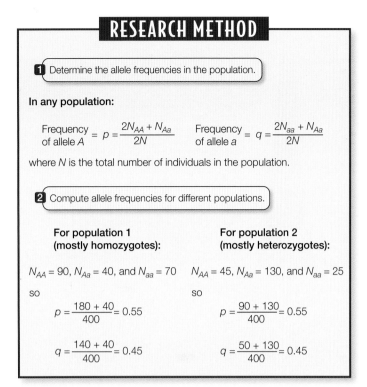

22.6 Calculating Allele Frequencies The gene pool and allele frequencies are the same in the two different populations, but the alleles are distributed differently between heterozygous and homozygous genotypes. In all cases, $p + q$ must equal 1.

type frequencies are calculated as the number of individuals that have a given genotype divided by the total number of individuals in the population. In population 1 in Figure 22.6, the genotype frequencies are 0.45 *AA*, 0.20 *Aa*, and 0.35 *aa*.

The frequencies of different alleles at each locus and the frequencies of different genotypes in a Mendelian population describe that population's **genetic structure**. Allele frequencies measure the amount of genetic variation in a population; genotype frequencies show how a population's genetic variation is distributed among its members. With these measurements, it becomes possible to consider how the genetic structure of a population changes or remains the same over generations—that is, to measure evolutionary change.

The genetic structure of a population does not change over time if certain conditions exist

In 1908, the British mathematician Godfrey Hardy and the German physician Wilhelm Weinberg independently deduced the conditions that must prevail if the genetic structure of a population is to remain the same over time. Hardy's equations explain why dominant alleles do not necessarily replace recessive alleles in populations, as well as other features of the genetic structure of populations: *If an allele is not advantageous, its frequency remains constant from generation to generation;* its frequency will not increase even if the allele is dominant.

Hardy wrote his equations in response to a question posed to him by the geneticist Reginald C. Punnett at the Cambridge University faculty club. Punnett wondered why, even though the allele for brachydactyly (short, stubby fingers) is dominant and the allele for normal-length fingers is recessive, most people in Britain have normal-length fingers.

Hardy–Weinberg equilibrium is the cornerstone of population genetics. The equation describes a model situation in which allele frequencies do not change across generations and genotype frequencies can be predicted from allele frequencies (**Figure 22.7**). The principles of Hardy–Weinberg equilibrium apply to sexually reproducing organisms. Several conditions must be met for a population to be at Hardy–Weinberg equilibrium:

- *Mating is random.* Individuals do not preferentially choose mates with certain genotypes.
- *Population size is infinite.* The larger a population, the smaller will be the effect of *genetic drift*—random (chance) fluctuations in allele frequencies.
- *There is no gene flow.* There is no migration either into or out of the population.
- *There is no mutation.* There is no change to alleles *A* and *a*, and no new alleles are added to change the gene pool.
- *Natural selection does not affect the survival of particular genotypes.* There is no differential survival of individuals with different genotypes.

If these ideal conditions hold, two major consequences follow. First, the frequencies of alleles at a locus remain constant from generation to generation. Second, following one generation of random mating, the genotype frequencies occur in the following proportions:

Genotype	*AA*	*Aa*	*aa*
Frequency	p^2	$2pq$	q^2

Consider generation 1 in Figure 22.7, in which the frequency of the *A* allele (p) is 0.55. Because we assume that individuals select mates at random, without regard to their genotype, gametes carrying *A* or *a* combine at random—that is, as predicted by the frequencies p and q. The probability that a particular sperm or egg in this example will bear an *A* allele rather than an *a* allele is 0.55. In other words, 55 out of 100 randomly sampled sperm or eggs will bear an *A* allele. Because $q = 1 - p$, the probability that a sperm or egg will bear an *a* allele is $1 - 0.55 = 0.45$. (You may wish to review the discussion of probability in Section 10.1.)

To obtain the probability of two *A*-bearing gametes coming together at fertilization, we multiply the two independent probabilities of their occurring separately:

$$p \times p = p^2 = (0.55)^2 = 0.3025$$

Therefore, 0.3025, or 30.25 percent, of the offspring in the next generation will have the *AA* genotype. Similarly, the probability of bringing together two *a*-bearing gametes is

$$q \times q = q^2 = (0.45)^2 = 0.2025$$

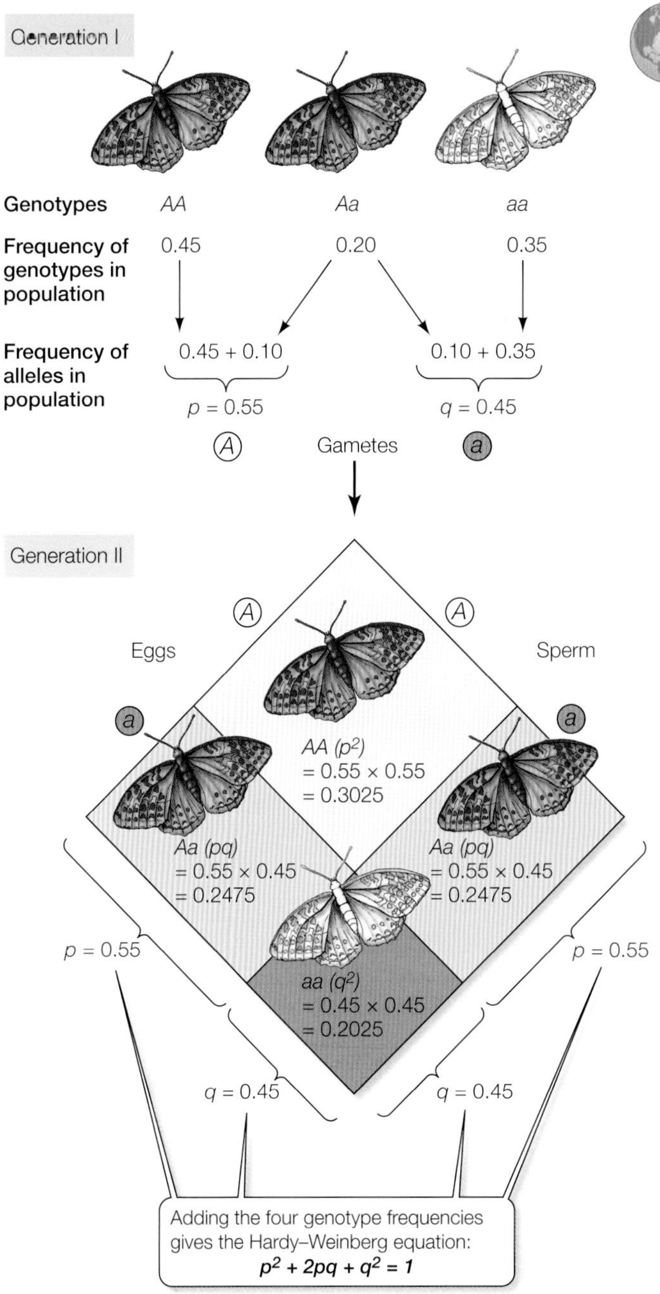

22.7 Calculating Hardy–Weinberg Genotype Frequencies The areas within the rectangle are proportional to the expected frequencies if mating is random with respect to genotype. Because there are two ways of producing a heterozygote, the probability of this event occurring is the sum of the two *Aa* squares. This example assumes that (1) the organism in question is diploid, (2) its generations do not overlap, (3) the gene under consideration has two alleles, and (4) allele frequencies are identical in males and females. Hardy–Weinberg equilibrium also applies if the gene has more than two alleles and generations overlap, but in those cases the mathematics is more complicated.

Thus 20.25 percent of the next generation will have the *aa* genotype.

Figure 22.7 also shows that there are two ways of producing a heterozygote: an *A* sperm may combine with an *a* egg, the probability of which is $p \times q$; or an *a* sperm may combine with an *A* egg, the probability of which is $q \times p$. Consequently, the overall probability of obtaining a heterozygote is $2pq$.

It is now easy to show that the allele frequencies p and q remain constant for each generation. If the frequency of *A* alleles in a randomly mating population is $p^2 + pq$, this frequency becomes $p^2 + p(1 - p) = p^2 + p - p^2 = p$. The original allele frequencies are unchanged, and the population is at *Hardy–Weinberg equilibrium*, as described by the **Hardy–Weinberg equation**:

$$p^2 + 2pq + q^2 = 1$$

If the genotype frequencies in the parental generation were to be altered (say, by emigration of a large number of *AA* individuals from the population), then the allele frequencies in the next generation would also be altered. However, based on the new allele frequencies, a single generation of random mating is sufficient to restore the genotype frequencies to Hardy–Weinberg (H-W) equilibrium, based on the new allele frequencies.

Deviations from Hardy–Weinberg equilibrium show that evolution is occurring

You may already have realized that populations in nature never meet the stringent conditions necessary to maintain them at H-W equilibrium. Why, then, is this model considered so important for the study of evolution? There are two reasons. First, the equation is useful for predicting the approximate genotype frequencies of a population from its allele frequencies.

Second, and crucially, the model describes the conditions that would result if there were *no* evolution in a population: the Hardy–Weinberg equation shows that *allele frequencies will remain the same from generation to generation unless some mechanism acts to change them*. Since the model's conditions are never met completely, allele frequencies in all populations do in fact deviate from Hardy–Weinberg equilibrium. In other words, *there are mechanisms acting to change allele frequencies, and these mechanisms drive evolution*. The patterns of deviation from Hardy–Weinberg equilibrium can help us identify specific mechanisms of evolutionary change.

22.1 RECAP

Laboratory experiments and the results of artificial selection demonstrate the existence of genetic variation in populations and provide evidence that natural selection can and does produce evolutionary change.

- Can you articulate the principle of natural selection? The two meanings of adaptation? See p. 489
- Do you see how calculating allele frequencies allows us to measure evolutionary change? See pp. 491–492 and Figure 22.6
- Why is the concept of Hardy–Weinberg equilibrium important even though the assumptions on which it is based are nearly always violated in nature? See pp. 492–493

We have briefly outlined Charles Darwin's vision of natural selection and adaptation and explained the mathematical basis of Hardy–Weinberg equilibrium and its importance for studying evolution. Now let's take a look at some of the forces that cause populations to deviate from equilibrium—the mechanisms of evolutionary change.

22.2 What Are the Mechanisms of Evolutionary Change?

Evolutionary mechanisms are forces that change the genetic structure of a population. Hardy–Weinberg equilibrium is a null hypothesis that assumes that those forces are absent. The known evolutionary mechanisms include mutation, gene flow, genetic drift, nonrandom mating, and natural selection. To understand evolutionary processes we need to discuss each of these mechanisms before considering natural selection in detail.

Mutations generate genetic variation

The origin of genetic variation is mutation. A mutation, as we saw in Section 12.6, is any change in an organism's DNA. Mutations appear to be random with respect to the adaptive needs of organisms. Most mutations are harmful to their bearers or are neutral, but if environmental conditions change, previously harmful or neutral alleles may become advantageous. In addition, mutations can restore to populations alleles that other evolutionary processes have removed. Thus mutations both create and help maintain genetic variation within populations.

Mutation rates are very low for most loci that have been studied. Rates as high as one mutation per locus in a thousand zygotes per generation are rare; one in a million is more typical. Nonetheless, these rates are sufficient to create considerable genetic variation because each of a large number of genes may mutate, chromosomal rearrangements may change many genes simultaneously, and populations often contain large numbers of individuals. For example, if the probability of a point mutation (an addition, subtraction, or substitution of a single base) were 10^{-9} per base pair per generation, then in each human gamete, the DNA of which contains 3×10^9 base pairs, there would be an average of three new point mutations ($3 \times 10^9 \times 10^{-9} = 3$). Therefore, each zygote would carry, on average, six new mutations. The current human population of about 6.5 billion people would be expected to carry about 40 billion new mutations that were not present one generation earlier.

One condition for Hardy–Weinberg equilibrium is that there be no mutation. Although this condition is never strictly met, the rate at which mutations arise at a single locus is usually so low that mutations by themselves result in only very small deviations from Hardy–Weinberg equilibrium. If large deviations are found, it is appropriate to dismiss mutation as the cause and to look for evidence of other evolutionary mechanisms acting on the population.

Gene flow may change allele frequencies

Few populations are completely isolated from other populations of the same species. Migration of individuals and movements of gametes between populations, referred to as **gene flow**, are common. If the arriving individuals or gametes survive and reproduce in their new location, they may add new alleles to the gene pool of the population, or they may change the frequencies of alleles already present if they come from a population with different allele frequencies. For a population to be at Hardy–Weinberg equilibrium, there must be no gene flow from populations with different allele frequencies.

Genetic drift may cause large changes in small populations

In small populations, **genetic drift**—random changes in allele frequencies—may produce large changes in allele frequencies from one generation to the next. Harmful alleles may increase in frequency and rare advantageous alleles may be lost. Even in large populations, genetic drift can influence the frequencies of alleles that do not influence the survival and reproductive rates of their bearers.

As an example, suppose we perform a cross of $Aa \times Aa$ fruit flies to produce an F_1 population in which $p = q = 0.5$ and in which the genotype frequencies are 0.25 *AA*, 0.50 *Aa*, and 0.25 *aa*. If we randomly select 4 individuals (= 8 copies of the gene) from the F_1 population to produce the F_2 generation, the allele frequencies in this small sample population may differ markedly from $p = q = 0.5$. If, for example, we happen by chance to draw 2 *AA* homozygotes and 2 heterozygotes (*Aa*), the allele frequencies in the sample will be $p = 0.75$ (6 out of 8) and $q = 0.25$ (2 out of 8). If we replicate this sampling experiment 1,000 times, one of the two alleles will be missing entirely from about 8 of the 1,000 sample populations.

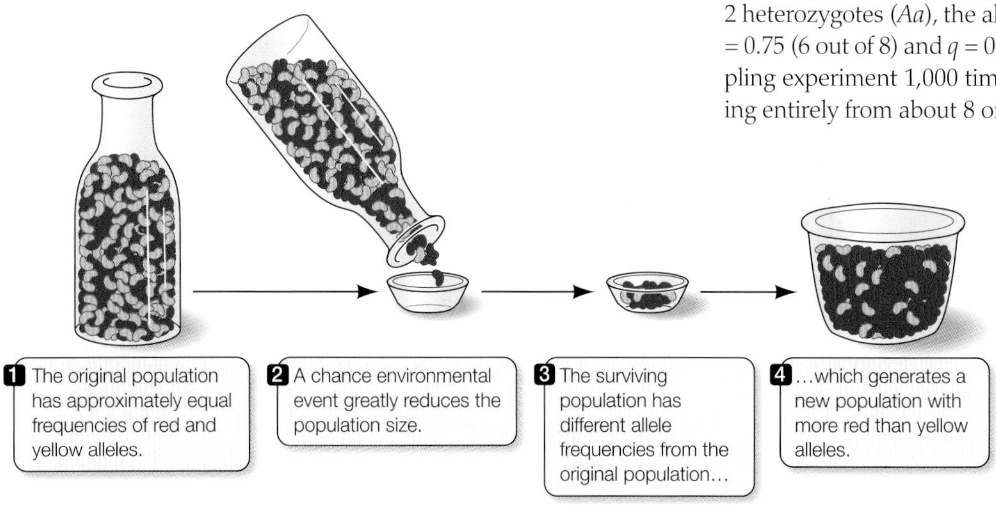

1 The original population has approximately equal frequencies of red and yellow alleles.
2 A chance environmental event greatly reduces the population size.
3 The surviving population has different allele frequencies from the original population…
4 …which generates a new population with more red than yellow alleles.

22.8 A Population Bottleneck Population bottlenecks occur when only a few individuals survive a random event, resulting in a shift in allele frequencies within the population.

The same principles operate when a population suffers large losses. Populations that are normally large may pass through occasional periods when only a small number of individuals survive. During these **population bottlenecks**, genetic variation can be reduced by genetic drift. How this works is illustrated in **Figure 22.8**, in which red and yellow beans represent two different alleles of a gene. Most of the "surviving" beans in the small sample taken from the original bean population are, just by chance, red, so the new population has a much higher frequency of red beans than the previous generation had. In a real population, the allele frequencies would be described as having "drifted."

A population forced through a bottleneck is likely to lose much of its genetic variation. Millions of greater prairie chickens lived in the prairies of North America when Europeans first arrived. As a result of both hunting and habitat destruction, the Illinois population of prairie chickens plummeted from about 100 million birds in 1900 to fewer than 50 individuals in the 1990s (**Figure 22.9A**).

22.9 Species with Low Genetic Variation (A) Prairie chickens in Illinois lost most of their genetic variation when the population crashed from millions to fewer than 100 individuals. (B) The California fan palm, whose range has been reduced to a small area of southern California and neighboring Mexico, has little genetic variation.

A comparison of DNA from birds collected in Illinois during the middle of the twentieth century with DNA from the surviving population in the 1990s showed that Illinois prairie chickens had lost most of their genetic diversity. The remaining population is experiencing low reproductive success. Similarly, the California fan palm (*Washingtonia filifera*) was once widespread in California and Mexico; today it is restricted to a few oases in extreme southern California and adjacent Mexico (**Figure 22.9B**). The species has little genetic variation: an average individual is heterozygous at only 0.009 of its loci.

South African cheetah populations have very low genetic variation, suggesting an extreme bottleneck occurred in the past. These animals are so close genetically that skin grafts from one cheetah to another are accepted as "self" and do not provoke an immune response.

Genetic drift can have similar effects when a few pioneering individuals colonize a new region. Because of its small size, the colonizing population is unlikely to have all the alleles found among members of its source population. The resulting change in genetic variation, called a **founder effect**, is equivalent to that in a large population reduced by a bottleneck. For example, the current population of the pitcher plant *Sarracenia purpurea* on a small island arose from a single individual that was planted there in 1912. Today the population has only one polymorphic locus in its entire genome.

Scientists were given an opportunity to study the genetic composition of founding populations when *Drosophila subobscura*, a well-studied species of fruit fly native to Europe, was discovered near Puerto Montt, Chile (in 1978), and at Port Townsend, Washington (in 1982). The *D. subobscura* founders probably reached Chile from Europe on a ship. A few flies carried north from Chile on another ship founded the North American population. In both South and North America, populations of the flies grew rapidly and expanded their ranges. Today in North America, *D. subobscura* ranges from British Columbia, Canada, to central California. In Chile it has spread across 15° of latitude (**Figure 22.10**).

European populations of *D. subobscura* have 80 chromosomal inversions (see Section 12.6), but the North and South American populations have only 20 such inversions—and they are the same 20 on both continents. North and South American populations also have lower allele diversity at certain enzyme-producing genes than European populations do. Only those alleles that have a frequency higher than 0.10 in European populations are present in the Americas. Thus, as expected for a small founding population, only a small part of the total genetic variation found in Europe reached the Americas. Geneticists estimate that somewhere between 4 and 100 flies founded the North and South American populations.

Nonrandom mating changes genotype frequencies

Mating patterns may alter genotype frequencies if individuals in a population choose other individuals of particular genotypes as mates (a phenomenon known as **nonrandom mating**). For exam-

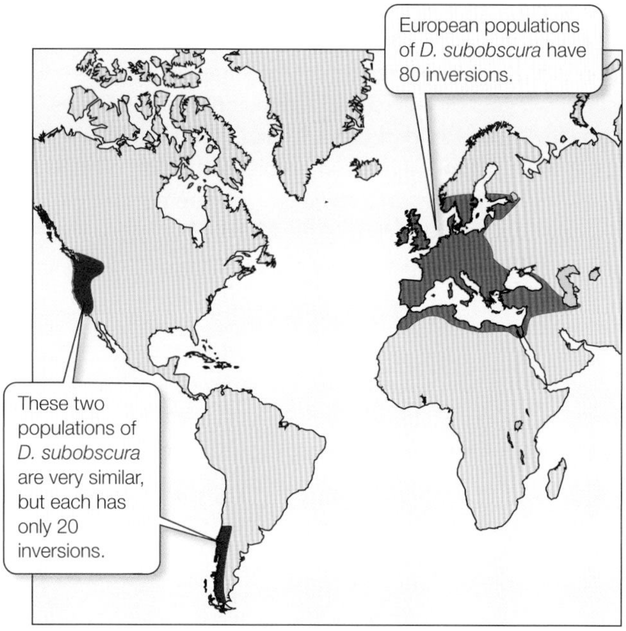

22.10 A Founder Effect Populations of the fruit fly *Drosophila subobscura* in North and South America contain less genetic variation than the European populations from which they came, as measured by the number of chromosome inversions in each population. Within two decades of arriving in the New World, *D. subobscura* populations had increased dramatically and spread widely in spite of their reduced genetic variation.

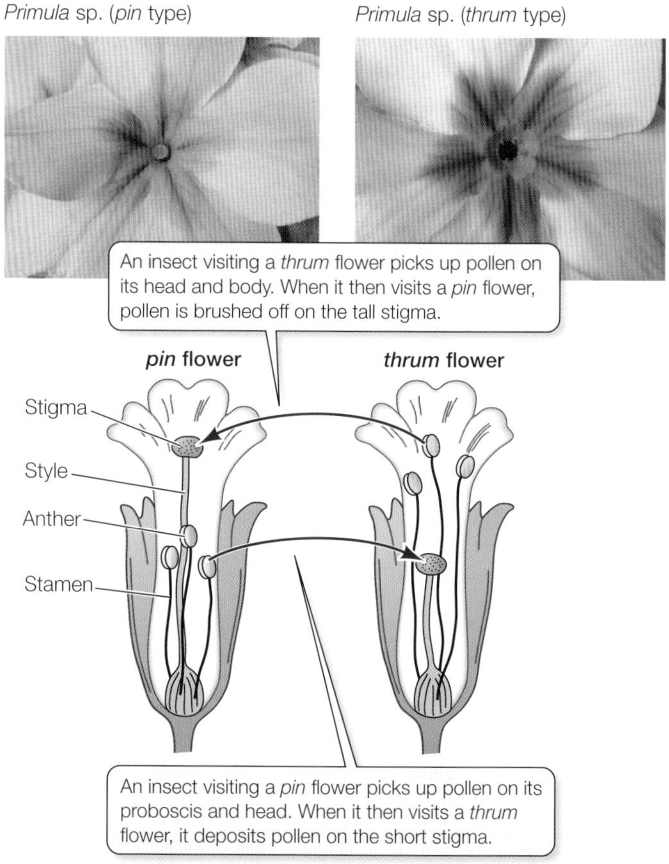

22.11 Flower Structure Fosters Nonrandom Mating Differing floral structure within the same plant species, as illustrated by this primrose, ensures that pollination usually occurs between individuals of different genotypes.

ple, if they mate preferentially with individuals of the same genotype, then homozygous genotypes will be overrepresented and heterozygous genotypes underrepresented, relative to Hardy–Weinberg expectations. Alternatively, individuals may mate primarily or exclusively with individuals of different genotypes.

Nonrandom mating is seen in some plant species, such as primroses (genus *Primula*), in which individual plants bear flowers of only one of two different types. One type, known as *pin*, has a long style (female reproductive organ) and short stamens (male reproductive organs). The other type, known as *thrum*, has a short style and long stamens (**Figure 22.11**). In many species with this reciprocal arrangement, pollen from one flower type can fertilize only flowers of the other type. Pollen grains from *pin* and *thrum* flowers are deposited on different parts of the bodies of insects that visit the flowers. When the insects visit other flowers, pollen grains from *pin* flowers are most likely to come into contact with stigmas of *thrum* flowers, and vice versa.

Self-fertilization (*selfing*), another form of nonrandom mating, is common in many groups of organisms, especially plants. Selfing reduces the frequencies of heterozygous individuals from Hardy–Weinberg equilibrium and increases the frequencies of homozygotes, but it does not change allele frequencies, and thus does not result in adaptation.

Sexual selection is a particularly important form of nonrandom mating that *does* change allele frequencies and often results in adaptations. We will discuss this important evolutionary mechanism in detail in the next section.

22.2 RECAP

Evolutionary mechanisms are forces that change the genetic structure of a population. The known evolutionary mechanisms include mutation, gene flow, genetic drift, nonrandom mating, and natural selection.

- Why do mutations, by themselves, result in only very small deviations from Hardy–Weinberg equilibrium? See p. 494
- Can you explain how genetic drift can cause large changes in small populations? See pp. 494–495 and Figure 22.8
- Why is it that some types of nonrandom mating alter genotype frequencies but do not change allele frequencies? See p. 496

The evolutionary mechanisms discussed so far influence the frequencies of alleles and genotypes in populations. Although all of these processes influence the course of biological evolution, none result in adaptations. For adaptation to occur, individuals that differ in heritable traits must survive and reproduce with different degrees of success. What evolutionary mechanisms have these effects?

22.3 What Evolutionary Mechanisms Result in Adaptation?

Adaptation occurs when some individuals in a population contribute more offspring to the next generation than others, such that *allele frequencies in the population change in a way that adapts individuals to the environment that influenced such reproductive success.* This is the mechanism that Darwin called "natural selection."

Although natural selection is formally defined as "the differential contribution of offspring to the next generation by various genotypes belonging to the same population," natural selection in fact acts on the *phenotype*—the physical features expressed by an organism with a given genotype—rather than acting directly on the genotype. The reproductive contribution of a phenotype to subsequent generations relative to the contributions of other phenotypes is called its **fitness**.

The absolute number of offspring produced by an individual does not influence the genetic structure of a population. Changes in absolute numbers of offspring are responsible for increases and decreases in the *size* of a population, but only changes in the *relative* success of different phenotypes within a population lead to changes in allele frequencies from one generation to another—that is, to adaptation. The fitness of individuals of a particular phenotype is a function of the probability of those individuals surviving multiplied by the average number of offspring they produce over their lifetimes. In other words, the *fitness of a phenotype is determined by the average rates of survival and reproduction of individuals with that phenotype.*

Natural selection produces variable results

To simplify our discussion until now, we have considered only characters influenced by alleles at a single locus. However, as we described in Section 10.3, most characters are influenced by alleles at more than one locus. Such characters are likely to show quantitative rather than qualitative variation. For example, the distribution of the body sizes of individuals in a population, a character that is influenced by genes at many loci as well as by the environment, is likely to resemble the bell-shaped curves shown in right-hand column of Figure 22.12.

Natural selection can act on characters with quantitative variation in any one of several different ways, producing quite different results:

- *Stabilizing selection* preserves the average characteristics of a population by favoring average individuals.
- *Directional selection* changes the characteristics of a population by favoring individuals that vary in one direction from the mean of the population.
- *Disruptive selection* changes the characteristics of a population by favoring individuals that vary in opposite directions from the mean of the population.

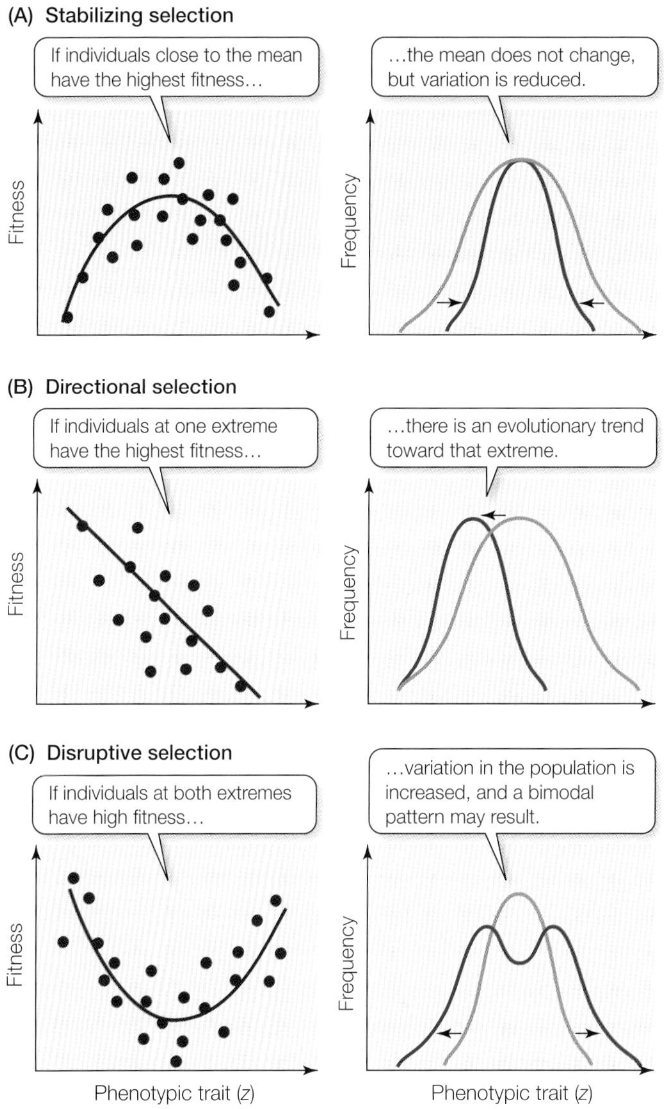

22.12 Natural Selection Can Operate on Quantitative Variation in Several Ways The graphs in the left-hand column show the fitnesses of individuals with different phenotypes of the same trait. Graphs at the right show the distribution of the phenotypes in the population before (light green) and after (dark green) the influence of selection.

STABILIZING SELECTION If the individuals in a population with both the smallest and the largest body sizes contribute relatively fewer offspring to the next generation than those closer to the average size do, then **stabilizing selection** is operating (**Figure 22.12A**). Stabilizing selection reduces variation in populations, but does not change the mean. Natural selection frequently acts in this way, countering increases in variation brought about by sexual recombination, mutation, or migration. Rates of evolution are typically very slow because natural selection is usually stabilizing. Stabilizing selection operates, for example, on human birth weight. Babies born lighter or heavier than the population mean die at higher rates than babies whose weights are close to the mean (**Figure 22.13**).

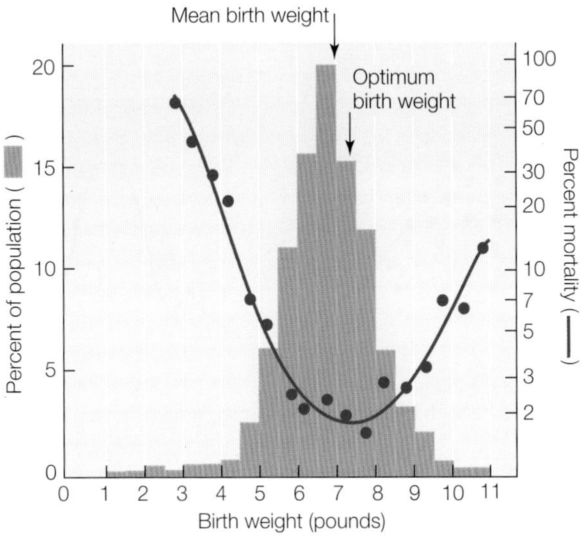

22.13 Human Birth Weight Is Influenced by Stabilizing Selection Babies that weigh more or less than average are more likely to die soon after birth than babies with weights close to the population mean.

DIRECTIONAL SELECTION If individuals at one extreme of a character distribution contribute relatively more offspring to the next generation than other individuals do, then the average value of that character in the population will shift toward that extreme. In this case, **directional selection** is operating. If directional selection operates over many generations, an *evolutionary trend* within the population is seen (**Figure 22.12B**). Directional evolutionary trends often continue for many generations, but they may be reversed when the environment changes and different phenotypes are favored, or they may be halted when an optimum phenotype is reached, or when trade-offs oppose further change. The character then falls under stabilizing selection.

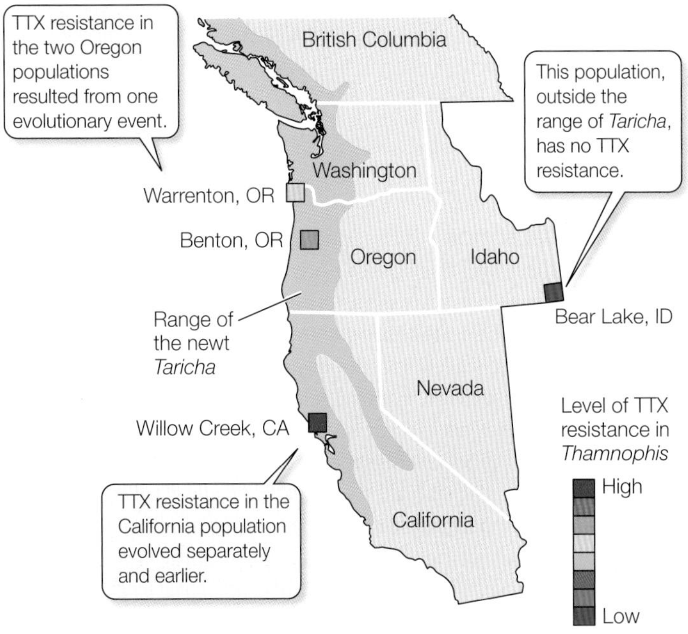

Directional selection produced the resistance to tetrodotoxin (TTX) in garter snakes discussed at the beginning of this chapter. TTX resistance evolved because, in some individual snakes, mutations in the allele (i.e., form of the gene) for the sodium channel proteins in their nerves and muscles resulted in a phenotype (i.e., physical protein) that could continue to function even when exposed to TTX. In areas where toxic newts abounded, such "mutant" snakes survived better than other individuals and left more offspring (who inherited their TTX-resistant allele for the sodium channel protein). TTX resistance has evolved independently several times within *Thamnophis sirtalis* populations in western North America via several different mutations in the same gene (**Figure 22.14**).

DISRUPTIVE SELECTION When **disruptive selection** operates, individuals at opposite extremes of a character distribution contribute more offspring to the next generation than do those close to the mean, increasing variation in the population (**Figure 22.12C**).

The strikingly bimodal (two-peaked) distribution of bill sizes in the black-bellied seedcracker (*Pyrenestes ostrinus*), a West African finch (**Figure 22.15**), illustrates how disruptive selection can influence populations in nature. The seeds of two types of sedges (marsh plants) are the most abundant food source for these finches during part of the year. Birds with large bills can readily crack the hard seeds of the sedge *Scleria verrucosa*. Birds with small bills can crack *S. verrucosa* seeds only with difficulty; however, they feed more efficiently on the soft seeds of *S. goossensii* than do birds with larger bills.

Young finches whose bills deviate markedly from the two predominant bill sizes do not survive as well as finches whose bills are close to one of the two sizes represented by the distribution peaks. Because there are few abundant food sources in the environment, and because the seeds of the two sedges do not overlap in hardness, birds with intermediate-sized bills are less efficient in using either one of the principal food sources. Disruptive selection therefore maintains a bimodal bill size distribution.

Sexual selection influences reproductive success

Sexual selection is a special type of natural selection that acts on characteristics that determine reproductive success. In *The Origin of Species*, Darwin devoted only a few pages to sexual selection, but he subsequently wrote a book about it—*The Descent of Man, and Selection in Relation to Sex* (1871). Sexual selection was Darwin's explanation for the evolution of apparently useless, and in some cases deleterious, but conspicuous characters, such as bright colors, long tails, horns, antlers, and elaborate courtship displays, in males of many species. He hypothesized that these features either improved the ability of their bearers to compete for access to mates (*intrasexual selection*) or made their bearers more attractive to members of the opposite sex (*intersexual selection*). The concept of sexual

22.14 Resistance to TTX Is Associated with the Presence of Newts Garter snakes (*Thamnophis sirtalis*) of the Pacific Coast have evolved resistance to the neurotoxin TTX, which is produced by a prey species found there, the newt *Taricha granulosa*. TTX resistance has evolved at least twice. The range of the newt is shown in blue.

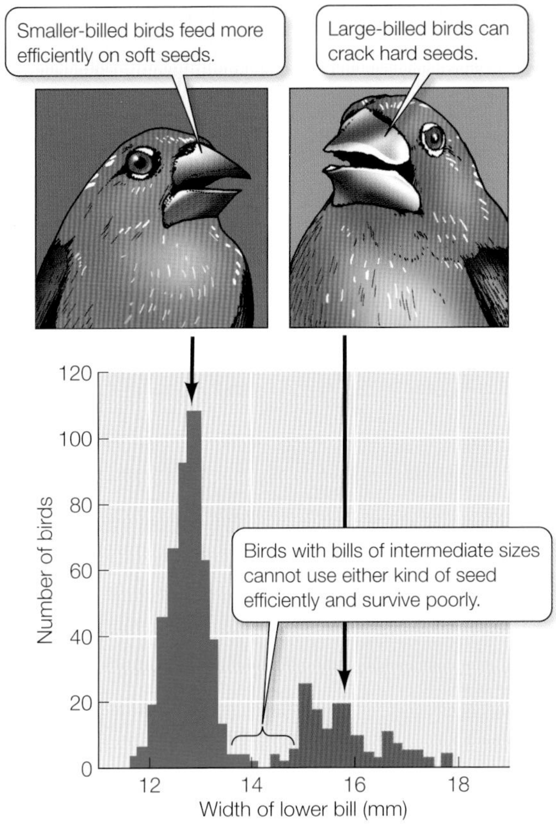

22.15 Disruptive Selection Results in a Bimodal Distribution The bimodal distribution of bill sizes in the black-bellied seedcracker of West Africa is a result of disruptive selection, which favors individuals with larger and smaller bill sizes over individuals with intermediate-sized bills.

selection was either ignored or questioned by Darwin's contemporaries, and for many decades afterward, but recent investigations have demonstrated its importance.

Darwin devoted an entire book to sexual selection because he recognized that, whereas natural selection typically favors traits that enhance the survival of their bearers or their descendants, sexual selection is primarily about success in reproduction. Of course, an animal must survive to reproduce, but if it survives and fails to reproduce, it makes no contribution to the next generation. Thus sexual selection may favor traits that enhance the bearer's chances of reproduction, but reduce its chances of survival. Such costly traits reliably demonstrate the quality of their possessors as mates because they enable the choosing sex (typically females) to distinguish between genuinely fit individuals and exaggerators. If females chose mates on the base of traits that could be easily faked, they would gain no fitness benefit.

One example of a trait that Darwin attributed to sexual selection is the remarkable tail of the male African long-tailed widowbird, which is longer than the bird's head and body combined. To demonstrate that sexual selection drove the evolution of widowbird tails, Malte Andersson, a behavioral ecologist at Gothenburg University, Sweden, captured some male widowbirds. He shortened the tails of some males by cutting them, and lengthened the tails of others by gluing on additional feathers. He cut and reglued the tail feathers of still other males, which served as controls. Male widowbirds normally select, and defend from other males, a territory where they perform courtship displays to attract females. Both short-tailed and long-tailed males successfully defended their display territories, indicating that a long tail does not confer an advantage in male–male competition. However, males with artificially elongated tails attracted about four times more females than did males with shortened tails (**Figure 22.16**).

Why do female widowbirds prefer males with long tails? The ability to grow and maintain a costly feature such as a long tail may indicate that the male bearing it is vigorous and healthy, even though the tail impairs the bird's ability to fly. Even though the manipulated males in Malte Andersson's investigation did not have to pay the price of growing and supporting (except briefly) artificially long tails, the hypothesis that having well-developed orna-

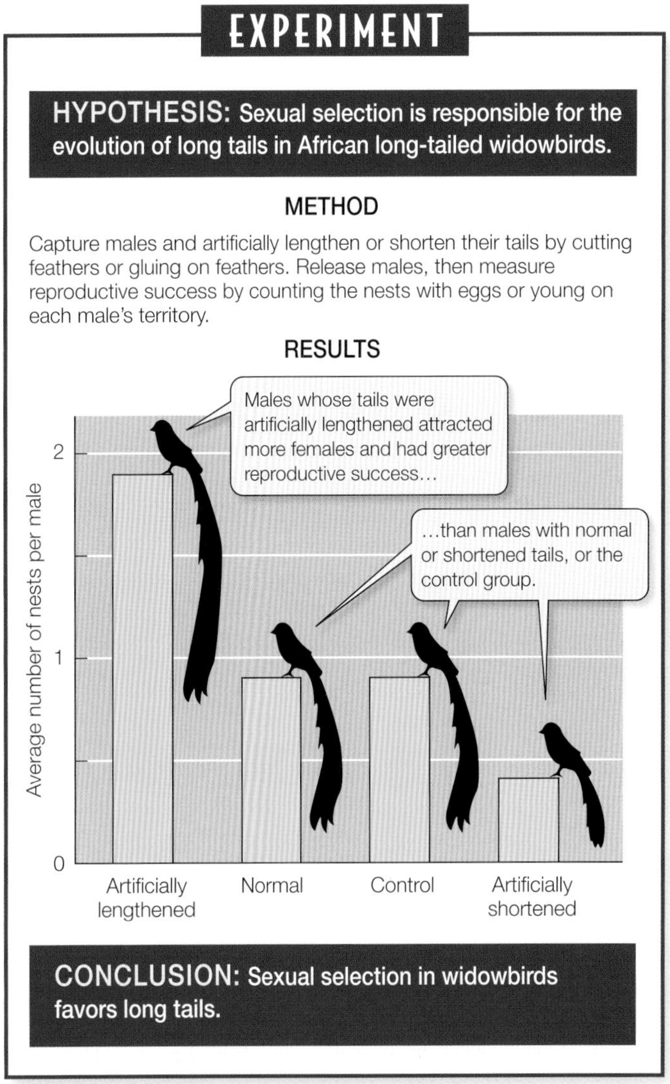

22.16 The Longer the Tail, the Better the Male Male widowbirds with shortened tails defended their display sites successfully, but attracted fewer females (and thus fathered fewer nests of eggs) than did males with normal and lengthened tails.

(A)

Taeniopygia guttata

22.17 Bright Bills Signal Good Health (A) Female zebra finches preferentially choose mates with the brightest bill color, thus choosing the healthiest males. (B) This laboratory experiment demonstrated that bright bill color in the male zebra finch indicates a healthy individual. FURTHER RESEARCH: How would you test this same hypothesis in the field? What would constitute experimental and control birds?

(B)

EXPERIMENT

HYPOTHESIS: Having a bright red bill signals good health in a male zebra finch.

METHOD

Provide carotenoids in drinking water for experimental, but not for control males. Challenge all males immunologically and measure response.

RESULTS

Experimental males responded more strongly to the immunological challenge. They also developed brighter bills than control males.

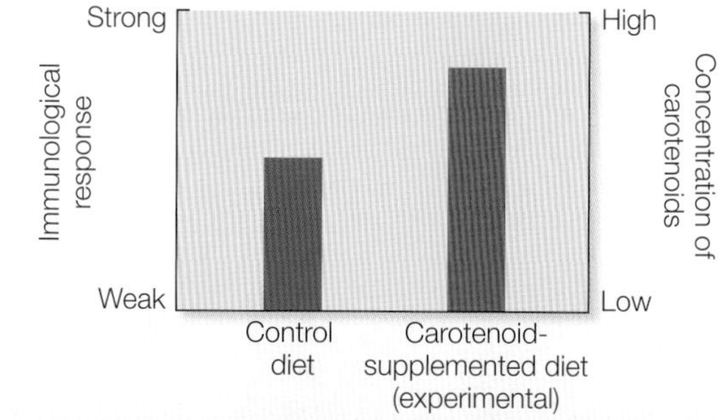

CONCLUSION: Males with high carotenoid levels have bright bills and are immunologically strong. Therefore, bill color is an indication of the health of a male zebra finch.

mental traits signals vigor and health has been tested experimentally with captive zebra finches.

The bright red bills of male zebra finches are the result of red and yellow carotenoid pigments. Zebra finches (and most other animals) cannot synthesize carotenoids and must obtain them from their food. In addition to influencing bill color, carotenoids are antioxidants and components of the immune system. Males in good health may need to allocate fewer carotenoids to immune function than males in poorer health. If so, then females can use the brightness of his bill to assess a male's health.

Tim Birkhead and his colleagues at Sheffield University manipulated blood levels of carotenoids in genetically similar male zebra finches by giving experimental males drinking water with added carotenoids; they gave control males only distilled water. All the males had access to the same food. After one month, the experimental males had higher levels of carotenoids in their blood, had much brighter bills than the control males, and were preferred by female zebra finches.

Next, the investigators challenged both groups of males immunologically by injecting phytohemagglutinin (PHA) into their wings. PHA induces a response by T lymphocytes (see Chapter 18) that results in an accumulation of white blood cells and thus a thickening of the skin at the injection site. Experimental males with enhanced carotenoid levels developed thicker skins because they responded more strongly to PHA than control males did, indicating that they had a stronger immune system (**Figure 22.17**).

This experiment showed that when a female chooses a male with a bright red bill, she probably gets a mate with a healthy immune system. Such males are less likely to become infected with parasites and diseases, so they are less likely to pass on infections to their mates. Healthier males are also better able to assist with parental care than are males with duller bills.

22.3 RECAP

Variation in genotype leads to variation in fitness. Fitness is the reproductive contribution of a phenotype to subsequent generations relative to the contributions of other phenotypes in a population. Natural selection acts on phenotypes with quantitative variation in several different ways that can result in adaptation.

- Do you understand the connection between genotype and phenotype, and why natural selection that acts on the phenotype results in changes in genotype frequencies? See p. 497
- Can you describe the differences between stabilizing, directional, and disruptive selection? See pp. 497–498 and Figure 22.12
- Can you explain why Darwin devoted an entire book to sexual selection? See pp. 498–499

Genetic drift, stabilizing selection, and directional selection all tend to reduce genetic variation within populations. Nevertheless, as we have seen, most populations have considerable genetic variation. What processes produce and maintain genetic variation within populations?

22.4 How Is Genetic Variation Maintained within Populations?

We have seen that genetic variation is the raw material on which mechanisms of evolution act. Several forces—neutral mutations, sexual recombination, and frequency-dependent selection—operate to maintain genetic variation in populations despite the action of other forces that reduce it (such as genetic drift and many types of selection). In addition, as we will show, genetic variation may be maintained over geographic space.

Neutral mutations may accumulate within populations

As we saw in Section 12.6, some mutations do not affect the function of the proteins encoded by the mutated genes. An allele that does not affect the fitness of an organism—that is, an allele that is no better or worse than alternative alleles at the same locus—is called a **neutral allele**. Neutral alleles are unaffected by natural selection. Even in large populations, neutral alleles may be lost or their frequencies may increase, purely by genetic drift. Neutral alleles tend to accumulate in a population over time, providing it with considerable genetic variation.

Much of the *phenotypic* variation that we are able to observe is not neutral. However, modern techniques enable us to measure neutral variation at the *molecular* level and provide the means by which to distinguish it from adaptive variation. In Section 24.2 we will see how these techniques enable us to discriminate among alleles and how variation in neutral molecular traits can be used to estimate rates of evolution.

Sexual recombination amplifies the number of possible genotypes

In asexually reproducing organisms, each new individual is genetically identical to its parent unless there has been a mutation. When organisms reproduce sexually, however, offspring differ from their parents, due to crossing over and independent assortment of chromosomes during meiosis as well as the combination of genetic material from two different gametes, as described in Chapter 9. Sexual recombination generates an endless variety of genotypic combinations that increases the evolutionary potential of populations—a long-term advantage of sex.

The evolution of the mechanisms of meiosis and sexual recombination were crucial events in the history of life. Exactly how these attributes arose is puzzling, however, because sex has at least three striking disadvantages in the short term:

- Recombination breaks up adaptive combinations of genes.
- Sex reduces the rate at which females pass genes on to their offspring.
- Dividing offspring into separate genders greatly reduces the overall reproductive rate.

To see why this last disadvantage exists, consider an asexual female that produces the same number of offspring as a sexual female. Let's assume that both females produce two offspring, but that the sexual female must produce 50 percent males. In the next generation, both asexual F_1 females will produce two more offspring, but there is only one sexual F_1 female to produce offspring. The evolutionary problem is to identify the advantages of sex that can overcome such short-term disadvantages.

A number of hypotheses have been proposed for the existence of sex, none of them mutually exclusive. One is that sexual recombination facilitates repair of damaged DNA, because breaks and other errors in DNA on one chromosome can be repaired by copying the intact sequence from the homologous chromosome.

Another advantage of sexual reproduction is that it permits the elimination of deleterious mutations. As we saw in Section 11.4, DNA replication is not perfect. Errors are introduced in every generation, and many or most of these errors result in lower fitness. Asexual organisms have no mechanism to eliminate deleterious mutations. Hermann J. Muller noted that the accumulation of mutations in a non-recombining genome is like a genetic ratchet. The mutations accumulate—"ratchet up"—at each replication: that is, a mutation occurs and is passed on when the genome replicates, then two new mutations occur in the next replication, so three mutations are passed on, and so on. Deleterious mutations cannot be eliminated except by the death of the lineage. This accumulation of deleterious mutations in asexual lineages is known as *Muller's ratchet*.

In sexual species, on the other hand, genetic recombination produces some individuals with more of these deleterious mutations, and some with fewer. The individuals with fewer deleterious mutations are more likely to survive. Therefore, sexual reproduction allows natural selection to eliminate deleterious mutations from the population over time.

Another explanation for the existence of sex is that the great variety of genetic combinations created each generation by sexual recombination may be especially valuable as a defense against pathogens and parasites. Most pathogens and parasites have much shorter life cycles than their hosts, and thus can potentially evolve counteradaptations to host defenses rapidly. Sexual recombination might give the host's defenses a chance to keep up.

Sexual recombination does not influence the frequencies of alleles; rather, *sexual recombination generates new combinations of alleles on which natural selection can act*. It expands variation in a character influenced by alleles at many loci by creating new genotypes. That is why artificial selection for bristle number in *Drosophila* (see Figure 22.5) resulted in flies with more bristles than any flies in the initial population had.

Frequency-dependent selection maintains genetic variation within populations

Natural selection often preserves variation as a polymorphism. A polymorphism may be maintained when the fitness of a genotype (or phenotype) depends on its frequency in a population. This phenomenon is known as **frequency-dependent selection**.

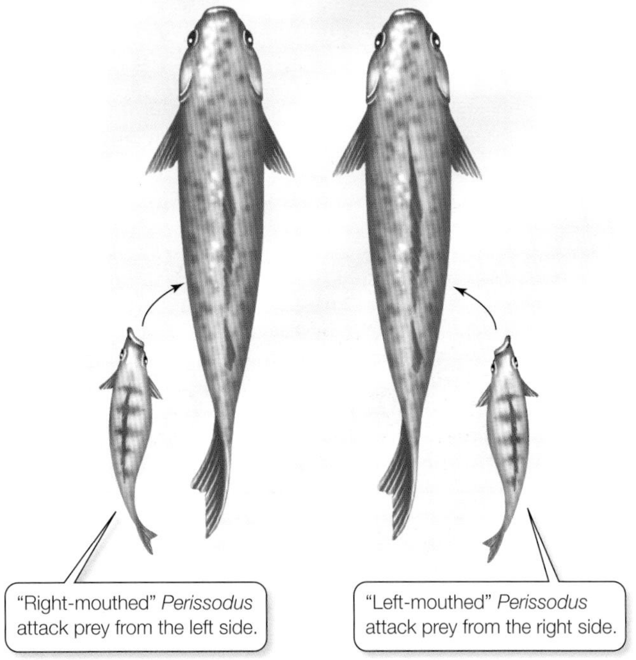

22.18 A Stable Polymorphism Frequency-dependent selection maintains equal proportions of left-mouthed and right-mouthed individuals of the scale-eating fish *Perissodus microlepis*.

A small fish that lives in Lake Tanganyika, in East Africa, provides an example of frequency-dependent selection. The mouth of this scale-eating fish, *Perissodus microlepis*, opens either to the right or to the left as a result of an asymmetrical jaw joint; the direction of opening is genetically determined (**Figure 22.18**). The scale-eater approaches its prey (another fish) from behind and dashes in to bite off several scales from its flank. "Right-mouthed" individuals always attack from the victim's left; "left-mouthed" individuals always attack from the victim's right. The distorted mouth enlarges the area of teeth in contact with the prey's flank, but only if the scale-eater attacks from the appropriate side.

Prey fish are alert to approaching scale-eaters, so attacks are more likely to be successful if the prey must watch both flanks. Vigilance by prey favors equal numbers of right-mouthed and left-mouthed scale-eaters, because if attacks from one side were more common than the other, prey fish would pay more attention to potential attacks from that side. Over an 11-year study of the scale-eaters in Lake Tanganyika, the polymorphism was found to be stable: the two forms of *P. microlepis* remained at about equal frequencies.

Environmental variation favors genetic variation

Environments vary greatly over time. A night is dramatically different from the preceding day. A cold, cloudy day differs from a clear, hot one. Day lengths and temperatures change seasonally. No single genotype is likely to perform well under all these conditions.

Colias butterflies of the Rocky Mountains live in environments where dawn temperatures often are too cold, and afternoon temperatures too hot, for the butterflies to fly. Populations of this butterfly are polymorphic for the enzyme phosphoglucose isomerase (PGI), which influences how well the butterfly flies at different temperatures. Some PGI genotypes can fly better during the cold hours of early morning; others perform better during midday heat. The optimal body temperature for flight is 35°C to 39°C, but some butterflies can fly with body temperatures as low as 29°C or as high as 40°C. During spells of unusually hot weather, heat-tolerant genotypes are favored; during spells of unusually cool weather, cold-tolerant genotypes are favored.

Heterozygous *Colias* butterflies can fly over a greater temperature range than homozygous individuals, which should give them an advantage in foraging and finding mates. Thus heterozygous males should be more successful in inseminating females than homozygous males. Ward Watt tested this prediction; His results are shown in **Figure 22.19**.

EXPERIMENT

HYPOTHESIS: Heterozygous male *Colias* butterflies should have better mating sucess than homozygous males because they can fly farther under a broader range of temperatures.

METHOD

Capture butterflies in the field, transfer them to the laboratory, and determine their genotypes. Allow females to lay eggs, and determine the genotypes of the offspring (and thus paternity and mating success) of males.

RESULTS

For both species, the proportion of heterozygous males that mated successfully was higher than the proportion of all males seeking females ("flying").

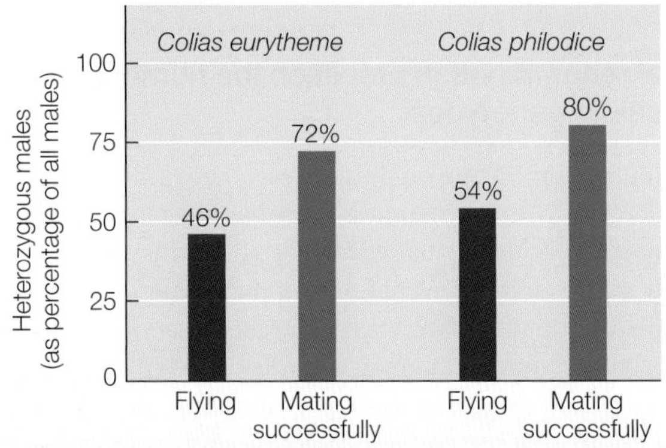

CONCLUSION: Heterozygous *Colias* males have a mating advantage over homozygous males.

22.19 A Heterozygote Mating Advantage Heterozygous male *Colias* butterflies can fly farther than homozygous males under a broader range of temperatures; thus heterozygotes are more successful in inseminating females.

Much genetic variation is maintained in geographically distinct subpopulations

Much of the genetic variation in large populations is preserved as differences among members living in different places (subpopulations). Subpopulations often vary genetically because they are subjected to different selective pressures in different environments. Environments may vary significantly over short distances. For example, in the Northern Hemisphere, temperature and soil moisture differ dramatically between north-facing and south-facing mountain slopes. In the Rocky Mountains of Colorado, the proportion of ponderosa pines (*Pinus ponderosa*) that are heterozygous for a particular peroxidase enzyme is particularly high on south-facing slopes, where temperatures fluctuate dramatically, often on a daily basis. This heterozygous genotype performs well over a broad range of temperatures. On north-facing slopes and at higher elevations, where temperatures are cooler and fluctuate less strikingly, a peroxidase homozygote, which has a lower optimum temperature, is much more frequent.

Plant species may also vary geographically in the chemicals they synthesize to defend themselves against herbivores. Some individuals of the white clover (*Trifolium repens*) produce the poisonous chemical cyanide. Poisonous individuals are less appealing to herbivores—particularly mice and slugs—than are nonpoisonous individuals. However, clover plants that produce cyanide are more likely to be killed by frost, because freezing damages cell membranes and releases cyanide into the plant's own tissues.

In European subpopulations of *Trifolium repens*, the frequency of cyanide-producing individuals increases gradually from north to south and from east to west (**Figure 22.20**). Poisonous plants make up a large proportion of clover subpopulations only in areas where winters are mild. Cyanide-producing individuals are rare where winters are cold, even though herbivores graze clovers heavily in those areas.

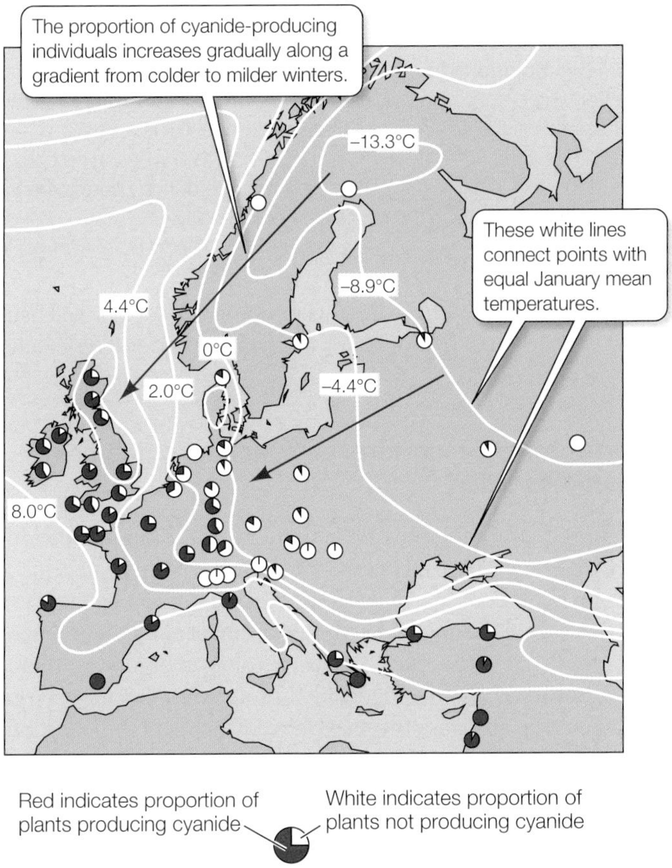

22.20 Geographic Variation in a Defensive Chemical The frequency of cyanide-producing individuals in each subpopulation of white clover (*Trifolium repens*) in Europe depends on the winter temperature of its environment.

22.4 RECAP

Neutral mutations, sexual recombination, and frequency-dependent selection all act to maintain considerable genetic variation within most populations. Variation is also maintained among geographically distinct, genetically variable subpopulations.

- Do you understand why sexual recombination is so prevalent in nature, despite having at least three short-term evolutionary disadvantages? See p. 501
- How does frequency-dependent selection act to maintain genetic variation within a population? See pp. 501–502

The mechanisms of evolution have produced a remarkable variety of organisms, some of which are adapted to most environments that exist on Earth. This natural variation, and the success of breeders attempting to produce desired traits in domesticated plants and animals, suggests that natural selection can favor most traits that might be adaptive. But is this impression correct?

22.5 What Are the Constraints on Evolution?

We would be mistaken to assume that natural selection always results in adaptive traits. There are obviously constraints on evolution. Lack of appropriate genetic variation, for example, could prevent the development of many potentially favorable traits. (That is, if the allele for a given trait does not exist in a population, that trait cannot evolve even if it would be highly favored by natural selection.) Evolutionary biologists have long debated whether lack of genetic variation has seriously constrained the rate and direction of evolutionary change, but nobody knows for sure how important this constraint has been. Evidence is accumulating, however, for other types of constraints on evolution—indeed, imagine how many different organisms might have evolved if there were none.

Earlier in the book we learned of constraints imposed on organisms by the dictates of physics and chemistry. The size of cells, for example, is constrained by the stringencies of surface-area-to-volume ratios (see Section 2.1). The ways in which proteins can fold are limited by the bonding capacities of their constituent molecules (see Section 3.2). And the energy transfers that fuel life must operate within the laws of thermodynamics (see Section 6.1). Keep in mind that evolution works within the boundaries of these universal constraints, as well those mentioned below.

Developmental processes constrain evolution

As we saw in Section 20.5, developmental constraints on evolution are paramount because all evolutionary innovations are modifications of previously existing structures. Human engineers seeking to power an airplane may be able to design a completely new type of engine (jet) that can replace a previous type (propeller), but evolutionary changes cannot happen that way.

A striking example of such developmental constraints is provided by the evolution of fish that spend most of their time on the sea bottom. One lineage, the bottom-dwelling skates and rays, share a common ancestor with sharks, whose bodies were already somewhat ventrally flattened, and whose skeletal frame is made of flexible cartilage. Skates and rays evolved a body plan that further flattened their bellies, allowing them to swim along the ocean floor (**Figure 22.21A**).

Plaice, sole, and flounder, on the other hand, are bottom-dwelling descendants of deep-bellied, laterally flattened ancestors with bony skeletons. The only way such a fish can lie flat is to flop over onto its side. Their ability to swim is thus curtailed, but their bodies can lie still and are well camouflaged. During development, the eyes of these flatfishes are grotesquely twisted so that both eyes are one side of the body (**Figure 22.21B**). Small shifts in the position of one eye probably helped ancestral flatfishes see better, eventually resulting in the body form found today.

(A) *Taeniura lymma*

(B) *Bothus lunatus*

22.21 Two Solutions to a Single Problem (A) Stingrays, whose ancestors were dorsoventrally flattened, lie on their bellies. Their bodies are symmetrical around the dorsal backbone. (B) Flounders, whose ancestors were laterally flattened, lie on their sides. (The backbone of this fish is at the left.) Their eyes migrate during development so that both eyes are on the same side of the body.

Trade-offs constrain evolution

As we saw in the case of the garter snakes described at the opening of this chapter, all adaptations impose both fitness costs and fitness benefits. For an adaptation to evolve, the fitness benefits it confers must exceed the fitness costs it imposes—in other words, the trade-off must be worthwhile. Garter snakes pay for their ability to eat rough-skinned newts by being unable to move rapidly after having eaten a newt. Sufficient benefits apparently exist to override that cost *only* in areas where newts are common and are potentially major items in the snakes' diet.

The costs of developing and maintaining certain conspicuous features used by males to compete with other males for access to females have been estimated using the comparative method. In some mammals, such as deer, lions, and baboons, one male controls reproductive access to many females. Species in which males have multiple mates are said to be *polygynous* (the term *monogamous* applies to species that have just one mate). Males of many polygynous species are significantly larger than the females and often bear large weapons (such as horns, antlers, and large canine teeth). Such dramatic differences between the sexes are known as *sexual dimorphism*.

In polygynous species, large size and weaponry are needed to defend a male's multiple mates against other males of the species. Developing and maintaining these structures is energetically costly, and males are often injured during their contests. Their life span is often considerably shorter than that of females of the same species. However, since during his "time at the top" a polygynous male enjoys access to multiple females, his reproductive success will usually be greater than that of males he chases away, or that females reject as "inferior."

Short-term and long-term evolutionary outcomes sometimes differ

The short-term changes in allele frequencies within populations that we have emphasized in this chapter are an important focus of study for evolutionary biologists. These changes can be observed directly, they can be manipulated experimentally, and they demonstrate the actual processes by which evolution occurs. By themselves, however, they do not enable us to predict—or, more properly, "postdict" (because they have already happened)—the kinds of long-term evolutionary changes described in Section 21.3.

Patterns of evolutionary change can be strongly influenced by events that occur so infrequently (a meteorite impact, for example) or so slowly (continental drift) that they are unlikely to be observed during short-term studies. In addition, the ways in which evolutionary processes act may change over time; even among the descendants of a single ancestral species, different lineages may evolve in different directions. Therefore, additional types of evidence, demonstrating the effects of rare and unusual events on trends in the fossil record, must be gathered if we wish to understand the course of evolution over billions of years. In subsequent chapters, we will discuss the methods that biologists use to study long-term evolutionary changes and infer the processes that led to them.

22.5 RECAP

Developmental processes constrain evolution because all evolutionary innovations are modifications of previously existing structures. An adaptation can evolve only if the fitness benefits it confers exceed the fitness costs it imposes.

- How do trade-offs constrain the evolution of adaptations? See p. 504
- Do you see why the presence of a great deal of genetic variation within a population could increase the chances that some members of the population would survive an unprecedented environmental change? Do you also see why there is no guarantee that this would be the case?

Human beings have not altered the mechanisms of evolution. Humans have, however, influenced evolutionary events.

22.6 How Have Humans Influenced Evolution?

Evolutionary processes are operating today just as they have been since life first appeared on Earth. Like all living things, humans are still evolving, but as humans have changed their environment, the selective forces affecting us have also changed, and now differ from those that dominated most of human prehistory. Today few humans are killed by predators, and deaths due to bad weather, although not infrequent, are rare compared with deaths caused by diseases, accidents, wars, and murders. Today, some of the differences in survival and reproductive success among humans are related to genes that confer defenses against diseases like malaria, as well as health conditions related to the stresses of modern industrial life, such as hypertension, diabetes, and AIDS.

> People in most parts of the world cannot digest lactose, the sugar in dairy products. Enzymes that digest lactose began to be selected for between 5,000 and 10,000 years ago when human populations in certain regions began to consume milk from their domesticated animals.

Many evolutionary changes are taking place all around us, and humans are influencing some of those changes. For example, our attempts to control populations of species we consider pests and to increase populations of those we consider desirable make us powerful agents of evolutionary change. In addition to producing the results we desire, these efforts often cause undesirable consequences, such as the evolution of resistance to antibiotics by pathogens and to pesticides by pests. Medicine and agriculture can respond creatively to the evolutionary changes they are causing only if their practitioners understand how and why those changes happen.

Ovis canadensis (trophy male)

Ovis canadensis (male)

22.22 Trophy Hunting Selects for Smaller Males Hunters preferentially kill large "trophy" rams rather than their less impressive counterparts.

Many current human-caused evolutionary changes are inadvertent. For example, sport hunters often seek large trophy animals to display on the walls of their homes. By preferentially shooting male mammals with particularly large ornaments, trophy hunters have inadvertently favored males with smaller ornaments. Trophy hunting is big business: one hunter paid more than a million Canadian dollars in 1998 and 1999 for permits to hunt trophy bighorn rams in Alberta. Since 1975, most rams (45 of 57) shot at Ram Mountain, Alberta, have been trophy rams: large individuals with large horns (**Figure 22.22**). The result has been a steady decline in the size of rams shot by hunters (and in the size of their horns). Nearly all of these rams were shot before they had achieved most of the reproductive success that would be expected over their lifetimes. Ironically, selectively harvesting trophy rams is making trophy animals scarce. Unrestricted trophy hunting of other species has similarly resulted in diminished populations of animals with the valued traits.

Humans have influenced evolution in a number of far-reaching ways. We have moved thousands of species around the globe and modified organisms using biotechnology. Human activities have changed the climate and greatly increased the rate of extinction of other species. We will continue to see examples of the effects of *Homo sapiens* on the evolution of other species throughout this book.

CHAPTER SUMMARY

22.1 What facts form the base of our understanding of evolution?

See Web/CD Tutorial 22.1

Charles Darwin attributed changes in species over time to the possession of advantageous traits by some individuals. He understood that *individuals* do not evolve, but *populations* evolve when individuals with different heritable genotypes survive and reproduce at different rates.

Adaptation refers both to characteristics of organisms and the way those characteristics are acquired via **natural selection**.

The sum of all copies of all alleles at all loci found in the population constitutes its **gene pool** and represents the **genetic variation** that results in different phenotypic traits upon which natural selection can act. Review Figure 22.3

Artificial selection and laboratory experiments demonstrate the existence of considerable genetic variation in most populations. Review Figure 22.5

Allele frequencies measure the amount of genetic variation in a population; **genotype frequencies** show how a population's genetic variation is distributed among its members. Review Figure 22.6

Hardy–Weinberg equilibrium predicts the allele frequencies in populations in the absence of evolution. Deviation from these frequencies indicates the work of evolutionary mechanisms. Review Figure 22.7, Web/CD Tutorial 22.2

22.2 What are the mechanisms of evolutionary change?

Migration of individuals between populations results in **gene flow**.

In small populations, **genetic drift**—the random loss of individuals and the alleles they possess—may produce large changes in allele frequencies from one generation to the next and greatly reduce genetic variation. Review Figure 22.8

Population bottlenecks occur when only a few individuals survive a random event, resulting in a drastic shift in allele frequencies within the population and the loss of variation. Similarly, a population established by a small number of individuals colonizing a new region may lose variation via a **founder effect**.

Nonrandom mating may result in genotype frequencies that deviate from Hardy–Weinberg equilibrium.

22.3 What evolutionary mechanisms result in adaptation?

See Web/CD Tutorial 22.3

Fitness is the reproductive contribution of a phenotype to subsequent generations relative to the contributions of other phenotypes.

Changes in numbers of offspring are responsible for changes in the absolute size of a population, but only changes in the relative success of different phenotypes within a population lead to changes in allele frequencies.

Natural selection can act on traits with quantitative variation in several different ways, resulting in **stabilizing**, **directional**, or **disruptive selection**. Review Figure 22.12

Sexual selection is primarily about success in reproduction, not about success in survival. Review Figures 22.16 and 22.17

22.4 How is genetic variation maintained within populations?

Although genetic drift, stabilizing selection, and directional selection all tend to reduce genetic variation within populations, most populations have considerable genetic variation.

Neutral mutations, sexual recombination, and frequency-dependent selection can maintain genetic variation within populations.

Neutral alleles do not affect the fitness of an organism, are not affected by natural selection, and may accumulate or be lost by genetic drift.

Sexual reproduction generates countless genotypic combinations that increase the evolutionary potential of populations despite short-term disadvantages.

A polymorphism may be maintained by **frequency-dependent selection** when the fitness of a genotype depends on its frequency in a population.

Genetic variation may be maintained by the existence of genetically distinct subpopulations over geographic space. Review Figure 22.20

22.5 What are the constraints on evolution?

Developmental processes constrain evolution because all evolutionary innovations are modifications of previously existing structures.

Most adaptations impose costs. An adaptation can evolve only if the benefits it confers exceed the costs it imposes, a situation that leads to **trade-offs**.

22.6 How have humans influenced evolution?

Humans have become major agents of evolution as they attempt to control pests and diseases, move species around the globe, and modify organisms via biotechnology. Human activities are changing the climate and have greatly increased the rates of extinction of other species.

SELF-QUIZ

1. Garter snakes that are resistant to TTX move more slowly than those that are susceptible to the toxin. Their reduced speed is an example of
   a. an adaptation.
   b. genetic drift.
   c. natural selection.
   d. a trade-off.
   e. none of the above
2. Which of the following is *not* true?
   a. Darwin and Wallace were both influenced by Malthus.
   b. Wallace proposed a theory of evolution by natural selection that was similar to Darwin's.
   c. Malthus claimed that because human population growth would outstrip any increases in food production, famine was a likely result.
   d. Darwin realized that all populations had the capacity for rapid increases in numbers.
   e. All of the above are true.
3. The phenotype of an organism is
   a. the type specimen of its species in a museum.
   b. its genetic constitution, which governs its traits.
   c. the chronological expression of its genes.
   d. the physical expression of its genotype.
   e. the form it achieves as an adult.

4. The appropriate unit for defining and measuring genetic variation is the
   a. cell.
   b. individual.
   c. population.
   d. community.
   e. ecosystem.
5. Which statement about allele frequencies is *not* true?
   a. The sum of all allele frequencies at a locus is always 1.
   b. If there are two alleles at a locus and we know the frequency of one of them, we can obtain the frequency of the other by subtraction.
   c. If an allele is missing from a population, its frequency in that population is 0.
   d. If two populations have the same gene pool for a locus, they will have the same proportion of homozygotes at that locus.
   e. If there is only one allele at a locus, its frequency is 1.
6. Which of the following is *not* required for a population at Hardy–Weinberg equilibrium ?
   a. There is no migration between populations.
   b. Natural selection is not acting on the alleles in the population.
   c. Mating is random.
   d. The frequency of one allele must be greater than 0.7.
   e. All of the above conditions must be met.
7. The fitness of a genotype is a function of the
   a. average rates of survival and reproduction of individuals with that genotype.
   b. individuals that have the highest rates of both survival and reproduction.
   c. individuals that have the highest rates of survival.
   d. individuals that have the highest rates of reproduction.
   e. average reproductive rate of individuals with that genotype.
8. Laboratory selection experiments with fruit flies have demonstrated that
   a. bristle number is not genetically controlled.
   b. bristle number is not genetically controlled, but changes in bristle number are caused by the environment in which the fly is raised.
   c. bristle number is genetically controlled, but there is little variation on which natural selection can act.
   d. bristle number is genetically controlled, but selection cannot result in flies having more bristles than any individual in the original population had.
   e. bristle number is genetically controlled, and selection can result in flies having more bristles than any individual in the original population had.
9. Disruptive selection maintains a bimodal distribution of bill size in the West African seedcracker because
   a. bills of intermediate shapes are difficult to form.
   b. the birds' two major food sources differ markedly in size and hardness.
   c. males use their large bills in displays.
   d. migrants introduce different bill sizes into the population each year.
   e. older birds need larger bills than younger birds.
10. Which of the following is *not* a reason why trade-offs constrain evolution?
   a. All adaptations impose both fitness costs and benefits.
   b. Sexual selection for large size and weaponry shortens the life span of their possessors.
   c. Changes in allele frequencies may be influenced by events that occur very infrequently.
   d. Ability to consume toxic prey may reduce mobility.
   e. Adaptations can evolve only if the fitness benefits they confer exceed the costs they impose.

FOR DISCUSSION

1. In what ways does artificial selection by humans differ from natural selection? Was Darwin wise to base so much of his argument for natural selection on the results of artificial selection?
2. In nature, mating among individuals in a population is never truly random, immigration and emigration are common, and natural selection is seldom totally absent. Why, then, is Hardy–Weinberg equilibrium, which is based on assumptions known generally to be false, so useful in our study of evolution? Can you think of other models in science that are based on false assumptions? How are such models used?
3. As far as we know, natural selection cannot adapt organisms to future events. Yet many organisms appear to respond to natural events before they happen. For example, many mammals go into hibernation while it is still quite warm. Similarly, many birds leave the temperate zone for their southern wintering grounds long before winter has arrived. How can such "anticipatory" behaviors evolve?
4. Populations of most of the thousands of species that have been introduced to areas where they were previously not found, including those that have become pests, began with a few individuals. They should therefore have begun with much less genetic variation than the parent populations have. If genetic variation is advantageous, why have so many of these species been successful in their new environments?
5. Why is it important that the ways in which males advertise their health and vigor to females reliably indicate their status?
6. How does selection on humans today differ from selection on humans in the preindustrial past (i.e., prior to 1700)?
7. As more humans live longer, many people face degenerative conditions such as Alzheimer's disease that (in most cases) are linked to advancing age. Assuming that some individuals may be genetically predisposed to successfully combat these conditions, is it likely that natural selection alone would act to favor such a predisposition in human populations? Why or why not?

FOR INVESTIGATION

During the past 50 years, more than 200 species of insects that attack crop plants have become highly resistant to DDT and other pesticides. Using your recently acquired knowledge of evolutionary processes, explain the rapid and widespread evolution of resistance. What proposals concerning pesticide use would you make in order to slow down the rate of evolution of resistance? Explain why you think your proposals could work and how you might test them.

CHAPTER 23 Species and Their Formation

Sex stimulates speciation (among other things)

Charles Darwin proposed that sexual selection was responsible for the evolution of such bright and conspicuous traits as the large antlers of male deer and the greatly enlarged tail feathers of male peacocks. Those traits are especially exaggerated in species in which individuals of one sex—usually males—compete for opportunities to mate with individuals of the opposite sex—usually females. Exaggerated traits evolve if they confer an advantage, either in competition among males for access to females, or in stimulating and attracting discriminating females.

In addition to leading to conspicuous male ornaments such as antlers and plumage, recent evidence suggests that sexual selection may also increase the rate at which new species form. Evidence for this effect of sexual selection comes from comparing the number of species found in sister clades, that is, clades that share a common ancestor. Because they share a common ancestor, sister clades have been evolving independently from one another for the same length of time. The rate at which species have formed in the two clades can be estimated by comparing the number of species in them today.

Birds with promiscuous mating systems offer some of the best examples of both of these effects of sexual selection. In many of these species males assemble in display grounds and females come there to choose with whom to copulate. After mating, the females of many species build their nests, lay their eggs, and raise their offspring with no help from the males. And in most of these species, the males have evolved bright plumage and often ornaments such as long tail feathers. In contrast, in monogamous species that form pair bonds and share the responsibilities of raising young, individuals of both sexes tend to have dull plumage and to look alike.

As an example of this phenomenon, there are about 320 species of hummingbirds in the Americas, all of which are promiscuous. However, there are only 103 species of swifts (sister clade to the hummingbirds), even though swifts are found worldwide. Male and female swifts form monogamous pair bonds and look alike.

Why does sexual selection stimulate the divergence of a lineage into many species? A likely reason is that random mutations result in different plumage elaborations in different parts of the range of a species. Thus, local populations of a species may develop unique male plumage patterns, each of which is favored by sexual selection in that area. Females may not respond positively to an immigrant male whose

A Male Hummingbird More than 300 species of hummingbirds are found in the Americas. Hummingbird males are promiscuous. They compete with other males for mates, and males with the most successful display will usually inseminate the most females and thus sire the most offspring.

CHAPTER OUTLINE

23.1 What Are Species?

23.2 How Do New Species Arise?

23.3 What Happens when Newly Formed Species Come Together?

23.4 Why Do Rates of Speciation Vary?

23.5 Why Do Adaptive Radiations Occur?

A Male Swift Swifts are the sister clade of hummingbirds. Swifts form pair bonds and the male and female, who look much alike, raise the young together. Thus the most successful males are those that help their mates raise the most offspring. There are only 103 species of swifts, even though they are found worldwide.

plumage differs from the pattern to which they are normally attracted.

The origin of new species—the splitting and divergence of a single lineage into distinct species—is one of the most important phenomena in biological science. Charles Darwin recognized its preeminence when he chose the title of his most famous book. But without the underlying knowledge supplied by modern genetics, Darwin was primarily viewing the consequences of speciation, not its underlying causes, and he recognized that as well. We are still looking for the answers to many questions about speciation, a process Darwin referred to as "the mystery of mysteries."

IN THIS CHAPTER we will describe what species are and discuss how Earth's millions of species came into being. We will examine the mechanisms by which a population splits into new species and how such separations are maintained. We will look at different factors that can make speciation a rapid or a very slow process. Finally, we will look at the conditions that give rise to the great species diversifications known as evolutionary radiations.

23.1 What Are Species?

The word *species* means, literally, "kinds." But how do biologists interpret "kinds"? Answering this question is complicated because species are the result of processes that unfold over time. In studying those processes, biologists pursue several different questions about species. How can we recognize and identify species? How do species form over time? How do species remain separate? Various definitions of species will emerge as we address these questions.

We can recognize and identify many species by their appearance

Someone who is knowledgeable about a group of organisms, such as lizards or birds, usually can distinguish the different species found in a particular area simply by looking at them. Standard field guides to birds, mammals, insects, and flowering plants are possible only because many species change little in appearance over large geographic distances. We can easily recognize male red-winged blackbirds from New York and from California, for example, as members of the same species (**Figure 23.1A**).

More than 200 years ago, the Swedish biologist Carolus Linnaeus originated the *binomial* system of Latinate nomenclature by which species are known today (see Section 1.2). Linnaeus described hundreds of species, and because he knew nothing about genetics or the mating behavior of the organisms he was naming, he classified them on the basis of their appearance; that is, he used a **morphological species concept**. Members of many of the groups that he classified as species by their appearance look alike because they share many of the alleles that code for their body structures. In many groups of organisms for which genetic data are unavailable, species are still recognized by their morphological traits.

But not all members of a species must look alike. For example, males, females, and young individuals may not resemble one another closely (**Figure 23.1B**). Scientists must use information other than appearance to decide whether such easily distinguished individuals are members of the same or different species. To understand the types of information they use, we need to understand how species form.

23.1 Members of the Same Species Look Alike—or Not
(A) Both of these male red-winged blackbirds are obviously members of the same species, even though one is from the eastern United States and the other is from California. (B) Because red-winged blackbirds are sexually dimorphic, the female of the species looks quite different from the male.

Species form over time

Evolutionary biologists think of species as branches on the tree of life. Each species has a history that starts at a speciation event and ends at either extinction or another speciation event, at which it produces two daughter species. This process is often gradual (**Figure 23.2**). **Speciation** is thus the process by which one species splits into two or more daughter species, which thereafter evolve as distinct lineages. The gradual nature of most speciation guarantees that in many cases, two populations at various stages in the process of becoming new species will exist. In these cases it is impractical to try to decide whether the individuals belong to species A or to species B. However, it *is* important to understand the processes that lead to the splitting of a single species into two species.

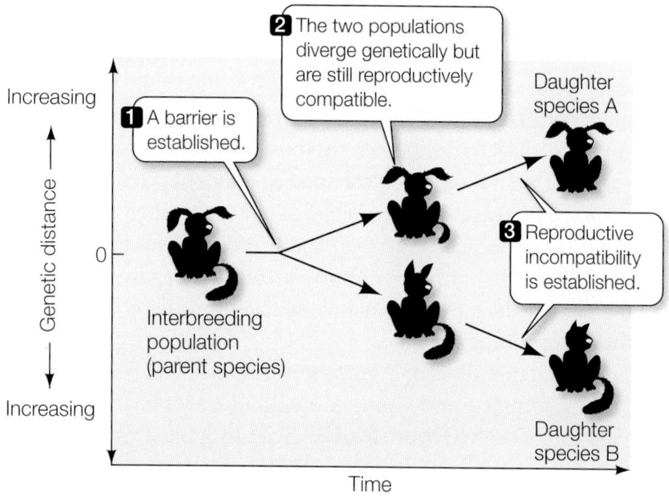

23.2 Speciation May Be a Gradual Process In this hypothetical example, genetic divergence between two separated populations begins before reproductive incompatibility evolves.

An important component of the process of speciation is the development of **reproductive isolation**. If individuals of a population mate with one another, but not with individuals of other populations, they constitute a distinct group within which genes recombine; that is, they are independent evolutionary units—separate branches on the tree of life. In 1940 Ernst Mayr proposed the following definition of species, known as the **biological species concept**: "Species are groups of actually or potentially interbreeding natural populations which are reproductively isolated from other such groups." The terms "actually" and "potentially" are important elements of the definition. "Actually" says that the individuals live in the same area and interbreed with one another. "Potentially" says that although the individuals do not live in the same area, and therefore cannot interbreed, other information suggests that they would do so if they did get together. This widely used species definition does not, however, apply to organisms that reproduce asexually.

Mayr's definition of species asserts that reproductive isolation is the most important criterion for identifying species, but how does reproductive isolation develop? In other words, how do species arise? We will devote much attention in this chapter to this important question.

23.1 RECAP

The biological species concept defines species in terms of reproductive isolation of groups. Speciation is usually a gradual process, and populations can exist at various stages in the process of becoming reproductively isolated from one another.

- Do you understand the importance of the biological species concept relative to the morphological species concept? See pp. 509–510
- Do you see why the morphological species concept is still a useful tool? See p. 510
- Do you see why the biological species concept is difficult to apply to asexually reproducing organisms? (Read Mayr's definition carefully.)

Although Charles Darwin titled his groundbreaking book *The Origin of Species,* it did not extensively discuss the processes of speciation. He devoted most of his attention to demonstrating that species are altered by natural selection over time. We will next discuss the many things that biologists have learned about "the mystery of mysteries" since Darwin's time.

23.2 How Do New Species Arise?

Not all evolutionary changes result in new species. A single lineage may change over time without giving rise to a new species. If evolutionary changes result in one species splitting into two or more daughter species, however, then speciation has occurred. Subsequently, within each isolated gene pool, allele and genotype frequencies may change as a result of the action of evolutionary mechanisms. If two populations are isolated from each other, and sufficient differences in their genetic structure accumulate during the period of isolation, then the two populations may not exchange genes when they come together again. As we will see, the amount of genetic difference that is needed to prevent gene exchange is highly variable.

Thus speciation requires that gene flow within a population whose members formerly exchanged genes be interrupted. Gene flow may be interrupted in two major ways, each of which characterizes a mode of speciation.

23.3 Allopatric Speciation Allopatric speciation may result when a population is divided into two separate populations by a physical barrier, such as rising sea levels.

Allopatric speciation requires almost complete genetic isolation

Speciation that results when a population is divided by a physical barrier is known as **allopatric speciation** (*allo,* "different"; *patris,* "country") or *geographic speciation* (**Figure 23.3**). Allopatric speciation is thought to be the dominant mode of speciation among most groups of organisms. The physical barrier that divides the range of a species may be a water body or a mountain range for terrestrial organisms, or dry land for aquatic organisms. Barriers can form when continents drift, sea levels rise, glaciers advance and retreat, and climates change. These processes continue to generate physical barriers today. The populations separated by such barriers are often, but not always, initially large. They evolve differences for a variety of reasons, including gene drift, but especially because the environments in which they live are, or become, different.

Allopatric speciation may also result when some members of a population cross an existing barrier and found a new, isolated population. The 14 species of finches of the Galápagos archipelago, 1,000 kilometers off the coast of Ecuador, were generated by allopatric speciation. Darwin's finches (as they are usually called, because Darwin was the first scientist to study them) arose in the Galápagos from a single South American species that colonized the islands. Today the 14 species of Darwin's finches differ strikingly from their closest mainland relative (**Figure 23.4**). The islands of the Galápagos archipelago are sufficiently distant from one another that finches infrequently disperse among them. In addition, environmental conditions differ among the islands. Some are relatively flat and arid; others have forested mountain slopes. Populations of finches on different islands have differentiated enough over millions of years that when occasional immigrants arrive from other islands, they either do not breed with the residents or, if they do, the resulting offspring often do not survive as well as those produced by pairs composed of island residents. The genetic distinctness and cohesiveness of the different finch species are thus maintained.

Many of the more than 800 species of the fruit fly genus *Drosophila* in the Hawaiian Islands are restricted to a single island. We know that these species are the descendants of new populations founded by individuals dispersing among the islands because the closest relative of a species on one island is often a species on a neighboring island rather than a species on the same island. Biologists who have studied the chromosomes of picture-winged species of *Drosophila* believe that speciation in this group of flies has resulted from at least 45 such founder events (**Figure 23.5**).

A physical barrier's effectiveness at preventing gene flow depends on the size and mobility of the species in question. The eight-lane highway that is an almost impenetrable barrier to a snail

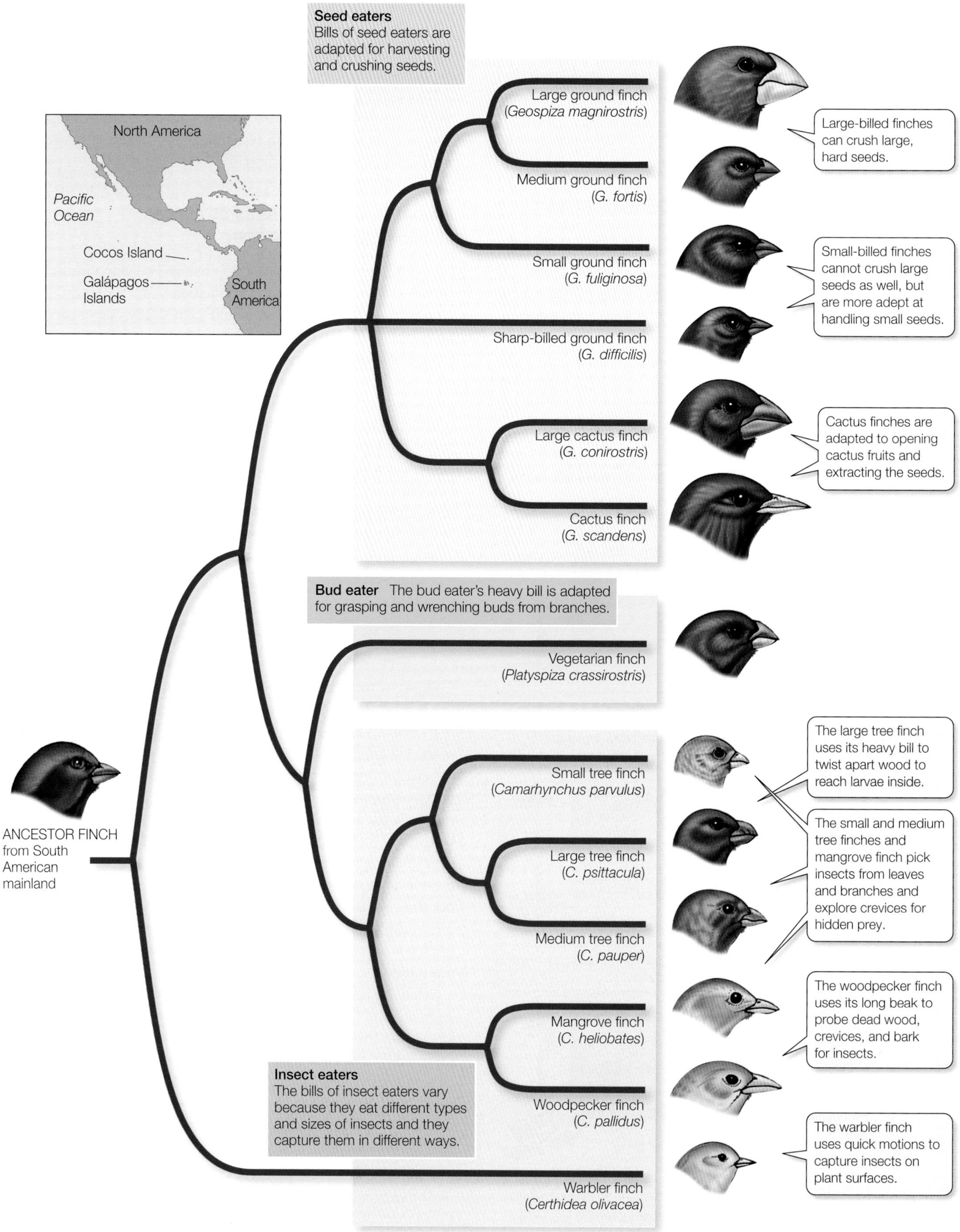
Seed eaters
Bills of seed eaters are adapted for harvesting and crushing seeds.
North America
Pacific Ocean
Cocos Island
Galápagos Islands
South America
Large ground finch (Geospiza magnirostris)
Medium ground finch (G. fortis)
Small ground finch (G. fuliginosa)
Sharp-billed ground finch (G. difficilis)
Large cactus finch (G. conirostris)
Cactus finch (G. scandens)
Large-billed finches can crush large, hard seeds.
Small-billed finches cannot crush large seeds as well, but are more adept at handling small seeds.
Cactus finches are adapted to opening cactus fruits and extracting the seeds.
Bud eater The bud eater's heavy bill is adapted for grasping and wrenching buds from branches.
Vegetarian finch (Platyspiza crassirostris)
ANCESTOR FINCH from South American mainland
Small tree finch (Camarhynchus parvulus)
Large tree finch (C. psittacula)
Medium tree finch (C. pauper)
Mangrove finch (C. heliobates)
Woodpecker finch (C. pallidus)
Warbler finch (Certhidea olivacea)
The large tree finch uses its heavy bill to twist apart wood to reach larvae inside.
The small and medium tree finches and mangrove finch pick insects from leaves and branches and explore crevices for hidden prey.
The woodpecker finch uses its long beak to probe dead wood, crevices, and bark for insects.
The warbler finch uses quick motions to capture insects on plant surfaces.
Insect eaters
The bills of insect eaters vary because they eat different types and sizes of insects and they capture them in different ways.

23.4 Allopatric Speciation among Darwin's Finches The descendants of the ancestral finch that colonized the Galápagos archipelago several million years ago evolved into 14 different species whose members are variously adapted to feed on seeds, buds, and insects. (The fourteenth species, not pictured here, lives in Cocos Island, farther north in the Pacific Ocean.)

is no barrier at all to a butterfly or a bird. Populations of wind-pollinated plants are isolated at the maximum distance pollen can be blown by the wind, but individual plants are effectively isolated at much shorter distances. Among animal-pollinated plants, the width of the barrier is the distance that animals can travel while carrying pollen or seeds. Even animals with great powers of dispersal are often reluctant to cross narrow strips of unsuitable habitat. For animals that cannot swim or fly, narrow water-filled gaps may be effective barriers. Gene flow can sometimes be interrupted, however, even in the absence of physical barriers.

Sympatric speciation occurs without physical barriers

Although physical isolation is usually required for speciation, under some circumstances speciation can occur without it. A partition of a gene pool without physical isolation is called **sympatric speciation** (*sym*, "together with"). But if speciation is usually a gradual process, how can reproductive isolation develop when individuals have frequent opportunities to mate with one another? What is required is some form of disruptive selection in which certain genotypes have high fitness on one or the other of two different resources, as we described for black-bellied seedcrackers (see Figure 22.15).

Disruptive selection in seedcrackers has not yet resulted in the formation of two species, but sympatric speciation via disruptive selection may be happening in a fruit fly (*Rhagoletis pomonella*) in New York State. Until the mid-1800s, *Rhagoletis* fruit flies courted, mated, and deposited their eggs only on hawthorn fruits. The larvae learned the odor of hawthorn as they fed on the fruits. When they emerged from their pupae as adults, they used this cue to locate other hawthorn plants on which to mate and lay their eggs. About 150 years ago, large commercial apple orchards were planted in the Hudson River valley. Apple trees are closely related to hawthorns. A few female *Rhagoletis* laid their eggs on apples, perhaps by mistake. Their larvae did not grow as well as the larvae on hawthorn fruits, but many did survive. These larvae recognized the odor of apples, so when they emerged as adults, they sought out apple trees, where they mated with other flies reared on apples.

Today there are two groups of *Rhagoletis pomonella* in the Hudson River valley that may be on the way to becoming distinct species (**Figure 23.6**). One feeds primarily on hawthorn fruits, the other on apples. The two incipient species are partly reproductively isolated because they mate primarily with individuals raised on the same fruit and because they emerge from their pupae at different times of the year. In addition, the apple-feeding flies have evolved so that they now grow more rapidly on apples than they originally did.

Sympatric speciation via ecological isolation (on different resources), as is happening in *Rhagoletis pomonella*, may be widespread among insects, many of which feed on only a single plant species, but the most common means of sympatric speciation is **polyploidy**, the production within an individual of duplicate sets of chromosomes. Polyploidy can arise either from chromosome duplication in a single species (**autopolyploidy**) or from the combining of the chromosomes of two different species (**allopolyploidy**).

An autopolyploid individual originates when (for example) cells that are normally diploid (with two sets of chromosomes) acciden-

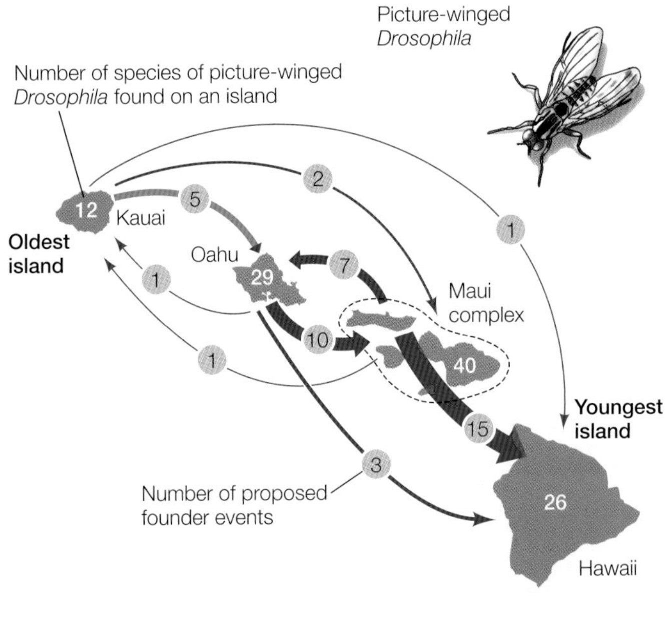

23.5 Founder Events Lead to Allopatric Speciation The large number of species of picture-winged *Drosophila* in the Hawaiian Islands is the result of founder events: the founding of new populations by individuals dispersing among the islands. The islands, which were formed in sequence as Earth's crust moved over a volcanic "hot spot," vary in age.

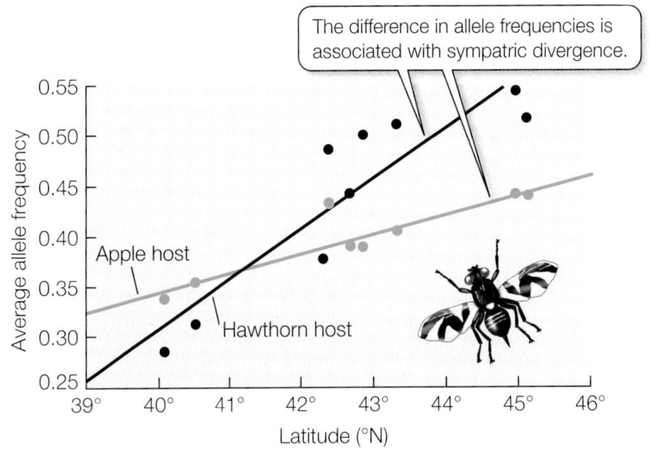

23.6 Sympatric Speciation May Be Underway in *Rhagoletis pomonella* Genetic differentiation (measured by the average frequencies of alleles at several loci) is increasing among *R. pomonella* flies who have developed preferences for different host plants.

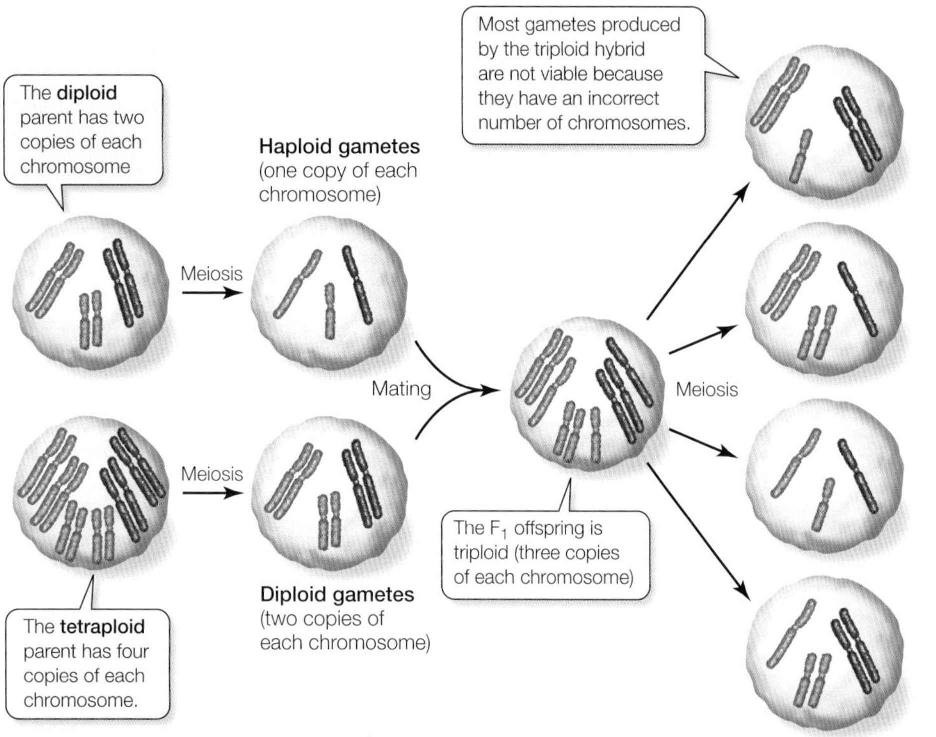

23.7 Tetraploids Are Soon Reproductively Isolated from Diploids Even if the triploid offspring of a diploid and a tetraploid parent survives and reaches sexual maturity, most of the gametes it produces have aneuploid numbers of chromosomes. Such triploid individuals are effectively sterile. (For simplicity, the diagram shows only three chromosomes; most species have many more than that.)

tally duplicate their chromosomes, resulting in a tetraploid individual (with four sets of chromosomes). Tetraploid and diploid plants of the same species are reproductively isolated because their hybrid offspring are triploid and are usually sterile. They cannot produce viable gametes because their chromosomes do not synapse correctly during meiosis (**Figure 23.7**). So a tetraploid plant cannot produce viable offspring by mating with a diploid individual—but it *can* do so if it self-fertilizes or mates with another tetraploid. Thus polyploidy can result in complete reproductive isolation in two generations—an important exception to the general rule that speciation is a gradual process.

Allopolyploids may also be produced when individuals of two different (but closely related) species interbreed, or **hybridize**. Allopolyploids are often fertile because each of the chromosomes has a nearly identical partner with which to pair during meiosis.

Speciation by polyploidy has been important in the evolution of plants. Botanists estimate that about 70 percent of flowering plant species and 95 percent of fern species are polyploids. Most of these arose as a result of hybridization between two species, followed by self-fertilization. New species arise by means of polyploidy much more easily among plants than among animals because plants of many species can reproduce by self-fertilization. In addition, if polyploidy arises in several offspring of a single parent, the siblings can fertilize one another.

How easily allopolyploidy can produce new species is illustrated by the salsifies (*Tragopogon*), relatives of sunflowers. Salsifies are weedy plants that thrive in disturbed areas around towns. People have inadvertently spread them around the world from their ancestral ranges in Eurasia. Three diploid species of salsify—*Tragopogon porrifolius*, *T. pratensis*, and *T. dubius*—were introduced into North America early in the twentieth century. Two tetraploid hybrids—*T. mirus* and *T. miscellus*—between the original three diploid species were first discovered in 1950. The hybrids have spread since their discovery and today are more widespread than their diploid parents (**Figure 23.8**).

Studies of their genetic material show that both salsify hybrids have formed more than once. Some populations of *T. miscellus*—a hybrid of *T. pratensis* and *T. dubius*—have the chloroplast genome of *T. pratensis*; other populations have the chloroplast genome of *T. dubius*. Genetic differences among local populations of *T. miscellus* show that this allopolyploid has formed independently at least 21 times. *T. mirus*, a hybrid of *T. porrifolius* and *T. dubius*, has formed 12 times. Scientists seldom know the dates and locations of species formation so well. *T. porri-*

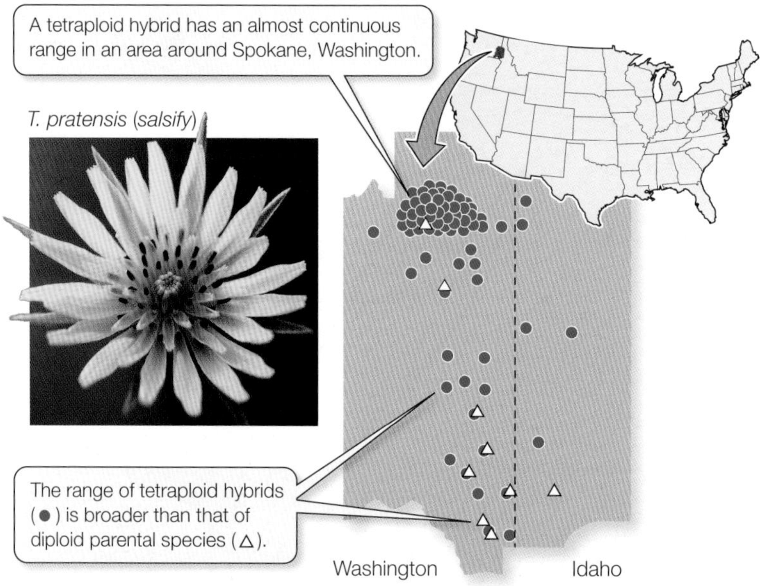

23.8 Polyploids May Outperform Their Parent Species *Tragopogon* species (salsifies) are members of the sunflower family. The map shows the distribution of the three diploid parent species and of the two tetraploid hybrid species of *Tragopogon* in eastern Washington and adjacent Idaho.

folius prefers wet, shady places; *T. dubius* prefers dry, sunny places. *T. mirus*, however, can grow in partly shaded environments where neither parent species does well. The success of these newly formed hybrid species of salsifies illustrates how so many species of flowering plants could have originated as polyploids.

23.2 RECAP

Allopatric speciation results from the separation of populations by geographic barriers; it is the dominant mode of speciation among most groups. Among plants and some animals, sympatric speciation via polyploidy has been frequent.

- Can you explain why an effective barrier to gene flow for one species may not effectively isolate another species? See pp. 511, 513
- How can speciation via polyploidy happen in two generations? See p. 514 and Figure 23.7

Polyploidy, as we have just seen, can result in a new species that is completely reproductively isolated from its parent species in two generations, but most populations separated by a physical barrier become reproductively isolated only very slowly. Let's see how reproductive isolation may become established once two populations have separated from each other.

23.3 What Happens when Newly Formed Species Come Together?

Once a barrier to gene flow is established, the separated populations may diverge genetically through the action of the evolutionary mechanisms we described in Section 22.2. Over many generations, differences may accumulate that reduce the probability that members of the two populations could mate and produce viable offspring. In this way, reproductive isolation can evolve as an incidental by-product of genetic changes in allopatric populations.

Partial reproductive isolation has evolved in this way in strains of *Phlox drummondii*. In 1835, Thomas Drummond, after whom this plant is named, collected seeds in Texas and distributed them to nurseries in Europe. Over the next 80 years, these nurseries established more than 200 true-breeding strains of this phlox, which differed in flower size and color and plant growth form. These plant breeders did not select directly for reproductive incompatibility between strains, but in subsequent experiments, in which the fertilization rates of flowers of various strains by pollen of other strains were measured and compared, it was discovered that reproductive compatibility between strains had been reduced from 14 to 50 percent, depending on the strains.

Geographic isolation does not necessarily lead to reproductive isolation, however, because genetic divergence does not cause reproductive isolation to appear as a byproduct. Therefore, populations that have been isolated from one another for millions of years may still be reproductively compatible. For example, American sycamores and European sycamores (also known as plane trees) have been geographically isolated from one another for at least 20 million years. Nevertheless, they are morphologically very similar (**Figure 23.9**), and they can form fertile hybrids, even though they never have an opportunity to do so in nature.

Reproductive incompatibility may arise by many different mechanisms, but it is useful to group them into two major types: prezygotic and postzygotic reproductive barriers.

Prezygotic barriers operate before fertilization

Several mechanisms that operate before fertilization—**prezygotic reproductive barriers**—may prevent individuals of different species or populations from interbreeding:

23.9 Geographically Separated, Morphologically Similar Although they have been separated by the Atlantic Ocean for at least 20 million years, American and European sycamores have diverged very little in appearance.

- *Habitat isolation.* Individuals of different species may select different habitats in the same general area in which to live or to mate. As a result, they may never come into contact during their respective mating periods. This is what is happening with *Rhagoletis* flies in the Hudson River valley.
- *Temporal isolation.* Many organisms have mating periods that are as short as a few hours or days. If the mating periods of two species do not overlap, they will be reproductively isolated by time, as the *Rhagoletis* flies are.
- *Mechanical isolation.* Differences in the sizes and shapes of reproductive organs may prevent the union of gametes from different species. Males of many insects, for example, have elaborate copulatory organs (penises) that may prevent them from inseminating females of other species.
- *Gametic isolation.* Sperm of one species may not attach to the eggs of another species because the eggs do not release the appropriate attractive chemicals, or the sperm may be unable to penetrate the egg because the two gametes are chemically incompatible.
- *Behavioral isolation.* Individuals of a species may reject, or fail to recognize, individuals of other species as mating partners. Behavioral isolation may have stimulated speciation in birds of paradise.

Sometimes the mate choice of one species is mediated by the behavior of individuals of other species. For example, whether two plant species hybridize may depend on the food preferences of their pollinators. The floral traits of plants can enhance reproductive isolation either by influencing which pollinators are attracted to the flowers or by altering where pollen is deposited on the bodies of pollinators.

The evolution of floral traits that generate reproductive isolation has been studied in columbines of the genus *Aquilegia*. Columbines have undergone recent and very rapid speciation. At the same time, they have evolved long floral nectar spurs—tubular outgrowths of petals that produce nectar at their tips. Animals pollinate these flowers while probing the spurs to collect nectar. The length of the spurs and the orientation of the flowers influence how efficiently pollinators can extract nectar. Two species, *Aquilegia formosa* and *A. pubescens*, that grow in the mountains of California can produce fertile hybrids. *A. formosa* has pendant (hanging) flowers and short spurs (**Figure 23.10A**); it is pollinated by hummingbirds. *A. pubescens* has upright flowers and long spurs (**Figure 23.10B**); it is pollinated by hawkmoths.

M. Fulton and S. Hodges tested discrimination among these flowers by hawkmoths by turning *A. formosa* flowers so that they were upright. Hawkmoths still visited mostly *A. pubescens* flowers (**Figure 23.10C**), probably because the flowers of the two species differ strongly in the color of light they reflect. Thus, although these two species are interfertile, hybrids rarely form in nature because the two species attract different pollinators.

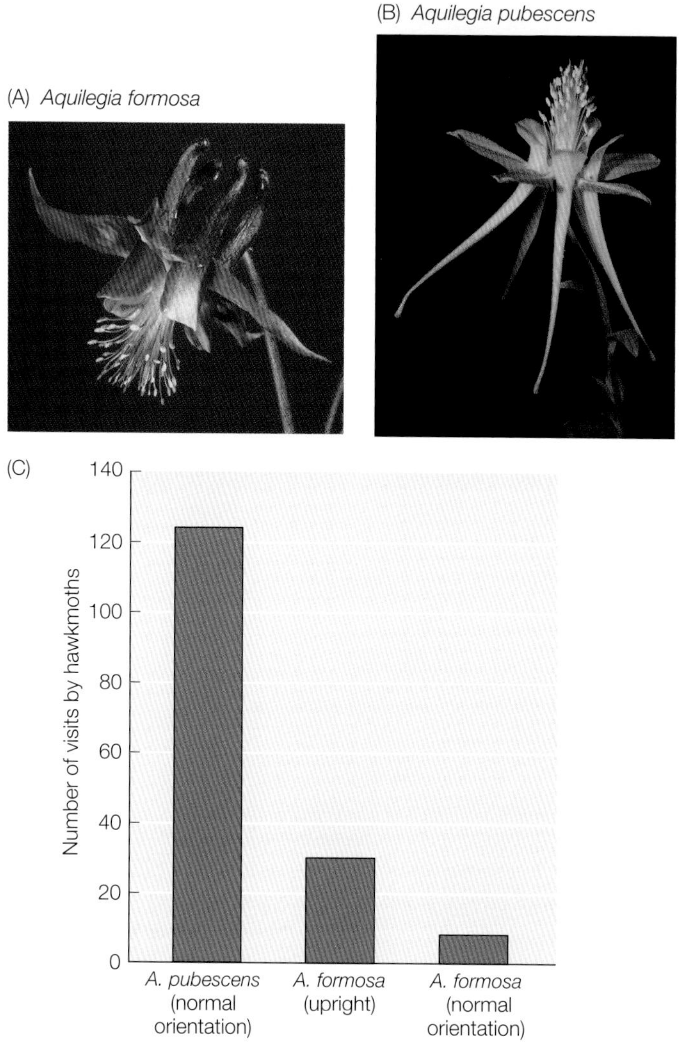

23.10 Hawkmoths Favor Flowers of One Columbine Species (A) Flowers of *Aquilegia formosa* are normally pendant. (B) Flowers of *A. pubescens* are normally upright. (C) The hawkmoths that pollinate *A. pubescens* distinguish between flowers of the two species, even when *A. formosa* flowers are experimentally modified to be upright.

Postzygotic barriers operate after fertilization

If individuals of two different populations lack complete prezygotic reproductive barriers, **postzygotic reproductive barriers** may still prevent gene exchange. Genetic differences that accumulate while the populations are isolated from each other may reduce the survival and reproduction of hybrid offspring in any of several ways:

- *Low hybrid zygote viability.* Hybrid zygotes may fail to mature normally, either dying during development or developing such severe abnormalities that they cannot mate as adults.
- *Low hybrid adult viability.* Hybrid offspring may simply survive less well than offspring resulting from matings within populations.
- *Hybrid infertility.* Hybrids may mature normally, but be infertile when they attempt to reproduce. For example, the offspring of matings between horses and donkeys—mules—are healthy, but they are sterile; they produce no descendants.

Although natural selection does not directly favor the evolution of postzygotic reproductive barriers, if hybrid offspring survive poorly, natural selection may favor the evolution of prezygotic barriers. This happens because individuals that mate with individuals of the

EXPERIMENT

HYPOTHESIS: ***Phlox drummondii*** **has red flowers only where it is sympatric with pink-flowered** ***P. cuspidata*** **because having red flowers decreases interspecific hybridization.**

METHOD

1. Introduce equal numbers of red- and pink-flowered *P. drummondii* individuals into an area with many pink-flowered *P. cuspidata*.
2. After the flowering season ends, assess the genetic composition of the seeds produced by *P. drummondii* plants of both colors.

RESULTS

Of the seeds produced by pink-flowered *P. drummondii*, 38% were hybrids with *P. cuspidata*. Only 13% of the seeds produced by red-flowered individuals were genetic hybrids.

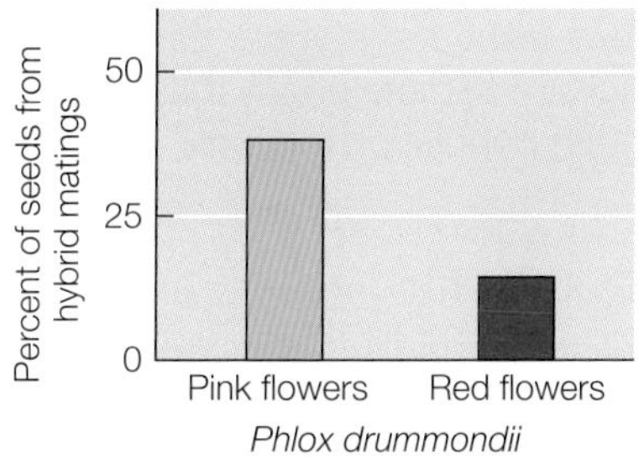

CONCLUSION: For ***Phlox drummondii*****, having flowers that differ in color from those of** ***P. cuspidata*** **reduces the amount of interspecific hybridization.**

23.11 Prezygotic Reproductive Barriers Most *Phlox drummondii* plants are pink, but in regions where they are sympatric with *P. cuspidata*, which is also pink, most individuals are red. An experiment by Donald Levin showed that the red color functions as a prezygotic reproductive barrier, because pollinators tend to visit flowers of only one color. FURTHER RESEARCH: This experiment did not address the probable reproductive advantages for individual *Phlox* plants of donating and receiving primarily intraspecific pollen. What experiments could be designed to measure such advantages?

other population will leave fewer surviving descendants than individuals that mate only within their own population. Such strengthening of prezygotic barriers is known as **reinforcement**.

Donald Levin of the University of Texas noticed that individuals of *Phlox drummondii* over most of the range of the species in Texas have pink flowers. However, where *P. drummondii* is sympatric with the pink-flowered *P. cuspidata*, they have red flowers. No other species of *Phlox* has red flowers. Levin performed an experiment whose results showed that reinforcement might explain the evolution of red flowers where the two species are sympatric (**Figure 23.11**).

Reinforcement can also be detected by using the comparative method. If reinforcement is occurring, then sympatric pairs of closely related species should evolve prezygotic reproductive barriers more rapidly than allopatric pairs of species. An investigation of related sympatric and allopatric species of *Agrodiaetus* butterflies is one of many studies that have demonstrated reinforcement. The colors of the wings of males, which females use to choose mates, have diverged much faster in sympatric than in allopatric populations.

Many closely related species in nature form hybrids in areas where their ranges overlap, and they may continue to do so for many years. Let's examine what happens when reproductive barriers do not completely prevent individuals from different populations from mating and producing offspring.

Hybrid zones may form if reproductive isolation is incomplete

If contact is reestablished between formerly isolated populations before complete reproductive isolation has developed, members of the two populations may interbreed. Three outcomes of such interbreeding are possible:

- If hybrid offspring are as fit as those resulting from matings within each population, hybrids may spread through both populations and reproduce with other individuals. The gene pools are then combined, and no new species result from the period of isolation.
- If hybrid offspring are less fit, complete reproductive isolation may evolve as reinforcement strengthens prezygotic reproductive barriers.
- Even if hybrid offspring are at some disadvantage, a narrow **hybrid zone** may exist if reinforcement does not happen, or the zone may persist for a long time while reinforcement may be developing.

Hybrid zones are excellent natural laboratories for the study of speciation. When a hybrid zone first forms, most hybrids are offspring of crosses between purebred individuals of the two species. However, subsequent generations include a variety of individuals with different proportions of their genes derived from the original two populations. Thus hybrid zones contain recombinant individuals resulting from many generations of hybridization. Detailed genetic studies can tell us much about why hybrid zones may be narrow and stable for long periods of time.

The hybrid zone between two species of European toads of the genus *Bombina* has been studied intensively. The fire-bellied toad (*B. bombina*) lives in eastern Europe. The closely related yellow-bellied toad (*B. variegata*) lives in western and southern Europe. The ranges of the two species meet in a narrow zone stretching 4,800 kilometers from eastern Germany to the Black Sea (**Figure 23.12**). Hybrids between the two species suffer from a range of defects, many of which are lethal. Those that survive often have skeletal abnormalities, such as misshapen mouths, ribs that are fused to vertebrae, and a reduced number of vertebrae.

By following the fates of thousands of toads from the hybrid zone, investigators found that a hybrid toad is on average only half as fit as a purebred individual. The hybrid zone is narrow because there is strong selection against hybrids, and because adult toads do not move over long distances. It has persisted for hundreds of years, however, because many purebred individuals move only a

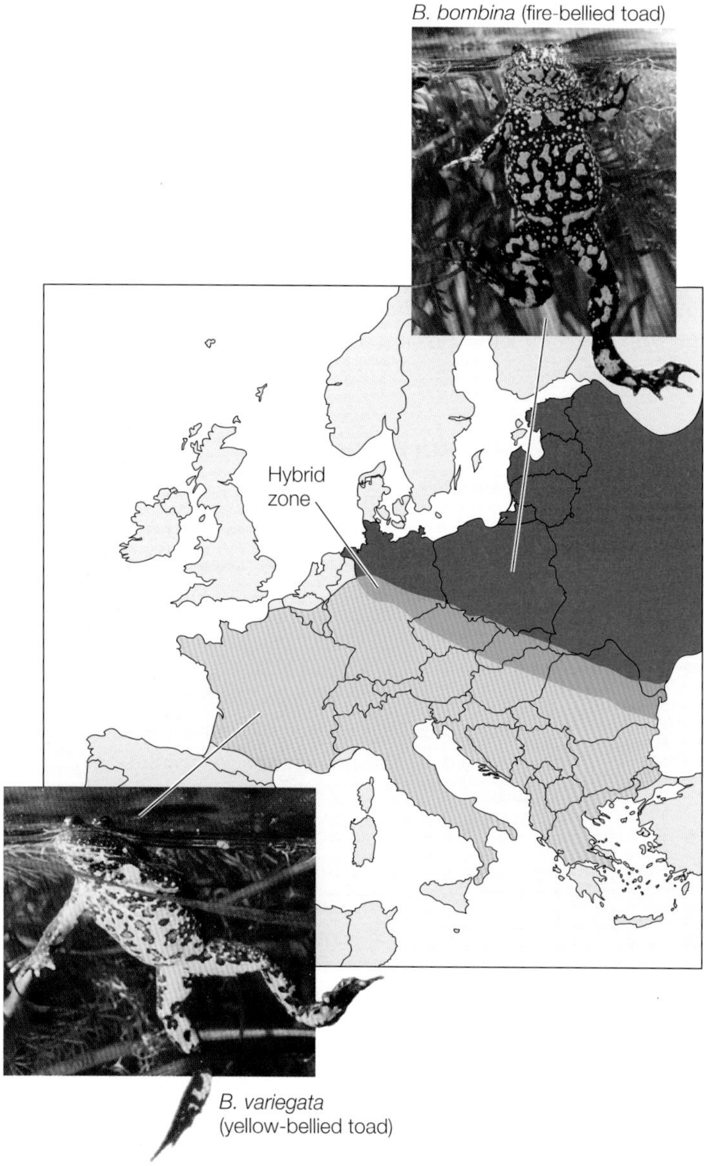

23.12 Hybrid Zones May Be Long and Narrow The narrow zone in Europe where fire-bellied toads meet and hybridize with yellow-bellied toads stretches across Europe. This hybrid zone has been stable for hundreds of years, but has never expanded, and no reinforcement has evolved.

short distance into the zone. They have not previously encountered individuals of the other species, so there has been no opportunity for reinforcement to evolve.

Human activity may result in new hybrid zones. One example is found among the many protead plants that are adapted to the dry, nutrient-poor soils of southwestern Australia. In undisturbed sand plain vegetation in this region, two related species of protead shrubs, *Banksia hookeriana* and *Banksia prionotes,* do not hybridize because their flowering seasons do not overlap. However, where human disturbances have disrupted their habitat, their flowering seasons expand and overlap, and fully fertile hybrids between the two species are now common.

The protead plant group is both distinctive and ancient, dating from the Mesozoic. Proteads were isolated on Gondwana when Pangaea broke apart, and no members of the group are native to the Northern Hemisphere. Modern proteads are most species-rich in southwest Australia (≈550 species) and the Cape region of South Africa (≈320 species)

23.3 RECAP

Reproduction isolation may result from prezygotic or postzygotic reproductive barriers. If reproductive isolation is incomplete, hybrid zones may form when previously separated populations come into contact.

- Can you describe the various kinds of prezygotic and postzygotic reproductive barriers? See pp. 515–516
- Why is reinforcement of prezygotic barriers likely if hybrid offspring survive more poorly than offspring produced by within-population matings? See p. 517

Some groups of organisms have many species; others have only a few. Hundreds of species of *Drosophila* evolved rapidly in the Hawaiian Islands, but worldwide there is only one species of horseshoe crab, even though its ancestry dates back more than 300 million years. Why do different groups of organisms have such different rates of speciation?

23.4 Why Do Rates of Speciation Vary?

Rates of speciation vary greatly because many factors influence the likelihood that a lineage will split to form two or more species. The larger the number of species in a group, the larger the number of opportunities for new species to form. For speciation by polyploidy, the more species in a group, the more species are available to hybridize with one another. For allopatric speciation, the larger the number of different species living in an area, the larger the number of species whose ranges will be bisected by a given physical barrier. For all these reasons, "the rich get richer": groups that are already species-rich are likely to speciate faster than species-poor groups.

Speciation rates are likely to be higher in species with poor dispersal abilities than in species with good dispersal abilities, because even a narrow barriers can be effective in dividing a species whose members are highly sedentary. The Hawaiian Islands have about a thousand species of land snails, many of which are restricted to a single valley. Because snails move only short distances, the high ridges that separate the valleys are effective barriers to their dispersal.

Populations of species that have specialized diets are more likely to diverge than are populations that have generalized diets. To investigate the effects of diet on rates of speciation, C. Mitter and colleagues compared species richness in some closely related groups of true bugs (hemipterans). The common ancestor of these

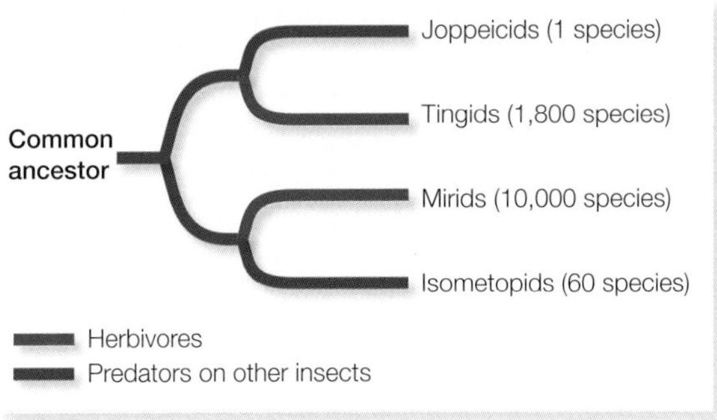

23.13 Dietary Shifts Can Promote Speciation Herbivorous groups of hemipteran insects have speciated several times faster than closely related predatory groups.

groups was a predator that fed on other insects, but a dietary shift to herbivory (eating plants) evolved at least twice in the groups under study. The herbivorous groups have many more species than those that are predatory (**Figure 23.13**). Herbivorous bugs typically specialize on one or a few closely related species of plants, whereas predatory bugs tend to feed on many different species of insects.

Speciation rates in plants are faster in animal-pollinated than wind-pollinated plants. Animal-pollinated groups have, on average, 2.4 times as many species as related groups pollinated by wind. Among animal-pollinated plants, speciation rates are correlated with pollinator specialization. In columbines (*Aquilegia*), the rate of evolution of new species has been about three times faster in lineages that have long nectar spurs as in closely related lineages that lack spurs. Why do spurs increase the speciation rate? Apparently it is because having longer spurs restricts the number of pollinator species that visit the flowers, thus increasing opportunities for reproductive isolation (see Figure 23.10).

The mechanisms of sexual selection (see Section 22.3) also appear to result in increased rates of speciation. Some of the most striking examples of sexual selection are found in birds with promiscuous mating systems.

Bird-watchers travel thousands of miles to Papua New Guinea to witness the mating displays of male birds of paradise, which have long, brightly colored tail feathers and are very distinct in appearance from the females (*sexual dimorphism*). In many of these species, males assemble at display grounds, and females come there to choose a male with whom to copulate. After mating, the females leave the display grounds, build their nests, lay their eggs, and feed their offspring with no help from the males. The males remain to court more females. There are 33 species of birds of paradise (**Figure 23.14A**).

The closest relatives of the birds of paradise are the manucodes. Male and female manucodes differ only slightly in size and plumage. They form monogamous pair bonds and both sexes contribute to raising the young. There are only 5 species of manucodes (**Figure 23.14B**).

Animals with complex sexually selected behaviors are likely to form new species at a high rate because they make sophisticated discriminations among potential mating partners. They distinguish members of their own species from members of other species, and they make subtle discriminations among members of their own species on the basis of size, shape, appearance, and behavior (see Figures 22.16 and 22.17). Such discriminations can greatly influence which individuals are most successful in mating and producing offspring, and may lead to rapid evolution of differences among populations.

(A) *Paradisaea minor*

(B) *Manucodia comrii*

23.14 Sexual Selection in Birds Can Lead to Higher Speciation Rates (A) Birds of paradise and (B) manucodes are closely related bird groups of the South Pacific. However, speciation rates are much higher among the sexually dimorphic, polygynous birds of paradise (33 species) than among manucodes (5 species).

23.4 RECAP

The existing species richness of a group, dispersal ability, dietary specialization, type of pollination, and sexual selection are among the many factors that influence speciation rates.

- Do you understand why speciation rates tend to be higher in groups that already have a lot of species? See p. 518
- Do you understand why pollinator specialization in plants and sexual selection in animals typically stimulate speciation? See p. 519

The fossil record reveals that, at certain times in certain groups, speciation rates have been much higher than extinction rates, resulting in the proliferation of a large number of daughter species. Let's examine why such evolutionary radiations happen.

23.5 Why Do Adaptive Radiations Occur?

The proliferation of a large number of daughter species from a single ancestor is called an **evolutionary radiation**. If the rapid proliferation of species results in an array of species that live in a variety of environments and differ in the characteristics they use to exploit those environments, the radiation is said to be *adaptive*. If a radiation is not accompanied by any observed ecological differentiation among the species, it is said to be nonadaptive.

An **adaptive radiation** begins when genetic differentiation between populations evolves in response to differences in the environments they inhabit and the resources they use. Such differentiation is likely to occur in environments with abundant resources. A population is likely to encounter underutilized resources when it colonizes a new environment that contains relatively few species. That is why, as we saw in Chapter 21, adaptive radiations have frequently followed mass extinctions. Similarly, islands lack many groups of organisms found on the mainland, so the ecological opportunities that exist on islands may stimulate rapid evolutionary changes when a new species colonizes them. Water barriers also restrict gene flow among the islands in an archipelago, so populations on the different islands may evolve adaptations to their local environments.

Remarkable adaptive radiations have occurred in the Hawaiian Islands. The native biota of the Hawaiian Islands includes 1,000 species of flowering plants, 10,000 species of insects, 1,000 land snails, and more than 100 bird species. However, there were no amphibians, no terrestrial reptiles, and only one native terrestrial mammal—a bat—on the islands until humans introduced additional species. The 10,000 known native species of insects on Hawaii are believed to have evolved from only about 400 immigrant species; only 7 immigrant species are believed to account for all the native Hawaiian land birds. Similarly, as we saw earlier in this chapter, an adaptive radiation in the Galápagos archipelago resulted in the 14 species of Darwin's finches, which differ strikingly in their bill sizes and shapes and, consequently, in the food resources they use (see Figure 23.4).

The Hawaiian archipelago is Earth's most isolated group of islands. The Hawaiian Islands lie 4,000 kilometers from the nearest continental land mass and 1,600 kilometers from the nearest group of islands.

23.15 Rapid Evolution among Hawaiian Silverswords The Hawaiian silverswords, three closely related genera of the sunflower family, are believed to have descended from a single common ancestor (a plant similar to the tarweed) that colonized Hawaii from the Pacific coast of North America. The four plants shown here are more closely related than they appear to be based on their morphology.

Madia sativa (tarweed)

Argyroxiphium sandwicense

Wilkesia hobdyi

Dubautia menziesii

Adaptive radiations have been frequent among plants on the Hawaiian Islands. More than 90 percent of the 1,000 plant species on the Hawaiian Islands are *endemic*—that is, they are found nowhere else. Several groups of flowering plants have more diverse forms and life histories on the islands, and live in a wider variety of habitats, than do their close relatives on the mainland. An outstanding example is the 28 species of Hawaiian sunflowers called silverswords (genera *Argyroxiphium*, *Dubautia*, and *Wilkesia*). Chloroplast DNA sequences show that these species share a relatively recent common ancestor with a species of tarweed from the Pacific coast of North America (**Figure 23.15**). Whereas all mainland tarweeds are small, upright herbs (nonwoody plants), the silverswords include prostrate and upright herbs, shrubs, trees, and vines. Silversword species occupy nearly all the habitats of the Hawaiian Islands, from sea level to above timberline in the mountains. Despite their extraordinary morphological diversification, the silverswords are genetically very similar.

The island silverswords are more diverse in size and shape than the mainland tarweeds because the original colonizers arrived on islands that had very few plant species. In particular, there were few trees and shrubs, because such large-seeded plants rarely disperse to oceanic islands. Trees and shrubs have evolved from nonwoody ancestors on many oceanic islands. On the mainland, however, tarweeds live in ecological communities that contain many tree and shrub species in lineages with long evolutionary histories. In those environments, opportunities to exploit the "tree" way of life had already been preempted.

Adaptive radiations are common on islands but are not confined to them. Genetic analyses of the 1,563 species of ice plants, nearly all of which are endemic to southern Africa, show that the group radiated in that region within the last 9 million years. Ice plants are *succulents*, plants whose cells utilize the processes of crassulacean acid metabolism (CAM; see Section 8.4) to withstand extremely dry conditions. Their prominence in their desert home has given the region its name—the Succulent Karoo.

23.5 RECAP

Evolutionary radiation is the rapid proliferation of species from a single ancestor.

- Can you describe the difference between adaptive and nonadaptive radiation? See p. 520
- Do you understand why adaptive radiations are particularly common on islands? See pp. 520–521

The processes we have discussed in this chapter, operating over billions of years, have produced a world in which life is organized into millions of species, each adapted to live in a particular environment and to use environmental resources in a particular way. How these millions of species are organized into ecological communities is explored in Part Nine of this book.

CHAPTER SUMMARY

23.1 What are species?

Speciation is the process by which one species splits into two or more daughter species, which thereafter evolve as distinct lineages. Review Figure 23.2

The **morphological species concept** distinguishes species on the basis of physical similarities.

The **biological species concept** distinguishes species on the basis of **reproductive isolation**; that is, a species are considered to be made up of populations that can interbreed with each other. Asexual species cannot be defined with the biological species concept.

23.2 How do new species arise?

See Web/CD Tutorial 23.1

Speciation requires that gene flow within a population that once exchanged genes be interrupted.

Allopatric speciation, which results when populations are separated by a physical barrier, is the dominant mode of speciation. This type of speciation may follow from founder events, in which some members of a population cross a barrier and found a new, isolated population. Review Figure 23.3, Web/CD Tutorial 23.2

Sympatric speciation results when the genomes of two groups diverge in the absence of physical isolation. It can result when groups are ecologically isolated (i.e., depend on different resources). Review Figure 23.6

In plants, sympatric speciation can occur within two generations via **polyploidy**, an increase in the number of chromosomes. Polyploidy may arise from chromosome duplications within a species (autopolyploidy) or from **hybridization** that results in combining the chromosomes of two species (**allopolyploidy**). Review Figure 23.7

23.3 What happens when newly formed species come together?

Prezygotic barriers to reproduction operate before fertilization; **postzygotic reproductive barriers** operate after fertilization. Prezygotic barriers may be favored by natural selection if postzygotic barriers are incomplete.

Hybrid zones may form when previously separated populations come into contact if reproductive isolation is incomplete.

23.4 Why do rates of speciation vary?

Species richness, dispersal ability, dietary specialization, type of pollination, and sexual selection all influence speciation rates. Review Figure 23.13

23.5 Why do adaptive radiations occur?

An **evolutionary radiation**, a rapid proliferation of species from a common ancestor, may result in an array of species that live in a variety of environments. Radiation may also be stimulated by sexual selection.

Adaptive radiation, during which daughter species become ecologically differentiated, is likely to occur in environments where there is an array of underutilized resources.

See Web/CD Activity 23.1 for a concept review of this chapter.

SELF-QUIZ

1. The biological species concept defines a species as a group of
   a. actually interbreeding natural populations that are reproductively isolated from other such groups.
   b. potentially interbreeding natural populations that are reproductively isolated from other such groups.
   c. actually or potentially interbreeding natural populations that are reproductively isolated from other such groups.
   d. actually or potentially interbreeding natural populations that are reproductively connected to other such groups.
   e. actually interbreeding natural populations that are reproductively connected to other such groups.
2. Which of the following is *not* a condition that favors allopatric speciation?
   a. Continents drift apart and separate previously connected lineages.
   b. A mountain range separates formerly connected populations.
   c. Different environments on two sides of a barrier cause populations to diverge.
   d. The range of a species is separated by loss of intermediate habitat.
   e. Tetraploid individuals arise in one part of the range of a species.
3. Finches speciated in the Galápagos Islands because
   a. the Galápagos Islands are not far from the mainland.
   b. the Galápagos Islands are arid.
   c. the Galápagos Islands are small.
   d. the islands of the Galápagos archipelago are sufficiently isolated from one another that there is little migration among them.
   e. the islands of the Galápagos archipelago are close enough to one another that there is considerable migration among them.
4. Which of the following is *not* a potential prezygotic reproductive barrier?
   a. Temporal segregation of breeding seasons
   b. Differences in chemicals that attract mates
   c. Hybrid infertility
   d. Spatial segregation of mating sites
   e. Sperm that cannot penetrate an egg
5. A common means of sympatric speciation is
   a. polyploidy.
   b. hybrid infertility.
   c. temporal segregation of breeding seasons.
   d. spatial segregation of mating sites.
   e. imposition of a geographic barrier.
6. Narrow hybrid zones may persist for long times because
   a. hybrids are always at a disadvantage.
   b. hybrids have an advantage only in narrow zones.
   c. hybrid individuals never move far from their birthplaces.
   d. individuals that move into the zone have not previously encountered individuals of the other species, so reinforcement of reproductive barriers has not occurred.
   e. Narrow hybrid zones are artifacts because biologists generally restrict their studies to contact zones between species.
7. Which statement about speciation is *not* true?
   a. It always takes thousands of years.
   b. Reproductive isolation may develop slowly between diverging lineages.
   c. Among animals, it usually requires a physical barrier.
   d. Among plants, it often happens as a result of polyploidy.
   e. It has produced the millions of species living today.
8. Speciation is often rapid within groups whose species have complex behavior because
   a. individuals of such species make fine discriminations among potential mating partners.
   b. such species have short generation times.
   c. such species have high reproductive rates.
   d. such species have complex relationships with their environments.
   e. such species are particularly abundant.
9. Evolutionary radiations
   a. often happen on continents, but rarely on island archipelagoes.
   b. characterize birds and plants, but not other groups of organisms.
   c. have happened on continents as well as on islands.
   d. require major reorganizations of the genome.
   e. never happen in species-poor environments.
10. Speciation is an important component of evolution because it
   a. generates the variation on which natural selection acts.
   b. generates the variation on which genetic drift and mutations act.
   c. enabled Charles Darwin to perceive the mechanisms of evolution.
   d. generates the high extinction rates that drive evolutionary change.
   e. has resulted in a world with millions of species, each adapted for a particular way of life.

FOR DISCUSSION

1. The North American snow goose has two distinct color forms, blue and white. Matings between the two color forms are common. However, blue individuals pair with blue individuals and white individuals pair with white individuals much more frequently than would be expected by chance. Suppose that 75 percent of all mated pairs consisted of two individuals of the same color. What would you conclude about speciation processes in these geese? If 95 percent of pairs were the same color? If 100 percent of pairs were the same color?
2. Suppose pairs of snow geese of mixed colors were found only in a narrow zone within the broad Arctic breeding range of the geese. Would your answer to Question 1 remain the same? Would your answer change if mixed-color pairs were widely distributed across the breeding range of the geese?
3. Although many butterfly species are divided into local populations among which there is little gene flow, these species often show relatively little morphological variation among populations. Describe the studies you would conduct to determine what maintains this morphological similarity.
4. Evolutionary radiations are common and easily studied on oceanic islands, but in what types of *mainland* situations would you expect to find major evolutionary radiations? Why?
5. Fruit flies of the genus *Drosophila* are distributed worldwide, but 30–40 percent of the species in the genus are found on the Hawaiian Islands. What might account for this distribution pattern?
6. Evolutionary radiations take place when speciation rates exceed extinction rates. What factors can cause extinction rates to exceed

speciation rates in a clade? Name some clades in which human activities are increasing extinction rates without increasing speciation rates.

7. If it is true that natural selection does not directly favor lower viability of hybrids, why is it that hybrid individuals so often have lowered viability?

FOR INVESTIGATION

The experiment in Figure 23.10 changed only the orientation of the flowers. Although the flowers in the experiment were similarly oriented, they still differed in color, and probably in odor as well. What experiments could you design to determine the traits that the bees use to distinguish among the flowers of different *Aquilegia* species?

CHAPTER 24 The Evolution of Genes and Genomes

Molecular evolution and the conquest of polio

Prior to the late 1950s, thousands of children every year died or were left paralyzed by poliomyelitis (polio). This disease is caused by several strains of poliovirus, which infect humans through the mouth and multiply in the intestines. There are no symptoms in most infected individuals, but sometimes the virus invades the nervous system, resulting in rapid paralysis.

In 1955, Jonas Salk developed a vaccine against this nightmare disease. The vaccine, IPV (*i*nactivated *p*olio *v*accine), was based on killed virus. An injection of IPV produces antibodies to poliovirus in the blood (*serum immunity*) and prevents the spread of poliovirus into the nervous system. IPV does not prevent infection of the intestine, however, so infected individuals can still spread the virus to others. Thus, even though IPV prevented many cases of polio, the disease persisted, especially in developing countries where it was difficult to vaccinate everyone.

In 1958, Albert Sabin introduced an alternative: a live-virus vaccine that could be administered orally and provided a local immune reaction in the intestine as well as serum immunity. Sabin's discovery required selection of mutant strains of poliovirus, called *attenuated viruses*, that did not cause the disease.

Sabin could not have known the molecular details back in 1958, but today we know a small number of nucleotide substitutions present in the attenuated strains prevent them from infecting the nervous system. Antibodies produced in the lining of the intestine in response to attenuated viruses also prevent the multiplication of wild-type poliovirus, and the wild-type cannot infect the intestine of an immunized individual; thus Sabin's vaccine—called OPV, for *o*ral *p*olio *v*accine—prevents person-to-person transmission. Moreover, the attenuated strains of OPVs can spread to other individuals (especially in regions with poor sanitation), so an entire local population can become protected, even if not everyone receives the vaccine.

But OPVs have one major disadvantage: the attenuated strains continue to evolve. Very rarely, an attenuated virus undergoes a simple back substitution that results in its reversion to a virulent strain. Therefore, in regions where poliovirus has been virtually eliminated, the IPV vaccine is once again preferred: since it is based on a dead virus, it

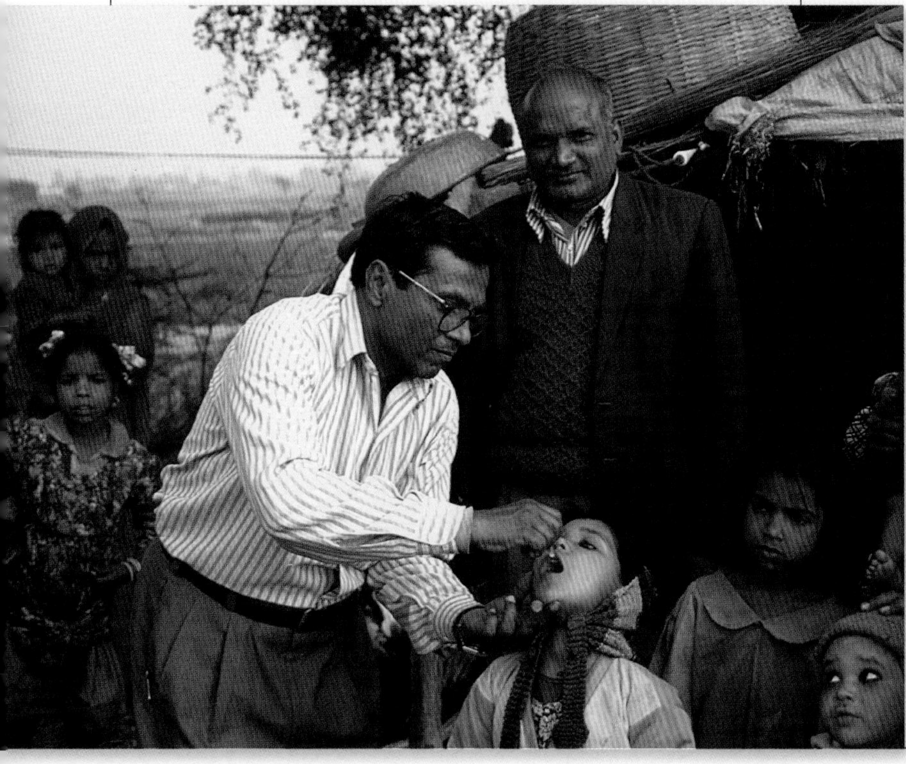

Combating Paralysis Mutant strains of poliovirus were the key to developing an effective vaccine that is helping to eliminate polio worldwide. Here a government health official administers the oral vaccine to children in rural India.

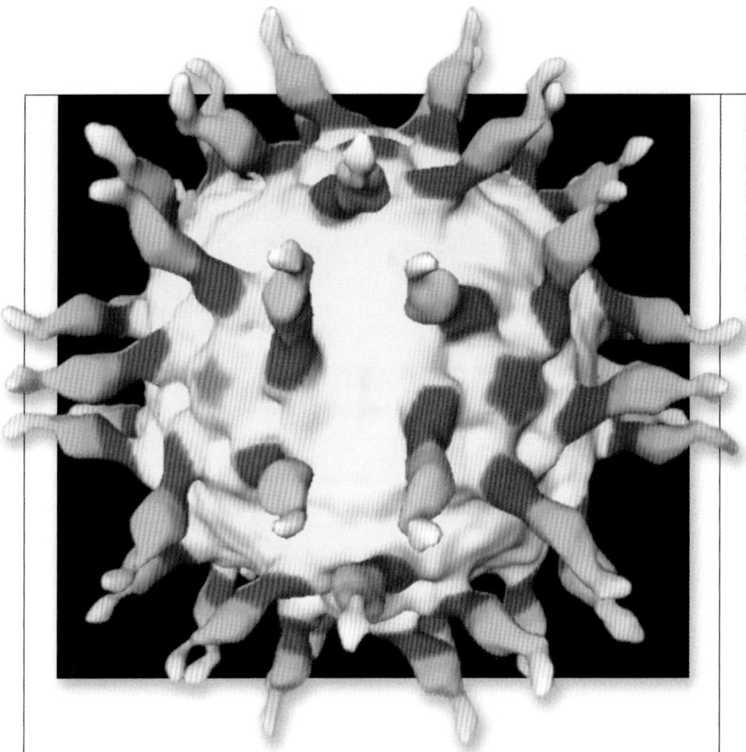

Killing a Killer Poliovirus infects its host by binding to receptors on the host cell surfaces. The binding sites on this inactivated virus are shown in red; antibodies have inactivated the sites and "killed" the virus. Injection of a vaccine made from inactivated poliovirus (IPV) stops the virus from affecting the nervous system.

cannot evolve. And now that entire poliovirus genomes have been sequenced, the molecular basis of attenuation is known—which is useful in selecting attenuated strains that are less subject to reversion to use in OPVs.

Today, when a polio outbreak occurs anywhere in the world, DNA sequencing can be used to determine the evolutionary relationships of the viruses involved and the preferred vaccination strategy. This medical success story is a result of our growing understanding of molecular evolution.

IN THIS CHAPTER we will see how molecular biologists use nucleic acids and proteins to infer both the patterns and the causes of molecular evolution. We will explore how the functions of molecules change, where new genes come from, and how genomes change in size. Finally, we will see how knowledge of the patterns of molecular evolution can help answer other biological questions.

CHAPTER OUTLINE

24.1 What Can Genomes Reveal about Evolution?

An organism's **genome** is the full set of genes it contains, as well as any noncoding regions of the DNA (or, in the case of some viruses, RNA). Most of the genes of eukaryotic organisms are found on chromosomes in the nucleus, but genes are also present in chloroplasts and mitochondria. In organisms that reproduce sexually, both males and females transmit nuclear genes, but mitochondrial and chloroplast genes usually are transmitted only via the cytoplasm of eggs, as we saw in Section 10.5.

Genomes must be replicated to be transmitted from parents to offspring. DNA replication does not occur without error, however. Mistakes in DNA replication—mutations—provide the raw material for evolutionary change. Mutations are essential for the long-term survival of life, for without the genetic variation they provide, organisms could not evolve in response to changes in their environment.

A particular copy of a gene will not be passed on to successive generations unless an individual with that copy survives and reproduces. Therefore, the capacity to cooperate with different combinations of other genes is likely to increase a particular allele's probability of fixation in a population. Moreover, the degree and timing of a gene's expression are affected by its location in the genome. For these reasons, the genes of an individual organism can be viewed as interacting members of a group, among which there are divisions of labor, but also strong interdependencies.

A genome, then, is not simply a random collection of genes in random order along chromosomes. Rather, it is a complex set of integrated genes and their regulatory sequences, as well as vast stretches of noncoding DNA that apparently have little direct function. The positions of genes, as well as their sequences, are subject to evolutionary change, as are the extent and location of noncoding DNA. All of these changes can affect the phenotype of an organism. Biologists have now sequenced the complete genomes of a large number of organisms (including humans), and this information is helping us to understand how and why organisms differ, how they function, and how they have evolved.

Evolution of genomes results in biological diversity

The field of **molecular evolution** investigates the mechanisms and consequences of the evolution of macromolecules. Molecular evolutionists study relationships between the structures of genes and proteins and the functions of organisms. They also use molecular variation to reconstruct evolutionary history and study the mechanisms and consequences of evolution. The molecules of special interest to molecular evolutionists are nucleic acids (DNA and RNA) and proteins. Students of this field ask questions such as, How do proteins acquire new functions? Why are the genomes of different organisms so variable in size? How has enlargement of genomes been accomplished? They also investigate the evolution of particular nucleic acids and proteins and use their findings to reconstruct the evolutionary histories of genes and the organisms that carry them. In this way, molecular evolutionists seek to understand the molecular basis for the biological diversity that we now observe in the world around us.

The evolution of nucleic acids and proteins depends on genetic variation introduced by mutations (for a review of the various kinds of mutations, refer to Section 12.6). One of several ways in which genes evolve is by means of **nucleotide substitutions**. In genes that encode proteins, some of these nucleotide substitutions can result in **amino acid replacements** in the encoded proteins. Changes in the amino acid sequence of a protein can change the charge, as well as the secondary (two-dimensional) and tertiary (three-dimensional) structure, of the protein. All of these phenotypic changes affect the way the protein functions in the organism.

Evolutionary changes in genes and proteins can be identified by comparing the nucleotide or amino acid sequences among different organisms. The longer two sequences have been evolving separately, the more differences they accumulate (although different genes in the same species evolve at different rates). Evolutionary analysis by such comparisons is essential for determining the timing of evolutionary changes in molecular characters, and knowing the timing of such changes is usually the first step in inferring their causes. Conversely, knowledge of the pattern and rate of evolutionary change in a given macromolecule is useful in reconstructing the evolutionary history of groups of organisms.

Before biologists can compare genes or proteins across different organisms, they must have a method for identifying homologous parts of these molecules. As we will see in Section 25.1, any features shared by two or more species that have been inherited from a common ancestor are said to be *homologous*. For example, the forelimbs of all mammals are homologous, although they differ greatly in their form and function (consider the fins of whales and the arms of humans). The concept of homology extends down to the level of particular nucleotide positions in genes. Therefore, one of the first steps in studying the evolution of genes or proteins is to align homologous positions in the nucleotide or amino acid sequences of interest.

Genes and proteins are compared through sequence alignment

Once the DNA or amino acid sequences of molecules from different organisms have been determined, they can be compared. Homologous positions can be identified only if we first pinpoint the locations of deletions and insertions that have occurred in the molecules of interest in the time since the organisms diverged from a common ancestor. A simple hypothetical example illustrates this **sequence alignment** technique. In **Figure 24.1** we compare two amino acid sequences (1 and 2) from homologous proteins in different organisms. The two sequences at first appear to differ in the number and identity of their amino acids, but if we insert a gap after the first amino acid in sequence 2 (after leucine), similarities in the two sequences become evident. This gap indicates the occurrence of one of two evolutionary events: an insertion of an amino acid in the longer protein, or a deletion of an amino acid

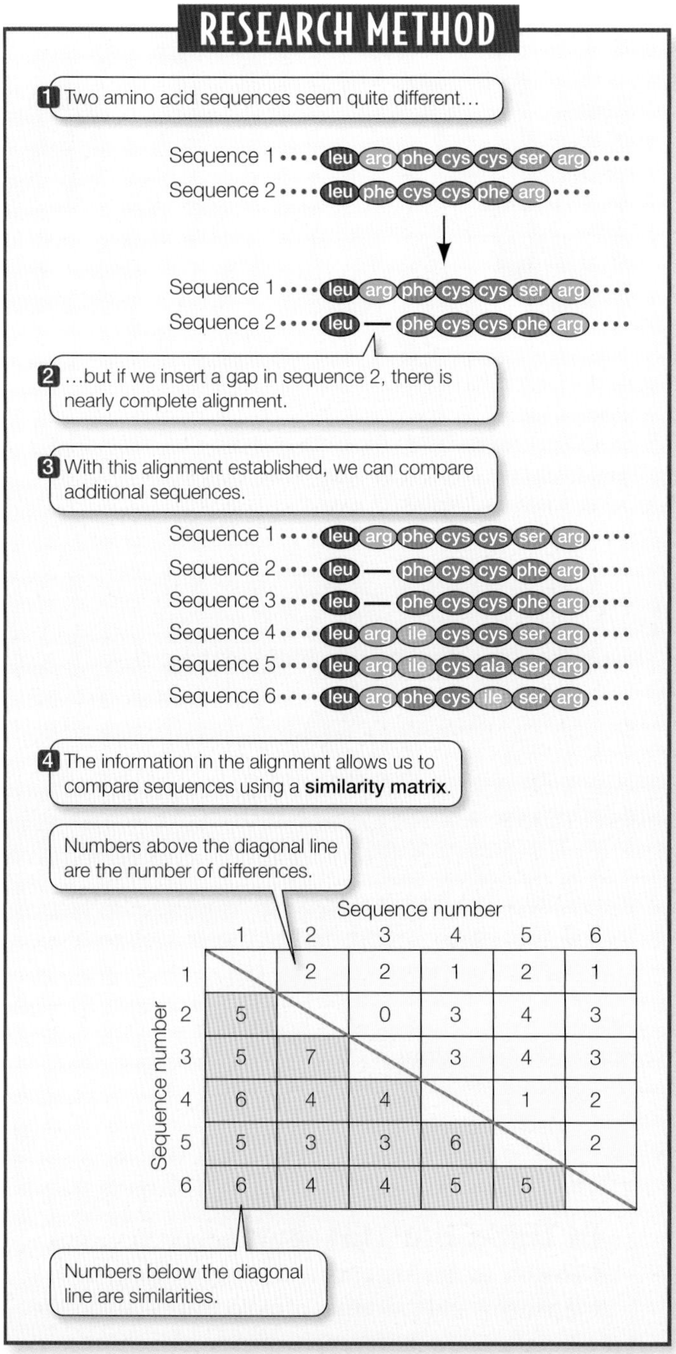

24.1 Amino Acid Sequence Alignment Insertion of a gap (—) allows us to align two homologous amino acid squences so that we can compare them. Once the alignment is established, sequences from more organisms can be added and compared. A similarity matrix sums similarities and differences between each pair of organisms.

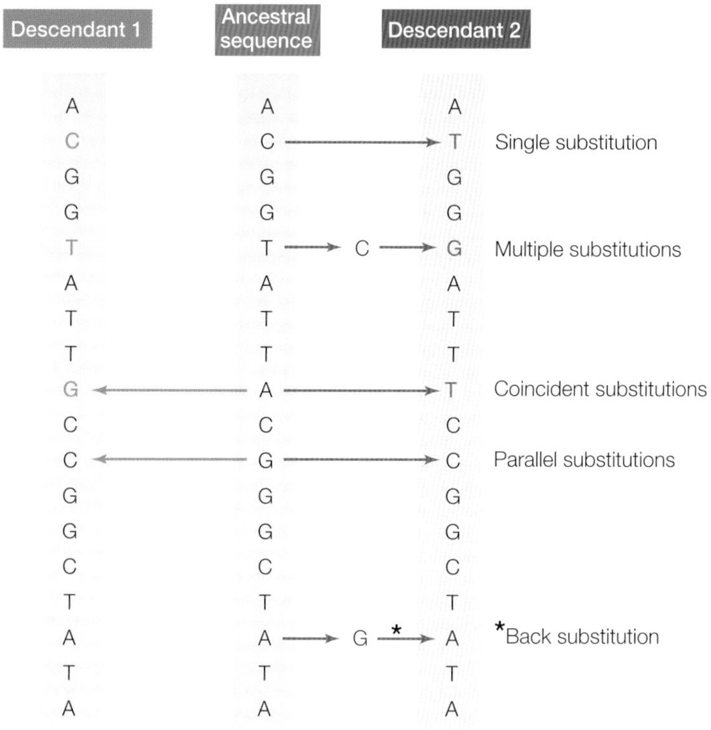

24.2 Multiple Substitutions Are Not Reflected in Pairwise Sequence Comparisons An ancestral sequence (center) gives rise to two descendant sequences through a series of substitutions. Although the two descendant sequences exhibit only three nucleotide differences (colored letters), these three differences resulted from a total of nine substitutions (arrows).

in the shorter protein. Having adjusted for this gap, we can see that the two sequences differ by only one amino acid at position 6 (serine or phenylalanine). Adding a single gap—that is, identifying a deletion or an insertion—*aligns* these sequences. Longer sequences and those that have diverged more extensively require more elaborate adjustments based on explicit models (computer algorithms) for the relative costs of deletions, insertions, and particular amino acid replacements.

Having aligned the sequences, we can compare them by counting the number of nucleotides or amino acids that differ between them. If we add more sequences to our original example and sum the number of similar and different amino acids in each pair of sequences, we can construct a **similarity matrix**, which gives us a measure of the minimum number of changes that have occurred during the divergence between each pair of organisms (see Figure 24.1).

Models of sequence evolution are used to calculate evolutionary divergence

The sequence comparison procedure illustrated in Figure 24.1 gives a simple count of the minimum number of changes between two species. In the context of two aligned DNA sequences, we can count the number of differences at homologous nucleotide positions, and this count indicates the minimum number of nucleotide substitutions that must have occurred between the two sequences.

However, this simple count of differences almost certainly underestimates the number of substitutions that have actually occurred since the sequences diverged from a common ancestor. When more than one substitution occurs between the ancestor and the descendants, as illustrated in **Figure 24.2**, the number of changes is not captured by a simple count of differences in a similarity matrix. Several phenomena can account for this, including:

- *Multiple substitutions:* More than one change occurs at a given position between the ancestor and a descendant.
- *Coincident substitutions:* At a given position, different substitutions occur between the ancestor and each descendant.
- *Parallel substitutions:* The same substitution occurs independently between the ancestor and each descendant.
- *Back substitutions:* In a variation on multiple substitutions, after a change at a given position, a subsequent substitution changes the position back to the ancestral state.

To correct for undercounting of substitutions, molecular evolutionists have developed mathematical models that describe how DNA (and protein) sequences evolve. These models take into account the rates of change from each nucleotide to another; for example, *transitions* (changes between two purines or between two pyrimidines) are more frequent than *transversions* (changes between a purine and pyrimidine). They also include parameters such as the different rates of substitution across different parts of a gene and the proportions of each nucleotide present in a given sequence. These parameters are estimated for a particular set of sequences, and then the models are used to correct for multiple substitutions, coincident substitutions, parallel substitutions, and back substitutions. The end result is a revised estimate of the total number of substitutions likely to have occurred between two sequences, which is almost always greater than the observed number of differences.

As sequence information becomes available for more and more proteins in an ever-expanding database, sequence alignments can be extended across multiple homologous sequences, and the minimum number of insertions, deletions, and substitutions can be summed across homologous genes or proteins of an entire group of organisms. **Figure 24.3** shows this kind of aligned data for cytochrome *c* sequences in a wide variety of animals, plants, and fungi. This type of information is used extensively in determining the evolutionary relationships among species.

Experimental studies examine molecular evolution directly

Although molecular evolutionists are often interested in naturally evolved sequences and proteins, molecular and phenotypic evolution can also be observed directly in the laboratory. Increasingly, evolutionary biologists are studying evolution experimentally. Because substitution rates are related to generation rate rather than to absolute time, most of these experiments use unicellular organisms or viruses with short generations. Viruses, bacteria, and unicellular eukaryotes (such as the yeasts) can be cultured in large populations in the laboratory, and many of these organisms can evolve

Tuna

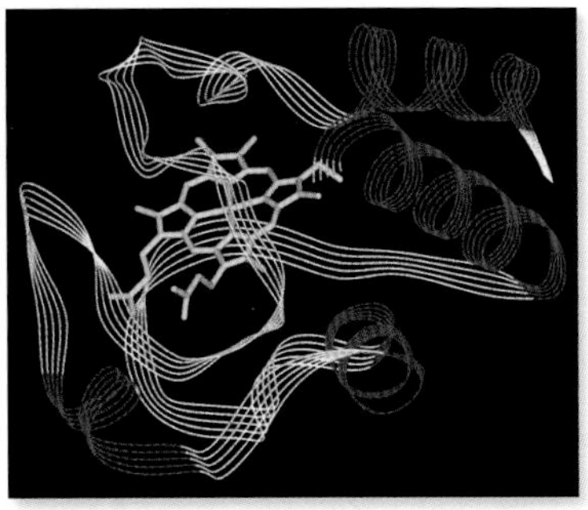

Rice

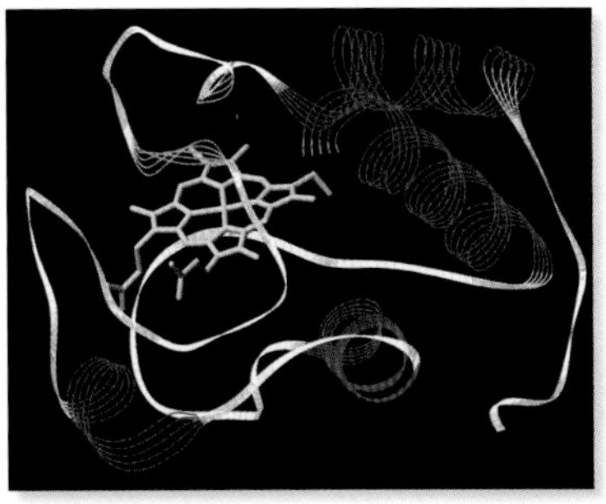

The number 1 indicates an invariant position in the cytochrome *c* molecule (i.e., all the organisms have the same amino acid in this position). Such a position is probably under strong stabilizing selection.

Amino acids at positions marked by red arrowheads have side chains that interact with the heme group.

Position in sequence	1				5					10					15					20					25					30				
														▼			▼	▼																
Number of amino acids at the position	1	3	5	5	5	1	3	3	4	1	4	3	2	1	3	3	1	1	2	4	3	4	2	3	4	2	1	4	1	1	2	1	5	1
Human, chimpanzee	G	D	V	E	K	G	K	K	I	F	I	M	K	C	S	Q	C	H	T	V	E	K	G	G	K	H	K	T	G	P	N	L	H	G
Rhesus monkey	G	D	V	E	K	G	K	K	I	F	I	M	K	C	S	Q	C	H	T	V	E	K	G	G	K	H	K	T	G	P	N	L	H	G
Horse	G	D	V	E	K	G	K	K	I	F	V	Q	K	C	A	Q	C	H	T	V	E	K	G	G	K	H	K	T	G	P	N	L	H	G
Donkey	G	D	V	E	K	G	K	K	I	F	V	Q	K	C	A	Q	C	H	T	V	E	K	G	G	K	H	K	T	G	P	N	L	H	G
Cow, pig, sheep	G	D	V	E	K	G	K	K	I	F	V	Q	K	C	A	Q	C	H	T	V	E	K	G	G	K	H	K	T	G	P	N	L	H	G
Dog	G	D	V	E	K	G	K	K	I	F	V	Q	K	C	A	Q	C	H	T	V	E	K	G	G	K	H	K	T	G	P	N	L	H	G
Rabbit	G	D	V	E	K	G	K	K	I	F	V	Q	K	C	A	Q	C	H	T	V	E	K	G	G	K	H	K	T	G	P	N	L	H	G
Gray whale	G	D	V	E	K	G	K	K	I	F	V	Q	K	C	A	Q	C	H	T	V	E	K	G	G	K	H	K	T	G	P	N	L	H	G
Gray kangaroo	G	D	V	E	K	G	K	K	I	F	V	Q	K	C	A	Q	C	H	T	V	E	K	G	G	K	H	K	T	G	P	N	L	N	G
Chicken, turkey	G	D	I	E	K	G	K	K	I	F	V	Q	K	C	S	Q	C	H	T	V	E	K	G	G	K	H	K	T	G	P	N	L	H	G
Pigeon	G	D	I	E	K	G	K	K	I	F	V	Q	K	C	S	Q	C	H	T	V	E	K	G	G	K	H	K	T	G	P	N	L	H	G
Pekin duck	G	D	V	E	K	G	K	K	I	F	V	Q	K	C	S	Q	C	H	T	V	E	K	G	G	K	H	K	T	G	P	N	L	H	G
Snapping turtle	G	D	V	E	K	G	K	K	I	F	V	Q	K	C	A	Q	C	H	T	V	E	K	G	G	K	H	K	T	G	P	N	L	N	G
Rattlesnake	G	D	V	E	K	G	K	K	I	F	T	M	K	C	S	Q	C	H	T	V	E	K	G	G	K	H	K	T	G	P	N	L	H	G
Bullfrog	G	D	V	E	K	G	K	K	I	F	V	Q	K	C	A	Q	C	H	T	C	E	K	G	G	K	H	K	V	G	P	N	L	Y	G
Tuna	G	D	V	A	K	G	K	K	T	F	V	Q	K	C	A	Q	C	H	T	V	E	N	G	G	K	H	K	V	G	P	N	L	W	G
Dogfish	G	D	V	E	K	G	K	K	V	F	V	Q	K	C	A	Q	C	H	T	V	E	N	G	G	K	H	K	T	G	P	N	L	S	G
Samia cynthia (moth)	G	N	A	E	N	G	K	K	I	F	V	Q	R	C	A	Q	C	H	T	V	E	A	G	G	K	H	K	V	G	P	N	L	H	G
Tobacco hornworm moth	G	N	A	D	N	G	K	K	I	F	V	Q	R	C	A	Q	C	H	T	V	E	A	G	G	K	H	K	V	G	P	N	L	H	G
Screwworm fly	G	D	V	E	K	G	K	K	I	F	V	Q	R	C	A	Q	C	H	T	V	E	A	G	G	K	H	K	V	G	P	N	L	H	G
Drosophila (fruit fly)	G	D	V	E	K	G	K	K	L	F	V	Q	R	C	A	Q	C	H	T	V	E	A	G	G	K	H	K	V	G	P	N	L	H	G
Baker's yeast	G	S	A	K	K	G	A	T	L	F	K	T	R	C	E	L	C	H	T	V	E	K	G	G	P	H	K	V	G	P	N	L	H	G
Candida krusei (yeast)	G	S	A	K	K	G	A	T	L	F	K	T	R	C	A	E	C	H	T	I	E	A	G	G	P	H	K	V	G	P	N	L	H	G
Neurospora crassa (mold)	G	D	S	K	K	G	A	N	L	F	K	T	R	C	A	E	C	H		E			N	L	T	Q	K	I	G	P	A	L	H	G
Wheat	G	N	P	D	A	G	A	K	I	F	K	T	K	C	A	Q	C	H	T	V	D	A	G	A		H	K	Q	G	P	N	L	H	G
Sunflower	G	D	P	T	T	G	A	K	I	F	K	T	K	C	A	Q	C	H	T	V	E	K	G	A		H	K	Q	G	P	N	L	N	G
Mung bean	G	D	S	K	S	G	E	K	I	F	K	T	K	C	A	Q	C	H	T	V	D	K	G	A		H	K	Q	G	P	N	L	N	G
Rice	G	N	P	K	A	G	E	K	I	F	K	T	K	C	A	Q	C	H	T	V	D	K	G	A		H	K	Q	G	P	N	L	N	G
Sesame	G	D	V	K	S	G	E	K	I	F	K	T	K	C	A	Q	C	H	T	V	D	K	G	A		H	K	Q	G	P	N	L	N	G

Gaps indicate insertion and/or deletion events.

Acidic side chains
- D Aspartic acid
- E Glutamic acid

Basic side chains
- H Histidine
- K Lysine
- R Arginine

Hydrophobic side chains
- F Phenylalanine
- I Isoleucine
- L Leucine
- M Methionine
- V Valine
- Y Tyrosine
- W Tryptophan
- A Alanine

Other
- C Cysteine
- P Proline
- Q Glutamine
- N Asparagine
- S Serine
- T Threonine
- G Glycine

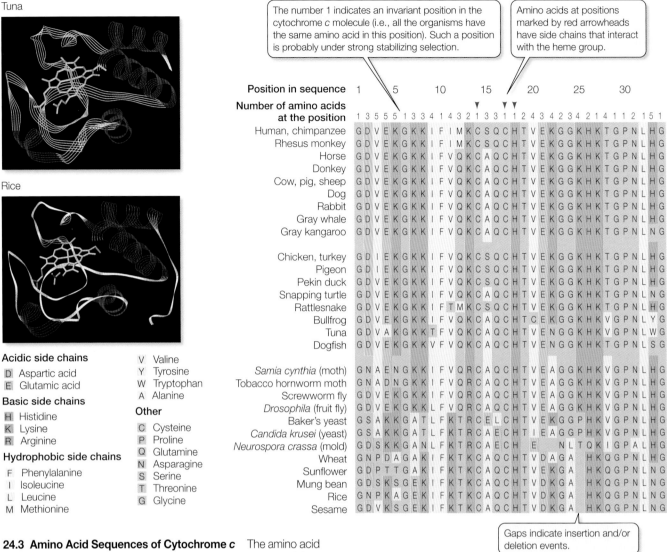

24.3 Amino Acid Sequences of Cytochrome *c* The amino acid sequences shown in the table were obtained from analyses of the enzyme cytochrome *c* from 33 species of plants, fungi, and animals. Note the lack of variation across the sequences at positions 70–80, suggesting that this region is under strong stabilizing selection, and changing its amino acid sequence would impair the protein's function. The computer graphics at the upper left are created from these sequences and show the three-dimensional structures of tuna and rice cytochrome *c*. Alpha helixes are in red, and the molecule's heme group is shown in yellow.

rapidly. In the case of some RNA viruses, the natural substitution rate may be as high as 10^{-3} substitutions per position per generation. Therefore, in a virus of a few thousand nucleotides, one or more substitutions are expected (on average) every generation, and these changes can easily be determined by sequencing the entire genome (because of its small size). Generation time may be only tens of minutes (rather than years or decades, as in humans), so biologists can directly observe substantial molecular evolution in a controlled population over the course of days, weeks, or months.

An example of an experimental evolutionary study is shown in **Figure 24.4**. Paul Rainey and Michael Travisano wanted to examine a potential cause of adaptive radiations, which are a major source of biological diversity. For instance, near the beginning of the Cenozoic era, mammals rapidly diversified into species as diverse as elephants, moles, whales, and bats. While Rainey and Travisano clearly couldn't experimentally manipulate mammals over many millions of years, they could test the idea that heterogenous environments lead to adaptive radiation by experimentally manipulating a lineage of bacteria.

Rainey and Travisano inoculated a number of flasks containing culture medium with the same strain of the bacterium *Pseudomonas fluorescens*. They then shook some of the cultures to maintain a constantly uniform environment, and left others alone (static cultures), allowing them to develop a spatially distinct structure. For instance, in the static cultures the environment on the surface film of the medium differed from that on the walls of the flasks and from parts of the culture not touching any surfaces.

When the cultures were started, the ancestral phenotype of the bacterium produced a smooth colony, which the investigators called a "smooth morph." Within just a few days, however, the static cultures consistently and independently developed two other

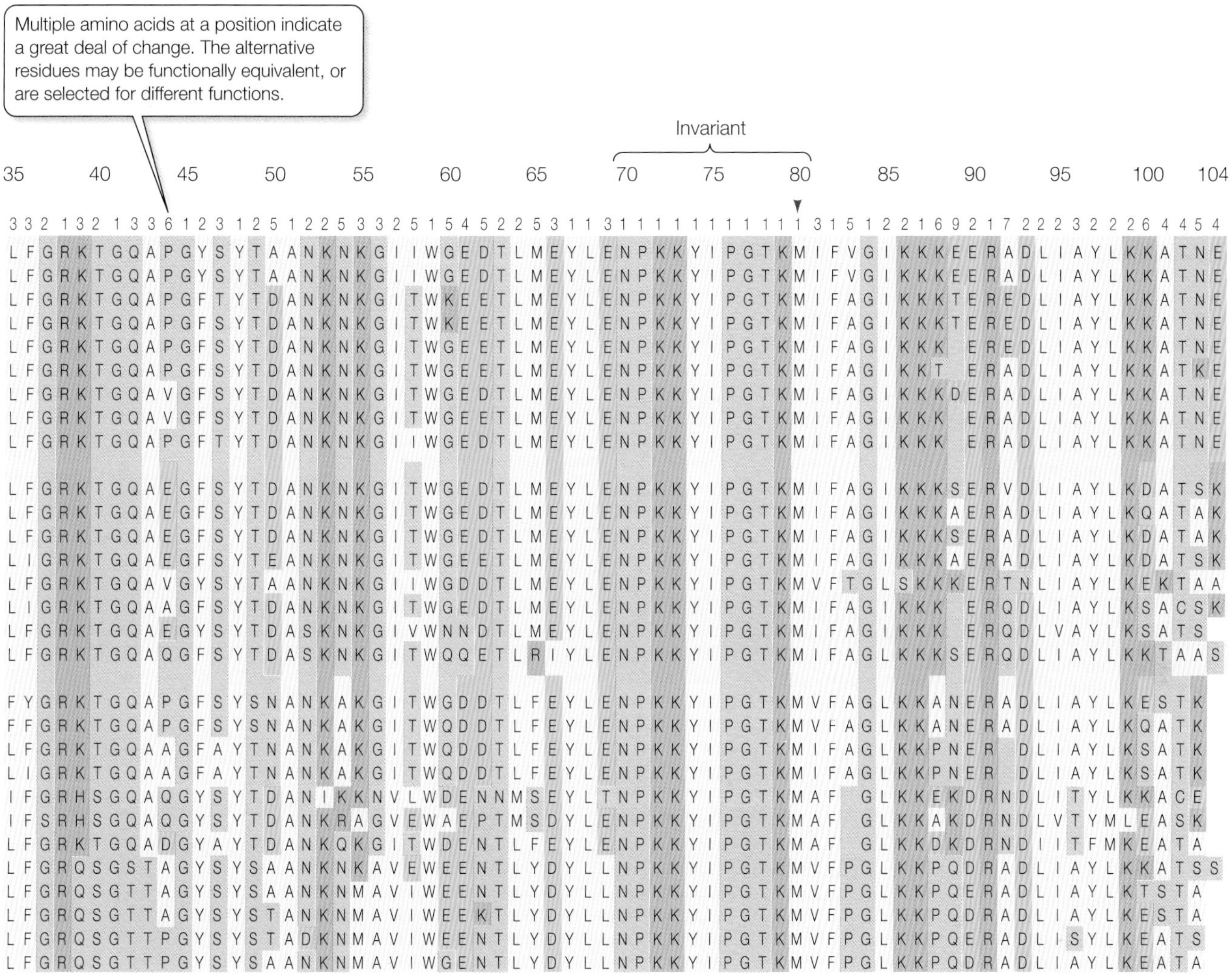

morphs: a "wrinkly spreader" and a "fuzzy spreader." The researchers determined that the two new morphs had a genetic basis and were adaptively superior in some of the environments of the static cultures. For example, the "wrinkly spreader" cells adhered firmly to one another as well as to surfaces, and were thus able to form a mat across the surface of the medium, where they could compete successfully for oxygen.

Molecular investigations of the genomes of these morphs showed that they had evolved repeatedly, and that many different substitutions could produce the same phenotypes. In contrast, the homogeneous shaken cultures showed no evolution. The same mutations occurred in the shaken cultures, but did not persist in those populations because the novel phenotypes that they produced were selectively disadvantageous under the "shaken" environmental conditions.

Experimental molecular evolutionary studies are used for a wide variety of purposes and have greatly expanded the ability of evolutionary biologists to test evolutionary concepts and principles. Biologists now routinely study evolution in the laboratory and, as we will see later in this chapter, use in vitro evolutionary techniques to produce novel molecules to perform new functions for industrial and pharmaceutical uses.

24.1 RECAP

A genome is the sum of all the genetic material of an organism—including both sequences coding for functional genes, and noncoding sequences. The genomes of all organisms evolve over time.

- Can you explain the relationship between a nucleotide substitution and an amino acid replacement? See p. 526
- Can you describe how biologists align nucleotide and amino acid sequences they wish to compare and how they estimate the number of changes that have occurred between pairs of aligned sequences? See p. 526–527 and Figure 24.1

Molecular evolutionists can directly observe the evolution of genomes over time, and they can compare the genomes of different organisms and reconstruct the changes that have occurred during their evolution. Let's turn now to the question of how genomes change and examine some of the consequences of those changes.

24.4 A Heterogeneous Environment Spurs Adaptive Radiation Rainey and Travisano's studies used a rapidly reproducing prokaryote species to model adaptive radiation. Their experiments indicated that phenotypic change is enhanced in a heterogeneous environment. The uniform environment of the shaken cultures showed no evolution. Molecular analysis revealed that same genetic mutations occurred in the shaken cultures, but did not persist because the novel phenotypes they produced were selectively disadvantageous under homogeneous environmental conditions. FURTHER RESEARCH: Once they have arisen in a heterogeneous environment, would the three evolved phenotypes compete successfully in a homogeneous environment?

EXPERIMENT

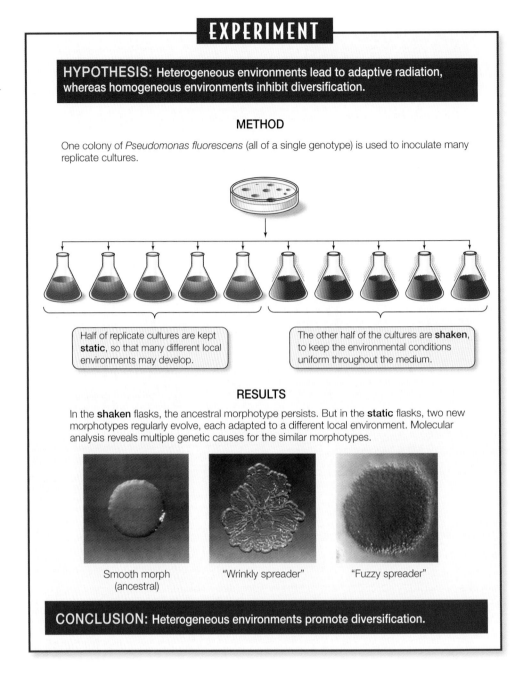

HYPOTHESIS: Heterogeneous environments lead to adaptive radiation, whereas homogeneous environments inhibit diversification.

METHOD

One colony of *Pseudomonas fluorescens* (all of a single genotype) is used to inoculate many replicate cultures.

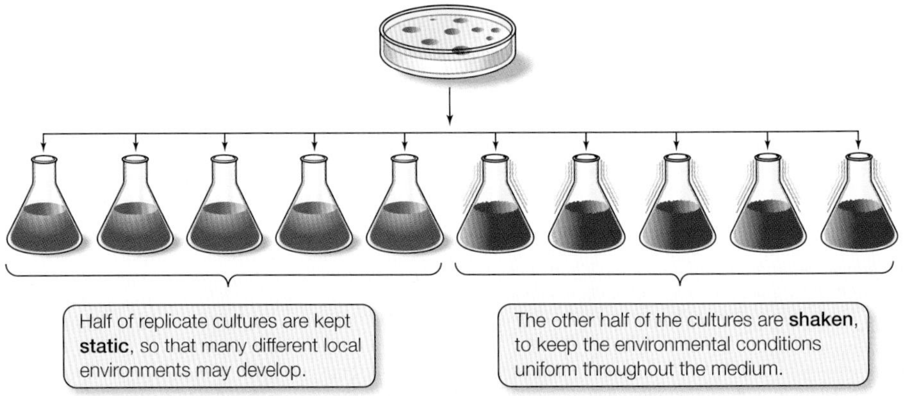

Half of replicate cultures are kept **static**, so that many different local environments may develop.
The other half of the cultures are **shaken**, to keep the environmental conditions uniform throughout the medium.

RESULTS

In the **shaken** flasks, the ancestral morphotype persists. But in the **static** flasks, two new morphotypes regularly evolve, each adapted to a different local environment. Molecular analysis reveals multiple genetic causes for the similar morphotypes.

CONCLUSION: Heterogeneous environments promote diversification.

24.2 What Are the Mechanisms of Molecular Evolution?

A *mutation*, as we saw in Chapter 12, is any change in the genetic material. A nucleotide substitution is one type of mutation. Many nucleotide substitutions have no effect on phenotype, even if the change occurs in a gene that encodes a protein, because most amino acids are specified by more than one codon (see Figure 12.6). A substitution that does not change the amino acid that is specified is known as a **synonymous** or **silent substitution** (**Figure 24.5A**). Synonymous substitutions do not affect the functioning of a protein (and hence the organism) and are therefore unlikely to be influenced by natural selection.

A nucleotide substitution that *does* change the amino acid sequence encoded by a gene is known as a **nonsynonymous substitution** (**Figure 24.5B**). In general, nonsynonymous substitutions are likely to be deleterious to the organism. But not every amino acid replacement alters a protein's shape and charge (and hence its functional properties). Therefore, some nonsynonymous substitutions may also be selectively neutral, or nearly so. Conversely, an amino acid replacement that confers an advantage to the organism would result in positive selection for the corresponding nonsynonymous substitution.

Enough analyses of mammalian genes have been performed to show that the rate of nonsynonymous nucleotide substitutions varies from nearly zero to about 3×10^{-9} substitutions per position per year. Synonymous substitutions in the protein-coding regions of genes have occurred about five times more rapidly than nonsynonymous substitutions. In other words, substitution rates are highest at nucleotide positions that *do not change the amino acid be-*

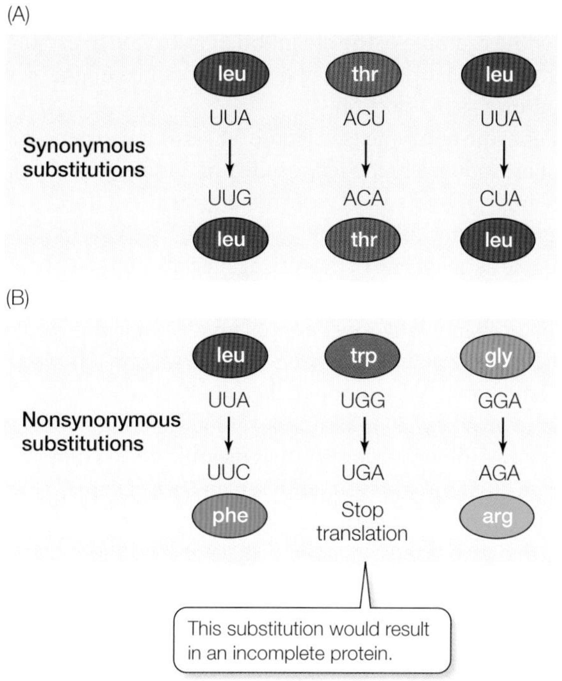

24.5 When One Nucleotide Does or Doesn't Make a Difference (A) Synonymous substitutions do not change the amino acid specified and do not affect protein function; such substitutions are less likely to be subject to natural selection. (B) Nonsynonymous substitutions do change the amino acid sequence and are likely to have an effect (often deleterious) on protein function; such substitutions are targets for natural selection.

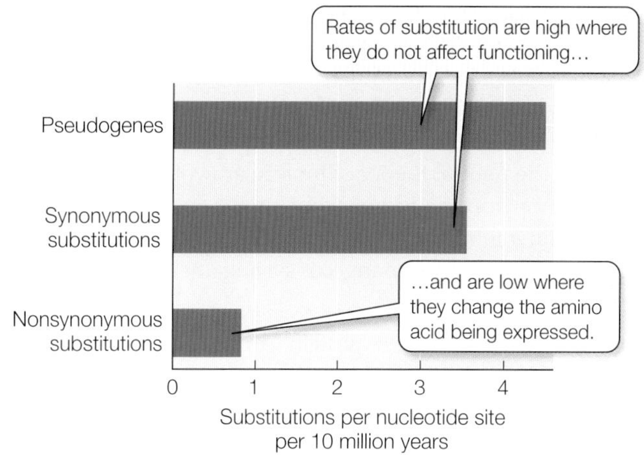

24.6 Rates of Substitution Differ Rates of nonsynonymous substitution typically are much lower than rates of synonymous substitution and the substitution rate in pseudogenes. This pattern reflects differing levels of functional constraints.

ing expressed (**Figure 24.6**). The rate of substitution is even higher in **pseudogenes**, which are duplicate copies of genes that have undergone one or more mutations that eliminate their ability to be expressed.

As we saw in Chapter 22, most natural populations of organisms harbor much more genetic variation than we would expect if genetic variation were influenced by natural selection alone. This discovery, combined with the knowledge that many mutations do not change molecular function, stimulated the development of the neutral theory of molecular evolution.

Much of evolution is neutral

In 1968, Motoo Kimura proposed the *neutral theory of molecular evolution*. Kimura suggested that, at the molecular level, the majority of mutations we observe in populations are selectively neutral; that is, they confer neither an advantage nor a disadvantage on their bearers. Therefore, these neutral mutations accumulate through genetic drift, rather than through positive directional selection (as described in Section 22.3).

The rate of fixation of neutral mutations by genetic drift is independent of population size. To see why this is so, consider a population of size N and a neutral mutation rate μ (mu) per gamete per generation at a locus. The number of new mutations would, on average, be $\mu \times 2N$, because $2N$ gene copies are available to mutate in a population of diploid organisms. According to genetic drift theory, the probability that a mutation will be fixed by drift alone is its frequency, p, which equals $1/(2N)$ for a newly arisen mutation. Therefore, the number of neutral mutations that arise per generation that are likely to become fixed in a given population is $2N\mu \times 1/(2N) = \mu$. Therefore, the rate of fixation of neutral mutations depends only on the neutral mutation rate (μ), and is independent of population size (N). A given mutation is more likely to appear in a large population than in a small one, but given that it does appear, it is more likely to become fixed in a small population. These two influences of population size exactly cancel each other out.

Therefore, the rate of fixation of neutral mutations is equal to the mutation rate. So if most mutations of macromolecules are neutral, and if the underlying mutation rate is constant, macromolecules evolving in different populations should diverge from one another at a constant rate. Empirically, the rate of evolution of particular genes and proteins is often relatively constant over time, and in effect can be used as a "molecular clock." (More will be said about this in Section 25.2, where we will see how molecular clocks can be used to calculate evolutionary divergence times between species.)

Although much of the genetic variation we observe in populations is the result of neutral evolution, that does not mean that most mutations have no effect on the organism. Many mutations are never observed in populations because they are lethal or strongly detrimental to the organism and are thus removed from the population through natural selection. Similarly, mutations that confer a selective advantage tend to be quickly fixed in populations, so they do not result in variation at the population level either. Nonetheless, if we compare homologous proteins from different populations or species, some amino acid positions will stay constant under stabilizing selection, others will vary through neutral genetic drift, and still others will vary as a result of positive selection for change. How can these evolutionary processes be distinguished?

Positive and stabilizing selection can be detected in the genome

As we have just seen, substitutions in a protein-coding gene can be either synonymous or nonsynonymous, depending on whether they change the resulting amino acid sequence of the protein. According to the neutral theory of molecular evolution, the relative rates of synonymous and nonsynonymous substitutions are expected to differ in regions of genes that are evolving neutrally, under positive selection for change, or staying unchanged under stabilizing selection.

- If a given amino acid in a protein can be one of many alternatives (without changing the protein's function), then an amino acid replacement is *neutral* with respect to the fitness of an organism. In this case, the rates of synonymous and nonsynonymous substitutions in the corresponding DNA sequences are expected to be very similar, so the ratio of the two rates would be close to 1.
- If a given amino acid position is under *strong stabilizing selection*, then the rate of synonymous substitutions in the corresponding DNA sequences is expected to be much higher than the rate of nonsynonymous substitutions.
- If a given amino acid position is under *strong selection for change*, the rate of nonsynonymous substitutions is expected to exceed the rate of synonymous substitutions in the corresponding DNA sequences.

By comparing the gene sequences that encode proteins from many species, the history and timing of synonymous and nonsynonymous substitutions can be determined (look back at Figure 24.2 for an example). This information can be mapped on a *phylogenetic tree*, a diagram of evolutionary relationships (the construction of phylogenetic trees will be discussed in more detail in Chapter 25).

Genes, or regions of genes, that are evolving under neutral, stabilizing, or positive selection can be identified by comparing the nature and rates of substitutions across the phylogenetic tree. Let's consider the example of the evolution of lysozyme to explore how and why particular positions of a gene sequence might be under different modes of selection.

The enzyme lysozyme (see Figure 3.8) is found in almost all animals. It is produced in the tears, saliva, and milk of mammals and in the whites of bird eggs. Lysozyme digests the cell walls of bacteria, rupturing and killing them. As a result, lysozyme plays an important role as a first line of defense against invading bacteria. Most animals defend themselves against bacteria by digesting them, which is probably why most animals have lysozyme. Some animals, however, also use lysozyme in the digestion of food.

Among mammals, a mode of digestion called *foregut fermentation* has evolved twice. In mammals with this mode of digestion, the foregut—the posterior esophagus and/or the stomach—has been converted into a chamber in which bacteria break down ingested plant matter by fermentation. Foregut fermenters can obtain nutrients from the otherwise indigestible cellulose that makes up a large proportion of the plant body. Foregut fermentation evolved independently in ruminants (a group of hoofed mammals that includes cattle) and in certain leaf-eating monkeys, such as langurs (**Figure 24.7A**). We know that these evolutionary events were independent because both langurs and ruminants have close relatives that are not foregut fermenters.

In both foregut-fermenting lineages, the enzyme lysozyme has been modified to play a new, nondefensive role. This lysozyme ruptures some of the bacteria that live in the foregut, releasing nutrients metabolized by the bacteria, which the mammal then absorbs. How many changes in the lysozyme molecule were needed to allow it to perform this function amid the digestive enzymes and acidic conditions of the mammalian foregut? To answer this question, molecular evolutionists compared the lysozyme-coding sequences in foregut fermenters with those of several of their nonfermenting relatives. They determined which amino acids differed and which were shared among the species (**Table 24.1**), and they also determined the rates of synonymous and nonsynonymous substitutions in the lysozyme genes across the evolutionary history of the sampled species.

For many of the amino acid positions of lysozyme, the rate of synonymous substitutions (in the corresponding gene) is much higher than the rate of nonsynonymous substitutions. This ob-

(A) *Presbytis entellus*

(B) *Opisthocomus hoazin*

24.7 Convergent Molecular Evolution Foregut-fermenting mammals such as the Hanuman langur (A) have been evolving independently from the hoatzin (B) for hundreds of millions of years, but they have independently evolved similar modifications to the enzyme lysozyme.

TABLE 24.1

Similarity Matrix for Lysozyme in Mammals

SPECIES	LANGUR	BABOON	HUMAN	RAT	CATTLE	HORSE
Langur*		14	18	38	32	65
Baboon	0		14	33	39	65
Human	0	1		37	41	64
Rat	0	1	0		55	64
Cattle*	5	0	0	0		71
Horse	0	0	0	0	1	

Shown above the diagonal line is the number of amino acid sequence *differences* between the two species being compared; below the line are the number of changes uniquely *shared* by the two species.

Asterisks (*) indicate foregut-fermenting species.

servation indicates that many of the amino acids that make up lysozyme are evolving under stabilizing selection. In other words, there is selection against change in the protein at these positions, and the observed amino acids must therefore be critical for lysozyme function. At other positions, several different amino acids function equally well, and the corresponding regions of the genes have similar rates of synonymous and nonsynonymous substitutions. The most striking finding is that amino acid replacements in lysozyme happened at a much higher rate in the lineage leading to langurs than in any other primates. The high rate of nonsynonymous substitutions in the langur lysozyme gene shows that lysozyme went through a period of rapid change in adapting to the stomachs of langurs. Moreover, the lysozymes of langurs and cattle share five amino acid replacements, all of which lie on the surface of the lysozyme molecule, well away from the enzyme's active site. Several of these shared replacements involve changes from arginine to lysine, which makes the proteins more resistant to attack by the stomach enzyme pepsin. By understanding the functional significance of amino acid replacements, molecular evolutionists can explain the observed changes in amino acid sequences in terms of changes in the functioning of the protein.

A large body of fossil, morphological, and molecular evidence shows that langurs and cattle do not share a recent common ancestor. However, langur and ruminant lysozymes share several amino acids that neither mammal shares with the lysozymes of its own closer relatives. The lysozymes of these two mammals have converged at some amino acid positions despite their very different ancestry. (We will see other examples of *convergent evolution* under similar selection pressures in Chapter 25.) The amino acids they share give these lysozymes the ability to lyse the bacteria that ferment plant material in the foregut.

An even more remarkable story emerges in the case of lysozyme in the crop of the hoatzin, a unique leaf-eating South American bird and the only known avian foregut fermenter (**Figure 24.7B**). Many birds have an enlarged esophageal chamber called a *crop*. Hoatzins have a crop that contains bacteria and acts as a fermenting chamber. Many of the amino acid replacements that occurred in the adaptation of hoatzin crop lysozyme are identical to the changes that evolved in ruminants and langurs. Thus, even though the hoatzin and the foregut-fermenting mammals have not shared a common ancestor in hundreds of millions of years, they have all evolved similar adaptations in their lysozymes that enable them to recover nutrients from their fermenting bacteria in a highly acidic environment.

Genome size and organization also evolve

We know that genome size varies tremendously among organisms. Across broad taxonomic categories, there is some correlation between genome size and organismal complexity. The genome of the tiny bacterium *Mycoplasma genitalium* has only 470 genes. *Rickettsia prowazekii*, the bacterium that causes typhus, has 634 genes. *Homo sapiens*, on the other hand, has about 23,000 protein-coding genes. **Figure 24.8** shows the relative number of genes for several prokaryotic and eukaryotic organisms.

Additional genomic comparisons reveal, however, that a larger genome does not always indicate greater complexity. It is not surprising that more complex genetic instructions are needed for building and maintaining a large, multicellular organism than a small, single-celled bacterium. What is surprising is that some organisms, such as lungfishes, some salamanders, and lilies, have about 40 times as much DNA as humans do. Structurally, a lungfish or a lily is not 40 times more complex as a human. So why does genome size vary so much?

Differences in genome size are not so great if we take into account only the portion of DNA that actually encodes RNAs or proteins. The organisms with the largest total amounts of nuclear DNA (some ferns and flowering plants) have 80,000 times as much DNA as the bacteria with the smallest genomes, but no species has more than about 100 times as many protein-coding genes as a bacterium. Therefore, much of the variation in genome size lies not in the number of functional genes, but in the amount of noncoding DNA (**Figure 24.9**).

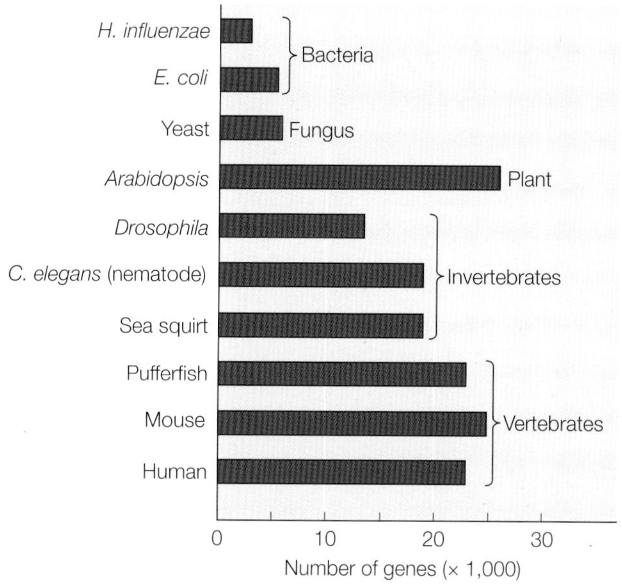

24.8 Genome Size Varies Widely The number of genes in the genome has been measured or estimated in a variety of organisms, ranging from single-celled prokaryotes to vertebrates.

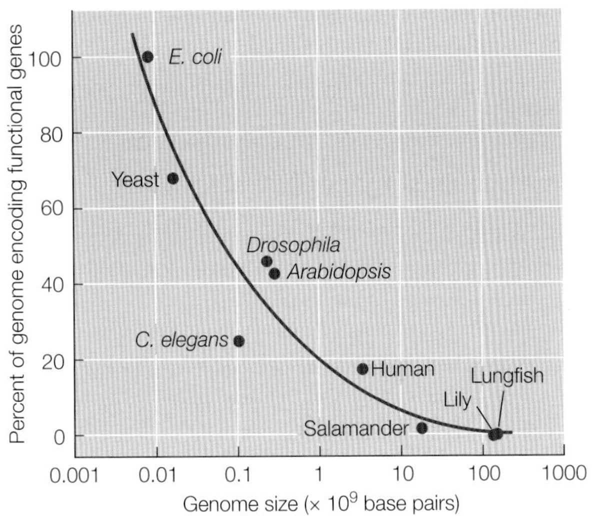

24.9 A Large Proportion of DNA Is Noncoding Most of the DNA of bacteria and yeasts encodes RNAs or proteins, but a large percentage of the DNA of multicellular species is noncoding.

Why do the cells of most organisms have so much noncoding DNA? Does this noncoding DNA have a function, or is it "junk"? Although much of this DNA does not appear to have a direct function, it can alter the expression of the surrounding genes. The degree or timing of gene expression can be changed dramatically depending on the gene's position relative to noncoding sequences. Other regions of noncoding DNA consist of pseudogenes that are simply carried in the genome because the cost of doing so is very small. These pseudogenes may become the raw material for the evolution of new genes with novel functions. Still other noncoding sequences consist of parasitic transposable elements that spread through populations because they reproduce faster than the host genome.

Investigators can use retrotransposons to estimate the rates at which species lose DNA. Retrotransposons are transposable elements that copy themselves through an RNA intermediate, as we saw in Section 14.1. The most common type of retrotransposon carries duplicated sequences at each end, called long terminal repeats (LTRs). Occasionally, LTRs recombine in the host genome, so that the DNA between them is excised. When this happens, one recombined LTR is left behind. The number of such "orphaned" LTRs in a genome is a measure of how many retrotransposons have been lost. By comparing the number of LTRs in the genomes of Hawaiian crickets (*Laupala*) and fruit flies (*Drosophila*), investigators found that *Laupala* loses DNA more than 40 times more slowly than *Drosophila*. As a result, the genome of *Laupala* is 11 times larger than that of *Drosophila*.

Why do species differ so greatly in the rate at which they gain or lose apparently functionless DNA? One hypothesis is that genome size is related to the rate at which the organism develops, which may be under selection pressure. Large genomes can slow down the rate of development and thus alter the relative timing of expression of particular genes. As we illustrated and discussed in Section 20.2, such changes in the timing of gene expression (*heterochrony*) can produce major changes in phenotype. Thus, although some noncoding DNA sequences may have no direct function, they may still affect the development of the organism.

Another hypothesis is that the proportion of noncoding DNA is related primarily to population size. Noncoding sequences that are only slightly deleterious are likely to be purged by selection primarily in species with large population sizes. In species with small populations, the effects of genetic drift can overwhelm selection against noncoding sequences that have small deleterious consequences. Therefore, selection against the accumulation of noncoding sequences is most effective in species with large populations, so such species (such as bacteria or yeasts) have relatively little noncoding DNA compared with species with small populations (see Figure 24.9).

New functions can arise by gene duplication

Gene duplication is another way in which proteins can acquire new functions. When a gene is duplicated, one copy of that gene is potentially freed from having to perform its original function. The identical copies of a duplicated gene can have any one of four different fates:

- Both copies of the gene may retain their original function (which can result in a change in the amount of gene product that is produced by the organism).
- Both copies of the gene may retain the ability to produce the original gene product, but the expression of the genes may diverge in different tissues or at different times of development.
- One copy of the gene may be incapacitated by the accumulation of deleterious substitutions and become a functionless pseudogene.
- One copy of the gene may retain its original function, while the second copy accumulates enough substitutions that it can perform a different function.

How often do gene duplications arise, and which of these four outcomes is most likely? Investigators have found that rates of gene duplication are fast enough for a yeast or *Drosophila* population to acquire several hundred duplicate genes over the course of a million years. They have also found that most of the duplicated genes in these organisms are very young. Many extra genes are lost from a genome within 10 million years (which is rapid on an evolutionary time scale).

Many gene duplications affect only one or a few genes at a time, but entire genomes are often duplicated in *polyploid* organisms (including many plants). When all the genes are duplicated, there are massive opportunities for new functions to evolve. That is exactly what appears to have happened in the evolution of vertebrates. The genomes of most jawed vertebrates appear to have four ancient copies of many major genes, which leads biologists to believe that two genome-wide duplication events occurred in the ancestor of these species. These duplications have allowed considerable specialization of individual vertebrate genes, many of which are now highly tissue-specific in their expression.

A novel function that evolved as a result of gene duplication is the ability of some fish—electric eels, among others—to produce electric signals. This function has evolved independently several times in different species, always through similar changes in duplicated sodium channel genes.

Although many extra genes disappear rapidly, some duplication events lead to the evolution of genes with new functions. Several successive rounds of duplication and mutation may result in a *gene family*, a group of homologous genes with related functions, often arrayed in tandem along a chromosome. An example of this process is provided by the *globin gene family* (see Figure 14.8) The globins were among the first proteins to be sequenced and compared. Comparisons of their amino acid sequences strongly suggest that the different globins arose via gene duplications. These comparisons can also tell us how long the globins have been evolving separately because differences among these proteins have accumulated with time.

Hemoglobin, a tetramer (four-subunit molecule) consisting of two α-globin and two β-globin polypeptide chains, carries oxygen in blood. Myoglobin, a monomer, is the primary oxygen storage protein in muscle. Myoglobin's affinity for O_2 is much higher than that of hemoglobin, but hemoglobin has evolved to be more diversified in its role. Hemoglobin binds O_2 in the lungs or gills, where the O_2 concentration is relatively high, transports it to deep body tissues, where the O_2 concentration is low, and releases it in those tissues. With its more complex tetrameric structure, hemoglobin is able to carry four molecules of O_2, as well as hydrogen ions and carbon dioxide, in the blood.

To estimate the time of the globin gene duplication that gave rise to the α- and β-globin gene clusters, we can create a *gene tree*, a phylogenetic tree based on the gene sequences that encode the various globins (**Figure 24.10**). The rate of molecular evolution of globin genes has been estimated from other studies, using the divergence times of groups of vertebrates that are well documented in the fossil record. These studies indicate an average rate of divergence for globin genes of about 1 substitution every 2 million years. Applying this rate to the gene tree, the two globin gene clusters are estimated to have split about 450 million years ago.

Some gene families evolve through concerted evolution

Although the members of the globin gene family have diversified in form and function, the members of many other gene families do not evolve independently of one another. For instance, almost all organisms have many (up to thousands of) copies of the ribosomal RNA genes. Ribosomal RNA is the principal structural element of the ribosome, and as such, has a primary role in protein synthesis. Every living species needs to synthesize proteins, often in large amounts (especially during early development). Having many copies of the ribosomal RNA genes ensures that organisms can rapidly produce many ribosomes and thereby maintain a high rate of protein synthesis.

Like all portions of the genome, ribosomal RNA genes evolve, and differences accumulate in the ribosomal RNA genes of different species. But within any one species, the multiple copies of ribosomal RNA genes are very similar, both structurally and functionally. This similarity makes sense, because ideally every ribosome within a species should synthesize proteins in the same way. In other words, the multiple copies of these genes within a species

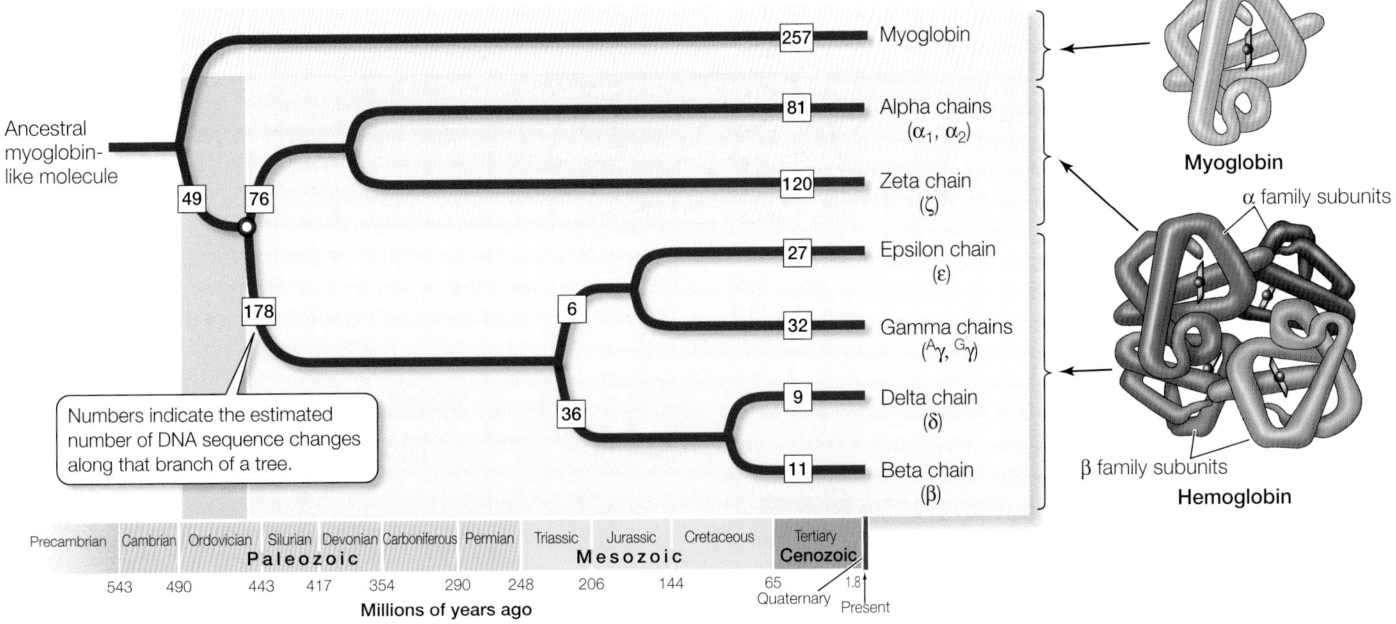

24.10 A Globin Family Gene Tree This gene tree suggests that the α-globin and β-globin gene clusters diverged about 450 million years ago (open circle), soon after the origin of the vertebrates.

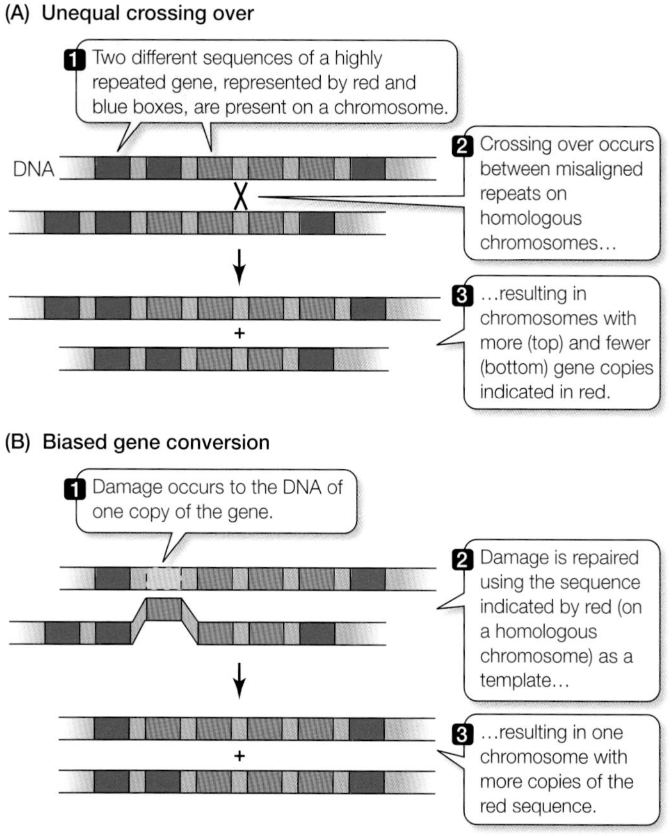

24.11 Concerted Evolution Two mechanisms can produce concerted evolution of highly repeated genes. (A) Unequal crossing over results in deletions and duplications of a repeated gene. (B) Biased gene conversion can rapidly spread a mutation across multiple copies of a repeated gene.

are evolving in concert with one another, a phenomenon called **concerted evolution**.

How does concerted evolution occur? There must be one or more mechanisms to cause a substitution in one copy to spread to other copies within a species so that all of the copies remain similar. In fact, two different mechanisms appear to be responsible for concerted evolution. The first of these is *unequal crossing over*. When DNA is replicated during meiosis in a diploid species, the homologous chromosome pairs align and recombine by crossing over (see Section 9.5).

However, in the case of highly repeated genes, it is easy for genes to become displaced in alignment, since so many copies of the same genes are present in the repeats (**Figure 24.11A**). The end result is that one chromosome will gain extra copies of the repeat and the other chromosome will have fewer copies of the repeat. If a new substitution arises in one copy of the repeat, it can spread to new copies (or be eliminated) through unequal crossing over. Thus, over time, a novel substitution will either become fixed or lost entirely from the repeat. In either case, all the copies of the repeat will remain very similar to one another.

The second mechanism that produces concerted evolution is *biased gene conversion*. This mechanism can be much faster than unequal crossing over, and has been shown to be the primary mechanism for concerted evolution of ribosomal RNA genes. DNA strands break often, and are repaired by the DNA repair systems of cells (see Section 11.4). At many times during the cell cycle, the ribosomal RNA genes are clustered together in close proximity. If damage occurs to one of the genes, another copy of the RNA gene on another chromosome may be used to repair the damaged copy, and the sequence that is used as a template can thereby replace the original sequence (**Figure 24.11B**). In many cases, this repair system appears to be biased in favor of using particular sequences as templates for repair, and thus the favored sequence rapidly spreads across all the copies of the gene. In this way, changes may appear in a single copy and then rapidly spread to all the other copies.

Regardless of the mechanism responsible, the net result of concerted evolution is that the copies of a highly repeated gene do not evolve independently of one another. Mutations still occur, but once they arise in one copy, they either spread rapidly across all the copies or are lost from the genome completely. This process allows the products of each copy to remain similar through time in both sequence and function.

24.2 RECAP

By examining the relative rates of synonymous and nonsynonymous substitutions in genes across evolutionary history, biologists can distinguish the evolutionary mechanisms acting on individual genes. This knowledge, in turn, can lead to an understanding of protein function.

- Can you describe how the ratio of synonymous to nonsynonymous substitutions can be used to determine whether a gene is evolving neutrally, under positive selection, or under stabilizing selection? See pp. 530–532 and Figure 24.6
- Can you contrast two hypotheses for the wide diversity of genome sizes among different organisms? See pp. 533–534
- What are four possible outcomes of gene duplication? See p. 534

We have seen how the principles and methods of molecular evolution have opened new vistas in evolutionary science. Now let's consider some of the practical applications of this field.

24.3 What Are Some Applications of Molecular Evolution?

Our understanding of molecular biology has helped reveal how biological molecules function as well as how they diversify. Such knowledge allows scientists to create new molecules with novel functions in the laboratory, and to understand and treat disease.

Molecular sequence data are used to determine the evolutionary history of genes

A **gene tree** shows the evolutionary relationships of a single gene in different species or of the members of a gene family (as in Figure 24.10). The methods for constructing a gene tree are the same as those that will be presented in Section 25.2 for building phylogenetic trees. The process involves identifying differences between genes and using those differences to reconstruct the evolutionary history of the genes. Gene trees are often used to infer species phylogenetic trees, but the two types of trees are not necessarily equivalent. Processes such as gene duplication can give rise to differences between the phylogenetic trees of genes and species. From a gene tree, biologists can reconstruct the history and timing of gene duplication events and learn how gene diversification has resulted in the evolution of new protein functions.

All of the genes of a particular gene family have similar sequences because they have a common ancestry. As we will see in Chapter 25, features that are similar as a result of common ancestry are referred to as *homologs* of one another. However, when discussing gene trees, we usually need to distinguish between two forms of homology. Genes found in different species, and whose divergence we can trace to the speciation events that gave rise to various species, are called **orthologs**. Genes in the same or different species that are related through gene duplication events are called **paralogs**. When we examine a gene tree, the questions we wish to address determine whether we should compare orthologous or paralogous genes. If we wish to reconstruct the evolutionary history of the species that contain the genes, then our comparison should be restricted to orthologs (because they will reflect the history of speciation events). On the other hand, if we are interested in the changes in function that have resulted from gene duplication events, then the appropriate comparison is among paralogs (because they will reflect the history of gene duplication events). If our focus is on the diversification of a gene family through both processes, then we will want to include both paralogs and orthologs in our analysis.

Figure 24.12 depicts a gene tree for the members of a gene family called *engrailed* (its members encode transcription factors that regulate development). At least three gene duplications have occurred in this family, resulting in up to four different *engrailed* genes in some vertebrates (such as the zebrafish). All of the *engrailed* genes (*En*) are homologs because they have a common ancestor. Gene duplication events have generated paralogous *engrailed* genes in some lineages of vertebrates. We could compare the orthologous sequences of the *En1* group of genes to reconstruct the history of the bony vertebrates (i.e., all the vertebrates in Figure 24.12 except the lamprey), or we could use the orthologous sequences of the *En2* group of genes and expect the same answer (because

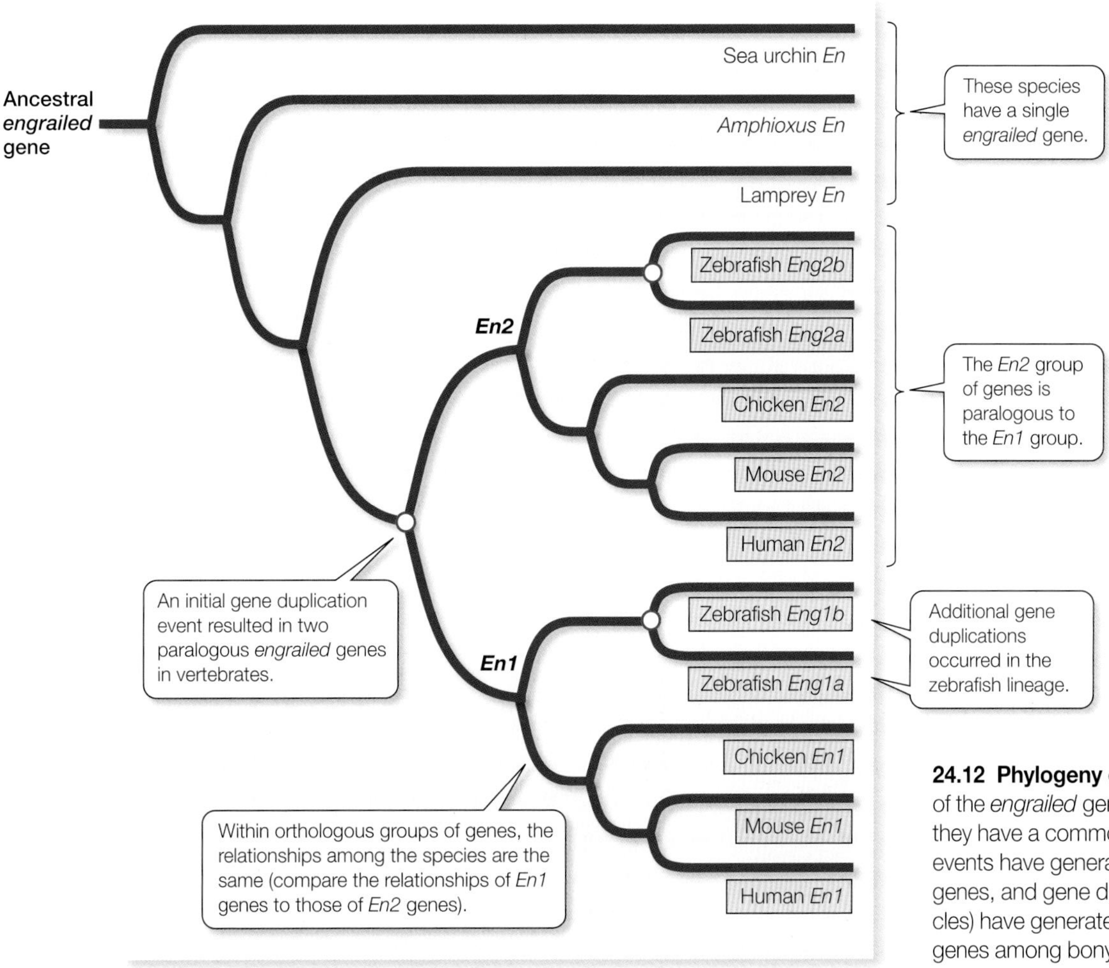

24.12 Phylogeny of the *engrailed* Genes All of the *engrailed* genes are homologous because they have a common ancestor. Speciation events have generated orthologous *engrailed* genes, and gene duplication events (open circles) have generated paralogous *engrailed* genes among bony vertebrates.

there is only one history of the underlying speciation events). All bony vertebrates have both of these groups of *engrailed* genes because the two groups arose from a gene duplication event in the common ancestor of bony vertebrates. If we wanted to focus on the diversification that occurred as a result of this gene duplication, then the appropriate comparison would be between the paralogous genes of the *En1* versus *En2* groups.

Gene evolution is used to study protein function

Earlier in this chapter we discussed the ways in which biologists can detect regions of genes that are under positive selection for change. What are the practical uses of this information? Consider the evolution of the family of gated sodium channel genes. Sodium channels have many functions, including the control of nerve impulses in the nervous system. Sodium channels can become blocked by various toxins, such as the tetrodotoxin that is present in newts (see the opening of Chapter 22), pufferfish, and many other animals. If a human eats those tissues of a pufferfish that contain tetrodotoxin, they can become paralyzed and die, because the tetrodotoxin blocks sodium channels and prevents nerves and muscles from functioning.

> Despite its potential toxicity, pufferfish sushi (fugu) is considered a delicacy in some cultures. Only those tissues with minimal amounts of tetrodotoxin can be consumed safely, and errors in preparation lead quickly to the death of the consumer. The danger in its consumption is part of this sushi's appeal.

Pufferfish themselves have sodium channels; why doesn't the tetrodotoxin cause paralysis in the pufferfish? The sodium channels of pufferfish (and other animals that sequester tetrodotoxin) have evolved to become resistant to the toxin. Nucleotide substitutions in the pufferfish genome have resulted in changes to the proteins that make up sodium channels, and those changes prevent tetrodotoxin from binding to the sodium channel pore and blocking it.

Many other changes that have nothing to do with the evolution of tetrodotoxin resistance have occurred in these genes as well. Biologists who study the function of sodium channels can learn a great deal about how the channels work (and about neurological diseases that are caused by mutations in the sodium channel genes) by understanding which changes have been selected for tetrodotoxin resistance. They can do this by comparing the rates of synonymous and nonsynonymous substitutions across the genes in various lineages that have evolved tetrodotoxin resistance (including the garter snakes mentioned in Chapter 22). In a similar manner, molecular evolutionary principles are used to understand function and diversification of function in many other proteins.

As biologists studied the relationship between selection, evolution, and function in macromolecules, they realized that molecular evolution could be used in a controlled laboratory environment to produce new molecules with novel and useful functions. Thus were born the applications of in vitro evolution.

In vitro evolution produces new molecules

Living organisms produce thousands of compounds that humans have found useful. The search for such naturally occurring compounds, which can be used for pharmaceutical, agricultural, or industrial purposes, has been termed *bioprospecting*. These compounds are the result of millions of years of molecular evolution across millions of species of living organisms. Yet biologists can imagine molecules that could have evolved but have not, lacking the right combination of selection pressures and opportunities.

For instance, we might like to have a molecule that binds a particular environmental contaminant so that it can be easily isolated and extracted from the environment. But if the environmental contaminant is synthetic (not produced naturally), then it is unlikely that any living organism would have evolved a molecule with the function we desire. This problem was the inspiration for the field of **in vitro evolution**, in which new molecules are produced in the laboratory to perform novel and useful functions.

The principles of in vitro evolution are based on the principles of molecular evolution that we have learned from the natural world. Consider the evolution of a new RNA molecule that was produced in the laboratory using some of the techniques described in Chapter 16. This molecule's intended function was to join two other RNA molecules (acting as a ribozyme with a function similar to that of the naturally occurring DNA ligase described in Section 11.3, but for RNA molecules). The process started with a large pool of random RNA sequences (10^{15} different sequences, each about 300 nucleotides long), which were then selected for any ligase activity (**Figure 24.13A**). None were effective ribozymes for ligase activity, but some were very slightly better than others. The best of the ribozymes were selected and reverse-transcribed into cDNA (using the enzyme reverse transcriptase).

The cDNA molecules were then amplified using the polymerase chain reaction (see Figure 11.23). PCR amplification is not perfect, and it introduced many new substitutions into the pool of sequences. These sequences were then transcribed back into RNA molecules using RNA polymerase, and the process was repeated. The ligase activity of the RNAs quickly evolved, and after 10 rounds of in vitro evolution, it had increased by about 7 million times (**Figure 24.13B**). Similar techniques have since been used to create a wide variety of molecules with novel enzymatic and binding functions.

Molecular evolution is used to study and combat diseases

We began this chapter with a discussion of some of the ways in which molecular evolution has proved critical for the global eradication of polio. Polio is just one of many diseases that can be controlled only by understanding and applying molecular evolution-

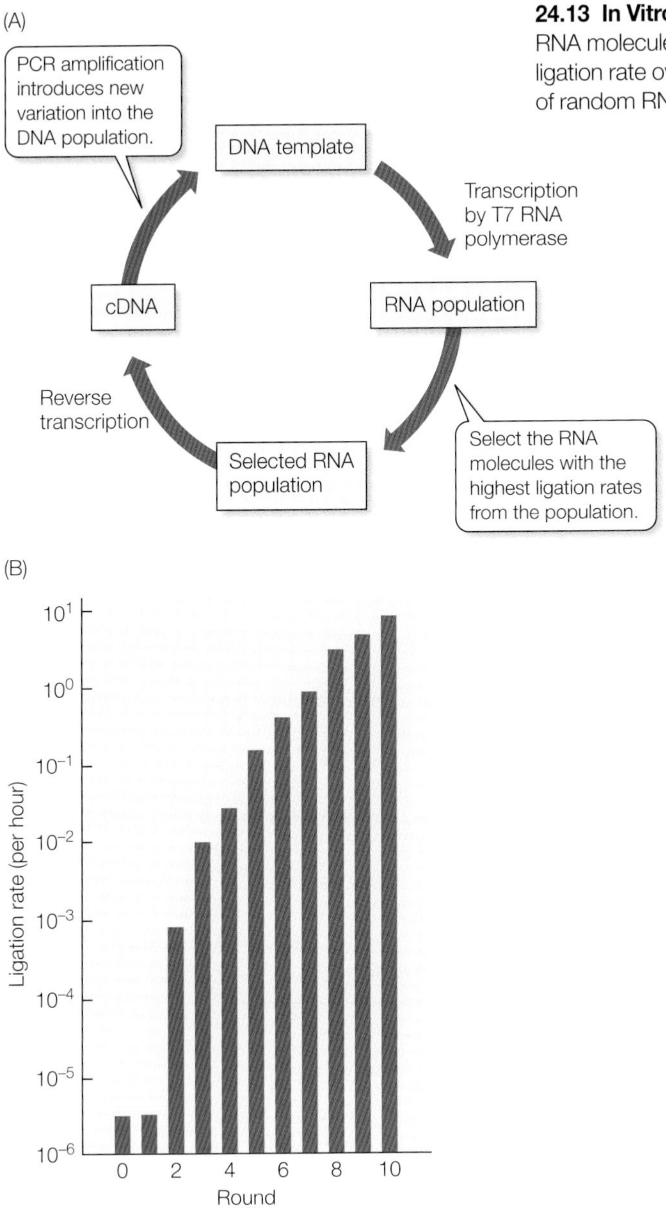

24.13 In Vitro Evolution of a Ribozyme (A) Methodology for in vitro evolution of new RNA molecules with the novel ability to ligate other RNA molecules. (B) Improvement of ligation rate over 10 rounds of in vitro evolution; the experiment started with a large pool of random RNA sequences.

ary principles. Many of humankind's most problematic diseases are caused by living, evolving organisms, and those organisms present a moving target. Their control depends on techniques that can track their evolution through time.

During the last century, transportation advances have allowed humans to move around the world with unprecedented speed and with increasing frequency. Unfortunately, this mobility has allowed pathogens to be transmitted among human populations at much higher rates, which has led to the global emergence of many "new" diseases. Most of these emerging diseases are caused by viruses, and virtually all new viral diseases have been identified by evolutionary comparison of their genomes with those of known viruses. In recent years, for example, rodent-borne hantaviruses have been identified as the source of widespread respiratory illnesses, and the virus (and its host) that causes Sudden Acute Respiratory Syndrome (SARS) has been identified using evolutionary comparisons of genes. Studies of the origins, the timing of emergence, and the global diversity of the human immunodeficiency virus (HIV) all depend on the principles of molecular evolution, as do the efforts to develop a vaccine for HIV. Global patterns of flu change as the influenza virus evolves each year, and new vaccines must be developed to track the evolution of these viruses.

In the future, molecular evolution will become even more critical to the identification of human (and other) diseases. Once biologists have collected data on the genomes of enough organisms, it will be possible to identify an infection by sequencing a portion of the infecting organism's genome and then comparing this sequence with other sequences on an evolutionary tree. At present, it is difficult to identify many common viral infections (those that cause "colds," for instance). As genomic databases and evolutionary trees increase, however, automated methods of sequencing and rapid phylogenetic comparison of the sequences will allow us to identify and treat a much wider array of human illnesses.

24.3 RECAP

Molecular evolutionary studies have provided biologists with new tools to understand the functions of macromolecules and how those functions can change through time. Molecular evolution is used to develop synthetic molecules for industrial and pharmaceutical uses and to identify and combat human diseases.

- Can you explain why a biologist might limit a particular investigation to orthologous (as opposed to paralogous) genes? See pp. 537–538
- Do you understand how gene evolution can be used to study protein function? See p. 538
- Can you describe the process of in vitro evolution? See p. 538 and Figure 24.13

Now that we have discussed how organisms and biological molecules evolve, we are ready to consider how evolutionary history is studied and why evolution is the principal organizing framework for biology. Chapter 25 describes some of the principles and uses of phylogenetic analysis: the study of evolutionary relationships.

CHAPTER SUMMARY

24.1 What can genomes reveal about evolution?

See Web/CD Activity 24.1

The field of **molecular evolution** concerns relationships between the structures of genes and proteins and the functions of organisms.

A **genome** is an organism's full set of genes and noncoding DNA. In eukaryotes, the genome includes genetic material in the nucleus of the cell as well as in mitochondria and chloroplasts (where present).

Nucleotide substitutions may or may not result in **amino acid replacements** in the encoded proteins.

The estimated number of substitutions between sequences can be calculated from a **similarity matrix** using models of sequence evolution that account for changes that cannot be observed directly. Review Figure 24.1, Web/CD Activity 24.2

The concept of homology (similarity that results from common ancestry) extends down to the level of particular positions in nucleotide or amino acid sequences. **Sequence alignments** from different organisms allow us to compare the sequences and identify homologous positions. Review Figure 24.3

24.2 What are the mechanisms of molecular evolution?

Nonsynonymous substitutions of nucleotides result in amino acid replacements in proteins, but **synonymous substitutions** do not. Review Figure 24.5

Rates of synonymous substitution are typically higher than rates of nonsynonymous substitution in protein-coding genes (a result of stabilizing selection). Review Figure 24.6

Much of the molecular change in nucleotide sequences is a result of neutral evolution. The rate of fixation of neutral mutations is independent of population size and is equal to the mutation rate.

Positive selection for change in a protein-coding gene may be detected by a higher rate of nonsynonymous than synonymous substitutions.

Genome size evolves by the addition or deletion of genes and noncoding DNA. The total size of genomes varies much more widely across living organisms than does the number of functional genes. Review Figures 24.8 and 24.9

Even though many noncoding regions of the genome may not have direct functions, these regions can affect the phenotype of an organism by influencing gene expression. Functionless **pseudogenes** can serve as the raw material for the evolution of new genes.

Gene duplications can result in increased production of the gene's product, in pseudogenes, or in new gene functions.

Some highly repeated genes evolve by **concerted evolution**: multiple copies within an organism maintain high similarity, while the genes continue to diverge between species.

24.3 What are some applications of molecular evolution?

See Web/CD Activity 24.3

Gene trees describe the evolutionary history of particular genes or gene families.

Orthologs are genes that are related through speciation events, whereas **paralogs** are genes that are related through gene duplication events. Review Figure 24.12

Protein function can be studied by examining gene evolution. Detection of positive selection can be used to identify molecular changes that have resulted in functional changes.

In vitro evolution is used to produce synthetic molecules with particular desired functions.

Many diseases are identified, studied, and combated through molecular evolutionary investigations.

SELF-QUIZ

1. A higher rate of synonymous than nonsynonymous substitutions in a protein-coding gene is expected under
   a. stabilizing selection.
   b. positive selection.
   c. neutral evolution.
   d. concerted evolution.
   e. none of the above
2. Before nucleotide and amino acid sequences can be compared in an evolutionary framework, they must be aligned in order to account for
   a. deletions and insertions.
   b. selection and neutrality.
   c. parallelisms and convergences.
   d. gene families.
   e. all of the above
3. Models of nucleotide sequence evolution, developed by biologists to estimate sequence divergence, include parameters that account for
   a. substitution rates between different nucleotides.
   b. differences in substitution rates across different positions in a gene.
   c. differences in nucleotide frequencies.
   d. all of the above
   e. none of the above
4. The rate of fixation of neutral mutations is
   a. independent of population size.
   b. higher in small populations than in large populations.
   c. higher in large populations than in small populations.
   d. slower than the rate of fixation of deleterious mutations.
   e. none of the above
5. Genome size differs widely among different species. What is the greatest contributing cause for these differences?
   a. The number of protein-coding genes
   b. The amount of noncoding DNA
   c. The number of duplicated genes
   d. The degree of concerted evolution
   e. The amount of positive selection for change in protein-coding genes
6. Which of the following is *not* true of concerted evolution?
   a. Concerted evolution refers to the nonindependent evolution of some repeated genes within a species.
   b. Unequal crossing over may produce concerted evolution.

c. Biased gene conversion may produce concerted evolution.
d. Ribosomal RNA genes are an example of a gene family that has undergone concerted evolution.
e. Concerted evolution results in divergence of members of a gene family within an organism.

7. When a gene is duplicated, which of the following outcomes may occur?
a. Production of the gene's product may increase.
b. The two copies may become expressed in different tissues.
c. One copy of the gene may accumulate deleterious substitutions and become functionless.
d. The two copies may diverge and acquire different functions.
e. All of the above

8. Paralogous genes are genes that trace back to a common
a. speciation event.
b. substitution event.
c. insertion event.
d. deletion event.
e. duplication event.

9. Which of the following is true of in vitro evolution?
a. In vitro evolution refers to bioprospecting for naturally occurring macromolecules.
b. In vitro evolution can produce new molecular sequences not known from nature.
c. In vitro evolution can only produce new proteins.
d. In vitro evolution only selects for changes that were present in the starting pool of molecules, and does not introduce any new mutations.
e. All of the above

10. Which of the following is true of the use of molecular evolutionary studies of human disease?
a. Molecular evolutionary studies are useful for identifying many diseases.
b. Molecular evolutionary studies are often used to determine the origin of emerging diseases.
c. Molecular evolutionary studies are important for developing vaccines against diseases.
d. Molecular evolutionary studies are used to determine whether outbreaks of polio are the result of naturally occurring viruses or viruses that have evolved from attenuated viruses.
e. All of the above

FOR DISCUSSION

1. Rates of evolutionary change differ among different molecules, and different species differ widely in generation times and population sizes. How does this variation limit how and in what ways we can use the concept of a molecular clock to help us answer questions about the evolution of both molecules and organisms?
2. One hypothesis proposed to explain the existence of large amounts of noncoding DNA is that the cost of maintaining that DNA is so small that natural selection is too weak to reduce it. How could you test this hypothesis against the hypothesis that genome size is functionally related to developmental rate?
3. If fossil evidence and molecular evidence disagree on the date of a major lineage split, which of the two kinds of evidence would you favor? Why?
4. Soon scientists will be able to produce and release into the wild genetically modified mosquitoes that are unable to harbor and transmit malarial parasites. What ethical issues need to be discussed before such releases are permitted?

FOR INVESTIGATION

Many groups of organisms that inhabit caves have evolved to become eyeless. For instance, although surface-dwelling crayfishes have functional eyes, several species that are restricted to underground habitats lack eyes. Opsins are a group of light-sensitive proteins known to have an important function in vision; opsin genes are expressed in eye tissues. Although cave-dwelling crayfishes lack eyes, opsin genes are still present in their genomes. Two alternative hypotheses are (1) the opsin genes are no longer experiencing stabilizing selection (because there is no longer selection for function in vision); or (2) the opsin genes are experiencing selection for a function other than vision. How would you investigate these alternatives using the sequences of the opsin genes in various species of crayfishes?

CHAPTER 25 Reconstructing and Using Phylogenies

Phylogenetic trees in the courtroom

Transmitting HIV, while irresponsible, is not usually prosecuted as a crime. But in one true-crime case, a woman we'll call "April" went to the police immediately upon learning she was HIV-positive. April believed she was the victim of an attempted murder by "Victor," a physician and her former boyfriend, who had repeatedly threatened violence when she tried to break up with him. April's contention was that Victor, under the pretense of administering vitamin therapy, had injected her with blood from one of his HIV-infected patients.

Police investigators discovered that Victor had drawn blood from one of his HIV-positive patients just before giving April the injection. This blood draw had no clinical purpose, and Victor had tried to hide the records of it. The police were convinced that he might indeed have committed the alleged crime.

The district attorney, however, had to show that April's HIV infection had come from Victor's patient, and from no other source. To reconstruct the history of the infection, the district attorney turned to *phylogenetic analysis*—the study of the evolutionary relationships among a group of organisms.

The district attorney's task was complicated by the nature of HIV. HIV is a retrovirus, in which poor repair of replication errors leads to a very high rate of evolution. Once a person is infected with HIV, the virus not only replicates quickly, but evolves quickly, so that the infected individual is soon host to a genetically diverse population of viruses. Thus, when one person transmits HIV to another, typically very few viral particles (often only one) initiate the infective event. But the person who is the source of the infection may be host to a large, genetically diverse population of viruses—not just the variant they transmit to the recipient.

Enter molecular phylogeny. Samples of HIV from an infected individual can be sequenced to trace their evolutionary lineages back to the originally transmitted virus. The virus that is passed to the recipient will be very closely related to some of the viruses in the source individual and more distantly related to others. A reconstruction of the evolutionary history of the viruses in both individuals is needed to reveal not only whether the two individuals' viruses are closely related, but also who infected whom.

To prove attempted murder, the district attorney needed to demonstrate that April's HIV was more closely related to that of Victor's patient than

Human Immunodeficiency Virus Human immunodeficiency virus (HIV) is the cause of acquired immunodeficiency syndrome, or AIDS. The biology of this retrovirus was discussed extensively in Chapter 13; however, to combat AIDS it is also essential to understand the phylogeny of HIV.

A Source of the Virus AIDS is a zoonotic disease, the virus having been transferred to humans from another animal. Phylogenetic analyses of immunodeficiency viruses show that humans acquired HIV-1 from chimpanzees (see Figure 25.8). Other forms of the virus have been passed to humans by different simians.

CHAPTER OUTLINE

to other HIV variants in her community. Samples of HIV were isolated from the blood of the patient, from April, and from other HIV-positive individuals in the community. Phylogenetic analysis revealed that April's HIV was indeed closely related to a subset of the patient's HIV, and more distantly related to the other HIV sources in the community. Given this fact along with the other evidence in the case, Victor was convicted of attempted murder.

IN THIS CHAPTER we will examine the field of systematics, the scientific study of the diversity of life. We will see how phylogenetic methods are used to reconstruct evolutionary history and to study diversity across genes, populations, species, and larger groups of organisms. We will see how systematists reconstruct the past and predict the course of evolution. We end the chapter with a look at taxonomy, the theory and practice of classifying organisms.

25.1 What Is Phylogeny?

A **phylogeny** is a description of the evolutionary history of relationships among organisms (or their parts). A **phylogenetic tree** is a diagram that portrays a reconstruction of that history. Phylogenetic trees are commonly used to depict the evolutionary history of species, populations, and genes. Each split (or *node*) in a phylogenetic tree represents a point at which lineages diverged in the past. In the case of species, these nodes represent past speciation events, when one lineage split into two. Thus a phylogenetic tree can be used to trace the evolutionary relationships from the ancient common ancestor of a group of species, through the various speciation events when lineages split, up to the present populations of the organisms.

A phylogenetic tree may portray the evolutionary history of all life forms; of a major evolutionary group (such as the insects); of a small group of closely related species; or in some cases, even the history of individuals, populations, or genes within a species. The common ancestor of all the organisms in the tree forms the *root* of the tree. The phylogenetic trees in this book depict time flowing from left (earliest) to right (most recent) (**Figure 25.1A**); it is equally common practice to draw trees with the earliest times at the bottom.

The timing of separations between lineages of organisms is shown by the positions of nodes on a time or divergence axis; this axis may have an explicit scale or simply show the relative timing of divergence events. In this book, the positions of nodes along the horizontal (time) axis have meaning, but the vertical distance between the branches does not. Vertical distances are adjusted for legibility and clarity of presentation; they do not correlate with the degree of similarity or difference between groups. Note too that lineages can be rotated around nodes in the tree, so the vertical order of taxa is also largely arbitrary (**Figure 25.1B**).

Any group of species that we designate or name is called a **taxon** (plural *taxa*). Some examples of familiar taxa include humans, primates, mammals, and vertebrates (note that in this series, each taxon in the list is also a member of the next, more inclusive taxon). Any taxon that consists of all the evolutionary descendants of a common ancestor is called a **clade**. Clades can be identified by picking any point on a phylogenetic tree and then tracing all the descendant lineages to the tips of the terminal branches. Two species that are each other's closest rela-

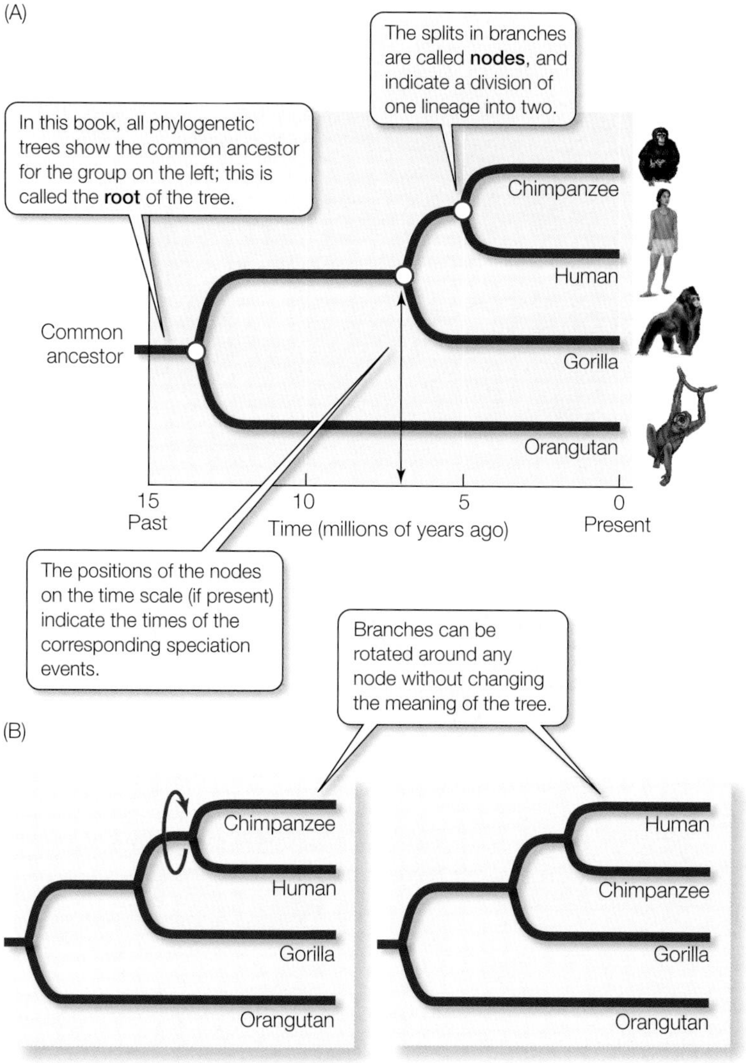

25.1 How to Read a Phylogenetic Tree (A) A phylogenetic tree displays the evolutionary relationships among organisms. Such trees can be produced with time scales, as shown here, or with no indication of time. If no time scale is shown, then the branch lengths show relative rather than absolute times of divergence. (B) Lineages can be rotated around a given node, so the vertical order of taxa is also largely arbitrary.

tives are called **sister species**; similarly, two clades that are each other's closest relatives are called **sister clades**.

Before the 1980s, phylogenetic trees were mostly seen in the literature on evolutionary biology, especially in **systematics**, the study of biodiversity. But today, in almost any scientific journal in the life sciences published in the last few years, you are likely to find articles that contain phylogenetic trees. Trees are widely used in studies of molecular biology, biomedicine, physiology, behavior, ecology, and virtually all other fields of biology. Why have phylogenetic studies become so important to biology?

All of life is connected through evolutionary history

In biology, we study life at all levels of organization—from genes, cells, organisms, populations, and species to the major divisions of life. In most cases, however, no individual gene or organism (or other unit of study) is exactly like any other gene or organism that we investigate.

Consider the individuals in your biology class. We recognize each person as an individual human, but we know that no two are exactly alike. If we knew everyone's family tree in detail, we could predict the genetic similarity of any pair of students. If we had this information, we would find that more closely related students would have many more traits in common (such as susceptibility or resistance to diseases). In a similar manner, biologists use phylogenies to make comparisons and predictions about shared traits across genes, populations, and species.

One of the greatest unifying concepts in biology is that all of life is connected through its evolutionary history. The complete, nearly 4-billion-year evolutionary history of life is known as the "Tree of Life." Biologists estimate that there are tens of millions of species on Earth. Of these, only 1.7 million have been formally described and named. New species are being discovered and named, and phylogenetic analyses reviewed and revised, all the time, but our knowledge of the Tree of Life is far from complete, even for known species. Nonetheless, knowledge of evolutionary relationships is essential for making comparisons in biology, so biologists build phylogenies for particular groups of interest as the need arises.

Any claim of an evolutionary association between a trait and a group of organisms is actually a statement about when during the history of the group the trait first arose, and about the maintenance of the trait since its first appearance. For example, the statement that the cytoskeleton is a trait possessed by all eukaryotes is an assertion that the cytoskeleton is an ancestral trait of eukaryotes that has been maintained during the subsequent evolution of all surviving eukaryote lineages.

Comparisons among species require an evolutionary perspective

When biologists make comparisons among species, they observe traits that differ within the group of interest, and try to ascertain when these traits evolved. In many cases, investigators are interested in how the evolution of a trait is dependent on environmental conditions or selective pressures. For instance, phylogenetic analyses have been used to discover changes in the genome of HIV that confer resistance to particular drug treatments. The association of a particular genetic change in HIV with a particular treatment provides a hypothesis about resistance that can be tested experimentally.

Any features shared by two or more species that have been inherited from a common ancestor are said to be **homologous**. As we noted in Section 24.1, homologous features may be any heritable traits, including DNA sequences, protein structures, anatomical structures, and even some behavior patterns. Traits that are shared by most or all of the organisms in a group of interest are likely to have been inherited from a common ancestor. For example, all living vertebrates have a vertebral column, all known fossil vertebrates had a vertebral column, and all vertebrates are descended from the same common ancestor. Therefore, the vertebral column is judged to be homologous in all vertebrates.

A trait that differs from its ancestral form is called a **derived trait**. Conversely, a trait that was present in the ancestor of a group is known as an **ancestral trait** for that group. Derived traits that are shared among a group of organisms, and are viewed as evidence of the common ancestry of the group, are called **synapomorphies** (*syn* means "shared," *apo* means "derived," and *morphy* refers to the "form" of a trait). Thus the vertebral column is considered a synapomorphy of the vertebrates.

Not all similar traits are evidence of relatedness, however. Similar traits in unrelated groups of organisms can develop for either of the following reasons:

- Independently evolved traits subjected to similar selection pressures may become superficially similar; this phenomenon is called **convergent evolution**. For example, although the bones of the wings of bats and birds are homologous, having been inherited from a common ancestor, the wings of bats and the wings of birds are not homologous because they evolved independently from the forelimbs of different nonflying ancestors (**Figure 25.2**).
- A character may revert from a derived state back to an ancestral state. Such a change is called an **evolutionary reversal**. For example, most frogs lack teeth in the lower jaw, but the ancestor of frogs did have such teeth. Teeth have been regained in the lower jaw of one frog genus, *Amphignathodon*, and thus represent an evolutionary reversal.

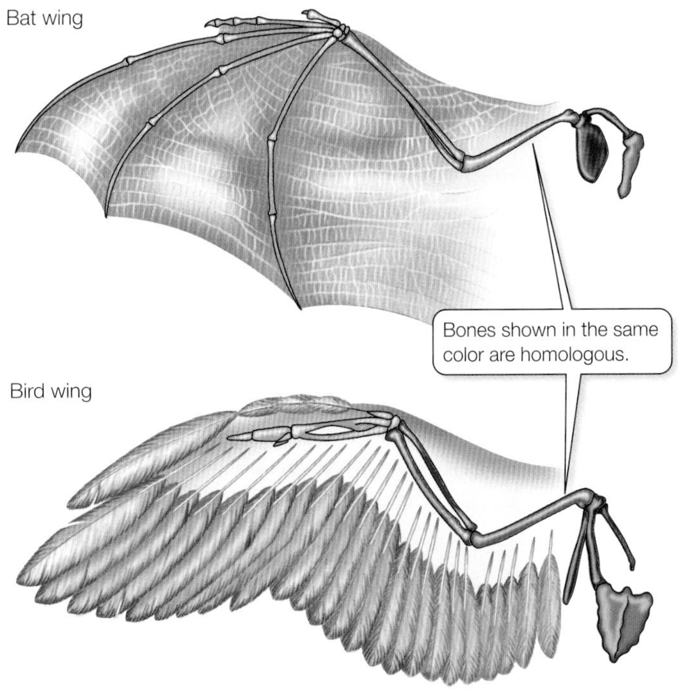

25.2 The Bones Are Homologous; the Wings Are Not The supporting bone structures of both bat wings and bird wings are derived from a common four-limbed ancestor and are thus homologous. However, the wings themselves—an adaptation for flight—evolved independently in the two groups.

Convergent evolution and evolutionary reversals generate traits that are similar for reasons other than inheritance from a common ancestor. Such traits are called *homoplastic traits* or **homoplasies**.

A particular trait may be ancestral or derived, depending on our point of reference in a phylogeny. For example, all birds have feathers, which are highly modified scales. We infer from this that feathers were present in the common ancestor of modern birds. Therefore, we consider the presence of feathers to be an *ancestral* trait for birds. However, feathers are not present in any other living vertebrates (or in any other animal species, for that matter). If we were reconstructing a phylogeny of all living vertebrates, the presence of feathers would be a *derived* trait that is found only among birds (and thus a synapomorphy of the birds).

25.1 RECAP

A phylogeny is a description of evolutionary relationships: how a group of genes, populations, or species have evolved from a common ancestor. All living organisms share a common ancestor and are related through the phylogenetic Tree of Life.

- Do you understand the different elements of a phylogenetic tree? See p. 543 and Figure 25.1
- Can you explain the difference between an ancestral and a derived trait? See p. 545
- Do you see how similar traits might arise in species that are not descended from a common ancestor? See p. 545 and Figure 25.2

Phylogenetic analyses have become increasingly important to many types of biological research in recent years, and they are the basis for the comparative nature of biology. But for the most part, evolutionary history cannot be observed directly. How, then, do biologists reconstruct the past?

25.2 How Are Phylogenetic Trees Constructed?

To illustrate how a phylogenetic tree is constructed, let's consider eight vertebrate animals: a lamprey, perch, pigeon, chimpanzee, salamander, lizard, mouse, and crocodile. We will assume initially that a given derived trait evolved only once during the evolution of these animals (that is, there has been no convergent evolution), and that no derived traits were lost from any of the descendant groups (there has been no evolutionary reversal). For simplicity, we have selected traits that are either present (+) or absent (–) (**Table 25.1**).

TABLE 25.1

Eight Vertebrates Ordered According to Unique Shared Derived Traits

TAXON	DERIVED TRAIT[a]							
	JAWS	LUNGS	CLAWS OR NAILS	GIZZARD	FEATHERS	FUR	MAMMARY GLANDS	KERATINOUS SCALES
Lamprey (outgroup)	–	–	–	–	–	–	–	–
Perch	+	–	–	–	–	–	–	–
Salamander	+	+	–	–	–	–	–	–
Lizard	+	+	+	–	–	–	–	+
Crocodile	+	+	+	+	–	–	–	+
Pigeon	+	+	+	+	+	–	–	+
Mouse	+	+	+	–	–	+	+	–
Chimpanzee	+	+	+	–	–	+	+	–

[a]A plus sign indicates the trait is present, a minus sign that it is absent.

As we will see in Chapter 33, a group of jawless fishes called the lampreys is thought to have separated from the lineage leading to the other vertebrates before the jaw arose. Therefore, we will choose the lamprey as the *outgroup* for our analysis. An outgroup can be any species or group of species outside the group of interest; it follows, then, that we call the group of primary interest the *ingroup*. The outgroup is used to determine which traits of the ingroup are derived (evolved in the ingroup) and which are ancestral (evolved before the origin of the ingroup). In this case, derived traits are those that have been acquired by other members of the vertebrate lineage since they separated from the lamprey, whereas any trait that is present in both the lamprey and the other vertebrates is judged to be ancestral. The root of the tree is determined by the relationship of the ingroup to the outgroup.

We begin by noting that the chimpanzee and mouse share two derived traits: mammary glands and fur. Those traits are absent in both the outgroup and the other species of the ingroup. Therefore, we infer that mammary glands and fur are derived traits that evolved in a common ancestor of chimpanzees and mice after that lineage separated from the ones leading to the other vertebrates. In other words, we provisionally assume that mammary glands and fur evolved only once among the animals in our ingroup. These characters are thus synapomorphies that unite chimpanzees and mice (as well as all other mammals, but we have not included other mammalian species in this example).

By the same reasoning, we can infer that the other shared derived traits are synapomorphies for the various groups in which they are expressed. For instance, keratinous scales are a synapomorphy of the crocodile, pigeon, and lizard, meaning that we infer that these species inherited this trait from a common ancestor.

The pigeon has one unique trait in our list: feathers. As before, we provisionally assume that feathers evolved only once, in the ancestor of birds. Feathers are a synapomorphy of birds, but since we only have one bird in this example, the presence of feathers does not provide any clues about the relationships among the eight species of vertebrates we have sampled. On the other hand, gizzards are found in birds and crocodilians, and so this trait is evidence of the close relationship between birds and crocodilians.

By combining information about the various synapomorphies, we can construct a phylogenetic tree. We infer, for example, that mice and chimpanzees, the only two animals that share fur and mammary glands in our example, share a more recent common ancestor with each other than they do with pigeons and crocodiles. Otherwise, we would need to assume that the ancestors of pigeons and crocodiles also had fur and mammary glands, but subsequently lost them—unnecessary additional assumptions.

Figure 25.3 shows a phylogenetic tree for these eight vertebrates, based on the traits we used and the assumption that each derived trait evolved only once. This particular phylogenetic tree was easy to construct because the animals and traits we used fulfilled the assumptions that derived traits appeared only once in the tree and that they were never lost after they appeared. Had we included a snake in the group, our second assumption would have been violated, because we know that the lizard ancestors of snakes had limbs that were subsequently lost (along with their claws). We would need to examine additional traits to determine that the lineage leading to snakes separated from the one leading to lizards long after the lineage leading to lizards separated from the others. In fact, the analysis of a number of traits shows that snakes evolved from burrowing lizards that became adapted to a subterranean existence.

Parsimony provides the simplest explanation for phylogenetic data

The phylogenetic tree shown in Figure 25.3 is based on only a very small sample of traits. Typically, biologists construct phylogenetic trees using hundreds or thousands of traits. With larger data sets, we would expect to observe some traits that have changed more than once, and thus we would expect to see some convergence and evolutionary reversal. How do we determine which traits are synapomorphies and which are homoplasies?

One approach is to use the **parsimony** principle. In its most general form, the parsimony principle states that the preferred explanation of the observed data is the simplest explanation. Its application to the reconstruction of phylogenies means minimizing the

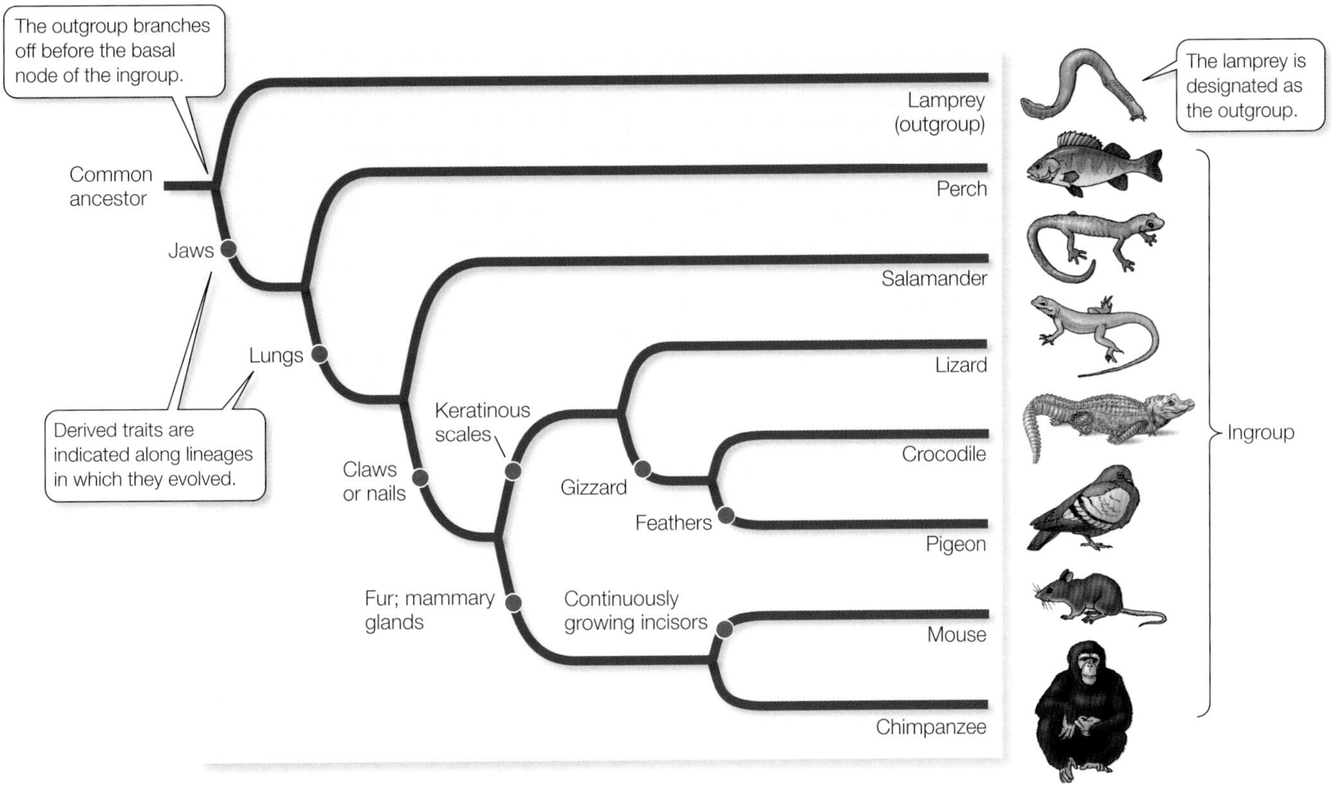

25.3 Inferring a Phylogenetic Tree This phylogenetic tree was constructed from the information given in Table 25.1 using the parsimony principle. Each clade in the tree is supported by at least one shared derived trait, or synapomorphy.

number of evolutionary changes that need to be assumed over all characters in all groups in the tree. In other words, the best hypothesis under the parsimony principle is one that requires the fewest homoplasies. This application of parsimony is a specific case of a general principle of logic called Occam's razor: the idea that the best explanation is the one that fits the data best and makes the fewest assumptions.

Using the parsimony principle is appropriate not because all evolutionary changes occurred parsimoniously, but because it is logical to adopt the simplest explanation that can account for the observed data. More complicated explanations are accepted only when the evidence requires them. Phylogenetic trees represent our best estimates about evolutionary relationships. They are continually modified as additional evidence becomes available.

Phylogenies are reconstructed from many sources of data

Naturalists constructed various forms of phylogenetic trees as far back as classical times. Tree construction has been revolutionized, however, by the advent of computer software for trait analysis and tree construction, allowing us to consider amounts of data far more vast than could ever before be processed.

Any trait that is genetically determined, and therefore heritable, can be used in a phylogenetic analysis. Evolutionary relationships can be revealed through studies of morphology, development, the fossil record, behavioral traits, and molecular traits such as DNA and protein sequences. All of these are used by modern systematists, and computer-driven tree construction and gene sequencing taken together have led to an explosion of new approaches to phylogeny.

> Although the construction of phylogenetic trees is conceptually straightforward, the computations involved with large sets of traits and taxa can be challenging. For perspective, the number of possible phylogenetic trees for just 50 species greatly exceeds the number of atoms in the universe!

MORPHOLOGY An important source of phylogenetic information is *morphology*—that is, the presence, size, shape, and other attributes of body parts. Since living organisms have been studied for centuries, we have a wealth of recorded morphological data as well as extensive museum and herbarium collections of organisms whose traits can be measured. New technological tools, such as the electron microscope and computed tomography (CT) scans, enable systematists to examine and analyze the structures of organisms at much finer scales than was formerly possible. Most species of living organisms have been described and are known primarily from morphological data, so morphology provides the most comprehensive data set available for many taxa. The features of morphology that are important for phylogenetic analysis are often specific to a particular group of organisms. For example, the presence, development, shape, and size of various features of the skeletal system are important for the study of vertebrate phylogeny, whereas floral structures are important for studying the relationships among flowering plants.

Despite the usefulness of morphological approaches to phylogenetic analysis, they have some limitations. Some taxa exhibit very little morphological diversity, despite great species diversity. At the other extreme, there are few morphological traits that can be compared between very distantly related species (consider bacteria and vertebrates, for instance). Some morphological variation has an environmental (rather than a genetic) basis and so must be excluded from phylogenetic analyses. Therefore, although morphology provides an important source of information on phylogeny, additional sources of information are often necessary.

DEVELOPMENT Observations of similarities in developmental patterns may also reveal evolutionary relationships. Similarities in early developmental stages may be lost during later development. For example, the larvae of marine creatures called sea squirts have a rod in the back—the *notochord*—that disappears as they develop into adults. All vertebrate animals also have a notochord at some time during their development (**Figure 25.4**). This shared structure is one of the reasons for inferring that sea squirts are more closely related to vertebrates than would be suspected if only adult sea squirts were examined.

PALEONTOLOGY The fossil record is another important source of information on evolutionary history. Fossils show us where and when organisms lived in the past and give us an idea of what they looked like. Fossils provide important evidence that helps us distinguish ancestral from derived traits. The fossil record can also reveal when lineages diverged and began their independent evolutionary histories. Furthermore, in groups with few species that have survived to the present, information on extinct species is often critical to an understanding of the large divergences between the surviving species. The fossil record does have limitations, however. Few or no fossils have been found for some groups whose phylogenies we may wish to determine, and the fossil record for many groups is fragmentary.

BEHAVIOR Some behavioral traits are culturally transmitted and others are inherited. If a particular behavior is culturally transmitted, then it may not accurately reflect evolutionary relationships (but may nonetheless reflect cultural connections). Bird songs, for instance, are often learned and may be inappropriate traits for phylogenetic analysis. Frog calls, on the other hand, are genetically determined and appear to be acceptable sources of information for reconstructing phylogenies.

MOLECULAR DATA All heritable variation is encoded in DNA, and so the complete genome of an organism contains an enormous set of traits (the individual nucleotide bases of DNA) that can be used in phylogenetic analyses. In recent years, DNA sequences have become one of the most widely used sources for constructing phylogenetic trees. Comparisons of nucleotide sequences are not limited to the DNA in the cell nucleus. Eukaryotes have genes in their mitochondria as well as in their nuclei; plant cells also have genes in their chloroplasts. The chloroplast genome (cpDNA), which is used extensively in phylogenetic studies of plants, has changed slowly over evolutionary time, so it is often used to study relatively ancient phylogenetic relationships. Most animal mitochondrial DNA (mtDNA) has changed more rapidly, so mitochondrial genes have been used extensively for studies of evolutionary relationships among closely related animal species. Many nuclear gene sequences are also commonly analyzed, and now that a number of whole genomes have been sequenced, they are also used to construct phylogenetic trees. Information on gene products (such as the amino acid sequences of proteins) is also widely used for phylogenetic analyses, as we discussed in Chapter 24.

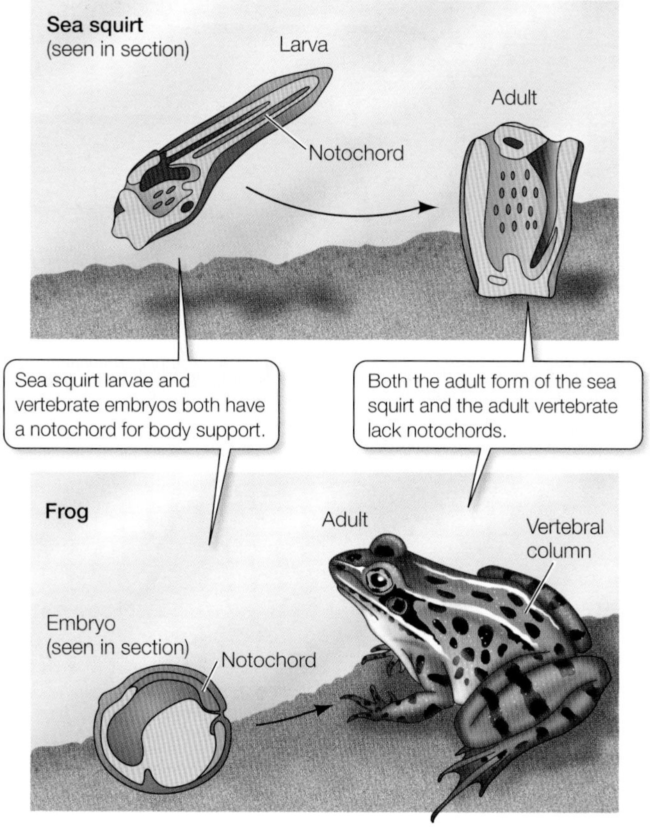

25.4 A Larva Reveals Evolutionary Relationships Sea squirt larvae, but not adults, have a well-developed notochord (orange) that reveals their evolutionary relationship to vertebrates, all of which have a notochord at some time during their life cycle. In adult vertebrates, the vertebral column replaces the notochord as the support structure.

Mathematical models expand the power of phylogenetic reconstruction

As biologists began to use DNA sequences to infer phylogenies in the 1970s and 1980s, they developed explicit mathematical models describing how DNA sequences change over time. These models account for multiple changes at a given position in a DNA sequence, and they also take into account different rates of change at different positions in a gene, at different positions in a codon, and among different nucleotides (see Section 24.1). For example, transitions (changes between two purines or between two pyrimidines) are usually more likely than are transversions (changes between a purine and pyrimidine).

Such mathematical models can be used to compute **maximum likelihood** solutions for phylogenetic estimation. A likelihood score of a tree is based on the probability of the observed data evolving on the specified tree, given an explicit mathematical model of evolution for the characters. The maximum likelihood solution, then, is the most likely tree given the observed data. Maximum likelihood methods can be used for any kind of characters, but they are most often used with molecular data, for which explicit mathematical models of evolutionary change are easier to develop. The principal advantages to maximum likelihood analyses are that they incorporate more information about evolutionary change than do parsimony methods, and they are easier to treat in a statistical framework. The principal disadvantages are that they are computationally intensive and require explicit models of evolutionary change (which may not be available for some kinds of character change).

The accuracy of phylogenetic methods can be tested

If phylogenetic trees represent reconstructions of past events, and if many of these events occurred before any humans were around to witness them, how can we test the accuracy of phylogenetic methods? Biologists have conducted experiments both in living organisms and with computer simulations that have demonstrated the effectiveness and accuracy of phylogenetic methods.

In one experiment that was designed to test the accuracy of phylogenetic analysis, a single viral culture of bacteriophage T7 was used as a starting point, and lineages were allowed to evolve from this ancestral virus in the laboratory (**Figure 25.5**). The initial culture was split into two separate lineages, one of which became the ingroup for analysis, and the other became the outgroup for rooting the tree. In the ingroup, the lineages were split in two after every 400 generations, and samples of the virus were saved for analysis at each branching point. The lineages were allowed to

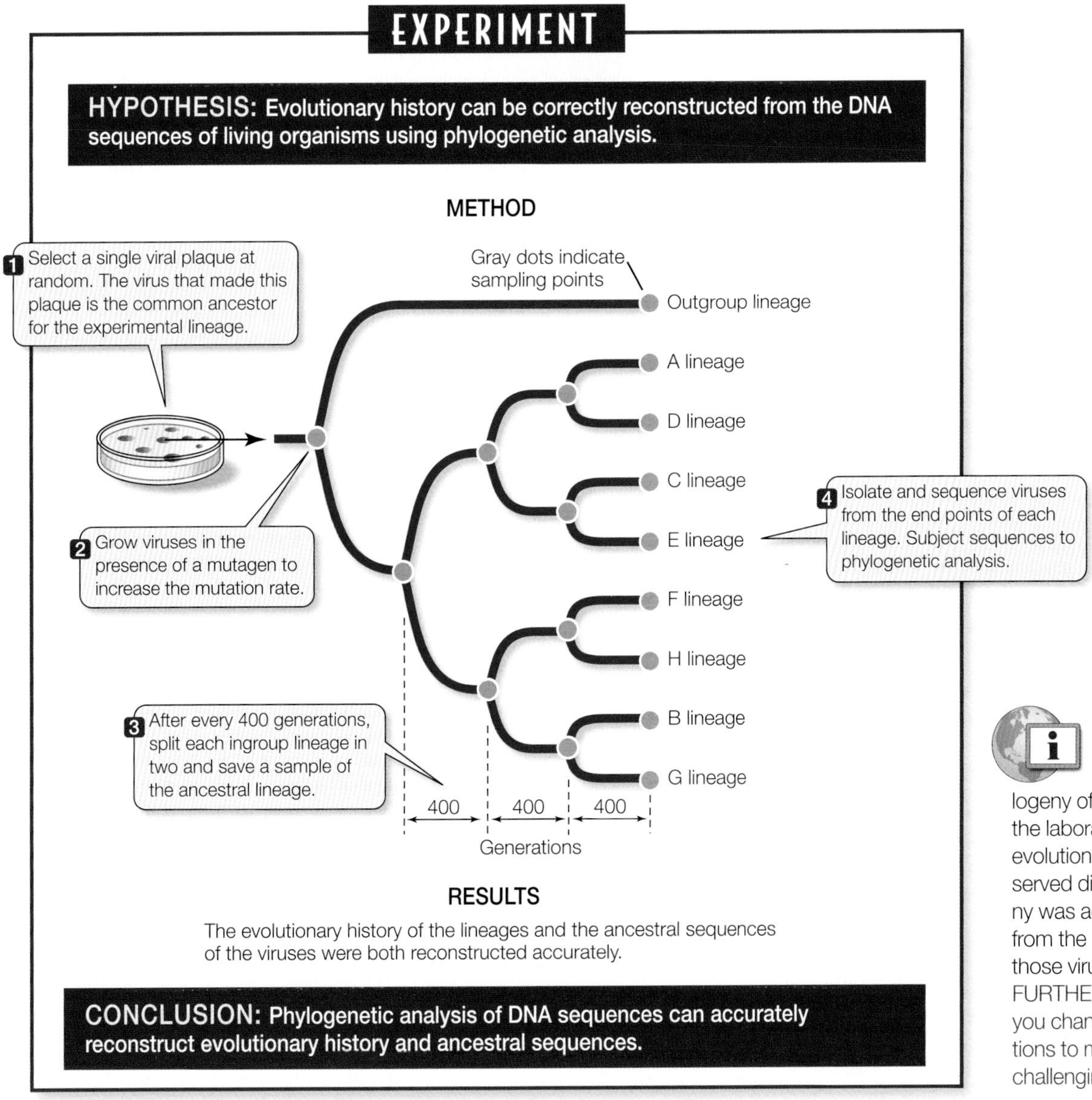

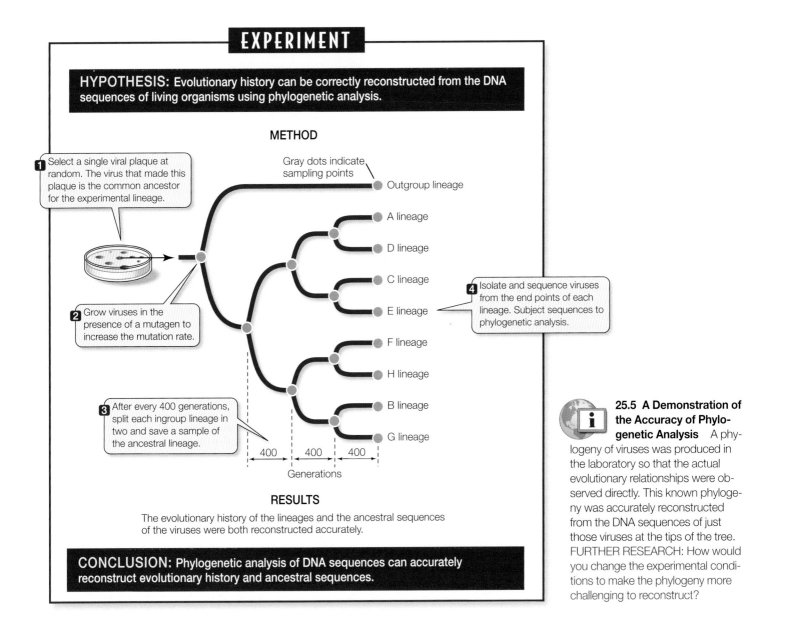

25.5 A Demonstration of the Accuracy of Phylogenetic Analysis A phylogeny of viruses was produced in the laboratory so that the actual evolutionary relationships were observed directly. This known phylogeny was accurately reconstructed from the DNA sequences of just those viruses at the tips of the tree. FURTHER RESEARCH: How would you change the experimental conditions to make the phylogeny more challenging to reconstruct?

evolve until there were eight lineages in the ingroup and one in the outgroup. Mutagens were added to the viral cultures to increase the mutation rate so that the amount of change and the degree of homoplasy would be typical of the organisms analyzed in average phylogenetic analyses. The investigators then sequenced samples from the end points of the eight lineages, as well as from the ancestors at the branching points. They then gave the sequences from the end points of the lineages to other investigators to analyze, without revealing the known history of the lineages or the sequences of the ancestral viruses.

After the phylogenetic analysis had been completed, the investigators asked two questions: Did phylogenetic methods reconstruct the known history correctly, and were the sequences of the ancestral viruses reconstructed accurately? The answer in both cases was yes: the branching order of the lineages was reconstructed exactly as it had occurred, over 98 percent of the nucleotide positions of the ancestral viruses were reconstructed correctly, and 100 percent of the amino acid changes in the viral proteins were reconstructed correctly.

The experiment shown in Figure 25.5 demonstrated that phylogenetic analysis was accurate under the conditions tested, but it did not examine all possible conditions. Other experimental studies have taken other factors into account, such as the sensitivity of phylogenetic analysis to convergent environments and highly variable rates of evolutionary change. In addition, computer simulations based on evolutionary models have been used extensively to study the effectiveness of phylogenetic analysis. These studies, too, have confirmed the accuracy of phylogenetic methods and have also been used to refine those methods and extend them to new applications.

Ancestral states can be reconstructed

In addition to inferring the evolutionary relationships among lineages, biologists can use phylogenetic methods to reconstruct the morphology, behavior, or nucleotide and amino acid sequences of ancestral species (as was demonstrated for the ancestral sequences of T7 in the experiment shown in Figure 25.5). For instance, a phylogenetic analysis was used to reconstruct an opsin protein in the ancestral archosaur (the last common ancestor of birds, dinosaurs, and crocodiles). Opsins are pigment proteins involved in vision; different opsins (with different amino acid sequences) are excited by different wavelengths of light. Knowledge of the opsin sequence in the ancestral archosaur would provide clues about the animal's visual capabilities and therefore about some of its probable behaviors. The investigators used phylogenetic analysis of opsin from living vertebrates to estimate the amino acid sequence of the pigment that existed in the ancestral archosaur. A protein with this same sequence was then constructed in the laboratory. The investigators tested the reconstructed opsin and found a significant shift toward the red end of the spectrum in the light sensitivity of this protein compared with most modern opsins. Modern species that exhibit similar sensitivity are adapted for nocturnal vision, so the investigators inferred that the ancestral archosaur might have been active at night. Thus, reminiscent of the movie *Jurassic Park*, phylogenetic analyses are being used to reconstruct extinct species, one protein at a time!

(A)

Harpagochromis sp.

Ptyochromis sp.

(B)

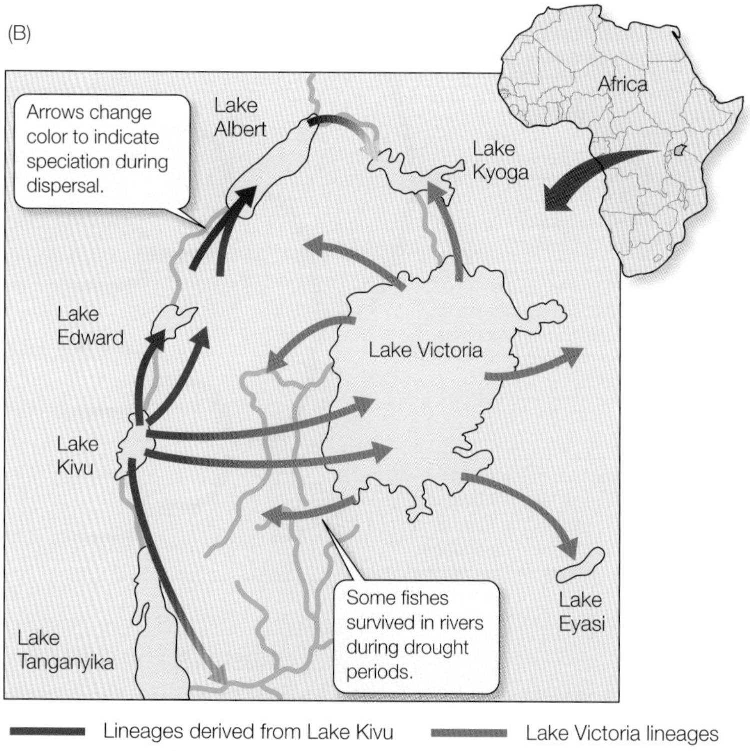

25.6 Origins of the Cichlid Fishes of Lake Victoria (A) These photographs show only two of the hundreds of cichlid species found in Lake Victoria. (B) Phylogenetic analysis suggests that cichlids colonized Lake Victoria by the routes shown on the map.

Molecular clocks add a dimension of time

For many applications, biologists want to know not only the order in which evolutionary lineages split, but also the timing of those splits. In 1965, Emile Zuckerkandl and Linus Pauling hypothesized that rates of molecular change were constant enough that they could be used to predict evolutionary divergence times—an idea that has become known as the *molecular clock hypothesis*.

Of course, different genes evolve at different rates, and there are also differences in evolutionary rates among species related to differing generation times, environments, efficiencies of DNA repair systems, and other biological factors. Nonetheless, among closely related species, a given gene usually evolves at a reasonably constant rate and can be used as a metric to gauge the time of divergence for a particular split in the phylogeny. These **molecular clocks** must be calibrated using independent data, such as the fossil record, known times of divergence, or biogeographic dates (such as the dates for separations of continents). Using such calibrations, times of divergence have been estimated for many groups of species.

Studies of cichlid fishes provide an example of the use of molecular clocks. The spectacular evolutionary radiation that produced more than 500 species in one group of cichlid fishes in Lake Victoria, in eastern Africa, was initially assumed to have occurred over a period of about 750,000 years, the presumed age of the lake basin. Recently discovered geological data suggest, however, that Lake Victoria dried up between 15,600 and 14,700 years ago. Biologists judged that the hundreds of morphologically diverse cichlids in Lake Victoria could not have evolved in such a short time, so they considered other hypotheses. One hypothesis assumed that the lake did not dry up completely. Another postulated that some of the fish species survived in rivers, from which they subsequently recolonized the lake. But how could they test these possibilities? Phylogenetic analyses based on molecular data helped resolve the problem.

Using mitochondrial DNA sequences from each of 300 species, investigators constructed a phylogenetic tree of the cichlid fishes of Lake Victoria and other lakes in the region. Their phylogenetic tree suggests that the ancestors of the Lake Victoria cichlids came from the geologically much older Lake Kivu. Today, Lake Kivu is home to only 15 species of cichlids, but the phylogenetic tree suggests that fishes from Lake Kivu colonized Lake Victoria on two different occasions (**Figure 25.6**). A molecular clock analysis also indicated that some of the cichlid lineages that are found only in Lake Victoria, and which therefore probably evolved there, split at least 100,000 years ago. Thus the inferred phylogeny of these fishes strongly suggests that Lake Victoria did not completely dry up about 15,000 years ago, and that many fish species survived in rivers and in pockets of water that remained in the deepest part of the lake throughout the most recent dry period.

Biologists once thought that HIV-1 moved from chimpanzees to humans shortly before AIDS became recognized as a disease in the early 1980s. However, molecular clock analysis of the main HIV-1 group of viruses demonstrated that HIV-1 in humans dates back at least to the 1920s or 1930s, and that other primate immunodeficiency viruses have moved into human populations several times over the past century.

25.2 RECAP

Phylogenetic trees can be constructed by using the parsimony principle to find the simplest explanation for the evolution of traits. Maximum likelihood methods incorporate more explicit models of evolutionary change to reconstruct evolutionary history.

- Do you understand how a phylogenetic tree is constructed? See pp. 547–548 and Figure 25.3
- Is there a way to test whether phylogenetic trees provide accurate reconstructions of evolutionary history? See pp. 549–550 and Figure 25.5
- How do molecular clocks add a time dimension to phylogenetic trees? See p. 551

Biologists in many fields now routinely reconstruct phylogenetic relationships. Let's examine some of the many uses of these phylogenetic trees.

25.3 How Do Biologists Use Phylogenetic Trees?

Information about the evolutionary relationships among organisms is useful to scientists investigating a wide variety of biological questions. In this section we will illustrate how phylogenetic trees can be used to ask questions about the past, to compare organisms in the present, and to make predictions about the future.

Phylogenies help us reconstruct the past

Most flowering plants reproduce by mating with another individual—a process called *outcrossing*. Many outcrossing species have mechanisms to prevent self-fertilization, and so are referred to as *self-incompatible*. Individuals of some species, however, regularly fertilize themselves with their own pollen; they are termed *selfing* species, which of course requires that they be *self-compatible*. How can we tell how often self-compatibility has evolved in a group of plants? We can do so by conducting a phylogenetic analysis of outcrossing and selfing species and testing the species for self-compatibility.

The evolution of fertilization mechanisms was examined in *Linanthus* (a genus in the phlox family), a group of plants with a diversity of breeding systems and pollination mechanisms. The outcrossing species of *Linanthus* have long petals and are pollinated by long-tongued flies; these species are self-incompatible. The self-compatible species, in contrast, all have short petals. The investigators reconstructed a phylogeny for 12 species in the genus using nuclear ribosomal DNA sequences (**Figure 25.7**). They determined whether each species was self-compatible by artificially pollinating flowers with the plant's own pollen or with pollen from other individuals and observing whether viable seeds formed.

Several lines of evidence suggest that self-incompatibility is the ancestral state in *Linanthus*. Multiple origins of self-incompatibil-

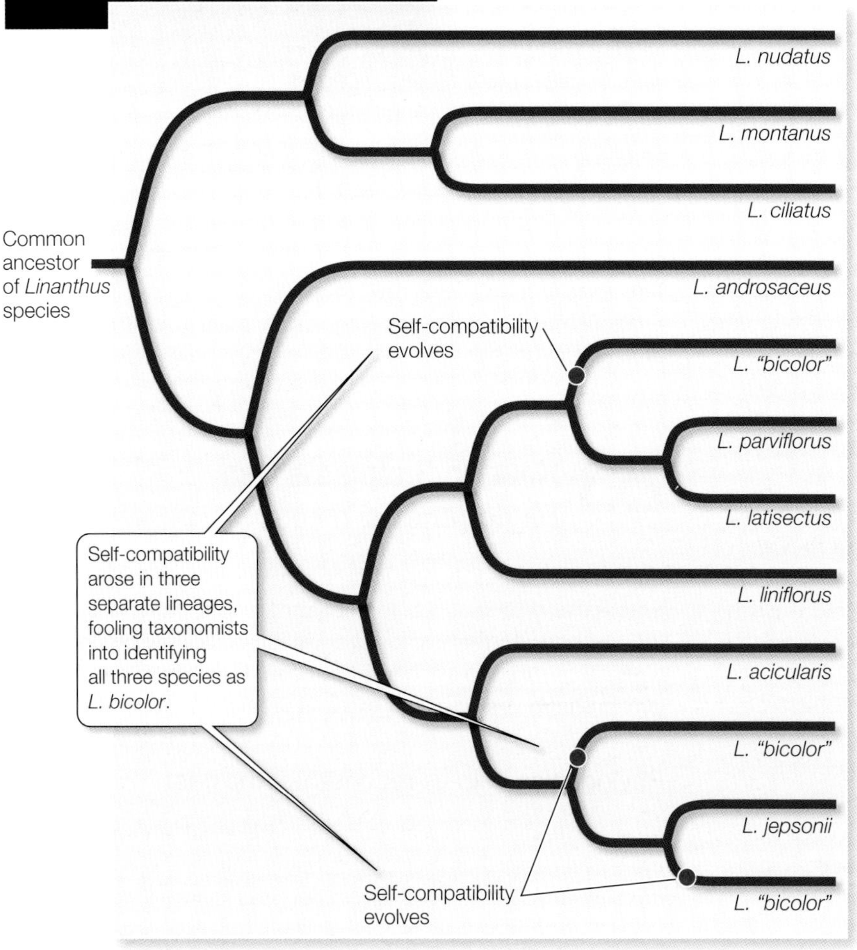

25.7 Phylogeny of a Section of the Plant Genus *Linanthus* Self-compatibility apparently evolved three times in this group. Because the form of the flowers converged in the three selfing lineages, taxonomists mistakenly thought that they were all varieties of a single species.

ity have not been found in any other flowering plant family. Self-incompatibility depends on physiological mechanisms in both the pollen and the stigma (the female organ on which pollen lands) and requires the presence of at least three different alleles. Therefore, a change from self-incompatibility to self-compatibility would be easier than the reverse change. In addition, in all self-incompatible species of *Linanthus*, the site of pollen rejection is the stigma, even though sites of pollen rejection vary greatly among other plant families.

Assuming that self-incompatibility is the ancestral state, the reconstructed phylogeny suggests that self-compatibility has evolved three times within this group of *Linanthus* (see Figure 25.7). The change to self-compatibility has been accompanied by the evolution of reduced petal size. Interestingly, the striking similarity of the flowers in the self-compatible groups once led to their being classified as members of a single species. The phylogenetic analysis using ribosomal DNA showed them to be members of three distinct lineages, however.

Reconstructing the past is important for understanding many biological processes. In the case of *zoonotic diseases* (diseases caused by infectious organisms that have been transferred to humans from another animal host), it is important to understand when, where, and how the disease first entered human populations. Human immunodeficiency virus (HIV) is the cause of such a zoonotic disease: acquired immunodeficiency syndrome, or AIDS. A phylogenetic analysis of immunodeficiency viruses shows that humans acquired these viruses from two different hosts: HIV-1 from chimpanzees, and HIV-2 from sooty mangabeys (**Figure 25.8**).

HIV-1 is the common form of the virus in human populations in central Africa, where chimpanzees are hunted for food, and HIV-2 is the common form of the virus in human populations in western Africa, where sooty mangabeys are hunted for food. Thus it seems likely that these viruses entered human populations through hunters who cut themselves while skinning chimpanzees and sooty mangabeys. The relatively recent global pandemic of AIDS occurred when these infections in local African populations rapidly spread through human populations around the world.

Phylogenies allow us to compare and contrast living organisms

Male swordtails (a group of fishes within the genus *Xiphophorus*) have a long, colorful extension of the tail (**Figure 25.9A**). The reproductive success of male swordtails is closely associated with these appendages. Males with long swords are more likely to mate successfully than are males with short swords (an example of *sexual selection*; see Chapters 22 and 23). Several explanations have been advanced for the evolution of these structures, including the hypothesis that the sword simply exploits a preexisting bias in the sensory system of the females. This *sensory exploitation hypothesis* suggests that female swordtails had a preference for males with long tails even before the tails evolved (perhaps because females assess the size of males by their total length, including the tail). Long tails became a sexually selected trait because of the preexisting preference of the females.

A phylogeny was used to identify the closest relatives of swordtails that had split from their lineage before the evolution of swords. These turned out to be the platyfishes, another group of *Xiphophorus* (**Figure 25.9B**). To test the sensory exploitation hypothesis, researchers attached artificial swordlike structures to the tails of some male platyfishes. Even though male platyfishes do not normally have such structures, female platyfishes preferred the males with the artificial swords, thus providing support for the hypothesis that

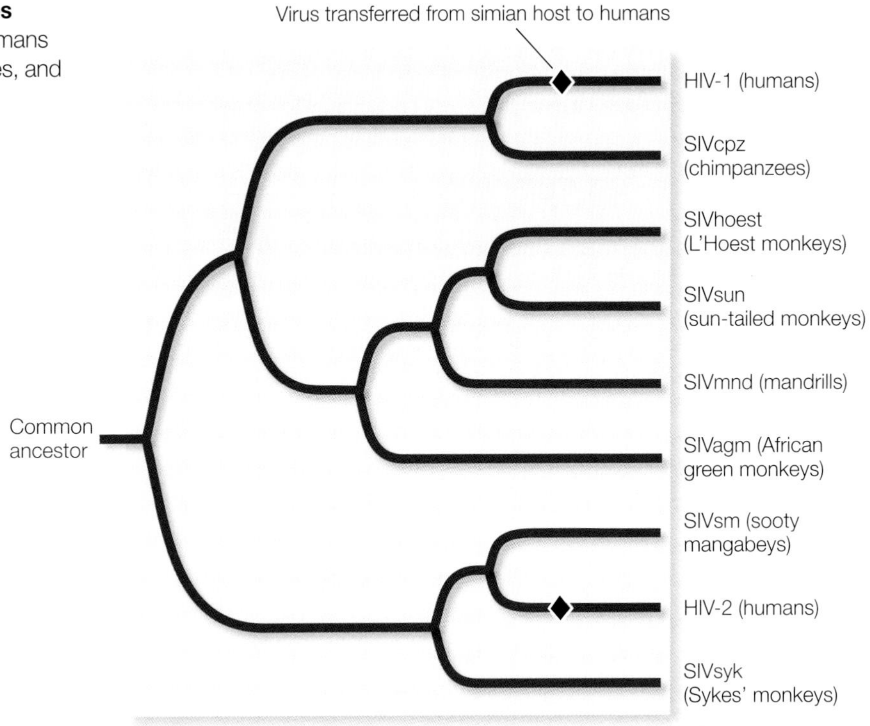

25.8 Phylogenetic Tree of Immunodeficiency Viruses Immunodeficiency viruses have been transmitted to humans from two different simian hosts (HIV-1 from chimpanzees, and HIV-2 from sooty mangabeys).

female *Xiphophorus* had a preexisting sensory bias that favored swords even before the trait evolved.

Biologists use phylogenies to predict the future

Influenza kills many people every year, and each year, many susceptible individuals receive a flu vaccine to reduce their chance of contracting the disease. Why do we need a new flu vaccine every year, when a single dose of vaccine for many other diseases provides many years of protection? The answer is that the rate of evolution of the influenza virus is sufficiently high that the flu viruses in circulation each year are often substantially different from the flu viruses that circulated in previous years.

A phylogenetic analysis of strains of influenza virus indicates that there is strong selection by the human immune system against most strains. Only those strains with the greatest number of amino acid replacements in particular positions of the protein hemagglutinin (a protein on the surface of influenza viruses that is recognized by the human immune system; **Figure 25.10**) are likely to leave descendant lineages in subsequent years. Thus, by conducting a phylogenetic analysis of hemagglutinin

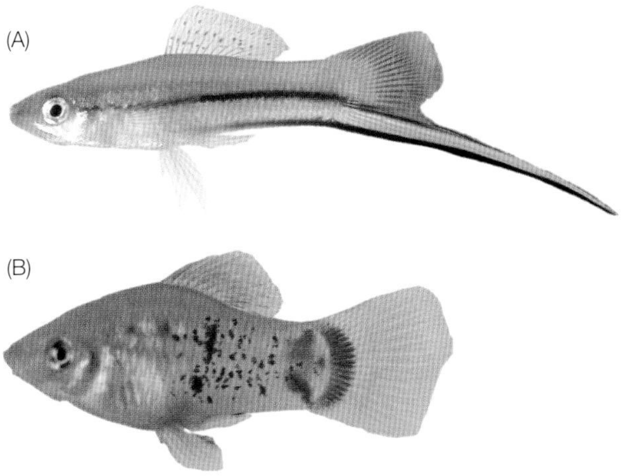

25.9 The Origin of a Sexually Selected Trait in the Fish Genus *Xiphophorus* (A) A male swordtail, showing the large tail that evolved through sexual selection. (B) A male platyfish, a related species. Phylogenetic analysis reveals that the platyfish split from the swordtails before the evolution of the sword. But experimental analysis demonstrates that female platyfish prefer males with artificial swords. This finding supports the idea that the sword of swordtails evolved as a result of a preexisting preference in the females.

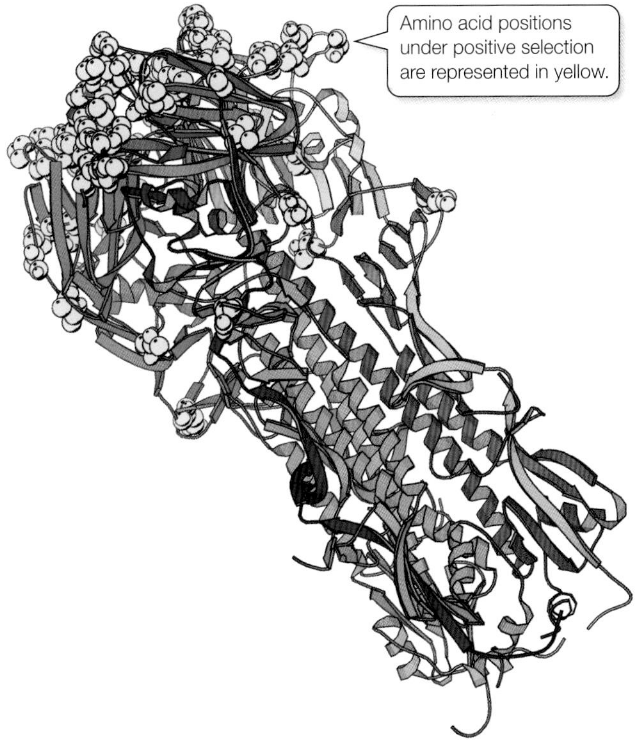

25.10 Model of Hemagglutinin, a Surface Protein of Influenza The yellow balloon-shaped structures represent amino acids in the protein that are under strong selection for change. Changes in these amino acids produce proteins that are more likely to escape detection by the human immune system.

sequences, biologists can predict which of the currently circulating strains of influenza virus are most likely to survive and leave descendants in the future. This information can help to determine which strains of influenza virus are the best candidates for developing flu vaccine for upcoming flu seasons.

25.3 RECAP

Phylogenetic trees are used to reconstruct the past history of lineages, to determine when and where traits arose, and to make relevant biological comparisons among genes, populations, and species. In some cases they can even be used to make predictions about future evolution.

- Can you explain how phylogenetic trees can help us determine the number of times a particular trait evolved? See pp. 551–552 and Figure 25.7
- Do you understand how phylogenetic analysis can help produce more effective flu vaccines? See p. 553

As we have now seen, all of life is connected through evolutionary history, and the relationships among organisms provide a natural basis for making biological comparisons. For these reasons, biologists use phylogenetic relationships as the basis for organizing life into a coherent classification system.

25.4 How Does Phylogeny Relate to Classification?

The biological classification system in widespread use today is derived from a system developed by the Swedish biologist Carolus Linnaeus and has been used since the mid-1700s. Linnaeus's system, referred to as **binomial nomenclature**, allows scientists throughout the world to refer unambiguously to the same organisms by the same names (**Figure 25.11**).

Linnaeus gave each species two names, one identifying the species itself and the other the genus to which it belongs. A **genus** (plural, *genera*; adjectival form *generic*) is a group of closely related species. Optionally, the name of the taxonomist who first proposed the species name may be added at the end. Thus *Homo sapiens* Linnaeus is the name of the modern human species. *Homo* is the genus to which the species belongs, and *sapiens* identifies the particular species in the genus *Homo*; Linnaeus proposed the species name *Homo sapiens*. You can think of the generic name *Homo* as equivalent to your surname and the specific name *sapiens* as equivalent to your first name. The generic name is always capitalized; the name identifying the species always lowercased. Both names are italicized, whereas common names of organisms are not. Rather than repeating a generic name when it is used several times in the same discussion, biologists often spell it out only once and abbreviate it to the initial letter thereafter (for example, *D. melanogaster* is the abbreviated form of *Drosophila melanogaster*).

As we noted earlier, any group of organisms that is treated as a unit in a biological classification system, such as the genus *Drosophila*, or all insects, is called a *taxon*. In the Linnaean system, species and genera are further grouped into a hierarchical system of higher taxonomic categories. The taxon above the genus in the Linnaean system is the **family**. The names of animal families end in the suffix "-idae." Thus Formicidae is the family that contains all ant species, and the family Hominidae contains humans and our recent fossil relatives, as well as our closest living relatives, the chimpanzees and gorillas. Family names are based on the name of a member genus; Formicidae is based on the genus *Formica*, and Hominidae is based on *Homo*. Plant classification follows the same procedures, except that the suffix "-aceae" is used with family names instead of "-idae." Thus Rosaceae is the family that includes the genus of roses (*Rosa*) and its close relatives. Families, in turn, are grouped into **orders**, orders into **classes**, and classes into **phyla** (singular *phylum*), and phyla into **kingdoms**. However, the application of these various ranked levels of classification is subjective,

(A) *Campanula rotundifolia*

(B) *Endymion non-scriptus*

(C) *Mertensia virginica*

25.11 Many Different Plants Are Called Bluebells All three of these unrelated plant species are commonly called "bluebells." Binomial nomenclature allows us to communicate exactly what is being described. (A) *Campanula rotundifolia*, found on the North American Great Plains, belongs to a larger group of bellflowers. (B) *Endymion non-scriptus*, the English bluebell, is related to hyacinths. (C) *Mertensia virginica*, the Virginia bluebell, belongs in a very different group of plants known as borages.

and they are used largely for convenience. Although families are always grouped within orders, orders within classes, and so forth, there is nothing that makes a family in one group equivalent (in age, for instance) to a family in another group.

Linnaeus developed his system before evolutionary thought had become widespread in biology, but he still recognized the overwhelming hierarchy of life. Although Linnaeus did not know the basis of this hierarchy, biologists today universally recognize the common Tree of Life as the organizational basis for biological classification. Today, some biologists de-emphasize the use of ranked classifications, since it is a largely subjective decision whether a particular taxon is considered, say, an order or a class. But regardless of whether modern biologists explicitly use these ranked categories, they use evolutionary relationships as the basis for recognizing all biological taxa.

Phylogeny is the basis for modern biological classification

Biological classification systems are used to express relationships among organisms. The kind of relationship we wish to express influences which features we use to classify organisms. If, for instance, we were interested in a system that would help us decide what plants and animals were desirable as food, we might devise a classification based on tastiness, ease of capture, and the type of edible parts each organism possessed. Early Hindu classifications of organisms were designed according to these criteria. Biologists do not use such systems in formal classification today, but those systems served the needs of the people who developed them.

Taxonomists today use biological classifications to express the evolutionary relationships of organisms. Therefore, taxa in biological classifications are expected to be **monophyletic**, meaning that the taxon contains an ancestor and all descendants of that ancestor, and no other organisms (**Figure 25.12**). In other words, it is a historical group of related species, or a complete branch on the Tree of Life (also known as a *clade*). Although biologists seek to name monophyletic taxa, the detailed phylogenetic information needed to do so is not always available. A group that does not include its common ancestor is called a **polyphyletic** group. A group that does not include all the descendants of a common ancestor is called a **paraphyletic** group.

A true monophyletic group (or clade) can be removed from a phylogenetic tree by a single "cut" in the tree, as shown in Figure 25.12. Note that there are many monophyletic groups on any phylogenetic tree, and that these groups are successively smaller subsets of larger monophyletic groups. This hierarchy of biological taxa, with all of life as the most inclusive taxon and many smaller taxa within larger taxa, down to the individual species, is the modern basis for biological classification.

Virtually all taxonomists now agree that polyphyletic and paraphyletic groups are inappropriate as taxonomic units. The classifications used today still contain such groups because some organisms have not been evaluated phylogenetically. As mistakes in prior classifications are detected, taxonomic names are revised and polyphyletic and paraphyletic groups are eliminated from the classifications.

Several codes of biological nomenclature govern the use of scientific names

Several sets of explicit rules govern the use of scientific names. Biologists around the world follow these rules voluntarily to facilitate communication and dialogue. Although there may be dozens of common names for an organism in many different languages, the rules of biological nomenclature are designed so that there is only one correct scientific name for any single recognized taxon and (ideally) a given scientific name applies only to a single taxon (that is, each scientific name is unique). Sometimes the

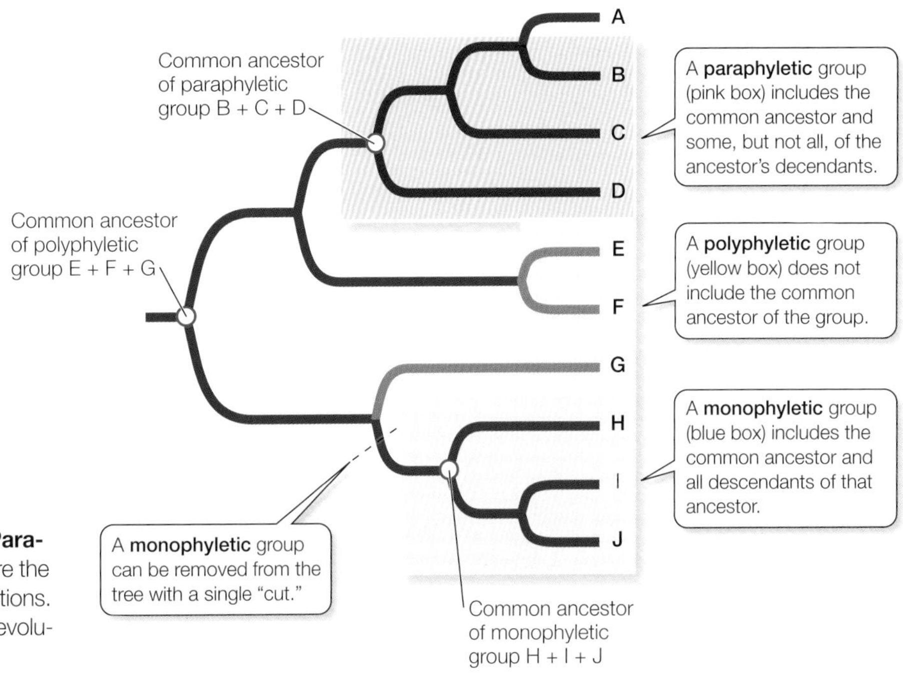

25.12 Monophyletic, Polyphyletic, and Paraphyletic Groups Monophyletic groups are the basis of biological taxa in modern classifications. Polyphyletic and paraphyletic groups do not reflect evolutionary history.

same species is named more than once (when more than one taxonomist has taken up the task); the rules specify that the valid name is the first name proposed. If the same name is inadvertently given to two different species, then a replacement name must be given to the species that was named second.

Because of the historical separation of the fields of zoology, botany (including, originally, the study of fungi), and microbiology, different sets of taxonomic rules were developed for each of these groups. Yet another set of rules for classifying viruses emerged later. This has resulted in many duplicated names between groups that are governed by different sets of rules: *Drosophila*, for instance, is both a genus of fruit flies and a genus of fungi, and there are species within both groups that have identical names. Until recently these duplicated names caused little confusion, since traditionally biologists who studied fruit flies were unlikely to read the literature on fungi (and vice versa). Today, however, given the use of large, universal biological databases (such as GenBank, which includes DNA sequences from across all life), it is increasingly important that each taxon have a unique name. Taxonomists are now working to develop common sets of rules that can be applied across all living organisms.

25.4 RECAP

Biologists organize and classify life by identifying and naming monophyletic groups. Several sets of rules govern the use of scientific names so that each species and higher taxon can be identified and named unambiguously.

- Can you explain the difference between monophyletic, paraphyletic, and polyphyletic groups? See p. 555 and Figure 25.12
- Do you understand why biologists prefer monophyletic groups for use in formal classifications? See p. 555

Now that we have seen how evolution occurs and how it can be studied, we are ready to consider the products of evolutionary history. The vastness and diversity of the living world are what first excite many future biologists and convince them to study biology. In Part Six we will examine the great diversity of life and describe some of the reasons that biologists turn their attention to particular groups of organisms.

CHAPTER SUMMARY

25.1 What is phylogeny?

A **phylogeny** is a description of the history of descent of a group of organisms from their common ancestor. Groups of evolutionarily related species are represented as related branches in a **phylogenetic tree**. Review Figure 25.1

A group of species that consists of all the evolutionary descendants of a common ancestor is called a **clade**. Named clades and species are called **taxa**.

Homologies are similar traits that have been inherited from a common ancestor. Review Figure 25.2

A **derived** trait is one that differs from its form in the common ancestor of a lineage. Derived traits that support the common relationship among a group of species are called **synapomorphies**.

Similar traits may occur among species that do not result from common ancestry. **Convergent evolution** and **evolutionary reversals** can give rise to such traits, which are called **homoplasies**.

25.2 How are phylogenetic trees constructed?

See Web/CD Activity 25.1

Phylogenies can be inferred from synapomorphies using the principle of **parsimony**. Review Figure 25.3

Sources of phylogenetic information include morphology, patterns of development, the fossil record, behavioral traits, and molecular traits such as DNA and protein sequences.

Phylogenies can also be inferred by finding **maximum likelihood** solutions, using explicit evolutionary models of character change.

Biologists can use phylogenetic trees to reconstruct ancestral states. See Web/CD Tutorial 25.1

Phylogenetic trees may include estimates of times of divergence of lineages, as determined by a **molecular clock** analysis.

25.3 How do biologists use phylogenetic trees?

Phylogenetic trees are used to reconstruct the past and understand the origin of traits. Review Figure 25.7

Phylogenetic trees are used to make appropriate evolutionary comparisons among living organisms.

Phylogenetic trees can sometimes be used to predict future evolution.

25.4 How does phylogeny relate to classification?

Taxonomists organize biological diversity on the basis of evolutionary history.

Taxa in modern classifications are expected to be **monophyletic** groups. **Paraphyletic** and **polyphyletic** groups are not considered appropriate taxonomic units. Review Figure 25.12, Web/CD Activity 25.2

Several sets of rules govern the use of scientific names, with the goal of providing unique and universal names for biological taxa.

SELF-QUIZ

1. A clade is
   a. a type of phylogenetic tree.
   b. a group of evolutionarily related species that share a common ancestor.
   c. a tool for constructing phylogenetic trees.
   d. an extinct species.
   e. an ancestral species.
2. Phylogenetic trees may be reconstructed for
   a. genes.
   b. species.
   c. major evolutionary groups.
   d. viruses.
   e. All of the above
3. A shared derived trait, used as the basis for inferring a monophyletic group, is called
   a. a synapomorphy.
   b. a homoplasy.
   c. a parallel trait.
   d. a convergent trait.
   e. a phylogeny.
4. The parsimony principle can be used to infer phylogenies because
   a. evolution is nearly always parsimonious.
   b. it is logical to adopt the simplest hypothesis capable of explaining the known facts.
   c. once a trait changes, it never reverses condition.
   d. all species have an equal probability of evolving.
   e. closely related species are always very similar to one another.
5. Convergent evolution and evolutionary reversal are two sources of
   a. homology.
   b. parsimony.
   c. synapomorphy.
   d. monophyly.
   e. homoplasy.
6. Which of the following are commonly used to infer phylogenetic relationships among plants but not among animals?
   a. Nuclear genes
   b. Chloroplast genes
   c. Mitochondrial genes
   d. Ribosomal RNA genes
   e. Protein-coding genes
7. Which of the following is *not* true of maximum likelihood or parsimony methods for inferring phylogeny?
   a. The maximum likelihood method requires an explicit model of evolutionary character change.
   b. The parsimony method is computationally easier than the maximum likelihood method.
   c. The maximum likelihood method is easier to treat in a statistical framework.
   d. The maximum likelihood method is most often used with molecular data.
   e. Parsimony is usually used to infer time on a phylogenetic tree.
8. Taxonomists strive to include taxa in biological classifications that are
   a. monophyletic.
   b. paraphyletic.
   c polyphyletic.
   d. homoplastic.
   e monomorphic.
9. Which of the following groups have separate sets of rules for nomenclature?
   a. Animals
   b. Plants and fungi
   c. Bacteria
   d. Viruses
   e. All of the above
10. If two scientific names are proposed for the same species, how do taxonomists decide which name should be used?
    a. The name that provides the most accurate description of the organism is used.
    b. The name that was proposed most recently is used.
    c. The name that was used in the most recent taxonomic revision is used.
    d. The first name to be proposed is used, unless that name was previously used for another species.
    e. Taxonomists use whichever name they prefer.

FOR DISCUSSION

1. Why are taxonomists concerned with identifying species that share a single common ancestor?
2. How are fossils used to identify ancestral and derived traits of organisms?
3. The parsimony principle is often used to reconstruct phylogenetic trees. Given that evolutionary processes are not always parsimonious, why is it used as a guiding principle?
4. A student of the evolution of frogs has proposed a strikingly new classification of frogs based on an analysis of a few mitochondrial genes from about 10 percent of frog species. Should frog taxonomists immediately accept the new classification? Why or why not?
5. Linnaeus developed his system of classification before Darwin developed his theory of evolution by natural selection, and most classifications of organisms initially were proposed by nonevolutionists. Yet many parts of these classifications are still used today, with minor modifications, by most evolutionary taxonomists. Why?
6. Classification systems summarize much information about organisms and enable us to remember the traits of many organisms. From our general knowledge, how many traits can you associate with the following names: conifer, fern, bird, mammal?

FOR INVESTIGATION

West Nile virus causes mortality in many species of birds, and can cause fatal encephalitis (inflammation of the brain) in humans and horses. It was first isolated in Africa (where it is thought to be endemic) in the 1930s, and by the 1990s it had been found throughout much of Eurasia. West Nile virus was not found in North America until 1999, but since then, it has quickly spread across most of the United States. The genome of West Nile virus evolves quickly. How could you use phylogenetic analysis to investigate the origin of the West Nile virus that was introduced into North America in 1999?

PART SIX
The Evolution of Diversity

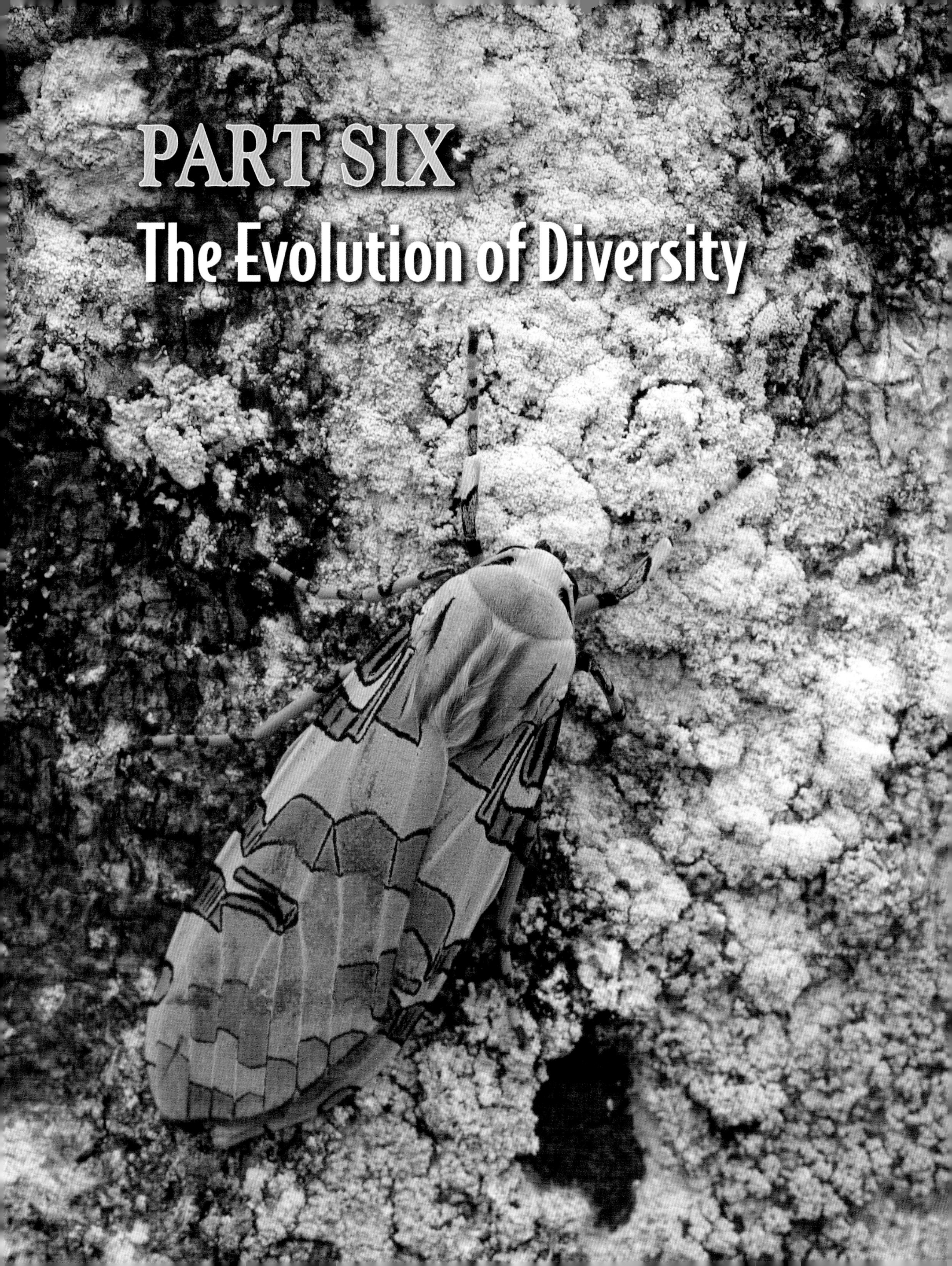

CHAPTER 26 Bacteria and Archaea: The Prokaryotic Domains

Life on the Red Planet?

The ancient Phoenicians called it the "river of fire." Today, Spanish astrobiologist Ricardo Amils Pibernat calls Spain's Río Tinto ("painted river") a possible model for the scene of the origin of life that may have existed on Mars. The Río Tinto wends its way through a huge deposit of iron pyrite—"fool's gold." The river's deep color comes about because prokaryotes in the river and in the acidic soil from which it arises convert iron pyrite into sulfuric acid and dissolved iron.

The Río Tinto has a pH of 2 and exceptionally high concentrations of heavy metals, especially iron. Oxygen concentrations in the river and in its source soil are extremely low. Amils believes this soil resembles the kind of environment in which life could have arisen on Mars. Whatever the truth of that speculation, the Río Tinto represents one of the most unusual habitats for life on Earth.

Is there life—presumably microscopic—on Mars today? Most scientists are skeptical about the possibility. However, at a meeting in 2005, Italian space scientist Vittorio Formisano reported some suggestive findings. In particular, there appears to be formaldehyde, a breakdown product of methane, in the Martian atmosphere.

In order to account for the observed concentration of formaldehyde in the Martian atmosphere, some 2.5 million tons of methane would have to be produced every year. Formisano argues that there are only three explanations for such a high rate of methane production: solar radiation-induced chemical reactions at the planetary surface; chemical reactions within the planet; or biochemical reactions in living organisms. There is no geological or chemical evidence for the first two explanations. Could living microorganisms similar to the many known methane-producing prokaryotes on Earth be responsible?

Just how likely is it that populations of simple organisms could be living in the Martian crust? We know there are enormous numbers of prokaryotes living deep in Earth's subsurface in rocks, mines, oil reservoirs, ice sheets, and sediments as much as 90 meters below the ocean floor. Some of these populations have been living and metabolizing underground for millions of years.

Prokaryotes are relatively simple in structure, but don't make the mistake of underestimating their capabilities. The prokaryotes are by far the most numer-

Earth or Ancient Mars? Spain's Río Tinto owes its rusty red color—and its extreme acidity—to the action of prokaryotes on iron pyrite-rich soil.

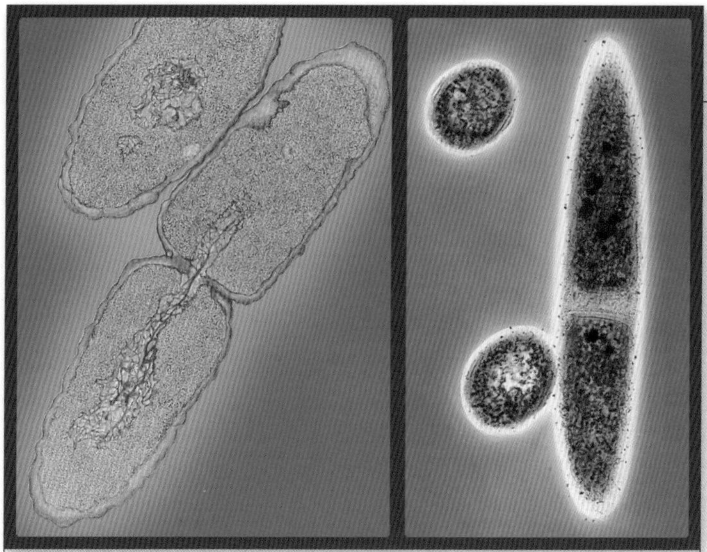

Very Different Prokaryotes *Salmonella typhimurium* (left) is a member of the domain Bacteria; *Methanospirillum hungatii* (right) is classified in the Archaea, and is probably more closely related to the eukaryotes than to the bacteria. The cells in both images are dividing, but are not shown to the same scale; the archaeal cells are actually about one-tenth the size of the *Salmonella* cells.

CHAPTER OUTLINE

26.1 **How** Did the Living World Begin to Diversify?

26.2 **Where** Are Prokaryotes Found?

26.3 **What** Are Some Keys to the Success of Prokaryotes?

26.4 **How** Can We Determine Prokaryote Phylogeny?

26.5 **What** Are the Major Known Groups of Prokaryotes?

26.6 **How** Do Prokaryotes Affect Their Environments?

ous organisms on Earth, where they are found in more different habitats than are the eukaryotes. There are more prokaryotes living on and inside your body than you have human cells. The prokaryotes are masters of metabolic ingenuity, having developed more ways to obtain energy from the environment than the eukaryotes have. And they have been around longer than other organisms.

Late in the twentieth century, it became apparent to many microbiologists that certain prokaryotes differ so fundamentally from others in their metabolic processes that they should be considered members of a distinct evolutionary lineage. Furthermore, the two prokaryotic lineages diverged very early in life's evolution. We refer to these lineages as the domains Bacteria and Archaea.

IN THIS CHAPTER we will discuss the distribution of prokaryotes and examine their remarkable metabolic diversity. We will describe the impediments to the resolution of evolutionary relationships among the prokaryotes and survey the surprising diversity of organisms within each domain. Finally, we will discuss the effects of prokaryotes on their environments.

26.1 How Did the Living World Begin to Diversify?

What does it mean to be *different*? You and the person nearest you look very different—certainly you appear more different than the two cells shown above. But the two of you are members of the same species, while these two tiny organisms that look so much alike actually are classified in entirely separate domains. Still, you—in the domain Eukarya—and those two prokaryotes in the domains Bacteria and Archaea have a lot in common. All three of you:

- conduct glycolysis
- replicate DNA semiconservatively
- have DNA that encodes polypeptides
- produce those polypeptides by transcription and translation using the same genetic code
- have plasma membranes and ribosomes in abundance

Despite these commonalities among life's domains, there are also major differences. Let's first distinguish between the Eukarya and the two prokaryotic domains. Note that "domain" is a subjective term used for the largest groups of life. There is no objective definition of a domain, any more than of a kingdom or a family.

The three domains differ in significant ways

Prokaryotic cells differ from eukaryotic cells in three important ways:

- Prokaryotic cells lack a cytoskeleton, and in the absence of organized cytoskeletal proteins, they lack mitosis. Prokaryotic cells divide by their own method, *binary fission*, after replicating their DNA (see Figure 9.2).
- The organization and replication of the genetic material differs. The DNA of the prokaryotic cell is not organized within a membrane-enclosed nucleus. DNA molecules in prokaryotes (both bacteria and archaea) are usually circu-

TABLE 26.1

The Three Domains of Life on Earth

CHARACTERISTIC	DOMAIN BACTERIA	DOMAIN ARCHAEA	DOMAIN EUKARYA
Membrane-enclosed nucleus	Absent	Absent	*Present*
Membrane-enclosed organelles	Absent	Absent	*Present*
Peptidoglycan in cell wall	*Present*	Absent	Absent
Membrane lipids	Ester-linked	*Ether-linked*	Ester-linked
	Unbranched	*Branched*	Unbranched
Ribosomes[a]	70S	70S	*80S*
Initiator tRNA	*Formylmethionine*	Methionine	Methionine
Operons	Yes	Yes	*No*
Plasmids	Yes	Yes	*Rare*
RNA polymerases	One	One[b]	*Three*
Ribosomes sensitive to chloramphenicol and streptomycin	*Yes*	No	No
Ribosomes sensitive to diphtheria toxin	*No*	Yes	Yes
Some are methanogens	No	*Yes*	No
Some fix nitrogen	Yes	Yes	*No*
Some conduct chlorophyll-based photosynthesis	Yes	*No*	Yes

[a]70S ribosomes are smaller than 80S ribosomes.

[b]Archaeal RNA polymerase is similar to eukaryotic polymerases.

lar; in the best-studied prokaryotes, there is a single chromosome, but there are often plasmids as well (see Section 13.3).

- Prokaryotes have none of the membrane-enclosed cytoplasmic organelles that modern eukaryotes have—mitochondria, Golgi apparatus, and others. However, the cytoplasm of a prokaryotic cell may contain a variety of infoldings of the plasma membrane and photosynthetic membrane systems not found in eukaryotes.

A glance at **Table 26.1** will show you that there are also major differences (most of which cannot be seen even under the microscope) between the two prokaryotic domains. In some ways the archaea are more like eukayrotes; in other ways they are more like bacteria. The architectures of prokaryotic and eukaryotic cells were compared in Chapter 4. The basic unit of the archaea and bacteria is the prokaryotic cell, which contains a full complement of genetic and protein-synthesizing systems, including DNA, RNA, and all the enzymes needed to transcribe and translate the genetic information into proteins. The prokaryotic cell also contains at least one system for generating the ATP it needs.

Genetic studies clearly indicate that all three domains had a single common ancestor, and that the present-day Archaea share a more recent common ancestor with the Eukarya than they do with the Bacteria (**Figure 26.1**). To treat all the prokaryotes as a single domain within a two-domain classification of organisms would result in a domain that is paraphyletic. That is, a single domain "Prokaryotes" would not include all the descendants of their common ancestor. (See Section 25.4 for a discussion of paraphyletic groups.)

The common ancestor of all three domains had DNA as its genetic material; its machinery for transcription and translation produced RNAs and proteins, respectively. It probably had a circular chromosome.

The Archaea, Bacteria, and Eukarya are all products of billions of years of mutation, natural selection, and genetic drift, and they are all well adapted to present-day environments. None is "primitive." The common ancestor of the Archaea and the Eukarya probably lived more than 2 billion years ago, and the common ancestor of the Archaea, the Eukarya, and the Bacteria probably lived more than 3 billion years ago. The earliest prokaryotic fossils date back at least 3.5 billion years, and those ancient fossils indicate that there was considerable diversity among the prokaryotes even during the earliest days of life.

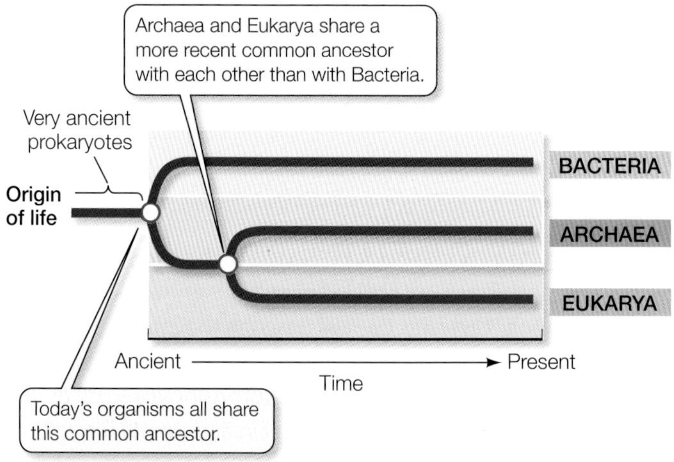

26.1 The Three Domains of the Living World Most biologists believe that all three domains share a common prokaryotic ancestor.

26.1 RECAP

Prokaryotic bacteria and archaea are as different from each other as they are from the eukaryotes. Eukaryotes share a more recent common ancestor with archaea than with bacteria.

- Do you understand the principal differences between the prokaryotes and the eukaryotes? See pp. 561–562 and Table 26.1
- Can you explain why we do not group the Bacteria and the Archaea together in a single domain? See p. 562 and Figure 26.1

The prokaryotes were alone on Earth for a very long time, adapting to new environments and to changes in existing environments. They have survived to this day—and in massive numbers. Prokaryotes are around us everywhere.

26.2 Where Are Prokaryotes Found?

Although we cannot see them with the naked eye, the prokaryotes are the most successful of all creatures on Earth, if success is measured by numbers of individuals. Bacteria and archaea in the oceans number more than 3×10^{28}. This stunning number is perhaps 100 million times as great as the number of stars in the visible universe.

> The bacteria living in a single human intestinal tract outnumber all the humans who have ever lived.

Prokaryotes are found in every type of environment on the planet, from the coldest to the hottest, from the most acidic to the most alkaline and the saltiest. Some live where oxygen is abundant, others where there is no oxygen at all. They have established themselves at the bottom of the seas, in rocks more than 2 kilometers deep in solid crust, in the clouds, and on and inside other organisms, large and small. Their effects on our environment are diverse and profound.

Three shapes are particularly common among the bacteria: spheres, rods, and curved or helical forms (**Figure 26.2**). Among their other shapes are long filaments and branched filaments. A spherical bacterium is called a *coccus* (plural *cocci*). Cocci may live singly or may associate in two- or three-dimensional arrays as chains, plates, blocks, or clusters of cells. A rod-shaped bacterium is called a *bacillus* (plural *bacilli*). Helical forms (like a corkscrew) are the third main bacterial shape. Bacilli and helical forms may be single, they may form chains, or they may gather in regular clusters. Less is known about the shapes of archaea because many of these organisms have never been seen; they are known only from samples of DNA from the environment, as we will describe in Section 26.4.

Prokaryotes are almost all unicellular, although some multicellular ones are known. Associations such as chains or clusters do not signify multicellularity because each individual cell is fully viable and independent. These associations arise as cells adhere to one another after reproducing by binary fission. Associations in the form of chains are called **filaments**. Some filaments become enclosed within delicate tubular sheaths.

Prokaryotes generally form complex communities

Prokaryotic cells and their associations do not usually live in isolation. Rather, they live in communities of many different species of organisms, often including microscopic eukaryotes. (Microscopic organisms are sometimes collectively referred to as *microbes*.) Some microbial communities form layers in sediments, and others form clumps a meter or more in diameter. While some microbial communities are harmful to humans, others provide us with important services. They help us digest our food, they break down municipal waste, and they recycle organic matter in the environment.

Many microbial communities tend to form dense **biofilms**. Upon contacting a solid surface, the cells lay down a gel-like polysaccharide matrix that then traps other cells, forming a biofilm (**Figure**

26.2 Bacterial Cell Shapes (A) These acid-producing cocci grow in the mammalian gut. (B) The bacillus *E. coli*, a resident of the human gut, is one of the most thoroughly studied organisms on Earth. (C) This helical bacterium belongs to a genus of human pathogens that cause leptospirosis, an infection spread by contaminated water. This particular strain was isolated in 1915 from the blood of a soldier serving in World War I.

(A) *Enterococcus* sp. (B) *Escherichia coli* (C) *Leptospira interrogans*

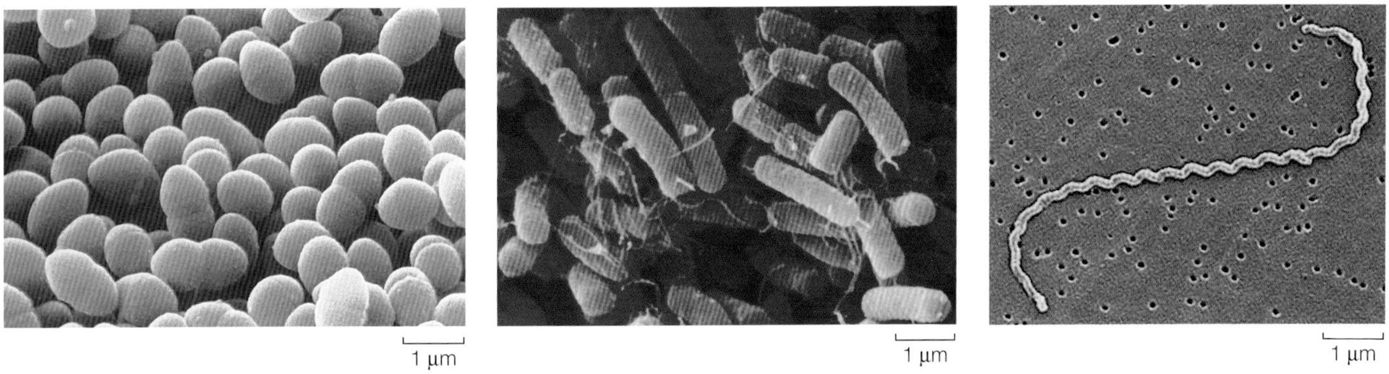

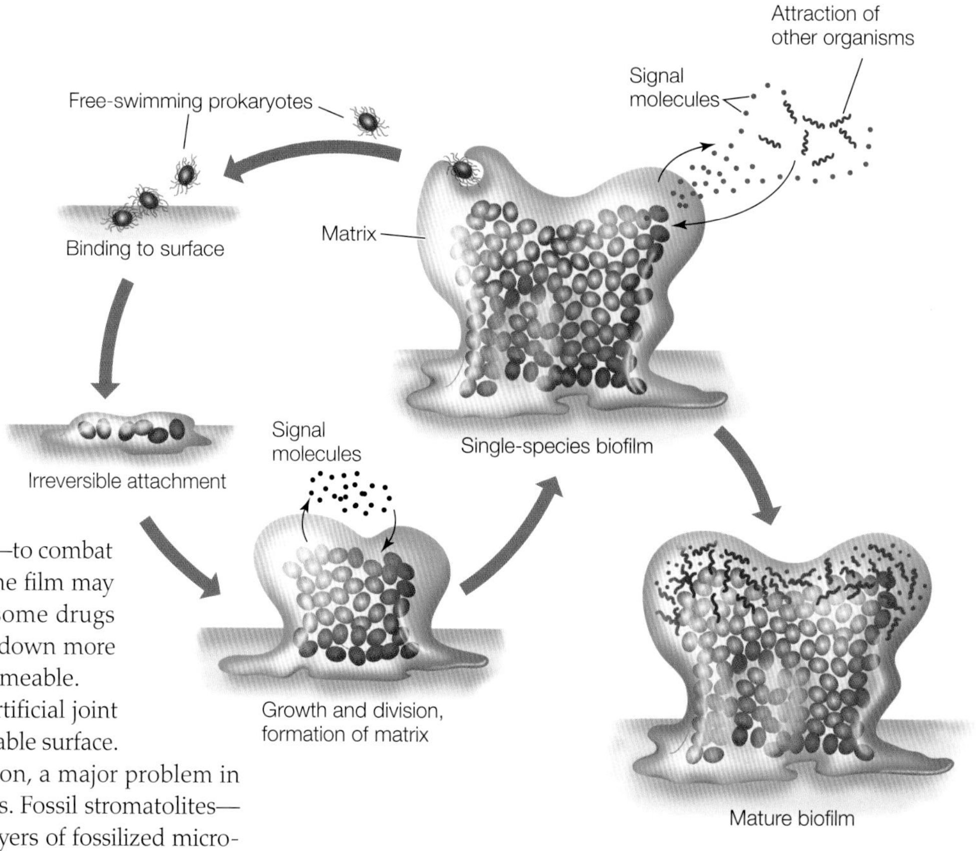

26.3 Forming a Biofilm Free-swimming microorganisms such as bacteria and archaea readily attach themselves to surfaces and form films stabilized and protected by a surrounding matrix. Once the population size is great enough, the developing biofilm can send chemical signals that attract other microorganisms.

26.3). Once a biofilm forms, it is difficult to kill the cells. Pathogenic (disease-causing) bacteria are hard for the immune system—and modern medicine—to combat once they form a biofilm. For example, the film may be impermeable to antibiotics. Worse, some drugs stimulate the bacteria in a biofilm to lay down more matrix, making the film even more impermeable.

Biofilms form on contact lenses, on artificial joint replacements, and on just about any available surface. They foul metal pipes and cause corrosion, a major problem in steam-driven electricity generation plants. Fossil stromatolites—large, rocky structures with alternating layers of fossilized microbial biofilm and calcium carbonate—are the oldest remnants of early life on Earth (see Figure 21.4). Stromatolites still form today in some parts of the world.

Biofilms are the object of much current research. For example, some biologists are studying the chemical signals used by bacteria in biofilms to communicate with one another. By blocking the signals that lead to the production of the matrix polysaccharides, they may be able to prevent biofilms from forming.

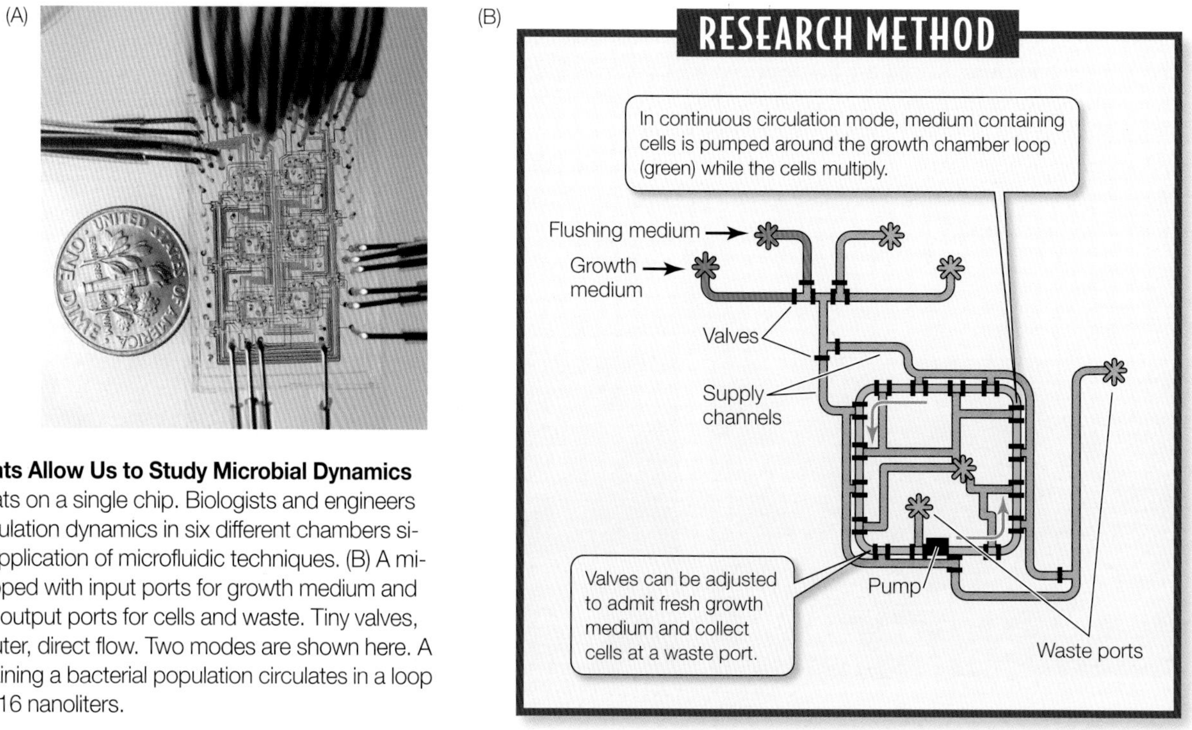

26.4 Microchemostats Allow Us to Study Microbial Dynamics (A) Six microchemostats on a single chip. Biologists and engineers monitor microbial population dynamics in six different chambers simultaneously by the application of microfluidic techniques. (B) A microchemostat is equipped with input ports for growth medium and flushing medium, and output ports for cells and waste. Tiny valves, controlled by a computer, direct flow. Two modes are shown here. A growth medium containing a bacterial population circulates in a loop with a total volume of 16 nanoliters.

That stain on your teeth is a biofilm called dental plaque, a coating of bacteria and hard matrix on and between the teeth. A softer bacterial biofilm forms on your tongue if you don't brush it regularly. Both biofilms can cause bad breath.

A team of bioengineers and chemical engineers recently devised a sophisticated technique that enables them to monitor biofilm development in extremely small populations of bacteria, cell by cell. They developed a tiny chip housing six separate growth chambers, or "microchemostats" (**Figure 26.4A**). The techniques of *microfluidics* use microscopic tubes and computer-controlled valves to direct fluid flow through complex "plumbing circuits" in the growth chambers (**Figure 26.4B**).

26.2 RECAP

Prokaryotes have established themselves everywhere on Earth. They often live in communities called biofilms.

- Can you describe the three most common forms of bacterial cells? See p. 563 and Figure 26.2
- Do you understand how biofilms form and why they are of special interest to researchers? See p. 563 and Figure 26.3

The prokaryotes are not only the most numerous living organisms, but also the most widely dispersed. To what do they owe their spectacular success?

26.3 What Are Some Keys to the Success of Prokaryotes?

Features of the prokaryotes that have contributed to their success include unique cell surfaces and modes of locomotion, communication, nutrition, and reproduction. These characteristics vary from group to group and even within groups of prokaryotes.

Prokaryotes have distinctive cell walls

Many prokaryotes have a thick and relatively stiff cell wall. This cell wall is quite different from those of land plants and algae, which contain cellulose and other polysaccharides, and from those of fungi, which contain chitin. Almost all bacteria have cell walls containing **peptidoglycan** (a polymer of amino sugars). Archaeal cell walls are of differing types, but most contain significant amounts of protein. One group of archaea has *pseudopeptidoglycan* in its cell wall; as you have probably already guessed from the prefix *pseudo*, pseudopeptidoglycan is similar to, but distinct from, the peptidoglycan of bacteria. The monomers making up pseudopeptidoglycan differ from and are differently linked than those of peptidoglycan. Peptidoglycan is a substance unique to bacteria; its absence from the walls of archaea is a key difference between the two prokaryotic domains.

To appreciate the complexity of some bacterial cell walls, consider the reactions of bacteria to a simple staining process. The **Gram stain** separates most types of bacteria into two distinct groups, Gram-positive and Gram-negative. A smear of cells on a microscope slide is soaked in a violet dye and treated with iodine; it is then washed with alcohol and counterstained with safranine (a red dye). **Gram-positive** bacteria retain the violet dye and appear blue to purple (**Figure 26.5A**). The alcohol washes the vi-

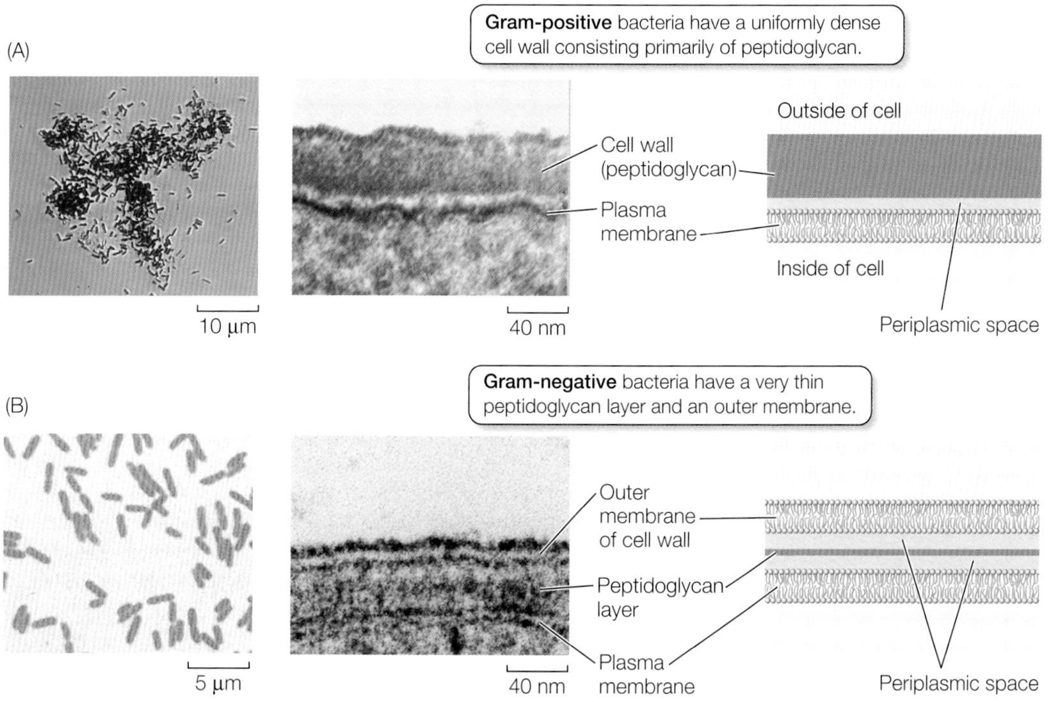

26.5 The Gram Stain and the Bacterial Cell Wall When treated with Gram stain, the cell wall components of different bacteria react in one of two ways. (A) Gram-positive bacteria have a thick peptidoglycan cell wall that retains the violet dye and appears deep blue or purple. (B) Gram-negative bacteria have a thin peptidoglycan layer that does not retain the violet dye, but picks up the counterstain and appears pink to red.

olet stain out of **Gram-negative** cells; these cells then pick up the safranine counterstain and appear pink to red (**Figure 26.5B**).

For many bacteria, the Gram-staining results correlate roughly with the structure of the cell wall. A Gram-negative cell wall usually has a thin peptidoglycan layer, and outside the peptidoglycan layer the cell is surrounded by a second, outer membrane quite distinct in chemical makeup from the plasma membrane (see Figure 26.5B). Between the inner (plasma) and outer membranes of Gram-negative bacteria is a *periplasmic space*. This space contains proteins that are important in digesting some materials, transporting others, and detecting chemical gradients in the environment.

A Gram-positive cell wall usually has about five times as much peptidoglycan as a Gram-negative wall. This thick peptidoglycan layer is a meshwork that may serve some of the same purposes as the periplasmic space of the Gram-negative cell wall.

The consequences of the different features of prokaryotic cell walls are numerous and relate to the disease-causing characteristics of some bacteria. Indeed, the cell wall is a favorite target in medical combat against pathogenic bacteria because it has no counterpart in eukaryotic cells. Antibiotics such as penicillin and ampicillin, as well as other agents that specifically interfere with the synthesis of peptidoglycan-containing cell walls, tend to have little, if any, effect on the cells of humans and other eukaryotes.

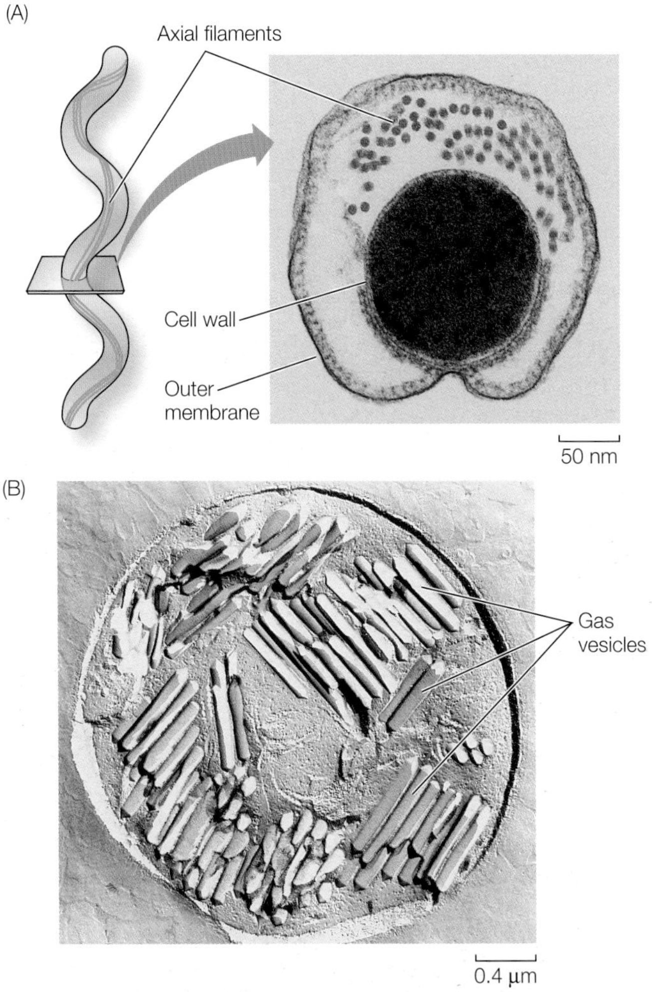

26.6 Structures Associated with Prokaryote Motility (A) A spirochete from the gut of a termite, seen in cross section, shows the axial filaments used to produce a corkscrew-like motion. (B) Gas vesicles in a cyanobacterium, visualized by the freeze-fracture technique.

Prokaryotes have distinctive modes of locomotion

Although many prokaryotes cannot move, others are *motile*. These organisms move by one of several means. Some helical bacteria, called *spirochetes*, use a corkscrew-like motion made possible by modified flagella, called *axial filaments*, running along the axis of the cell beneath the outer membrane (**Figure 26.6A**). Many cyanobacteria and a few other groups of bacteria use various poorly understood gliding mechanisms, including rolling. Various aquatic prokaryotes, including some cyanobacteria, can move slowly up and down in the water by adjusting the amount of gas in *gas vesicles* (**Figure 26.6B**). By far the most common type of locomotion in prokaryotes, however, is that driven by flagella.

Prokaryotic **flagella** are slender filaments that extend singly or in tufts from one or both ends of the cell or are randomly distributed all around it (**Figure 26.7**). A prokaryotic flagellum consists of a single fibril made of the protein *flagellin*, projecting from the cell surface, plus a hook and basal body responsible for motion (see Figure 4.5). In contrast, the flagellum of eukaryotes is enclosed by the plasma membrane and usually contains a circle of nine pairs of microtubules surrounding two central microtubules, all containing the protein tubulin, along with many other associated proteins. The prokaryotic flagellum rotates about its base, rather than beating as a eukaryotic flagellum or cilium does.

Prokaryotes reproduce asexually, but genetic recombination can occur

Prokaryotes reproduce by **binary fission**, an asexual process. Recall, however, that there are also processes—transformation, conjugation, and transduction—that allow the exchange of genetic information between some prokaryotes quite apart from reproduction (see Chapter 13).

Some prokaryotes multiply very rapidly. One of the fastest is the bacterium *Escherichia coli*, which under optimal conditions has a generation time of about 20 minutes. The shortest known prokaryote generation times are about 10 minutes. Generation times of 1 to 3 hours are common for others; some extend to days. Bacteria living deep in Earth's crust may suspend their growth for more than a century without dividing and then multiply for a few days before suspending growth again.

Some prokaryotes communicate

Prokaryotes can communicate with one another and with other organisms. One communication channel that they employ is chemical. Another is physical, with light as the medium.

Some bacteria release chemical signals—substances that are sensed by other bacteria of the same species. They can announce their availability for conjugation, for example, by means of such signals. They can also monitor the size of their population. As the number of bacteria in a particular region increases, the concentration of a chemical signal builds up. When the bacteria sense that

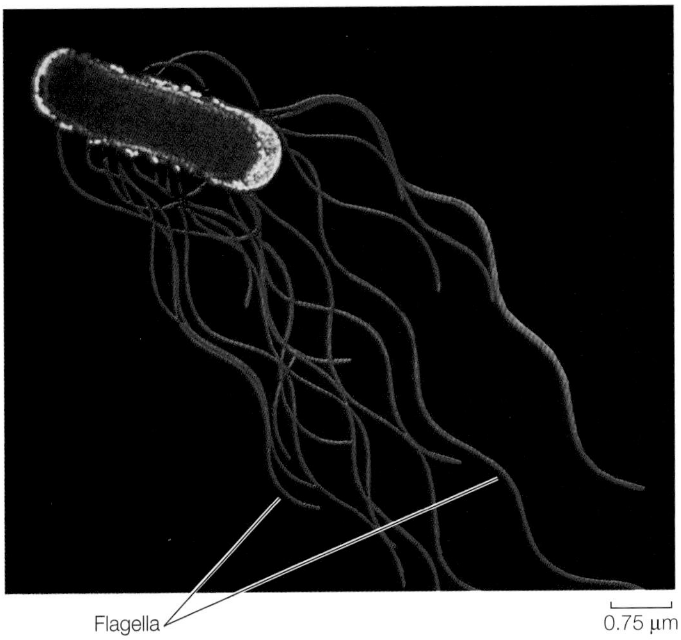

26.7 Some Prokaryotes Use Flagella for Locomotion Multiple flagella propel this *Salmonella* bacillus.

their population has become sufficiently large, they can commence activities that smaller numbers could not manage, such as forming a biofilm (see Figure 26.3). This "counting" technique is called **quorum sensing**.

Like many other organisms, such as fireflies, some bacteria can emit light by a process called **bioluminescence**. A complex, enzyme-catalyzed reaction requiring ATP causes the emission of light, but not heat. Often such bacteria luminesce only when a quorum has been sensed.

How could bioluminescence be useful to a prokaryote? One fairly well understood case is that of some bacteria of the genus *Vibrio*. These bacteria can live freely, but thrive when inside the guts of fish. Inside the fish, they may attach to food particles and be expelled as waste along with the particulate matter. Reproducing on the particles, their population increases until the glowing particle attracts another fish, which ingests it along with the bacteria—giving them a new home and food source for a while. In this case, *Vibrio* is communicating with another species and enhancing its own nutritional status. *Vibrio* in the Indian Ocean, off the eastern coast of Africa, provide the only instance in which bioluminescence is visible from space. These bacteria become concentrated over an area of several thousand square kilometers (**Figure 26.8**).

Some fish have "headlights"—colonies of bioluminescent bacteria that glow together. Other fish are attracted to the light and then eaten by the host fish. This arrangement, profitable to the host, also probably provides the bacteria with nutrients released as the fish eats.

Some soil-dwelling bacteria bioluminesce, producing eerily glowing patches of ground at night. What is the adaptive value of this particular case of bacterial bioluminescence? We're not certain, but one hypothesis links the phenomenon to the time in evolutionary history when most organisms went extinct because their metabolisms could not deal with increasing levels of oxygen in the atmosphere. The distant ancestor of bioluminescent soil bacteria may have avoided oxidative reactions (and subsequent death) by dissipating excess energy, emitting it as light.

Prokaryotes have amazingly diverse metabolic pathways

The members of the two prokaryotic domains outdo all other groups in metabolic diversity. Eukaryotes, while much more diverse in size and shape, draw on fewer metabolic mechanisms for their energy needs. In fact, much of the energy metabolism of

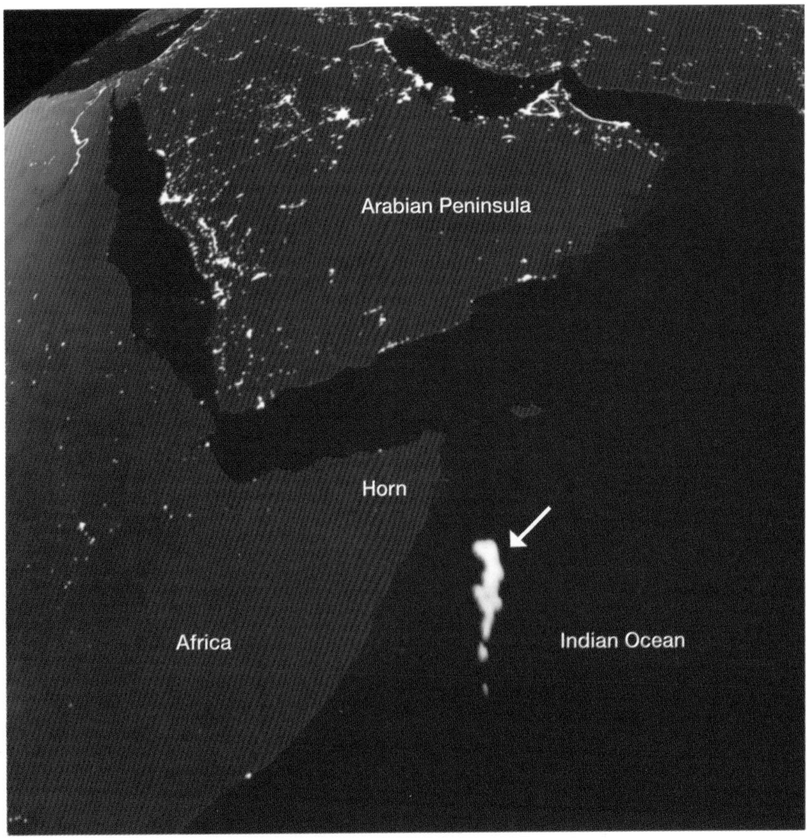

26.8 Bioluminescent Bacteria Seen from Space In this satellite photo, legions of *Vibrio harveyi* form a glowing patch (arrow) some thousands of square kilometers in area in the Indian Ocean, off the Horn of Africa. Compare their blue glow with the white light of cities in eastern Africa and the Middle East.

eukaryotes is carried out in organelles—mitochondria and chloroplasts—that are descended from bacteria, as we saw in Section 4.5.

The long evolutionary history of the bacteria and archaea, during which they have had time to explore a wide variety of habitats, has led to the extraordinary diversity of their metabolic "lifestyles"—their use or nonuse of oxygen, their energy sources, their sources of carbon atoms, and the materials they release as waste products.

ANAEROBIC VERSUS AEROBIC METABOLISM Some prokaryotes can live only by anaerobic metabolism because molecular oxygen is poisonous to them. These oxygen-sensitive organisms are called **obligate anaerobes**.

Other prokaryotes can shift their metabolism between anaerobic and aerobic modes (see Chapter 7) and thus are called **facultative anaerobes**. Many facultative anaerobes alternate between anaerobic metabolism (such as fermentation) and cellular respiration as conditions dictate. **Aerotolerant anaerobes** cannot conduct cellular respiration, but are not damaged by oxygen when it is present. By definition, an anaerobe does not use oxygen as an electron acceptor for its respiration.

At the other extreme from the obligate anaerobes, some prokaryotes are **obligate aerobes**, unable to survive for extended periods in the *absence* of oxygen. They require oxygen for cellular respiration.

NUTRITIONAL CATEGORIES Biologists recognize four broad nutritional categories of organisms: photoautotrophs, photoheterotrophs, chemolithotrophs, and chemoheterotrophs. Prokaryotes are represented in all four groups (**Table 26.2**).

Photoautotrophs perform photosynthesis. They use light as their source of energy and carbon dioxide as their source of carbon. Like the photosynthetic eukaryotes, the cyanobacteria, a group of photoautotrophic bacteria, use chlorophyll *a* as their key photosynthetic pigment and produce oxygen as a by-product of noncyclic electron transport (see Section 8.1).

In contrast, the other photosynthetic bacteria use *bacteriochlorophyll* as their key photosynthetic pigment, and they do not release oxygen gas. Some of these photosynthesizers produce particles of pure sulfur instead because hydrogen sulfide (H_2S), rather than H_2O, is their electron donor for photophosphorylation. Bacteriochlorophyll absorbs light of longer wavelengths than the chlorophyll used by all other photosynthesizing organisms. As a result, bacteria using this pigment can grow in water beneath fairly dense layers of algae, using light of wavelengths that are not absorbed by the algae (**Figure 26.9**).

Photoheterotrophs use light as their source of energy, but must obtain their carbon atoms from organic compounds made by other organisms. They use compounds such as carbohydrates, fatty acids, and alcohols as their organic "food." For example, compounds released from plant roots (as in rice paddies) or from decomposing cyanobacteria in hot springs are taken up by photoheterotrophs and metabolized to form building blocks for other compounds; sunlight provides the necessary ATP through photophosphorylation (see Section 8.2). The purple nonsulfur bacteria, among others, are photoheterotrophs.

Chemolithotrophs (also called chemoautotrophs) obtain their energy by oxidizing inorganic substances, and they use some of that energy to fix carbon dioxide. Some chemolithotrophs use reactions identical to those of the typical photosynthetic cycle (see Chapter 8), but others use other pathways to fix carbon dioxide. Some bacteria oxidize ammonia or nitrite ions to form nitrate ions. Others oxidize hydrogen gas, hydrogen sulfide, sulfur, and other materials. Many archaea are chemolithotrophs.

Deep-sea hydrothermal vent ecosystems are based on chemolithotrophic prokaryotes that are incorporated into large communities of crabs, mollusks, and giant worms, all living at a depth of 2,500 meters, below any hint of light from the sun. These bacteria obtain energy by oxidizing hydrogen sulfide and other substances released in the near-boiling water that flows from volcanic vents in the ocean floor.

TABLE 26.2

How Organisms Obtain Their Energy and Carbon

NUTRITIONAL CATEGORY	ENERGY SOURCE	CARBON SOURCE
Photoautotrophs (found in all three domains)	Light	Carbon dioxide
Photoheterotrophs (some bacteria)	Light	Organic compounds
Chemolithotrophs (some bacteria, many archaea)	Inorganic substances	Carbon dioxide
Chemoheterotrophs (found in all three domains)	Organic compounds	Organic compounds

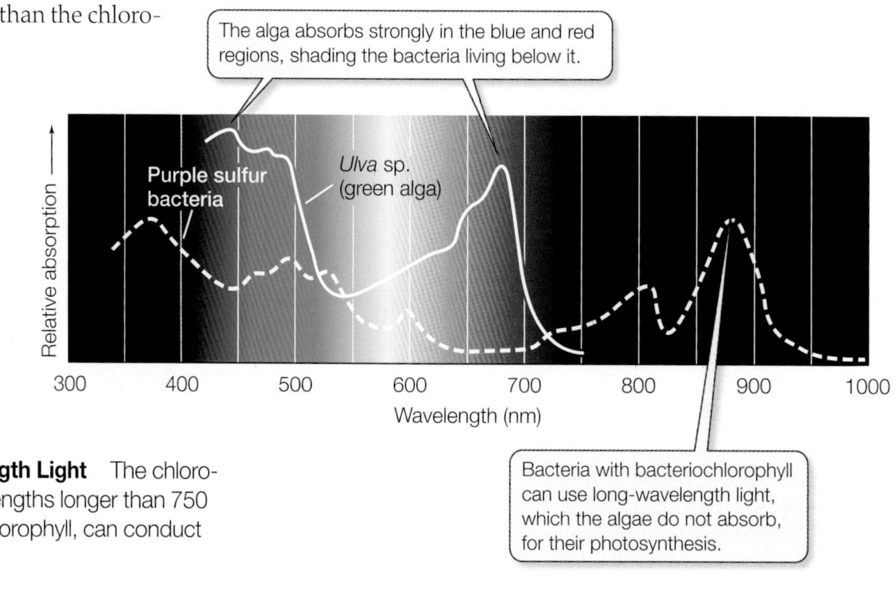

26.9 Bacteriochlorophyll Absorbs Long-Wavelength Light The chlorophyll in *Ulva*, a green alga, absorbs no light of wavelengths longer than 750 nm. Purple sulfur bacteria, which contain bacteriochlorophyll, can conduct photosynthesis using longer wavelengths.

Finally, **chemoheterotrophs** obtain both energy and carbon atoms from one or more complex organic compounds. Most known bacteria and archaea are chemoheterotrophs—as are all animals and fungi and many protists.

NITROGEN AND SULFUR METABOLISM Key metabolic reactions in many prokaryotes involve nitrogen or sulfur. For example, some bacteria carry out respiratory electron transport without using oxygen as an electron acceptor. These organisms use oxidized inorganic ions such as nitrate, nitrite, or sulfate as electron acceptors. Examples include the **denitrifiers**, bacteria that release nitrogen to the atmosphere as nitrogen gas (N_2). These normally aerobic bacteria, mostly species of the genera *Bacillus* and *Pseudomonas*, use nitrate (NO_3^-) as an electron acceptor in place of oxygen if they are kept under anaerobic conditions:

$$2\,NO_3^- + 10\,e^- + 12\,H^+ \rightarrow N_2 + 6\,H_2O$$

Nitrogen fixers convert atmospheric nitrogen gas into a chemical form usable by the nitrogen fixers themselves as well as by other organisms. They convert nitrogen gas into ammonia:

$$N_2 + 6\,H \rightarrow 2\,NH_3$$

All organisms require nitrogen for their proteins, nucleic acids, and other important compounds. Nitrogen fixation is thus vital to life as we know it; this all-important biochemical process is carried out by a wide variety of archaea and bacteria (including cyanobacteria), but by no other organisms. We'll discuss this vital process in detail in Chapter 36.

Ammonia is oxidized to nitrate in soil and in seawater by chemolithotrophic bacteria called **nitrifiers**. Bacteria of two genera, *Nitrosomonas* and *Nitrosococcus*, convert ammonia to nitrite ions (NO_2^-), and *Nitrobacter* oxidizes nitrite to nitrate (NO_3^-).

What do the nitrifiers get out of these reactions? Their metabolism is powered by the energy released by the oxidation of ammonia or nitrite. For example, by passing the electrons from nitrite through an electron transport chain, *Nitrobacter* can make ATP, and using some of this ATP, it can also make NADH. With this ATP and NADH, the bacterium can convert CO_2 and H_2O to glucose.

26.3 RECAP

Prokaryotes have distinctive cell walls and modes of locomotion, communication, reproduction, and nutrition.

- Can you describe bacterial cell wall architecture? See p. 565 and Figure 26.5
- How are the four nutritional categories of prokaryotes distinguished? See p. 568 and Table 26.2
- Can you see why nitrogen metabolism in the prokaryotes is vital to other organisms? See p. 569

We noted earlier that only very recently have scientists appreciated the huge distinctions between Bacteria and Archaea. How do researchers approach the classification of organisms they can't even see?

26.4 How Can We Determine Prokaryote Phylogeny?

As detailed in Chapter 25, there are three primary motivations for classification schemes: to identify unknown organisms, to reveal evolutionary relationships, and to provide universal names. Classifying bacteria and archaea is of particular importance to humans, because scientists and medical technologists must be able to identify bacteria quickly and accurately—when the bacteria are pathogenic, lives may depend on it. In addition, many emerging biotechnologies (see Chapter 16) depend on a thorough knowledge of prokaryote biochemistry, and understanding an organism's phylogeny can contribute to such knowledge.

Size complicates the study of prokaryote phylogeny

Until recently, taxonomists based their classification schemes for the prokaryotes on readily observable phenotypic characters such as color, motility, nutritional requirements, antibiotic sensitivity, and reaction to the Gram stain. Although such schemes have facilitated the identification of prokaryotes, they have not provided insights into how these organisms evolved—a question of great interest to microbiologists and to all students of evolution. The prokaryotes and the microbial eukaryotes (protists; Chapter 27) have presented major challenges to those who attempted phylogenetic classifications.

Nobody had *seen* an individual prokaryote until about 300 years ago. Prokaryotes are so small that they were invisible before the invention of the first simple microscope. Even under the best light microscopes, we still don't learn much about them.

When biologists learned how to grow bacteria in pure culture on nutrient media, some useful information began to flow in. A great deal was learned about the genetics, nutrition, and metabolism of prokaryotes. With little but cell shape, size, and nutritional requirements for criteria, however, microbiologists were unable to deduce classification schemes that made sense in evolutionary terms. Only recently have systematists had the right tools for tackling this task.

The nucleotide sequences of prokaryotes reveal their evolutionary relationships

Analyses of the nucleotide sequences of ribosomal RNA have provided us with the first apparently reliable measures of evolutionary distance among taxonomic groups. Ribosomal RNA (rRNA) is particularly useful for evolutionary studies of living organisms for several reasons:

- rRNA is evolutionarily ancient.
- No living organism lacks rRNA.
- rRNA plays the same role in translation in all organisms.
- rRNA has evolved slowly enough that sequence similarities between groups of organisms are easily found.

Comparisons of short stretches of rRNA from a great many organisms have revealed recognizable base sequences that are characteristic of particular taxonomic groups. More recent investigations have focused on longer stretches of DNA, even entire genes (especially rRNA genes).

These data are very helpful, but things aren't as simple as we might wish. When biologists examined more genes, contradictions began to appear and new questions arose. In some groups of prokaryotes, analyses of different nucleotide sequences suggested different phylogenetic patterns. How could such a situation have arisen?

Lateral gene transfer may complicate phylogenetic studies

It is now clear that, from early in evolution to the present day, genes have been moving among many prokaryotic species by **lateral gene transfer**. As we have seen, a gene from one species can become incorporated into the genome of another. For example, the 1,869 genes of *Thermotoga maritima*, a bacterium that can survive extremely high temperatures, have all been sequenced. In comparing the sequences of *T. maritima* genes against sequences of genes for the same proteins in other species, it was found that nearly 20 percent of this bacterium's genes have their closest match not with other bacterial species, but with archaeal species.

Mechanisms of lateral gene transfer include transfer by plasmids and viruses and uptake of DNA by transformation. Such transfers are well documented, not just between species in the prokaryotic domains, but between prokaryotes and eukaryotes.

A gene that has been transferred will be inherited by the recipient's progeny and in time will be recognized as part of the normal genome of its descendants. When phylogenies are inferred using gene trees, the presence of a laterally transferred gene can result in mistaken assumptions about relationships (**Figure 26.10A,B**). Biologists are still assessing the extent of lateral gene transfer among prokaryotes and its implications for phylogeny, especially at the early stages of evolution.

It is unclear whether lateral gene transfer has seriously complicated our attempts to resolve the tree of prokaryotic life. Recent work suggests that it has not—while it complicates studies within some individual species, it need not present problems at higher levels. It is now possible to make nucleotide sequence comparisons involving entire genomes, and many scientists feel this work will reveal a *stable core* of crucial genes that are uncomplicated by lateral gene transfer. Gene trees using this stable core should reveal more accurate phylogenetic relationships (**Figure 26.10C**). The problem remains that only a very small proportion of the prokaryotic world has been described and studied.

The great majority of prokaryote species have never been studied

Some prokaryotes have defied all attempts to grow them in pure culture, causing biologists to wonder how many species, and possibly even important clades, we might be missing. A window onto this problem was opened with the introduction of a new way to look at nucleic acid sequences. Unable to work with the whole genome of a single species, biologists instead examine sequences in individual genes collected from a random sample of the environment.

Norman Pace, of the University of Colorado, isolated individual rRNA gene sequences from extracts of environmental samples such as soil or seawater. Comparison of such sequences with previously known ones revealed an extraordinary number of new sequences, implying that they came from previously unrecognized species. Biologists have described only about 5,000 species of bacteria and archaea. The results of Pace's studies strongly suggest that there may be as many as half a million prokaryote species out there. This finding presents a great challenge, but also a great opportu-

26.10 Lateral Gene Transfer Complicates Phylogenetic Relationships (A) The phylogeny of four hypothetical species is shown as a shaded tree. (B) In a tree based on gene *x*, the true relationship is obscured. (C) Many systematists studying prokaryote phylogeny believe that eventually we will establish a "stable core" of prokaryote genes that is resistant to lateral transfer. Such a tree, however, remains in the future, awaiting the sequencing of many more prokaryotic genomes.

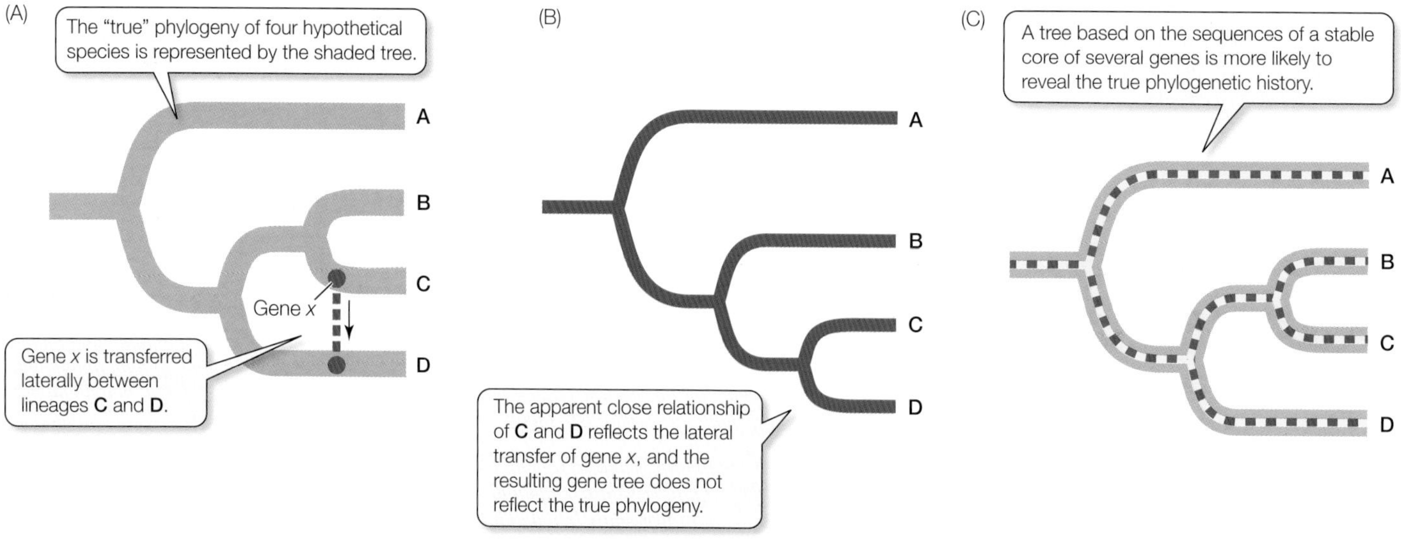

nity to explore prokaryotic diversity and better understand the phylogenetic tree of life.

Mutations are a major source of prokaryotic variation

Assuming that the prokaryote groups we are about to describe do indeed represent clades, these groups are amazingly complex. A single clade of bacteria or archaea may contain extraordinarily diverse species; on the other hand, a species in one group may be phenotypically almost indistinguishable from one or many species in another group. What are the sources of these phylogenetic patterns?

Although prokaryotes can acquire new alleles by transformation, transduction, or conjugation, the most important sources of genetic variation in populations of prokaryotes are probably mutation and genetic drift (described in Section 22.2). Mutations, especially recessive mutations, are slow to make their presence felt in populations of humans and other diploid organisms. In contrast, a mutation in a prokaryote, which is haploid, has immediate consequences for that organism. If it is not lethal, it will be transmitted to and expressed in the organism's daughter cells—and in their daughter cells, and so on. Thus a beneficial mutant allele spreads rapidly.

The rapid multiplication of many prokaryotes, coupled with mutation, natural selection, genetic drift, and lateral gene transfer, allows rapid phenotypic changes within their populations. Important changes, such as loss of sensitivity to an antibiotic, can occur over broad geographic areas in just a few years. Think how many significant metabolic changes could have occurred over even modest time spans, let alone over the entire history of life on Earth. When we introduce the proteobacteria, the largest group of bacteria, you will see that its different subgroups have easily and rapidly adopted and abandoned metabolic pathways under selective pressure from their environments.

26.4 RECAP

The study of prokaryote phylogeny is complicated by the organisms' small size, our inability to grow some of them in pure culture, and lateral gene transfer. Nucleotide sequences are especially helpful in distinguishing prokaryote clades.

- How did biologists classify bacteria before it became possible to determine nucleotide sequences? See p. 569
- Can you explain why rRNA is useful for evolutionary studies? See p. 569
- Do you see how lateral gene transfer could complicate evolutionary studies? See p. 570 and Figure 26.10

In spite of the difficulties described here, biologists have identified many clades of prokaryotes. We identify the characteristics of some of them in the next section.

26.5 What Are the Major Known Groups of Prokaryotes?

The Bacteria are the better-studied of the two prokaryote domains, and here we use a widely accepted classification scheme that enjoys considerable support from nucleotide sequence data. More than a dozen clades have been proposed under this scheme; we will describe just of a few of them here. We pay the closest attention to six groups: the spirochetes, chlamydias, high-GC Gram-positives, cyanobacteria, low-GC Gram-positives, and proteobacteria (**Figure 26.11**). First, however, we'll mention one property that is shared by members of three other groups.

Three of the bacterial groups once thought to have branched out earliest during bacterial evolution are all **thermophiles** (heat-lovers), as are the most ancient of the archaea. This observation supported the hypothesis that the first living organisms were thermophiles that appeared in much hotter environments than are predominant today. Recent nucleic acid-based evidence suggests, however, that those clades may have arisen more recently than did the spirochetes and chlamydias.

Spirochetes move by means of axial filaments

Spirochetes are Gram-negative, motile, chemoheterotrophic bacteria characterized by unique structures called axial filaments, which are modified flagella running through the periplasmic space (see Figure 26.6A). The cell body is a long cylinder coiled into a he-

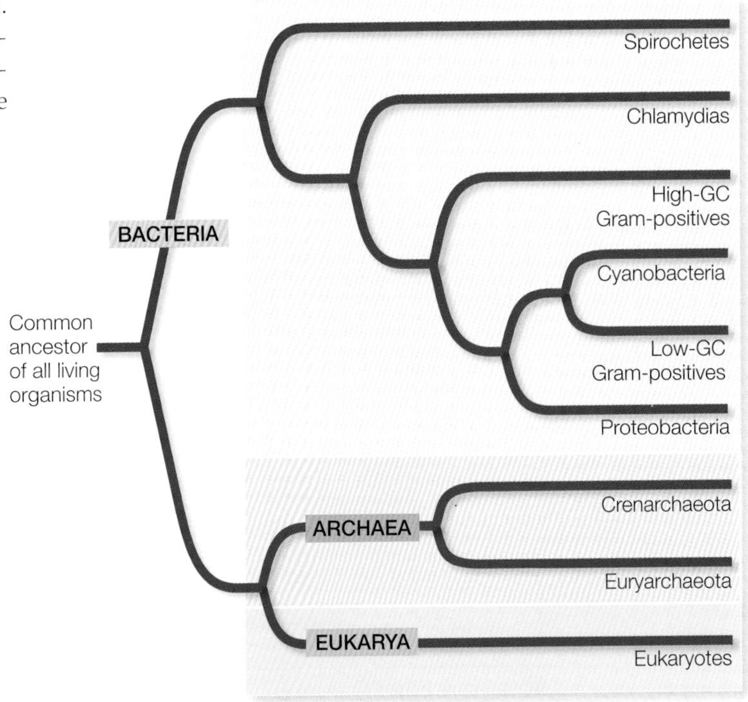

26.11 Two Domains: A Brief Overview This abridged summary classification of the domains Bacteria and Archaea shows their relationships to each other and to the Eukarya. The relationships among the many clades of bacteria, not all of which are listed here, are incompletely resolved at this time.

26.12 A Spirochete This corkscrew-shaped bacterium causes syphilis in humans.

lix (**Figure 26.12**). The axial filaments begin at either end of the cell and overlap in the middle, and there are typical protein motors where they are attached to the cell wall. These structures rotate, as they do in other prokaryotic flagella (see Figure 4.5) Many spirochetes live in humans as parasites; a few are pathogens, including those that cause syphilis and Lyme disease. Others live free in mud or water.

Chlamydias are extremely small parasites

Chlamydias are among the smallest bacteria (0.2–1.5 μm in diameter). They can live only as parasites within the cells of other organisms. It was once believed that this obligate parasitism resulted from an inability of chlamydias to produce ATP—that chlamydias were "energy parasites." Genomic sequencing results from the end of the twentieth century indicate, however, that chlamydias have the genetic capability to produce at least some ATP. They can augment this capacity by using an enzyme called a translocase, which allows them to take up ATP from the cytoplasm of their host in exchange for ADP from their own cells.

These tiny Gram-negative cocci are unique prokaryotes because of their complex life cycle, which involves two different forms of cells, *elementary bodies* and *reticulate bodies* (**Figure 26.13**). In humans, various strains of chlamydias cause eye infections (especially trachoma), sexually transmitted diseases, and some forms of pneumonia.

Some high-GC Gram-positives are valuable sources of antibiotics

High-GC Gram-positives, also known as *actinobacteria*, derive their name from the relatively high G+C/A+T ratio of their DNA. They develop an elaborately branched system of filaments (**Figure 26.14**). The shapes of these bacteria closely resemble the filamentous growth habit of fungi at a reduced scale. Some high-GC

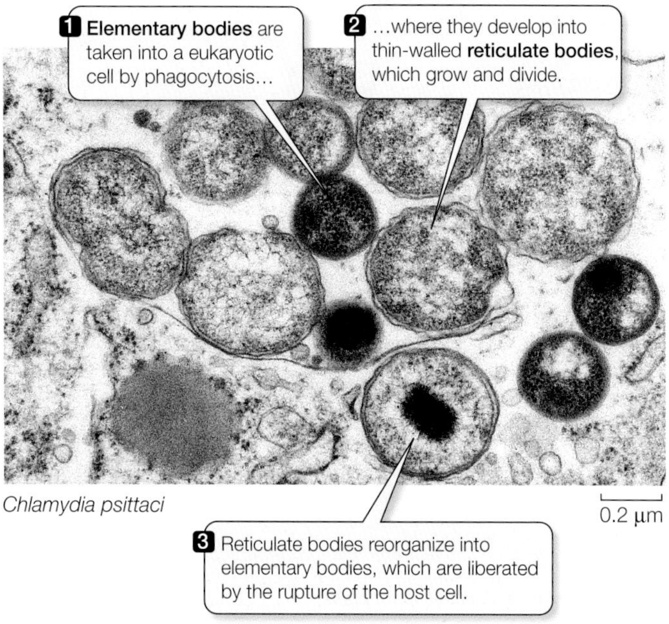

1 Elementary bodies are taken into a eukaryotic cell by phagocytosis...
2 ...where they develop into thin-walled reticulate bodies, which grow and divide.
Chlamydia psittaci
0.2 μm
3 Reticulate bodies reorganize into elementary bodies, which are liberated by the rupture of the host cell.

26.13 Chlamydias Change Form during Their Life Cycle Elementary bodies and reticulate bodies are the two major phases of the chlamydia life cycle.

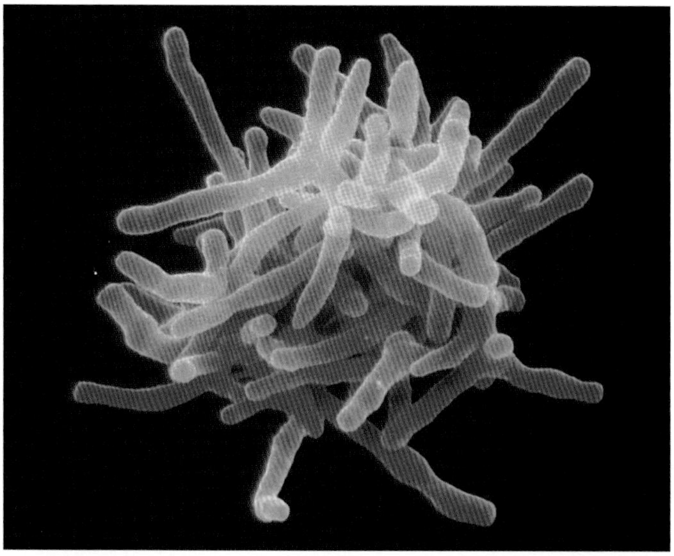

26.14 Filaments of a High-GC Gram-Positive The branching filaments seen in this scanning electron micrograph are typical of this medically important bacterial group.

Gram-positives reproduce by forming chains of spores at the tips of the filaments. In species that do not form spores, the branched, filamentous growth ceases, and the structure breaks up into typical cocci or bacilli, which then reproduce by binary fission.

The high-GC Gram-positives include several medically important bacteria. *Mycobacterium tuberculosis* causes tuberculosis, which kills 3 million people each year. Genetic data suggest that this bacterium arose 3 million years ago in East Africa, making it the most ancient known human bacterial affliction. *Streptomyces* produce streptomycin as well as hundreds of other antibiotics. We derive most of our antibiotics from members of the high-GC Gram-positives.

Cyanobacteria are important photoautotrophs

Cyanobacteria, sometimes called *blue-green bacteria* because of their pigmentation, are photoautotrophs that require only water, nitrogen gas, oxygen, a few mineral elements, light, and carbon dioxide to survive. They use chlorophyll *a* for photosynthesis and release oxygen gas; many species also fix nitrogen. Their photosynthesis was the basis of the "oxygen revolution" that transformed Earth's atmosphere (see Section 21.2).

Cyanobacteria carry out the same type of photosynthesis that is characteristic of eukaryotic photosynthesizers. They contain elaborate and highly organized internal membrane systems called *photosynthetic lamellae* or *thylakoids*. The chloroplasts of photosynthetic eukaryotes are derived from an endosymbiotic cyanobacterium.

Cyanobacteria may live free as single cells or associate in colonies. Depending on the species and on growth conditions, colonies of cyanobacteria may range from flat sheets one cell thick to filaments to spherical balls of cells.

Some filamentous colonies of cyanobacteria differentiate into three cell types: vegetative cells, spores, and heterocysts (**Figure 26.15**). **Vegetative cells** photosynthesize, **spores** are resting stages that can survive harsh environmental conditions and eventually develop into new filaments, and **heterocysts** are cells specialized for nitrogen fixation. All of the known cyanobacteria with heterocysts fix nitrogen. Heterocysts also have a role in reproduction: when filaments break apart to reproduce, the heterocyst may serve as a breaking point.

Not all low-GC Gram-positives are Gram-positive

The **low-GC Gram-positives**, as their name suggests, have a lower G+C/A+T ratio than do the high-GC Gram-positives. They are also sometimes called *firmicutes*. Some of the low-GC Gram-positives are in fact Gram-negative, and some have no cell wall at all. Nonetheless, this group is a clade.

Some low-GC Gram-positives produce **endospores** (**Figure 26.16**)—heat-resistant resting structures—when a key nutrient such as nitrogen or carbon becomes scarce. The bacterium replicates its DNA and encapsulates one copy, along with some of its cytoplasm, in a tough cell wall heavily thickened with peptidogly-

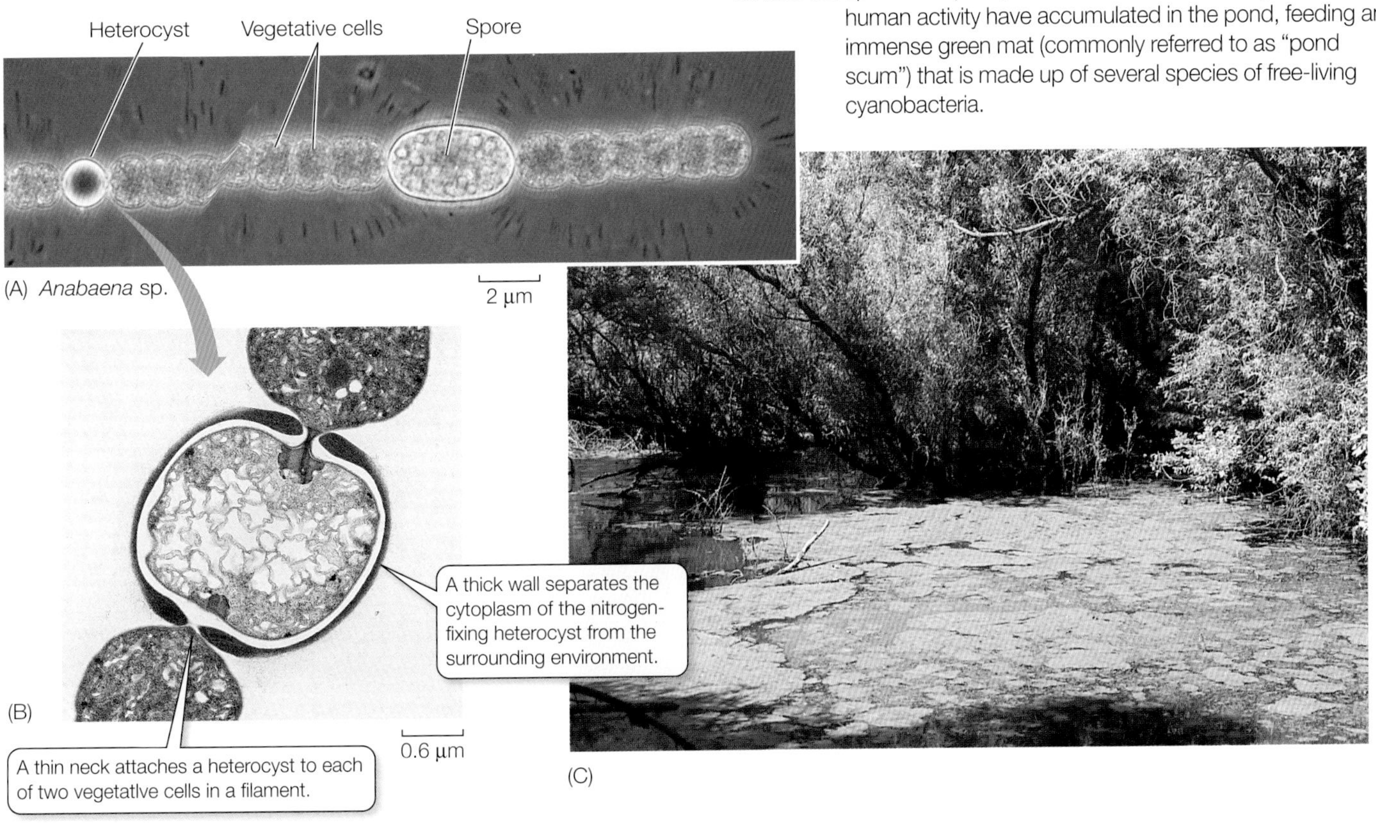

26.15 Cyanobacteria (A) *Anabaena* is a genus of cyanobacteria that form filamentous colonies containing three cell types. (B) Heterocysts are specialized for nitrogen fixation and serve as a breaking point when filaments reproduce. (C) Cyanobacteria appear in enormous numbers in some environments. This California pond has experienced eutrophication: phosphorus and other nutrients generated by human activity have accumulated in the pond, feeding an immense green mat (commonly referred to as "pond scum") that is made up of several species of free-living cyanobacteria.

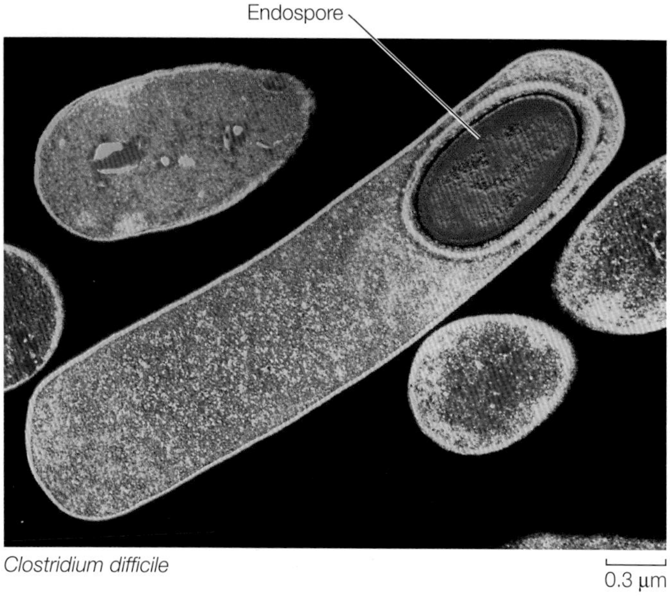

26.16 A Structure for Waiting Out Bad Times This low-GC Gram-positive bacterium, which can cause severe colitis in humans, produces endospores as resistant resting structures.

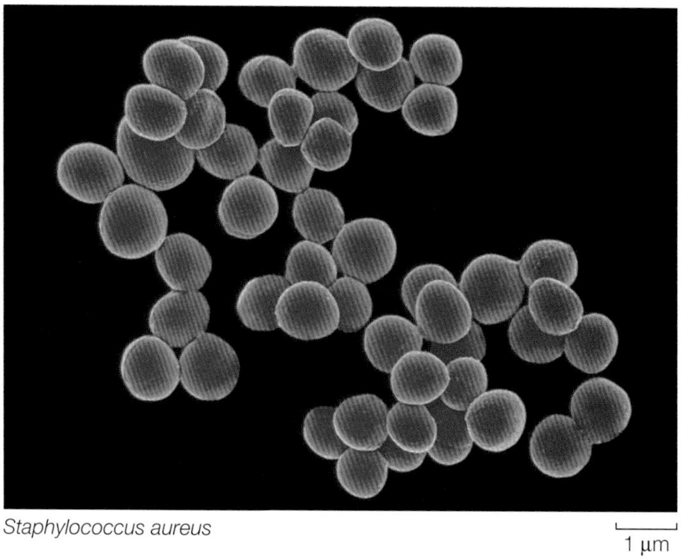

26.17 Low-GC Gram-Positives "Grape clusters" are the usual arrangement of staphylococci.

can and surrounded by a spore coat. The parent cell then breaks down, releasing the endospore. Endospore production is not a reproductive process; the endospore merely replaces the parent cell. The endospore, however, can survive harsh environmental conditions that would kill the parent cell, such as high or low temperatures or drought, because it is *dormant*—its normal activity is suspended. Later, if it encounters favorable conditions, the endospore becomes metabolically active and divides, forming new cells like the parent.

Dormant endospores of *Bacillus anthracis*, the causal agent of anthrax, germinate upon sensing specific molecules in the cytoplasm of blood cells called macrophages. Endospores of nonpathogenic *Bacillus* species do not germinate in this environment. Some endospores can be reactivated after more than a thousand years of dormancy. There are credible claims of reactivation of *Bacillus* endospores after millions of years. Members of this endospore-forming group of low-GC Gram-positives include the many species of *Clostridium* and *Bacillus*. The toxins produced by *C. botulinum* are among the most poisonous ever discovered; the lethal dose for humans is about one-millionth of a gram (1 μg).

The genus *Staphylococcus*—the staphylococci—includes low-GC Gram-positives that are abundant on the human body surface; they are responsible for boils and many other skin problems (**Figure 26.17**). *Staphylococcus aureus* is the best-known human pathogen in this genus; it is found in 20 to 40 percent of normal adults (and in 50 to 70 percent of hospitalized adults). It can cause respiratory, intestinal, and wound infections in addition to skin diseases.

Another interesting group of low-GC Gram-positives, the **mycoplasmas**, lack cell walls, although some have a stiffening material outside the plasma membrane. Some of them are the smallest cellular creatures ever discovered—they are even smaller than chlamydias (**Figure 26.18**). The smallest mycoplasmas capable of multiplication have a diameter of about 0.2 μm. They are small in another crucial sense as well: they have less than half as much DNA as most other prokaryotes—but they still can grow autonomously. It has been speculated that the amount of DNA in a mycoplasma may be the minimum amount required to encode the essential properties of a living cell.

The Proteobacteria are a large and diverse group

By far the largest group of bacteria, in terms of numbers of described species, is the **proteobacteria**, sometimes referred to as the *purple bacteria*. Among the proteobacteria are many species of Gram-negative, bacteriochlorophyll-containing, sulfur-using photoautotrophs. However, the proteobacteria also include dramatically diverse bacteria that bear no resemblance to those species

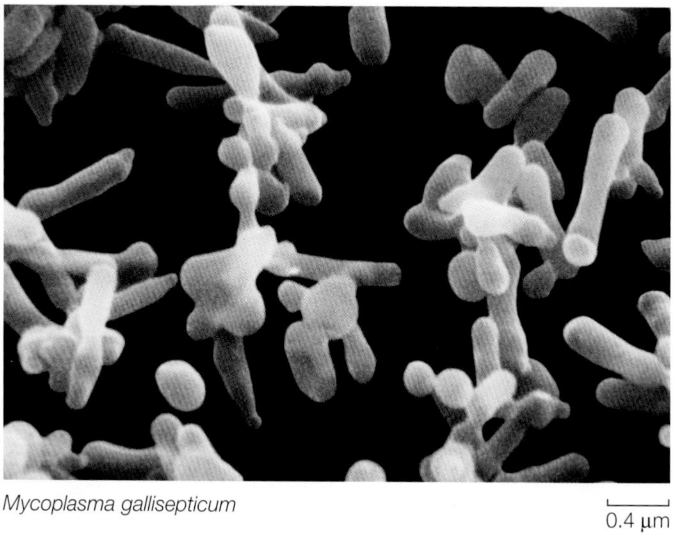

26.18 The Tiniest Living Cells Containing only about one-fifth as much DNA as *E. coli*, mycoplasmas are the smallest known bacteria.

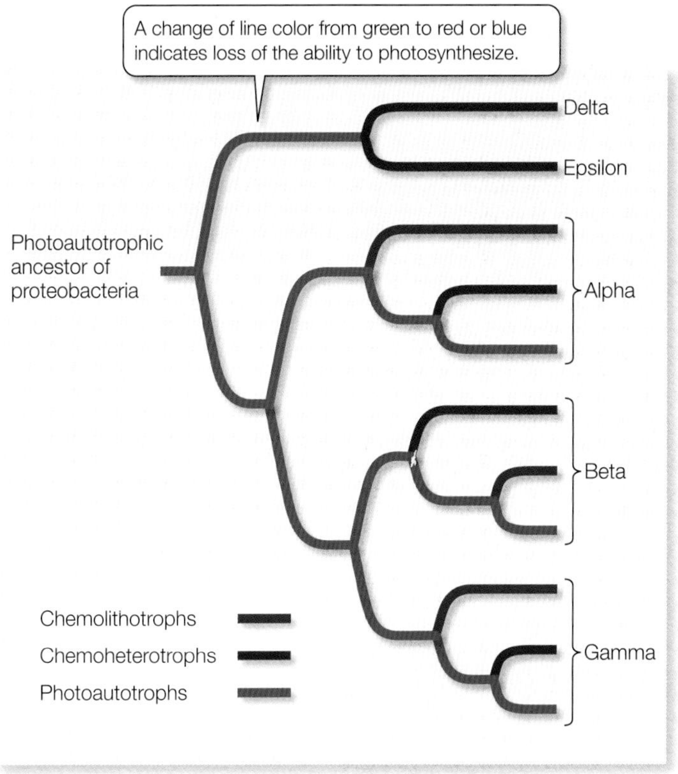

26.19 Modes of Nutrition in the Proteobacteria The common ancestor of all proteobacteria was probably a photoautotroph. As they encountered new environments, the delta and epsilon proteobacteria lost the ability to photosynthesize. In the other three groups, some evolutionary lineages became chemolithotrophs or chemoheterotrophs.

in phenotype. The mitochondria of eukaryotes were derived from a proteobacterium by endosymbiosis, as we'll see in Section 27.2.

No characteristic demonstrates the diversity of the proteobacteria more clearly than their metabolic pathways (**Figure 26.19**). There are five groups of proteobacteria: alpha, beta, gamma, delta, and epsilon. The common ancestor of all the proteobacteria was a photoautotroph. Early in evolution, two of the groups of proteobacteria lost their ability to photosynthesize and have been chemoheterotrophs ever since. The other three groups still have photoautotrophic members, but in *each* group, some evolutionary lines have abandoned photoautotrophy and taken up other modes of nutrition. There are chemolithotrophs and chemoheterotrophs in all three groups. Why? One possibility is that each of the trends shown in Figure 26.19 was an evolutionary response to selective pressures encountered as these bacteria colonized new habitats that presented new challenges and opportunities. Lateral gene transfer may have played a role in these responses.

Among the proteobacteria are some nitrogen-fixing genera, such as *Rhizobium* (see Figure 36.7), and other bacteria that contribute to the global nitrogen and sulfur cycles. *Escherichia coli*, one of the most studied organisms on Earth, is a proteobacterium. So, too, are many of the most famous human pathogens, such as *Yersinia pestis* (plague), *Vibrio cholerae* (cholera), and *Salmonella typhimurium* (gastrointestinal disease).

26.20 A Crown Gall This colorful tumor (the large mass at the left) growing on the stem of a geranium plant is caused by the proteobacterium *Agrobacterium tumefaciens*.

Fungi cause most plant diseases, and viruses cause others, but about 200 known plant diseases are of bacterial origin. *Crown gall*, with its characteristic tumors (**Figure 26.20**), is one of the most striking. The causal agent of crown gall is *Agrobacterium tumefaciens*, a proteobacterium that harbors a plasmid used in recombinant DNA studies as a vehicle for inserting genes into new plant hosts.

We have discussed six clades of bacteria in some detail, but other bacterial clades are well known, and there probably are dozens more waiting to be discovered. This conservative estimate is based on the fact that many bacteria have never been cultured in the laboratory.

Archaea differ in several important ways from bacteria

Archaea are well known for living in extreme habitats such as those with high salinity (salt content), low oxygen concentrations, high temperatures, or high or low pH (**Figure 26.21**). However, many archaea live in habitats that are not extreme. Perhaps the largest number of archaea live in the ocean depths.

One current classification scheme divides Archaea into two principal groups, **Euryarchaeota** and **Crenarchaeota**. Less is known about two more recently discovered groups, **Korarchaeota** and **Nanoarchaeota**. In fact, we know relatively little about the phylogeny of archaea, in part because the study of archaea is still in its early stages. Two characteristics shared by all archaea are the absence of peptidoglycan in their cell walls and the presence of lipids of distinctive composition in their cell membranes (see Table 26.1). Their separation from the Bacteria and Eukarya was supported when biologists sequenced the first archaeal genome. It consisted of 1,738 genes, more than half of which were unlike any genes ever found in the other two domains.

The unusual lipids in the membranes of archaea are found in all archaea, and in no bacteria or eukaryotes. Most bacterial and

EXPERIMENT

HYPOTHESIS: Some prokaryotes can grow and multiply at temperatures above 120°C.

METHOD

1. Seal samples of unidentified microorganisms taken from the vicinity of a thermal vent in tubes with medium containing Fe^{3+} as an electron acceptor. Control tubes contain the electron acceptor but no cells.
2. Hold experimental and control tubes for 10 hours in a sterilizer at 121°C. Reduction of the Fe^{3+} produces Fe^{2+} as magnetite, indicating the presence of living cells (left-hand photo).
3. In a second experiment, isolate and test for growth at various temperatures.

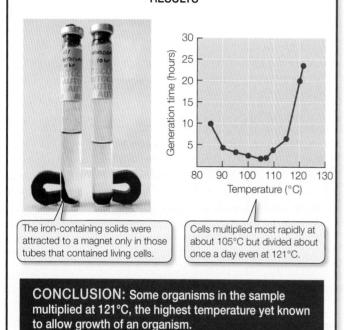

RESULTS

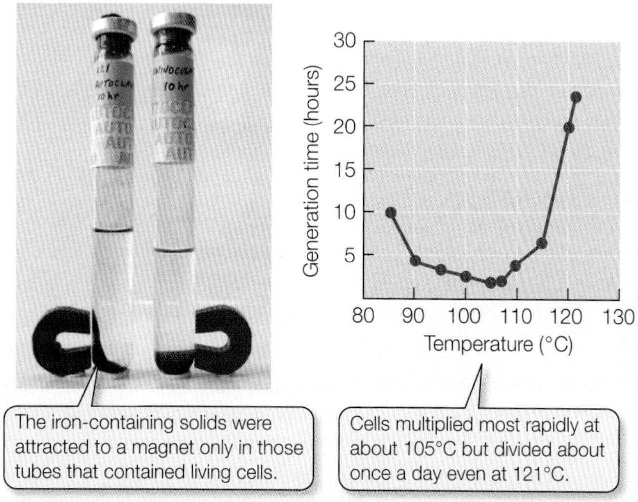

CONCLUSION: Some organisms in the sample multiplied at 121°C, the highest temperature yet known to allow growth of an organism.

26.21 What Is the Highest Temperature an Organism Can Tolerate? Kazem Kashefi and Derek Lovley isolated an unidentified prokaryotic organism from water samples taken near a hydrothermal vent in the northeastern Pacific Ocean. Tests to establish growth and metabolism were run at temperatures from 85°C to 130°C in sealed tubes with Fe^{3+} as an electron acceptor (as Fe_2O_3, or common rust).* The organism continued growing at temperatures as high as 121°C—a sterilizing temperature, known to destroy all previously described microorganisms. Genes from the surviving organism (dubbed "Strain 121") were sequenced, and comparison of those sequences with genes from known archaean species indicated that Strain 121 is probably also an archaean. FURTHER RESEARCH: Strain 121 did not grow during a 2-hour exposure to 130°C, but it did not die, either. How would you demonstrate that it was still alive?

*Some prokaryotes metabolize Fe^{3+} to Fe^{2+}, converting rust to magnetite. Living, growing cells thus produce a material that is attracted to magnets, whereas dead cells do not.

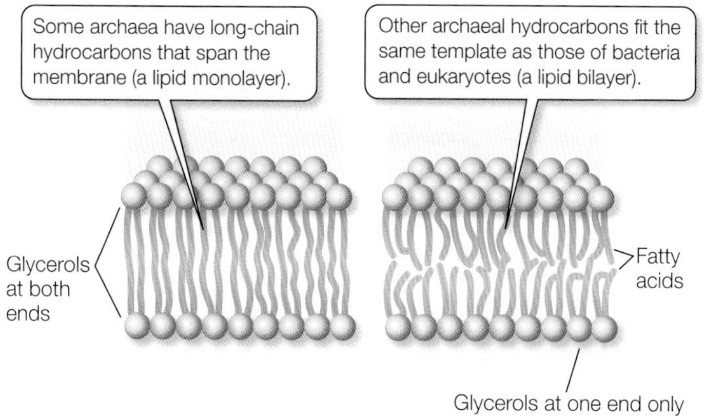

26.22 Membrane Architecture in Archaea The long-chain hydrocarbons of many archaeal membranes are branched, and may have glycerol at both ends. This lipid monolayer structure (on the left) still fits into a biological membrane.

eukaryotic membrane lipids contain unbranched long-chain fatty acids connected to glycerol by ester linkages:

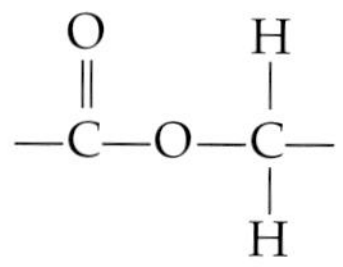

In contrast, some archaeal membrane lipids contain long-chain hydrocarbons connected to glycerol by *ether linkages*:

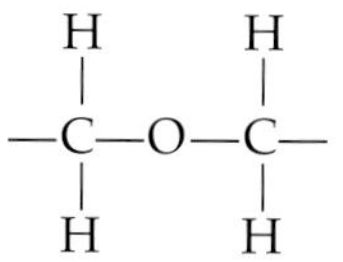

In addition, the long-chain hydrocarbons of the archaea are branched. One class of these lipids, with hydrocarbon chains 40 carbon atoms in length, contains glycerol at *both* ends of the hydrocarbons (**Figure 26.22**). This *lipid monolayer* structure, unique to Archaea, still fits in a biological membrane because the lipids are twice as long as the typical lipids in the bilayers of other membranes. Lipid monolayers and bilayers are both found among the archaea. The effects, if any, of these structural features on membrane performance are unknown. In spite of this striking difference in their membrane lipids, the membranes seen in all three domains have similar overall structures, dimensions, and functions.

Many Crenarchaeota live in hot, acidic places

Most known Crenarchaeota are both thermophilic (heat-loving) and **acidophilic** (acid-loving). Members of the genus *Sulfolobus* live in hot sulfur springs at temperatures of 70°C –75°C. They die of "cold" at 55°C (131°F). Hot sulfur springs are also extremely acidic. *Sulfolobus* grows best in the range from pH 2 to pH 3, but it readily tolerates pH values as low as 0.9. One species of the genus *Ferroplasma* lives at a pH near 0. Some acidophilic hyperthermophiles maintain an internal pH near 7 (neutral) in spite of their

26.23 Some Would Call It Hell; These Archaea Call It Home Masses of heat- and acid-loving archaea form an orange mat inside a volcanic vent on the island of Kyushu, Japan. Sulfurous residue is visible at the edges of the archaeal mat.

26.24 Extreme Halophiles Commercial seawater evaporating ponds, such as these in San Francisco Bay, are attractive homes for salt-loving archaea, which are easily visible here because of their carotenoid pigments.

acidic environment. These and other hyperthermophiles thrive where very few other organisms can even survive (**Figure 26.23**).

The Euryarchaeota live in many surprising places

Some species of Euryarchaeota share the property of producing methane (CH_4) by reducing carbon dioxide. All of these **methanogens** are obligate anaerobes, and methane production is the key step in their energy metabolism. Comparison of rRNA nucleotide sequences has revealed a close evolutionary relationship among all these methanogens, which were previously assigned to several unrelated bacterial groups.

Methanogens release approximately 2 billion tons of methane gas into Earth's atmosphere each year, accounting for 80 to 90 percent of the methane in the atmosphere, including that associated with mammalian belching. Approximately a third of this methane comes from methanogens living in the guts of grazing herbivores such as cattle, sheep, and deer. Methane is increasing in Earth's atmosphere by about 1 percent per year and is a contributor to the greenhouse effect. Part of the increase is due to increases in cattle and rice farming and the methanogens associated with both.

One methanogen, *Methanopyrus*, lives on the ocean bottom near boiling hot hydrothermal vents. *Methanopyrus* can survive and grow at 110°C. It grows best at 98°C and not at all at temperatures below 84°C.

Another group of Euryarchaeota, the **extreme halophiles** (salt lovers), lives exclusively in very salty environments. Because they contain pink carotenoid pigments, they can be easily seen under some circumstances (**Figure 26.24**). Halophiles grow in the Dead Sea and in brines of all types: pickled fish may sometimes show reddish pink spots that are colonies of halophilic archaea. Few other organisms can live in the saltiest of the homes that the extreme halophiles occupy; most would "dry" to death, losing too much water to the hypertonic environment. Extreme halophiles have been found in lakes with pH values as high as 11.5—the most alkaline environment inhabited by living organisms, and almost as alkaline as household ammonia.

Some of the extreme halophiles have a unique system for trapping light energy and using it to form ATP—without using any form of chlorophyll—when oxygen is in short supply. They use the pigment *retinal* (also found in the vertebrate eye) combined with a protein to form a light-absorbing molecule called *bacteriorhodopsin*, and they form ATP by a chemiosmotic mechanism of the kind described in Figure 7.13.

Another member of the Euryarchaeota, *Thermoplasma*, has no cell wall. It is thermophilic and acidophilic, its metabolism is aerobic, and it lives in coal deposits. It has the smallest genome among the archaea, and perhaps the smallest (along with the mycoplasmas) of any free-living organism—1,100,000 base pairs.

Korarchaeota and Nanoarchaeota are less well known

The Korarchaeota are known only by evidence derived from DNA isolated directly from hot springs. No korarchaeote has been successfully grown in pure culture.

Another archaean has been discovered at a deep-sea hydrothermal vent off the coast of Iceland. It is the first representative of a group christened Nanoarchaeota because of their minute size. This organism lives attached to cells of *Ignicoccus*, a crenarchaeote. Because of their association, the two species can be grown together in culture (**Figure 26.25**).

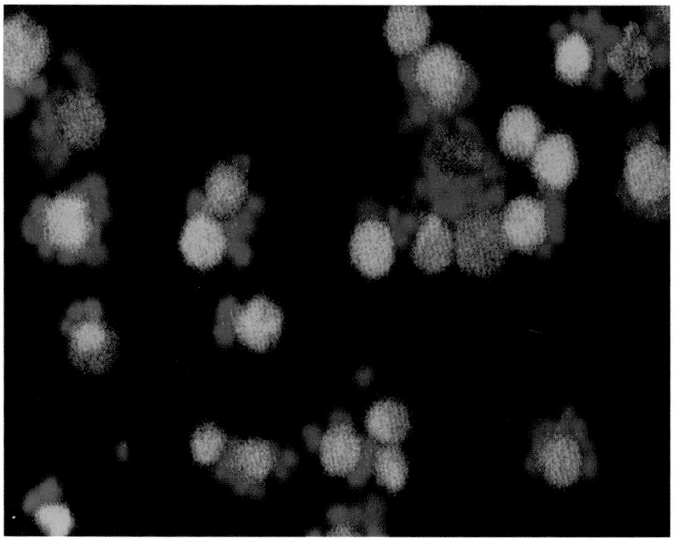

26.25 A Nanoarchaeote Growing in Mixed Culture with a Crenarchaeote *Nanoarchaeum equitans* (red), discovered living near deep-ocean hydrothermal vents, is the only representative of the nanoarchaeote group so far discovered. This tiny organism lives attached to cells of the crenarchaeote *Ignicoccus* (green). For this confocal laser micrograph, the two species were visually differentiated by fluorescent dye "tags" that are specific to their separate gene sequences.

26.5 RECAP

The relationships among the groups of Bacteria and Archaea are only partially understood. Each group is diverse in form and metabolism.

- Can you explain how metabolic diversity could have become so great in the proteobacteria? See p. 575 and Figure 26.19
- What makes the membranes of archaea unique? See p. 576 and Figure 26.22

Because prokaryotes have so many different metabolic and nutritional capabilities, and because they can live in so many environments, it is reasonable to expect that they affect their environments in many ways. As we are about to see, prokaryotes directly affect humans—in ways both beneficial and harmful.

26.6 How Do Prokaryotes Affect Their Environments?

Prokaryotes live in and exploit all kinds of environments and are part of all ecosystems. In this section we'll examine the roles of prokaryotes that live in soils, in water, and even in other organisms, where they may exist in a neutral, beneficial, or parasitic relationship with their host's tissues. The roles of some prokaryotes living in extreme environments have yet to be determined.

Remember that in spite of our frequent mention of prokaryotes as human pathogens, only a small minority of the known prokaryotic species are pathogenic. Many more prokaryotes play positive roles in our lives and in the biosphere. We make direct use of many bacteria and a few archaea in such diverse applications as cheese production, sewage treatment, and the industrial production of an amazing variety of antibiotics, vitamins, organic solvents, and other chemicals.

Prokaryotes are important players in element cycling

Many prokaryotes are **decomposers**, organisms that metabolize organic compounds in dead organisms and other organic material and return the products to the environment as inorganic substances. Prokaryotes, along with fungi, return tremendous quantities of organic carbon to the atmosphere as carbon dioxide, thus carrying out a key step in the carbon cycle. Prokaryotic decomposers also return inorganic nitrogen and sulfur to the environment.

Animals depend on plants and other photosynthetic organisms for their food, directly or indirectly. But plants depend on other organisms—prokaryotes—for their own nutrition. The extent and diversity of life on Earth would not be possible without nitrogen fixation by prokaryotes. Nitrifiers are crucial to the biosphere because they convert the products of nitrogen fixation into nitrate ions, the form of nitrogen most easily used by many plants (see Figure 36.8). Plants, in turn, are the source of nitrogen compounds for animals and fungi. Denitrifiers also play a key role in keeping the nitrogen cycle going. Without denitrifiers, which convert nitrate ions back into nitrogen gas, all forms of nitrogen would leach from the soil and end up in lakes and oceans, making life on land impossible. Other prokaryotes—both bacteria and archaea—contribute to a similar cycle of sulfur.

In the ancient past, the cyanobacteria had an equally dramatic effect on life: their photosynthesis generated oxygen, converting Earth's atmosphere from an anaerobic to an aerobic environment. A major result was the wholesale loss of obligate anaerobic species that could not tolerate the O_2 generated by the cyanobacteria. Only those anaerobes that were able to adapt to aerobic conditions or colonize environments that remained anaerobic (such as very wet ones) survived. However, this transformation to aerobic environments made possible the evolution of cellular respiration and the subsequent explosion of eukaryotic life.

Ten trillion tons of methane gas lie deep under the ocean floor. Is it possible we could experience a disastrous "big burp" as this gas escapes to the atmosphere? Fortunately, legions of archaea also live below the ocean bottom, where they metabolize the methane as it rises. Virtually none of the gas gets even as far as the ocean deeps.

Prokaryotes live on and in other organisms

Prokaryotes work together with eukaryotes in many ways. As we have seen, mitochondria and chloroplasts are descended from what were once free-living bacteria. Much later in evolutionary history, some plants became associated with bacteria to form cooperative nitrogen-fixing nodules on their roots (see Figure 36.9).

Many animals harbor a variety of bacteria and archaea in their digestive tracts. Cattle depend on prokaryotes to perform important steps in digestion. Like most animals, cattle cannot produce cellulase, the enzyme needed to start the digestion of the cellulose that makes up the bulk of their plant food. However, bacteria living in a special section of the gut, called the rumen, produce enough cellulase to process the daily diet for the cattle.

Humans use some of the metabolic products—especially vitamins B_{12} and K—of bacteria living in our large intestine. These and other bacteria and archaea line our intestines with a dense biofilm that is in intimate contact with the lining of the gut. This biofilm facilitates nutrient transfer from the intestine into the body and induces immunity to the gut contents. The biofilm in the gut is a major part of an "organ" consisting of prokaryotes that is essential to our health. Its makeup varies from time to time and from region to region of the intestinal tract, and it has a complex ecology that scientists have just begun to explore in detail.

We are heavily populated, inside and out, by bacteria. Although very few of them are agents of disease, popular notions of bacteria as "germs" arouse our curiosity about those few.

A small minority of bacteria are pathogens

The late nineteenth century was a productive era in the history of medicine—a time during which bacteriologists, chemists, and physicians proved that many diseases are caused by microbial agents. During this time the German physician Robert Koch laid down a set of four rules for establishing that a particular microorganism causes a particular disease:

- The microorganism is always found in individuals with the disease.
- The microorganism can be taken from the host and grown in pure culture.
- A sample of the culture produces the disease when injected into a new, healthy host.
- The newly infected host yields a new, pure culture of microorganisms identical to those obtained in the second step.

These rules, called **Koch's postulates**, were very important in a time when it was not widely understood that microorganisms cause disease. Although medical science today has more powerful diagnostic tools, the postulates remain useful on occasion. For example, physicians were taken aback in the 1990s when stomach ulcers—long accepted and treated as the result of excess stomach acid—were proven by Koch's postulates to be caused by the bacterium *Helicobacter pylori* (see Figure 50.14).

Only a tiny percentage of all prokaryotes are **pathogens** (disease-producing organisms), and of those that are known, all are in the domain Bacteria. For an organism to be a successful pathogen, it must overcome several hurdles:

- It must arrive at the body surface of a potential host.
- It must enter the host's body.
- It must evade the host's defenses.
- It must multiply inside the host.
- It must infect a new host.

Failure to overcome any of these hurdles ends the reproductive career of a pathogenic organism. However, in spite of the many defenses available to potential hosts (see Chapter 18), some bacteria are very successful pathogens.

For the host, the consequences of a bacterial infection depend on several factors. One is the **invasiveness** of the pathogen—its ability to multiply within the body of the host. Another is its **toxigenicity**—its ability to produce chemical substances (*toxins*) that are harmful to the tissues of the host. *Corynebacterium diphtheriae*, the agent that causes diphtheria, has low invasiveness and multiplies only in the throat, but its toxigenicity is so great that the entire body is affected. In contrast, *Bacillus anthracis*, which causes anthrax (a disease primarily of cattle and sheep, but which is also sometimes fatal in humans), has low toxigenicity, but an invasiveness so great that the entire bloodstream ultimately teems with the bacteria. Both are low-GC Gram-positives.

There are two general types of bacterial toxins: exotoxins and endotoxins. **Endotoxins** are released when certain Gram-negative bacteria grow or lyse (burst). These toxins are lipopolysaccharides (complexes consisting of a polysaccharide and a lipid component) that form part of the outer bacterial membrane (see Figure 26.5). Endotoxins are rarely fatal; they normally cause fever, vomiting, and diarrhea. Among the endotoxin producers are some strains of the gamma proteobacteria *Salmonella* and *Escherichia*.

Exotoxins are usually soluble proteins released by living, multiplying bacteria, and they may travel throughout the host's body. They are highly toxic—often fatal—to the host, but do not produce fevers. Exotoxin-induced human diseases include tetanus (from *Clostridium tetani*), botulism (from *Clostridium botulinum*), cholera (from *Vibrio cholerae*), and plague (from *Yersinia pestis*). Anthrax results from three exotoxins produced by *Bacillus anthracis*.

Pathogenic bacteria are often surprisingly difficult to combat, even with today's arsenal of antibiotics. One source of this difficulty is the ability of prokaryotes to form biofilms.

26.6 RECAP

Prokaryotes play key roles in the cycling of Earth's elements. While many prokaryotes are beneficial and even necessary to other forms of life, some are pathogens.

- Can you describe the roles of bacteria in the nitrogen cycle? See p. 569 and p. 578
- Do you understand the challenges facing a pathogen? See p. 579

CHAPTER SUMMARY

26.1 How did the living world begin to diversify?

See Web/CD Tutorial 26.1

Two of life's three domains, **Bacteria** and **Archaea**, are prokaryotic. They are distinguished from Eukarya in several ways, including the cell's lack of membrane-enclosed organelles. Review Table 26.1

Archaea and Eukarya share a common ancestor not shared by Bacteria. The common ancestor of all three domains probably lived more than 3 billion years ago, and the common ancestor of the Archaea and Eukarya at least 2 billion years ago. Review Figure 26.1

26.2 Where are prokaryotes found?

Prokaryotes are the most numerous organisms on Earth. They occupy an enormous variety of habitats, including inside other organisms and deep in Earth's crust.

The three most common bacterial body forms are **cocci** (spheres), **bacilli** (rods), and **helices** (spirals). The cells of some bacteria aggregate, forming **filaments** and other structures.

Prokaryotes form complex communities, some of which become dense films called **biofilms**. Review Figure 26.3

26.3 What are some keys to the success of prokaryotes?

Most prokaryotes have cell walls. Almost all bacterial cell walls contain **peptidoglycan**. Review Figure 26.5, Web/CD Activity 26.1

Bacteria can be classified into two groups by the **Gram stain**.

Prokaryotes move by a variety of means, including axial filaments, gas vesicles, and flagella.

Prokaryotes undergo genetic recombination but reproduce asexually.

Prokaryote metabolism is very diverse. Some prokaryotes are anaerobic, others are aerobic, and yet others can shift between these modes. Prokaryotes are classified as **photoautotrophs**, **photoheterotrophs**, **chemolithotrophs**, or **chemoheterotrophs**. Review Table 26.2

The metabolic pathways of some prokaryotes involve sulfur or nitrogen. **Nitrogen fixers** convert nitrogen gas into a form organisms can metabolize.

26.4 How can we determine prokaryote phylogeny?

Early attempts to classify prokaryotes were hampered by their small size and difficulties growing them in pure culture. Phylogenetic classification of prokaryotes is now based on rRNA and other nucleotide sequences.

Lateral gene transfer has occurred throughout evolutionary history, but it may not complicate the elucidation of prokaryote phylogeny. Review Figure 26.10

Only a tiny percentage of all prokaryote species have been described.

26.5 What are the major known groups of prokaryotes?

Several clades of prokaryotes have been recognized. The members of a prokaryote clade often differ profoundly from one another. Review Figures 26.11 and 26.19

Of the clades of Bacteria, the **proteobacteria** embrace the largest number of species. Other important groups include the **cyanobacteria**, the **spirochetes**, the **chlamydias**, and the **low-GC Gram-positives**. Some **high-GC Gram positives** produce important antibiotics.

The cell walls of archaeans lack peptidoglycan, and archaeal membrane lipids differ from those of bacteria and eukaryotes.

The best-studied groups of Archaea are the **Euryarchaeota** and the **Crenarchaeota**.

26.6 How do prokaryotes affect their environments?

Prokaryotes play key roles in the cycling of elements such as nitrogen, oxygen, sulfur, and carbon. One such role is as **decomposers** of dead organisms.

Nitrogen-fixing bacteria fix the nitrogen needed by all other organisms. Nitrifiers convert that nitrogen into forms that can be used by plants, and denitrifiers ensure that nitrogen is returned to the atmosphere.

Oxygen production by early photosynthetic cyanobacteria reconfigured Earth's atmosphere, which made aerobic forms of life possible.

Prokaryotes inhabiting the guts of many animals help them digest their food.

Koch's postulates establish the criteria by which an organism may be classified as a **pathogen**. Relatively few bacteria—and no archaea—are known to be pathogens.

SELF-QUIZ

1. Most prokaryotes
   a. are agents of disease.
   b. lack ribosomes.
   c. evolved from the most ancient eukaryotes.
   d. lack a cell wall.
   e. are chemoheterotrophs.
2. The division of the living world into three domains
   a. is strictly arbitrary.
   b. is based on the morphological differences between archaea and bacteria.
   c. emphasizes the greater importance of eukaryotes.
   d. was proposed by the early microscopists.
   e. is strongly supported by data on rRNA sequences.
3. Which statement about the archaeal genome is true?
   a. It is much more similar to the bacterial genome than to eukaryotic genomes.
   b. Many of its genes are genes that are never observed in bacteria or eukaryotes.
   c. It is much smaller than the bacterial genome.
   d. It is housed in the nucleus.
   e. No archaeal genome has yet been sequenced.
4. Which statement about nitrogen metabolism is *not* true?
   a. Certain prokaryotes reduce atmospheric N_2 to ammonia.
   b. Some nitrifiers are soil bacteria.
   c. Denitrifiers are obligate anaerobes.
   d. Nitrifiers obtain energy by oxidizing ammonia and nitrite.
   e. Without the nitrifiers, terrestrial organisms would lack a nitrogen supply.

5. All photosynthetic bacteria
   a. use chlorophyll *a* as their photosynthetic pigment.
   b. use bacteriochlorophyll as their photosynthetic pigment.
   c. release oxygen gas.
   d. produce particles of sulfur.
   e. are photoautotrophs.
6. Gram-negative bacteria
   a. appear blue to purple following Gram staining.
   b. are the most abundant of the bacterial groups.
   c. are all either bacilli or cocci.
   d. contain no peptidoglycan in their cell walls.
   e. are all photosynthetic.
7. Endospores
   a. are produced by viruses.
   b. are reproductive structures.
   c. are very delicate and easily killed.
   d. are resting structures.
   e. lack cell walls.
8. Chlamydias
   a. are among the smallest archaea.
   b. live on the surface of human skin.
   c. are never pathogenic to humans.
   d. are Gram-negative.
   e. have a very simple life cycle.
9. Which statement about mycoplasmas is *not* true?
   a. They lack cell walls.
   b. They are the smallest known cellular organisms.
   c. They contain the same amount of DNA as do other prokaryotes.
   d. They cannot be killed with penicillin.
   e. Some are pathogens.
10. Archaea
   a. have cytoskeletons.
   b. have distinctive lipids in their plasma membranes.
   c. survive only at moderate temperatures and near neutrality.
   d. all produce methane.
   e. have substantial amounts of peptidoglycan in their cell walls.

FOR DISCUSSION

1. Why do systematic biologists find rRNA sequence data more useful than data on metabolism or cell structure for classifying prokaryotes?
2. Why does lateral gene transfer make it so difficult to arrive at agreement on prokaryote phylogeny?
3. Differentiate among the members of the following sets of related terms:
   a. prokaryotic/eukaryotic
   b. obligate anaerobe/facultative anaerobe/obligate aerobe
   c. photoautotroph/photoheterotroph/chemolithotroph/chemoheterotroph
   d. Gram-positive/Gram-negative
4. Why are the endospores of low-GC Gram-positives not considered to be reproductive structures?
5. Originally, the cyanobacteria were called "blue-green algae" and were not grouped with the bacteria. Suggest several reasons for this (abandoned) tendency to separate the cyanobacteria from the bacteria. Why are the cyanobacteria now grouped with the other bacteria?
6. The high-GC Gram-positives are of great commercial interest. Why?
7. Thermophiles are of great interest to molecular biologists and biochemists. Why? What practical concerns might motivate that interest?
8. How can biologists discuss the Korarchaeota when they have never seen one?

FOR INVESTIGATION

Kashefi and Lovley were able to grow an unnamed archaean at temperatures over 120°C only because they used Fe^{3+} as an electron acceptor—no other electron acceptor that they tried allowed growth (see Figure 26.21). How might you explore the same or other high-temperature environments for other hyperthermophilic organisms not detected by Kashefi and Lovley using Fe^{3+}?

CHAPTER 27 The Origin and Diversification of the Eukaryotes

A tale of three trypanosomes

Among the most deadly organisms on Earth are the trypanosomes, single-celled microscopic organisms that cause several grim diseases, mainly in developing countries. In central Africa, tsetse flies carry *Trypanosoma brucei* and its relatives, which cause African sleeping sickness. There is no vaccine to prevent the infection, and only one drug is currently available to treat it. That drug—melarsoprol—kills about 5 percent of the patients who take it and is without effect on another 30 percent, but it is the only treatment available. There are 300,000 to 500,000 cases of sleeping sickness and more than 50,000 deaths from the disease each year.

Assassin bugs carry another trypanosome, *Trypanosoma cruzi*, which causes Chagas' disease. This disease affects 16 to 18 million people, primarily in Central and South America. Again, there is no vaccine and no effective drug, and 20,000 to 50,000 people die from Chagas' disease each year. Still another trypanosome, *Leishmania major*, causes a family of often fatal human diseases collectively called leishmaniasis. This organism is transmitted by sand flies. There are about 2 million cases and an estimated 60,000 deaths from leishmaniasis each year, and—you guessed it—there is no vaccine and no good treatment.

Research and development of a new medicine typically requires about a billion dollars and a dozen years. The income from sales must compensate the developers and yield enough profit to make production a viable commercial enterprise. Trypanosome diseases mostly strike the poorest of the poor in developing countries, and so offer little financial incentive for drug and vaccine developers.

International health organizations rank the trypanosome diseases among the "most neglected diseases." Their enormous toll is barely mentioned in the Western news media, in stark contrast to the copious coverage given to any proposed new cancer treatment or the latest brouhaha over a flu vaccine.

Some developing countries, such as India, Cuba, Brazil, and South Africa, have the technical expertise and manufacturing capacity for pharmaceutical research and production. But even with such expertise, preventing and treating trypanosome diseases presents obstacles. Trypanosomes can evade recognition and destruction by the human immune system, vaccines, and drugs by constantly changing the cell surface recognition molecules of their own or infected host cells.

In addition, unlike the prokaryotic bacteria, these unicellular microbes are eukaryotes—

Thugs in a Huddle Sometimes trypanosomes like these *Leishmania major* form clusters held together by a tangle of mucilage secreted around their flagella; nobody is sure yet why they behave this way.

A Giant among Microbial Eukaryotes "Microbial" certainly does not describe *Macrocystis pyrifera*, the giant kelp, which can reach 60 meters in length. Despite this, they are often placed with the microbial eukaryotes in a large, paraphyletic grouping.

their cells are similar to ours, and remedies that attack them often damage our own cells as well. Trypanosomes are one group among the legions of *microbial eukaryotes*. Several different microbial eukaryote lineages were the ancestors of all multicellular life. Indeed, some of these organisms are multicellular themselves, and hardly microbial. They are a diverse lot, and difficult to categorize.

IN THIS CHAPTER we look at some of the many effects that microbial eukaryotes have on their environment. We then describe the origin and early diversification of the eukaryotes and the complexity achieved by some single cells. We then explore some of the diversity of microbial eukaryote body forms and adaptations and present the developing current view of the evolutionary relationships among some of the microbial eukaryotes.

CHAPTER OUTLINE

27.1 How Do Microbial Eukaryotes Affect the World Around Them?

Many modern members of the Eukarya (such as trees, mushrooms, and dogs, not to mention human beings) are familiar to us. We have no problem recognizing these organisms as land plants, fungi, and animals, respectively. However, trypanosomes and a dazzling assortment of other eukaryotes—mostly microscopic organisms—do not fit into any of these three groups. Eukaryotes that are neither land plants, animals, nor fungi have traditionally been grouped into a category called **protists**, and for the sake of clarity and convenience, other chapters in this book have used that term. In this chapter, however, we will refer to them as **microbial eukaryotes** (even though not all of them are actually microbial) to emphasize that these organisms do not constitute a clade, but are *paraphyletic* (see Figure 25.12).

The microbial eukaryotes are extremely diverse, and their effects on other organisms and on the physical environment are almost as diverse. Some microbial eukaryotes are food for marine animals, while others poison the sea; some are packaged as nutritional supplements, and some are pathogens; the remains of some form the sands of many modern beaches, and others are a major source of today's ever more expensive crude oil.

The phylogeny and morphology of the microbial eukaryotes both illustrate their diversity

The true phylogeny of these organisms is still the subject of research and debate, but we do know that *the microbial eukaryotes are not a clade*. Some groups of microbial eukaryotes are more closely related to the animals than they are to other microbial eukaryotes, while others are closely related to the land plants (**Table 27.1**; see also Figure 27.17). Some microbial eukaryotes are motile, while others do not move; some are photosynthetic, others heterotrophic; most are unicellular, but some are multicellular. Most are microscopic, but a few are huge: giant kelps, for example, sometimes achieve lengths greater than that of a football field (see above left). Many unicellular eukaryotes are constituents of the **plankton** (free-floating microscopic aquatic organisms). Photosynthetic members of the plankton are called **phytoplankton**.

TABLE 27.1

Major Eukaryote Clades

CLADE	ATTRIBUTES	EXAMPLE (GENUS)
Chromalveolates		
Haptophytes	Unicellular, often with calcium carbonate scales	*Emiliania*
Alveolates	Sac-like structures beneath plasma membrane	
Apicomplexans	Apical complex for penetration of host	*Plasmodium*
Dinoflagellates	Pigments give golden-brown color	*Gonyaulax*
Ciliates	Cilia; two types of nuclei	*Paramecium*
Stramenopiles	Hairy and smooth flagella	
Brown algae	Multicellular; marine; photosynthetic	*Macrocystis*
Diatoms	Unicellular; photosynthetic; two-part cell walls	*Thalassiosira*
Oomycetes	Mostly coenocytic; heterotrophic	*Saprolegnia*
Plantae		
Glaucophytes	Peptidoglycan in chloroplasts	*Cyanophora*
Red algae	No flagella; chlorophyll *a* and *c*; phycoerythrin	*Chondrus*
Chlorophytes	Chlorophyll *a* and *b*	*Ulva*
*Land plants (Chs. 28–29)	Chlorophyll *a* and *b*; protected embryo	*Ginkgo*
Charophytes	Chlorophyll *a* and *b*; mitotic spindle oriented as in land plants	*Chara*
Excavates		
Diplomonads	No mitochondria; two nuclei; flagella	*Giardia*
Parabasalids	No mitochondria; flagella and undulating membrane	*Trichomonas*
Heteroloboseans	Can transform between amoeboid and flagellate stages	*Naegleria*
Euglenids	Flagella; spiral strips of protein support cell surface	*Euglena*
Kinetoplastids	Kinetoplast within mitochondrion	*Trypanosoma*
Rhizaria		
Cercozoans	Threadlike pseudopods	*Cercomonas*
Foraminiferans	Long, branched pseudopods; calcium carbonate shells	*Globigerina*
Radiolarians	Glassy endoskeleton; thin, stiff pseudopods	*Astrolithium*
Unikonts		
Opisthokonts	Single, posterior flagellum	
*Fungi (Ch. 30)	Heterotrophs that feed by absorption	*Penicillium*
Choanoflagellates	Resemble sponge cells; heterotrophic; with flagella	*Choanoeca*
*Animals (Chs. 31–33)	Heterotrophs that feed by ingestion	*Drosophila*
Amoebozoans	Amoebas with lobe-shaped pseudopods	
Loboseans	Feed individually	*Amoeba*
Plasmodial slime molds	Form coenocytic feeding bodies	*Physarum*
Cellular slime molds	Cells retain their identity in pseudoplasmodium	*Dictyostelium*

*Clades marked with an asterisk are made up of multicellular organisms and are discussed in the chapters indicated. All other groups listed are treated here as *microbial eukaryotes* (often known as *protists*).

Phytoplankton are the primary producers of the marine food web

A single microbial eukaryote clade, the *diatoms,* is responsible for about a fifth of all of the photosynthetic carbon fixation on Earth—about the same amount of photosynthesis performed by all of Earth's rainforests. These spectacular unicellular organisms (**Figure 27.1**) are the predominant members of the phytoplankton, but other microbial eukaryote clades also include important phytoplanktonic species that contribute heavily to global photosynthesis. Like green plants on land, the phytoplankton serve as a gateway for energy from the sun into the living world; in other words, they are *primary producers*. In turn, they are eaten by heterotrophs, including animals and other microbial eukaryotes. Those consumers are, in turn, eaten by other consumers. Most aquatic heterotrophs

27.1 Architecture in Miniature: A Photosynthetic Diatom This artificially colored scanning electron micrograph shows the intricate patterning of diatom cell walls. These spectacular unicellular eukaryotes are photosynthetic and dominate the aquatic phytoplankton community.

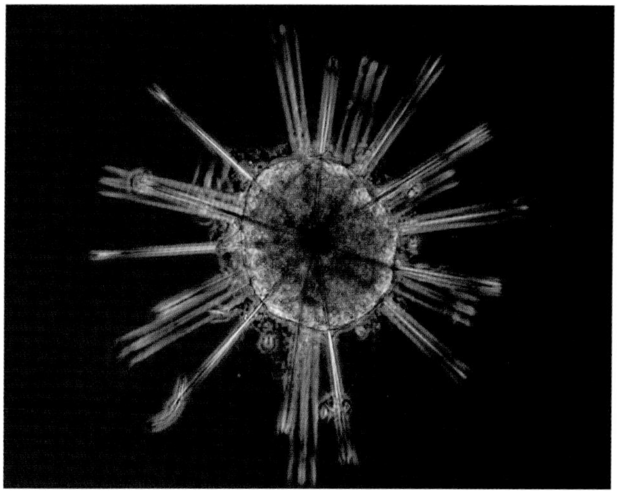

27.2 Two Microbial Eukaryotes in an Endosymbiotic Relationship Photosynthetic dinoflagellates (see Figure 27.18) are living as endosymbionts within this radiolarian, providing organic nutrients for the radiolarian and imparting the golden brown pigmentation seen at the center of its glassy skeleton.

(with the exception of some species existing in the deep ocean) depend on the photosynthesis performed by phytoplankton.

Some microbial eukaryotes are endosymbionts

Endosymbiosis is the condition in which two organisms live together, one inside the other (see Figure 4.15C). Endosymbiosis is very common among the microbial eukaryotes, many of which live within the cells of animals. Members of the *dinoflagellates* are common microbial eukaryote symbionts in both animals and in other microbial eukaryotes; most but not all dinoflagellate endosymbiont species are photosynthetic. Many *radiolarians*, for example, harbor photosynthetic endosymbionts (**Figure 27.2**). As a result, the radiolarians, which are not photosynthetic themselves, appear greenish or golden, depending on the type of endosymbiont they contain. This arrangement is often mutually beneficial: the radiolarian can make use of the organic nutrients produced by its photosynthetic guest, and the guest may in turn make use of metabolites made by the host or receive physical protection. In some cases, however, the guest is exploited for its photosynthetic products while receiving no benefit itself.

Some dinoflagellates live endosymbiotically in the cells of corals, contributing products of their photosynthesis to the partnership. The importance to the coral is demonstrated when the dinoflagellates are attacked by certain bacteria; the coral is ultimately damaged or destroyed when its nutrient supply is reduced.

Some microbial eukaryotes are deadly

The best-known pathogenic microbial eukaryotes are members of the genus *Plasmodium*, a highly specialized group of *apicomplexans* that spend part of their life cycle as parasites within human red blood cells, where they are the cause of malaria (**Figure 27.3**). In terms of the number of people affected, malaria is one of the world's three most serious infectious diseases, and it kills more than a million people each year. Every 30 seconds malaria kills someone somewhere—usually in sub-Saharan Africa, although malaria occurs in more than 100 countries. About 600 million people suffer from this disease.

Female mosquitoes of the genus *Anopheles* transmit *Plasmodium* to humans. In other words, *Anopheles* is the *vector* for malaria. The parasite enters the human circulatory system when an infected *Anopheles* mosquito penetrates the human skin in search of blood. The parasites find their way to cells in the liver and the lymphatic system, change their form, multiply, and reenter the bloodstream, attacking red blood cells.

The parasites multiply inside the red blood cells, which then burst, releasing new swarms of parasites. If another *Anopheles* bites the victim, the mosquito takes in *Plasmodium* cells along with blood. Some of the ingested cells are gametes that formed in human cells. The gametes unite in the mosquito, forming zygotes that lodge in the mosquito's gut, divide several times, and move into its salivary glands, from which they can be passed on to another human host. Thus *Plasmodium* is an extracellular parasite in the mosquito vector and an intracellular parasite in the human host.

Plasmodium has proved to be a singularly difficult pathogen to attack. The complex *Plasmodium* life cycle is best broken by the removal of stagnant water, in which mosquitoes breed. The use of insecticides to reduce the *Anopheles* population can be effective, but their benefits must be weighed against the ecological, economic, and health risks posed by the insecticides themselves.

The genomes of one malarial parasite, *Plasmodium falciparum*, and one of its vectors, *Anopheles gambiae*, have been sequenced and published. These advances should lead to a better understand-

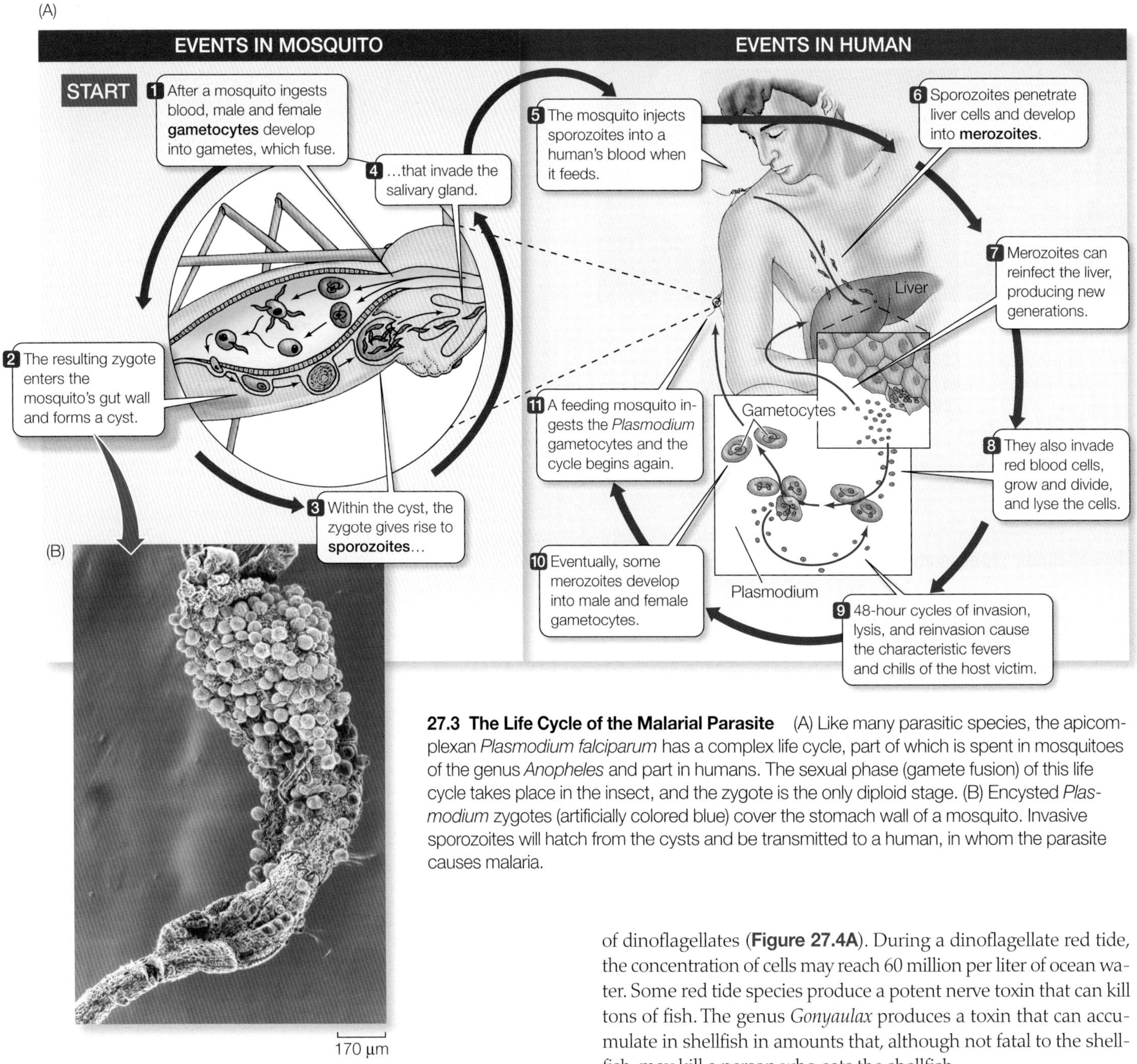

27.3 The Life Cycle of the Malarial Parasite (A) Like many parasitic species, the apicomplexan *Plasmodium falciparum* has a complex life cycle, part of which is spent in mosquitoes of the genus *Anopheles* and part in humans. The sexual phase (gamete fusion) of this life cycle takes place in the insect, and the zygote is the only diploid stage. (B) Encysted *Plasmodium* zygotes (artificially colored blue) cover the stomach wall of a mosquito. Invasive sporozoites will hatch from the cysts and be transmitted to a human, in whom the parasite causes malaria.

ing of the biology of malaria and to the possible development of drugs, vaccines, or other means of dealing with this pathogen or its insect vectors. In the opening of Chapter 30 we describe a novel treatment of mosquito netting with fungi that attack mosquitoes, first reported in 2005.

Some *kinetoplastids* are human pathogens, such as the trypanosomes discussed at the opening of this chapter. Recall from the opening of this chapter that trypanosomes cause sleeping sickness, leishmaniasis, and Chagas' disease. The genomes of all three of these trypanosomes were sequenced in 2005.

Some *chromalveolates*, including diatoms, dinoflagellates, and haptophytes, reproduce in enormous numbers in warm and somewhat stagnant waters. The result can be a "red tide," so called because of the reddish color of the sea that results from the pigments of dinoflagellates (**Figure 27.4A**). During a dinoflagellate red tide, the concentration of cells may reach 60 million per liter of ocean water. Some red tide species produce a potent nerve toxin that can kill tons of fish. The genus *Gonyaulax* produces a toxin that can accumulate in shellfish in amounts that, although not fatal to the shellfish, may kill a person who eats the shellfish.

The haptophyte *Emiliania huxleyi* is one of the smallest unicellular eukaryotes, but it can form tremendous blooms in ocean waters. This *coccolithophore* ("sphere of stone") has an armored coating that makes the surface water more reflective (**Figure 27.4B**). This reflectivity cools the deeper layers of water below the bloom by reducing the amount of sunlight that penetrates. At the same time, it is possible that *E. huxleyi* contributes to global warming, because its metabolism increases the amount of dissolved CO_2 in ocean waters.

We continue to rely on the products of ancient marine microbial eukaryotes

Diatoms are lovely to look at, as we saw in Figure 27.1, but their importance to us goes far beyond aesthetics. They store oil as an

27.4 Chromalveolates Can Bloom in the Oceans (A) By reproducing in astronomical numbers, the dinoflagellate *Gonyaulax tamarensis* can cause toxic red tides, such as this one along the coast of Baja California. (B) Massive blooms of this coccolithophore, a tiny haptophyte, can reduce the amount of sunlight that is able to penetrate to the waters below.

energy reserve and to help them float at the correct depth in the ocean. Over millions of years, diatoms have died and sunk to the ocean floor, ultimately undergoing chemical changes and becoming a major source of petroleum and natural gas, two of our most important energy supplies and political concerns.

Other marine microbial eukaryotes have also contributed to today's world. Some *foraminiferans,* for example, secrete cells of calcium carbonate. After they reproduce (by mitosis and cytokinesis), the daughter cells abandon the parent shell and make new shells of their own. The discarded shells of ancient foraminiferans make up extensive limestone deposits in various parts of the world, forming a layer hundreds to thousands of meters deep over millions of square kilometers of ocean bottom. Foraminiferan shells also make up the sand of some beaches. A single gram of such sand may contain as many as 50,000 foraminiferan shells and shell fragments.

The shells of individual foraminiferans are easily preserved as fossils in marine sediments. The shells of foraminiferan species have distinctive shapes (**Figure 27.5**), and each geological period has a distinctive assemblage of foraminiferan species. For this reason, and because they are so abundant, the remains of foraminiferans are especially valuable in classifying and dating sedimentary rocks, as well as in oil prospecting. Studies of foraminiferan fossil shells are also used in determining the global temperatures prevalent at the time of their existence.

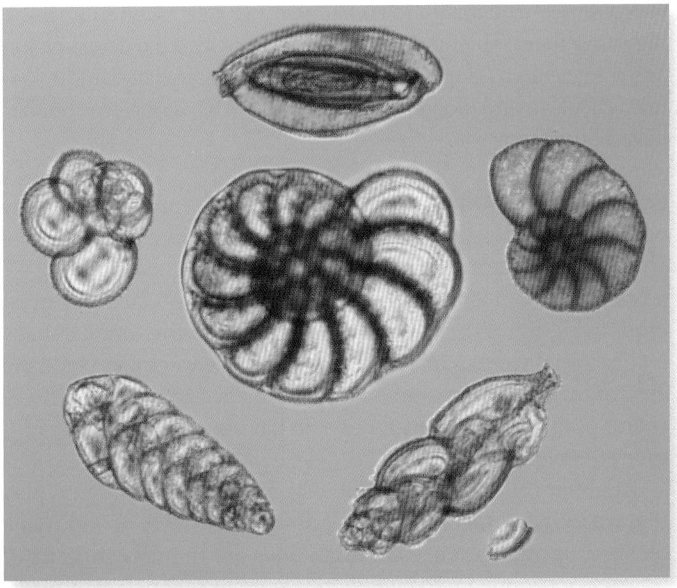

27.5 Foraminiferan Shells Are Building Blocks Foraminiferan shells are made of protein hardened with calcium carbonate. Over millions of years their remains have formed limestone deposits and sandy beaches. Several species are shown in this micrograph.

27.1 RECAP

Many different groups of species are treated here as "microbial eukaryotes," also known as "protists." Most but not all of these organisms are unicellular. These organisms have many effects, both positive and negative, on other organisms and on global ecosystems. Some species are primary producers, many are endosymbionts, and some are pathogens.

- Do you appreciate the full extent to which these organisms are *not* monophyletic? See p. 583 and Table 27.1; glance ahead to Figure 27.17
- Describe the role of female mosquitoes of the genus *Anopheles* as the vector for malaria. See p. 585 and Figure 27.3

This section has presented a very brief overview of the many diverse types of microbial eukaryotes. Indeed, perhaps the only thing all these organisms clearly have in common is that they are all eukaryotes. As we work to appreciate their origins and diversity, we are also working to understand the origin of the eukaryotic cell itself.

27.2 How Did the Eukaryotic Cell Arise?

The eukaryotic cell differs in many ways from the prokaryotic cell. Given the nature of evolutionary processes, these many differences cannot all have arisen simultaneously. We can make some reasonable inferences about the most important events that led to the evolution of a new cell type, bearing in mind that the global environment underwent an enormous change—from anaerobic to aerobic—during the course of these events (see Section 21.2). Keep in mind as you read that these inferences, although reasonable and grounded, are still conjectural; the hypothesis we pursue here is one of a few under current consideration. We present it as a framework for thinking about this challenging problem.

The modern eukaryotic cell arose in several steps

Several events preceded the origin of the modern eukaryotic cell:

- The origin of a flexible cell surface
- The origin of a cytoskeleton
- The origin of a nuclear envelope
- The appearance of digestive vesicles, or *vacuoles*
- The endosymbiotic acquisition of certain organelles

RAMIFICATIONS OF A FLEXIBLE CELL SURFACE Many ancient fossil prokaryotes look like rods, and we presume that they, like most present-day prokaryotic cells, had firm cell walls. The first step toward the eukaryotic condition was the loss of the cell wall by an ancestral prokaryotic cell. This wall-less condition is present in some present-day prokaryotes, although many others have developed new types of cell walls. Let's consider the possibilities open to a flexible cell without a wall.

First, think of cell size. As a cell grows larger, its surface area-to-volume ratio decreases (see Figure 4.2). Unless the surface area can be increased, the cell volume will reach an upper limit. If the cell's surface is flexible, it can fold inward and elaborate itself, creating more surface area for gas and nutrient exchange (**Figure 27.6**).

With a surface flexible enough to allow infolding, the cell can exchange materials with its environment rapidly enough to sustain a larger volume and more rapid metabolism. Furthermore, a flexible surface can pinch off bits of the environment, bringing them into the cell by endocytosis (**Figure 27.7, steps 1–3**).

CHANGES IN CELL STRUCTURE AND FUNCTION Other early steps in the evolution of the eukaryotic cell are likely to have included three advances: the appearance of a cytoskeleton; the formation of ribosome-studded internal membranes, some of which surrounded the DNA; and the evolution of digestive vesicles (**Figure 27.7, steps 3–7**).

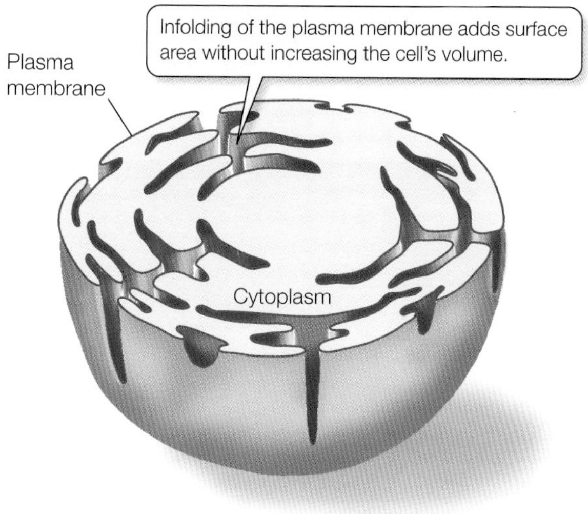

27.6 Membrane Infolding The loss of the rigid prokaryotic cell wall may have allowed the plasma membrane to fold inward and create more surface area.

A cytoskeleton made up of microfilaments and microtubules would support the cell and allow it to manage changes in shape, to distribute daughter chromosomes, and to move materials from one part of the now much larger cell to other parts. The presence of microtubules in the cytoskeleton could have evolved in some cells to give rise to the characteristic eukaryotic flagellum. The origin of the cytoskeleton is becoming clearer, as homologs of the genes that encode many cytoskeletal proteins have been found in modern prokaryotes.

The DNA of a prokaryotic cell is attached to a site on its plasma membrane. If that region of the plasma membrane were to fold into the cell, the first step would be taken toward the evolution of a *nucleus*, a primary feature of the eukaryotic cell.

From an intermediate kind of cell, the next step was probably *phagocytosis*—the ability to consume other cells by engulfing and digesting them. The first true eukaryote possessed a cytoskeleton and a nuclear envelope. It may have had an associated endoplasmic reticulum and Golgi apparatus, and perhaps one or more flagella of the eukaryotic type.

ENDOSYMBIOSIS AND ORGANELLES While the processes already outlined were taking place, the cyanobacteria were busy generating oxygen gas as a product of photosynthesis. The increasing O_2 levels in the atmosphere had disastrous consequences for most other living things because most organisms of the time (archaea and bacteria) were unable to tolerate the newly aerobic, oxidizing environment. But some prokaryotes managed to cope with these changes, and—fortunately for us—so did some of the new phagocytic eukaryotes.

At about this time endosymbiosis might have come into play (**Figure 27.7, steps 8 and 9**). Recall that the theory of endosymbiosis proposes that certain organelles are the descendants of prokaryotes engulfed, but not digested, by ancient eukaryotic cells (see Section 4.5). A crucial endosymbiotic event in the history of the Eukarya was the incorporation of a proteobacterium that evolved into the mitochondrion. Initially, the new organelle's primary function was probably to detoxify O_2 by reducing it to water.

Later, this reduction became coupled with the formation of ATP—respiration. Upon completion of this step, the basic modern eukaryotic cell was complete.

Some important eukaryotes are the result of yet another endosymbiotic step, the incorporation of a prokaryote related to today's cyanobacteria, which became the chloroplast.

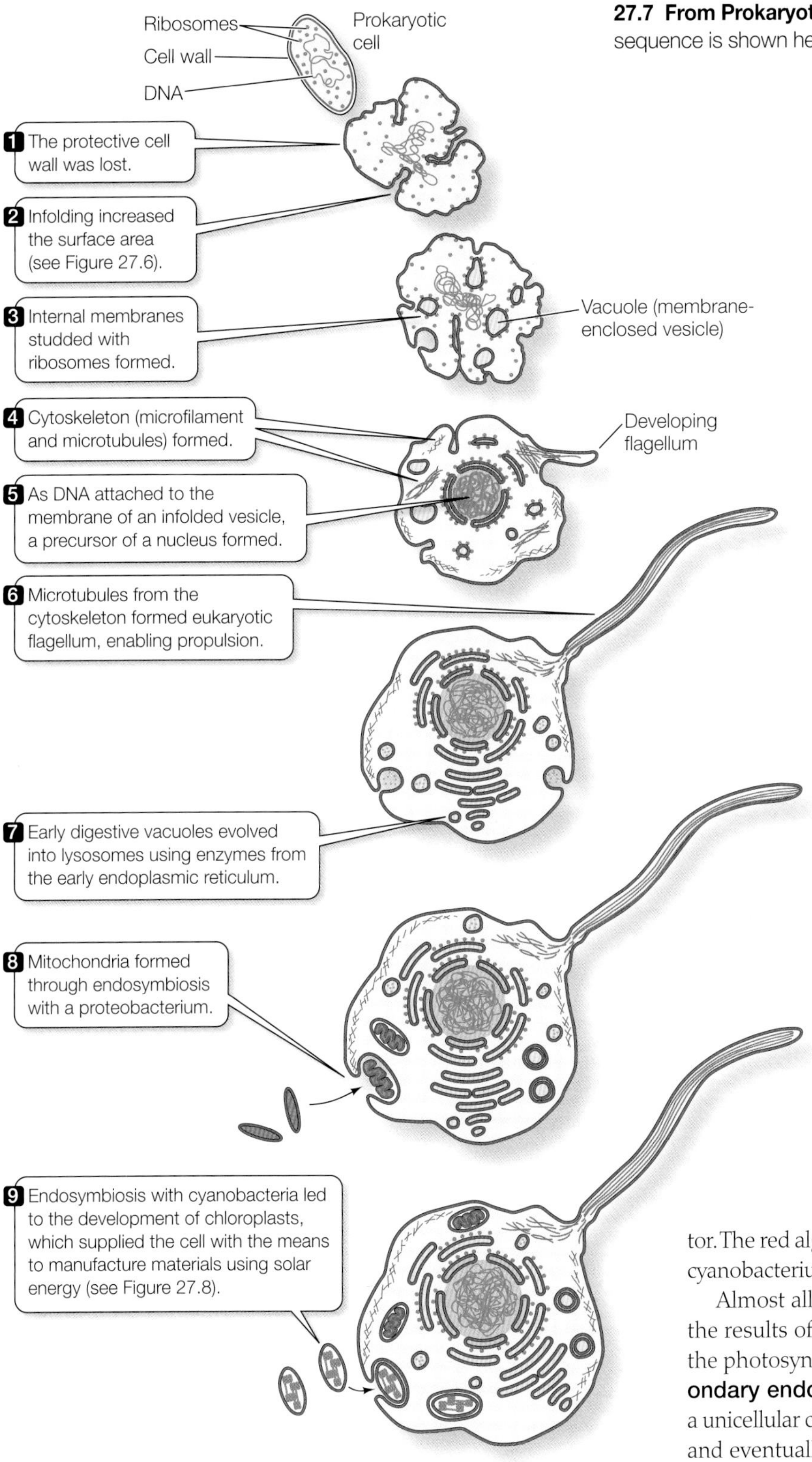

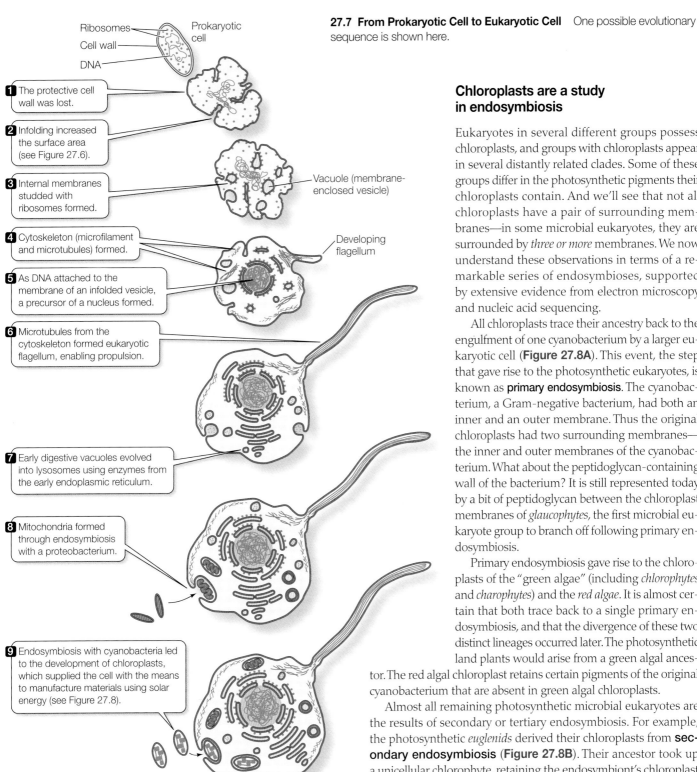

27.7 From Prokaryotic Cell to Eukaryotic Cell One possible evolutionary sequence is shown here.

Chloroplasts are a study in endosymbiosis

Eukaryotes in several different groups possess chloroplasts, and groups with chloroplasts appear in several distantly related clades. Some of these groups differ in the photosynthetic pigments their chloroplasts contain. And we'll see that not all chloroplasts have a pair of surrounding membranes—in some microbial eukaryotes, they are surrounded by *three or more* membranes. We now understand these observations in terms of a remarkable series of endosymbioses, supported by extensive evidence from electron microscopy and nucleic acid sequencing.

All chloroplasts trace their ancestry back to the engulfment of one cyanobacterium by a larger eukaryotic cell (**Figure 27.8A**). This event, the step that gave rise to the photosynthetic eukaryotes, is known as **primary endosymbiosis**. The cyanobacterium, a Gram-negative bacterium, had both an inner and an outer membrane. Thus the original chloroplasts had two surrounding membranes—the inner and outer membranes of the cyanobacterium. What about the peptidoglycan-containing wall of the bacterium? It is still represented today by a bit of peptidoglycan between the chloroplast membranes of *glaucophytes*, the first microbial eukaryote group to branch off following primary endosymbiosis.

Primary endosymbiosis gave rise to the chloroplasts of the "green algae" (including *chlorophytes* and *charophytes*) and the *red algae*. It is almost certain that both trace back to a single primary endosymbiosis, and that the divergence of these two distinct lineages occurred later. The photosynthetic land plants would arise from a green algal ancestor. The red algal chloroplast retains certain pigments of the original cyanobacterium that are absent in green algal chloroplasts.

Almost all remaining photosynthetic microbial eukaryotes are the results of secondary or tertiary endosymbiosis. For example, the photosynthetic *euglenids* derived their chloroplasts from **secondary endosymbiosis** (**Figure 27.8B**). Their ancestor took up a unicellular chlorophyte, retaining the endosymbiont's chloroplast and eventually losing the rest of its constituents. This history explains why the photosynthetic euglenids have the same photosynthetic pigments as the chlorophytes and land plants. It also accounts for the third membrane of the euglenoid chloroplast, which is derived from the euglenid's plasma membrane (as a result of endocytosis). Other evidence for secondary endosymbiosis comes from the observation that certain chloroplast products of secondary endosymbiosis contain traces of the nuclei of the cells that were engulfed.

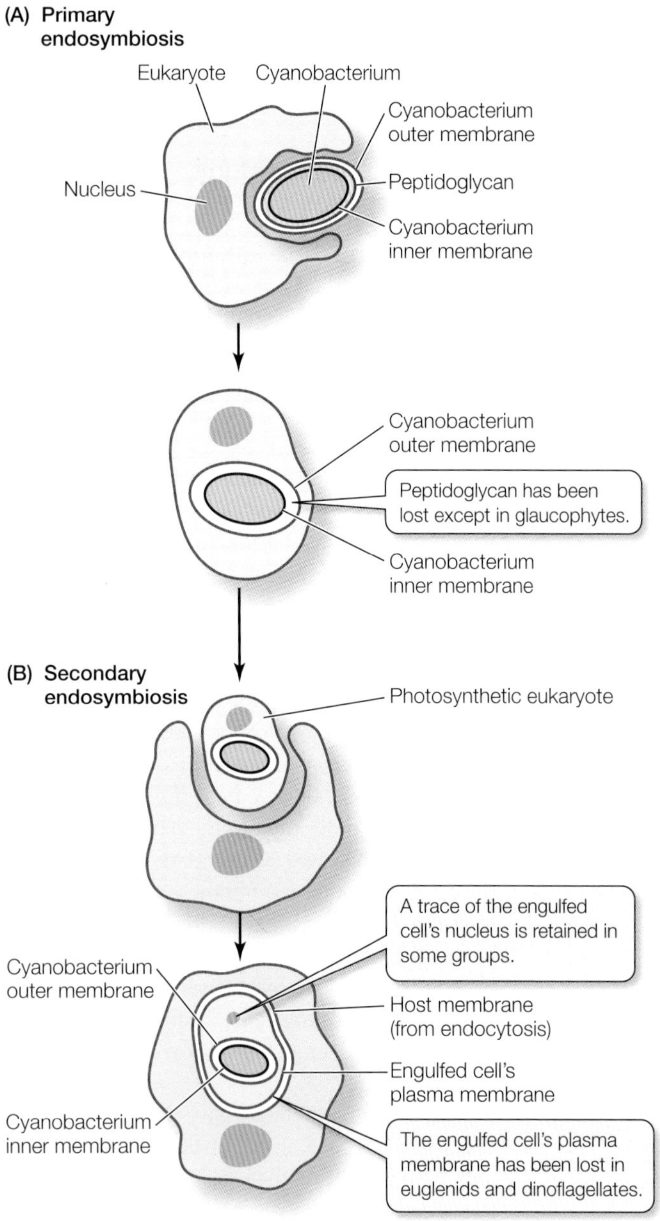

27.8 Endosymbiotic Events in the Family Tree of Chloroplasts (A) A single instance of primary endosymbiosis ultimately gave rise to all of today's chloroplasts. A eukaryotic cell engulfed a cyanobacterium, but did not digest it. (B) Secondary endosymbiosis—the uptake and retention of a chloroplast-containing cell by another eukaryotic cell—took place a few times. Tertiary endosymbiosis and sequential secondary endosymbiosis have also occurred.

Members of another photosynthetic microbial eukaryote group have chloroplasts derived by secondary endosymbiosis with a unicellular red alga. The chloroplasts of cryptophytes (one of the clades of chromalveolates) contain reduced red algal nuclei and appear to be sister to all other chromalveolate chloroplasts. The red clade of the chloroplasts may have been involved in only one secondary endosymbiosis, but the green clade of chloroplasts appears to have been involved in more than one secondary endosymbiosis.

Some dinoflagellates took up other partners by *tertiary endosymbiosis*. For example, one dinoflagellate apparently lost its chloroplast and took up a haptophyte (itself the result of secondary endosymbiosis). The result is the dinoflagellate *Karenia brevis*. There has been at least one case of *sequential secondary endosymbiosis*—another dinoflagellate lost its red algal chloroplast and engulfed a chlorophyte, thus becoming a new dinoflagellate species with a green algal chloroplast with two membranes. The actual number of endosymbiotic events in the course of evolution was small, but the results were diverse and profound.

Although euglenoid chloroplasts are descendants of a chlorophyte and haptophyte chloroplasts are descendants of a red alga, this does not mean that euglenids themselves are descendants of a chlorophyte, nor are haptophytes themselves descendants of a red alga. The ancestors that took up green or red algae in secondary endosymbioses had their own evolutionary histories. Thus the nuclear and chloroplast genomes of these organisms have different histories.

We cannot yet account for the presence of some prokaryotic genes in eukaryotes

Several uncertainties remain about the origins of eukaryotic cells. Lateral gene transfer complicates the study of eukaryote origins, just as it complicates the study of relationships among prokaryotes. And it seems unlikely that lateral gene transfer could have been extensive enough to account for the increasing numbers of genes of bacterial origin that are being found in eukaryotes by ongoing genetic analyses.

An endosymbiotic origin of mitochondria and chloroplasts accounts for the presence of bacterial genes encoding enzymes for energy metabolism (respiration and photosynthesis) in eukaryotes, but it does not explain the presence of some other bacterial genes. The eukaryotic genome clearly is a mixture of genes with different origins. A recent suggestion is that the Eukarya might have arisen from the mutualistic fusion of a Gram-negative bacterium and an archaean.

Many interesting ideas about eukaryotic origins await additional data and analysis. We can expect that these questions and others will eventually yield to additional research.

27.2 RECAP

The modern eukaryotic cell arose from an ancestral prokaryote in several steps, one of which involved endosymbiosis. The exact origins of eukaryotic cells are uncertain.

- Can you explain the importance of a flexible cell surface in the origin of eukaryotes? See p. 588 and Figure 27.6
- Can you identify some of the probable events involved in the evolution of the eukaryotic cell from a prokaryotic cell? See p. 588 and Figure 27.7
- Do you understand the difference between the origin of chloroplasts by primary and secondary endosymbiosis? See p. 589 and Figure 27.8

Having considered some of the known and suspected steps that led from the prokaryotic to the eukaryotic condition, let's now see what use the microbial eukaryotes made of their new features.

27.3 How Did the Microbial Eukaryotes Diversify?

The eukaryotic cell possesses some very useful features. The cytoskeleton allows for various means of locomotion, and also manages the controlled movement of cellular constituents (notably the mitotic and meiotic chromosomes). The specialized organelles of eukaryotes support a variety of activities. Given these tools, the microbial eukaryotes have been able to explore many environments and have exploited a variety of nutrient sources.

Microbial eukaryotes have different lifestyles

Most microbial eukaryotes are aquatic. Some live in marine environments, others in fresh water, and still others in the body fluids of other organisms. Many unicellular aquatic eukaryotes are plankton floating freely in the water. The *slime molds* inhabit damp soil and the moist, decaying bark of rotting trees. Other microbial eukaryotes also live in soil water, and some of them contribute to the global nitrogen cycle by preying on soil bacteria and recycling their nitrogen compounds into nitrates.

Some microbial eukaryotes are photosynthetic autotrophs, some are heterotrophs, and some switch with ease between the autotrophic and heterotrophic modes of nutrition. Some of the heterotrophs ingest their food; others, including many parasites, absorb nutrients from their environment.

Certain microbial eukaryotes, formerly classified as animals, are sometimes referred to as *protozoans*, although biologists increasingly believe that term lumps together too many phylogenetically distant groups. Most protozoans are ingestive heterotrophs. Similarly, there are several kinds of photosynthetic microbial eukaryotes that some biologists still refer to as *algae* (singular *alga*). Although these two terms are useful in some contexts, they do not correspond with our current understanding of phylogeny, and we generally avoid them in this chapter, except as parts of descriptive names such as "red algae."

Microbial eukaryotes have diverse means of locomotion

Although a few microbial eukaryote groups consist entirely of nonmotile organisms, most groups include cells that move, either by amoeboid motion, by ciliary action, or by means of flagella. Each of these types of motion is based on activities of the cytoskeleton.

In *amoeboid motion*, the cell forms **pseudopods** ("false feet") that are extensions of its constantly changing cell shape. Cells such as the one shown in **Figure 27.9** simply extend a pseudopod and then flow into it. Regions of the cytoplasm alternate between a more liquid state and a stiffer state, and a network of cytoskeletal microfilaments squeezes the more liquid cytoplasm forward.

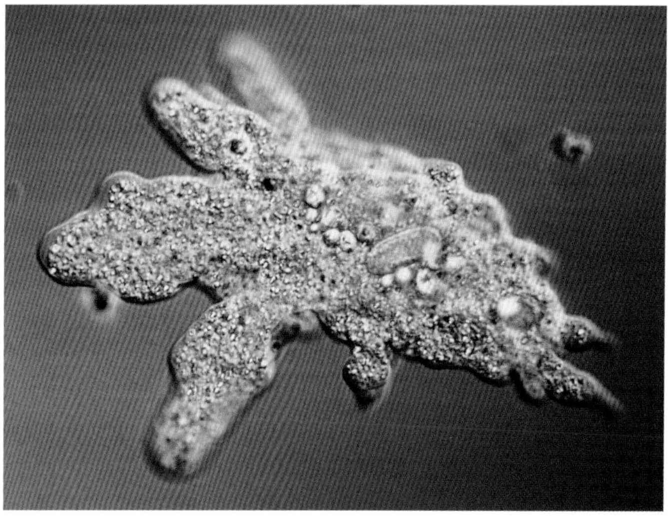

27.9 An Amoeba The flowing pseudopods are constantly changing shape as the amoeba moves and feeds.

As shown in Figure 27.7, the proteins of the eukaryotic cytoskeleton form microtubules that allowed the evolution of different means of locomotion. *Cilia* are tiny, hairlike organelles that beat in a coordinated fashion to move the cell forward or backward. Some ciliated organisms can change direction rapidly in response to their environment. A eukaryotic *flagellum* moves like a whip; some flagella *push* the cell forward, others *pull* the cell forward. Cilia and eukaryotic flagella are identical in cross section, with a "9 + 2" arrangement of microtubules (see Figure 4.22); they differ only in length.

Microbial eukaryotes employ vacuoles in several ways

Most unicellular organisms are microscopic. As we noted above, an important reason that cells are small is that they need enough membrane surface area in relation to their volume to support the exchange of materials required for their existence. Many relatively large unicellular eukaryotes minimize this problem by having membrane-enclosed *vacuoles* of various types that increase their effective surface area.

Organisms living in fresh water are hypertonic to their environment (see Section 5.3). Many freshwater microbial eukaryotes such as *Paramecium* address this problem by means of specialized vacuoles that excrete the excess water they constantly take in by osmosis. Members of several groups have such **contractile vacuoles**. The excess water collects in the contractile vacuole, which then expels the water from the cell (**Figure 27.10**).

A second important type of vacuole found in *Paramecium* and many other microbial eukaryotes is the **food vacuole**. These organisms engulf solid food by endocytosis, forming a vacuole within which the food is digested (**Figure 27.11**). Smaller vesicles containing digested food pinch away from the food vacuole and enter the cytoplasm. These tiny vesicles provide a large surface area across which the products of digestion may be absorbed by the rest of the cell.

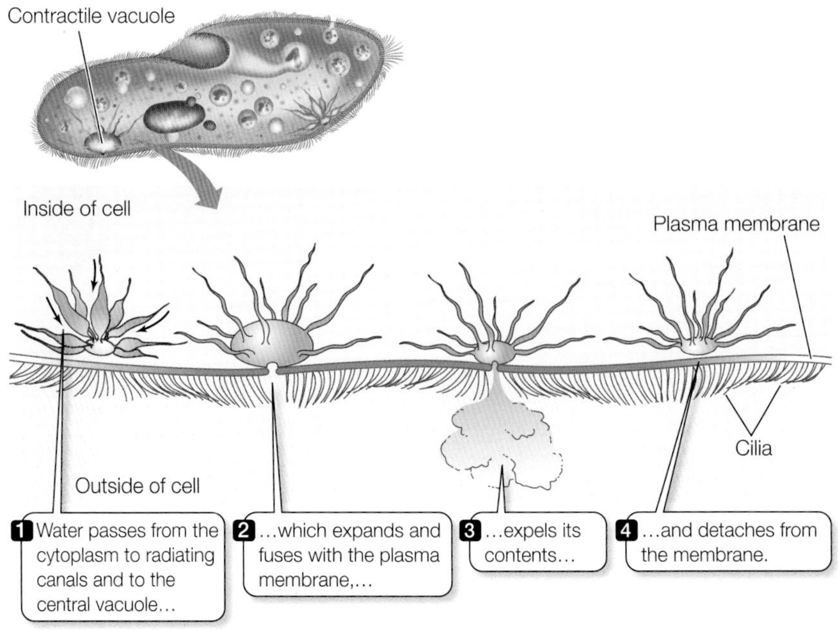

27.10 Contractile Vacuoles Bail Out Excess Water Contractile vacuoles remove the water that constantly enters freshwater microbial eukaryotes by osmosis.

The cell surfaces of microbial eukaryotes are diverse

A few microbial eukaryotes, such as the amoeba shown in Figure 27.9, are surrounded by only a plasma membrane, but most have stiffer surfaces that maintain the structural integrity of the cell. Many have cell walls, which are often complex in structure, outside the plasma membrane. Other microbial eukaryotes that lack cell walls have a variety of ways of strengthening their surfaces.

Paramecium has proteins in its cell surface—known as a *pellicle* in this genus—that make it flexible but resilient. Other groups have external "shells," which the organism either produces itself, as foraminiferans do (see Figure 27.5), or makes from bits of sand and thickenings immediately beneath the plasma membrane, as some amoebas do (**Figure 27.12A**). The complex cell walls of diatoms are glassy, based on silicon (**Figure 27.12B**; see also Figure 27.1). Biologists recently measured, at a microscopic scale, the forces needed to break single, living diatoms. They discovered that the glassy cell walls are exceptionally strong. Evolution of these walls by natural selection may have given diatoms an enhanced defense against predators, and thus an edge over competitors.

27.3 RECAP

Microbial eukaryotes are diverse in their habitat, nutrition, locomotion, and body form.

- Can you explain the roles of the cytoskeleton in the locomotion of microbial eukaryotes? See p. 591
- Do you understand the operation of contractile and food vacuoles in *Paramecium*? See p. 591 and Figures 27.10 and 27.11

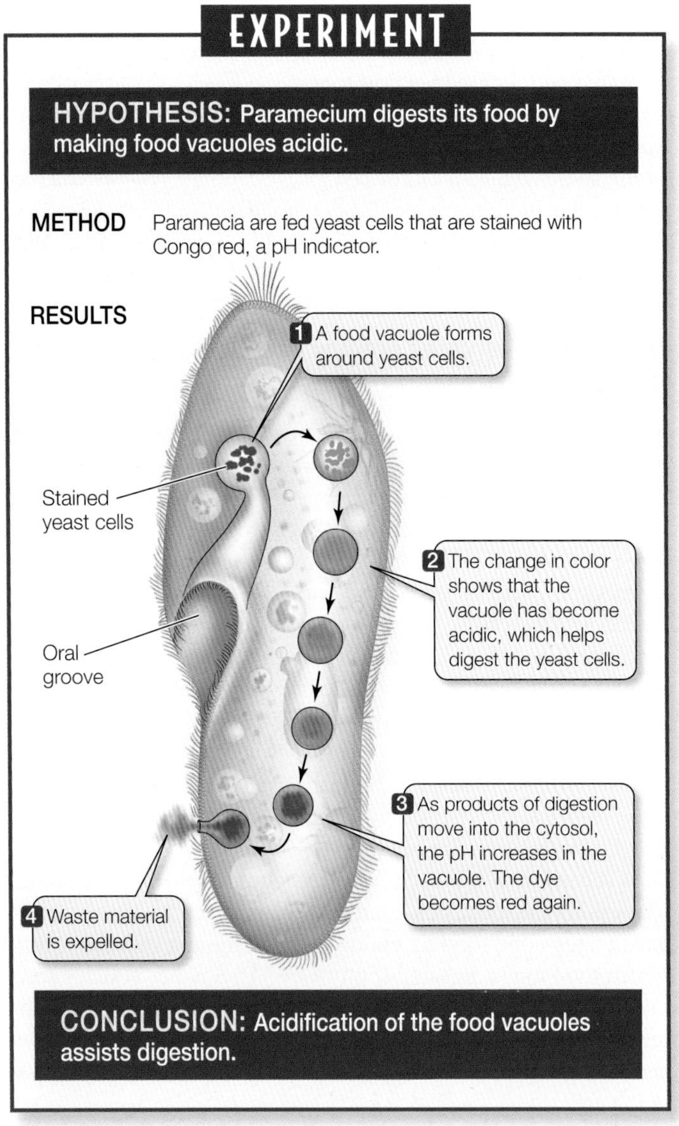

27.11 Food Vacuoles Handle Digestion and Excretion An experiment with *Paramecium* demonstrates the function of food vacuoles. *Paramecium* ingests food by way of the oral groove shown at the left. The dye Congo red turns green at an acidic pH and red at neutral or basic pH. Acid has a digestive function in other organisms (as in the human stomach), so its presence in food vacuoles suggests that digestion is taking place there. FURTHER RESEARCH: How might you determine whether food vacuoles in *Paramecium* contain digestive enzymes?

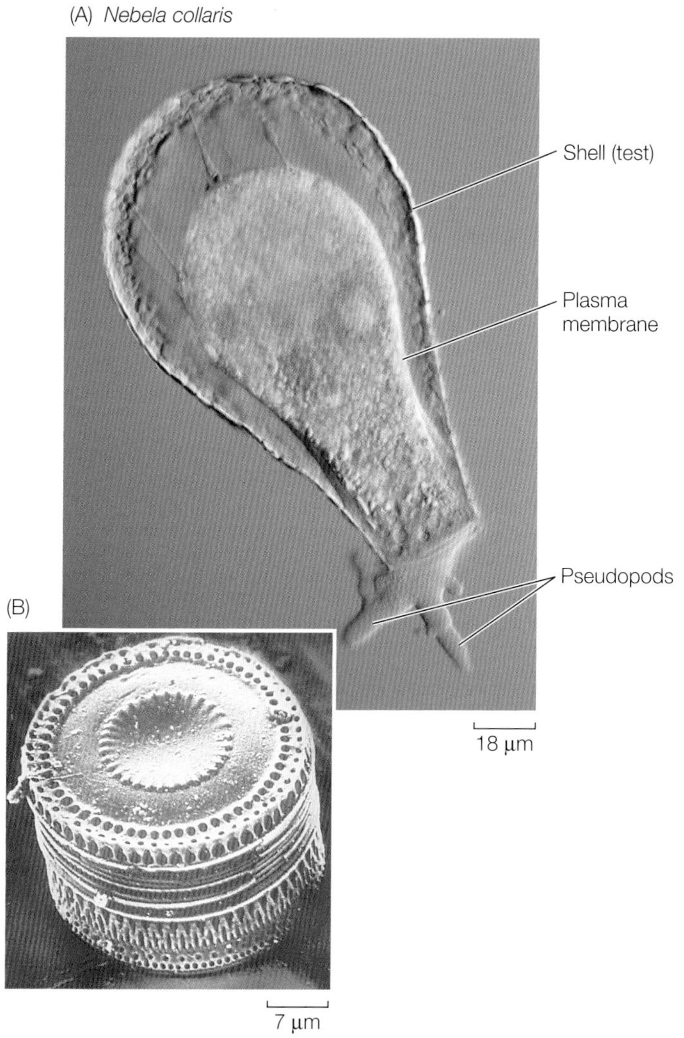

27.12 Cell Surfaces in the Microbial Eukaryotes (A) This testate amoeba has built a lightbulb-shaped shell, or test, by gluing sand grains together. Its pseudopods extend through the single aperture in the test (compare with Figure 27.9). (B) The artificial color added to this scanning electron micrograph shows the intricate patterning of the silicate cell surface of a diatom.

The diversity of body form, habitat, nutrition and locomotion found among the microbial eukaryotes reflects the diversity of avenues pursued during the early evolution of eukaryotes. The diverse reproductive modes and life cycles that have evolved among the microbial eukaryotes are a case in point.

27.4 How Do Microbial Eukaryotes Reproduce?

Although most microbial eukaryotes practice both asexual and sexual reproduction, some groups lack sexual reproduction. As we will see, some microbial eukaryotes separate the acts of sex and reproduction, so that the two are not directly linked.

The asexual reproductive processes found among the microbial eukaryotes include

- *Binary fission*: equal splitting of the cell, with mitosis followed by cytokinesis
- *Multiple fission*: splitting into more than two cells
- *Budding*: the outgrowth of a new cell from the surface of an old one
- *Spores*: the formation of specialized cells that are capable of developing into new organisms

Sexual reproduction in microbial eukaryotes takes various forms. In some microbial eukaryotes, as in animals, the gametes are the only haploid cells. In others, the zygote is the only diploid cell. In still others, by contrast, both diploid and haploid cells undergo mitosis, giving rise to *alternation of generations*.

Some microbial eukaryotes have reproduction without sex, and sex without reproduction

Like all groups in the *ciliate* clade, members of the genus *Paramecium* possess two types of nuclei, commonly a single *macronucleus* and, within the same cell, from one to several *micronuclei*. The micronuclei, which are typical eukaryotic nuclei, are essential for genetic recombination. The macronucleus is derived from micronuclei. Each macronucleus contains many copies of the genetic information, packaged in units containing very few genes each. The macronuclear DNA is transcribed and translated to regulate the life of the cell. In asexual reproduction, all of the nuclei are copied before the cell divides.

Paramecia also have an elaborate sexual behavior called **conjugation**, in which two paramecia line up tightly against each other and fuse in the oral groove region of the body. Nuclear material is extensively reorganized and exchanged over the next several hours (**Figure 27.13**). As a result of this process, each cell ends up with two haploid micronuclei, one of its own and one from the other cell, which fuse to form a new diploid micronucleus. A new macronucleus develops from the micronucleus through a series of dramatic chromosomal rearrangements. The exchange of nuclei is fully reciprocal—each of the two paramecia gives and receives an equal amount of DNA. The two organisms then separate and go their own ways, each equipped with new combinations of alleles.

Conjugation in *Paramecium* is a *sexual* process of genetic recombination, but it is not a *reproductive* process. The same two cells that begin the process are there at the end, and no new cells are created. As a rule, each asexual clone of paramecia must periodically conjugate. Experiments have shown that if some species are not permitted to conjugate, the clones can live through no more than approximately 350 cell divisions before they die out.

Many microbial eukaryote life cycles feature alternation of generations

What do we mean by **alternation of generations**, the type of life cycle found in many multicellular microbial eukaryotes and all land plants? In alternation of generations, a multicellular, diploid, spore-producing organism gives rise to a multicellular, haploid, gamete-producing organism. When two haploid gametes fuse (a process

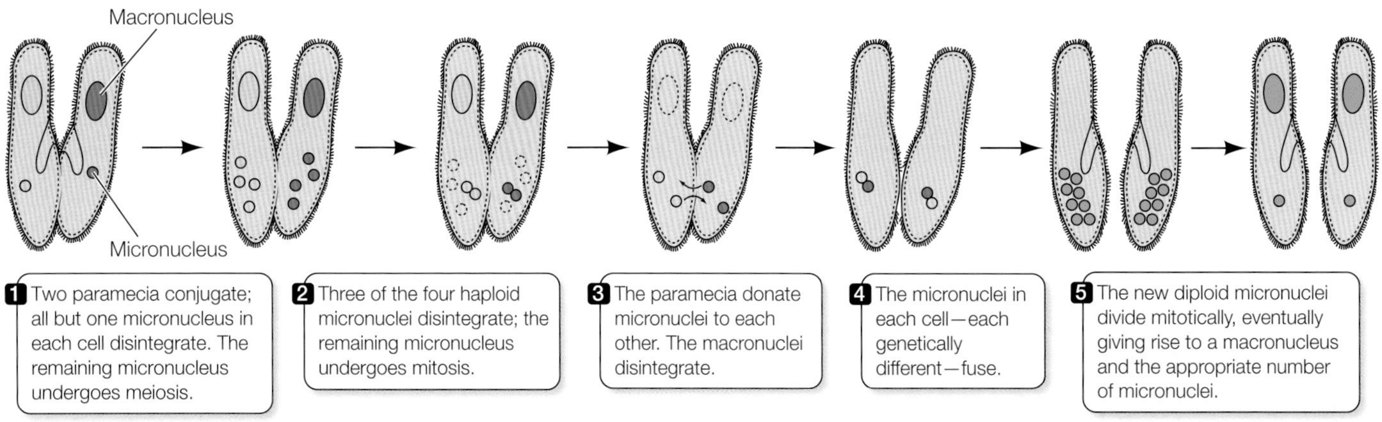

Macronucleus
Micronucleus
1 Two paramecia conjugate; all but one micronucleus in each cell disintegrate. The remaining micronucleus undergoes meiosis.
2 Three of the four haploid micronuclei disintegrate; the remaining micronucleus undergoes mitosis.
3 The paramecia donate micronuclei to each other. The macronuclei disintegrate.
4 The micronuclei in each cell—each genetically different—fuse.
5 The new diploid micronuclei divide mitotically, eventually giving rise to a macronucleus and the appropriate number of micronuclei.

27.13 Paramecia Achieve Genetic Recombination by Conjugating The exchange of micronuclei by conjugating *Paramecium* individuals results in genetic recombination. After conjugation, the cells separate and continue their lives as two individuals.

called *fertilization*, or *syngamy*), a diploid organism is formed (**Figure 27.14**). The haploid organism, the diploid organism, or both may also reproduce asexually.

The two alternating generations (spore-producing and gamete-producing) differ genetically (one has diploid cells and the other has haploid cells), but they may or may not differ morphologically. In **heteromorphic** alternation of generations, the two generations differ morphologically; in **isomorphic** alternation of generations, they do not, despite their genetic difference.

Examples of both heteromorphic and isomorphic alternation of generations are found in both brown algae and green algae. In discussing the life cycles of land plants and multicellular photosynthetic microbial eukaryotes, we will use the terms **sporophyte** ("spore plant") and **gametophyte** ("gamete plant") to refer to the multicellular diploid and haploid generations, respectively.

Gametes are not produced by meiosis because the gametophyte generation is already haploid. Instead, specialized cells of the diploid sporophyte, called **sporocytes**, divide meiotically to produce four haploid spores. The spores may eventually germinate and divide mitotically to produce multicellular haploid gametophytes, which produce gametes by mitosis and cytokinesis.

Gametes, unlike spores, can produce new organisms only by fusing with other gametes. The fusion of two gametes produces a diploid zygote, which then undergoes mitotic divisions to produce a diploid organism: the sporophyte generation. The sporocytes of the sporophyte generation then undergo meiosis and produce haploid spores, starting the cycle anew.

Chlorophytes provide examples of several life cycles

We can use different chlorophyte species to summarize the major features of microbial eukaryote life cycles. Let's begin with the sea lettuce *Ulva lactuca*. Like many chlorophytes, sea lettuce exhibits alternation of generations. The diploid sporophyte of this common multicellular seashore organism is a broad sheet only two cells thick. Some of its cells (sporocytes) differentiate and undergo meiosis and cytokinesis, producing motile haploid spores. These *zoospores* swim away, each propelled by four flagella, and some eventually find a suitable place to settle. The zoospores then lose their flagella and begin to divide mitotically, producing a thin filament that develops into a broad sheet only two cells thick. The gametophyte thus produced looks just like the sporophyte—in other words, *Ulva lactuca* has an isomorphic life cycle (**Figure 27.15**).

In most species of *Ulva*, the female and male gametes are also structurally indistinguishable, making those species **isogamous**—having gametes of identical appearance. Other chlorophytes, including some other species of *Ulva*, are **anisogamous**—having female gametes that are distinctly larger than the male gametes.

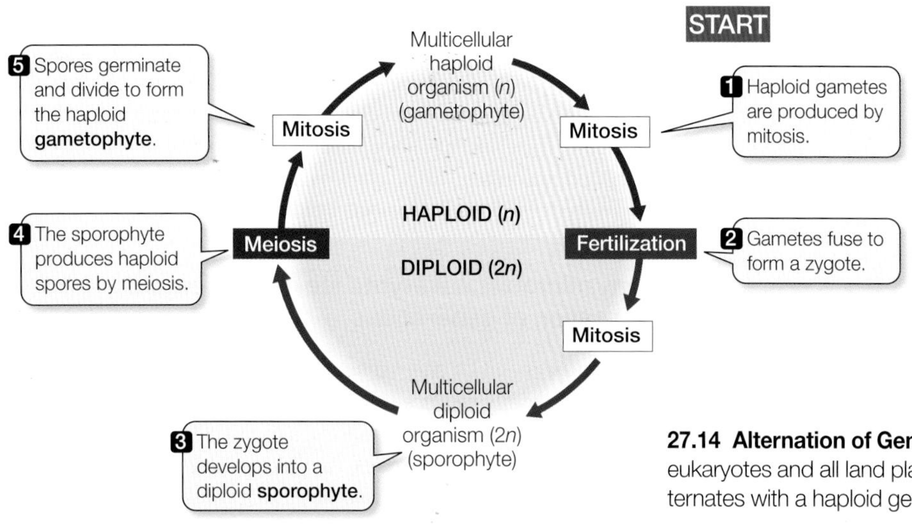

27.14 Alternation of Generations In many multicellular photosynthetic eukaryotes and all land plants, a diploid generation that produces spores alternates with a haploid generation that produces gametes.

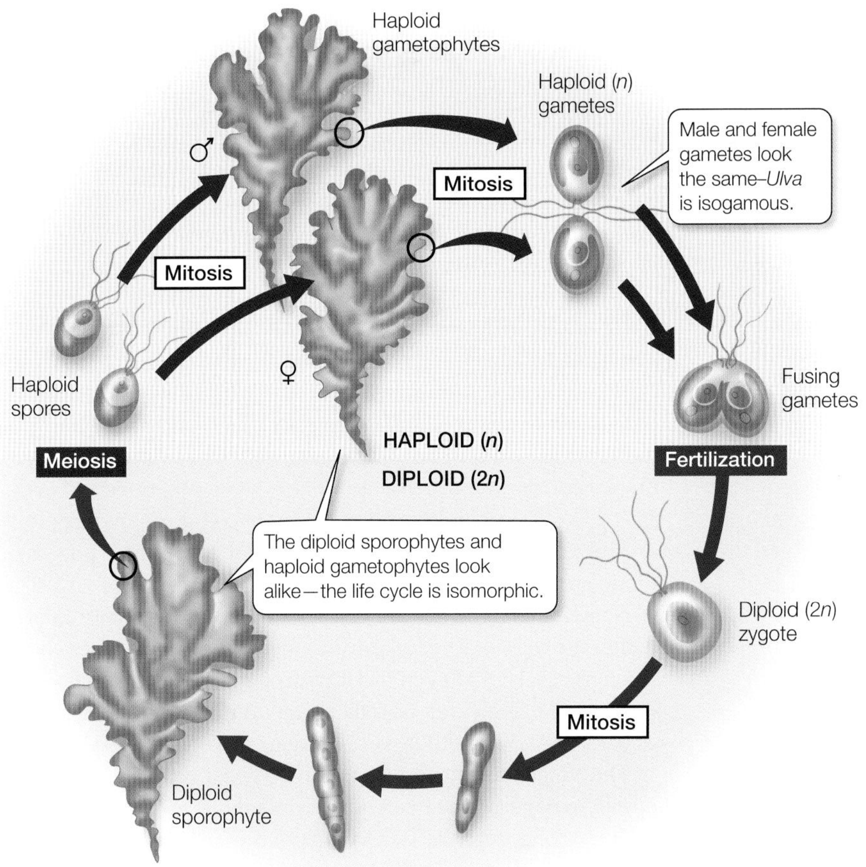

27.15 An Isomorphic Life Cycle The sexual life cycle of *Ulva lactuca* (sea lettuce) is an example of isomorphic alternation of generations.

The life cycles of many other chlorophytes do not feature alternation of generations. Some chlorophytes have a **haplontic** life cycle, in which a multicellular haploid individual produces gametes that fuse to form a zygote. The zygote functions directly as a sporocyte, undergoing meiosis to produce spores, which in turn produce a new haploid individual. In the entire haplontic life cycle, only one cell—the zygote—is diploid. The filamentous organisms of the genus *Ulothrix* are examples of haplontic chlorophytes (**Figure 27.16**).

Some other chlorophytes have a **diplontic** life cycle like that of many animals. In a diplontic life cycle, meiosis of diploid sporocytes produces haploid gametes directly; the gametes fuse, and the resulting diploid zygote divides mitotically to form a new multicellular diploid sporophyte. In such organisms, all cells except the gametes are diploid. Between these two extremes are chlorophytes in which the gametophyte and sporophyte generations are both multicellular, but one generation (usually the sporophyte) is much larger and more prominent than the other.

The life cycles of some microbial eukaryotes require more than one host species

The three trypanosome diseases discussed at the opening of this chapter share a striking feature with malaria: in each case, the eukaryote pathogen completes only part of its life cycle in the human host, and the other part in an insect (see Figure 27.3). Many other microbial eukaryote life cycles require the participation of two different host species.

What could be the advantage of a life cycle with two hosts? This remains an intriguing question. It may be relevant that in the human pathogens described above, the sexual phase of the organism's life cycle—the fusion of gametes into a zygote—takes place in the insect vector. Could this imply that the human host is nothing but a copying machine for the products of sexual reproduction in the vector?

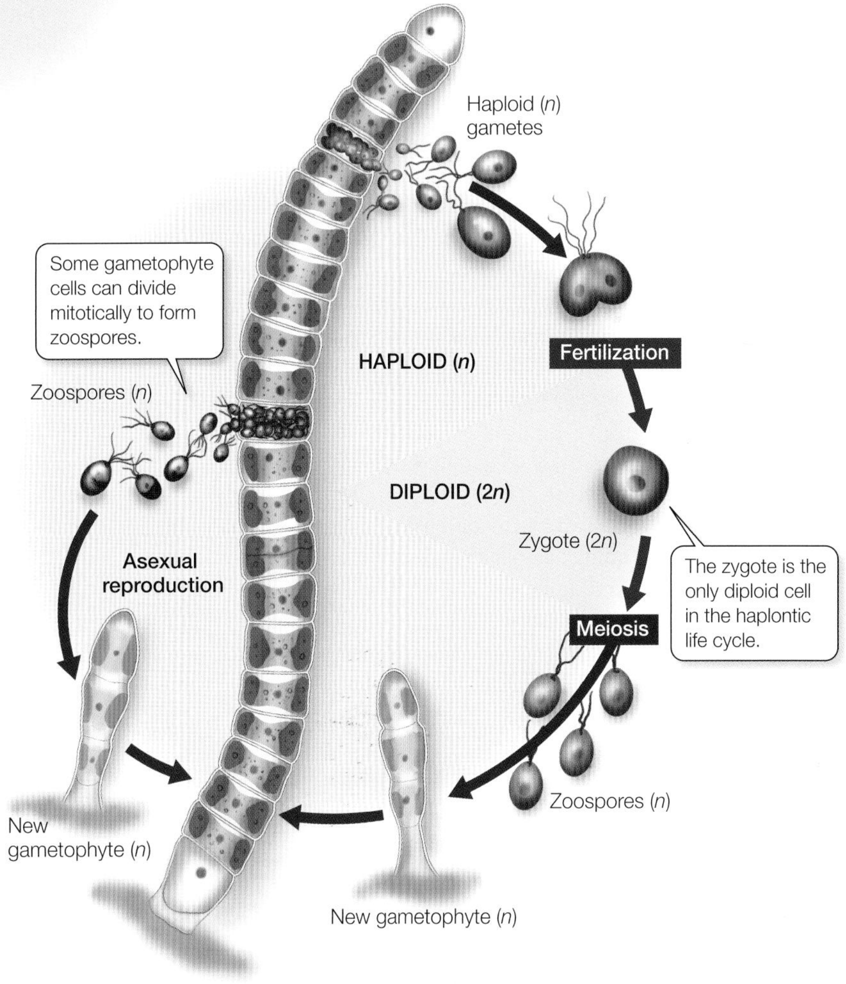

27.16 A Haplontic Life Cycle In the life cycle of *Ulothrix*, a filamentous, multicellular haploid gametophyte generation alternates with a diploid sporophyte generation consisting of a single cell (the zygote). *Ulothrix* gametophytes can also reproduce asexually.

27.4 RECAP

Microbial eukaryotes reproduce both asexually and sexually, and their life cycles are diverse.

- Why is conjugation between paramecia considered a sexual process, but not a reproductive process? See p. 593 and Figure 27.13
- Can you explain the difference between the diplontic human life cycle and a life cycle with alternation of generations? See p. 595

The success of the microbial eukaryotes' diverse adaptations for nutrition, locomotion, and reproduction is evident from the abundance and diversity of eukaryotes living today. In the next section we will survey that diversity.

27.5 What Are the Major Groups of Eukaryotes?

Biologists used to classify the microbial eukaryotes on the basis of features such as those we have described in Sections 27.3 and 27.4. However, scientists using the advanced technologies of electron microscopy and gene sequencing have revealed many new patterns of evolutionary relatedness. New molecular biological techniques, such as rRNA gene sequencing (see Section 26.4), are making it possible to explore evolutionary relationships among the microbial eukaryotes in ever greater detail and with greater confidence. Today we recognize great diversity within many microbial eukaryote clades, whose members have explored a great variety of lifestyles.

The phylogeny of microbial eukaryotes is an area of exciting, challenging research. Their marvelous diversity of body forms and nutritional lifestyles alone justify study of these organisms, and questions about how the multicellular eukaryotic groups (land plants, fungi, and animals) originated from the microbial groups stimulate further interest.

Most eukaryotes can be divided into five major groups: chromalveolates (including alveolates and stramenopiles), Plantae, excavates, Rhizaria, and unikonts (opisthokonts and amoebozoans) (**Figure 27.17**; also see Table 27.1). As we will see, some of these groups consist of organisms with very diverse body plans.

CHROMALVEOLATES

We'll begin our tour of microbial eukaryote groups with the **chromalveolates**, a group that includes the haptophytes, the cryptophytes, and two other large clades: the alveolates and the stramenopiles. The **haptophytes** are unicellular organisms with flagella; many are "armored" with elaborate scales (see Figure 27.4B). The role of the **cryptophytes** in the story of chloroplast evolution was described in Section 27.2.

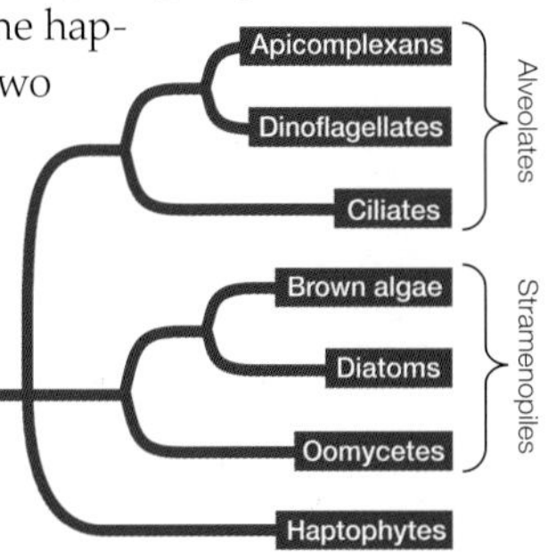

Alveolates have sacs under their plasma membrane

The synapomorphy that characterizes the **alveolate** clade is the possession of sacs called *alveoli* just below their plasma membranes. The alveoli may play a role in supporting the cell surface. These organisms are all unicellular, but they are diverse in body form. The alveolate groups we'll consider in detail here are the dinoflagellates, apicomplexans, and ciliates.

DINOFLAGELLATES ARE UNICELLULAR MARINE ORGANISMS WITH TWO FLAGELLA Most **dinoflagellates** are marine organisms. A distinctive mixture of photosynthetic and accessory pigments gives their chloroplasts a golden brown color. (The endosymbiotic events that gave rise to dinoflagellates with different numbers of membranes surrounding their chloroplasts were described in Section 27.2.) The dinoflagellates are of great ecological, evolutionary, and morphological interest. They are important primary photosynthetic producers of organic matter in the oceans.

Some dinoflagellates are photosynthetic endosymbionts living within the cells of other organisms, including various invertebrates (such as corals) and even other marine microbial eukaryotes (see Figure 27.2). Some dinoflagellates are nonphotosynthetic and live as parasites within other marine organisms.

Dinoflagellates have a distinctive appearance. They generally have two flagella, one in an equatorial groove around the cell, the other starting near the same point as the first and passing down a longitudinal groove before extending into the surrounding medium (**Figure 27.18**). Some dinoflagellates, notably *Pfiesteria piscicida*, can take on different forms, including amoeboid ones, depending on environmental conditions. It has been claimed that *P. piscicida* can occur in at least two dozen distinct forms, although this claim is highly controversial. In any case, this remarkable dinoflagellate is harmful to fish and can, when present in great numbers, both stun and feed on them.

ALL APICOMPLEXANS ARE PARASITES Exclusively parasitic organisms, the **apicomplexans** derive their name from the *apical complex*, a mass of organelles contained within the *apical* end (the tip) of a cell. These organelles help the apicomplexan invade its host's tissues. For example, the apical complex enables the merozoites and sporozoites of *Plasmodium*, the causal agent of malaria, to enter their target cells in the human body (see Figure 27.3).

Like many obligate parasites, apicomplexans have elaborate life cycles featuring asexual and sexual reproduction by a series of very dissimilar life stages. Often these stages are associated with two different types of host organisms, as is the case with *Plasmodium*.

The apicomplexan *Toxoplasma* alternates between cats and rats to complete its life cycle. A rat infected with *Toxoplasma* loses its fear of cats, making it more likely to be eaten by, and thus transfer the parasite to, a cat.

Apicomplexans lack contractile vacuoles. They contain a much-reduced, nonfunctional chloroplast (derived, like all chromalveolate chloroplasts, from secondary endosymbiosis of a red alga). This chloroplast might be a target for a future antimalarial drug.

27.17 Major Eukaryote Groups in an Evolutionary Context Most of the groups shown here are clades. The microbial eukaryotes themselves do not constitute a clade. The dashed lines indicate clades for which the evidence is weak or disputed. The root of the tree is uncertain.

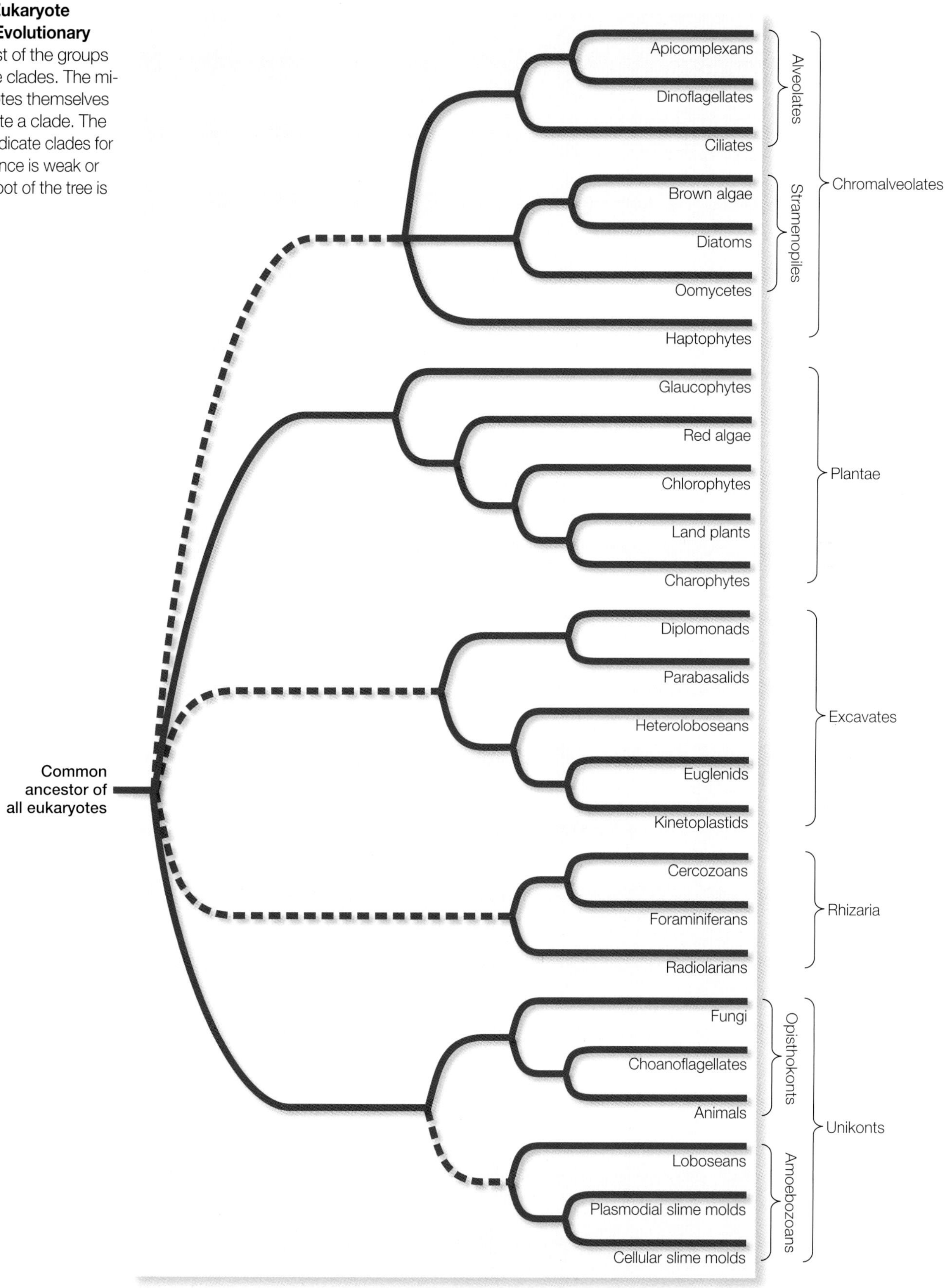

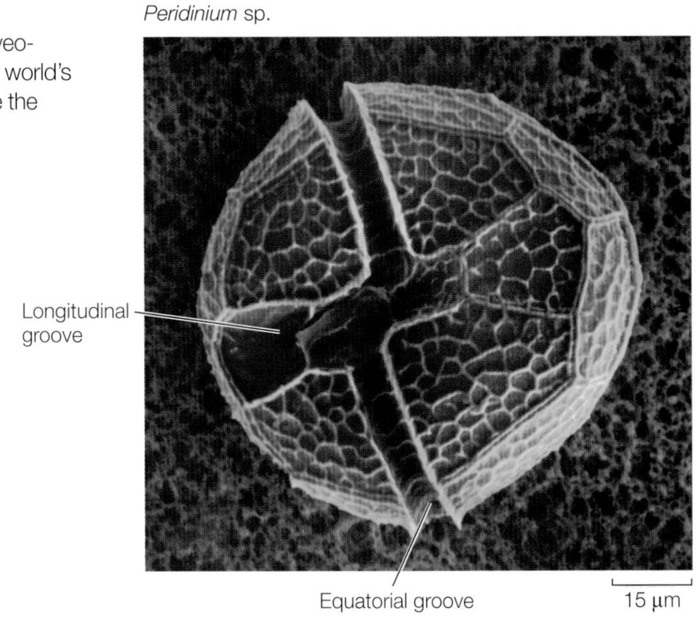

27.18 A Dinoflagellate The dinoflagellates are an important group of alveolates. Most of them are photosynthetic and are a crucial component of the world's phytoplankton. They are often endosymbiotic (see Figure 27.2) and can be the agents of deadly ocean "blooms" (see Figure 27.4A).

CILIATES HAVE TWO TYPES OF NUCLEI The **ciliates** are so named because they characteristically have numerous hairlike cilia shorter than, but otherwise identical to, eukaryotic flagella. This group is noteworthy for its diversity and ecological importance (**Figure 27.19**). Almost all ciliates are heterotrophic (although a few contain photosynthetic endosymbionts), and they are much more complex in body form than are most other unicellular eukaryotes. The definitive characteristic of ciliates is the possession of two types of nuclei (see Figure 27.13).

Paramecium, a frequently studied ciliate genus, exemplifies the complex structure and behavior of ciliates (**Figure 27.20**). The slipper-shaped cell is covered by an elaborate pellicle, a structure composed principally of an outer membrane and an inner layer of closely packed, membrane-enclosed sacs (the alveoli) that surround the bases of the cilia. Defensive organelles called *trichocysts* are also present in the pellicle. In response to a threat, a microscopic explosion expels the trichocysts in a few milliseconds, and they emerge as sharp darts, driven forward at the tip of a long, expanding filament.

The cilia provide a form of locomotion that is generally more precise than locomotion by flagella or pseudopods. A paramecium can coordinate the beating of its cilia to propel itself either forward or backward in a spiraling manner. It can also back off swiftly when it encounters a barrier or a negative stimulus. The coordination of ciliary beating is probably the result of a differential distribution of ion channels in the plasma membrane near the two ends of the cell.

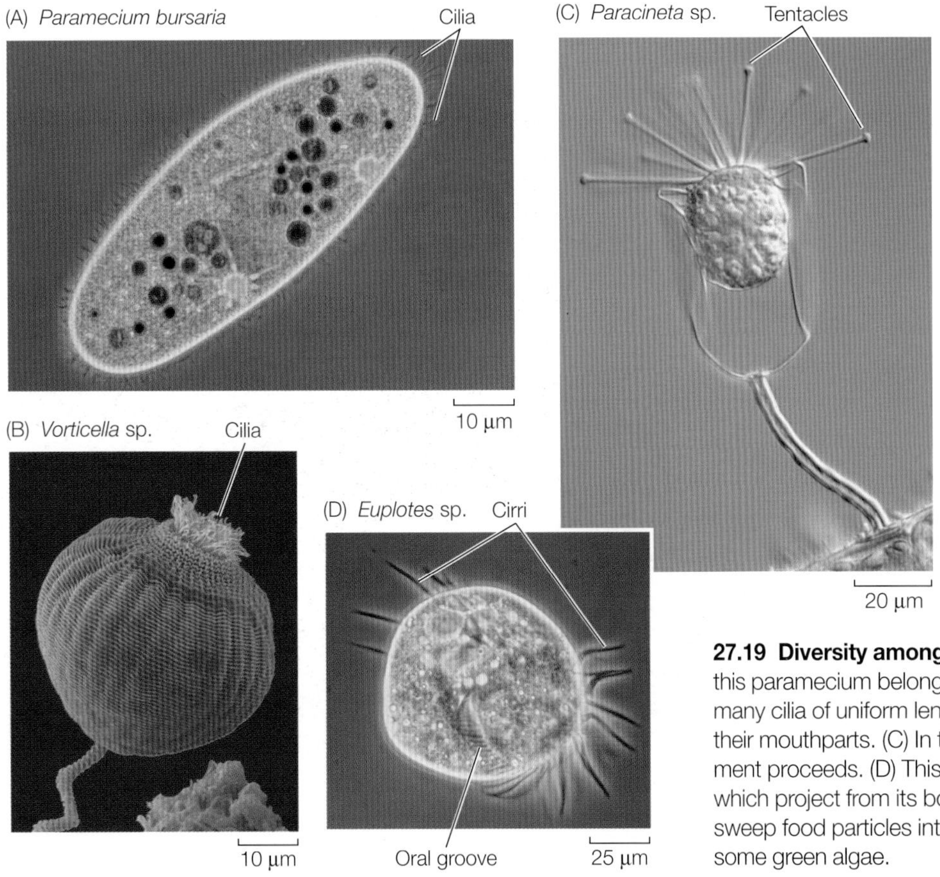

27.19 Diversity among the Ciliates (A) A free-swimming organism, this paramecium belongs to a ciliate group whose members have many cilia of uniform length. (B) Members of this group have cilia on their mouthparts. (C) In this group, tentacles replace cilia as development proceeds. (D) This ciliate "walks" on bundled cilia, called cirri, which project from its body. Other cilia are grouped into flat sheets that sweep food particles into its oral groove; this individual has ingested some green algae.

Stramenopiles have two unequal flagella, one with hairs

The synapomorphy that defines the **stramenopiles** is the possession of rows of tubular hairs on the longer of their two flagella. Some stramenopiles lack flagella, but they are descended from ancestors that possessed flagella. The stramenopiles include the diatoms and the brown algae, which are photosynthetic, and the oomycetes and slime nets, which are not. Most golden algae are photosynthetic, but nearly all of them become heterotrophic when light intensity is limiting or when there is a plentiful food supply; some even feed on diatoms or bacteria. The slime nets, formerly thought to be close relatives of the slime molds,

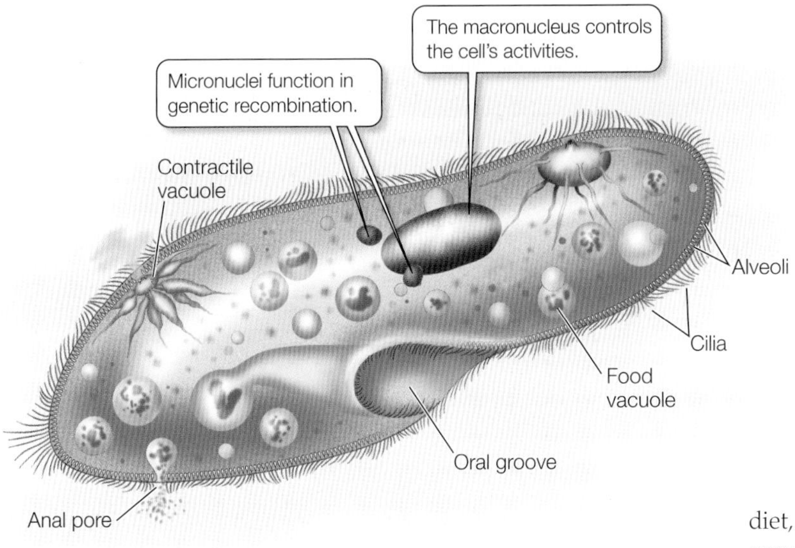

27.20 Anatomy of *Paramecium* This diagram shows the complex structure of a typical paramecium.

are unicellular organisms that produce networks of filaments along which the cells move.

DIATOMS ABOUND IN THE OCEANS **Diatoms** are unicellular organisms, although some species associate in filaments. Many have sufficient carotenoids in their chloroplasts to give them a yellow or brownish color. All make *chrysolaminarin* (a carbohydrate) and oils as photosynthetic storage products. Diatoms lack flagella except in male gametes.

Architectural magnificence on a microscopic scale is the hallmark of the diatoms. As mentioned earlier, almost all diatoms deposit silicon in their cell walls. The cell wall is constructed in two pieces, with the top overlapping the bottom like the top and bottom of a petri plate (see Figure 27.1). The silicon-impregnated walls have intricate patterns unique to each species (**Figure 27.21**). Despite their remarkable morphological diversity, however, all diatoms are symmetrical—either bilaterally (with "right" and "left" halves) or radially (with the type of symmetry possessed by a circle).

Diatoms reproduce both sexually and asexually. Asexual reproduction is by binary fission and is somewhat constrained by the stiff, silica-containing cell wall. Both the top and the bottom of the "petri plate" become tops of new "plates" without changing appreciably in size; as a result, the new cell made from the former bottom is smaller than the parent cell. If this process continued indefinitely, one cell line would simply vanish, but sexual reproduction largely solves this potential problem. Gametes are formed, shed their cell walls, and fuse. The resulting zygote then increases substantially in size before a new cell wall is laid down.

Diatoms are found in all the oceans and are frequently present in great numbers, making them major photosynthetic producers in coastal waters. Diatoms are also common in fresh water and even occur on the wet surfaces of terrestrial mosses.

Patches of ocean are occasionally populated by dense plankton blooms dominated by diatoms. Surprisingly, these diatom blooms are not heavily grazed by the copepods (tiny planktonic crustaceans) that are their usual predators. Thus much of the diatom population dies and sinks to the seafloor ungrazed. Later, as the bloom dissipates, the copepod population increases.

Why doesn't the copepod population peak and take advantage of the apparent riches of the diatom bloom? It had been suspected that this "failure" might result from a slow response by the copepod life cycle. Recent experimental data suggest, however, that the explanation lies in the effects of the diatoms on the reproductive success of the copepods when diatoms make up too great a proportion of their diet (**Figure 27.22**). While diatoms constitute a good food source for copepods when they make up a moderate proportion of the diet, they are toxic in large quantities because they release a toxic compound when they are crushed.

Because the silicon-containing walls of dead diatom cells resist decomposition, certain sedimentary rocks are composed almost entirely of diatom skeletons that sank to the seafloor over time. Diatomaceous earth, which is obtained from such rocks, has many industrial uses, such as insulation, filtration, and metal polishing. It has also been used as an "Earth-friendly" insecticide that clogs the tracheae (breathing structures) of insects.

THE BROWN ALGAE ARE MULTICELLULAR All the **brown algae** are multicellular, and some are large: giant kelps, such as those of the genus *Macrocystis*, may be up to 60 meters long (see p. 583). The brown algae obtain their namesake color from the carotenoid *fucoxanthin*, which is abundant in their chloroplasts. The combination of this yellow-orange pigment with the green of chlorophylls *a* and *c* yields a brownish tinge.

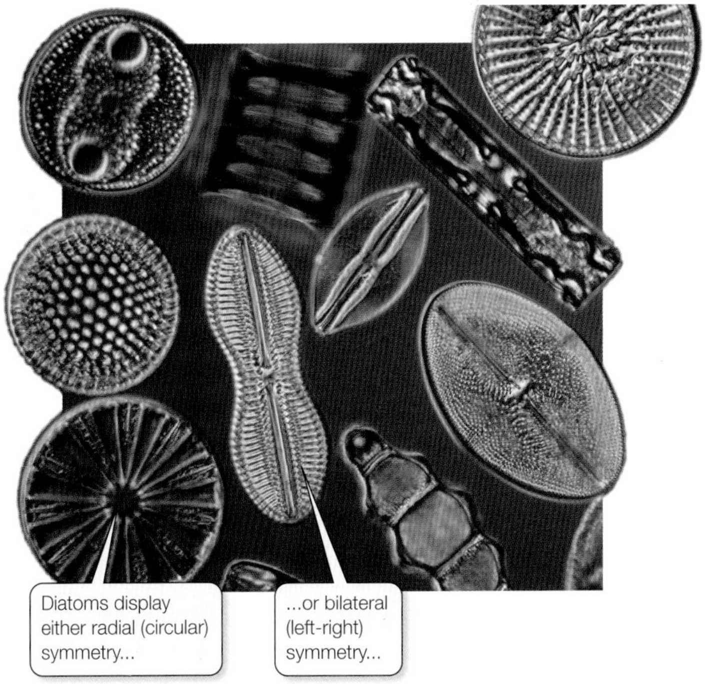

27.21 Diatom Diversity Diatoms exhibit a splendid variety of species-specific forms.

EXPERIMENT

HYPOTHESIS: Copepods fail to flourish during a diatom bloom because of the toxicity of the diatoms.

METHOD

1. Female copepods are fed for 10 days on either the diatom *Skeletonema costatum* or a nontoxic diet of dinoflagellates (control).
2. Eggs and newly hatched larvae are counted each day.

RESULTS

Although almost all eggs of the control group hatched each day, fewer and fewer of the eggs from the diatom-fed females hatched as the experiment progressed.

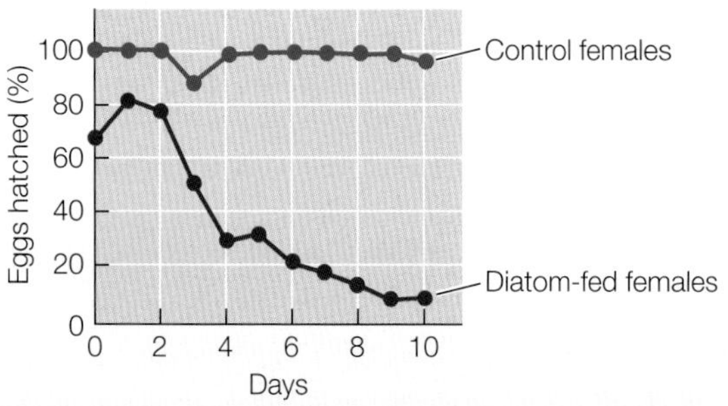

CONCLUSION: Toxicity is a plausible explanation for the failure of copepods to flourish during a diatom bloom.

27.22 Why Don't Copepods Flourish During Diatom Blooms? Adrianna Ianora and her colleagues tested the hypothesis that diatoms impair the reproduction of copepods when they constitute too great a proportion of the diet. FURTHER RESEARCH: How might you further test this hypothesis?

The brown algae are almost exclusively marine. They are composed either of branched filaments (**Figure 27.23A,B**) or of leaflike growths (**Figure 27.23C**). Some float in the open ocean; the most famous example is the genus *Sargassum*, which forms dense mats in the Sargasso Sea in the mid-Atlantic. Most brown algae, however, are attached to rocks near the shore. A few thrive only where they are regularly exposed to heavy surf; a notable example is the sea palm *Postelsia palmaeformis* of the Pacific coast. All of the attached forms develop a specialized structure, called a *holdfast*, that literally glues them to the rocks (**Figure 27.23D**). The "glue" of the holdfast is *alginic acid*, a gummy polymer of sugar acids found in the walls of many brown algal cells. In addition to its function in holdfasts, alginic acid cements algal cells and filaments together, and is harvested and used by humans as an emulsifier in ice cream, cosmetics, and other products.

Some brown algae differentiate extensively into specialized organs. Some, like the sea palm, have stemlike stalks and leaflike blades. Some develop gas-filled cavities or bladders that serve as floats. In addition to organ differentiation, the larger brown algae also exhibit considerable tissue differentiation. Most of the giant kelps have photosynthetic filaments only in the outermost regions of their stalks and blades. Within the stalks and blades lie filaments of tubular cells that closely resemble the nutrient-conducting tissue of land plants. Called *trumpet cells* because they have flaring ends, these tubes rapidly conduct the products of photosynthesis through the body of the organism.

The specialized flotation bladders of brown algae often contain as much as 5 percent carbon monoxide—a concentration high enough to kill a human.

THE OOMYCETES INCLUDE WATER MOLDS AND THEIR RELATIVES A nonphotosynthetic stramenopile group called the **oomycetes** consists in large part of the water molds and their terrestrial relatives, such as the downy mildews. Water molds are filamentous and stationary, and they are *absorptive heterotrophs*—that is, they secrete enzymes that digest large food molecules into smaller molecules that the water mold can absorb. If you have seen a whitish, cottony mold growing on dead fish or dead insects in water, it was probably a water mold of the common genus *Saprolegnia* (**Figure 27.24**).

Don't be confused by the *mycete* in the name of this group. That term means "fungus," and it is there because these organisms were once classified as fungi. However, we now know that the oomycetes are unrelated to the fungi.

Some oomycetes are **coenocytes**: that is, they have many nuclei enclosed in a single plasma membrane. Their filaments have no cross-walls to separate the many nuclei into discrete cells. Their cytoplasm is continuous throughout the body of the organism, and there is no single structural unit with a single nucleus, except in certain reproductive stages. A distinguishing feature of the oomycetes is their flagellated reproductive cells. Oomycetes are diploid throughout most of their life cycle and have cellulose in their cell walls.

The water molds, such as *Saprolegnia*, are all aquatic and **saprobic** (they feed on dead organic matter). Some other oomycetes are terrestrial. Although most of the terrestrial oomycetes are harmless or helpful decomposers of dead matter, a few are serious plant parasites that attack crops such as avocados, grapes, and potatoes.

Although their presumed chromalveolate ancestors had chloroplasts and were photosynthetic, the oomycetes lack chloroplasts. The next major group we'll consider, the Plantae, is predominantly photosynthetic.

Plantae

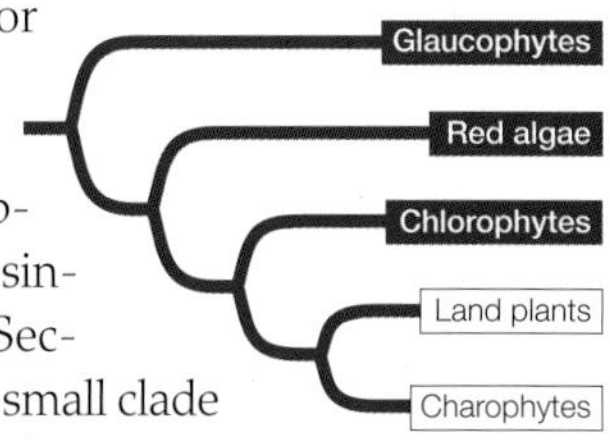

The **Plantae** consists of several major clades, including glaucophytes, red algae, chlorophytes, charophytes, and the land plants, all of which probably trace their chloroplasts back to a single incidence of endosymbiosis (see Section 27.2). It is for this reason that the small clade known as the **glaucophytes**, unicellular organisms that live in fresh water, is of great interest to students of evolution.

27.23 Brown Algae (A) Channeled wrack seaweed illustrates the filamentous growth form of brown algae. (B) Filaments of the microscopic brown alga *Ectocarpus* seen through a light microscope. (C) Sea palms exemplify the leaflike growth form. Sea palms grow in the intertidal zone, where they take a tremendous pounding by the surf. (D) Sea palms and many other brown algal species are "glued" to the substratum by tough, branched structures called holdfasts.

The glaucophytes were likely the first group to diverge after the primary endosymbiosis event. Their chloroplast is unique in containing a small amount of peptidoglycan between its inner and outer membranes—the same arrangement as that found in cyanobacteria (see Figure 27.8A). The presence of peptidoglycan, the characteristic cell wall component of bacteria, suggests that the glaucophytes resemble the common ancestor of all Plantae.

27.24 An Oomycete The filaments of a water mold radiate from the carcass of an insect.

Red algae have a distinctive accessory photosynthetic pigment

Almost all **red algae** are multicellular (**Figure 27.25**). Their characteristic color is a result of the accessory photosynthetic pigment *phycoerythrin*, which is found in relatively large amounts in the chloroplasts of many species. In addition to phycoerythrin, red algae contain phycocyanin, carotenoids, and chlorophyll *a*.

The red algae include species that grow in the shallowest tide pools as well as the photosynthesizers found deepest in the ocean (as deep as 260 meters if nutrient conditions are right and the water is clear enough to permit light to penetrate). Very few red algae inhabit fresh water. Most grow attached to a substratum by a holdfast.

In a sense, the red algae are misnamed. They have the capacity to change the relative amounts of their various photosynthetic pigments depending on the light conditions where they are growing. Thus the leaflike *Chondrus crispus*, a common North Atlantic red alga, may appear bright green when it is growing at or near the surface of the water and deep red when growing at greater depths. The ratio of pigments present depends to a remarkable degree on the intensity of the light that reaches the alga. In deep water,

27.25 Red Algae (A) Under the light microscope, reproductive structures can be seen in this red alga. (B) A coralline red alga grows along the coast of central Oregon.

where the light is dimmest, the alga accumulates large amounts of phycoerythrin. The algae in deep water have as much chlorophyll as the green ones near the surface, but the accumulated phycoerythrin makes them look red.

In addition to being the only photosynthetic eukaryotes with phycoerythrin among their pigments, the red algae have two other distinctive characteristics:

- They store the products of photosynthesis in the form of *floridean starch*, which is composed of very small, branched chains of approximately 15 glucose monomers.
- They produce no motile, flagellated cells at any stage of their life cycle. The male gametes lack cell walls and are slightly amoeboid; the female gametes are completely immobile.

Some red algal species enhance the formation of coral reefs (see Figure 27.25B). Like coral animals, they possess the biochemical machinery for secreting calcium carbonate, which they deposit both in and around their cell walls. After the deaths of corals and algae, the calcium carbonate persists, sometimes forming substantial rocky masses.

Some red algae produce large amounts of mucilaginous polysaccharide substances, which contain the sugar galactose with a sulfate group attached. This material readily forms solid gels and is the source of agar, a substance widely used in the laboratory for making a solid aqueous medium on which tissue cultures and many microorganisms can be grown.

A red alga became the ancestor of the distinctive chloroplasts of the photosynthetic chromalveolates by secondary endosymbiosis, as we saw in Section 27.2.

Chlorophytes, charophytes, and land plants contain chlorophylls *a* and *b*

One major clade of "green algae" consists of the **chlorophytes**. A sister group to the chlorophytes contains another green algal clade (the **charophytes**, or **charales**) along with the *land plants* (see Section 28.1). The green algae share several characters that distinguish them from other microbial eukaryotes: like the land plants, they contain chlorophylls *a* and *b*, and their reserve of photosynthetic products is stored as starch in chloroplasts. Through secondary endosymbiosis, a chlorophyte became the chloroplast of the euglenids.

There are more than 17,000 species of chlorophytes. Most are aquatic—some are marine, but more are freshwater forms—but others are terrestrial, living in moist environments. The chlorophytes range in size from microscopic unicellular forms to multicellular forms many centimeters in length.

CHLOROPHYTES VARY IN SHAPE AND CELLULAR ORGANIZATION We find among the chlorophytes an incredible variety in shape and body form. *Chlamydomonas* is an example of the simplest type: unicellular and flagellated.

Surprisingly large and well-formed colonies of cells are found in such freshwater groups as the genus *Volvox* (**Figure 27.26A**). The cells in these colonies are not differentiated into specialized tissues and organs, as in land plants and animals, but the colonies show vividly how the preliminary step of this great evolutionary innovation might have been taken. In *Volvox*, the origins of cell specialization can be seen in certain cells within the colony that are specialized for reproduction.

While *Volvox* is colonial and spherical, *Oedogonium* is multicellular and filamentous, and each of its cells has only one nucleus. *Cladophora* is multicellular, but each cell is multinucleate. *Bryopsis* is tubular and coenocytic, forming cross-walls only when reproductive structures form. *Acetabularia* is a single, giant uninucleate cell a few centimeters long that becomes multinucleate only at the end of its reproductive stage. *Ulva lactuca* is a thin, membranous sheet a few centimeters across; its distinctive appearance justifies its common name: sea lettuce (**Figure 27.26B**).

As we mentioned above, the chlorophytes are the largest clade of green algae, but there are other green algal clades as well. Those clades are branches of a clade that also includes the land plants, which will be described in the next chapter.

Excavates

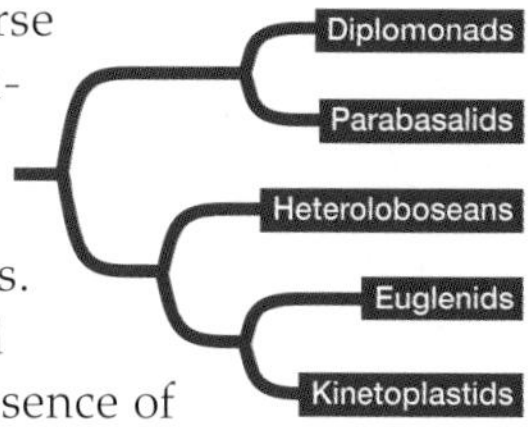

The **excavates** includes several diverse clades, several of which lack mitochondria. This absence of mitochondria once led to the view that these groups might be the basal clades of the eukaryotes. However, the trait seems to be a derived condition, judging in part from the presence of

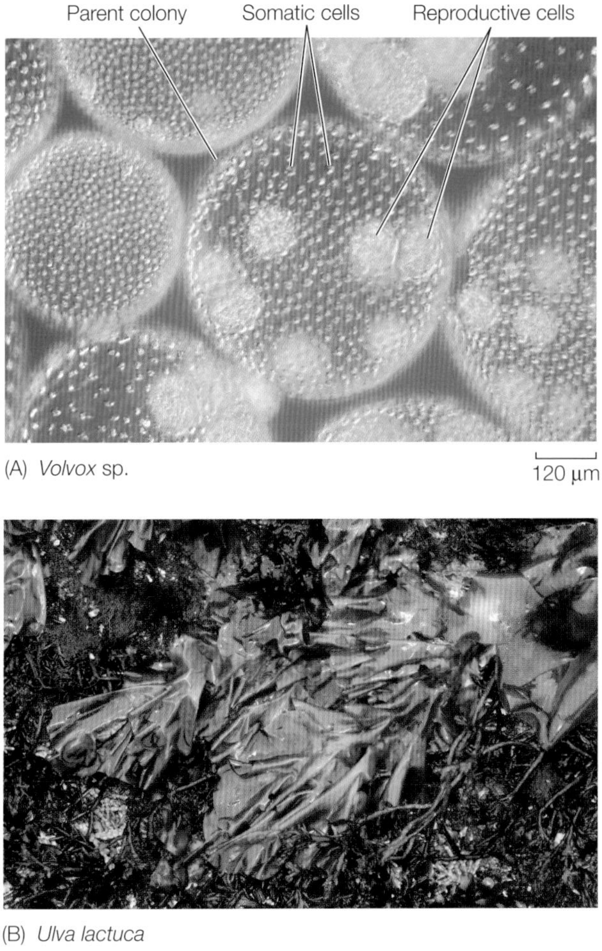

27.26 Chlorophytes (A) *Volvox* colonies are precisely spaced arrangements of cells. Specialized reproductive cells produce daughter colonies, which will eventually release new individuals. (B) A stand of sea lettuce exposed by low tide.

nuclear genes normally associated with mitochondria. Ancestors of these organisms probably possessed mitochondria that were lost or reduced in the course of evolution. The existence of such organisms today shows that eukaryotic life is feasible without mitochondria, and for that reason, these groups are the focus of much attention.

Diplomonads and parabasalids are excavates that lack mitochondria

The **diplomonads** and **parabasalids**, all unicellular, lack mitochondria. *Giardia lamblia*, a diplomonad, is a familiar parasite that contaminates water supplies and causes the intestinal disease giardiasis (**Figure 27.27A**). This tiny organism contains two nuclei bounded by nuclear envelopes, and it has a cytoskeleton and multiple flagella.

Trichomonas vaginalis is a parabasalid responsible for a sexually transmitted disease in humans (**Figure 27.27B**). Infection of the male urethra, where it may occur without symptoms, is less common than infection of the vagina. In addition to flagella and a cytoskeleton, the parabasalids have undulating membranes that also contribute to the cell's locomotion.

Heteroloboseans alternate between amoeboid forms and forms with flagella

The amoeboid body form appears in several microbial eukaryote groups—including the loboseans and heteroloboseans—that are only distantly related to one another. These groups belong, respectively, to the unikonts and excavates. Amoebas of the free-living heterolobosean genus *Naegleria*, some of which can enter humans and cause a fatal disease of the nervous system, usually have a two-stage life cycle, in which one stage has amoeboid cells and the other flagellated cells.

Euglenids and kinetoplastids have distinctive mitochondria and flagella

The euglenids and kinetoplastids, both of which are excavate clades, together constitute a clade of unicellular organisms with flagella. Their mitochondria contain distinctive, disc-shaped cristae,

27.27 Some Excavate Groups Lack Mitochondria (A) *Giardia*, a diplomonad, has flagella and two nuclei. (B) *Trichomonas*, a parabasalid, has flagella and undulating membranes. Neither of these organisms possesses mitochondria.

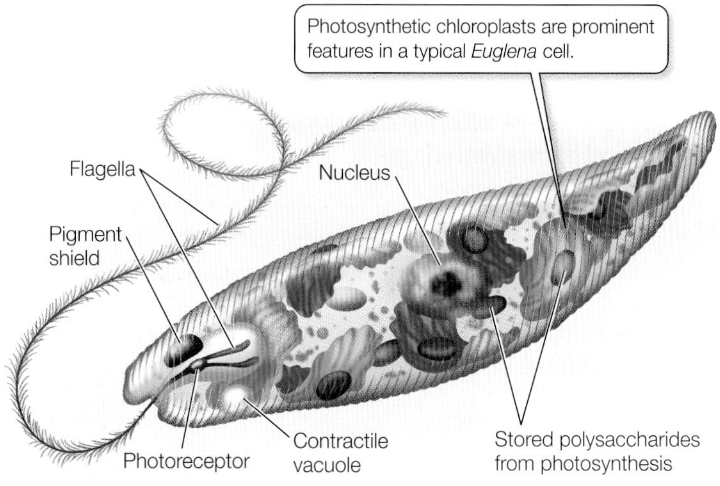

27.28 A Photosynthetic Euglenid Several *Euglena* species possess flagella. In this species, the second flagellum is rudimentary.

and their flagella contain a crystalline rod not found in other organisms. They reproduce asexually by binary fission.

EUGLENIDS HAVE SUPPORTING STRIPS OF PROTEIN The **euglenids** possess flagella arising from a pocket at the anterior end of the cell. Spiraling strips of proteins under their plasma membranes control the cell's shape. Some members of the group are photosynthetic. Euglenids used to be claimed by the zoologists as animals and by the botanists as plants.

Figure 27.28 depicts a cell of the genus *Euglena*. Like most other euglenids, this common freshwater organism has a complex cell structure. It propels itself through the water with the longer of its two flagella, which may also serve as an anchor to hold the organism in place. The second flagellum is often rudimentary.

Euglenids have very diverse nutritional requirements. Many species are always heterotrophic. Other species are fully autotrophic in sunlight, using chloroplasts to synthesize organic compounds through photosynthesis. The chloroplasts of euglenids are surrounded by three membranes as a result of secondary endosymbiosis (see Figure 27.8B). When kept in the dark, these euglenids lose their photosynthetic pigment and begin to feed exclusively on dissolved organic material in the water around them. Such a "bleached" *Euglena* resynthesizes its photosynthetic pigment when it is returned to the light and becomes autotrophic again. But *Euglena* cells treated with certain antibiotics or mutagens lose their photosynthetic pigment completely; neither they nor their descendants are ever autotrophs again. However, those descendants function well as heterotrophs.

KINETOPLASTIDS HAVE MITOCHONDRIA THAT EDIT THEIR OWN RNA The **kinetoplastids** are unicellular parasites with two flagella and a single, large mitochondrion. That mitochondrion contains a *kinetoplast*—a unique structure housing multiple, circular DNA molecules and associated proteins. Some of these DNA molecules encode "guides" that edit messenger RNA within the mitochondrion.

The trypanosomes discussed at the opening of this chapter are kinetoplastids. Recall that they are able to change their cell surface recognition molecules frequently, which allows them to evade our best attempts to kill them and eradicate the diseases they cause (**Table 27.2**)

Rhizaria

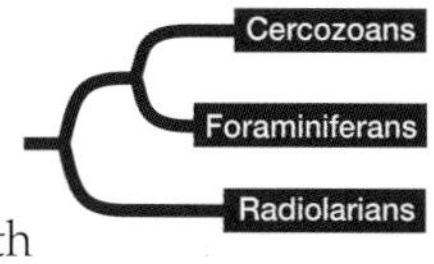

Three closely related groups of **Rhizaria** are unicellular aquatic eukaryotes. Foraminiferans, radiolarians, and cercozoans typically have long, thin pseudopodia that contrast with the broader, lobelike pseudopodia of the familiar amoebas. These groups have contributed to ocean sediments, some of which have become terrestrial features in the course of geological history.

The **cercozoans** are a diverse group, with many forms and habitats. Some are amoeboid, while others have flagella. Some are aquatic; others live in soil. One group of cercozoans possesses chloroplasts derived from a green alga by secondary endosymbiosis—and that chloroplast contains a trace of the green alga's nucleus.

Foraminiferans have created vast limestone deposits

Foraminiferans secrete external shells of calcium carbonate (see Figure 27.5). Some foraminiferans live as plankton, and many others live at the bottom of the sea. Living foraminiferans have been

TABLE 27.2

A Comparison of Three Kinetoplastid Trypanosomes

	TRYPANOSOMA BRUCEI	*TRYPANOSOMA CRUZI*	*LEISHMANIA MAJOR*
Human disease	Sleeping sickness	Chagas' disease	Leishmaniasis
Insect vector	Tsetse fly	Assassin bug	Sand fly
Vaccine or effective cure	None	None	None
Strategy for survival	Changes surface recognition molecules frequently	Causes changes in surface recognition molecules on host cell	Reduces effectiveness of macrophage hosts
Site in human body	Bloodstream; attacks nerve tissue in final stages	Enters cells, especially muscle cells	Enters cells, primarily macrophages
Deaths per year	>50,000	43,000	60,000 (?)

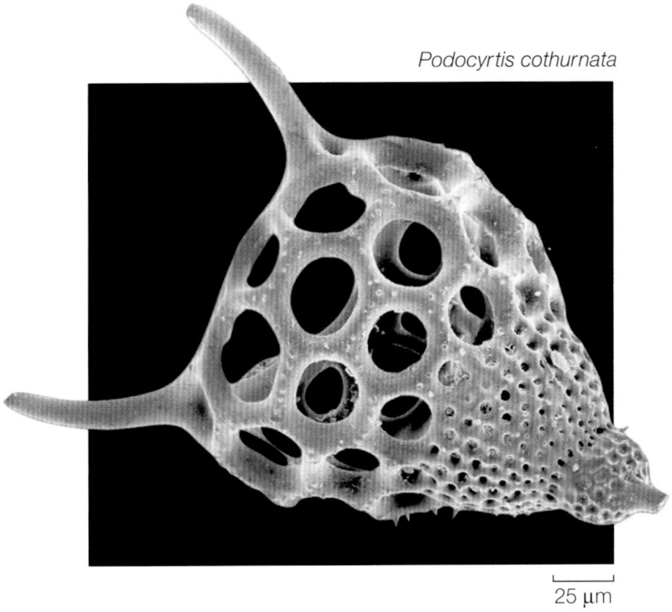

27.29 A Radiolarian's Glass House Radiolarians secrete intricate glassy skeletons such as the one shown here. A living radiolarian is shown in Figure 27.2.

found at the deepest point in the world's oceans—10,896 meters down in the Challenger Deep, in the western Pacific. At that depth, they cannot secrete a normal shell because the surrounding water is too poor in calcium carbonate.

Long, threadlike, branched pseudopods reach out through numerous microscopic pores in the shell and interconnect to create a sticky net, which planktonic foraminifera use to catch smaller plankton. The pseudopods provide locomotion in some species.

Radiolarians have thin, stiff pseudopods

The **radiolarians** are recognizable by their thin, stiff pseudopods, which are reinforced by microtubules. These pseudopods play important roles:

- They greatly increase the surface area of the cell for exchange of materials with the environment.
- They help the cell float in its marine environment.

Found exclusively in marine environments, radiolarians may be the most beautiful of all microorganisms (see Figure 27.2). Almost all radiolarian species secrete glassy *endoskeletons* (internal skeletons). A central capsule lies within the cytoplasm. The skeletons of the different species are as varied as snowflakes, and many have elaborate geometric designs (**Figure 27.29**). A few radiolarians are among the largest of the unicellular eukaryotes, measuring several millimeters across.

Unikonts

We now consider a large clade that may be close to the root of the eukaryote tree: the **unikonts**, eukaryotes whose flagella, if present, are single. (The name *unikont* derives from "single cone.") The animals and fungi are both believed to have arisen from a common ancestor within the **opisthokont** clade. Recent extensive work with multiple gene phylogenies (see Section 26.4) have led many scientists to the conclusion that the *amoebozoans* are sister to the opisthokonts.

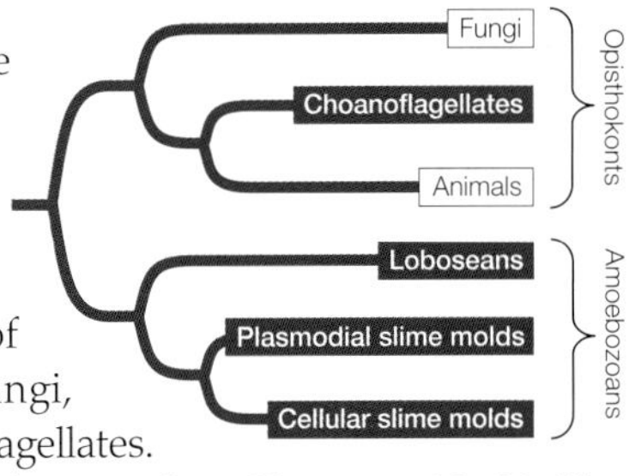

The synapomorphy of the opisthokonts is that their flagellum, if present, is posterior, as in animal sperm. The flagella of all other eukaryotes are anterior. Major groups of opisthokonts include the fungi, the animals, and the choanoflagellates. Fungi and animals will be discussed in Chapters 30–33. The choanoflagellates, or collar flagellates, are sister to the animals, and the animal–choanoflagellate clade is sister to the fungi.

Some choanoflagellates are colonial (**Figure 27.30A**). They bear a striking resemblance to the most characteristic type of cell found in the sponges (compare **Figure 27.30B** with Figure 31.7).

Amoebozoans use lobe-shaped pseudopods for locomotion

The lobe-shaped pseudopods used by **amoebozoans** are a hallmark of the amoeboid body plan. The amoebozoan pseudopod differs in form and function from the slender pseudopods of Rhizaria. We'll consider three amoebozoan groups here: the loboseans and two clades of slime molds.

LOBOSEAN CELLS LIVE INDEPENDENTLY OF ONE ANOTHER A **lobosean**, such as the *Amoeba proteus* shown in Figure 27.9, consists of a single cell. Unlike the cells of the slime molds, loboseans

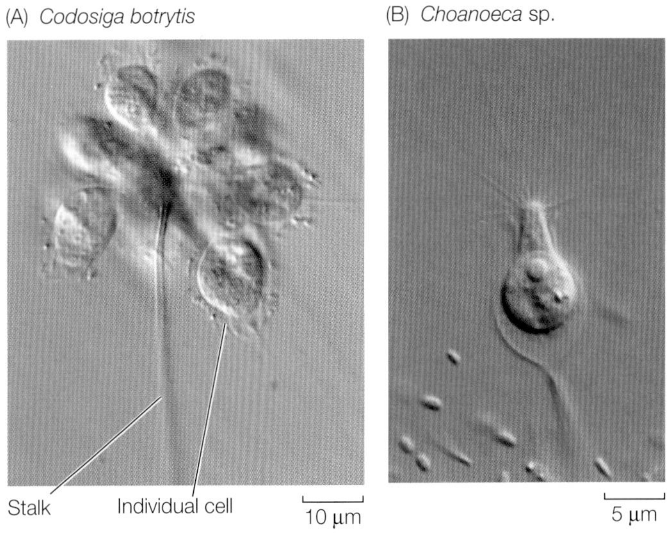

27.30 A Link to the Animals Choanoflagellates are sister to the animals. (A) The formation of colonies by unicellular organisms, as in this choanoflagellate species, is one route to the evolution of multicellularity. (B) A solitary choanoflagellate illustrates the similarity of this microbial eukaryote group to a cell type present in the multicellular sponges (see Figure 31.7).

do not aggregate. A lobosean feeds on small organisms and particles of organic matter by phagocytosis, engulfing them with its pseudopods. Many are adapted for life on the bottoms of lakes, ponds, and other bodies of water. Their creeping locomotion and their manner of engulfing food particles fit them for life close to a relatively rich supply of sedentary organisms or organic particles. Most loboseans exist as predators, parasites, or scavengers.

Some loboseans are shelled, living in casings of sand grains glued together (see Figure 27.12A). Others have shells secreted by the organism itself.

SLIME MOLDS RELEASE SPORES FROM ERECT FRUITING BODIES The two major groups of *slime molds* share only general characteristics. All are motile, all ingest particulate food by endocytosis, and all form spores on erect structures called *fruiting bodies*. They undergo striking changes in organization during their life cycles, and one stage consists of isolated cells that take up food particles by endocytosis. Some slime molds may cover areas of 1 meter or more in diameter while in their less aggregated stage. Such a large slime mold may weigh more than 50 grams. Slime molds of both types favor cool, moist habitats, primarily in forests. They range from colorless to brilliantly yellow and orange.

PLASMODIAL SLIME MOLDS FORM MULTINUCLEATE MASSES If the nucleus of an amoeba began rapid mitotic division, accompanied by a tremendous increase in cytoplasm and organelles but no cytokinesis, the resulting organism might resemble the **plasmodial slime molds**. During its vegetative (feeding) phase, a plasmodial slime mold is a wall-less mass of cytoplasm with numerous diploid nuclei. This mass streams very slowly over its substratum in a remarkable network of strands called a *plasmodium*.* The plasmodium of such a slime mold is another example of a coenocyte, with many nuclei enclosed in a single plasma membrane. The outer cytoplasm of the plasmodium (closest to the environment) is normally less fluid than the interior cytoplasm and thus provides some structural rigidity (**Figure 27.31A**).

Plasmodial slime molds provide a dramatic example of movement by **cytoplasmic streaming**. The outer cytoplasmic region of the plasmodium becomes more fluid in places, and cytoplasm rushes into those areas, stretching the plasmodium. This streaming somehow reverses its direction every few minutes as cytoplasm rushes into a new area and drains away from an older one, moving the plasmodium over its substratum. Sometimes an entire wave of plasmodium moves across the substratum, leaving strands behind. Microfilaments and a contractile protein called *myxomyosin* interact to produce the streaming movement. As it moves, the plasmodium engulfs food particles by endocytosis—predominantly bacteria, yeasts, spores of fungi, and other small organisms, as well as decaying animal and plant remains.

A plasmodial slime mold can grow almost indefinitely in its plasmodial stage, as long as the food supply is adequate and other conditions, such as moisture and pH, are favorable. However, one of two things can happen if conditions become unfavorable. First, the plasmodium can form an irregular mass of hardened cell-like components called a *sclerotium*. This resting structure rapidly becomes a plasmodium again when favorable conditions are restored.

Alternatively, the plasmodium can transform itself into spore-bearing fruiting structures (**Figure 27.31B**). These stalked or branched structures rise from heaped masses of plasmodium. They derive their rigidity from walls that form and thicken between their nuclei. The diploid nuclei of the plasmodium divide by meiosis as the fruiting structure develops. One or more knobs, called *sporangia*, develop on the end of the stalk. Within a sporangium, haploid nuclei become surrounded by walls and form spores. Eventually, as the fruiting body dries, it sheds its spores.

*Do not confuse the plasmodium of a plasmodial slime mold with the genus *Plasmodium*, the apicomplexan that is the cause of malaria.

27.31 Plasmodial Slime Molds (A) Plasmodia of the yellow slime mold *Physarum* cover a rock in Nova Scotia. (B) The fruiting structures of *Physarum*.

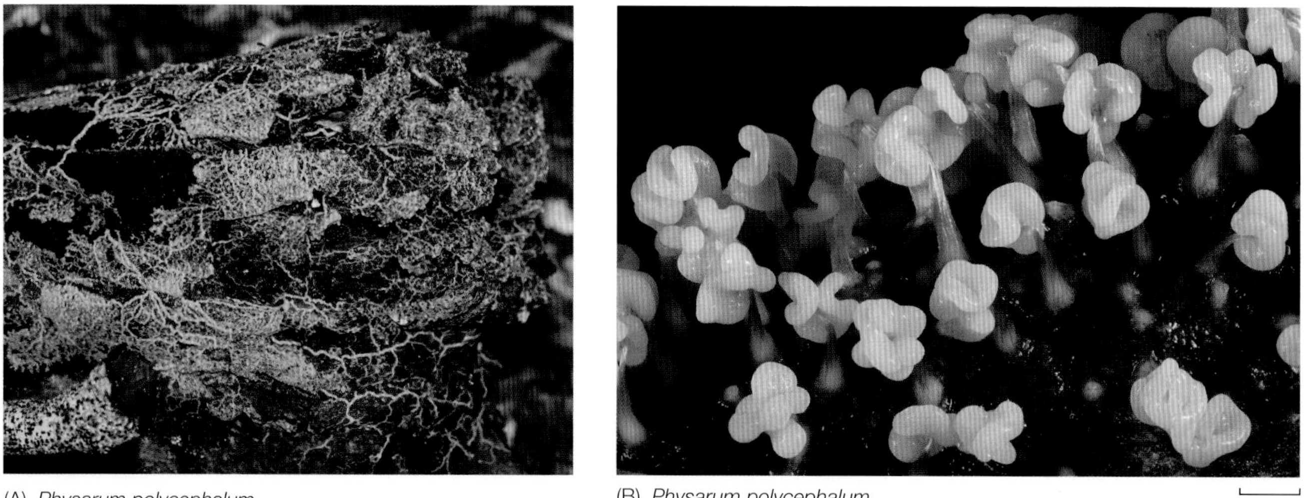

(A) *Physarum polycephalum*

(B) *Physarum polycephalum*

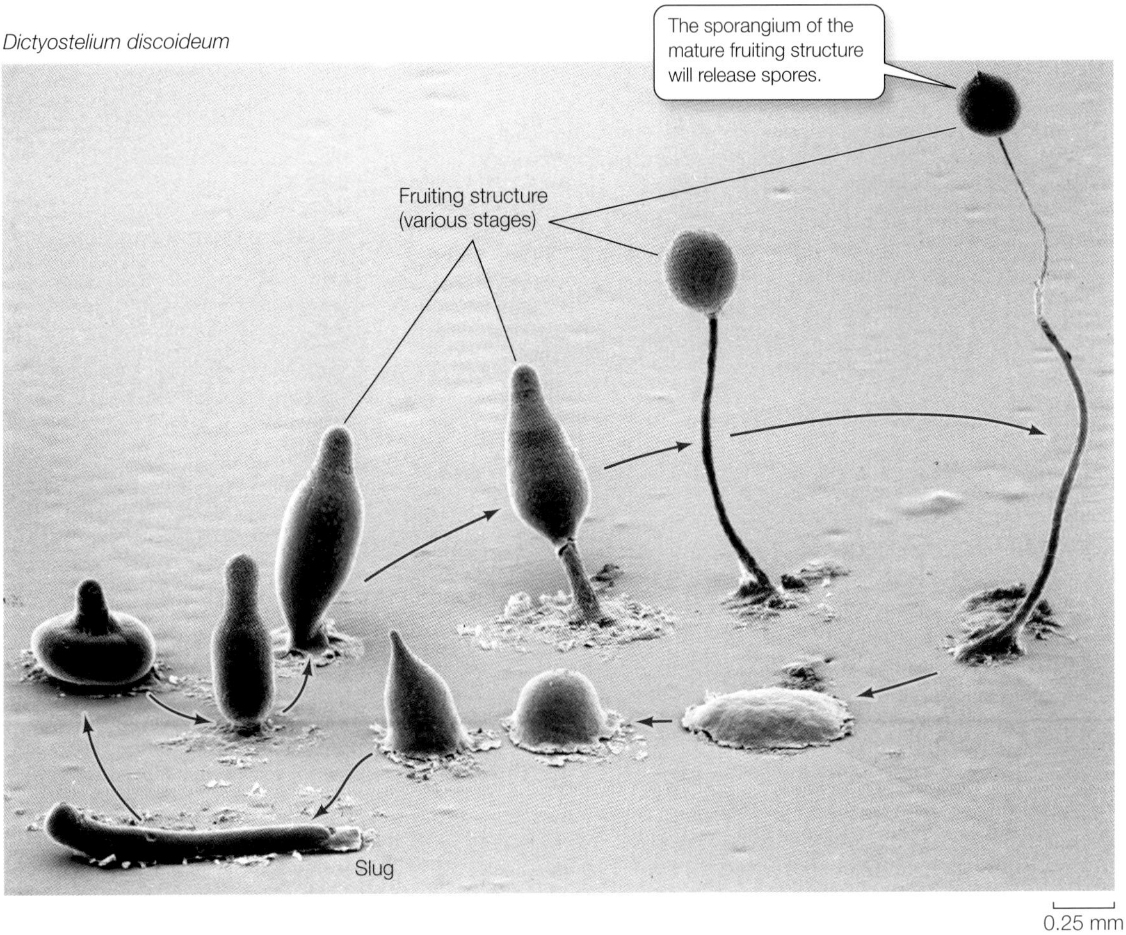

27.32 A Cellular Slime Mold The life cycle of the slime mold *Dictyostelium* is shown here in a composite micrograph.

The spores germinate into wall-less, haploid cells called *swarm cells*, which can either divide mitotically to produce more haploid swarm cells or function as gametes. Swarm cells can live as separate individual cells that move by means of flagella or pseudopods, or they can become walled and resistant resting cysts when conditions are unfavorable; when conditions improve again, the cysts release swarm cells. Two swarm cells can also fuse to form a diploid zygote, which divides by mitosis (but without a wall forming between the nuclei) and thus forms a new, coenocytic plasmodium.

CELLS RETAIN THEIR IDENTITY IN THE CELLULAR SLIME MOLDS Whereas the plasmodium is the basic vegetative (feeding, nonreproductive) unit of the plasmodial slime molds, an amoeboid cell is the vegetative unit of the **cellular slime molds**. Large numbers of cells called *myxamoebas*, which have single haploid nuclei, engulf bacteria and other food particles by endocytosis and reproduce by mitosis and fission. This simple life cycle stage, consisting of swarms of independent, isolated cells, can persist indefinitely as long as food and moisture are available.

When conditions become unfavorable, however, the cellular slime molds aggregate and form fruiting structures, as do their plasmodial counterparts. The individual myxamoebas aggregate into a mass called a *slug* or *pseudoplasmodium* (**Figure 27.32**). Unlike the true plasmodium of the plasmodial slime molds, this structure is not simply a giant sheet of cytoplasm with many nuclei; the individual myxamoebas retain their plasma membranes and, therefore, their identity.

A slug may migrate over its substratum for several hours before becoming motionless and reorganizing to construct a delicate, stalked fruiting structure. Cells at the top of the fruiting structure develop into thick-walled spores, which are eventually released. Later, under favorable conditions, the spores germinate, releasing myxamoebas.

The cycle from myxamoebas through slug and spores to new myxamoebas is asexual. Cellular slime molds also have a sexual cycle, in which two myxamoebas fuse. The product of this fusion develops into a spherical structure that ultimately germinates, releasing new haploid myxamoebas.

In subsequent chapters we will explore the three classic groups of multicellular eukaryotes that arose within the opisthokonts. Chapters 28 and 29 describe the rise and diversification of the land plants, Chapter 30 presents the fungi, and Chapters 31–33 describe the animals. All three of these groups arose from microbial eukaryote ancestors.

CHAPTER SUMMARY

27.1 How do microbial eukaryotes affect the world around them?

The **microbial eukaryotes** (also called **protists**) embrace several diverse groups of mostly unicellular eukaryotic organisms. They constitute a paraphyletic group, not a clade.

The diatoms, part of the **plankton**, are responsible for up to a fifth of all carbon fixation on Earth. **Phytoplankton** are the primary producers in the marine environment.

Endosymbiosis is common among microbial eukaryotes and often helpful to both partners. Pathogenic microbial eukaryotes include species of *Plasmodium* and trypanosomes. Review Figure 27.3

27.2 How did the eukaryotic cell arise?

Several events led to the evolution of the modern eukaryotic cell from an ancestral prokaryote. Probable early events include the loss of the cell wall and infolding of the plasma membrane. Review Figure 27.6

The development of a cytoskeleton gave the evolving cell increasing control over its shape and distribution of daughter chromosomes. Review Figure 27.7

Some organelles were acquired by endosymbiosis. Mitochondria evolved from a proteobacterium. **Primary endosymbiosis** of a eukaryote and a cyanobacterium gave rise to the chloroplasts, beginning with those of glaucophytes, red algae, and green algae. Review Figure 27.8A, Web/CD Tutorial 27.1

Secondary endosymbiosis of eukaryotes with other eukaryotes already equipped with chloroplasts gave rise to the chloroplasts of euglenids, stramenopiles, and other groups. Review Figure 27.8B

27.3 How did the microbial eukaryotes diversify?

Some microbial eukaryotes are photosynthetic autotrophs, some are heterotrophs, and some are both.

The cytoskeleton allows for various means of locomotion. Most microbial eukaryotes are motile, moving by amoeboid motion with **pseudopods** or by means of cilia or flagella.

Some microbial eukaryote cells contain **contractile vacuoles** that pump out excess water or **food vacuoles** where food is digested. Review Figures 27.10 and 27.11, Web/CD Tutorial 27.2

Many microbial eukaryotes have protective cell surfaces such as cell walls, external "shells," or shells constructed from sand.

27.4 How do microbial eukaryotes reproduce?

Most microbial eukaryotes reproduce both asexually and sexually.

Conjugation in paramecia is a sexual process but not a reproductive one. Review Figure 27.13

Alternation of generations is a feature of many microbial eukaryote life cycles, which include a multicellular diploid phase and a multicellular haploid phase. Review Figure 27.14

Alternation of generations may be **heteromorphic** or **isomorphic**. The alternating generations are the (diploid) **sporophyte** and (haploid) **gametophyte**. Specialized cells of the sporophyte, called **sporocytes**, divide meiotically to produce haploid spores.

Depending on whether their gametes appear identical or dissimilar, species are termed **isogamous** or **anisogamous**.

In a **haplontic** life cycle, the zygote is the only diploid cell. In a **diplontic** life cycle, the gametes are the only haploid cells. Review Figures 27.15 and 27.16, Web/CD Activities 27.1 and 27.2

Some microbial eukaryote life cycles involve more than one host species.

27.5 What are the major groups of eukaryotes?

Most eukaryotes can be divided into five major groups: chromalveolates, Plantae, excavates, Rhizaria, and unikonts. Review Figure 27.17

The **chromalveolates** include the haptophytes, cryptophytes, alveolates, and stramenopiles. The **cryptophytes** represent a key link in the story of the evolution of chloroplasts.

Alveolates are unicellular organisms with sacs (alveoli) beneath their plasma membranes. Alveolate clades include the marine **dinoflagellates**; the parasitic **apicomplexans**; and diverse, highly motile **ciliates.** Web/CD Activity 27.3

Stramenopiles typically have two flagella of unequal length, the longer bearing rows of tubular hairs. Included among the stramenopiles are the unicellular **diatoms**, the multicellular **brown algae**, and the nonphotosynthetic **oomycetes** including the water molds and downy mildews.

Plantae is a clade containing several other clades, including **glaucophytes**, **red algae**, **chlorophytes**, **land plants**, and **charophytes**. All are photosynthetic and contain chloroplasts. Glaucophyte chloroplasts contain peptidoglycan between their inner and outer membranes.

The **excavates** include the diplomonads, parabasalids, heteroloboseans, euglenids, and kinetoplastids. The **diplomonads** and **parabasalids** lack mitochondria, having apparently lost them during their evolution. **Heteroloboseans** are amoebas with a two-stage life cycle. **Euglenids** are often photosynthetic and have anterior flagella and strips of protein that support their cell surface. **Kinetoplastids** have a single, large mitochondrion, in which mitochondrial messenger RNA is edited.

Rhizaria are unicellular and aquatic; most are amoeboid. This group includes the **foraminiferans** whose shells have contributed to great limestone deposits; the **radiolarians** with thin, stiff pseudopods and glassy endoskeletons; and the **cercozoans**, which take many forms and live in diverse habitats.

The **unikonts** encompass organisms with single flagella on their flagellated cells (if any). They can be divided into two subgroups: the opisthokonts and the amoebozoans. In the **opisthokonts**, the flagellum (if any) is posterior. The opisthokont subgroups are the fungi, the choanoflagellates, and the animals. **Choanoflagellates** resemble the cells of sponges and are sister to the animal clade.

The **amoebozoans** move by means of lobe-shaped pseudopodia. They comprise the loboseans, plasmodial slime molds, and cellular slime molds. A **lobosean** consists of a single cell; these cells do not aggregate. **Plasmodial slime molds** are amoebozoans whose feeding phase is coenocytic. In the feeding phase, movement is by **cytoplasmic streaming**. In **cellular slime molds**, the individual cells maintain their identity at all times, but aggregate to form fruiting bodies.

SELF-QUIZ

1. Microbial eukaryotes with flagella
   a. appear in several clades.
   b. are all algae.
   c. all have pseudopods.
   d. are all colonial.
   e. are never pathogenic.
2. Which statement about eukaryotic phytoplankton is *not* true?
   a. Some are important primary producers.
   b. Some contributed to the formation of petroleum.
   c. Some form toxic "red tides."
   d. Some are food for marine animals.
   e. They constitute a clade.
3. Apicomplexans
   a. possess flagella.
   b. possess a glassy shell.
   c. are all parasitic.
   d. are algae.
   e. include the trypanosomes that cause sleeping sickness.
4. The ciliates
   a. move by means of flagella.
   b. use amoeboid movement.
   c. include *Plasmodium,* the agent of malaria.
   d. possess both a macronucleus and micronuclei.
   e. are autotrophic.
5. The chloroplasts of photosynthetic microbial eukaryotes
   a. are structurally identical.
   b. gave rise to mitochondria.
   c. are all descended from a once free-living cyanobacterium.
   d. all have exactly two surrounding membranes.
   e. are all descended from a once free-living red alga.
6. Which statement about the brown algae is *not* true?
   a. They are all multicellular.
   b. They use the same photosynthetic pigments as do land plants.
   c. They are almost exclusively marine.
   d. A few are many meters in length.
   e. They are stramenopiles.
7. Which statement about the chlorophytes is *not* true?
   a. They use the same photosynthetic pigments as do land plants.
   b. Some are unicellular.
   c. Some are multicellular.
   d. All are microscopic in size.
   e. They display a great diversity of life cycles.
8. The red algae
   a. are mostly unicellular.
   b. are mostly marine.
   c. owe their red color to a special form of chlorophyll.
   d. have flagella on their gametes.
   e. are all heterotrophic.
9. The plasmodial slime molds
   a. form a plasmodium that is a coenocyte.
   b. lack fruiting bodies.
   c. consist of large numbers of myxamoebas.
   d. consist at times of a mass called a pseudoplasmodium.
   e. possess flagella.
10. The cellular slime molds
    a. possess apical complexes.
    b. lack fruiting bodies.
    c. form a plasmodium that is a coenocyte.
    d. have haploid myxamoebas.
    e. possess flagella.

FOR DISCUSSION

1. For each type of organism below, give a single characteristic that may be used to differentiate it from the other, related organism(s) named in parentheses.
   a. Foraminiferans (radiolarians)
   b. *Euglena* (*Volvox*)
   c. *Trypanosoma* (*Giardia*)
   d. Plasmodial slime molds (cellular slime molds)
2. In what sense are sex and reproduction independent of each other in the ciliates? What does that suggest about the role of sex in biology?
3. Why are dinoflagellates and apicomplexans placed in one group of microbial eukaryotes and brown algae and oomycetes in another?
4. Unlike many microbial eukaryotes, apicomplexans lack contractile vacuoles. Why don't apicomplexans need a contractile vacuole?
5. Giant seaweeds (mostly brown algae) have "floats" that aid in keeping their fronds suspended at or near the surface of the water. Why is it important that the fronds be suspended in this way?
6. Why are algal pigments so much more diverse than those of land plants?
7. Consider the chloroplasts of chlorophytes, euglenids, and red algae. For each of these groups, indicate how many membranes surround their chloroplasts, and offer a reasonable explanation in each case. Why do some dinoflagellates have more membranes around their chloroplasts than other dinoflagellates?

FOR INVESTIGATION

During a bloom of diatoms, their predators, the copepods, do not show a corresponding increase in population. The copepod population rises only later, after the diatom bloom has declined. When the diatoms are at a normal (non-bloom) population level, what proportion of the copepods' diet do they constitute?

CHAPTER 28 Plants without Seeds: From Sea to Land

What surprises lurk in a rock?

John William Dawson, later Sir William Dawson, was an outspoken critic of Charles Darwin and evolutionary theory. Ironically, he made one of the first contributions to our understanding of the early evolution of plants. While surveying the geology of Nova Scotia, Dawson, a Canadian scientist, found a puzzling fossil fragment on the Gaspé Peninsula. It was 1859—the very year in which Darwin published *The Origin of Species*.

What Dawson found was the remains of a remarkable plant, which he named *Psilophyton*, meaning "naked plant." The fossilized plant appeared to have no roots and no leaves, and its stem grew both below and above the ground. The aboveground part of the stem, which branched and ended in spore cases, was about 50 centimeters tall. Could this fossil plant represent one of the stages in the momentous transition of plants from an aquatic existence to the forests that cover the land today? Dawson presented *Psilophyton* and other finds in a major lecture delivered in 1870, but his audience was unreceptive, with his fellow scientists chuckling that *Psilophyton* grew nowhere but in Dawson's imagination. His published drawing of the fossil plant was regarded as an amusing curiosity.

Decades later, Dawson's interpretation of *Psilophyton* was vindicated by the work of two British botanists who studied plant fossils. The year 1915 found Robert Kidston and William Lang in the hills near Rhynie, Scotland, where they discovered evidence that a marsh had existed there 400 million years ago. In the intervening hundreds of millions of years, that Devonian marsh had become a flinty, fossil-laden rock called chert. Although the rock now lay in a hilly site some 50 kilometers from the sea, its original location must have been near the shore.

The most startling feature of the Rhynie chert was the great abundance of fossils of a small plant with spore cases but neither roots nor leaves—a plant, in fact, that looked strikingly like Dawson's "imaginary" *Psilophyton*! Kidston and Lang described their find in detail and gave it the genus name *Rhynia* in honor of the site of its discovery.

Rhynia is just one of several ancient groups of plants that lack seeds and other features of more modern plants. The glory

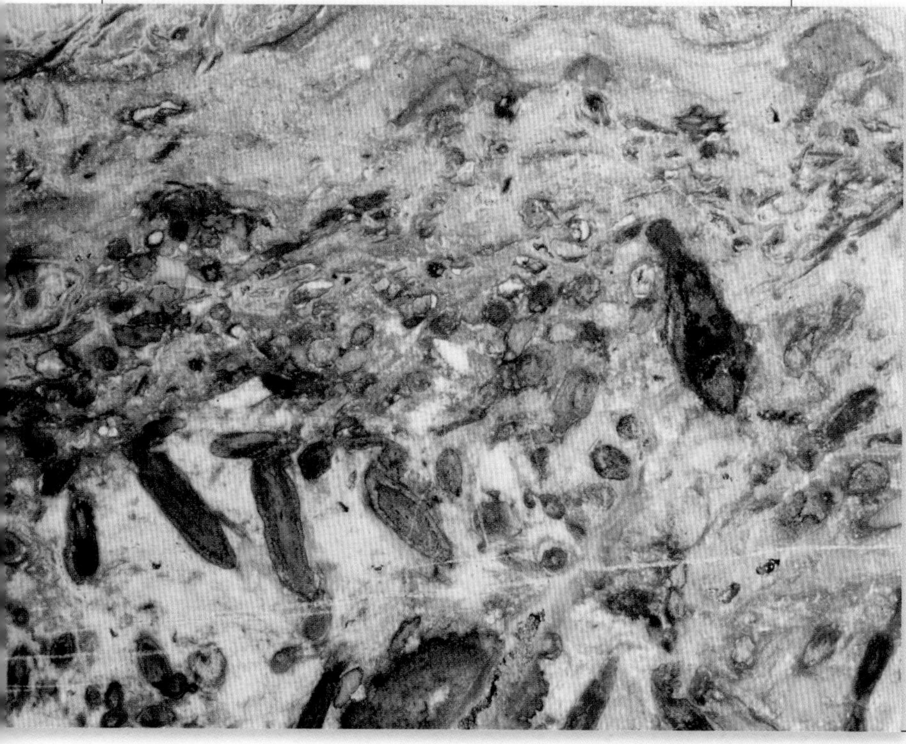

Making It on Land A sample of Rhynie chert shows fragments of fossilized *Rhynia gwynne-vaughani*, a vascular plant abundant during the Devonian. These early land plants were 30–60 centimeters tall and had no leaves and no roots. The dark, spindle-shaped objects seen against the lightest rock are glancing cuts through stems; the circles and ovals are stem cross-sections.

Green Mansions Seedless plants—mosses and ferns—dominate the understory of this pristine rainforest on the island of Maui, Hawaii. The only seed plant we see is the branching tree in the background.

days of the seedless plants are long past, but many of them—most notably the mosses and ferns—are still abundant. We rely on these surviving plants, as well as on fossils, to help us understand the evolution of plants from aquatic algae growing at the edge of a sea or marsh. Much of the fossilized history, and the great bulk of the mass, of seedless plants was preserved as coal, which today is an economically important substance that we burn as fuel and to produce electricity. What a library of botanical history the fossils in the world's coal beds provide!

IN THIS CHAPTER we will see how plants invaded the land, how land plants evolved, and how plant clades diversified, resulting in plants ever better equipped to face the challenges of terrestrial environments. Our descriptions here will concentrate on those land plants that lack seeds. Chapter 29 completes our survey of plants by considering the seed plants, which dominate the terrestrial scene today.

CHAPTER OUTLINE

28.1 How Did the Land Plants Arise?

The question "Where did plants come from?" embraces two questions: "What were the ancestors of plants?" and "Where did those ancestors live?" We will consider both questions in this section.

The **land plants** are monophyletic: all land plants descend from a single common ancestor and form a branch of the evolutionary tree of life. One of the key shared derived traits, or synapomorphies, of the land plants is development from an embryo protected by tissues of the parent plant. For this reason, land plants are sometimes referred to as **embryophytes** (*phyton*, "plant"). Land plants retain the derived features they share with the "green algae" described in Chapter 27: the use of chlorophylls *a* and *b* in photosynthesis, and the use of starch as a photosynthetic storage product. Both land plants and green algae have cellulose in their cell walls.

There are several ways to define "plant" and still refer to a clade (**Figure 28.1**). For example, if the land plants are included with a certain paraphyletic group of green algae, we define a monophyletic group (the **streptophytes**) with several synapomorphies, including the retention of the egg in the parent body. The addition of the remainder of the green algae to the streptophytes gives another clade, commonly called the **green plants**, with synapomorphies including the possession of chlorophyll *b*. Other groups, such as stramenopiles and red algae, are also called "plants" by some people. Green plants, streptophytes, and land plants have each been called "the plant kingdom" by different authorities. There are no objective criteria for defining a kingdom (or any other taxonomic rank); definitions of plant groups are thus somewhat subjective. In this book we use the unmodified common name "plants" to refer to the embryophytes, or monophyletic land plants.

There are ten major groups of land plants

The land plants of today fall naturally into ten major clades (**Table 28.1**). Members of seven of those clades possess well-developed vascular systems that transport materials throughout the plant body. We call these seven groups, collectively, the **vascular plants**, or *tracheophytes*, because they all possess conducting cells called **tracheids**. Together, the seven groups of vascular plants constitute a clade.

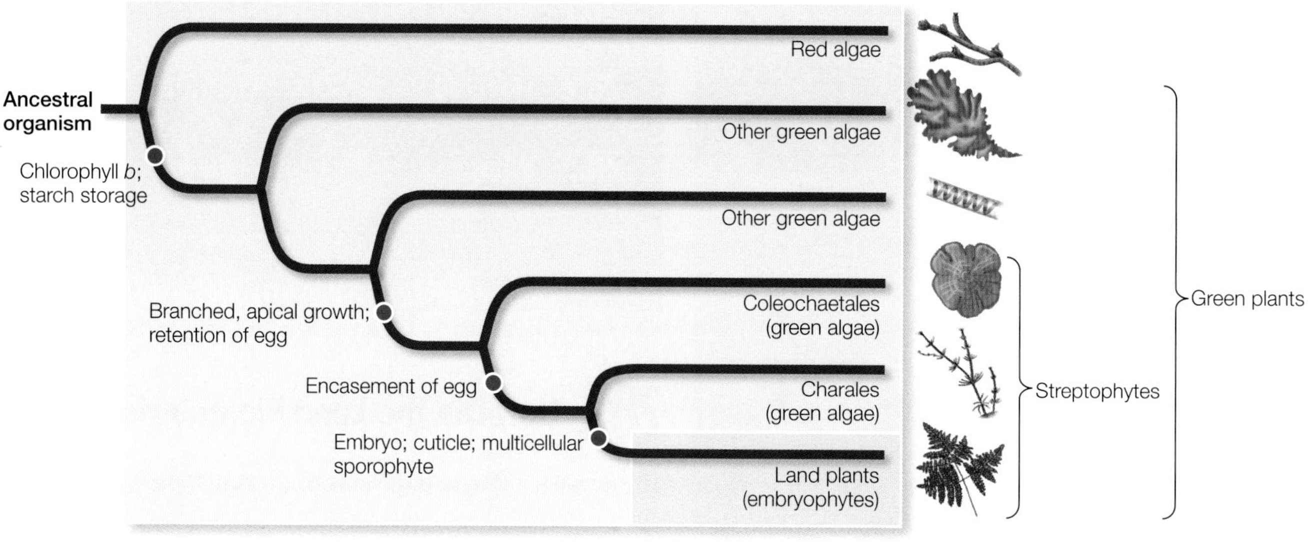

28.1 What Is a Plant? There are various definitions of the term "plant." In this book, we use the most restrictive definition: plants as embryophytes. To include the green algae, we use the broader term "green plants."

The remaining three clades (liverworts, hornworts, and mosses) lack tracheids. These three groups are sometimes collectively called *bryophytes*, but in this text we reserve that term for their most familiar members, the mosses. Instead, we refer to them as **nonvascular plants**, but note that, collectively, *these three groups are not a clade*. Some nonvascular plants have conducting cells, but none have tracheids.

Where did the plants come from? To determine which living organisms are most closely related to plants, biologists considered several of the synapomorphies of land plants and looked for their origins in various other groups.

The land plants arose from a green algal clade

Several synapomorphies, backed by clear-cut evidence from molecular studies, indicate that the sister clade of the plants is a group of aquatic green algae called the **Charales** (see Figure 28.1):

- Plasmodesmata joining the cytoplasm of adjacent cells (see Figure 15.20)
- Branching, apical (tip) growth
- Retention of the egg in the parent organism

TABLE 28.1

Classification of Land Plants

GROUP	COMMON NAME	CHARACTERISTICS
NONVASCULAR PLANTS		
Hepatophyta	Liverworts	No filamentous stage; gametophyte flat
Anthocerophyta	Hornworts	Embedded archegonia; sporophyte grows basally (from the ground)
Bryophyta	Mosses	Filamentous stage; sporophyte grows apically (from the tip)
VASCULAR PLANTS		
Lycophyta	Club mosses and allies	Microphylls in spirals; sporangia in leaf axils
Pteridophyta	Horsetails, whisk ferns, ferns	Differentiation between main stem and side branches (overtopping growth)
SEED PLANTS		
Gymnosperms		
Cycadophyta	Cycads	Compound leaves; swimming sperm; seeds on modified leaves
Ginkgophyta	Ginkgo	Deciduous; fan-shaped leaves; swimming sperm
Gnetophyta	Gnetophytes	Vessels in vascular tissue; opposite, simple leaves
Coniferophyta	Conifers	Seeds in cones; needle-like or scale-like leaves
Angiosperms	Flowering plants	Endosperm; carpels; gametophytes much reduced; seeds within fruit

Note: No extinct groups are included in this classification.

(A) *Chara* sp. (stonewort)

(B) *Coleochaete* sp.

28.2 The Closest Relatives of Land Plants The land plants probably evolved from a common ancestor shared with the Charales, a green algal group. (A) Molecular evidence seems to favor stoneworts of the genus *Chara* as sister group to the plants. (B) Evidence from morphology indicates that the clade including this coleochaete alga is closely related to the land plants.

Stoneworts of the genus *Chara* are members of the Charales that resemble plants in terms of their rRNA and DNA sequences, peroxisome contents, mechanics of mitosis and cytokinesis, and chloroplast structure (**Figure 28.2A**). Strong evidence from morphology-based cladistic analysis suggests that the Coleochaetales, a group of green algae that includes the genus *Coleochaete* (**Figure 28.2B**), are also fairly closely related to the land plants. *Coleochaete*-like algae have several features found in plants, such as a flattened form in contrast to the branched-filament form of *Chara*.

28.1 RECAP

Land plants are photosynthetic organisms that develop from embryos protected by parent plant tissue.

- Can you explain the different possible uses of the term "plant"? See p. 611 and Figure 28.1
- Do you recognize the key difference between the vascular plants and the other three clades of land plants? See pp. 611–612 and Table 28.1
- What evidence supports the phylogenetic relationship between land plants and Charales? See pp. 612–613

The green algal ancestors of the plants lived at the margins of ponds or marshes, ringing them with a green mat. It was from such a marginal habitat, which was sometimes wet and sometimes dry, that early plants made the transition onto land.

28.2 How Did Plants Colonize and Thrive on Land?

Land plants, or their immediate ancestors in those ancient green mats, first appeared in the terrestrial environment between 400 and 500 million years ago. How did they survive in an environment that differed so dramatically from the aquatic environment of their ancestors? While the water essential for life is everywhere in the aquatic environment, it is hard to obtain and retain water in the terrestrial environment.

Adaptations to life on land distinguish land plants from green algae

No longer bathed in fluid, organisms on land faced potentially lethal desiccation (drying). Large terrestrial organisms had to develop ways to transport water to body parts distant from the source. And whereas water provides aquatic organisms with support against gravity, a plant living on land must either have some other support system or sprawl unsupported on the ground. A land plant must also use different mechanisms for dispersing its gametes and progeny than its aquatic relatives, which can simply release them into the water. Survival on land required numerous adaptations. The first colonists—the *nonvascular plants*—met these challenges.

Most of the characteristics that distinguish land plants from green algae are evolutionary adaptations to life on land:

- The *cuticle*, a waxy covering that retards water loss (desiccation)
- *Gametangia*, cases that enclose plant gametes and prevent them from drying out
- *Embryos*, which are young plants contained within a protective structure
- Certain *pigments* that afford protection against the mutagenic ultraviolet radiation that bathes the terrestrial environment
- Thick *spore walls* containing *sporopollenin*, a polymer that protects the spores from desiccation and resists decay
- A *mutually beneficial association with a fungus* that promotes nutrient uptake from the soil

The **cuticle** may be the most important—and earliest—of these features. Composed of several unique waxy lipids (see Section 3.4) that coat the leaves and stems of land plants, the cuticle has several functions, the most obvious and important of which is to keep water (which the plant's ancestors obtained from their environment) from evaporating away from the plant body.

The lotus is revered as a symbol of purity. Its leaves are never dirty, even when they emerge from muddy water. The specific pattern of wax deposition in the lotus cuticle allows water to roll off the leaf surface, pushing dirt particles away. This "lotus effect" keeps the plant clean (and thus better able to trap light for photosynthesis).

Ancient plants also helped themselves and their descendants by contributing to the formation of soil. Acid secreted by plants helps break down rock, and the organic compounds produced by the breakdown of dead plants contribute to soil structure. Such effects are repeated today as plants grow in new areas.

The nonvascular plants usually live where water is available

The nonvascular plants—today's liverworts, hornworts, and mosses—are thought to be similar to the earliest land plants. Most of these plants grow in dense mats, usually in moist habitats (**Figure 28.3**). Even the largest nonvascular plants are only about half a meter tall, and most are only a few centimeters tall or long. Why have they not evolved to be taller? The probable answer is that they lack an efficient vascular system for conducting water and minerals from the soil to distant parts of the plant body.

The nonvascular plants lack the leaves, stems, and roots that characterize the vascular plants, although they have structures analogous to each. Their growth pattern allows water to move through the mats of plants by capillary action. They have leaflike structures that readily catch and hold any water that splashes onto them. They are small enough that minerals can be distributed throughout their bodies by diffusion. As in all land plants, layers of maternal tissue protect their embryos from desiccation. They also have a cuticle, although it is often very thin (or even absent in some species) and thus is not highly effective in retarding water loss.

Most nonvascular plants live on the soil or on other plants, but some grow on bare rock, dead and fallen tree trunks, and even on buildings. The ability to grow on such marginal surfaces results from a mutualistic association with the glomeromycetes, a clade of fungi. The earliest land plants were already colonized by these fungi; the association dates back at least 460 million years. The fungal association probably promoted the absorption of water and minerals, especially phosphorus, from the first "soils."

Nonvascular plants are widely distributed over six continents and even exist (albeit very locally) on the coast of the seventh, Antarctica. They are successful and well adapted to their environments. Most are terrestrial. Some live in wetlands. Although a few nonvascular plant species live in fresh water, these aquatic forms are descended from terrestrial ones. There are no marine nonvascular plants.

Land plants and green algae differ not just in structure, but with respect to their life cycles. Differences in the life cycles of land plants and those of their ancestors are also a function of the relative dependence of plants on a water supply.

Life cycles of land plants feature alternation of generations

A universal feature of the life cycles of land plants is alternation of generations (**Figure 28.4**). Recall from Section 27.4 the two hallmarks of alternation of generations:

- The life cycle includes both multicellular diploid individuals and multicellular haploid individuals.
- Gametes are produced by mitosis, not by meiosis. Meiosis produces spores that develop into multicellular haploid individuals.

If we begin looking at the plant life cycle at the single-cell stage—the diploid zygote—then the first phase of the cycle is the formation, by mitosis and cytokinesis, of a multicellular embryo, which eventually grows into a mature diploid plant. This multicellular diploid plant is the **sporophyte** ("spore plant").

Cells contained in **sporangia** (singular *sporangium*, "spore vessel") of the sporophyte undergo meiosis to produce haploid, uni-

28.3 Mosses Form Dense Mats Dense moss forms hummocks in a valley on New Zealand's South Island.

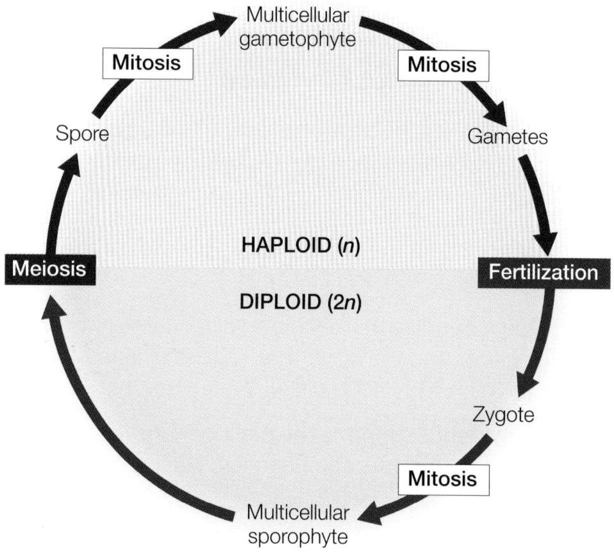

28.4 Alternation of Generations in Plants A diploid sporophyte generation that produces spores by meiosis alternates with a haploid gametophyte generation that produces gametes by mitosis.

cellular spores. By mitosis and cytokinesis, a spore forms a haploid plant. This multicellular haploid plant, called the **gametophyte** ("gamete plant"), produces haploid gametes by mitosis. The fusion of two gametes (*syngamy*, or *fertilization*) forms a single diploid cell—the zygote—and the cycle is repeated (**Figure 28.5**).

The *sporophyte generation* extends from the zygote through the adult multicellular diploid plant; the *gametophyte generation* extends from the spore through the adult multicellular haploid plant to the gametes. The transitions between the generations are accomplished by fertilization and meiosis. In all plants, the sporophyte and gametophyte differ genetically: the sporophyte has diploid cells, and the gametophyte has haploid cells.

Reduction of the gametophyte generation is a major theme in plant evolution. In the nonvascular plants, the gametophyte is larger, longer-lived, and more self-sufficient than the sporophyte. However, in those groups that appeared later in plant evolution, the sporophyte generation is the larger, longer-lived, and more self-sufficient one. In the seed plants, this evolutionary trend has led to a condition in which water is not required for the sperm to reach the egg.

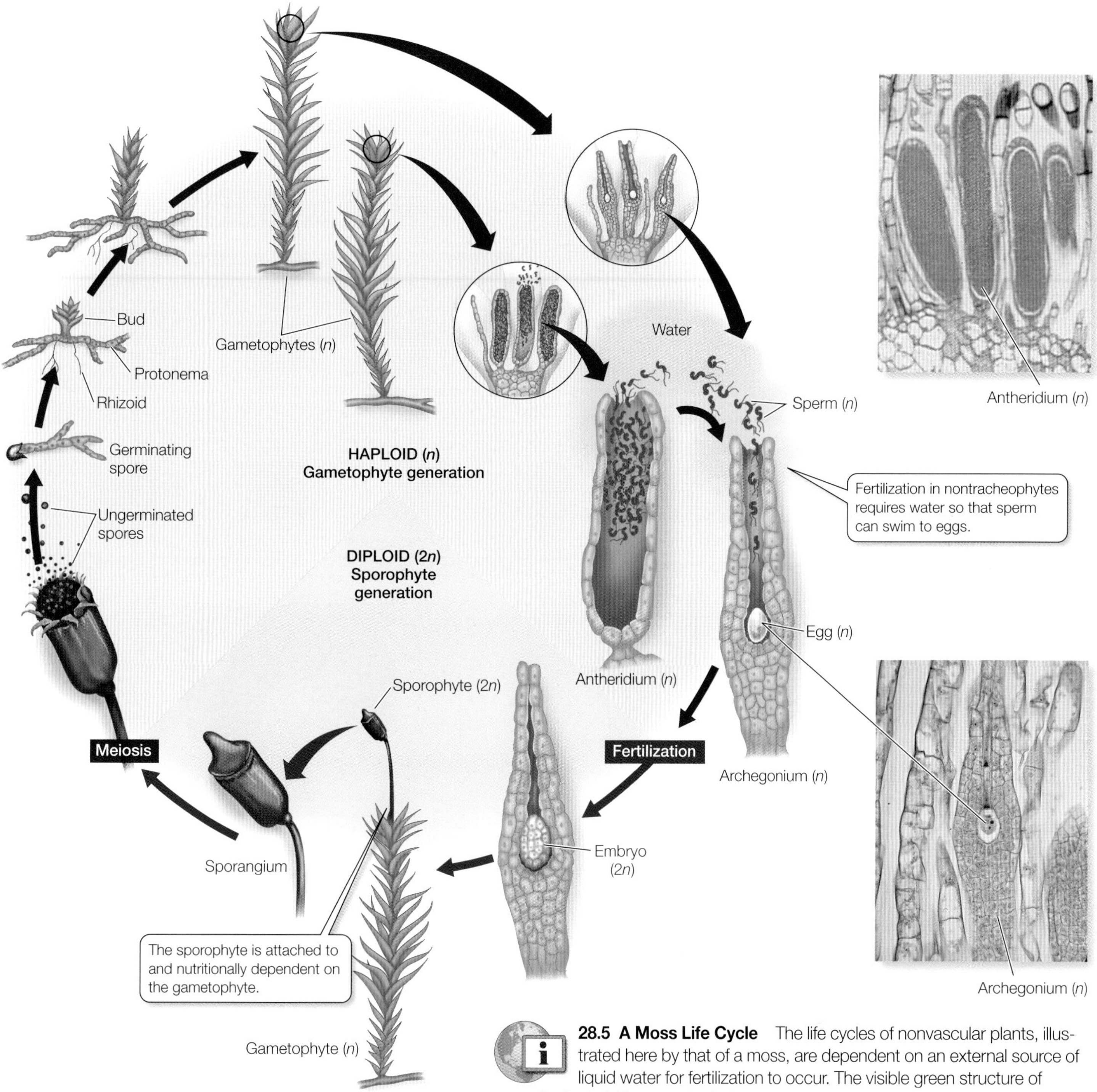

28.5 A Moss Life Cycle The life cycles of nonvascular plants, illustrated here by that of a moss, are dependent on an external source of liquid water for fertilization to occur. The visible green structure of such plants is the gametophyte.

The sporophytes of nonvascular plants are dependent on gametophytes

In nonvascular plants, the conspicuous green structure visible to the naked eye is the gametophyte (see Figure 28.5), in contrast to the familiar forms of vascular plants, such as ferns and seed plants, which are sporophytes. The gametophyte of liverworts, hornworts, and mosses is photosynthetic and is therefore nutritionally independent, whereas the sporophyte may or may not be photosynthetic, but is always nutritionally dependent on the gametophyte and remains permanently attached to it.

A nonvascular plant sporophyte produces unicellular haploid spores as products of meiosis within a sporangium. A spore germinates, giving rise to a multicellular haploid gametophyte whose cells contain chloroplasts and are thus photosynthetic. Eventually gametes form within specialized sex organs, the **gametangia**. The **archegonium** is a multicellular, flask-shaped female sex organ with a long neck and a swollen base, which produces a single egg. The **antheridium** is a male sex organ in which sperm, each bearing two flagella, are produced in large numbers (see the insets in Figure 28.5).

Once released from the antheridium, the sperm must swim or be splashed by raindrops to a nearby archegonium on the same or a neighboring plant. The sperm are aided in this task by chemical attractants released by the egg or the archegonium. Before sperm can enter the archegonium, certain cells in the neck of the archegonium must break down, leaving a water-filled canal through which the sperm swim to complete their journey. Note that *all of these events require liquid water.*

On arrival at the egg, the nucleus of a sperm fuses with the egg nucleus to form a diploid zygote. Mitotic divisions of the zygote produce a multicellular, diploid sporophyte embryo. The base of the archegonium grows to protect the embryo during its early development. Eventually, the developing sporophyte elongates sufficiently to break out of the archegonium, but it remains connected to the gametophyte by a "foot" that is embedded in the parent tissue and absorbs water and nutrients from it. In liverworts, hornworts, and mosses, the sporophyte remains attached to the gametophyte throughout its life. The sporophyte produces a sporangium, within which meiotic divisions produce spores and thus the next gametophyte generation.

28.2 RECAP

Terrestrial organisms must obtain and retain water for their life processes, for support against gravity, and for reproduction.

- Describe several adaptations of plants to the terrestrial environment. See p. 613
- Can you explain the cycle of alternation of generations? See pp. 614–615 and Figure 28.5
- Why do some plants require liquid water for fertilization? See p. 616
- In the nonvascular plants, how is the sporophyte dependent on the gametophyte? See p. 616

Further adaptations to the terrestrial environment appeared as plants continued to evolve. One of the most important of these later adaptations was the appearance of vascular tissues.

28.3 What Features Distinguish the Vascular Plants?

We now know that nonvascular plants evolved tens of millions of years before the earliest vascular plants, even though vascular plants—which fossilize more readily because of the chemical makeup of their vascular tissues—appear earlier in the fossil record. Persuasive DNA and structure-based evidence for the earlier appearance of nonvascular plants was further supported by recent experimental evidence suggesting that ancient microfossils predating the first appearance of nonvascular plants are actually fragments of ancient liverworts (**Figure 28.6**).

EXPERIMENT

HYPOTHESIS: Ancient microfossils, predating the earliest known nonvascular plant fossils, could be fragments of ancient liverworts.

METHOD

1. Investigators allowed liverworts to rot in soil or subjected them to high-temperature acid treatment, then examined the degraded material by light and scanning electron microscopy.
2. They compared images of the degraded material with those of microfossils from ancient rocks.

RESULTS

Sheets of decay-resistant cells from the lower surface of the degraded liverworts resembled some cell-sheet microfossils, showing rosette-like groupings of cells around "pores."

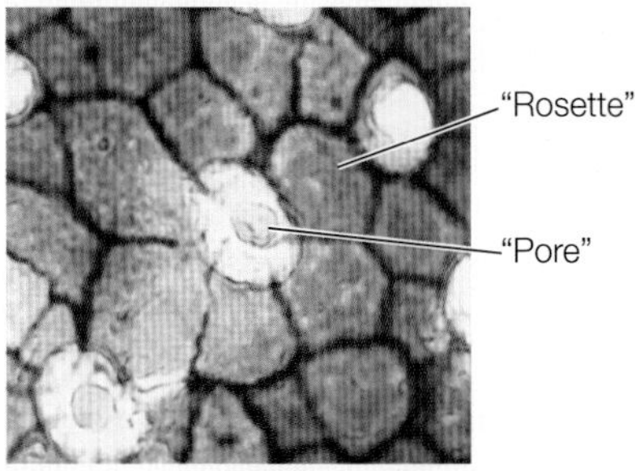

Resistant fragments of liverwort rhizoids resembled some tubular microfossils.

CONCLUSION: The ancient microfossils may be fragments of liverworts.

28.6 Mimicking a Microfossil? Linda Graham and her collaborators studied two species of decay-resistant liverworts. These plants yielded degradation fragments closely similar in appearance to microfossils characteristic of rocks from the Cambrian through the Devonian, suggesting that those microfossils were also from liverworts.

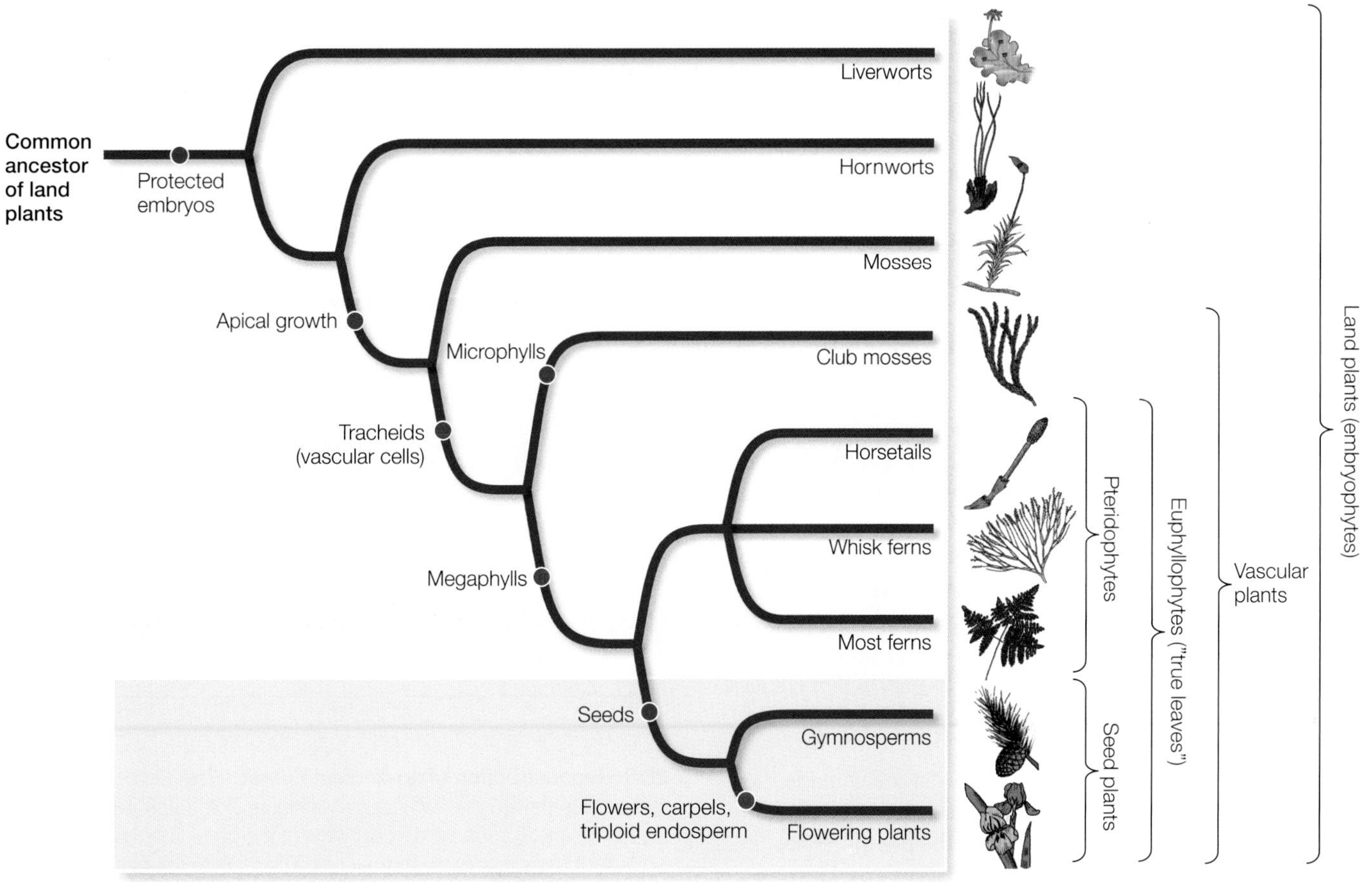

28.7 The Evolution of Today's Plants Three key characteristics that emerged during plant evolution—protected embryos, vascular tissues, and seeds—are adaptations to life in a terrestrial environment.

The first land plants were nonvascular, lacking both water-conducting and food-conducting tissue. The first true vascular plants, possessing specialized tracheids, arose later (**Figure 28.7**).

Vascular tissues transport water and dissolved materials

Vascular plants differ from the other land plants in crucial ways, one of which is the possession of a well-developed **vascular system** consisting of tissues specialized for the transport of materials from one part of the plant to another. One type of vascular tissue, the **xylem**, conducts water and minerals from the soil to aerial parts of the plant; because some of its cell walls are stiffened by a substance called *lignin*, xylem also provides support in the terrestrial environment. The other type of vascular tissue, the **phloem**, conducts the products of photosynthesis from sites where they are produced or released to sites where they are used or stored.

Familiar vascular plants include the club mosses, ferns, conifers, and angiosperms (flowering plants). Although they are an extraordinarily large and diverse group, the vascular plants can be said to have been launched by a single evolutionary event. Sometime during the Paleozoic era, probably well before the Silurian period (440 mya), the sporophyte generation of a now long-extinct plant produced a new cell type, the tracheid. The tracheid is the principal water-conducting element of the xylem in all vascular plants except the angiosperms, and even in the angiosperms, tracheids persist alongside a more specialized and efficient system of vessels and fibers derived from them.

The evolution of tracheids had two important consequences. First, it provided a pathway for transport of water and mineral nutrients from a source of supply to regions of need in the plant body. Second, the stiff cell walls of tracheids provided something almost completely lacking—and unnecessary—in the largely aquatic green algae: rigid structural support. Support is important in a terrestrial environment because it allows plants to grow upward as they compete for sunlight to power photosynthesis. A taller plant can receive direct sunlight and photosynthesize more readily than a shorter plant, whose leaves may be shaded by the taller one. Increased height also improves the dispersal of spores. Thus the tracheid set the stage for the complete and permanent invasion of land by plants.

The vascular plants featured another evolutionary novelty: a branching, independent sporophyte. A branching sporophyte can produce more spores than an unbranched body, and it can develop in complex ways. The sporophyte of a vascular plant is nutritionally independent of the gametophyte at maturity. Among the vascular plants, the sporophyte is the large and obvious plant that one normally notices in nature, in contrast to the sporophyte of nonvascular plants, which is attached to, dependent on, and usually much smaller than the gametophyte.

The present-day evolutionary descendants of the early vascular plants belong to six major groups (see Figure 28.7). Two types of life cycles are seen in the vascular plants: one that involves seeds and another that does not. The life cycles of the clades that include the club mosses and the ferns and their relatives, the horsetails and whisk ferns, do not involve seeds. We will describe these nonseed vascular plant groups in detail after taking a closer look at vascular plant evolution. The major groups of seed plants will be described in Chapter 29.

Vascular plants have been evolving for almost half a billion years

The evolution of an effective cuticle and of protective layers for the gametangia (archegonia and antheridia) helped make the first vascular plants successful, as did the initial absence of herbivores (plant-eating animals) on land. By the late Silurian period, vascular plants were being preserved as fossils that we can study today. During the Silurian, the largest vascular plants were only a few centimeters tall, yet fossils uncovered in Wales in 2004 gave clear evidence for the earliest known wildfire, which burned vigorously even in the Silurian atmosphere, which had 14 percent less oxygen than today's. The small plants must have been abundant to sustain fire in such an atmosphere.

Two groups of vascular plants that still exist today made their first appearances during the Devonian period (409–354 mya): the lycophytes (club mosses and relatives) and the pteridophytes (including horsetails and ferns). Their proliferation made the terrestrial environment more hospitable to animals. Amphibians and insects arrived on land at about the time the plants became established.

Trees of various kinds appeared in the Devonian period and dominated the landscape of the Carboniferous period. Mighty forests of lycophytes up to 40 meters tall, along with horsetails and tree ferns, flourished in the tropical swamps of what would become North America and Europe (**Figure 28.8**). Plant parts from those forests sank in the swamps and were gradually covered by sediment. Over millions of years, as the buried plant material was subjected to intense pressure and elevated temperatures, coal formed. Today that coal provides over half our electricity—and contributes to air pollution and global warming. The deposits, although huge, are not infinite, and they are not being renewed.

Coal comes from the remains of Carboniferous plants, but what about the other two great "fossil fuels"—petroleum and natural gas? They come from the remains of plankton that lived in ancient oceans.

28.8 Reconstruction of an Ancient Forest This Carboniferous forest once thrived in what is now Michigan. The "trees" to the left and in the background are lycophytes of the genus *Lepidodendron*; abundant ferns are visible to the right. The plant in the foreground is a relative of the horsetails.

In the subsequent Permian period, the continents came together to form a single gigantic land mass called Pangaea. The continental interior became warmer and drier, but late in the period glaciation was extensive. The 200-million-year reign of the lycophyte–fern forests came to an end as they were replaced by forests of seed plants (gymnosperms), which were prevalent until a different group of seed plants (angiosperms) overtook the landscape less than 80 million years ago.

The earliest vascular plants lacked roots and leaves

The earliest known vascular plants belonged to the now-extinct group called **rhyniophytes**. The rhyniophytes were one of a very few types of vascular plants in the Silurian period. The landscape at that time probably consisted of bare ground, with stands of rhyniophytes in low-lying moist areas. Early versions of the structural features of all the other vascular plant groups appeared in the rhyniophytes of that time. These shared features strengthen the case for the origin of all vascular plants from a common nonvascular plant ancestor.

At the beginning of this chapter we described the discovery of some important fossils in Devonian rocks near Rhynie, Scotland. The preservation of these plants was remarkable, considering that the rocks were more than 395 million years old. These fossil plants had a simple vascular system of phloem and xylem, but not all had the tracheids characteristic of today's vascular plants.

These plants also lacked roots. Like most modern ferns and lycophytes, they were apparently anchored in the soil by horizontal portions of stem, called **rhizomes**, which bore water-absorbing unicellular filaments called **rhizoids**. These rhizomes also bore aerial branches, and sporangia—homologous to the sporangia of mosses—were found at the tips of those branches. Their branching pattern was *dichotomous*; that is, the apex (tip) of the shoot divided to produce two equivalent new branches, each pair diverging at approximately the same angle from the original stem (**Figure 28.9**). Scattered fragments of such plants had been found earlier, but never in such profusion or so well preserved as those discovered by Kidston and Lang.

Although they were apparently ancestral to the other vascular plant groups, the rhyniophytes themselves are long gone. None of their fossils appear anywhere after the Devonian period.

The vascular plants branched out

A new group of vascular plants—the **lycophytes** (club mosses and their relatives)—also appeared in the Silurian period. Another—the **pteridophytes** (ferns and fern allies)—appeared during the Devonian period. These two groups, both still with us today, arose from rhyniophyte-like ancestors. These new groups featured specializations not found in the rhyniophytes, including true roots, true leaves, and a differentiation between two types of spores. The pteridophytes and seed plants constitute a clade called the **euphyllophytes**.

An important synapomorphy of the euphyllophytes is **overtopping** growth, a pattern in which one branch differentiates from and grows beyond the others. Overtopping growth would have given these plants an advantage in the competition for light for photosynthesis, enabling them to shade their dichotomously growing competitors. And, as we'll see, the overtopping growth of the euphyllophytes enabled a new type of leaf to evolve.

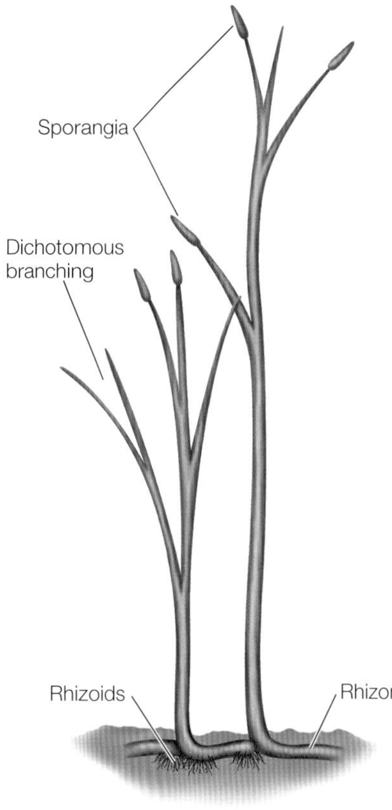

28.9 An Ancient Vascular Plant Relative This extinct plant, *Aglaophyton major* (a rhyniophyte), lacked roots and leaves. It had a central column of xylem running through its stems, but no true tracheids. The rhizome is a horizontal underground stem, not a root. The dichotomously branching aerial stems were less than 50 centimeters tall, and some were topped by sporangia. Other very similar rhyniophytes, such as *Rhynia*, did have tracheids.

Roots may have evolved from branches

The rhyniophytes had only rhizoids arising from a rhizome with which to gather water and minerals. How, then, did subsequent groups of vascular plants come to have the complex roots we see today? Roots probably arose separately in the lycophytes and euphyllophytes.

It is probable that roots had their evolutionary origins as a branch, either of a rhizome or of the aboveground portion of a stem. That branch presumably penetrated the soil and branched further. The underground portion could anchor the plant firmly, and even in this primitive condition, it could absorb water and minerals. The discovery of several fossil plants from the Devonian period, all having horizontal stems (rhizomes) with both underground and aerial branches, supported this hypothesis.

Underground and aboveground branches, growing in sharply different environments, were subjected to very different selection pressures during the succeeding millions of years. Thus the two parts of the plant body—the aboveground shoot system and the underground root system—diverged in structure and evolved distinct internal and external anatomies. In spite of these differences, scientists believe that the root and shoot systems of vascular plants are homologous—that they were once part of the same organ.

Pteridophytes and seed plants have true leaves

Thus far we have used the term "leaf" rather loosely. In the strictest sense, a *leaf* is a flattened photosynthetic structure emerging laterally from a stem or branch and possessing true vascular tissue. Using this precise definition as we take a closer look at true leaves in the vascular plants, we see that there are two different types of leaves, very likely of different evolutionary origins.

The first leaf type, the **microphyll**, is usually small and only rarely has more than a single vascular strand, at least in plants alive today. Club mosses (lycophytes), of which only a few genera survive, have such simple leaves. Some biologists believe that microphylls had their evolutionary origins as sterile sporangia (**Figure 28.10A**). The principal characteristic of this type of leaf is a vascular strand that departs from the vascular system of the stem in such a way that the structure of the stem's vascular system is scarcely disturbed. This was true even in the lycophyte trees of the Carboniferous period, many of which had leaves many centimeters long.

The other leaf type is found in pteridophytes and seed plants. This larger, more complex leaf is called a **megaphyll**. The megaphyll is thought to have arisen from the flattening of a dichotomously branching stem system with overtopping growth. This change was followed by the development of photosynthetic tissue between the members of overtopped groups of branches (**Figure 28.10B**), which had the advantage of increasing the photosynthetic surface area of those branches. Megaphylls evolved more than once, in different clades of euphyllophytes with overtopping growth.

The first megaphylls, which were very small, appeared in the Devonian period. We might expect that evolution should have led swiftly to the appearance of more and larger megaphylls because of their greater photosynthetic capacity. However, it took some 50 million years, until the Carboniferous period, for large megaphylls to become common. Why should this have been so, especially given that other advances in plant structure were taking place during that time?

According to one theory, the high concentration of CO_2 in the atmosphere during the Devonian period restricted the development of the tiny pores, called *stomata*, that allow a leaf to take up CO_2 for use in photosynthesis. With more CO_2 available, fewer stomata were needed. Today, when stomata are open, they allow water vapor to escape the leaf and CO_2 to enter. In the Devonian, larger leaves would have absorbed heat from sunlight, but would

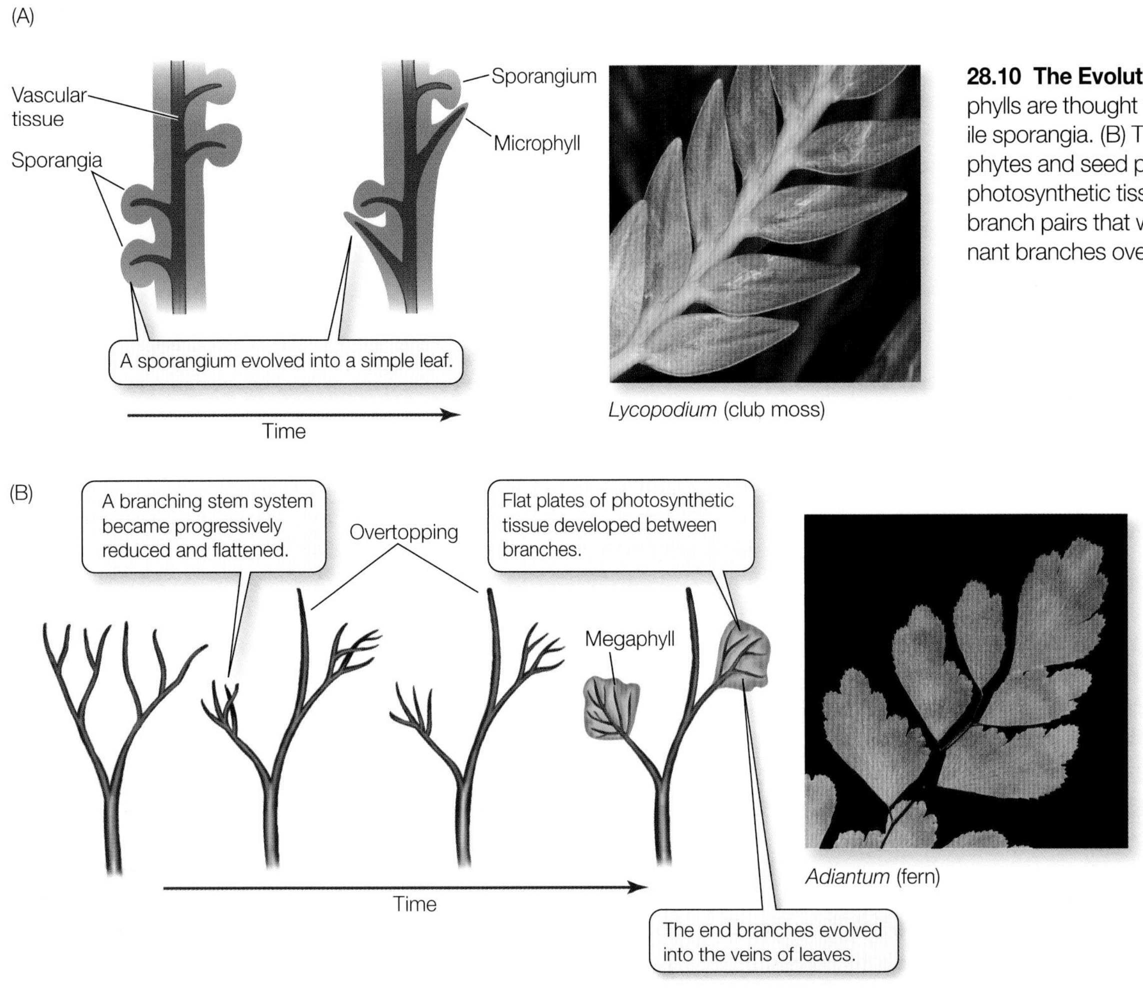

28.10 The Evolution of Leaves (A) Microphylls are thought to have evolved from sterile sporangia. (B) The megaphylls of pteridophytes and seed plants may have arisen as photosynthetic tissue developed between branch pairs that were "left behind" as dominant branches overtopped them.

EXPERIMENT

HYPOTHESIS: High concentrations of CO_2 in the Devonian atmosphere delayed the increase in leaf size.

METHOD

1. Scientists analyzed 300 plant fossils from the Devonian and Carboniferous periods and calculated the sizes of the leaves.
2. They compared the pattern of change in leaf size over time with that of change in atmosphere CO_2 concentrations.

RESULTS

Among these fossils, a rise in leaf size corresponded in time with a decline in the estimated concentration of CO_2 in the atmosphere.

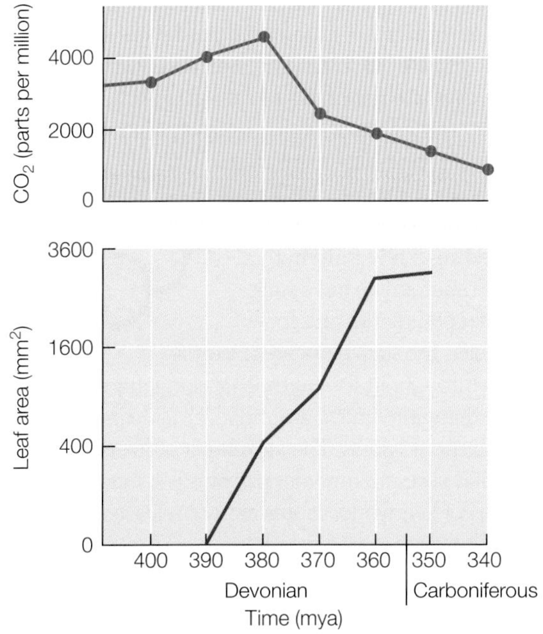

CONCLUSION: Leaf sizes increased as CO_2 concentrations decreased.

28.11 Decreasing CO_2 Levels and the Evolution of Megaphylls How can biologists test hypotheses relating to events that happened hundreds of millions of years ago? C. P. Osborne and colleagues gathered and measured the leaves of plant fossils from the Devonian and Carboniferous periods. They then compared the sizes of the leaves with estimates of atmospheric CO_2 concentrations under which the plants had lived. FURTHER RESEARCH: What sort of experiment would you do to determine the effects of stomata on overheating of fern leaves today?

have been unable to lose heat fast enough by evaporation of water through their limited number of stomata. The resulting overheating would have been lethal. Recent research has supported this theory, indicating that larger megaphylls evolved only as CO_2 concentrations dropped over millions of years (**Figure 28.11**).

Heterospory appeared among the vascular plants

In the most ancient of the present-day vascular plants, the gametophyte and the sporophyte are independent, and both are usually photosynthetic. Spores produced by the sporophyte are of a single type and develop into a single type of gametophyte that bears both female and male reproductive organs. The female organ is a multicellular archegonium containing a single egg. The male organ is an antheridium, producing many sperm. Such plants, which bear a single type of spore, are said to be **homosporous** (**Figure 28.12A**).

A system with two distinct types of spores evolved somewhat later. Plants of this type are said to be **heterosporous** (**Figure 28.12B**). In heterospory, one type of spore—the **megaspore**—develops into a specifically female gametophyte (a **megagametophyte**) that produces only eggs. The other type, the **microspore**, is smaller and develops into a male gametophyte (a **microgametophyte**) that produces only sperm. The sporophyte produces megaspores in small numbers in **megasporangia**, and microspores in large numbers in **microsporangia**. Heterospory affects not only the spores and the gametophyte, but also the sporophyte itself, which must develop two types of sporangia.

The most ancient vascular plants were all homosporous, but heterospory evidently evolved several times in the early descendants of the rhyniophytes. The fact that heterospory evolved repeatedly suggests that it affords selective advantages. Subsequent evolution in the land plants featured ever greater specialization of the heterosporous condition. All seed plants are heterosporous.

The nonseed vascular plants, like the nonvascular plants, require liquid water at a key stage in their life cycles: sperm reach eggs only by means of liquid water. As we will see in the next chapter, the seed plants have taken the next evolutionary step and can unite sperm and egg without relying on water.

28.3 RECAP

A new type of cell, the tracheid, marked the origin of the vascular plants. Later evolutionary events included the appearance of roots and leaves.

- How do the vascular tissues xylem and phloem serve the vascular plants? See p. 617
- Do you understand the difference between the two leaf types, microphylls and megaphylls? See p. 620 and Figure 28.10
- Can you explain the concept of heterospory? See p. 621 and Figure 28.12

The liverworts, hornworts, mosses, lycophytes, and pteridophytes have come a long way from the aquatic environment to meet the challenges of life on dry land. Let's look at the diversity within these groups.

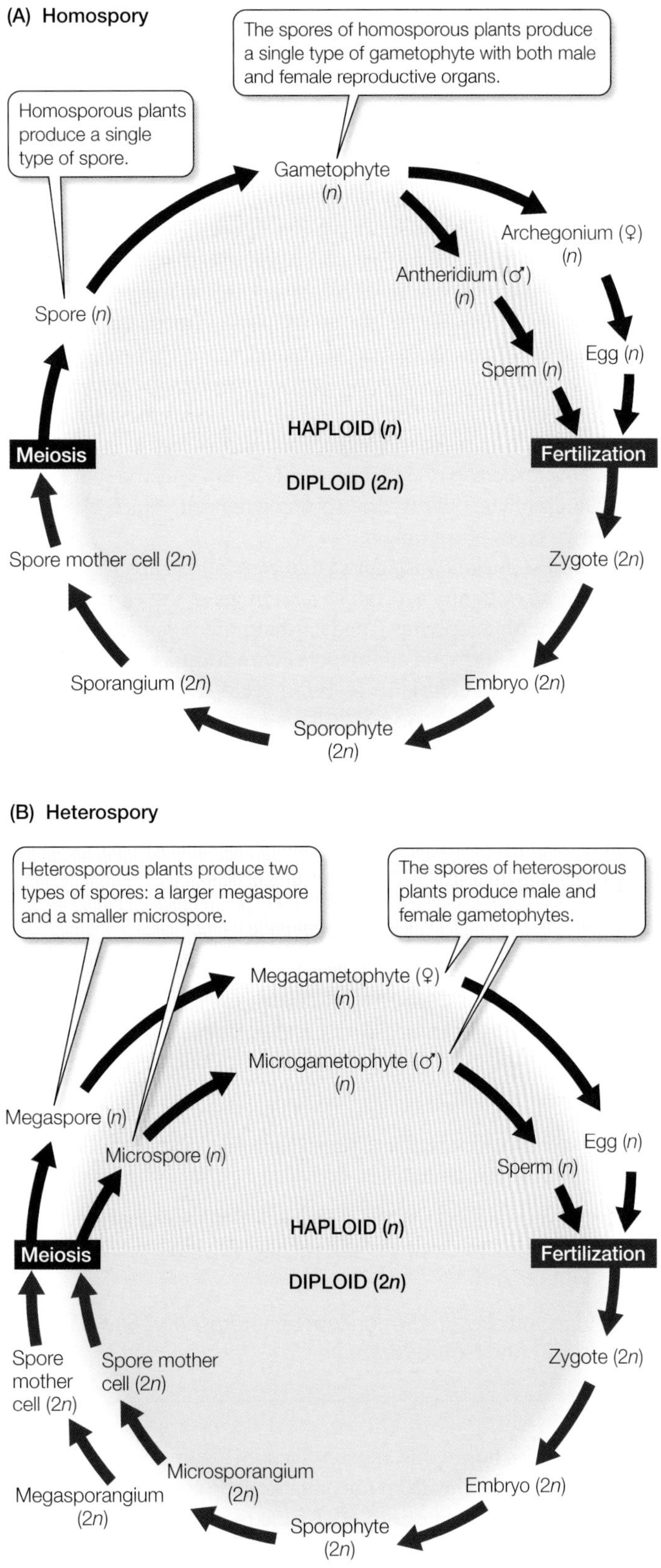

28.12 Homospory and Heterospory (A) Homosporous plants bear a single type of spore. Each gametophyte has two types of sex organs, antheridia (male) and archegonia (female). (B) Heterosporous plants bear two types of spores that develop into distinctly male and female gametophytes.

28.4 What Are the Major Clades of Seedless Plants?

Three clades of land plants lack tracheids; these nonvascular plants are the liverworts, hornworts, and mosses. Seedless vascular plants include three clades—club mosses, horsetails and whisk ferns—and the ferns, which are not a clade. The structure and growth pattern of the sporophyte differ among the three nonvascular land plant groups.

Liverworts may be the most ancient surviving plant clade

There are about 9,000 species of **liverworts** (**Hepatophyta**). Most liverworts have leafy gametophytes (**Figure 28.13A**). Some have *thalloid* gametophytes—green, leaflike layers that lie flat on the ground (**Figure 28.13B**). The simplest liverwort gametophytes, however, are flat plates of cells, a centimeter or so long, that produce antheridia or archegonia on their upper surfaces and rhizoids on their lower surfaces.

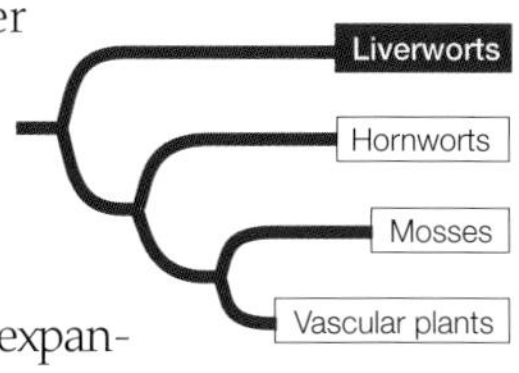

Liverwort sporophytes are shorter than those of mosses and hornworts, rarely exceeding a few millimeters. The liverwort sporophyte has a stalk that connects sporangium and foot. In most species, the stalk elongates by expansion of cells throughout its length. This elongation raises the sporangium above ground level, allowing the spores to be dispersed more widely. The sporangia of liverworts are simple: a globular sporangium wall surrounds a mass of spores. In some species of liverworts, spores are not released by the sporophyte until the surrounding sporangium wall rots. In other liverworts, however, the spores are thrown from the sporangium by structures that shorten and compress a "spring" as they dry out. When the stress becomes sufficient, the compressed spring snaps back to its resting position, throwing spores in all directions.

Among the most familiar thalloid liverworts are species of the genus *Marchantia*. *Marchantia* is easily recognized by the characteristic structures on which its male and female gametophytes bear their antheridia and archegonia (**Figure 28.13C**). Like most liverworts, *Marchantia* also reproduces asexually by simple fragmentation of the gametophyte. *Marchantia* and some other liverworts and mosses also reproduce asexually by means of *gemmae* (singular *gemma*), which are lens-shaped clumps of cells. In a few liverworts, the gemmae are loosely held in structures called *gemmae cups*, which promote dispersal of the gemmae by raindrops (see Figure 28.13C).

Hornworts have stomata, distinctive chloroplasts, and sporophytes without stalks

The group **Anthocerophyta** comprises the approximately 100 species of **hornworts**, so named because their sporophytes look like little horns (**Figure 28.14**). Hornworts appear at first glance to be liverworts with very simple gametophytes. Their gametophytes are flat plates of cells a few cells thick.

These cups contain gemmae—small, lens-shaped outgrowths of the plant body, each capable of developing into a new plant.
The banana-like structures bear archegonia.

(A) *Bazzania trilobata* (B) *Marchantia* sp. (C) *Marchantia* sp.

28.13 Liverwort Structures Liverworts display various characteristic structures. (A) The gametophyte of a leafy liverwort. (B) Gametophytes of a thalloid liverwort. (C) This thalloid liverwort bears archegonia in the structures that look like bunches of bananas. It also bears gemmae cups containing gemmae.

The hornworts, along with the mosses and vascular plants, share an advance over the liverwort clade in their adaptation to life on land: they have stomata. Stomata may be a shared derived trait of hornworts and all other land plants except liverworts, although hornwort stomata do not close, as do those in mosses and vascular plants, and may have evolved independently.

Hornworts have two characteristics that distinguish them from both liverworts and mosses. First, the cells of hornworts each contain a single large, platelike chloroplast, whereas the cells of the other two groups contain numerous small, lens-shaped chloroplasts. Second, of the sporophytes in all three of these groups, those of the hornworts come closest to being capable of growth without a set limit. Liverwort and moss sporophytes have a stalk that stops growing as the sporangium matures, so elongation of the sporophyte is strictly limited. The hornwort sporophyte, however, has no stalk. Instead, a basal region of the sporangium remains capable of indefinite cell division, continuously producing new spore-bearing tissue above. The sporophytes of some hornworts growing in mild and continuously moist conditions can become as tall as 20 centimeters. Eventually the sporophyte's growth is limited by the lack of a transport system.

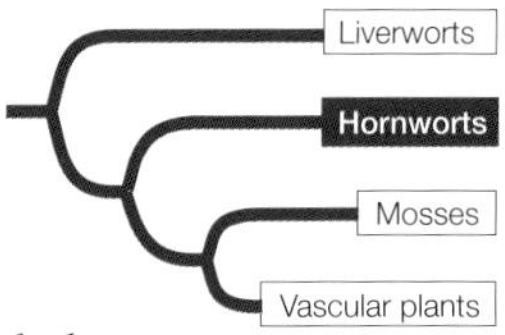
Liverworts
Hornworts
Mosses
Vascular plants

To support their metabolism, the hornworts need access to nitrogen. Hornworts have internal cavities filled with mucilage; these cavities are often populated by cyanobacteria that convert atmospheric nitrogen gas into a form usable by their host plant.

We have presented the hornworts as sister to the clade consisting of mosses and vascular plants, but this is only one possible interpretation of the current data. The exact evolutionary status of the hornworts is still unclear, and in some analyses they are placed as the sister group to the remaining land plants.

Water and sugar transport mechanisms emerged in the mosses

The most familiar of the land plants without tracheids are the **mosses** (**Bryophyta**). There are about 15,000 species of mosses—more than of liverworts and hornworts combined—and these hardy little plants are found in almost every terrestrial environment. They are often found on damp, cool ground, where they form thick mats (see Figure 28.3). The mosses are probably sister to the vascular plants (see Figure 28.7).

In mosses, the gametophyte begins its development following spore germination as a branched, filamentous structure called a *protonema* (see Figure 28.5). Although the protonema looks a bit like a filamentous green alga, it is unique to the mosses. Some of the filaments contain chloroplasts and are photosynthetic; others,

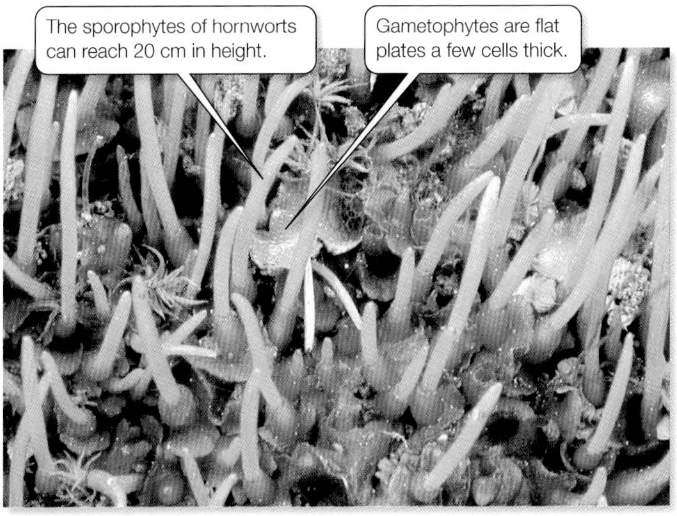
The sporophytes of hornworts can reach 20 cm in height.
Gametophytes are flat plates a few cells thick.

Anthoceros sp.

28.14 A Hornwort The sporophytes of many hornworts resemble little horns.

called rhizoids, are nonphotosynthetic and anchor the protonema to the substratum. After a period of linear growth, cells close to the tips of the photosynthetic filaments divide rapidly in three dimensions to form *buds*. The buds eventually develop a distinct tip, or apex, and produce the familiar leafy moss shoot with leaflike structures arranged spirally. These leafy shoots produce antheridia or archegonia (see Figure 28.5).

Sporophyte development in most mosses follows a precise pattern, resulting ultimately in the formation of an absorptive foot anchored to the gametophyte, a stalk, and, at the tip, a swollen sporangium. In contrast to liverworts and hornworts, the sporophytes of mosses and vascular plants grow by **apical cell division**, in which a region at the growing tip provides an organized pattern of cell division, elongation, and differentiation. This growth pattern allows extensive and sturdy vertical growth of sporophytes. Apical cell division is a synapomorphy of mosses and vascular plants.

Some moss gametophytes are so large that they could not transport enough water solely by diffusion. Gametophytes and sporophytes of many mosses contain a type of cell called a *hydroid*, which dies and leaves a tiny channel through which water can travel. The hydroid may be the progenitor of the tracheid, the characteristic water-conducting cell of the vascular plants, but it lacks lignin and the cell wall structure found in tracheids. The possession of hydroids and of a limited system for transport of sugar by some mosses (via cells called *leptoids*) shows that the old term "nonvascular plant" is somewhat misleading when applied to mosses. Despite their simple system of internal transport, however, the mosses are not vascular plants because they lack true xylem and phloem.

Mosses of the genus *Sphagnum* often grow in swampy places, where the plants begin to decompose in the water after they die. Rapidly growing upper layers of moss compress the deeper-lying, decomposing layers. Partially decomposed plant matter is called *peat*. In some parts of the world, people derive the majority of their fuel from peat bogs (**Figure 28.15**). *Sphagnum*-dominated peatlands cover an area approximately half as large as the United States—more than 1 percent of Earth's surface. Long ago, continued compression of peat composed primarily of other nonseed plants gave rise to coal.

28.15 Harvesting Peat from a Bog A farmer extracts peat, formed from decomposing *Sphagnum* mosses. Peat bogs are an important source of fuel in some areas of the world, such as the southern Ireland location shown here.

Some vascular plants have vascular tissue but not seeds

The earliest vascular plant clades that survive to this day do not have seeds. These plants have a large, independent sporophyte and a small gametophyte that is independent of the sporophyte. The gametophytes of the surviving nonseed vascular plants are rarely more than 1 or 2 centimeters long and are short-lived, whereas their sporophytes are often highly visible and long-lived; the sporophyte of a tree fern, for example, may be 15 or 20 meters tall and may live for many years.

The most prominent resting stage in the life cycle of the seedless vascular plants is the single-celled spore. A spore may "rest" for some time before developing further. This feature makes their life cycle similar to those of the fungi, the green algae, and the nonvascular plants, but not, as we will see in the next chapter, to that of the seed plants. Nonseed vascular plants must have an aqueous environment for at least one stage of their life cycle because fertilization is accomplished by flagellated, swimming sperm.

The ferns are the most abundant and diverse group of seedless vascular plants today, but the club mosses and horsetails were once dominant elements of Earth's vegetation. A fourth group, the whisk ferns, contains only two genera. Let's look at the characteristics of these four groups and at some of the evolutionary advances that appeared in them.

The club mosses are sister to the other vascular plants

The **club mosses** and their relatives, the spike mosses and quillworts (together called lycophytes), diverged earlier than all other living vascular plants; that is, the remaining vascular plants share an ancestor that was not ancestral to the lycophytes. There are relatively few surviving species of lycophytes— just over 1,200.

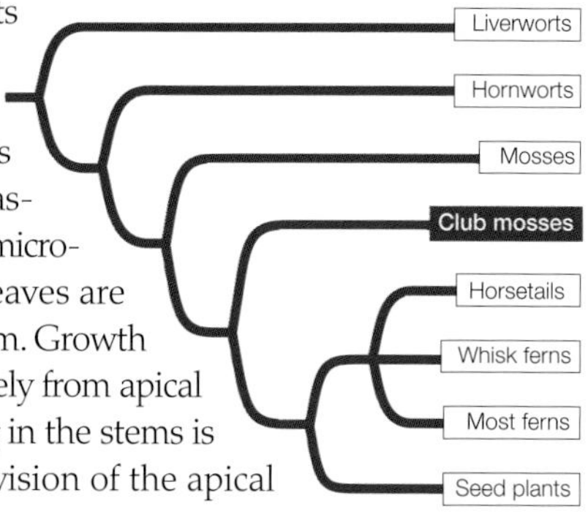

The lycophytes have roots that branch dichotomously. The arrangement of vascular tissue in their stems is simpler than in the other vascular plants. They bear only microphylls, and these simple leaves are arranged spirally on the stem. Growth in club mosses comes entirely from apical cell division, and branching in the stems is also dichotomous, by a division of the apical cluster of dividing cells.

The sporangia of many club mosses are aggregated in conelike structures called *strobili* (singular *strobilus*; **Figure 28.16**). A strobilus is a cluster of spore-bearing leaves inserted on an axis (linear supporting structure). Other club mosses lack strobili and bear their sporangia on (or adjacent to) the upper surfaces of leaves called *sporophylls*. This placement contrasts with the terminal spo-

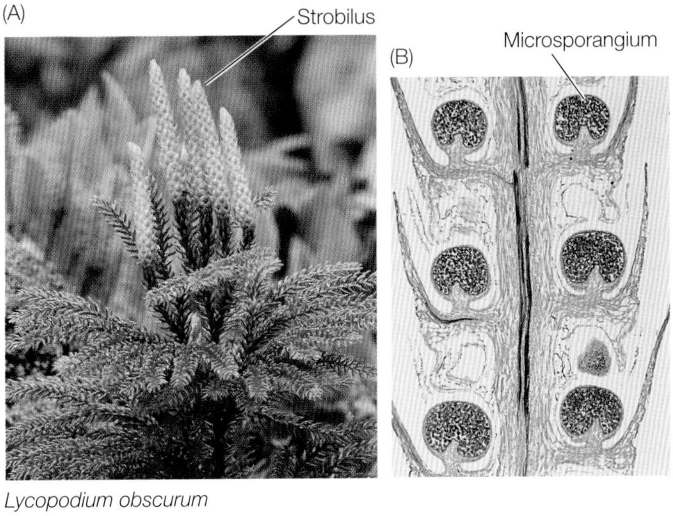

28.16 Club Mosses (A) Strobili are visible at the tips of this club moss. Club mosses have microphylls arranged spirally on their stems. (B) A thin section through a strobilus of a club moss, showing microsporangia.

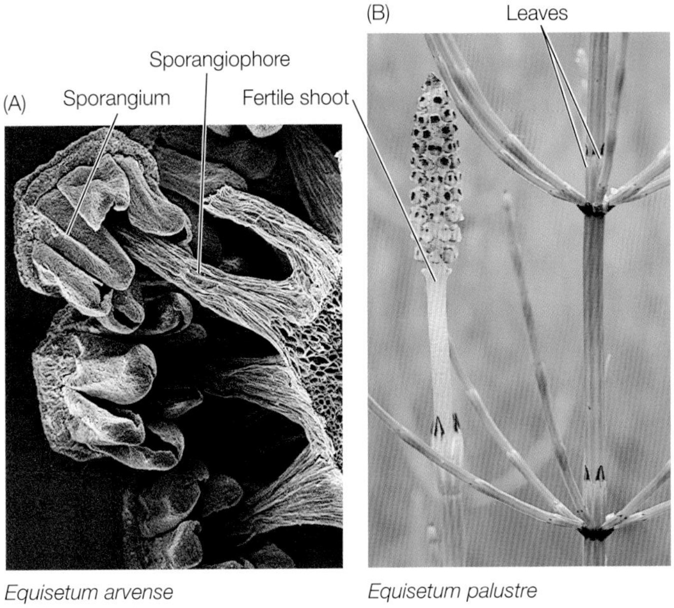

28.17 Horsetails (A) Sporangia and sporangiophores of a horsetail. (B) Vegetative and fertile shoots of the marsh horsetail. Reduced megaphylls can be seen in whorls on the stem of the vegetative shoot on the right. The sporangia on the fertile shoot (left) are ready to disperse their spores.

rangia of the rhyniophytes. There are both homosporous species and heterosporous species of club mosses.

Although they are only a minor element of present-day vegetation, the lycophytes are one of two groups that appear to have been the dominant vegetation during the Carboniferous period. One type of coal (cannel coal) is formed almost entirely from fossilized spores of the tree lycophyte *Lepidodendron*—which gives us an idea of the abundance of this genus in the forests of that time (see Figure 28.8). Other major elements of Carboniferous vegetation included horsetails and ferns.

Horsetails, whisk ferns, and ferns constitute a clade

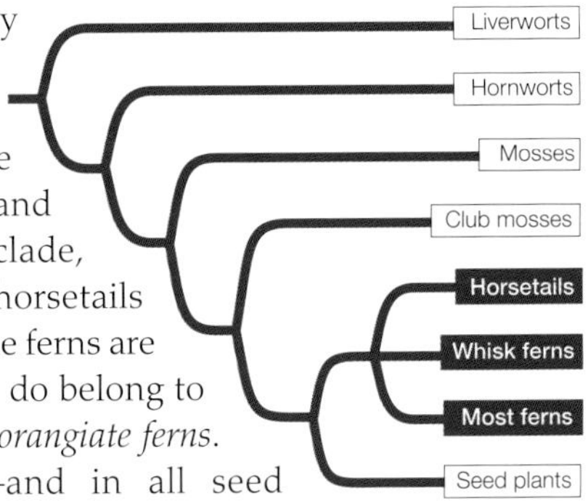

Once thought to be only distantly related, the horsetails, whisk ferns, and ferns form a clade, the pteridophytes, or "ferns and fern allies." Within that clade, the whisk ferns and the horsetails are both monophyletic; the ferns are not. However, most ferns do belong to a single clade, the *leptosporangiate ferns*. In the pteridophytes—and in all seed plants—there is differentiation between the main stem and side branches (overtopping). This pattern contrasts with the dichotomous branching characteristic of the lycophytes and rhyniophytes (see Figure 28.9).

HORSETAILS The **horsetails** are represented by only about 15 present-day species. All are in a single genus, *Equisetum*. These plants are sometimes called "scouring rushes" because silica deposits found in their cell walls made them useful for cleaning. They have true roots that branch irregularly. Their sporangia curve back toward the stem on the ends of short stalks called *sporangiophores* (**Figure 28.17A**). Horsetails have a large sporophyte and a small gametophyte, both independent.

The small leaves of horsetails are reduced megaphylls and form in distinct whorls (circles) around the stem (**Figure 28.17B**). Growth in horsetails originates to a large extent from discs of dividing cells just above each whorl of leaves, so each segment of the stem grows from its base. Such basal growth is uncommon in plants, although it is found in the grasses, a major group of flowering plants, as well as in the hornworts.

WHISK FERNS There once was some disagreement about whether the rhyniophytes are entirely extinct. The confusion arose because of the existence today of about 15 species in two genera of rootless, spore-bearing plants, *Psilotum* and *Tmesipteris*, collectively called the **whisk ferns**. *Psilotum flaccidum* (**Figure 28.18**) has only minute scales instead of true leaves, but plants of the genus *Tmesipteris* have flattened photosynthetic organs—reduced megaphylls—with well-developed vascular tissue. Are these two genera the living relics of the rhyniophytes, or do they have more recent origins?

Psilotum and *Tmesipteris* once were thought to be evolutionarily ancient descendants of anatomically simple ancestors. That hypothesis was weakened by an enormous hole in the fossil record between the rhyniophytes that apparently became extinct more than 300 million years ago, and *Psilotum* and *Tmesipteris*, which are modern plants. DNA sequence data finally settled the question in favor of a more modern origin of the whisk ferns from fernlike ancestors. The whisk ferns are a clade of highly specialized plants that evolved fairly recently from anatomically more complex ancestors by loss or reduction of megaphylls and true roots. Whisk fern gametophytes live below the surface of the ground and lack chlorophyll. They depend on fungal partners for their nutrition.

Psilotum flaccidum

28.18 A Whisk Fern *Psilotum* was once considered by some to be a surviving rhyniophyte and by others to be a fern. It is now included in the pteridophytes, and it is widespread in the tropics and subtropics.

FERNS The **ferns** constitute a group that first appeared during the Devonian period and today consists of more than 12,000 species. The ferns are not monophyletic, although the clade of leptosporangiate ferns includes about 97 percent of fern species. The leptosporangiate ferns differ from other ferns in having sporangia with walls only one cell thick, borne on a stalk.

The sporophytes of the ferns, like those of the seed plants, have true roots, stems, and leaves. Ferns are characterized by large leaves with branching vascular strands (**Figure 28.19A**). During its development, the fern leaf unfurls from a tightly coiled "fiddlehead" (**Figure 28.19B**). Some fern leaves become climbing organs and may grow to be as much as 30 meters long. A few species have small leaves as a result of evolutionary reduction, but even these small leaves have more than one vascular strand, and are thus megaphylls (**Figure 28.19C**).

THE FERN LIFE CYCLE In all ferns, *spore mother cells* inside the sporangia undergo meiosis to form haploid spores. Once shed, the spores may be blown great distances by the wind and eventually germinate to form independent gametophytes far from the parent sporophyte. A case in point: Old World climbing fern, *Lygodium microphyllum*, is currently spreading disastrously through the Florida Everglades, choking off the growth of other plants. This rapid spread is testimony to the effectiveness of windborne spores. Another example is the remarkable diversity of ferns that have spread through the isolated Hawaiian Islands.

Fern gametophytes have the potential to produce both antheridia and archegonia, although not necessarily at the same time or on the same gametophyte. Sperm swim through water to archegonia—often to those on other gametophytes—where they unite with an egg. The resulting zygote develops into a new sporophyte embryo. The young sporophyte sprouts a root and can thus grow independently of the gametophyte. In the alternating generations of a fern, the gametophyte is small, delicate, and short-lived, but the sporophyte can be very large and can sometimes survive for hundreds of years (**Figure 28.20**).

Because they require water for the transport of the male gametes to the female gametes, most ferns inhabit shaded, moist woodlands and swamps. Tree ferns can reach heights of 20 meters.

28.19 Fern Leaves Take Many Forms (A) The leaves of northern maidenhair fern form a pattern in this photograph. (B) The "fiddlehead" (developing leaf) of a common forest fern will unfurl and expand to give rise to a complex adult leaf such as those in (A). (C) The leaves of two species of water ferns.

Tree ferns are not as rigid as woody plants and they have poorly developed root systems. Thus they do not grow in sites exposed directly to strong winds, but rather in ravines or beneath trees in forests. The sporangia of ferns are found on the undersurfaces of the leaves, sometimes covering the whole undersurface and sometimes only at the edges. In most species the sporangia are found in clusters called *sori* (singular *sorus*) (see the inset in Figure 28.20).

Most ferns are homosporous. However, two groups of aquatic ferns, the Marsileaceae and the Salviniaceae (see Figure 28.19C), are derived from a common ancestor that evolved heterospory. The megaspores and microspores of these plants (which germinate to produce female and male gametophytes, respectively) are produced in different sporangia (megasporangia and microsporangia), and the microspores are always much smaller and greater in number than the megaspores.

A few genera of ferns produce a tuberous, fleshy gametophyte instead of the characteristic flattened, photosynthetic structure produced by most ferns. These tuberous gametophytes depend on a mutualistic fungus for nutrition; in some genera, even the sporophyte embryo must become associated with the fungus before its development can proceed. In Section 30.2 we will see that there are many other important plant–fungus mutualisms.

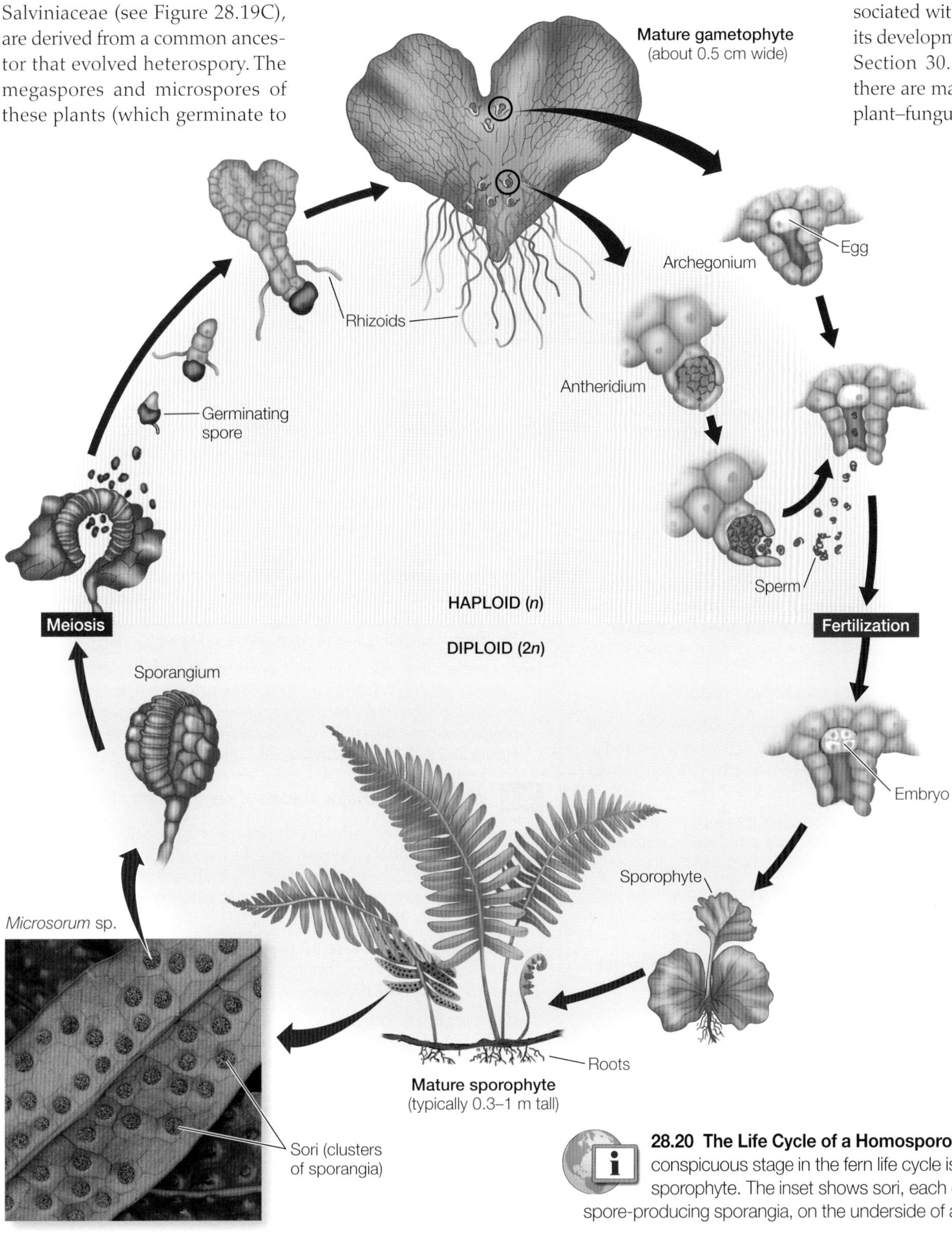

28.20 The Life Cycle of a Homosporous Fern The most conspicuous stage in the fern life cycle is the mature diploid sporophyte. The inset shows sori, each containing many spore-producing sporangia, on the underside of a leaf of a fern.

28.4 RECAP

Three clades of land plants lack true vascular systems (liverworts, hornworts, and mosses). The seedless vascular plants are the club mosses, horsetails, whisk ferns, and ferns. The ferns are not a clade.

- What is the difference between the branching patterns of lycophytes and pteridophytes? See p. 625
- Why was it once thought that whisk ferns were close relatives of the rhyniophytes? See p. 625
- Why do most ferns live in shady, moist areas? See p. 626

The seedless vascular plants, and especially the ferns, were long considered an evolutionary *cul-de-sac*—that is, a group with great diversity in the fossil record but less diversity in the present. However, recent DNA-based research has suggested that the diversification of today's ferns took place much more recently than previously thought. The expansion of flowering plants and their dominance of forests actually predates the diversification of extant ferns, which presumably took advantage of the new environments created by those forests.

All the vascular plants we have discussed thus far disperse themselves by spores. In the next chapter we will discuss the plants that dominate most of Earth's vegetation today—the seed plants, whose seeds afford new sporophytes protection unavailable to those of the other vascular plants. Even without such protection, ancient vascular plants were so successful and abundant that they formed vast forests, some of the remains of which are with us to this day as coal. Their modern representatives have come a long way from the plants discovered by Dawson and by Kidston and Lang.

CHAPTER SUMMARY

28.1 How did the land plants arise?

Land plants, sometimes referred to as **embryophytes**, are photosynthetic eukaryotes that develop from embryos protected by parental tissue. Review Figure 28.1

Streptophytes include the land plants and certain green algae. **Green plants** include the streptophytes and the remaining green algae.

Land plants arose from an aquatic green algal ancestor related to today's **Charales**.

There are ten major groups of living land plants. Seven groups (the **vascular plants**) have well developed water-conducting tissues with cells including **tracheids**; three groups (the **nonvascular plants**) do not. Review Table 28.1

28.2 How did plants colonize and thrive on land?

The acquisition of a **cuticle**, gametangia, a protected embryo, protective pigments, thick spore walls with a protective polymer, and a mutualistic association with a fungus are all adaptations to terrestrial life.

All land plant life cycles feature alternation of generations, in which a multicellular **sporophyte** alternates with a multicellular **gametophyte**. Review Figure 28.4

Spores form in **sporangia**; gametes form in **gametangia**. In nonvascular plants the female and male gametangia are, respectively, an **archegonium** and an **antheridium**.

The sporophyte of liverworts, hornworts, and mosses is smaller than the gametophyte and depends on it for water and nutrition. Review Figure 28.5, Web/CD Tutorial 28.1

28.3 What features distinguish the vascular plants?

A **vascular system**, consisting of **xylem** and **phloem**, conducts water, minerals, and products of photosynthesis through the bodies of vascular plants. Review Figure 28.7

In vascular plants, the sporophyte is larger than and independent of the gametophyte.

The **rhyniophytes**, the earliest vascular plants, are known to us only in fossil form. They lacked roots and leaves but possessed **rhizomes** and **rhizoids**. Review Figure 28.9

Lycophytes (club mosses and relatives) and **pteridophytes** (ferns and allies) appeared later. **Euphyllophytes** include the pteridophytes and seed plants.

Roots may have evolved from rhizomes or from branches. **Microphylls** probably evolved from sterile sporangia, and **megaphylls** may have resulted from the flattening and reduction of an **overtopping**, branching stem system. Review Figure 28.10

Many seedless vascular plants are **homosporous**, but **heterospory**—the production of distinct **megaspores** and **microspores**—evolved several times. Megaspores develop into **megagametophytes**; microspores develop into **microgametophytes**. Review Figure 28.12, Web/CD Activities 28.1 and 28.2

28.4 What are the major clades of seedless plants?

The nonvascular plant clades are the **liverworts** (Hepatophyta), the **hornworts** (Anthocerophyta), and the **mosses** (Bryophyta). The nonseed vascular plant groups are the **club mosses** and their relatives and the pteridophytes (**horsetails**, **whisk ferns**, and **ferns**). Review Figure 28.7

Hornworts, mosses, and vascular plants all have surface pores (stomata) in their leaves. In mosses and vascular plants, the sporophytes grow by **apical cell division**.

The ferns are not a clade, although 97 percent of fern species do constitute a clade called the leptosporangiate ferns. Ferns have megaphylls with branching vascular strands. See Web/CD Activity 28.3

SELF-QUIZ

1. Land plants differ from photosynthetic protists in that only the plants
   a. are photosynthetic.
   b. are multicellular.
   c. possess chloroplasts.
   d. have multicellular embryos protected by the parent.
   e. are eukaryotic.
2. Which statement about alternation of generations in land plants is *not* true?
   a. The gametophyte and sporophyte differ in appearance.
   b. Meiosis occurs in sporangia.
   c. Gametes are always produced by meiosis.
   d. The zygote is the first cell of the sporophyte generation.
   e. The gametophyte and sporophyte differ in chromosome number.
3. Which statement is not evidence for the origin of plants from the green algae?
   a. Some green algae have multicellular sporophytes and multicellular gametophytes.
   b. Both plants and green algae have cellulose in their cell walls.
   c. The two groups have the same photosynthetic pigments.
   d. Both plants and green algae produce starch as their principal storage carbohydrate.
   e. All green algae produce large, stationary eggs.
4. Liverworts, hornworts, and mosses
   a. lack a sporophyte generation.
   b. grow in dense masses, allowing capillary movement of water.
   c. possess xylem and phloem.
   d. possess true leaves.
   e. possess true roots.
5. Which statement is *not* true of the mosses?
   a. The sporophyte is dependent on the gametophyte.
   b. Sperm are produced in archegonia.
   c. There are more species of mosses than of liverworts and hornworts combined.
   d. The sporophyte grows by apical cell division.
   e. Mosses are probably sister to the vascular plants.
6. Megaphylls
   a. probably evolved only once.
   b. are found in all the vascular plant groups.
   c. probably arose from sterile sporangia.
   d. are the characteristic leaves of club mosses.
   e. are the characteristic leaves of horsetails and ferns.
7. The rhyniophytes
   a. lacked tracheids.
   b. possessed true roots.
   c. possessed sporangia at the tips of stems.
   d. possessed leaves.
   e. lacked branching stems.
8. Club mosses and horsetails
   a. have larger gametophytes than sporophytes.
   b. possess small leaves.
   c. are represented today primarily by trees.
   d. have never been a dominant part of the vegetation.
   e. produce fruits.
9. Which statement about ferns is *not* true?
   a. The sporophyte is larger than the gametophyte.
   b. Most are heterosporous.
   c. The young sporophyte can grow independently of the gametophyte.
   d. The leaf is a megaphyll.
   e. The gametophytes produce archegonia and antheridia.
10. The leptosporangiate ferns
    a. are not a monophyletic group.
    b. have sporangia with walls more than one cell thick.
    c. constitute a minority of all ferns.
    d. are pteridophytes.
    e. produce seeds.

FOR DISCUSSION

1. Mosses and ferns share a common trait that makes water droplets a necessity for sexual reproduction. What is that trait?
2. Are the mosses well adapted to terrestrial life? Justify your answer.
3. Ferns display a dominant sporophyte generation (with large leaves). Describe the major advance in anatomy that enables most ferns to grow much larger than mosses.
4. What features distinguish club mosses from horsetails? What features distinguish these groups from rhyniophytes? From ferns?
5. Why did some botanists once believe that the whisk ferns should be classified together with the rhyniophytes?
6. Contrast microphylls with megaphylls in terms of structure, evolutionary origin, and occurrence among plants.

FOR INVESTIGATION

Osborne's findings on the evolution of megaphylls (see Figure 28.11) support the concept that large megaphylls became common only after the atmospheric CO_2 level had dropped, so that more stomata were produced, allowing water to evaporate and cool larger leaves. How might you extend that work to confirm the involvement of temperature as a factor limiting leaf size?

CHAPTER 29 The Evolution of Seed Plants

A seed from Biblical times germinates in 2005

The Judean date was once much prized. The Prophet Muhammad admired its nutritional and medicinal properties; in the Koran it is associated with heaven and described as a symbol of goodness. The Judean date was the source of the "honey" in the Biblical "land of milk and honey." Today that ancient strain of Judean date is gone. Or is it?

Around 2,000 years ago, at the beginning of the Common Era, a seed developed in a fruit on a Judean date palm. The fruit that contained that seed found its way to a storeroom in the fortress Masada in Judea. In 73 C.E. a band of 960 Jewish Zealots involved in a religious revolt against Rome fled to this refuge with their families. Roman legions followed, and the ensuing siege lasted more than two years. In the end, rather than be killed or enslaved by the Roman soldiers, the Zealots are said to have killed themselves and their families in a dramatic mass suicide.

Twenty centuries later, archeologists working in Masada discovered the seed of that overlooked Judean date and confirmed its age. The previous record for seed survival and germination was 1,200 years, held by lotus seeds that recently germinated under the care of scientists in China. But botanist Elaine Solowey succeeded in making the 2,000-year-old date seed germinate! The resulting seedling has continued to thrive and grow. Perhaps the ancient Egyptians were right to place date seeds in the tombs of the Pharaohs as symbols of immortality.

Seeds are important structures for the evolutionary survival of plants. They protect the plant embryo within from environmental extremes through what may be a very long and stressful resting period—in the case of the Judean date, many centuries in a harsh desert. Seeds of the coconut palm remain dormant for years as they float across vast expanses of ocean, finally washing up on a distant shore, where they germinate and grow. Such hardiness is one of the properties that have contributed to making seed plants the predominant plants on Earth. All of today's forests are dominated by seed plants.

A Refuge As a precaution, King Herod of Judea fortified Masada and stocked it with water and food (including Judean dates).

The Hardy Seed This coconut seed has arrived on a beach, where it germinated successfully. The evolution of seeds was a major factor in the eventual dominance of the seed plants.

So will the seedling growing under Dr. Solowey's care serve as the parent of a new population of Judean dates, thus resurrecting that genotype from extinction? Unfortunately, it cannot, because date trees are of two different sexes. It will be many years before we learn the sex of this plant (if it survives). And regardless of its sex, it cannot reproduce alone.

In fact many seed plant species do not need a partner in order to reproduce sexually, but the sexuality of date palms and other plants was recognized from the earliest days of agriculture. The life-giving attributes of the seed plants are indispensable, and humans have always sought to understand and augment plant reproductive cycles.

IN THIS CHAPTER we will first describe the defining characteristics of the seed plants as a group. Then we will describe the flowers and fruits that are characteristic of their most dominant group, the flowering plants or angiosperms. We will examine some of the unsolved problems in seed plant evolution, concluding with a survey of the diversity of living seed plants.

CHAPTER OUTLINE

29.1 How Did Seed Plants Become Today's Dominant Vegetation?

By the late Devonian period, more than 360 million years ago, Earth was home to a great variety of land plants, many of which we discussed in the previous chapter. These plants shared the hot, humid terrestrial environment with insects, spiders, centipedes, and fishlike amphibians. The plants and animals affected one another, acting as agents of natural selection.

Late in this period an innovation appeared: some plants developed extensively thickened woody stems, which resulted from the proliferation of xylem. This type of growth in the diameter of stems and roots is called **secondary growth**. The first plants with this adaptation were seedless vascular plants called *progymnosperms*, all species of which are now extinct.

The seed plants are the most recent group of vascular plants to appear. The earliest fossil evidence of seed plants is found in Devonian rocks. Like the progymnosperms, these *seed ferns* were woody. They possessed fernlike foliage, but had seeds attached to their leaves. By the Carboniferous period, new lineages of seed plants had evolved (**Figure 29.1**).

The several clades of seed ferns are known only as fossils. Two of those clades are basal to the surviving seed plants, which fall into two groups, the **gymnosperms** (such as pines and cycads) and the **angiosperms** (flowering plants). There are four living groups of gymnosperms and one of angiosperms (**Figure 29.2**). The phylogenetic relationships among these five groups have not yet been resolved; we will discuss some of the issues involved later in this chapter. All living gymnosperms and many angiosperms show secondary growth. The life cycles of all seed plants also share major features, as we are about to see.

Features of the seed plant life cycle protect gametes and embryos

In Section 28.2, we noted a trend in plant evolution: the sporophyte became less dependent on the gametophyte, which became smaller in relation to the sporophyte. This trend continued with the appearance of the seed plants, whose gametophyte generation is reduced even further than it is in the ferns

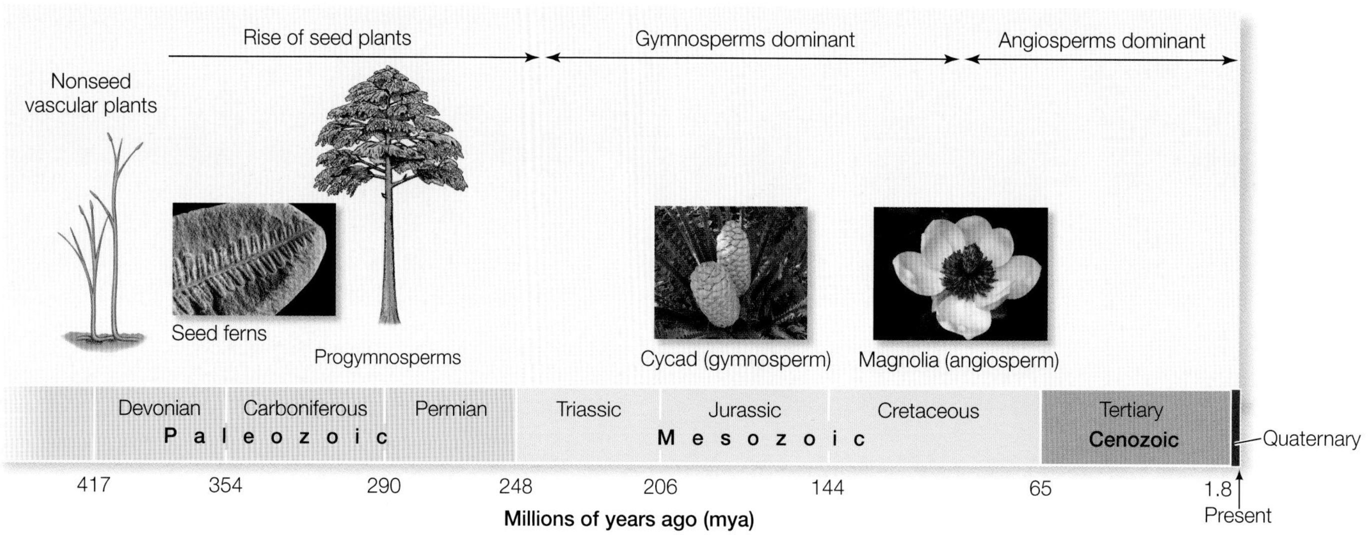

29.1 Highlights in the History of Seed Plants Woody growth evolved in the seedless progymnosperms. The now-extinct seed ferns had woody growth, fernlike foliage, and seeds attached to their leaves. New lineages of seed plants arose during the Carboniferous.

(**Figure 29.3**). The haploid gametophyte develops partly or entirely while attached to and nutritionally dependent on the diploid sporophyte.

Among the seed plants, only the earliest groups of gymnosperms (and their few survivors) had swimming sperm. Later groups of gymnosperms and the angiosperms evolved other means of bringing eggs and sperm together. The culmination of this striking evolutionary trend in seed plants was independence from the liquid water that earlier plants needed for sexual reproduction. This gave the seed plants a major advantage as they spread over the terrestrial environment.

Seed plants are heterosporous (see Figure 28.12B); that is, they produce two types of spores, one that becomes the male gametophyte and one that becomes the female gametophyte. They form separate megasporangia and microsporangia on structures that are grouped on short axes, such as the cones and strobili of conifers and the flowers of angiosperms.

As in other land plants, the spores of seed plants are produced by meiosis within the sporangia, but in seed plants, the megaspores are not shed. Instead, they develop into female gametophytes within the megasporangia. These megagametophytes are dependent on the sporophyte for food and water.

In most seed plant species, only one of the meiotic products in a megasporangium survives. The surviving haploid nucleus divides mitotically, and the resulting cells divide again to produce a multicellular female gametophyte. This megagametophyte is retained within the megasporangium, where it matures and produces an egg by mitosis. The megagametophyte, in turn, houses the early development of the next sporophyte generation when the egg is fertilized. The megasporangium is surrounded by sterile sporophytic structures, which form an **integument** that protects the megasporangium and its contents. Together, the megasporangium and integument constitute the **ovule**, which develops into a seed.

Within the microsporangium, the meiotic products are microspores, which divide mitotically within the spore wall one or a few times to form a male gametophyte called a **pollen grain**. Pollen

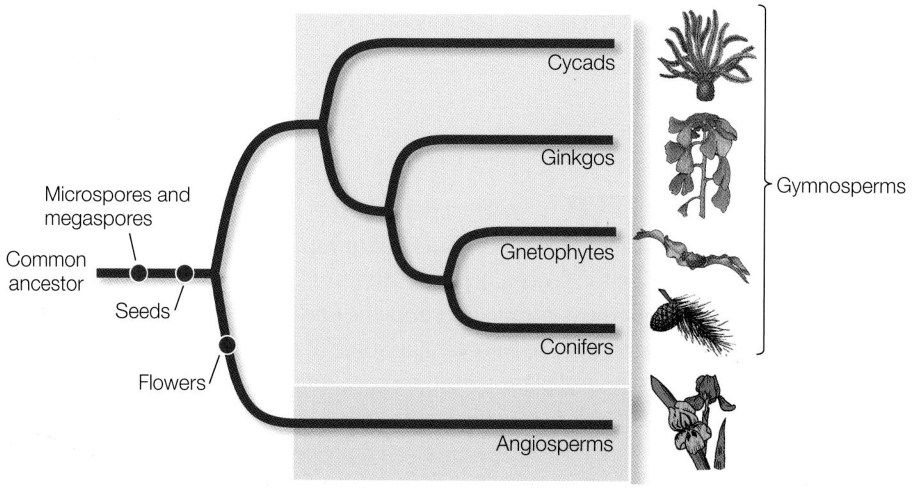

29.2 The Major Groups of Living Seed Plants There are four groups of gymnosperms and one of angiosperms. Their exact evolutionary relationship is still uncertain, but this cladogram represents one current interpretation.

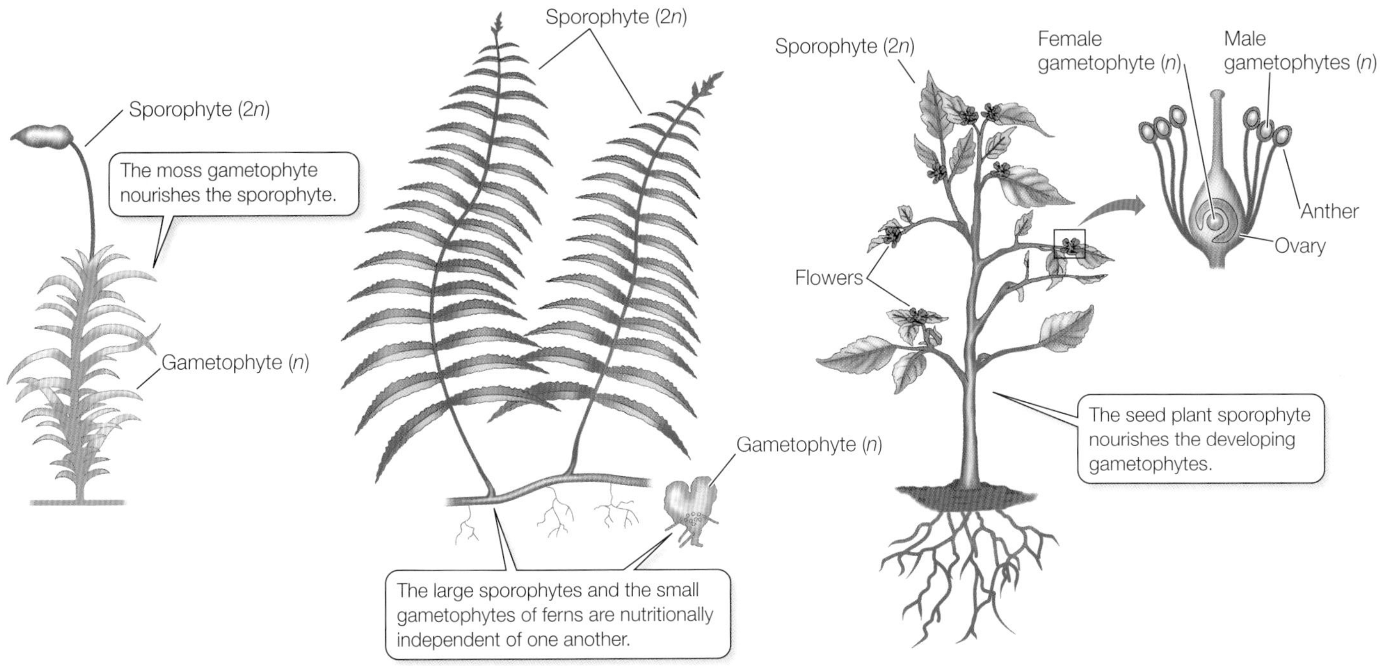

29.3 The Relationship between Sporophyte and Gametophyte Has Evolved In the course of plant evolution, the gametophyte has been reduced and the sporophyte has become more prominent.

grains are released from the microsporangium to be distributed by wind, an insect, a bird, or a plant breeder (**Figure 29.4**). The wall of the pollen grain contains *sporopollenin,* the most chemically resistant biological compound known, which protects the pollen grain against dehydration and chemical damage—another advantage in terms of survival in the terrestrial environment. Recall that sporopollenin in spore walls also contributed to the successful colonization of the terrestrial environment by the earliest land plants.

The arrival of a pollen grain at an appropriate landing point, close to a female gametophyte on a sporophyte of the same species, is called **pollination**. A pollen grain that reaches this point develops further. It produces a slender **pollen tube** that elongates and digests its way through the sporophytic tissue toward the megagametophyte. When the tip of the pollen tube reaches the megagametophyte, sperm are released from the tube, and fertilization occurs.

The resulting diploid zygote divides repeatedly, forming an embryonic sporophyte. After a period of embryonic development, growth is temporarily suspended (the embryo enters a *dormant* stage). The end product at this stage is a multicellular **seed**.

The seed is a complex, well-protected package

A seed may contain tissues from three generations. A *seed coat* develops from the tissues of the diploid sporophyte parent that surround the megasporangium (the integument). Within the megasporangium is the haploid female gametophytic tissue from the next generation, which contains a supply of nutrients for the developing embryo. (This tissue is fairly extensive in most gymnosperm seeds. In angiosperm seeds its place is taken by a tissue called endosperm, which we will describe below.) In the center

Betula pendula

29.4 Pollen Grains Pollen grains are the male gametophytes of seed plants. The pollen of this silver birch is dispersed by the wind, and grains may land near the female gametophytes of the same or other silver birch trees.

of the seed is the third generation, the embryo of the new diploid sporophyte.

The seed of a gymnosperm or an angiosperm is a well-protected resting stage. The seeds of some species may remain *viable* (capable of growth and development) for many years, germinating only when conditions are favorable for the growth of the sporophyte, as did the date seed from Biblical times mentioned at the beginning of this chapter. In contrast, the embryos of nonseed plants develop directly into sporophytes, which either survive or die, depending on environmental conditions; there is no dormant stage in the life cycle.

During the dormant stage, the seed coat protects the embryo from excessive drying and may also protect it against potential predators that would otherwise eat the embryo and its nutrient reserves. Many seeds have structural adaptations that promote their dispersal by wind or, more often, by animals. When the young sporophyte resumes growth, it draws on the food reserves in the seed. The possession of seeds is a major reason for the enormous evolutionary success of the seed plants, which are the dominant life forms of Earth's modern terrestrial flora in most areas. But there is another reason for their dominance: secondary growth.

A change in anatomy enabled seed plants to grow to great heights

The most ancient seed plants produced **wood**—extensively proliferated xylem—which gave them the support to grow taller than other plants around them, thus capturing more light for photosynthesis. The younger portion of wood is well adapted for water transport, while older wood becomes clogged with resins or other materials. Although no longer functional in transport, the older wood continues to provide support for the plant.

Not all seed plants are woody. In the course of seed plant evolution, many seed plants lost the woody growth habit; however, other advantageous attributes helped them become established in an astonishing variety of places.

29.1 RECAP

Pollen, seeds, and wood are major evolutionary innovations of the seed plants. Protection of the gametes and embryos is a hallmark of the seed plants.

- Can you distinguish between the roles of the megagametophyte and the pollen grain? See p. 632
- Do you understand the importance of pollen in freeing seed plants from dependence on liquid water? See p. 633
- What are some of the advantages afforded by seeds? By wood? See pp. 633–634

The seed ferns and progymnosperms have long been extinct, but the surviving seed plants have been remarkable successes. Let's look at the most ancient seed plants that survive today: the gymnosperms.

29.2 What Are the Major Groups of Gymnosperms?

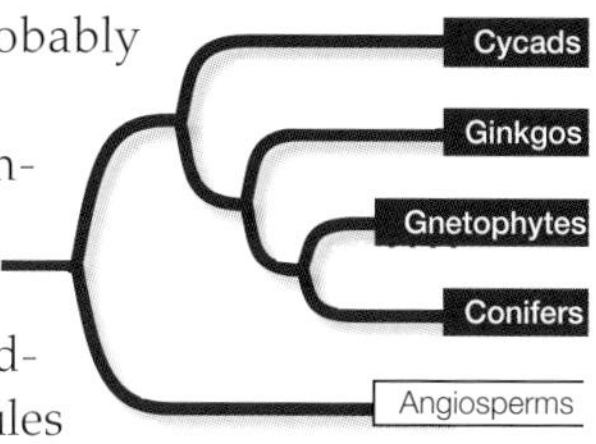

The extant gymnosperms are probably a clade, although that has not been established beyond a doubt. The gymnosperms are seed plants that do not form flowers. Gymnosperms derive their name (which means "naked-seeded") from the fact that their ovules and seeds are not protected by ovary or fruit tissue. Although there are probably fewer than 850 species of living gymnosperms, these plants are second only to the angiosperms in their dominance of the terrestrial environment.

The four major groups of living gymnosperms bear little superficial resemblance to one another:

- The **cycads** (*Cycadophyta*) are palmlike plants of the tropics and subtropics, growing as tall as 20 meters (**Figure 29.5A**). Of the present-day gymnosperms, the cycads are probably the earliest-diverging clade. There are 140 species of cycads. Their tissues are often highly toxic to humans.
- **Ginkgos** (*Ginkgophyta*), which were common during the Mesozoic era, are represented today by a single genus and species, *Ginkgo biloba*, the maidenhair tree (**Figure 29.5B**). There are both male (microsporangiate) and female (megasporangiate) maidenhair trees. The difference is determined by X and Y sex chromosomes, as in humans; few other plants have sex chromosomes.
- **Gnetophytes** (*Gnetophyta*) number about 90 species in three very different genera, which share certain characteristics analogous to ones found in the angiosperms, as we will see. One of the gnetophytes is *Welwitschia* (**Figure 29.5C**), a long-lived desert plant with just two straplike leaves that sprawl on the sand and can grow as long as 3 meters.
- **Conifers** (*Coniferophyta*) are by far the most abundant of the gymnosperms. There are about 600 species of these cone-bearing plants, including the pines and redwoods (**Figure 29.5D**).

All living gymnosperms except for the gnetophytes have only tracheids as water-conducting and support cells within the xylem. In angiosperms, cells called vessel elements and fibers, which are specialized for water conduction and support, respectively, are found alongside tracheids. While the gymnosperm water transport and support system may thus seem somewhat less efficient than that of the angiosperms, it serves some of the largest trees known. The coastal redwoods of California are the tallest gymnosperms; the largest are well over 100 meters tall.

During the Permian, as environments became warmer and dryer, the conifers and cycads flourished. Gymnosperm forests changed over time as the gymnosperm groups evolved. Gymnosperms dominated the Mesozoic era, during which the continents drifted apart and dinosaurs strode the Earth. They were the principal trees in all forests until less than 100 million years ago, and even down to the present day, conifers are the dominant trees in many forests.

(A) *Encephalartos villosus*

(B) *Ginkgo biloba*

(C) *Welwitschia mirabilis*

(D) *Sequoiadendron giganteum*

29.5 Diversity among the Gymnosperms (A) Many cycads have growth forms that resemble both ferns and palms, but are not closely related to either. (B) The characteristic fleshy seed coat and broad leaves of the maidenhair tree. (C) A gnetophyte growing in the Namib Desert of Africa. Straplike leaves grow throughout the life of the plant, breaking and splitting as they grow. (D) Conifers, like this giant sequoia growing in Sequoia National Park, California, dominate many modern forests.

The oldest living thing on Earth is a gymnosperm in California—a bristlecone pine called "Methuselah"—that began its life about 4,800 years ago, at about the time the ancient Egyptians were just starting to develop writing.

The relationship between gnetophytes and conifers is a subject of continuing research

Although the cycads, ginkgo, and gnetophytes are all clades, the status of the conifers is uncertain. There is evidence favoring a hypothesis—the "gnetifer" hypothesis—that the gnetophytes and conifers are sister clades (**Figure 29.6A**). However, there is also evidence favoring an alternative hypothesis—the "gnepine" hypothesis—that the gnetophytes are nested within a paraphyletic conifer group (**Figure 29.6B**). Why should this confusion exist, and how will it be resolved?

DNA data make it clear that the gnetophytes and conifers are closely related—but how? Most studies based on chloroplast and mitochondrial genes have supported the gnepine hypothesis. However, other studies based on rDNA (DNA that codes for rRNA) have favored the gnetifer hypothesis, as have some studies of chloroplast genes. An important source of confusion may be the use of different (usually small) sets of species for comparison; another is the use of different DNA sources.

It is widely agreed that what is needed are studies including more species and more genes. It may also be helpful to use new groups as outgroups (see Section 25.2) in future analyses. It is reasonable to expect that such studies will clarify the situation.

Whether or not they are a clade, the conifers are the most abundant gymnosperms. Let's look at them in more detail.

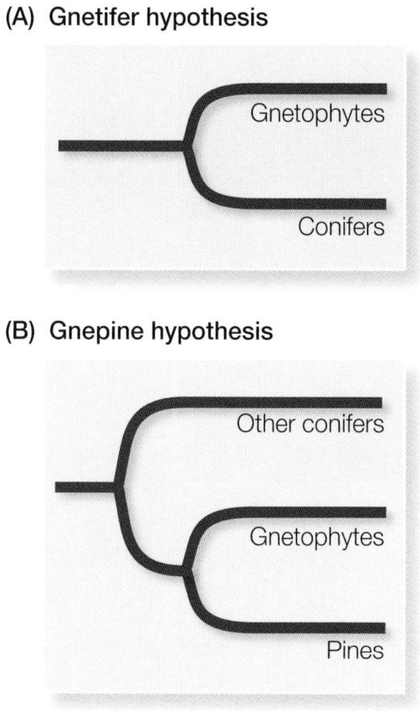

29.6 Two Interpretations of Conifer Phylogeny (A) According to the gnetifer hypothesis, the conifers are a clade. (B) According to the gnepine hypothesis, the conifers are paraphyletic.

Conifers have cones but no motile gametes

The great Douglas fir and cedar forests of the northwestern United States and the massive boreal forests of pine, fir, and spruce found in northern regions of Eurasia and North America, as well as on the upper slopes of mountain ranges everywhere, rank among the great vegetation formations of the world. All these trees belong to one group of gymnosperms, Coniferophyta—the conifers, or cone-bearers. A **cone** is a short axis (a modified stem) bearing a tight cluster of scales, which are reduced *branches* specialized for reproduction (**Figure 29.7A**). A **strobilus** is a conelike cluster of scales that are modified *leaves* inserted on an axis (**Figure 29.7B**). Megaspores, and thus seeds, are produced in cones, and microspores are produced in strobili. Cones are usually much larger than strobili.

We will use the life cycle of a pine to illustrate reproduction in gymnosperms (**Figure 29.8**). The production of male gametophytes in the form of pollen grains frees the plant completely from its dependence on liquid water for fertilization. Wind, rather than water, assists conifer pollen grains in their first stage of travel from the strobilus to the female gametophyte inside a cone (see Figure 29.4). The pollen tube provides the sperm with the means for the last stage of travel by elongating and digesting its way through maternal sporophytic tissue. When it reaches the female gametophyte, it releases two sperm, one of which degenerates after the other unites with an egg. Union of sperm and egg results in a zygote; mitotic divisions and further development of the zygote result in an embryo.

The megasporangium, in which the female gametophyte will form, is enclosed in a layer of sporophytic tissue—the integument—that will eventually develop into the seed coat that protects the embryo. The integument, the megasporangium inside it, and the tissue attaching it to the maternal sporophyte constitute the ovule. The pollen grain enters through a small opening in the integument at the tip of the ovule, the **micropyle**.

Most conifer ovules are borne exposed on the upper surfaces of the modified branches that form the scales of the cone. The only protection of the ovules comes from the scales, which are tightly pressed against one another within the cone. Some pines, such as the lodgepole pine, have such tightly closed cones that only fire suffices to split them open and release the seeds.

About half of all conifer species have soft, fleshy fruitlike tissues associated with their seeds; examples are the fleshy cones or "berries" of juniper and yew. Animals may eat these tissues and then disperse the seeds in their feces, often carrying them considerable distances from the parent plant.

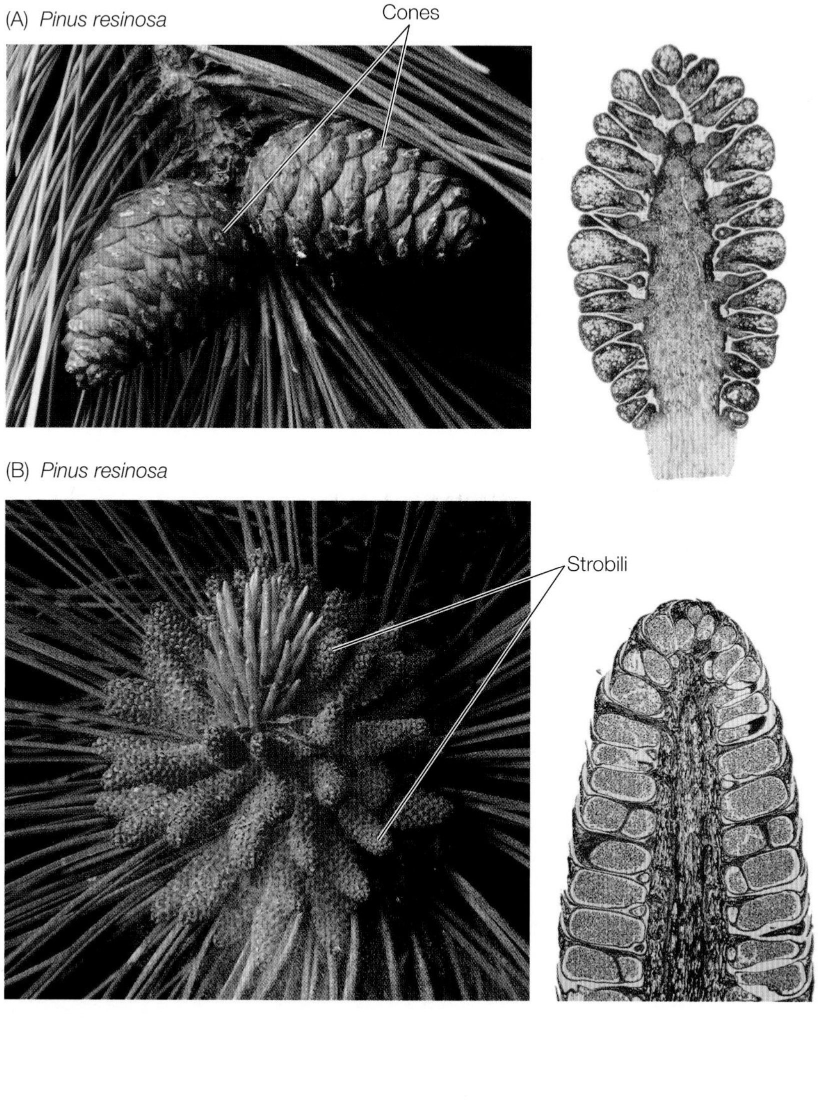

29.7 Cones and Strobili (A) The scales of cones are modified branches. (B) The spore-bearing structures in strobili are modified leaves.

29.8 The Life Cycle of a Pine Tree In conifers and other gymnosperms, the gametophytes are microscopically small and nutritionally dependent on the sporophyte generation.

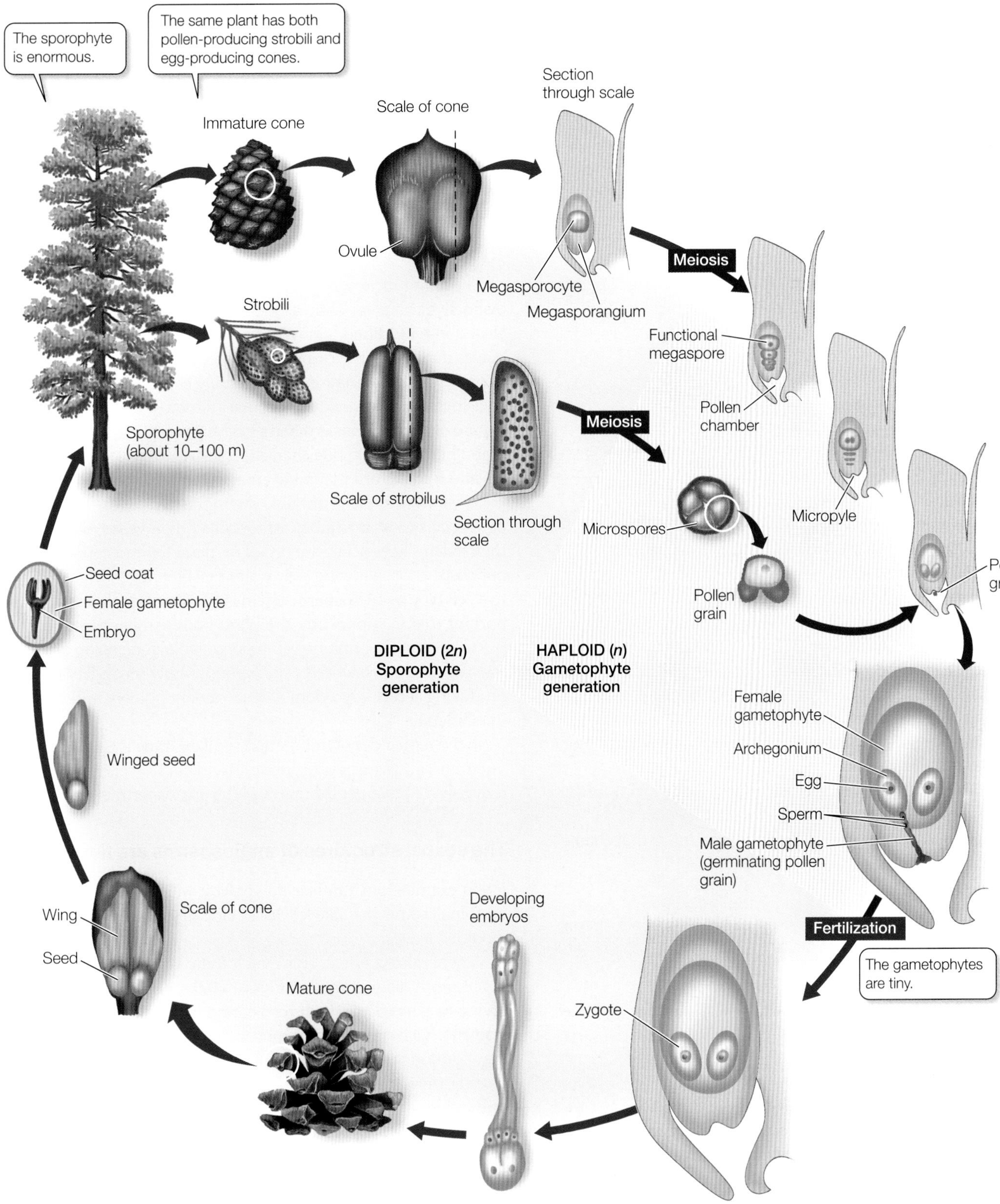

29.2 RECAP

The four groups of gymnosperms are woody and have naked seeds. Their phylogenetic relationships are still uncertain.

- What distinguishes the gnetifer from the gnepine hypothesis? See p. 635 and Figure 29.6
- Can you explain the difference between a cone and a strobilus? See p. 636 and Figure 29.8
- Do you understand the role of the integument? See pp. 632 and 636

Juniper and yew "berries" are not true fruits, which are characteristic of the plant group that is dominant today: the angiosperms. How else do angiosperms differ from gymnosperms?

29.3 What Features Distinguish the Angiosperms?

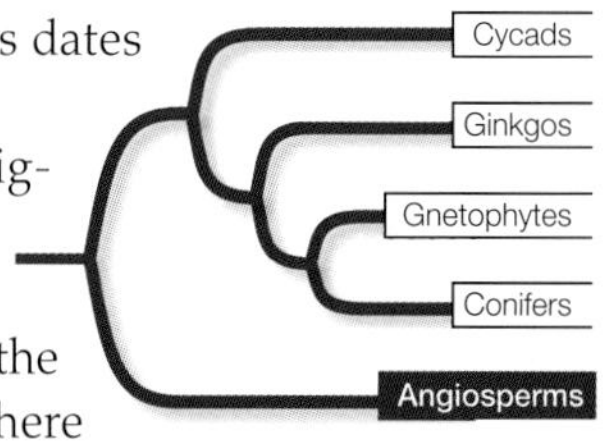

The oldest evidence of angiosperms dates back to the early Cretaceous period, about 140 million years ago (see Figure 29.1). The angiosperms radiated explosively and, over a period of only about 60 million years, became the dominant plant life of the planet. There are more than a quarter million species of angiosperms today.

The female gametophyte of the angiosperms is even more reduced than that of the gymnosperms, usually consisting of just seven cells. Thus the angiosperms represent the current extreme of the trend we have traced throughout the evolution of the vascular plants: the sporophyte generation becomes larger and more independent of the gametophyte, while the gametophyte generation becomes smaller and more dependent on the sporophyte. What else sets the angiosperms apart from other plants?

The major synapomorphies (shared derived traits) that characterize the angiosperms include:

- Double fertilization
- Production of a triploid nutritive tissue called the endosperm
- Ovules and seeds enclosed in a carpel
- Flowers
- Fruits
- Xylem with vessel elements and fibers
- Phloem with companion cells

In the angiosperms, pollination consists of the arrival of a microgametophyte—a pollen grain—on a receptive surface in a flower. As in the gymnosperms, pollination is just the first in a series of events that result in the formation of a seed. The next is the growth of a pollen tube extending to the megagametophyte. The third event is a fertilization process which, in detail, is unique to the angiosperms.

Double fertilization was long considered the single most reliable distinguishing characteristic of the angiosperms. *Two* male gametes, contained within a single microgametophyte, participate in fertilization events within the megagametophyte of an angiosperm. The nucleus of one sperm combines with that of the egg to produce a diploid zygote, the first cell of the sporophyte generation. In most angiosperms, the other sperm nucleus combines with two other haploid nuclei of the female gametophyte to form a *triploid* ($3n$) nucleus (see Figure 29.14 below). That nucleus, in turn, divides to form a triploid tissue, the **endosperm**, that nourishes the embryonic sporophyte during its early development. This process, in which two fertilization events take place, is known as **double fertilization**.

Double fertilization occurs in nearly all present-day angiosperms. We are not sure when and how it evolved because there is no known fossil evidence on this point. It may have first resulted in two diploid embryos, as it does in the three existing genera of the gnetophytes..

The name *angiosperm* ("enclosed seed") is drawn from another distinctive character of these plants: the ovules and seeds are enclosed in a modified leaf called a **carpel**. Besides protecting the ovules and seeds, the carpel often interacts with incoming pollen to prevent self-pollination, thus favoring cross-pollination and increasing genetic diversity. Of course, the most evident diagnostic feature of angiosperms is that they have **flowers**. Production of **fruits** is another of their unique characteristics. As we will see, both flowers and fruits afford major advantages to angiosperms.

Most angiosperms are also distinguished by the possession of specialized water-transporting cells called **vessel elements** in their xylem. These cells are broad in diameter and connect without obstruction, allowing easy water movement. A second distinctive cell type in angiosperm xylem is the **fiber**, which plays an important role in supporting the plant body. Angiosperm phloem possesses another unique cell type, called a **companion cell**. Like the gymnosperms, woody angiosperms show secondary growth, producing secondary xylem and secondary phloem and growing in diameter.

In the remainder of this section we'll examine the structure and function of flowers, evolutionary trends in flower structure, the functions of pollen and fruits, and the angiosperm life cycle.

The sexual structures of angiosperms are flowers

If you examine any familiar flower, you will notice that the outer parts look somewhat like leaves. In fact, all the parts of a flower *are* modified leaves.

A generalized flower (for which there is no exact counterpart in nature) is diagrammed in **Figure 29.9** for the purpose of identifying its parts. The structures bearing microsporangia are called **stamens**. Each stamen is composed of a **filament** bearing an **anther** that contains pollen-producing microsporangia. The structures bearing megasporangia are the carpels. A structure composed of one carpel or two or more fused carpels is called a **pistil**. The swollen base of the pistil, containing one or more ovules (each containing a megasporangium surrounded by its protective integument), is called the **ovary**. The apical stalk of the pistil is the **style**, and the terminal surface that receives pollen grains is the **stigma**.

In addition, a flower often has several specialized sterile (non-spore-bearing) leaves. The inner ones are called **petals** (collectively, the **corolla**) and the outer ones **sepals** (collectively, the

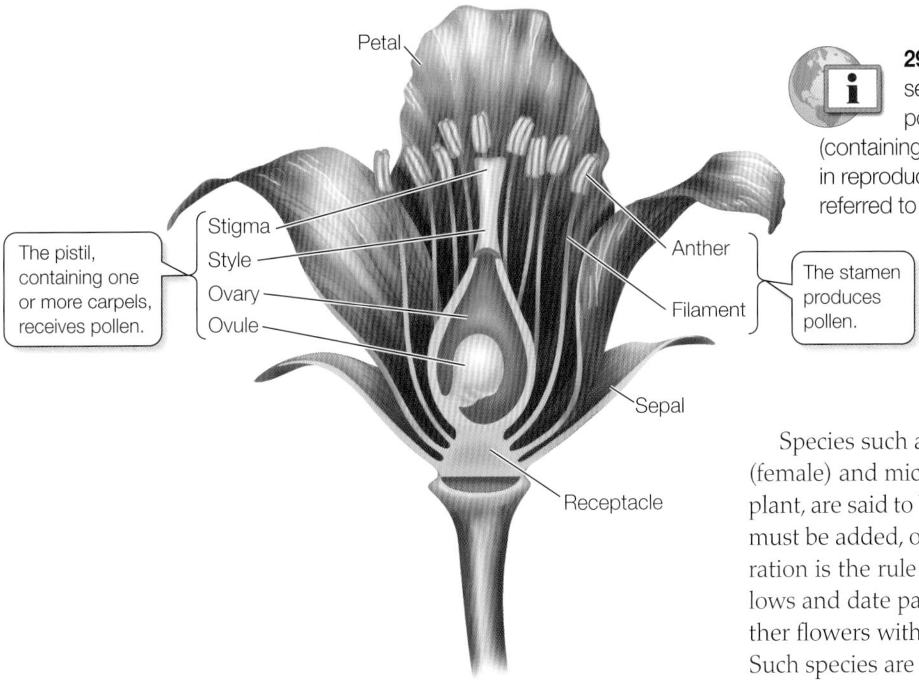

29.9 A Generalized Flower Not all flowers possess all the structures shown here, but they must possess a stamen (bearing microsporangia), a pistil (containing megasporangia), or both in order to play their role in reproduction. Flowers that have both, as this one does, are referred to as perfect.

calyx). The corolla and calyx, which can be quite showy, often play roles in attracting animal pollinators to the flower. The calyx more commonly protects the immature flower in bud. From base to apex, the sepals, petals, stamens, and carpels (which are referred to as the *floral organs*; see Figure 19.15) are usually positioned in circular arrangements or whorls and attached to a central stalk called the **receptacle**.

The generalized flower shown in Figure 29.9 has both megasporangia and microsporangia; such flowers are referred to as **perfect**. Many angiosperms produce two types of flowers, one with only megasporangia and the other with only microsporangia. Consequently, either the stamens or the carpels are nonfunctional or absent in a given flower, and the flower is referred to as **imperfect**.

Species such as corn or birch, in which both megasporangiate (female) and microsporangiate (male) flowers occur on the same plant, are said to be **monoecious** (meaning "one-housed"—but, it must be added, one house with separate rooms). Complete separation is the rule in some other angiosperm species, such as willows and date palms; in these species, a given plant produces either flowers with stamens or flowers with pistils, but never both. Such species are said to be **dioecious** ("two-housed").

Flowers come in an astonishing variety of forms, as you will realize if you think of some of the flowers you recognize. The generalized flower shown in Figure 29.9 has distinct petals and sepals arranged in distinct whorls. In nature, however, petals and sepals sometimes are indistinguishable. Such appendages are called **tepals** (see Figure 29.11A). In other flowers, petals, sepals, or tepals are completely absent.

Flowers may be single, or they may be grouped together to form an **inflorescence**. Different families of flowering plants have their own, characteristic types of inflorescences, such as the compound umbels of the carrot family, the heads of the aster family, and the spikes of many grasses (**Figure 29.10**).

(A) *Aegopodium podagraria*

(B) *Helianthus annuus*

(C) *Pennisetum setaceum*

29.10 Inflorescences (A) The inflorescence of bishop's goutweed (a member of the carrot family) is a compound umbel. Each umbel bears flowers on stalks that arise from a common center. (B) Sunflowers are members of the aster family; their inflorescence is a head. In a head, each of the long, petal-like structures is a ray flower; the central portion of the head consists of dozens to hundreds of disk flowers. (C) Grasses such as this fountain grass have inflorescences called spikes.

29.11 Flower Form and Evolution (A) A magnolia flower shows the major features of early flowers: it is radially symmetrical, and the individual tepals, carpels, and stamens are separate, numerous, and attached at their bases. (B) Orchids, such as this ladyslipper, have a bilaterally symmetrical structure that evolved much later.

(A) *Magnolia watsonii*

(B) *Paphiopedilum maudiae*

Flower structure has evolved over time

The flowers of the most basal clades of angiosperms have a large and variable number of tepals (or sepals and petals), carpels, and stamens (**Figure 29.11A**). Evolutionary change within the angiosperms has included some striking modifications of this early condition: reductions in the number of each type of floral organ to a fixed number, differentiation of petals from sepals, and changes in symmetry from radial (as in a lily or magnolia) to bilateral (as in a sweet pea or orchid), often accompanied by an extensive fusion of parts (**Figure 29.11B**).

According to one theory, the first carpels to evolve were modified leaves, folded but incompletely closed, and thus differing from the scales of the gymnosperms, which are modified branches. In the groups of angiosperms that evolved later, the carpels fused and became progressively more buried in receptacle tissue (**Figure 29.12A**). In the flowers of the most recent groups, the other floral organs are attached at the very top of the ovary, rather than at the bottom as in Figure 29.9. The stamens of the most ancient flowers may have appeared leaflike (**Figure 29.12B**), little resembling those of the generalized flower in Figure 29.9.

Why do so many flowers have pistils with long styles and anthers with long filaments? Natural selection has favored length in both of these structures, probably because length increases the likelihood of successful pollination. Long filaments may bring the anthers into contact with insect bodies, or they may place the anthers in a better position to catch the wind. Similar arguments apply to long styles.

A perfect flower represents a compromise of sorts. In attracting a pollinating bird or insect, the plant is attending to both its female and male functions with a single flower type, whereas plants with imperfect flowers must create that attraction twice—once for each type of flower. On the other hand, the perfect flower can favor self-pollination, which is usually disadvantageous. Another potential problem is that the female and male functions might interfere with each other—for example, the stigma might be so placed as to

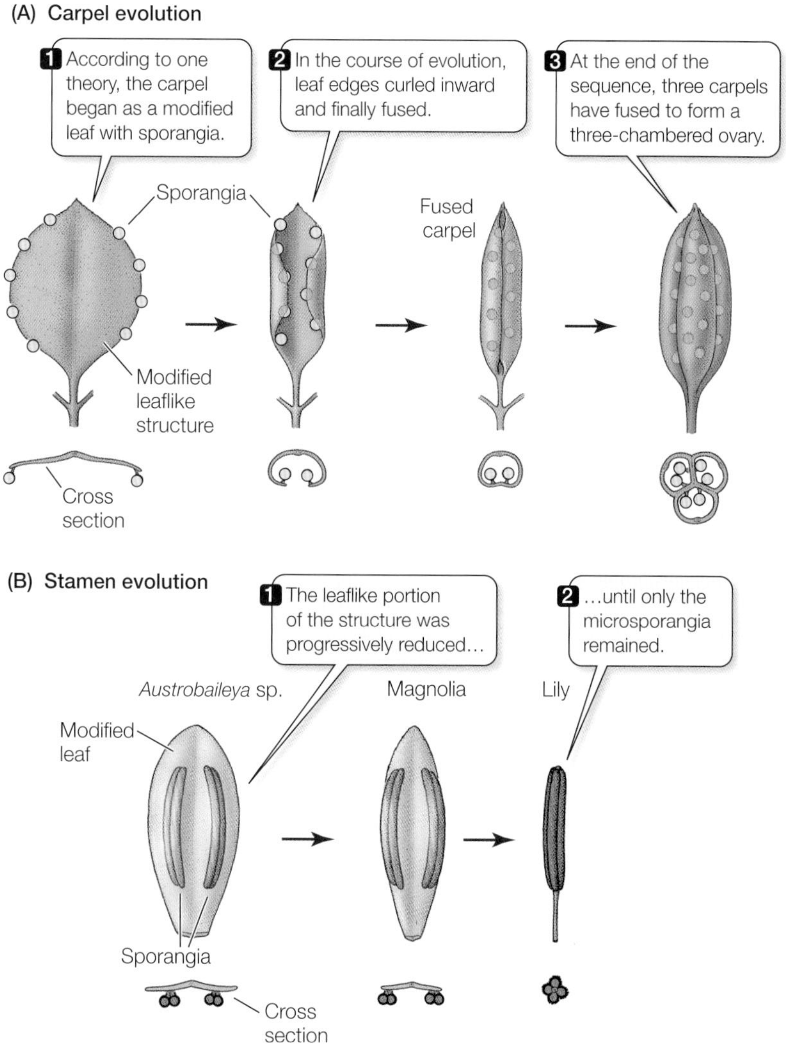

29.12 Carpels and Stamens Evolved from Leaflike Structures (A) Possible stages in the evolution of a carpel from a more leaflike structure. (B) The stamens of three modern plants show the various stages in the evolution of that organ. It is *not* implied that these species evolved one from another; they simply illustrate the structures.

(A)

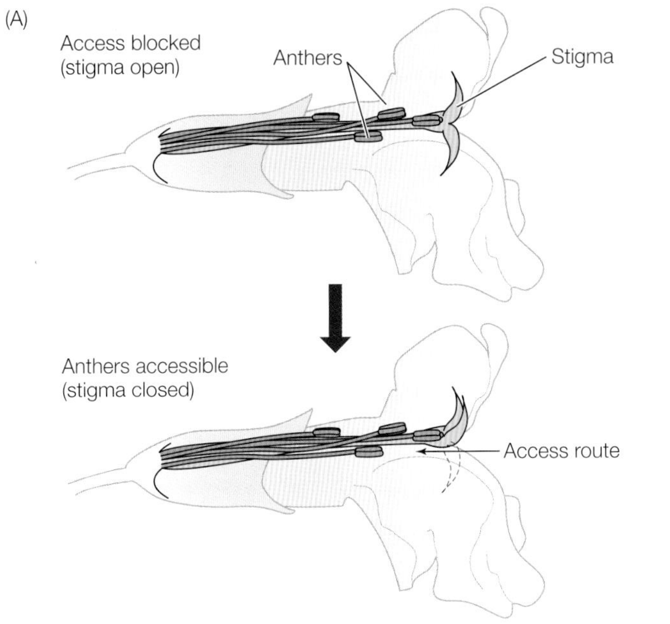

(B)

EXPERIMENT

HYPOTHESIS: Stigma responses in the bush monkeyflower favor the export of pollen.

METHOD

1. Set up experimental arrays of monkeyflowers such that only one flower in each array can donate pollen.
2. Pollen donors in some arrays are normal controls (untouched open stigmas); stigmas of other donors are artificially closed; a third group of donor stigmas are permanently propped open.
3. After hummingbirds visit the arrays, count the grains from each donor on the stigma of the next flower visited.

RESULTS

Almost twice as much pollen was exported from control flowers (i.e., those whose stigmas functioned normally) as from those whose stigmas were experimentally propped open. Experimentally closing the stigmas resulted in even greater pollen dispersal.

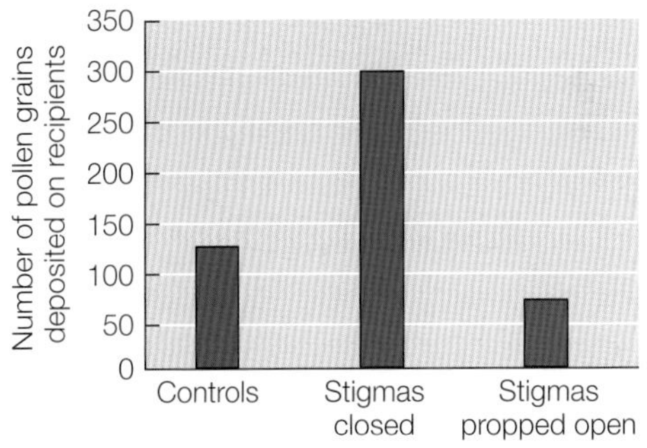

CONCLUSION: Stigma responses enhance the male function of the flower (pollen dispersal) once its female function (pollen deposition) has been performed.

29.13 Stigma Behavior Increases Pollen Export in Monkeyflowers (A) Initially, the stigma of the bush monkeyflower is open, blocking access to the anthers. A hummingbird's touch as it deposits pollen on the stigma causes one lobe of the stigma to retract, creating a path to the anthers. (B) Elizabeth Fetscher explored the reproductive consequences of stigma retraction in this unusual flower. FURTHER RESEARCH: How might you test how this mechanism affects self-pollination of the flower?

make it difficult for pollinators to reach the anthers, thus reducing the export of pollen to other flowers.

Might there be a way around these problems? One solution is seen in the bush monkeyflower (*Mimulus aurantiacus*), which is pollinated by hummingbirds and has a stigma that initially serves as a screen, hiding the anthers. Once a hummingbird touches the stigma, one of the stigma's two lobes folds, providing access by subsequent pollinators to the previously screened anthers (**Figure 29.13A**). The first bird can transfer pollen to the stigma, eventually leading to fertilization. Later visitors can pick up pollen from the anthers, fulfilling the flower's male function. The experiment that revealed the function of this mechanism is described in **Figure 29.13B**.

Angiosperms have coevolved with animals

Whereas many gymnosperms are wind-pollinated, most angiosperms are animal-pollinated. Many flowers entice animals to visit them by providing food rewards. Some flowers produce a sugary fluid called nectar, and the pollen grains themselves may serve as food for animals. In the process of visiting flowers to obtain nectar or pollen, animals often carry pollen from one flower to another or from one plant to another. Thus, in its quest for food, the animal contributes to the genetic diversity of the plant population. Insects, especially bees, are among the most important pollinators; birds and some species of bats also play major roles as pollinators.

For more than 130 million years, angiosperms and their animal pollinators have coevolved in the terrestrial environment. The animals have affected the evolution of the plants, and the plants have affected the evolution of the animals. Flower structure has become incredibly diverse under these selection pressures. Some of the products of coevolution are highly specific; for example, some yucca species are pollinated by only one species of moth. Pollination by just one or a few animal species provides a plant species with a reliable mechanism for transferring pollen from one of its members to another.

Most plant–pollinator interactions are much less specific; that is, many different animal species pollinate the same plant species, and the same animal species pollinate many different plant species. However, even these less specific interactions have developed some specialization. Bird-pollinated flowers are often red and odorless. Many insect-pollinated flowers have characteristic odors, and bee-pollinated flowers may have conspicuous markings, or *nectar guides*, that may be evident only in the ultraviolet region of the spectrum, where bees have better vision than in the red region.

29.14 The Life Cycle of an Angiosperm The formation of a triploid endosperm distinguishes the angiosperms from the gymnosperms. One sperm nucleus fertilizes the egg to form the zygote, while the other combines with the two polar nuclei to form the endosperm.

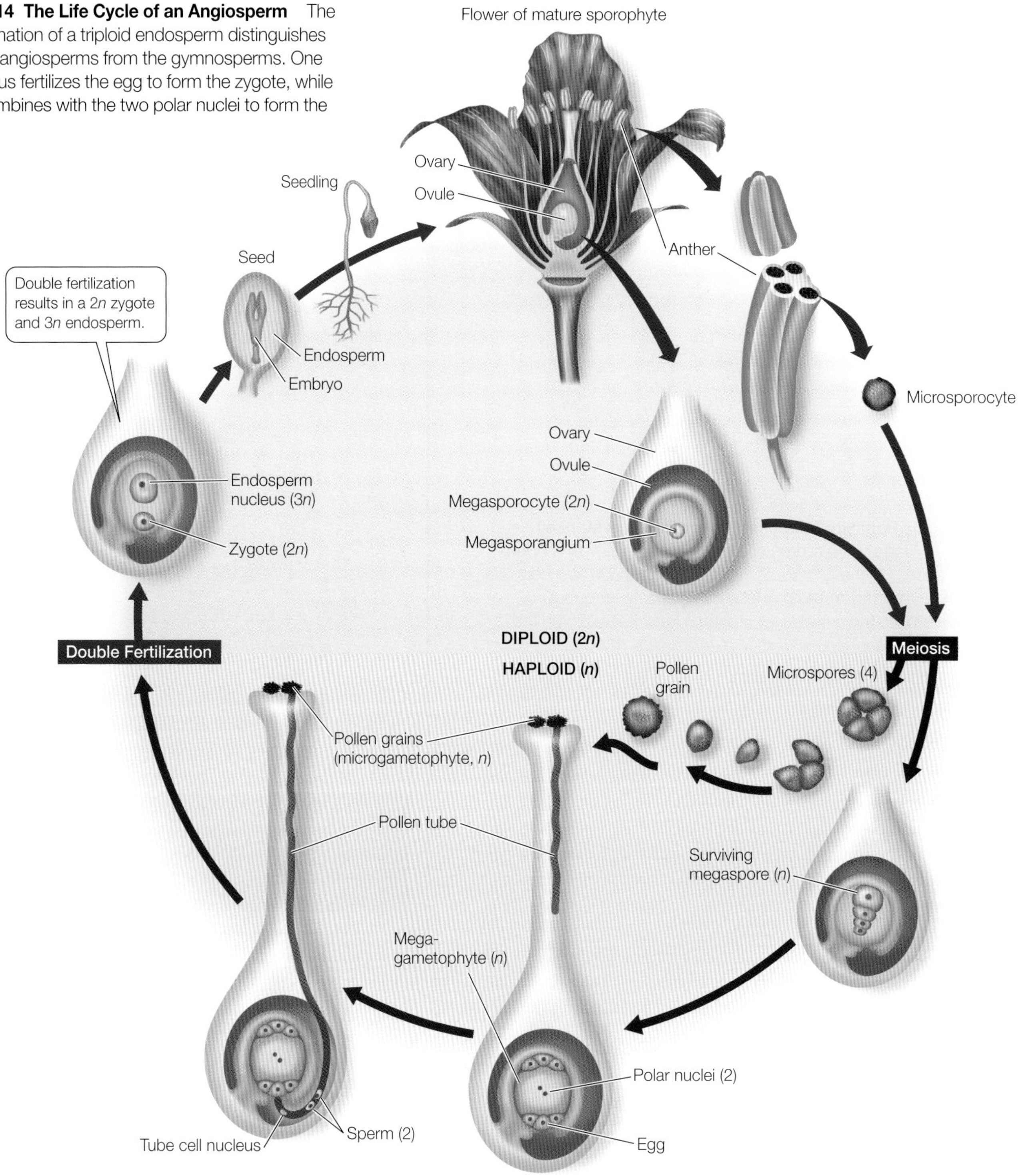

The angiosperm life cycle features double fertilization

The life cycle of the angiosperms is summarized in **Figure 29.14**. The angiosperm life cycle will be considered in detail in Chapter 38, but let's look at it briefly here and compare it with the conifer life cycle in Figure 29.8.

Like all seed plants, angiosperms are heterosporous. As we have seen, their ovules are contained within carpels, rather than being exposed on the surfaces of scales, as in most gymnosperms. The male gametophytes, as in the gymnosperms, are pollen grains.

The ovule develops into a seed containing the products of the double fertilization that characterizes angiosperms: a diploid zygote and a triploid endosperm. The endosperm serves as storage tissue for starch or lipids, proteins, and other substances that will be needed by the developing embryo.

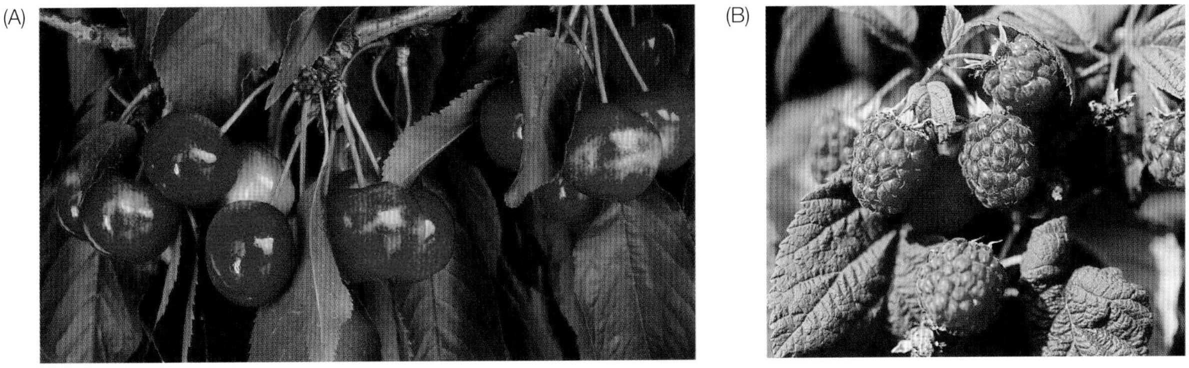

29.15 Fruits Come in Many Forms and Flavors (A) A simple fruit (sour cherry). (B) An aggregate fruit (raspberry). (C) A multiple fruit (pineapple). (D) An accessory fruit (strawberry).

The zygote develops into an embryo, which consists of an embryonic axis (the "backbone" that will become a stem and a root) and one or two **cotyledons**, or seed leaves. The cotyledons have different fates in different plants. In many, they serve as absorptive organs that take up and digest the endosperm. In others, they enlarge and become photosynthetic when the seed germinates. Often they play both roles.

Angiosperms produce fruits

The ovary of a flower (together with the seeds it contains) develops into a fruit after fertilization. The fruit protects the seeds and can also promote seed dispersal by becoming attached to or being eaten by an animal. A fruit may consist only of the mature ovary and its seeds, or it may include other parts of the flower or structures associated with it. A *simple fruit*, such as a cherry (**Figure 29.15A**), is one that develops from a single carpel or several united carpels. A raspberry is an example of an *aggregate fruit* (**Figure 29.15B**)—one that develops from several separate carpels of a single flower. Pineapples and figs are examples of *multiple fruits* (**Figure 29.16C**), formed from a cluster of flowers (an inflorescence). Fruits derived from parts in addition to the carpel and seeds are called *accessory fruits* (**Figure 29.15D**); examples are apples, pears, and strawberries. The development, ripening, and dispersal of fruits will be considered in Chapters 37 and 38.

29.3 RECAP

The synapomorphies of angiosperms include double fertilization, triploid endosperm, flowers, fruit, and distinctive cells in their xylem and phloem.

- Can you distinguish between pollination and fertilization?
- Can you give examples of how animals have affected the evolution of the angiosperms? See p. 641
- Do you understand the roles of the two sperm in double fertilization? See p. 642 and Figure 29.14
- Do you understand the difference between seeds and fruits, and the difference in their roles? See p. 643

We have been considering the shared characteristics of the angiosperms. What are the differences that separate the various angiosperm clades, and where did the angiosperms come from?

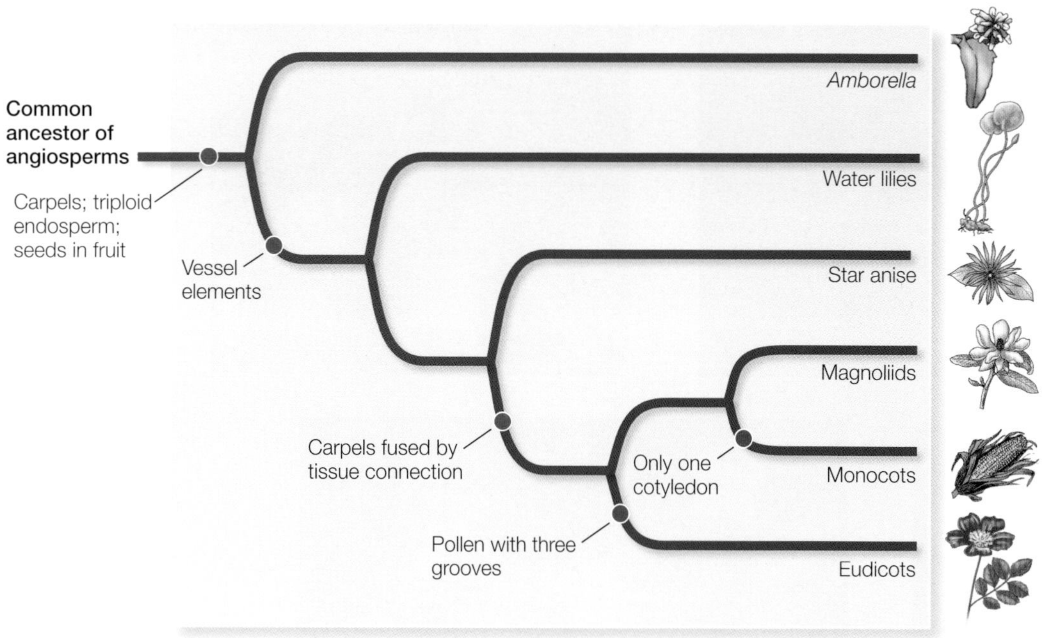

29.16 Evolutionary Relationships among the Angiosperms This diagram is a conservative interpretation of current data on relationships among the clades.

(A) *Amborella trichopoda* (B) *Nymphaea* sp. (C) *Illicium floridanum*

(D) *Piper nigrum* (E) *Aristolochia grandiflora* (F) *Persea* sp.

29.17 Monocots and Eudicots Are Not the Only Surviving Angiosperms (A) *Amborella*, a shrub, is the closest living relative of the first angiosperms; its clade is sister to the remaining extant angiosperms. (B) The water lily clade is the next most basal clade after *Amborella*'s. (C) Star anise and its relatives belong to another basal clade. (D–F) The largest clade other than the monocots and eudicots is the magnoliid complex, represented here by (D) a black pepper, (E) Dutchman's pipe, and (F) an avocado tree. The magnolia in Figure 29.11A is another magnoliid.

29.4 How Did the Angiosperms Originate and Diversify?

The relationships among certain angiosperm clades are controversial, but one conservative interpretation of those relationships is shown in **Figure 29.16**. Two large clades include the great majority of angiosperm species: the **monocots** and the **eudicots**. The monocots are so called because they have a single embryonic cotyledon; the eudicots have two. (We will describe other differences between these two groups in Chapter 34).

Some familiar angiosperms belong to clades other than the monocots and eudicots (**Figure 29.17**). These clades include the water lilies, star anise and its relatives, and the magnoliid complex. The magnoliids are less numerous than the monocots and eudicots, but they include many familiar and useful plants such as magnolias, avocados, cinnamon, and black pepper.

Although the structure of grasses is not favorable for fossilization, recently unearthed fossils in India have established that several grass clades were abundant at least 66 million years ago—much earlier than we once thought. These grasses were preserved in the fossilized dung of titanosaurs, among the heaviest of the dinosaurs.

The monocots (**Figure 29.18**) include grasses, cattails, lilies, orchids, and palms. The eudicots (**Figure 29.19**) include the vast majority of familiar seed plants, including most herbs (i.e., nonwoody plants), vines, trees, and shrubs. Among them are such diverse plants as oaks, willows, beans, snapdragons, and sunflowers.

The basal angiosperm clade is a matter of controversy

Which angiosperms constitute the basal clade of flowering plants is a matter of great controversy. Two leading candidates were the magnolia family (see Figures 29.17D–F) and another family, the Chloranthaceae, whose flowers are much simpler than those of the magnolias. At the close of the twentieth century, however, an impressive convergence of evidence led to the conclusion that the most basal living angiosperm belongs to neither of those families, but rather to a clade that today consists of a single species of the genus *Amborella* (see Figure 29.17A). This woody shrub, with cream-colored flowers, lives only on New Caledonia, an island in the South Pacific. Its 5 to 8 carpels are in a spiral arrangement, and it has 30 to 100 stamens. The xylem of *Amborella* lacks vessel elements, which appeared later in angiosperm evolution. The characteristics of *Amborella* give us an idea of what the first angiosperms might have been like. But are there extinct angiosperms that may represent still more ancient lineages?

In 2002, Chinese and American botanists examined fossils of two species of a 125-million-year-old aquatic genus, *Archaefructus* (see Figure 21.18). Their studies established an extinct group, Archaefructaceae, that is posited to be the sister taxon of all other angiosperms. The flower of these plants had its ovules enclosed in carpels, as in all angiosperms. The flower had neither petals nor sepals, however, and its carpels and stamens were arranged spirally around elongated shoots. This arrangement of carpels and stamens is seen today in the magnolias.

(A) *Phoenix dactylifera*

29.18 Monocots (A) Palms are among the few monocot trees. Date palms are a major food source in some areas of the world. (B) Grasses such as this cultivated wheat and the fountain grass in Figure 29.10C are monocots. (C) Monocots also include popular garden flowers such as these lilies. Orchids (Figure 29.11B) are highly sought-after monocot flowers.

(B) *Triticum* sp.

(C) *Lilium* sp.

(A) *Opuntia* sp.

(B) *Cornus florida*

(C) *Rosa rugosa*

29.19 Eudicots (A) The cactus family is a large group of eudicots, with about 1,500 species in the Americas. This cactus bears yellow flowers for a brief period of the year. (B) The flowering dogwood is a small eudicot tree. (C) Climbing Cape Cod roses are members of the eudicot family Rosaceae, as are the familiar roses from your local florist.

The origin of the angiosperms remains a mystery

We have learned a lot about evolution within the angiosperm clade. But how did the angiosperms first arise? Are the angiosperms sister to any single gymnosperm group? Charles Darwin himself brooded long and hard over the origin of the angiosperms. The problem vexed him so deeply that he referred to it as "an abominable mystery." A few years ago, it seemed that we were on the verge of solving that mystery. We do not yet have the answer, however.

Botanists were long intrigued by certain similarities between the gnetophytes and the angiosperms. One similarity is the presence of vessel elements in the xylem of both groups. Another is the occurrence of two fertilization events in both. Other similarities that gnetophytes and flowering plants shared with certain other groups known only as fossils led to the "anthophyte hypothesis," which groups all of these plants in a single clade (called anthophytes) and leaves the remaining gymnosperms as a paraphyletic group. Support for the anthophyte hypothesis comes from both morphological and molecular evidence.

Other molecular evidence, however, places the gnetophytes firmly within the gymnosperms. This alternative hypothesis leaves the extant gymnosperms as a sister clade to the angiosperms. It is possible that the vessel elements and "double fertilization" of the gnetophytes are simply analogous, rather than homologous, to those of the angiosperms. Reexamination of certain fossils has led some botanists to doubt their phylogenetic relationship to the angiosperms, further discrediting the anthophyte hypothesis. Figure 29.2 assumes that the anthophyte hypothesis is invalid, and it accepts the gnetifer hypothesis (see Figure 29.6A). Both hypotheses are still the subject of intense research, however.

How will we resolve these uncertainties? What morphological characters should be selected as important, or should they all be treated as equally important? Are all molecular differences and similarities significant, or are some of them incidental? Which fossils should be chosen for comparisons? Molecular biologists and paleobotanists are focusing their efforts on gathering new data and determining how best to interpret them. We eagerly await their findings.

29.4 RECAP

The largest angiosperm clades are the monocots and the eudicots. We still have questions about some aspects of angiosperm phylogeny.

- Can you describe the relationship between *Amborella* and the other angiosperms? See p. 645 and Figure 29.16
- Why is *Archaefructus* considered to have been an angiosperm? See p. 645

The remarkable diversity of the seed plants has been shaped in part by the different environments in which these and other plants have evolved. In turn, land plants—and seed plants in particular—affect their environments.

29.5 How Do Plants Support Our World?

Plants make profound contributions to **ecosystem services**—processes by which the environment maintains resources that benefit humans. These benefits include the effects of plants on soils, water, the atmosphere, and the climate. As we'll see in Section 36.3, plants play important roles in soil formation and in renewing the fertility of soils. Plant roots help hold soil in place, resisting erosion by wind and water. Plants store water in their bodies. They also moderate local climate in various ways, such as by increasing humidity, providing shade, and blocking wind.

Plants are **primary producers**; that is, their photosynthesis traps energy and carbon, making those resources available not only for their own needs, but also for the herbivores and omnivores that consume them, for the carnivores and omnivores that eat the herbivores, and for the prokaryotes and fungi that complete the food chain.

Seed plants are our primary food source

Twelve seed plant species stand between our species and starvation: rice, coconut, wheat, corn (also called maize), potato, sweet potato, cassava (also called tapioca or manioc), sugarcane, sugar beet, soybean, common bean, and banana. Other seed plants are cultivated for food, but none rank with these twelve in importance.

Indeed, more than half of the world's human population derives the bulk of its food energy from the seeds of a single plant: rice, *Oryza sativa*. Rice is particularly important in the diets of people in the Far East, where it has been cultivated for nearly 5,000 years. People also use rice straw in many ways, such as thatching for roofs, food and bedding for livestock, and clothing. Rice hulls, too, have many uses, ranging from fertilizer to fuel.

Let's look at what another one of these dozen plants provides. In some cultures the coconut palm (*Cocos nucifera*, **Figure 29.20**) is called the Tree of Life because every aboveground part of the plant has value to humans. People use the stem (the trunk) of this tropical coastal lowland monocot tree as lumber. They dry the sap from its trunk for use as a sugar, or they ferment it to drink. They use the leaves to thatch their homes and to make hats and baskets. They eat the apical bud at the top of the trunk in salads.

The coconut fruit serves many purposes. The hard shell can be used as a container or burnt as fuel. The fibrous middle layer, or coir, of the fruit wall can be made into mats and rope. The seed of the coconut palm contains both "milk" and "meat." Because the refreshing and delicious coconut milk contains no bacteria or other pathogens, it is particularly important wherever the water is not fit for drinking. Millions of people get most of their protein from coconut meat. Much coconut meat is dried and marketed as copra, from which coconut oil is pressed. Coconut oil is the most widely used vegetable oil in the world; it is used in the manufacture of a range of products from hydraulic brake fluid to synthetic rubber and, although nutritionally poor, as food. Ground copra serves as fertilizer and as food for livestock.

Seed plants have been sources of medicines since ancient times

One likely candidate for the oldest human profession is that of medicine man or shaman—a person who cures others with medicines derived from seed plants. It is claimed that a legendary Chinese emperor around 2700 B.C.E. knew some 365 medicinal plants. We use many medicines derived from molds, lichens, and actinobacteria as well as synthetic ones. However, we still rely on seed plants for many of our medicines, just a few of which are shown in **Table 29.1**.

How are plant-based medicines discovered? These days many are found by systematic testing of tremendous numbers of plants from all over the world, a process that began in the 1960s. One example is taxol, an important anticancer drug. Among the myriad plant samples that had been tested by 1962, extracts of the bark of Pacific yew (*Taxus brevifolia*) showed antitumor activity in tests against rodent tumors. The active ingredient, taxol, was isolated in 1971 and finally tested against human cancers in 1977. After another 16 years, the U.S. Food and Drug Administration approved it for human use, and taxol is now widely used in treating breast and ovarian cancers as well as several other types of cancers.

This type of widespread screening of plant samples eventually was deemphasized in favor of a purely chemical approach. Using

29.20 The Tree of Life Fruits of the coconut palm await harvest on a plantation in the South Pacific.

TABLE 29.1

Some Medicinal Plants and Their Products

PRODUCT	PLANT SOURCE	MEDICAL APPLICATION
Atropine	Belladonna	Dilating pupils for eye examination
Bromelain	Pineapple stem	Controlling tissue inflammation
Digitalin	Foxglove	Strengthening heart muscle contraction
Ephedrine	*Ephedra*	Easing nasal congestion
Menthol	Japanese mint	Relief of coughing
Morphine	Opium poppy	Relief of pain
Quinine	Cinchona bark	Treatment of malaria
Taxol	Pacific yew	Treatment of ovarian and breast cancers
Tubocurarine	Curare plant	As muscle relaxant in surgery
Vincristine	Periwinkle	Treatment of leukemia and lymphoma

automation and miniaturization, pharmaceutical laboratories generate vast numbers of compounds that are screened just as plant materials were screened in the search for taxol and other plant-based medicines. Now, however, the plant screening is getting renewed interest. Both approaches are based on trial and error.

The other leading source of medicinal plants is work by *ethnobotanists*, who study how and why people use and view plants in their local environments. This work proceeds all over the globe today. An older example is the discovery of quinine as a treatment for malaria. In 1630, Spanish priests in Peru successfully used the bark of the local cinchona tree as a cure for malaria. They were aware that the Peruvians had been using the bark to treat fevers. Word of the successful treatment made its way to Europe, and cinchona bark became the standard treatment for malaria. The active ingredient, quinine, was finally identified in 1820.

CHAPTER SUMMARY

29.1 How did seed plants become today's dominant vegetation?

Only seed plants show **secondary growth**, producing wood.

The surviving groups of seed plants are the **gymnosperms** and **angiosperms**.

All seed plants are heterosporous, and their gametophytes are much smaller than, and dependent on, their sporophytes. Review Figure 29.3

An **ovule** consists of the seed plant megagametophyte and the **integument** which protects it. The ovule develops into a seed.

Pollen grains, the microgametophytes, do not require liquid water to perform their functions. Following **pollination**, a **pollen tube** emerges from the pollen grain and elongates to deliver gametes to the megagametophyte. Pollen grains are enclosed in highly resistant sporopollenin walls.

Seeds are well protected, and they are often capable of long periods of dormancy, germinating when conditions are favorable.

29.2 What are the major groups of gymnosperms?

The extant gymnosperms may be a clade, as are at least three of the four gymnosperm groups. Review Figures 29.2 and 29.6

The gymnosperm clades are the **cycads**, **ginkgos**, and **gnetophytes**. **Conifers** are the most abundant gymnosperms.

The megaspores of pines are produced in **cones**, and microspores are produced in **strobili**. Pollen reaches the megasporangium by way of the **micropyle**, an opening in the integument of the ovule. Review Figures 29.7 and 29.8, See Web/CD Activity 29.1 and Tutorial 29.1

29.3 What features distinguish the angiosperms?

Only angiosperms have **flowers** and **fruits**. Review Figure 29.9, Web/CD Activity 29.2

The ovules and seeds of angiosperms are enclosed in and protected by **carpels**. Review Figure 29.12

Angiosperms have **double fertilization**, resulting in the production of a zygote and a triploid **endosperm**. Review Figure 29.14, Web/CD Tutorial 29.2

The xylem and phloem of angiosperms are more complex and efficient than those of the gymnosperms. **Vessel elements**, **fibers**, and **companion cells** contribute to the efficiency.

The floral organs, from the apex to the base of the flower, are the **pistil**, **stamens**, **petals**, and **sepals**. Stamens bear microsporangia in **anthers**. The pistil (consisting of one or more carpels) includes an **ovary** containing ovules. The **stigma** is the receptive surface of the pistil. The floral organs are born on the **receptacle**.

A flower with both megasporangia and microsporangia is **perfect**; all other flowers are **imperfect**. Flowers may be grouped to form an **inflorescence**. Most flowers are pollinated by animals.

A **monoecious** species has megasporangiate and microsporangiate flowers on the same plant. A **dioecious** species is one in which megasporangiate and microsporangiate flowers never occur on the same plant.

29.4 How did the angiosperms originate and diversify?

The most abundant angiosperm clades are the **monocots** and the **eudicots**. The relationships of these to the other extant angiosperm clades are not fully resolved. Review Figure 29.16

The most basal living angiosperm appears to be a single species in the genus *Amborella*. The most basal angiosperm clade appears to be the extinct Archaefructaceae. We do not know what group is sister to the angiosperms.

29.5 How do plants support our world?

Plants provide **ecosystem services** that affect soils, water, the air, and the climate.

Plants are **primary producers**, providing food for the entire terrestrial food web.

Plants provide many important medicines. Review Table 29.1

SELF-QUIZ

1. Which of the following statements about seed plants is true?
   a. The phylogenetic relationships among the major groups have been well established.
   b. The sporophyte generation is more reduced than in the ferns.
   c. The gametophytes are independent of the sporophytes.
   d. All seed plant species are heterosporous.
   e. The zygote divides repeatedly to form the gametophyte.
2. The gymnosperms
   a. dominate all land masses today.
   b. have never dominated land masses.
   c. have active secondary growth.
   d. all have vessel elements.
   e. lack sporangia.
3. Conifers
   a. produce ovules in strobili and pollen in cones.
   b. depend on liquid water for fertilization.
   c. have triploid endosperm.
   d. have pollen tubes that release two sperm.
   e. have vessel elements.
4. Angiosperms
   a. have ovules and seeds enclosed in a carpel.
   b. produce triploid endosperm by the union of two eggs and one sperm.
   c. lack secondary growth.
   d. bear two kinds of cones.
   e. all have perfect flowers.
5. Which statement about flowers is *not* true?
   a. Pollen is produced in the anthers.
   b. Pollen is received on the stigma.
   c. An inflorescence is a cluster of flowers.
   d. A species having female and male flowers on the same plant is dioecious.
   e. A flower with both megasporangia and microsporangia is said to be perfect.
6. Which statement about fruits is *not* true?
   a. They develop from ovaries.
   b. They may include other parts of the flower.
   c. A multiple fruit develops from several carpels of a single flower.
   d. They are produced only by angiosperms.
   e. A cherry is a simple fruit.
7. Which statement is *not* true of angiosperm pollen?
   a. It is the male gamete.
   b. It is haploid.
   c. It produces a long tube.
   d. It interacts with the carpel.
   e. It is produced in microsporangia.
8. Which statement is *not* true of carpels?
   a. They are thought to have evolved from leaves.
   b. They bear megasporangia.
   c. They may fuse to form a pistil.
   d. They are floral organs.
   e. They were absent in *Archaefructus*.
9. *Amborella*
   a. was the first flowering plant.
   b. belongs to the first gymnosperm clade.
   c. belongs to the oldest angiosperm clade still extant.
   d. is a eudicot.
   e. has vessel elements in its xylem.
10. The eudicots
    a. include many herbs, vines, shrubs, and trees.
    b. and the monocots are the only extant angiosperm clades.
    c. are not a clade.
    d. include the magnolias.
    e. include orchids and palm trees.

FOR DISCUSSION

1. In most seed plant species, only one of the products of meiosis in the megasporangium survives. How might this be advantageous?
2. Suggest an explanation for the great success of the angiosperms in occupying terrestrial habitats.
3. In many locales, large gymnosperms predominate over large angiosperms. Under what conditions might gymnosperms have the advantage, and why?
4. Not all flowers possess all of the following floral organs: sepals, petals, stamens, and carpels. Which floral organ or organs do you think might be found in the flowers that have the smallest number of floral organ types? Discuss the possibilities, both for a single flower and for a species.
5. The problem of the origin of the angiosperms has long been "an abominable mystery," as Charles Darwin once put it. Scientists still do not know the nearest relatives of the angiosperms. It has often been suggested (correctly or incorrectly) that the gnetophytes are sister to the angiosperms. What pieces of evidence suggested this connection?

FOR INVESTIGATION

The flower of a particular species of orchid has a long, spurlike tube into which a pollinating insect can insert its proboscis to suck nectar. The spurs are of different lengths in different habitats, apparently correlated with the lengths of the probosces of the local pollinators. How might you test the hypothesis that this correlation increases reproductive success, in terms of pollen transfer to the flower?

CHAPTER 30 Fungi: Recyclers, Pathogens, Parasites, and Plant Partners

A fungus battles witchweed

About 300 million Africans in 25 countries are suffering because of the invasion of crops by witchweed (*Striga*), a parasitic flowering plant. This parasite has attacked more than two-thirds of the sorghum, corn, and millet crops in sub-Saharan Africa, doing damage estimated at U.S.$7 billion each year.

A team of Canadian scientists set out to find a biological solution to the *Striga* problem. Their strategy was to look for an organism that would destroy witchweed in the fields. They succeeded in isolating a strain of a fungus—the mold *Fusarium oxysporum*—that has several outstanding properties. First, it grows on *Striga*, killing a high percentage of these parasitic plants. Second, it is not toxic to humans, nor does it attack the crop plants on which *Striga* is growing. In subsequent fieldwork in Mali, the scientists established techniques for proper application of *F. oxysporum* to witchweed. Now farmers apply the fungus to their crops and are rewarded by greatly increased crop yields as *Striga* is held in check.

It may be possible to repeat the *Striga* story—the use of a fungus to wipe out a particular type of flowering plant—in a very different context. A different strain of *F. oxysporum* preferentially attacks coca plants (the source of cocaine). A controversial proposal to use *F. oxysporum* to wipe out the coca plantations of Andean South America and in other parts of the world has been proposed. Some naturally occurring strains of *F. oxysporum* attack important crops in various parts of the world. What can we do about that? Recent work in India has identified two *other* species of fungi that inhibit the development of *F. oxysporum* growing on safflower when their spores are sprayed on the plants, allowing the crop to grow better. Thus we are using fungi to battle fungi! Fortunately, these other fungi are not harmful to safflower or any other crop plants that have been tested.

In another promising development, research teams reported in 2005 that two fungi, *Beauveria bassiana* and *Metarhizium anisopliae*, killed malaria-carrying mosquitoes when applied to mosquito netting. Certain fungi are already in use against other insect pests, notably termites and aphids. And in research conducted thus far, insect pests have not developed resistance to fungi as they can to DDT and other pesticides.

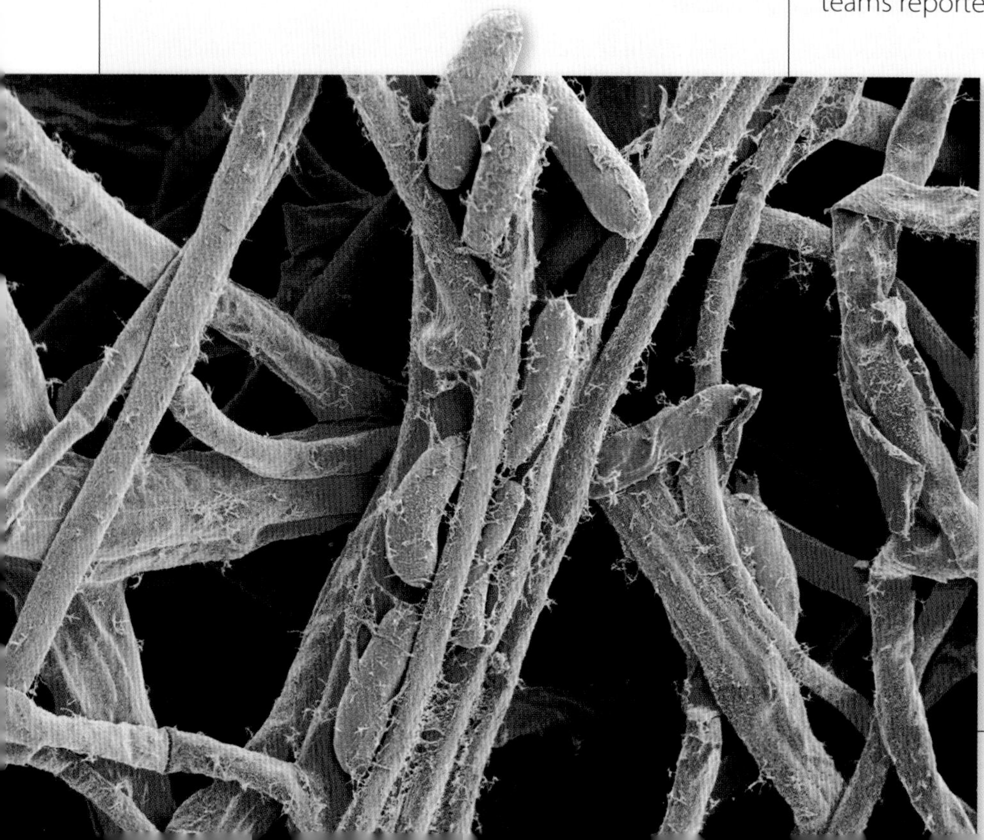

Pathogenic Fungus, Parasitic Plant The fungus *Fusarium oxysporum* is a potent pathogen of witchweed (*Striga*), a parasitic plant that attacks crops. The fungal spores are shown in blue; the fungal filaments are in tan. Both colors were added to enhance this electron micrograph.

An Alien Meal The tropical fungus whose fruiting body is growing on the stalk projecting from this ant's carcass developed internally in the host, from a spore ingested by the ant.

Of course, fungi don't need our help to find organisms to grow on. Fungal spores ingested by animals such as ants can germinate in their new host's gut. They develop into multicellular individuals that absorb nutrients from their unwitting hosts. Eventually they kill their hosts, producing new spores to infect other organisms as they do so.

Fungi interact with other organisms in many different ways, some of which are helpful and some harmful to the other organism. As we begin our study, recall that the fungi and the animals are descended from a common ancestor, and thus molds and mushrooms are more closely related to you than they are to the flowers we admired in Chapter 29.

IN THIS CHAPTER we will see how fungi differ from other eukaryotes in some very interesting ways. We will explore the diversity of body forms, reproductive structures, and life cycles that have evolved among the five groups of fungi, as well as examining the mutually beneficial associations of certain fungi with other organisms.

CHAPTER OUTLINE

30.1 How Do Fungi Thrive in Virtually Every Environment?

Modern fungi are believed to have evolved from a unicellular protistan ancestor that had a flagellum. The probable common ancestor of the animals was also a flagellated microbial eukaryote that may have been similar to the existing choanoflagellates (see Figure 27.30). Current evidence suggests that today's choanoflagellates, fungi, and animals share a single common ancestor, and thus the three lineages are often grouped together as the *opisthokonts.* The synapomorphies that distinguish the fungi among the opisthokonts are absorptive heterotrophy, and the presence of chitin in their cell walls (**Figure 30.1**).

The fungi live by **absorptive nutrition**: they secrete digestive enzymes that break down large food molecules in the environment, then absorb the breakdown products through the plasma membranes of their cells. Absorptive heterotrophy is successful in virtually every conceivable environment. Many fungi are **saprobes** that absorb nutrients from dead organic matter. Others are **parasites** that absorb nutrients from living hosts, such as the ant shown at the opening of this chapter. Still others are **mutualists** that live in intimate associations with other organisms that benefit both partners.

We will discuss five major groups of fungi: chytrids, zygomycetes, ascomycetes, basidiomycetes, and glomeromycetes. The first four of these groups were originally defined by their methods and structures for sexual reproduction and also, to a lesser extent, by other morphological differences. More recently, biologists have turned to evidence from DNA analyses, which established the glomeromycetes as a group independent of the zygomycetes. The chytrids and zygomycetes appear to be paraphyletic, although their phylogenetic relationships are not yet well understood. The other three groups are clades (**Figure 30.2**). The chytrids are aquatic, but the other groups are terrestrial.

The body of a multicellular fungus is composed of hyphae

Most fungi are multicellular, but single-celled species are found in most of the fungal groups. Unicellular members of all but the basal group of fungi (the chytrids) are called **yeasts** (**Figure 30.3**).

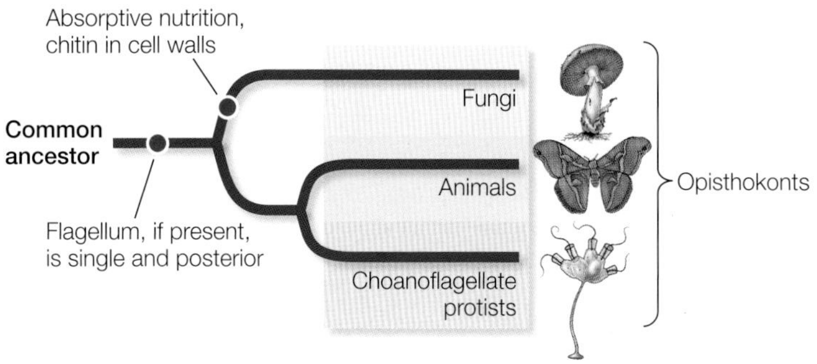

30.1 Fungi in Evolutionary Context Shared derived traits (synapomorphies) distinguish the fungi and other opisthokonts from one another and from the rest of the tree of life.

There are no unicellular glomeromycetes. Yeasts live in liquid or moist environments and absorb nutrients directly across their cell surfaces.

The body of a multicellular fungus is called a **mycelium** (plural, *mycelia*). It is composed of rapidly growing individual tubular filaments called **hyphae** (singular, *hypha*). The cell walls of the hyphae are greatly strengthened by microscopic fibrils of *chitin*, a nitrogen-containing polysaccharide. Hyphae may be subdivided into cell-like units by *incomplete* cross-walls called **septa** (singular, *septum*). Septa do not completely close off compartments in the hyphae; gaps in the septa known as *pores* allow organelles—sometimes even nuclei—to move in a controlled way between cells (**Figure 30.4**). Hyphae that have septa are referred to as **septate**; hyphae that lack septa—but may contain hundreds of nuclei—are referred to as **coenocytic**. The coenocytic condition results from repeated nuclear divisions unaccompanied by cytokinesis.

Certain modified hyphae, called *rhizoids*, anchor some fungi to their substratum (the dead organism or other matter on which they feed). These rhizoids are not homologous to the rhizoids of plants because they are not specialized to absorb nutrients and water. Parasitic fungi may possess modified hyphae that take up nutrients from their host.

The total hyphal growth of a mycelium (not the growth of an individual hypha) may exceed 1 kilometer per day. The hyphae may be widely dispersed to forage for nutrients over a large area, or they may clump together in a cottony mass to exploit a rich nutrient source. In some members of some fungal groups, when sexual spores are produced, the mycelium becomes reorganized into a reproductive *fruiting body* such as a mushroom.

Fungi are in intimate contact with their environment

The filamentous hyphae of a fungus give it a unique relationship with its physical environment. The fungal mycelium has an enormous surface area-to-volume ratio compared with that of most large multicellular organisms. This large ratio is a marvelous adaptation for absorptive nutrition. Throughout the mycelium (except in fruiting structures), all the hyphae are very close to their environmental food source.

There is a downside to the great surface area-to-volume ratio of the mycelium, however: it results in a tendency to lose water rapidly in a dry environment. Thus fungi are most common in moist environments. You have probably observed the tendency of molds, toadstools, and other fungi to appear in damp places.

Another characteristic of some fungi is their tolerance for highly hypertonic environments (those with a solute concentration higher than their own; see Figure 5.9). Many fungi are more resistant than bacteria to damage in hypertonic surroundings. Jelly in the refrigerator, for example, will not become a growth medium for bacteria because it is too hypertonic to those organisms, but it may eventually harbor mold colonies. Their presence in the refrigerator illustrates another trait of many fungi: tolerance of temperature extremes. Many fungi tolerate temperatures as low as –6°C, and some tolerate temperatures over 50°C.

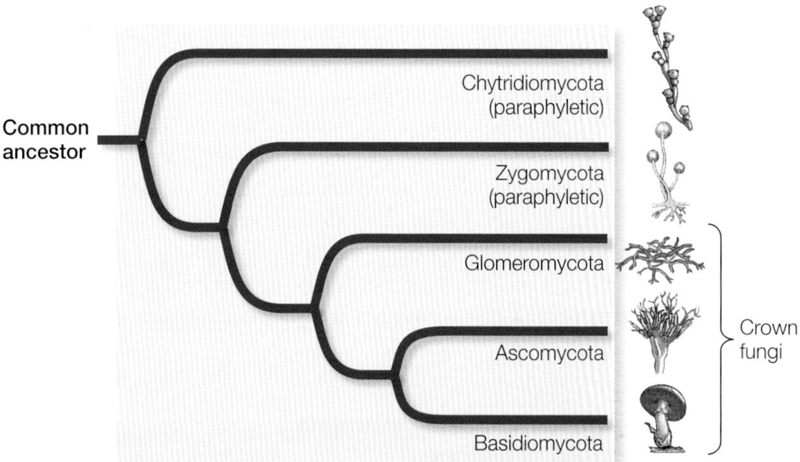

30.2 Phylogeny of the Fungi Five groups are recognized among the fungi. The chytrids and the zygomycetes are paraphyletic. Each of the other groups is a clade, and together the three groups comprise the crown fungi, which is also a clade.

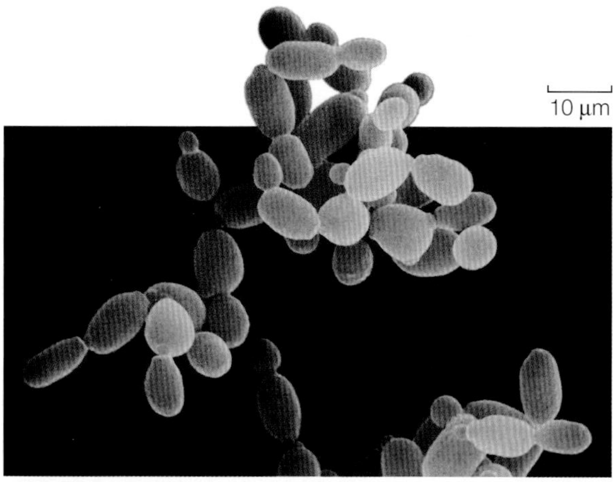

Saccharomyces sp.

30.3 Yeasts Are Unicellular Fungi Unicellular zygomycetes, ascomycetes, and basidiomycetes are known as yeasts. Many yeasts reproduce by budding—mitosis followed by asymmetrical cell division—as those shown here are doing.

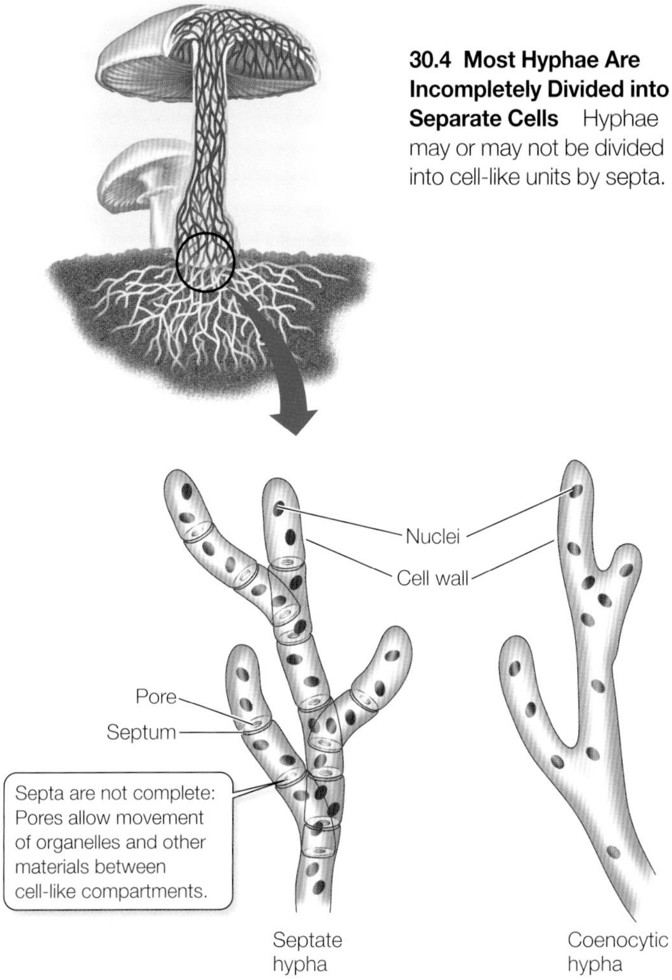

30.4 Most Hyphae Are Incompletely Divided into Separate Cells Hyphae may or may not be divided into cell-like units by septa.

How large can a fungus get? One fungus in Michigan covers 15 hectares (37 acres). At the surface, only isolated clumps of mushrooms are visible. But, growing underground, the vast mycelium of the fungus weighs more than a blue whale. We know from exhaustive analysis of its DNA that this giant fungus is a single individual.

Fungi exploit many nutrient sources

Different types of fungi find different sources of nutrients. While the majority are saprobes, obtaining their energy, carbon, and nitrogen directly from dead organic matter, many form mutualistic associations with other organisms. Others are parasitic—even predatory.

Because many saprobic fungi are able to grow on artificial media, we can perform experiments to determine their exact nutritional requirements. Sugars are their favored source of carbon. Most fungi obtain nitrogen from proteins or the products of protein breakdown. Many fungi can use nitrate (NO_3^-) or ammonium (NH_4^+) ions as their sole source of nitrogen. No known fungus can get its nitrogen directly from nitrogen gas, as can some bacteria and plant–bacteria associations (see Section 36.4). Nutritional studies also reveal that most fungi are unable to synthesize certain vitamins and must absorb them from their environment. On the other hand, fungi can synthesize some vitamins that animals cannot. Like all organisms, fungi also require some mineral elements.

PARASITIC FUNGI Nutrition in the parasitic fungi is particularly interesting to biologists. *Facultative* parasites can attack living organisms but can also be grown by themselves on artificial media. *Obligate* parasites cannot be grown on any available medium; they can grow only on their specific living host, usually a plant species. Because their growth is limited to a living host, such fungi must have specialized nutritional requirements.

The filamentous structure of fungal hyphae is especially well suited to a life of absorbing nutrients from plants. The slender hyphae of a parasitic fungus can invade a plant through stomata, through wounds, or in some cases, by direct penetration of epidermal cells (**Figure 30.5A**). Once inside the plant, the hyphae form a mycelium. Some hyphae produce **haustoria**, branching projections that push into living plant cells, absorbing the nutrients within those cells. The haustoria do not break through the plant cell plasma membranes; they simply press into the cells, with the membrane fitting them like a glove (**Figure 30.5B**). Fruiting structures may form, either within the plant body or on its surface. This type of symbiosis is not usually lethal to the plant.

Some parasitic fungi not only derive nutrition from the bodies of other organisms, but also sicken or kill those hosts. Such a fungus is called a *pathogen*.

SOME PARASITIC FUNGI ARE PATHOGENS Although most human diseases are caused by bacteria or viruses, fungal pathogens are a major cause of death among people with compromised immune systems. Most people with AIDS die of fungal diseases, such as the pneumonia caused by *Pneumocystis carinii* or incurable diarrhea caused by other fungi. *Candida albicans* and certain other yeasts also cause severe diseases, such as esophagitis (which impairs swallowing), in individuals with AIDS and in individuals taking immunosuppressive drugs. Fungal diseases are a growing international health problem, requiring vigorous research. Our limited understanding of the basic biology of these fungi still hampers our ability to treat the diseases they cause. Various fungi cause other, less threatening human diseases, such as ringworm and athlete's foot.

Fungi are by far the most important plant pathogens, causing crop losses amounting to billions of dollars. Bacteria and viruses are less important as plant pathogens. Major fungal diseases of crop plants include black stem rust of wheat and other diseases of wheat, corn, and oats. The agent of black stem rust is *Puccinia graminis*, whose complicated life cycle we will discuss later in the chapter. In an epidemic in 1935, *P. graminis* was responsible for the loss of about one-fourth of the entire wheat crop in Canada and the United States. However, as we saw at the beginning of this chapter, pathogenic fungi that kill certain weed species can be a boon to agriculture.

PREDATORY FUNGI Some fungi have adaptations that enable them to function as active predators, trapping nearby microscopic protists or animals. The most common predatory strategy seen in fungi is to secrete sticky substances from the hyphae so that passing organisms stick tightly to them. The hyphae then quickly invade

30.5 Attacks on a Leaf (A) The white structures in the micrograph are hyphae of the parasitic fungus *Blumeria graminis* growing on the dark surface of the leaf of a grass. (B) Haustoria are fungal hyphae that push into the living cells of plants, from which they absorb nutrients.

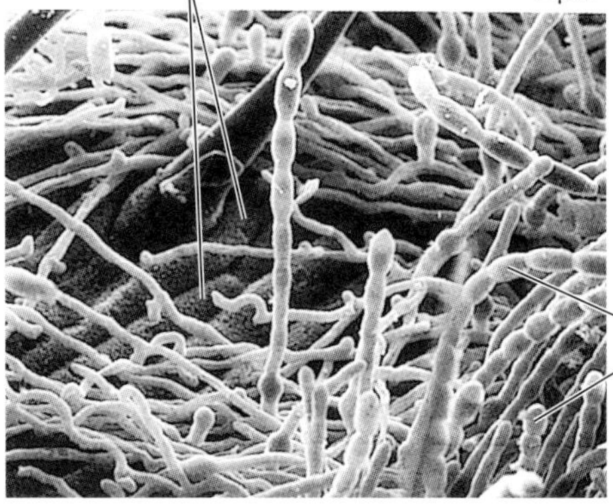

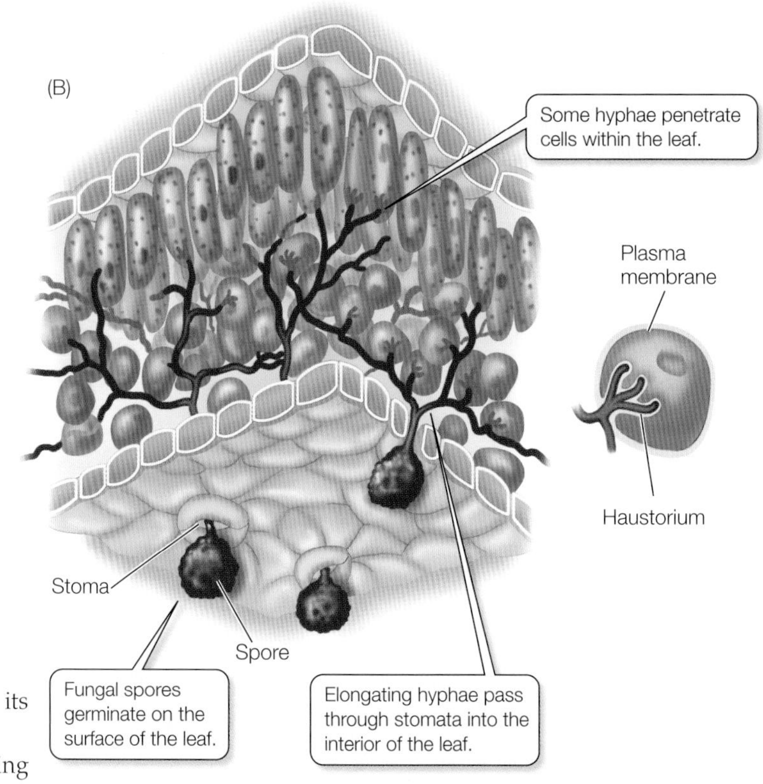

the prey, growing and branching within it, spreading through its body, absorbing nutrients, and eventually killing it.

A more dramatic adaptation for predation is the constricting ring formed by some species of *Arthrobotrys*, *Dactylaria*, and *Dactylella* (**Figure 30.6**). All of these fungi grow in soil. When nematodes (tiny roundworms) are present in the soil, these fungi form three-celled rings with a diameter that just fits a nematode. A nematode crawling through one of these rings stimulates the fungus, causing the cells of the ring to swell and trap the worm. Fungal hyphae quickly invade and digest the unlucky victim.

Fungi balance nutrition and reproduction

What happens if a fungus faces a dwindling food supply? A common strategy is to reproduce rapidly and abundantly. Even when conditions are good, fungi produce great quantities of spores. But the rate of spore production commonly goes up when nutrient supplies go down.

Not only are fungal spores abundant in number, but they are extremely tiny and easily spread by wind or water (**Figure 30.7**). This virtually assures that the individual that produced them will have many progeny, scattered over sometimes great distances. No wonder we find fungi just about everywhere.

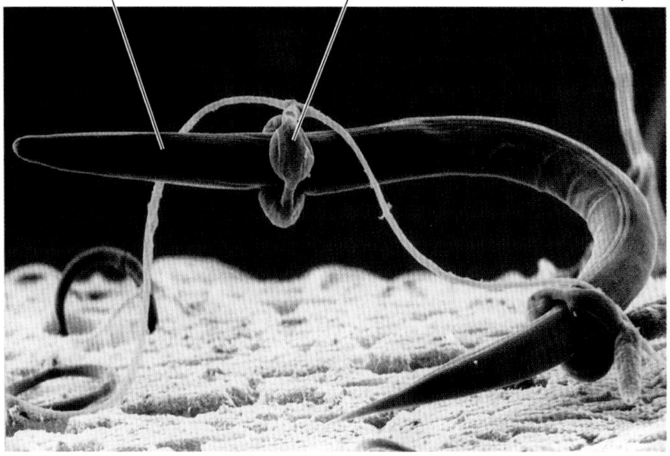

30.6 Some Fungi Are Predators A nematode is trapped in a constricting ring of the soil-dwelling fungus *Arthrobotrys anchonia*.

30.1 RECAP

The rapid growth and large surface area-to-volume ratio of fungal hyphae allow fungi to practice absorptive nutrition efficiently in moist environments.

- Can you explain the relationship between fungal structure and absorptive nutrition? See p. 652
- Can you distinguish among the nutritional modes of saprobes, parasites, and mutualists? See p. 653

The thought of all those fungal spores blowing around can be alarming. But fungi are not always a threat to the organisms around them. There are ways in which they benefit other organisms, and even the biosphere as a whole.

Lycoperdon sp.

30.7 Spores Galore Puffballs disperse trillions of spores in great bursts. Very few of the spores, however, travel very far; some 99 percent of the spores will fall within 100 m of the parent puffball.

30.2 How Are Fungi Beneficial to Other Organisms?

Without the fungi, our planet would be very different. Picture Earth with only a few stunted plants and with watery environments choked with the remains of dead organisms. The colonization of the terrestrial environment was made possible in large part by associations fungi formed with other organisms. Fungi also do much of Earth's garbage disposal. Fungi that absorb nutrients from dead organisms not only help clean up the landscape and form soil, but also play a key role in recycling mineral elements.

Saprobic fungi dispose of Earth's garbage and contribute to the planetary carbon cycle

Saprobic fungi, along with bacteria, are the major decomposers on Earth, contributing to decay and thus to the recycling of the elements used by living things. In forests, for example, the mycelia of fungi absorb nutrients from fallen trees, thus decomposing their wood. Fungi are the principal decomposers of cellulose and lignin, the main components of plant cell walls (most bacteria cannot break down these materials). Other fungi produce enzymes that decompose keratin and thus break down animal structures such as hair and nails.

Were it not for the fungi, great quantities of carbon atoms would have remained trapped on forest floors and elsewhere almost forever—the carbon cycle would have failed. Instead, that carbon is returned to the atmosphere as respiratory CO_2, available for photosynthesis by plants. There was in fact a time when there was a decline in the population of saprobic fungi. During the Carboniferous period, plants in tropical swamps died and began to form peat (see Section 28.4). Peat formation led to acidification of the swamps; in turn, the acidity drastically reduced the fungal population. The result? With the decomposers largely absent, the dead plant material remained on the floor of the swamp and over time was converted to peat and, eventually, coal.

The fungi rose to the occasion during the Permian, a quarter of a billion years ago, when a meteorite crashed to Earth, triggering a planet-wide extinction event (see Section 21.3). The fossil record shows that even though 96 percent of all species became extinct, fungi flourished, demonstrating both their hardiness and their role in recycling the elements in the dead bodies of plants and animals.

As decomposers, the saprobes are essential to life as a whole. There are also many fungi that interact more specifically with other organisms to play key roles as mutualists.

Mutualistic relationships are beneficial to both partners

Certain kinds of relationships between fungi and other organisms have nutritional consequences for the fungal partner. Two of these relationships are highly specific and are **symbiotic** (the partners live in close, permanent contact with one another) as well as **mutualistic** (the relationships benefit both partners).

Lichens are associations of a fungus with a cyanobacterium, a unicellular photosynthetic alga, or both. **Mycorrhizae** (singular *mycorrhiza*) are associations between fungi and the roots of plants. In these associations, the fungus obtains organic compounds from its photosynthetic partner, but provides it with minerals and water in return, so that the partner's nutrition is also promoted. In fact, many plants could not grow at all without their fungal partners.

Lichens can grow where plants cannot

A lichen is not a single organism, but rather a meshwork of two radically different organisms: a fungus and a photosynthetic microorganism. Together the organisms constituting a lichen can survive some of the harshest environments on Earth. The biota of Antarctica, for example, features more than a hundred times as many species of lichens as of plants.

In spite of this hardiness, lichens are very sensitive to air pollution because they are unable to excrete toxic substances that they absorb. Hence they are not common in industrialized cities. Because of their sensitivity, lichens are good biological indicators of air pollution.

The fungal components of most lichens belong to the clade called ascomycetes, or sac fungi, which also includes various cup fungi, yeasts, and molds such as the *Fusarium* mentioned at the beginning of the chapter. The photosynthetic component of a lichen is most often a unicellular green alga, but may be a cyanobacterium, or may include both. Relatively little experimental work has focused on lichens, perhaps because they grow so slowly—typically less than 1 centimeter per year.

(A)

Caloplaca flavescens

(B)

Xanthoria sp.

(C)
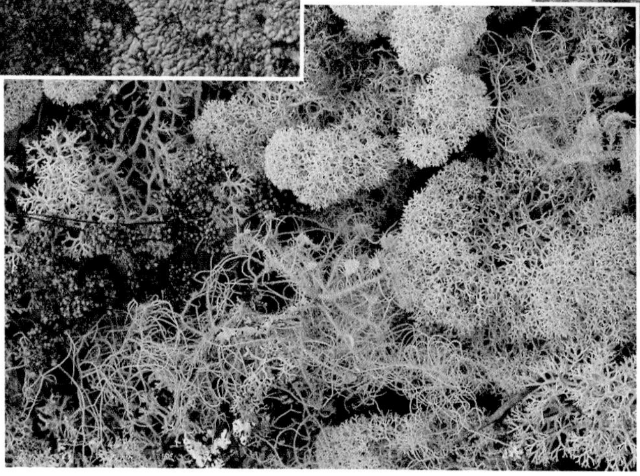

30.8 Lichen Body Forms
Lichens fall into three principal classes based on their body form. (A) These crustose lichens are growing on otherwise bare rock. (B) Foliose lichens have a leafy appearance. (C) A miniature jungle of "shrubby" fruticose lichens.

There are more than 15,000 "species" of lichens, each of which is assigned the name of its fungal component. These fungal components may constitute as many as 20 percent of all fungal species. Some of these fungi are able to grow independently without a photosynthetic partner, but others have not been observed in nature other than in a lichen association. Lichens are found in all sorts of exposed habitats: on tree bark, on open soil, and on bare rock. Reindeer moss (actually not a moss at all, but the lichen *Cladonia subtenuis*) covers vast areas in Arctic, sub-Arctic, and boreal regions, where it is an important part of the diets of reindeer and other large mammals. Lichens come in various forms and colors. *Crustose* (crustlike) lichens look like colored powder dusted over their substratum (**Figure 30.8A**); *foliose* (leafy) and *fruticose* (shrubby) lichens may have complex forms (**Figure 30.8B,C**).

The most widely held interpretation of the lichen relationship is that it is a mutually beneficial symbiosis. The hyphae of the fungal mycelium are tightly pressed against the algae or cyanobacteria and sometimes even invade them. The bacterial or algal cells not only survive these indignities, but continue their growth and photosynthesis. In fact, the algal cells in a lichen "leak" photosynthetic products at a greater rate than do similar cells growing on their own. On the other hand, photosynthetic cells from lichens grow more rapidly on their own than when associated with a fungus. On this basis, we could consider lichen fungi to be parasitic on their photosynthetic partners. However, in many places where lichens grow, the photosynthetic cells would not grow at all on their own.

Lichens can reproduce simply by fragmentation of the vegetative body, which is called the *thallus*, or by means of specialized structures called *soredia* (singular *soredium*). Soredia consist of one or a few photosynthetic cells surrounded by fungal hyphae. The soredia become detached from the lichen, are dispersed by air currents, and upon arriving at a favorable location, develop into a new lichen. Alternatively, the fungal partner may go through its sexual cycle, producing haploid spores. When these spores are discharged, however, they disperse alone, unaccompanied by the photosynthetic partner, and thus may not be capable of reestablishing the lichen association, or even of surviving on their own.

Visible in a cross section of a typical foliose lichen are a tight upper region of fungal hyphae, a layer of cyanobacteria or algae, a looser hyphal layer, and finally hyphal rhizoids that attach the whole structure to its substratum (**Figure 30.9**). The meshwork of fungal hyphae takes up some nutrients needed by the photosynthetic cells and provides a suitably moist environment for them by holding water tenaciously. The fungi derive fixed carbon from the photosynthetic products of the algal or cyanobacterial cells.

Lichens are often the first colonists on new areas of bare rock. They get most of the nutrients they need from the air and rainwater, augmented by minerals absorbed from dust. A lichen begins to grow shortly after a rain, as it begins to dry. As it grows, the lichen acidifies its environment slightly, and this acidity contributes to the slow breakdown of rocks, an early step in soil formation. After further drying, the lichen's photosynthesis ceases. The water content of the lichen may drop to less than 10 percent of its dry weight, at which point it becomes highly insensitive to extremes of temperature.

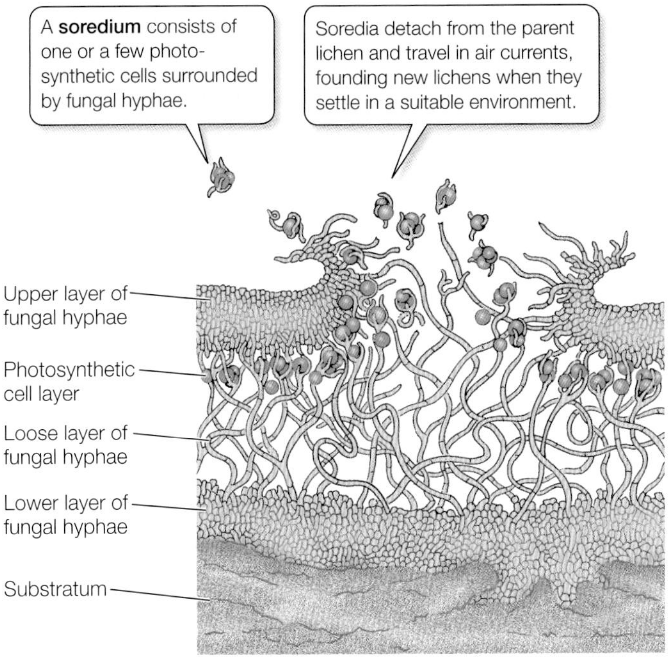

30.9 Lichen Anatomy Cross section showing the layers of a foliose lichen and the release of soredia.

Mycorrhizae are essential to most plants

Almost all vascular plants require a symbiotic association with fungi. Unassisted, the root hairs of such plants do not take up enough water or minerals to sustain growth. However, their roots usually do become infected with fungi, forming an association called a mycorrhiza. Mycorrhizae are of two types, based on whether or not the fungal hyphae penetrate the plant cells.

In *ectomycorrhizae*, the fungus wraps around the root, and its mass is often as great as that of the root itself (**Figure 30.10A**). The fungal hyphae wrap around individual cells in the root, but do not penetrate them. An extensive web of hyphae penetrates the soil in the area around the root, so that up to 25 percent of the soil volume near the root may be fungal hyphae. The hyphae attached to the root increase the surface area for the absorption of water and minerals, and the mass of the mycorrhiza in the soil, like a sponge, holds water efficiently in the neighborhood of the root. Infected roots characteristically branch extensively and become swollen and club-shaped, and they lack root hairs.

In *arbuscular mycorrhizae* the fungal hyphae enter the root and penetrate the cell wall of the root cells, forming arbuscular (tree-like) structures inside the cell wall, but outside the plasma membrane. These structures, like the haustoria of parasitic fungi, become the primary site of exchange between plant and fungus (**Figure 30.10B**). As with the ectomycorrhizae, the fungus forms a vast web of hyphae leading from the root surface into the surrounding soil.

The mycorrhizal association is important to both partners. The fungus obtains needed organic compounds, such as sugars and amino acids, from the plant. In return, the fungus, because of its very high surface area-to-volume ratio and its ability to penetrate the fine structure of the soil, greatly increases the plant's ability to absorb water and minerals (especially phosphorus). The fungus may also provide the plant with certain growth hormones and may protect it against attack by disease-causing microorganisms. Plants that have active arbuscular mycorrhizae typically are a deeper green and may resist drought and temperature extremes better than plants of the same species that have little mycorrhizal development. Attempts to introduce some plant species to new areas have failed until a bit of soil from the native area (presumably containing the fungus necessary to establish mycorrhizae) was provided. Trees without ectomycorrhizae will not grow well in the absence of abundant nutrients and water, so the health of our forests depends on the presence of ectomycorrhizal fungi.

The partnership between plant and fungus results in a plant that is better adapted for life on land. It has been suggested that the evolution of mycorrhizae was the single most important step in the colonization of the terrestrial environment by living things. Fossils of mycorrhizal structures 460 million years old have been found. Some liverworts, which are among the most ancient terrestrial plants (see Section 28.4), form mycorrhizae.

30.10 Mycorrhizal Associations (A) Ectomycorrhizal fungi wrap themselves around a plant root, increasing the area available for absorption of water and minerals. (B) Hyphae of arbuscular mycorrhizal fungi infect the root internally and penetrate the root cells, branching within the cells and forming tree-like (arbuscular) structures that provide the plant with nutrients. Hyphae fill much of the cell outside the nucleus.

(A)

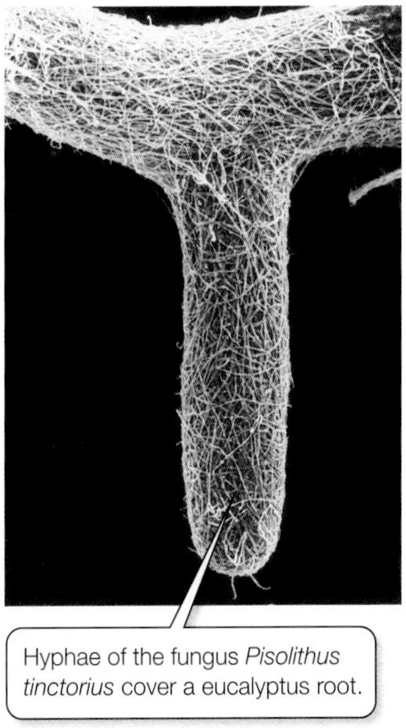

Hyphae of the fungus *Pisolithus tinctorius* cover a eucalyptus root.

(B)

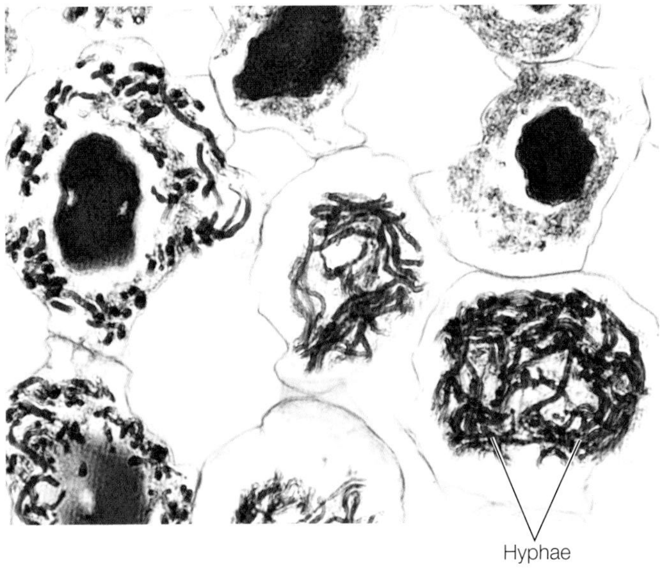

Certain plants that live in nitrogen-poor habitats, such as cranberry bushes and orchids, invariably have mycorrhizae. Orchid seeds will not germinate in nature unless they are already infected by the fungus that will form their mycorrhizae. Plants that lack chlorophyll always have mycorrhizae, which they often share with the roots of green, photosynthetic plants. In effect, these plants without chlorophyll are feeding on nearby green plants, using the fungus as a bridge.

Biologists had long suspected that roots secrete a chemical signal that enables fungi to find and invade them to form arbuscular mycorrhizae. This was proved to be correct in 2005 when researchers succeeded in isolating the signaling compound. Might the compound also be used by parasitic plants to attack their host plants? Indeed it is. *Striga*, discussed at the beginning of this chapter, turns out to be one of the parasitic plants that use exactly this signal. Thus, in attracting its helper fungus, a plant may also attract a dangerous parasite.

Endophytic fungi protect some plants from pathogens, herbivores, and stress

In a tropical jungle, 10,000 or more fungal spores land on a single leaf each day. Some are plant pathogens, some do not attack the plant at all, and some invade the plant in a beneficial way. Fungi that live within aboveground parts of plants are called **endophytic fungi**. Recent research has shown that endophytic fungi are abundant in plants in all terrestrial environments.

Grasses with endophytic fungi are more resistant to pathogens and to insect and mammalian herbivores than are grasses lacking endophytes. The fungi produce alkaloids (nitrogen-containing compounds) that are toxic to animals. The alkaloids do not harm the host plant; in fact, some plants produce alkaloids (such as nicotine) themselves. The fungal alkaloids also increase the ability of host plants to resist stress of various types, including drought (water shortage) and salty soils. Such resistance is useful in agriculture.

The role, if any, of endophytic fungi in most trees is unclear. They may simply occupy space within leaves, without conferring any benefit but also without doing harm.

Some fungi are food for the ants that farm them

In another kind of association, some leaf-cutting ants "farm" fungi, feeding the fungi and later eating the specialized fruiting bodies (called *gongylidia*) the fungi produce. The ants collect leaves and flower petals, chew them into small bits, and "plant" bits of fungal mycelium on their surfaces. The fungi in this "garden" secrete enzymes that digest the plant material, breaking it down into products that the fungi absorb for their own nutrition. The fungi then produce the gongylidia that are harvested and eaten by the ants.

An established garden is not a mixture of fungi; rather, it contains a single clone of fungus. When ants bring other fungi to the garden, the newcomers are killed by substances contained in the ants' feces. Do these substances derive from the ants, or from the fungal garden itself, which thus denies the ants the possible benefits of farming a genetically more diverse garden? A simple experiment performed in 2005 showed that it is indeed the fungus that produces the substances that keep potential competitors from joining the garden and enjoying the food supplied by the ants (**Figure 30.11**).

EXPERIMENT

HYPOTHESIS: Fungi, not ants, produce the substance that keeps "foreign" fungi from growing in the garden.

METHOD

1. Collect samples of distinct fungus gardens and ants from the field and establish colonies in the laboratory.
2. Some ants from each of two colonies (1 and 2) are forced to feed for 10 days on fungi obtained from the other colony. Other ants continue to feed on their "home" gardens.
3. At the beginning and end of the 10-day period, place fecal droplets of ants on fungi from their own garden or from the "alien"garden.
4. Test for incompatibility reactions (avoidance of droplet by fungus, toxicity to fungus).

RESULTS

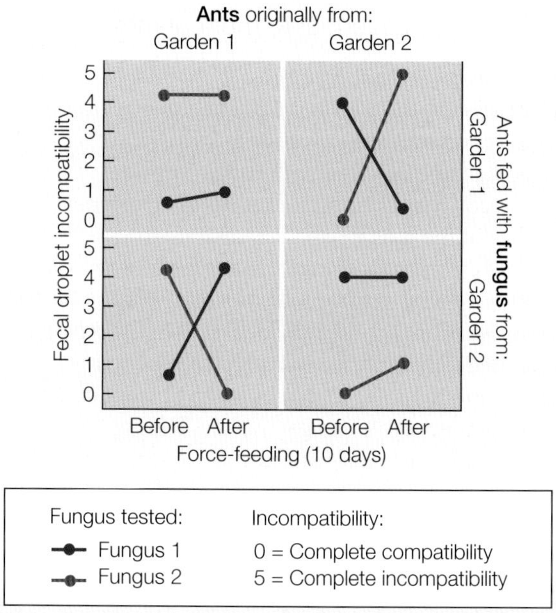

CONCLUSION: Foreign fungi are deterred by substances produced by the fungus an ant eats, not by the ants themselves.

30.11 Keeping Fungal Interlopers Away Only a single clone of a single fungal species is maintained in the fungal garden of leaf-cutting ants. What keeps these gardens from becoming contaminated by other fungi? An experiment performed by Michael Poulsen and Jacobus Boomsma explored this question. FURTHER RESEARCH: Some termites raise fungal gardens, but the fungi feed on the insects' fecal matter rather than on plant material. How would you determine whether a similar system to avoid contamination by incompatible fungal species is maintained in the termite-fungus symbiosis?

30.2 RECAP

Fungi form many beneficial associations with other organisms. Mycorrhizal fungi are essential for the survival of most plant species.

- What is the role of fungi in the planetary carbon cycle? See p. 655
- Can you explain the nature and benefits of the lichen association? See pp. 655–656
- Do you understand why plants grow better in the presence of mycorrhizal fungi? See p. 657

Most lichen fungi belong to one particular fungal clade, and arbuscular mycorrhizal fungi are members of another. One of the most important criteria for assigning fungi to taxonomic groups, before molecular techniques made things simpler and more certain, was the nature of their life cycles.

30.3 How Do Fungal Life Cycles Differ from One Another?

Different fungal groups have different life cycles. One group features alternation of generations, a type of life cycle found in all plants and some protists. Other groups have an unusual life cycle that is unique to the fungi, featuring a stage called a *dikaryon*. We will contrast these cycles after surveying the reproductive modes of the various fungal groups. Let's begin by briefly reviewing asexual and sexual reproduction in general terms.

Fungi reproduce both sexually and asexually

Both asexual and sexual reproduction are common among the fungi (**Figure 30.12**). Asexual reproduction takes several forms:

- The production of (usually) haploid spores within structures called *sporangia*
- The production of naked spores (not enclosed in sporangia) at the tips of hyphae; such spores are called *conidia* (from the Greek *konis*, "dust")
- Cell division by unicellular fungi—either a relatively equal division (called *fission*) or an asymmetrical division in which a small daughter cell is produced (called *budding*)
- Simple breakage of the mycelium

Asexual reproduction in fungi can be spectacular in terms of quantity. A 2.5-centimeter colony of *Penicillium*, the mold that produces the antibiotic penicillin, can produce as many as 400 million conidia. The air we breathe contains as many as 10,000 fungal spores per cubic meter.

Sexual reproduction in many fungi features an interesting twist. There is often no morphological distinction between female and male structures, or between female and male individuals. Rather, there is a genetically determined distinction between two *or more* **mating types**. Individuals of the same mating type cannot mate with one another, but they can mate with individuals of another mating type within the same species. This distinction prevents self-

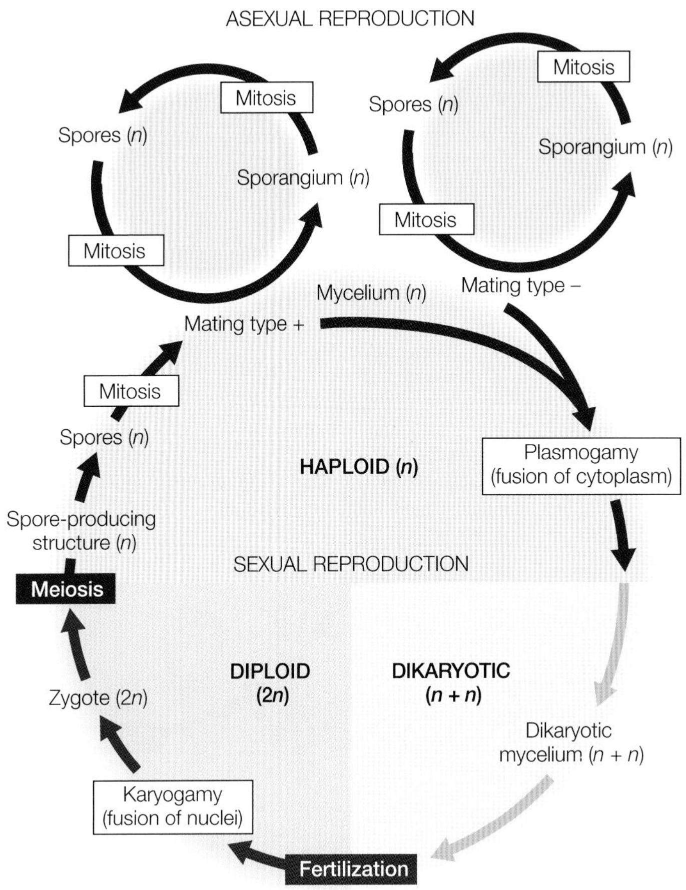

30.12 Asexual and Sexual Reproduction in a Fungal Life Cycle Environmental conditions may determine which mode of reproduction takes place at a given time.

fertilization. Individuals of different mating types differ genetically from one another, but are often visually and behaviorally indistinguishable. Many protists also have mating type systems.

Fungi reproduce sexually when hyphae (or, in one group, motile cells) of different mating types meet and fuse (**Figure 30.13**). In many fungi, the zygote nuclei formed by sexual reproduction are the only diploid nuclei in the life cycle. These nuclei undergo meiosis, producing haploid nuclei that become incorporated into spores. Haploid fungal spores, whether produced sexually in this manner or asexually, germinate, and their nuclei divide mitotically to produce hyphae. This type of life cycle, called a *haplontic* life cycle, is also characteristic of many microbial eukaryotes.

The alternation between multicellular haploid (n) and multicellular diploid ($2n$) generations that evolved in plants and certain protist groups (see Section 28.2) is found in the chytrids and certain yeasts (**Figure 30.13A**). Alternation of generations does not appear in the other groups. As one might expect, these basal fungi, which are aquatic, possess flagellated gametes; they also have flagellated spores. Flagella have been lost in the terrestrial fungi.

What are the consequences of alternation of generations in the chytrids? It is possible that the multicellular haploid organisms serve as a "filter" for harmful mutations. A haploid individual with such a mutation would die, and the mutant allele would not be passed to

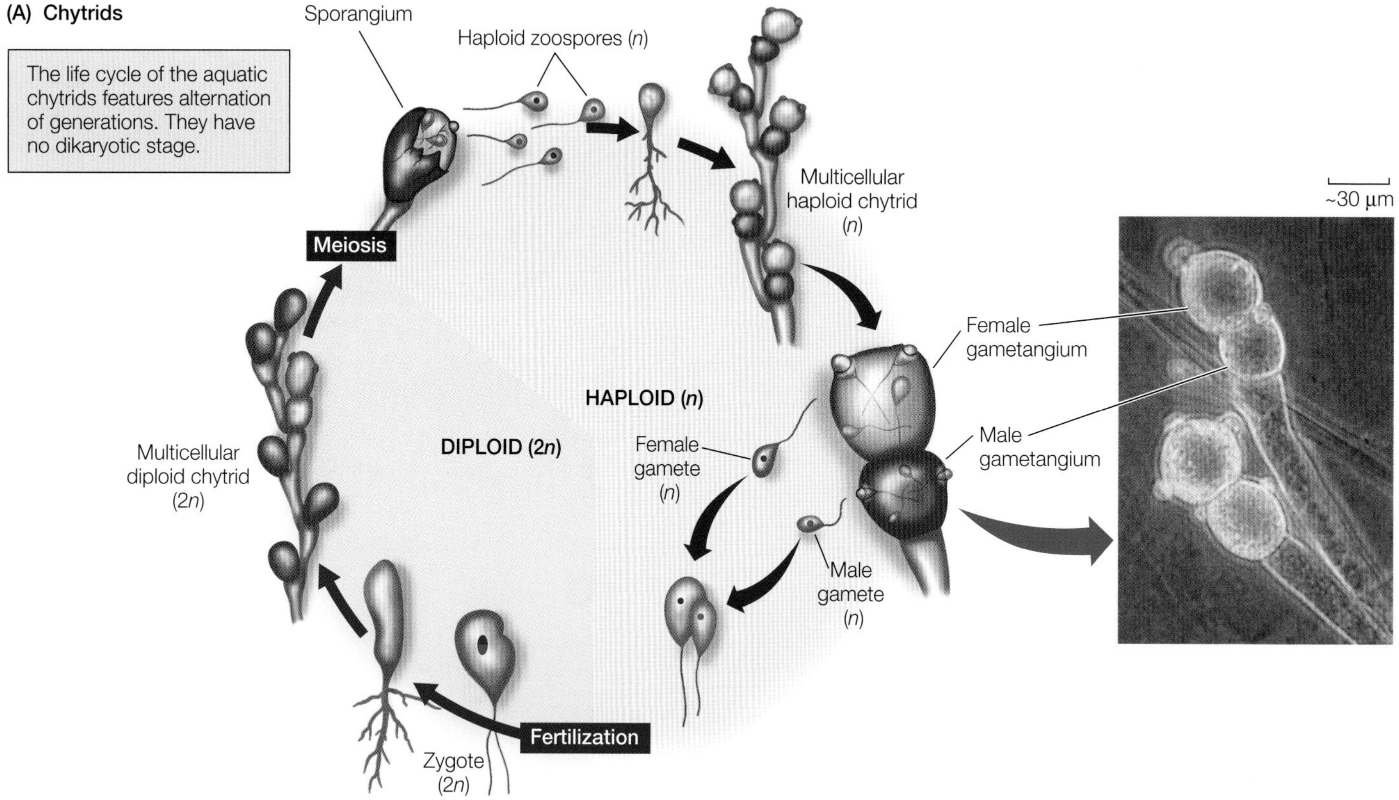

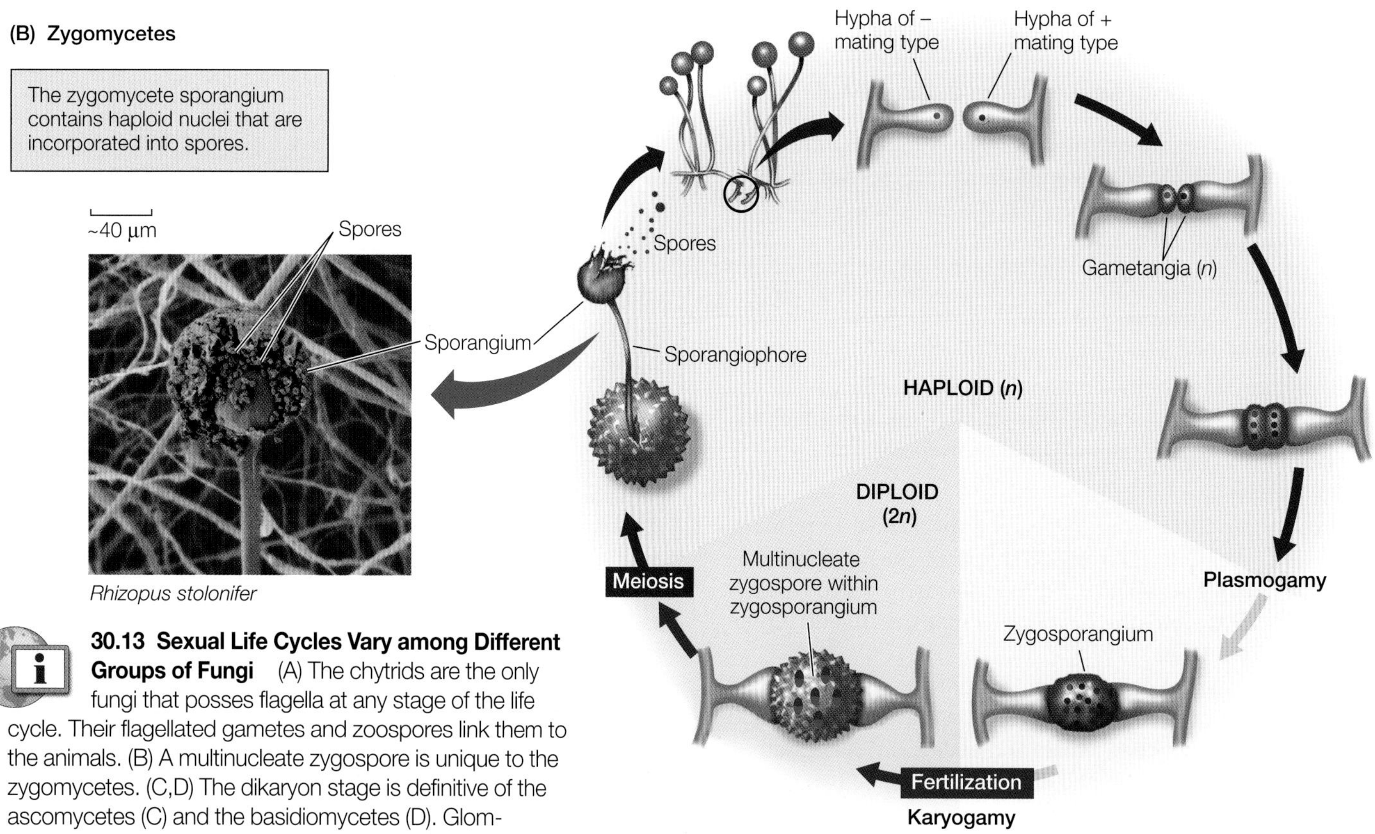

30.13 Sexual Life Cycles Vary among Different Groups of Fungi (A) The chytrids are the only fungi that posses flagella at any stage of the life cycle. Their flagellated gametes and zoospores link them to the animals. (B) A multinucleate zygospore is unique to the zygomycetes. (C,D) The dikaryon stage is definitive of the ascomycetes (C) and the basidiomycetes (D). Glomeromycetes (the mycorrhizal fungi) reproduce asexually only and are not depicted here.

(C) Ascomycetes

The products of meiosis in ascomycetes are borne in a microscopic sac called an ascus. The fleshy fruiting bodies consist of both dikaryotic and haploid hyphae.

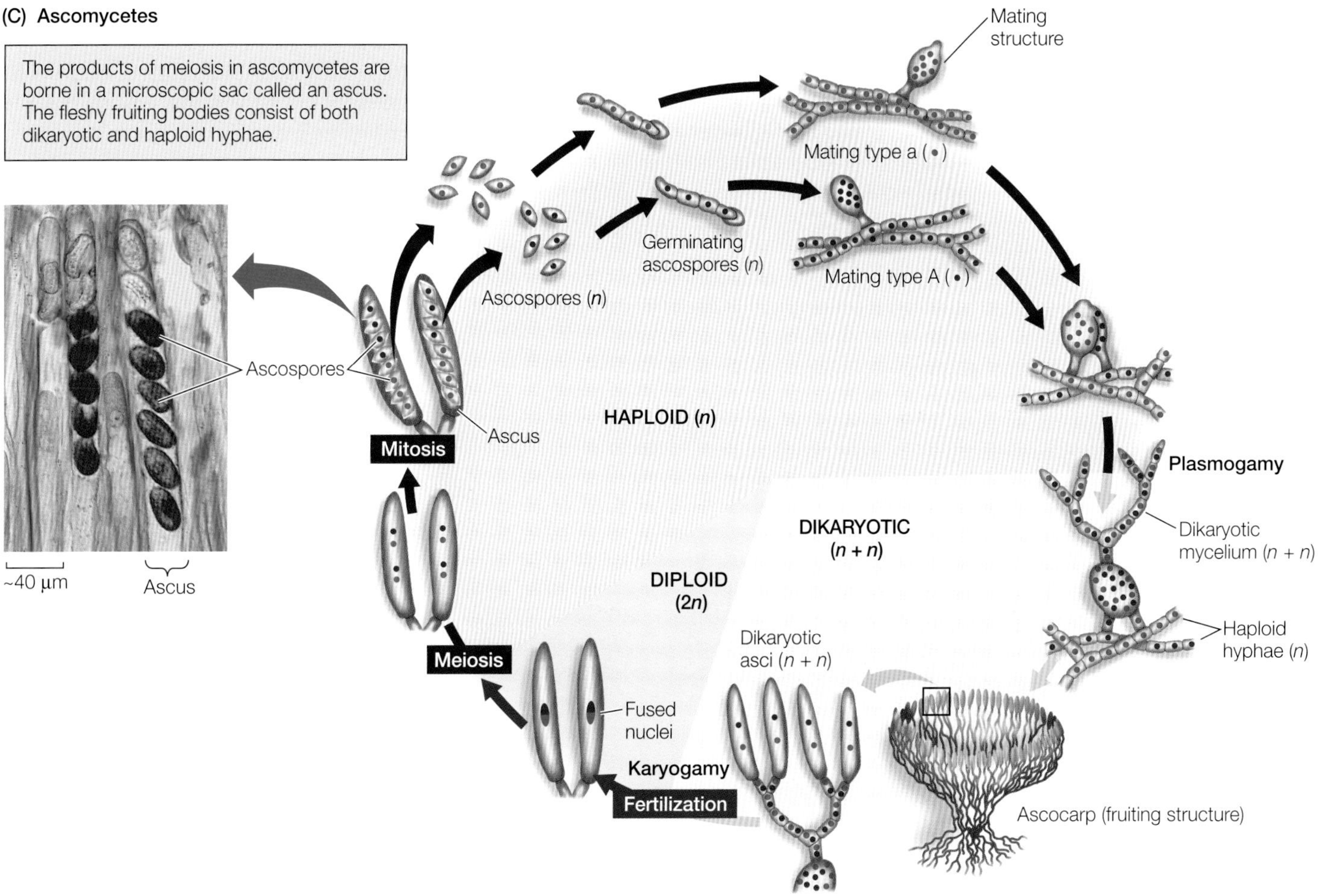

(D) Basidiomycetes

In basidiomycetes the products of meiosis are borne exposed on pedestals called basidia. Fruiting bodies consist solely of dikaryotic hyphae, and the dikaryotic phase can last a long time.

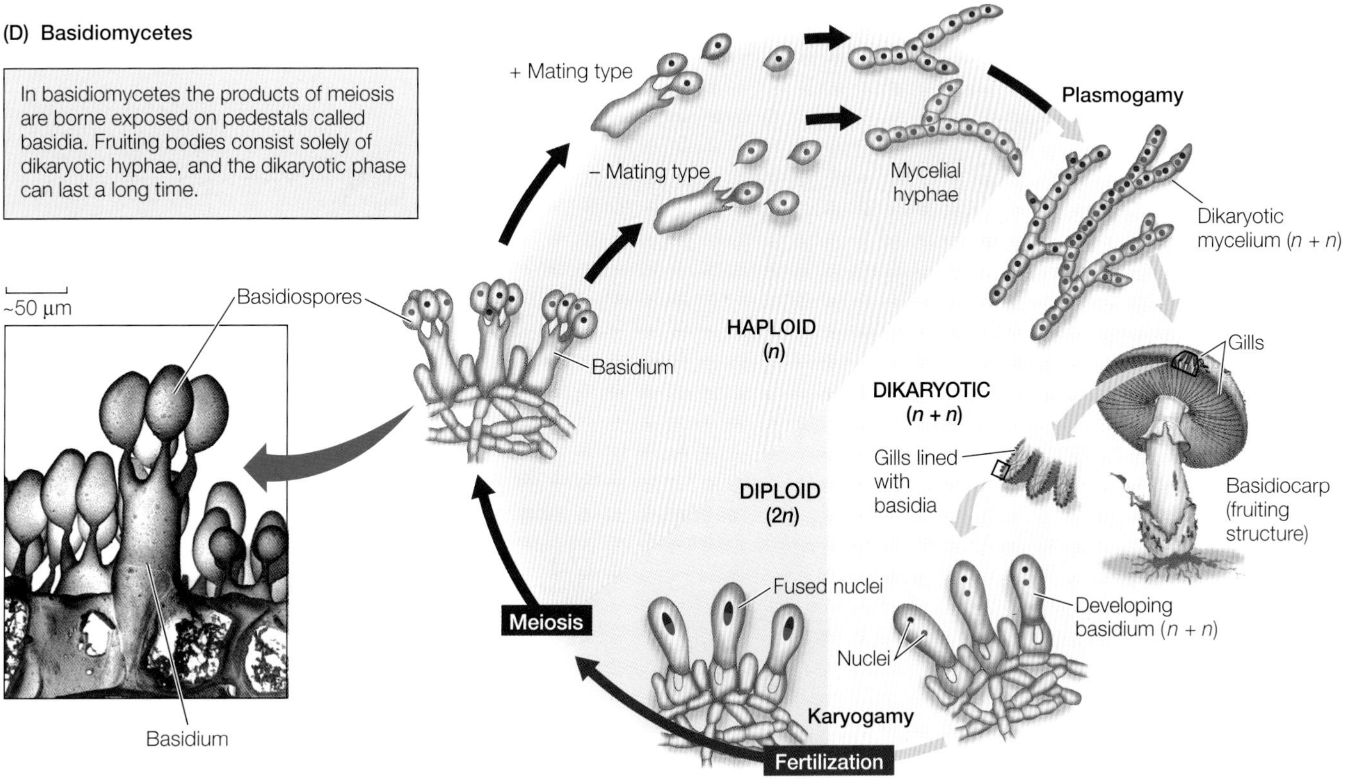

progeny. In these basal fungi, the multicellular diploid stage includes a resistant structure capable of withstanding freezing or drying.

Although the terrestrial fungi grow in moist places, their gamete nuclei are not motile and are not released into the environment (**Figure 30.13B–D**). Instead, the cytoplasms of two individuals of different mating types fuse (**plasmogamy**) before their nuclei fuse (**karyogamy**). Therefore, liquid water is not required for fertilization.

The dikaryotic condition is unique to the fungi

Certain hyphae of some terrestrial ascomycetes and basidiomycetes have a nuclear configuration other than the familiar haploid or diploid states (**Figures 30.13C,D**). In these fungi, sexual reproduction begins in an unusual way: karyogamy occurs long after plasmogamy, so that *two genetically different haploid nuclei coexist and divide within the same hypha*. Such a hypha is called a **dikaryon** ("two nuclei"). Because the two nuclei differ genetically, such a hypha is also called a *heterokaryon* ("different nuclei").

Eventually, specialized fruiting structures form, within which the pairs of genetically dissimilar nuclei—one from each parent—fuse, giving rise to zygotes long after the original "mating." The diploid zygote nucleus undergoes meiosis, producing four haploid nuclei. The mitotic descendants of those nuclei become spores, which give rise to the next generation of hyphae.

This type of life cycle displays several unusual features. First, there are no gamete *cells*, only gamete *nuclei*. Second, there is never any true diploid tissue, although for a long period the genes of both parents are present in the dikaryon and can be expressed. In effect, the hypha is neither diploid ($2n$) nor haploid (n); rather, it is *dikaryotic* ($n + n$). A harmful recessive mutation in one nucleus may be compensated for by a normal allele on the same chromosome in the other nucleus. Dikaryosis is perhaps the most distinctive of the genetic peculiarities of the fungi.

The dikaryotic condition often lasts for months or even years. Basidiomycetes have an elegant mechanism that ensures that the dikaryotic condition is maintained as new cells are formed. One consequence of a sustained dikaryotic condition is an increased opportunity for multiple fusions of hyphae of different mating types before fruiting bodies are formed. This could allow for more genetic recombination than with just two mating types.

The life cycles of some parasitic fungi require two hosts

Some parasitic fungi are very specific about what host organism they use as a source of nutrition, and some even use different hosts for different stages of their life cycle. One of the most striking examples of a fungal life cycle that involves two different hosts is that of *Puccinia graminis*, the agent of black stem rust of wheat mentioned earlier in the chapter (see p. 653).

Let's examine a year in the life cycle of *P. graminis* (**Figure 30.14**; the figure is keyed to the text by parenthetical numbers). During the summer, dikaryotic hyphae of *P. graminis* proliferate in the stem and leaf tissues of wheat plants. These dikaryotic hyphae produce great quantities of dikaryotic summer spores called *uredospores* **(1)**. The one-celled, orange uredospores are scattered by the wind and infect other wheat plants, on which the summer hyphae then proliferate.

Dark brown winter spores—*teliospores*—begin to appear on special hyphae in late summer. Each teliospore consists initially of two dikaryotic cells. The two haploid nuclei in each cell fuse to form a single diploid ($2n$) nucleus. These are the only diploid cells in the entire life cycle **(2)**. Both cells have thick walls and usually survive

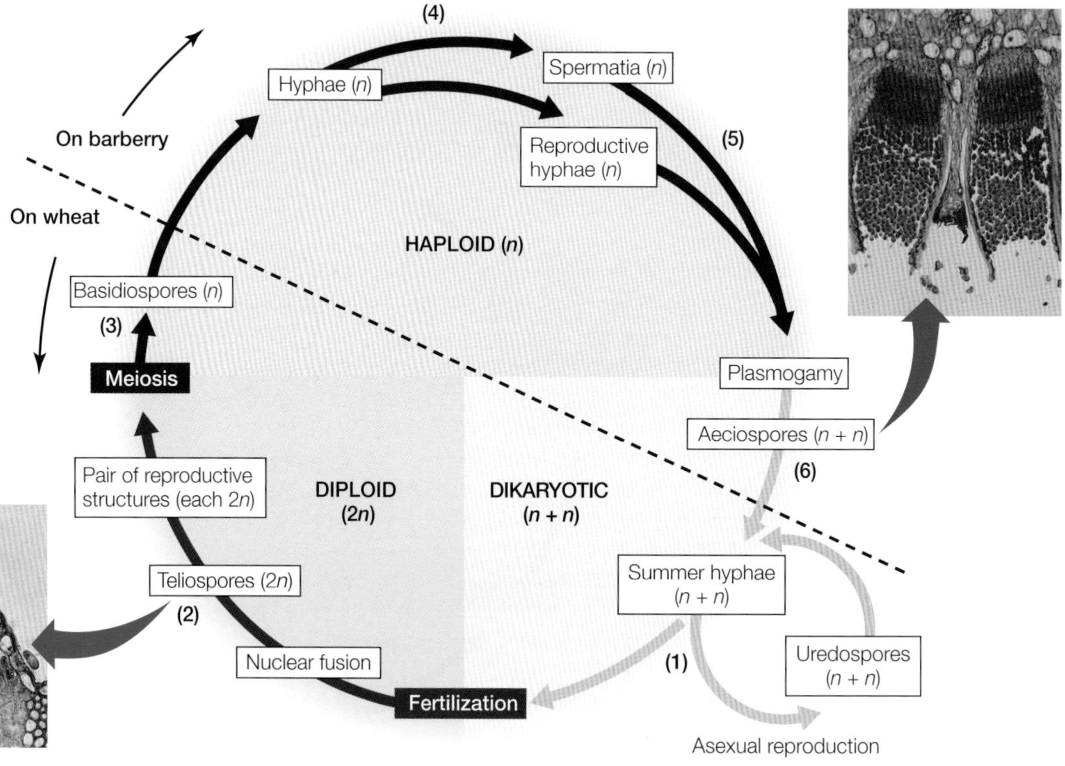

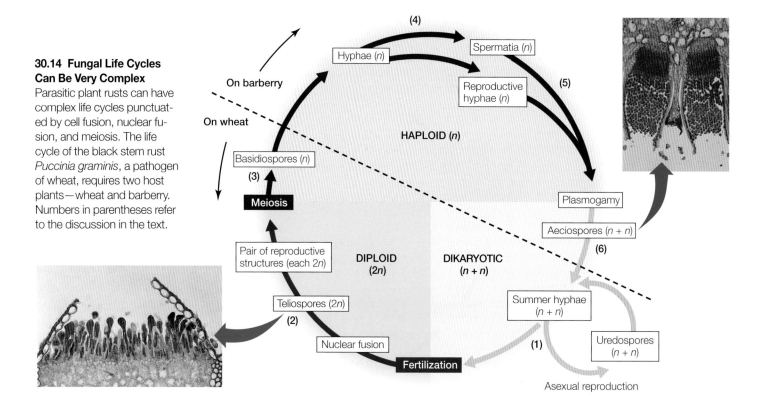

30.14 Fungal Life Cycles Can Be Very Complex Parasitic plant rusts can have complex life cycles punctuated by cell fusion, nuclear fusion, and meiosis. The life cycle of the black stem rust *Puccinia graminis*, a pathogen of wheat, requires two host plants—wheat and barberry. Numbers in parentheses refer to the discussion in the text.

freezing. The teliospores remain dormant until spring, when they germinate. Each of the two cells develops into a reproductive structure, within which the nucleus divides meiotically to produce four haploid *basidiospores* **(3)**.

Basidiospores are one of two mating types: plus (+) or minus (–). They are carried by the wind, and if they land on a leaf of the common barberry plant, they germinate and produce haploid hyphae (+ or –, depending on the mating type of the germinating basidiospore). These hyphae invade the barberry leaf, forming flask-shaped structures on the upper surface of the leaf, within which some of the hyphae pinch off tiny, colorless, haploid *spermatia* from their tips **(4)**.

The hyphae also form another type of structure, called an *aecial primordium*, near the lower surface of the leaf. Thus the leaf contains "flasks" on the upper surface and aecial primordia near the lower one, and these structures are connected by hyphae. Insects attracted to a sweetish liquid produced in the flasks carry spermatia from one flask to another, initiating the next step in the cycle.

A spermatium of one mating type fuses with a receptive hypha of the other type within a flask. The nucleus of the spermatium repeatedly divides mitotically, and the products move through the hyphae into immature aecial primordia, where they produce dikaryotic cells with two haploid nuclei, one of each mating type. These cells develop into *aeciospores*, each containing two unlike nuclei **(5)**. The aeciospores are scattered by the wind, and some land on wheat plants, where they continue the life cycle **(6)**. When the dikaryotic aeciospores germinate on wheat, they produce the summer hyphae with two nuclei in each cell **(1)**.

The different types of spores produced during the life cycle of *P. graminis* play very different roles. Consider first the stages on the wheat host. Windborne uredospores are the primary agents for spreading the rust from wheat plant to wheat plant and to other fields. Resistant teliospores allow the rust to survive the harsh winter, but contribute little to the spreading of the rust. Basidiospores spread the rust from wheat to barberry plants. On the barberry host, spermatia and receptive hyphae initiate the sexual cycle of the rust. Finally, aeciospores spread the rust from barberry to wheat plants. The adaptive value of this division of the life cycle between two host plants is still a matter of speculation. Life cycles including two hosts have evolved many times in many groups of organisms—recall, for example, the parasitic protist that causes malaria (see Figure 27.3).

"Imperfect fungi" lack a sexual stage

As we have just seen, mechanisms of sexual reproduction help distinguish members of the groups of fungi from one another. But many fungi, including both saprobes and parasites, appear to lack sexual stages entirely; presumably these stages have been lost during the evolution of these species or have not yet been observed. Classifying these fungi used to be difficult, but biologists can now assign most of them to one of the five groups on the basis of their DNA sequences.

Fungi that have not yet been placed in any of the existing groups are pooled together in a polyphyletic group called **deuteromycetes**, informally known as "imperfect fungi." Thus the deuteromycete group is a holding area for species whose status is yet to be resolved. At present, about 25,000 species are classified as imperfect fungi.

If sexual structures are found on a fungus classified as a deuteromycete, that fungus is reassigned to the appropriate group. That happened, for example, to a fungus that produces plant growth hormones called gibberellins (see Section 37.1). Originally classified as the deuteromycete *Fusarium moniliforme*, this fungus was later found to produce a reproductive structure characteristic of an ascomycete, whereupon it was renamed and transferred to that group.

30.3 RECAP

Fungi have a variety of life cycles, including one requiring multiple hosts.

- Do you understand the concept of mating types? See p. 659
- Can you explain the phenomenon of dikaryosis in terms of plasmogamy and karyogamy? See p. 662

In examining the most important properties of the fungi as a clade, we have often drawn examples from specific groups. Now let's briefly consider the diversity of the fungi.

30.4 How Do We Tell the Fungal Groups Apart?

In this section we'll examine representative species from each of the five major groups of fungi—chytrids, zygomycetes, glomeromycetes, ascomycetes, and basidiomycetes (**Table 30.1**). The chytrids and the zygomycetes are not monophyletic, but the last three groups are all clades; furthermore, together they form a clade called the **crown fungi**.

Chytrids are the only fungi with flagella

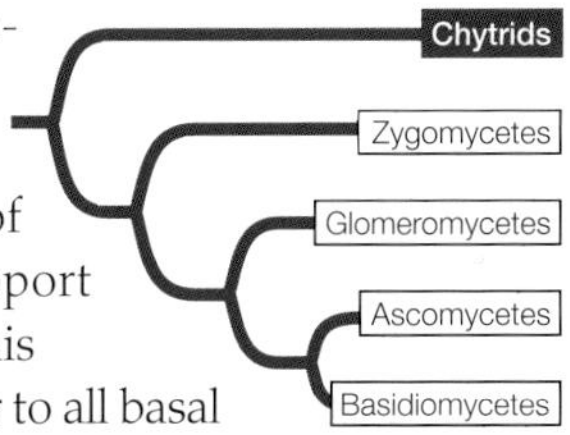

The **chytrids** are aquatic microorganisms once classified with the protists. However, morphological evidence (cell walls that consist primarily of chitin) and molecular evidence support their classification as basal fungi. In this book we use the term "chytrid" to refer to all basal fungi, but some mycologists reserve the term to apply to a particular clade within that group. There are fewer than 1,000 described species of chytrids.

Like the animals, the chytrids possess flagellated gametes. The retention of this character reflects the aquatic environment in which fungi first evolved. Chytrids are the only fungi that have flagella at any life cycle stage.

Chytrids are either parasitic (on organisms such as algae, mosquito larvae, and nematodes) or saprobic. Chytrids in the compound stomachs of foregut-fermenting animals such as cattle may be an exception, living in a mutualistic association with their hosts. Most chytrids live in freshwater habitats or in moist soil, but some are marine. Some chytrids are unicellular, others have rhizoids and still others have coenocytic hyphae (**Figure 30.15**). Chytrids reproduce both sexually and asexually, but they do not have a dikaryon stage.

TABLE 30.1

A Classification of the Fungi

GROUP	COMMON NAME	FEATURES	EXAMPLES
Chytridiomycetes	Chytrids	Aquatic; zoospores have flagella	*Allomyces*
Zygomycetes	Conjugating fungi	Zygosporangium; no regularly occurring septa; usually no fleshy fruiting body	*Rhizopus*
Glomeromycetes	Mycorrhizal fungi	Form arbuscular mycorrhizae on plant roots	*Glomus*
Ascomycetes	Sac fungi	Ascus; perforated septa	*Neurospora*
Basidiomycetes	Club fungi	Basidium; perforated septa	*Armillariella*

Allomyces, a well-studied genus of chytrids, displays alternation of generations. A haploid *zoospore* (a spore with flagella) comes to rest on dead plant or animal material in water and germinates to form a small, multicellular haploid mycelium. That mycelium produces female and male *gametangia* (gamete cases; see Figure 30.13A). *Mitosis* in the gametangia results in the formation of haploid gametes, each with a single nucleus.

Both female and male gametes have flagella. The motile female gamete produces a *pheromone*, a chemical signal that attracts the swimming male gamete. The two gametes fuse, and then their nuclei fuse to form a diploid zygote. Mitosis and cytokinesis in the zygote gives rise to a small, multicellular, diploid organism, which produces numerous diploid flagellated zoospores. These diploid zoospores disperse and germinate to form more diploid organisms. Eventually, the diploid organism produces thick-walled resting sporangia that can survive unfavorable conditions such as dry weather or freezing. Nuclei in the resting sporangia eventually undergo meiosis, giving rise to haploid zoospores that are released into the water and begin the cycle anew.

Zygomycetes reproduce sexually by fusion of two gametangia

Most **zygomycetes** ("conjugating fungi") are terrestrial, living on soil as saprobes or as parasites on insects, spiders, and other animals. They produce no cells with flagella, and only one diploid cell—the zygote—appears in the entire life cycle. Their hyphae are coenocytic. The mycelium of a zygomycete spreads over its substratum, growing forward by means of vegetative hyphae. Most zygomycetes do not form a fleshy fruiting structure; rather, the hyphae spread in an apparently random fashion, with occasional stalked **sporangiophores** reaching up into the air (**Figure 30.16**). These reproductive structures may bear one or many sporangia.

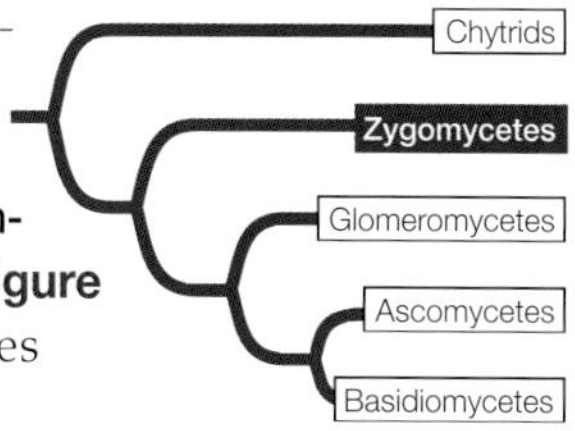

More than 700 species of zygomycetes have been described. A zygomycete that you may have seen is *Rhizopus stolonifer*, the black bread mold. *Rhizopus* reproduces asexually by producing many stalked sporangiophores, each bearing a single

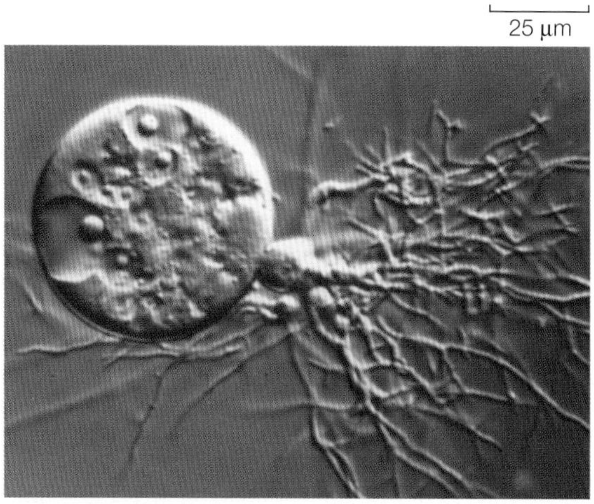

Chytriomyces hyalinus

30.15 A Chytrid Branched rhizoids emerge from the fruiting body of a mature chytrid.

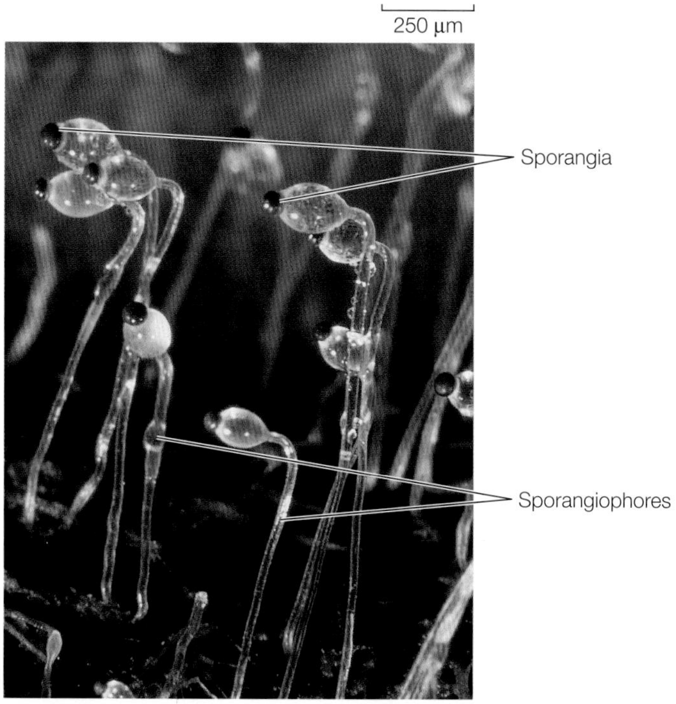

Pilobolus sp.

30.16 Zygomycetes Produce Sporangiophores These transparent structures are sporangiophores (spore-bearing hyphae) growing on decomposing animal dung. Sporangiophores grow toward the light and end in tiny sporangia, which these filamentous structures can eject as far as 2 meters. Animals ingest those sporangia that land on grass and then disseminate the spores in their feces.

sporangium containing hundreds of minute spores (see Figure 30.13B). As in other filamentous fungi, the spore-forming structure is separated from the rest of the hypha by a wall.

The zygomycete *Rhizopus microsporus* has been known as a pathogen of rice. However, it isn't the fungus that kills the rice, but a bacterial symbiont living inside the fungal cells. The bacterium produces rhizoxin, a rice-killing toxin that also happens to be a potent antitumor agent

Zygomycetes reproduce sexually when adjacent hyphae of two different mating types release pheromones, which cause them to grow toward each other. These hyphae produce gametangia, which fuse to form a **zygosporangium** (see Figure 30.13B). Sometime later, the gamete nuclei now contained within the zygosporangium fuse to form a single multinucleate *zygospore*. The zygosporangium develops a thick, multilayered wall that protects the zygospore. The highly resistant zygospore may remain dormant for months before its nuclei undergo meiosis and a sporangiophore sprouts. The sporangium contains the products of meiosis: haploid nuclei that are incorporated into spores. These spores disperse and germinate to form a new generation of haploid hyphae.

Glomeromycetes form arbuscular mycorrhizae

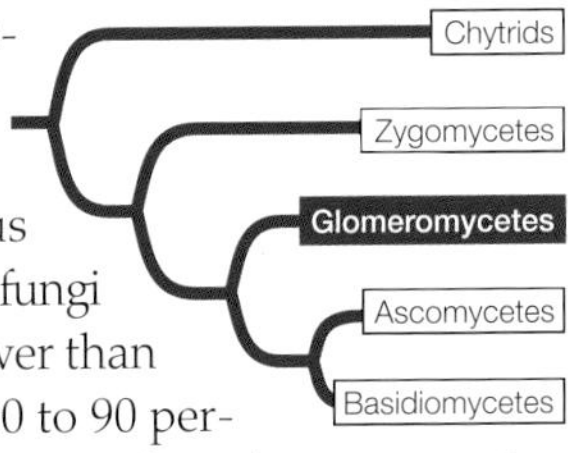

The **glomeromycetes** are strictly terrestrial. They associate with plant roots to form arbuscular mycorrhizae, so they are crucial to the plant world and thus to us (see Figure 30.10B). About half the fungi found in soils are glomeromycetes. Fewer than 200 species have been described, but 80 to 90 percent of all plants have associations with them. These fungi were originally classified among the zygomycetes, but molecular systematic studies have shown that they are a distinct clade, sister to the ascomycetes and basidiomycetes. The glomeromycetes, ascomycetes, and basidiomycetes constitute the clade known as the crown fungi.

The hyphae of glomeromycetes are coenocytic and do not form interconnected mycelia. These fungi use glucose from their plant partners as their primary energy source. They then convert the glucose to other, fungus-specific sugars that cannot return to the plant. Glomeromycetes reproduce asexually; there is no evidence that they reproduce sexually.

The next two fungal clades that we'll discuss are related groups with many similarities, including a dikaryon stage and septate hyphae. A key feature distinguishing between them is whether the sexual spores are borne inside a sac (in the ascomycetes) or on a pedestal (in the basidiomycetes).

The sexual reproductive structure of ascomycetes is the ascus

The **ascomycetes** are a large and diverse group of fungi found in marine, freshwater, and terrestrial habitats. There are approximately 60,000 known species of ascomycetes, about half of which are the fungal partners in lichens. Ascomycete hyphae are segmented by more or less regularly spaced septa. A pore in each septum permits extensive movement of cytoplasm and organelles (including nuclei) from one segment to the next.

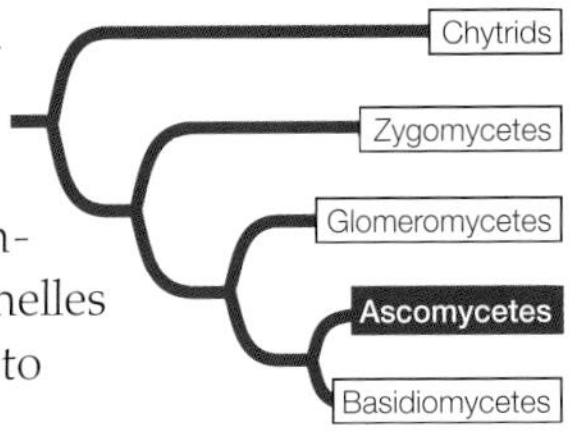

The ascomycetes are distinguished by the production of sacs called **asci** (singular *ascus*), which contain sexually produced *ascospores* (see Figure 30.13C). The ascus is the characteristic sexual reproductive structure of the ascomycetes.

The ascomycetes can be divided into two broad groups, depending on whether the asci are contained within a specialized fruiting structure. Species that have this fruiting structure, the **ascocarp**, are collectively called **euascomycetes** ("true ascomycetes"); those without ascocarps are called **hemiascomycetes** ("half ascomycetes").

Most hemiascomycetes are microscopic, and many species are unicellular. Perhaps the best known are the ascomycete yeasts, especially baker's or brewer's yeast (*Saccharomyces cerevisiae*; see Figure 30.3). These yeasts are among the most important domesticated fungi. *S. cerevisiae* metabolizes glucose obtained from its environment to ethanol and carbon dioxide by fermentation. It forms carbon dioxide bubbles in bread dough and gives baked bread its light texture. Although they are baked away in bread making, the ethanol and carbon dioxide are both retained when yeast ferments grain into beer. Other yeasts live on fruits such as figs and grapes and play an important role in the making of wine.

Hemiascomycete yeasts reproduce asexually either by fission (splitting in half after mitosis) or by budding. Sexual reproduction takes place when two adjacent haploid cells of opposite mating types fuse. In some species, the resulting zygote buds to form a diploid cell population. In others, the zygote nucleus undergoes meiosis immediately; when this happens, the entire cell becomes an ascus. Depending on whether the products of meiosis then undergo mitosis, a yeast ascus contains either eight or four ascospores. The ascospores germinate to become haploid cells. Hemiascomycetes have no dikaryon stage.

The euascomycetes include the cup fungi (**Figure 30.17**). In most of these organisms the ascocarps are cup-shaped and can be as large as several centimeters across. The inner surfaces of the cups, which are covered with a mixture of vegetative hyphae and asci, produce huge numbers of spores. The edible ascocarps of some species, including morels and truffles, are regarded by humans as gourmet delicacies.

Truffles grow underground in a mutualistic association with the roots of oak trees. Europeans traditionally used pigs to find truffles because some truffles secrete a substance that has an odor similar to a pig's sex pheromone.

The euascomycetes also include many of the filamentous fungi known as molds. Many euascomycetes are parasites on flowering plants. Chestnut blight and Dutch elm disease are both caused by euascomycetes. Between its introduction to the United States

(A) *Morchella esculenta* (B) *Sarcoscypha coccinea*

30.17 Two Cup Fungi (A) Morels, which have a spongelike ascocarp and a subtle flavor, are considered a delicacy by humans. (B) These brilliant red cups are the ascocarps of another cup fungus.

in the 1890s and 1940, the chestnut blight fungus destroyed the American chestnut as a commercial species. Before the blight, this species accounted for over half the trees in the great forests of the eastern United States. Another familiar story is that of the American elm, once considered the ideal street tree. Sometime before 1930, the Dutch elm disease fungus (first discovered in the Netherlands) was introduced into the United States on infected elm logs from Europe. Spreading rapidly—sometimes by way of connected root systems—the fungus destroyed great numbers of American elm trees.

Other euascomycete plant pathogens include the powdery mildews that infect cereal grains, lilacs, and roses, among many other plants. Mildews can be a serious problem to farmers and gardeners, and a great deal of research has focused on ways to control these agricultural pests.

Brown molds of the genus *Aspergillus* are important in some human diets. *A. tamarii* acts on soybeans in the production of soy sauce, and *A. oryzae* is used in brewing the Japanese alcoholic beverage sake. Some species of *Aspergillus* that grow on grains and on nuts such as peanuts and pecans produce extremely carcinogenic (cancer-inducing) compounds called *aflatoxins*. In the United States, moldy grain infected with *Aspergillus* is thrown out; In Africa, where food is scarcer, the grain gets eaten, moldy or not, and causes severe health problems.

The witchcraft trials that took place in Salem, Massachusetts in 1692 have been linked to the mold *Claviceps purpurea* (whose common name is ergot), a parasite of rye. The bizarre behavior of the "possessed" young women who accused their neighbors of witchcraft may have been a case of "ergotism"—the powerful hallucinogenic effects of toxins produced by *C. purpurea* that can result when a person eats bread made with contaminated rye flour.

Penicillium is a genus of green molds, of which some species produce the antibiotic penicillin, presumably for defense against competing bacteria. Two species, *P. camembertii* and *P. roquefortii*, are the organisms responsible for the characteristic strong flavors of Camembert and Roquefort cheeses, respectively.

The euascomycetes reproduce asexually by means of conidia that form at the tips of specialized hyphae (**Figure 30.18**). Small chains of conidia are produced by the millions and can survive for weeks in nature. The conidia are what give molds their characteristic colors. *Fusarium oxysporum*, the plant pathogen mentioned at the beginning of this chapter, is a euascomycete with no known sexual stage. It produces conidia in abundance.

The sexual reproductive cycle of euascomycetes includes the formation of a dikaryon, although this stage is relatively brief. Most euascomycetes form mating structures, some "female" and some "male" (see Figure 30.13C). Nuclei from a male structure on one hypha enter a female mating structure on a hypha of a compatible mating type. Dikaryotic *ascogenous* (ascus-forming) hyphae develop from the now dikaryotic female mating structure. The introduced nuclei divide simultaneously with the host nuclei. Eventually asci form at the tips of the ascogenous hyphae. Only with the formation of asci do the nuclei finally fuse. Both nuclear fusion and the subsequent meiosis of the resulting diploid nucleus take place within individual asci. The meiotic products are incorporated into ascospores that are ultimately shed by the ascus to begin the new haploid generation.

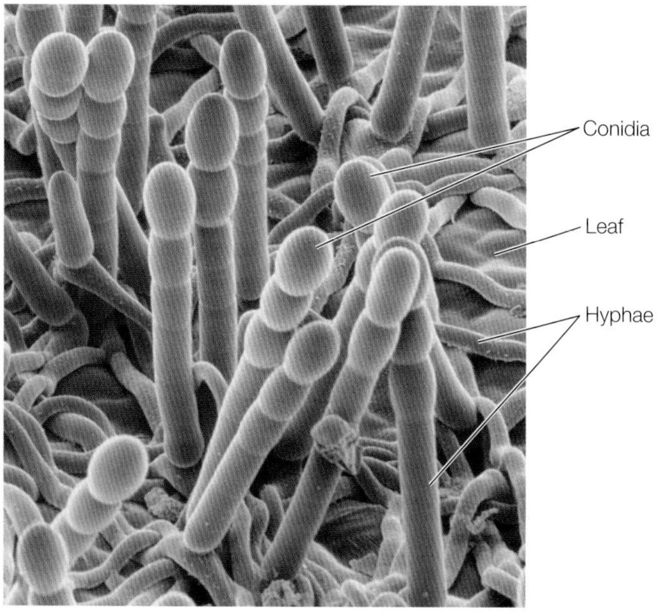

Erysiphe sp.

30.18 Conidia Chains of conidia are developing at the tips of specialized hyphae arising from this powdery mildew growing on a leaf.

(A) *Amanita muscaria*

(B) *Laetiporus sulphureus*

30.19 Basidiomycete Fruiting Structures The basidiocarps of the basidiomycetes are probably the most familiar structures produced by fungi. (A) These mushrooms were produced by a member of a highly poisonous genus, *Amanita*, that forms mycorrhizal relationships with trees. (B) This edible bracket fungus is parasitizing a tree.

The sexual reproductive structure of basidiomycetes is a basidium

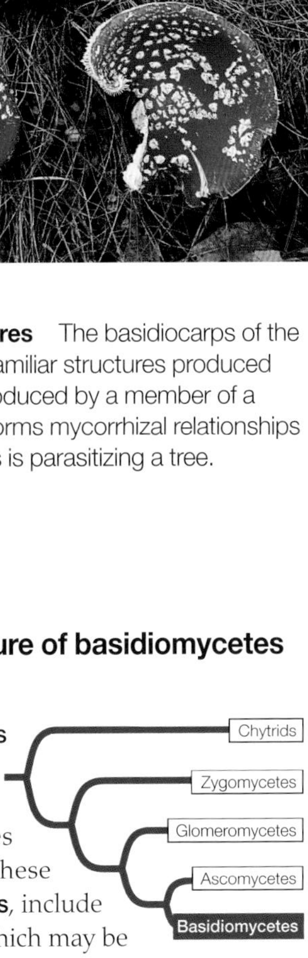

About 25,000 species of **basidiomycetes** (club fungi) have been described. Basidiomycetes produce some of the most spectacular fruiting structures found anywhere among the fungi. These fruiting structures, called **basidiocarps**, include puffballs (see Figure 30.7), some of which may be more than half a meter in diameter, mushrooms of all kinds, and the bracket fungi often encountered on trees and fallen logs in a damp forest. There are more than 3,250 species of mushrooms, including the familiar *Agaricus bisporus* you may enjoy on your pizza as well as poisonous species, such as members of the genus *Amanita* (**Figure 30.19A**). Bracket fungi (**Figure 30.19B**) do great damage to cut lumber and stands of timber. Some of the most damaging plant pathogens are basidiomycetes, including the rust fungi and the smut fungi that parasitize cereal grains. In contrast, other basidiomycetes contribute to the survival of plants as fungal partners in ectomycorrhizae.

Basidiomycete hyphae characteristically have septa with small, distinctive pores. The **basidium** (plural *basidia*), a swollen cell at the tip of a hypha, is the characteristic sexual reproductive structure of the basidiomycetes. It is the site of nuclear fusion and meiosis. Thus the basidium plays the same role in the basidiomycetes as the ascus does in the ascomycetes and the zygosporangium does in the zygomycetes.

As seen in Figure 30.13D, after nuclei fuse in the basidium, the resulting diploid nucleus undergoes meiosis, and the four resulting haploid nuclei are incorporated into haploid *basidiospores*, which form on tiny stalks on the outside of the basidium. A single basidiocarp of the common bracket fungus *Ganoderma applanatum* can produce as many as 4.5 *trillion* basidiospores in one growing season. Basidiospores typically are forcibly discharged from their basidia and then germinate, giving rise to haploid hyphae. As these hyphae grow, haploid hyphae of different mating types meet and fuse, forming dikaryotic hyphae, each cell of which contains two nuclei, one from each parent hypha. The dikaryotic mycelium grows and eventually, when triggered by rain or another environmental cue, produces a basidiocarp. The dikaryon stage may persist for years—some basidiomycetes live for decades or even centuries. This pattern contrasts with the life cycle of the ascomycetes, in which the dikaryon is found only in the stages leading up to formation of the asci.

30.4 RECAP

Five major taxonomic groups of fungi can be distinguished on the basis of differences in their life cycles and reproductive structures.

- What feature of the chytrids suggests an aquatic ancestor to the fungi? See p. 663 and Figure 30.13A
- What happens inside a zygosporangium? See p. 665 and Figure 30.13B
- What distinguishes the fruiting bodies of ascomycetes from those of basidiomycetes? See p. 667 and Figure 30.13C and D

Whether living on their own or in symbiotic associations, fungi have spread successfully over much of Earth since their origin from a protist ancestor. That ancestor also gave rise to the choanoflagellates and the animals, as we will describe in Chapter 31.

CHAPTER SUMMARY

30.1 How do fungi thrive in virtually every environment?

Fungi are heterotrophic organisms with **absorptive nutrition** and with chitin in their cell walls. Fungi have various nutritional modes: some are **saprobes**, others are **parasites**, and some are **mutualists**. **Yeasts** are unicellular fungi. See Web/CD Activity 30.1

The body of a multicellular fungus is a **mycelium**—a meshwork of **hyphae** that may be **septate** (having **septa**) or **coenocytic**. Review Figure 30.4

The cells of fungi live in intimate association with their surroundings, and are well adapted for absorptive nutrition in many environments. Many are parasitic plant pathogens, harvesting nutrients from plant cells by means of **haustoria**. Review Figure 30.5

Fungi are tolerant to hypertonic environments, and many are tolerant to low or high temperatures.

30.2 How are fungi beneficial to other organisms?

Saprobic fungi, as decomposers, make crucial contributions to the recycling of elements. Certain fungi have relationships with other organisms that are both **symbiotic** and **mutualistic**.

Some fungi associate with cyanobacteria or green algae to form **lichens**, which contribute to soil formation. Review Figure 30.9

Mycorrhizae are mutualistic associations of fungi with plant roots. They improve a plant's ability to take up nutrients. Review Figure 30.10

Endophytic fungi protect their plant hosts.

30.3 How do fungal life cycles differ from one another?

See Web/CD Tutorial 30.1

Most fungi reproduce sexually, although several modes of asexual reproduction are used by fungi. In many fungi, reproductive pairing is determined by two or more morphologically indistinguishable **mating types** rather than by two sexes.

Chytrids and some yeasts are the only fungi whose life cycle includes alternation of generations.

In the sexual reproduction of all other fungal groups, hyphae fuse, allowing "gamete" nuclei to be transferred. In the ascomycetes and basidiomycetes, **plasmogamy** precedes **karyogamy**, with the result that a **dikaryon** is formed. This $(n + n)$ dikaryotic condition is unique to the fungi. Review Figures 30.13C and D, Web/CD Activity 30.2

The **deuteromycetes** are a polyphyletic group of fungi for which no sexual stage has been observed and which have not yet been assigned to one of the other five fungal groups.

30.4 How do we tell the fungal groups apart?

There are five major fungal groups, of which the chytrids and the zygomycetes are paraphyletic. The glomeromycetes, ascomycetes, and basidiomycetes are each a clade, and together they form the **crown fungi** clade. Review Figure 30.2 and Table 30.1

The **chytrids**, a paraphyletic group of aquatic fungi, have flagellated gametes. Their life cycle displays alternation of generations. Review Figure 30.13A

The **zygomycetes** are a paraphyletic group with coenocytic hyphae. Gametangia fuse to form a **zygosporangium** within which a zygospore (a single cell containing many diploid nuclei) develops. Review Figure 30.13B

Glomeromycetes form arbuscular mycorrhizae with plant roots. They reproduce only asexually. Their hyphae are coenocytic.

Ascomycetes have septate hyphae; their sexual reproductive structures are **asci**. Many ascomycetes are partners in lichen associations. Most **hemiascomycetes** are yeasts. **Euascomycetes** produce fleshy fruiting bodies called **ascocarps**. The dikaryon stage in the ascomycete life cycle is relatively brief. Review Figure 30.13C

Basidiomycetes have septate hyphae. Their fruiting bodies are called **basidiocarps**, and their sexual reproductive structures are **basidia**. The dikaryon stage may last for years. Review Figure 30.13D

SELF-QUIZ

1. Which statement about fungi is *not* true?
   a. A multicellular fungus has a body called a mycelium.
   b. Hyphae are composed of individual mycelia.
   c. Many fungi tolerate highly hypertonic environments.
   d. Many fungi tolerate low temperatures.
   e. Some fungi are anchored to their substrate by rhizoids.
2. The absorptive nutrition of fungi is aided by
   a. dikaryon formation.
   b. spore formation.
   c. the fact that they are all parasites.
   d. their large surface area-to-volume ratio.
   e. their possession of chloroplasts.
3. Which statement about fungal nutrition is *not* true?
   a. Some fungi are active predators.
   b. Some fungi form mutualistic associations with other organisms.
   c. All fungi require mineral nutrients.
   d. Fungi can make some of the compounds that are vitamins for animals.
   e. Facultative parasites can grow only on their specific hosts.
4. Which statement about dikaryosis is *not* true?
   a. The cytoplasm of two cells fuses before their nuclei fuse.
   b. The two haploid nuclei are genetically different.
   c. The two nuclei are of the same mating type.
   d. The dikaryon stage ends when the two nuclei fuse.
   e. Not all fungi have a dikaryon stage.
5. Reproductive structures consisting of one or more photosynthetic cells surrounded by fungal hyphae are called
   a. ascospores.
   b. basidiospores.
   c. conidia.
   d. soredia.
   e. gametes.
6. The zygomycetes
   a. have hyphae without regularly occurring septa.
   b. produce motile gametes.
   c. form fleshy fruiting bodies.
   d. are haploid throughout their life cycle.
   e. have sexual reproductive structures similar to those of the ascomycetes.

7. Which statement about ascomycetes is *not* true?
   a. They include yeasts.
   b. They form reproductive structures called asci.
   c. Their hyphae are segmented by septa.
   d. Many of their species have a dikaryotic state.
   e. All have fruiting structures called ascocarps.
8. The basidiomycetes
   a. often produce fleshy fruiting structures.
   b. have hyphae without septa.
   c. have no sexual stage.
   d. produce basidia within basidiospores.
   e. form diploid basidiospores.
9. The deuteromycetes
   a. have distinctive sexual stages.
   b. are all parasitic.
   c. have "lost" some members to other fungal groups.
   d. include the ascomycetes.
   e. are never components of lichens.
10. Which statement about lichens is *not* true?
   a. They can reproduce by fragmentation of the vegetative body.
   b. They are often the first colonists in a new area.
   c. They render their environment more basic (alkaline).
   d. They contribute to soil formation.
   e. They may contain less than 10 percent water by weight.

FOR DISCUSSION

1. You are shown an object that looks superficially like a pale green mushroom. Describe at least three criteria (including anatomical and chemical traits) that would enable you to tell whether the object is a piece of a plant or a piece of a fungus.
2. Differentiate among the members of the following pairs of related terms:
   a. hypha/mycelium
   b. euascomycete/hemiascomycete
   c. ascus/basidium
   d. ectomycorrhiza/arbuscular mycorrhiza
3. For each type of organism listed below, give a single characteristic that may be used to differentiate it from the other, related organism(s) in parentheses.
   a. zygomycete (ascomycete)
   b. basidiomycete (deuteromycete)
   c. ascomycete (basidiomycete)
   d. baker's yeast (*Penicillium*)
4. Many fungi are dikaryotic during part of their life cycle. Why are dikaryons described as $n + n$ instead of $2n$?
5. If all the fungi on Earth were suddenly to die, how would the surviving organisms be affected? Be thorough and specific in your answer.
6. How might the first mycorrhizae have arisen?
7. What attributes might account for the ability of lichens to withstand the intensely cold environment of Antarctica? Be specific in your answer.
8. What factors must be taken into account in using fungi to combat agricultural pests?

FOR INVESTIGATION

Recall Poulsen and Boomsma's studies on how the fungi in ant "gardens" prevent the growth of competing fungi (see Figure 30.11). How would you test the hypothesis that a green mold that you found on an orange prevented the growth of other molds?

CHAPTER 31 Animal Origins and the Evolution of Body Plans

Dinosaur embryos illuminate evolution

The images of dinosaurs that are vividly etched in most people's minds include the frightening bipedal carnivores such as *Tyrannosaurus rex* on the one hand, and, on the other, the massive quadrupedal herbivores—the largest land animals ever to walk the planet. In fact, the fossil record indicates that all of the earliest dinosaur species were probably bipedal.

Quadrupedal dinosaurs achieved sizes far too great to be supported by hindlimbs alone, but paleontologists have long believed that these four-legged giants evolved from smaller, bipedal species. The recent discovery of the fossilized eggs of one of the earliest known dinosaur species has shed new light on this hypothesis.

Massospondylus was a bipedal, plant-eating dinosaur that lived in Africa about 190 million years ago. Fully grown adults of this genus measured about 28 meters from head to tail. Within the fossilized eggs of *Massospondylus carinatus*, researchers discovered remarkably well preserved, nearly fully developed embryos about 10 centimeters long.

Embryos reveal the form of an animal when it is born or hatched—and this form can be very different from that of the adult. The *M. carinatus* embryos revealed an animal that was born with a short tail, a neck that it carried horizontally, long forelimbs, a large head, and no teeth. The hatchlings would have been awkward little animals that moved around on all four limbs. To achieve the bipedal adult form, the necks and the hindlimbs of the young animals would need to have grown more rapidly than their front legs and heads.

By measuring *Massospondylus* fossils of various sizes, researchers determined that these dinosaurs walked on all fours while young, but walked and ran on their hindlimbs as adults. These findings suggest that the giant quadrupedal dinosaurs could have evolved from bipedal ancestors by means of rather straightforward changes in their growth pattern: If forelimb growth rate shifted slightly so that the forelimbs continued to grow as rapidly as the hindlimbs, an individual would retain the quadrupedal juvenile body form into adulthood.

The discovery of mature dinosaur embryos is just one example showing how the rapidly expanding fossil record is yielding valuable information about the forms of ancient animals and how they may have changed during their life cy-

Fossilized Embryos Help Explain Evolution This remarkably preserved embryo of *Massospondylus carinatus* revealed to paleontologists that the young of this bipedal dinosaur walked on all fours.

A Giant Quadruped Brachiosaurs lived around 130 million years ago, during the late Jurassic. This individual could have weighed as much as 80 tons. Its front legs were markedly longer than its hind legs—the reverse of the bipedal condition, but one that could have evolved from bipedal ancestors whose hatchlings were quadrupedal.

cle. Mutations in the genes governing development can result in dramatic changes in the forms of adult animals. Developmental and genetic knowledge, combined with fossil evidence, can help explain how a few fundamental body plans could have been modified to yield the remarkable variety of animal forms that we will describe in this and the following two chapters.

IN THIS CHAPTER we first enumerate the evidence that has led biologists to conclude that the animals are monophyletic and present a current phylogenetic tree of animals. We then describe how the diverse animal forms are based on modifications of a few key features within a small array of body plans. We will discuss how animals obtain food and describe their varied life cycles—how they are born, grow, disperse, and reproduce. Finally, we describe the members of several clades of structurally "simpler" animals.

CHAPTER OUTLINE

31.1 What Evidence Indicates the Animals Are Monophyletic?

We have no trouble identifying frogs, lizards, birds, and dogs as animals, but things are not always so simple. For example, many aquatic animals, such as sea anemones, look like plants. Indeed, many of these species were thought to be plants when they were first described. Sponges, for example, were not recognized as animals until 1765.

What traits distinguish the animals from the other groups of organisms?

- In contrast to the Bacteria, Archaea, and most microbial eukaryotes (see Chapters 26 and 27), all animals are *multicellular*. Animal life cycles feature complex patterns of *development* from a single-celled zygote into a multicellular adult.
- In contrast to most plants (see Chapters 28 and 29), all animals are *heterotrophs*. Animals are able to synthesize very few organic molecules from inorganic chemicals, so they must take in nutrients from their environment.
- The fungi (see Chapter 30) are also heterotrophs. In contrast to the fungi, however, animals use *internal* processes to break down materials from their environment into the organic molecules they need most. Most animals *ingest* food into an internal *gut* that is continuous with the outside environment, in which digestion takes place.
- In contrast to plants, most animals can *move*. Animals must move to find food or bring food to them. Animals have specialized *muscle* tissues that allow them to move, and many animal body plans are specialized for movement.

Animal monophyly is supported by gene sequences and morphology

The most convincing evidence that all the organisms considered to be animals share a common ancestor comes from their many shared derived molecular and morphological traits, most of which we have described in earlier chapters:

- Many gene sequences, such as the ribosomal RNA genes (see Section 26.4), support the monophyly of animals.

- Animals display similarities in the organization and function of their Hox genes (see Chapter 20).
- Animals have unique types of junctions between their cells (tight junctions, desmosomes, and gap junctions; see Figure 5.7).
- Animals have a common set of *extracellular matrix* molecules, including collagen and proteoglycans (see Figure 4.25).

Although there are animals in a few clades that lack one or another of these *synapomorphies*, these species apparently once possessed the traits and lost them during their later evolution.

The ancestor of the animal clade was probably a colonial flagellated protist similar to existing colonial choanoflagellates (see Figure 27.30A). The most reasonable current scenario postulates a choanoflagellate lineage in which certain cells within the colony began to be specialized—some for movement, others for nutrition, others for reproduction, and so on. Once this *functional specialization* had begun, cells could have continued to differentiate. Coordination among groups of cells could have improved by means of specific regulatory molecules that guided the differentiation and migration of cells in developing embryos. Such coordinated groups of cells eventually evolved into the larger and more complex organisms that we call animals.

More than a million animal species have been named and described, and there are doubtless millions of living species that have yet to be named. The synapomorphies that indicate animal monophyly cannot be used to infer evolutionary relationships among animals, because nearly all animals have them. Clues to the evolutionary relationships among animal groups thus must be sought in *derived traits* that are found in some groups but not in others. Such characteristics can be found in fossils, in patterns of embryonic development, in the morphology and physiology of living animals, in the structure of animal molecules, and in the genomes of animals (for example, in mitochondrial and ribosomal RNA genes).

Developmental patterns show evolutionary relationships among animals

Differences in patterns of embryonic development traditionally provided some of the most important clues to animal phylogeny, although analyses of gene sequences are now showing that some developmental patterns are more evolutionarily labile than previously thought. We will discuss the details of animal development in Chapter 43; here we will describe the basic developmental patterns that vary among the animal groups shown in **Figure 31.1**.

The first few cell divisions of a zygote are known as *cleavage*. In general, the number of cells in the embryo doubles with each cleavage. A number of different **cleavage patterns** exist among animals.

As will be described in Section 43.1, cleavage patterns are influenced by the configuration of the *yolk*, the nutritive material that

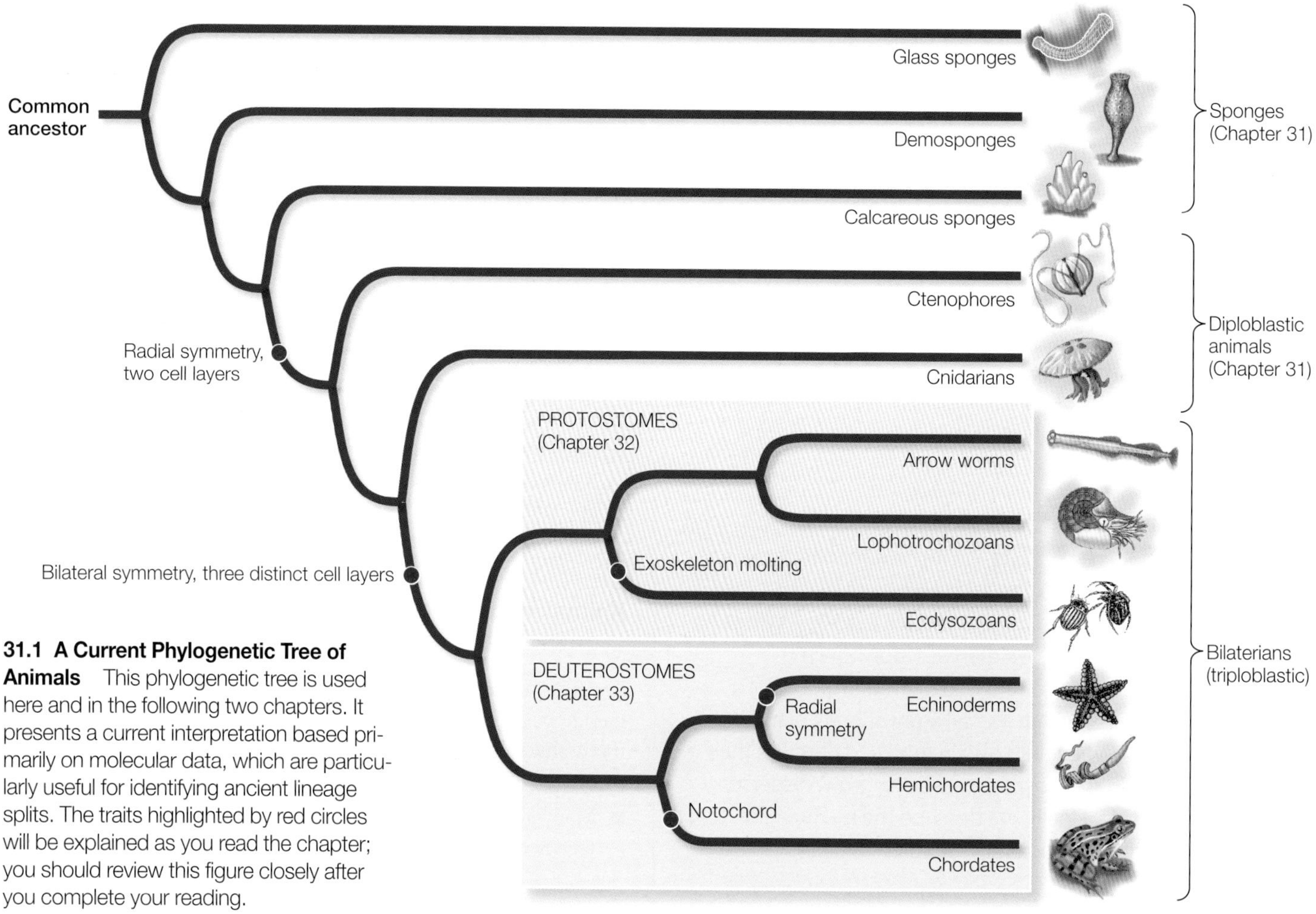

31.1 A Current Phylogenetic Tree of Animals This phylogenetic tree is used here and in the following two chapters. It presents a current interpretation based primarily on molecular data, which are particularly useful for identifying ancient lineage splits. The traits highlighted by red circles will be explained as you read the chapter; you should review this figure closely after you complete your reading.

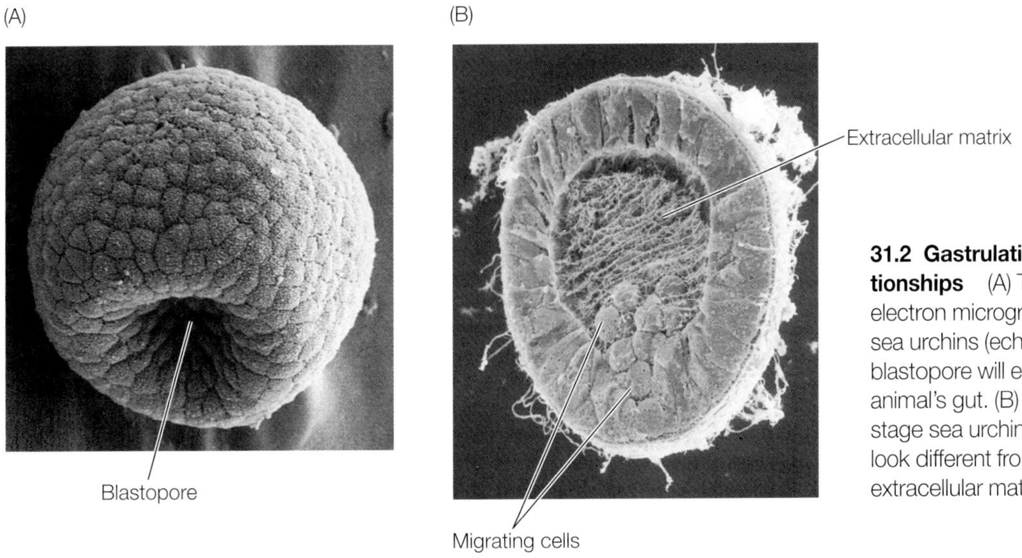

31.2 Gastrulation Illuminates Evolutionary Relationships (A) The blastopore is clear in this scanning electron micrograph of a sea urchin gastrula. Because sea urchins (echinoderms) are deuterostomes, this blastopore will eventually become the anal end of the animal's gut. (B) In this cross section through a later-stage sea urchin gastrula, the cells are beginning to look different from one another. The molecules of the extracellular matrix guide cell movement.

nourishes the growing embryo. In reptiles, for example, the presence of a large body of a cellular yolk within the fertilized egg creates an *incomplete* cleavage pattern in which the dividing cells form an embryo on top of the yolk mass (see Figure 43.3C). In echinoderms such as sea urchins, small yolk particles are evenly distributed throughout the egg cytoplasm, so cleavage is *complete*, with the fertilized egg cell dividing in an even pattern known as *radial cleavage* (see Figure 43.3A). Radial cleavage is the ancestral condition for eumetazoans, so it is found among many protostomes and diploblastic animals as well as deuterostomes. *Spiral cleavage*, a complicated derived permutation of radial cleavage, is found among many lophotrochozoans, such as earthworms and clams. Lophotrochozoans with spiral cleavage are thus sometimes known as *spiralians*. The early branches of the ecdysozoans have radial cleavage, although most ecdysozoans have an idiosyncratic cleavage pattern that is neither radial nor spiral in organization.

During the early development of most animals, distinct layers of cells form. These cell layers differentiate into specific organs and organ systems as development continues. The embryos of **diploblastic** animals have only two of these cell layers: an outer *ectoderm* and an inner *endoderm*. The embryos of **triploblastic** animals have, in addition to ectoderm and endoderm, a third distinct cell layer, the *mesoderm*, which lies between the ectoderm and the endoderm. The existence of three cell layers is a synapomorphy of triploblastic animals, whereas the paraphyletic diploblastic animals (ctenophores and cnidarians) exhibit the ancestral condition.

During early development in many animals, a hollow ball one cell thick indents to form a cup-shaped structure. This process is known as *gastrulation*. The opening of the cavity formed by this indentation is called the *blastopore* (**Figure 31.2**). The pattern of development after formation of the blastopore has been used to divide the triploblastic animals into two major groups. Among members of the first group, the **protostomes** (Greek, "mouth first"), the mouth arises from the blastopore; the anus forms later. This appears to be the derived condition. Among the **deuterostomes** ("mouth second"), the blastopore becomes the anus; the mouth forms later. This is thought to be the ancestral condition. We now know that the developmental patterns of animals are more varied than suggested by this simple dichotomy, but the protostomes and deuterostomes are still recognized as distinct animal clades based upon sequence similarities of their genes.

> The embryologist Lewis Wolpert once stated that "It is not birth, marriage or death, but gastrulation, which is truly the most important time in your life." Aside from its significance to the individual organism, the process of gastrulation is central to our understanding of evolutionary relationships among the animals.

31.1 RECAP

The animals are thought to be monophyletic because they share many derived traits, including multicellularity, mobility, and a heterotrophic lifestyle based on the ingestion of outside nutrients. Evolutionary relationships among animals are inferred from fossils and from molecular and developmental traits that are shared by different groups of animals.

- Do you understand why the traits biologists use to determine evolutionary relationships among animals must differ from those they use to infer that all animals share a common ancestor? See p. 672
- Can you describe the differences between diploblastic and triploblastic embryos, and between protostomes and deuterostomes? See p. 673

We devote Chapter 32 to the protostomes and Chapter 33 to the deuterostomes. Later in this chapter we will describe several groups of animals with relatively simple structural organization. Millions of species of triploblastic animals with complex structures eventually evolved from these ancestral states.

31.2 What Are the Features of Animal Body Plans?

The general structure of an animal, the arrangement of its organ systems, and the integrated functioning of its parts are referred to as its **body plan**. Although animal body plans are extremely varied, they can be seen as variations on four key features:

- The *symmetry* of the body
- The structure of the *body cavity*
- The *segmentation* of the body
- *External appendages* that move the body

All of these features affect the way in which an animal moves and interacts with its environment. These four attributes vary across a number of basic animal body plans.

As we learned in Chapter 20, the *regulatory genes* that govern the development of body symmetry, body cavities, segmentation, and appendages are widely shared among the different animal groups. Thus we might expect animals to share body plans.

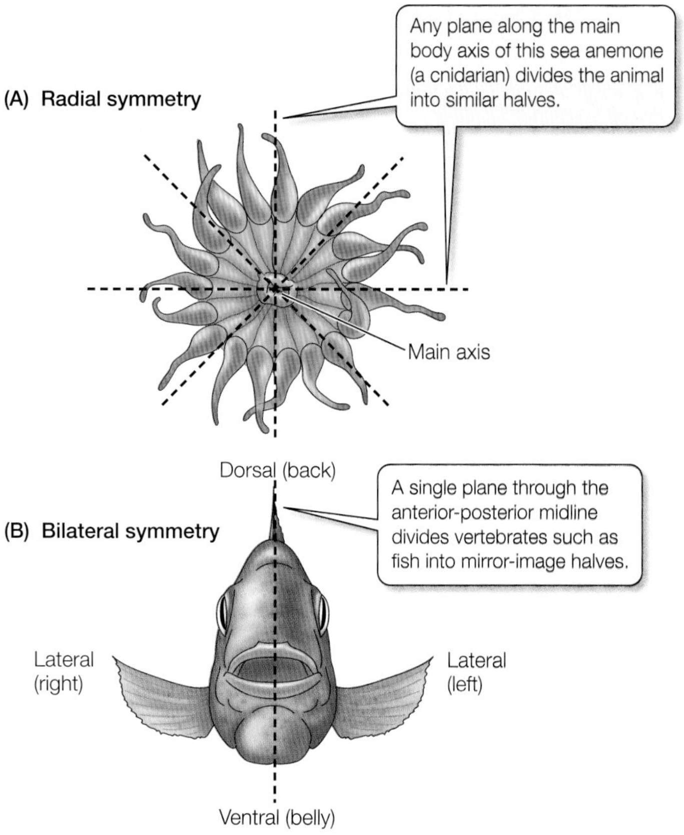

31.3 Body Symmetry Most animals are either radially or bilaterally symmetrical.

Most animals are symmetrical

The overall shape of an animal can be described by its **symmetry**. An animal is said to be *symmetrical* if it can be divided along at least one plane into similar halves. Animals that have no plane of symmetry are said to be *asymmetrical*. Many sponges are asymmetrical, but most other animals have some kind of symmetry, which is governed by the expression of regulatory genes.

The simplest form of symmetry is **spherical symmetry**, in which body parts radiate out from a central point. An infinite number of planes passing through the central point can divide a spherically symmetrical organism into similar halves. Spherical symmetry is widespread among unicellular protists, but most animals possess other forms of symmetry.

In organisms with **radial symmetry**, there is one main axis around which body parts are arranged. Two animal groups—ctenophores and cnidarians—are composed primarily of radially symmetrical animals (**Figure 31.3A**). A perfectly radially symmetrical animal can be divided into similar halves by any plane that contains the main axis. However, most radially symmetrical animals—including the adults of echinoderms such as sea stars and sand dollars—are slightly modified so that fewer planes can divide them into identical halves. Many radially symmetrical animals are sessile (sedentary). Others move slowly, but can move equally well in any direction.

Bilateral symmetry is characteristic of animals that move in one direction. A bilaterally symmetrical animal can be divided into mirror-image (left and right) halves by a single plane that passes through the midline of its body (**Figure 31.3B**). This plane runs from the tip, or **anterior** of the body to its tail, or **posterior**.

A plane at right angles to the midline divides the body into two dissimilar sides. The back of a bilaterally symmetrical animal is its **dorsal** surface; the belly, which contains the mouth, is its **ventral** surface.

Bilateral symmetry is strongly correlated with **cephalization**, which is the concentration of sensory organs and nervous tissues in a head at the anterior end of the animal. Cephalization has been evolutionarily favored because the anterior end of a bilaterally symmetrical animal typically encounters new environments first.

The structure of the body cavity influences movement

Animals can be divided into three types—*acoelomate, pseudocoelomate*, and *coelomate*— based on the presence and structure of an internal, fluid-filled **body cavity**. The structure of an animal's body cavity strongly influences the ways in which it can move.

Acoelomate animals such as flatworms lack an enclosed, fluid-filled body cavity. Instead, the space between the gut (derived from endoderm) and the muscular body wall (derived from mesoderm) is filled with masses of cells called *mesenchyme* (**Figure 31.4A**). These animals typically move by beating cilia.

Body cavities come in two types. Both types lie between the ectoderm and the endoderm; they are differentiated by their relationship to the mesoderm.

- **Pseudocoelomate** animals have a body cavity called a *pseudocoel*, a fluid-filled space in which many of the internal organs are suspended. A pseudocoel is enclosed by muscles (mesoderm) only on its outside; there is no inner layer of mesoderm surrounding the internal organs (**Figure 31.4B**).
- **Coelomate** animals have a *coelom*, a body cavity that develops within the mesoderm. It is lined with a layer of muscular tissue called the *peritoneum*, which also surrounds the internal

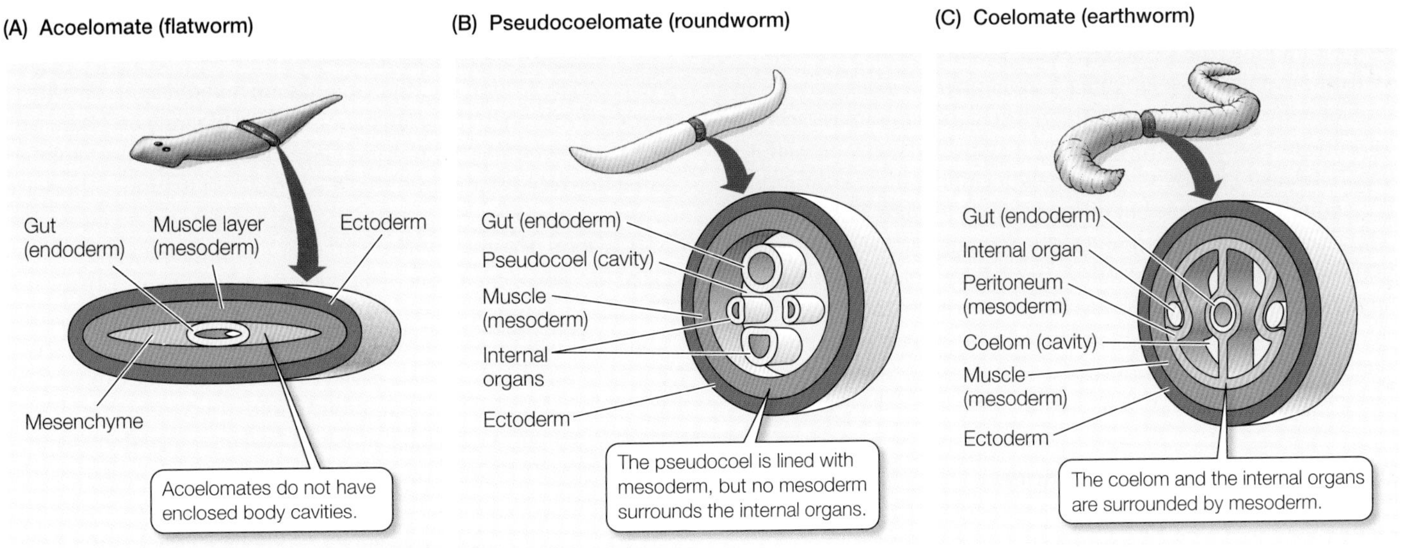

31.4 Animal Body Cavities (A) Acoelomates do not have enclosed body cavities. (B) Pseudocoelomates have a body cavity enclosed by only one layer of mesoderm, which lies outside the cavity. (C) Coelomates have a peritoneum surrounding the internal organs.

organs. The coelom is thus enclosed on both the inside and the outside by mesoderm (**Figure 31.4C**). A coelomate animal has better control over the movement of the fluids in its body cavity than a pseudocoelomate animal does.

The body cavities of many animals function as **hydrostatic skeletons**. Fluids are relatively incompressible, so when the muscles surrounding them contract, they move to another part of the cavity. If the body tissues around the cavity are flexible, fluids squeezed out of one region can cause some other region to expand. The moving fluids can thus move specific body parts. (You can see how a hydrostatic skeleton works by watching a snail emerge from its shell.) An animal with both *circular muscles* (encircling the body cavity) and *longitudinal muscles* (running along the length of the body) has even greater control over its movement.

Although the hydrostatic function of fluid-filled body cavities is important, most animals also have hard skeletons that provide protection and facilitate movement. Muscles are attached to those firm structures, which may be inside the animal or on its outer surface (in the form of a shell or cuticle).

Segmentation improves control of movement

Many animals have bodies that are divided into segments. **Segmentation** facilitates specialization of different body regions. Segmentation also allows an animal to alter the shape of its body in complex ways and to control its movements precisely. If an animal's body is segmented, muscles in each individual segment can change the shape of that segment independently of the others. In only a few segmented animals is the body cavity separated into discrete compartments, but even partly separated compartments allow better control of movement. As we will see, segmen-

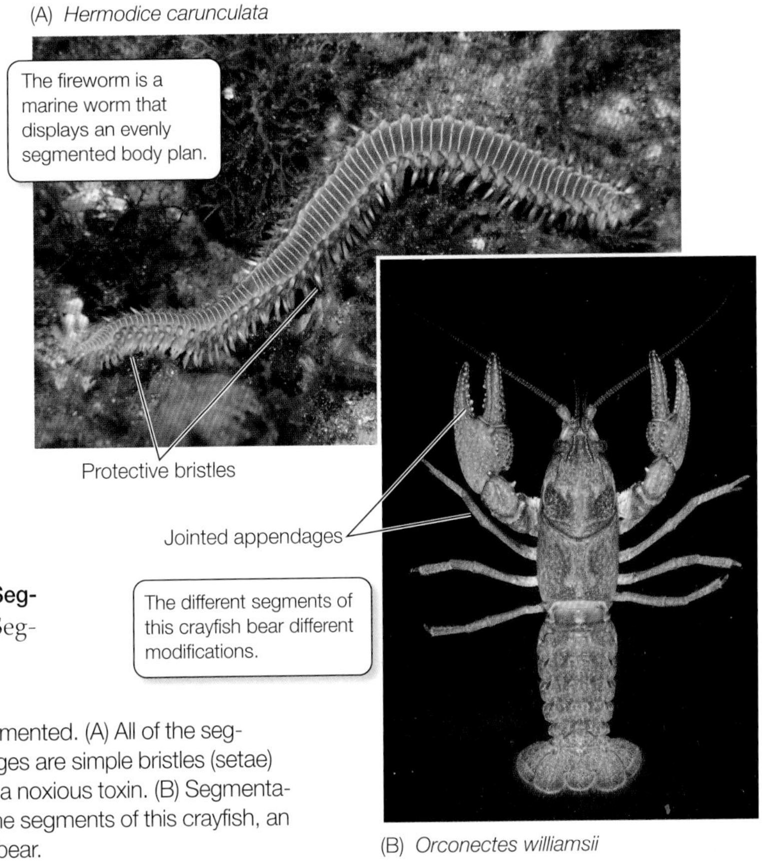

31.5 Segmentation The body cavities of many animals are segmented. (A) All of the segments of this marine fireworm, an annelid, are similar. Its appendages are simple bristles (setae) that, in this animal, serve mainly for protection—the setae contain a noxious toxin. (B) Segmentation allows the evolution of differentiation among the segments. The segments of this crayfish, an arthropod, differ in their form, function, and the appendages they bear.

tation evolved independently several different times in both the protostomes and the deuterostomes.

In some animals, segments are not apparent externally (as in the segmented vertebrae of vertebrates). In other animals, such as the annelids, similar body segments are repeated many times (**Figure 31.5A**); and in yet other animals, including most arthropods, the segments are visible but differ strikingly (**Figure 31.5B**). As we will describe in the next chapter, the dramatic evolutionary radiation of the arthropods (including the insects, spiders, centipedes, and crustaceans) was based on changes in a segmented body plan that features muscles attached to the inner surface of an external skeleton, along with a variety of external appendages that move these animals.

Appendages enhance locomotion

Getting around under their own steam is important to many animals. It allows them to obtain food, to avoid predators, and to find mates. Even some sessile species, such as sea anemones, have larval stages that use cilia to swim, thus increasing the animal's chances of finding a suitable habitat to settle on.

Appendages that project externally from the body greatly enhance an animal's ability to move around. Many echinoderms, including sea urchins and sea stars, have myriad *tube feet* that allow them to move slowly across the substratum. Highly controlled, rapid movement is greatly enhanced in animals whose appendages have become modified into specialized *limbs*. In two animal groups, the arthropods and the vertebrates, the presence of *jointed limbs* has been a prominent factor in their evolutionary success (see Figure 31.5B).

In several independent instances—among the arthropod insects, the pterosaurs, the birds, and the bats—body plans emerged in which limbs were modified into wings, allowing animals to take to the air (see Figure 20.12).

31.2 RECAP

The body plans of animals are all variations on patterns of symmetry, body cavities, segmentation, and appendages. All four can have bearing on movement and locomotion, which are important aspects of the animal way of life.

- Can you describe the main types of symmetry found in animals? Do you see how an animal's symmetry can influence the way it moves? See p. 674 and Figure 31.3
- Can you explain several ways in which body cavities and segmentation improve control over movement? See pp. 674–676

Many of the modifications to their body plans involve ways of finding, capturing, and processing food. Evolutionary changes in the symmetry, body cavities, appendages, and segmentation of animals have played a key role in enabling them to obtain food from their environments, as well as helping them avoid becoming food for other animals.

31.3 How Do Animals Get Their Food?

We noted in Section 31.1 that animals are heterotrophs, or "other-feeders." Although there are many animals that rely on photosynthetic *endosymbionts* for nutrition (see Figure 4.15B), most animals must actively obtain an outside source of nutrition, otherwise known as food.

The food of animals includes most other members of their own clade as well as members of all other groups of living organisms. The need to locate food has favored the evolution of sensory structures that can provide animals with detailed information about their environment, as well as nervous systems that can receive, process, and coordinate that information.

To acquire food, animals must expend energy, either to move through the environment to where food is located or to move the environment and the food it contains to them. Animals that can move from one place to another are **motile**; animals that stay in one place are **sessile**.

The feeding strategies that animals use fall into a few broad categories:

- *Filter feeders* capture small organisms delivered to them by their environment.
- *Herbivores* eat plants or parts of plants.
- *Predators* capture and eat other animals that typically are relatively large in relation to themselves.
- *Parasites* live in or on other organisms from which they obtain energy and nutrients.
- *Detritivores* actively feed on dead organic material.

Each of these modes of feeding can be found in many different animal groups, and none of them is limited to a single group. In addition, individuals of some species may employ more than one feeding strategy, and some animals employ entirely different feeding strategies at different points in their life cycle. The constant and ongoing need to obtain food, the variety of nutrient sources available in any given environment, and the necessity of competing with other animals to get food means that a variety of feeding strategies can be found among all the major animal groups.

Filter feeders capture small prey

Air and water often contain small organisms and organic molecules that are potential food for animals. Moving air and water may carry those items to an animal that positions itself in a good location. These **filter feeders** then use some kind of straining device to filter the food from the environment. Many sessile aquatic animals rely on water currents to bring prey to them (**Figure 31.6A**).

Baleen whales—including the blue whale, the largest animal on Earth—rely on some of the smallest organisms for their nutrition. These whales filter tons of tiny, often microscopic zooplankton out of the ocean water that passes through comblike baleen plates in their upper jaws.

31.6 Filter Feeding Strategies (A) Sessile marine filter feeders such as this "Christmas tree worm," a polychaete, allow the ocean currents to bring their food—plankton—to them. (B) The greater flamingo of South America is a motile filter feeder, using its appendages (legs) to stir up mud as it wades through ocean lagoons and salty lakes. The birds then use their beaks to strain small organisms out of the muddy mixture.

Motile filter feeders bring the nutrient-containing medium to them. The serrated beak of the flamingo, for example, filters small organisms out of the muddy mixture it picks up as it wades through shallow water (**Figure 31.6B**).

Some sessile filter feeders expend energy to move water past their food-capturing devices. Sponges, for example, bring water into their bodies by beating the flagella of their specialized feeding cells, called **choanocytes** (**Figure 31.7**). It is these flagellated feeding cells of sponges that link the animals with choanoflagellate protists and the fungi. The choanoflagellates and the animals are most closely related to the fungi, and these three groups form a clade known as the *opisthokonts* (see p. 605).

Herbivores eat plants

Animals that eat plants are referred to as **herbivores**. An individual plant has many different structures—leaves, wood, sap, flowers, fruits, nectar, and seeds—that animals can consume. Not surprisingly, then, many different kinds of herbivores may feed on a single kind of plant, consuming different parts of the plant or eating the same part in different ways (**Figure 31.8**). An individual animal that is captured by a predator is likely to die, but herbivores often feed on plants without killing them.

Animals do not need to expend energy subduing and killing plants. However, they do need to digest them. This can be difficult for terrestrial herbivores, because the dominant land plants tend to have many different kinds of tissues, many of which are tough or fibrous. Plant tissues may also contain chemicals that must be detoxified before they can be ingested. Herbivorous animals typically have long, complex guts to accomplish the tasks involved in digesting plants (see Section 50.2).

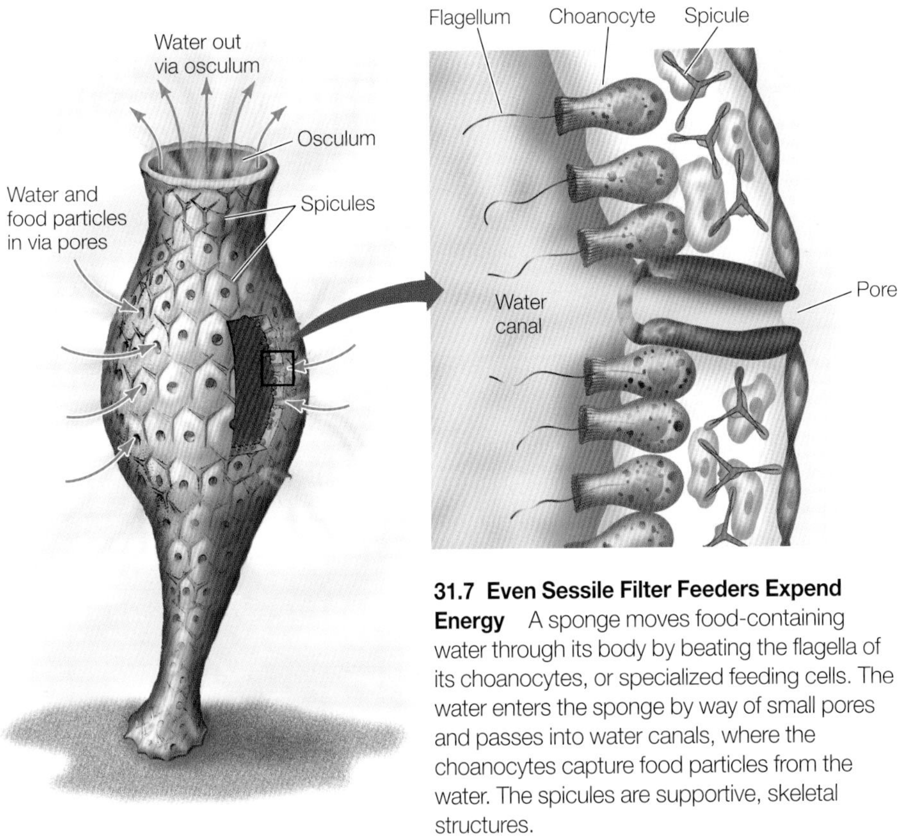

31.7 Even Sessile Filter Feeders Expend Energy A sponge moves food-containing water through its body by beating the flagella of its choanocytes, or specialized feeding cells. The water enters the sponge by way of small pores and passes into water canals, where the choanocytes capture food particles from the water. The spicules are supportive, skeletal structures.

31.8 A Single Plant Can Feed Many Different Herbivores
Many species of insects feed on a single species of willow, each species consuming different tissues in different ways.

Chrysomela knabi (leaf beetle, adult)

Nematus sp. (willow sawfly, larvae)

Papilio sp. (tiger swallowtail butterfly, larva)

Salix sericea (silky willow)

Plagiodera versicolora (leaf beetle, larva)

Predators capture and subdue large prey

Predators possess features that enable them to capture and subdue relatively large animals (referred to as their **prey**). Many vertebrate predators have sensitive sensory organs that enable them to locate prey as well as sharp teeth or claws that allow them to capture and subdue large prey (**Figure 31.9**). Predators may stalk and pursue their prey, or wait (often camouflaged) for their prey to come to them.

Another weapon of predators (as well as of prey; see page 486) is toxins. We are all aware of the dangers of encountering the toxins of a venomous snake. Cnidarians (jellyfish and their relatives), which are among the simplest of animals, use toxins to capture and subdue prey that are much larger and more complex than themselves. Their tentacles are covered with specialized cells that contain stinging organelles called **nematocysts**, which inject toxins into their prey (**Figure 31.10**).

31.9 Tooth and Claw (A) The teeth of this Kodiak brown bear are adapted to an omnivorous diet that includes fish. (B) The appendages (legs and wings) of the bald eagle, along with its strong beak, are adapted to the life of a predatory hunter.

(A) *Ursus arctos*

(B) *Haliaeetus leucocephalus*

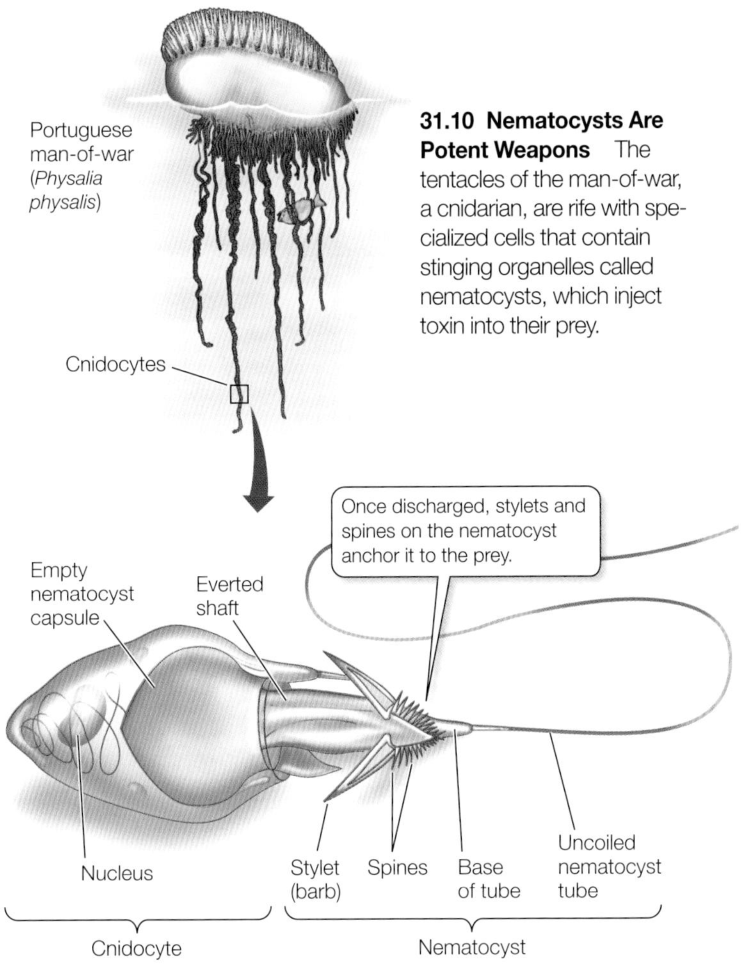

31.10 Nematocysts Are Potent Weapons The tentacles of the man-of-war, a cnidarian, are rife with specialized cells that contain stinging organelles called nematocysts, which inject toxin into their prey.

Cone-shelled snails capture large prey by injecting them with toxins from a highly modified tooth. To date more than 80 U.S. patents have been awarded for various medical uses of cone shell toxins.

Many animals, such as raccoons and humans, eat both plants and other animals. These animals are called **omnivores**. There are also many animals whose diet differs at different life stages, such as the many songbirds that eat fruit or seeds as adults but feed their young on insects.

Parasites live in or on other organisms

Animals that live in or on another organism—called a *host*—and obtain their nutrients by consuming parts of that organism are called **parasites**. Most animal parasites are much smaller than their hosts, and many parasites can consume parts of their host without killing it. However, they first must overcome the host's defenses. Parasites often have complex life cycles, that rely on multiple hosts, as we detail in the next section.

Parasites that live inside their hosts are called *endoparasites* and are often morphologically very simple. They can often function without a digestive system because they absorb food directly from the host's gut or bodily tissues. Many flatworms are endoparasites of humans and other mammals, as will be described in Chapter 32.

Parasites that live outside their hosts—*ectoparasites*—are generally more morphologically complex than endoparasites. They have digestive tracts and mouthparts that enable them to pierce the host's tissues or suck on their body fluids. Fleas and ticks are widely known ectoparasitic arthropods that many humans have unfortunately experienced.

31.3 RECAP

Animals have many ways of acquiring food. Filter feeders strain food particles from the water or air. Herbivores have digestive adaptations that allow them to eat plants, while predators are physically adapted to capture and subdue other animals (prey) and consume them. Parasites obtain their nutrition from a host organism.

- Can you describe the different types of adaptations that are necessary for animals that eat plants as opposed to the adaptations needed for a predatory lifestyle? See pp. 677–678
- Looking at the brief overview of some of the diverse feeding modes of animals presented here, and using whatever you already know about different animals, how useful do you think feeding behavior would be as a criterion for grouping animals into broader categories?

As an animal grows from a single cell into a larger, more complex adult, its body structure, its diet, and the environment in which it lives may all change. In the next section we describe some animal life cycles and discuss why they are so varied.

31.4 How Do Animal Life Cycles Differ?

The **life cycle** of an animal encompasses its embryonic development, birth, growth to maturity, reproduction, and death. During its life an individual animal ingests food, grows, interacts with other individuals of the same and other species, and reproduces.

In some groups of animals, newborns are similar in many ways to adults (a pattern called *direct development*). Newborns of most species, however, differ dramatically from adults. An immature life cycle stage that has a form different from that of the adult is called a **larva** (plural *larvae*). Some of the most striking life cycle changes are found among insects such as beetles, flies, moths, butterflies, and bees, which undergo radical changes (called *metamorphosis*) between their larval and adult stages (**Figure 31.11**). In these animals, one stage may be specialized for feeding and the other for reproduction. Adults of most moth species, for example, do not eat. Alternatively, individuals of all life cycle stages may eat, but what they eat may change. Butterfly larvae, known as *caterpillars*, eat leaves and flowers; most adult butterflies eat only nectar. Having different life cycle stages that are specialized for different activities may increase the efficiency with which the animal performs particular tasks.

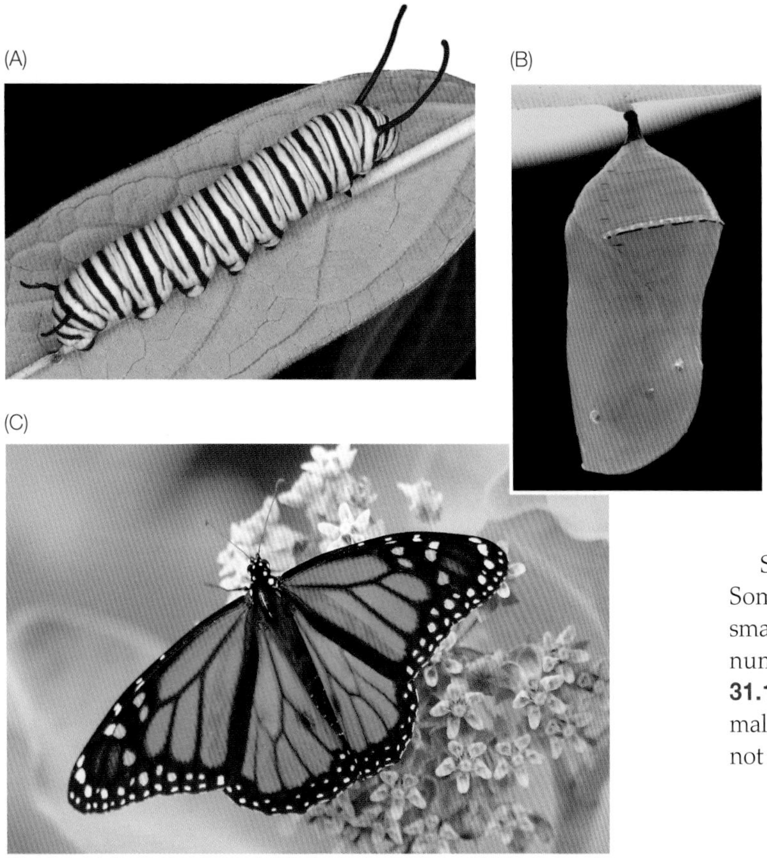

31.11 A Life Cycle with Metamorphosis (A) The larval stage (caterpillar) of the monarch butterfly, *Danaus plexippus*, is specialized for feeding. (B) The pupa is the stage during which the transformation to the adult form occurs. (C) The adult butterfly is specialized for dispersal and reproduction.

All life cycles have at least one dispersal stage

At some time during its life, an animal moves, or is moved, so that it does not die exactly where it was born. Such movement is called **dispersal**.

Animals that are sessile as adults typically disperse as eggs or larvae. This pattern is common among sessile marine animals, most of which discharge their small eggs and sperm into the water, where fertilization takes place. A larva soon hatches and floats freely in the plankton as it filters small prey from the water.

Many animals that live on the seafloor, including polychaete worms and mollusks, have a common larval form, the **trochophore** (**Figure 31.12A**). Some other marine animals, such as crustaceans, have a different, bilaterally symmetrical larval form, called a **nauplius** (**Figure 31.12B**). Both types of larvae feed for some time in the plankton before settling on a substratum and transforming into adults. Larvae that feed in, and are dispersed by the movement of, water probably evolved in these diverse groups of animals because they are all filter feeders on small organisms that are widely dispersed in the water column.

Most animals that are motile as adults disperse when they are mature. A caterpillar, for example, may spend its entire larval stage feeding on a single plant, but after its metamorphosis into a flying adult—a butterfly—it may fly to and lay eggs on other plants located far from the one where it spent its caterpillar days. In some species, individuals disperse during several different life cycle stages.

No life cycle can maximize all benefits

The common saying, "Jack of all trades, master of none," suggests why there are constraints on the evolution of life cycles. The characteristics of an animal in any one life cycle stage may improve its performance in one activity, but reduce its performance in another—a situation known as a **trade-off**. An animal that is good at filtering small food particles from the water, for example, probably cannot capture large prey. Similarly, energy devoted to building protective structures such as shells cannot be used for growth.

Some major trade-offs can be seen in animal reproduction. Some animals produce large numbers of small eggs, each with a small energy store (**Figure 31.13A**). Other animals produce a small number of large eggs, each with a large energy store (**Figure 31.13B**). With a fixed amount of available energy, a female animal can produce many small eggs or a few large eggs, but she cannot do both.

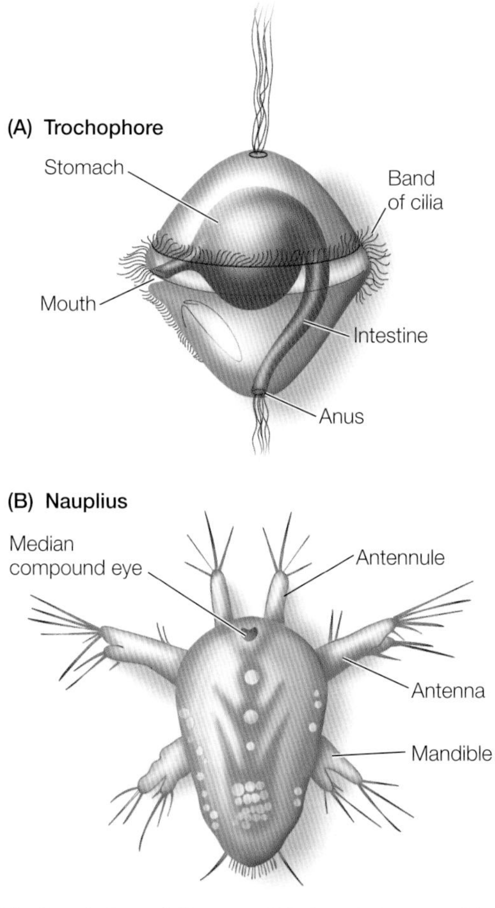

31.12 Planktonic Larval Forms of Marine Animals (A) The trochophore ("wheel-bearer") is a distinctive larval form found in several marine animal clades with spiral cleavage, most notably the polychaete worms and the mollusks. (B) This nauplius larva will mature into a crustacean with a segmented body and jointed appendages.

(A) *Rana temporaria*

(B) *Pygoscelis antarctica*

31.13 Many Small or Few Large Allocation of energy to eggs requires trade-offs. (A) This frog has divided her reproductive energy among a large number of small eggs. (B) This penguin has invested all her reproductive energy in one large egg.

The larger the energy store in an egg, the longer an offspring can develop before it must either find its own food or be fed by its parents. Birds of all species lay relatively small numbers of relatively large eggs, but incubation periods vary. In some species, eggs hatch when the young are still helpless (**Figure 31.14A**). Such *altricial* young must be fed and cared for until they can feed themselves; parents can provide for only a small number of altricial offspring. On the other hand, some bird species incubate their eggs longer, and the hatchlings are developed to a point that they are able to forage for themselves almost immediately (**Figure 31.14B**). The young of such species are called *precocial*.

Parasite life cycles evolve to facilitate dispersal and overcome host defenses

Animals that live as internal parasites are bathed in the nutritious tissues of their host or in the digested food that fills their host's digestive tract. Thus they may not need to exert much energy to obtain food, but to survive, they must overcome the host's defenses. Furthermore, either they or their offspring must disperse to new hosts while their host is still living, because they die when their host dies.

(A) *Parus caeruleus*

(B) *Branta canadensis*

31.14 Helpless or Independent (A) The altricial young of the blue tit are essentially helpless when they hatch. Their parents feed and care for them for several weeks. (B) Canada geese hatchlings are precocial. They are ready to swim and feed independently almost immediately after hatching.

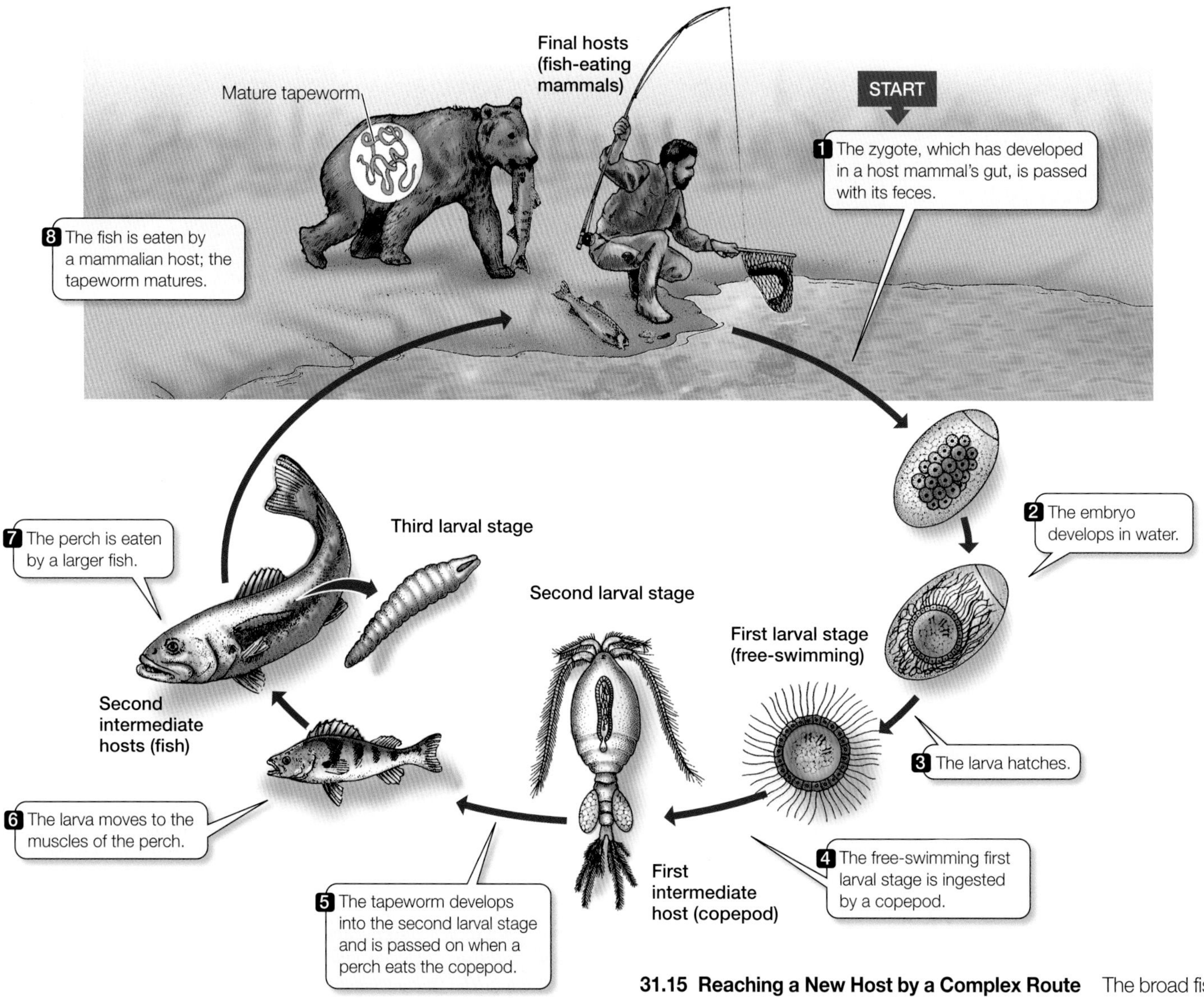

31.15 Reaching a New Host by a Complex Route The broad fish tapeworm, *Diphyllobothrium latum*, must pass through the bodies of a copepod (a type of crustacean) and a fish before it can reinfect its primary host, a mammal. Such complex life cycles assist the parasite's colonization of new host individuals, but they also provide opportunities for humans to break the cycle with hygienic measures.

The fertilized eggs of some parasites are voided with the host's feces and later ingested directly by other host individuals. Most parasite species, however, have complex life cycles involving one or more intermediate hosts and several larval stages (**Figure 31.15**). Some intermediate hosts transport individual parasites directly between other hosts. Others house and support the parasite until another host ingests it. Thus complex life cycles may facilitate the transfer of individual parasites among hosts.

31.4 RECAP

Different life cycle stages of animals may differ in form and be specialized for different activities.

- How do trade-offs constrain the evolution of life cycles? See p. 680
- Can you explain why parasites often have complicated life cycles? See pp. 681–682 and Figure 31.15

31.5 What Are the Major Groups of Animals?

The millions of species of animals display variations in body symmetry and body cavity structure and have a host of different life cycles, patterns of development, and survival strategies. In the remainder of this chapter and in the following two chapters we will become acquainted with the major animal groups and learn how the general characteristics described in this chapter apply to each of them.

Table 31.1 provides a summary of the living members of the major animal groups. The **Bilateria** is a large monophyletic group embracing all animals other than sponges, ctenophores, and cnidarians. The traits that support the monophyly of Bilateria are

TABLE 31.1

Summary of Living Members of the Major Groups of Animals

	APPROXIMATE NUMBER OF LIVING SPECIES DESCRIBED	MAJOR GROUPS
Glass sponges	500	
Demosponges	7,000	
Calcareous sponges	500	
Ctenophores	100	
Cnidarians	11,000	Anthozoans: Corals, sea anemones
		Hydrozoans: Hydras and hydroids
		Scyphozoans: Jellyfishes
PROTOSTOMES		
Arrow worms	100	
Lophotrochozoans		
Ectoprocts	4,500	
Flatworms	25,000	Free-living flatworms; flukes and tapeworms (all parasitic); monogeneans (ectoparasites of fishes)
Rotifers	1,800	
Ribbon worms	1,000	
Phoronids	20	
Brachiopods	335	
Annelids	16,500	Polychaetes (all marine)
		Clitellates: Earthworms, freshwater worms, leeches
Mollusks	95,000	Monoplacophorans
		Chitons
		Bivalves: Clams, oysters, mussels
		Gastropods: Snails, slugs, limpets
		Cephalopods: Squids, octopuses, nautiloids

	APPROXIMATE NUMBER OF LIVING SPECIES DESCRIBED	MAJOR GROUPS
Ecdysozoans		
Kinorhynchs	150	
Loriciferans	100	
Priapulids	16	
Horsehair worms	320	
Nematodes	25,000	
Onychophorans	150	
Tardigrades	800	
Arthropods:		
Crustaceans	50,000	Crabs, shrimps, lobsters, barnacles, copepods
Hexapods	1,000,000	Insects and relatives
Myriapods	14,000	Millipedes, centipedes
Chelicerates	89,000	Horseshoe crabs, arachnids (scorpions, harvestmen, spiders, mites, ticks)
DEUTEROSTOMES		
Echinoderms	7,000	Crinoids (sea lilies and feather stars); brittle stars; sea stars; sea daisies; sea urchins; sea cucumbers
Hemichordates	95	Acorn worms and pterobranchs
Urochordates	3,000	Ascidians (sea squirts)
Cephalochordates	30	Lancelets
Vertebrates	52,000	Hagfish; lampreys
		Cartilaginous fishes
		Ray-finned fishes
		Coelacanths
		Amphibians
		Reptiles (including birds)
		Mammals

bilateral symmetry, three cell layers, and the presence of at least seven Hox genes (see Chapters 19 and 20).

The bilaterian animals comprise the two major categories mentioned earlier in this chapter, and are classified as either **protostomes** or **deuterostomes** (see Figure 31.1). These two groups have been evolving separately for over 500 million years, since the early Cambrian or late Precambrian. We will describe the protostomes in Chapter 32 and the deuterostomes in Chapter 33.

The remainder of this chapter describes those animal groups that are not bilaterians. The simplest animals, the sponges, have no cell layers and no organs. Sponges are not a clade, but the name is used for three groups that exhibit the ancestral body organization of animals. All other animals, including the bilatarians, are known as **eumetazoans**. They have obvious body symmetry, a gut, a nervous system, special types of cell junctions, and well-organized tissues in distinct cell layers (although there have been secondary losses of some of these structures in some eumetazoans). Sponges lack all these features.

Sponges are loosely organized animals

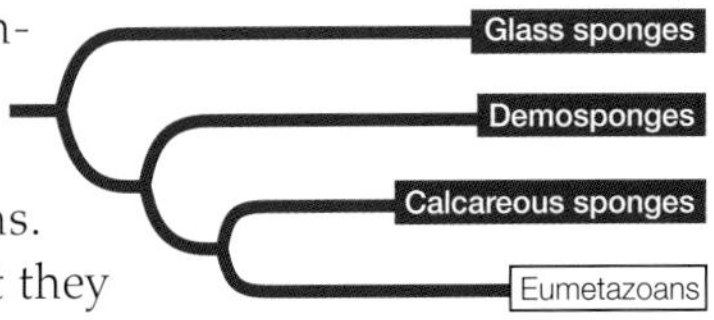

Sponges are the simplest of animals. They have some specialized cells, but no distinct cell layers and no true organs. Early naturalists thought that they were plants because they lacked body symmetry.

Sponges have hard skeletal elements called **spicules**, which may be small and simple or large and complex. Recent analyses of ribosomal RNA genes suggest that there are three major groups of sponges, which are paraphyletic with respect to the remaining animals. Members of two groups (*glass sponges* and *demosponges*) have skeletons composed of silicaceous spicules made of hydrated silicon dioxide (**Figure 31.16A,B**). These spicules are remarkable in having greater flexibility and toughness than synthetic glass rods of similar length. Members of the third group, the *calcareous sponges*, take their name from their calcium carbonate skeletons (**Figure 31.16C**). It is the latter group that is most closely related to the eumetazoans.

The body plan of sponges of all three groups—even large ones, which may reach a meter or more in length—is an aggregation of cells built around a water canal system. Water, along with any food particles it contains, enters the sponge by way of small pores and passes into the water canals, where choanocytes capture food particles (see Figures 31.7 and 31.16A).

Sponges have been collected for thousands of years. The sponge industry peaked in 1938, when more than 3.6 million kilograms were harvested and marketed. Today most commercial "sponges" are made of synthetic materials created in laboratories.

A skeleton of simple or branching spicules, and often a complex network of elastic fibers, supports the bodies of most sponges. Sponges also have an extracellular matrix, composed of collagen, adhesive glycoproteins, and other molecules, that holds the cells together. Most species are filter feeders; a few species are carnivores that trap prey on hook-shaped spicules that protrude from the body surface.

Most of the 8,000 species of sponges are marine animals; only about 50 species live in fresh water. Sponges come in a wide variety of sizes and shapes that are adapted to different movement patterns of water. Sponges living in intertidal or shallow subtidal environments with strong wave action are firmly attached to the substratum. Most sponges that live in slowly flowing water are flattened and are oriented at right angles to the direction of current flow. They intercept water and the prey it contains as it flows past them.

(A) *Xestospongia testudinaria*

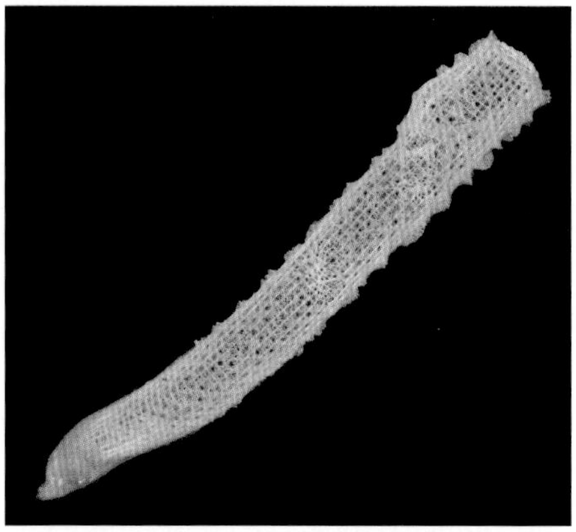

(B) *Euplectella aspergillum*

(C) *Leucilla nuttingi*

31.16 There Are Three Groups of Sponges (A) The great majority of sponge species are demosponges such as these Pacific barrel sponges. The system of pores and water canals "typical" of the sponge body plan is apparent. (B) The supporting structures of both demosponges and glass sponges are silicaceous spicules, seen here in the skeleton of a glass sponge. (C) The skeletons of calcareous sponges are made of calcium carbonate. Calcareous sponges are more closely related to the eumetazoans than to the other two groups of sponges.

Sponges reproduce both sexually and asexually. In most species, a single individual produces both eggs and sperm, but individuals do not self-fertilize. Water currents carry sperm from one individual to another. Asexual reproduction is by budding and fragmentation.

Ctenophores are radially symmetrical and diploblastic

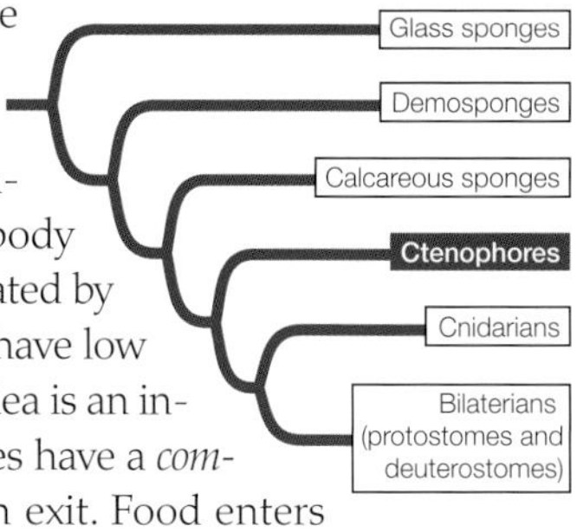

The **ctenophores**, also known as the comb jellies, lack most of the Hox genes possessed by all other eumetazoans. Ctenophores have a radially symmetrical, diploblastic body plan, with the two cell layers separated by a thick, gelatinous **mesoglea**. They have low metabolic rates because the mesoglea is an inert extracellular matrix. Ctenophores have a *complete gut*, with an entrance and an exit. Food enters through a mouth, and wastes are eliminated through two anal pores.

Ctenophores have eight comblike rows of fused plates of cilia, called **ctenes** (**Figure 31.17**). A ctenophore moves through the water by beating these cilia rather than by muscular contractions. Its feeding tentacles are covered with cells that discharge adhesive material when they contact prey. After capturing its prey, a ctenophore retracts its tentacles to bring the food to its mouth. In some species, the entire surface of the body is coated with sticky mucus that captures prey. All of the 100 known species of ctenophores eat small planktonic organisms. They are common in open seas.

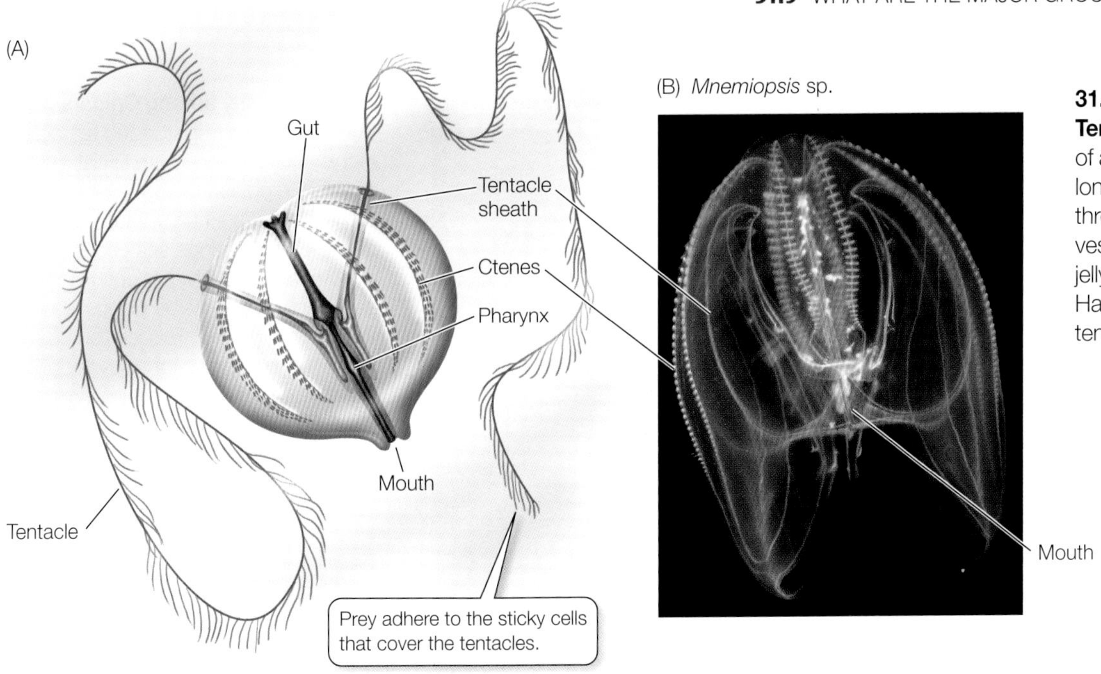

31.17 Comb Jellies Feed with Tentacles (A) The body plan of a typical ctenophore. The long, sticky tentacles sweep through the water, efficiently harvesting small prey. (B) This comb jelly photographed in Sydney Harbor, Australia, has short tentacles.

Ctenophore life cycles are uncomplicated. Gametes are released into the body cavity and then discharged through the mouth or the anal pores. Fertilization takes place in open seawater. In nearly all species, the fertilized egg develops directly into a miniature ctenophore that gradually grows into an adult.

Cnidarians are specialized carnivores

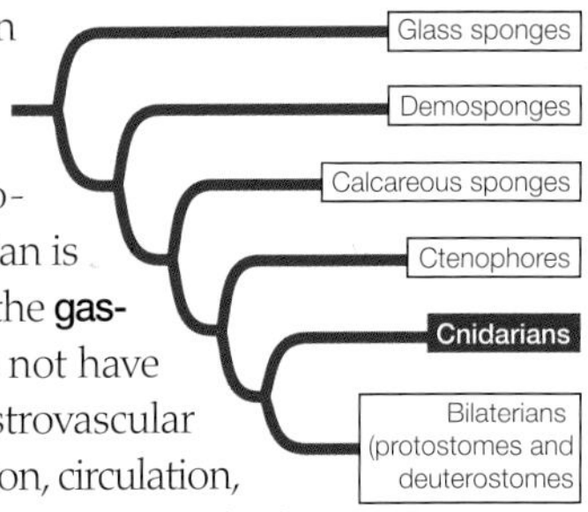

One branch of the next split in the animal lineage led to the **cnidarians** (jellyfishes, sea anemones, corals, and hydrozoans). The mouth of a cnidarian is connected to a blind sac called the **gastrovascular cavity** (thus it does not have a complete gut). The gastrovascular cavity functions in digestion, circulation, and gas exchange, and it also acts as a hydrostatic skeleton. The single opening serves as both mouth and anus.

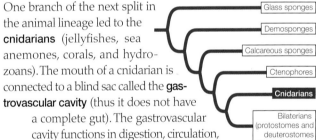

The life cycle of most cnidarians has two distinct stages, one sessile and the other motile (**Figure 31.18**). In the sessile **polyp** stage, a cylindrical stalk is attached to the substratum. Individual polyps may reproduce asexually by budding, thereby forming a colony. The motile **medusa** (plural *medusae*) is a free-swimming stage shaped like a bell or an umbrella. It typically floats with its mouth and feeding tentacles facing downward. Medusae of many species produce eggs and sperm and release them into the water. A fertilized egg develops into a free-swimming, ciliated larva called a **planula**, which eventually settles to the bottom and develops into a polyp.

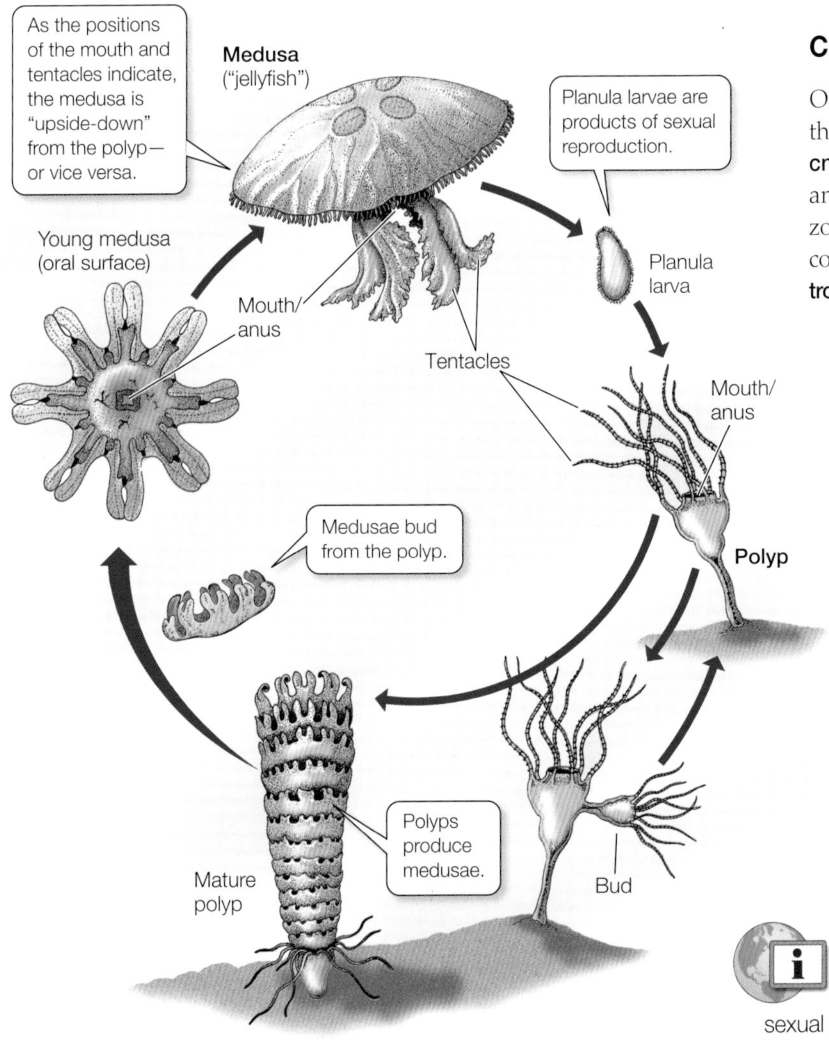

31.18 The Cnidarian Life Cycle Has Two Stages The life cycle of a scyphozoan (jellyfish) exemplifies the typical cnidarian body forms: the sessile, asexual polyp; and the motile, sexual medusa.

(A) *Anthopleura elegantissima*

(B) *Ptilosarcus gurneyi*

(C) *Phyllorhyza punctata*

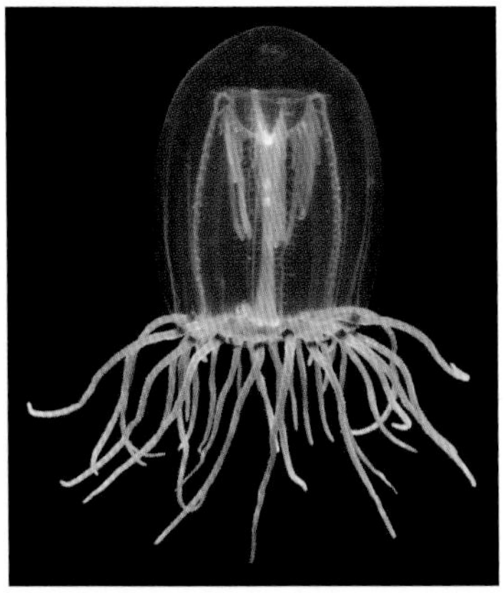

(D) *Polyorchis penicillatus*

31.19 Diversity among Cnidarians (A) The nematocyst-studded tentacles of this sea anemone from British Columbia are poised to capture large prey carried to the animal by water movement. (B) The orange sea pen is a colonial cnidarian that lives in soft bottom sediments and projects polyps above the substratum. (C) This jellyfish illustrates the complexity of a scyphozoan medusa. (D) The internal structure of the medusa of a North Atlantic colonial hydrozoan is visible here.

Cnidarians have epithelial cells with muscle fibers whose contractions enable the animals to move, as well as simple *nerve nets* that integrate their body activities. They also have specialized structural molecules (collagen, actin, and myosin) and Hox genes. They are specialized carnivores, using the toxin in their nematocysts to capture relatively large and complex prey (see Figure 31.10). Some cnidarians, including corals and anemones, gain additional nutrition from photosynthetic protists that live in their tissues. Cnidarians, like ctenophores, are largely made up of inert mesoglea. They have low metabolic rates and can survive in environments where they encounter prey only infrequently.

Of the roughly 11,000 cnidarian species living today, all but a few live in the oceans (**Figure 31.19**). The smallest cnidarians can hardly be seen without a microscope; the largest known jellyfish is 2.5 meters in diameter. We will describe three clades of cnidarians that have many species: the scyphozoans, anthozoans, and hydrozoans.

SCYPHOZOANS The several hundred species of scyphozoans are all marine. The mesoglea of their medusae is thick and firm, giving rise to their common name—jellyfishes or sea jellies. The medusa rather than the polyp dominates the life cycle of scyphozoans. An individual medusa is male or female, releasing eggs or sperm into the open sea. The fertilized egg develops into a small planula larva that quickly settles on a substratum and develops into a small polyp. This polyp feeds and grows and may produce additional polyps by budding. After a period of growth, the polyp begins to bud off small medusae, which feed, grow, and transform themselves into adult medusae (see Figures 31.18 and 31.19C).

ANTHOZOANS Members of the **anthozoan** clade include sea anemones, sea pens, and corals. Sea anemones (**Figure 31.19A**), all of which are solitary, are widespread in both warm and cold ocean waters. Sea pens (**Figure 31.19B**), by contrast, are colonial. Each colony consists of at least two different kinds of polyps. The primary polyp has a lower portion anchored in the bottom sediment and a branched upper portion, which projects above the substratum. Along the upper portion, the primary polyp produces smaller secondary polyps by budding. Some of these secondary polyps differentiate into feeding polyps; others circulate water through the colony.

Corals also are sessile and colonial. The polyps of most corals form a skeleton by secreting a matrix of organic molecules on which they deposit calcium carbonate, which forms the eventual skeleton of the coral colony. As the colony grows, old polyps die but their calcium carbonate skeletons remain. The living members form a layer on top of a growing bank of skeletal remains, eventually forming chains of islands and reefs (**Figure 31.20A**). The common names of coral groups—horn corals, brain corals, staghorn corals, and organ pipe corals, among others—describe their appearance (**Figure 31.20B**).

The Great Barrier Reef along the northeastern coast of Australia is a system of coral formations more than 2,000 km long—about the distance from New York City to St. Louis. A single coral reef in the Red Sea has been calculated to contain more material than all the buildings in the major cities of North America combined.

(A)

(B) *Diploria labyrinthiformis*

31.20 Corals (A) Many different coral species form an Indian Ocean reef in Chumbe Coral Park off the coast of Zanzibar. (B) The very descriptive common name of this Caribbean coral is "brain coral." The Latinate species name reflects its maze-like, or "labyrinthine," appearance.

31.5 RECAP

The bilaterian animals fall into two major clades, the protostomes and the deuterostomes. The nonbilateran animals—the sponges, ctenophores, and cnidarians—have relatively simple structures and correspondingly simple feeding strategies.

- Do you understand why sponges are considered to be animals, even though they lack the complex body structures found among the other animal groups? See p. 683 and Figure 31.7
- What are some major features of the cnidarian clade? See pp. 685–686 and Figure 31.18

Corals flourish in clear, nutrient-poor tropical waters. They can grow rapidly in such environments because unicellular photosynthetic protists live endosymbiotically within their cells. These protists provide the corals with products of photosynthesis; the corals, in turn, provide the protists with nutrients and a place to live. This endosymbiotic relationship explains why reef-forming corals are restricted to clear surface waters, where light levels are high enough to support photosynthesis.

Coral reefs throughout the world are threatened both by global warming, which is raising the temperatures of shallow tropical ocean waters (see Figure 57.10), and by polluted runoff from development on adjacent shorelines. An overabundance of nitrogen in the runoff gives an advantage to algae, which overgrow and eventually smother the corals.

HYDROZOANS Hydrozoans have diverse life cycles. The polyp typically dominates the life cycle, but some species have only medusae; others have only polyps. Most hydrozoans are colonial. A single planula larva eventually gives rise to a colony of many polyps, all interconnected and sharing a continuous gastrovascular cavity (**Figure 31.21**). Within such a colony some polyps have tentacles with many nematocysts; they capture prey for the colony. Others lack tentacles and are unable to feed, but are specialized for the production of medusae. Still others are fingerlike and defend the colony with their nematocysts.

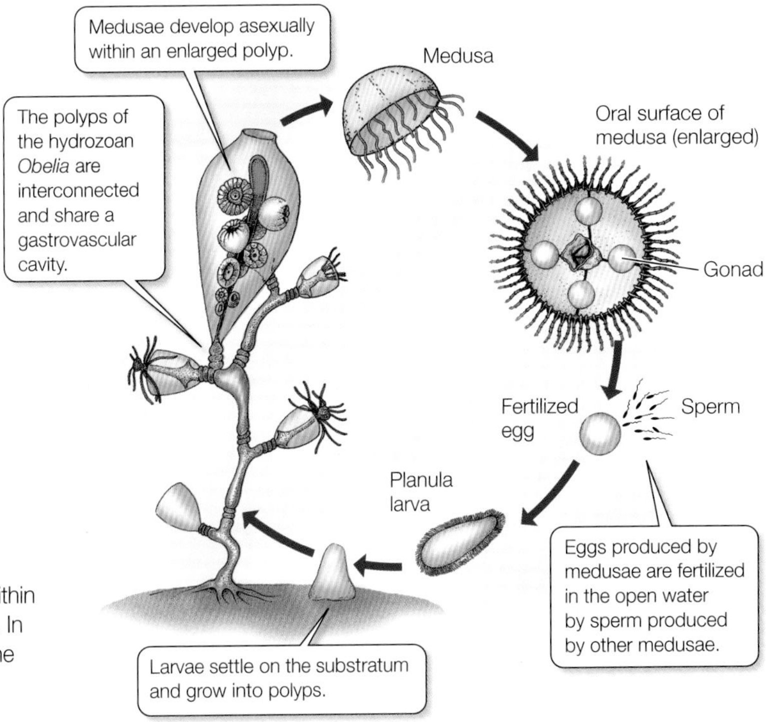

31.21 Hydrozoans Often Have Colonial Polyps The polyps within a hydrozoan colony may differentiate to perform specialized tasks. In the species whose life cycle is diagrammed here, the medusa is the sexual reproductive stage, producing eggs and sperm in organs called gonads.

CHAPTER SUMMARY

31.1 What evidence indicates the animals are monophyletic?

Most **animals** share a set of derived traits not found in other groups of organisms. These traits include similarities in their ribosomal RNA and Hox genes, cell junctions, and an extracellular matrix.

Patterns of embryonic development provide important clues to the evolutionary relationships among animals. **Diploblastic** animals develop two embryonic cell layers; **triploblastic** animals develop three.

Differences in patterns of early development also characterize two major clades of triploblastic animals, the **protostomes** and **deuterostomes**.

31.2 What are the features of animal body plans?

Animal **body plans** can be described in terms of **symmetry**, **body cavity** structure, **segmentation**, and **appendages**.

A few animals have **spherical** symmetry, but most animals have either **radial** or **bilateral** symmetry. Review Figure 31.3

Most animals with radial symmetry move slowly or not at all, while animals with bilateral symmetry are able to move more rapidly. Many bilaterally symmetrical animals are **cephalized**, with sensory and nervous tissues in an anterior head.

On the basis of their body cavity structure, animals can be described as **acoelomates**, **pseudocoelomates**, or **coelomates**. Review Figure 31.4

Segmentation, which takes many forms, improves control of movement, especially if the animal also has appendages.

31.3 How do animals get their food?

Motile animals can move to find food; **sessile** animals stay in one place and capture food that is brought to them.

Filter feeders filter small organisms and organic molecules from their environment.

Predators have features that enable them to capture and subdue large animal prey.

Herbivores consume plants, usually without killing them.

Parasites live in or on other organisms and obtain nutrition from these host individuals.

31.4 How do animal life cycles differ?

The stages of an animal's **life cycle** may be specialized for different activities.

An immature stage that is dramatically different from the adult stage is called a **larva**.

All life cycles have at least one **dispersal** stage, so that the animal does not die in the same place where it was born. Many sessile marine animals can be grouped by the presence of one of two distinct dispersal stages, the **trochophore** larva and the **nauplius** larva.

Parasites have complex life cycles that may involve one or more hosts and several larval stages. Review Figure 31.15

A characteristic of an animal or a life cycle stage may improve its performance in one activity, but reduce its performance in another, a situation known as a **trade-off**.

31.5 What are the major groups of animals?

All animals other than sponges, ctenophores, and cnidarians belong to a large monophyletic group called the **Bilateria**. The clade **Eumetazoa** embraces all animals other than sponges.

Sponges are simple animals that lack cell layers and true organs. They have skeletons made up of silicaceous or calcareous **spicules**. They create water currents and capture food with flagellated feeding cells called **choanocytes**. Choanocytes are an evolutionary link between the animals and choanoflagellate protists. Review Figure 31.7

The **ctenophores** and **cnidarians** are diploblastic, radially symmetrical animals.

The two cell layers of ctenophores are separated by an inert extracellular matrix called **mesoglea**. They move by beating fused plates of cilia called **ctenes**. Review Figure 31.17

The life cycle of cnidarians has two distinct stages: a sessile **polyp** stage and a motile **medusa**. A fertilized egg develops into a free-swimming **planula** larva, which settles to the bottom and develops into a polyp. Review Figures 31.18 and 31.21, Web/CD Tutorial 31.1

See Web/CD Activities 31.1 and 31.2 for a concept review of this chapter.

SELF-QUIZ

1. The body plan of an animal is
   a. its general structure.
   b. the integrated functioning of its parts.
   c. its general structure and the integrated functioning of its parts.
   d. its general structure and its evolutionary history.
   e. the integrated functioning of its parts and its evolutionary history.
2. A bilaterally symmetrical animal can be divided into mirror images by
   a. any plane through the midline of its body.
   b. any plane from its anterior to its posterior end.
   c. any plane from its dorsal to its ventral surface.
   d. any plane through the midline of its body from its anterior to its posterior end.
   e. a single plane through the midline of its body from its dorsal to its ventral surface.
3. Among protostomes, cleavage of the fertilized egg is
   a. delayed while the egg continues to mature.
   b. always radial.
   c. spiral in some species and radial in others.
   d. triploblastic.
   e. diploblastic.
4. Many parasites evolved complex life cycles because
   a. they are too simple to disperse readily.
   b. they are poor at recognizing new hosts.
   c. they were driven to it by host defenses.
   d. complex life cycles increase the probability of a parasite's transfer to a new host.
   e. their ancestors had complex life cycles and they simply retained them.

5. Animals are believed to be monophyletic because
   a. their presumed ancestor has been found in the fossil record.
   b. they all have the same number of Hox genes.
   c. they all share a set of derived traits not found in other clades.
   d. they all lack ribosomal RNAs.
   e. they all lack extracellular matrix molecules.
6. In the common ancestor of the protostomes and deuterostomes, the pattern of early cleavage was
   a. spiral.
   b. radial.
   c. biradial.
   d. deterministic.
   e. haphazard.
7. A fluid-filled body cavity can function as a hydrostatic skeleton because
   a. fluids are moderately compressible.
   b. fluids are highly compressible.
   c. fluids are relatively incompressible.
   d. fluids have the same density as body tissues.
   e. fluids can be moved by ciliary action.
8. Which of the following is *not* a feature that enables some animals to capture large prey?
   a. Sharp teeth
   b. Claws
   c. Toxins
   d. A filtering device
   e. Tentacles with stinging cells
9. The sponge body plan is characterized by
   a. a mouth and digestive cavity but no muscles or nerves.
   b. muscles and nerves but no mouth or digestive cavity.
   c. a mouth, digestive cavity, and spicules.
   d. muscles and spicules but no digestive cavity or nerves.
   e. the lack of a mouth, digestive cavity, muscles, or nerves.
10. Cnidarians have the ability to
    a. live in both salt and fresh water.
    b. move rapidly in the water column.
    c. capture and consume large numbers of small prey.
    d. survive where food is scarce because of their low metabolic rate.
    e. capture large prey and to move rapidly.

FOR DISCUSSION

1. Differentiate among the members of each of the following sets of related terms:
   a. radial symmetry/bilateral symmetry
   b. protostome/deuterostome
   c. diploblastic/triploblastic
   d. coelomate/pseudocoelomate/acoelomate
2. In this chapter we listed some of the traits shared by all animals that convince most biologists that all animals are descendants of a single common ancestral lineage. In your opinion, which of these traits provides the most compelling evidence that animals are monophyletic? If morphological and molecular data do not agree, should one type of evidence be given greater weight? If so, which one?
3. Describe some features that allow animals to capture prey that are larger and more complex than they themselves are.
4. Why is bilateral symmetry strongly associated with cephalization, the concentration of sense organs in an anterior head?
5. How does a slow metabolic rate enable an animal to live in an unproductive environment?

FOR INVESTIGATION

The discovery of fossils of mature dinosaur embryos yielded valuable information about how these animals changed over the course of their lives. What further investigations might be carried out using information from fossils to tell us what modern developmental biology theories must explain?

CHAPTER 32 Protostome Animals

Tiny parasites exert mind control

Most people have never heard of a strepsipteran, and even those who *have* heard of strepsipterans probably have never seen one. They don't know what they are missing! These tiny insects—there are some 600 different species of them—parasitize hundreds of other insect species, including bees, wasps, ants, grasshoppers, and cockroaches. Males and females of most strepsipteran species parasitize the same host species, although in one clade males parasitize ants while females parasitize grasshoppers.

Strepsipteran males and females are often so different that even determining that they are members of the same species requires a DNA analysis, and they have some of the strangest life cycles of any animal. Once grown to maturity within their hosts (whose internal organs they consume), the males of most species emerge looking like a "typical" insect. The females also consume the host from within, but they usually remain inside the host. A mature female extrudes her head and parts of her body from the body of her host. The extruded body parts contain an opening that receives sperm from a male. Much later, this opening becomes an exit for the strepsipteran larvae. The host insects are left dead or severely damaged and produce no offspring of their own.

Strepsipterans dramatically change the behavior of their hosts in ways that help them complete their life cycles at the expense of their hosts' reproduction. For example, when wasps—a typical host—are parasitized by strepsipterans, the parasites generate signals that induce the wasps to leave their nest and form a mating aggregation. This aggregation, however, serves the strepsipterans, not the wasps. Once the wasps aggregate, the male strepsipterans emerge from their hosts to search for and mate with the females whose heads are now poking out of the bodies of other wasps.

Adult male strepsipterans live only a few hours, during which they must find a female and mate. Because the protruding part of a female's body is barely visible, the males have unusually large eyes, and about 75 percent of their brain cells are allocated to vision. This sensory system serves a single purpose: to help the male find a female.

Same Parasite, Different Lifestyles Strepsipterans are parasitic insects that grow to maturity within host insects. Most male strepsipterans look insect-like upon reaching maturity and leave their host to find and mate with a very different-appearing female strepsipteran, who remains inside her host.

A Host Insect Strepsipterans parasitize many different insect species; wasps of the genus *Polistes* (paper wasps) are common hosts. Molecular signals generated by the parasites can dramatically affect the behavior of the wasps.

Strepsipterans and their hosts are all insects, and insects account for more than half of the described species of protostomes. Other protostome groups, such as mollusks, nematodes, crabs, spiders, and ticks, are also species-rich. Many of the protostome species are parasites. Parasites often live within their hosts and absorb nutrients through their body walls. Some parasites, including the strepsipterans, can more technically be described as *parasitoids*, which consume the host's tissues as they develop from eggs laid on or within the host's body and often grow to be almost as large as their hosts.

IN THIS CHAPTER we will describe the characteristics of protostome animals and describe the members of two major protostome clades, the lophotrochozoans and the ecdysozoans. We will give particular attention to the arthropods, an incredibly species-rich group of ecdysozoans with rigid exoskeletons and jointed appendages.

CHAPTER OUTLINE

32.1 What Is a Protostome?

Some time after the origin of the diplobastic radiate animals, (the cnidarians and ctenophores), a third embryological germ layer—the mesoderm—arose. This feature is found in the two great triploblastic animal clades, the protostomes and the deuterostomes. If we were to judge solely on the basis of numbers, both of species and of individuals, the protostomes would emerge as by far the more successful of the two groups.

As described in Chapter 31, the name protostome means "mouth first," and was applied because in most of these species the embryonic blastopore becomes the mouth; this is in contrast to deuterostome animals, in which the blastopore becomes the anal opening of the digestive tract (see Figure 31.2). This trait is not universally shared, however; for example, no blastopore forms during the early development of insects.

The protostomes are extremely varied, but they are all bilaterally symmetrical animals whose bodies exhibit two major derived traits:

- An anterior *brain* that surrounds the entrance to the digestive tract
- A ventral *nervous system* consisting of paired or fused longitudinal nerve cords

Other aspects of protostome body organization can be quite different from group to group (**Table 32.1**). There are several coelomate groups as well as several groups of pseudocoelomates; one important clade, the flatworms, are acoeolomate (see Figure 31.4). In two of the most prominent clades, the coelom has been secondarily modified:

- The *arthropods* have lost the ancestral coelom over the course of evolution. Their internal body cavity has become a *hemocoel*, or "blood chamber," in which fluid from an open (i.e., no blood vessels) circulatory system bathes the internal organs.
- The *mollusks* are also generally characterized by an open circulatory system and have some of the attributes of the hemocoel, but they retain vestiges of an enclosed coelom around their major organs.

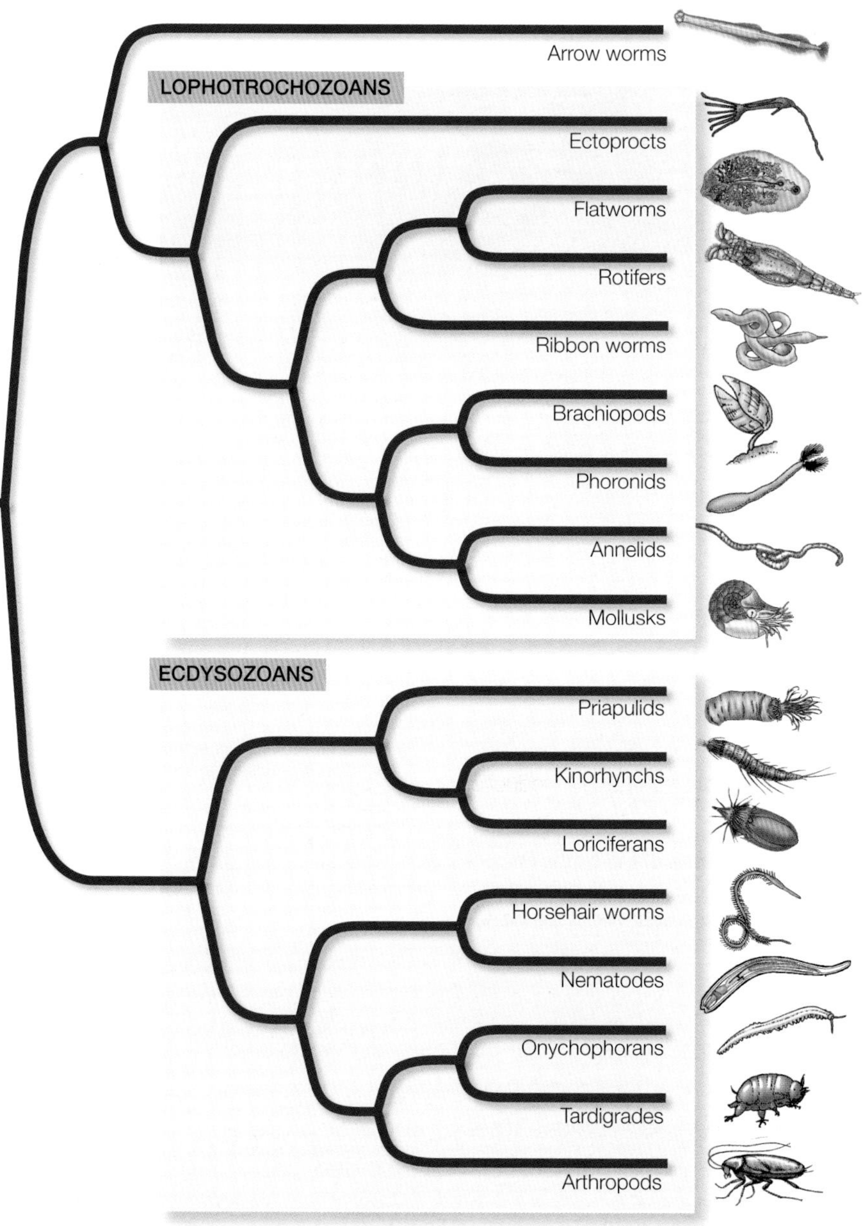

32.1 A Current Phylogenetic Tree of Protostomes Two major lineages, the lophotrochozoans and the ecdysozoans, dominate the tree. Many small clades are not included. The position of the arrow worms is somewhat uncertain; another possibility is that they are the sister group of all other protostomes, not just the lophotrochozoans.

In ancient times, the protostomes split into two major clades—*lophotrochozoans* and *ecdysozoans*—that have been evolving independently ever since (**Figure 32.1**).

Trochophores, lophophores, and spiral cleavage evolved among the lophotrochozoans

The skeletons of most species of **lophotrochozoans** are internal, which means these animals grow by adding to the size of their skeletal elements. Some of them use cilia for locomotion; many groups have a type of free-living larva known as a *trochophore* that moves by beating a band of cilia (see Figure 31.12A).

Several distantly related groups of lophotrochozoans (including ectoprocts, brachiopods, and phoronids) have a **lophophore**, a circular or U-shaped ridge around the mouth that bears one or two rows of ciliated, hollow tentacles (**Figure 32.2**). This complex structure is an organ for both food collection and gas exchange. Biologists once grouped these taxa together as *lophophorates*, but it is now clear that they are not each other's closest relatives. The lophophore appears to have evolved independently at least twice, or else it is an ancestral feature of lophotrochozoans and has been lost in many groups. Nearly all animals with a lophophore are sessile as adults. They use the tentacles and cilia of the lophophore to capture small floating organisms from the water. Other sessile lophotrochozoans use less well developed tentacles for the same purpose.

As discussed in the previous chapter, some lophotrochozoans (including flatworms, ribbon worms, annelids, and mollusks) exhibit a derived form of cleavage in early development known as *spiral cleavage*. Some biologists group these taxa

TABLE 32.1

Anatomical Characteristics of Some Major Protostome Groups[a]

GROUP	BODY CAVITY	DIGESTIVE TRACT	CIRCULATORY SYSTEM
Arrow worms	Coelom	Complete	None
LOPHOTROCHOZOANS			
Flatworms	None	Dead-end sac	None
Rotifers	Pseudocoelom	Complete	None
Ectoprocts	Coelom	Complete	None
Brachiopods	Coelom	Complete in most	Open
Phoronids	Coelom	Complete	Closed
Ribbon worms	Coelom	Complete	Closed
Annelids	Coelom	Complete	Closed or open
Mollusks	Reduced coelom	Complete	Open except in cephalopods
ECDYSOZOANS			
Horsehair worms	Pseudocoelom	Greatly reduced	None
Nematodes	Pseudoceolom	Complete	None
Arthropods	Hemocoel	Complete	Open

[a]Note that all protostomes have bilateral symmetry.

Lophopus crystallinus

32.2 Ectoprocts Use Their Lophophore to Feed The extended lophophore dominates the anatomy of the colonial ectoprocts. This species inhabits freshwater, although most ectoprocts are marine. Ectoproct colonies can grow to contain over a million individuals, all stemming from the asexual reproduction of the colony's founder.

together as *spiralians,* although phylogenetic analyses of gene sequences do not support spiralians as monophyletic.

Members of several of the groups with spiral cleavage are *wormlike*, which means they are bilaterally symmetrical, legless, soft-bodied, and at least several times longer than they are wide. A wormlike body form enables animals to burrow efficiently through muddy and sandy marine sediment or soil. However, as we will describe later in this chapter, the *mollusks*—the most species-rich of the groups with spiral cleavage—have a very different body plan.

Ecdysozoans must shed their exoskeletons

Ecdysozoans have an external skeleton, or **exoskeleton**, which is a nonliving covering secreted by the underlying *epidermis* (the outermost cell layer). The exoskeleton provides these animals with both protection and support. Once formed, however, an exoskeleton cannot grow. How, then, can ecdysozoans increase in size? They do so by shedding, or **molting**, the exoskeleton and replacing it with a new, larger one. This molting process gives the clade its name (from the Greek word *ecdysis*, "to get out of").

A recently discovered fossil of a Cambrian soft-bodied arthropod, preserved in the process of molting, shows that molting evolved more than 500 million years ago (**Figure 32.3A**). An increasingly rich array of molecular and genetic evidence, including a set of Hox genes shared by all ecdysozoans, suggests that they have a single common ancestor. Thus molting of an exoskeleton is a trait that may have evolved only once during animal evolution.

Before an ecdysozoan molts, a new exoskeleton is already forming underneath the old one. Once the old exoskeleton is shed, the new one expands and hardens. But until it has hardened, the animal is vulnerable to its enemies, both because its outer surface is easy to penetrate and because an animal with a soft exoskeleton can move slowly or not at all (**Figure 32.3B**).

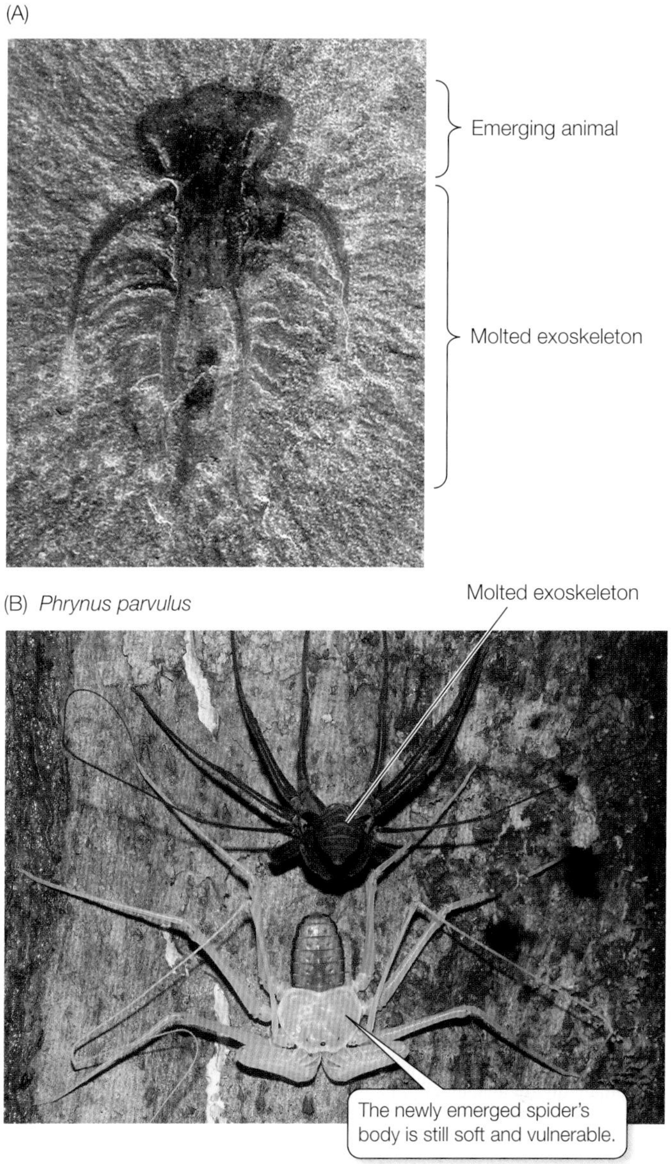

32.3 Molting: Past and Present (A) This 500 million-year-old fossil from the Cambrian captured an individual of a long-extinct arthropod species in the process of molting and shows that the molting process is an evolutionarily ancient trait. (B) This whip spider has just emerged from its discarded exoskelton and will be highly vulnerable until its new cuticle has hardened.

Some ecdysozoans have wormlike bodies covered by exoskeletons that are relatively thin and flexible. Such an exoskeleton, called a **cuticle**, offers the animal some protection, but provides only modest body support. A thin cuticle allows the exchange of gases, minerals, and water across the body surface, but restricts the animal to moist habitats. Many species of ecdysozoans with thin cuticles live in marine sediments from which they obtain prey, either by ingesting sediments and extracting organic material from them, or by capturing larger prey using a toothed *pharynx* (a muscular organ at the anterior end of the digestive tract). Some freshwater species absorb nutrients directly through their thin cuticles, as do parasitic species that live within their hosts. Many wormlike ecdysozoans are predators, eating protists and small animals.

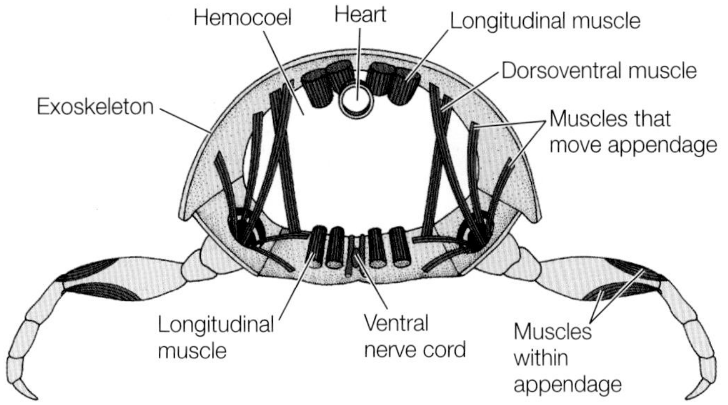

32.4 Arthropod Skeletons Are Rigid and Jointed This cross section through a thoracic segment of a generalized arthropod illustrates the arthropod body plan, which is characterized by a rigid exoskeleton with jointed appendages.

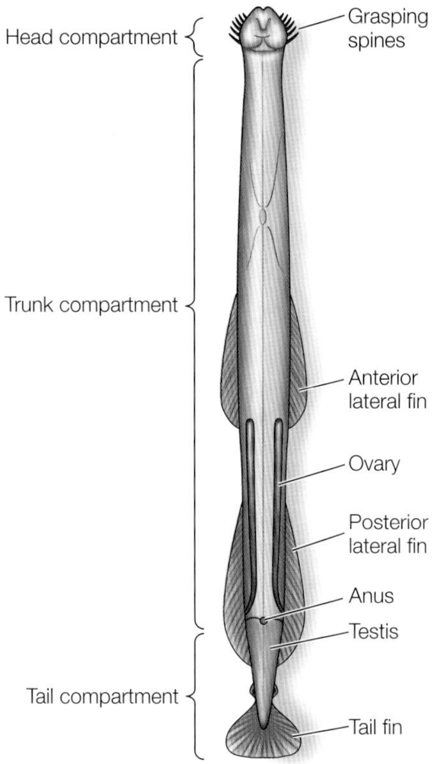

32.5 An Arrow Worm Arrow worms have a three-part body plan. Their fins and grasping spines are adaptations for a predatory lifestyle. Individuals are hermaphroditic, producing both eggs and sperm (ovary and testis).

The exoskeletons of other ecdysozoans are thickened by the incorporation of layers of protein and a strong, waterproof polysaccharide called **chitin**. An animal with a rigid, chitin-reinforced body covering can neither move in a wormlike manner nor use cilia for locomotion. A hard exoskeleton also impedes the passage of oxygen and nutrients into the animal, presenting new challenges in other areas besides growth. Therefore, ecdysozoans with hard exoskeletons evolved new mechanisms of locomotion and gas exchange.

To move rapidly, an animal with a rigid exoskeleton must have extensions of the body that can be manipulated by muscles. Such *appendages* evolved in the late Precambrian, leading to the **arthropod** ("jointed foot") clade. Arthropod appendages exist in an amazing variety of forms. They serve many functions, including walking and swimming, gas exchange, food capture and manipulation, copulation, and sensory perception. Arthropods grasp food with their mouths and associated appendages and digest it internally. Their muscles are attached to the inside of the exoskeleton. Each segment has muscles that operate that segment and the appendages attached to it (**Figure 32.4**).

The arthropod exoskeleton has had a profound influence on the evolution of these animals. Encasement within a rigid body covering provides support for walking on dry land, and the waterproofing provided by chitin keeps the animal from dehydrating in dry air. Aquatic arthropods were, in short, excellent candidates to invade terrestrial environments. As we will see, they did so several times.

Arrow worms retain some ancestral developmental features

Nearly all triploblastic animal groups can be readily classified as being either protostomes or deuterostomes, but the evolutionary relationships of one small group, the **arrow worms**, were debated for many years. The early development of arrow worms looks similar to that of deuterostomes, although it is now known that they merely retain developmental features that are ancestral to triploblastic animals in general. Recent studies of gene sequences clearly identify them as protostomes. There is still some question as to whether they are the closest relatives of the lophotrochozoans (as shown in Figure 32.1), or possibly the sister group of all other protostomes.

The arrow worm body plan is based on a coelom divided into three compartments: head, trunk, and tail (**Figure 32.5**). Their bodies are transparent or translucent. Most arrow worms swim in the open sea. A few species live on the seafloor. Their abundance as fossils indicates that they were common more than 500 million years ago. The 100 or so living species of arrow worms are small enough—ranging from 3 mm to less than 12 cm in length—that their gas exchange and waste excretion requirements are met by diffusion through the body surface. They lack a circulatory system; wastes and nutrients are moved around the body in the coelomic fluid, which is propelled by cilia that line the coelom. There is no distinct larval stage. Miniature adults hatch directly from eggs that are fertilized internally following elaborate courtship between two hermaphroditic individuals.

Arrow worms are stabilized in the water by means of one or two pairs of lateral fins and a tail fin. They are major predators of planktonic organisms in the open ocean, ranging in size from small protists to young fish as large as the arrow worms themselves. An arrow worm typically lies motionless in the water until water movement signals the approach of prey. The arrow worm then darts forward and grasps the prey with the stiff spines adjacent to its mouth.

32.1 RECAP

The shared derived traits of protostomes include an anterior brain and a ventral nervous system. Many lophotrochozoans grow by adding to the size of their skeletal elements, whereas ecdysozoans, which have an exoskeleton, must molt periodically in order to grow.

- How does an animal's body covering influence the way it breathes, feeds, and moves? See pp. 693–694
- What features made arthropods well adapted for colonizing terrestrial environments? See p. 694

32.2 What Are the Major Groups of Lophotrochozoans?

Lophotrochozoans come in a variety of sizes and shapes, ranging from relatively simple animals with a blind gut (that is, a gut with only one opening) and no internal transport system to animals with a complete gut (having separate entrance and exit openings) and a complex internal transport system. They include some highly species-rich groups, such as flatworms, annelids, and mollusks. A number of these groups exhibit wormlike bodies, but the lophotrochozoans encompass a wide diversity of morphologies, including several groups with external shells. Some lophotrochozoan groups have only recently been discovered by biologists.

Ectoprocts live in colonies

The 4,500 species of **ectoprocts** (also called bryozoans, or "moss animals") are colonial animals that live in a "house" made of material secreted by the external body wall. Almost all ectoprocts are marine, although a few species occur in fresh or brackish water. An ectoproct colony consists of many small (1–2 mm) individuals connected by strands of tissue along which nutrients can be moved (**Figure 32.6**). The colony is created by the asexual reproduction of its founding member, and a single colony may contain as many as 2 million individuals. Rocks in coastal regions in many parts of the world are covered with luxuriant growths of ectoprocts. Some ectoprocts create miniature reefs in shallow waters. In some species, the individual colony members are differentially specialized for feeding, reproduction, defense, or support.

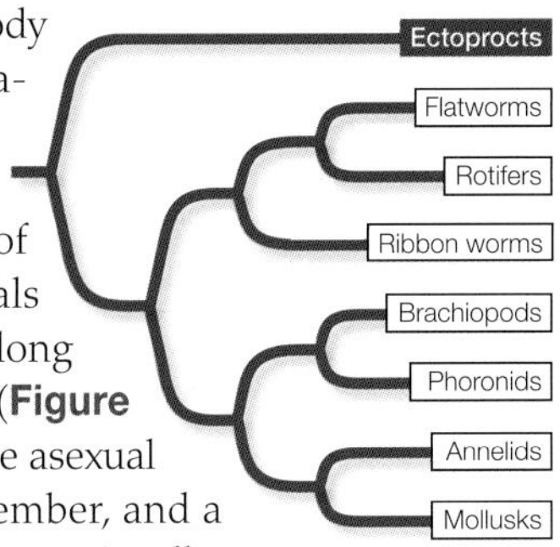

Sertella septentrionalis

32.6 An Ectoproct Colony The membranous orange tissue of this ectoproct colony connects and supplies nutrients to thousands of individual animals (see Figure 32.2).

Individual ectoprocts are able to oscillate and rotate their lophophore to increase contact with prey. They can also retract it into their "house" (see Figure 32.2). Ectoprocts also can reproduce sexually by releasing sperm into the water, which carries them to other individuals. Eggs are fertilized internally; developing embryos are brooded before they exit as larvae to seek suitable sites for attachment to the substratum.

Flatworms, rotifers, and ribbon worms are structurally diverse relatives

The flatworms, rotifers, and ribbon worms are structurally a diverse group, and only recently have their relationships to one another been elucidated. Before the availability of gene sequences for phylogenetic analysis, biologists considered the structure of the body cavity to be an important feature in animal classification. However, this monophyletic group of animals includes subgroups that are acoelomate (flatworms), pseudocoelomate (rotifers), and coelomate (ribbon worms). Thus, systematists today understand that the forms of body cavities have undergone considerable convergence in the course of animal evolution (see Table 32.1).

Flatworms lack specialized organs for transporting oxygen to their internal tissues. Their lack of a gas transport system dictates that each cell must be near a body surface, a requirement met by the dorsoventrally flattened body form. The digestive tract of a flatworm consists of a mouth opening into a blind sac. The sac is often highly branched, however, forming intricate patterns that increase the surface area available for the absorption of nutrients. Some small free-living flatworms are cephalized, with a head bearing chemoreceptor organs, two simple eyes, and a tiny brain composed of anterior thickenings of the longitudinal nerve cords. Free-living flatworms glide over surfaces, powered by broad bands of cilia (**Figure 32.7A**).

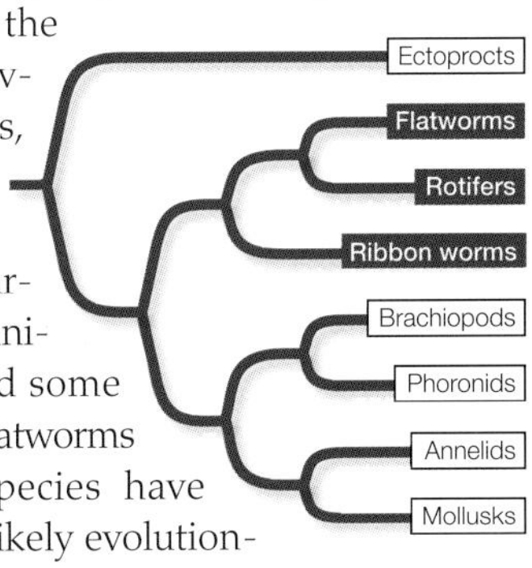

Most flatworms are internal parasites; others feed externally on animal tissues (living or dead), and some graze on plants. Although many flatworms are free-living, many other species have evolved to become parasites. A likely evolution-

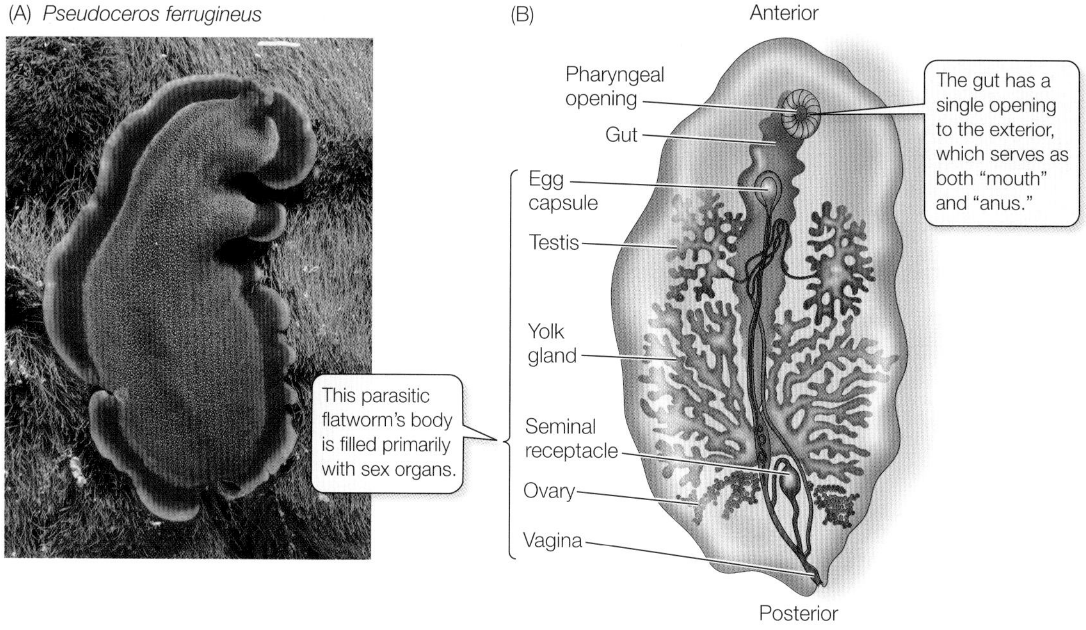

32.7 Flatworms May Live Freely or Parasitically (A) Some flatworm species, such as this Pacific marine flatworm, are free-living. (B) The fluke diagrammed here lives parasitically in the gut of sea urchins and is representative of parasitic flatworms. Because their hosts provide all the nutrition they need, these internal parasites do not require elaborate feeding or digestive organs and can devote most of their bodies to reproduction.

ary transition was from feeding on dead organisms to feeding on the body surfaces of dying hosts to invading and consuming parts of healthy hosts. Most of the 25,000 species of living flatworms are *tapeworms* and *flukes*; members of these two groups are internal parasites, particularly of vertebrates (**Figure 32.7B**). They absorb digested food from the digestive tracts of their hosts; many of them lack digestive tracts of their own. Some cause serious human diseases, such as schistosomiasis. Members of another flatworm group, the *monogeneans*, are external parasites of fishes and other aquatic vertebrates. The *turbellarians* include most of the free-living species.

Most **rotifers** are tiny (50–500 μm long)—smaller than some ciliate protists—but they have specialized internal organs (**Figure 32.8**). A complete gut passes from an anterior mouth to a posterior anus; the body cavity is a pseudocoel that functions as a hydrostatic skeleton. Most rotifers propel themselves through the water by means of rapidly beating cilia rather than by muscular contraction.

The most distinctive organ of rotifers is a conspicuous ciliated organ called the *corona*, which surmounts the head of many species. Coordinated beating of the cilia sweeps particles of organic matter from the water into the animal's mouth and down to a complicated structure called the *mastax*, in which the food is ground into small pieces. By contracting the muscles around the pseudocoel, a few rotifer species that prey on protists and small animals can protrude the mastax through the mouth and seize small objects with it. Males and females are found in some species, but the bdelloid rotifers have only females, which produce eggs that develop without being fertilized by a male. This group is the only known case of a group of animals that appear to have existed for millions of years without the benefits of sexual reproduction.

Most of the 1,800 known species of rotifers live in fresh water. Members of a few species rest on the surfaces of mosses or lichens in a desiccated, inactive state until it rains. When rain falls, they absorb water and become mobile, feeding in the films of water that temporarily cover the plants. Most rotifers live no longer than a week or two.

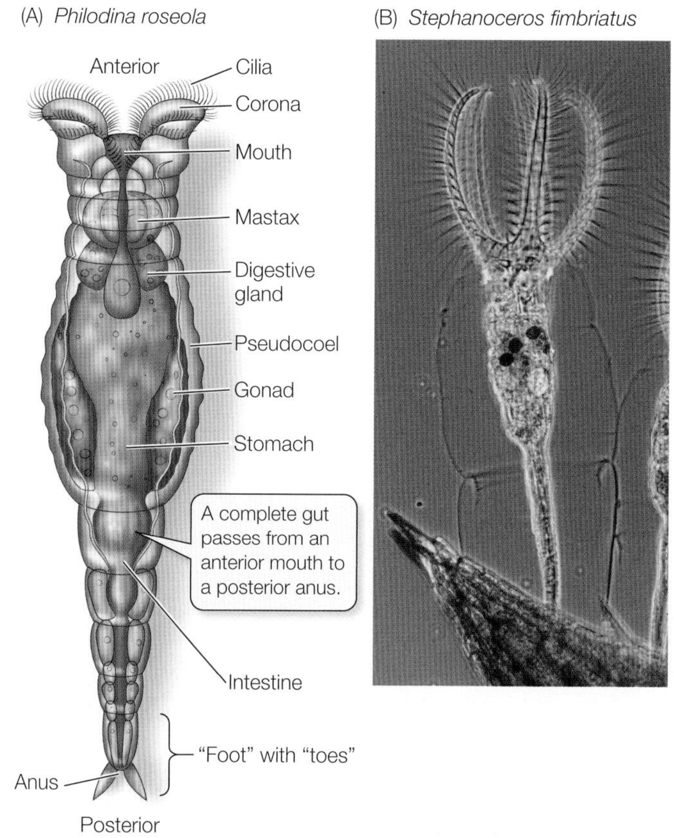

32.8 Rotifers (A) The rotifer diagrammed here reflects the general structure of many rotifers. (B) A micrograph reveals the internal complexity of a rotifer that has ingested photosynthetic protists.

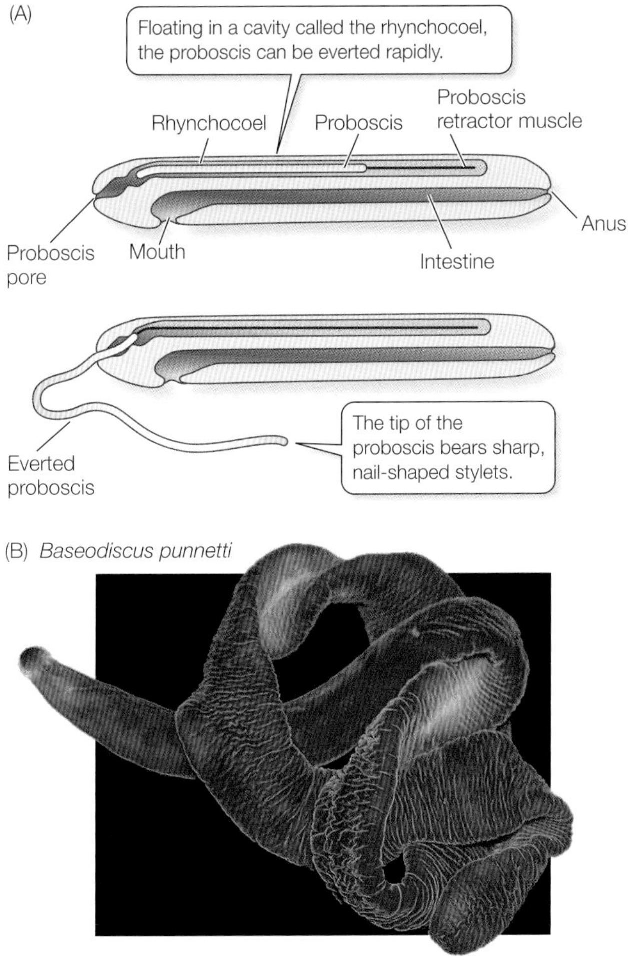

32.9 Ribbon Worms (A) The proboscis is the ribbon worm's feeding organ. (B) This deep-water nemertean is an impressive length despite the fact that its proboscis is not visible.

Ribbon worms (nemerteans) have simple nervous and excretory systems similar to those of flatworms. Unlike flatworms, however, they have a complete digestive tract with a mouth at one end and an anus at the other. Small ribbon worms move slowly by beating their cilia. Larger ones employ waves of muscle contraction to move over the surface of sediments or to burrow into them.

Within the body of nearly all of the 1,000 species of ribbon worms is a fluid-filled cavity called the *rhynchocoel*, within which lies a hollow, muscular *proboscis*. The proboscis, which is the worm's feeding organ, may extend much of the length of the body. Contraction of the muscles surrounding the rhynchocoel causes the proboscis to exit explosively through an anterior pore (**Figure 32.9A**). The proboscis may be armed with sharp stylets that pierce prey and discharge paralysis-causing toxins into the wound.

Ribbon worms are largely marine, although there are species that live in freshwater or on land. Most species are less than 20 cm in length, but individuals of some species reach 20 *meters* or more. Some genera feature species that are conspicuous and brightly colored (**Figure 32.9B**).

Phoronids and brachiopods use lophophores to extract food from the water

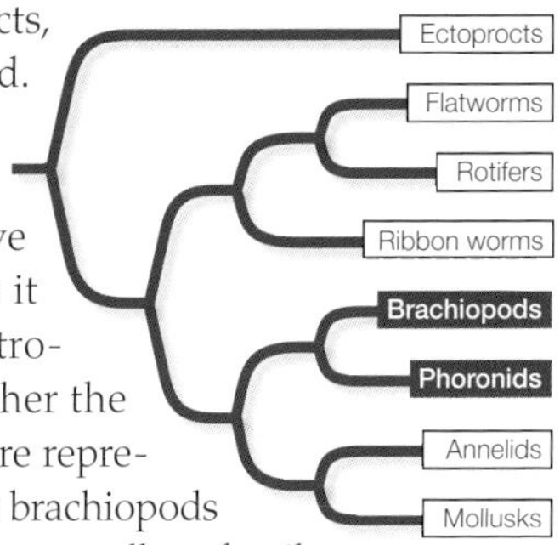

Earlier we discussed the ectoprocts, which use a lophophore to feed. Phoronids and brachiopods also feed using a lophophore, but this structure appears to have evolved independently (or else it has been lost in other lophotrochozoan groups). Although neither the phoronids nor the brachiopods are represented by many living species, the brachiopods (which have shells, and so leave an excellent fossil record) are known to have been much more abundant during the Paleozoic and Mesozoic eras.

The 20 known species of **phoronids** are small (5–25 cm in length), sessile worms that live in muddy or sandy sediments or attached to rocky substrata. Phoronids are found in marine waters, from the intertidal zone to about 400 meters deep. They secrete tubes made of chitin, inside which they live (**Figure 32.10**). Their cilia drive water into the top of the lophophore, and the water exits through the narrow spaces between the tentacles. Suspended food particles are caught and transported to the mouth by ciliary action. In most species, eggs are released into the water, where they are fertilized, but some species produce large eggs that are fertilized internally and retained in the parent's body, where they are brooded until they hatch.

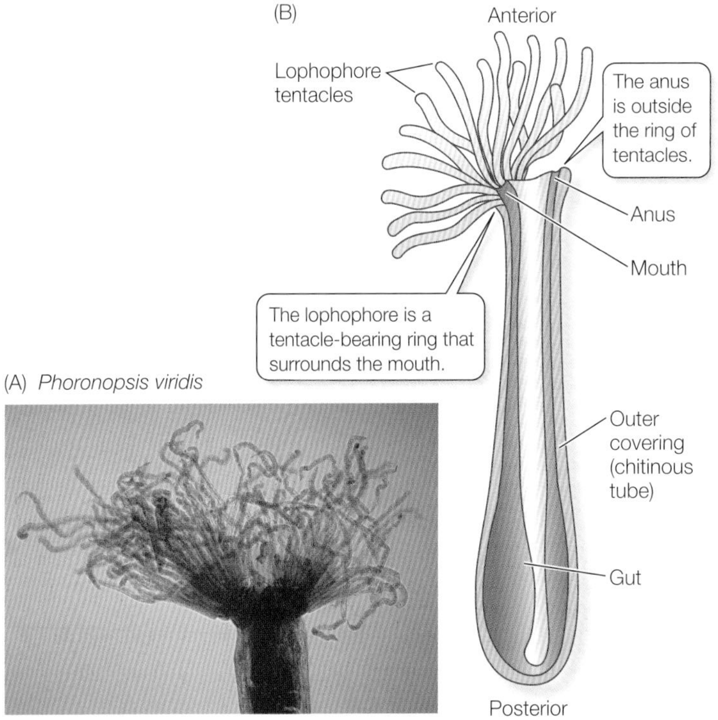

32.10 Phoronids Have Impressive Lophophores (A) The photograph shows the tentacles of a phoronid's lophophore. (B) The phoronid gut is U-shaped, as seen in brown in this generalized diagram.

32.11 A Brachiopod The lophophore of this North Pacific brachiopod can be seen between the valves of its shell.

Brachiopods are solitary marine animals. They have rigid shells that are divided into two parts connected by a ligament (**Figure 32.11**). The two halves can be pulled shut to protect the soft body. Brachiopods superficially resemble bivalve mollusks, but shells have evolved independently in the two groups. The two halves of the brachiopod shell are dorsal and ventral, rather than lateral, as in bivalves. The lophophore is located within the shell. The beating of cilia on the lophophore draws water into the slightly opened shell. Food is trapped in the lophophore and directed to a ridge, along which it is transferred to the mouth. Most brachiopods are 4–6 centimeters long, but they can be as long as 9 centimeters.

Brachiopods live attached to a solid substratum or embedded in soft sediments. Most species are attached by means of a short, flexible stalk that holds the animal above the substratum. Gases are exchanged across body surfaces, especially the tentacles of the lophophore. Most brachiopods release their gametes into the water, where they are fertilized. The larvae remain among the plankton for only a few days before they settle and develop into adults.

Brachiopods reached their peak abundance and diversity in Paleozoic and Mesozoic times. More than 26,000 fossil species have been described. Only about 335 species survive, but they are common in some marine environments.

The annelids and the mollusks are sister groups

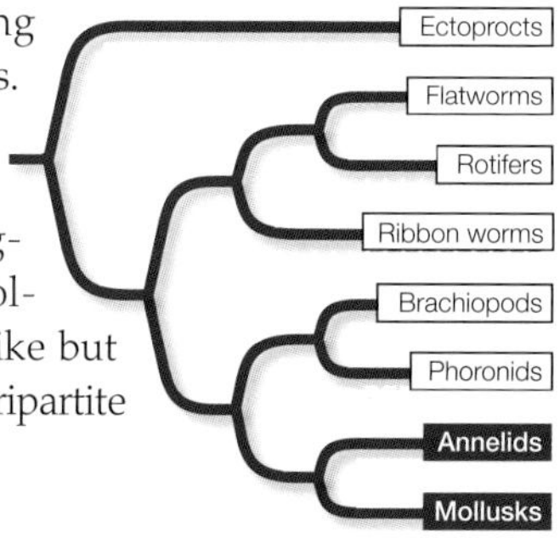

Annelids and mollusks are among the most familiar lophotrochozoans. Annelids are another wormlike group of animals, but unlike other worms, they are clearly segmented. The many species of mollusks, in contrast, are not wormlike but display modifications of a unique tripartite body plan.

Annelids have segmented bodies

The bodies of **annelids** are clearly segmented. Segmentation, as we saw in Section 31.2, allows an animal to move different parts of the body independently of one another, giving it much better control of its movement. The earliest segmented worms, preserved as fossils from the middle Cambrian, were burrowing marine annelids.

In most annelids, the coelom in each segment is isolated from those in other segments (**Figure 32.12**). A separate nerve center called a *ganglion* controls each segment; nerve cords that connect the ganglia coordinate their functioning. Most annelids lack a rigid external protective covering; instead, they have a thin, permeable body wall that serves as a general surface for gas exchange. Thus most annelids are restricted to moist environments because they lose body water rapidly in dry air. The approximately 16,500 described species live in marine, freshwater, and moist terrestrial environments.

POLYCHAETES More than half of all annelid species are *polychaetes* ("many hairs"), most of which are marine animals. Many polychaetes live in burrows in soft sediments. Most of them have one or more pairs of eyes and one or more pairs of tentacles, with which they filter prey from the surrounding water, at the an-

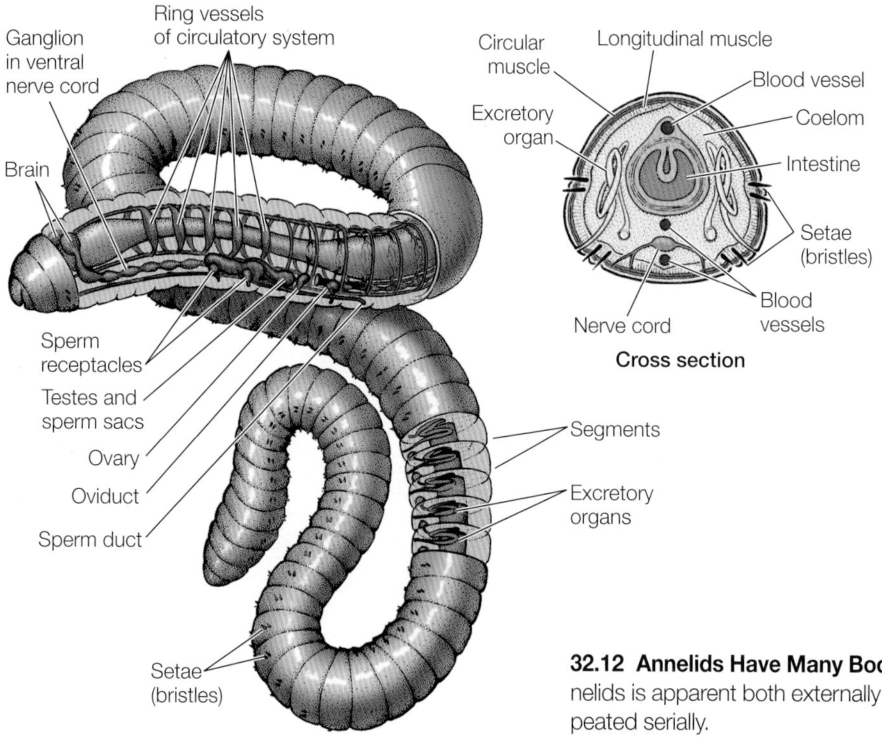

32.12 Annelids Have Many Body Segments The segmented structure of the annelids is apparent both externally and internally. Many organs of this earthworm are repeated serially.

32.13 Diversity among the Annelids
(A) "Cluster dusters" or "social feathers" are sessile marine polychaetes that grow in masses. They filter food from the water with their elegant tentacles. (B) These pogonophorans live around hydrothermal vents deep in the ocean. Their tentacles can be seen protruding from their chitinous tubes. (C) Earthworms are hermaphroditic; when they copulate, each individual donates and receives sperm. (D) The medicinal leech has been a tool of physicians and healers for many centuries. Even today leeches have uses in modern clinical practice.

(A) *Bispira brunnea*

(B) *Riftia* sp.

(C) *Lumbricus terrestris*

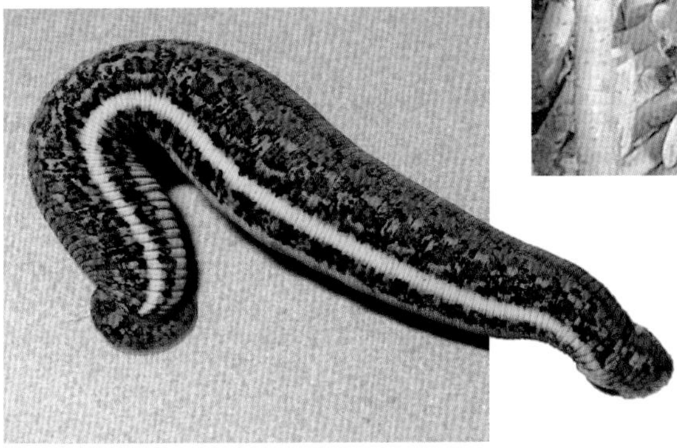

(D) *Hirudo medicinalis*

terior end of the body (**Figure 32.13A**; see also Figure 31.6A). In the marine species, the body wall of most segments extends laterally as a series of thin outgrowths called *parapodia*. The parapodia function in gas exchange, and some species use them to move. Stiff bristles called *setae* protrude from each parapodium, forming temporary attachments to the substratum that prevent the animal from slipping backward when its muscles contract. Recent molecular studies suggest that polychaetes are probably paraphyletic with respect to the remaining annelids.

Members of one clade of polychaetes, the *pogonophorans*, lost their digestive tract (they have no mouth or gut). They are burrowing animals with a crown of tentacles through which gases are exchanged. Pogonophorans secrete tubes made of chitin and other substances, in which they live (**Figure 32.13B**). The pogonophoran coelom consists of an anterior compartment, into which the tentacles can be withdrawn, and a long, subdivided cavity that extends much of the length of the body. The posterior end of the body is segmented. Pogonophorans take up dissolved organic matter at high rates from the sediments in which they live or the surrounding water. This uptake is facilitated by endosymbiotic bacteria that live in a specialized organ known as the *trophosome*. Developmentally, the trophosome develops from the embryonic gut, so the organ has retained its nutritional function, although the process of feeding has changed considerably.

Pogonophorans were not discovered until the twentieth century, when deep-sea explorers found them living many thousands of meters below the ocean surface. In these deep oceanic sediments, they may reach densities of many thousands per square meter. About 160 species have been described. The largest and most remarkable pogonophorans are 2 meters or more in length and live near deep-sea *hydrothermal vents*—volcanic openings in the seafloor through which hot, sulfide-rich water pours. The tissues of these species harbor endosymbiotic bacteria that fix carbon using energy obtained from the oxidation of hydrogen sulfide (H_2S).

CLITELLATES Most of the approximately 3,000 described species of the other major annelid group, the *clitellates*, live in freshwater or terrestrial environments. The clitellates are likely phylogenetically nested within the polychaetes, although the exact relationships are not yet clear. There are two major groups of clitellates, the oligochaetes and the leeches.

Oligochaetes ("few hairs") have no parapodia, eyes, or anterior tentacles, and they have relatively few setae. Earthworms—the most familiar oligochaetes—burrow in and ingest soil, from which they extract food particles. All oligochaetes are *hermaphroditic*; that is, each individual is both male and female. Sperm are exchanged simultaneously between two copulating individuals (**Figure 32.13C**). Eggs are laid in a cocoon outside the adult's body. The cocoon is shed, and when development is complete, miniature worms emerge and immediately begin independent life.

Leeches, like oligochaetes, lack parapodia and tentacles. The coelom of leeches is not divided into compartments; the coelomic space is largely filled with undifferentiated tissue. Groups of segments at each end of the body are modified to form suckers, which serve as temporary anchors that aid the leech in movement. With its posterior sucker attached to a substratum, the leech extends its body by contracting its circular muscles. The anterior sucker is then attached, the posterior one detached, and the leech shortens itself by contracting its longitudinal muscles. Leeches live in freshwater or terrestrial habitats.

A leech makes an incision in its host, from which blood flows. It can ingest so much blood in a single feeding that its body may enlarge several-fold. The leech secretes an anticoagulant into the wound that keeps the host's blood flowing. For centuries, medical practitioners have employed leeches to draw blood to treat diseases they believed were caused by an excess of blood or by "bad blood." Although most leeching practices (such as inserting a leech in a person's throat to alleviate swollen tonsils) have been abandoned, *Hirudo medicinalis* (the medicinal leech; **Figure 32.13D**) is used today to reduce fluid pressure and prevent blood clotting in damaged tissues, to eliminate pools of coagulated blood, and to prevent scarring. The anticoagulants of certain other leech species that also contain anesthetics and blood vessel dilators are being studied for possible medical uses.

Mollusks have undergone a dramatic evolutionary radiation

The ancestors of modern **mollusks** were probably unsegmented, wormlike animals. However, the origin of a tripartite body organization in this lineage set the stage for one of the most dramatic of animal evolutionary radiations.

The molluscan "body plan" has three major components: a foot, a visceral mass, and a mantle (**Figure 32.14**).

- The *foot* is a large, muscular structure that originally was both an organ of locomotion and a support for the internal organs. In squids and octopuses, the foot has been modified to form arms and tentacles borne on a head with complex sense organs. In other groups, such as clams, the foot is a burrowing organ. In some groups the foot is greatly reduced.
- The heart and the digestive, excretory, and reproductive organs are concentrated in a centralized, internal *visceral mass*.
- The *mantle* is a fold of tissue that covers the organs of the visceral mass The mantle secretes the hard, calcareous shell that is typical of many mollusks.

In most mollusks, the mantle extends beyond the visceral mass to form a *mantle cavity*. Within this cavity lie *gills* that are used for gas exchange. When cilia on the gills beat, they create a current of water. The tissue of the gills, which is highly *vascularized* (contains many blood vessels), takes up oxygen from the water and releases carbon dioxide. Many mollusk species use their gills as filter-feeding devices, whereas others feed using a rasping structure known as the *radula* to scrape algae from rocks. In some mollusks, such as the marine cone snails, the radula has been modified into a drill or poison dart.

Molluscan blood vessels do not form a closed system. Blood and other fluids empty into a large, fluid-filled *hemocoel*, through which fluids move around the animal and deliver oxygen to the internal organs. Eventually, the fluids re-enter the blood vessels and are moved by a heart.

Monoplacophorans were the most abundant mollusks during the Cambrian period, 500 million years ago, but only a few species

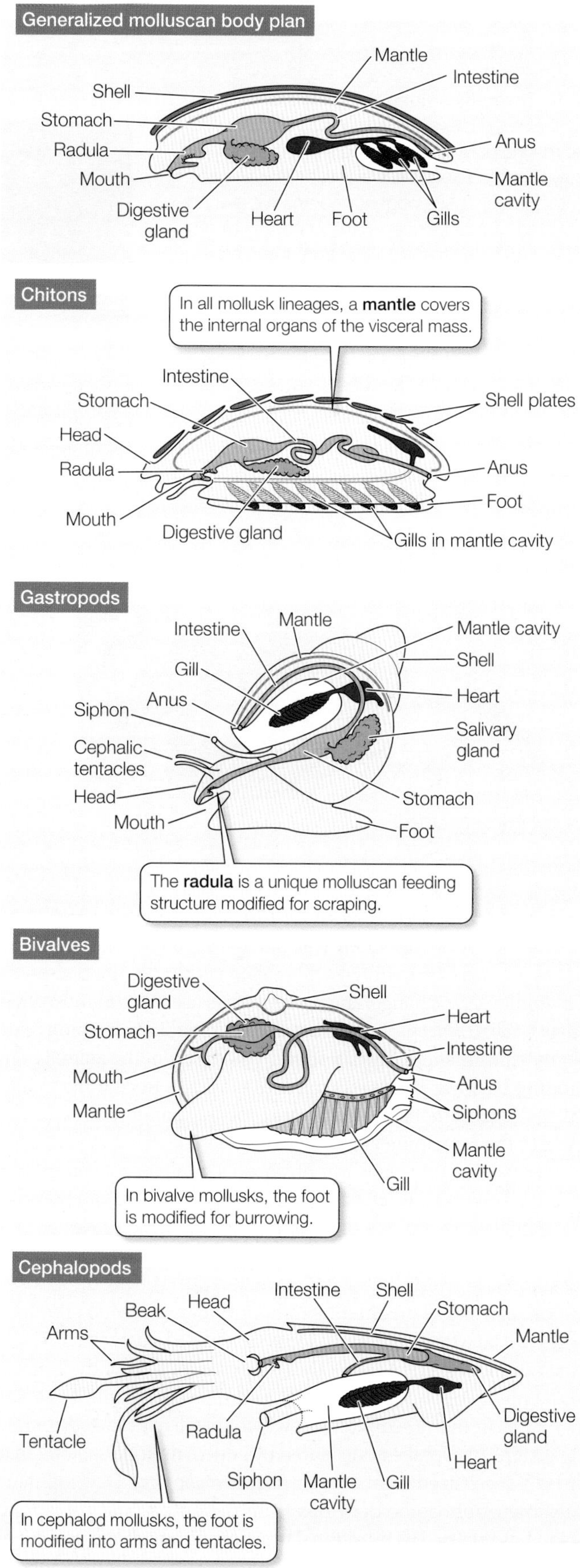

32.14 Molluscan Body Plans The mollusks are all variations on a general body plan that includes three major components: a foot, a visceral mass of internal organs, and a mantle. The mantle may secrete a calcareous shell, as in the gastropods and bivalves.

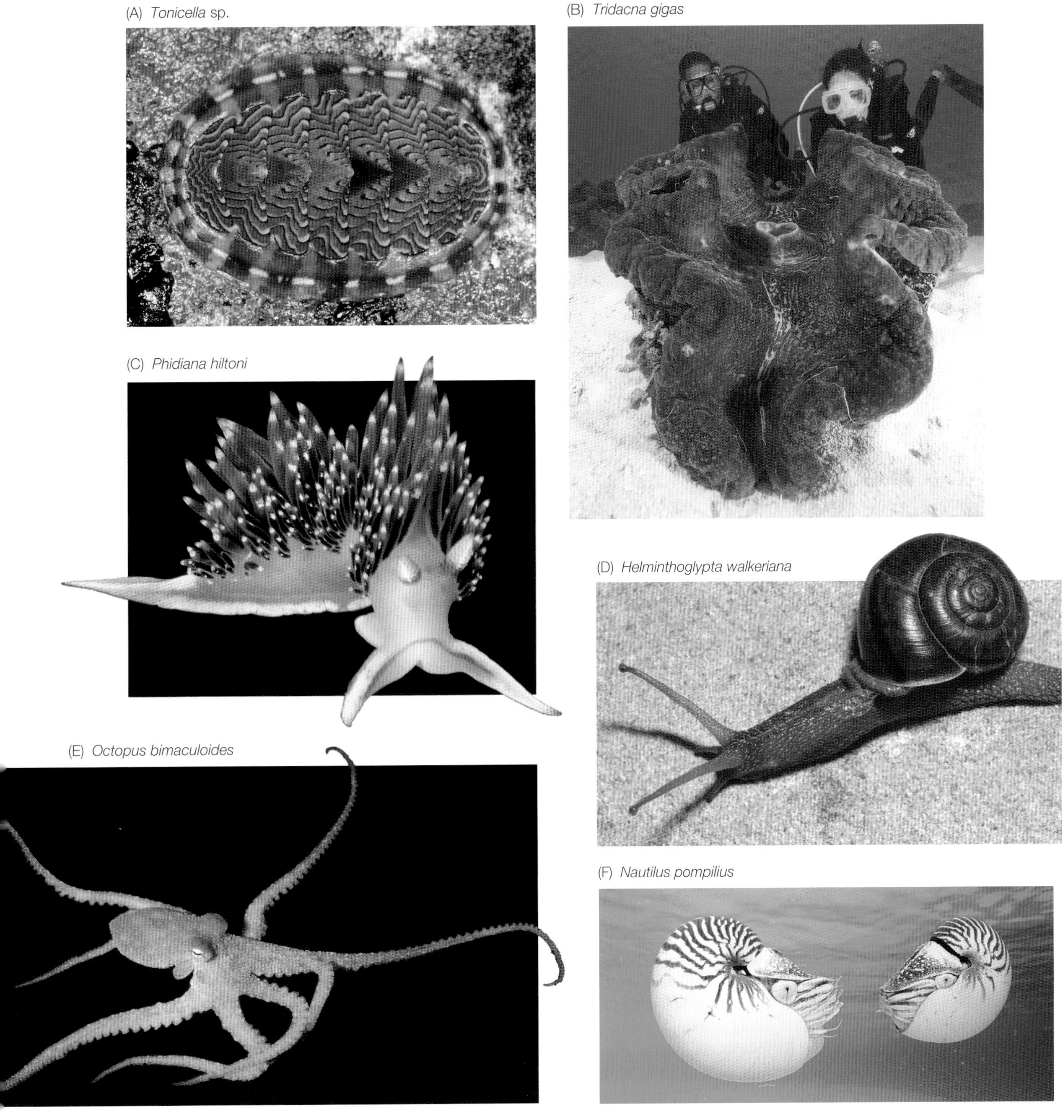

32.15 Diversity among the Mollusks (A) Chitons are common in the intertidal zones of temperate-zone coasts. (B) The giant clam of Indonesia is among the largest of the bivalve mollusks. (C) This shell-less nudibranch ("naked gills"), or sea slug, is colored to signal its toxicity. (D) Land snails are shelled, terrestrial gastropods. (E) Cephalopods such as the octopus are active predators. (F) The boundaries of chambers are clearly visible on the outer surfaces of these shelled cephalopods, known as chambered nautiluses.

survive today. They look something like the gastropod limpets, but in contrast to all other living mollusks, the gas exchange organs, muscles, and excretory pores of monoplacophorans are repeated over the length of the body.

The four major molluscan clades living today—chitons, bivalves, gastropods, and cephalopods—are illustrated in Figure 32.15. The species shown here are a tiny sample of the 95,000 living species of mollusks.

CHITONS Multiple gills and shell plates characterize the **chitons**, but most of their body organs are not iterated (**Figure 32.15A**). The chiton body is bilaterally symmetrical, and the internal organs, particularly the digestive and nervous systems, are relatively simple. Most chitons are marine herbivores that scrape algae from rocks with their sharp radulae. An adult chiton spends most of its life clinging tightly to rock surfaces with its large, muscular, mucus-covered foot. It moves slowly by means of rippling waves of muscular contraction in the foot. Fertilization in most chitons takes place in the water, but in a few species fertilization is internal and embryos are brooded within the body.

BIVALVES Clams, oysters, scallops, and mussels are all familiar **bivalves**. They are found in both marine and freshwater environments. Bivalves have very small heads and hinged, two-part shells that extend over the sides of the body as well as the top (**Figure 32.15B**). Many clams use the foot to burrow into mud and sand. Bivalves feed by taking in water through an opening called an *incurrent siphon* and filtering food from the water with their large gills, which are also the main sites of gas exchange. Water and gametes exit through the *excurrent siphon*. Fertilization takes place in open water in most species.

GASTROPODS **Gastropods** are the most species-rich and widely distributed mollusks. Snails, whelks, limpets, slugs, nudibranchs (sea slugs), and abalones are all gastropods. Most species move by gliding on the muscular foot, but in a few species—the sea butterflies and heteropods—the foot is a swimming organ, with which the animal moves through open ocean waters. The nudibranchs, or sea slugs, have lost their protective shells over the course of evolution; their sometimes brilliant coloration is *aposematic*; that is, bright colors often serve to warn potential predators of toxicity (**Figure 32.15C**). Other nudibranch species exhibit camouflaged coloration.

Shelled gastropods have one-piece shells. The only mollusks that live in terrestrial environments—land snails and slugs—are gastropods (**Figure 32.15D**). In these terrestrial species, the mantle tissue is modified into a highly vascularized lung.

CEPHALOPODS The **cephalopods**—squids, octopuses, and nautiluses—first appeared near the beginning of the Cambrian period. By the Ordovician period a wide variety of types were present.

In the cephalopods, the excurrent siphon is modified to allow the animal to control the water content of the mantle cavity. The modification of the mantle into a device for forcibly ejecting water from the cavity through the siphon enables these animals to move rapidly by "jet propulsion" through the water. With their greatly enhanced mobility, cephalopods became the major predators in the open waters of the Devonian oceans. They remain important marine predators today.

Cephalopods capture and subdue their prey with their tentacles; octopuses also use their tentacles to move over the substratum (**Figure 32.15E**). As is typical of active, rapidly moving predators, cephalopods have a head with complex sensory organs, most notably eyes that are comparable to those of vertebrates in their ability to resolve images. The head is closely associated with a large, branched foot that bears the tentacles and a siphon. The large, muscular mantle provides a solid external supporting structure. The gills hang in the mantle cavity. Many cephalopods have elaborate courtship behavior, which may involve striking color changes.

Many early cephalopods had chambered shells divided by partitions penetrated by tubes through which gases and liquids could be moved to control their buoyancy. Nautiloids (genus *Nautilus*) are the only cephalopods with such external chambered shells that survive today (**Figure 32.15F**).

32.2 RECAP

Lophotrochozoans include animals with diverse body types. Wormlike forms include some flatworms, ribbon worms, phoronids, and annelids. There has been convergent evolution of lophophores (in ectoprocts versus brachiopods and phoronids) and in external two-part shell coverings (in brachiopods versus bivalve mollusks).

- How can flatworms survive without a gas transport system? See p. 695 and Figure 32.7
- Do you understand why most annelids are restricted to moist environments? See p. 698
- Describe how the basic body plan of mollusks has been modified to yield a wide diversity of animals. See p. 700 and Figure 32.14

The second of the two major protostome clades, the ecdysozoans, contains the vast majority of the Earth's animal species. Let's see what aspects of their body plan led to this massive amount of diversity.

32.3 What Are the Major Groups of Ecdysozoans?

Many ecdysozoans are wormlike in form, although the *arthropods* have limblike appendages. In this section we will look at the two clades of wormlike ecdysozoans: the priapulids, kinorhynchs, and loriciferans in one group, and the horsehair worms and nematodes in the other. Section 32.4 will describe the most familiar and diverse ecdysozoans—the arthropods and their relatives—and the many forms their appendages take in the different arthropod groups.

Several marine groups have relatively few species

Members of several species-poor clades of wormlike marine ecdysozoans—priapulids, kinorhynchs, and loriciferans—have relatively thin cuticles that are molted periodically as the animals grow to full size. In 2004, embryos of a fossil species in this group were discovered in sediments laid down in China about 500 million years ago. These remarkable discoveries show that the ancestors of these animals developed directly from an egg to the adult form, as their modern descendants do.

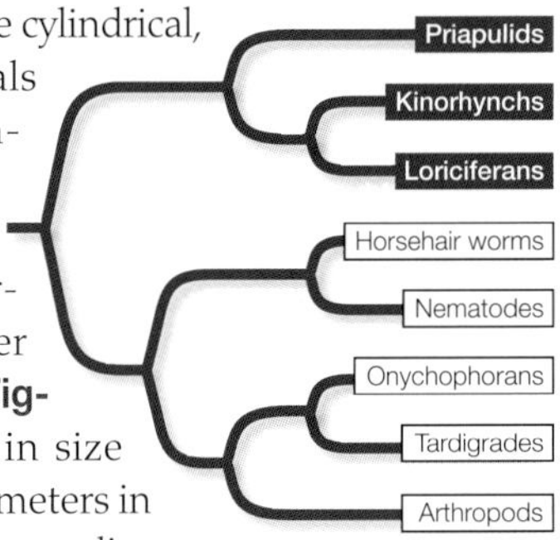

The 16 species of **priapulids** are cylindrical, unsegmented, wormlike animals with a three-part body plan consisting of a proboscis, trunk, and caudal appendage ("tail"). It should be clear from their appearance why they were named after the Greek fertility god Priapus (**Figure 32.16A**). Priapulids range in size from half a millimeter to 20 centimeters in length. They burrow in fine marine sediments and prey on soft-bodied invertebrates such as polychaetes. They capture prey with a toothed, muscular pharynx that they evert through the mouth and then withdraw into the body together with the grasped prey. Fertilization is external, and most species have a larval form that also lives in the mud.

About 150 species of **kinorhynchs** have been described. They live in marine sands and muds and are virtually microscopic; no kinorhynch species is larger than 1 millimeter in length. Their bodies are divided into 13 segments, each with a separate cuticular plate (**Figure 32.16B**). These plates are periodically molted during growth. Kinorhynchs feed by ingesting sediments through their retractable proboscis (*kinorhynch* means "movable snout"). They then digest the organic material found in the sediment, which may include living algae as well as dead matter. Kinorhynchs have no distinct larval stage; fertilized eggs develop directly into juveniles, which emerge from their egg cases with 11 of the 13 body segments already formed.

Loriciferans are also minute animals less than a millimeter in length. They were not discovered until 1983. About 100 living species are known to exist, although many of these are still being described. The body is divided into a head, neck, thorax, and abdomen and is covered by six plates, from which the loriciferans get their name (Latin *lorica*, "corset"). The plates around the base of the neck bear anteriorly directed spines of unknown function (**Figure 32.16C**). Loriciferans live in coarse marine sediments. Little is known about what they eat, but some species apparently eat bacteria.

Nematodes and their relatives are abundant and diverse

About 320 species of the unsegmented **horsehair worms** have been described. As their name implies, these animals are extremely thin in diameter; horsehair worms range from a few millimeters up to a meter in length (**Figure 32.17**). Most adult worms live in fresh water, among leaf litter and algal mats near the edges of streams and ponds. A few species live in damp soil. The larvae are internal parasites of terrestrial and aquatic insects and freshwater crayfish.

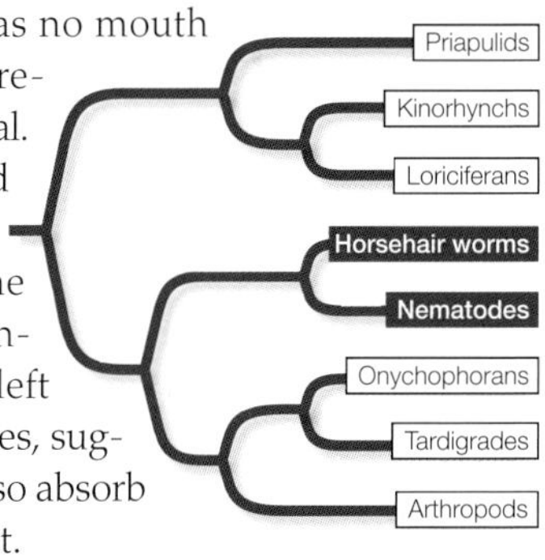

An adult horsehair worm has no mouth opening, and its gut is greatly reduced and probably nonfunctional. Some of these worms may feed only as larvae, absorbing nutrients from their hosts across the body wall. Others, however, continue to grow after they have left their hosts, shedding their cuticles, suggesting that adult worms may also absorb nutrients from their environment.

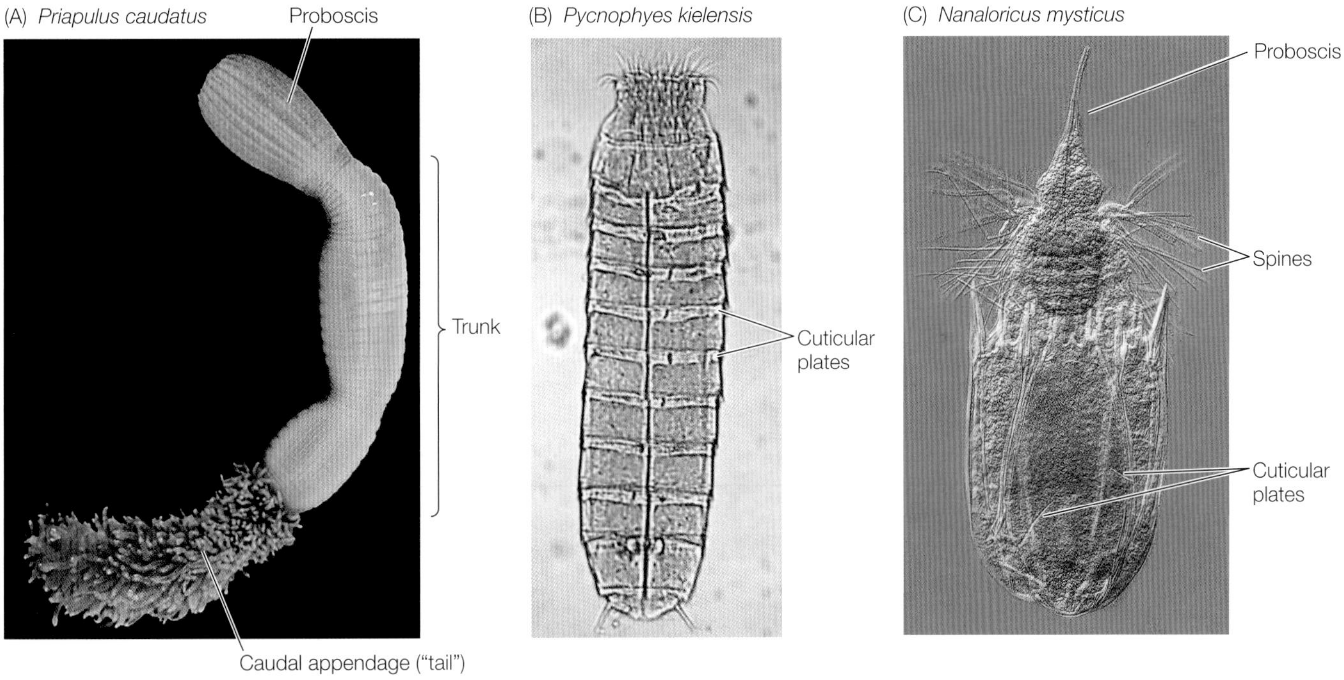

32.16 Wormlike Marine Ecdysozoans (A) A priapulid. Priapulids are marine worms that live, usually in burrows, on the ocean floor. (B) Kinorhynchs are virtually microscopic. The cuticular plates that cover their bodies are periodically molted. (C) Six cuticular plates form a "corset" around the loriciferan body.

Paragordius sp.

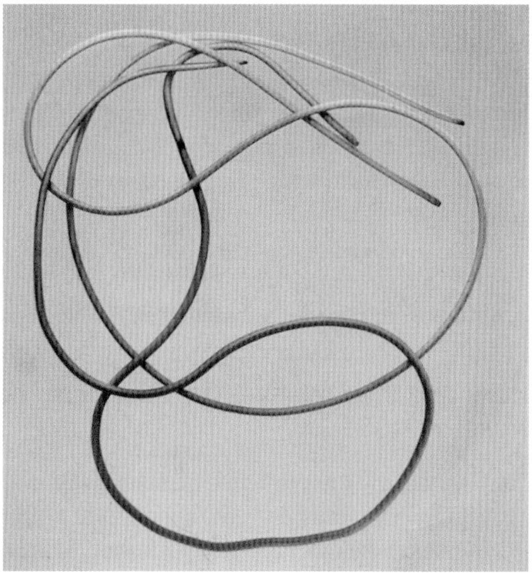

32.17 A Horsehair Worm Horsehair worms get their name from their hairlike or threadlike shape. They can grow to be up to a meter long.

Nematodes (roundworms) have a thick, multilayered cuticle that gives their unsegmented body its shape (**Figure 32.18**). As a nematode grows, it sheds its cuticle four times. Nematodes exchange oxygen and nutrients with their environment through both the cuticle and the gut, which is only one cell layer thick. Materials are moved through the gut by rhythmic contraction of a highly muscular organ, the *pharynx*, at the worm's anterior end. Nematodes move by contracting their longitudinal muscles.

Nematodes are one of the most abundant and universally distributed of all animal groups. Many are microscopic; the largest known nematode, which reaches a length of 9 meters, is a parasite in the placentas of female sperm whales. About 25,000 species have been described, but the actual number of living species may be more than a million. Countless nematodes live as scavengers in the upper layers of the soil, on the bottoms of lakes and streams, and in marine sediments. The topsoil of rich farmland may contain from 3 to 9 billion nematodes per acre. A single rotting apple may contain as many as 90,000 individuals.

The zoologist Ralph Buchsbaum wrote that "If all the matter in the universe except the nematodes were swept away, our world would still be dimly recognizable ... we should find its mountains, hills, vales, rivers, lakes, and oceans represented by a film of nematodes. ... The location of the various plants and animals would still be decipherable."

One soil-inhabiting nematode, *Caenorhabitis elegans*, serves as a "model organism" in the laboratories of geneticists and developmental biologists. It is ideal for such research because it is easy to cultivate, matures in three days, and has a fixed number of body cells. Its genome has been completely mapped.

Many nematodes are predators, feeding on protists and small animals (including other roundworms). Most significant to humans, however, are the many species that parasitize plants and animals. The nematodes that are parasites of humans (causing serious tropical diseases such as trichinosis, filariasis, and elephantiasis), domestic animals, and economically important plants have been studied intensively in an effort to find ways of controlling them.

The structure of parasitic nematodes is similar to that of free-living species, but the life cycles of many parasitic species have special stages that facilitate the transfer of individuals among hosts. *Trichinella spiralis*, the species that causes the human disease trichi-

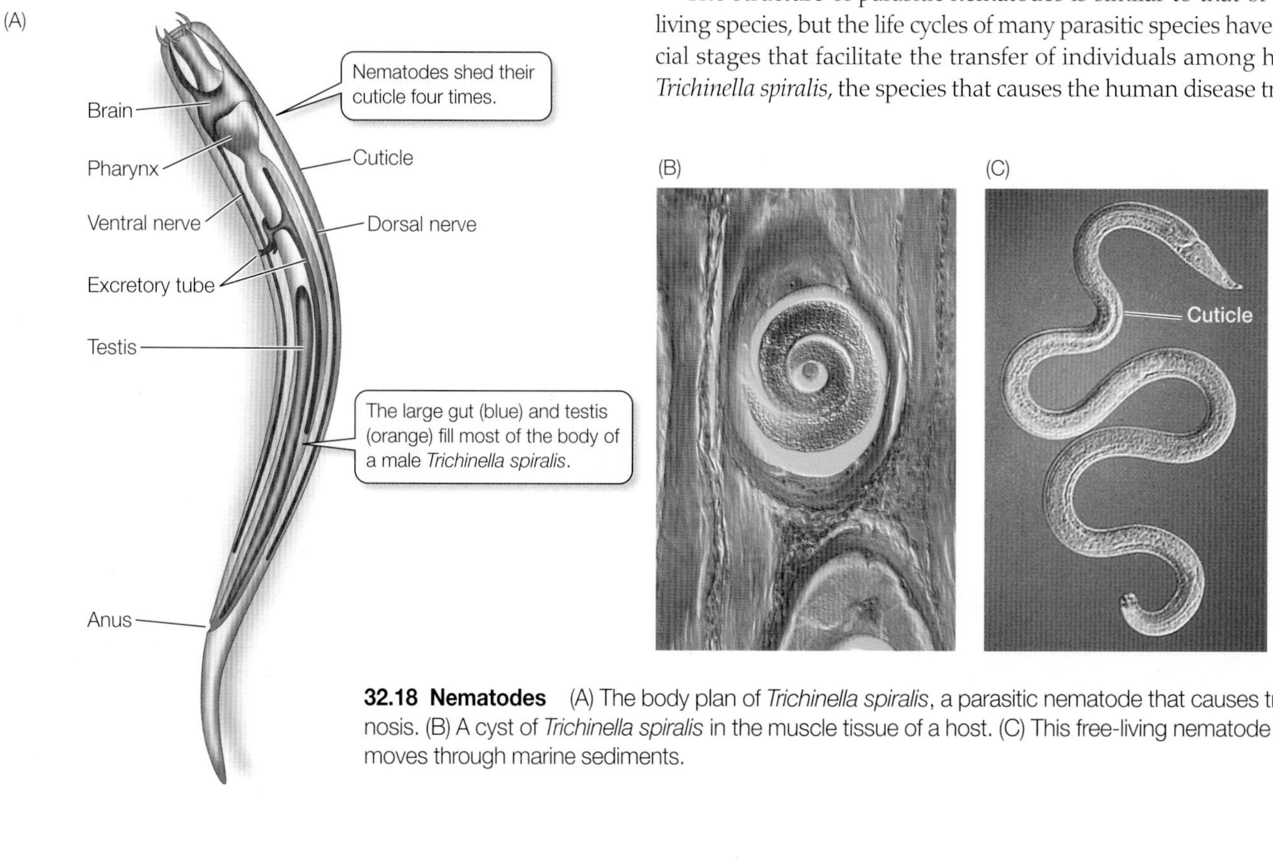

32.18 Nematodes (A) The body plan of *Trichinella spiralis*, a parasitic nematode that causes trichinosis. (B) A cyst of *Trichinella spiralis* in the muscle tissue of a host. (C) This free-living nematode moves through marine sediments.

nosis, has a relatively simple life cycle. A person may become infected by eating the flesh of an animal (usually a pig) containing larvae of *Trichinella* encysted in its muscles (see Figure 32.18B). The larvae are activated in the person's digestive tract, emerge from their cysts, and attach to the intestinal wall, where they feed. Later, they bore through the intestinal wall and are carried in the bloodstream to muscles, where they form new cysts. If present in great numbers, these cysts can cause severe pain or death.

32.3 RECAP

The priapulids, kinorhynchs, and loriciferans are relatively small, poorly known groups of wormlike marine ecdysozoans. The horsehair worms and the nematodes are also wormlike, but the biology of some species has been studied extensively. The nematodes are among the most abundant and species-rich groups on Earth.

- Can you describe at least three different ways in which nematodes have significant impact on humans? See pp. 704–705

We now turn to the animals that not only dominate the ecdysozoan clade, but are also the most numerous animals on Earth.

32.4 Why Do Arthropods Dominate Earth's Fauna?

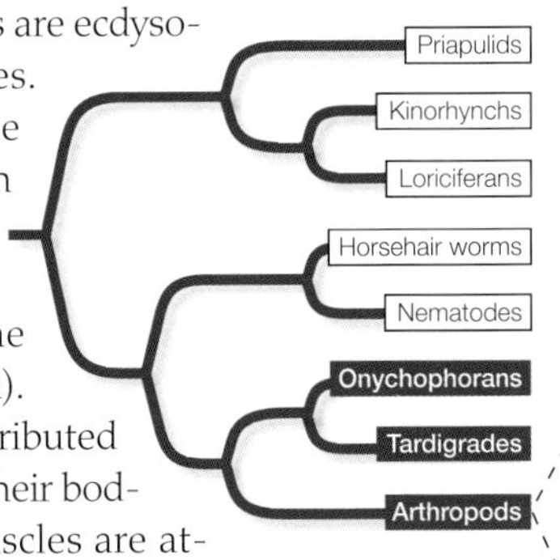

The arthropods and their relatives are ecdysozoans with limblike appendages. Collectively, arthropods are the dominant animals on Earth, both in numbers of species (over a million described) and numbers of individuals (estimated at some 10^{18} individuals, or a billion billion).

Several key features have contributed to the success of the arthropods. Their bodies are segmented, and their muscles are attached to the inside of their rigid exoskeletons. Each segment has muscles that operate that segment and the jointed appendages attached to it (see Figure 32.4). The jointed appendages permit complex movement patterns, and different appendages are specialized for different functions. Encasement of the body within a rigid exoskeleton provides the animal with support for walking in the water or on dry land and provides some protection against predators. The waterproofing provided by chitin keeps the animal from dehydrating in dry air.

Representatives of the four major arthropod groups living today are all species-rich: the crustaceans (including shrimps, crabs, and barnacles), hexapods (insects and their relatives), myriapods (millipedes and centipedes), and chelicerates (including the arachnids—spiders, scorpions, mites and their relatives). The phylogenetic relationships among arthropod groups are currently being reexamined in the light of a wealth of new information, much of it based on gene sequences (see Chapter 24). As shown in **Figure 32.19,** most of the current uncertainty concerns the relationship of the myriapods to the other three groups. However, it appears clear that the arthropods constitute a monophyletic group, and most analyses support the close relationship between hexapods and crustaceans (the lower two trees in Figure 32.19).

Arthropods evolved from ancestors with simple, unjointed appendages. The exact forms of those ancestors are unknown, but some arthropod relatives with segmented bodies and unjointed appendages survive today. Before we describe the modern arthropods, we will discuss those arthropod relatives as well as an early arthropod clade that went extinct, but left an important fossil record.

Arthropod relatives have fleshy, unjointed appendages

Until fairly recently, biologists debated whether the **onychophorans** (velvet worms) were more closely related to annelids or arthropods, but molecular evidence clearly links them to the latter. In-

32.19 The Placement of Myriapods Among Arthropods Is Under Study Although many aspects of arthropod phylogeny are well understood, the placement of the myriapods is still debated among systematists. These three trees show three possible placements of the myriapods; most analyses favor one of the lower two trees.

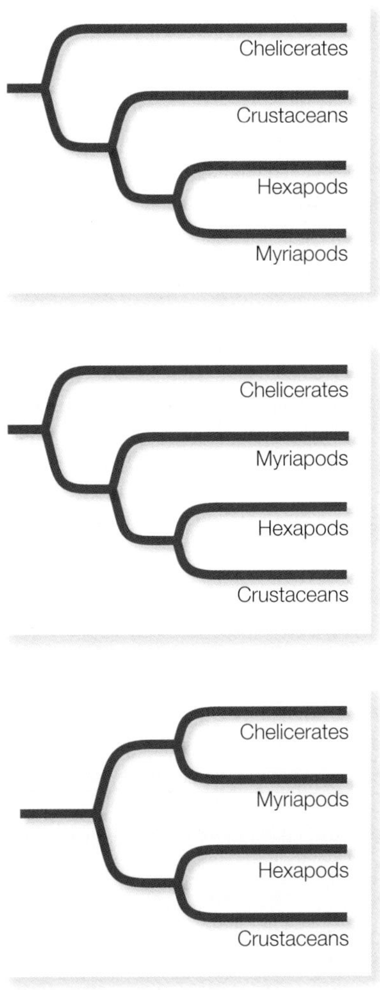

(A) *Peripatus* sp.

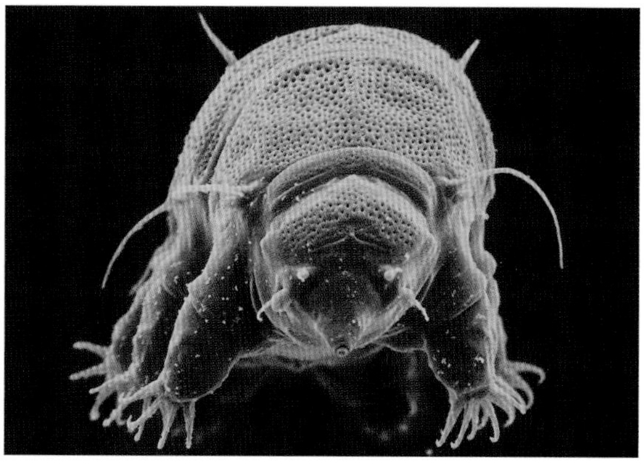

(B) *Echiniscus springer*

32.20 Arthropod Relatives with Unjointed Appendages (A) Onychophorans, also called velvet worms, have unjointed legs and use the body cavity as a hydrostatic skeleton. (B) The appendages and general anatomy of tardigrades superficially resemble those of onychophorans.

deed, with their soft, fleshy, unjointed, claw-bearing legs, onychophorans may be similar in appearance to the ancestors of arthropods (**Figure 32.20A**). The 150 species of onychophorans live in leaf litter in humid tropical environments. They have soft, segmented bodies that are covered by a thin, flexible cuticle that contains chitin. They use their fluid-filled body cavities as hydrostatic skeletons. Fertilization is internal, and the large, yolky eggs are brooded within the body of the female.

Tardigrades (water bears) also have fleshy, unjointed legs and use their fluid-filled body cavities as hydrostatic skeletons (**Figure 32.20B**). Tardigrades are extremely small (0.1–0.5 mm in length) and lack both circulatory systems and gas exchange organs. The 800 known extant species live in marine sands and on temporary water films on plants. When these films dry out, the animals also lose water and shrink to small, barrel-shaped objects that can survive for at least a decade in a dormant state. Tardigardes have been found to have densities as high as 2 million per square meter of moss.

Jointed legs first appeared in the trilobites

The **trilobites** flourished in Cambrian and Ordovician seas, but they disappeared in the great Permian extinction at the close of the Paleozoic era (251 mya). Because they had heavy exoskeletons that readily fossilized, they left behind an abundant record of their existence (**Figure 32.21**). About 10,000 species have been described.

The body segmentation and appendages of trilobites followed a relatively simple, repetitive plan, but some of their jointed appendages were modified for different functions. This specialization of appendages is a theme in the continuing evolution of the arthropods.

The jointed appendages of arthropods gave the clade its name, from the Greek words *arthron*, "joint," and *podos*, "foot" or "limb." A related derivation can be discerned that describes the widespread mammalian affliction *arthritis* (inflamed joints).

Crustaceans are diverse and abundant

Crustaceans are the dominant marine arthropods today. The most familiar crustaceans are the shrimps, lobsters, crayfishes, and crabs (all *decapods*; **Figure 32.22A**) and the sow bugs (*isopods*; **Figure 32.22B**). *Krill* are small but very abundant oceanic crustaceans that are important food items for a variety of large vertebrates, including baleen whales. Also included in the crustacean clade are a variety of small *copepod* species, many of which superficially resemble shrimps (**Figure 32.22C**). Copepods are also abundant in the open oceans. Additional species-rich groups of crustaceans include *amphipods* and *ostracods* (both found in freshwater and marine environments).

Barnacles are unusual crustaceans that are sessile as adults (**Figure 32.22D**). Adult barnacles look more like mollusks than like other crustaceans, but, as the zoologist Louis Agassiz remarked more than a century ago, a barnacle is "nothing more than a little shrimp-like animal, standing on its head in a limestone house and kicking food into its mouth."

Odontochile rugosa

32.21 A Trilobite The relatively simple, repetitive segments of the now-extinct trilobites are illustrated by a fossil trilobite from the shallow seas of the Devonian period, some 400 million years ago.

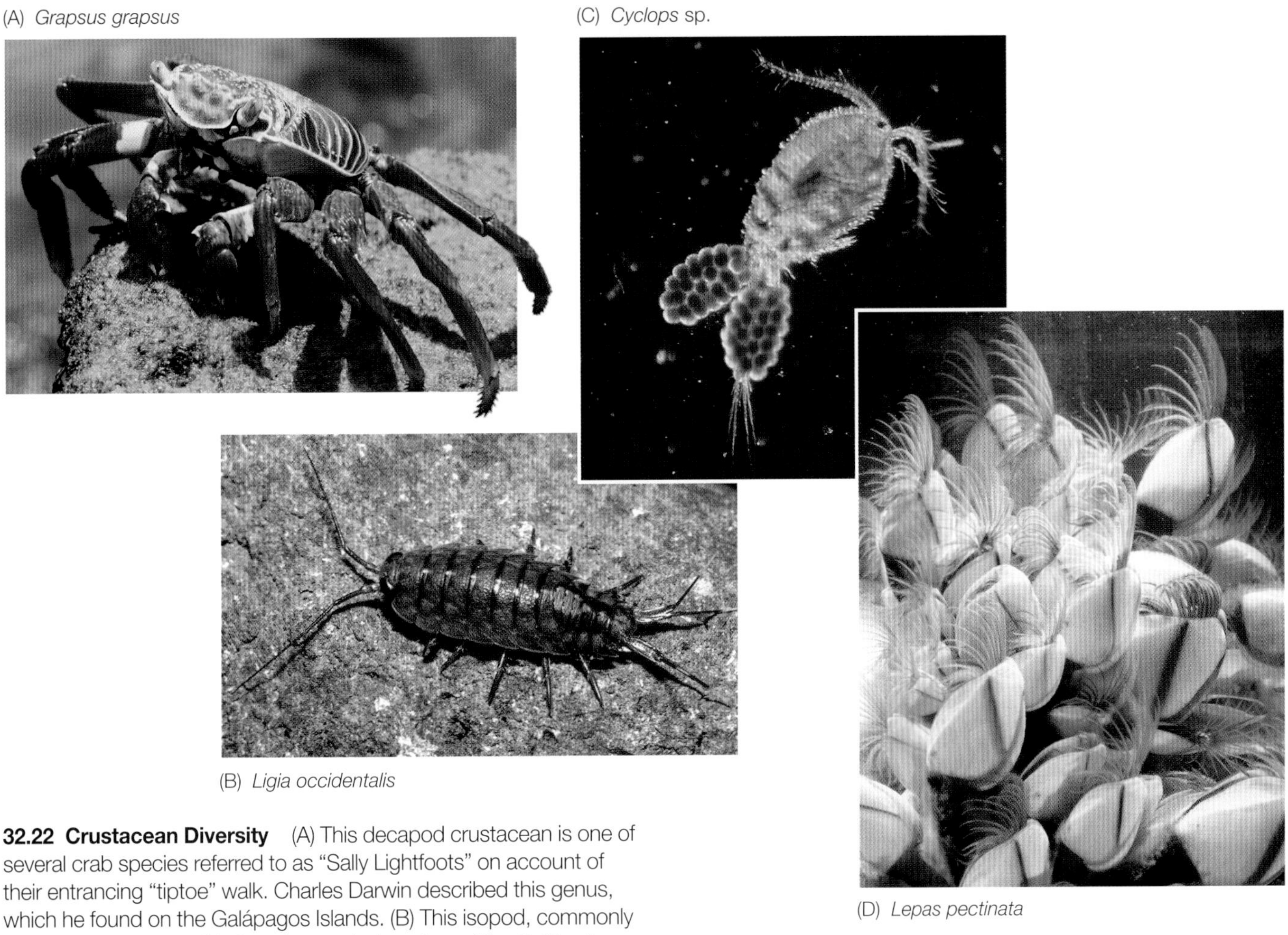

32.22 Crustacean Diversity (A) This decapod crustacean is one of several crab species referred to as "Sally Lightfoots" on account of their entrancing "tiptoe" walk. Charles Darwin described this genus, which he found on the Galápagos Islands. (B) This isopod, commonly called the "rock louse," is found on rocky beaches of the California coast. (C) This microscopic freshwater copepod is only about 30 μm long. (D) Gooseneck barnacles attach to a substratum by their muscular stalks and feed by protruding and retracting their feeding appendages.

Most of the 50,000 described species of crustaceans have a body that is divided into three regions: *head, thorax,* and *abdomen* (**Figure 32.23**). The segments of the head are fused together, and the head bears five pairs of appendages. Each of the multiple thoracic and abdominal segments usually bears one pair of appendages. The appendages on different parts of the body are specialized for different functions, such as gas exchange, chewing, sensing, walking, and swimming. In some cases, the appendages are branched, with different branches serving different functions. In many species, a fold of the exoskeleton, the *carapace*, extends dorsally and laterally back from the head to cover and protect some of the other segments.

The fertilized eggs of most crustacean species are attached to the outside of the female's body, where they remain during their early development. At hatching, the young of some species are released as larvae; those of other species are released as juveniles that are similar in form to the adults. Still other species release eggs into the water or attach them to an object in the environment. The typical crustacean larva, called a *nauplius*, has three pairs of appendages and one central eye (see Figure 31.12B). In many crustaceans, the nauplius larva develops within the egg before it hatches.

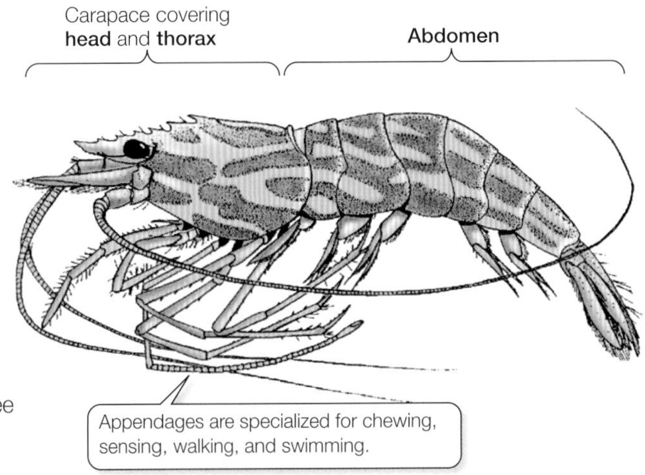

32.23 Crustacean Structure The bodies of crustaceans are divided into three regions—the head, thorax, and abdomen. Each region bears specialized appendages, and a thin, shell-like carapace covers the head and thorax.

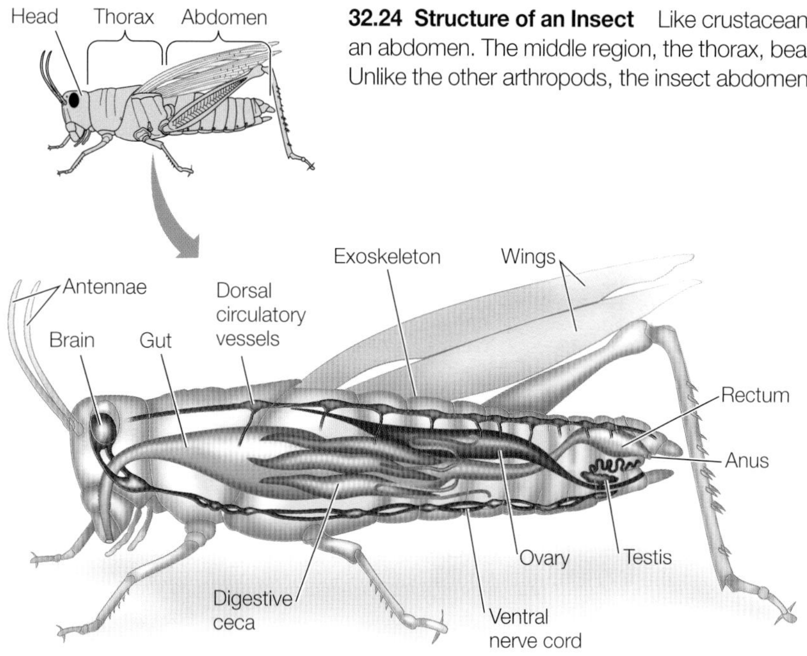

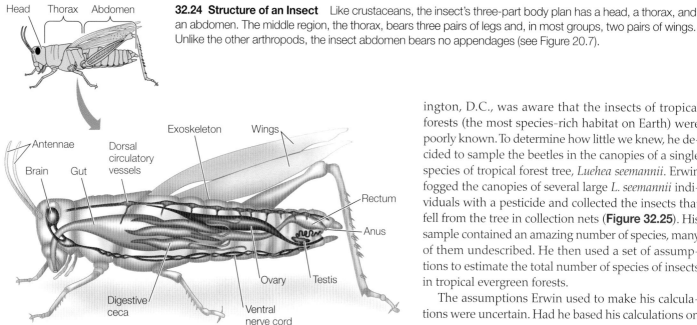

32.24 Structure of an Insect Like crustaceans, the insect's three-part body plan has a head, a thorax, and an abdomen. The middle region, the thorax, bears three pairs of legs and, in most groups, two pairs of wings. Unlike the other arthropods, the insect abdomen bears no appendages (see Figure 20.7).

Insects are the dominant terrestrial arthropods

During the Devonian period, more than 400 million years ago, some arthropods colonized terrestrial environments. Of the several groups that successfully colonized the land, none is more prominent today than the six-legged *hexapods*: the **insects** and their relatives. Insects are abundant and diverse in terrestrial and freshwater environments; only a few live in the oceans.

Insects, like crustaceans, have bodies with three regions—head, thorax, and abdomen. In addition, insects have a unique mechanism for gas exchange in air: a system of air sacs and tubular channels called *tracheae* (singular *trachea*) that extend from external openings called *spiracles* inward to tissues throughout the body. They have a single pair of antennae on the head and three pairs of legs attached to the thorax (**Figure 32.24**). Unlike the other arthropods, insects have no appendages growing from their abdominal segments.

Insects use nearly all species of plants and many species of animals as food. Herbivorous insects can consume massive amounts of plant matter. Gypsy moth caterpillars, for example, have been known to denude entire forests of their hardwood tree leaves, and the cyclical depradations wrought by locust hordes can be of biblical proportions. Many insects are predatory, feeding on small animals or on other insects. *Detrivorous* insects such as dung beetles are important in the recycling of chemicals through ecosystems (see Section 56.1). Some insects are internal parasites of plants or animals; others suck their host's blood or consume surface body tissues.

As recently as 1982, most biologists thought that about half of existing insect species had been described, but today they think that the approximately 1,000,000 species of described insects are a small fraction of the total number of living species. Why did they change their minds?

A simple but important field study provided the stimulus. One entomologist, Terry Erwin of the Smithsonian Institution in Washington, D.C., was aware that the insects of tropical forests (the most species-rich habitat on Earth) were poorly known. To determine how little we knew, he decided to sample the beetles in the canopies of a single species of tropical forest tree, *Luehea seemannii*. Erwin fogged the canopies of several large *L. seemannii* individuals with a pesticide and collected the insects that fell from the tree in collection nets (**Figure 32.25**). His sample contained an amazing number of species, many of them undescribed. He then used a set of assumptions to estimate the total number of species of insects in tropical evergreen forests.

The assumptions Erwin used to make his calculations were uncertain. Had he based his calculations on

32.25 Erwin's Research Project In an important study, entomologist Terry Erwin fogged tree canopies in the Central American rainforest with insecticide. Using the number of different species represented among the fallen insects as a base, he applied extrapolatory mathematical equations to estimate the total number of insect species on Earth. FURTHER RESEARCH: What kinds of experiments would you perform to refine Erwin's estimates of the number of species of insects on Earth?

different assumptions, his estimates of the number of species of insects could have changed dramatically. For example, if he overestimated the proportion of species that are specialists, his calculations would have overestimated the total number of species; but if he underestimated the proportion of specialists, the reverse would be true. Erwin assumed that he had adequately sampled the insect fauna of *Luehea seemannii* trees, but had he fogged the canopies of more individual trees, especially from different areas, he might have found many more species, which would mean his estimate was too low. But despite its uncertainties, Erwin's pioneering study alerted biologists to the fact that we live on a poorly known planet, most of whose species have yet to be named and described.

Terry Erwin's field within entomology (the study of insects) is *coleoptology*—the study of beetles. This specialty is not as esoteric as it might sound; beetles account for nearly 40 percent of the known insect species. Noted Scottish evolutionary biologist J. B. S. Haldane, when asked in 1951 what his studies revealed about the Creator, is said to have replied, "An inordinate fondness for beetles."

TABLE 32.2

The major insect groups[a]

GROUP	APPROXIMATE NUMBER OF DESCRIBED LIVING SPECIES
Jumping bristletails (*Archaeognatha*)	300
Silverfish (*Thysanura*)	370
PTERYGOTE (WINGED) INSECTS (*PTERYGOTA*)	
Mayflies (*Ephemeroptera*)	2,000
Dragonflies and damselflies (*Odonata*)	5,000
Neopterans (*Neoptera*)[b]	
Ice-crawlers (*Grylloblattodea*)	25
Gladiators (*Mantophasmatodea*)	15
Stoneflies (*Plecoptera*)	1,700
Webspinners (*Embioptera*)	300
Angel insects (*Zoraptera*)	30
Earwigs (*Dermaptera*)	1,800
Grasshoppers and crickets (*Orthoptera*)	20,000
Stick insects (*Phasmida*)	3,000
Cockroaches (*Blattodea*)	3,500
Termites (*Isoptera*)	2,750
Mantids (*Mantodea*)	2,300
Booklice and barklice (*Psocoptera*)	3,000
Thrips (*Thysanoptera*)	5,000
Lice (*Phthiraptera*)	3,100
True bugs, cicadas, aphids, leafhoppers (*Hemiptera*)	80,000
Holometabolous neopterans (*Holometabola*)[c]	
Ants, bees, wasps (*Hymenoptera*)	125,000
Beetles (*Coleoptera*)	375,000
Strepsipterans (*Strepsiptera*)	600
Lacewings, ant lions, dobsonflies (*Neuropterida*)	4,700
Scorpionflies (*Mecoptera*)	600
Fleas (*Siphonaptera*)	2,400
True flies (*Diptera*)	120,000
Caddisflies (*Trichoptera*)	5,000
Butterflies and moths (*Lepidoptera*)	250,000

[a] The hexapod relatives of insects include the springtails (*Collembola*; 3,000 spp.), two-pronged bristletails (*Diplura*; 600 spp.), and proturans (*Protura*; 10 spp.). All are wingless and have internal mouthparts.

[b] Neopteran insects can tuck their wings close to their bodies

[c] Holometabolous insects are neopterans that undergo complete metamorphosis.

Entire textbooks—and hefty ones at that—have been written about the described groups of insects. Here and in **Table 32.2** we present a brief outline of the major groups covered by these detailed entomology texts.

The wingless relatives of the insects—springtails (**Figure 32.26**), two-pronged bristletails, and proturans—are probably the most similar of living forms to insect ancestors. These relatives of insects have a simple life cycle; they hatch from eggs as miniature adults. They differ from insects in having internal mouth parts. Springtails can be extremely abundant (up to 200,000 per square meter) in soil, leaf litter, and on vegetation, and are the most abundant hexapods in the world.

Insects can be distinguished from other hexapods by their external mouthparts and paired antennae that contain a sensory receptor called Johnston's organ. Two groups of insects—the jumping bristletails and silverfish—are wingless and have simple life cycles, like the springtails and other close insect relatives. The remaining insects are the *pterygote insects*. Pterygotes have two pairs of wings, except in some groups where the wings have been secondarily lost.

Sminthurides aquaticus

32.26 Wingless Hexapods The wingless hexapods, such as this springtail, have a simple life cycle. They hatch looking like miniature adults, then grow by successive molts of the cuticle.

Hatchling pterygote insects do not look like adults, and they undergo substantial changes at each molt. The immature stages of insects between molts are called **instars**. As we saw in Section 31.4, a substantial change that occurs between one developmental stage and another is called **metamorphosis**. If the changes between its instars are gradual, an insect is said to have **incomplete metamorphosis**. If the change between at least some instars is dramatic, an insect is said to have **complete metamorphosis**.

The most familiar examples of species with complete metamorphosis are butterflies (see Figure 31.11). The wormlike butterfly larva, called a *caterpillar*, transforms itself during a specialized phase, called the **pupa**, in which many larval tissues are broken down and the adult form develops. In many insects with complete metamorphosis, the different life stages are specialized for living in different environments and using different food sources. In many species, the larvae are adapted for feeding and growing; the adults are specialized for reproduction and dispersal.

Pterygote insects were the first animals in evolutionary history to achieve the ability to fly (although members of several groups of winged insects—including parasitic lice and fleas, as well some beetles and the worker individuals in many ant groups—subsequently lost their wings and their ability to fly). Flight opened up many new lifestyles and feeding opportunities that only the insects could exploit, and it is almost certainly one of the reasons for the remarkable numbers of insect species and individuals, and for their unparalleled evolutionary success.

The adults of most flying insects have two pairs of stiff, membranous wings attached to the thorax. True flies, however, have only one pair of wings. In beetles, one pair of wings—the forewings—forms heavy, hardened wing covers. Flying insects are important pollinators of flowering plants.

Two groups of pterygote insects, the mayflies and dragonflies (**Figure 32.27A**), cannot fold their wings against their bodies. This is the ancestral condition for pterygote insects, and these two groups are not closely related to one another. All members of these groups have predatory aquatic larvae that transform themselves into flying adults after they crawl out of the water. Many of them are excellent flyers. Dragonflies (and their relatives the damselflies) are active predators as adults, but adult mayflies lack functional digestive tracts; adult mayflies live only about a day, just long enough to mate and lay eggs.

All other pterygote insects—the *neopterans*—can tuck their wings out of the way upon landing and crawl into crevices and other tight places. Many neopteran groups have an incomplete metamorphosis, so hatchlings of these insects are sufficiently similar in form to adults to be recognizable. Examples include the grasshoppers (**Figure 32.27B**), roaches, mantids, stick insects, termites, stoneflies, earwigs, thrips, true bugs (**Figure 32.27C**), aphids, cicadas, and leafhoppers. They acquire adult organ systems, such as wings and compound eyes, gradually through several juvenile instars. Remarkably, a new major group, the mantophasmatodeans, was described only recently, in 2002 (**Figure 32.27D**). These small insects are common in the Cape Region of South Africa, an area of exceptional species richness and endemism for many animal and plant groups.

Over 80 percent of all insects are a subgroup of the neopterans called the *holometabolous insects* (see Table 32.2), which have complete metamorphosis. The many species of beetles comprise almost half of this group (**Figure 32.27E**). Also included are lacewings and their relatives; caddisflies; butterflies and moths (**Figure 32.27F**); sawflies; true flies (**Figure 32.27G**); and bees, wasps, and ants, some species of which display unique and highly specialized social behaviors (**Figure 32.27H**).

Molecular data suggest that the lineage leading to the insects separated from the lineage leading to modern crustaceans about 450 million years ago, about the time of the appearance of the first land plants. These ancestral hexapods penetrated a terrestrial environment that lacked any other similar organisms, which in part accounts for their remarkable success. But the success of the insects is also due to their wings. Homologous genes control the development of insect wings and crustacean appendages, suggesting that that the insect wing evolved from a dorsal branch of a crustacean-like limb (**Figure 32.28**). The dorsal limb branch of crustaceans is used for gas exchange. Thus the insect wing probably evolved from a gill-like structure that had a gas exchange function.

32.27 The Diversity of Winged Insects (A) Unlike most flying insects, this dragonfly cannot fold its wings over its back. (B) The Mexican bush katydid represents an orthopteran. (C) Dock bugs are "true" bugs (hemipterans) specialized for feeding on a plants of the genus *Rumex*. (D) This mantophasmatodean represents a recently discovered insect group found only in southern Africa. (E) A predatory diving beetle (a coleopteran). (F) The California dogface butterfly is a lepidopteran. (G) Flies such as this green bottle fly are dipterans. (H) Many hymenopteran genera, such as the honeybees seen here on their hive, are social insects.

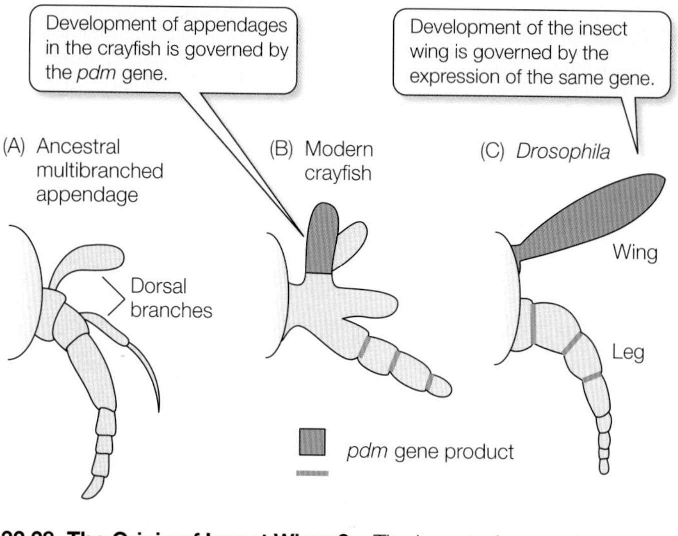

32.28 The Origin of Insect Wings? The insect wing may have evolved from an ancestral appendage similar to that of modern crustaceans. (A) A diagram of the ancestral, multibranched arthropod limb. (B,C) The *pdm* gene, a Hox gene, is expressed throughout the dorsal limb branch and walking leg of the thoracic limb of a crayfish (B) and in the wings and legs of *Drosophila* (C).

(A) *Libellula luctuosa*

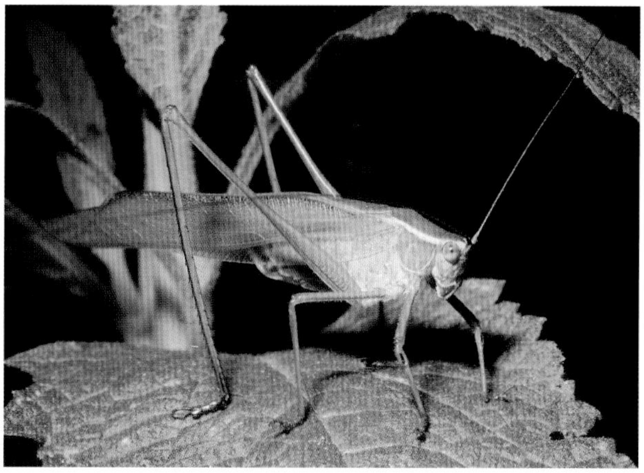
(B) *Scudderia mexicana*

(C) *Coreus marginatus*

(D) *Mantophasma zephyra*

(E) *Dytiscus marginalis*

(F) *Colias eurydice*

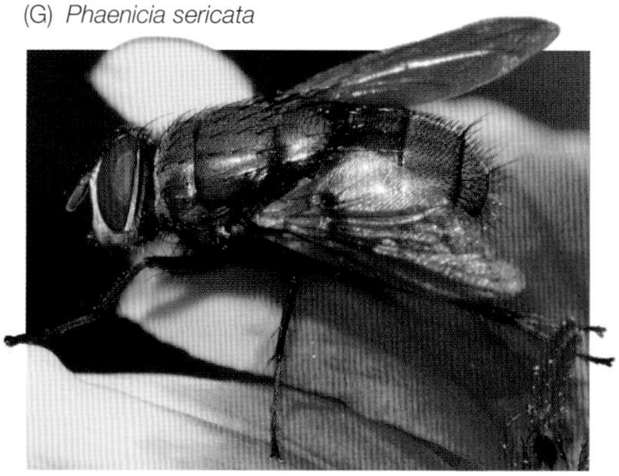
(G) *Phaenicia sericata*

(H) *Apis mellifera*

Myriapods have many legs

As we have seen, insects and most crustaceans have a three-part body plan, with a head, thorax, and abdomen. Members of two other arthropod groups, the myriapods and chelicerates, have a body plan with just two regions: a head and a trunk.

The **myriapods** comprise the centipedes, millipedes, and two other groups. Centipedes and millipedes have a well-formed head and a long, flexible, segmented trunk that bears many pairs of legs (**Figure 32.29**). Centipedes, which have one pair of legs per segment, prey on insects and other small animals. In millipedes, two adjacent segments are fused so that each fused segment has two pairs of legs. Millipedes scavenge and eat plants. More than 3,000 species of centipedes and 11,000 species of millipedes have been described; many more species probably remain unknown. Although most myriapods are less than a few centimeters long, some tropical species are ten times that size.

Most chelicerates have four pairs of legs

In the two-part body plan of **chelicerates**, the head bears two pairs of appendages modified to form mouthparts. In addition, many chelicerates have four pairs of walking legs. The 89,000 described species are usually placed in three clades: pycnogonids, horseshoe crabs, and arachnids.

The *pycnogonids*, or sea spiders, are a poorly known group of about 1,000 marine species (**Figure 32.30A**). Most are small, with leg spans less than 1 centimeter, but some deep-sea species have leg spans up to 60 centimeters. A few pycnogonids eat algae, but most are carnivorous, eating a variety of small invertebrates.

There are four living species of *horseshoe crabs*, but many close relatives are known only from fossils. Horseshoe crabs, which have changed very little morphologically during their long fossil history, have a large horseshoe-shaped covering over most of the body. They are common in shallow waters along the eastern coasts of North America and the southern and eastern coasts of Asia, where

(A) *Scolopendra gigantea*

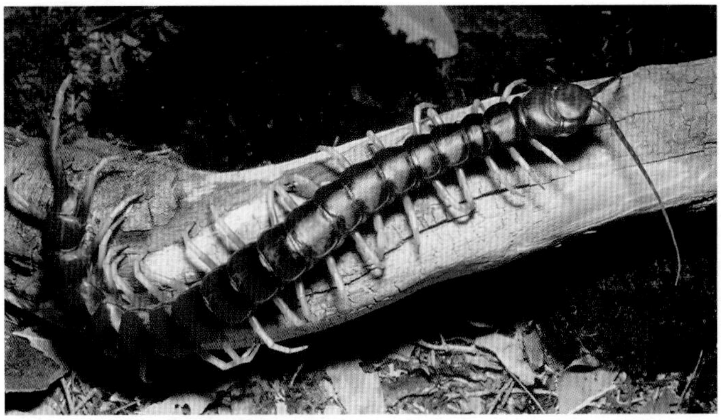

(B) *Sigmoria trimaculata*

32.29 Myriapods (A) Centipedes have powerful jaws for capturing active prey and one pair of legs per segment. (B) Millipedes, which are scavengers and plant eaters, have smaller jaws and legs. They have two pairs of legs per segment.

(A) *Colossendeis megalonyx*

(B) *Limulus polyphemus*

32.30 Some Small Groups of Chelicerates (A) Although they are not spiders, it is easy to see why sea spiders were given their common name. (B) This spawning aggregation of horseshoe crabs was photographed on a sandy beach of the Delaware Bay. Horseshoe crabs are an ancient group that has changed very little in morphology over time; such species are sometimes referred to as "living fossils."

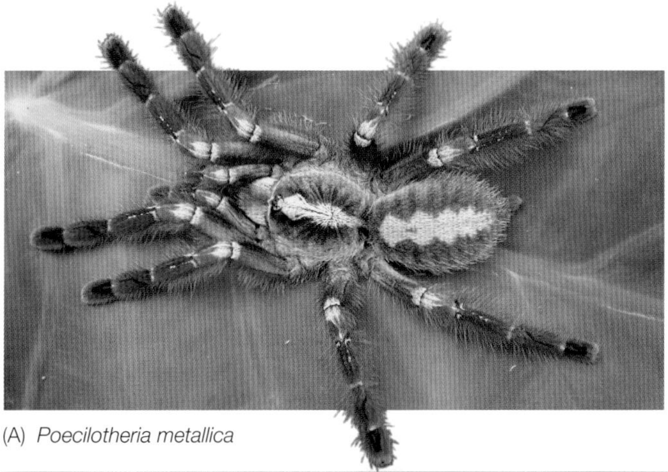

(A) *Poecilotheria metallica*

(B) *Pseudouroctonus minimus*

(C) *Hadrobunus maculosus*

(D) *Brevipalpus phoenicis*

32.31 Arachnid Diversity (A) The name "tarantula" encompasses several hundred species of hairy, ground-dwelling spiders, some of which can grow to the size of a dinner plate. Their venomous bite, although painful, is not usually deadly to humans. (B) Scorpions are nocturnal predators. (C) Harvestmen, also called daddy longlegs, are scavengers. (D) Mites are blood-sucking, external parasites.

they scavenge and prey on bottom-dwelling animals. Periodically they crawl into the intertidal zone in large numbers to mate and lay eggs (**Figure 32.30B**).

Arachnids are abundant in terrestrial environments. Most arachnids have a simple life cycle in which miniature adults hatch from internally fertilized eggs and begin independent lives almost immediately. Some arachnids retain their eggs during development and give birth to live young.

The most species-rich and abundant arachnids are the spiders, scorpions, harvestmen, mites, and ticks (**Figure 32.31**). The 50,000 described species of mites and ticks live in soil, leaf litter, mosses, and lichens, under bark, and as parasites of plants and animals. Mites are vectors for wheat and rye mosaic viruses; they cause mange in domestic animals and skin irritation in humans.

Spiders (of which 38,000 species have been described) are important terrestrial predators. Some have excellent vision that enables them to chase and seize their prey. Others spin elaborate webs made of protein threads in which they snare prey. The threads are produced by modified abdominal appendages connected to internal glands that secrete the proteins, which dry on contact with air. The webs of different groups of spiders are strikingly varied, and this variation enables the spiders to position their snares in many different environments.

32.4 RECAP

All arthropods have segmented bodies. Muscles in each segment operate that segment and the appendages attached to it. Their jointed, specialized appendages permit complex patterns of movement, including the ability of the insects to fly.

- What features have contributed to making arthropods the dominant animals on Earth, both in number of species and number of individuals? Can you think of some possible reasons why there are so many species of insects?
- Do you understand the difference between incomplete and complete metamorphosis? See p. 710

The protostomes encompass the vast majority of the Earth's animal species, so it is not surprising that the different protostome groups display a huge variety of different characteristics.

An Overview of Protostome Evolution

The myriad protostome species encompass a staggering number of different life forms. Consider how the following aspects of protostome evolution have contributed to this enormous diversity:

- The evolution of *segmentation* permitted some protostome groups to move different parts of the body independently of one another. Species in some groups gradually evolved the ability to move rapidly over and through the substratum, through water, and through air.
- *Complex life cycles* with dramatic changes in form between one stage and another allow individuals of different stages to specialize on different resources.
- *Parasitism* has evolved repeatedly, and many protostome groups parasitize multicellular plants and animals.
- The evolution of *diverse feeding structures* allowed protostomes to specialize on many different food sources. Early protostomes were mostly filter-feeders because dissolved organic matter and very small organisms were the most readily available food sources in the oceans. Although filter-feeding remains common, the ability to move rapidly on land and through water and air allowed the evolution of predation and herbivory.
- Predation was a major selection pressure favoring the development of *hard external body coverings* (exoskeletons). Such coverings evolved independently in many lophotrochozoan and ecdysozoan groups. In addition to providing protection, they became key elements in the development of new systems of locomotion.
- *Better locomotion* permitted prey to escape from predators, but also allowed predators to pursue their prey more effectively. Thus the evolution of animals has been, and continues to be, a complex "arms race" among predators and prey.

Many major evolutionary trends among the protostomes are shared by the deuterostomes, which includes the chordates, the group to which humans belong.

CHAPTER SUMMARY

32.1 What is a protostome?

Protostomes ("mouth first") have an anterior brain that surrounds the entrance to the digestive tract and a ventral nervous system. The major protostome clades are grouped into the **lophotrochozoans** and the **ecdysozoans**. Review Figure 32.1

Lophotrochozoans include a wide diversity of animals. Within this group, **lophophores**, free-living **trochophore** larvae, and **spiral cleavage** evolved.

Ecdysozoans have an **exoskeleton**, which they must **molt** in order to grow. Some ecdysozoans have a relatively thin exoskeleton, called a **cuticle**. Others, especially the **arthropods**, have a rigid exoskeleton reinforced with **chitin**. New mechanisms of locomotion and gas exchange evolved in these animals. Review Figure 32.4

32.2 What are the major groups of lophotrochozoans?

Lophotrochozoans range from relatively simple animals having only one entrance to the digestive tract and no oxygen transport system to animals with complete digestive tracts and complex internal transport systems.

Lophophores, wormlike body forms, and external shells are each found in multiple unrelated groups of lophotrochozoans.

The most species-rich groups of lophotrochozoans are the **flatworms**, **annelids**, and **mollusks**.

Segmentation of the body first evolved among the annelids. Review Figure 32.12

Mollusks underwent a dramatic evolutionary radiation based on a body plan consisting of three major components: a **foot**, a **mantle**, and a **visceral mass**. The five major molluscan clades—the **monoplacophorans**, **chitons**, **bivalves**, **gastropods**, and **cephalopods**—demonstrate the diversity that evolved from this three-part body plan. Review Figure 32.14

32.3 What are the major groups of ecdysozoans?

Many groups of ecdysozoans are wormlike in form. Members of several species-poor clades of marine wormlike ecdysozoans (**priapulids**, **kinorhynchs**, and **loriciferans**) have thin cuticles.

Horsehair worms are extremely thin; many are internal parasites as larvae.

Nematodes, or roundworms, have thick, multilayered cuticles. Nematodes are one of the most abundant and universally distributed of all animal groups. Review Figure 32.17

One major ecdysozoan clade, the **arthropods** and their relatives, have evolved limblike appendages. Collectively, the arthropods are the dominant animals on Earth, both in number of species and number of individuals.

32.4 Why do arthropods dominate Earth's fauna?

Encasement within a rigid exoskeleton provides arthropods with support for walking in the water and on dry land as well as some protection from predators. The waterproofing provided by chitin keeps arthropods from dehydrating in dry air.

Jointed appendages permit complex movement patterns. Each arthropod segment has muscles attached to the inside of the exoskeleton that operate that segment and the appendages attached to it.

Two groups of arthropod relatives, the **onychophorans** and the **tardigrades**, have simple, unjointed appendages. The **trilobites** were early marine arthropods that disappeared in the Permian extinction.

Crustaceans are the dominant marine arthropods. Their segmented bodies are divided into three regions (head, thorax, abdomen) with different, specialized appendages in each region. Review Figure 32.23

Hexapods—insects and their relatives—are the dominant terrestrial arthropods. They have the same three body regions as crustaceans, but no appendages form in their abdominal segments. Wings and the ability to fly first evolved among the insects, allowing them to exploit new lifestyles. Review Figure 32.24

The bodies of **myriapods** have only two regions, a head, and a long trunk with many segments, each of which carries appendages. **Chelicerates** also have a two-part body plan; most chelicerates have four pairs of walking legs.

See Web/CD Activities 32.1 and 32.2 for a concept review of this chapter.

SELF-QUIZ

1. Members of which groups have lophophores?
   a. Phoronids, brachiopods, and nematodes
   b. Phoronids, brachiopods, and ectoprocts
   c. Brachiopods, ectoprocts, and flatworms
   d. Phoronids, rotifers, and ectoprocts
   e. Rotifers, ectoprocts, and brachiopods
2. Which of the following is *not* part of the molluscan body plan?
   a. Mantle
   b. Foot
   c. Radula
   d. Visceral mass
   e. Jointed skeleton
3. Nautiluses control their buoyancy by
   a. adjusting salt concentrations in their blood.
   b. forcibly expelling water from the mantle.
   c. pumping water and gases in and out of internal chambers.
   d. using the complex sense organs in their heads.
   e. swimming rapidly.
4. The outer covering of ecdysozoans
   a. is always hard and rigid.
   b. is always thin and flexible.
   c. is present at some stage in the life cycle but not always among adults.
   d. ranges from very thin to hard and rigid.
   e. prevents the animals from changing their shapes.
5. Nematodes are abundant and diverse because
   a. they are both parasitic and free-living and eat a wide variety of foods.
   b. they are able to molt their exoskeletons.
   c. their thick cuticle enables them to move in complex ways.
   d. their body cavity is a pseudocoelom.
   e. their segmented bodies enable them to live in many different places.
6. The arthropod exoskeleton is composed of a
   a. mixture of several kinds of polysaccharides.
   b. mixture of several kinds of proteins.
   c. single complex polysaccharide called chitin.
   d. single complex protein called arthropodin.
   e. mixture of layers of proteins and a polysaccharide called chitin.
7. Which groups are arthropod relatives with unjointed legs?
   a. Trilobites and onychophorans
   b. Onychophorans and tardigrades
   c. Trilobites and tardigrades
   d. Onychophorans and chelicerates
   e. Tardigrades and chelicerates
8. The body plan of insects is composed of which of the three following regions?
   a. Head, abdomen, and trachea
   b. Head, abdomen, and cephalothorax
   c. Cephalothorax, abdomen, and trachea
   d. Head, thorax, and abdomen
   e. Abdomen, trachea, and mantle
9. Insects whose hatchlings are sufficiently similar in form to adults to be recognizable are said to have
   a. instars.
   b. neopterous development.
   c. accelerated development.
   d. incomplete metamorphosis.
   e. complete metamorphosis.
10. Factors that may have contributed to the remarkable evolutionary diversification of insects include
    a. the terrestrial environments penetrated by insects lacked any other similar organisms.
    b. the ability to fly.
    c. complete metamorphosis.
    d. a new mechanism for delivering oxygen to their internal tissues.
    e. all of the above

FOR DISCUSSION

1. Segmentation has arisen several times during animal evolution. What advantages does segmentation provide? Given these advantages, why do so many unsegmented animals survive? Why have some animals lost their segmentation?
2. Major structural novelties have arisen only infrequently during the course of evolution. Which of the features of protostomes do you think are major evolutionary novelties? What criteria do you use to judge whether a feature is a major as opposed to a minor novelty?
3. There are more described and named species of insects than of all other animal groups combined. However, only a very few species of insects live in marine environments, and those species are restricted to the intertidal zone or the ocean surface. What factors may have contributed to this lack of success in the oceans?

FOR INVESTIGATION

If you were given funding to carry out studies to improve estimates of the number of species of insects on Earth, how would you spend it? Would you do the same things if your objective were to determine the number of species of nematodes? Mites?

CHAPTER 33 Deuterostome Animals

Hobbits of Flores Island

In 2004, archeological workers on the Indonesian island of Flores unearthed the skeleton of an adult hominid female who stood less than a meter tall, weighed about 20 kilograms, and had a brain the size of a chimpanzee's. More fossils of these diminutive hominids were subsequently discovered on Flores, and radioactive dating indicated that some of them were shockingly recent—a mere 18,000 years old.

Although not all anthropologists are convinced, many experts think that the remains are those of a new species, dubbed *H. floresiensis,* and that this species is more closely related to *Homo erectus*, an extinct hominid species, than it is to *Homo sapiens*. *H. floresiensis* is thought to have evolved from *H. erectus* ancestors that arrived on Flores at least 840,000 years ago, at a time when the island lacked any other flightless mammals except rodents and pygmy elephants. The fossils tell us that these "hobbits" hunted pygmy elephants, which may have been their major food source.

Assuming that the majority opinion is correct, the discovery of a small hominid species surviving until relatively recent times stimulates several questions. How did their ancestors reach Flores? Why did they evolve to be so small? How did they manage to avoid extermination by the much larger modern humans who spread across Indonesia and reached Australia at least 46,000 years ago?

A probable scenario is that ancestral *H. erectus* colonized Flores during a period of glacial expansion, when sea levels were about 150 meters lower than they are today. At that time, Java and Bali would have been part of the Asian mainland. Even so, to reach Flores, the colonists had to cross three water gaps. These gaps were narrow enough that an island on the other side would have been visible, and reachable on a simple raft or canoe.

That the "hobbits" evolved small size is not surprising; biogeographers have observed that when large mammals colonize small islands, they typically evolve smaller size. Pygmy hippos, buffaloes, ground sloths, elephants, deer, and other mammals have all evolved on islands. In part this may be because islands typically lack both the resources to sustain large animals and the kinds of predators that feed on smaller animals. *Homo*

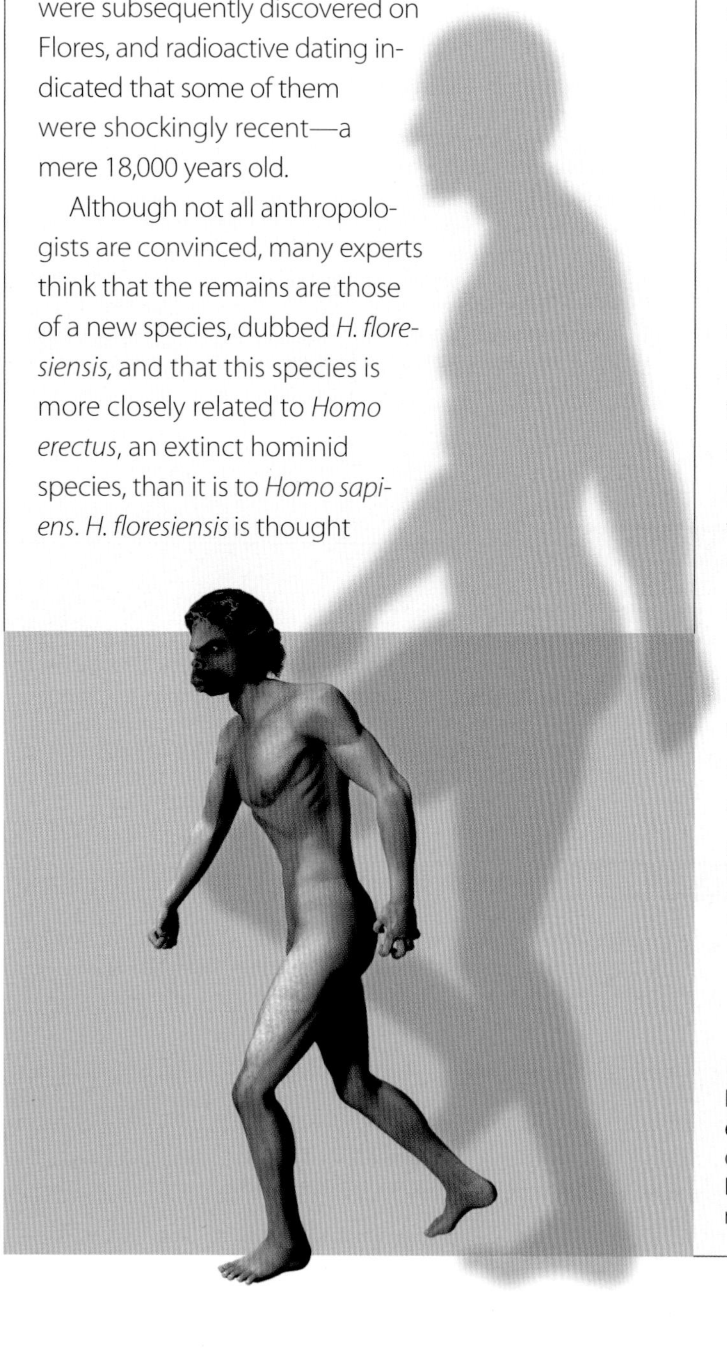

Hominids Are Deuterostomes In this artist's rendition, the extinct hominid *Homo floresiensis*, at less than a meter tall, is dwarfed by the silhouette of a modern human, *Homo sapiens*. Modern and extinct hominid species are vertebrates and thus members of the deuterostome clade.

Different Deuterostomes A sea star rests on a colony of tunicates ("sea squirts"); both of these species are deuterostomes. Although their morphology is not like that of any vertebrate, both animals share several synapomorphies with the vertebrates.

floresiensis and the pygmy elephants they hunted would be examples of this well-known phenomenon.

The fact that the *H. floresiensis* lineage survived was probably due in large part to the three water gaps, which would have fluctuated in size with glacial fluctuations. Later hominid expansion across Indonesia to New Guinea could have occurred via a more easily traversed northerly route, leaving the "hobbits" isolated and untouched for many millennia.

The study of our own history—the evolution of *Homo sapiens*—has its own field, anthropology. But it is also the legitimate study of biologists, as we will see when we view humans in the light of their position in the realm of the deuterostomes.

IN THIS CHAPTER we introduce the deuterostomes and describe the major deuterostome groups—echinoderms, hemichordates, and chordates. We will then discuss the features that allowed some chordates to colonize dry land and will trace the evolution of those terrestrial vertebrates, taking an especially close look at the primate lineage that includes our own species.

CHAPTER OUTLINE

33.1 What is a Deuterostome?

It may surprise you to learn that you and a sea urchin are both deuterostomes. Adult sea stars, sea urchins, and sea cucumbers—the most familiar echinoderms—look so different from adult vertebrates (fish, frogs, lizards, birds, and mammals) that it may be difficult to believe all these animals are closely related to one another. The evidence that deuterostomes all share a common ancestor that is different from the common ancestor of the protostomes is provided by their early developmental patterns and by phylogenetic analysis of gene sequences, neither of which is apparent in the forms of the adult animals.

Three early developmental patterns characterize the deuterostomes:

- Radial cleavage
- Formation of the mouth at the opposite end of the embryo from the blastopore (the pattern that gives the deuterostomes their name; see Figure 31.2)
- Development of a coelom from mesodermal pockets that bud off from the cavity of the gastrula rather than by splitting of the mesoderm, as occurs among protostomes

The first two of these features represent the ancestral condition for bilaterian animals in general, and as we saw in the previous chapter, some protostomes also have radial cleavage and other developmental similarities to deuterostomes. (In fact, some of the groups we now consider protostomes were once thought to be deuterostomes because of their retained ancestral developmental similarities with echinoderms and chordates.) Therefore, although developmental features that were once thought to be uniquely derived in deuterostomes have historically been important for hypotheses about the group's monophyly, these features are now known to be ancestral and therefore not indicative of monophyly. Nonetheless, we still recognize the close evolutionary relationships of echinoderms, hemichordates, and chordates (the groups that now compose the deuterostomes) because phylogenetic analyses of DNA sequences of many slowly evolving genes support their common ancestry.

There are many fewer species of deuterostomes than of protostomes (see Table 31.1), but we have a special interest in deuterostomes, in part because we are members of that clade.

33.1 A Current Phylogenetic Tree of the Deuterostomes There are many fewer species of deuterostomes than of protostomes.

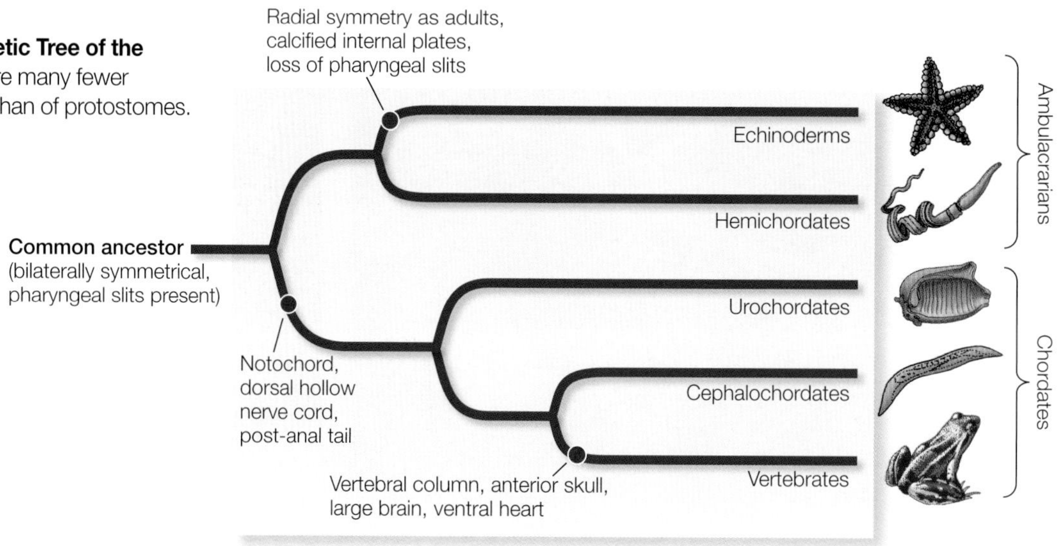

We are also interested in them because they include the largest living animals; these large deuterostomes strongly influence the characteristics of ecosystems.

Living deuterostomes include three major clades (**Figure 33.1**):

- *Echinoderms*: sea stars, sea urchins, and their relatives
- *Hemichordates*: acorn worms and pterobranchs
- *Chordates*: sea squirts, lancelets, and vertebrates

All deuterostomes are triploblastic, coelomate animals (see Figure 31.4) with internal skeletons. Some species have segmented bodies, but the segments are less obvious than those of annelids and arthropods.

Scientists are learning much about the ancestors of modern deuterostomes from recently discovered fossils of several early forms. The most important finds have come from 520-million-year-old fossil beds in China. The *homalozoans* had a skeleton similar to that of a modern echinoderm, but they had pharyngeal slits and bilateral symmetry. The *vetulicosystids*, which were first discovered in 2002, also had pharyngeal slits. Many fossils of a third kind, the *yunnanozoans*, were discovered in China's Yunnan Province. These well-preserved animals had a large mouth, six pairs of external gills, and a segmented posterior body section bearing a light cuticle (**Figure 33.2**).

The features of these fossil animals support the findings from phylogenetic analyses of living species in showing that the earliest deuterostomes were bilaterally symmetrical, segmented animals with a pharynx that had slits through which water flowed (see Figure 33.1). Echinoderms evolved their adult forms with unique pentaradial symmetry much later, whereas other deuterostomes retained the ancestral bilateral symmetry.

Yunnanozoon lividum

33.2 Ancestral Deuterostomes Had External Gills The extinct yunnanozoans may be ancestral to all deuterostomes. This fossil, which dates from the Cambrian, shows the six pairs of external gills and segmented posterior body that characterized these animals.

33.2 What Are the Major Groups of Echinoderms and Hemichordates?

About 13,000 species of echinoderms in 23 major groups have been described from their fossil remains. They are probably only a small fraction of those that actually lived. Only 6 of the 23 groups known from fossils are represented by species that survive today; many clades became extinct during the periodic mass extinctions that have occurred throughout Earth's history. Nearly all of the 7,000 extant species of echinoderms live only in marine environments. There are only 95 known living species of hemichordates.

The echinoderms and hemichordates (together known as *ambulacrarians*) have a bilaterally symmetrical, ciliated larva (**Figure 33.3**). Adult hemichordates also are bilaterally symmetrical. Echinoderms, however, undergo a radical change in form as they develop into adults, changing from a bilaterally symmetrical larva to an adult with **pentaradial symmetry** (symmetry in five or multiples of five). As is typical of animals with radial symmetry, echinoderms have no head, and they move slowly and equally well in

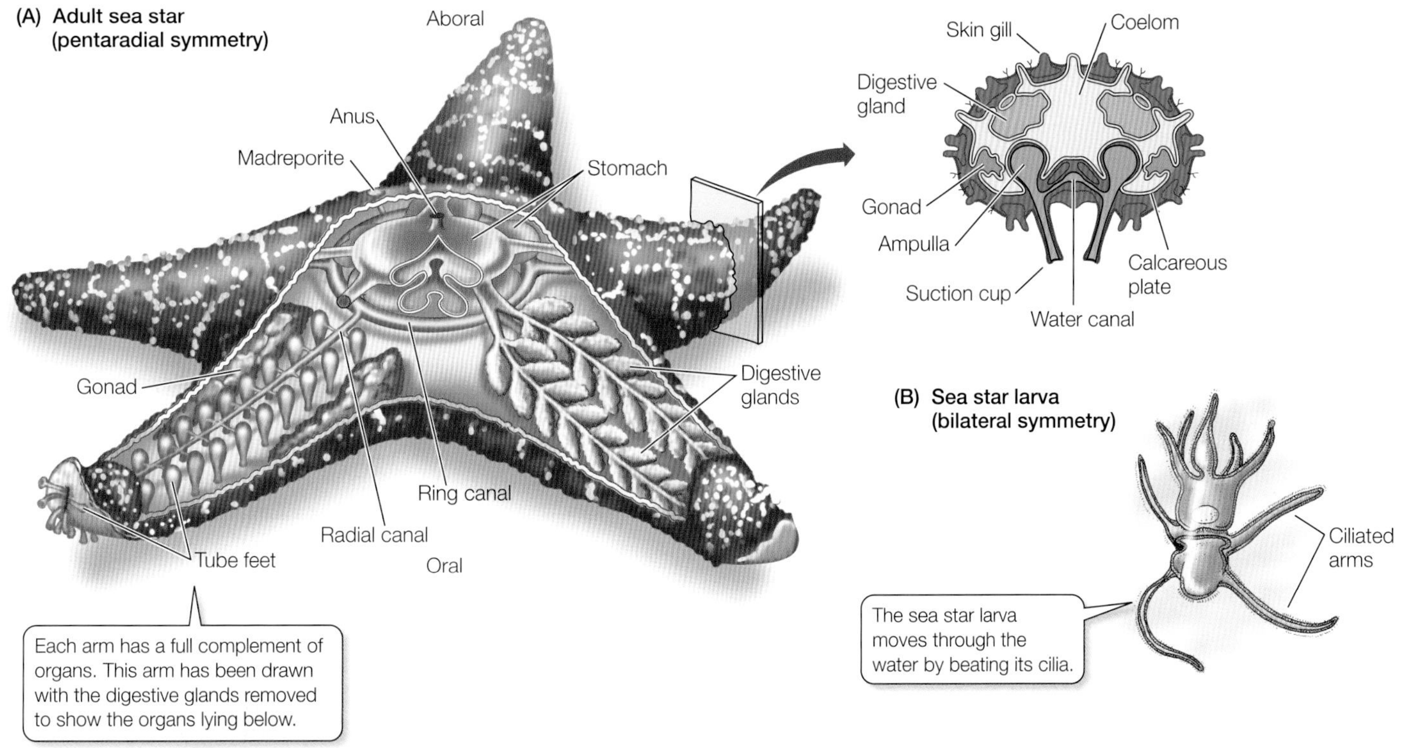

33.3 Evolutionary Innovations of Echinoderms (A) A dorsal view of a sea star displays the canals and tube feet of the echinoderm water vascular system, as well as the calcified internal skeleton. (B) The ciliated larva of a sea star has bilateral symmetry.

many directions. Rather than having an anterior-posterior (head-tail) and dorsal-ventral (back-belly) body organization, echinoderms have an *oral* side (containing the mouth) and an opposite *aboral* side (containing the anus).

Echinoderms have a water vascular system

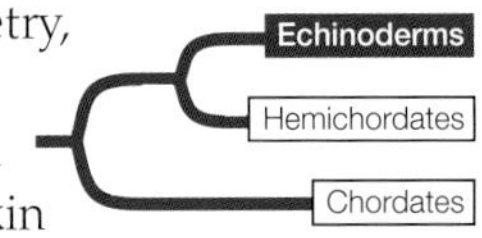

In addition to having pentaradial symmetry, adult echinoderms have two unique structural features. One is a system of calcified internal plates covered by thin layers of skin and some muscles. The calcified plates of most echinoderms are thick, and they fuse inside the entire body, forming an *internal skeleton*. The other unique feature is a **water vascular system**, a network of water-filled canals leading to extensions called **tube feet**. This system functions in gas exchange, locomotion, and feeding (see Figure 33.3). Seawater enters the system through a perforated structure called a *madreporite*. A calcified canal leads from the madreporite to another canal that rings the *esophagus* (the tube leading from the mouth to the stomach). Other canals radiate from this *ring canal*, extending through the arms (in species that have arms) and connecting with the tube feet. Echinoderms use their tube feet in many different ways to move and to capture prey. These structural innovations have been modified in many ways to result in a striking array of very different animals.

Members of one major extant echinoderm clade, the *crinoids* (sea lilies and feather stars), were more abundant and species-rich 300–500 million years ago than they are today. Most of the 80 living species of sea lilies attach to a substratum by means of a flexible stalk consisting of a stack of calcareous discs. The main body of the animal is a cup-shaped structure that contains a tubular digestive system. Five to several hundred arms, usually in multiples of five, extend outward from the cup. The jointed calcareous plates of the arms enable them to bend (**Figure 33.4A**).

Feather stars are similar to sea lilies, but they grasp the substratum with their flexible appendages (**Figure 33.4B**). They can walk on the tips of their arms or swim by rhythmically beating them. Feather stars feed in much the same manner as sea lilies. About 600 living species have been described.

Most surviving echinoderm species—including sea urchins, sea cucumbers, sea stars, and brittle stars—are not crinoids. Sea urchins are hemispherical in shape and lack arms (**Figure 33.4C**). They are covered with spines that are attached to the underlying skeleton via ball-and-socket joints. These joints enable the spines to be moved so that they can converge toward a point that has been touched. The spines come in varied sizes and shapes; a few produce toxic substances. They provide effective protection for the urchin, as many a scuba diver has found out the hard way. Sand dollars are flattened and disc-shaped relatives of sea urchins.

The sea cucumbers also lack arms, and their bodies are oriented in an atypical manner for an echinoderm (**Figure 33.4D**). The mouth is anterior and the anus is posterior, not oral and aboral as in other echinoderms. Sea cucumbers use most of their tube feet primarily for attaching to the substratum rather than for moving.

Sea stars are the most familiar echinoderms (**Figure 33.4E**). Their gonads and digestive organs are located in the arms, as seen in Figure 33.3. Their tube feet serve as organs of locomotion, gas exchange, and attachment. Each tube foot of a sea star consists of an internal *ampulla* connected by a muscular tube to an external suction cup that can stick to the substratum. The tube foot is

(A) *Nominus novus*

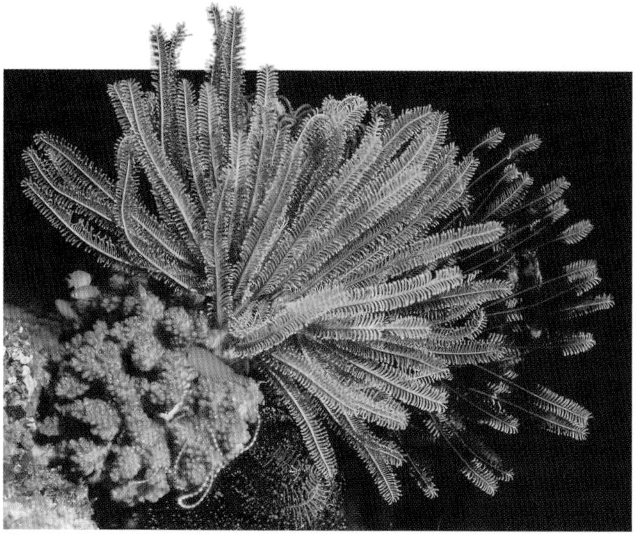

(B) *Oxycomanthus bennetti*

(C) *Strongylocentrotus purpuratus*

(D) *Bohadschia argus*

(E) *Henricia leviuscula*

(F) *Ophiothrix spiculata*

33.4 Diversity among the Echinoderms (A) Sea lilies can have up to several hundred arms, usually in multiples of five. (B) The flexible arms of the golden feather star are clearly visible. (C) Purple sea urchins are important grazers on algae in the intertidal zone of the Pacific Coast of North America. (D) This sea cucumber lives on rocky substrata in the seas around Papua New Guinea. (E) Sea stars are important predators on bivalve mollusks such as mussels and clams. Suction tips on its tube feet allow this animal to grasp both shells of the bivalve and pull them apart. (F) The arms of the brittle star are composed of hard but flexible plates.

moved by expansion and contraction of the circular and longitudinal muscles of the tube. Brittle stars are similar in structure to sea stars, but their flexible arms are composed of jointed hard plates (**Figure 33.4F**).

Sea daisies were discovered only in 1986; little is known about them. They have tiny, disc-shaped bodies with a ring of marginal spines, and two ring canals, but no arms. Sea daisies live on rotting wood in ocean waters. They apparently eat prokaryotes, which they digest outside their bodies and absorb either through a membrane that covers the oral surface or via a shallow, saclike stomach. Recent molecular data suggest that they are greatly modified sea stars.

Echinoderms use their tube feet in a great variety of ways to capture prey. Sea lilies, for example, feed by orienting their arms in passing water currents. Food particles then strike and stick to the tube feet, which are covered with mucus-secreting glands. The tube feet transfer these particles to grooves in the arms, where ciliary action carries the food to the mouth. Sea cucumbers capture food with their anterior tube feet, which are modified into large, feathery, sticky tentacles that can be protruded from the mouth (see Figure 33.4D). Periodically, a sea cucumber withdraws the tentacles, wipes off the material that has adhered to them, and digests it.

Many sea stars use their tube feet to capture large prey such as polychaetes, gastropod and bivalve mollusks, small crustaceans such as crabs, and fish. With hundreds of tube feet acting simultaneously, a sea star can grasp a bivalve in its arms, anchor the arms with its tube feet, and, by steady contraction of the muscles in its arms, gradually exhaust the muscles the bivalve uses to keep its shell closed (see Figure 33.4E). To feed on a bivalve, a sea star can push its stomach out through its mouth and then through the narrow space between the two halves of the bivalve's shell. The sea star's stomach then secretes enzymes that digest the prey.

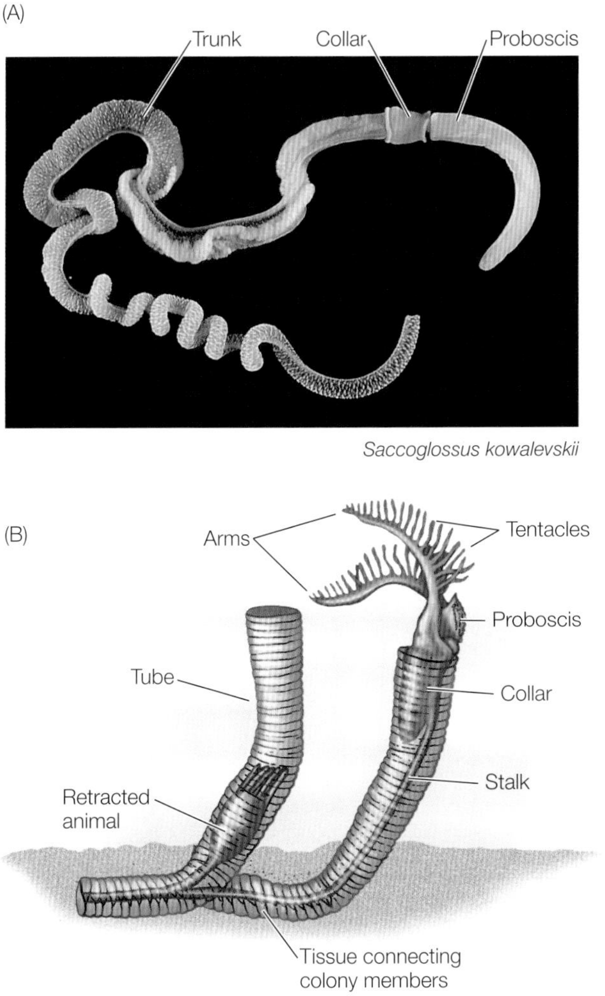

33.5 Hemichordates (A) The proboscis of an acorn worm is modified for burrowing. (B) The structure of a colonial pterobranch.

Members of other echinoderm groups do not use their tube feet to capture food. Most sea urchins eat algae, which they scrape from rocks with a complex rasping structure. Most of the 2,000 species of brittle stars ingest particles from the upper layers of sediments and assimilate the organic material from them, although some species filter suspended food particles from the water and others capture small animals.

Hemichordates have a three-part body plan

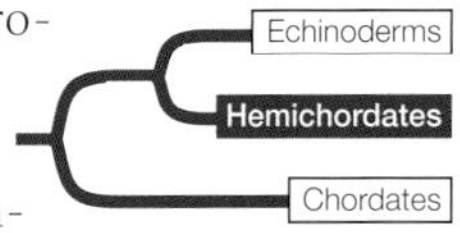

Hemichordates—acorn worms and pterobranchs—have a three-part body plan, consisting of a *proboscis*, a *collar* (which bears the mouth), and a *trunk* (which contains the other body parts). The 75 known species of *acorn worms* range up to 2 meters in length (**Figure 33.5A**). They live in burrows in muddy and sandy marine sediments. The digestive tract of an acorn worm consists of a mouth behind which are a muscular *pharynx* and an *intestine*. The pharynx opens to the outside through a number of *pharyngeal slits* through which water can exit. Highly vascularized tissue surrounding the pharyngeal slits serves as a gas exchange apparatus. Acorn worms breathe by pumping water into the mouth and out through the pharyngeal slits. They capture prey with the large proboscis, which is coated with sticky mucus to which small organisms in the sediment stick. The mucus and its attached prey are conveyed by cilia to the mouth. In the esophagus, the food-laden mucus is compacted into a ropelike mass that is moved through the digestive tract by ciliary action.

The 20 living species of *pterobranchs* are sedentary marine animals up to 12 millimeters in length that live in a tube secreted by the proboscis. Some species are solitary; others form colonies of individuals joined together (**Figure 33.5B**). Behind the proboscis is a collar with 1–9 pairs of arms. The arms bear long tentacles that capture prey and function in gas exchange.

33.2 RECAP

Echinoderms have an internal skeleton of calcified plates and a unique water vascular system. Hemichordates have a bilaterally symmetrical body divided into three parts: proboscis, collar, and trunk.

- What two unique structural features characterize the echinoderms? See p. 719
- Can you describe some of the ways that echinoderms use their tube feet to obtain food? See p. 721
- How do hemichordates obtain food? See p. 721

We have described the deuterostome groups that are most distantly related to us. Now we turn our attention to the chordates, the clade to which humans belong.

33.3 What New Features Evolved in the Chordates?

It is not obvious from examining adult animals that echinoderms and chordates share a common ancestor. For the same reason, the evolutionary relationships among some chordate groups are not immediately apparent. The features that reveal the evolutionary relationships both among the chordates and between chordates and echinoderms are primarily seen in the larvae—in other words, it is during the early developmental stages that their evolutionary relationships are evident.

There are three chordate clades: the **urochordates**, the **cephalochordates**, and the **vertebrates**. There are about 3,000 species of urochordates and 50,000 species of vertebrates, but only about 30 species of cephalochordates.

Adult chordates vary greatly in form, but all chordates display the following derived structures at some stage in their development (**Figure 33.6**):

- A dorsal, hollow nerve cord
- A tail that extends beyond the anus
- A dorsal supporting rod called the *notochord*

The **notochord** is the most distinctive derived chordate trait. It is composed of a core of large cells with turgid fluid-filled vacuoles, which make it rigid but flexible. In the urochordates the notochord is lost during metamorphosis to the adult stage. In most vertebrate species, it is replaced by skeletal structures that provide support for the body.

The ancestral pharyngeal slits (not a derived feature of this group) are present at some developmental stage of chordates, although they are often lost in adults. The *pharynx*, which develops around the pharyngeal slits, functioned in chordate ancestors as the site for oxygen uptake and the elimination of carbon dioxide and water (as in acorn worms). The pharynx is much enlarged in some chordate species (as in the *pharyngeal basket* of the lancelet in **Figure 33.6**), but has been lost in others.

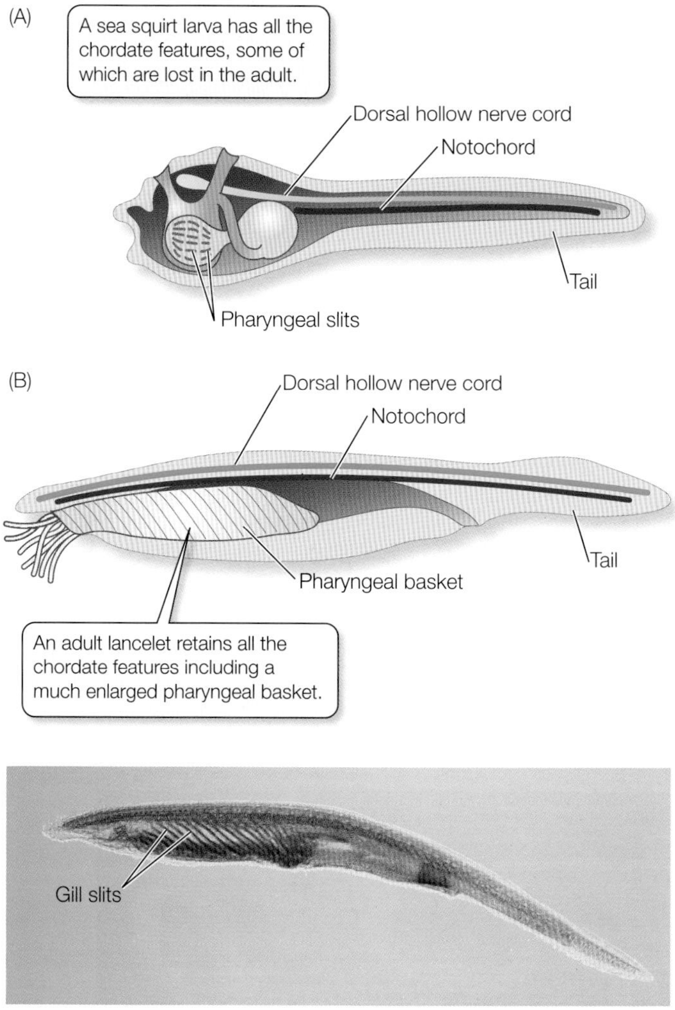

33.6 The Key Features of Chordates Are Most Apparent in Early Developmental Stages The pharyngeal slits of both the urochordate sea squirt and the cephalochordate lancelet develop into pharyngeal baskets. (A) The sea squirt larva (but not the adult) has all three chordate features. (B) The pharyngeal basket of the adult lancelet features external gill slits.

Adults of most urochordates and cephalochordates are sessile

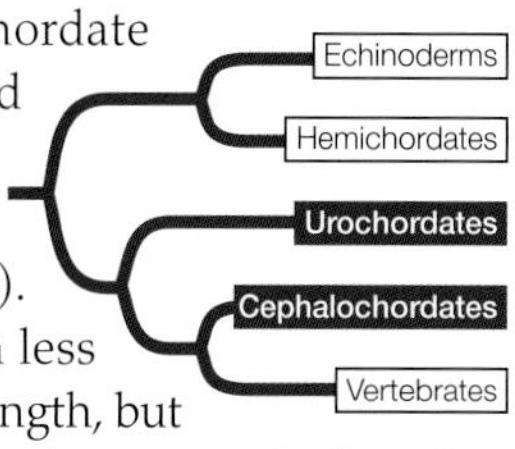

All members of the three major urochordate groups—the ascidians, thaliaceans, and larvaceans—are marine animals. More than 90 percent of the known species of urochordates are *ascidians* (sea squirts). Individual ascidians range in size from less than 1 millimeter to 60 centimeters in length, but some form colonies by asexual budding from a single founder. These colonies may measure several meters across. The baglike body of an adult ascidian is enclosed in a tough tunic, leading to its alternate name of "tunicate" (**Figure 33.7A**). The tunic is composed of proteins and a complex polysaccharide secreted by epidermal cells. The ascidian pharynx is enlarged into a *pharyngeal basket* that filters prey from the water passing through it.

In addition to its pharyngeal slits, an ascidian larva has a dorsal, hollow nerve cord and a notochord that is restricted to the tail region (see Figure 33.6). Bands of muscle that surround the notochord provide support for the body. After a short time swimming in the plankton, the larvae of most species settle on the seafloor and transform into sessile adults. As Darwin realized, the swimming, tadpole-like larvae suggest a close evolutionary relationship between ascidians and vertebrates (see Figure 25.4).

Thaliaceans (salps) can live singly or in chainlike colonies up to several meters long (**Figure 33.7B**). Salps float in tropical and subtropical oceans at all depths down to 1,500 meters. *Larvaceans* are solitary planktonic animals that retain their notochords and nerve cords throughout their lives. Most larvaceans are less than 5 millimeters long, but some species that live near the bottom of deep ocean waters build delicate casings of slime that may be more than a meter wide. They snare sinking organic particles (their pri-

(A) *Rhopalaea* sp.

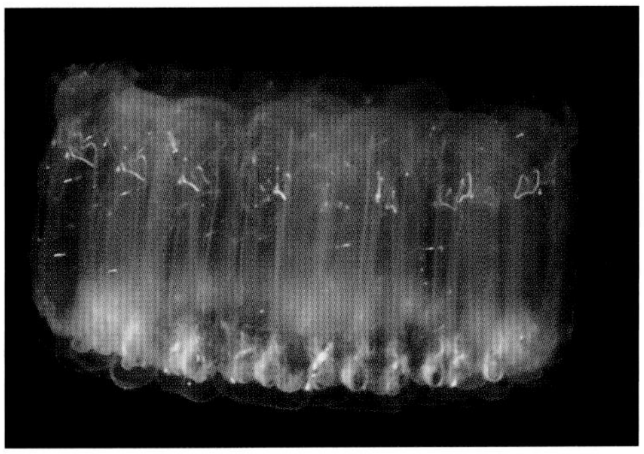

(B) *Pegea socia*

33.7 Adult Urochordates (A) The iridescent tunic is clearly visible in these transparent ascidians ("sea squirts," also known as tunicates). Two different species of the same genus appear in this photograph. (B) A chainlike colony of salps floats in tropical waters.

mary food source) with these slimy "houses." When the old "house" gets clogged, they build new a one.

The 30 species of cephalochordates, or *lancelets*, are small animals that rarely exceed 5 centimeters in length. The notochord, used in burrowing, extends the entire length of the body throughout their lives (see Figure 33.6B). Lancelets are found in shallow marine and brackish waters worldwide. Most of the time they lie covered in sand with their heads protruding above the sediment, but they can swim. They extract prey from the water with their pharyngeal baskets. During the reproductive season, the gonads of the males and females greatly enlarge. At spawning, the walls of the gonads rupture, releasing eggs and sperm into the water column, where fertilization takes place.

A new dorsal supporting structure replaces the notochord in vertebrates

In one chordate group, a new dorsal supporting structure evolved. The **vertebrates** take their name from the jointed, dorsal **vertebral column** that replaces the notochord during early development as their primary supporting structure.

The lineage that led to the vertebrates is thought to have evolved in an estuarine environment (where freshwater meets salt water). Modern vertebrates have since radiated into marine, freshwater, and terrestrial environments worldwide (**Figure 33.8**).

33.8 Vertebrates Have Colonized a Wide Diversity of Environments This current phylogenetic hypothesis shows the distribution of the major vertebrate groups over marine, freshwater, terrestrial, and estuarine (where fresh and salt water meet) environments. Representatives of the amniote vertebrates can be found in all four of these environments, as well as in the air (birds and bats).

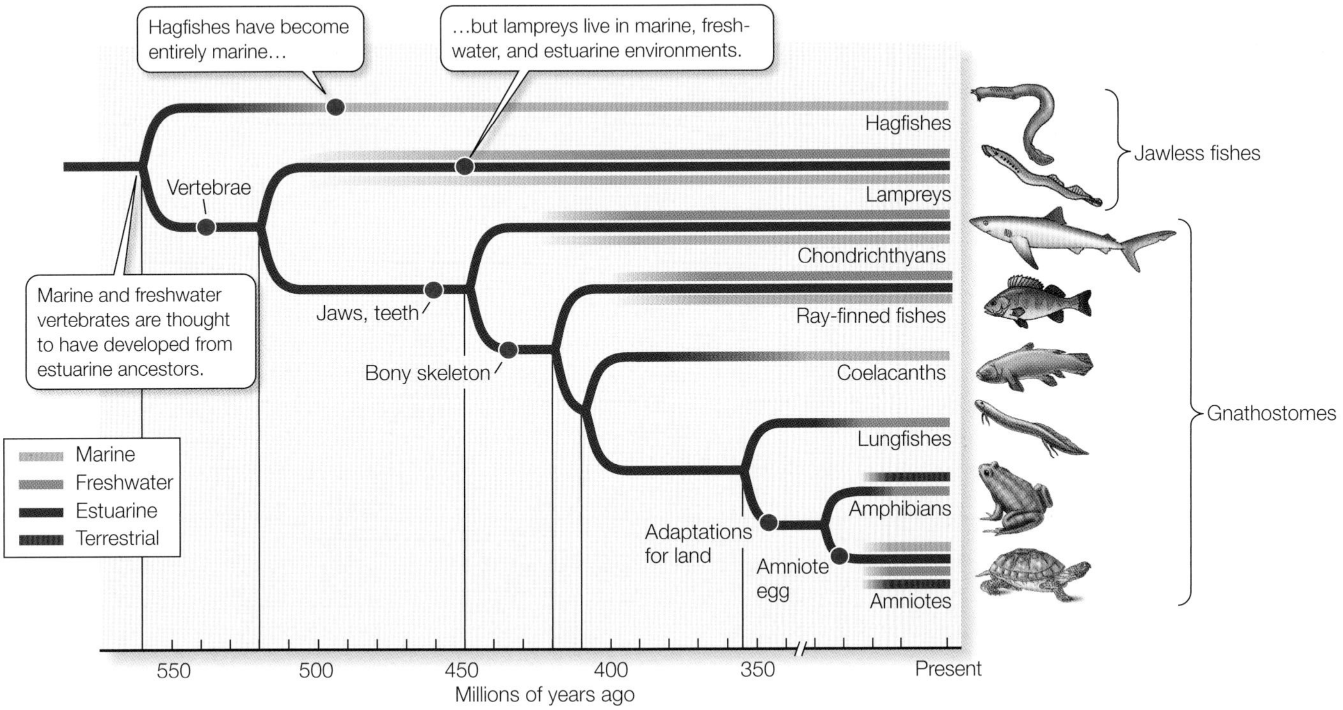

(A) *Eptatretus stouti*

(B) *Petromyzon marinus*

33.9 Modern Jawless Fishes (A) Hagfish burrow in the ocean mud, from which they extract small prey. They are also scavengers on dead or dying fish. Hagfish have degenerate eyes, which has led to their being miscalled "blind eels." (B) Some lampreys can live in either fresh or salt water, although all lampreys breed in freshwater. Many species are ectoparasites that attach to the bodies of living fish and use their large, jawless mouths to suck blood and flesh.

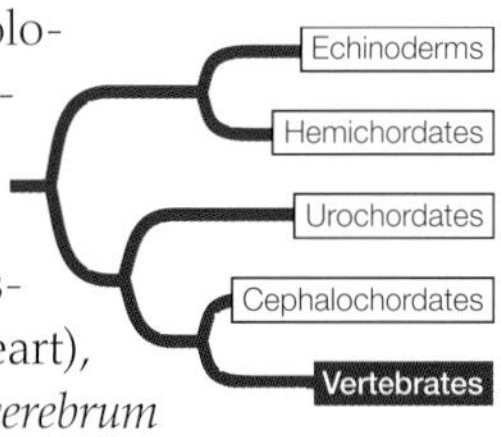

The *hagfishes* are thought by many biologists to be the sister group to the remaining vertebrates (as shown in Figure 33.8). Hagfishes (**Figure 33.9A**) have a weak circulatory system, with three small accessory hearts (rather than a single, large heart), a partial *cranium* or skull (containing no *cerebrum* or *cerebellum*, two main regions of the brain of other vertebrates), and no jaws or stomach. They also lack vertebrae, and have a skeleton composed of cartilage. Thus, some biologists do not consider hagfishes vertebrates, but instead call the hagfishes plus the vertebrates the *craniates*. On the other hand, recent analyses of some gene sequences suggest that hagfishes may be more closely related to the vertebrate lampreys (**Figure 33.9B**), in which case the two groups are placed together in the vertebrates as *cyclostomes* ("circle mouths"). If this latter hypothesis is correct, then hagfishes must have secondarily lost many of the major vertebrate morphological features during their evolution.

The 58 known species of hagfishes are unusual marine animals that produce copious quantities of slime as a defense. They are virtually blind, and rely largely on the four pairs of sensory tentacles around their mouths to detect food. Although they have no jaws, they have a tongue-like structure in their mouths that is equipped with tooth-like rasps, which hagfishes can use to tear apart dead organisms and to capture their principal prey (polychaete worms). They have direct development (no larvae), and individuals may actually change sex from year to year (from male to female and vice versa).

The nearly 50 species of lampreys either live in freshwater, or live in coastal salt water and then move into freshwater to breed. Although the lampreys and hagfishes may look superficially similar in body form (with elongate eel-like bodies and no paired fins), they are otherwise very different in their biology.

Lampreys have a complete braincase and true (although rudimentary) vertebrae, all cartilaginous rather than bony. Lampreys undergo a complete metamorphosis from filter-feeding larvae known as *ammocoetes*, which are morphologically quite similar in general structure to adult lancelets. The adults of many species of lampreys are parasitic, although several lineages of lampreys evolved to become non-feeding as adults. These non-feeding adult lampreys survive only a few weeks to breed after metamorphosis. In the species that are parasitic as adults, the round mouth is a rasping and sucking organ (see Figure 33.9B) that is used to attach to their prey and rasp at the flesh. One predatory species, the sea lamprey, began spreading through man-made canals into the Great Lakes in 1835, and contributed to enormous losses of commercial fisheries there until control measures were introduced in the middle of the 1900s.

The vertebrate body plan can support large animals

Vertebrates are characterized by several key features:

- A rigid internal *skeleton* supported by the vertebral column
- An anterior *skull* with a large brain
- Internal organs *suspended in a coelom*
- A well-developed *circulatory system*, driven by contractions of a ventral *heart*

This body plan is exemplified by the bony fish diagrammed in **Figure 33.10**. Fishes underwent many millenia of evolution in Earth's watery environments before the first vertebrate colonization of land and remain the most species-rich vertebrate group.

Many kinds of jawless fishes were found in the seas, estuaries, and fresh waters of the Devonian period. However, hagfishes and lampreys are the only jawless fishes that survived beyond the Devonian. During that period, the *gnathostomes* ("jaw mouths") evolved jaws via modifications of the skeletal arches that supported the gills (**Figure 33.11A**). Jaws greatly improved feeding efficiency, and an animal with jaws can grasp, subdue, and swallow large prey.

The earliest jaws were simple, but the evolution of *teeth* made predators more effective (**Figure 33.11B**). In predators, teeth function crucially both in grasping and in breaking up prey. In both

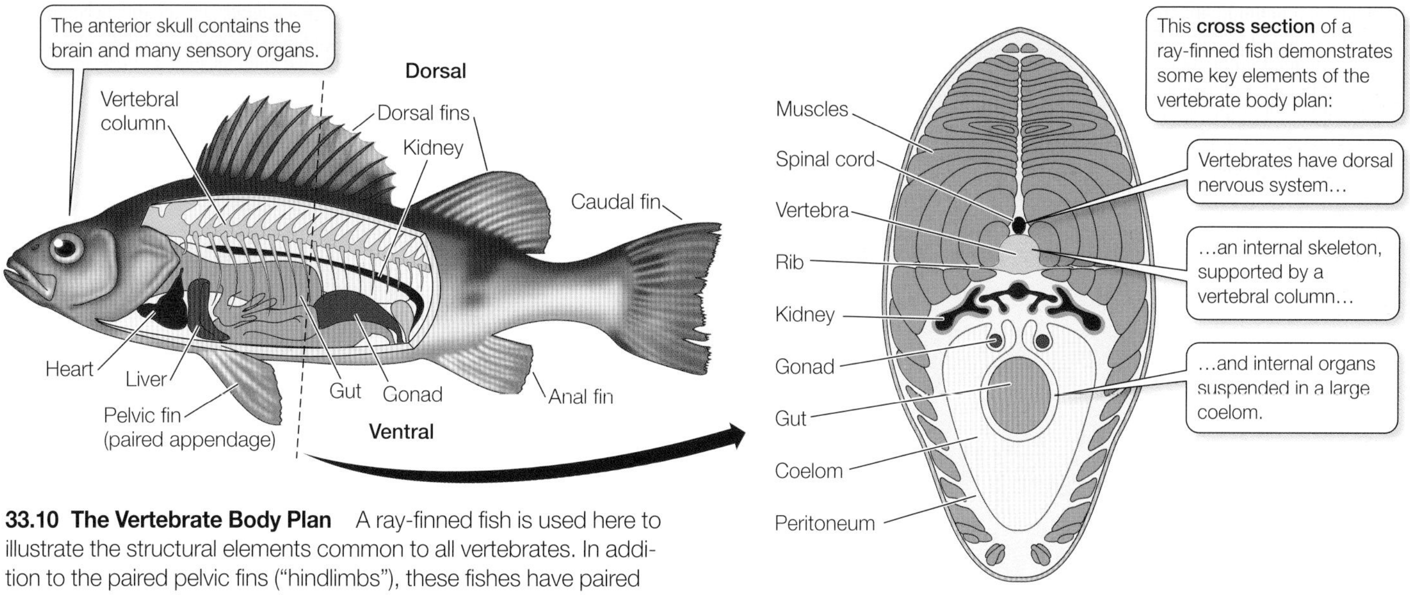

33.10 The Vertebrate Body Plan A ray-finned fish is used here to illustrate the structural elements common to all vertebrates. In addition to the paired pelvic fins ("hindlimbs"), these fishes have paired pectoral fins ("forelimbs") on the sides of their bodies (not seen in this cutaway view).

predators and herbivores, teeth enable their possessors to chew both soft and hard body parts of their food (see Figure 50.7). Chewing also aids chemical digestion and improves the animal's ability to extract nutrients from its food, as will be described in Chapter 50. Vertebrates are remarkable in the diversity of their jaws and teeth.

Fins and swim bladders improved stability and control over locomotion

Jawed fishes stabilize their position in water, as well as propel themselves through water, using their *fins*. Most fishes have a pair of pectoral fins just behind the gill slits, and a pair of pelvic fins anterior to the anal region (see Figure 33.10). Median dorsal and anal fins stabilize the fish as it moves, or may be used for propulsion

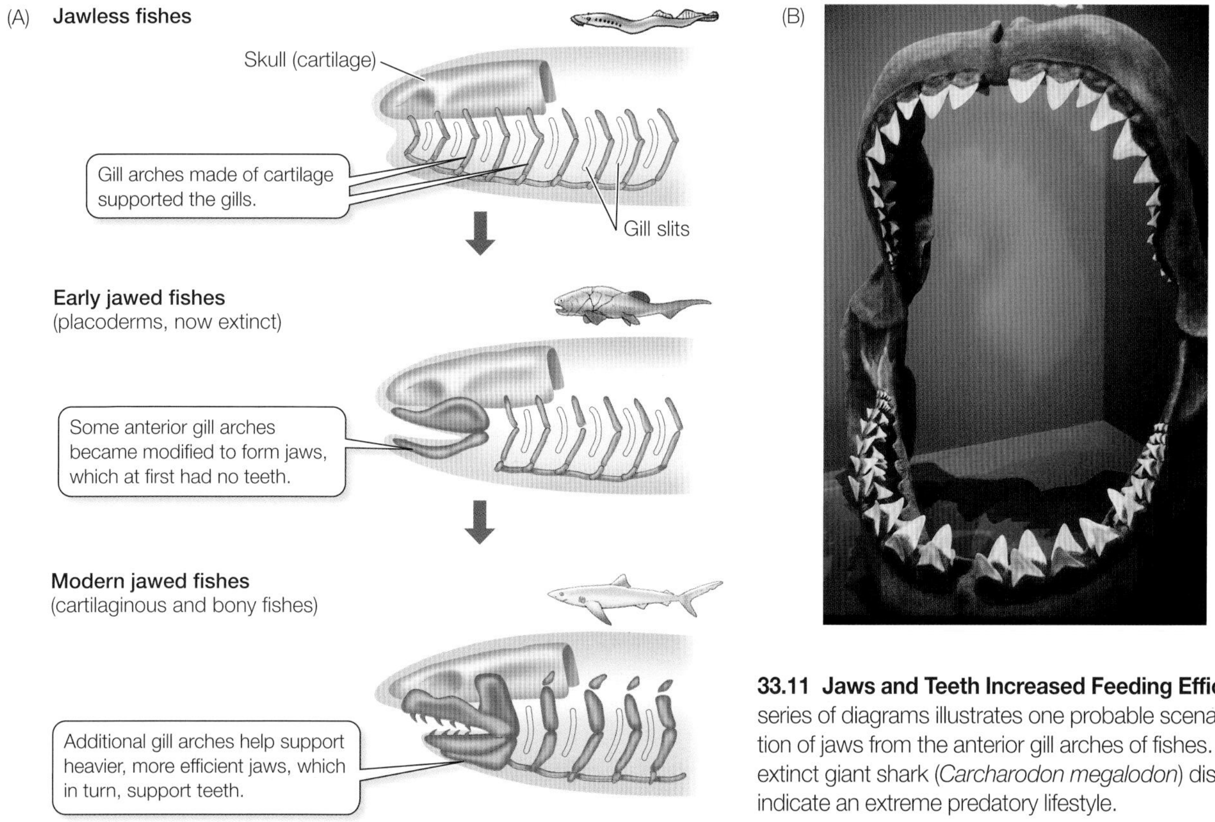

33.11 Jaws and Teeth Increased Feeding Efficiency (A) This series of diagrams illustrates one probable scenario for the evolution of jaws from the anterior gill arches of fishes. (B) Jaws of the extinct giant shark (*Carcharodon megalodon*) display the teeth that indicate an extreme predatory lifestyle.

in some species. In many fishes, the caudal fins help propel the animal and enable it to turn rapidly.

Several groups of finned fishes became abundant during the Devonian. Among them were **chondrichthyans**—sharks, skates and rays (820 known species), and chimaeras (30 known species). Like hagfishes and lampreys, these fishes have a skeleton composed entirely of a firm but pliable material called *cartilage*. Their skin is flexible and leathery, sometimes bearing scales that give it the consistency of sandpaper. Sharks move forward by means of lateral undulations of their bodies and caudal (tail) fins (**Figure 33.12A**). Skates and rays propel themselves by means of vertical undulating movements of their greatly enlarged pectoral fins (**Figure 33.12B**).

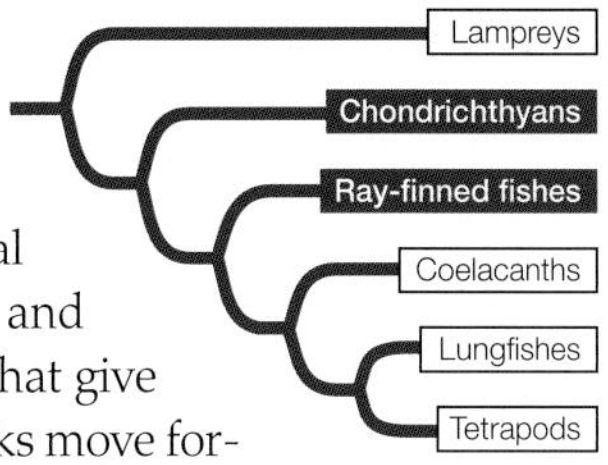

Most sharks are predators, but some feed by straining plankton from the water. Most skates and rays live on the ocean floor, where they feed on mollusks and other animals buried in the sediments. Nearly all cartilaginous fishes live in the oceans, but a few are estuarine or migrate into lakes and rivers. One group of stingrays is found in river systems of South America. The less familiar chimaeras (**Figure 33.12C**) live in deep-sea or cold waters.

In some early fishes, gas-filled sacs supplemented the gas exchange function of the gills by giving the animals access to atmospheric oxygen. These features enabled those fishes to live where oxygen was periodically in short supply, as it often is in freshwater environments. These lunglike sacs evolved into *swim bladders*, which are organs of buoyancy. By adjusting the amount of gas in its swim bladder, a fish can control the depth at which it remains suspended in the water while expending very little energy to maintain its position.

Ray-finned fishes, and most remaining groups of vertebrates, have internal skeletons of calcified, rigid *bone* rather than flexible cartilage. The outer body surface of most species of ray-finned fishes is covered with flat, thin, lightweight scales that provide some protection or enhance movement through the water. The gills of ray-finned fishes open into a single chamber covered by a hard flap, called an *operculum*. Movement of the operculum improves the flow of water over the gills, where gas exchange takes place.

Ray-finned fishes radiated during the Tertiary into about 24,000 species, encompassing a remarkable variety of sizes, shapes, and lifestyles (**Figure 33.13**). The smallest are less than 1 centimeter long as adults; the largest weigh as much as 900 kilograms. Ray-finned fishes exploit nearly all types of aquatic food sources. In the oceans they filter plankton from the water, rasp algae from rocks, eat corals and other soft-bodied colonial animals, dig animals from soft sediments, and prey on virtually all kinds of other fishes. In fresh water they eat plankton, devour insects, eat fruits that fall into the water in flooded forests, and prey on other aquatic vertebrates and, occasionally, terrestrial vertebrates. Many fishes are solitary, but in open water others form large aggregations called *schools*. Many fishes perform complicated behaviors to maintain schools, build nests, court and choose mates, and care for their young.

Although ray-finned fishes can readily control their position in open water using their fins and swim bladders, their eggs tend to sink. Some species produce small eggs that are buoyant enough to complete their development in the open water, but most marine

33.12 Chondrichthyans (A) Most sharks, such as this whitetip reef shark, are active marine predators. (B) Skates and rays, represented here by an eagle ray, feed on the ocean bottom. Their modified pectoral fins are used for propulsion, and their other fins have secondarily become vestigial. (C) A chimaera, or ratfish. These deep-sea fish often possess modified dorsal fins that contain toxins.

(A) *Sphyraena barracuda*

(B) *Chromis punctipinnis*

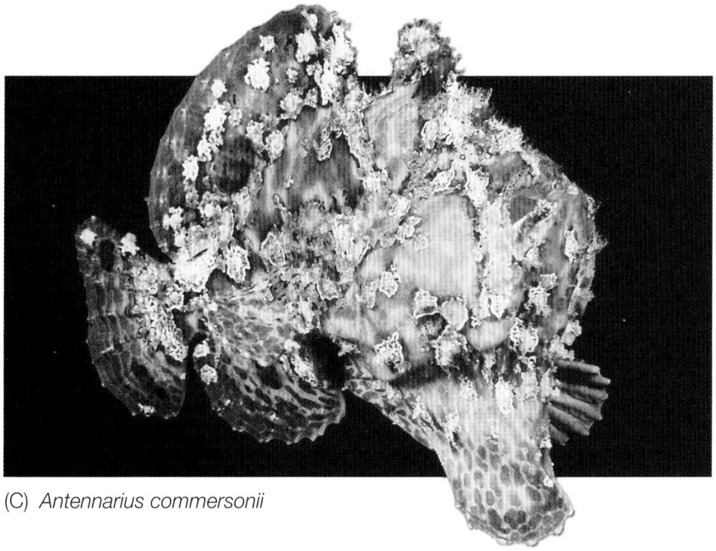

(C) *Antennarius commersonii*

(D) *Phycodurus eques*

33.13 Diverse Ray-Finned Fishes (A) The barracuda has the large teeth and powerful jaws of a predator. (B) The blacksmith fish of North America's Pacific coastal waters displays the "typical" body form of fishes that school in the open ocean. (C) Commerson's frogfish can change its color over a range from pale yellow to orange-brown to deep red, thus enhancing its camouflage abilities. (D) This leafy sea dragon is difficult to see when it hides in vegetation. It is a larger relative of the more familiar seahorse.

fishes move to food-rich shallow waters to lay their eggs. That is why coastal waters and estuaries are so important in the life cycles of many marine species. Some, such as salmon, abandon salt water when they breed, ascending rivers to spawn in freshwater streams and lakes.

How do the larvae of coral reef fish that hatch out in the open ocean find their way back to the reefs? Perhaps their relatives call them home. Biologists built artificial reefs off the coast of Australia and played recordings of calls of reef fishes on half of them. Many more young fish were attracted to the noisy reefs than to the quiet ones.

33.3 RECAP

Chordates are characterized by a dorsal hollow nerve chord, a post-anal tail, and a dorsal supporting rod called a notochord (although all of these features are not found at all life stages). Specialized structures for support (such as vertebrae), locomotion (such as fins), and feeding (such as jaws and teeth) evolved among the vertebrates, which allowed them to colonize and adapt to most of Earth's environments.

- Can you remember the synapomorphies that characterize the chordates and those that characterize the vertebrates? See pp. 722–724 and Figures 33.6 and 33.10
- Can you describe the ways that hagfishes differ from lampreys in their morphology and life history? Do you understand why some biologists do not consider the hagfishes to be vertebrates? See p. 724
- There are more species of fishes than of any other single vertebrate group. Why do you think this is so? Can you describe the adaptations that differentiate the major groups of fishes? See pp. 724–726

In some fishes, the lunglike sacs that gave rise to swim bladders became specialized for another purpose: breathing air. That adaptation set the stage for the vertebrates to move onto the land.

33.4 How Did Vertebrates Colonize the Land?

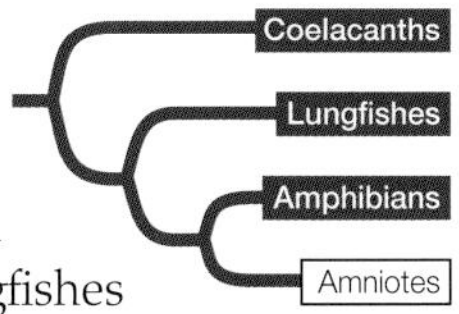

The evolution of lunglike sacs in fishes set the stage for the invasion of the land. Some early ray-finned fishes probably used those structures to supplement their gills when oxygen levels in the water were low, as lungfishes and many groups of ray-finned fishes do today. But with their unjointed fins, those fishes could only flop around on land, as do most modern fishes. Changes in the structure of the fins first allowed some fishes to support themselves better in shallow water and, later, to move better on land.

Jointed fins enhanced support for fishes

Jointed fins evolved in the ancestor of the **sarcopterygians,** which includes coelacanths, lungfishes, and tetrapods. The coelacanths flourished from the Devonian until about 65 million years ago, when they were thought to have become extinct. However, in 1938, a commercial fisherman caught a living coelacanth off South Africa. Since that time, hundreds of individuals of this extraordinary fish, *Latimeria chalumnae*, have been collected. A second species, *L. menadoensis*, was discovered in 1998 off the Indonesian island of Sulawesi. *Latimeria*, a predator of other fish, reaches a length of about 1.8 meters and weighs up to 82 kilograms (**Figure 33.14A**). Its skeleton is mostly composed of cartilage, not bone. A cartilaginous skeleton is a derived feature in this clade because they had bony ancestors.

Lungfishes, which also have jointed fins, were important predators in shallow-water habitats in the Devonian, but most lineages died out. The six surviving species live in stagnant swamps and muddy waters in South America, Africa, and Australia (**Figure 33.14B**). Lungfishes have lungs derived from the lunglike sacs of their ancestors as well as gills. When ponds dry up, individuals of most species can burrow deep into the mud and survive for many months in an inactive state while breathing air.

It is believed that some early aquatic sarcopterygians began to use terrestrial food sources, became more fully adapted to life on land, and eventually evolved to become ancestral **tetrapods** (four-legged vertebrates).

How was the transition from an animal that swam in water to one that walked on land accomplished? Early in 2006, scientists reported the discovery of a Devonian fossil they believe represents an intermediate between the finned appendages of fishes and the limbs of terrestrial tetrapods (**Figure 33.14C**). It appears that limbs able to prop up a large fish with the front-to-rear movement necessary for walking may have evolved while the animals still lived in water.

Amphibians adapted to life on land

During the Devonian, the first tetrapods arose from an aquatic ancestor. In this lineage, stubby, jointed fins evolved into walking legs. The basic elements of those legs have remained throughout the evolution of terrestrial vertebrates, although they have changed considerably in their form.

Most modern **amphibians** are confined to moist environments because they lose water rapidly through the skin when exposed to dry air. In addition, their eggs are enclosed within delicate membranous envelopes that cannot prevent water loss in dry conditions. In many temperate-zone amphibian species, adults live mostly on land, but they return to fresh water to lay their eggs,

(A) *Latimeria chalumnae*

(B) *Neoceratodus forsteri*

(C) *Tiktaalik roseae*

33.14 The Closest Relatives of Tetrapods (A) This coelacanth, found in deep waters of the Indian Ocean, represents one of two surviving species of a group that was once thought to be extinct. (B) All surviving lungfish species live in the Southern Hemisphere. (C) A recently discovered fossil provides further insight into the evolution of limbs from pectoral fins.

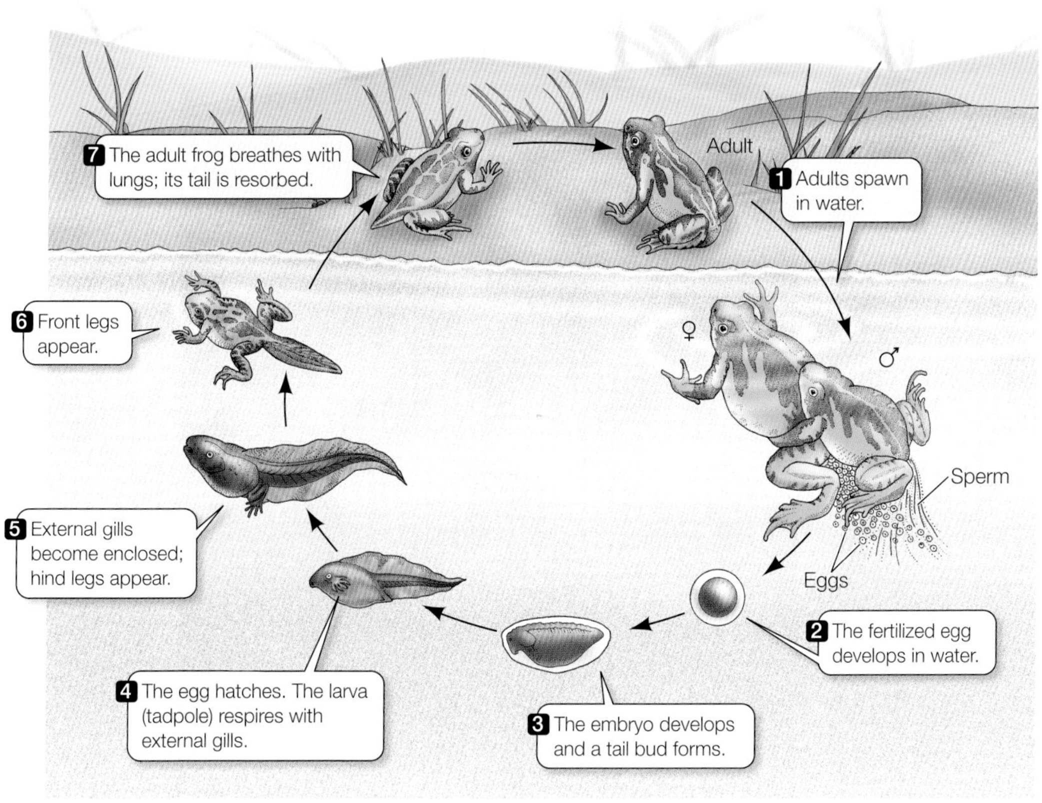

33.15 In and Out of the Water Most early stages in the life cycle of temperate-zone amphibians take place in water. The aquatic tadpole is transformed into a terrestrial adult through metamorphosis.

which are usually fertilized external to the body (**Figure 33.15**). The fertilized eggs give rise to larvae that live in water until they undergo metamorphosis to become terrestrial adults. However, many amphibians (especially those in tropical and subtropical areas) have evolved a wide diversity of additional reproductive modes and types of parental care. Internal fertilization evolved several times among the amphibians. Many species develop directly into adult-like forms from fertilized eggs laid on land or carried by the parents. Other species of amphibians are entirely aquatic, never leaving the water at any stage of their lives, and many of these species retain a larval-like morphology.

The more than 6,000 known species of amphibians living on Earth today belong to three major groups: the wormlike, limbless, tropical, burrowing or aquatic *caecilians* (**Figure 33.16A**), the tailless frogs and toads (collectively called *anurans;* **Figure 33.16B**), and the tailed *salamanders* (**Figure 33.16C,D**).

Anurans are most diverse in wet tropical and warm temperate regions, although a few are found at very high latitudes. There are far more anurans than any other amphibians, with over 5,300 described species and many more being discovered every year. Some anurans have tough skins and other adaptations that enable them to live for long periods in very dry deserts, whereas others live in moist terrestrial and arboreal environments. Some species are completely aquatic as adults. All anurans have a very short vertebral column, with a strongly modified pelvic region that is modified for leaping, hopping, or propelling their bodies through water by kicking their hind legs.

Salamanders are most diverse in temperate regions of the Northern Hemisphere, but many species are also found in cool, moist environments in mountains of Central America. Many salamanders live in rotting logs or moist soil. One major group has lost lungs, and these species exchange gases entirely through the skin and mouth lining—body parts that all amphibians use in addition to their lungs. Through *paedomorphosis* (retention of the juvenile state; see Chapter 20), a completely aquatic lifestyle has evolved several times among the salamanders (see Figure 33.16D). Most species of salamanders have internal fertilization, which is usually achieved through the transfer of a small jelly-like, sperm-containing capsule (called a *spermatophore*).

Many amphibians have complex social behaviors. Most male anurans utter loud, species-specific calls to attract females of their own species (and sometimes to defend breeding territories), and they compete for access to females that arrive at the breeding sites. Many amphibians lay large numbers of eggs, which they abandon once they are deposited and fertilized. Some species, however, lay only a few eggs, which are fertilized and then guarded in a nest, or carried on the backs, in the vocal pouches, or even in the stomachs of one of the parents. A few species of frogs, salamanders, and caecilians are *viviparous*, meaning that they give birth to well-developed young that have received nutrition from the female during gestation.

Amphibians are the focus of much attention today because populations of many species are declining rapidly, especially in mountainous regions of western North America, Central and South America, and northeastern Australia. Scientists are investigating several hypotheses to account for amphibian population declines, including the adverse effects of habitat alteration by humans, increased solar radiation caused by destruction of the Earth's ozone layer, pollution from urban and industrial areas and airborne agricultural pesticides and herbicides, and the spread of a pathogenic chytrid fungus that attacks amphibians. Scientists have documented the spread of the chytrid fungus through Central Amer-

(A) *Dermophis mexicanus*

(B) *Bufo periglenes*

(C) *Gyrinophilus porphyriticus*

(D) *Necturus* sp.

33.16 Diversity among the Amphibians (A) Burrowing caecilians superficially look more like worms than amphibians. (B) Male golden toads in the Monteverde cloud forest of Costa Rica. This species has recently become extinct, one of many amphibian species to do so in the past few decades. (C) An adult spring salamander. (D) The mudpuppy is a salamander that has developed an entirely aquatic life; it has no terrestrial portion of its life cycle.

ica, where many species of amphibians have become extinct, including Costa Rica's golden toad (see Figure 33.16B).

Amniotes colonized dry environments

Several key innovations contributed to the ability of members of one clade of tetrapods to exploit a wide range of terrestrial habitats. The animals that evolved these water-conserving traits are called **amniotes**.

The **amniote egg** (which gives the group its name) is relatively impermeable to water and allows the embryo to develop in a contained aqueous environment (**Figure 33.17**). The leathery or brittle, calcium-impregnated shell of the amniote egg retards evaporation of the fluids inside, but permits passage of oxygen and carbon dioxide. The egg also stores large quantities of food in the form of *yolk*, permitting the embryo to attain a relatively advanced state of development before it hatches. Within the shell are *extraembryonic membranes* that protect the embryo from desiccation and assist its gas exchange and excretion of waste nitrogen.

In several different groups of amniotes, the amniote egg became modified, allowing the embryo to grow inside (and receive nutrition from) the mother. For instance, the mammalian egg lost its shell and yolk while the functions of the extraembryonic membranes were retained and expanded; we will examine the roles of these membranes in detail in Chapter 43.

Other innovations were found in the organs of terrestrial adults. A tough, impermeable skin, covered with scales or modifications

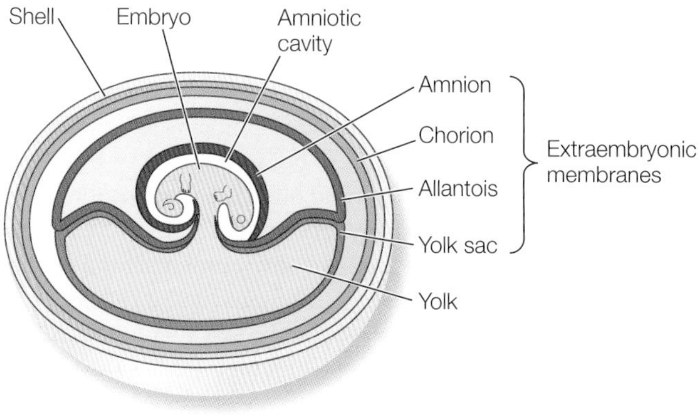

33.17 An Egg for Dry Places The evolution of the amniote egg, with its water-retaining shell, four extraembryonic membranes, and embryo-nourishing yolk, was a major step in the colonization of the terrestrial environment.

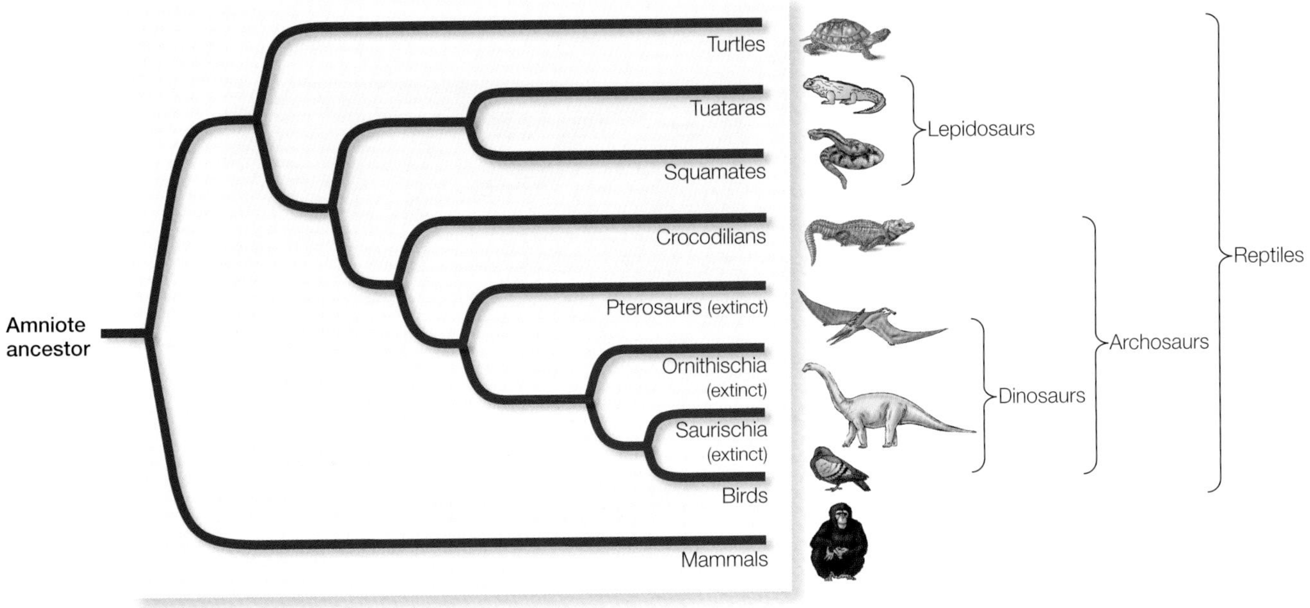

33.18 A Current Phylogenetic Tree of Amniotes This tree of amniote relationships shows the primary split between mammals and reptiles. The reptile portion of the tree shows a lineage that led to the turtles, another to the lepidosaurs (snakes, lizards, and tuataras), and a third branch that includes all the archosaurs (crocodiles, several extinct groups, and the birds).

of scales such as hair and feathers, greatly reduced water loss. And, perhaps most importantly, adaptations to the vertebrate excretory organs, the kidneys, allowed amniotes to excrete concentrated urine, ridding the body of waste nitrogen without losing a large amount of water in the process (see Chapter 51).

During the Carboniferous, amniotes split into two major groups, the mammals and reptiles (**Figure 33.18**). More than 17,200 species of **reptiles** exist today, over half of which are *birds*. Birds are the only living members of the otherwise extinct *dinosaurs*, the dominant terrestrial predators of the Mesozoic.

Reptiles adapted to life in many habitats

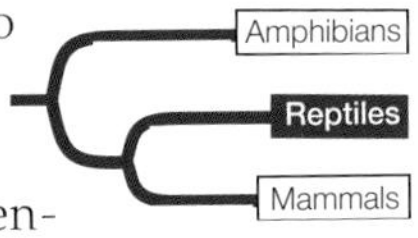

The lineage leading to modern reptiles began to diverge from other amniotes about 250 million years ago. One reptilian group that has changed very little over the intervening millennia is the *turtles*. The dorsal and ventral bony plates of turtles form a shell into which the head and limbs can be withdrawn in many species (**Figure 33.19A**). The dorsal shell is an expansion of the ribs, and it is a mystery how the pectoral girdles evolved to be inside the ribs of turtles, unlike any other vertebrates. Most turtles live in aquatic environments, but several groups, such as tortoises and box turtles, are terrestrial. Sea turtles spend their entire lives at sea except when they come ashore to lay eggs. Human exploitation of sea turtles and their eggs has resulted in worldwide declines of these species, all of which are now endangered. A few species of turtles are strict herbivores or carnivores, but most species are omnivores that eat a variety of aquatic and terrestrial plants and animals.

The *lepidosaurs* constitute the second-most species-rich clade of living reptiles. This group is composed of the *squamates* (lizards, snakes, and amphisbaenians—the latter a group of mostly legless, wormlike, burrowing reptiles with greatly reduced eyes) and the *tuataras*, which superficially resemble lizards, but differ from them in tooth attachment and several internal anatomical features. Many species related to the tuataras lived during the Mesozoic era, but today only two species, restricted to a few islands off the coast of New Zealand, survive (**Figure 33.19B**).

The skin of a lepidosaur is covered with horny scales that greatly reduce loss of water from the body surface. These scales, however, make the skin unavailable as an organ of gas exchange. Gases are exchanged almost entirely via the lungs, which are proportionally much larger in surface area than those of amphibians. A lepidosaur forces air into and out of its lungs by bellows-like movements of its ribs. The lepidosaur heart is divided into *chambers* that partially separate oxygenated blood from the lungs from deoxygenated blood returning from the body. With this type of heart, lepidosaurs can generate high blood pressure and can sustain relatively high levels of metabolism.

Most lizards are insectivores, but some are herbivores; a few prey on other vertebrates. The largest lizard, which grows as long as 3 meters, is the predaceous Komodo dragon of the East Indies. Most lizards walk on four limbs (**Figure 33.19C**), although limblessness has evolved repeatedly in the group, especially in burrowing and grassland species. One major group of limbless squamates are known as snakes (**Figure 33.19D**). All snakes are carnivores; many can swallow objects much larger than themselves. Several snake groups evolved venom glands and the ability to inject venom rapidly into their prey.

Crocodilians and birds share their ancestry with the dinosaurs

Another reptilian group, the *archosaurs*, includes the crocodilians, dinosaurs, and birds. *Dinosaurs* rose to prominence about 215 million years ago and dominated terrestrial environments for about 150 million years; only one group of dinosaurs, the *birds*, survived the mass extinction event at the Cretaceous–Tertiary boundary.

(A) *Chelonia mydas*

(B) *Sphenodon punctatus*

(C) *Chlamydosaurus kingii*

(D) *Diadophis punctatus*

33.19 Reptilian Diversity (A) Green sea turtles are widely distributed in tropical oceans. (B) This tuatara represents one of only two surviving species in a lineage that separated from lizards long ago. (C) An Australian frilled lizard offers a threat display. (D) The ringneck snake of North America is nonvenomous. It has a bright orange underbelly, which it coils in an oddly characteristic way that seems to discourage predators.

During the Mesozoic, most terrestrial animals more than a meter in length were dinosaurs. Many were agile and could run rapidly; they had special muscles that enabled the lungs to be filled and emptied while the limbs moved. We can infer the existence of such muscles in dinosaurs from the structure of the vertebral column in fossils. Some of the largest dinosaurs weighed as much as 80 tons (see the opening of Chapter 31).

Until recently, scientists assumed that the non-avian dinosaurs grew slowly. However, analyses of growth lines in fossil bones indicate that *Tyrannosaurus rex* reached full size in 15 to 18 years—much faster than the 25 to 35 years it takes a smaller African elephant to achieve full size.

Modern *crocodilians*—crocodiles, caimans, gharials, and alligators—are confined to tropical and warm temperate environments (**Figure 33.20A**). Crocodilians spend much of their time in water, but they build nests on land or on floating piles of vegetation. The eggs are warmed by heat generated by decaying organic matter that the female places in the nest. Typically, the female guards the eggs until hatching; and she often facilitates hatching. In some species, the female continues to guard and communicate with her offspring after they hatch. All crocodilians are carnivorous. They eat vertebrates of all kinds, including large mammals.

Biologists have long accepted the phylogenetic position of birds among the reptiles, although birds clearly have many unique, derived morphological features. In addition to the strong morphological evidence for the placement of birds among the reptiles, fossil and molecular data emerging over the last few decades have provided definitive supporting evidence. Birds are thought to have emerged among the *theropods*, a group of predatory dinosaurs that share such traits as bipedal stance, hollow bones, a *furcula* ("wishbone"), elongated metatarsals with three-fingered feet, elongated forelimbs with three-fingered hands, and a pelvis that points backwards.

The living bird species fall into two major groups that diverged during the late Cretaceous, about 80–90 million years ago, from a flying ancestor. The few modern descendants of one lineage include a group of secondarily flightless and weakly flying birds, some of which are very large. This group, called the *palaeognaths*, includes the South and Central American tinamous and several large flightless birds of the southern continents—the rhea, emu, kiwi, cassowary, and the world's largest bird, the ostrich (**Figure 33.20B**). The second lineage (*neognaths*) has left a much larger number of descendants, most of which have retained the ability to fly.

(A) *Crocodylus niloticus*

(B) *Struthio camelus*

33.20 Archosaurs (A) The Nile crocodile. Crocodiles, alligators, and their relatives live in tropical and warm temperate climates. (B) Birds are the other living archosaur group, represented here by the winged but flightless ostrich.

The evolution of feathers allowed birds to fly

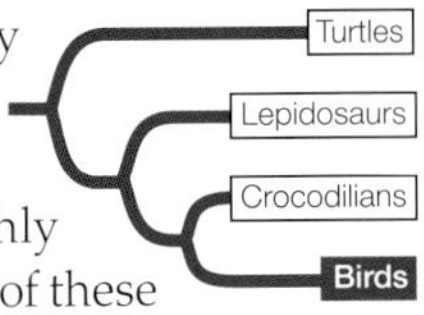

Fossil dinosaurs discovered recently in early Cretaceous deposits in Liaoning Province, in northeastern China, show that the scales of some small predatory dinosaurs were highly modified to form *feathers*. The feathers of one of these dinosaurs, *Microraptor gui*, were structurally similar to those of modern birds (**Figure 33.21A**).

During the Mesozoic era, about 175 million years ago, a lineage of theropods gave rise to the birds. The oldest known avian fossil, *Archaeopteryx*, which lived about 150 million years ago, had teeth, but was covered with feathers that are virtually identical to those of modern birds (**Figure 33.21B**). It also had well-developed wings, a long tail, and a furcula, or "wishbone," to which some of the flight muscles were probably attached. *Archaeopteryx* had clawed fingers on its forelimbs, but it also had typical perching bird claws on its hind limbs. It probably lived in trees and shrubs and used the fingers to assist it in clambering over branches.

The evolution of feathers was a major force for diversification. Feathers are lightweight but are strong and structurally complex (**Figure 33.22**). The large quills that support wing feathers arise from the skin of the forelimbs to create the flying surfaces. Other strong feathers sprout like a fan from the shortened tail and serve as stabilizers during flight. The feathers that cover the body, along with an underlying layer of down feathers, provide birds with insulation that helps them to survive in virtually all of Earth's climates.

The bones of theropod dinosaurs, including birds, are hollow with internal struts that increase their strength. Hollow bones would have made early theropods lighter and more mobile; later they facilitated the evolution of flight. The sternum (breastbone) of flying birds forms a large, vertical keel to which the flight muscles are attached.

Flight is metabolically expensive. A flying bird consumes energy at a rate about 15–20 times faster than a running lizard of the same weight. Because birds have such high metabolic rates, they generate large amounts of heat. They control the rate of heat loss using their feathers, which may be held close to the body or elevated to alter the amount of insulation they provide. Another adaptation to the needs of flight is found in the lungs of birds, which allow air to flow through unidirectionally rather than pumping air in and out (see Section 48.2).

There are about 9,600 species of living birds, which range in size from the 150-kilogram ostrich to a tiny hummingbird weighing

(A)

(B)

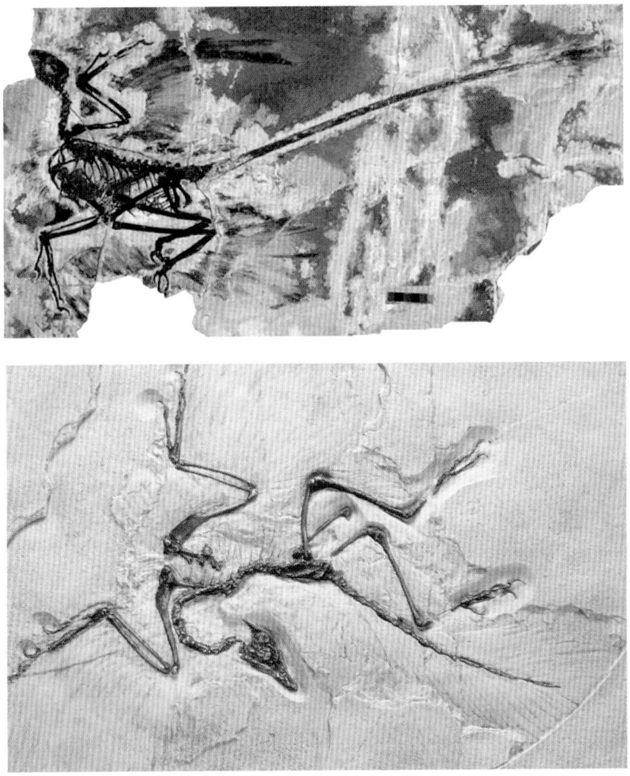

33.21 Mesozoic Bird Fossils Fossil remains demonstrate the evolution of birds from other dinosaurs. (A) *Microraptor gui*, a feathered dinosaur from the early Cretaceous (about 140 mya). (B) *Archaeopteryx* is the oldest known birdlike fossil.

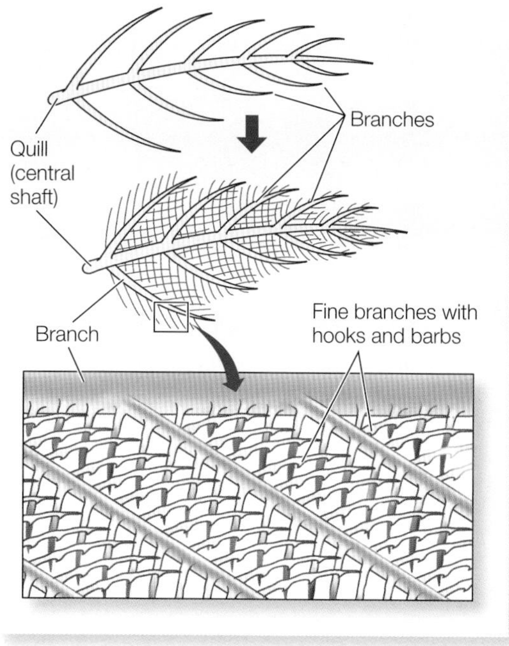

33.22 A Major Evolutionary Breakthrough Early feathers may have been a simple central shaft, or quill, with side branches. A pattern of fine branches with interlocking hooks and barbs creates a strong, lightweight surface that allows flight.

only 2 grams. The teeth so prominent among other dinosaurs were secondarily lost in the ancestral birds, but birds nonetheless eat almost all types of animal and plant material. Insects and fruits are the most important dietary items for terrestrial species. Birds also eat seeds, nectar and pollen, leaves and buds, carrion, and other vertebrates. By eating the fruits and seeds of plants, birds serve as major agents of seed dispersal. Representatives of some of the major groups of birds are shown in **Figure 33.23**.

Mammals radiated after the extinction of dinosaurs

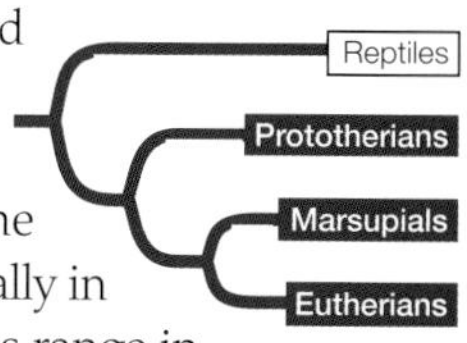

Small and medium-sized **mammals** coexisted with the dinosaurs for millions of years. After the non-avian dinosaurs disappeared during the mass extinction at the close of the Mesozoic era, mammals increased dramatically in numbers, diversity, and size. Today mammals range in size from tiny shrews and bats weighing only about 2 grams to the blue whale, the largest animal on Earth, which measures up to 33

(A) *Aix galericulata*

(B) *Frigata minor*

(C) *Tyto alba*

33.23 Diversity among the Birds (A) This bright-plumaged male Mandarin duck is a member of the group that includes ducks, geese, and swans. (B) Frigate birds are among the ocean-going birds that are often found miles from shore. This male is in dis play mode, with an inflated pouch that advertises his strength to females. (C) This barn owl is a nighttime predator that can find prey using its sensitive auditory "sonar" system. (D) Perching, or passeriform, birds such as this male northern cardinal are the most species-rich of all bird group.

(D) *Cardinalis cardinalis*

meters long and can weigh as much as 160,000 kilograms. Mammals have far fewer, but more highly differentiated, teeth than fishes, amphibians, and reptiles do. Differences among mammals in the number, type, and arrangement of teeth reflect their varied diets (see Figure 50.7).

Four key features distinguish the mammals:

- *Sweat glands,* whose secretions cool the animals as they evaporate.
- *Mammary glands* in female mammals secrete a nutritive fluid (milk) on which newborn individuals feed.
- *Hair,* which gives a protective and insulating covering.
- A *four-chambered heart* that completely separates the oxygenated blood coming from the lungs from the deoxygenated blood returning from the body (this last characteristic is convergent with the archosaurs, including modern birds and crocodiles).

Mammalian eggs are fertilized within the female's body, and the embryos undergo a period of development within the female's body in an organ called the *uterus* prior to being born. Most mammals have a covering of hair (fur), which is luxuriant in some species but has been greatly reduced in others, including the cetaceans (whales and dolphins) and humans. Thick layers of insulating fat (blubber) replace hair as a heat retention mechanism in the cetaceans; humans learned to use clothing for this purpose when they dispersed from warm tropical areas.

The approximately 5,000 species of living mammals are divided into two major groups. The three species of *prototherians* are found only in Australia and New Guinea. These mammals, the duck-billed platypus and two species of echidnas, differ from other mammals in lacking a placenta, laying eggs, and having sprawling legs (**Figure 33.24**). Prototherians supply milk for their young, but they have no nipples on their mammary glands; the milk simply oozes out and is lapped off the fur by the offspring.

(A) *Tachyglossus aculeata*

(B) *Ornithorhynchus anatinus*

33.24 Prototherians (A) The short-beaked echidna is one of the two surviving species of echidnas. (B) The duck-billed platypus is another surviving prototherian species.

Most mammals are therians

Members of the *therian* clade are further subdivided into two more groups, marsupials and eutherians. Females of most species of *marsupials* have a ventral pouch in which they carry and feed their offspring (see Figure 33.25A). Gestation (pregnancy) in marsupials is brief; the young are born tiny but with well-developed forelimbs, with which they climb to the pouch. They attach to a nipple, but cannot suck. The mother ejects milk into the tiny offspring until it grow large enough to suckle. Once her offspring have left the uterus, a female marsupial may become sexually receptive again. She can then carry fertilized eggs capable of initiating development and replacing the offspring in her pouch should something happen to them.

At one time marsupials were found on all continents, but the approximately 330 living species are now restricted to the Australian region (**Figure 33.25A,B**) and the Americas (especially South America; **Figure 33.25C**). One marsupial species, the Virginia opossum, is widely distributed in North America. Marsupials radiated to become herbivores, insectivores, and carnivores, but no marsupials live in the oceans. None can fly, although some *arboreal* (tree-dwelling) marsupials are gliders. The largest living marsupials are the kangaroos of Australia, which can weigh up to 90 kilograms. Much larger marsupials existed in Australia until humans exterminated them soon after reaching that continent about 40,000 years ago (see Figure 57.1).

Most bat species are predators, primarily of insects. A number of bats feed on fruit, however, and some fruit-eating species became specialized fluid feeders, able to pierce the skin of fruit and feed on the juice. It is probably from such juice-drinking ancestors that the three species of notorious "vampire" bats arose, able to pierce the skin and feed on the blood of other mammals.

The largest group of therians are the *eutherians*. Eutherians are sometimes called *placental mammals*, but this name is inappropriate because some marsupials also have placentas. Eutherians are more developed at birth than are marsupials; no external pouch houses them after they are born.

The more than 4,500 species of eutherians belong to one of 20 major groups (**Table 33.1**). The largest group is the rodents, with over 2,000 species. Rodents are traditionally defined by the unique morphology of their teeth, which are adapted for gnawing through substances such as wood. The next largest group comprises the approximately 1,000 bat species—the flying mammals. The bats are followed by the moles and shrews, with more than 400 species. The relationships of the major groups of mammals to

33.25 Marsupials (A) Australia's eastern gray kangaroos are among the largest living marsupials. This female carries her young offspring in the characteristic marsupial pouch. (B) The carnivorous Tasmanian devil is found only in Tasmania, an island off Australia's southern coast. (C) This arboreal opossum is a South American marsupial species.

(B) *Sarcophilus harrisii*
(A) *Macropus giganteus*
(C) *Caluromys philander*

one another have been difficult to determine, because most of the major groups diverged in a short period of time during an explosive adaptive radiation.

Eutherians are extremely varied in their form and ecology (**Figure 33.26**). The extinction of the non-avian dinosaurs at the end of the Cretaceous may have made it possible for them to diversify and radiate into a large range of ecological *niches.* Many eutherian species grew to become quite large in size, and some assumed the role of dominant terrestrial predators previously occupied by the large dinosaurs. Among these predators, social hunting behavior evolved in a number of species, including members of the canid (wolf/dog), felid (cat), and primate lineages.

Grazing and *browsing* by members of several eutherian groups helped transform the terrestrial landscape. Herds of grazing herbivores feed on open grasslands, while browsers feed on shrubs and trees. The effects of herbivores on plant life favored the evolution of the spines, tough leaves, and difficult-to-eat growth

TABLE 33.1

Major groups of living eutherian mammals

GROUP	APPROXIMATE NUMBER OF LIVING SPECIES	EXAMPLES
Gnawing mammals (*Rodentia*)	>2,000	Rats, mice, squirrels, woodchucks, ground squirrels, beaver, capybara
Flying mammals (*Chiroptera*)	1,000	Bats
Soricomorph insectivores (*Soricomorpha*)	430	Shrews, moles
Even-toed hoofed mammals and cetaceans (*Cetartiodactyla*)	300	Deer, sheep, goats, cattle, antelope, giraffes, camels, swine, hippopotamus, cetaceans (whales, dolphins)
Carnivores (*Carnivora*)	280	Wolves, dogs, bears, cats, weasels, pinnipeds (seals, sea lions, walruses)
Primates (*Primates*)	235	Lemurs, monkeys, apes, humans
Lagomorphs (*Lagomorpha*)	80	Rabbits, hares, pikas
African insectivores (*Afrosoricida*)	30	Tenrecs and golden moles
Spiny insectivores (*Erinaceomorpha*)	24	Hedgehogs
Armored mammals (*Cingulata*)	21	Armadillos
Tree shrews (*Scandentia*)	20	Tree shrews
Odd-toed hoofed mammals (*Perissodactyla*)	20	Horses, zebras, tapirs, rhinoceros
Long-nosed insectivores (*Macroscelidea*)	15	Elephant shrews
Pilosans (*Pilosa*)	10	Anteaters and sloths
Pholidotans (*Pholidota*)	8	Pangolins
Sirenians (*Sirenia*)	5	Manatees, dugongs
Hyracoids (*Hyracoidea*)	4	Hyraxes, dassies
Elephants (*Proboscidea*)	3	African and Indian elephants
Dermopterans (*Dermoptera*)	2	Flying lemurs
Aardvark (*Tubulidentata*)	1	Aardvark

33.26 Diversity among the Eutherians (A) The Cape ground squirrel of South Africa is one of many species of small, diurnal rodents. (B) Virtually all bat species are nocturnal. Many predatory bat species find their prey via a unique sound wave system similar to sonar. (C) These Hawaiian spinner dolphins are a type of cetacean, a cetartiodactyl group that returned to the marine environment. (D) Large hoofed mammals are important herbivores in terrestrial environments. Although this bull is grazing by himself, caribou are usually found in huge herds.

forms found in many plants. In turn, adaptations to the teeth and digestive systems of many herbivore lineages allowed these species to consume many plants despite such defenses—a striking example of coevolution. A large animal can survive on food of lower quality than a small animal can, and large size evolved in several of the grazing and browsing animals (see Figure 33.26D). The most striking examples of large body size, of course, were found among the giant herbivorous dinosaurs (see pp. 670–671). The evolution of large herbivores, in turn, favored the evolution of large carnivores able to attack and overpower them.

Several lineages of terrestrial eutherians subsequently returned to the aquatic environments their ancestors had left behind. The completely aquatic marine cetaceans—whales and dolphins—evolved from artiodactyl ancestors (whales are closely related to the hippopotamus). The seals, sea lions, and walruses also returned to the marine environment and their limbs became modified into flippers. Weasel-like otters retain their limbs but have also returned to aquatic environments, colonizing both fresh and salt water. The manatees and dugongs colonized estuaries and shallow seas.

33.4 RECAP

The vertebrate colonization of dry land was facilitated by the evolution of an impermeable body covering, efficient kidneys, and the amniote egg—a structure that resists desiccation and provides an aqueous internal environment in which the embryo grows. Major amniote groups include turtles, lepidosaurs, archosaurs (crocodilians and birds), and mammals.

- In the not-too-distant past, the idea that birds were reptiles met with extreme skepticism. Do you understand how fossils, morphology, and molecular evidence support the position of birds vis-à-vis the reptiles? See pp. 732–733
- In reviewing the discussion of the various vertebrate groups, can you identify several reasons why tooth structure is such an important area of study?

The evolutionary history of one eutherian group—the primates—is of special interest to us because it includes the human lineage. The primates have been the subject of extensive research, both physical and molecular. Let's take a closer look at the characteristics and evolutionary history of the primates.

33.5 What Traits Characterize the Primates?

The eutherian **primates** underwent extensive evolutionary radiation from an ancestral small, arboreal, insectivorous mammal. A nearly complete fossil of an early primate, *Carpolestes*, was found in Wyoming and dated at 56 million years ago; it had grasping feet with an opposable big toe that had a nail rather than a claw. Grasp-

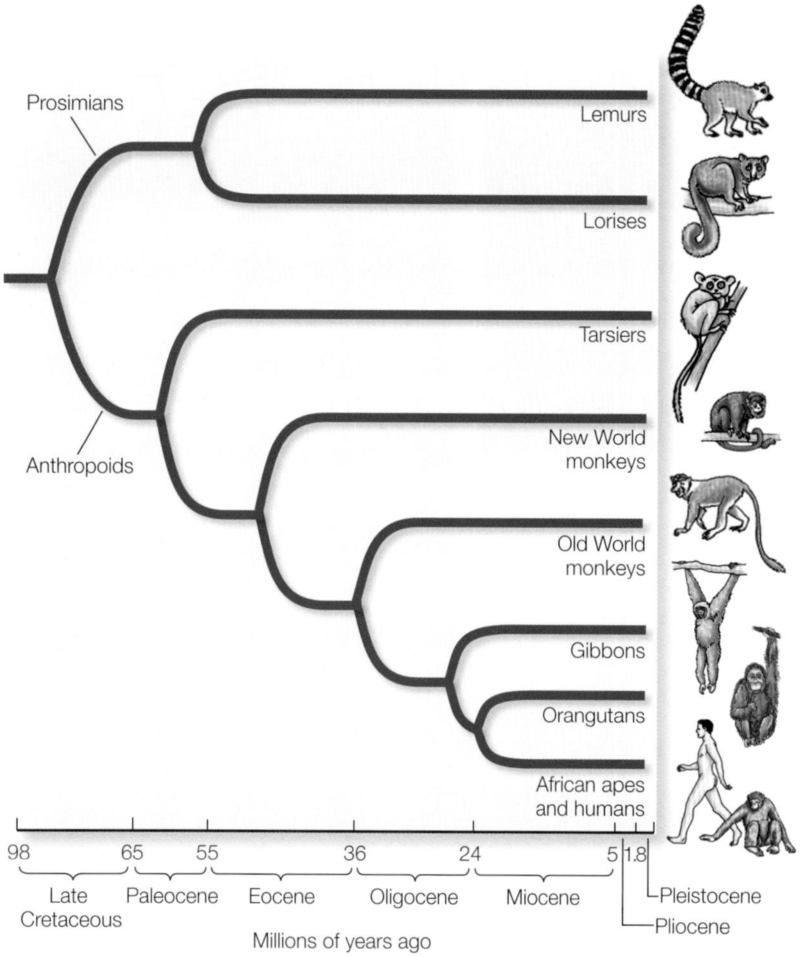

33.27 A Current Phylogenetic Tree of the Primates The phylogeny of primates is among the best studied of any major group of mammals. This tree is based on evidence from many genes, morphology, and fossils.

ing limbs with opposable digits are one of the major adaptations to arboreal life that distinguish primates from other mammals.

Early in their evolutionary history, the primates split into two main clades, the prosimians and the anthropoids (**Figure 33.27**). *Prosimians*—lemurs, pottos, and lorises—once lived on all continents, but today they are restricted to Africa, Madagascar, and tropical Asia. All mainland prosimian species are arboreal and nocturnal (**Figure 33.28**). On the island of Madagascar, however, the site of a remarkable radiation of lemurs, there are also diurnal and terrestrial species.

A second primate lineage, the *anthropoids*—tarsiers, Old World monkeys, New World monkeys, apes, and humans—appeared about 65 million years ago in Africa or Asia. New World monkeys diverged from Old World monkeys at a slightly later date, but early enough that they might have reached South America from Africa when those two continents were still close to each other. All New World monkeys are arboreal (**Figure 33.29A**). Many of them have long, prehensile tails with which they can grasp branches. Many Old World monkeys are arboreal as well, but a number of species are terrestrial (**Figure 33.29B**). No Old World primate has a prehensile tail.

About 35 million years ago, a lineage that led to the modern apes separated from the Old World monkeys. Between 22 and 5.5 million years ago, dozens of species of apes lived in Europe, Asia, and Africa. The Asian apes—gibbons and orangutans (**Figure 33.30A,B**)—descended from two of these ape lineages. Another extinct genus, *Dryopithecus*, is the sister group of the modern African apes—gorillas and chimpanzees (**Figure 33.30C,D**)—and of humans.

Human ancestors evolved bipedal locomotion

About 6 million years ago in Africa, a lineage split occurred that would lead to the chimpanzees on the one hand, and to the *hominid* clade that includes modern humans and their ancestors on the other.

The earliest protohominids, known as *ardipithecines*, had distinct morphological adaptations for *bipedal locomotion* (walking on two legs). Bipedal locomotion frees the forelimbs to manipulate objects and to carry them while walking. It also elevates the eyes, enabling the animal to see over tall vegetation to spot predators and prey. Bipedal locomotion is also energetically more economical than quadrupedal locomotion. All three advantages were probably important for the ardipithecines and their descendants, the australopithecines.

The first *australopithecine* skull was found in South Africa in 1924. Since then australopithecine fossils have been found at many sites in Africa. The most complete fossil skeleton of an australopithecine yet found was discovered

Eulemur coronatus

33.28 A Prosimian The crowned lemur is one of the many lemur species found in Madagascar, where they are part of a unique assemblage of endemic plants and animals.

33.29 Monkeys (A) The spider monkeys of Central America are typical of the New World monkeys, all of which are arboreal. (B) Although many Old World monkey species are arboreal, these drills are among the many terrestrial groups.

in Ethiopia in 1974. The skeleton was approximately 3.5 million years old and was that of a young female who has since become known to the world as "Lucy." Lucy was assigned to the species *Australopithecus afarensis*, and her discovery captured worldwide interest. Fossil remains of more than 100 *A. afarensis* individuals have since been discovered, and there have been recent discoveries of fossils of other australopithecines who lived in Africa 4–5 million years ago.

Experts disagree over how many species are represented by australopithecine fossils, but it is clear that at least two distinct types lived together over much of eastern Africa several million years ago (**Figure 33.31**). The larger type (about 40 kilograms) is represented by at least two species (*Paranthropus robustus* and *P. boisei*), both of which died out about 1.5 million years ago. The smaller type, represented by *A. afarensis*, probably gave rise to the genus *Homo*.

Early members of the genus *Homo* lived contemporaneously with australopithecines in Africa for perhaps half a million years. Some

33.30 Apes (A) Gibbons are the smallest of the apes. They are found in Asia; Indonesia is home to this Borneo gibbon. (B) Orangutans live in the forests of Sumatra and Borneo. (C) Gorillas, the largest apes, are restricted to humid African forests. This male is a lowland gorilla. (D) Chimpanzees, our closest relatives, are found in forested regions of Africa.

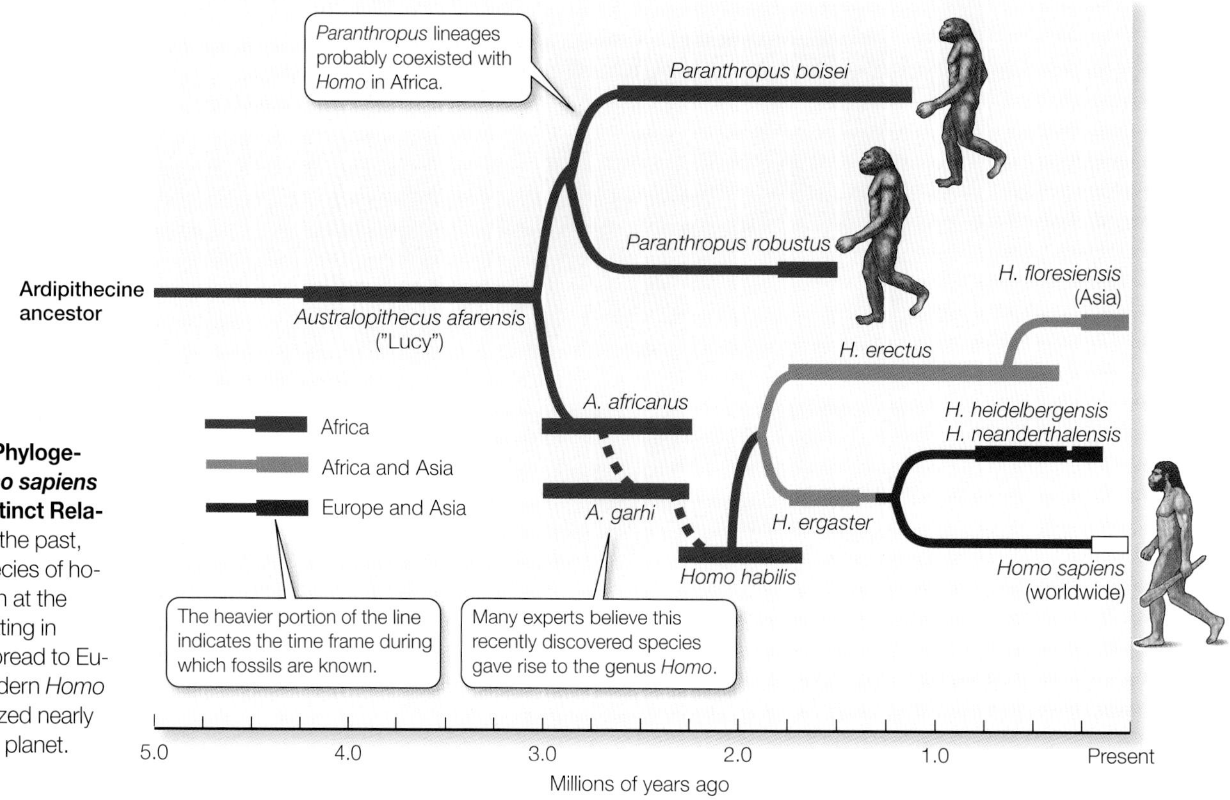

33.31 A Current Phylogenetic Tree of *Homo sapiens* and Our Close Extinct Relatives At times in the past, more than one species of hominid lived on Earth at the same time. Originating in Africa, hominids spread to Europe and Asia; modern *Homo sapiens* has colonized nearly every corner of the planet.

2-million-year-old fossils of an extinct species called *H. habilis* were discovered in the Olduvai Gorge, Tanzania. Other fossils of *H. habilis* have been found in Kenya and Ethiopia. Associated with the fossils are tools that these early hominids used to obtain food.

Another extinct hominid species, *Homo erectus*, evolved in Africa about 1.6 million years ago. Soon thereafter it had spread as far as eastern Asia. Members of *H. erectus* were as large as modern people, but their brains were smaller and they had unusually thick skulls. The cranium, which had thick, bony walls, may have been an adaptation to protect the brain, ears, and eyes from impacts caused by a fall or a blow from a blunt object. What would have been the source of such blows? Fighting with other *H. erectus* males is a highly probable answer.

Homo erectus used fire for cooking and for hunting large animals, and made characteristic stone tools that have been found in many parts of the Old World. Although *H. erectus* survived in Eurasia until about 250,000 years ago, by about 200,000 years ago only *Homo sapiens* was found in most tropical regions. However, as described at the opening of this chapter, a *H. erectus* descendant, *Homo floresiensis*, survived until as recently as 18,000 years ago.

Human brains became larger as jaws became smaller

In the hominid lineage leading to *Homo sapiens*, the brain increased rapidly in size, reaching its modern size by about 160,000 years ago. At the same time, the powerful jaw muscles found in living apes and the australopithecines dramatically decreased in size. These two changes were simultaneous, suggesting that they might have been functionally correlated. A mutation in a regulatory gene that is expressed only in the head may have removed a barrier that had previously prevented this remodeling of the human cranium.

The striking enlargement of brains relative to body size in the hominid lineage was probably favored by an increasingly complex social life. Any features that allowed group members to communicate more effectively with one another would have been valuable in cooperative hunting and gathering and for improving one's status in the complex social interactions that must have characterized early human societies, just as they do in ours today.

Several *Homo* species existed during the mid-Pleistocene epoch, from about 1.5 million to about 300,000 years ago. All were skilled hunters of large mammals, but plants were important components of their diets. During this period another distinctly human trait emerged: rituals and a concept of life after death. Deceased individuals were buried with tools and clothing, supplies for their presumed existence in the next world.

One species, *Homo neanderthalensis*, was widespread in Europe and Asia between about 75,000 and 30,000 years ago. Neanderthals were short, stocky, and powerfully built. Their massive skulls housed brains somewhat larger than our own. They manufactured a variety of tools and hunted large mammals, which they probably ambushed and subdued in close combat. For a short time, their range overlapped that of the *H. sapiens* known as Cro-Magnons, but then the Neanderthals abruptly disappeared. Many scientists believe that they were exterminated by the Cro-Magnons after the latter moved out of Africa into the range of Neanderthals.

Cro-Magnon people made and used a variety of sophisticated tools. They created the remarkable paintings of large mammals, many of them showing scenes of hunting, found in European caves. The animals depicted were characteristic of the cold steppes and grasslands that occupied much of Europe during periods of glacial expansion. Cro-Magnon people spread across Asia, reaching North America perhaps as early as 20,000 years ago, although the date of their arrival in the New World is still uncertain. Within a few thousand years, they had spread southward through North America to the southern tip of South America.

Humans developed complex language and culture

As our ancestors evolved larger brains, their behavioral capabilities increased, especially the capacity for language. Most animal communication consists of a limited number of signals, which refer mostly to immediate circumstances and are associated with charged emotional states induced by those circumstances. Human language is far richer in its symbolic character than other animal vocalizations. Our words can refer to past and future times and to distant places. We are capable of learning thousands of words, many of them referring to abstract concepts. We can rearrange words to form sentences with complex meanings.

The expanded mental abilities of humans enabled the development of a complex **culture**, in which knowledge and traditions are passed along from one generation to the next by teaching and observation. Cultures can change rapidly because genetic changes are not necessary for a cultural trait to spread through a population. On the other hand, cultural norms are not transferred automatically, but must be deliberately taught to each generation.

Cultural transmission greatly facilitated the development and use of domestic plants and animals and the resultant conversion of most human societies from ones in which food was obtained by hunting and gathering to ones in which *pastoralism* (herding large animals) and *agriculture* provided most of the food. The development of agriculture led to an increasingly sedentary life, the growth of cities, greatly expanded food supplies, rapid increases in the human population, and the appearance of occupational specializations, such as artisans, shamans, and teachers.

33.5 RECAP

Grasping limbs with opposable digits distinguish primates from other mammals. Human ancestors developed bipedal locomotion and large brains.

- What are some major trends in primate evolution?
- Can you describe the differences between Old World and New World monkeys? See p. 738
- Do you understand how cultural evolution differs from genetic evolution? See p. 741

CHAPTER SUMMARY

33.1 What is a deuterostome?

Deuterostomes vary greatly in adult form, but because they share distinctive patterns of early development, they are judged to be monophyletic. There are many fewer species of deuterostomes than of protostomes, but many deuterostomes are large and ecologically important. Review Figure 33.1, Web/CD Activity 33.1

33.2 What are the major groups of echinoderms and hemichordates?

Echinoderms and hemichordates both have a bilaterally symmetrical larva. Adult echinoderms, however, have **pentaradial symmetry**. Echinoderms have an internal skeleton of calcified plates and a unique **water vascular system** connected to extensions called **tube feet**. Review Figure 33.3

Hemichordates are bilaterally symmetrical and have a three-part body that is divided into a proboscis, collar, and trunk. They include the acorn worms and the pterobranchs. Review Figure 33.5

33.3 What new features evolved in the chordates?

Chordates fall into three major subgroups: **urochordates**, **cephalochordates**, and **vertebrates**.

At some stage in their development, all chordates have a dorsal, hollow nerve cord, a post-anal tail, and a **notochord**. Review Figure 33.6

Urochordates include the ascidians (sea squirts) and salps. Cephalochordates are the lancelets, which live buried in the sand of shallow marine and brackish waters.

The vertebrate body plan is characterized by a rigid internal skeleton, which is supported by a **vertebral column** that replaces the notochord, and an anterior skull with a large brain. Review Figure 33.10

The evolution of jaws from gill arches enabled individuals to grasp large prey and, together with teeth, cut them into small pieces. Review Figure 33.11

Chondrichthyans have skeletons of cartilage; almost all are marine. The skeletons of ray-finned fishes are made of bone, and they have colonized all aquatic environments.

33.4 How did vertebrates colonize the land?

Lungs and jointed appendages enabled vertebrates to colonize the land. The earliest tetrapod vertebrates were the **amphibians**. Most modern amphibians are confined to moist environments because they and their eggs lose water rapidly. Review Figure 33.16, Web/CD Tutorial 33.1

An impermeable skin, efficient kidneys and an egg that could resist desiccation evolved in the **amniotes** (reptiles and mammals). Review Figure 33.17, Web/CD Activity 33.2

The major reptile groups are the turtles; the lepidosaurs (tuataras, lizards, snakes, and amphisbaenians); and the archosaurs (crocodilians and birds). Review Figure 33.18

Mammals are unique among animals in supplying their young with a nutritive fluid (milk) secreted by **mammary glands**. There are two mammalian clades: the three **protherian** species, and the species-rich **therian** clade, which is further subdivided into the **marsupials** and the **eutherians**. Review Table 33.1

33.5 What traits characterize the primates?

Grasping limbs with opposable digits distinguish **primates** from other mammals. The **prosimian** clade includes the lemurs and lorises; the **anthropoid** clade includes monkeys, apes, and humans. Review Figure 33.27

Hominid ancestors were terrestrial primates who developed efficient bipedal locomotion. In the lineage leading to modern *Homo sapiens*, brains became larger as jaws became smaller; the two events may have been functionally linked. Review Figure 33.31

See Web/CD Activity 33.3 for a concept review of this chapter.

SELF-QUIZ

1. Which of the following deuterostome groups have a three-part body plan?
   a. Acorn worms and urochordates
   b. Acorn worms and pterobranchs
   c. Pterobranchs and urochordates
   d. Pterobranchs and lancelets
   e. Urochordates and lancelets
2. The structure used by adult urochordates to capture food is a
   a. pharyngeal basket.
   b. proboscis.
   c. lophophore.
   d. mucus net.
   e. radula.
3. The pharyngeal gill slits of chordate ancestors functioned as sites for
   a. uptake of oxygen only.
   b. release of carbon dioxide only.
   c. both uptake of oxygen and release of carbon dioxide.
   d. removal of small prey from the water.
   e. forcible expulsion of water to move the animal.
4. The key to the vertebrate body plan is a
   a. rigid internal skeleton supported by a vertebral column.
   b. vertebral column to which internal organs are attached.
   c. vertebral column to which two pairs of appendages are attached.
   d. vertebral column to which a pharyngeal basket is attached.
   e. pharyngeal basket and two pairs of appendages.
5. In most fishes, lunglike sacs evolved into
   a. pharyngeal gill slits.
   b. true lungs.
   c. coelomic cavities.
   d. swim bladders.
   e. none of the above
6. Most temperate-zone amphibians return to water to lay their eggs because
   a. water is isotonic to egg fluids.
   b. adults must be in water while they guard their eggs.
   c. there are fewer predators in water than on land.
   d. amphibians need water to produce their eggs.
   e. amphibian eggs quickly lose water and desiccate if their surroundings are dry.
7. The horny scales that cover the skin of reptiles prevent them from
   a. using their skin as an organ of gas exchange.
   b. sustaining high levels of metabolic activity.
   c. laying their eggs in water.
   d. flying.
   e. crawling into small spaces.
8. Which statement about bird feathers is *not* true?
   a. They are highly modified reptilian scales.
   b. They provide insulation for the body.
   c. They exist in two layers.
   d. They help birds fly.
   e. They are important sites of gas exchange.
9. Prototherians differ from other mammals in that they
   a. do not produce milk.
   b. lack body hair.
   c. lay eggs.
   d. live in Australia.
   e. have a pouch in which the young are raised.
10. Bipedalism is believed to have evolved in the human lineage because bipedal locomotion is
    a. more efficient than quadrupedal locomotion.
    b. more efficient than quadrupedal locomotion, and it frees the forelimbs to manipulate objects.
    c. less efficient than quadrupedal locomotion, but it frees the forelimbs to manipulate objects.
    d. less efficient than quadrupedal locomotion, but bipedal animals can run faster.
    e. less efficient than quadrupedal locomotion, but natural selection does not act to improve efficiency.

FOR DISCUSSION

1. In what animal groups has the ability to fly evolved? How do the structures used for flying differ among these animals?
2. Extracting suspended food from the water is a common mode of feeding among animals. Which groups contain species that extract prey from the air? Why is this mode of obtaining food so much less common than extracting prey from the water?
3. What risks and benefits are posed by large size?
4. Amphibians have survived and prospered for many millions of years, but today many species are disappearing, and populations of others are declining seriously. What features of amphibian life histories might make them especially vulnerable to the kinds of environmental changes now happening on Earth?
5. The body plan of most vertebrates is based on four appendages. What are the varied forms that these appendages take, and how are they used?
6. Compare the ways in which different animal lineages colonized the land. How were those ways influenced by the body plans of animals in the different groups?

FOR INVESTIGATION

A mutation in the gene that encodes the myosin heavy chain (see Chapter 47) decreased the size of the jaw muscles in human ancestors. This mutation may have enabled a restructuring of the cranium and larger brain size. How could we determine when this mutation arose in the phylogenetic history of the primates?

PART NINE

Ecology

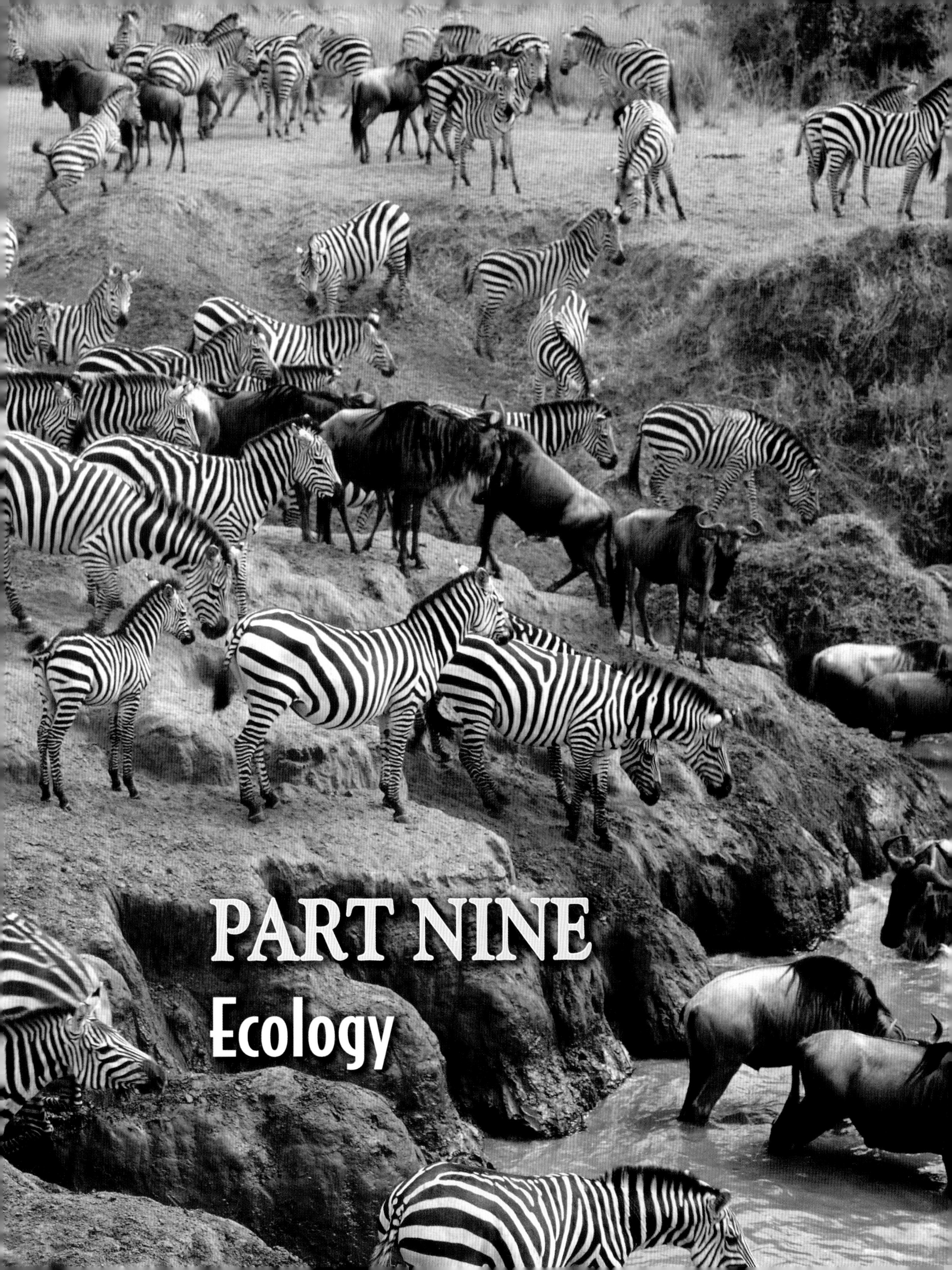

CHAPTER 52 Ecology and the Distribution of Life

Shelter from the storm

The city of New Orleans lies along the Mississippi River at the point where the river spreads into a magnificent, low-lying delta of swamps and wetlands. More than half the city is below sea level. An elaborate flood control system of dams and levees constructed by the Army Corps of Engineers holds back both the river and Lake Pontchartrain, a large, brackish body of water to the north of the city.

On August 29, 2005, Hurricane Katrina made landfall just to the east of New Orleans. At first it seemed that the storm, once a terrifying category 5 hurricane, had dissipated enough to spare the city the worst. But over a 24-hour period, the combination of strong winds, heavy rain, and storm surges breached some of the levees, with catastrophic effects. More than 1,800 people died. Many thousands lost their homes. The 2000 U.S. census figures put the population of New Orleans at 485,000; figures collected in June of 2006—nine months after Katrina—estimated the city's population at about 225,000.

Katrina's devastating impact was due in part to a situation that had been developing for decades. The upstream dams that protect New Orleans also prevent the Mississippi River from depositing the sediments that sustained the surrounding wetlands for centuries. The oil and natural gas industry has cut thousands of small canals through the delta wetlands in order to lay pipelines and drilling rigs, and the extraction of oil and gas from beneath the land has caused it to sink. Increased dredging of shipping lanes and rising sea levels due to global warming contribute to a rise in salinity, killing off many of the great cypress tree swamps. It all adds up to the loss of over 80 percent (more than 1.2 *million* acres) of protective wetlands between 1930 and 2005. At the time Katrina made landfall, the greatly reduced coastal wetlands could not protect New Orleans. Storm surges raced along the paths carved by canals and shipping lanes to breach the levees, inundating about 80 percent of the city.

Coastal wetlands provide many services besides protection. Louisiana's wetlands provide winter habitat for about 70 percent of the migrating birds in the huge Mississippi Valley. They are also the spawning grounds for marine organisms, some of which are commercially

Approach of a Mighty Storm A satellite weather photo shows Katrina at full strength, as a category 5 hurricane approaching the U.S. Gulf Coast. Although the storm weakened somewhat before making landfall, the city of New Orleans and other coastal towns suffered devastating damage and destruction.

The Sheltering Bayou Bald cypress (*Taxodium distichum*) growing in freshwater swamps are a distinctive and vital part of the Louisiana bayou. Their destruction by encroaching salt water from the Gulf of Mexico is endangering the region.

valuable. The shrimping industry of that region has long been renowned, and the Gulf Coast contributes about 30 percent by weight of the total commercial fish harvest in the continental U.S.

Around the world, valuable, indeed, indispensable, coastal wetlands and other ecosystems are being destroyed or altered at a frightening rate. Some of the services provided by these systems can be replaced—usually at tremendous expense—by human technology, but some of them are irreplaceable. Ecologists work to find ways to manage Earth's ecosystems and to guarantee the sustained flow of their benefits to future generations.

IN THIS CHAPTER we define the field of ecology and the kinds of questions ecologists try to answer. Then we describe the distribution of Earth's climates and the forces that create them. We explain some of the ways in which climates and other features of the physical environment influence where different organisms are found. We conclude by describing the major biomes and biogeographic regions that reflect the distribution of life on Earth.

CHAPTER OUTLINE

52.1 What Is Ecology?

Ecology is the scientific study of the rich and varied interactions between organisms and their environment. Ecologists study these interactions at many levels. As we will see in the next chapter, behavioral interactions among individuals of the same species may give rise to elaborate social systems. Organisms also interact with individuals of different species (for example, as predators, prey, mutualists, and competitors) and with their physical environment. These interactions, in turn, influence the structure of **communities** (systems embracing all the organisms living together in the same area), **ecosystems** (systems embracing all organisms in an area plus their physical environment), and the **biosphere** (the system that embraces all regions of the planet where organisms live).

The term **environment**, as used by ecologists, encompasses both **abiotic** (physical and chemical) factors, such as water, mineral nutrients, light, temperature, and wind, and **biotic** factors (living organisms). Interactions between organisms and their environment are two-way processes: organisms both influence and are influenced by their environment. Indeed, dealing with environmental changes caused by our own species is one of the major challenges facing humanity today. For this reason, ecologists are often asked to help analyze the causes of environmental problems and to assist in finding solutions.

There are many reasons to care about ecology. Our lives are enriched by the fascinating interactions among species that we encounter whenever we venture outside. We watch a butterfly pollinating flowers (**Figure 52.1**) and want to know more about how species interact with one another. We are curious about the consequences of those interactions.

Beyond simple curiosity, however, information and insights from ecological science are needed to solve many practical problems. Humans depend on vibrant, functioning ecosystems to provide many goods and services—including the essential services of clean air and water. An understanding of ecology allows us to manage ecosystems so as to maintain the availability of those goods and services. An understanding of ecology allows us to grow food, control pests and diseases, and deal with natural disasters such as floods without causing a cascade of other unanticipated consequences.

52.1 An Ecological Interaction The pollination of flowers by butterflies is mutualism that benefits both participants.

Ecological information can help us solve practical problems only if we know how and why those problems arise. Thus ecologists must become familiar with various environments and understand how organisms adapt to them. This chapter will focus on the big picture: Earth's climates and general patterns in the distribution of life. In subsequent chapters we will look in greater detail at the ways in which individual organisms interact, the dynamics of their populations, and the ecological communities in which they live.

52.1 RECAP

Ecology is the scientific investigation of the interactions among organisms and between organisms and their physical environment.

- What are the components of the environment as defined by ecologists? See p. 1113
- At what levels do ecologists study ecological systems? See p. 1113
- Do you understand why understanding ecology can be considered essential to human survival? See p. 1113

Climate is one of the abiotic factors that determines what kinds of organisms can survive and reproduce in a particular place. We begin our study of ecology with an overview of Earth's climates.

52.2 How Are Climates Distributed on Earth?

The **climate** of a region is the average of the atmospheric conditions (temperature, precipitation, and wind direction and velocity) found there over the long term. *Weather* is the short-term state of those conditions. In other words, climate is what you expect; weather is what you get! Climates vary greatly from place to place on Earth, primarily because different places receive different amounts of solar energy. In this section we will examine how these differences in solar energy input determine atmospheric and oceanic circulation patterns—the factors that most strongly influence climates.

Solar energy drives global climates

The differences in air temperature among different places on Earth are largely determined by differences in solar energy input. Every place on Earth receives the same total number of hours of sunlight each year—an average of 12 hours per day—but not the same amount of solar energy. The rate at which solar energy arrives on Earth per unit of Earth's surface depends primarily on the angle of sunlight. If the sun is low in the sky, a given amount of solar energy is spread over a larger area (and is thus less intense) than if the sun is directly overhead. In addition, when the sun is low in the sky, its light must pass through more of Earth's atmosphere, so more of its energy is absorbed and reflected before it reaches the ground. Thus higher latitudes (closer to the poles) receive less solar energy than latitudes closer to the equator. On average, mean annual air temperature decreases about 0.4°C for every degree of latitude (about 110 kilometers) at sea level. In addition, higher latitudes experience greater variation in both day length and the angle of arriving solar energy over the course of a year, leading to more seasonal variation in temperature.

Air temperature also decreases with elevation. As a parcel of air rises, it expands (its molecules move farther apart), its pressure and temperature drop, and it releases moisture. When a parcel of air descends, it is compressed, its pressure rises, its temperature increases, and it takes up moisture.

Global air circulation patterns result from the global variation in solar energy input that we have just described and from the spinning of Earth on its axis (**Figure 52.2**). Air rises when it is heated by the sun, so warm air rises in the tropics, which receive the greatest solar energy input. This rising air is replaced by air that flows in toward the equator from the north and south. The coming together of these air masses produces the **intertropical convergence zone**. Cool air cannot hold as much moisture as warm air, so heavy rains fall in the intertropical convergence zone as the rising air cools and releases its moisture. The intertropical convergence zone shifts latitudinally with the seasons, following the shift in the zone of greatest solar energy input. This shift results in predictable rainy and dry seasons in tropical and subtropical regions.

The air that moves into the intertropical convergence zone to replace the rising air is replaced, in turn, by air from aloft that descends at roughly 30° north and south latitudes after having traveled away from the equator high in the atmosphere. This air cooled and lost its moisture while it rose at the equator. It now descends, warms, and takes up, rather than releases, moisture. Many of Earth's deserts, such as the Sahara and the Australian deserts, are located at these latitudes where dry air descends.

At about 60° north and south latitudes, air rises again and moves either toward or away from the equator. At the poles, where there is little solar energy input, air descends. These movements of air masses are largely responsible for global wind patterns.

The spinning of Earth on its axis also influences surface winds because Earth's velocity is rapid at the equator, where its diameter is greatest, but relatively slow close to the poles. A stationary air

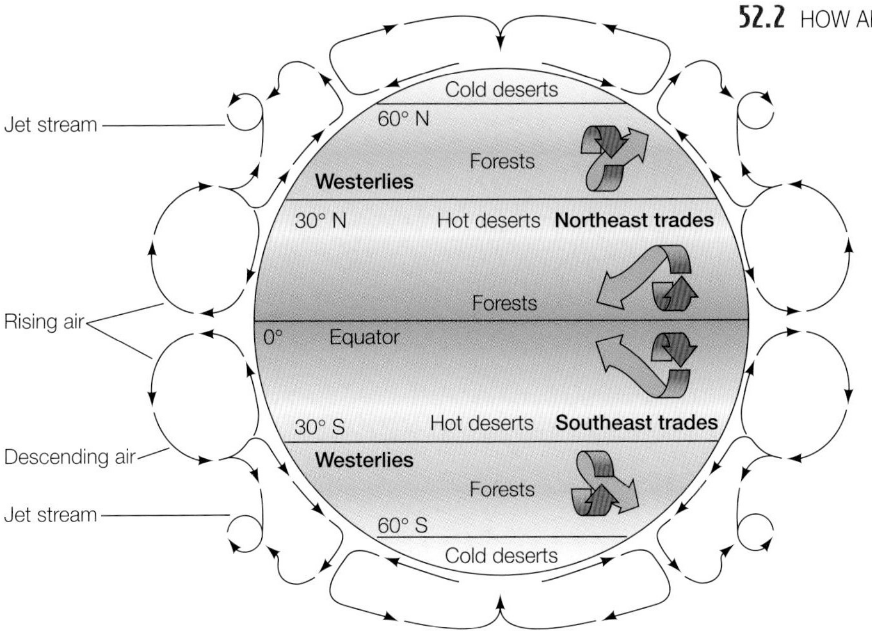

52.2 The Circulation of Earth's Atmosphere If we could stand outside Earth and observe air movements in its atmosphere, we would see vertical air circulation patterns similar to those indicated by the black and red arrows and surface winds similar to those shown by the blue arrows.

mass has the same velocity as Earth at the same latitude. As an air mass moves toward the equator, it confronts an increasingly faster spin, and its rotational movement is slower than that of Earth beneath it. Conversely, as an air mass moves poleward, it confronts an increasingly slower spin, and it speeds up relative to Earth beneath it. Therefore, air masses moving latitudinally are deflected to the right in the Northern Hemisphere and to the left in the Southern Hemisphere. The air masses moving toward the equator from the north and south veer to become the northeast and southeast *trade winds*, respectively. Air masses moving away from the equator also veer and become the *westerly* winds that prevail at mid-latitudes.

When these prevailing winds bring air masses into contact with a mountain range, the air rises to pass over the mountains, cooling as it does so. Thus clouds frequently form on the windward side of mountains (the side facing into the winds) and release moisture as rain or snow. On the leeward side (opposite from the direction of the winds) of mountains, the now-dry air descends, warms, and once again picks up moisture. This pattern often results in a dry area called a **rain shadow** on the leeward sides of mountain ranges (**Figure 52.3**).

Global oceanic circulation is driven by wind patterns

The global air circulation patterns that we have just described drive the circulation patterns of surface ocean waters, known as *currents* (**Figure 52.4**). The trade winds blowing toward the equator from the northeast and southeast cause water to converge at the equator and move westward until it encounters a continental land mass. At that point the water splits, with some of it moving north and some of it moving south along continental shores. This poleward movement of ocean water that has been warmed in the tropics transfers large amounts of heat to high latitudes. As these currents move toward the poles, the water, driven by the winds, veers right in the Northern Hemisphere and left in the Southern Hemisphere. Thus water flowing toward the poles turns eastward until it encounters another continent and is deflected laterally along its shores. In both hemispheres, water flows toward the equator along the western sides of continents, continuing to veer right or left until it meets at the equator and flows westward again.

Organisms must adapt to changes in their environment

Some kinds of changes in an organism's environment, such as the approach of a fire, storm, or predator, require immediate responses; other changes allow time for a more gradual response. Many plants adapt to hot conditions by reducing water loss and avoid overheating by shifting the position of their leaves during the day so that they intercept sunlight early and late in the day, but do not overheat at midday. Lizards bask in sunshine in the morning to raise their body temperature, but move into the shade when it gets too hot.

In addition to short-term changes in behavior, the evolution of many morphological and physiological features has enabled organisms to function in a variable physical environment. Many such features are described in Parts Seven and Eight of this book.

Few individuals die exactly where they were born; at some time during their lives, most organisms move, or are moved, to a new place, a phenomenon known as **dispersal**. Individuals may leave the site of their birth to find a better place to reproduce. Others may seek new places to live when local conditions deteriorate.

If repeated seasonal changes alter an environment in predictable ways, organisms may evolve life cycles that appear to anticipate those changes. **Migration** is one response to such cycli-

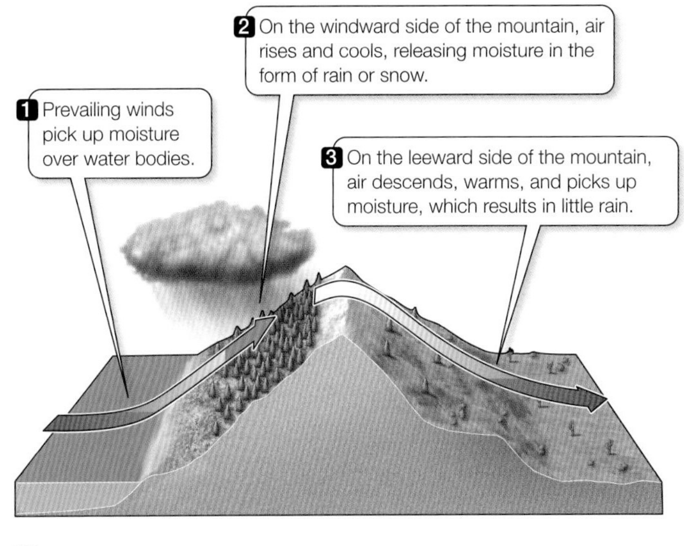

52.3 A Rain Shadow Average annual rainfall tends to be lower on the leeward side of a mountain range than on the windward side.

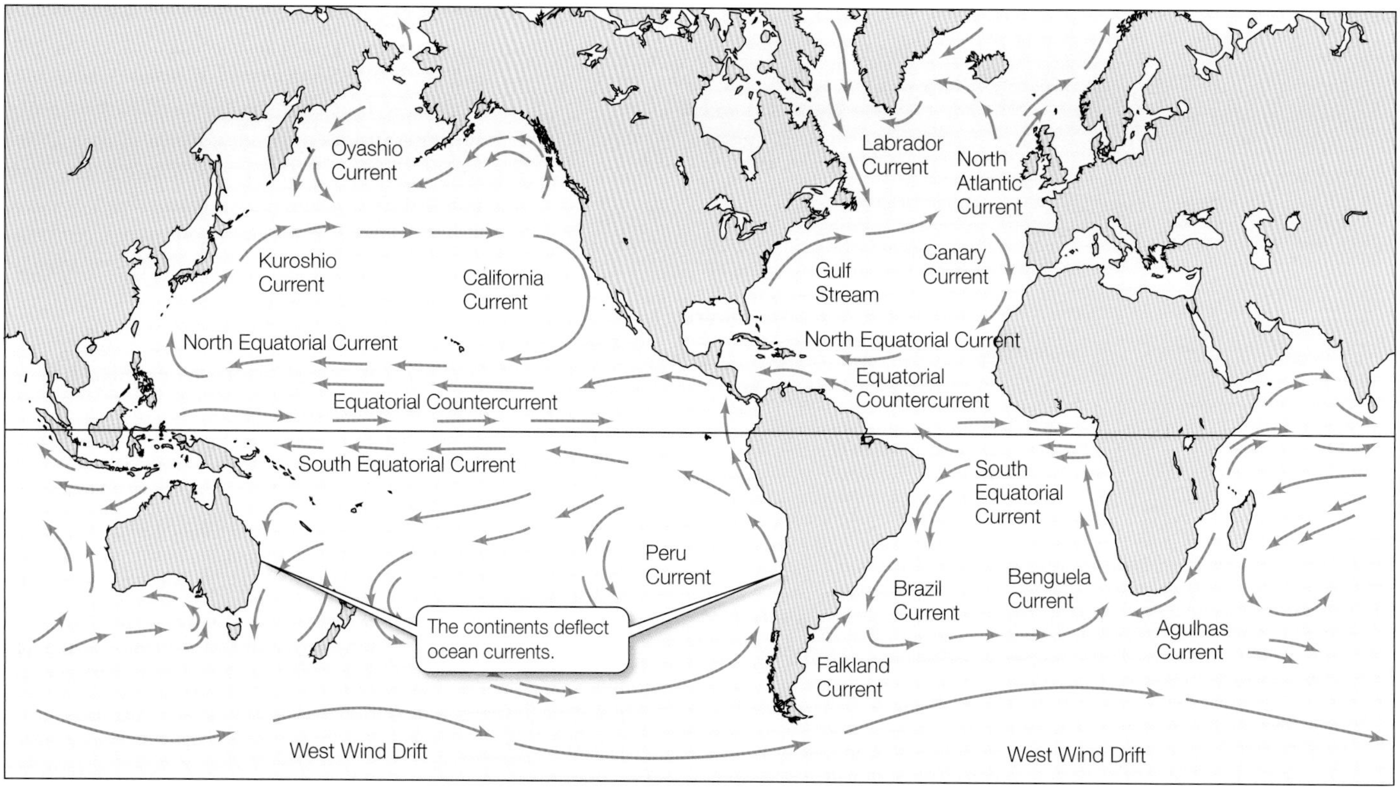

52.4 Global Oceanic Circulation To see that surface ocean currents are driven primarily by winds, compare the currents shown here with the prevailing winds shown in Figure 52.2.

cal environmental changes. Other animals enter a resting state (estivation, hibernation, or diapause) before adverse conditions materialize. They remain in that state until environmental signals indicate that conditions have improved.

Humans have clearly influenced Earth's climate, but such effects were thought to be only recent. However, climate models built by scientists at the University of Colorado suggest that if the early human occupants of Australia had not extensively burned that continent's forests, Australia's climate would be much wetter today than it actually is.

Most changes in the physical environment, both short-term and long-term, happen independently of anything that organisms do. Glaciers advance and retreat, storms brew, heat waves and cold fronts come and go, uninfluenced by the organisms that must deal with them. Other environmental changes, however, have been, and continue to be, influenced by the activities of organisms. For example, as Chapter 21 describes, organisms generated Earth's oxygenated atmosphere and produced Earth's soils. As we will see in the following chapters, many significant changes in the environment to which organisms must adapt are caused by other organisms, and many characteristics of organisms have been determined to a large degree by the history of such interactions.

52.2 RECAP

Differences in solar energy input create patterns of atmospheric circulation, and these prevailing winds affect oceanic circulation patterns. Organisms adapt to both short- and long-term climatic changes in their environment.

- Why does mean annual air temperature decrease with both latitude and altitude? See p. 1114
- Do you understand how variations in solar energy drive global air circulation patterns? See p. 1115 and Figure 52.2
- How do global air circulation patterns drive ocean currents? See p. 1115 and Figure 52.4

The great variability of the effects of Earth's climate has given rise to many different assemblages of organisms. Ecologists have found it useful to classify these assemblages into different ecosystem types. Depending on what parts of the system they intend to study, ecologists may classify ecosystems into biomes or biogeographic regions.

52.3 What Is a Biome?

When ecologists identify ecosystems, how do they decide where to locate the boundaries between different ecosystem types? The growth forms of the dominant plants of a particular environment, by determining the structure of the vegetation and by modifying the climate near the ground, strongly influence the lives of the other organisms that live there. A **biome** is a terrestrial environment defined by the growth forms of its plants. Common biomes include forests, grassland, desert, and tundra (**Figure 52.5**).

The distributions of plants are strongly influenced by annual patterns of temperature and rainfall. In some biomes, such as temperate deciduous forest, precipitation is relatively constant throughout the year, but temperature varies strikingly between summer and winter. In other biomes, both temperature and precipitation change seasonally. In still other biomes, temperatures are nearly constant, but rainfall varies seasonally. In the tropics, where seasonal temperature fluctuations are small, annual cycles are dominated by wet and dry seasons. Tropical biome types are determined primarily by the length of the dry season.

It is easiest to grasp the similarities and differences between biomes by means of a combination of photographs and graphs of temperature, precipitation, and biological activity, supplemented by a few words that describe other attributes of those biomes. In the following pages, each biome is represented by a map showing its locations and two photographs that illustrate either the biome at different times of year or representatives of the biome in different places on Earth. One set of graphs plots seasonal patterns of temperature and precipitation at a site in the biome. Other graphs show activity patterns of different kinds of organisms during the year. (For high-latitude biomes, patterns in the Southern Hemisphere are 6 months out of phase with those shown, which represent the Northern Hemisphere.) Levels of biological activity, shown by the width of the horizontal bars, change either because resident organisms become more or less active (produce leaves, come out of hibernation, hatch, or reproduce) or because organisms migrate into and out of the biome at different times of the year. A small box describes the growth forms of the plants that dominate the vegetation in the biome and its pattern of **species richness** (the number of species present in its communities).

These descriptions of biomes are very general and cannot begin to describe the variation that exists within each biome. For example, the temperate deciduous forest biome contains lakes, rivers, low-growing vegetation on cliffs, and grassy areas recovering from fire or other types of disturbance, as well as forests. In addition, the boundaries between biomes are somewhat arbitrary. Although sometimes an abrupt change can be seen in a landscape, more often one biome gradually merges into another. For example, the boundary between a forest and a grassland may not be distinct; instead, the spaces between trees may gradually increase, allowing more grasses to grow among them. Nevertheless, it is useful to recognize the major biomes of the world.

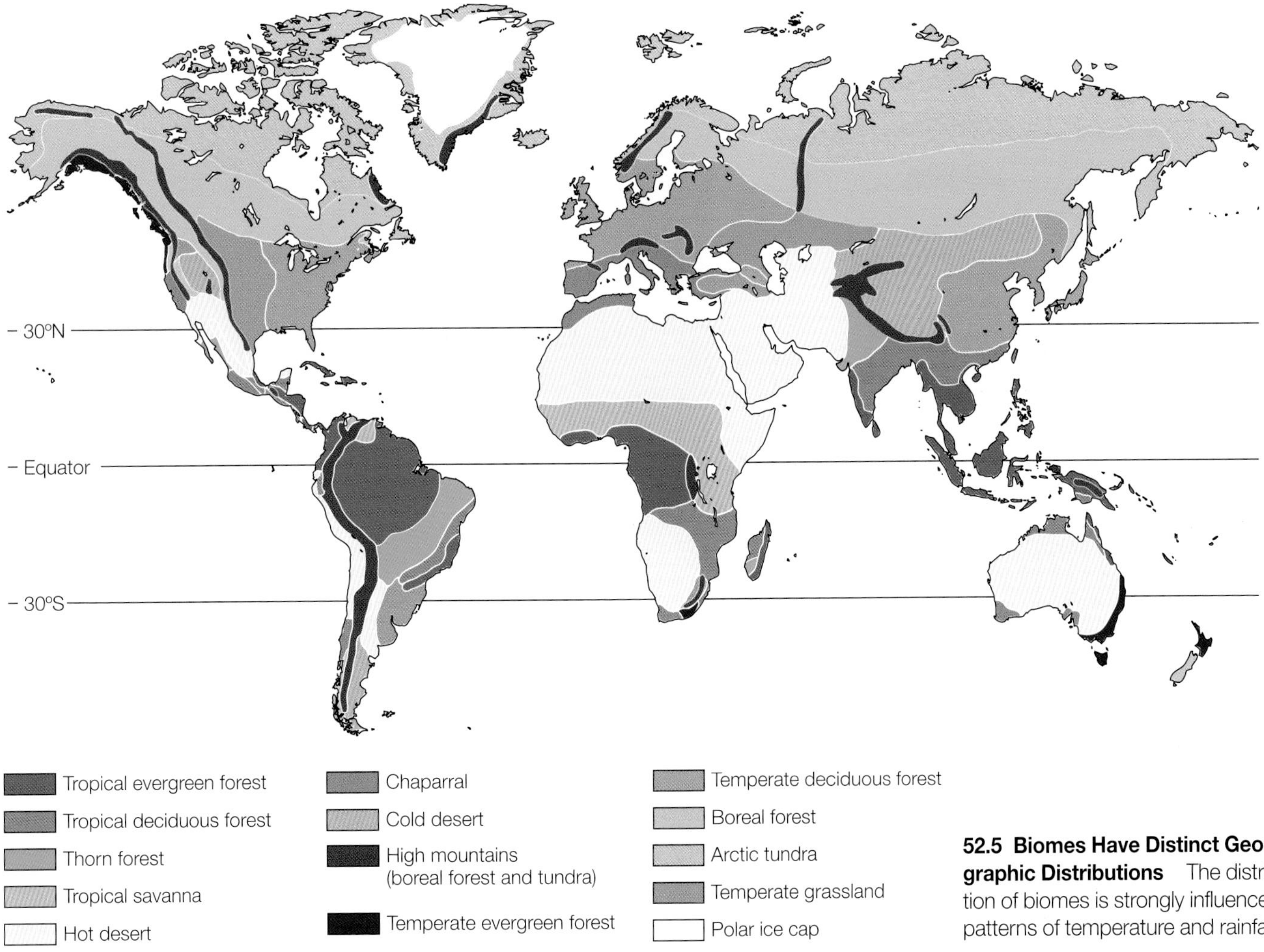

52.5 Biomes Have Distinct Geographic Distributions The distribution of biomes is strongly influenced by patterns of temperature and rainfall.

TUNDRA

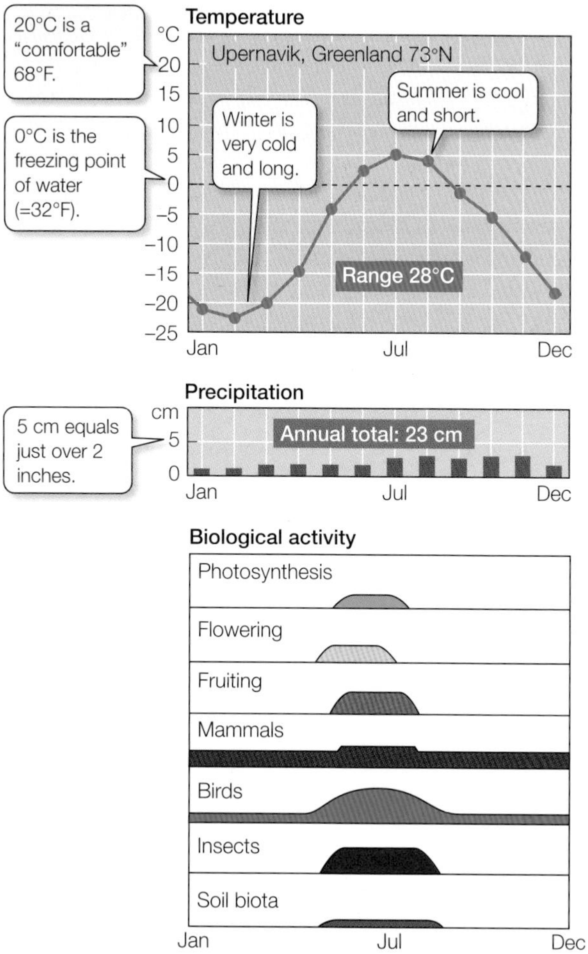

Community composition

Dominant plants
Perennial herbs and small shrubs

Species richness
Plants: Low; higher in tropical alpine
Animals: Low; many birds migrate in for summer; a few species of insects abundant in summer

Soil biota
Few species

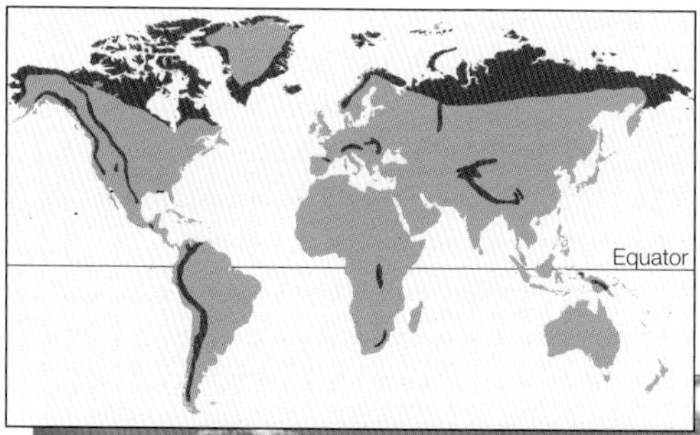

Arctic tundra, Greenland

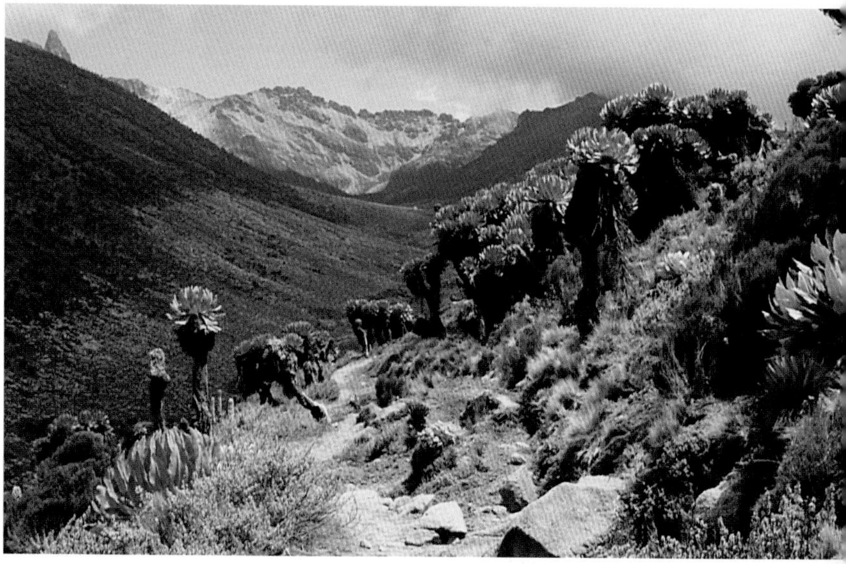

Tropical alpine tundra, Teleki Valley, Mt. Kenya, Kenya

Tundra is found at high latitudes and in high mountains

The **tundra** biome is found in the Arctic and at high elevations in mountains at all latitudes. In *Arctic tundra,* the vegetation, which consists of low-growing perennial plants, is underlain by *permafrost*—soil whose water is permanently frozen. The top few centimeters of soil thaw during the short summers, when the sun may be above the horizon 24 hours a day. Even though there is little precipitation, lowland Arctic tundra is very wet because water cannot drain down through the permafrost. Plants grow for only a few months each year. Most Arctic tundra animals either migrate into the area only for the summer or are dormant for most of the year.

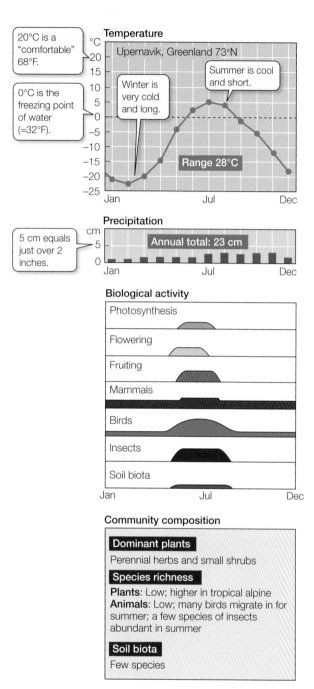

Tropical alpine tundra is not underlain by permafrost, so photosynthesis and most other biological activities continue (albeit slowly) throughout the entire year. As the photo of alpine vegetation on Mt. Kenya shows, more plant growth forms are present in tropical alpine tundra than in Arctic tundra vegetation.

BOREAL FOREST and TEMPERATE EVERGREEN FOREST

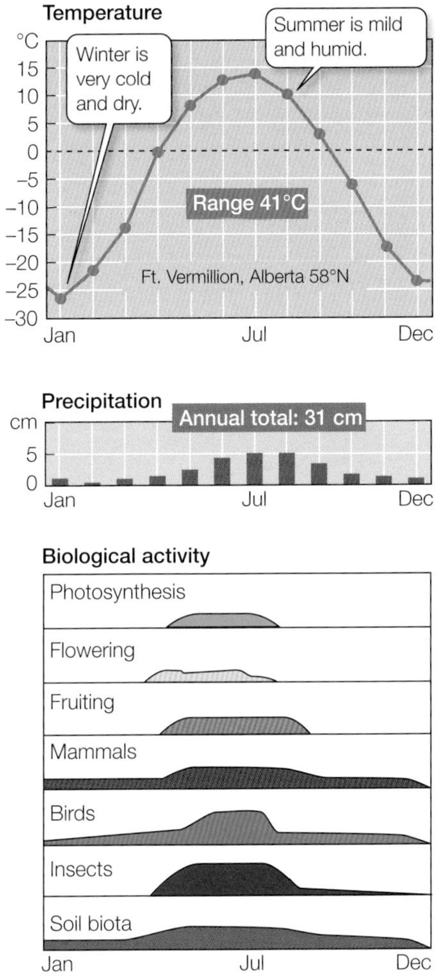

Community composition

Dominant plants
Trees, shrubs, and perennial herbs

Species richness
Plants: Low in trees, higher in understory
Animals: Low, but with summer peaks in migratory birds

Soil biota
Very rich in deep litter layer

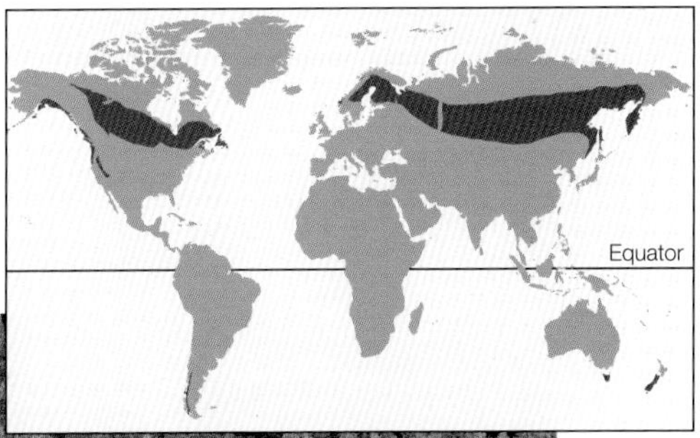

Northern boreal forest, Gunnison National Forest, Colorado

Southern boreal forest, Fiordland National Park, New Zealand

Evergreen trees dominate most boreal forests

The **boreal forest** biome is found toward the equator from Arctic tundra and at lower elevations on temperate-zone mountains. Boreal forest winters are long and very cold; summers are short (although often warm). The shortness of the summer favors trees with evergreen leaves because these trees are ready to photosynthesize as soon as temperatures warm in spring.

The boreal forests of the Northern Hemisphere are dominated by evergreen coniferous gymnosperms. In the Southern Hemisphere the dominant trees are southern beeches (*Nothofagus*), some of which are evergreen. **Temperate evergreen forests** also grow along the western coasts of continents at middle to high latitudes in both hemispheres, where winters are mild but very wet and summers are cool and dry. These forests are home to Earth's tallest trees.

Boreal forests have only a few tree species. The dominant mammals, such as moose and hares, eat leaves. The seeds in the cones of conifers support a fauna of rodents, birds, and insects.

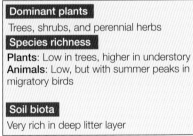

TEMPERATE DECIDUOUS FOREST

A Rhode Island forest in summer and... ...in winter

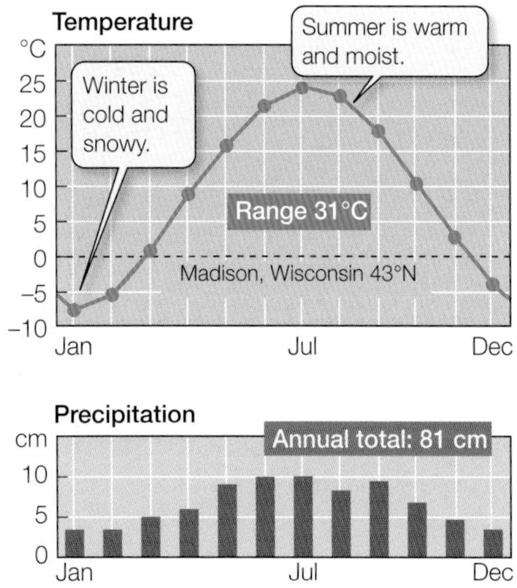

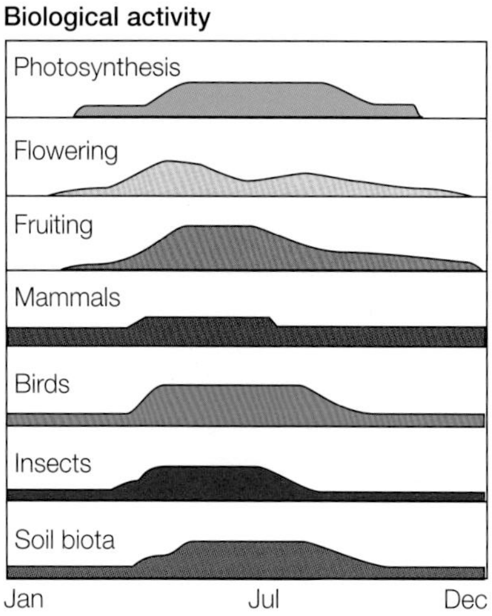

Community composition

Dominant plants
Trees and shrubs

Species richness
Plants: Many tree species in southeastern U.S. and eastern Asia, rich shrub layer
Animals: Rich; many migrant birds, richest amphibian communities on Earth, rich summer insect fauna

Soil biota
Rich

Temperate deciduous forests change with the seasons

The **temperate deciduous forest** biome is found in eastern North America, eastern Asia, and Europe. Temperatures in these regions fluctuate dramatically between summer and winter. Precipitation is relatively evenly distributed throughout the year.

Deciduous trees, which dominate these forests, lose their leaves during the cold winters and produce leaves that photosynthesize rapidly during the warm, moist summers. Many more tree species live here than in boreal forests. The temperate forests richest in species are in the southern Appalachian Mountains of the United States and in eastern China and Japan—areas that were not covered by glaciers during the Pleistocene. Many genera of plants and animals are shared among the three geographically separate regions with a deciduous forest biome.

TEMPERATE GRASSLANDS

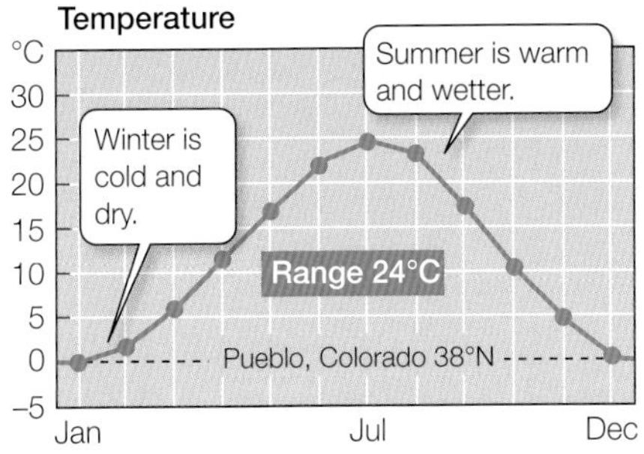

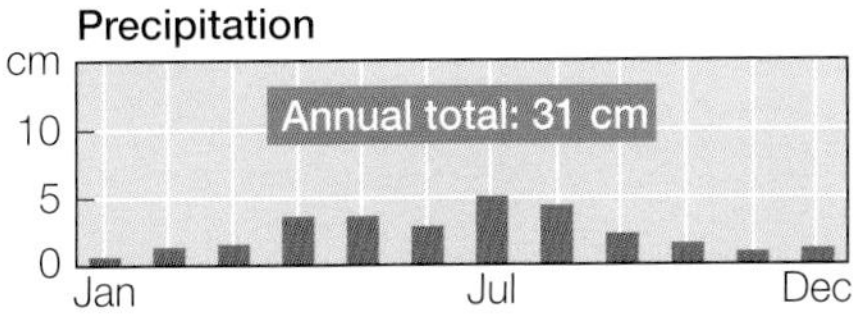

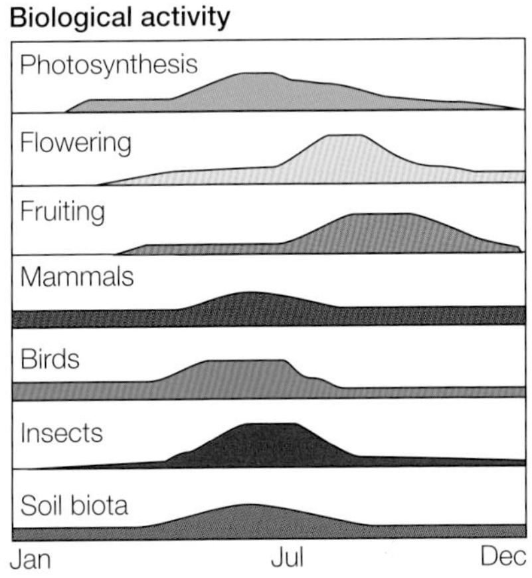

Community composition

Dominant plants
Perennial grasses and forbs

Species richness
Plants: Fairly high
Animals: Relatively few birds because of simple structure; mammals fairly rich

Soil biota
Rich

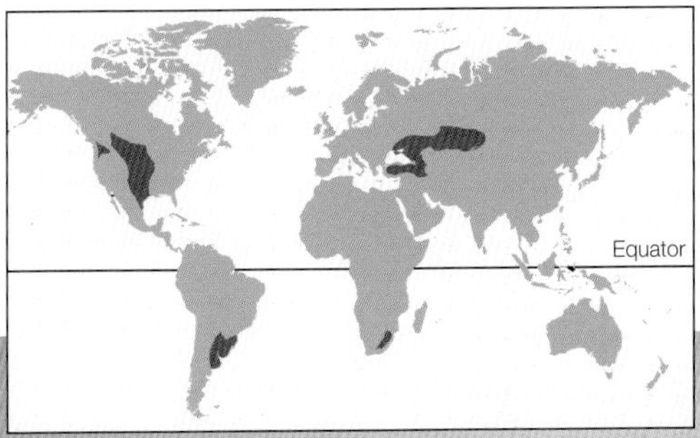

Nebraska prairie in spring

The Veldt, Natal, South Africa

Temperate grasslands are widespread

The **temperate grassland** biome is found in many parts of the world, all of which are relatively dry for much of the year. Most grasslands, such as the pampas of Argentina, the veldt of South Africa, and the Great Plains of North America, have hot summers and relatively cold winters. Most of this biome has been converted to agriculture. In some grasslands, most of the precipitation falls in winter (California grasslands); in others, the majority falls in summer (Great Plains, Russian steppe).

Grassland vegetation is structurally simple, but it is rich in species of perennial grasses, sedges, and *forbs* (herbaceous plants other than grasses). Grasslands are often a riot of color when forbs are in bloom. Grassland plants are adapted to grazing and fire. They store much of their energy underground and quickly resprout after they are burned or grazed.

COLD DESERT

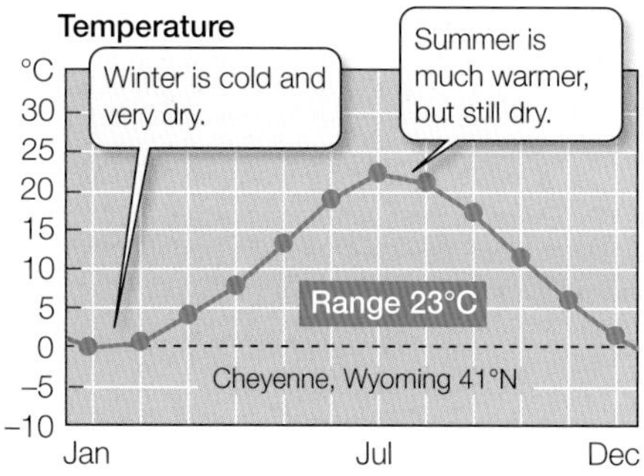

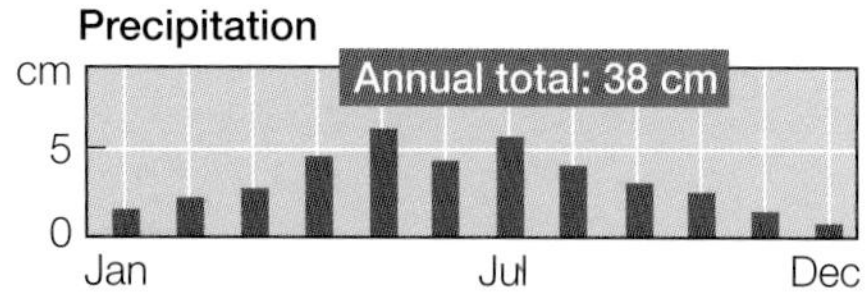

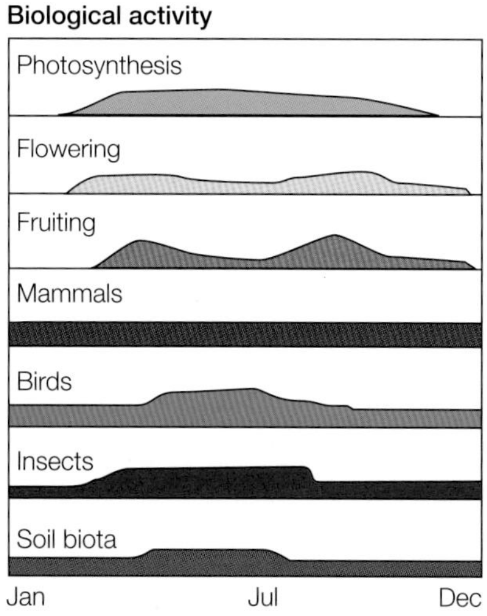

Community composition

Dominant plants
Low-growing shrubs and herbaceous plants

Species richness
Plants: Few species
Animals: Rich in seed-eating birds, ants, and rodents; low in all other taxa

Soil biota
Poor in species

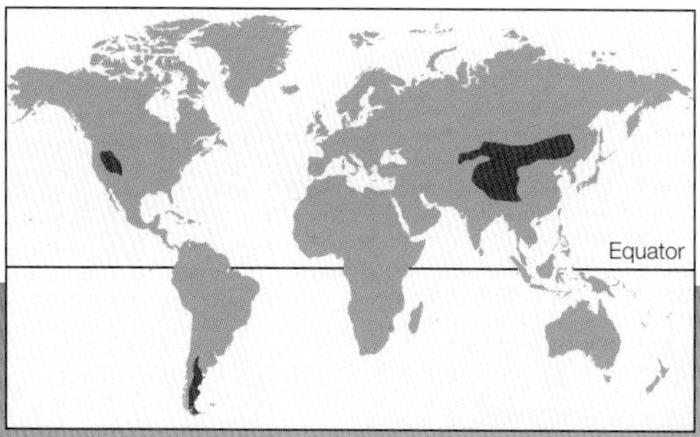

Sagebrush steppe near Mono Lake, California

Patagonia, Argentina

Cold deserts are high and dry

The **cold desert** biome is found in dry regions at middle to high latitudes, especially in the interiors of large continents in the rain shadows of mountain ranges. Seasonal changes in temperature are great.

Cold deserts are dominated by a few species of low-growing shrubs. The surface layers of the soil are recharged with moisture in winter, and plant growth is concentrated in spring. Because soils dry rapidly in spring, annual productivity is low. Cold deserts are relatively poor in species of most taxonomic groups, but the plants of this biome tend to produce large numbers of seeds, supporting many species of seed-eating birds, ants, and rodents.

HOT DESERT

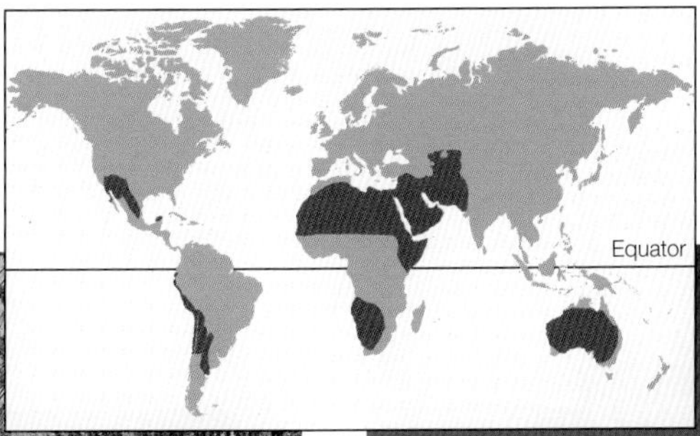

Anza Borrego Desert, California

Simpson Desert, Australia, following rain

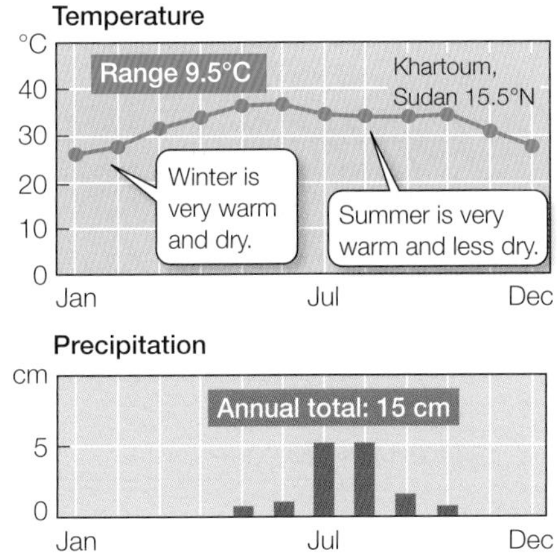

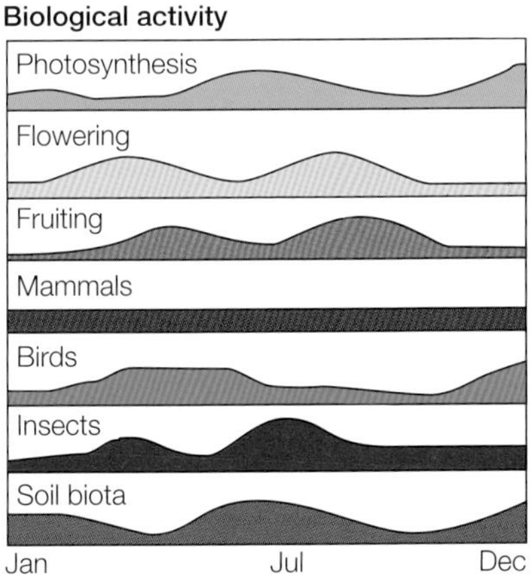

Community composition

Dominant plants

Many different growth forms

Species richness

Plants: Moderately rich; many annuals
Animals: Very rich in rodents; richest bee communities on Earth; very rich in reptiles and butterflies

Soil biota

Poor in species

Hot deserts form around 30° latitude

The **hot desert** biome is found in two belts, centered around 30° north and 30° south latitudes, where air descends, warms, and picks up moisture. Hot deserts receive most of their scarce rainfall in summer, but they also receive winter rains from storms that form over the mid-latitude oceans. The driest large regions, where summer and winter rains rarely penetrate, are in the center of Australia and the middle of the Sahara Desert of Africa.

Except in the driest regions, hot deserts have richer and structurally more diverse vegetation than cold deserts. Succulent plants (such as cacti) that store water in their expandable stems are conspicuous in some hot deserts. When rain falls, annual plants germinate and grow in abundance. Pollination and dispersal of fruits by animals are common. A rich fauna of rodents, termites, ants, lizards, and snakes is found in hot deserts.

CHAPARRAL

Southwest Australia

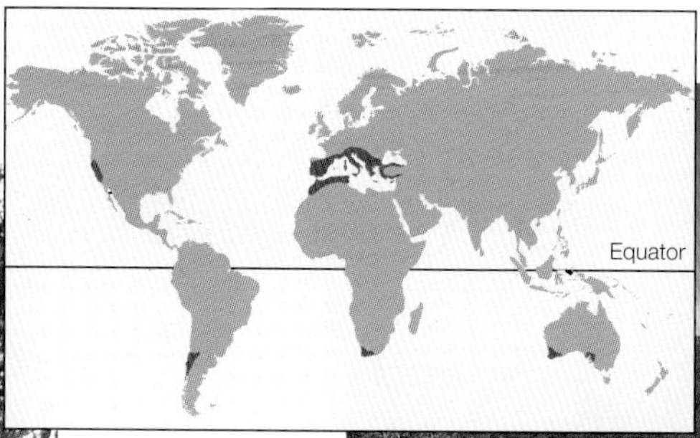

Santa Barbara County, California

Temperature

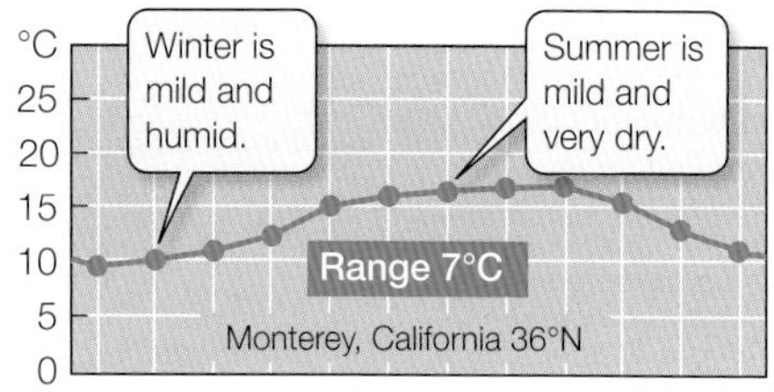

Precipitation

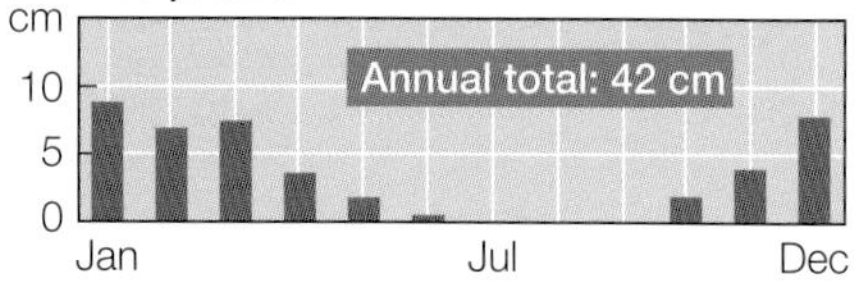

Biological activity

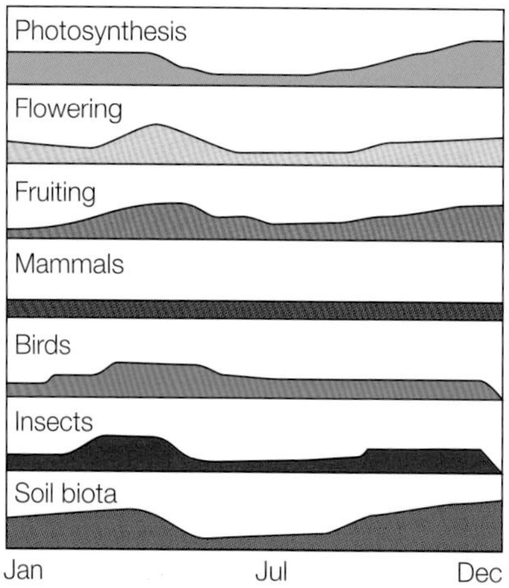

Community composition

Dominant plants

Low-growing shrubs and herbaceous plants

Species richness

Plants: Extremely high in South Africa and Australia
Animals: Rich in rodents and reptiles; very rich in insects, especially bees

Soil biota

Moderately rich

The chaparral climate is dry and pleasant

The **chaparral** biome is found on the western sides of continents at mid-latitudes (around 30°), where cool ocean currents flow offshore. Winters in this biome are cool and wet; summers are warm and dry. Such climates are found in the Mediterranean region of Europe, coastal California, central Chile, extreme southern Africa, and southwestern Australia.

The dominant plants of chaparral vegetation are low-growing shrubs and trees with tough, evergreen leaves. The shrubs carry out most of their growth and photosynthesis in early spring, when insects are active and birds breed. Annual plants are abundant and produce copious seeds that fall onto the soil. This biome thus supports large populations of small rodents, most of which store seeds in underground burrows. Chaparral vegetation is naturally adapted to survive periodic fires. Many shrubs of Northern Hemisphere chaparral produce bird-dispersed fruits that ripen in the late fall, when large numbers of migrant birds arrive from the north.

THORN FOREST and TROPICAL SAVANNA

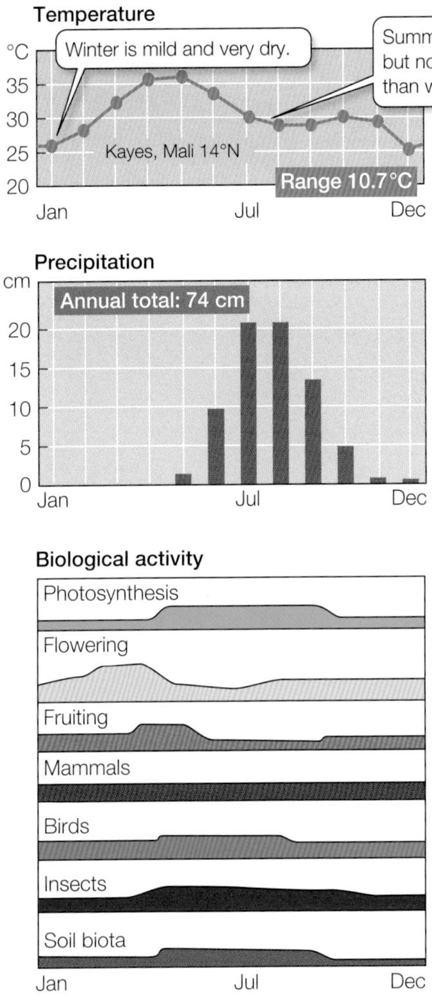

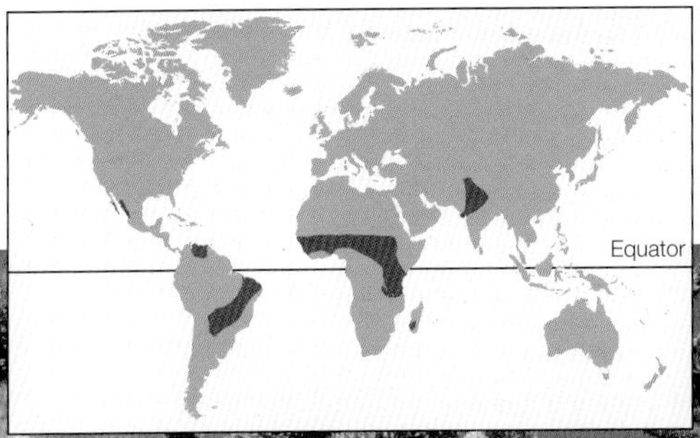

Thorn forest in Madagascar

Community composition

Dominant plants
Shrubs and small trees; grasses

Species richness
Plants: Moderate in thorn forest; low in savanna
Animals: Rich mammal faunas; moderately rich in birds, reptiles, and insects

Soil biota
Rich

KwaZulu-Natal, South Africa

Thorn forests and tropical savannas have similar climates

The **thorn forest** biome is found on the equatorial sides of hot deserts. The climate is semiarid; little or no rain falls during winter, but rainfall may be heavy during summer. Thorn forests contain many plants similar to those found in hot deserts. The dominant plants are spiny shrubs and small trees, many of which drop their leaves during the long, dry winter. Members of the genus *Acacia* are common in thorn forests worldwide.

The dry tropical and subtropical regions of Africa, South America, and Australia have extensive areas of the **tropical savanna** biome—expanses of grasses and grasslike plants with scattered trees. The largest tropical savannas are found in central and eastern Africa, where the biome supports huge numbers of grazing and browsing mammals and many large carnivores that prey on them. The grazers and browsers maintain the savannas. If savanna vegetation is not grazed, browsed, or burned, it typically reverts to dense thorn forest.

TROPICAL DECIDUOUS FOREST

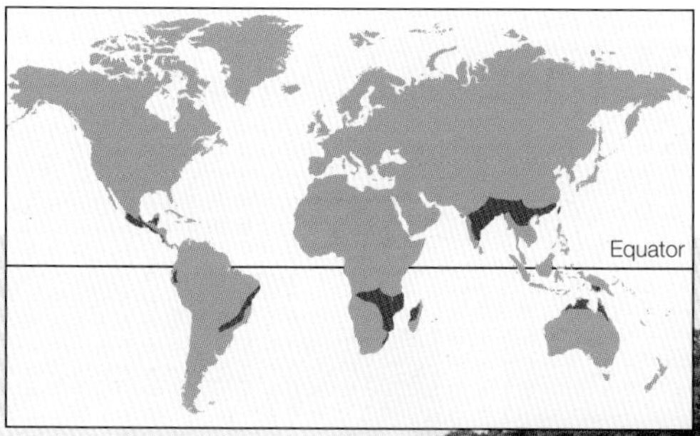

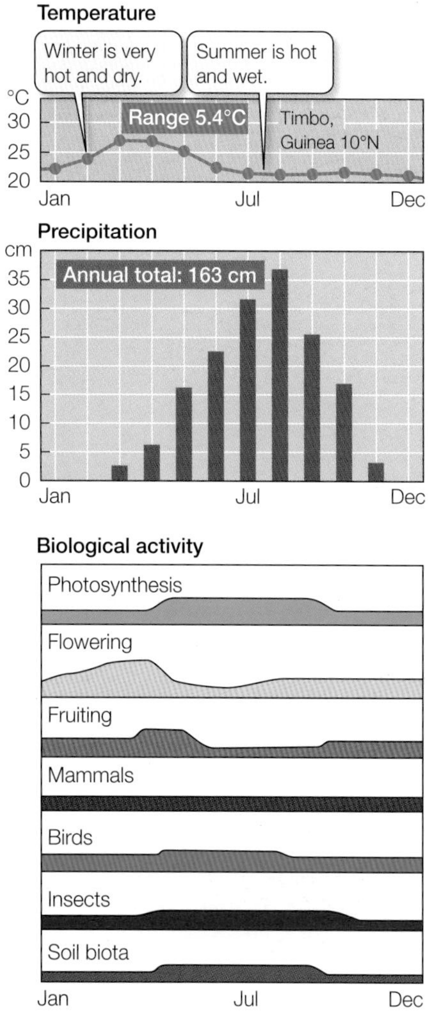

Palo Verde National Park, Costa Rica, in the rainy season…

Community composition

Dominant plants

Deciduous trees

Species richness

Plants: Moderately rich in tree species
Animals: Rich mammal, bird, reptile, and amphibian communities; rich in insects

Soil biota

Rich, but poorly known

…and in the dry season

Tropical deciduous forests occur in hot lowlands

As the length of the rainy season increases toward the equator, the **tropical deciduous forest** biome replaces thorn forests. Tropical deciduous forests have taller trees and fewer succulent plants than thorn forests, and they are much richer in plant and animal species. Most of the trees, except for those growing along rivers, lose their leaves during the long, hot dry season. Many of them flower while they are leafless, and most species are pollinated by animals. During the hot rainy season, biological activity is intense.

The soils of the tropical deciduous forest biome are some of the best soils in the tropics for agriculture because they contain more nutrients than the soils of wetter areas. As a result, most tropical deciduous forests worldwide have been cleared for agriculture and cattle grazing. Restoration efforts are under way on several continents.

TROPICAL EVERGREEN FOREST

The exterior of lowland wet forest...

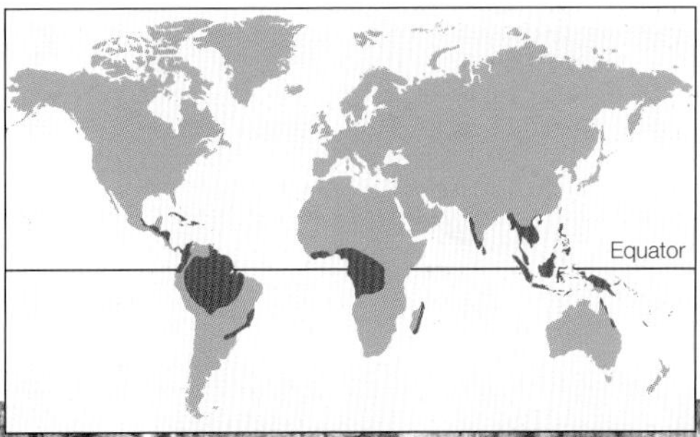

...and its interior, Cocha Cashu, Peru

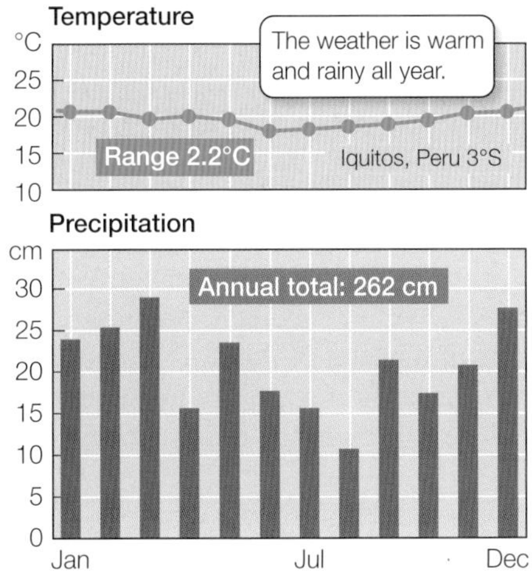

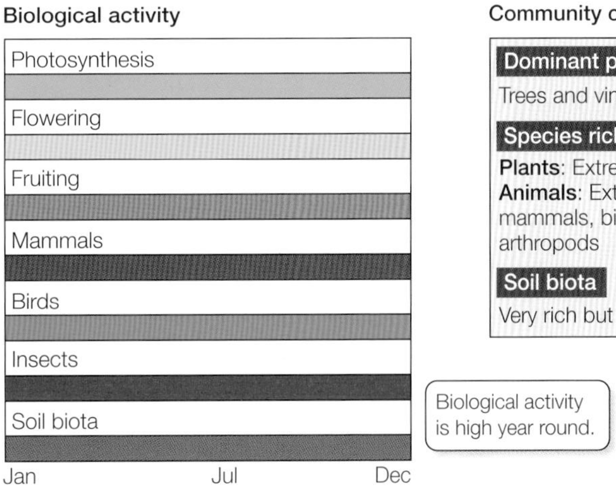

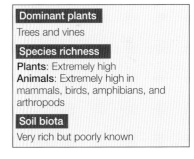

Community composition

Dominant plants
Trees and vines

Species richness
Plants: Extremely high
Animals: Extremely high in mammals, birds, amphibians, and arthropods

Soil biota
Very rich but poorly known

Tropical evergreen forests are species-rich

The **tropical evergreen forest** biome is found in equatorial regions where total rainfall exceeds 250 centimeters annually and the dry season lasts no longer than 2 or 3 months. It is the richest of all biomes in numbers of species of both plants and animals, with up to 500 species of trees per square kilometer. Along with their immense species richness, tropical evergreen forests have the highest overall productivity of all ecological communities. However, most mineral nutrients are tied up in the vegetation. The soils usually cannot support agriculture without massive applications of fertilizers.

On the slopes of tropical mountains, trees are shorter than lowland tropical trees. Their leaves are smaller, and there are more *epiphytes* (plants that grow on other plants, deriving their nutrients and moisture from air and water rather than soil).

52.6 Australian Deserts Differ from Those on Other Continents Extensive areas in Australia are dominated by a tough, fibrous grass, called spinifex (*Triodia*) that burns readily when green.

The distribution of biomes is not determined solely by climate

From the previous discussion, you might conclude that only climate determines the distribution of biomes and the characteristics of the plants that dominate them. Climate is important, but other factors, particularly soil fertility and fire, also exert strong influences on the structure of the vegetation in an area. For example, the vegetation of Australian deserts grows on extremely nutrient-poor soils. The leaves of those plants are so difficult to digest that herbivores consume little of them. Thus flammable litter accumulates rapidly under the trees, and intense fires periodically sweep across the landscape. As a result, succulent plants, which are easily killed by fires, are not found in Australia, although they are common in deserts on other continents (see p. 1123). Australian deserts have distinctive vegetation (**Figure 52.6**).

52.3 RECAP

Ecologists recognize a number of large ecological units called biomes, which are based on the growth forms of the dominant vegetation. The distribution of terrestrial biomes is determined primarily by temperature and precipitation, but is also influenced by soil fertility and fire.

Climate explains why chaparral is confined to the western coasts of continents at mid-latitudes, but it cannot explain why the chaparral in California and the chaparral in South Africa do not share any species in common. To explain the distribution of species on Earth, we need to study past as well as current physical environments.

52.4 What Is a Biogeographic Region?

Until European naturalists traveled the globe, they had no way of knowing what organisms were found elsewhere. Alfred Russel Wallace, who along with Charles Darwin put forth the idea that natural selection could account for the evolution of life on Earth (see Section 22.1), was one of those global travelers. He returned to England in the spring of 1862 after having spent seven years traveling in the Malay Archipelago. During his travels he noticed some remarkable patterns in the distributions of organisms across the archipelago. For example, he described the dramatically different birds that inhabited Bali and Lombok, two adjacent islands:

> In Bali we have barbets, fruit-thrushes and woodpeckers; on passing over to Lombock these are seen no more, but we have an abundance of cockatoos, honeysuckers, and brush-turkeys, which are equally unknown in Bali, or any island further west. The strait here is fifteen miles wide, so that we may pass in two hours from one great division of the earth to another, differing as essentially in their animal life as Europe does from America.

Wallace pointed out that these striking biological differences could not be explained by climate or geology, because in those respects, Bali and Lombok were essentially identical.

Wallace saw that he could draw a line through the Malay Archipelago that would divide it into two distinct halves based on the distributions of plant and animal species. He correctly deduced that these dramatic differences in flora and fauna were related to the depth of the channel separating Bali and Lombok. This channel is so deep that it remained a barrier to the movement of terrestrial animals even during the Pleistocene glaciations, when sea levels dropped more than 100 meters and Java and other islands to the west were connected to the Asian mainland (**Figure 52.7**).

With these insights, Wallace established the conceptual foundations of **biogeography**, the scientific study of the patterns of distribution of populations, species, and ecological communities across Earth. He pointed out that the geological history of regions profoundly influences the kinds of organisms found in them. Today we call the line he drew through the Malay Archipelago "Wallace's line."

The flora, fauna, and microorganisms of different parts of the world—that is, their *biota*—differ enough to allow us to divide Earth into several major **biogeographic regions** (**Figure 52.8**). Biogeographic regions are based on the taxonomic composition of the organisms living in them. Their boundaries are set where species compositions change dramatically over short distances. Wallace's line, for example, separates the Malay Archipelago into two distinct biogeographic regions. The biotas of biogeographic regions differ because oceans, mountains, deserts, and other barriers restrict the dispersal of organisms between them. Although organisms do disperse between adjacent biogeographic regions, such interchanges have not been frequent enough or massive enough to eliminate the striking differences that have resulted from speciation and extinction within each region. Most species are confined to a single biogeographic region.

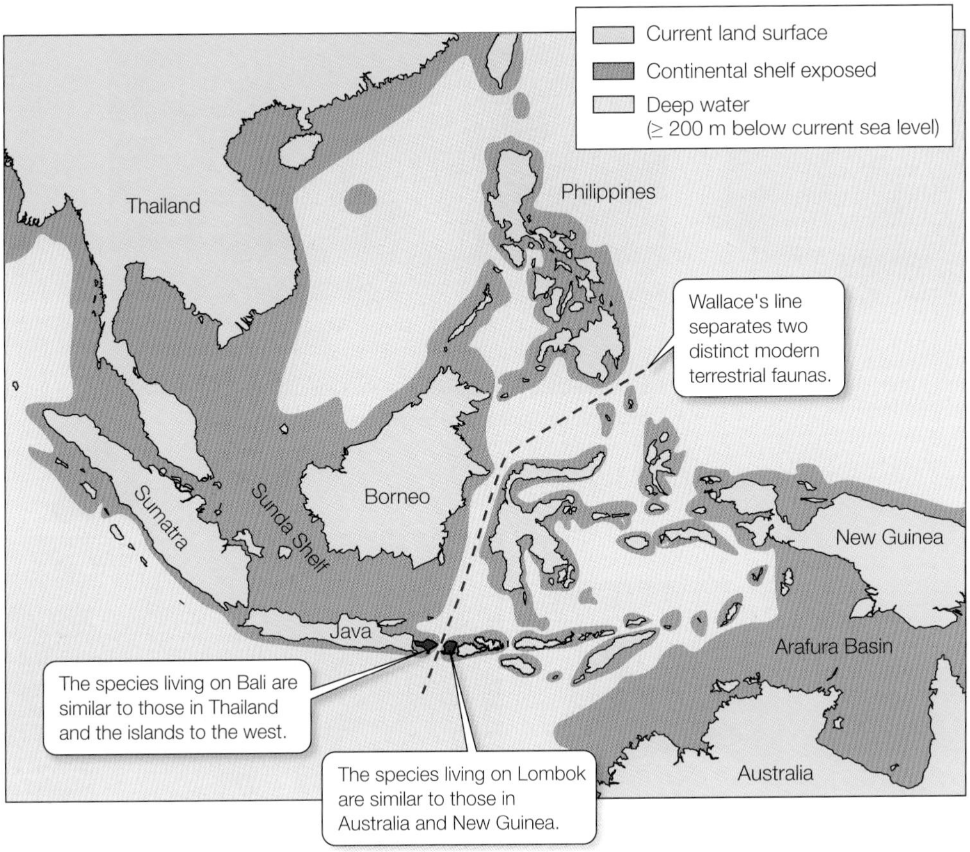

52.7 The Malay Archipelago during the Most Recent Glacial Maximum Wallace's line, which divides two distinct geographic regions, corresponds to a deep-water channel between the islands of Bali and Lombok. This channel is deep enough to have blocked the movement of terrestrial organisms even during the glaciations of the Pleistocene era when the sea level was 100 meters lower than it is today.

Today the most widespread terrestrial animal species is *Homo sapiens*, but there are also a few bird species—for example, the great egret, osprey, peregrine falcon, and barn owl—that are found on all continents except Antarctica.

A species found only within a certain region is said to be **endemic** to that region. Remote islands typically have distinctive endemic biotas because water barriers greatly restrict immigration. For example, nearly all the species of vascular plants and vertebrates of Madagascar, a large island off the eastern coast of Africa, are endemic to that island (**Figure 52.9**). Madagascar by itself could be considered a biogeographic region, but because dozens of islands would qualify as biogeographic regions on the basis of the distinctiveness of their biotas, islands are not called biogeographic regions.

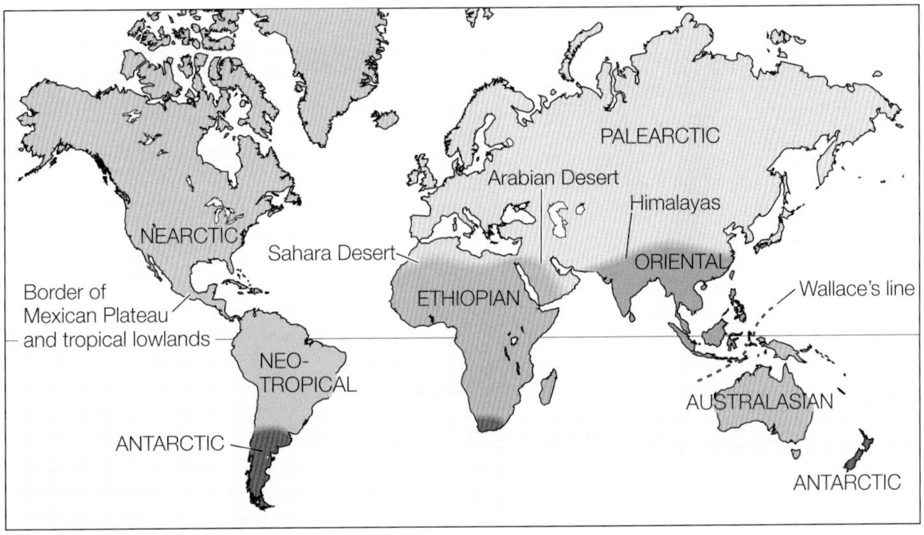

52.8 Major Biogeographic Regions The biotas of Earth's major biogeographic regions differ strikingly from one another. These regions are separated by topographic, climatic, or aquatic barriers to dispersal that cause their biotas to differ from one another.

Three scientific advances changed the field of biogeography

For many decades after biogeographers determined that the biotas of the major biogeographic regions are strikingly different from one another, they devoted their efforts to assembling information on the distributions of organisms. They speculated about the causes of the distributional patterns they found, but the field remained primarily descriptive. Biogeographers did gain insights from the study of fossils, which can show how long a taxon has been present in an area and whether its members formerly lived in areas where they are no longer found. But it took three scientific ad-

Chamaeleo parsonii

Hemicentetes semispinosus (yellow-streaked tenrec)

Cryptoprocta ferox (fossa)

Lemur catta (ring-tailed lemur) climbing *Alluaudia procera* (Madagascar ocotillo)

Adansonia grandidieri (giant baobob tree)

Pachypodium rosulatum (elephant's foot)

52.9 Madagascar Abounds with Endemic Species The majority of the plant and vertebrate species found on the island of Madagascar are found nowhere else on Earth.

vances in the latter part of the twentieth century to change biogeography into the dynamic multidisciplinary field it is today:

- The acceptance of the theory of *continental drift*
- The development of *phylogenetic taxonomy*
- The development of the theory of *island biogeography*

Let's see how each of these advances contributed to the development of biogeography.

CONTINENTAL DRIFT Prior to the mid-nineteenth century, most people believed that Earth was young and had changed little over time. Carolus Linnaeus, the naturalist who initiated the classification of species some 250 years ago, believed that all organisms had been created in one place (which he called Paradise), from which they later dispersed. Indeed, because most people believed that the continents were fixed in their positions, the only way to account for the current distributions of organisms was to invoke massive dispersal.

The notion that the continents might have moved was not seriously considered until 1912, when Alfred Wegener put forward the concept of continental drift, proposing that the continents had changed position over time. When Wegener proposed this idea, few scientists took him seriously, but as Section 21.2 described, geological evidence and plausible mechanisms that could move continents were eventually discovered.

About 280 million years ago, the continents were united to form a single land mass, Pangaea (see Figure 21.15). The continents then began to separate from one another, but by the Triassic period, when they were still very close to one another (about 245 mya), many groups of terrestrial and freshwater organisms, such as insects, freshwater fishes, frogs, and vascular plants, had already evolved. The ancestors of some organisms that live on widely separated continents today were probably present on those land masses when they were part of Pangaea.

PHYLOGENETIC TAXONOMY As discussed in Chapter 25, taxonomists have developed powerful methods of reconstructing phylogenetic relationships among organisms. Biogeographers have adapted these methods to help them answer biogeographic questions. Biogeographers can transform phylogenetic trees into **area phylogenies** by replacing the names of the taxa on a tree with the names of the places where those taxa live or lived. For example, we know from the fossil record that the earliest ancestors of horses evolved in North America, but an area phylogeny of horses makes it possible to follow their evolution and dispersal much further. It suggests that the ancestors of today's horses dispersed from North America to Asia, and then from Asia to Africa, and further, that the speciation of zebras took place entirely in Africa (**Figure 52.10**).

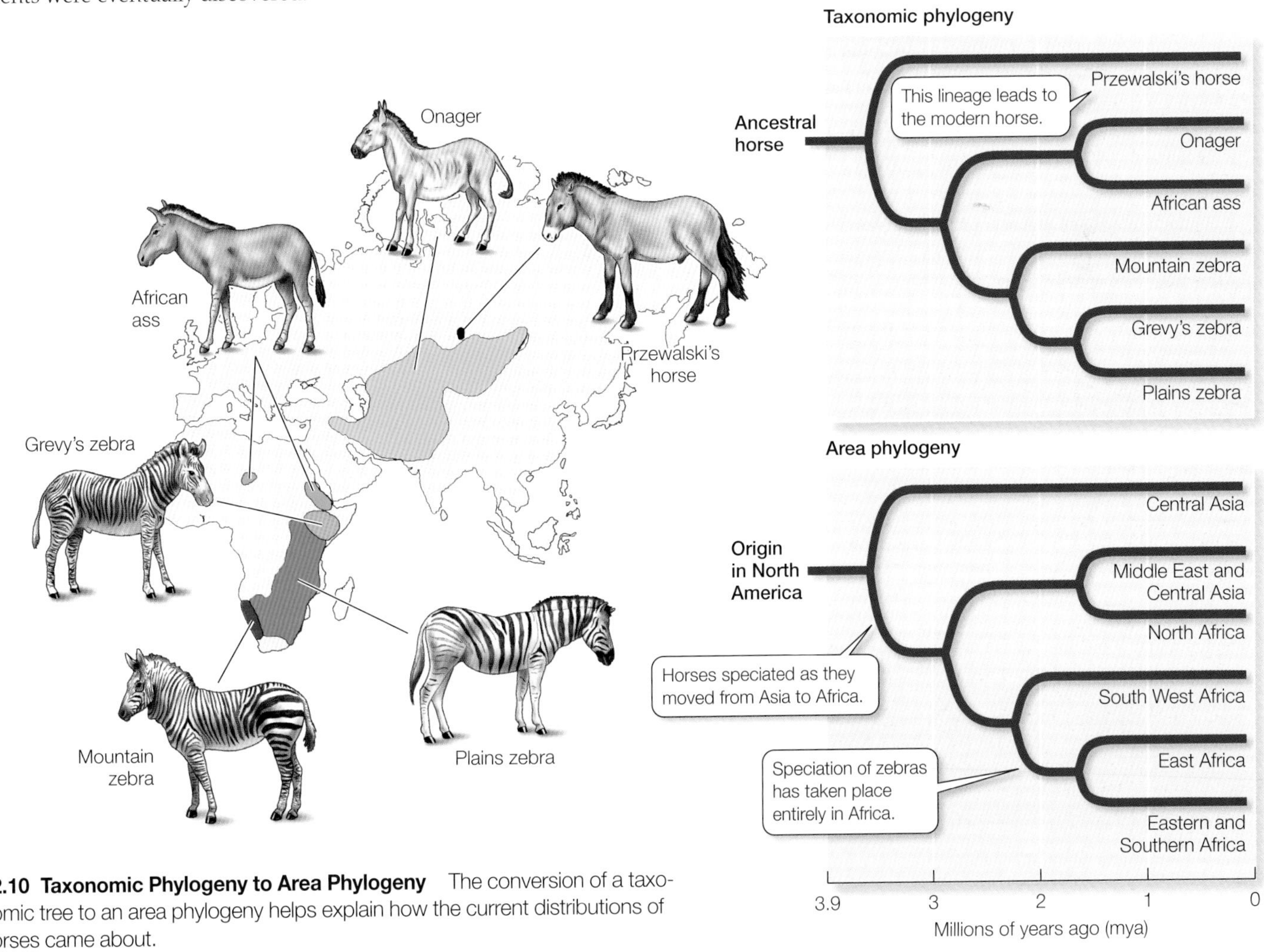

52.10 Taxonomic Phylogeny to Area Phylogeny The conversion of a taxonomic tree to an area phylogeny helps explain how the current distributions of horses came about.

ISLAND BIOGEOGRAPHY Robert MacArthur and Edward O. Wilson developed the theory of **island biogeography** to help explain why oceanic islands always have fewer species than a mainland area of equivalent size. They based their theory on just two processes: the immigration of new species to an island and the extinction of species already present on that island. They modeled the unfolding of these two processes over periods of only a few hundred years or less, during which they assumed that no speciation took place.

Imagine a newly formed oceanic island that receives colonists from a mainland area. The list of species on the mainland that might possibly colonize the island constitutes the **species pool**. The first colonists to arrive on the island are all "new" species because no species live there initially. As the number of species on the island increases, a larger fraction of colonizing individuals will be members of species already present. Therefore, even if the same number of species arrive as before, the rate of arrival of new species decreases, until it reaches zero when the island has all the species in the species pool.

Now consider extinction rates. At first there will be only a few species on the island, and their populations may grow large. As more species arrive and their populations increase, the resources of the island will be divided among more species. Therefore, the average population size of each species will become smaller as the number of species increases. The smaller a population, the more likely it is to become extinct (see Section 54.4). In addition, the number of species that can possibly become extinct increases as species accumulate on the island. Furthermore, new arrivals on the island may include pathogens and predators that increase the probability of extinction for other species. For all these reasons, the rate of extinction increases as the number of species on the island increases.

Because the rate of arrival of new species decreases and the extinction rate increases as the number of species increases, eventually the number of species on the island should reach an equilibrium at which the rates of arrival and extinction are equal (**Figure 52.11A**). If there are more species than the equilibrium number, extinctions should exceed arrivals, and species richness should decline. If there are fewer species than the equilibrium number, arrivals should exceed extinctions, and species richness should increase. The equilibrium is dynamic because if either rate fluctuates, as they generally do, the equilibrium number of species shifts up or down.

MacArthur and Wilson's model can also be used to predict how species richness should differ among islands of different sizes and at different distances from the mainland. We expect extinction rates to be higher on small islands than on large islands because species' populations are, on average, smaller there. Similarly, we expect fewer immigrants to reach distant islands. **Figure 52.11B** gives hypothetical relative species numbers for islands of different sizes and distances from the mainland. As you can see, the number of species should be highest for islands that are relatively large and relatively close to the mainland (such as Madagascar). These principles can be applied not only to oceanic islands, but also to "habitat islands" in terrestrial landscapes, as we will see in Section 57.2.

The most important contribution of the theory of island biogeography was to demonstrate that biogeography could actually have an experimental component. Immigration and extinction rates can be measured and, sometimes, manipulated to test hypotheses. In addition, major disturbances, which can serve as "natural experiments," sometimes permit scientists to estimate colonization and extinction rates. In August 1883, Krakatau, an island in the Sunda Strait between Sumatra and Java, was devastated by a series of volcanic eruptions that destroyed all life on the island's surface. After the lava cooled, plants and animals from Sumatra to

52.11 The Theory of Island Biogeography (A) The rate of arrival of new species and the rate of extinction of species already present determine the equilibrium number of species on an island. (B) These rates are affected by the size of the island and its distance from the mainland.

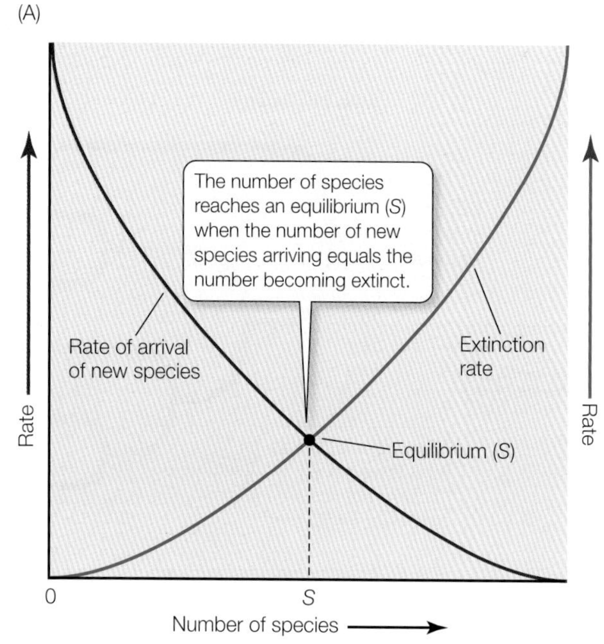

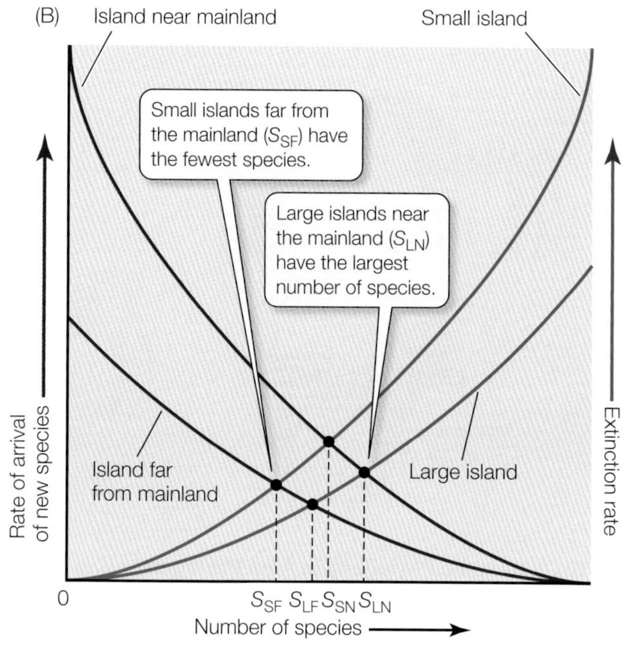

TABLE 52.1

Number of Species of Resident Land Birds on Krakatau

PERIOD	NUMBER OF SPECIES	EXTINCTIONS	COLONIZATIONS
1908	13		
1908–1919		2	17
1919–1921	28		
1921–1933		3	4
1933–1934	29		
1934–1951		3	7
1951	33		
1952–1984		4	7
1984–1996	36		

the west and Java to the east rapidly colonized Krakatau. By 1933, the island was again covered with a tropical evergreen forest, and 271 species of plants and 27 species of resident land birds were found there.

During the 1920s, as a forest canopy was developing on Krakatau, both birds and plants colonized the island at a high rate (**Table 52.1**). Birds probably brought the seeds of many plants because, between 1908 and 1934, both the percentage (from 20% to 25%) and the absolute number (from 21 to 54) of plant species with bird-dispersed seeds increased. Today the numbers of plant and bird species are not increasing as fast as they did during the 1920s, but colonizations and extinctions continue, as predicted by the theory of island biogeography.

Experiments can also be conducted to test components of the theory. To test the hypothesis that there is an equilibrium number of species on islands, Daniel Simberloff and Edward O. Wilson of Harvard University defaunated (eliminated all animals from) tiny islets of red mangroves in the Florida Keys and monitored their recolonization (**Figure 52.12**). The islands were rapidly recolonized, and all of them eventually supported roughly the same number of species that they had before defaunation. In addition, the rate of recolonization was slowest on the most remote island.

The role of experimentation in biogeography is necessarily limited, however, because the processes that drive biogeographic patterns are fully expressed only after long time spans, usually extending over millions of years. The striking patterns that Alfred Russel Wallace observed had their origins deep in time. Let's now consider some of the long-term processes that result in recognizably different biogeographic regions.

EXPERIMENT

HYPOTHESIS: Defaunated islands will be rapidly recolonized, eventually achieving about the same number of species that they had prior to defaunation.

METHOD

Erect scaffolding and tent to enclose islets. Fumigate small islets with a chemical (methyl bromide) that kills arthropods but does not harm plants. Periodically monitor recolonizations and extinctions of arthropods on the islands.

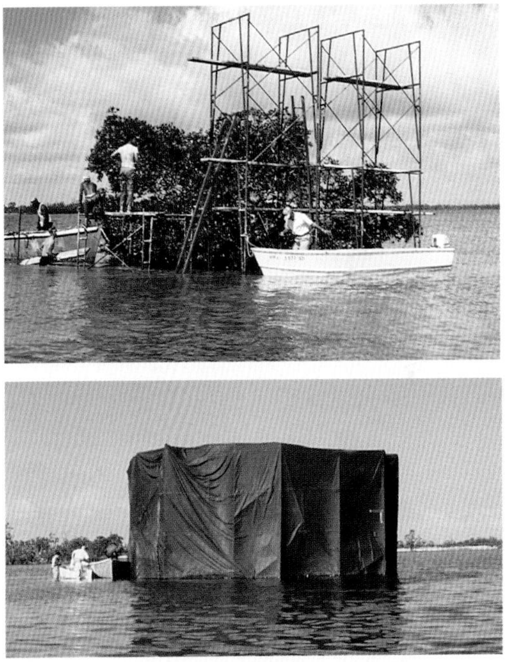

RESULTS

Recolonization was rapid, turnover rates were high, and the rate of recolonization was slowest on the most remote island.

CONCLUSION: An island can support a certain equilibrium number of species.

52.12 Experimental Island Defaunation Simberloff and Wilson surrounded several small islands with scaffolding so that they could be covered and fumigated to remove all arthropods. The researchers then counted and monitored arthropods as they recolonized the islands. FURTHER RESEARCH: Although the results of Wilson and Simberloff's defaunation experiment were dramatic, the experiment lasted only a short time, and only arthropods were counted. What experiments could you conduct to assess some of the longer-term predictions of the theory of island biogeography?

A single barrier may split the ranges of many species

Two major processes—vicariance and dispersal—generate biogeographic patterns. The appearance of a physical barrier that splits the range of a species is called a **vicariant event**. A vicariant event divides the population of a species into two or more discontinuous populations, even though no individuals have dispersed to new areas. If, however, members of a species cross an already existing barrier and establish a new population, the species' discontinuous range is considered the result of dispersal.

By studying a clade, a biogeographer may discover evidence suggesting that the distribution of an ancestral species was influenced by a vicariant event, such as a change in sea level or mountain building (see Figure 23.3). If that inference is correct, it is reasonable to assume that other ancestral species would have been affected by the same event and that similar distribution patterns would be seen in other clades. Differences in distribution patterns among clades may indicate that they responded differently to the same vicariant events, that they diverged at different times, or that the clades have had very different dispersal histories. By analyzing such similarities and differences, biogeographers seek to discover the relative roles of vicariant events and dispersal in determining today's distribution patterns.

Biotic interchange follows fusion of land masses

When formerly separated land masses fuse, as may happen through sea level changes or continental drift, two different biotas can merge. Many species of both biotas are likely to disperse into the region they had not previously inhabited. This phenomenon is called **biotic interchange**. Such a massive colonization of new areas by mammals occurred when the Central American land bridge formed about 4 million years ago. This land bridge connected North and South America for the first time in about 65 million years. While the two continents were separated, their mammals evolved independently of one another because terrestrial mammals (with the exception of bats) are poor dispersers across water barriers. South America evolved a distinctive mammalian fauna dominated by marsupials, primates, edentates (armadillos, sloths, ground sloths, and anteaters), and caviomorph rodents (porcupines, capybaras, pacas, agoutis, guinea pigs, and chinchillas).

Many species of mammals dispersed across the newly established land bridge. Only a few South American species—the porcupine, nine-banded armadillo, and Virginia opossum—became established north of the tropical forests of Mexico (**Figure 52.13A**). However, many North America mammals—rabbits, mice, foxes, bears, raccoons, weasels, cats, tapirs, peccaries, camels, and deer—successfully colonized South America. The North American invasion apparently caused the extinction of several kinds of large marsupial carnivores and the large herbivores that were their prey. Subsequently, the northern invaders gave rise to new species that today exist only in South America (**Figure 52.13B**).

The exchange of mammals between North and South America did not eliminate the differences between the mammalian faunas of the two continents. North American visitors to South America today still enjoy seeing a rich array of mammals that are unfamiliar to them, most of which are endemic to South America or penetrate only a short distance into Central America. But without knowing the long-term history of the region, we would not be aware of the many recently extinct South American species, nor could we know which of the species there today are recent arrivals.

(A)

Dasypus novemcinctus

Erethizon dorsatum

52.13 North and South America Exchanged Mammalian Faunas (A) The nine-banded armadillo and the porcupine are among only a handful of species that colonized North America from South America. (B) Some species that exist today only in South America are descended from ancestors who migrated from North America, including the Patagonian fox, the Chacoan peccary, and the vicuña (a member of the camel family).

(B)

Pseudalopex griseus

Catagonus wagneri

Vicugna vicugna

Both vicariance and dispersal influence most biogeographic patterns

If both vicariance and dispersal influence distribution patterns, how can we determine their relative importance in particular cases? When several hypotheses can explain a pattern, scientists typically prefer the most parsimonious one—the one that requires the smallest number of unobserved events to account for it. We saw how the parsimony principle is used in the reconstruction of phylogenies in Section 25.2. To see how it is applied to biogeography, consider the distribution of the New Zealand flightless weevil *Lyperobius huttoni*, a species that is found in the mountains of South Island and on sea cliffs at the extreme southwestern corner of North Island (**Figure 52.14**). If you knew only its current distribution and the current positions of the two islands, you might surmise that, even though this weevil cannot fly, it had somehow managed to cross Cook Strait, the 25-kilometer body of water that separates the two islands.

More than 60 other animal and plant species, however, including other species of flightless insects, live on both sides of Cook Strait. Although organisms do cross physical barriers, it is unlikely that all of these species made the same ocean crossing. In fact, we do not need to make that assumption. Geological evidence indicates that the present-day southwestern tip of North Island was formerly united with South Island. Therefore, none of the 60 species need have made a water crossing. A single vicariant event—the separation of the northern tip of South Island from the remainder of the island by the newly formed Cook Strait—could have split all of the distributions.

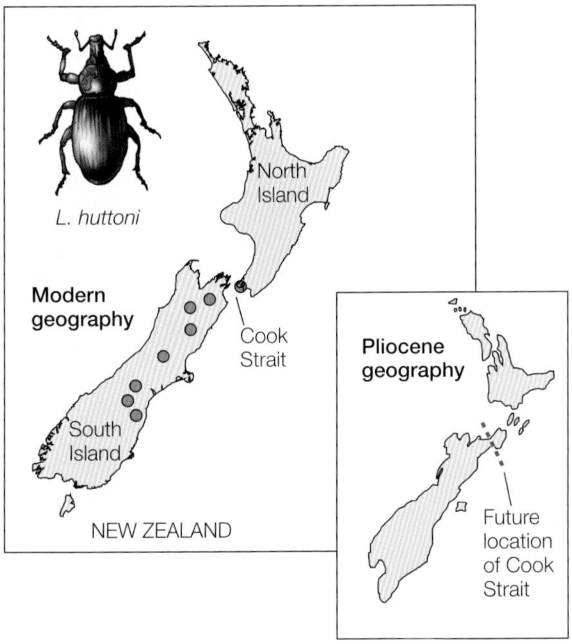

52.14 A Vicariant Distribution Explained Blue circles indicate the current distribution of the weevil *Lyperobius huttoni*. A comparison of the present New Zealand geography with that of the Pliocene, when the southern part of today's North island was part of South Island, suggests that a vicariant event—a physical split separating populations—explains this distribution.

52.4 RECAP

Biogeographic regions are based on the taxonomic similarities of the organisms living in them. Biotas of biogeographic regions differ because barriers restrict the dispersal of organisms between them.

- What determines the boundaries of Earth's major biogeographic regions? See p. 1128 and Figure 52.8
- Do you understand how the concepts of continental drift, phylogenetic taxonomy, and island biogeography have transformed and energized the field of biogeography? See pp. 1131–1133
- How do vicariance and dispersal interact to generate major biogeographic patterns? See pp. 1133–1135

Now that we have seen how species are distributed over Earth's land masses, let's take a look at the other three-quarters of the planet—the blue world of water.

52.5 How Is Life Distributed in Aquatic Environments?

Most of Earth's aquatic environment is saltwater oceans and seas. However, the small percentage of the watery world that constitutes freshwater lakes, rivers, ponds, and streams hosts a significant number of ecosystems and species.

Currents create biogeographic regions in the oceans

Earth's oceans form one large, interconnected water mass interrupted only partly by continents. Therefore, we might not expect to find biogeographic regions in the oceans, but we would be wrong. Currents generate striking physical and biological discontinuities, dividing the oceans into distinct regions.

Figure 52.4 depicted the great circular patterns of the world's ocean currents. Even organisms with limited swimming abilities can travel long distances simply by floating with the ocean currents. Nevertheless, most marine organisms have restricted ranges. Why is this true?

The oceans may be connected, but water temperatures, salinities, and food supplies all change spatially. Surviving under these varied conditions requires different physiological tolerances and morphological attributes. Changes in water temperatures, for example, can be barriers to dispersal because many marine organisms function well in only a relatively narrow range of temperatures. The main biogeographic divisions of the ocean coincide with regions where the surface water temperatures and salinities change relatively abruptly as a result of horizontal and vertical currents (**Figure 52.15**).

Primary food production by photosynthesis also varies across the oceans. Temperature changes, in combination with seasonal changes in the amount of sunlight, determine the seasons of maximum photosynthesis. Different species of marine algae photosyn-

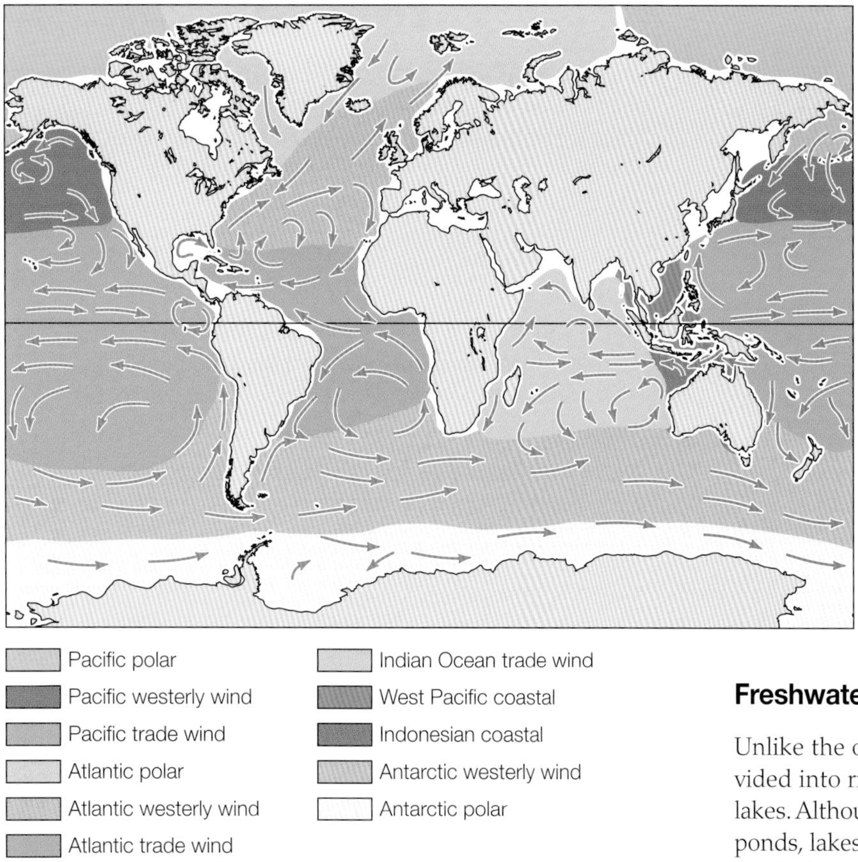

52.15 Oceanic Biogeographic Regions Are Determined by Ocean Currents The arrows represent ocean currents. Different biogeographic regions, in which photosynthesis is maximized at different seasons, are indicated by different colors.

thesize either in summer or in winter, but not during both seasons. Mangrove forests grow in estuaries in tropical regions, corals generate complex structures around tropical islands, and large algae form beds along many coasts. But over most of the oceans, where the dominant photosynthesizers are unicellular organisms, water temperature, salinity, and currents determine the physical environment to which organisms must adjust.

Deep ocean waters are barriers to the dispersal of marine organisms that live only in shallow water. The distance that ocean currents carry eggs and larvae of many marine organisms is determined in large part by the time it takes for larvae to metamorphose into sedentary adults. Relatively few species have eggs and larvae that survive long enough to disperse across wide areas of deep water. As a result, the richness of shallow-water species in the intertidal and subtidal zones of isolated islands in the Pacific Ocean decreases with distance from New Guinea (**Figure 52.16**).

Freshwater environments may be rich in species

Unlike the oceans, the world's freshwater environments are divided into river basins and into thousands of relatively isolated lakes. Although only about 2.5 percent of Earth's water is found in ponds, lakes, and streams, about 10 percent of all aquatic species live in these freshwater ecosystems. Prominent among freshwater taxa are the more than 25,000 species of insects that have at least one aquatic stage in their life cycle. Many freshwater insects are capable of dispersing across terrestrial barriers and are found over several continents. Typically, eggs and juveniles are aquatic; the adults have wings. Some of these insects, such as dragonflies, are powerful flyers, but mayflies and some other species are weak flyers, desiccate rapidly in air, and live no longer than a few days. As you would expect, oceanic islands have few, if any, species of these weak flyers.

Discontinuities in the ranges of aquatic organisms are also seen because most animals that live in the oceans cannot survive in fresh water, and vice versa. For example, most freshwater fishes are unable to live in salt water. They can disperse only within the connected rivers and lakes of a river basin or in brackish coastal waters. Most clades of freshwater fishes that cannot tolerate salt water are restricted to a single continent. Those groups with species distributed on both sides of major saltwater barriers are believed to be ancient clades whose ancestors were distributed widely in Laurasia or Gondwana, the two supercontinents that formed when Pangaea broke up some 250 million years ago (see Figure 21.17).

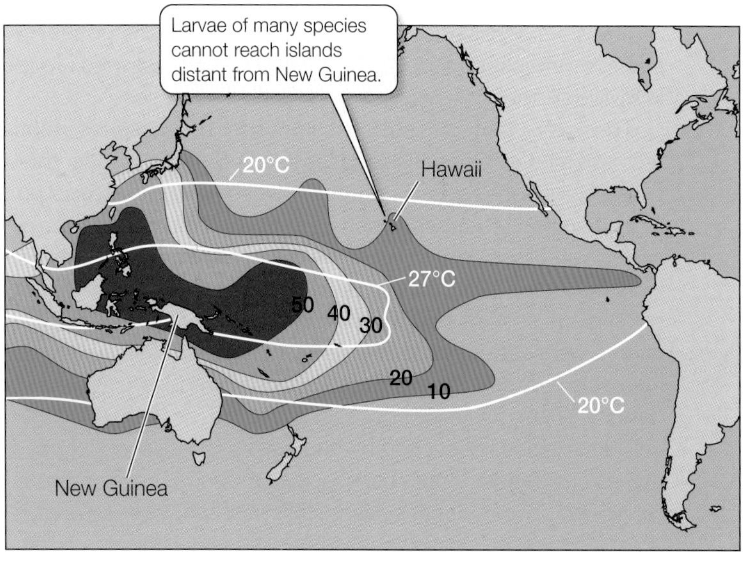

52.16 Generic Richness of Reef-Building Corals Declines with Distance from New Guinea The colored zones represent areas with equal numbers of coral genera. The 20°C and 27°C mean annual temperature isotherms are also shown.

52.5 RECAP

Despite being connected, the oceans are divided into distinct regions by currents that generate striking physical and biological discontinuities. Fresh waters, in contrast, are divided into river basins and thousands of relatively isolated lakes.

- Why do the main biogeographic divisions of the ocean coincide with regions where the surface water temperatures and salinities change relatively abruptly? See p. 1135 and Figure 52.15
- How do organisms that live in fresh water disperse among watersheds? See p. 1136

Now that we have described how the physical environment influences the distribution of species and ecosystems on Earth, we are ready to delve into the details of the interactions among organisms and their environments. In Chapter 53 we will consider the behavior of individual organisms and the choices they make that enable them to survive and reproduce in their environments. In Chapter 54 we will consider the dynamics of populations. In Chapter 55 we will discuss how populations of different species interact and how these interactions influence ecological communities. The final two chapters in the book recall global ecological patterns and emphasize the importance of ecology both to humanity and to our planet as a whole.

CHAPTER SUMMARY

52.1 What is ecology?

Ecology is the field of science that investigates interactions among organisms and between organisms and the physical environment.

An organism's **environment** encompasses both **abiotic** (physical and chemical) and **biotic** factors (other living organisms).

Ecologists study interactions on many levels, from the level of an individual organism's interactions with members of its own species to the functioning of **communities**, **ecosystems**, and the entire **biosphere**.

52.2 How are climates distributed on Earth?

See Web/CD Tutorial 52.1

The **climate** of a region is the average of the atmospheric conditions (temperature, precipitation, and wind direction and velocity) found there over time.

The rate at which solar energy arrives on Earth per unit of Earth's surface depends primarily on the angle at which sunlight arrives.

Mean annual air temperature decreases with both latitude and altitude.

Global atmospheric circulation patterns drive ocean currents and strongly influence global climates. Review Figures 52.2 and 52.4

Organisms respond behaviorally to short-term environmental changes.

Morphological and physiological features have evolved in organisms in response to long-term changes in the physical environment.

52.3 What is a biome?

See Web/CD Tutorial 52.2

Biomes are large ecological units that are classified by the structure of the dominant vegetation. Review Figure 52.5

Many biomes are recognized, including Arctic and alpine **tundra**, **boreal forest**, **temperate evergreen forest**, **temperate deciduous forest**, **temperate grassland**, **hot** and **cold deserts**, **chaparral**, **thorn forest** and **tropical savanna**, **tropical deciduous forest**, and **tropical evergreen forest**.

The distribution of biomes is determined primarily by climate, but other factors, such as soil fertiltity and fire, also influence patterns of vegetation.

52.4 What is a biogeographic region?

Biogeography is the scientific study of the distributional patterns of populations, species, and ecological communities. **Biogeographic regions** are based on the taxonomic similarities of the organisms living within them. Review Figure 52.8, Web/CD Activity 52.1

A species that is found only within a certain region is said to be **endemic** to that region.

Three scientific advances have played a role in explaining the distributions of species: the understanding of continental drift; the development of phylogenetic taxonomy; and the theory of island biogeography.

Area phylogenies are phylogenetic trees that show where species originated and where they now live. Review Figure 52.10

The theory of **island biogeography** predits an equlibrium number of species on an island based on rates of immigration of new species and rates of extinction of species already present. Review Figure 52.11, Web/CD Tutorial 52.3

Both vicariant events and dispersal generate biogeographic patterns. When formerly separated land masses come together, many species from both biotas may disperse into the other region, a phenomenon known as **biotic interchange**.

52.5 How is life distributed in aquatic environments?

The oceans are divided into distinct biogeographic regions by currents that generate striking physical and biological discontinuities. The main biogeographic divisions of the oceans coincide with regions where the surface water temperatures and salinities change relatively abruptly. Review Figure 52.15

Fresh waters are divided into river basins and thousands of relatively isolated lakes. Most organisms that live in fresh water cannot survive in the oceans, and vice versa.

SELF-QUIZ

1. Ecosystems are composed of
   a. all the organisms that live in an area.
   b. the organisms that most strongly influence the environment of other organisms in an area.
   c. all the organisms that live in an area plus the physical environment.
   d. the geology, soils, weather, and climate of an area.
   e. Different ecosystems have different components.
2. A biome is
   a. a large ecological unit based on the growth forms of the dominant plants.
   b. a large ecological unit based on the life forms of the dominant animals.
   c. a large ecological unit based on the regional climate.
   d. a large ecological unit based on both topography and climate.
   e. a large ecological unit based on biogeochemical cycles.
3. The rate at which solar energy arrives per unit of Earth's surface depends primarily on
   a. the angle of the sun's rays.
   b. the moisture content of the air.
   c. the amount of cloud cover.
   d. the strength of the winds.
   e. day length.
4. When a region lies under the intertropical convergence zone,
   a. it experiences unusually strong winds.
   b. it is in the middle of the dry season
   c. it is in the middle of the wet season.
   d. the winds are not unusually strong but they come from multiple directions.
   e. day lengths change rapidly.
5. Biogeography as a science began when
   a. nineteenth-century naturalists first noted intercontinental differences in the distributions of organisms.
   b. Europeans went to the Middle East during the Crusades.
   c. phylogenetic methods were developed.
   d. the fact of continental drift was accepted.
   e. Charles Darwin proposed the theory of natural selection.
6. Vicariant events
   a. are infrequent in nature.
   b. were common in the past but are rare today.
   c. separate species ranges in the absence of dispersal.
   d. were rare in the past but are common today.
   e. caused most of today's discontinuous distributions.
7. Marine biogeographic regions exist even though the oceans are all connected because
   a. the rate of photosynthesis is low in the oceans.
   b. ocean currents keeps organisms close to where they were born.
   c. most taxa of marine organisms evolved before the oceans were separated by continental drift.
   d. water temperatures and salinities often change abruptly where ocean currents meet.
   e. oceanic circulation is too slow to carry marine organisms from one ocean to another.
8. A parsimonious interpretation of a distribution pattern is one that
   a. requires the smallest number of undocumented vicariant events.
   b. requires the smallest number of undocumented dispersal events.
   c. requires the smallest total number of undocumented vicariant plus dispersal events.
   d. accords with the phylogeny of a lineage.
   e. accounts for centers of endemism.
9. The only major biogeographic region that today is completely isolated by water from other biogeographic regions is
   a. Greenland.
   b. Africa.
   c. South America.
   d. Australasia.
   e. North America.
10. According to the theory of island biogeography, equilibrium species richness is reached when
    a. immigration rates of new species and extinction rates of species are equal.
    b. immigration rates of all species and extinction rates of species are equal.
    c. the rate of vicariant events equals the rate of dispersal.
    d. the rate of island formation equals the rate of island loss.
    e. No equilibrium number of species exists according to the theory.

FOR DISCUSSION

1. Horses evolved in North America, but subsequently became extinct there. They survived to modern times only in Africa and Asia. In the absence of a fossil record, we would probably infer that horses originated in the Old World. Today, the Hawaiian Islands have by far the greatest number of species of fruit flies (*Drosophila*). Would you conclude that the genus *Drosophila* originally evolved in Hawaii and spread to other regions? Under what circumstances do you think it is safe to conclude that a group of organisms evolved close to where the greatest number of species live today?
2. Processes in nature do not always conform to the parsimony principle. Why, then, do biogeographers often use the parsimony principle to infer geographic histories of species and lineages?
3. A well-known legend states that Saint Patrick drove the snakes out of Ireland. Give some alternative explanations, based on sound biogeographic principles, for the absence of indigenous snakes in that country.
4. Most of the world's flightless birds are either nocturnal and secretive (such as the kiwi of New Zealand) or large, swift, and powerful (such as the ostrich of Africa). The exceptions are found primarily on islands. Many of these island species have become extinct with the arrival of humans and their domestic animals. What special conditions on islands might permit the survival of flightless birds? Why has human colonization so often resulted in the extinction of such birds? The power of flight has been lost secondarily in representatives of many groups of birds and insects; what are some possible evolutionary advantages of flightlessness that might offset its obvious disadvantages?
5. MacArthur and Wilson's theory of island biogeography incorporates almost nothing about the biology of species. What traits of species should be incorporated into more realistic models of rates of colonization and extinction of species on islands?
6. A legislator introduces a controversial bill into the U.S. Congress that would ban all introductions of exotic species to the Hawaiian Islands. Would you vote in favor of this bill if you were in Congress? Why or why not?

FOR INVESTIGATION

Alfred Russel Wallace observed that dramatically different birds inhabited Bali and Lombok, even though the strait that separates them is only 15 miles wide. But most birds can easily fly 15 miles, and can probably see the other island across the water. What kinds of data, in addition to those gathered by Wallace, could be obtained to determine why birds do not fly across the strait or, if indeed they do cross the water, why they fail to colonize the island on the other side?

CHAPTER 53 Behavior and Behavioral Ecology

Monkey see, monkey do

Scientists studying a troop of Japanese macaques often fed the monkeys by throwing pieces of sweet potato onto the beach from a passing boat. The monkeys tried to brush the sand off the potato pieces, but they were still gritty. One day a young female macaque took her sweet potatoes to the water and began washing them. Soon her siblings and other juveniles in her playgroup were imitating her new behavior. Next their mothers began washing their potatoes. None of the adult males imitated this behavior, but young males learned the behavior from their mothers and their siblings.

The scientists were fascinated by the way in which the creative, insightful behavior of one juvenile female spread throughout the population, so they presented the monkeys with a new challenge: they threw grains of wheat onto the beach. Picking the wheat out of the sand was tedious and difficult. The same juvenile female came up with a solution: she carried handfuls of sand and wheat to the water and threw them in. The sand sank, but the wheat floated, enabling her to skim the grain off the surface and eat it. This behavior spread throughout the population just as washing sweet potatoes had—first to other juveniles, then to mothers, and then from mothers to their male and female offspring, but not to adult males.

The macaques of that troop now routinely wash their food. They play in the water, which they never did before, and they have added some marine food items to their diet. Clearly, this population of monkeys has acquired a *culture*: a set of behaviors shared by members of the population and transmitted through learning.

Animals also display many elaborate behaviors that they do not have to learn. Web spinning by spiders, for example, requires no prior experience. In many species of spiders, when juvenile spiders hatch, their mother is already dead (remember *Charlotte's Web* by E. B. White?). They have no experience with their mother's web. Yet when they construct their own web, they do it perfectly, without the benefit of experience or a model to copy. When generations do not overlap, we can rule out parental guidance as a factor in the acquisition of a behavior.

Web spinning requires thousands of movements performed in just the right sequence. For any given spider species, most of that sequence is *stereotypic*—

Shared Learned Behaviors Become a Culture In the space of a single generation, a population of Japanese macaques (*Macaca fuscata*) learned and transmitted a set of behaviors that included washing food, playing in the water, and eating marine food items—a new "culture" of water-related behaviors.

Spiders Are Born Web Designers Each spider performs a stereotypic sequence of movements that results in a species-specific web design. Spiders such as this orb weaver are born with this ability, with no need to learn from experience or to model their movements after those of a parent.

CHAPTER OUTLINE

performed almost exactly the same way every time. Different spider species spin webs of different designs, using different sequences of movements. Yet every spider knows how to spin its web at birth; the behavior is part of its genetic inheritance.

The relationship between learned and inherited behavior has been a fascinating one for psychologists, sociologists, and philosophers as well as for biologists. In addition, the behavioral responses of individuals, both learned and inherited, are the foundation of much of ecology. The interactions, densities, and distributions of populations, and the effects of these population interactions on ecosystems, may all change as a result of the behavioral responses of individual animals.

IN THIS CHAPTER we will see how biologists identify the hereditary and experiential underpinnings of behavior. We will consider how genes and environment interact to shape the development of both the behavior of individuals and the long-term evolution of behavior. We will discuss several types of animal behaviors: how animals respond to changes in the environment, decide where to carry out their activities, select the resources they need (food, water, shelter, nest sites), respond to predators and competitors, and associate with other members of their own species.

53.1 What Questions Do Biologists Ask about Behavior?

Niko Tinbergen, one of the founders of **ethology**—the study of animal behavior from an evolutionary perspective—pointed out that to gain a complete understanding of any form of behavior, we need to ask several questions about it. Some questions focus on describing behavior patterns and how and when individuals perform them. They also address the *proximate* mechanisms that underlie behavior—the neuronal, hormonal, and anatomical mechanisms that we described in Part Eight of this book. Other questions concern the origins and means of acquisition of specific behaviors, especially on the relative roles of genes and experience. Most behaviors result from complex interactions between inherited anatomical and physiological mechanisms and the ability to modify behavior as a result of experience.

Other questions concern the *ultimate* causes of behavior—the selection pressures that shaped its evolution. In what ways does the behavior contribute to the fitness of the organism that possesses it (see Section 22.3)?

The behavior of the macaques described at the beginning of this chapter stimulates many questions. How did one young female macaque first think of washing sweet potatoes? Did the innovative young female differ genetically from the other monkeys, and if so, how was she different? Why were females more likely than males to learn the behavior? Why didn't any adult males learn the new behavior? The observations also stimulate questions about how changes in behavior influence the functioning of populations. How do animals incorporate new foods into their diets? How do individuals benefit from being members of social groups? How do differences in the benefits individuals receive from group membership influence how groups form and why they remain cohesive?

For many animals, much of their behavior is like the web-spinning behavior of spiders—unlearned and highly stereotypic (i.e., always exactly the same). Stereotypic behavior is often *species-specific,* in that most individuals of a given species perform the behavior in the same way. We can identify different species of spiders by the patterns of their webs. Why are the patterns so different? We know that a spider's genes code for the proteins of the silk from which its web is spun; but how do genes encode the limb movements necessary to produce the web? The questions raised by animal behavior are endless and fascinating.

53.1 RECAP

The study of animal behavior focuses on descriptive, mechanistic, and ultimate questions about behavior.

- Do you understand the difference between proximate and ultimate causes of a behavior? See p. 1141
- Can you explain what we mean when we say that a behavior is stereotypic and species-specific? See p. 1141

Distinguishing the various genetic and environmental influences on behavior is not always straightforward. The next section introduces some experimental techniques that can help us make this distinction. Then we'll investigate how inheritance and learning interact to shape certain behaviors.

53.2 How Do Genes and Environment Interact to Shape Behavior?

The observation that a given behavior is stereotypic and requires little or no learning tells biologists little about the relative roles of genes and experience in the development of behavior. An animal may fail to perform even a "genetically controlled behavior" if the environmental conditions needed to stimulate it are absent. Conversely, all individuals may behave in the same way not because of their genes but because they all imitated the same teacher.

Genes do not encode behaviors. Rather, gene products such as enzymes can affect behavior by setting in motion a series of gene-environment interactions that underlie the development of proximate mechanisms that enable individuals to make certain behavioral responses. We next describe some of the methods biologists use to determine how genes affect behavior and how genes and experience interact to influence how behaviors develop and when they are performed.

Experiments can distinguish between genetic and environmental influences on behavior

Experiments that either carefully control the environment to eliminate opportunities for learning or that modify the organism's genome can help us distinguish between genetic and environmental influences on the development of behavior. Two experimental approaches are especially useful to biologists in evaluating how genes and experience interact to shape behaviors:

- In a *deprivation experiment*, investigators rear a young animal so that it is deprived of all experience relevant to the behavior under study. If it still exhibits the behavior, we may assume that the behavior can develop without opportunities to learn it.
- In *genetic experiments*, investigators alter the genomes of organisms by interbreeding closely related species, by comparing individuals that differ in only one or a few genes, or by knocking out or inserting specific genes to determine how these manipulations affect their behavior.

DEPRIVATION EXPERIMENTS A simple deprivation experiment can yield useful information. A newborn tree squirrel was reared in isolation, on a liquid diet, and in a cage without soil or other particulate matter. When the juvenile squirrel was given a nut, it put the nut in its mouth and ran around the cage. Eventually it made stereotypic digging movements in the corner of its cage, placed the nut in the imaginary hole, went through the motions of refilling the hole, and ended by tamping the nonexistent soil with its nose. The squirrel had never handled a food object and had never experienced soil, yet it fully expressed the stereotypic behavior of a squirrel burying a nut. This experiment showed that heredity underlies the food-storing behavior of this tree squirrel species, but the behavior was expressed only when the environment provided conditions that stimulated the behavior (the presence of a nut).

SELECTIVE BREEDING *Selective breeding* is a means of genetic manipulation that has been in use since plants and animals were first domesticated. Selective breeding has been used extensively to select for both anatomical traits and behavior, as can be seen in many breeds of dogs, such as retrievers, pointers, shepherds, and bloodhounds. Thus, although its approach is applied and not based in theoretical science, it does provide insights about the effect genetic constitution has on behavior.

53.1 The Mallard Courtship Display The courtship display of the male mallard duck contains ten elements. The displays of closely related duck species contain some of the same ten elements, but have other elements not displayed by mallards. The elements of the courtship display and their sequence are species-specific and act to discourage mating between species.

(A)

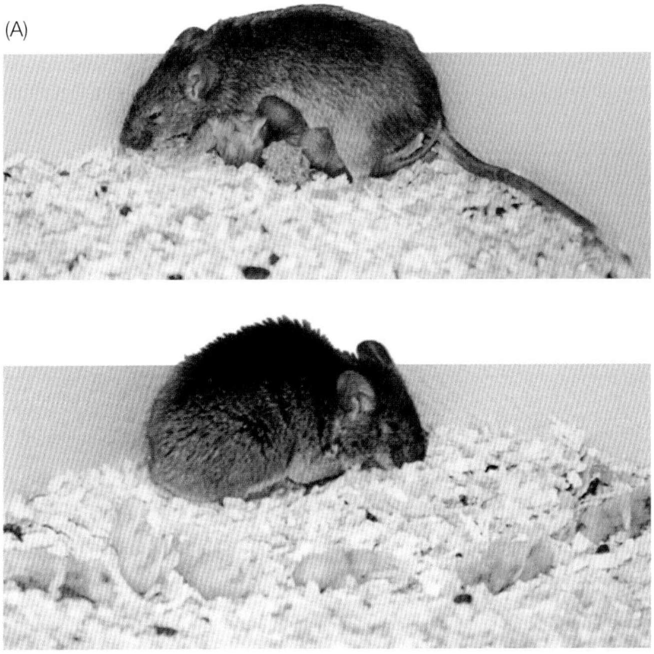

(B)

EXPERIMENT

HYPOTHESIS: Absence of the *fosB* gene changes the maternal behavior of female mice.

METHOD

1. Inactivate ("knock out") the *fosB* gene in a strain of female mice. When mature, allow them to mate and give birth at the same time wild-type females of the same strain do so.
2. Immediately upon birth, separate mouse pups from mothers. Place three newborn pups in opposite corners of a cage with their mother.
3. Count number of pups each mother retrieves within 20 minutes.

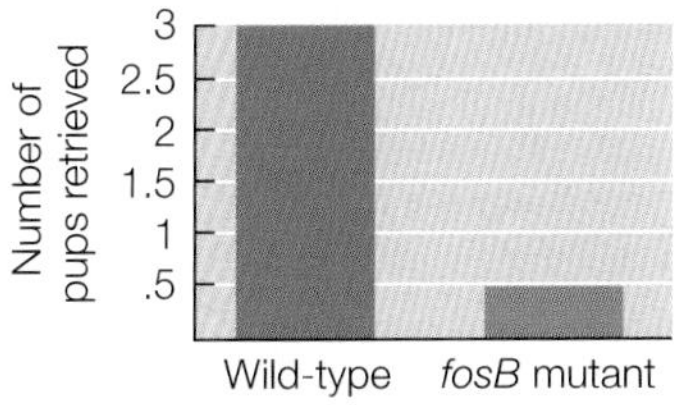

RESULTS

Normal females retrieved all of their pups within 20 minutes, whereas *fosB* mutant females retrieved an average of only 0.5 pups.

CONCLUSION: The protein product of the *fosB* gene is needed in order for female mice to display normal maternal behavior.

53.2 A Single Gene Affects Maternal Behavior in Mice (A) A wild-type female mouse gathers her pups together and crouches over them (top), but a female lacking the *fosB* gene product (bottom) does not exhibit these behaviors; her pups can be seen scattered in the foreground of the photograph. (B) An experiment confirms that *fosB* knockout mice do not exhibit a basic maternal behavior.

INTERBREEDING Konrad Lorenz, one of the pioneers in ethology, used interbreeding (hybridization) of duck species to investigate the hereditary basis of their elaborate courtship displays. Some duck species, such as mallards, teals, pintails, and gadwalls, are closely related to one another and can interbreed, but they rarely do so in nature. Each male duck performs a carefully choreographed water ballet that is typical of his species (**Figure 53.1**). A female is not likely to accept his advances unless the entire display is successfully and correctly completed, and she uses his skill in performing the display to judge his quality.

When Lorenz crossbred these duck species, the hybrid offspring expressed some elements of each parent's courtship display, but in new combinations. Furthermore, Lorenz observed that the hybrids sometimes exhibited display elements that were not in the repertoire of either parent species, but were characteristic of other species. Lorenz's hybridization studies clearly demonstrated that the stereotypic motor patterns of the courtship displays are inherited. The observation that females are not interested in males performing hybrid displays is evidence that *sexual selection* (see Section 22.3) has shaped these genetically determined behaviors.

GENE KNOCKOUT EXPERIMENTS Gene mutations that affect behavior have been studied in fruit flies and rodents using gene knockout and gene silencing techniques such as those described in Section 16.5. One striking example of the influence of a single gene mutation on a complex behavior is provided by experiments with house mice.

Female mice in which the *fosB* gene is active gather their pups together, keep them warm, and nurse them. Females carrying a mutation that inactivates ("knocks out") the *fosB* gene appear normal, but they differ from other mice in the way they treat their newborn pups. After giving birth, they inspect their pups but then ignore them.

Investigators displaced pups from the litters of both normal and *fosB* mutant females and recorded how many pups each mother retrieved. Normal females retrieved all of their pups within 20 minutes, whereas mutant females almost never retrieved their pups (**Figure 53.2**).

How does the *fosB* gene influence maternal behavior? The answer is not definite, but it would seem that the protein encoded by *fosB* is involved in stimulating neural changes in the hypothalamus of the mother's brain, possibly in response to olfactory (odor) molecules she encounters upon her initial inspection of the pups. These altered neural connections apparently play a part in motivating the mother to retrieve and care for her pups; the neural changes do not occur if the *fosB* gene is inactivated.

Genetic control of behavior is adaptive under many conditions

As we saw at the opening of this chapter, the ability to learn and to modify behavior as a result of experience can be highly adaptive. So why are behavior patterns in many species so strongly influenced by genes? One answer to this question has already been pointed out. Without role models and opportunities for learning—as in species with nonoverlapping generations—individuals might fail to acquire the appropriate behavior, or acquire inappropriate behavior, if genes did not exert strong influences on the development of the behavior.

Inherited behavior is also adaptive when mistakes are costly or dangerous. Mating with a member of the wrong species is a costly mistake, and in an environment in which incorrect as well as correct models exist, learning the wrong pattern of courtship behavior would be possible (see the story of whooping cranes and sandhill cranes at the start of Chapter 57).

Inheritance of behavior patterns used to avoid predators or capture dangerous prey is obviously adaptive. These situations allow no room for mistakes: if the behavior is not performed promptly and accurately the first time, the animal will probably not get a second chance.

Kangaroo rats are small, nocturnal desert rodents with powerful jumping legs. They can avoid rattlesnakes in total darkness because as soon as a rattlesnake begins to strike, they hear it moving through the air and jump out of the way. This jumping behavior does not need to be learned.

Even if it is not expressed during a deprivation experiment or in genetically modified individuals, a behavior may nonetheless be genetically influenced. Some behaviors are expressed only under certain conditions. The deprived squirrel described earlier, for example, did not exhibit digging and burying behaviors until it had been given a nut; that is, the nut triggered the behavior. The nut acts as a **releaser**—an object, event, or condition required to elicit a behavior. The sight of red feathers on the breasts of adult male European robins, for example, may release aggressive behavior in other males. During the breeding season, when males are defending breeding territories, the sight of an adult male robin stimulates another male robin to sing, perform aggressive displays, and attack the intruder. An immature male robin, whose feathers are brown, does not elicit this aggressive behavior. A tuft of red feathers on a stick, however, is a sufficient releaser for male aggressive behavior, even though it looks nothing like a real robin. The response to the releaser may depend on the motivational state of the animal; a male robin does not respond to red feathers when he is not on his territory and in breeding condition.

Even though genetic control of behavior is adaptive under many conditions, complete stereotypy may not be. Thus spiders adjust some details of their webs to accommodate the geometry of the structures to which they anchor their webs. As discussed in Section 20.4, the genes that govern development often allow organ-

53.3 Spatial Learning by a Wasp Tinbergen's classic experiment showed that a female digger wasp learns the positions of objects in her environment.

isms to adjust their forms and behavior to the particular environment in which they develop.

For example, many animals select nesting sites with specific features, but no two sites are identical. To relocate its nest, an individual must learn and remember the particular features around it. By placing pine cones near the entrance of a female digger wasp's nest, Niko Tinbergen demonstrated that the wasp used features in her environment to learn the location of her nest. After the wasp left her nest, he moved the cones a short distance away. Upon returning, the wasp oriented to the displaced cones and could not find her nest entrance (**Figure 53.3**).

Imprinting takes place at a specific point in development

Some types of learning take place only at a specific time in the animal's development, called a **critical period**. In a type of learning called **imprinting**, an animal learns a set of stimuli during a limited critical period. The classic example of a behavior learned by imprinting is the recognition of offspring by their parents and of parents by their offspring, which is essential if parents and their young are to find one another in a crowded situation such as a colony or a herd. Individual recognition must often be learned quickly because the opportunity to do so may arise only once. Konrad Lorenz demonstrated that young goslings imprint on their parents at birth by making sure that he was the first thing they saw upon hatching. They imprinted on him and followed him around thereafter as if he were their parent goose (**Figure 53.4A**).

Imprinting requires only a brief exposure, but its effects are strong and can last a long time. Emperor penguins reproduce during the coldest, darkest time of the year in Antarctica. The parents walk up to 150 kilometers inland to form a dense colony where a female lays her egg. Her mate incubates the egg while the mother walks back to the ocean to feed. By the time she returns, the chick has hatched. She takes over its care and feeding, and the father walks back to the ocean to feed. Before he gets back, the mother must leave to avoid starvation. Thus, after being away for weeks, the father has to find his chick in a crowded, milling colony of chicks, all calling for their parents (**Figure 53.4B**). Yet he finds it by recognizing its call, which he learned before he left to feed.

The critical period for imprinting may be determined by a brief developmental or hormonal state. For example, if a mother goat does not nuzzle and lick her newborn within 10 minutes after its birth, she will not recognize it as her own offspring later. The high levels of the hormone oxytocin in the mother's circulatory system at the time she gives birth and olfactory cues from the newborn kid determine the critical period.

Some behaviors result from intricate interactions between inheritance and learning

Male songbirds use a species-specific song in territorial displays and courtship. For many species, such as the white-crowned sparrow, learning is an essential step in the acquisition of this song. If eggs of white-crowned sparrows are hatched in an incubator and the young male birds are reared in isolation, they will not sing the typical species-specific song when they mature. White-crowned sparrows cannot produce their species-specific song unless they hear it as nestlings (**Figure 53.5**). But even though a male white-crowned sparrow must hear the song of his own species as a nestling to sing it as an adult, he does not sing for almost a year. Instead, the sounds he heard as a nestling form a song memory in his nervous system.

As the young male sparrow approaches sexual maturity the following spring, he begins to sing. Eventually he matches his stored song memory through trial and error. If a bird that has heard his species-specific song as a nestling is deafened before he begins to sing, he will not develop his species-specific song (Experiment 2 in Figure 53.5). He must be able to hear himself to match his song memory. If he is deafened *after* he sings his correct species-specific song, however, he will continue to sing like a normal bird. Thus there are two critical periods for song learning: the first in the nestling stage, the second as the bird approaches sexual maturity.

Although a nestling male white-crowned sparrow in the wild is likely to hear the songs of his father and neighboring male white-crowned sparrows, he is also likely to hear the songs of many other species of birds. Deprivation experiments have demonstrated that young male sparrows do not learn the songs of other

(A)

(B) *Aptenodytes forsteri*

53.4 Imprinting Helps Parents and Offspring Recognize One Another (A) Graylag geese that imprinted on Konrad Lorenz as hatchlings followed him everywhere he went. (B) A male emperor penguin must quickly find his own chick among many others.

EXPERIMENT 1

HYPOTHESIS: Learning is essential for song acquisition in white-crowned sparrows.

METHOD

Raise young sparrows in the presence of an adult male sparrow singing. Record song of these control birds when they mature and plot as a sonogram.

RESULTS

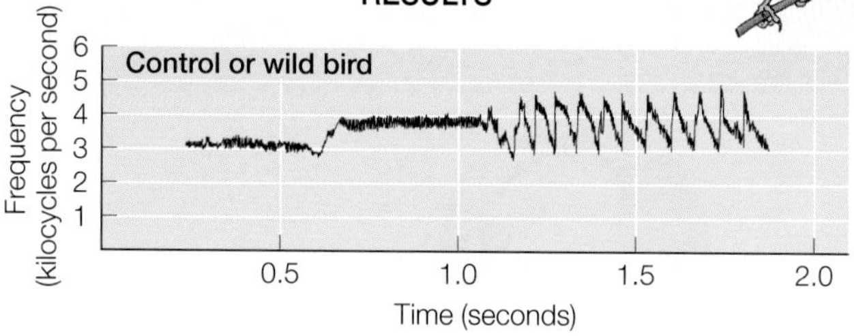

METHOD

Hatch eggs in an incubator and rear the birds in isolation. Record and plot their song. Compare to the control birds' song.

RESULTS

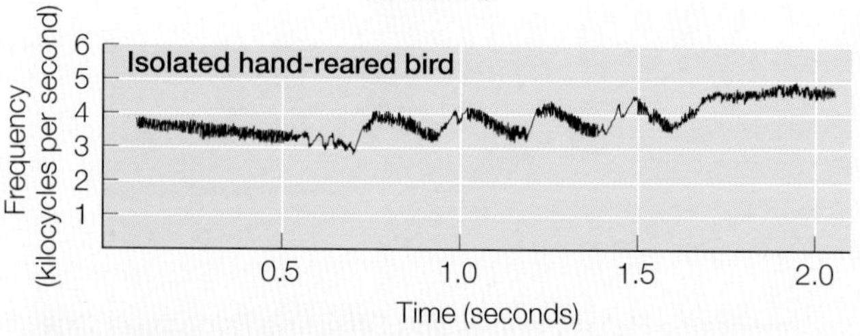

CONCLUSION: White-crowned sparrows that do not hear adult song as nestlings do not express their species-specific song when they mature.

EXPERIMENT 2

HYPOTHESIS: Maturing white-crowned sparrows require auditory feedback to learn to express their species-specific song.

METHOD

Deafen subadult birds that have heard the song of adult males when they were nestlings. Record their song and plot as a sonogram.

RESULTS

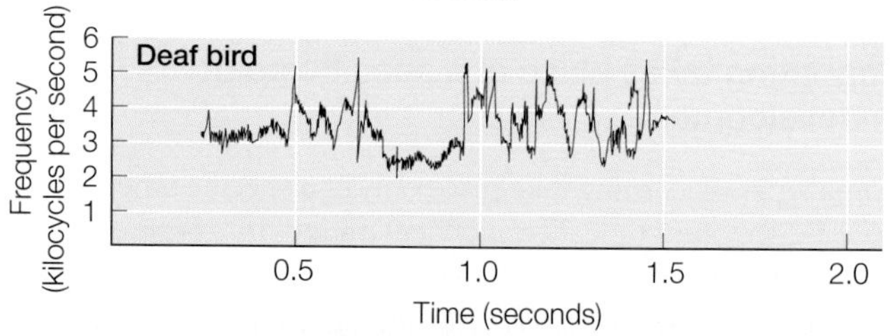

CONCLUSION: Even if white-crowned sparrows have formed a song memory, they need auditory feedback to learn to match it.

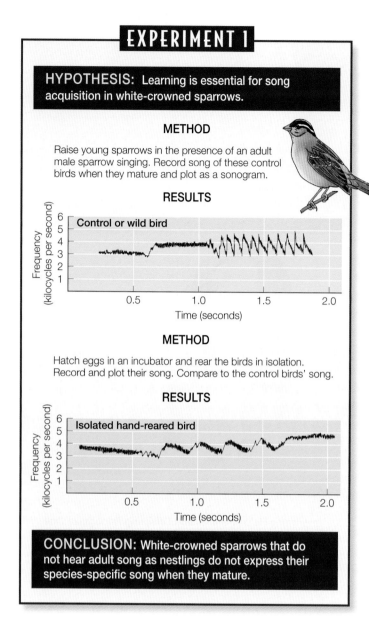

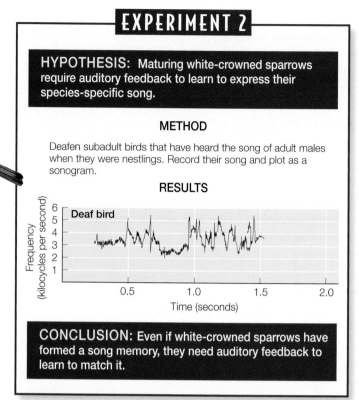

53.5 Two Critical Periods for Song Learning To sing his species-specific song as an adult, a male white-crowned sparrow must acquire a song memory by hearing the song as a nestling, and must be able to hear himself as he attempts to match his singing to that song memory.

species, even if they hear them many times, but exposure to just a few songs of their own species is sufficient for imprinting. Thus, although the sparrows must learn their songs, their genes make it very difficult for them to learn the songs of other species.

Hormones influence behavior at genetically determined times

In multicellular organisms, all behavior depends on the nervous system for initiation, coordination, and execution. Frequently, however, the hormones of the endocrine system, through controlling influences on development and on the physiological state of the animal (see Chapter 41), determine when a particular behavior is performed, as well as when certain behaviors can be learned.

We have already seen that learning is essential for the acquisition of song by some birds. Both male and female songbirds hear their species-specific song as nestlings, but only the males of most migratory species sing as adults. Male birds use song to claim and advertise a breeding territory, compete with other males, and declare dominance. They also use song to attract females, which suggests that the females recognize the song of their species even if they do not sing it. What is responsible for this difference in the learning and expression of song in male and female songbirds?

After leaving the nest where they heard their fathers' or neighbors' songs, young songbirds from temperate and Arctic habitats migrate and associate with other species in mixed-species flocks. During this time they do not sing, and they do not hear their species-specific song again until the following spring. As spring approaches and the days become longer, the young male's testes begin to grow and mature. As his testosterone level rises, he begins to sing, matching his own vocalizations to his imprinted memory of his father's song.

Why don't the females of these species sing? Can't they learn the vocalization patterns of their species-specific song? Do they lack the muscular or nervous system capabilities necessary to sing? Or do they simply lack the hormonal stimulus for developing the behavior? To answer these questions, investigators injected female songbirds with testosterone in the spring. In response to these injections, the females developed their species-specific song and sang just as the males did. Apparently females form a memory of their species-specific song when they are nestlings and have the capability to express it, but they normally lack the hormonal stimulation.

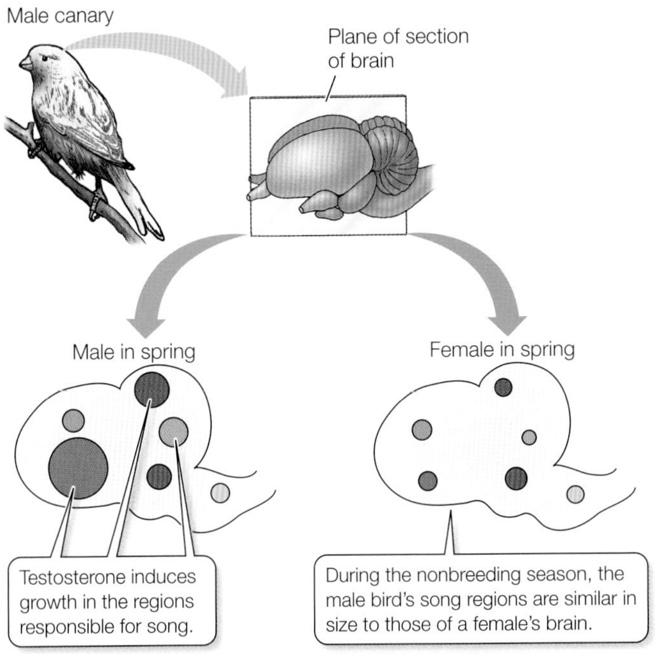

53.6 Effects of Testosterone on Bird Brains In spring, rising testosterone levels in the male cause the song regions of the brain to develop. The size of each circle is proportional to the volume of the brain occupied by that region.

How does testosterone cause a male songbird to sing? A remarkable study revealed that each spring, an increase in testosterone causes certain parts of the male's brain necessary for learning and developing song to grow larger (**Figure 53.6**). Individual neurons in those regions of the brain increase in size and grow longer extensions, and the number of neurons in those regions increases. Such research on the neurobiology of bird song has revealed that, contrary to the widely held belief that new neurons are not produced in the brains of adult vertebrates, hormones can control behavior by changing brain structure as well as brain function, both developmentally and seasonally.

53.2 RECAP

Some behaviors are strongly influenced by genes, resulting in highly stereotypic behavior patterns. Genes also play a role in shaping what individuals learn. Much adaptive behavior results from interactions between heredity and learning, often modified by hormones.

- What kinds of experiments can show whether a behavior is strongly influenced by genes? See pp. 1142–1143 and Figure 53.2
- What adaptive advantages does imprinting offer? See p. 1145
- Can you explain how inheritance and learning interact when songbirds learn their species-specific songs? See pp. 1145–1146 and Figure 53.5

Let's look now at the many decisions about how and when to behave in a certain way that animals must make during their lives, and that influence their fitness.

53.3 How Do Behavioral Responses to the Environment Influence Fitness?

Any animal that reaches old age has made decisions throughout its life as it grew to maturity, reproduced, and experienced the effects of aging. It is likely to have chosen where to settle, how long to stay there, and when, if ever, to leave. It will also have found its way around the environment and selected places for specific activities, such as resting and nesting. It will have decided which things to eat from among the rich array of potential food sources in its immediate environment. Most animals also choose with whom to associate and for what purposes.

All of these decisions are made in an environment that varies dramatically in space and in time. For example, in the Sierra Nevada of California, you can drive from a hot, dry sagebrush desert to alpine snowfields in a half hour. In the woodlands of Pennsylvania it can be hot and humid in the summer and below freezing in the winter. In the marine environment, only a few kilometers may separate the rocky intertidal zone from the dark, deep ocean floor. Given this great variation, how do animals find a good place to live and decide where to go if the local environment deteriorates?

Choosing where to live influences survival and reproductive success

Selecting a place in which to live is one of the most important decisions an individual makes. The environment in which an organism lives is called its **habitat**. Once a habitat is chosen, an animal seeks its food, resting places, nest sites, and escape routes within that habitat. A challenge to behavioral ecologists is to discover what information animals use to make their habitat choices, and how that information relates to environmental qualities that influence their fitness. Some of the ways in which organisms make their choices are surprisingly simple. The cues most organisms use to select suitable habitats have a common feature: *they are good predictors of conditions suitable for future survival and reproduction*.

The red abalone, a marine mollusk, sticks to rocks and does not move very far as an adult. Abalones, however, shed their eggs and sperm into the water. The fertilized eggs and the larvae that hatch from them drift and swim in the open ocean for about 8 days before they drop to the ocean floor, choose places to settle, and develop into adults. The best place for a red abalone is on coralline algae, its major source of food. Red abalone larvae recognize coralline algae by a chemical the algae produce. In the laboratory, abalone larvae will settle on any surface on which these molecules have been placed, but in nature, only coralline algae produce them. By using this simple chemosensory cue, the larvae always settle on a surface that has the potential of supplying the food necessary for their future survival and reproductive success.

Visual information can also provide useful cues about the quality of a habitat. Many animals use the presence of already settled individuals as an indication that the habitat may be good. While collared flycatchers are on their breeding grounds in spring, they regularly peer into the nests of other individuals. Observing this behavior, the Swedish and French biologists Blandine Doligez, Etienne Danchin, and Jean Clobert hypothesized that the flycatchers were assessing the quality of the habitat by seeing how well their neighbors were doing. To test this hypothesis, they created some areas with super-sized broods—normally an indication of abundant food—by taking young birds from some nests and adding them to nests in another area. The next year, flycatchers preferentially settled in the areas where broods had been artificially enlarged (**Figure 53.7**).

Some highly social animals actually "vote" on the quality of habitats. For example, honeybees, live in large colonial nests that consist of vertical sheets of hexagonal cells that they build out of wax. Some of the cells are used to raise larvae, some are used to store honey, and some are used to store pollen, a source of protein. When a bee colony outgrows its hive space, the queen leaves it, along with a huge retinue of worker bees—the whole group is called a *swarm*. The queen is not a very good flyer, so the swarm does not go far before coming to rest as a huge mass of bees tightly clustered around the queen. From this cluster, individual worker bees fly out to search for a cavity that would be a good site for a new colony. When a worker (a scout) finds a cavity, she explores it by walking around inside. She then returns to the swarm and dances on its surface. Her dance conveys information to other bees in the swarm about the distance and direction to the potential nest site she has discovered (we will see how she does so in Section 53.4). The vigor of her dance reflects the quality of the nest site. In response to her dance, other worker bees fly to the site and assess it. When they return to the swarm, they also dance to communicate information about the site and recruit more workers to visit it. Different workers typically provide information to the swarm about several potential sites at the same time. Eventually the site that excites the most workers is chosen, and the swarm takes flight again to move to the new location.

EXPERIMENT

HYPOTHESIS: Collared flycatchers use information on the reproductive success of already established individuals to decide where to settle.

METHOD

Transfer nestlings from nests in one area to nests in another area to create an area with super-sized broods and another area with under-sized broods. Leave a third area unaltered to serve as a control. Observe settlement the following year.

RESULTS

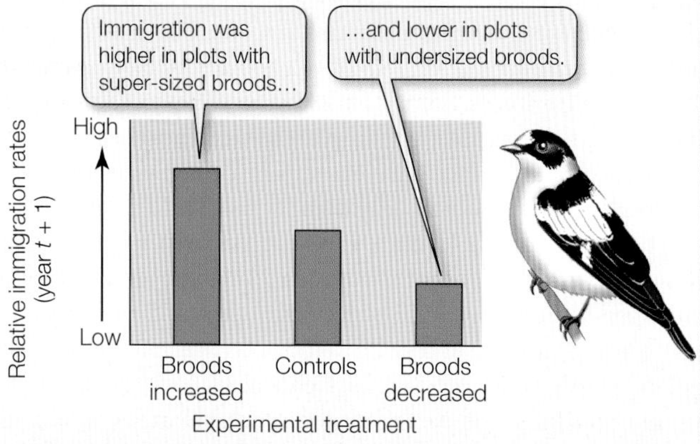

CONCLUSION: Collared flycatchers use brood size of already settled individuals to assess how good an area is for breeding.

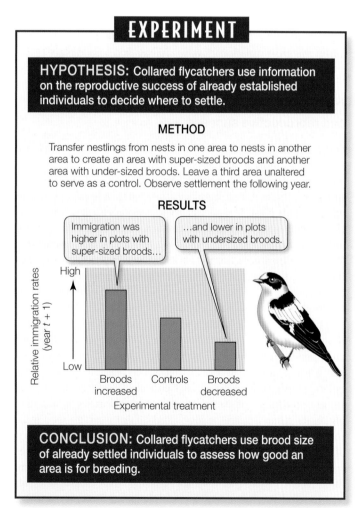

53.7 Flycatchers Use Neighbors' Success to Assess Habitat Quality Collared flycatchers (*Ficedula albicollis*) settled at higher densities in areas where experimenters had artificially enlarged the broods of other flycatchers.

Defending a territory has benefits and costs

When high-quality habitats are in limited supply, animals may compete for access to them. Under these conditions, an animal may improve its fitness by establishing exclusive use of its chosen habitat. The most common way of doing so is to establish a **territory** from which it excludes *conspecifics* (other individuals of the same species)—and sometimes individuals of other species as well—by advertising that it owns the area and, if necessary, chasing others away. But advertising and chasing take time and energy that could have been used for other beneficial purposes, such as finding food and watching out for predators.

To understand the evolution of behaviors that involve these kinds of trade-offs, ecologists often use a cost–benefit approach. A **cost–benefit approach** assumes that an animal has only a limited amount of time and energy to devote to its activities. Animals cannot long perform behaviors whose total costs are greater than the sum of their benefits. A cost–benefit approach provides a framework that behavioral ecologists can use to make predictions, design experiments, and make observations that can explain why behavior patterns evolve as they do.

The benefits of a behavior are the improvements in survival and reproductive success that the animal achieves by performing the behavior. The total cost of any particular behavior typically has three components:

- **Energetic cost** is the difference between the energy the animal would have expended had it rested and the energy expended in performing the behavior.
- **Risk cost** is the increased chance of being injured or killed as a result of performing the behavior, compared with resting.
- **Opportunity cost** is the sum of the benefits the animal forfeits by not being able to perform other behaviors during the same time interval. An animal that devotes all of its time to foraging, for example, cannot achieve high reproductive success!

Michael Moore and Catherine Marler at Arizona State University performed an experiment to estimate the costs incurred by male lizards when defending a territory. Male Yarrow's spiny lizards defend territories that include the habitats of several females, from which they exclude conspecific males. They normally do so most vigorously during September and October, when females are most receptive to mating. To assess the costs of territorial behavior, the researchers inserted small capsules containing testosterone, a hormone that the lizards normally produce in the fall and that induces aggressive territorial behavior, beneath the skin of some males. They performed the experiment in June and July, a time of year when the lizards are normally only weakly territorial. They also captured and released control males, but the controls received no testosterone implants.

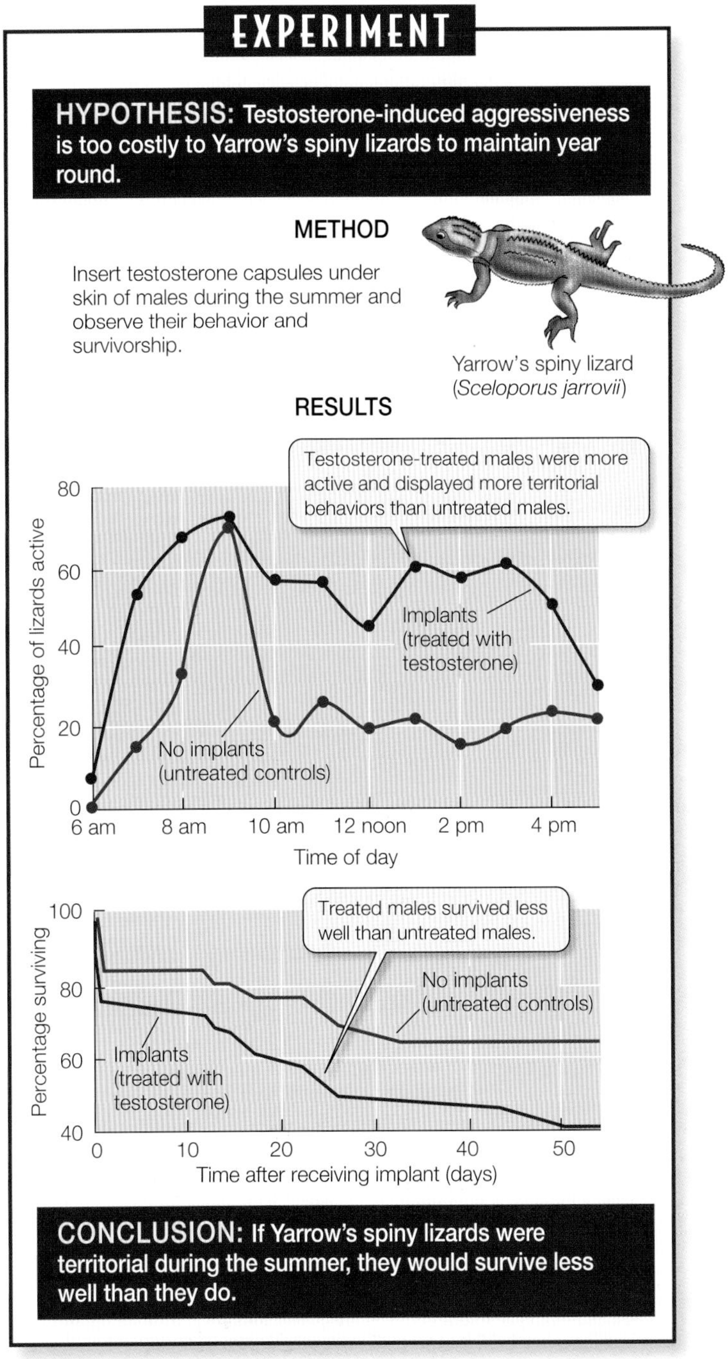

53.8 The Costs of Defending a Territory By using testosterone implants to increase territorial behavior, Moore and Marler measured the costs to male lizards of defending a territory during the summer months.

Males with implanted testosterone capsules spent more time patrolling their territories, performed more advertising displays, and expended about one-third more energy (an energetic cost) than control males. As a result, they had less time to feed (an opportunity cost), captured fewer insects, stored less energy, and died at a higher rate (a risk cost) (**Figure 53.8**). In June and July, when females are less receptive, these high costs of vigorous territorial defense probably outweigh the reproductive benefits of territoriality. Probably that is why the lizards normally wait to perform these behaviors until later in the summer.

Some animals defend all-purpose territories that include all of the resources the animal needs for nesting, mating, and finding food. Tigers defend territories that are many square kilometers in extent by leaving scent marks on their boundaries (**Figure 53.9A**). Many songbirds defend all-purpose territories by singing and threat postures.

The food supplies of some animals cannot be defended because they are widely distributed in space or are subject to fluctuations in availability. For example, the open ocean where seabirds feed cannot be defended. However, most seabirds nest on islands or on rocky cliffs, which offer protection from predators. Such nesting sites can be in very short supply and are vigorously defended. Some of these territories are no larger than the distance the bird can reach while sitting on its nest (**Figure 53.9B**).

Some animals defend territories that are used only for mating. When the breeding season arrives, males of some grouse species congregate on display grounds, where they defend very small areas (**Figure 53.9C**). Their vigorous displays show other males how strong they are and show the females their quality as mates. A female visits a display ground and selects the male with whom she will mate. She then leaves and raises her brood alone. The males often expend so much energy defending their little display areas that less exhausted males may eventually evict them.

Animals choose what foods to eat

Cost–benefit analyses have also been used to investigate the food choices animals make. When an animal *forages* (looks for food), how much time should it spend searching an area before moving to another site? When many different types of food are available, which ones should it eat, and which ones should it ignore? By applying cost–benefit approaches to feeding behavior, scientists have produced a body of knowledge known as **foraging theory**, which helps us understand the survival (ultimate) value of feeding choices. The primary benefits of foraging are the nutritional value of the food obtained—its energy, minerals, and vitamins. The costs of foraging are similar to those for territorial defense: energy expended, time lost for other activities that could enhance fitness, and the risk of increased exposure to predators.

As an example, let's consider some tests of the hypothesis that animals make choices among available prey in such a way as to maximize the rate at which they obtain energy. This is a plausible hypothesis because the more rapidly an animal captures food, the more time and energy it will have for other activities, such as reproduction or avoiding predators. To determine how an animal should choose food items to maximize its energy intake rate, behavioral ecologists can characterize each type of available food item in two

53.9 Animals Defend Territories of Different Sizes (A) To mark the boundaries of her extensive territory, this female tiger rubs the pheromone-producing scent glands on her face against a tree. Other tigers passing the spot will know that the area is "claimed," as well as knowing something about the individual who claimed it. (B) The nesting territories of some colonial birds are no larger than the distance the incubating bird can reach. (C) Some territorial displays are related to mating. Here two male black grouse defend a small territory in snowy Sweden; the winner will mate with nearby females.

ways: by the time it takes the animal to pursue, capture, and consume an individual item, and by the amount of energy an individual item contains. They can then rank the food types according to the amount of energy the animal gets relative to the time it spends pursuing, capturing, and handling the item. The most valuable food type is the one that yields the most energy per unit of time expended.

With this information, behavioral ecologists can determine the rate at which an animal would obtain energy given a particular foraging strategy. They can then compare alternative foraging strategies and determine the one that yields the highest rate of energy intake. Such calculations show that, if the most valuable food type is abundant enough, an animal gains the most energy per unit of time spent foraging by taking only the most valuable type and ignoring all others. However, as the abundance of the most valuable type decreases, an energy-maximizing animal adds less valuable types to its diet in order of the energy per unit of time that they yield.

Earl Werner and Donald Hall of Michigan State University performed laboratory experiments with bluegill sunfish to test the energy maximization hypothesis (**Figure 53.10**). In preparation for their experiments, they measured the energy content of water fleas of different sizes (the different food types), how much time bluegill sunfish (the foragers) needed to capture and eat these different food types, the energy they spent pursuing and capturing the different food types, and the rates at which they encountered the different food types under different food densities. Werner and Hall then stocked experimental environments with different densities and proportions of large, medium, and small water fleas. They made two predictions from the energy maximization hypothesis: first, that in an environment with abundant large water fleas, the fish would ignore smaller water fleas, and second, that in an environment stocked with low densities of all three sizes of water fleas, the fish

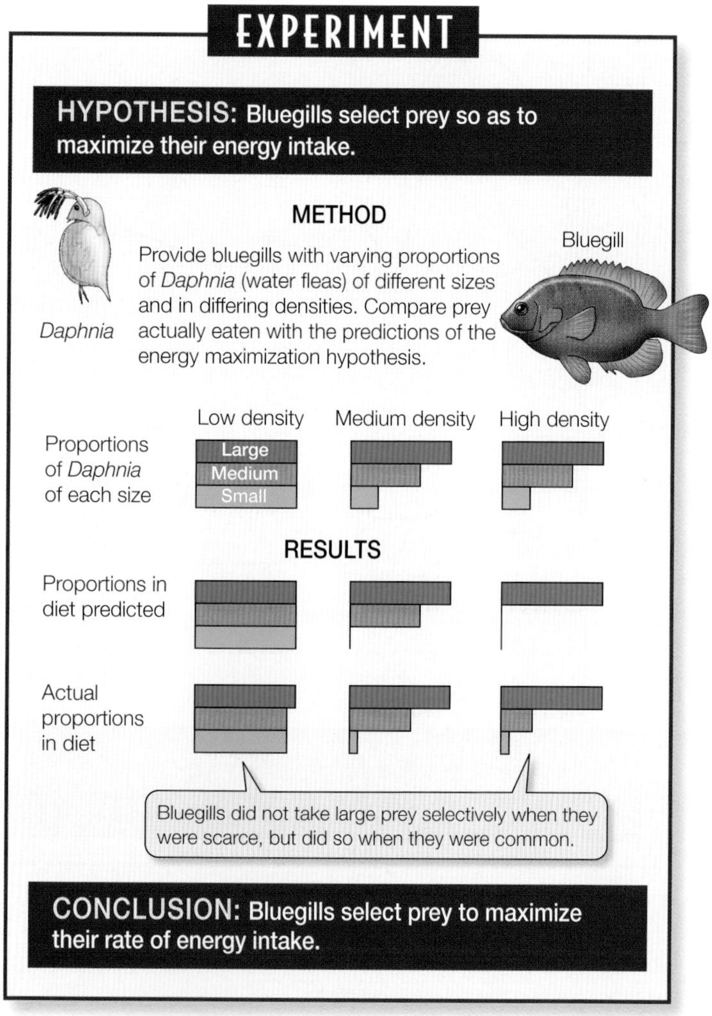

53.10 Bluegills Are Energy Maximizers The prey choices of bluegill sunfish (*Lepomis macrochirus*) were very similar to those predicted by the energy maximization hypothesis.

Ara chloroptera

53.11 Mineral Seekers Red-and-green macaws obtain mineral nutrients from a clay lick in the Amazon jungle of Peru.

would eat every water flea they encountered. The proportions of large, medium, and small water fleas taken by the fish under different conditions were close to those predicted by the hypothesis.

In the bluegill experiment, only the energy content of water fleas mattered, but minerals are also important to some foragers. Some animals incur large energetic costs and the risks of traveling great distances to obtain essential minerals. Many species of mammals and birds, particularly herbivores and seed eaters, get mineral nutrients by eating soil at particular sites where mineral-rich soil is exposed (**Figure 53.11**). Moose wade into streams and ponds to eat plants that contain more sodium than the terrestrial plants from which they get most of their energy.

Animals may ingest some foods for reasons other than the energy or nutrients they provide. For example, some species of frogs have poisons in their skins. Like many other animals that defend themselves from predators with toxic compounds, these frogs have bright, conspicuous coloration that warns potential predators of the dangers of eating them. Where do these frogs get their poisons? Some species get them by eating certain species of ants that have evolved poisons as their own predator defense mechanisms. The frogs can eat these ants because they are immune to the ants' poisons.

A striking example of the ingestion of non-nutritive foods is the use of spices by humans in food preparation. Spices must provide some benefit to humans because we value them so highly. Spices contain chemicals known to protect the plants that produce them against bacteria and fungi, so perhaps the use of spices in food preparation protects people from contaminated food. Zhi-He Liu and Hiroyuki Nakano of Hiroshima University in Japan tested this antimicrobial hypothesis by exposing food-borne bacteria to chemicals found in spices. Most of the commonly used spices inhibited the growth of more than one kind of food-borne bacteria (**Figure 53.12**).

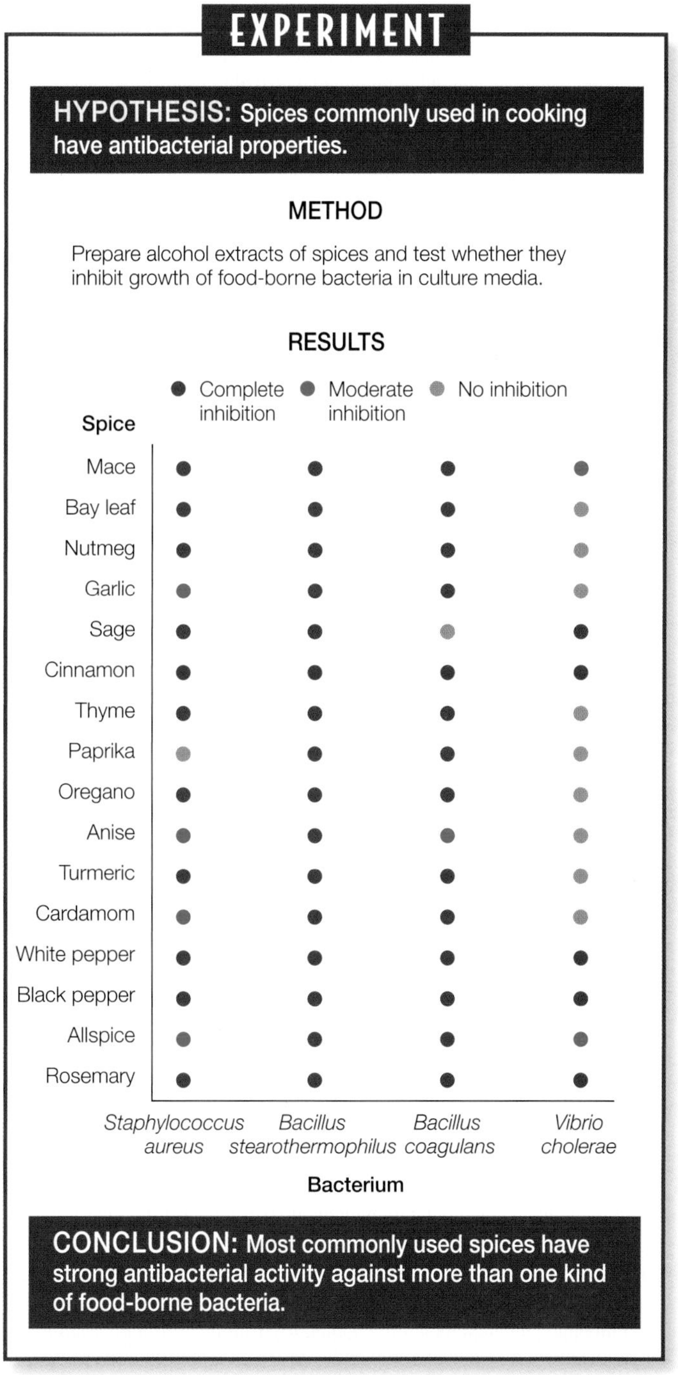
EXPERIMENT
HYPOTHESIS: Spices commonly used in cooking have antibacterial properties.
METHOD
Prepare alcohol extracts of spices and test whether they inhibit growth of food-borne bacteria in culture media.
RESULTS
Complete inhibition
Moderate inhibition
No inhibition
Spice
Mace
Bay leaf
Nutmeg
Garlic
Sage
Cinnamon
Thyme
Paprika
Oregano
Anise
Turmeric
Cardamom
White pepper
Black pepper
Allspice
Rosemary
Staphylococcus aureus
Bacillus stearothermophilus
Bacillus coagulans
Vibrio cholerae
Bacterium
CONCLUSION: Most commonly used spices have strong antibacterial activity against more than one kind of food-borne bacteria.

53.12 Most Spices Have Antimicrobial Activity Laboratory tests have shown that most commonly used spices can inhibit the growth of bacteria.

Although these tests support the antimicrobial hypothesis, they do not exclude the possibility that an alternative hypothesis might also explain the human fondness for spices. One alternative hypothesis is that spices disguise the smell and taste of spoiled foods. This hypothesis cannot be tested easily, but it predicts that spoiled foods could be safely consumed if aversions to their smell and taste could be overcome. We know, however, that toxins from foodborne bacteria kill thousands of people every year and debilitate millions more. Therefore, eating spoiled food by covering up its bad flavors would be a dangerous thing to do, even for a starving person. Natural selection is not likely to have favored people who ate rancid food, no matter how tasty.

An animal's choice of associates influences its fitness

Most animals do not lead solitary lives. They associate with other individuals for a variety of reasons. An important decision made by individuals of all sexually reproducing species is the choice of mating partners. These choices may be based on the inherent qualities of a potential mate, on the resources it controls (food, nest sites, escape places), or on a combination of the two.

The ways that males and females choose mating partners are often very different. Males of most species initiate courtship, seldom reject receptive females, and often fight for opportunities to mate with females. Females of most species seldom fight over males, and they often reject courting males. Why are these sexual roles so different?

The answer lies in part in the costs of producing sperm and eggs. Because sperm are small and cheap to produce, one male produces enough to father a very large number of offspring—usually many more than the number of eggs a female can produce or the number of young she can nourish. Therefore, males of many species can increase their reproductive success by mating with many females.

Eggs, on the other hand, are typically much larger than sperm and are expensive to produce. Consequently, a female usually cannot increase her reproductive output very much by increasing the number of males she mates with. The reproductive success of a female may depend on the quality of the genes she receives from her mate, the resources he controls, or the amount of assistance he provides in the care of their offspring. By their choices among males, females may cause the evolution of traits that reliably signal male quality, a component of sexual selection.

Males employ a variety of tactics to induce females to copulate with them. A male may use courtship behavior that signals in some way that he is in good health, that he is a good provider of parental care, that he can control resources, or that he has a good genotype. For example, males of some species of hangingflies court females by offering them a dead insect, a valuable resource for the female. A female hangingfly will mate with a male only if he provides her with food in this manner. The bigger the food item, the longer she copulates with him, and the more of her eggs he fertilizes (**Figure 53.13**).

Females can improve their reproductive success if they can correctly assess the genetic quality and health of potential mates, the quantity of parental care they may provide, and the quality of the resources they control. But how can females make such assessments when all males would benefit by attempting to signal that they excel in the favored traits? The answer is that by paying particular attention to those signals that males cannot fake, females have favored the evolution of "reliable" signals. Possession of a large dead insect reliably indicates that a male hangingfly is a good forager.

53.13 A Male Wins His Mate The male hangingfly on the right has just presented a moth to his mate, thus demonstrating his foraging skills. He also provides a material reward to his partner, who eats it while they copulate. The bigger the moth, the longer they copulate, and the more eggs he fertilizes.

Responses to the environment must be timed appropriately

Most environments are not constant, and therefore behaviors that are adaptive at one time may not be adaptive at another time. Earth turns on its axis once every 24 hours, generating daily cycles of light and dark, temperature, humidity, and tides. In addition, Earth is tilted on its axis, so the light–dark cycle, and hence the seasons, change as it revolves around the sun. These environmental changes have been discussed in detail as they pertain to plants (see especially Section 38.2) but such cycles profoundly influence the physiology and behavior of animals as well.

CIRCADIAN RHYTHMS CONTROL THE DAILY CYCLE OF BEHAVIOR If animals are kept in constant darkness, at a constant temperature, with food and water available all the time, they will still demonstrate daily cycles of activities such as sleeping, eating, drinking, and just about anything else that can be measured. The persistence of these daily cycles in the absence of environmental time cues suggests that animals have an endogenous (internal) clock. Although these daily cycles seldom are exactly 24 hours long, they are known as **circadian rhythms** (Latin *circa*, "about," and *dies*, "day").

As we learned in Chapter 38, any biological rhythm can be thought of as a series of cycles, and the length of one of those cycles is the *period* of the rhythm (see Figure 38.13). Any point on the cycle is a *phase* of that cycle: Hence, when two rhythms completely match, they are *in phase*, and if a rhythm is shifted (as in the resetting of a clock), it is *phase-advanced* or *phase-delayed*. Since the period of a circadian rhythm is not exactly 24 hours, it must be phase-advanced or phase-delayed each day to remain in phase with the daily cycle of the environment. In other words, the rhythm has to be **entrained** to the cycle of light and dark in the animal's environment.

When you fly across time zones, your circadian rhythm gets out of phase, resulting in "jet lag." Because your biorhythms can only be shifted by 30–60 minutes each day, it may take several days to re-entrain your internal clock to reflect the environmental time in your new location.

An animal kept in constant conditions will not be entrained to the cycle of the environment, and its circadian clock will run according to its natural period—it will be *free-running*. If its period is less than 24 hours, the animal will begin its activity a little earlier each day (**Figure 53.14**). The free-running circadian rhythm is under genetic control. Different species may have different average periods, and within a species, mutations can lead to different period lengths.

Under natural conditions, environmental time cues, such as the onset of light or dark, entrain the free-running rhythm to the light-dark cycle of the environment. In the laboratory, it is possible to entrain the circadian rhythms of free-running animals with short pulses of light or dark administered every 24 hours (bottom panel of Figure 53.14).

A variety of different sensory capabilities have evolved in response to daily cycles of light and dark. Animals that are active at night are *nocturnal* and have different sensory capabilities than those that are *diurnal*, or active during the day. Diurnal animals (including humans and most birds) tend to be highly visual, whereas nocturnal animals depend more on their abilities to hear and smell and use tactile information. Among animals that rely mostly on vision, adaptations have evolved to support diurnal versus nocturnal habits. Ground squirrels are diurnal and have retinas made up entirely of cone cells (visual receptors that can process light wavelengths into the perception of color but are not sensitive to low light levels). In contrast, flying squirrels are nocturnal and their retinas are composed entirely of rods (visual receptors that are highly sensitive to light stimuli but do not provide information about color; see the discussion of photoreceptors in Section 45.4).

PHOTOPERIOD AND CIRCANNUAL RHYTHMS CONTROL SEASONAL BEHAVIORS Seasonal changes in the environment present challenges to many species. Most animals reproduce most successfully if they time their reproductive behavior to coincide with the most favorable time of year for the survival of their offspring. It is advantageous for animals to be able to anticipate when to shift their reproductive systems into breeding mode, because it may take considerable time to prepare physiologically for reproduction.

For many species, a change in day length—the *photoperiod*—is a reliable indicator of seasonal changes to come. For some animals, however, change in day length is not a reliable seasonal cue. Hi-

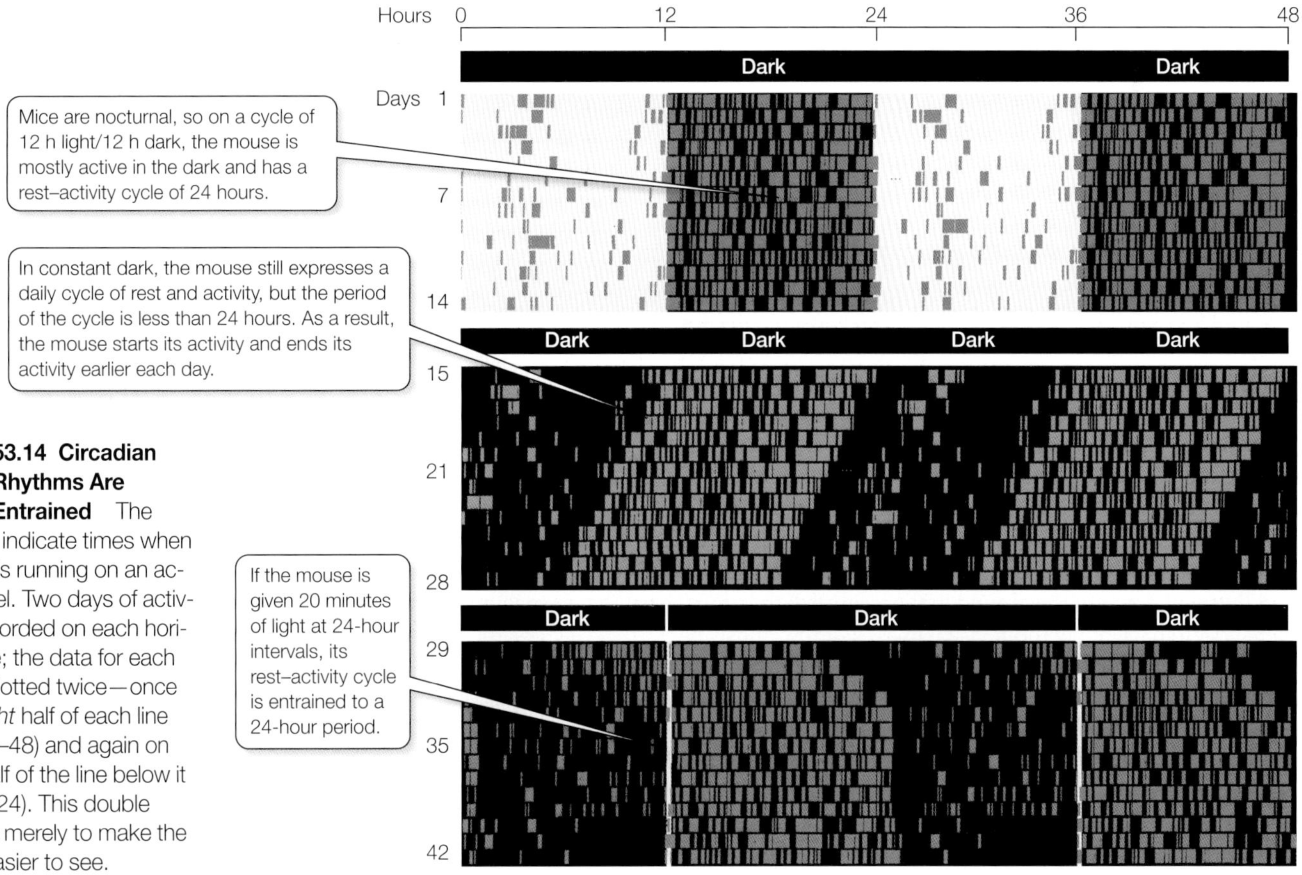

53.14 Circadian Rhythms Are Entrained The gray bars indicate times when a mouse is running on an activity wheel. Two days of activity are recorded on each horizontal line; the data for each day are plotted twice—once on the *right* half of each line (hours 24–48) and again on the *left* half of the line below it (hours 0–24). This double plotting is merely to make the pattern easier to see.

bernators, for example, spend long months in dark burrows underground, but have to be physiologically prepared to breed almost as soon as they emerge in the spring. A bird overwintering near the equator cannot use changes in photoperiod as a cue to time its migration north to the breeding grounds. Hibernators and equatorial migrants have *circannual rhythms*, built-in neural calendars that keep track of the time of year.

Arctic ground squirrels have a very short season for breeding and raising young. Females are ready to mate almost as soon as they end hibernation and may remain in estrus for only a day. Males must be prepared! They stop hibernating a month earlier, but remain in their dens, burning precious metabolic fuel. Why? Because cold testes do not produce sperm.

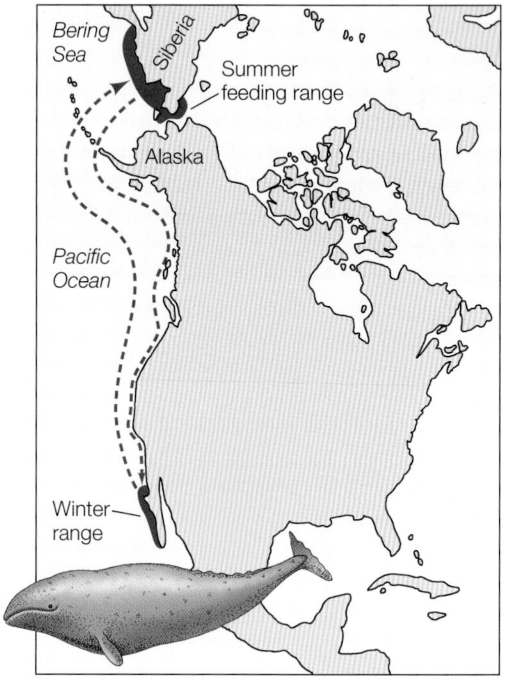

53.15 Piloting Gray whales migrate south in winter from the Bering Sea to the coast of Baja California, in part by following the western coast of North America.

Animals must find their way around their environment

To locate good habitats, find food and mates, and escape from predators and bad weather, an animal needs to be able to find its way around its environment. Within its local habitat, an animal can organize its behavior spatially by orienting to landmarks, as digger wasps do (see Figure 53.3). But what if its destination is a considerable distance away?

PILOTING: ORIENTATION BY LANDMARKS Most animals find their way by knowing and remembering the structure of their environment in a process called **piloting**. Gray whales, for example, migrate seasonally between the Bering Sea and the coastal lagoons of Mexico (**Figure 53.15**). They find their way in part by following the west coast of North America. Coastlines, mountain chains, rivers, water currents, and wind patterns can all serve as piloting cues. But some remarkable cases of long-distance orientation and movement cannot be explained by piloting.

HOMING: RETURN TO A SPECIFIC LOCATION The ability to return over long distances to a nest site, burrow, or other specific location is called **homing**. Homing can be accomplished by piloting in a known environment, but some animals perform much more sophisticated homing.

Pigeons, for example, can return home after being transported to remote sites where they have never been. They do not search around until they encounter familiar territory, but fly fairly directly home from the point of release. In one series of experiments, pigeons were fitted with frosted contact lenses so that they could see nothing but the degree of light and dark. These pigeons still homed and fluttered down to the ground in the vicinity of their loft. They were able to navigate without visual images of the landscape.

MIGRATION: TRAVELING GREAT DISTANCES For as long as humans have inhabited temperate latitudes, they must have been aware that whole populations of animals, especially birds, disappear and reappear seasonally—that is, they migrate. Not until the early nineteenth century, however, were patterns of migration established by marking individual birds with identification bands around their legs. Only when individuals could be unmistakably identified was it possible to show that the same birds and their offspring often returned to the same breeding grounds year after year, and that these same birds could be found during the nonbreeding season at locations hundreds or even thousands of kilometers from their breeding grounds.

Because many homing and migrating species are able to take direct routes to their destinations through environments they have never experienced, they must have mechanisms of navigation other than piloting. There are two systems of navigation:

- **Distance-and-direction navigation** requires knowing in what direction and how far away the destination is. With a compass to determine direction and a means of measuring distance, humans can navigate.
- **Bicoordinate navigation**, also known as *true navigation*, requires knowing the latitude and longitude (the map coordinates) of both the current position and the destination.

The behavior of many animals suggests that they are capable of bicoordinate navigation. Gray-headed albatrosses, for example, breed on oceanic islands. When a young gray-headed albatross first leaves its parents' nest, it flies widely over the southern oceans for 8 or 9 years before it reaches reproductive maturity (**Figure 53.16A**). At that time, it flies back to the island where it was raised to select a mate and build a nest (**Figure 53.16B**). How can it find a tiny island in such an enormous ocean after years of wandering? A circadian clock probably gives the albatross enough information about time of day and the position of the sun to determine their coordinates—much as sailors did in the days before global positioning satellites.

How do animals determine distance and direction? Two obvious means of determining direction are the sun and the stars. During the day, the sun can serve as a compass, as long as the time of day is known. As we have seen, animals can tell the time of day by means of their circadian clocks. Experiments in which the internal clocks of

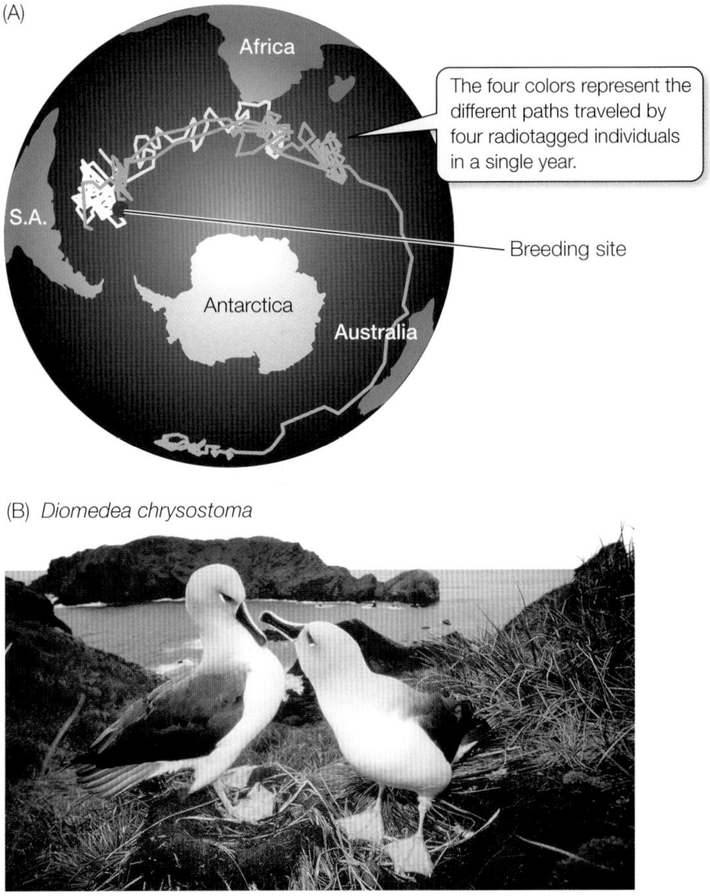

53.16 Coming Home (A) Gray-headed albatrosses are born on islands in the subantarctic oceans. Young birds roam widely over the southern oceans for 8 or 9 years. (B) Once reaching maturity, the birds return to the island on which they were born, where they mate and raise their own young. A courting couple is shown here.

animals are shifted have demonstrated that animals can use their circadian clocks to determine direction using the position of the sun.

Researchers placed pigeons in a circular cage that enabled them to see the sun and sky, but no other visual cues. Food bins were arranged around the sides of the cage, and the birds were trained to expect food in the bin at one particular direction—south, for example. After training, no matter what time they were fed, and even if the cage was rotated between feedings, the birds always went to the bin at the southern end of the cage for food, even if that bin contained no food (**Figure 53.17**).

Next the birds were placed in a room with a controlled light cycle, and their circadian rhythms were shifted by turning the lights on at midnight and off at noon. After about 2 weeks, the birds' circadian clocks had been advanced by 6 hours. Then the birds were returned to the circular cage under natural light conditions, with sunrise at 6:00 A.M. Because of the shift in their circadian rhythms, their endogenous clocks anticipated that the sun would come up at noon. Assuming that birds use the sun as a compass, if the birds expected to find food in the south bin, then at sunrise, they should have looked for food 90 degrees to the right of the direction of the sun. But because their circadian clocks were telling them it was noon, they looked for food in the direction of the sun—in the east bin. The 6-hour shift in their circadian clocks resulted in a 90-

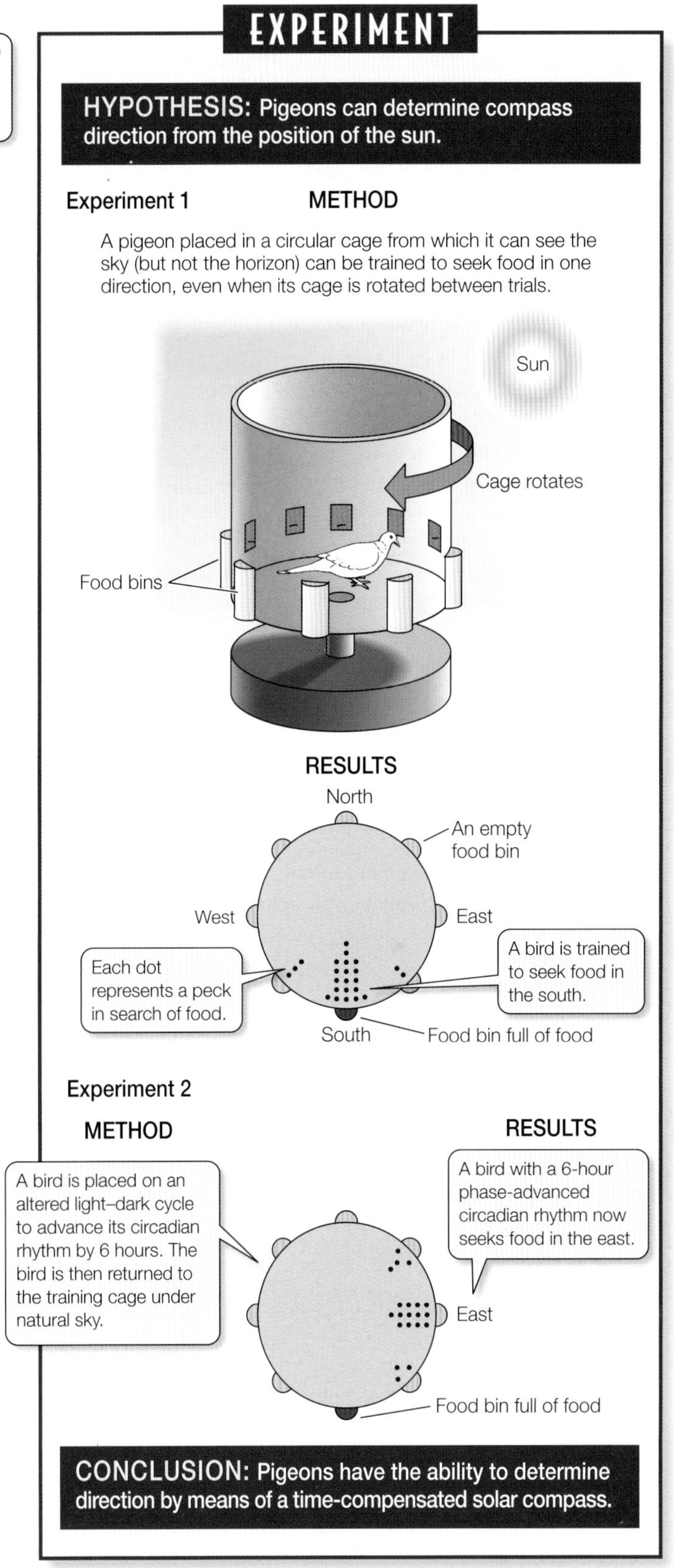

53.17 The Time-Compensated Solar Compass Pigeons whose circadian rhythms were phase shifted forward by 6 hours oriented as though the dawn sun was at its noon position. These results show that birds are capable of using their circadian clocks to determine direction from the position of the sun.

degree error in their orientation. Similar experiments on many species have shown that animals can orient by means of a *time-compensated solar compass*.

Many animals are normally active at night; in addition, many diurnal bird species migrate at night and thus cannot use the sun to determine direction. The stars offer two sources of information about direction: moving constellations and a fixed point. The positions of constellations change because Earth is rotating. With a star map and a clock, direction can be determined from any constellation. But one point that does not change position during the night is the point directly over the axis on which Earth turns. In the Northern Hemisphere, a star called Polaris, or the North Star, lies in that position and always indicates north.

Stephen Emlen at Cornell University raised young birds in a planetarium, in which star patterns are projected on the ceiling of a large, domed room. The star patterns in the planetarium could be slowly rotated to simulate the rotation of Earth. If the star patterns were rotated each night as the young birds matured, they were able to orient in the planetarium. Birds reared in the planetarium under a nonmoving sky moved around, but they did not orient themselves in any particular direction. These experiments showed that birds can learn to use star patterns for orientation.

Animals cannot use sun and star compasses when the sky is overcast, yet they still home and migrate under such conditions. Pigeons home as well on overcast days as on clear days, but this ability is severely impaired if small magnets are attached to their heads—evidence that the birds use a magnetic sense, although the neurophysiology of this sense is largely unknown. Although there is as yet no experimental evidence for these possibilities, birds could also use the plane of polarization of light, which can give directional information even under heavy cloud cover, and very low frequencies of sound, which can provide information about coastlines and mountain chains.

53.3 RECAP

Behavioral ecologists investigate the factors influencing how animals choose places to live, foods to eat, and mates, as well as how they find their way in changing environments. A cost–benefit analysis helps determine why animals make the choices they do.

- Describe three kinds of costs an animal may incur by defending a territory. See pp. 1148–1149 and Figure 53.8
- What prediction does the energy maximization hypothesis make about a foraging animal's choice of food types? See pp. 1149–1151 and Figure 53.10
- Can you explain why the ways that males and females choose mating partners are often very different? See p. 1152
- Describe some ways in which circadian and circannual rhythms improve an animal's fitness. See pp. 1152–1154

Many of the behaviors we have discussed in this section are responses to the physical environment, but responses to other animals of the same and different species are also an important aspect of animal behavior. Let's consider how individual animals behave in response to interactions with other animals of their own and different species and how those interactions have influenced the evolution of animal communication.

53.4 How Do Animals Communicate with One Another?

As individuals interact, they convey information; therefore, their behavior can evolve into systems of information exchange, or **communication**. The behaviors of individuals may become elaborated as **displays** or **signals**, however, only if the transmission of information benefits both the sender and the receiver. To understand why these conditions must be met, consider male courtship displays (see Figure 53.13). A courtship display may benefit a male if it improves his attractiveness to females, but a female will not benefit from responding to such displays unless they allow her to assess whether the male is of the right species, and whether he is strong, healthy, and has other attributes that will make him a good mate or father for her offspring. Female red deer, for example, pay attention to the loud vocalizations of competing stags (**Figure 53.18**) because only large and healthy males can sustain them. Weaker males lose these roaring contests.

Animals communicate in various sensory modes. These modes differ not only in how the communication is performed and in the properties of the signals, but also in what can be communicated. Let's look at some examples of different sensory modes that animals use to communicate with one another.

53.18 Strong Males Win Roaring Contests Weak male red deer cannot sustain long roaring bouts, so females can use the results of these contests to judge the quality of potential mates.

Visual signals are rapid and versatile

Visual signals are easy to produce, come in an endless variety, can be changed rapidly, and clearly indicate the position of the signaler. Most animals are sensitive to light and can therefore receive visual signals. However, the extreme directionality of visual signals means that the receiver must be looking directly at the signaler.

Visual communication is not useful at night or in environments that lack light, such as caves and the ocean depths. Some species have surmounted this constraint by evolving their own light-emitting mechanisms. Fireflies, for example, use an enzymatic mechanism to create flashes of light. By emitting flashes in species-specific patterns, fireflies can advertise for mates at night.

Fireflies also illustrate how some species can exploit the communication systems of other species. There are predatory species of fireflies that mimic the mating flashes of females of other species. When an eager suitor approaches the signaling "female," he is eaten. Thus deception can be part of animal communication systems, just as it is part of human use of language.

Chemical signals are durable

Molecules used for chemical communication between individuals of the same species are called **pheromones**. Because of the diversity of their molecular structures, pheromones can communicate very specific, information-rich messages. The mate attraction pheromone of the female Japanese silkworm moth, for example, enables male moths as far as several kilometers downwind to determine that a female of their species is sexually receptive (see Figure 45.3). By orienting to the wind direction and following the concentration gradient of the pheromone, they can find her.

Pheromone messages left in the environment by mammals who mark their territories with urine or other secretions can reveal a great deal of information about the signaler: species, individual identity, reproductive status, size (indicated by the height of the message), and how recently the animal has been in the area (indicated by the strength of the scent).

Pheromones remain in the environment for some time after they are released. In contrast, vocal and visual displays disappear as soon as the animal stops signaling or displaying. The durability of pheromones makes them useful for marking trails (as ants do), marking territories (as many mammals do), or indicating location (as with the moth sex attractant). Their durability, however, makes pheromones unsuitable for a rapid exchange of information.

Auditory signals communicate well over a distance

Compared with visual communication, auditory communication has both advantages and disadvantages. As is implied by the expression "A picture is worth a thousand words," auditory signals cannot convey complex information as rapidly as visual signals can. But auditory signals can be used at night and in dark environments. They can go around objects that would interfere with visual signals, so they can be transmitted in complex environments such as forests. They are often better than visual signals at getting the attention of a receiver because the receiver does not have to be focused on the signaler for the message to be received. Sounds are also useful for communicating over long distances. Even though the intensity of a sound decreases with distance from the source, loud sounds can be used to communicate over distances much greater than those possible with visual signals.

The complex songs of humpback whales, when produced at a depth of about 1,000 meters, can be heard hundreds of kilometers away. In this way, humpback whales locate one another across vast expanses of ocean.

Tactile signals can communicate complex messages

Communication by touch is common, although not always obvious. Animals in close contact use tactile interactions extensively, especially under conditions in which visual communication is difficult. The best-studied use of tactile communication is the dance of honeybees, first described by Karl von Frisch. When a forager bee finds food, she returns to the hive and communicates her discovery to her hivemates by dancing in the dark on the vertical surface of the honeycomb. Other bees follow and touch the dancer to interpret the message.

If the food source the forager has found is more than about 80 meters away from the hive, she performs a *waggle dance* (**Figure 53.19**), which conveys information about both the distance and the direction of the food source. The forager bee repeatedly traces out a figure-eight pattern as she runs on the honeycomb. She alternates half-circles to the left and right with vigorous wagging of her abdomen in the short, straight run between turns. The angle of the straight run indicates the direction of the food source relative to the direction of the sun. The speed of the dancing indicates the distance to the food source: the farther away it is, the slower the waggle run.

If the food she has found is less than 80 meters from the hive, the forager performs a *round dance*, running rapidly in a circle and reversing her direction after each circumference. The odor on her body and the round dance combine tactile and chemical cues: the odor indicates the flower to be looked for, and the dance communicates the fact that the food source is within 80 meters of the hive.

Electric signals can convey messages in murky water

Some species of fish generate electric fields in the water around them by emitting a series of electric pulses. These pulse trains can be used both for sensing objects in the immediate surroundings and for communication.

For example, each individual glass knife fish emits pulses at a different frequency, and the frequency each fish uses relates to its status in the population. Males emit lower frequencies than females; the most dominant male has the lowest frequency, and the most dominant female has the highest frequency. When a new individual is introduced into the group, the other individuals adjust their frequencies so that they do not overlap with the newcomer's, and the newcomer's signal indicates its position in the hierarchy. In their natural environment—the murky waters of tropical evergreen forests—these fish can determine the identity, sex, and social position of another fish by its electric signals. They can also locate prey by the distortions in the electric field they cause.

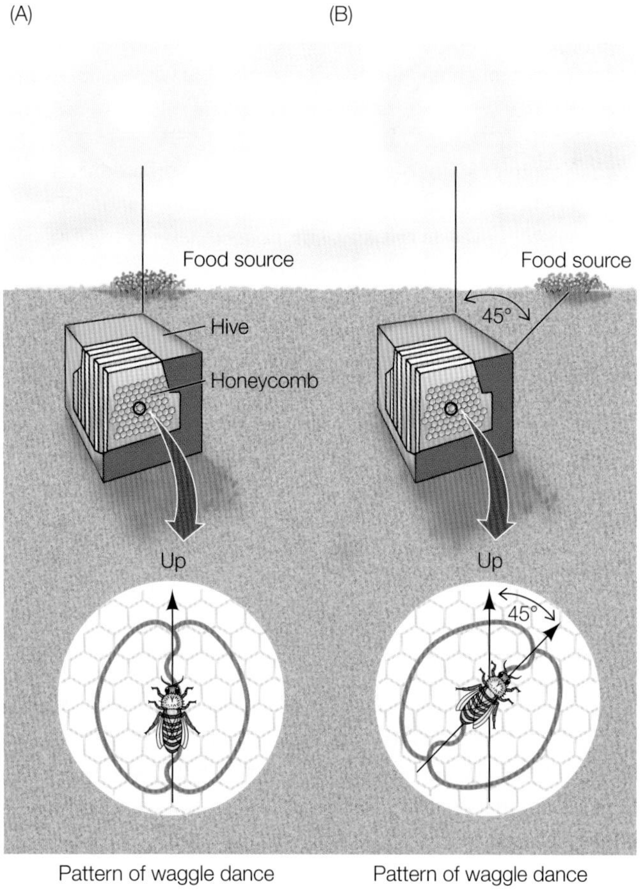

53.19 The Waggle Dance of the Honeybee (A) A honeybee runs straight up on the vertical surface of the honeycomb in the dark hive while wagging her abdomen to tell her hivemates that there is a food source in the direction of the sun. (B) When her waggle runs are at an angle from the vertical, the other bees know that the same angle separates the direction of the food source from the direction of the sun.

53.4 RECAP

A communication system can evolve only if both the sender and receiver of a signal benefit from the information provided by a signal.

- What are the advantages and disadvantages of chemical and visual signals with respect to information transfer? See p. 1157
- Do you understand how honeybee "dances" communicate the location of a good food source? See p. 1157 and Figure 53.19

The use of communication signals allows animals to function not only individually but as part of a pair or larger group. Many animals must associate with each other for mating, but some form groups that remain together for long time periods, eventually evolving into complex societies. Why do animals form such associations?

53.5 Why Do Animal Societies Evolve?

Social behavior evolves when, by cooperating, conspecific individuals can achieve, on average, higher rates of survival and reproduction than they would if they lived alone. The elaborate colonies of ants, bees, and wasps and the social groups of lions and primates are examples of complex social systems. How did these complex animal societies evolve? Because behavior leaves few traces in the fossil record, biologists must infer possible routes for the evolution of social systems by studying current patterns of social organization. Fortunately, many degrees of social system complexity exist among living species; the simpler systems suggest stages through which the more complex ones may have passed.

As we describe a few animal social systems, keep in mind that social systems are dynamic; individuals repeatedly communicate with one another and adjust their relationships. Their relationships change regularly because the costs and benefits experienced by individuals in a social system change with their age, sex, physiological condition, and status.

Also keep in mind that social systems are best understood not by asking how they benefit the species as a whole, but by asking how the individuals that join together benefit from doing so. All individuals must benefit to some degree from being a member of the group; otherwise they would leave it. The option of solitary living may confer few benefits, but it is likely to be better than living in a social group that offers no benefits or imposes high costs.

Group living confers benefits but also imposes costs

Living in groups may confer many types of benefits. It may improve hunting success or expand the range of food that can be captured. For example, by hunting in groups, our ancestors were able to kill large mammals they could not have subdued as individual hunters. These social humans could also defend their prey and themselves from other carnivores, and they could tell one another about the locations of food and enemies.

Many small birds forage in flocks. To test whether flocking provides protection against predators, an investigator released a trained goshawk near wood pigeons in England. The hawk was most successful when it attacked solitary pigeons. Its success in capturing a pigeon decreased as the number of pigeons in the flock increased (**Figure 53.20**). The larger the flock of pigeons, the sooner some individual in the flock spotted the hawk and flew away. This escape behavior stimulated other individuals in the flock to take flight as well.

Social behavior, however, has many costs as well as benefits. The pigeons in a flock interfere with one another's ability to find seeds. Social individuals may inhibit one another's attempts to reproduce or injure one another's offspring. An almost universal cost associated with group living is higher exposure to diseases and parasites. Long before the causes of human diseases were known, people knew that association with sick persons increased their chances of getting sick. Quarantine has been used to combat the spread of disease throughout recorded history. The diseases of wild animals are not well understood, but many of them are probably spread by close contact as well.

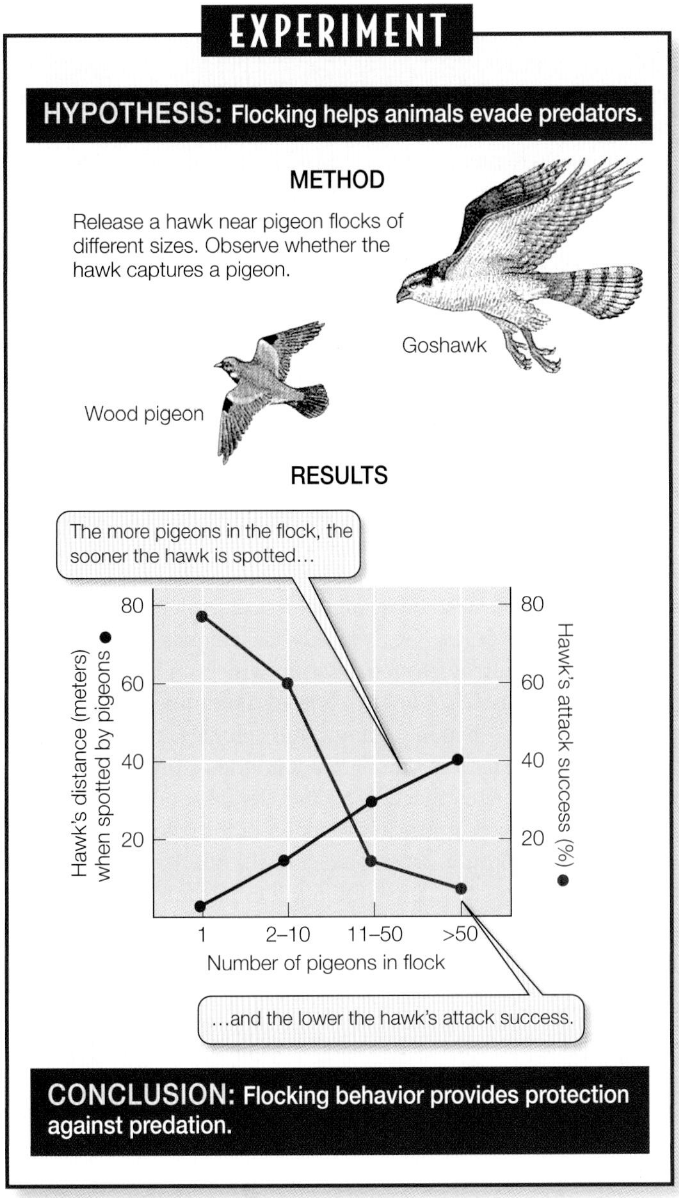

53.20 Groups Provide Protection from Predators The larger a flock of pigeons, the greater the distance at which they detect an approaching hawk, and the less likely the hawk is to succeed in capturing a pigeon.

Parental care can evolve into more complex social systems

The most widespread form of social system is an association of one or two parents with their immature, dependent offspring. Some family units include more adult individuals, typically older offspring that remain with their parents and help care for their younger siblings. Florida scrub jays, for example, live year-round on territories, each of which contains a breeding pair and up to six helpers that bring food to the nest. Nearly all helpers are offspring from the previous breeding season that remain with their parents.

Many mammalian species have also evolved social systems based on extended families. In simple mammalian social systems, solitary females or male–female pairs care for their young. As the period of parental care increases, older offspring may still be present when the next generation is born. They often help rear their younger siblings. In most social mammal species, female offspring remain in the group in which they were born, but males tend to leave, or are driven out. Therefore, among mammals, most helpers are females. In some cases, these related females remain a unit even after they begin having their own offspring. For example, the females in a pride of lions are related, and they care for each other's offspring. Males from the outside compete to take over prides and mate with the females.

Caring for young involves tremendous costs for parents and helpers. Animals that provide food for their young may sacrifice food for themselves, and protecting its young may put an animal in danger. It is easy to understand how natural selection favors parental care in spite of such costs when the benefit is the survival of the animal's own offspring. But natural selection can favor **altruistic** acts—behaviors that reduce the helper's reproductive chances, but increase the fitness of the helped individual. Why?

Altruism can evolve by contributing to an animal's inclusive fitness

Helping behaviors exhibited by parents toward their offspring are easily understood in terms of the close genetic relatedness between them. An animal's offspring contribute to its **individual fitness**. Genetic relatedness extends beyond the parent–offspring relationship, however. In diploid organisms, two offspring of the same parents share, on average, 50 percent of the same alleles, and an individual is likely to share 25 percent of its alleles with its sibling's offspring. Therefore, by helping its relatives, an individual can increase the representation of some of its own alleles in the population. Together, individual fitness plus the fitness gained by increasing the reproductive success of nondescendant kin determine the **inclusive fitness** of an individual. Occasional altruistic acts may eventually evolve into altruistic behavior patterns if their benefits (in terms of inclusive fitness) exceed their costs (in terms of decreases in the altruist's own reproductive success).

Many social groups consist of some individuals that are close relatives and others that are unrelated or distantly related. Individuals of some species recognize their relatives and adjust their behavior accordingly. White-fronted bee-eaters are African birds that nest colonially. Most breeding pairs are assisted by nonbreeding adults that help incubate eggs and feed nestlings. Nearly all of these helpers assist close relatives (**Figure 53.21**). When helpers have a choice of two nests at which to help, about 95 percent of the time they choose the nest with the young more closely related to them. Individual bee-eaters improve their inclusive fitness by helping because nests with helpers produce more fledglings than do nests without helpers. Helpers also increase their chances of inheriting the nesting territory when the owners die.

Species whose social groups include sterile individuals are said to be **eusocial**. This extreme form of social behavior has evolved in termites and many hymenopterans (ants, bees, and wasps). In these species, most females are nonreproductive *workers* that forage for the colony or defend it against predators. Workers may include *soldiers* with large, specialized defensive weapons (**Fig-**

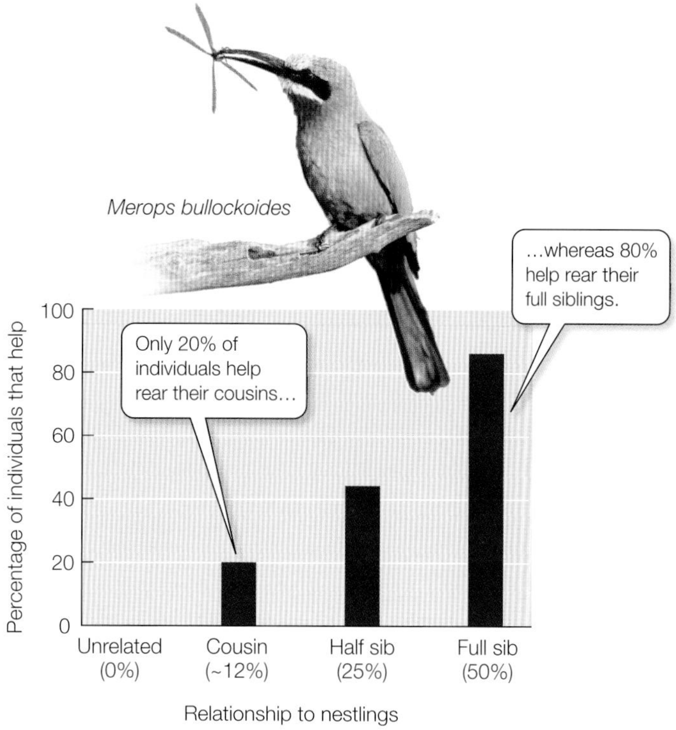

53.21 White-Fronted Bee-Eaters Are Discriminating Altruists Bee-eaters preferentially help close relatives rear their nestlings. The relationship axis gives the approximate percentage of alleles that would be identical between two relatives. Because some individuals help out more than one nest, the percentages add to more than 100 percent.

ure 53.22). They may be killed while defending the colony. Only a few females, known as *queens*, are fertile; they produce all the offspring of the colony.

Both genetic and environmental factors may facilitate the evolution of eusociality. The British evolutionary biologist W. D. Hamilton first suggested that eusociality evolved among the Hymenoptera because its members have an unusual sex determination system in which males are haploid but females are diploid. Among the Hymenoptera, a fertilized (diploid) egg hatches into a female; an unfertilized (haploid) egg hatches into a male.

If a female hymenopteran copulates with only one male, all the sperm she receives are identical because a haploid male has only one set of chromosomes, all of which are transmitted to every sperm cell. Therefore, a female's daughters share all of their father's genes. They also share, on average, half of the genes they receive from their mother. As a result, they share 75 percent of their alleles, on average, rather than the 50 percent they would share if both parents were diploid. If they reproduced, they would share only 50 percent of their alleles with their own female offspring. Since workers are more genetically similar to their sisters than they would be to their own offspring, they can potentially increase their fitness more by caring for their sisters than by producing and caring for their own offspring. If, however, a queen copulates with more than one male, workers may not be so closely related to all of their sisters.

Eusociality may also be favored if it is difficult or dangerous to establish new colonies. Nearly all eusocial animals construct elaborate nests or burrow systems within which their offspring are reared (see Figure 53.24 below). Naked mole-rats—the most eusocial mammals—live in underground colonies containing 70 to 80 individuals. Sterile workers maintain the colony's tunnel systems. Breeding is restricted to a single female and several males that live in a nest chamber in the center of the colony. Individuals attempting to found new colonies are at high risk of being captured by predators, and most founding events fail. These circumstances, which favor cooperation among founding individuals, may facilitate the evolution of eusociality.

Inbreeding—the mating of individuals who are closely related—increases genetic relatedness within a group. Even if two parents are unrelated, but each is the product of generations of intense inbreeding, all of their offspring may be genetically nearly identical. Such offspring would increase their fitness by helping to rear siblings. Genetic similarity generated by inbreeding could explain the evolution of eusociality among the many hymenopteran species in which queens mate with many males, and among termites and naked mole-rats, in which both sexes are diploid.

Eciton hamatum

53.22 Sterile Workers are Extreme Altruists Eusocial insect species contain classes of sterile worker individuals. The army ant soldier from Costa Rica protects her colonymates with her large, powerful jaws.

53.5 RECAP

Social behavior evolves when the benefits of group living outweigh its costs. Most complex social systems evolved from extended families.

- How does genetic relatedness among group members influence the evolution of altruistic behavior? See p. 1159 and Figure 53.21
- Why is eusociality particularly common in hymenopterans? See p. 1160

Animals with complex social systems often achieve remarkably dense populations. Why? In the next section we'll see how the ways in which animals interact with members of their own and other species influence what habitats and foods a species uses in nature, which species live together, and how abundant they are.

53.6 How Does Behavior Influence Populations and Communities?

Individuals use the behaviors we have described in this chapter to perform the activities they need to grow, reproduce, and avoid becoming food for other organisms. These behaviors have evolved primarily because they have provided benefits to the individuals that performed them. In addition, however, the ways in which animals make decisions about habitats, food, and associates influence the structure and functioning of populations, communities, and ecosystems.

Habitat and food selection influence the distributions of organisms

Although animals settle preferentially in good habitats, individuals that are already there may prevent them from using those habitats. If population densities are high, many individuals of a species may be forced to settle in poor habitats. And if prior residents do not exclude newcomers, their activities may lower the quality of the best habitats.

Foraging behavior can influence population dynamics in many ways. Some animals buffer fluctuations in food availability by storing food. The energy resources gained from food can be stored within an organism's body; animals typically store them as fat, and plants typically store them as starch. In cold and arid environments, where food does not decompose quickly, animals may store large amounts of food outside their bodies (**Figure 53.23**). Some food types, notably seeds, spores, wood, fungi, leaves, and nectar, are more readily stored than others, either because they already have a low moisture content or because they can be dried prior to storage. Storing food enables animals to live in environments where food would otherwise be unavailable seasonally.

How animals choose their food may influence the species composition and abundance of their prey. For example, fish influence the composition of communities of small planktonic animals in lakes by selectively capturing and eating the largest of these animals (as we saw in Figure 53.10). If predatory fish are absent, large plankton outcompete smaller plankton; consequently, large planktonic species dominate fishless lakes, but small planktonic species dominate lakes with plankton-eating fish.

Territoriality influences community structure

Individuals of most territorial species defend their territories only against other individuals of the same species, but some animals benefit by defending their territories against individuals of other species as well. Typically the winner of interspecific territorial contests is the species with the larger body size. Because individuals of larger species need to harvest energy at a higher rate than those of smaller species, the individuals of the dominant species may be able to survive and reproduce only in high-productivity habitats. If the dominant species prevents subordinate species from living in those habitats, it follows that subordinate species would occupy those habitats if the dominant individuals were removed. Removal of the subordinate species, however, should have little effect on the distribution of the dominant species.

Melanerpes formicivorus

53.23 Storing Energy for a Poor Season Acorn woodpeckers drill holes in dead trees in which they hoard acorns to sustain them over long dry or cold seasons. The birds may form social nesting groups, and members of the group will defend their cache tree against potential "robbers."

Stuart Pimm and Michael Rosenzweig tested this hypothesis with hummingbirds in southern Arizona. They created artificial "flower patches" by setting up an array of feeders. Some feeders contained artificial nectar rich in sucrose. Others contained a more dilute sucrose solution. Male blue-throated hummingbirds, which weigh an average of 8.3 grams, dominated male black-chinned hummingbirds, which weigh only 3.2 grams. When male blue-throats were absent, black-chin males fed almost exclusively at the rich feeders, but when male blue-throats were present, male black-chins fed extensively at the poorer-quality feeders. They were able to maintain their weight on the poorer-quality feeders, whereas male blue-throats could not.

Social animals may achieve great population densities

The population densities achieved by some social animals are impressive. For example, up to 94 percent of the individuals and 86 percent of the biomass of arthropods in the canopies of tropical evergreen forests are eusocial ants. Termites (also eusocial insects) are the primary consumers of plant tissues in the savannas of Africa and Australia. They live in and build large mounds, within which many other species of animals live (**Figure 53.24**). In parts of Australia, termite density may reach 1,000 colonies per hectare.

Ants and termites have achieved these remarkable abundances in part because their social organization allows them to exploit the services of other organisms in harvesting vital resources. The most abundant and productive ants and termites actively cultivate fungi that break down difficult-to-digest plant tissues, including wood. Some ants protect aphids and other insects that tap phloem fluids

53.24 Termite Mounds Are Large and Complex These immense Australian termite mounds are constructed over many years by millions of worker termites. Elaborate nests or burrows, which are very costly to construct and maintain, characterize nearly all eusocial animals.

of plants. Phloem is rich in carbohydrates but poor in proteins, so phloem suckers ingest more carbohydrates than they can use. They eject the excess in the form of sugar-rich anal drops (see Figure 35.13), which the ants eat. Because the ants can easily obtain enough carbohydrates from the anal drops, they need to get only proteins in other ways, such as by eating other insects. Moreover, with their high metabolic rates fueled by ready access to carbohydrates, the ants can expend the energy needed to drive other predatory insects away from their food sources. Ants strongly influence the community of insects in tropical forest canopies.

The population densities achieved by our own species serve as striking reminders of the consequences of social behavior (**Figure 53.25**). Social living enabled members of human groups to specialize in different activities. Among the benefits of specialization were the domestication of plants and animals and the cultivation of land. These innovations enabled our ancestors to increase the resources at their disposal dramatically. Those increases, in turn, stimulated rapid population growth up to the limit determined by the agricultural productivity that was possible with human- and animal-powered tools. Agricultural machines and artificial fertilizers, made possible by the tapping of fossil fuels, greatly increased agricultural productivity and removed that earlier limit. In addition, the development of modern medicine reduced mortality rates in human populations. Medicine and better hygiene have also allowed people to live in large numbers in areas where diseases formerly kept numbers very low. However, these successes are accompanied by many social and environmental problems, some of which we will discuss in subsequent chapters.

53.6 RECAP

The ways in which individuals choose habitats and foods influence interactions both within and among species. Social animals can achieve great population densities and can therefore be a major influence on communities and ecosystems.

- How does foraging behavior enable animals to live in environments with fluctuating food supplies? See p. 1161
- Can you explain why highly social animals often achieve high population densities? See pp. 1161–1162

53.25 Social Organization Allows Humans to Live at High Densities Densely populated human cities, such as Benidorm on Spain's Costa Blanca, serve as examples of how social organization allows our species to achieve and sustain extreme population densities.

Observations of the behavior of animals show us that the ways in which animals choose what to eat, where to seek food, and with whom to associate influence the sizes and distributions of populations of many species and how they interact in nature. These aspects of populations will form the focus of the next chapter.

CHAPTER SUMMARY

53.1 What questions do biologists ask about behavior?

Scientists who study behavior describe what they observe and then try to answer either proximate or ultimate questions about the behavior.

53.2 How do genes and environment interact to shape behavior?

Animals perform many stereotypic and species-specific behaviors without prior experience .

In a deprivation experiment, an animal is deprived of all experience relevant to the behavior under study so that its genetic component can be assessed.

In a genetic experiment, investigators are able to compare the behaviors of individuals that differ in only one or a few known genes. Review Figure 53.2

Inherited behaviors are often triggered by simple stimuli called **releasers**. Such behavior is adaptive when opportunities to learn are lacking and when mistakes are costly or even lethal.

Imprinting is a type of learning that takes place during a **critical period** in an animal's development.

Learning abilities require proximate mechanisms whose construction requires genetic information. Review Figure 53.5

Hormones can influence the development and expression of behavior patterns. Review Figure 53.6

53.3 How do behavioral responses to the environment influence fitness?

The cues most organisms use to select suitable **habitats** are good predictors for their future survival in those habitats.

Animals may establish and defend a **territory**, giving themselves exclusive use of that space.

A **cost–benefit approach** analyzes the total cost of any particular behavior in terms of **energetic**, **risk**, and **opportunity costs**. Review Figure 53.8, Web/CD Tutorial 53.1

Foraging theory helps behavioral ecologists understand the survival value of animals' food choices. When choosing food, animals often function as energy maximizers. Review Figure 53.10, Web/CD Tutorial 53.2

Individuals of all sexually reproducing species choose mating partners based on the inherent qualities of a potential mate, on the resources it controls, or on a combination of the two.

Circadian rhythms enable animals to anticipate daily changes in the environment. Photoperiods and circannual rhythms enable them to anticipate seasonal changes. Review Figure 53.14, Web/CD Tutorial 53.3

Piloting involves navigating by landmarks; **homing** is the ability to return home regardless of distance. **Migration** often requires long-distance navigation.

Humans can use both **distance-and-direction** and **bicoordinate navigation**. It is not known whether animals can use bicoordinate navigational techniques, but they can use the sun and stars to determine direction. Review Figure 53.17, Web/CD Tutorial 53.4

53.4 How do animals communicate with one another?

Interactions between individuals can evolve into **communication systems** if the transmission of information by **displays** or **signals** benefits both signaler and receiver.

Animals communicate in different sensory modes, which differ in their properties and in what can be communicated by using them.

Visual signals can convey complex information rapidly. Chemical signals such as **pheromones** can convey complex information and last a long time. Auditory signals can convey complex information over long distances in complex habitats. Tactile and electric signals can convey complex information rapidly where other kinds of signals are not possible. See Web/CD Activity 53.1

53.5 Why do animal societies evolve?

Group living inevitably imposes both costs and benefits. Social behavior evolves when cooperation among conspecifics results in higher rates of survival and reproduction than solitary individuals can achieve. Review Figure 53.20

The simplest social systems consist of one or two parents with their immature, dependent offspring. Other social systems include members of extended families.

An animal's own reproductive success contributes to its **individual fitness**. The reproductive success of its close relatives, with whom it shares alleles, contributes to its **inclusive fitness**. An animal that performs altruistic behavior reduces its own reproductive success but may gain inclusive fitness by increasing the fitness of its relatives. Review Figure 53.21

Eusocial species live in groups that include sterile individuals that help close relatives reproduce.

53.6 How does behavior influence populations and communities?

The ways in which animals select habitats and food and interact with individuals of their own and other species can influence the range of habitats and foods a species uses in nature, which species live together, and how abundant they are.

Animals with complex social systems often achieve remarkably high population densities.

See Web/CD Activity 53.2 for a concept review of this chapter.

SELF-QUIZ

1. In a deprivation experiment, young animals are reared
   a. with only young, naive individuals of their own species.
   b. with conspecifics of all ages, but in a limited environment.
   c. so that they have no experience relevant to the behavior being studied.
   d. in an environment that lacks the conditions that would stimulate them to perform the behavior.
   e. under conditions that deprive them of food.
2. Which of the following is *not* a component of the cost of performing a behavior?
   a. Its energetic cost
   b. The risk of being injured
   c. Its opportunity cost
   d. The risk of being attacked by a predator
   e. Its information cost
3. Birds that migrate at night
   a. inherit a star map.
   b. determine direction by knowing the time and the position in the sky of a constellation.
   c. orient to a particular point in the sky.
   d. imprint on one or more key constellations.
   e. determine distance, but not direction, from the stars.
4. If a bird is trained to seek food on the western side of a cage open to the sky, its circadian rhythm is then delayed by 6 hours, and it is returned to the open cage at noon in real time, it will seek food in the
   a. north.
   b. south.
   c. east.
   d. west.
5. To be able to pilot, an animal must
   a. have a time-compensated solar compass.
   b. orient to a fixed point in the night sky.
   c. know the distance between two points.
   d. know landmarks.
   e. know its longitude and latitude.
6. Which of the following statements about communication is true?
   a. Complex information can be conveyed most rapidly by pheromones.
   b. Visual signaling is advantageous in complex environments.
   c. Auditory communication always reveals the location of the signaler.
   d. An advantage of pheromones is that the message can persist through time.
   e. The dance of bees is an example of visual signaling.
7. One cost commonly associated with group living is
   a. increased risk of predation.
   b. interference with foraging.
   c. higher exposure to diseases and parasites.
   d. poorer access to mates.
   e. All of the above
8. The choice of a mating partner may be based on
   a. the inherent qualities of a potential mate.
   b. the resources held by a potential mate.
   c. both the inherent qualities of a potential mate and the resources it holds.
   d. the courtship display of a potential mate.
   e. All of the above
9. Altruistic behavior
   a. confers a benefit on the performer by inflicting some cost on some other individual.
   b. confers a benefit both on the performer and on some other individual.
   c. inflicts a cost both on the performer and on some other individual.
   d. confers a benefit on an individual other than the performer's offspring at some cost to the performer.
   e. inflicts a cost on the performer without benefiting any other individual.
10. An animal is said to be *eusocial* if
    a. group members interact very intensively.
    b. some group members produce many more offspring than others do.
    c. a dominance hierarchy exists among group members.
    d. young individuals remain in the group to help their parents rear other offspring.
    e. the social group contains sterile individuals.

FOR DISCUSSION

1. Cowbirds are nest parasites. A female cowbird lays her eggs in the nest of another bird species, which then incubates the eggs and raises the young. What do you think would characterize the acquisition of song in cowbirds? In a given area, cowbirds tend to parasitize the nests of particular bird species. How do you think female cowbirds learn this behavior? How would you test your hypothesis?
2. Male dogs lift a hind leg when they urinate; female dogs squat. If a male puppy receives an injection of estrogen when it is a newborn, it will never lift its leg to urinate for the rest of its life; it will always squat. How might this result be explained?
3. The short-tailed shearwater is a bird that winters in Antarctica and summers in the Arctic. What problems would this species have in using either the sun or the stars for navigation? What is the most likely means it uses to find its way to its summer and its winter feeding grounds?
4. Most hawks are solitary hunters. Swallows often hunt in groups. What are some plausible explanations for this difference? How could you test your ideas?
5. Among birds, males of species that mate with many females and perform communal courtship displays are usually much larger and more brightly colored than females, whereas among species that form monogamous pairs, males are usually similar to females in size, whether or not they are more brightly colored. What hypotheses can be advanced to explain this difference?
6. Many animals defend territories, but the sizes of the territories they defend and the resources those areas provide vary enormously. Why don't all animals defend the same type and size of territory?
7. Among vertebrates, helpers are individuals capable of reproducing, and most of them later breed on their own. Among eusocial insects, sterile castes have evolved repeatedly. What differences between vertebrates and insects might explain the failure of sterile castes to evolve in the former?

FOR INVESTIGATION

Experiments on animal migration show that animals can use a time-compensated solar compass or identification of a fixed point in the night sky as a basis for distance and direction navigation. However, the observations on the homing ability of seabirds such as the albatross indicate that animals might be capable of true or bicoordinate navigation. What experiments could you do to prove that animals have such abilities, and what experiments could you do to test hypotheses about the mechanisms they might use? At least two hypotheses might involve using the elevation of the sun or the angles of Earth's magnetic lines of force as a means of determining latitude.

CHAPTER 54 Population Ecology

Exotic moth controls exotic cactus

In the early 1830s, someone thought the South American prickly-pear cactus *Opuntia stricta* would make a nice ornamental ground cover in the semiarid climate of Australia. The cactus reacted well to its new home—by 1925 it had covered more than 25 million hectares of eastern Australia's valuable sheep-grazing land.

In 1926, another *exotic* (non-native) species was introduced in Australia, this time to control the burgeoning *Opuntia* population. The caterpillar larvae of the moth *Cactoblastis cactorum* feed on the cactus. About 2 million eggs of this moth, whose native origins are also in South America, were imported and dispersed among the Australian *Opuntia*. The strategy was extremely successful, and today the total numbers of both the cactus and the moth in Australia are constant and fairly low, although there are many local population *oscillations*.

A female *Cactoblastis* moth lays her eggs anchored to a cactus spine. Hundreds of caterpillars hatch and feed communally inside the leathery cactus pads, which the caterpillars reduce to a gooey green mess. *Cactoblastis* caterpillars can completely destroy *Opuntia* patches, but new patches arise in other places from seeds dispersed by birds. The new patches flourish until a female *Cactoblastis* finds them and lays large numbers of eggs on them. In Australia, both *Opuntia* and *Cactoblastis* are now distributed as scattered subpopulations among which individuals occasionally disperse.

Various *Opuntia* speces native to North and South America have been introduced to Europe and Africa, as well as to Hawaii and many Caribbean islands. In some of these places, opuntias have become invasive and *C. cactorum* has been introduced to control them. The moth was introduced onto the Caribbean island of Nevis, where it successfully controlled an invasive exotic *Opuntia* species, but also destroyed some native species of *Opuntia*. The moth reached Florida in 1989. Today it is also found in the southwestern United States, where it has become a pest on the native opuntias, many of which are threatened species. In Mexico, where opuntias are economically important producers of fruit, fodder, dyes, and medicinals, the moth is a serious pest.

Why has *Cactoblastis* been so successful in Australia, yet caused so many problems in North America? The answer is probably that

A Pest in Australia Humans introduced *Opuntia* into Australia, where there are no native cactus species. The exotic cactus quickly overran valuable grazing land, but was brought under control by importing eggs of the moth *Cactoblastis cactorum,* whose caterpillars feed voraciously on the fleshy cactus pads.

A Valuable Crop in Mexico In Mexico, many native species of *Opuntia* are economically valuable farm crops. The presence of the accidentally introduced *Cactoblastis* moth threatens some of the native cacti.

Australia has no native cactus species, and so far the moth has not attacked any other type of plant. In areas where there are other native species of *Opuntia*, the exotic moth is destructive.

Species introduced to regions beyond their original ranges often achieve much greater population densities than they had in their native ranges. This may happen because in the new region, no predators exist to attack them—which is why we often introduce predators and parasites in an attempt to control an invasive species. As we will see in this chapter, surprises often accompany our efforts to manage populations.

IN THIS CHAPTER we first explore how ecologists study populations and then describe some of the relationships between the life history traits of organisms and the dynamics of their populations. Next we will discuss the factors that influence population densities and explore how environmental variation over space affects population dynamics. Finally, we will show how ecological knowledge is used to manage populations that are of special interest or importance to humans.

CHAPTER OUTLINE

54.1 How Do Ecologists Study Populations?

Before we consider how ecologists study populations, we need to know how they define populations. A **population** consists of the individuals of a species within a given area. At any given moment, an individual organism occupies only one spot in space, and it is of a particular age and size. The members of a population, however, are distributed over space, and they differ in age and size. The age distribution of individuals in a population and the way those individuals are spread over the environment describe its **population structure**. Ecologists study population structure because the spatial distributions of individuals and their ages influence the stability of a population and affect how that population interacts with other species.

The number of individuals of a population per unit of area (or volume) is its **population density**. Population densities exert strong influences on the ways in which members of a population interact with one another and with populations of other species. Scientists working in agriculture, conservation, and medicine typically try to maintain or increase population densities of some species (crop plants, game animals, aesthetically attractive species, threatened or endangered species) and reduce the densities of others (agricultural pests, pathogens). To manage populations, we need to know what factors cause populations to grow or decrease in density and how those factors work.

The structure of a population changes continually because **demographic events**—births, deaths, immigration (movement of individuals into the area), and emigration (movement of individuals out of the area)—are common occurrences. Knowledge of when individuals are born and when they die provides a surprising amount of information about a population. The study of the birth, death, and movement rates that create *population dynamics* (changes in population structure and density) is known as *demography*.

Ecologists use a variety of devices to track individuals

In order to study populations, ecologists determine how many individuals are found in an area and where they are located. They also measure the rates at which individual are born, die, and move into and out of an area. How do they make those measurements?

Chapter 53 describes various behaviors that affect animal population dynamics. For example, sometimes individual animals increase their inclusive fitness by helping to care for close relatives. Individuals also change their locations by migrating or dispersing. To recognize and study such events, investigators need to be able to recognize and track individual animals. Although an experienced investigator may be able to distinguish individuals by their appearance—elephants by their ears or orcas by differences in the pattern of white blotches on their bodies (**Figure 54.1A**)—differences among individuals are often too minor to detect in the field. Most field studies of animal populations require tagging or marking individuals in some way.

No single form of marking can be used on all species. Birds are typically marked by colored bands on their legs (**Figure 54.1B**), butterflies by placing colored spots on their wings, bees by placing numbered tags on their bodies, and mammals by tags or dyeing their fur. Plants are marked by tags tied to their branches or, because they do not move, by stakes in the ground nearby.

Until recently, most field techniques for tracking animals depended on the ability of researchers to see them. Today's tracking devices are so sophisticated that they can record and transmit information not only about an animal's location, but also about its physiology (by measuring its heartbeat), feeding behavior (its stomach temperature), and social behavior (its vocalizations) many times per second. Microchips and other forms of electronic tagging are used on organisms of all sizes (**Figure 54.1C**). The latest devices can also record information about the animal's environment.

Molecular markers can be used to determine the movement of individuals over long distances. American redstarts (*Setophaga ruticilla*) breed in the deciduous forests of eastern North America. Following the energetically demanding breeding season, these small songbirds migrate south to the Caribbean and Central America for the winter. It is easy to determine where redstarts breed, but how might we determine where they molt? One way is to analyze the chemical composition of the feathers that the birds molt as they migrate south. Scientists can determine where they molted by evaluating hydrogen isotopes in feathers because there is a strong latitudinal gradient in stable hydrogen isotopes in precipitation. These hydrogen isotopes are incorporated into the tissues of plants, the animals that eat the plants (such as herbivorous insects), and eventually, the animals, such as redstarts, that eat the herbivores. Because feathers are metabolically inert after they have formed, the hydrogen isotopes in redstart feathers indicate the latitude at which the individuals grew those feathers (**Figure 54.2**). Most individuals molted their feathers close to the breeding ground.

Population densities can be estimated from samples

Because organisms and their environments differ, population densities must be measured in more than one way. Ecologists usually measure the densities of organisms in terrestrial environments as the number of individuals per unit of area. The number per unit of volume is generally a more useful measure for aquatic organisms. For species whose members differ markedly in size, as do most plants and some animals (such as mollusks, fishes, and non-avian reptiles), the percentage of ground covered or the total mass of individuals may be more useful measures of density than the number of individuals.

The most accurate way to determine the density and structure of a population is to count every individual and note its location. Ecologists studying populations of woody plants can often do this. In most field studies of animal populations, however, counting every individual would be impossible. Fortunately,

(A) *Orcinus orca*

(B) *Calidris alba* (C) *Odocoileus virginianus*

54.1 By Their Marks You May Know Them
(A) Many individual animals are distinguishable from each other by subtle differences in their size and markings. The slight differences in the white markings of these orcas are discernible when they are side by side. (B) Colored bands on the legs of this sanderling (a small shorebird) allow researchers to identify this individual. (C) A female white-tailed deer has an individually identifying ear tag. In addition, researchers have attached a collar that emits a radio signal so that the deer's movements can be monitored even when she cannot be seen.

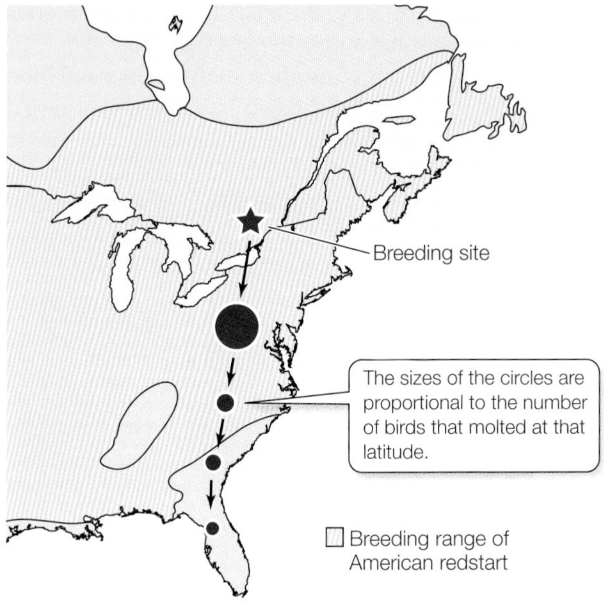

54.2 Hydrogen Isotopes Tell Where Migratory American Redstarts Molted Their Feathers Redstarts that breed in southern Ontario migrate to the Caribbean for the winter. The values of a stable hydrogen isotope in their tail feathers indicate where the feathers were molted.

statistical methods allow us to reliably estimate population densities from representative samples without locating and counting every individual.

Estimating population densities is easiest for sedentary organisms. Investigators need only count the number of individuals in a sample of representative habitats and extrapolate the counts to the entire ecosystem. Individuals may be counted on sampling plots of various shapes (square, rectangular, circular) or along linear transects across the habitat. By making repeated censuses, investigators can also determine trends in the size and distribution of the population.

Counting mobile organisms is much more difficult because individuals move into and out of census areas. Estimating the number of individuals in a population often involves capturing, marking, and then releasing a number of individuals. Later, after the marked individuals have had time to mix with the unmarked individuals in the population, another sample of individuals is taken. The proportion of individuals in the new sample that is marked can be used to estimate the size of the population using the formula

$$\frac{m_2}{n_2} = \frac{n_1}{N}$$

where

n_1 = the number of marked individuals in the first sample (all get marked)

n_2 = the total number of individuals in the second sample

m_2 = the number of marked individuals in the second sample

N = the estimated size of the total population

You may already have guessed that an estimate of total population size computed from such a capture–mark–recapture effort would be accurate only if the marked individuals had randomly mixed with the unmarked individuals and if marked and unmarked individuals were equally likely to survive and be captured. Estimates of total population size will be inaccurate if marked individuals either learn to avoid traps or become "trap-happy" (because, say, they get a quick meal), or if, as a result of being captured, individuals leave the study area. Ecologists have developed statistical techniques that can correct for these errors and improve the accuracy of population estimates.

Birth and death rates can be estimated from population density data

Ecologists use estimates of population densities to estimate the *rate* (number per unit of time) at which births, deaths, and movements take place in a population, and they study how these rates are influenced by environmental factors, life histories, and population densities.

The number of individuals in a population at any given time is equal to the number present at some time in the past, plus the number born between then and now, minus the number that died, plus the number that immigrated into the population, minus the number that emigrated from the population. That is, the number of individuals at a given time, N_1, is given by the equation

$$N_1 = N_0 + B - D + I - E$$

where

N_1 = the number of individuals at time 1

N_0 = the number of individuals at time 0

B = the number of individuals born between time 0 and time 1

D = the number that died between time 0 and time 1

I = the number that immigrated between time 0 and time 1

E = the number that emigrated between time 0 and time 1

If we make these counts over many time intervals, we can determine how a population's density changes over time.

A useful way to display information about birth and death rates in a population is to construct a **life table**. We can construct a life table by tracking a group of individuals born at the same time (called a **cohort**) and determining the number that are still alive at later dates (**survivorship**). Some life tables also include the number of offspring produced by the cohort (*fecundity*) during each time interval.

Life tables were first developed by the Romans nearly 2,000 years ago to determine how much money needed to be set aside to compensate families of soldiers who might be killed in battle. Today, life insurance companies use life tables ("actuarial tables") to determine how much to charge people of different ages for insurance policies.

TABLE 54.1

Life Table of the 1978 Cohort of the Cactus Finch (*Geospiza scandens*) on Isla Daphne

AGE IN YEARS (x)	NUMBER ALIVE	SURVIVORSHIP[a]	SURVIVAL RATE[b]	MORTALITY RATE[c]
0	210	1.000	0.434	0.566
1	91	0.434	0.857	0.143
2	78	0.371	0.898	0.102
3	70	0.333	0.928	0.072
4	65	0.309	0.955	0.045
5	62	0.295	0.678	0.322
6	42	0.200	0.548	0.452
7	23	0.109	0.652	0.348
8	15	0.071	0.933	0.067
9	14	0.067	0.786	0.214
10	11	0.052	0.909	0.091
11	10	0.048	0.400	0.600
12	4	0.019	0.750	0.250
13	3	0.014	0.996	

[a]Survivorship = the proportion of newborns who survive to age x.
[b]Survival rate = the proportion of individuals of age x who survive to age $x + 1$.
[c]Mortality rate = the proportion of individuals of age x who die before the age of $x + 1$.

Biologists can use life tables to predict future trends in populations. A life table based on an intensive study of the seed-eating cactus finch, carried out on Isla Daphne in the Galápagos Archipelago, is shown in **Table 54.1**. The data come from 210 birds that hatched in 1978 and were followed until 1991, at which time only 3 individuals were still alive. The table shows that the mortality rate for these birds was high during the first year of life. It then dropped dramatically for several years, then showed a general increase in later years. Mortality rates fluctuated among years because the survival of these birds depends on seed production, which is strongly correlated with rainfall. The Galápagos archipelago experiences both drought years and years of heavy rain. During drought years, plants produce few seeds, birds do not nest, and adult survival is poor. In years when rainfall is heavy, seed production is high, most birds breed several times, and adult survival is high. The survival rates in the table reflect these rainfall fluctuations. Reproductive rates (not shown in the table) also varied greatly from year to year, but females of all ages bred successfully during years of high rainfall.

Graphs are helpful for highlighting important changes in populations. Graphs of survivorship in relation to age show when individuals survive well and when they do not. Survivorship curves for many populations fall into one of three patterns. In some populations, most individuals survive for most of their potential life span and then die at about the same age. For example, because of intensive parental care and the availability of medical services, the survivorship of humans in the United States is high for many decades, but then declines rapidly in older individuals (**Figure 54.3A**). In a second pattern, which is characteristic of many songbirds, the probability of surviving to the next year is about the same over most of the life span once individuals are a few months old (**Figure 54.3B**). A third widespread survivorship pattern is found among organisms that produce a large number of offspring, each of which receives little investment of energy or parental care. In these species, high death rates of young individuals are followed by high survival rates during the middle part of the life span. *Spergula vernalis*, an annual plant that grows on sand dunes in Poland, illustrates this pattern (**Figure 54.3C**).

The age distribution of individuals in a population reveals much about the recent history of births and deaths in the population. The timing of births and deaths can influence age distributions for many years in populations of long-lived species. The human population of the United States is a good example. Between 1947 and 1964, the United States experienced what is known as the post-World War II baby boom. During these years, average family size grew from 2.5 to 3.8 children; an unprecedented 4.3 million babies were born in

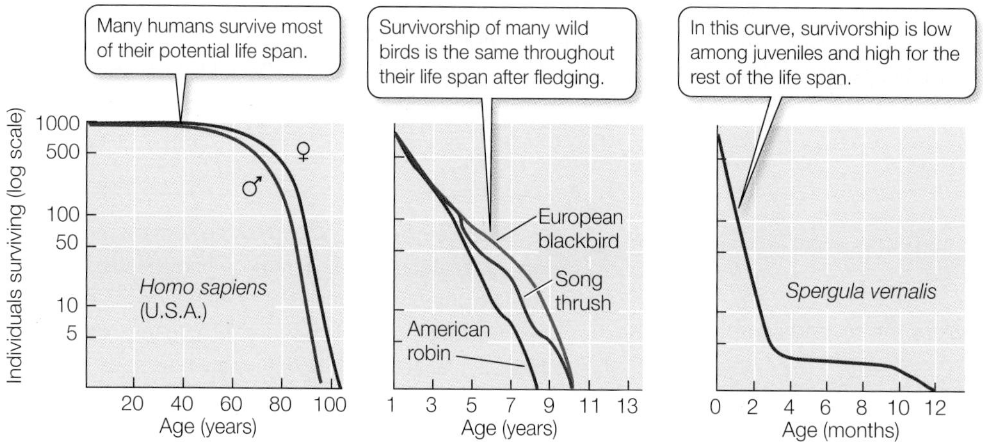

54.3 Survivorship Curves Three common survivorship curves. (A) For some species, mortality is highest at advanced ages. (B) For other species, the probability of survivorship is similar throughout much of the life span. (C) Still other species have high mortality rates at early ages, but survive well once past a critical point.

1957. Birth rates declined during the 1960s, but Americans born during the baby boom still constitute the dominant age class in the first part of the twenty-first century (**Figure 54.4**). "Baby boomers" became parents in the 1980s, producing another bulge in the age distribution—a "baby boom echo"—but they had, on average, fewer children than their parents did, so the bulge is not as large.

By summarizing information on when individuals are born and die, life tables help us understand why population densities change over time. Life table data can also be used to determine how heavily a population can be harvested and which age groups should be the focus of our efforts to save rare species.

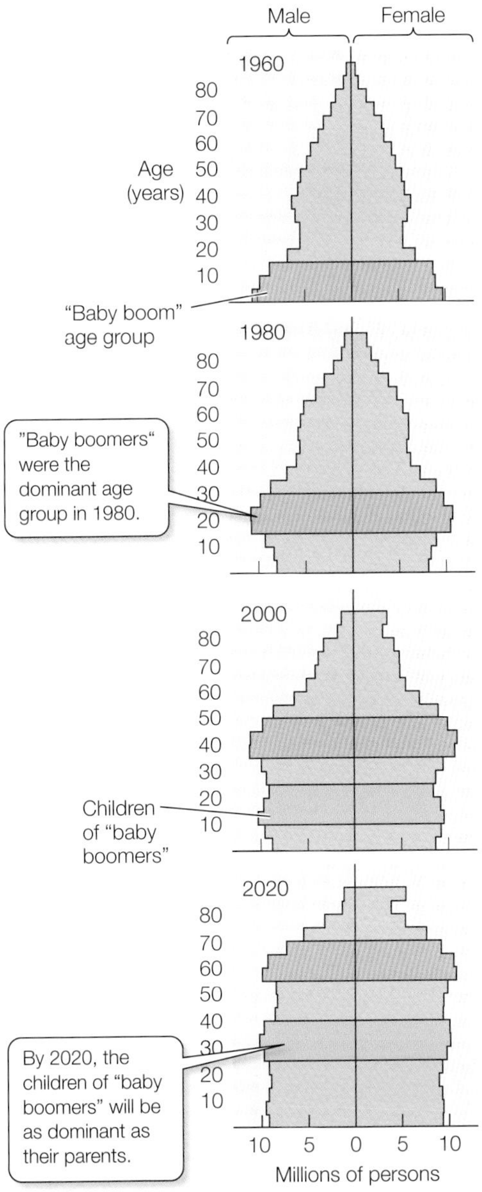

54.4 Age Distributions Change over Time The graphs shows age distributions for the human population of the United States from 1960 to 2020. The high birth rate during the "baby boom" has influenced the structure of the population over many decades.

54.1 RECAP

A population consists of the individuals of a species within a particular area. Population density is a measure of the number of individuals per unit of area or volume. Birth rates, death rates, and other aspects of population dynamics can be estimated from population data collected over time.

- Describe some ways in which population density can be measured. See pp. 1168–1169
- What kinds of information can life tables provide about a population? See pp. 1169–1170 and Table 54.1

During its life, an individual organism ingests nutrients, grows, interacts with other individuals of the same and other species, and usually moves or is moved so that it does not die exactly where it was born. In the next section we discuss the ecological significance of the different ways in which organisms allocate their time and energy to these activities.

54.2 How Do Ecological Conditions Affect Life Histories?

An organism's **life history** describes how it allocates its time and energy among the various activities that occupy its life. The life histories of different organisms vary dramatically. Some plants grow rapidly, produce large numbers of small seeds, and die soon thereafter. Other plants grow slowly, do not reproduce until they are many years old, produce only a few large seeds, and continue to reproduce for decades. Some animals, such as elephants and humans, usually give birth to a single offspring in each reproductive episode; others, such as oysters, produce thousands of eggs in one batch. Some organisms, such as agaves and salmon, usually reproduce only once and then die (**Figure 54.5**). Ecologists study life histories because they influence how populations grow and are distributed.

The life history of the black rockfish (*Sebastes melanops*), which lives off the Pacific coast of North America, offers an example of how life history traits influence the growth of a population that people would like to manage. Female rockfish continue to grow throughout their lives. Larger females are much more productive than smaller females because the number of eggs a female produces is proportional to her size. In addition, older, larger females produce eggs containing larger oil droplets. These droplets provide energy to newly hatched fish, giving them a head start in their independent lives (**Figure 54.6**). Larvae that hatch from eggs with large oil droplets grow faster and survive better than do larvae that hatch from eggs with small oil droplets. These facts have important implications for the harvesting of rockfish populations. Intensive bottom fishing off the Oregon coast from 1996 to 1999 reduced the average age of female rockfish from 9.5 to 6.5 years, so reproductive females were much smaller in 1999 than they were in 1996. This age reduction decreased the number of eggs produced by females in the population and reduced the average growth rates

(A) *Agave americana*

(B) *Oncorhyncus* sp.

54.5 Big Bang Reproduction (A) Agaves, also known as century plants, mobilize the energy stored during their long lives to produce a large flowering stalk with hundreds of flowers, literally reproducing themselves to death. (B) After years spent feeding and growing in the open sea, salmon fight their way up the river in which they hatched. There they spawn and soon die.

of larvae by about 50 percent. Thus maintaining productive populations of rockfish may require setting aside no-fishing zones where some females can grow to very large sizes without being harvested.

Ecological interactions influence the evolution of life histories. The influence of predation on the evolution of life history traits was tested in experiments on guppies (*Poecilia reticulata*). In Trinidad, guppies live in streams where they are preyed on by larger fish. But some streams have waterfalls that predatory fishes are unable to surmount. Guppies that live in the predator-free areas upstream from those waterfalls have low mortality rates; those that live below the falls have high mortality rates. David Reznick and his colleagues reared in the laboratory 240 guppies from high-predation and low-predation sites. They supplied some guppies from each group with plentiful food and others with limited food to match the variation that the fish would encounter in nature. In the laboratory, where no predators were present, they found that guppies from high-predation sites matured earlier, reproduced more frequently, and produced more offspring in each brood than guppies from low-predation sites, no matter how much food they received. Thus predation favored early and frequent reproduction, leading to change in the guppy genotype.

As these examples illustrate, a study of the population dynamics of an organism must take its life history characteristics into account. As we will see at the end of this chapter, life history information is vital for designing population management plans as well.

(A) *Sebastes melanops*

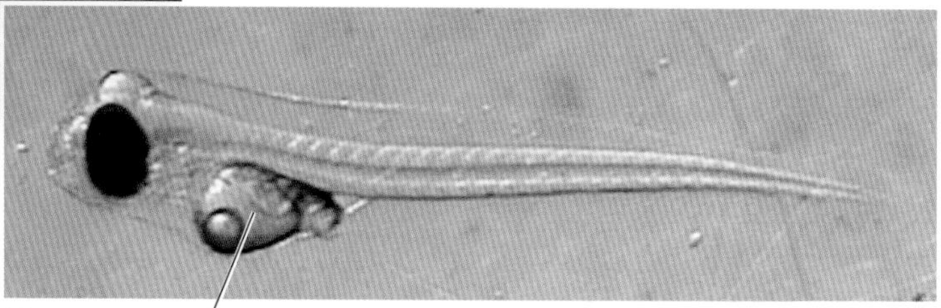

(B)

54.6 An Oil Droplet Is an Energy Kick Start
(A) Among black rockfish, the older, larger females are more reproductively successful, producing both more eggs and eggs with larger stores of nutritive oil. (B) The oil droplet beneath this rockfish larva provides it with nutrition to fuel its growth until it can feed on its own.

54.2 RECAP

The life history of an organism describes how it allocates its time and energy among growth, reproduction, and other activities.

- How can age-related differences in fecundity influence population dynamics? See pp. 1171–1172
- How do mortality patterns influence the evolution of life history traits? See p. 1172

A locally rare species may be abundant somewhere else; but some species exist at low population densities everywhere they are found. A species that is rare at a given time may be abundant at some later time, or vice versa. What factors determine population densities, and why do they vary so much?

54.3 What Factors Influence Population Densities?

Although many populations fluctuate markedly in density, none fluctuate as dramatically as is theoretically possible. Consider, for example, a single bacterium selected at random from the surface of this book. If all its descendants were able to grow and reproduce in an environment with unlimited resources, the population would grow explosively. In a month, this bacterial colony would weigh more than the visible universe and would be expanding outward at the speed of light. Similarly, a single pair of Atlantic cod and their descendants, reproducing at the maximum rate of which they are capable, would fill the Atlantic Ocean basin in 6 years if none of them died. What prevents such dramatic population growth from happening in nature?

All populations have the potential for exponential growth

All populations have the potential for explosive growth. As the number of individuals in a population increases, the number of new individuals added per unit of time accelerates, even if the rate of increase expressed on a per individual basis—called the *per capita growth rate*—remains constant. If births and deaths occur continuously and at constant rates, a graph of the population size over time forms a continuous upward curve (see Figure 54.7): this pattern, known as **exponential growth**, can be expressed as

Rate of increase in number of individuals =
(Average per capita birth rate − Average per capita death rate)
× Number of individuals

or, more concisely,

$$r = \frac{\Delta N}{\Delta t} = (b - d)N$$

where

r = the net reproductive rate

ΔN = the change in number of individuals

Δt = change in time

b = the average population per capita birth rate

d = the average population per capita death rate

The term $\Delta N/\Delta t$ thus conveys the rate of change in the size of the population over time. In other words, the difference between the average per capita birth rate in a population (b) and its average per capita death rate (d) is the *net reproductive rate* (r). (In these equations, b includes both births and immigrations, and d includes both deaths and emigrations.)

The highest possible value for the net reproductive rate—the rate at which the population would grow under optimal conditions—is called r_{max}, or the **intrinsic rate of increase**; r_{max} has a characteristic value for each species. The intrinsic rate of increase can be expressed as

$$\frac{\Delta N}{\Delta t} = r_{max}N$$

For very short periods, some populations may grow at rates close to the intrinsic rate of increase. For example, northern elephant seals were hunted nearly to extinction in the late nineteenth century. In 1890, only about 20 animals remained, confined to Isla Guadalupe off the northwestern coast of Mexico. Ample elephant seal habitat remained available, however, so once the hunting was stopped, the population began to increase rapidly (**Figure 54.7**). Elephant seals recolonized Año Nuevo Island near Santa Cruz, California, in 1960. During the 20 years following colonization, the population breeding on the island expanded exponentially.

Population growth is limited by resources and biotic interactions

No real population can maintain exponential growth for very long. As a population increases in density, environmental limits cause birth rates to drop and death rates to rise. The simplest way to picture the limits imposed by the environment is to assume that an environment can support no more than a certain number of individuals of any particular species per unit of area (or volume). This number, called the **environmental carrying capacity** (***K***), is determined by the availability of resources—such as food, nest sites, or shelter—as well as by disease, predators, and, in some cases, social interactions.

The growth of a population typically slows down as its density approaches the environmental carrying capacity because resource limitations and the activities of predators and pathogens lower birth rates and increase death rates. A graph of population size over time typically forms an S-shaped curve; this pattern is known as **logistic growth** (**Figure 54.8**). The simplest way to generate an S-shaped growth curve is to add to the equation for exponential growth a term that slows the population's growth as it approaches the carrying capacity. This term, expressed as $(K - N)/K$, implies

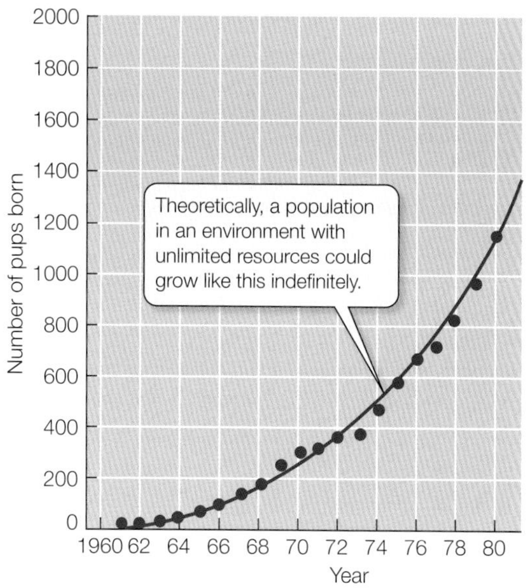

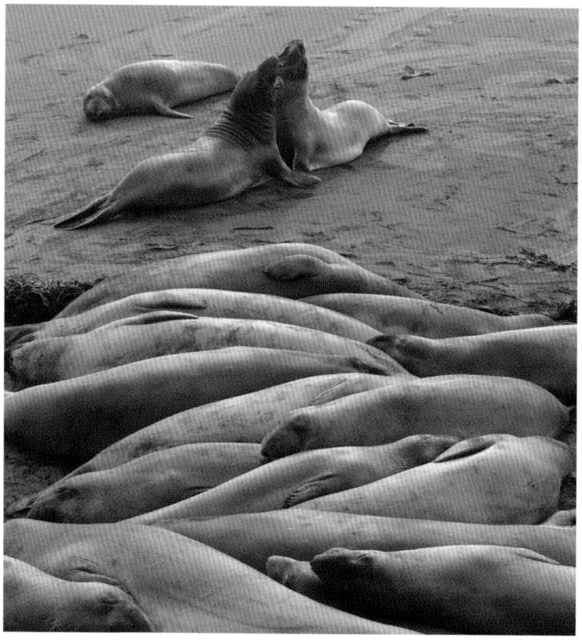

Mirounga angustirostris

54.7 Exponential Population Growth The elephant seal population on Año Nuevo Island, California grew exponentially between 1960 and 1980. Since no resource on Earth is unlimited, this pattern cannot continue for very long. The growth of the seal population on Año Nuevo has indeed slowed down.

that each individual added to the population depresses population growth by an equal amount:

$$\frac{\Delta N}{\Delta t} = r\left(\frac{K-N}{K}\right)N$$

Population growth stops when $N = K$ because then $(K - N) = 0$, so $(K - N)/K = 0$, and thus $\Delta N/\Delta t = 0$.

Population densities influence birth and death rates

Because each additional individual typically makes things worse for other members of the population in an environment with limited resources, per capita birth and death rates usually change together with changes in population density; that is, they are **density-dependent**. Birth and death rates may be density-dependent for several reasons:

- As a species increases in abundance, it may deplete its food supply, reducing the amount of food available to each individual. Poorer nutrition may then increase death rates and decrease birth rates.
- Predators may be attracted to areas with high densities of their prey. If predators capture a larger proportion of the prey than they did when the prey were scarce, the per capita death rate of the prey rises.
- Diseases can spread more easily in dense populations than in sparse populations.

Not all factors affecting population size act in a density-dependent way. A cold spell in winter or a hurricane that blows down most of the trees in its path may kill a large proportion of the individuals in a population regardless of its density. Factors that change per capita birth and death rates in a population independently of its density are said to be **density-independent**.

Fluctuations in the density of a population are determined by all of the density-dependent and density-independent factors acting on it. The combined action of these factors can be seen in a study of the dynamics of a population of song sparrows (*Melospiza melodia*) on Mandarte Island, off the coast of British Columbia, Canada. Over a period of 12 years, the number of song sparrows fluctuated between 4 and 72 breeding females and between 9 and 100 territorial males. Death rates were high during particularly cold, snowy winters, regardless of the density of the population. Several density-dependent factors also contributed to fluctuations in the density of the population. The number of breeding males, for example, was limited by territorial behavior: the larger the number of males, the larger the number that failed to gain a territory and lived as "floaters" with little chance of reproducing (**Figure 54.9A**). In addition, the larger the number of breeding females, the fewer offspring each female fledged (**Figure 54.9B**). And the more birds alive

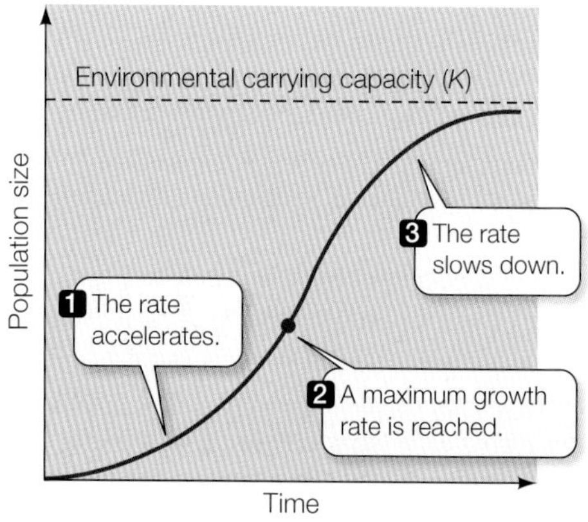

54.8 Logistic Population Growth Typically, a population in an environment with limited resources stops growing exponentially long before it reaches the environmental carrying capacity.

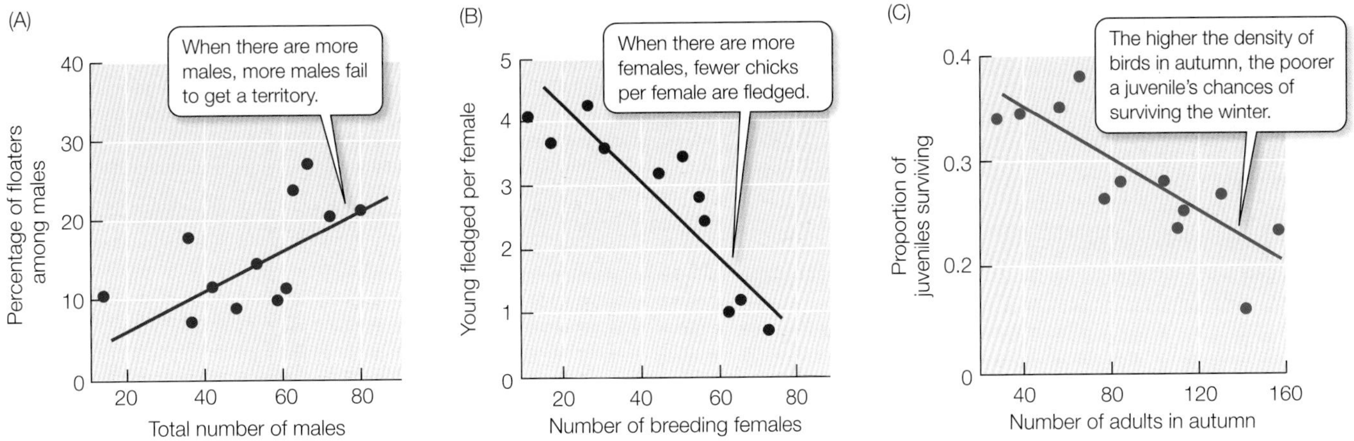

54.9 Regulation of an Island Population of Song Sparrows The size of the population of song sparrows (*Melospiza melodia*) on Mandarte Island, British Columbia, is determined in part by the severity of winter weather. In addition, the population is regulated by density-dependent factors, including (A) the territorial behavior of males, (B) the reproductive success of females, and (C) the survival of juveniles over the winter.

in autumn, the poorer were the chances that juveniles born in that year would survive the winter (**Figure 54.9C**). Thus the number of males and females breeding each year was influenced by both density-independent and density-dependent factors.

All populations fluctuate less than the theoretical maximum, but the sizes of some populations fluctuate remarkably little. In general, more stable populations are seen in species with long-lived individuals that have low reproductive rates than in short-lived species that have high reproductive rates. Small, short-lived individuals are generally more vulnerable to environmental changes than long-lived individuals. Insect population densities tend to fluctuate much more than those of birds and mammals, and population densities of annual plants fluctuate much more than those of trees.

Most fluctuations in the densities of populations are driven by changes in the biotic and abiotic environment that change the environmental carrying capacity for the species. Let's consider two examples.

EPISODIC REPRODUCTION GENERATES POPULATION FLUCTUATIONS

For most species, some years are better than others for reproductive success. In Lake Erie, 1944 was such an excellent year for whitefish reproduction that individuals born that year dominated whitefish catches in the lake for several years thereafter (**Figure 54.10A**). Similarly, most of the individuals found in a population of black cherry trees in a Wisconsin forest in 1971 had become established between 1931 and 1941 (**Figure 54.10B**). Population densities increase following years of good reproductive success, but they decrease following years of poor reproduction.

RESOURCE FLUCTUATIONS GENERATE CONSUMER FLUCTUATIONS

Densities of populations of species that depend on a single or just a few resources are likely to fluctuate more than those of species that use a greater variety of resources. Several species of birds and mammals that live in boreal forests feed on the seeds found in conifer cones. Most trees in these forests reproduce *synchronously* (all at the same time) and *episodically* (on an irregular basis). Over large areas, there are years of massive seed production and years of little or no seed production. Some birds (such as crossbills) wander over large areas, looking for places where cones have been produced. Other birds (such as jays and nutcrackers) and some mammals (squirrels) store seeds during years of high production, but they often suffer high mortality rates during years when the trees in their area produce few or no seeds.

Several factors explain why some species are more common than others

The processes that we have just discussed enable us to understand how populations grow, why they fluctuate in size, and why fluctuations in their densities are much less than would be theoretically possible. But they do not explain why some species are common whereas others are rare. Many factors determine why typical population densities vary so greatly among species, but four of them—resource abundance, the size of individuals, the length of time a species has lived in an area, and social organization—exert especially strong influences.

- *Species that use abundant resources generally reach higher population densities than species that use scarce resources.* Thus, on average, animals that eat plants are typically more common than animals that eat other animals.
- *Species with small body sizes generally reach higher population densities than species with large body sizes.* In general, population density decreases as body size increases, because small individuals require less energy to survive than large individuals.

The relationship between body size and population density is illustrated by a logarithmic plot of population density against body size for mammal species worldwide (**Figure 54.11**). Although the relationship is strong, the great scatter of the points on the graph shows that some small species use scarce resources and some large species use abundant resources.

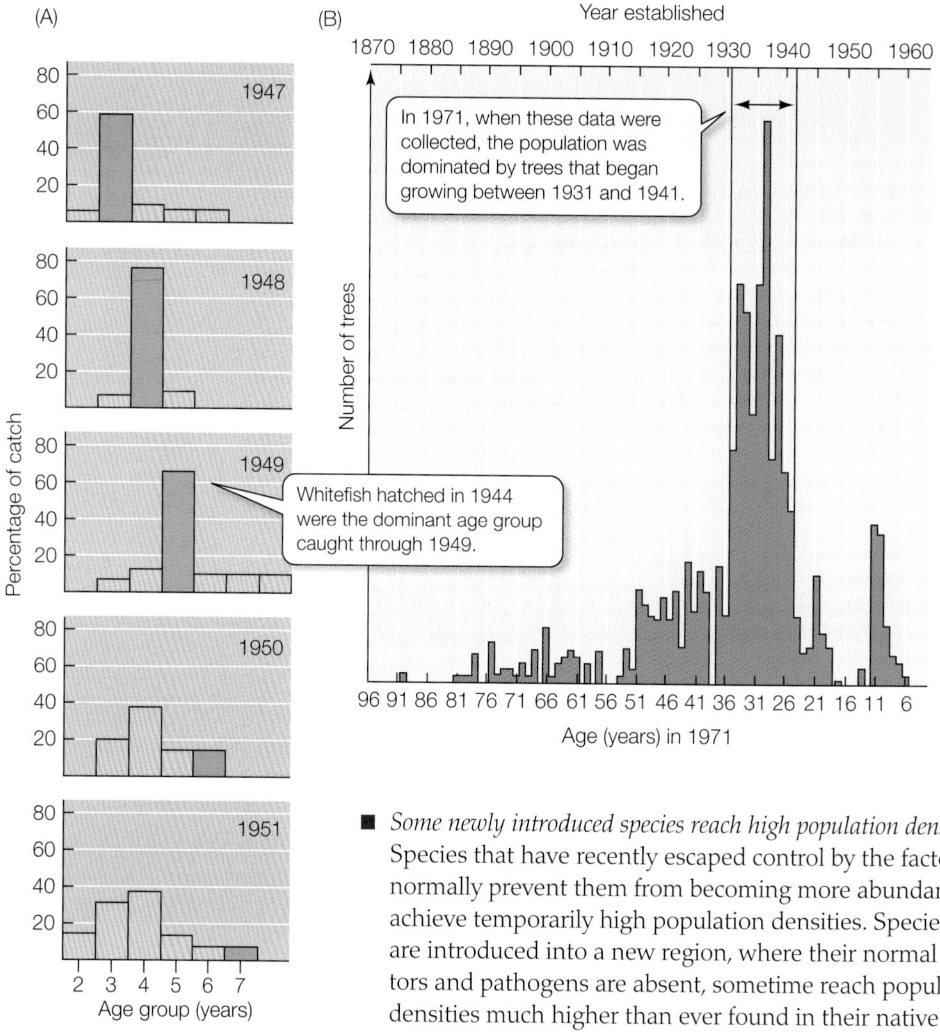

54.10 Individuals Born during Years of Good Reproduction May Dominate Populations (A) Whitefish born in 1944 dominated catches in Lake Erie for many years thereafter. (B) The population of black cherry trees (*Prunus serotina*) in a Wisconsin forest in 1971 was dominated by trees that became established between 1931 and 1941.

- *Some newly introduced species reach high population densities.* Species that have recently escaped control by the factors that normally prevent them from becoming more abundant may achieve temporarily high population densities. Species that are introduced into a new region, where their normal predators and pathogens are absent, sometime reach population densities much higher than ever found in their native ranges.

The zebra mussel (*Dreissena polymorpha*), whose larvae were carried from Europe in the ballast water of commercial cargo ships, became established in the Great Lakes in about 1985. Zebra mussels spread rapidly, and today they occupy much of the Great Lakes and the Mississippi River drainage (**Figure 54.12**). In some places these mussels have reached densities as high as 400,000 individuals per square meter; such densities are never found in Europe. Densities of zebra mussels in North America may decrease in the future if local predators and pathogens begin to attack them.

- *Complex social organization may facilitate high densities.* As we saw in Section 53.6, highly social species, including ants, termites, and humans, can achieve remarkably high population densities.

Important though these four factors are, they cannot explain many differences in abundances of species. For example, both Douglas firs and giant sequoias are large trees that use the same source

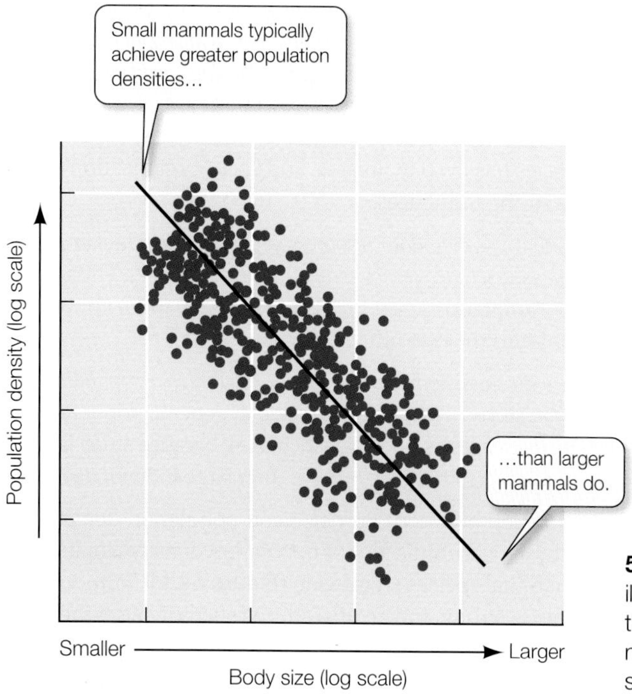

54.11 Population Density Decreases as Body Size Increases This trend is illustrated by a logarithmic plot (that is, each tick mark represents a number 10 times greater than the one before it) of population density against body size for mammals of different sizes. Each dot represents a different mammalian species, and the resulting slope (straight line) is determined algebraically.

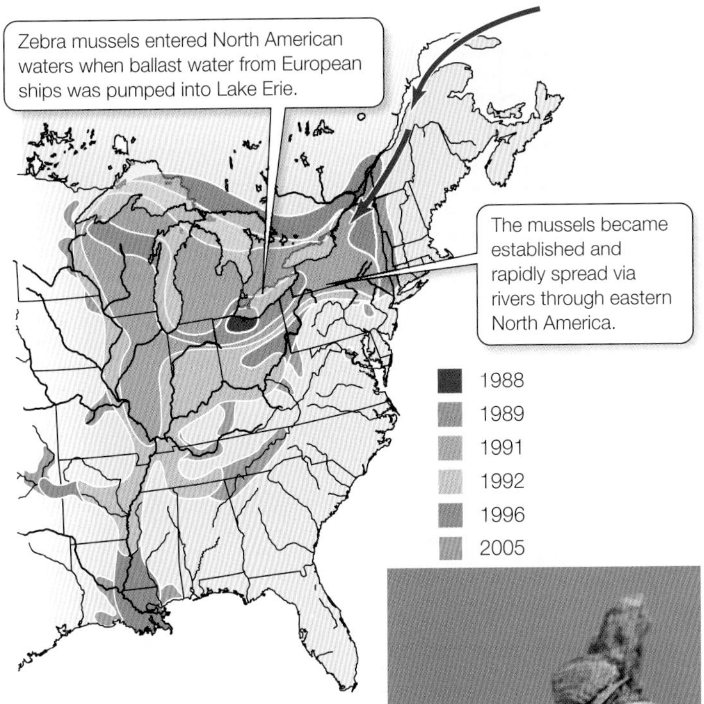

Dreissena polymorpha

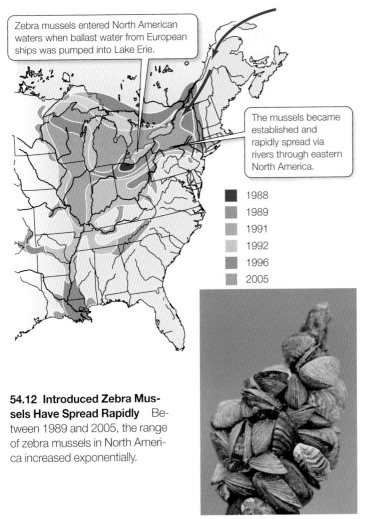

54.12 Introduced Zebra Mussels Have Spread Rapidly Between 1989 and 2005, the range of zebra mussels in North America increased exponentially.

Sequoiadendron giganteum

54.13 The Last Refuge The range of the giant sequoia has progressively shrunk over thousands of years, primarily because of changes in climate. Only a few groves of trees remain, scattered in the southern Sierra Nevada of California.

of energy (sunlight) and need the same nutrients. Yet Douglas firs are widespread and abundant in western North America, whereas giant sequoias are restricted to a few groves in the southern Sierra Nevada of California (**Figure 54.13**). Similarly, several species of desert pupfish are each restricted to a single spring in Death Valley, California, whereas smallmouth bass live in most of the rivers and lakes in eastern North America. To explain these differences, we also need to know about the origins and long-term histories of species.

As Chapter 23 describes, a new species can originate in several ways. A species that arises by polyploidy inevitably begins with a very small, local population. Many polyploid plant species that have formed only recently have not spread much beyond the site of their origin. They have small ranges, although their local population densities may be high. Similarly, species that arise through founder events typically begin their history with only a few individuals. In contrast, most species that arise from a vicariant event (see Section 52.4) begin with large populations and ranges. Finally, as a species declines toward extinction, as may be happening to giant sequoias, its range shrinks until it vanishes when the last individual dies.

Interactions with other species may also limit both the densities and ranges of species. We will explore the consequences of those interactions in Chapter 55.

54.3 RECAP

Many populations fluctuate markedly in density, but even the most dramatic fluctuations are much less than those that are theoretically possible. Population sizes are limited by the environmental carrying capacity, which is determined by biotic interactions and by the availability of resources.

- Why can populations grow exponentially only for short periods? See pp. 1173–1174
- How do density-dependent and density-independent factors interact to determine population densities? See pp. 1174-1175 and Figure 54.9
- How do resource availability, the size of individuals, the length of time a species has lived in an area, and social organization influence population densities? See pp. 1175–1177

All species, no matter how abundant, are found in only those habitats in which they can survive and reproduce well enough to persist over time. Yet a species is rarely found in all of the habitats that seem suitable for it. The next section explores why this is the case.

54.4 How Do Spatially Variable Environments Influence Population Dynamics?

Most natural history field guides display maps that show the geographic range over which a species is found. But you know that you will not find individuals of a species everywhere within the area indicated on the map. No species, not even the most abundant, is found everywhere within its mapped range. To know where to look for the species, you consult the text that describes the habitat in which the species lives.

Many populations live in separated habitat patches

Most populations are divided into separated, discrete *subpopulations* that live in distinct **habitat patches**—areas of a particular kind of environment that are surrounded by other kinds. The larger population to which such subpopulations belong is referred to as a **metapopulation**. Each subpopulation has a probability of "birth" (colonization of that habitat patch) and "death" (extinction in that patch). Growth occurs in each subpopulation in the ways we have just described, but because the subpopulations are much smaller than the metapopulation, local disturbances and random fluctuations in numbers of individuals are more likely to cause the extinction of a subpopulation than the extinction of an entire metapopulation. However, if individuals move frequently between subpopulations, immigrants may prevent declining subpopulations from becoming extinct, a process called the **rescue effect**.

The bay checkerspot butterfly (*Euphydryas editha bayensis*) provides a good illustration of the dynamics of metapopulations. The caterpillars (larvae) of this butterfly feed on only a few species of annual plants, which are restricted to outcrops of serpentine rock on hills south of San Francisco, California. The bay checkerspot has been studied for many years by Stanford University biologists. During drought years, most host plants die early in spring, before the caterpillars have developed far enough to be able to enter their summer resting stage. At least three butterfly subpopulations became extinct during a severe drought in 1975–1977. The largest patch of suitable butterfly habitat, Morgan Hill, typically supported thousands of butterflies (**Figure 54.14**). Until the population on Morgan Hill became extinct recently, it probably served as a source of individuals that dispersed to and recolonized small patches where the butterflies had become extinct.

In another study, ecologists manipulated the habitat of tiny arthropods (springtails—tiny hexapods without wings—and mites) to investigate the metapopulation dynamics of these animals. In one experiment, they created isolated patches of their habitat—mosses growing on rocks—by clearing moss from parts of the rock surface (**Figure 54.15, Experiment 1**). The number of species present in these patches declined about 40 percent within a year, with more rare species than common species disappearing from the patches. The experiment demonstrated that small populations were more likely to become extinct than large populations.

In a second experiment, the investigators created similar patches, but these patches were connected by narrow corridors of moss that were either intact or disrupted by a barrier only 10 millimeters wide

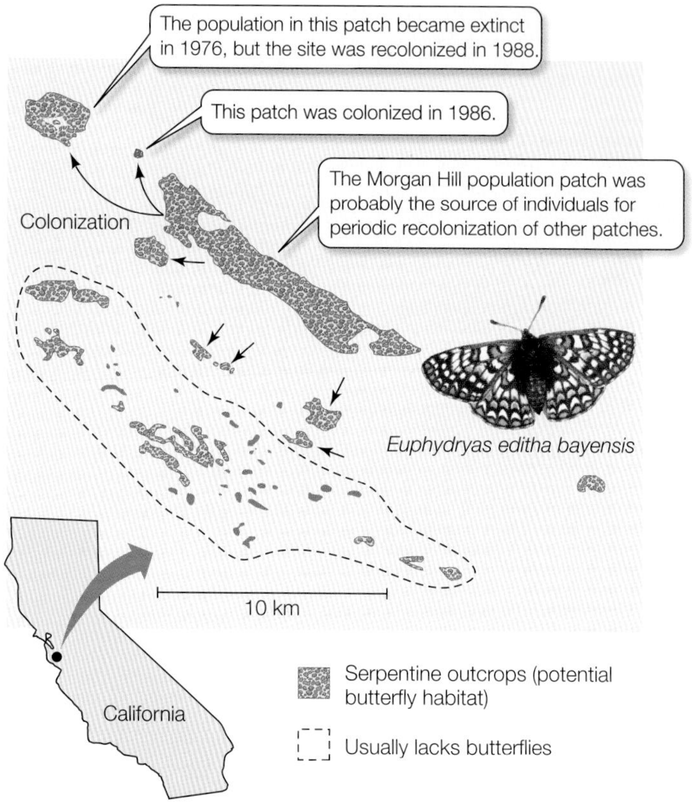

54.14 Metapopulation Dynamics The bay checkerspot butterfly population is divided into a number of subpopulations confined to patches of habitat (serpentine rock outcrops) that contain the food plants of its larvae. Extinction of these subpopulations is common. Indeed, no butterflies survive today in any of the habitat patches.

(**Figure 54.15, Experiment 2**). Moss patches connected by unbroken corridors contained more species of arthropods 6 months later than patches connected by discontinuous "pseudocorridors." Thus a gap between patches of only 10 millimeters was sufficient to reduce the rescue effect for these tiny organisms.

Distant events may influence local population densities

Field guides to birds have many maps that show both breeding and winter ranges of migratory species. In some cases the winter range is on another continent. Populations of migratory species may be influenced by events on both the breeding and wintering grounds, as well as in the places where the individuals stop to rest and feed during migration. To understand fluctuations in populations of migratory species, ecologists may need to study distant events.

A census of breeding birds conducted in Eastern Wood, in southeastern England, illustrates this point. Between 1950 and 1980, populations of some species increased while others decreased (**Figure 54.16**). The population of wood pigeons more than doubled, but the population of garden warblers decreased to zero in 1971; no more than two pairs have bred in the wood since then. The population of blue tits increased from just a few pairs to an average of more than 15 pairs. Why did these populations show such different dynamics?

No matter how intensively ecologists might have studied the birds of Eastern Wood, they could not have answered that question, because populations of two of these three species were

EXPERIMENT

HYPOTHESIS: Even small barriers to recolonization may reduce the number of species in a habitat patch.

METHOD

Moss growing on rocks was trimmed to form distinct habitat patches. The number of small organisms (mostly arthropods) living in the patches was observed over time.

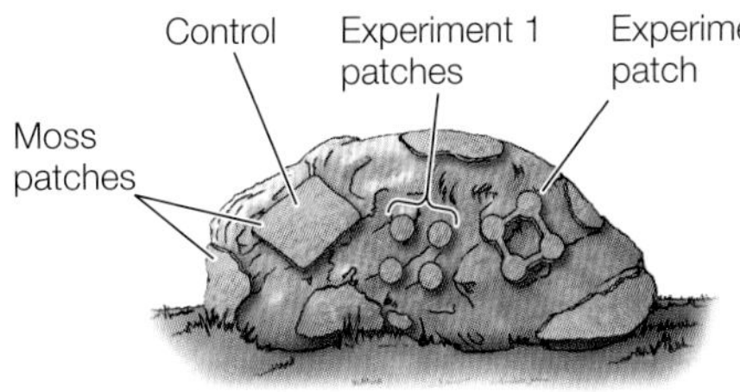

Experiment 1

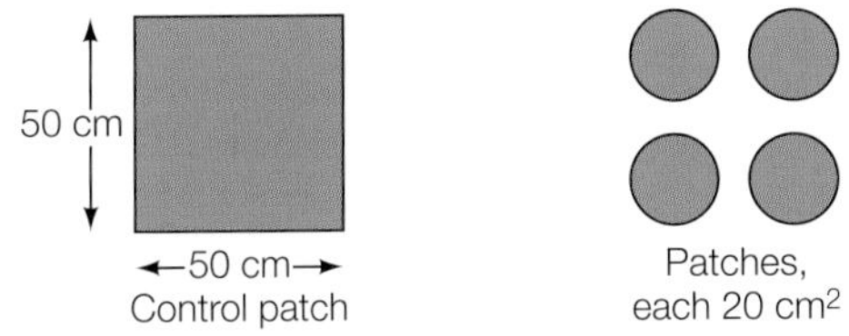

RESULTS

In patches, 40% of species became extinct after 1 year.

Experiment 2

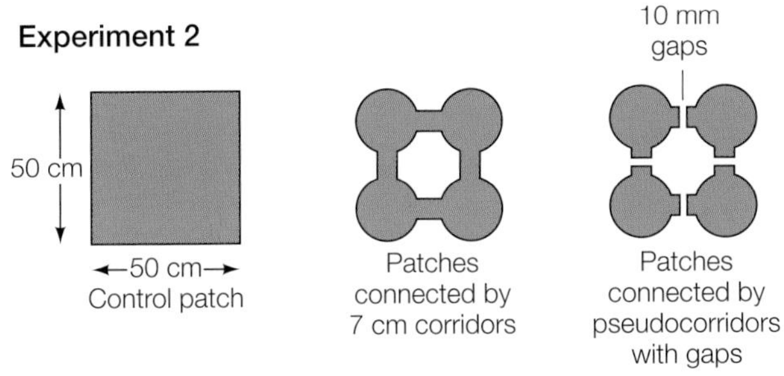

RESULTS:	14% of the species became extinct after 6 months.	41% of the species became extinct after 6 months.

CONCLUSION: Even small barriers to recolonization raised extinction rates in a metapopulation.

54.15 Narrow Barriers Suffice to Separate Arthropod Subpopulations Many species of small arthropods went extinct in isolated habitat patches. Barriers to dispersal as small as 10 millimeters prevented recolonization of patches. FURTHER RESEARCH: These experiments investigated the dispersal of very small organisms over very short distances over very short time spans. The number of species in the isolated patches was unlikely to have reached equilibrium during that time. How could the longer-term effects of imposing barriers to dispersal be investigated?

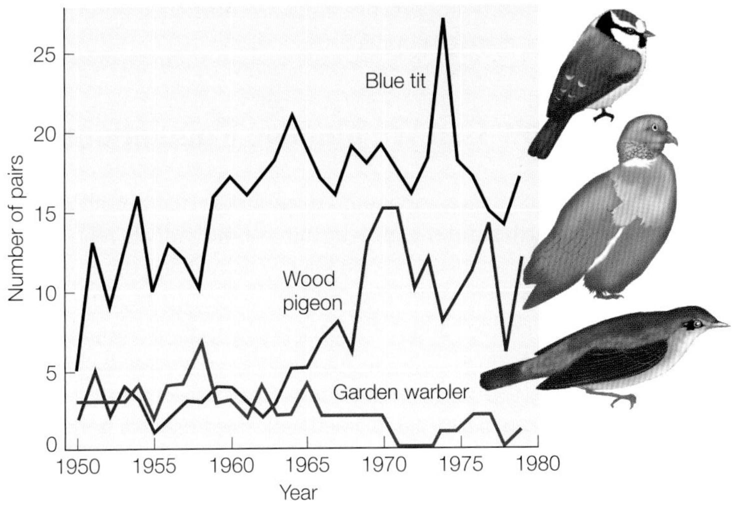

54.16 Populations May Be Influenced by Remote Events Populations of some birds increased in Eastern Wood, England, but others decreased. The wood pigeon (*Columba palumbus*) and garden warbler (*Sylvia borin*) population shifts were strongly influenced by different events that took place far from Eastern Wood. Only the blue tit (*Parus caeruleus*) population was affected most strongly by events within Eastern Wood itself.

strongly influenced by events outside Eastern Wood. Wood pigeons increased greatly over most of southern England during this 30-year period because of the widespread adoption of oilseed rape (the source of canola oil) as an agricultural crop. Rape fields provide wood pigeons with abundant winter food. Garden warblers decreased because their overwinter survival was poor due to a severe drought on their wintering grounds in West Africa.

The population of blue tits was influenced primarily by changes within Eastern Wood itself. Until the early 1950s, trees in Eastern Wood were periodically felled and sold for timber. After the cutting stopped, more holes, in which blue tits nest, became available in mature and dead trees.

54.4 RECAP

Most populations are divide into subpopulations that inhabit patches of suitable habitat.

- Why are many populations divided into subpopulations? See p. 1178 and Figure 54.14
- How can barriers to dispersal reduce species richness in a habitat patch? See p. 1178 and Figure 54.15
- How can distant events influence local population densities? See p. 1179

For many centuries, people have tried to reduce populations of species they consider undesirable and maintain or increase populations of desirable or useful species. Efforts to control and manage populations of organisms are more likely to be successful if they are based on knowledge of how those populations grow and what determines their densities. Let's see how such information can be used to manage populations.

54.5 How Can We Manage Populations?

A general principle of population dynamics is that both the total number of births and the growth rates of individuals tend to be highest when a population is well below its carrying capacity (see Figure 54.8). Therefore, if we wish to maximize the number of individuals that can be harvested from a population, we should manage the population so that it is far enough below carrying capacity to have high birth and growth rates. Hunting seasons for game birds and mammals are established with this objective in mind.

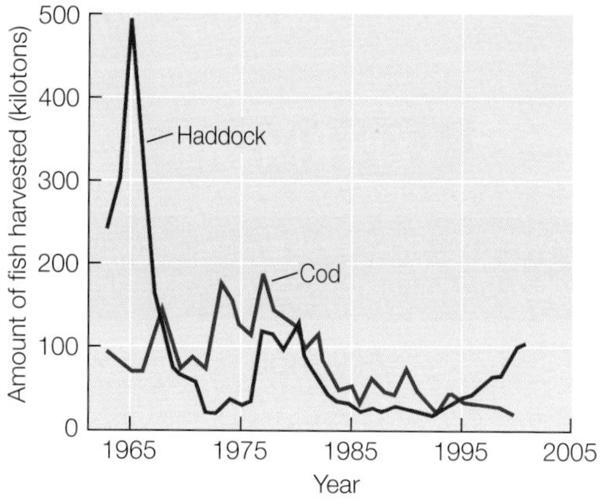

54.17 Overharvesting Can Reduce Fish Populations Populations of cod and haddock on Georges Bank have crashed due to overfishing.

Demographic traits determine sustainable harvest levels

Populations that have high reproductive capacities can persist even if harvest rates are high. In such populations (which include many species of fish), each female may produce thousands or millions of eggs. In these fast-reproducing populations, growth rates of individuals are often density-dependent. Therefore, if prereproductive individuals are harvested at a high rate, the remaining individuals may grow faster. Some fish populations can be harvested heavily on a sustained basis because a modest number of females can produce sufficient eggs to maintain the population.

Fish can, of course, be overharvested, as illustrated by the story of the black rockfish in Section 54.2. Many fish populations have been greatly reduced because so many individuals were harvested that the few surviving reproductive adults could not maintain the population. Georges Bank off the coast of New England—a source of cod, haddock, and other prime food fishes—was exploited so heavily during the twentieth century that many fish stocks were reduced to levels insufficient to support a commercial fishery (**Figure 54.17**). The haddock fishery has rebounded enough to support a valuable fishery today because commercial fishing ceased and was restarted only after the population had recovered. In contrast, managers reduced fishing pressure on cod only slowly; the cod population has failed to increase.

Mid-nineteenth-century New England fishing logs contain geographically specific catch records. Using these documents, scientists estimated that in 1852, the biomass of cod on the rich fishing bank south of Nova Scotia was about 1,260,000 metric tons. Today the cod biomass is less than 50,000 metric tons.

The whaling industry has also engaged in overharvesting. Twentieth-century whalers hunted the blue whale, Earth's largest animal, nearly to extinction. They then turned to smaller species of whales that were still numerous enough to support commercially viable whaling operations. Most whale populations have failed to recover.

Management of whale populations is difficult for two reasons. First, unlike most fish, whales reproduce at very low rates. They live for years before becoming reproductively mature, produce only one offspring at a time, and have long intervals between births. Thus many adult whales are needed to produce even a small number of offspring. Second, because whales are distributed widely throughout Earth's oceans, they are an international resource whose conservation and wise management depend on cooperative action by all whaling nations.

It is therefore not surprising that the International Whaling Commission, established as an international body to guide the recovery of whale populations, has had a contentious history. The Commission's member nations initially voted to ban all commercial harvesting of whales, but some members have lobbied to restore harvests of the species that are not endangered. Other member nations and most nongovernmental organizations continue to oppose resumption of commercial whaling under any circumstances. The behind-the-scenes machinations are complex and the recommendations of the Commission are not binding. A member nation can choose to continue commercial harvesting, as Norway has done, or conduct commercial whaling under the guise of conducting scientific research to better understand whale population dynamics, as Japan has done.

Market forces rather than international pressures may cause the end of commercial whaling. For example, the Japanese Government's Institute of Cetacean Research relies on selling whale meat to fund its "research" program, but fewer and fewer Japanese people eat whale meat. There is a glut of whale meat on the market despite government campaigns encouraging young people to eat whale meat and providing "whale burgers" for school lunches. Faced with a declining market, the Japanese government may no longer have reason to push for resumption of commercial whaling.

Demographic information is used to control populations

The same management principles apply if we wish to reduce the size of populations of undesirable species and keep them at low densities. At densities well below carrying capacity, populations typically have high birth rates and can therefore withstand higher death rates than they can when they are closer to carrying capacity. When population dynamics are influenced primarily by factors that operate in a density-dependent manner, killing part of a population typically reduces it to a density at which it reproduces at a

Bufo marinus

54.18 Biological Control Gone Awry The toad *Bufo marinus* not only failed to control destructive cane beetles in Australia, but increased dramatically in abundance and now threatens many species of Australian animals.

higher rate. A more effective approach to reducing such a population is to remove its resources, thereby lowering the carrying capacity of its environment. For example, we can rid our cities of rats more easily by making garbage unavailable (reducing the carrying capacity of the rats' environment) than by poisoning rats (which only increases their reproductive rate). However, this option is not useful as a means to control agricultural pests because a high density of the crop is the management objective.

As we saw at the opening of this chapter, humans may attempt to control populations of an undesirable species by introducing one of its predators; this worked for *Opuntia* and *Cactoblastis* in Australia. Sometimes such efforts are successful, but just as often they are not.

Sometimes an introduced predator or parasite simply fails to control its host; more serious consequences occur when a species introduced to control an exotic pest not only attacks the pest species but also destroys other species that are considered valuable. This happened with *Cactoblastis* in Mexico and the southwestern U.S. It is also what happened when the toad *Bufo marinus* was introduced into Australia from Central America (**Figure 54.18**), only to become one of Australia's worst ecological disasters.

The idea was for the toads to control the cane beetles that were attacking sugar cane fields in northern Australia. The toads were believed to have controlled cane beetles in Hawaii, so a number of them were released in Australian fields. Unfortunately, the toads could not reach the beetles, which stayed high on the upper stalks of the sugar cane plants, so they had no effect on the beetle population. But they had massive effects on other species.

All stages of the *Bufo marinus* life cycle are poisonous, and Australian snakes and mammals that eat them usually die. The toads grow fast and outcompete native amphibian species for resources. They have spread from northern Australia down the east coast, where they are threatening the native frog species. Currently the Australian government is spending millions of dollars in an attempt to genetically manipulate a pathogen that will kill *B. marinus* exclusively.

Can we manage our own population?

Managing our own population has become a matter of great concern because the size of the human population contributes to most of the environmental problems we are facing today, from pollution to extinctions of other species. For thousands of years, Earth's carrying capacity for humans was set at a low level by food and water supplies and disease. We saw in Section 53.6 how human social behavior and specialization have allowed us to develop technologies for increasing our resources and combating diseases. Our social behavior, the domestication of plants and animals, improved crops and farm yields, mining and use of fossil fuels, and the development of modern medicine have all contributed to a staggering increase in Earth's human population.

It took more than 10,000 years for Earth's population of *Homo sapiens* to reach 1 billion, an event that occurred in the late nineteenth century. In the subsequent 125 years, the human population has increased to 6.5 billion. The growth rate has slowed somewhat in the last two decades; the World Resources Institute estimates the current worldwide rate of population increase to be about 1.1 percent per year. But with a base of 6.5 billion, even a minimal growth rate means millions more individuals.

> If half the people alive today were to mysteriously vanish overnight, there would still be more than twice as many people on Earth as there were in 1900.

Earth's present carrying capacity for humans is set in part by the biosphere's ability to absorb the by-products—especially carbon dioxide—of our enormous consumption of fossil fuel energy; by water availability (in many areas); and by whether we are willing to cause the extinction of millions of other species to accommodate our increasing use of Earth's resources. We will explore the global cycles of these resources in Chapter 56, and we will enumerate some of the consequences of high human population densities and high per capita use of resources for the survival of other species in Chapter 57. Chapter 57 will also discuss the actions being taken to try and maintain Earth's biological diversity.

54.5 RECAP

Efforts to control and manage populations are more likely to be successful if they are based on knowledge of how those populations grow and what determines their densities. At densities well below carrying capacity, populations typically have high birth rates and can therefore withstand higher death rates than they can when they are closer to carrying capacity.

- What general ecological principles guide human efforts to manage other species?

CHAPTER SUMMARY

54.1 How do ecologists study populations?

A **population** consists of the individuals of a species within a particular area.

The distribution of the ages and locations of individuals in a population describes the **population structure**.

Population density is the number of individuals per unit of area or volume.

Demographic events such as births, deaths, immigration, and emigration affect the structure of a population.

Ecologists often mark and track individual animals in population studies.

Population sizes can be estimated from representative samples.

Life tables provide summaries of births, deaths, and other demographic events in a population. Review Table 54.1

A life table tracks a **cohort** of individuals born at the same time and records the **survivorship** of those individuals over time.

54.2 How do ecological conditions affect life histories?

The **life history** of an organism describes how it allocates its time and energy among growth, reproduction, and other activities.

Mortality rates can influence the evolution of life history characteristics.

54.3 What factors influence population densities?

Many populations fluctuate markedly in density, but even the most dramatic fluctuations are much less than those that are theoretically possible.

Populations can exhibit **exponential growth** for short periods, but eventually the **environmental carrying capacity** is approached, causing birth rates to drop and death rates to rise. Review Figure 54.7, Web/CD Activity 54.1 and Tutorial 54.1

Logistic growth is the pattern seen when the growth of a population slows as its density approaches the environmental carrying capacity. Review Figure 54.8, Web/CD Tutorial 54.2

Population densities are determined by the combined influences of **density-dependent** and **density-independent** factors. Review Figure 54.9

When and how species form influences their range and local population density.

Several factors—including resource abundance, the size of individuals in a population, the length of time a species has occupied an area, and social organization—exert strong influences on the densities achieved by populations of different species.

54.4 How do spatially variable environments influence population dynamics?

No species is found everywhere within its mapped range. Members of most species live as separate subpopulations within suitable **habitat patches**. See Web/CD Tutorial 54.3

A **metapopulation** consists of separate subpopulations among which some individuals move on a regular basis.

Extinction of a subpopulation may be prevented by immigration of individuals from another subpopulation; this process is known as the **rescue effect**.

Distant events may influence local population densities.

54.5 How can we manage populations?

To maximize the number of individuals that can be harvested from a population, the population should be kept well below carrying capacity.

Species that have high reproductive capacities can persist even if harvested at high rates.

Reducing the carrying capacity of the environment is a more effective way to reduce an unwanted population than killing its members.

Predators may be introduced to control populations of introduced species, but they may cause other problems.

Earth's carrying capacity for humans depends on our use of resources and the effects of our activities on other species.

SELF-QUIZ

1. The distribution of the ages of individuals in a population and the way those individuals are spread over the environment describes
   a. population dynamics.
   b. population regulation.
   c. population structure.
   d. subpopulation structure.
   e. biomass distribution.
2. The age distribution of a population is determined by
   a. the timing of births.
   b. the timing of deaths.
   c. the timing of both births and deaths.
   d. the rate at which the population is growing.
   e. all of the above
3. Which of the following is *not* a demographic event?
   a. Growth
   b. Birth
   c. Death
   d. Immigration
   e. Emigration
4. A group of individuals born at the same time is known as a
   a. deme.
   b. subpopulation.
   c. Mendelian population.
   d. cohort.
   e. taxon.
5. A population grows at a rate closest to its intrinsic rate of increase when
   a. its birth rates are the highest.
   b. its death rates are the lowest.
   c. environmental conditions are optimal.
   d. it is close to the environmental carrying capacity.
   e. it is well below the environmental carrying capacity.
6. The process by which immigrants prevent a subpopulation from becoming extinct is called the
   a. colonization effect.
   b. rescue effect.
   c. metapopulation effect.
   d. genetic drift effect.
   e. salvage effect.

7. Density-dependent factors have the greatest effect on population densities when
   a. only birth rates change in response to density.
   b. only death rates change in response to density.
   c. diseases spread in populations at all densities.
   d. both birth and death rates change in response to density.
   e. population densities fluctuate very little.
8. A metapopulation is
   a. an unusually large population.
   b. a population that is spread out over a very large area.
   c. a group of subpopulations among which some individuals move.
   d. a group of subpopulations that are isolated from one another.
   e. a group of subpopulations among which individuals move frequently.
9. The best way to reduce the population of an undesirable species in the long term is to
   a. reduce the carrying capacity of the environment for the species.
   b. selectively kill reproducing adults.
   c. selectively kill prereproductive individuals.
   d. attempt to kill individuals of all ages.
   e. sterilize individuals.
10. Populations that are most readily overharvested are characterized by having
   a. very long-lived adults.
   b. short prereproductive periods and many offspring.
   c. short prereproductive periods and few offspring.
   d. long prereproductive periods and few offspring.
   e. long prereproductive periods and many offspring.

FOR DISCUSSION

1. Most organisms whose populations we wish to manage for higher densities are long-lived and have low reproductive rates, whereas most organisms whose populations we attempt to reduce are short-lived, but have high reproductive rates. What is the significance of these differences for management strategies and the effectiveness of management practices?
2. In the mid-nineteenth century, the human population of Ireland was largely dependent on a single food crop, the potato. When a disease caused the potato crop to fail, the Irish population declined drastically for three reasons: (1) a large percentage of the population emigrated to the United States and other countries; (2) the average age of a woman at marriage increased from about 20 to about 30 years; and (3) many families starved to death rather than accept food from Britain. None of these social changes was planned at the national level, yet all contributed to adjusting the population size to the new carrying capacity. Discuss the ecological principles involved, using examples from other species. What would you have done had you been in charge of the national population policy for Ireland at that time?
3. Because some species introduced to control a pest have become pests themselves, some scientists argue that species introductions should not be used under any circumstances to control pests. Others argue that, provided they are properly researched and controlled, we should continue to use introductions as part of our set of tools for managing pest populations. Which view do you support, and why?

FOR INVESTIGATION

The species whose metapopulation dynamics were studied in the experiment described in Figure 54.15 were all tiny animals with limited dispersal abilities. How could experiments be designed to test the role of barriers to recolonization on the metapopulation dynamics of species such as birds, lizards, and mammals, which readily disperse across large areas? Would you expect the results of such experiments to be similar to those using small arthropods on rocks? Why or why not?

CHAPTER 55 Community Ecology

Host sweet host

Many plant species produce sweet nectar in their flowers. This floral nectar attracts pollinators—animals the plant relies on in order to reproduce. But plants in at least several hundred genera also produce nectar on their nonreproductive (vegetative) parts; this *extrafloral* nectar attracts ants. The plant provides the ants with the nectar as well as with other food rewards and, in some cases, nesting sites. For their part the ants patrol the plants and attack herbivores, pathogens, and competing plants.

Some of these ant-hosting plants are vitally dependent on the insects, which in turn live on and depend on a single host plant species. The phenomenon has been extensively studied among Central American thorn trees of the genus *Acacia* and ants of the genus *Pseudomyrmex*. Because both partners in this *mutualism* reproduce independently, the association must be reestablished in each subsequent generation. Individuals of competing species of ants that consume nectar but do not defend the plant may arrive on a young plant before its mutualistic ants have colonized that plant and can drive them away. How can a plant attract the ants that will help it while discouraging ants that will eat its nectar and then move on?

One way in which plants can attract the right insects is by controlling the composition of their nectars. The nectar produced by most plants contains sucrose and varying amounts of glucose and fructose. Sucrose is a particularly important food for most ant species, which produce an enzyme called invertase that cleaves sucrose into monomers that are easily transported through cell membranes. A surprising finding is that, although the nectars of acacia species that do not have intimate associations with ants contain sucrose, the nectar produced by several species of acacias that are defended by specialist ants does not contain sucrose.

Nectar that lacks sucrose is not attractive to generalist ants because they cannot digest its sugars. The specialist ants, on the other hand, readily eat and efficiently digest the acacia's sucrose-free nectar. Why? Because the nectar of the host plant has a trait that makes digestion of its nectar by specialist ants possible: it contains an enzyme that stimulates specific enzymatic activity in the guts of the specialist ants that enables them to digest the sugars found in the nectar.

Thus, the nectar produced by those acacia species that have intimate associations with ants differs chemically from the nectar produced by other acacias, even closely related species. In addition, the digestive enzymes of specialist ants differ from those of generalist

Home Sweet Home A worker acacia ant (*Pseudomyrmex flavicornis*) enters the hollowed-out thorn of a bullhorn acacia (*Acacia cornigera*) in Costa Rica. The hollow thorn provides a nesting site. The stinging ant protects the tree from many herbivores.

All You Can Eat The acacia ant *Pseudomyrmex ferrugineus* gathers carrot-like growths known as beltian bodies from *Acacia collinsii*. Beltian bodies have no known function beyond serving as food for the ant larvae; the tree gains no benefit beyond the presence of the protective ants.

ants. These differences imply that the mutually beneficial association between the acacias and the ants has existed for a very long time. During that time, traits evolved in both partners that benefit the specialist ants but exclude other species that would not help the plant.

All species interact with other species in various ways. Most of those associations are not as specialized as the ones between individual ant and acacia species, but they nevertheless influence the structure and dynamics of ecological communities. How they do so is the subject of this chapter.

IN THIS CHAPTER we will describe ecological communities, discuss the processes that determine community structure, and show how those processes interact in nature. We will also consider how disturbances affect communities and how ecological communities are assembled over time. We will conclude by considering factors that determine how many species can live together in ecological communities.

CHAPTER OUTLINE

55.1 What Are Ecological Communities?

Charles Darwin is remembered mostly for his contributions to evolutionary theory, but as the following quote from *The Origin of Species* shows, he was also a pioneering ecologist who understood the nature and complexity of the interactions among the species of organisms that live in a particular place.

> It is interesting to contemplate an entangled bank, clothed with many plants of many kinds, with birds singing on the bushes, with various insects flitting about, and with worms crawling through the damp earth, and to reflect that these elaborately constructed forms, so different from each other, and dependent on each other in so complex a manner, have all been produced by the laws acting around us.

The species that live and interact in an area constitute an ecological **community**. The "entangled bank" near Darwin's home was an ecological community that had obvious boundaries defined by adjacent crops, pastures, and gardens. But the organisms living in the bank were not confined within those boundaries. Some of the seeds that landed in the bank and grew into trees and shrubs came from parent plants living far away. The insects and birds Darwin observed must have flown into and out of the bank from a large area. To understand which species live in the bank and how they interact, he would have needed to know about such movements. Knowledge of how the bank had changed over time would also be relevant. Glaciers had covered the area 10,000 years earlier. The plant species Darwin observed had colonized Britain at different times over the thousands of years since the glaciers melted.

Communities are loose assemblages of species

Early in the twentieth century, two leading North American plant ecologists debated the nature of communities. Henry Gleason argued in 1926 that plant communities were loose assemblages of species, each of which was individually distributed according to its unique interactions with the physical environment. In contrast, in a paper published in 1936, Frederick Clements argued that plant communities were tightly inte-

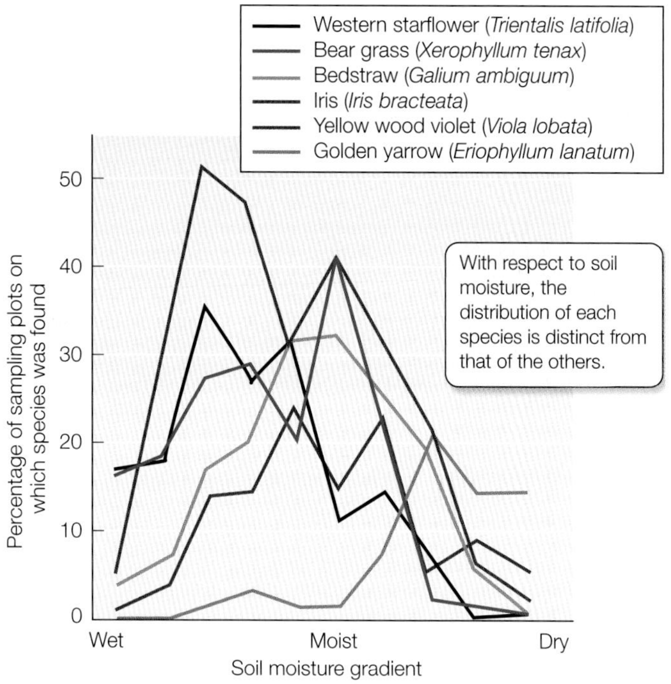

55.1 Plant Distributions along an Environmental Gradient The abundances of different plant species change gradually and individually along a soil moisture gradient in Oregon's Siskiyou Mountains.

grated "superorganisms," and that communities in similar environments would have the same species composition unless they had been recently disturbed.

The debate was resolved by detailed studies of the distributions of plants. Especially influential were analyses of the vegetation of the Siskiyou Mountains of Oregon carried out by Robert Whittaker, who showed that different combinations of plant species are found at different locations. Species enter and drop out of communities independently over environmental gradients (**Figure 55.1**). These and other results generally supported Gleason's view of the nature of communities. However, where environmental conditions change abruptly, as they do at the edges of lakes and streams, the ranges of many species may terminate at the same place.

Thus ecological communities are not assemblages of organisms that move together as units when environmental conditions change. Rather, each species has unique interactions with its biotic and abiotic environments. But if ecological communities are just loose assemblages of species, why do we care about them? We care about ecological communities in part because we wish to know how these assemblages of species, however loose they might be, function. But we also care about them because we too are members of those communities. We interact with many other species, and as we will see, those interactions affect human welfare in many ways.

The organisms in a community use diverse sources of energy

Most ecological communities contain hundreds of species that interact with one another and the physical environment in multitudinous ways, so trying to understand how communities function might seem to be an impossible task. Fortunately, we do not need to know all of the details to make considerable progress. We can understand a great deal by simply finding out who eats whom. The organisms in a community can be divided into trophic levels based on the source of their energy (**Table 55.1**). A **trophic level** consists of the organisms whose energy source has passed through the same number of steps to reach them. Plants and other photosynthetic organisms (*autotrophs*) get their energy directly from sunlight. Collectively, they constitute a trophic level called *photosynthesizers*, or **primary producers**. They produce the energy-rich organic molecules that nearly all other organisms consume.

In most ecological communities, all nonphotosynthetic organisms (*heterotrophs*) consume, either directly or indirectly, the energy-rich organic molecules produced by primary producers. Organisms that eat plants constitute a trophic level called *herbivores* or **primary consumers**. Organisms that eat herbivores are called **secondary consumers**. Those that eat secondary consumers are called *tertiary consumers*, and so on. Organisms that eat the dead bodies of organisms or their waste products are called *detritivores* or **decomposers**. Organisms that obtain their food from more than one trophic level are called *omnivores*. Because many species are omnivores, trophic levels are often not clearly distinct, but if we remember that boundaries between trophic levels are fuzzy, the con-

TABLE 55.1

The Major Trophic Levels

TROPHIC LEVEL	SOURCE OF ENERGY	EXAMPLES
Photosynthesizers (primary producers)	Solar energy	Green plants, photosynthetic bacteria and protists
Herbivores (primary consumers)	Tissues of primary producers	Termites, grasshoppers, gypsy moth larvae, anchovies, deer, geese, white-footed mice
Primary carnivores (secondary consumers)	Herbivores	Spiders, warblers, wolves, copepods
Secondary carnivores (tertiary consumers)	Primary carnivores	Tuna, falcons, killer whales
Omnivores	Several trophic levels	Humans, opossums, crabs, robins
Detritivores (decomposers)	Dead bodies and waste products of other organisms	Fungi, many bacteria, vultures, earthworms

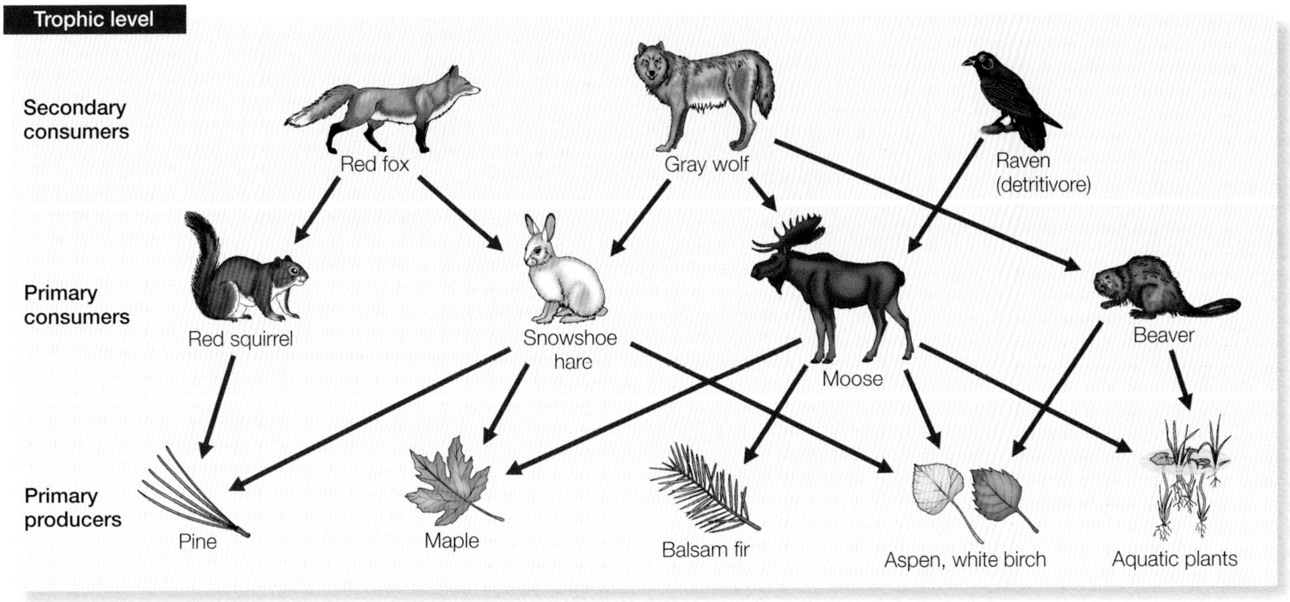

55.2 Food Webs Show Trophic Interactions in a Community This food web for Isle Royale National Park, located on a large island in Lake Superior, includes only large vertebrates and the plants on which they depend. Even with these restrictions, the web is complex. The arrows show who eats whom.

cept still provides a useful way of thinking about energy flow through communities.

A sequence of interactions in which a plant is eaten by an herbivore, which is in turn eaten by a secondary consumer, and so on, can be diagrammed as a **food chain**. Food chains are usually interconnected to make a **food web** because most species in a community eat and are eaten by more than one other species (**Figure 55.2**). Ecological communities contain so many species that it is impossible to show all of them in a food web. Even so, simplified diagrams of food webs can help us understand the trophic interactions among organisms in an ecosystem.

Despite their considerable differences, most communities have only three to five trophic levels. Why are there so few levels? Loss of energy between trophic levels is partly responsible. To show how energy decreases at each step as it flows from lower to higher trophic levels, ecologists construct diagrams that show the distribution of energy or **biomass** (the weight of living matter) at each trophic level in a community. Diagram such as those in **Figure 55.3** show the amount of energy or biomass that is available at a given time for organisms at the next trophic level.

Distributions of energy and biomass for a particular ecosystem usually have similar shapes. Variations in their dimensions depend on the nature of the dominant organisms at each trophic level and how they allocate their energy. In most terrestrial ecosystems, photosynthetic plants dominate, both in terms of the energy they represent and the biomass they contain. They store energy for long periods, some of it in difficult-to-digest forms (such as cellulose and lignin). In forests, the biomass at the primary producer level is mostly wood, which is rarely eaten unless the plant is diseased or otherwise weakened. In contrast, grassland plants produce few hard-to-digest woody tissues. Mammals may consume 30–40 percent of the annual aboveground grassland plant biomass; insects may consume an additional 5–15 percent. Soil organisms, primarily nematodes, may consume 6–40 percent of the belowground biomass in grasslands. Thus, relative to the biomass of plants, the biomass of herbivores is larger in grasslands than in forests (**Figure 55.3A,B**).

In most aquatic ecosystems, the dominant photosynthesizers are bacteria and protists. Those unicellular organisms have such high rates of cell division that a small biomass of photosynthesizers can feed a much larger biomass of herbivores, which grow and reproduce much more slowly. This pattern can result in an inverted distribution of biomass, even though the energy distribution for the same ecosystem has the typical shape (**Figure 55.3C**).

Much of the energy ingested by organisms is converted to biomass that is eventually consumed by decomposers, members of a trophic level not shown in Figure 55.3. **Detritivores** such as bacteria, fungi, worms, mites, and many insects, transform *detritus* (the dead remains and waste products of organisms) into free mineral nutrients that can again be taken up by plants. If there were no detritivores, most nutrients would eventually be tied up in dead bodies, where they would be unavailable to plants. Continued ecosystem productivity depends on the rapid decomposition of detritus.

Dense populations of detrivorous crabs live in the nutrient-poor hydrothermal vent communities of certain shallow ocean waters. During slack tide, when water currents cease, the sulfurous vent plumes asphyxiate vast numbers of copepods, which sink to the ocean floor. The crabs swarm out of cracks in the surrounding rocks and feed on the dead copepods.

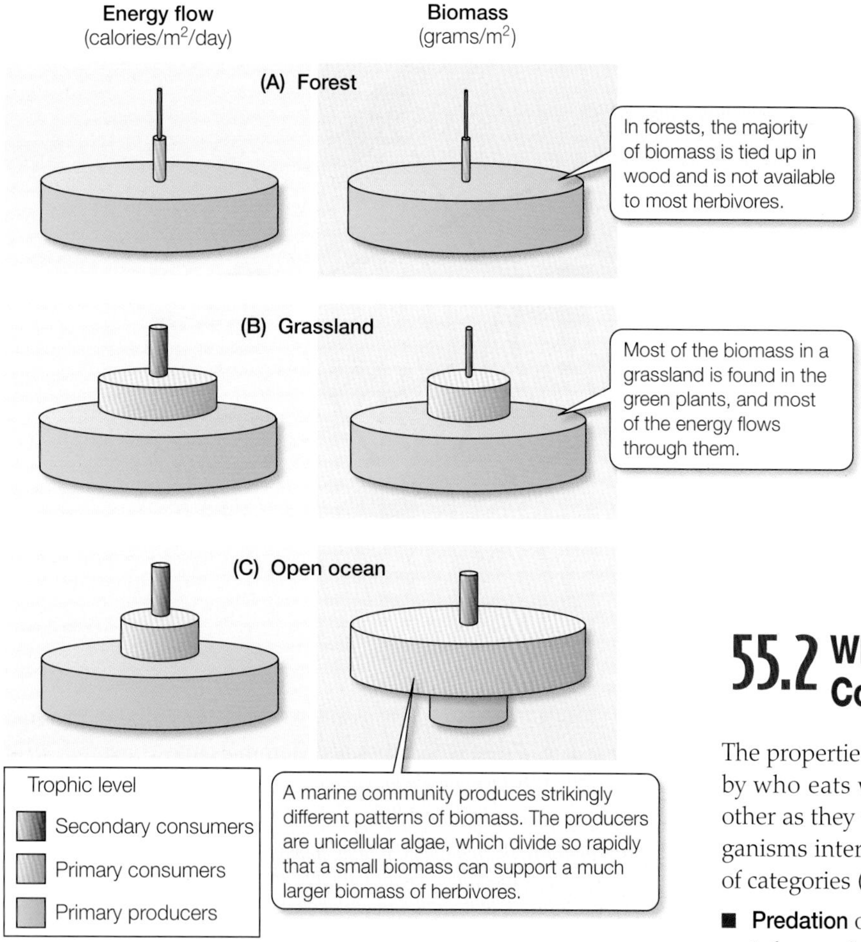

55.3 Diagrams of Biomass and Energy Distributions Energy diagrams (left column) allow ecologists to compare patterns of energy flow through trophic levels in different ecosystems. Biomass diagrams (right column) allow them to compare the amount of material present in living organisms at different trophic levels.

55.1 RECAP

The species that live and interact in an area constitute an ecological community. Although each species has unique interactions within the ecological community, the community as a whole can be studied on the basis of the distribution of energy and biomass within it.

- Do you understand how the concept of trophic levels is useful when describing ecological communities? See p. 1186 and Table 55.1
- What are food webs and what do they describe? See p. 1187 and Figure 55.2
- What is the primary cause of the differences between patterns of distribution of energy and biomass in communities? See p. 1187 and Figure 55.3

Energy and biomass distributions reveal some important things about ecological communities, but they do not tell us which processes most strongly influence the structure and dynamics of a community. In the next section we explore different types of interactions between species and how they influence the properties of communities as a whole.

55.2 What Processes Influence Community Structure?

The properties of ecological communities are influenced not only by who eats whom, but also by how organisms affect one another as they seek food. Fortunately, the many ways in which organisms interact with one another also fall into a small number of categories (**Table 55.2**):

- **Predation** or **parasitism**: interactions in which one participant is harmed, but the other benefits (+/– interactions)
- **Competition**: interactions in which two organisms use the same resources and those resources are insufficient to supply their combined needs (–/– interactions)
- **Mutualism**: interactions in which both participants benefit (+/+ interactions)
- **Commensalism**: interactions in which one participant benefits but the other is unaffected (+/0 interactions)
- **Amensalism**: interactions in which one participant is harmed but the other is unaffected (0/– interactions)

All five types of interactions, combined with the effects of the physical environment, influence the population densities of species. They may also restrict the range of environmental conditions under which species can persist. If there were no competitors, predators, or pathogens in their environment, most species would be able to persist under a broader array of abiotic conditions than they do in the presence of other species. On the other hand, the presence of mutualists may increase the range of physical conditions under which a species can persist.

Predation and competition might, at first glance, seem to be distinct processes, but they interact strongly because most organisms are preyed on by more than one other species, and because most predators include many other species in their diets. Competitors are often predators that share diets. Let's look more closely at all of these types of interactions to see under what conditions they operate and how they influence community dynamics.

TABLE 55.2

Types of Ecological Interactions

		EFFECT ON ORGANISM 2		
		HARM	BENEFIT	NO EFFECT
EFFECT OF ORGANISM 1	HARM	Competition (–/–)	Predation or parasitism (–/+)	Amensalism (–/0)
	BENEFIT	Predation or parasitism (+/–)	Mutualism (+/+)	Commensalism (+/0)
	NO EFFECT	Amensalism (0/–)	Commensalism (0/+)	—

Predation and parasitism are universal

Predation and parasitism are universal processes. Every species is eaten by at least one other species, and no species is entirely free of parasites and pathogens. *Parasites* are typically smaller than their hosts. They may live inside or outside the bodies of their hosts. Parasites often feed on their hosts without killing them. Some parasites are only slightly smaller than their hosts, but *microparasites*, such as pathogenic viruses, bacteria, and protists, are much smaller than their hosts. Multiple generations of microparasites may reside within a single host individual, and a host may harbor thousands or millions of them. As a result, parasite–host interactions differ in interesting ways from predator–prey interactions.

Predators are typically larger than and live outside the bodies of their prey. Herbivores are predators of plants; they may feed on many individual plants without killing them, whereas predators of animals typically kill their prey.

PREDATOR AND PREY POPULATIONS OFTEN OSCILLATE You might think that predators would simply reduce the sizes of their prey populations, but the consequences of predator–prey interactions are much more complex. Predators often do depress populations of their prey, but they also cause fluctuations in prey population *densities*. In part because of this aspect of predation, predator–prey interactions may actually result in increases in the prey populations.

Growth of a predator population nearly always lags behind growth in its prey populations. As a predator population grows, it may eat most of its prey populations. The predator population, which no longer has enough food, then crashes. Population oscillations among small mammals and their predators living at high latitudes, where there are only a few species of predators and prey, are the best-known examples of such population density fluctuations driven by predator–prey interactions. Populations of Arctic lemmings and their chief predators—snowy owls, jaegers, and Arctic foxes—oscillate with a 3- to 4-year periodicity. Populations of Canadian lynx and their principal prey, snowshoe hares, oscillate on a 9- to 11-year cycle (**Figure 55.4**).

For many years, ecologists thought that hare–lynx population oscillations were caused only by interactions between hares and lynx. Recently, Charles Krebs and his associates at the University of British Columbia performed experiments in Yukon Territory, Canada, to test the hypothesis that the lynx–hare oscillations are caused by fluctuations in the hares' food supply as well as by lynx predation. They enclosed some areas with fences through which hares, but not lynx, could pass, and they provided food in some of the enclosures. The results of the experiments showed that the oscillations are driven both by lynx predation and by interactions between hares and their food supply (**Figure 55.5**).

55.4 Hare and Lynx Populations Cycle in Nature The 9- to 11-year population cycle of the snowshoe hare and its major predator, the Canadian lynx, was revealed in records of the number of pelts that were sold to the Hudson's Bay Company by fur trappers.

EXPERIMENT

HYPOTHESIS: Population cycles of hares are influenced by both food supply and predators.

METHOD

1. Select 9 1-km^2 blocks of undisturbed coniferous forest.
2. In two of the blocks give the hares supplemental food year-round.
3. Erect an electric fence around two other blocks, with mesh large enough to allow hares, but not lynxes, to pass through.
4. Provide extra food in one of these enclosed blocks.
5. In two other blocks add fertilizer to increase food quality.
6. Use three other blocks as unmanipulated controls.

RESULTS

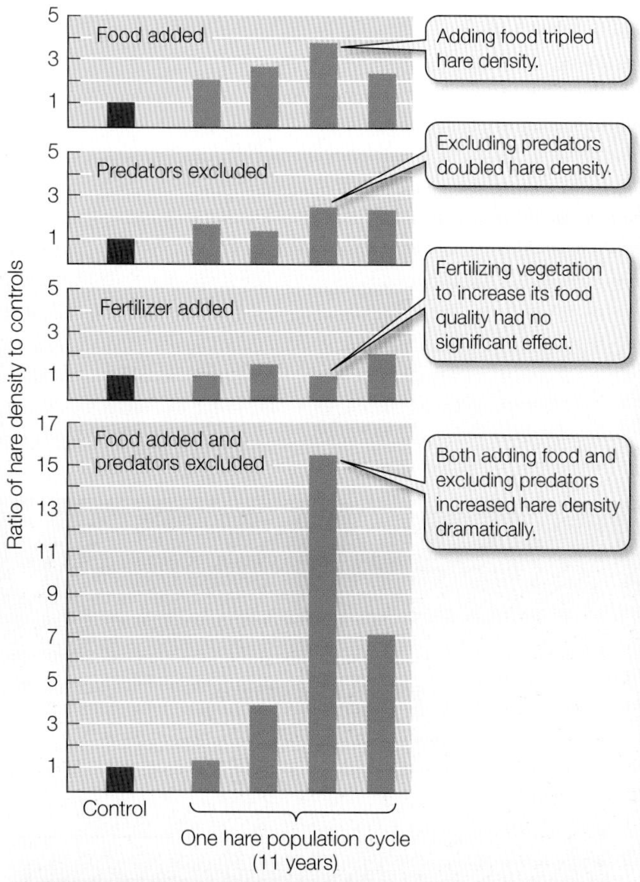
Food added
Adding food tripled hare density.
Predators excluded
Excluding predators doubled hare density.
Fertilizer added
Fertilizing vegetation to increase its food quality had no significant effect.
Food added and predators excluded
Both adding food and excluding predators increased hare density dramatically.
Ratio of hare density to controls
1
3
5
7
9
11
13
15
17
Control
One hare population cycle (11 years)

CONCLUSION: Population cycles of the snowshoe hare are influenced by their food supply as well as by interactions with their predators.

55.5 Prey Population Cycles May Have Multiple Causes Experiments showed that both food supply (but not food quality) and predation by lynx affected the population densities of snowshoe hares.

PREDATORS MAY RESTRICT SPECIES' RANGES Predators may also restrict the habitat and geographic distribution of their prey. The Australasian biogeographic region (see Figure 52.8) is home to a group of birds—megapodes, also called mound-builders—that do not incubate their eggs. Instead, they lay their eggs in a mound of decomposing vegetable material, where they are warmed by the heat of decomposition (**Figure 55.6**). The parent megapodes regularly visit the nest mound and, if necessary, add or remove decaying material to maintain the eggs at an appropriate temperature.

Megapodes are excellent dispersers. They have colonized many remote oceanic islands and are found on many islands west of Wallace's line (see Figure 52.7), but they are absent from all islands with Asian mammalian predators. Unattended eggs in large, conspicuous mounds of decomposing vegetation would be obvious to egg-eating mammals. Megapodes have survived only in regions where the primary predators are marsupials, few of which eat eggs.

MIMICRY EVOLVES IN RESPONSE TO PREDATION Predators do not capture prey individuals randomly. Prey individuals vary in ways that make them more or less susceptible to being captured. Therefore, prey species have evolved a rich variety of adaptations that make them more difficult to capture, subdue, and eat. Among these adaptations are toxic hairs and bristles, tough spines, noxious chemicals, camouflage, and mimicry of inedible objects or of larger and more dangerous organisms. Predators, in turn, have evolved to be more effective at overcoming such prey defenses.

Mimicry is the best-studied adaptation of prey to predation. A palatable species may mimic an unpalatable or noxious one—a process called **Batesian mimicry**—or two or more unpalatable or noxious species may converge to resemble one another—a process called **Müllerian mimicry**. Batesian mimicry works because a predator that captures an individual of an unpalatable or noxious species learns to avoid other prey individuals of similar appearance. However, if a predator captures a palatable mimic, it is re-

Leipoa ocellata

55.6 Megapode Distributions are Limited by Mainland Predators A megapode at its nest mound. Egg-eating mammalian predators quickly destroy the easily accessible eggs; thus megapodes are limited to areas where the only mammalian predators are marsupials, among whom egg-eating is rare.

warded with food. It learns to associate palatability with the appearance of that prey. As a result, individuals of unpalatable species are attacked more often than they would be if they had no Batesian mimics. Unpalatable individuals that differ from their mimics more than the average are less likely to be attacked by predators that have eaten a mimic. Therefore, directional selection (see Section 22.3) causes unpalatable species to evolve away from their mimics. Batesian mimicry systems can evolve and be maintained only if the mimic evolves toward an unpalatable species faster than the unpalatable species evolves away from it. Generally, this happens only if the mimic is less common than the unpalatable species.

All species in a Müllerian mimicry system benefit when inexperienced predators eat individuals of any of the species because the predators learn that that all species of similar appearance are unpalatable. Some of the most spectacular tropical butterflies are members of Müllerian mimicry systems (**Figure 55.7**), as are many kinds of bees and wasps.

HOSTS RESIST INFECTION BY MICROPARASITES For a microparasite population to persist in a host population, at least one new host individual, on average, must become infected with the microparasite before each infected host dies. Members of host populations involved in a microparasite–host interaction fall into three distinct classes: susceptible, infected, or recovered (and thus immune; see Chapter 18). Changes in the numbers of individuals in each class depend on births, deaths, infections, and development and loss of immunity.

A microparasite can readily invade a host population dominated by susceptible individuals, but as the infection spreads, fewer and fewer susceptible individuals remain. Eventually a point is reached at which infected individuals do not, on average, transmit the infection to at least one other individual. Then the infection dies out. As a result, rates of microparasite infection typically rise, then fall, and do not rise again until a sufficiently dense population of susceptible host individuals has reappeared.

A microparasite can be transferred from one host individual to another by direct body contact; via the breath, or body fluids, or waste products of infected individuals; by water; or by an animal vector. A single infected host can most readily infect a large number of other individuals if it continues to infect others for a long time, even after it has died. An infected host individual can most easily spread an infection to many other hosts if the microparasite is dispersed by water. Cholera, one of the most deadly human diseases, is caused by the bacterium *Vibrio cholerae*, which usually lives in the ocean where it infects copepods and other small planktonic animals, but also lives in fresh water. People ingest *V. cholerae* by drinking contaminated water. The bacterium produces a toxin that damages the salt balance mechanisms of the cells that line the small intestine, and infected persons defecate large quantities of the bacteria (see the opening of Chapter 5). A single infected person can release thousands of pathogenic bacteria into local waters and stimulate a new infection. In Bangladesh, where most people get their drinking water directly from rivers, filtering river water through folded cloth can cut infection rates by 50 percent (**Figure 55.8**).

Researchers recently found six viruses of primate origin in the blood of 930 people in Cameroon who had eaten freshly killed primates, or "bush meat." Logging of tropical forests, which increases hunters' access to primates and generates local markets for bush meat, is thus implicated in the transfer of potentially dangerous viruses to humans.

55.7 Batesian and Müllerian Mimicry Systems By converging in appearance, the unpalatable Müllerian mimics among these different species of Costa Rican butterflies and moths reinforce one another's ability to deter predators. The palatable Batesian mimics benefit because predators learn to associate these color patterns with unpalatability.

55.8 Filtering Water Can Help Combat Cholera The *Vibrio cholerae* bacterium is spread in contaminated water supplies. When drinking water is obtained from open supplies such as this one, even the simple act of filtering water through a cloth can lessen the rate of infection.

Competition is widespread because all species share resources

Nearly all species share at least part of their diet and use of other resources with other species, but resource sharing influences the abundances and distributions of species only if individuals reduce the ability of others to access resources, either by interfering with their activities—**interference competition**—or by reducing the available resources—**exploitation competition**.

Competition may occur among individuals of a single species or among individuals of different species. *Intraspecific* competition—competition among individuals of the same species—may result in reduced growth and reproductive rates for some individuals, may exclude some individuals from better habitats, and may cause the deaths of others. Intraspecific competition is a primary cause of the density-dependent birth and death rates that we discussed in Section 54.3. *Interspecific* competition—competition among individuals of different species—affects individuals in the same way, but in addition, a superior competitor can prevent all members of another species from using a habitat, a phenomenon called **competitive exclusion**.

COMPETITION MAY RESTRICT SPECIES' HABITAT USE Photosynthetic plants typically compete for space, which is why gardeners and farmers weed their crops. Occupation of space gives a plant access to sunlight (for which shoots compete) and to water and mineral nutrients (for which roots compete).

Competition among sessile animals may also restrict their habitat distribution. For example, two species of barnacles, *Balanus balanoides* and *Chthamalus stellatus*, compete for space in intertidal zones (**Figure 55.9**). The planktonic larvae of both species settle between high and low tide levels on the rocky shorelines of the North Atlantic Ocean and metamorphose into sessile adults. Curiously, however, the two populations end up occupying distinct settlement areas. Adult *Chthamalus* generally live higher in the intertidal zone than do adult *Balanus*, and there is little overlap be-

55.9 Competition Restricts the Intertidal Ranges of Barnacles Interspecific competition between *Balanus* and *Chthamalus* makes the zone each species occupies smaller than the zone it could occupy in the absence of the other species. The widths of the red and blue bars are proportional to the densities of the populations.

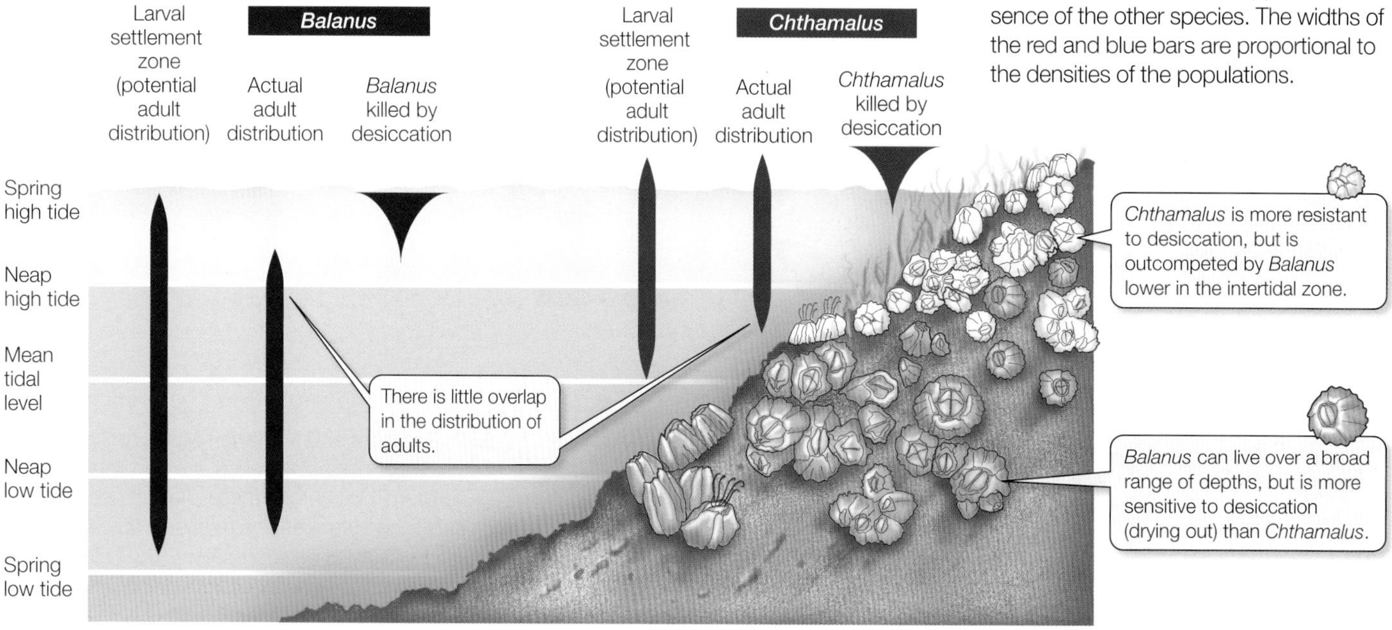

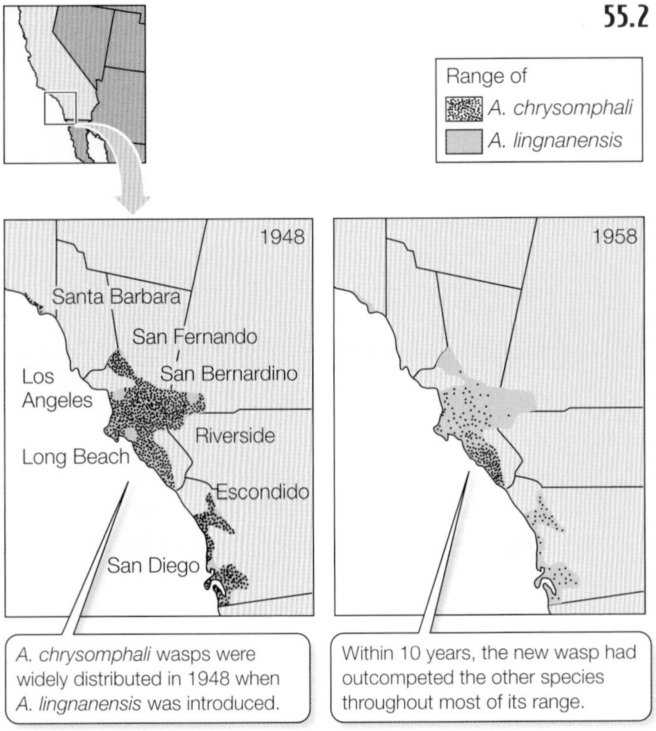

55.10 A Parasitoid Wasp Outcompetes Its Close Relative *Aphytis lingnanensis* displaced *A. chrysomphali* over most of its range within a decade of its introduction.

tween areas occupied by the two species. What explains their distinct distributions in the intertidal zone?

By experimentally removing one or the other species, Joseph Connell showed that the vertical range of adults of each species is greater in the absence of the other species. *Chthamalus* larvae normally settle in large numbers in the *Balanus* zone. If *Balanus* are absent, young *Chthamalus* survive and grow well in the *Balanus* zone, but if *Balanus* are present, they smother, crush, or undercut the *Chthamalus*. *Balanus* larvae also settle in the *Chthamalus* zone, but the young *Balanus* grow slowly there because they lose water rapidly when exposed to air, so *Chthamalus* outcompete *Balanus* in that zone. The result of the competitive interaction between the two species is a pattern of *intertidal zonation*, with *Chthamalus* growing above *Balanus*.

COMPETITION MAY RESTRICT SPECIES' RANGES A species can restrict the range of another species by reducing populations of shared prey to such low levels that the other species cannot persist. Consider the distributions of two species of parasitoid wasps that parasitize scale insects. The wasps, whose larvae burrow into, eat, and kill scale insects, were introduced to southern California to control outbreaks of scale insects that were damaging citrus orchards. The Mediterranean wasp *Aphytis chrysomphali* was introduced around 1900, but it failed to control the scale insects. Therefore, a close relative from China, *A. lingnanensis*, was introduced in 1948. *A. lingnanensis*, which has a higher reproductive rate, increased rapidly. Within a decade it had greatly reduced population densities of the scale insects and had displaced *A. chrysomphali* from most of its range in California (**Figure 55.10**).

Commensal and amensal interactions are widespread

Amensalisms are widespread and inevitable interactions. For example, herds of mammals drinking at a water hole may trample and kill many plants. There is no benefit to the mammals in trampling the plants; the destruction of the plants is unintentional but inevitable. Trees drop dead leaves and branches, often harming smaller plants and unsuspecting animals beneath them. Commensal and mutualistic relationships are also everywhere.

Consider a large herbivorous mammal such as a rhinoceros grazing on the African plains (**Figure 55.11**). As it forages, the mammal disturbs multitudes of insects in the grass, crushing some and inadvertently consuming others in an amensal relationship.

55.11 A Single Small Community Demonstrates Many Interactions On the African plain, large herbivores such as the black rhinoceros disturb insect communities. The cattle egret pounces on the displaced insects; neither the amensal displacement of the insects nor the commensal activity of the birds has any effect on the rhino. Oxpeckers (the small, dark bird on the rhino's back) remove ticks from the skin of many large African mammals. In this mutualism, the birds get food and the mammals get relief from parasites. The diagram at right reveals the ecological complexity of the simple scene.

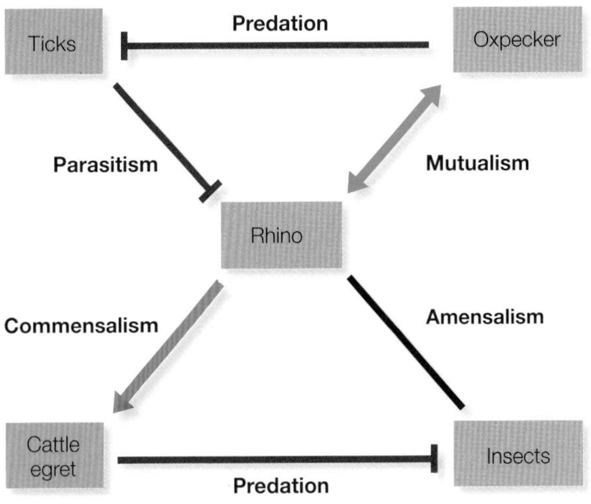

In addition, a commensal relationship has evolved between large herbivores and some bird species that are predators of insects. Birds such as cattle egrets typically forage on the ground near the mammals' heads and feet, where they capture insects the insects flushed by their hooves and mouths Cattle egrets foraging close to grazing mammals capture more food for less effort than egrets foraging away from grazing mammals. The benefit to the egrets is clear; the mammal neither gains nor loses. In another aspect of the community shown in Figure 55.11, a *mutualistic* relationship has evolved between the rhino and another bird species; birds known as oxpeckers pluck blood-sucking ticks from the skin of the grazing mammal. The bird gains a meal, and the mammal gains some protection from a parasite. Such mutualisms exist between many species and have been the subject of wide study.

Most organisms participate in mutualistic interactions

Mutualistic interactions exist between plants and microorganisms, between protists and fungi, among plants and insects, among animals, and among plants. Most plants have critical and beneficial associations with soil-inhabiting fungi called mycorrhizae, which enhance the plant's ability to extract minerals from the soil (see Figure 30.10). The mutualistic relationship between plants and nitrogen-fixing bacteria of the genus *Rhizobium* is the basis of much of life as we know it (see Section 36.4).

Examples of mutualisms between animals and protists abound. Corals and some other marine organisms gain most of their energy from photosynthetic protists living within their tissues. In exchange, they provide the protists with nutrients from the small planktonic animals they capture. Termites have protists in their guts that help them digest the cellulose in the wood they eat. The termites provide the protists with a suitable environment in which to live and an abundant supply of cellulose.

Terrestrial plants have many mutualistic associations with animals. As we saw at the opening of this chapter, many plants produce nectar on their vegetative parts that attracts ants, which provide the plants with protection from their predators and competitors. Experiments with Central American acacias have shown that trees deprived of their ants are heavily attacked by herbivores and grow poorly.

Many plants depend on animals to move their pollen and provide those animals with nutrient-rich rewards (**Figure 55.12A**). The plants benefit by having their pollen carried to other plants and by receiving pollen to fertilize their ovules. The animals benefit by obtaining food in the form of nectar and pollen. Movement to another plant of the same species is encouraged by the limited amount of nectar on any one plant and by the existence of similar rewards on other plants of the same species. But this arrangement has a trade-off for the plant: the energy and materials it uses to produce nectar and other rewards cannot be used for growth or reproduction.

Interactions between plants and their pollinators and seed dispersers are clearly mutualistic, but they are not purely mutualistic. Many seed dispersers are also seed predators that destroy some of the seeds they remove from plants. Some animals that visit flowers cut holes in the petals and gain access to the nectar without transferring any pollen. On the other hand, some plants exploit their pollinators. The flowers of certain orchids, for example, mimic female insects, enticing male insects to copulate with them (**Figure 55.12B**). The male insects neither sire any offspring nor obtain any reward, but they transfer pollen between flowers, benefiting the orchids.

55.12 Plant–Animal Mutualisms Are Important in Pollination (A) A bat of the genus *Brachyphylla* obtains nectar from an orchid plant in the West Indies. Some yellow pollen usually adheres to the bat's mouthparts and head and is then spread to other flowers. (B) In a twist, pollination of the *Ophrys scolopax* orchid is no longer a mutualism. The male bee (*Eucera longicornis*) is deceived by its odor and appearance into attempting copulation. The orchid will be pollinated, but the bee will receive no reward and has wasted valuable energy.

55.2 RECAP

Species interactions can be grouped into five categories: predator–prey (or parasite–host) interactions; competition; commensalisms; amensalisms; and mutualisms.

- Can you explain why predator and prey population densities often undergo oscillations? See p. 1189 and Figure 55.5
- Can you name two different ways in which organisms compete? See p. 1192
- Can you name at least two widespread kinds of mutualisms that are vital to sustain the plants that support all terrestrial life? See p. 1194

As illustrated in Figure 55.11, many interactions are going on at the same time in any given community. How do these multiple interactions influence the properties of ecological communities?

55.3 How Do Species Interactions Cause Trophic Cascades?

The interactions of a single predator species in a community can cause a progression of indirect effects across successively lower trophic levels. Such a pattern is called a **trophic cascade** and can be illustrated by the effects of the wolf population in the Lamar Valley of Yellowstone National Park.

One predator can affect many different species

The food web in Yellowstone National Park is extremely complex. Wolves in the park feed on pronghorn, mule deer, elk, bison, and bighorn sheep in addition to moose. They share these prey with coyotes, mountain lions, and grizzly and black bears. Despite the complexity of interactions in Yellowstone, wolves exert particularly strong effects on the Park's community structure and dynamics.

America's first national park, Yellowstone was established in 1872, but unrestricted hunting of mammals continued within and around the park for years after its establishment. In 1886, the U.S. Army assumed responsibility for protecting wildlife resources in the park, but the soldiers continued to kill mammalian predators (except for bears) in order to increase the populations of the large herbivores visitors wanted to see. Killing of predators continued when the National Park Service took over management of the park in 1918. By 1926, wolves had been extirpated from the park (**Figure 55.13A**).

Annual censuses of elk were initiated in Yellowstone in 1920. To prevent elk from exceeding the park's carrying capacity, the park service culled elk herds until 1968, when, in response to public pressure, elk kills were stopped. When the culling ended, the elk population rapidly increased (**Figure 55.13B**). When wolves were absent, the elk population browsed aspen trees so intensely that no young trees were recruited to the population after 1920 (**Figure 55.13C,D**). The elk also severely browsed streamside willows,

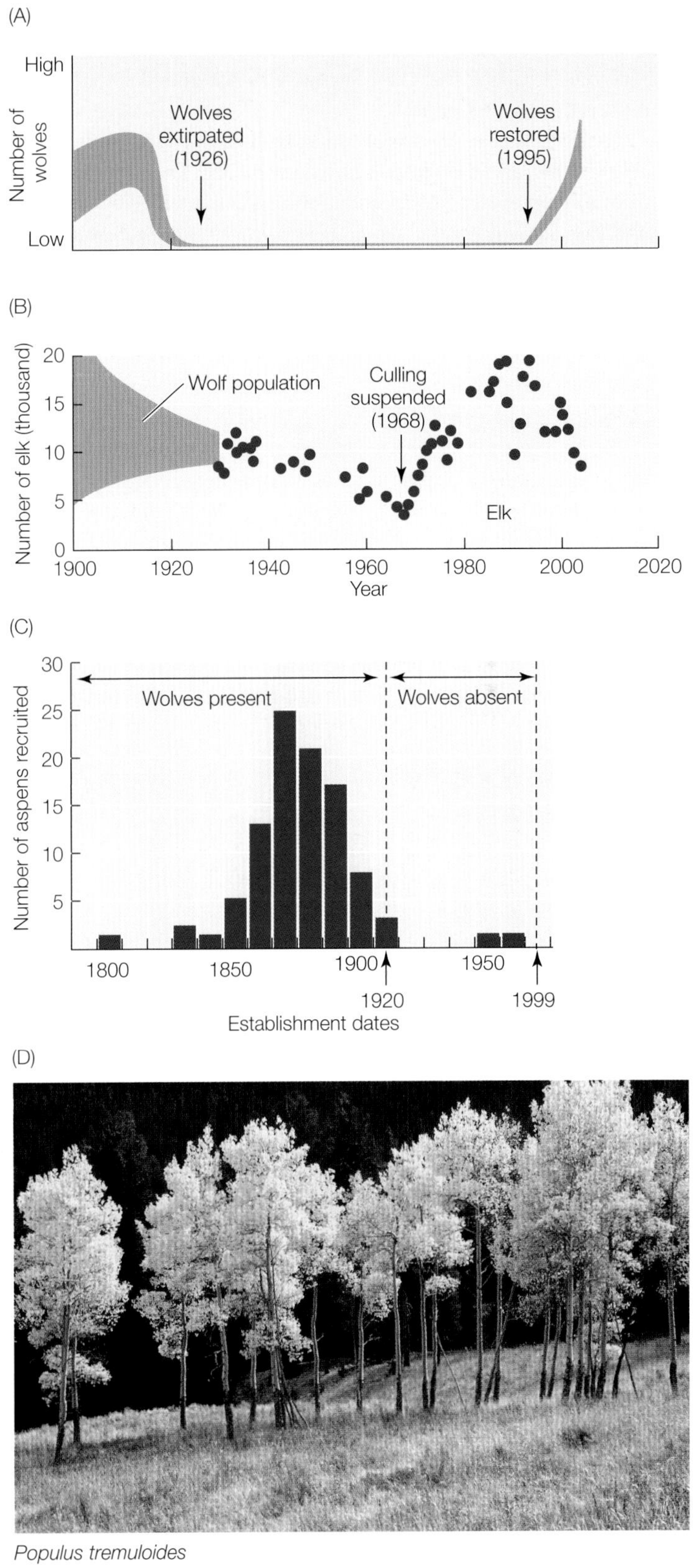

55.13 Wolves Initiated a Trophic Cascade (A) Wolves were eliminated from Yellowstone National Park in 1926 and reintroduced in 1995. (B) In the absence of wolves, culling controlled elk populations until 1968. When culling of elk ended in 1968, elk populations grew rapidly. (C) In the absence of wolves, the elk prevented recruitment of aspens. No young individuals are recruiting under these old trees (D).

with the result that beavers in Lamar Valley were nearly exterminated. (Aspen and willows grew well in areas from which elk were excluded, showing that their decline was not due to climate conditions during that period).

In 1995, after a 70-year absence, wolves were reintroduced to Yellowstone, and their population grew rapidly. The wolves preyed primarily on elk. The elk population of the Lamar Valley dropped and the elk avoided the aspen groves where they were very vulnerable to wolves. Young aspen began to grow, willows regrew along streams, and the number of beaver colonies increased from one in 1996 to seven in 2003. Thus the presence or absence of a single predator, the wolf, influenced not only populations of its prey, but also the structure of the vegetation and populations of other species that depend on it.

Trophic cascades may have effects across multiple and very different ecosystems because individuals of many species move from one habitat type to another. For example, larval dragonflies are aquatic predators that eat other insects and even small fish; the larvae are eaten in turn by larger fish. The surviving larvae metamorphose into adult dragonflies that eat flying insects.

Dragonfly larvae are abundant in ponds without fish; they are much less common in ponds with fish. Ecologists at the University of Florida studied 12 permanent ponds that differed primarily in whether they had fish (8 ponds) or lacked fish (4 ponds). They found that adult dragonflies (**Figure 55.14A**) were much more abundant around fish-free ponds than around ponds with fish (**Figure 55.14B**). Insect pollinators that were preyed on by the dragonflies were much *less* common around fish-free ponds. Flowers of Saint-John's-wort, the most common plant near the ponds, were visited much less frequently near fish-free ponds than near ponds with fish (**Figure 55.14C**). The plants near fish-free ponds also produced fewer seeds; by artificially pollinating some plants, the ecologists showed that they produced fewer seeds because they did not receive enough pollen. Thus the effects of fish predation in one habitat reached into another habitat where fish do not live.

Beavers cause trophic cascades both through what they eat and through what they construct. By preferentially cutting down some species of trees, they alter the composition of the vegetation. By building dams, they create meadows and ponds that are habitats for other species that would otherwise not be able to live in the area. Organisms that create structures are called **ecosystem engineers**.

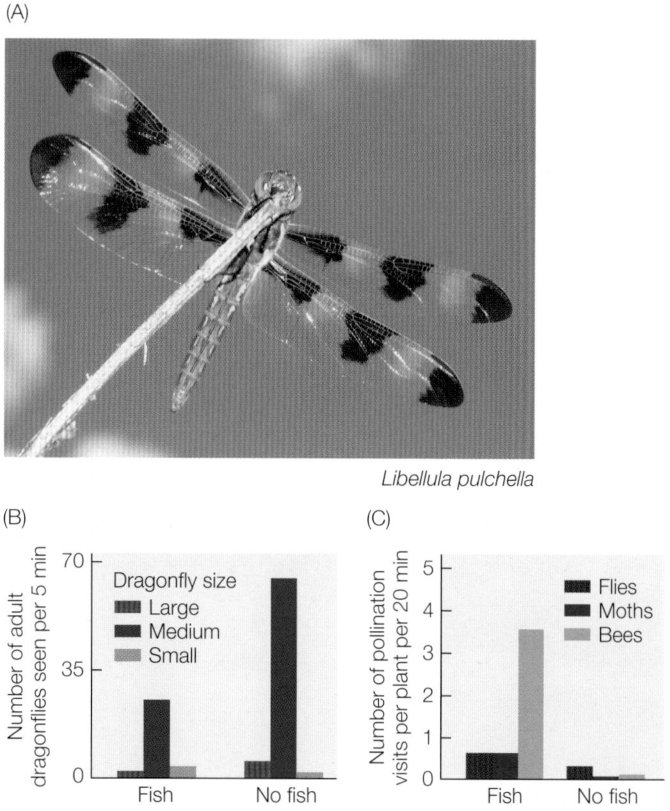

55.14 Trophic Cascades May Cross Habitats (A) Dragonflies affect the composition of communities in several habitats. (B) Adult dragonfly populations are greater near ponds without fish than near ponds with fish. (C) More pollinating insects are found around ponds with fish than around ponds without fish because fish reduce populations of dragonflies, which are major predators on the pollinators.

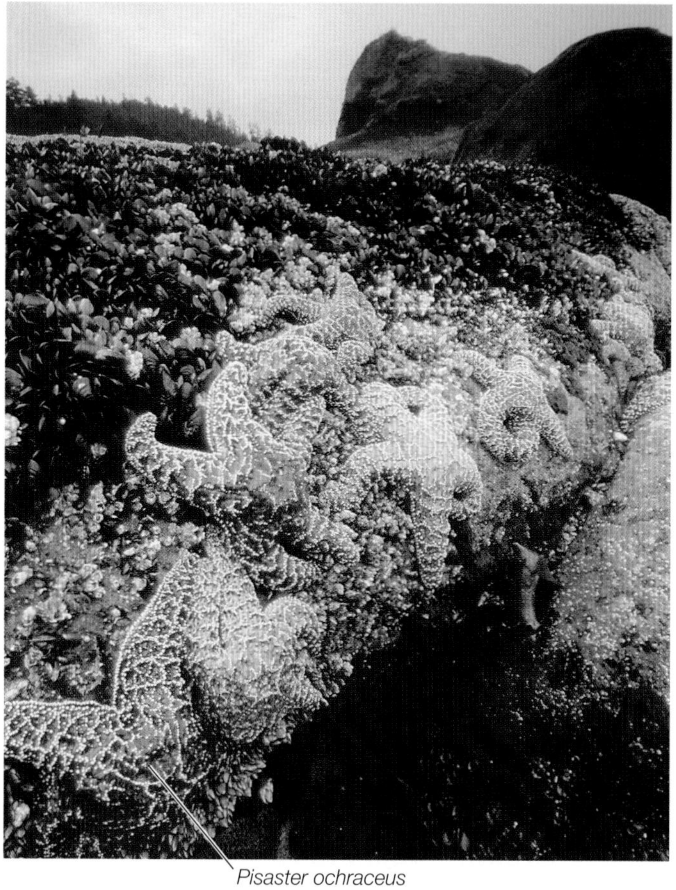

55.15 Some Sea Stars are Keystone Species Ochre sea stars (*Pisaster ochraceus*) have harvested all the mussels from the lower parts of these rocks on the Olympic Peninsula of Washington. By consuming mussels, *Pisaster* creates bare spaces on the rocks that are in turn occupied by a variety of other species.

Keystone species have wide-ranging effects

A species that exerts an influence out of proportion to its abundance is called a **keystone species**. Keystone species may influence both the species richness of communities and the flow of energy and materials through ecosystems. The sea star *Pisaster ochraceus*, which lives in rocky intertidal ecosystems on the Pacific coast of North America, is an example of a keystone species. Its preferred prey is the mussel *Mytilus californianus*. In the absence of sea stars, these mussels crowd out other competitors in a broad belt of the intertidal zone. By consuming mussels, *P. ochraceus* creates bare spaces that are taken over by a variety of other species (**Figure 55.15**).

Robert Paine of the University of Washington demonstrated the influence of *Pisaster* on species richness by removing sea stars from selected parts of the intertidal zone repeatedly over a 5-year period. Two major changes occurred in the areas from which he removed sea stars. First, the lower edge of the mussel bed extended farther down into the intertidal zone, showing that sea stars are able to eliminate mussels completely where they are covered with water most of the time. Second, and more dramatically, 28 species of animals and algae disappeared from the sea star removal zone. Eventually only *Mytilus*, the dominant competitor, occupied the entire substratum. Through its effect on competitive relationships, predation by *Pisaster* largely determines which species live in these rocky intertidal ecosystems.

Keystone species are not necessarily predators. A plant species that serves as food for many different animals can also be a keystone species, as has been suggested for fig tree species in tropical forests. Figs ripen their fruits during times of the year when fruit is otherwise scarce. The populations of dozens of fruit-eating species depend on figs when no other fruits are present in the forests.

55.3 RECAP

Some interspecific interactions result in a trophic cascade of indirect effects on species at lower trophic levels. Keystone species have an especially strong influence on the species richness of communities and on the flow of energy and materials through ecosystems.

- Describe an example of how a species causes indirect effects across many trophic levels. See pp. 1195–1196 and Figure 55.13
- What are keystone species, and how do they affect the environment of other species? See pp. 1196–1197

Now that we've seen some of the ways in which species interact among themselves to influence the structure of ecological communities, let's look at how abiotic physical disturbances affect what species live in a particular community, and how the assemblage of species in a community can change over time after a disturbance.

55.4 How Do Disturbances Affect Ecological Communities?

A **disturbance** is an event that changes the survival rate of one or more species in an ecological community. Disturbances may remove some species from a community, but may open up space and resources for other species. Keystone species generate disturbances, as we saw in the case of sea stars, but so do physical events. Logs carried by waves may crush algae and animals attached to rocks in an intertidal community. A windstorm may blow down trees, crushing shrubs and herbs. The effects of such disturbances are typically limited to small areas. Other kinds of disturbances, such as hurricanes and volcanic eruptions, may affect much larger areas. Small disturbances are much more common than large disturbances, but a few large events may cause most of the changes in a community. One hurricane, for example, may fell more trees than years of "normal" storms. The effects of disturbances also depend on how often they occur. If strong windstorms are frequent, for example, trees may never have the opportunity to grow tall.

A particular type of disturbance can have a variety of effects. For example, in 1988, massive fires burned one-third of Yellowstone National Park, but they created a mosaic that included unburned patches, areas where only herbs and shrubs were burned, and areas where all trees were consumed (**Figure 55.16**). How does such an ecological community recover after a disturbance?

Succession is a change in a community after a disturbance

A change in the composition of an ecological community following a disturbance is called **succession**. Ecologists divide succession into two major types: primary succession and secondary succession. **Primary succession** begins on sites that lack living organisms. **Secondary succession** begins on sites where some organisms have survived the most recent disturbance. The patterns and causes of ecological succession are varied, but the species that colonize a site soon after the disturbance often alter environmental conditions for other species that come after them.

A good example of primary succession is seen in the changes in plant communities that have followed the retreat of a glacier in Glacier Bay, Alaska, over the last 200 years. The melting and retreating glacier left a series of *moraines*—gravel deposits formed where the

55.16 Fires Create Mosaics of Burned and Unburned Patches
This view of Yellowstone National Park was taken a year after the massive forest fires of 1988.

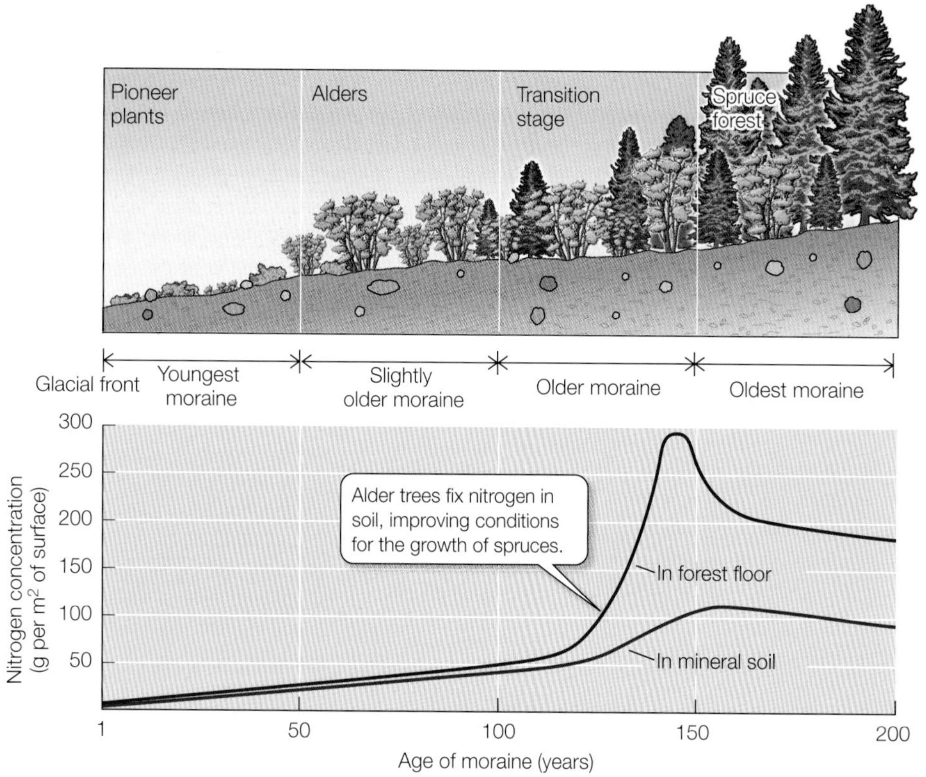

55.17 Primary Succession on a Glacial Moraine As the plant community occupying a glacial moraine at Glacier Bay, Alaska, changes from an assemblage of pioneer plants such as *Dryas* to a spruce forest, nitrogen accumulates in the mineral soil.

glacial front was stationary for a number of years. No human observer was present to measure changes over the entire 200-year period, but ecologists have inferred the temporal pattern of succession by studying plant communities on moraines of different ages. The youngest moraines, closest to the current glacial front, are populated with bacteria, fungi, and photosynthetic microorganisms. Slightly older moraines farther from the glacial front have lichens, mosses, and a few species of shallow-rooted herbs. Still farther from the glacial front, successively older moraines have shrubby willows, alders, and spruces.

By comparing moraines of different ages, ecologists deduced the pattern of plant succession and changes in soil nitrogen content on a glacial moraine (**Figure 55.17**). Succession is caused in part by changes in the soil brought about by the plants themselves. Nitrogen is virtually absent from glacial moraines, so the plants that grow best on recently formed moraines at Glacier Bay are the herbaceous plant *Dryas* and alder trees (*Alnus*), both of which have nitrogen-fixing bacteria in nodules on their roots (see Figure 36.7). Nitrogen fixation by *Dryas* and alders improves the soil so that spruces can grow. Spruces then outcompete and displace the alders and *Dryas*. If the local climate does not change dramatically, a forest community dominated by spruces is likely to persist for many centuries on old moraines at Glacier Bay.

Secondary succession may begin with the dead parts of organisms. The succession of fungal species that decompose pine needles in litter beneath Scots pines (*Pinus sylvestris*) is shown in **Figure 55.18**. New needles continuously fall from the pines, so the surface layer of litter is the youngest and deeper layers are progressively older. Decomposition begins when the first group of fungi starts attacking the needles soon after they fall. Each group of fungi derives its energy by decomposing certain compounds, convert-

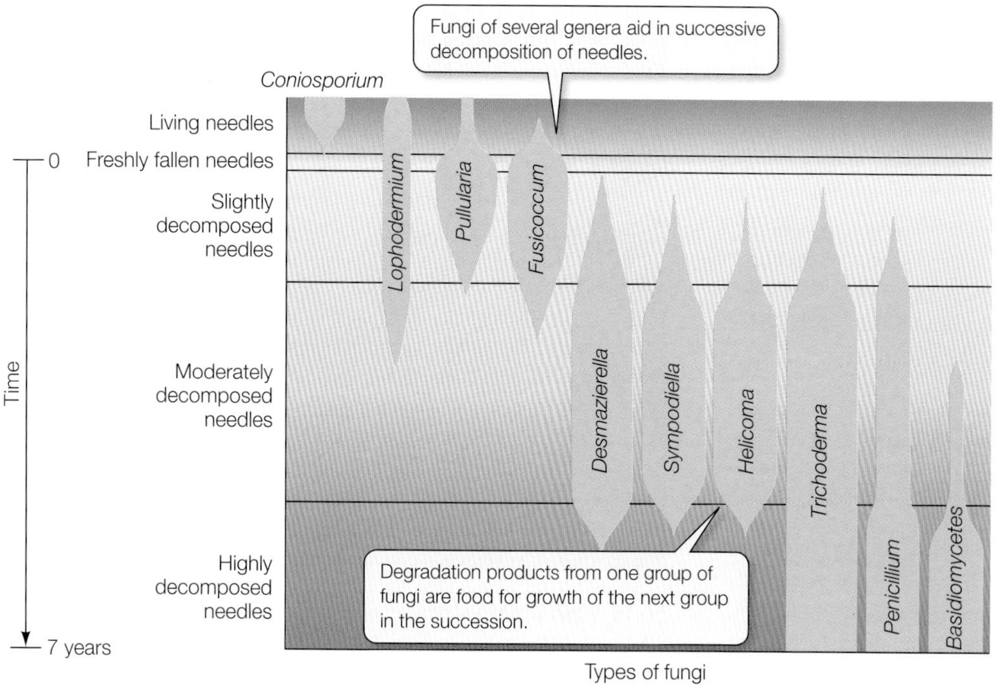

55.18 Secondary Succession on Pine Needles As indicated by the widths of the gray bars, the abundances of ten types of fungi in pine needle litter change over time as the needles are decomposed.

ing them to other compounds that are used by the next group of species. This process continues over about 7 years, by which time the last group of fungi—basidiomycetes—has decomposed the last remaining compounds.

Species richness is greatest at intermediate levels of disturbance

Although the consequences of various kinds of disturbances are highly variable, their results conform to a general pattern: communities with very high levels of disturbance and those with very low levels of disturbance have fewer species than communities subjected to intermediate levels of disturbance. The discovery of this general pattern generated the **intermediate disturbance hypothesis**, which explains the low species richness in areas with high disturbance levels by suggesting that only species with great dispersal abilities and high reproductive rates can persist in such areas. Conversely, the hypothesis explains the decline in species richness where disturbance levels are low by suggesting that competitively dominant species displace other species, as mussels did when sea stars were removed.

The intermediate disturbance hypothesis was tested using boulders in the intertidal zone on beaches in California. Wayne Sousa observed that boulders of intermediate size had more species of algae and barnacles attached to them than smaller boulders did. He hypothesized that this pattern existed because waves move small boulders more easily, more often, and farther than large boulders, not because small boulders are intrinsically unsuitable habitats for many species. When a wave moves a boulder, its motion crushes organisms living on the boulder's surface. To test his hypothesis, Sousa altered disturbance levels by gluing some small boulders to the substratum. Within 6 months, these boulders had more species than the unsecured small boulders (**Figure 55.19**). The results of the experiment supported the intermediate disturbance hypothesis and also refuted the hypothesis that species not normally found on small boulders were absent because small boulders were not suitable habitat for them.

The boulders in Sousa's experiments accumulated species rapidly. Most ecological succession progresses much more slowly, as happened at Glacier Bay.

EXPERIMENT

HYPOTHESIS: Small boulders have fewer species growing on them than larger boulders because they are subjected to high levels of disturbance.

METHOD

Sterilize a number of small boulders. Secure some of them to the natural substratum with glue. Leave other small boulders unsecured to serve as controls. Observe accumulation of species on the boulders over time.

RESULTS

Secured small boulders accumulated many more species than unsecured small boulders.

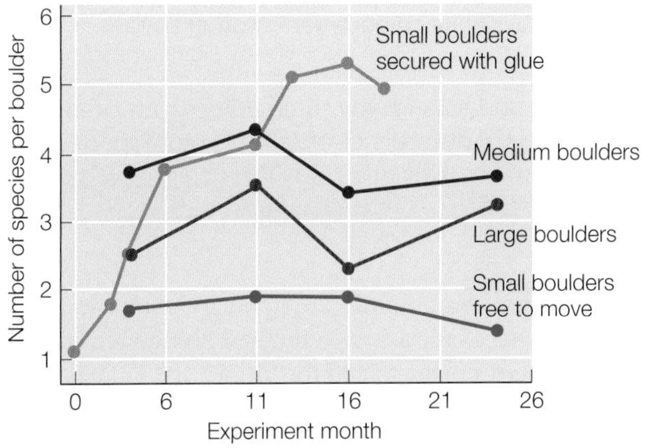

CONCLUSION: Small boulders have fewer species because the higher rates at which they are moved by waves prevent many species from surviving on them, not because they are unsuitable habitat for local species.

55.19 Species Richness Depends on Level of Disturbance By gluing boulders to the substratum, an ecologist showed that small boulders can support more species if they are not disturbed at high rates. Why they accumulated more species than larger boulders is unknown.

Both facilitation and inhibition influence succession

Succession at Glacier Bay illustrates how early colonizers of a changed environment can further change it in ways that *facilitate* the establishment of other species. Is this always the case, or might established species in some cases *inhibit* colonization by other species?

Wayne Sousa again turned to intertidal boulders to test the possible influence of inhibition on succession. He removed all organisms from some boulders and placed them in the intertidal zone. The first species to colonize these boulders were the green algae *Ulva* and *Enteromorpha*. These species were replaced much later by large brown algae, and finally by the red alga *Gigartina canaliculata*. To assess the role of inhibition during this succession, Sousa removed *Ulva* from some of the boulders. The boulders from which *Ulva* had been removed soon had a much greater population density of *Gigartina* than the others.

Inhibition clearly influenced the pattern of succession on intertidal boulders, but its effects lasted only a few months. Can inhibition have an effect lasting centuries or even millennia? Most ecological communities will reach an equilibrium number of species within a few decades to a few centuries after a disturbance. But the species living in some communities may have been assembled over periods of up to many millennia (as Darwin noted in describing an entangled bank at the start of this chapter).

As an example, we learned in Section 52.5 that South American ecological communities include many mammalian species whose ancestors arrived in South America from North America over the past 4 million years. North American communities, on the other hand, contain few mammalian species whose ancestors recently came to North America from South America (see Figure 52.13).

Such a pattern suggests that the effects of inhibition may persist for millions of years, but experiments cannot evaluate processes that take place over such long time spans. To evaluate such processes, scientists use the comparative method and examine patterns of species richness in ecological communities worldwide.

55.4 RECAP

Disturbances are events that change the survival rate of one or more species in an ecological community. Ecological succession—a process of change in community structure—typically follows a disturbance.

- Describe an example of primary succession. See pp. 1197–1198 and Figure 55.17
- Name some ways in which already-established species facilitate colonization by other species. See p. 1198
- Do you understand why communities with intermediate levels of disturbance often have more species than communities with either very low or high levels of disturbance? See p. 1199 and Figure 55.19

Some ecological communities are home to many species; other communities have only a few. In the next section we will discuss patterns of species richness and the factors that influence them.

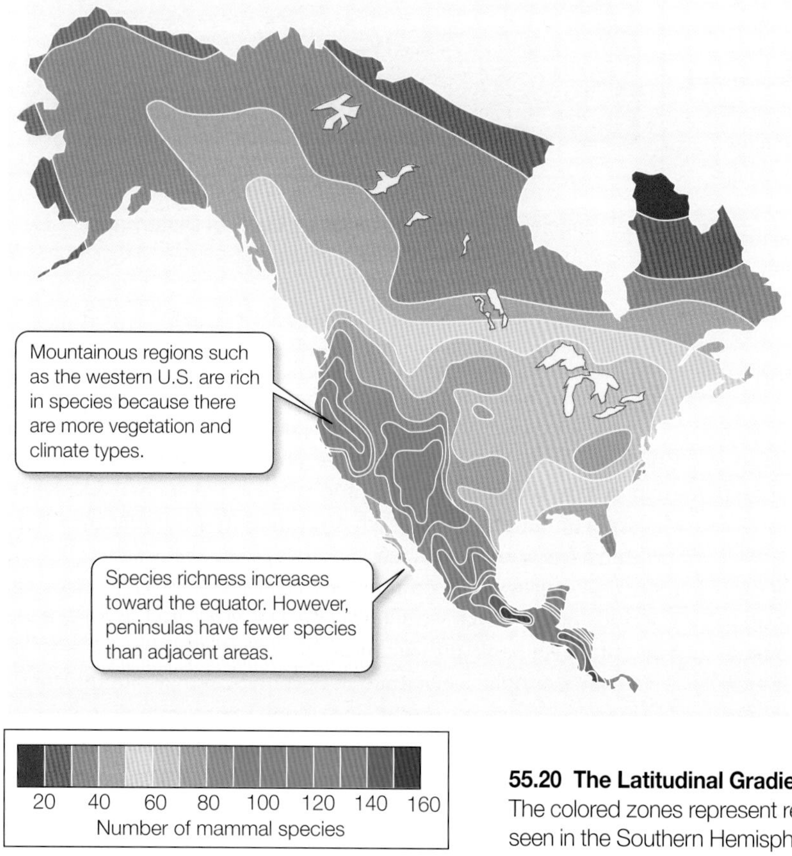

55.20 The Latitudinal Gradient of Species Richness of North American Mammals The colored zones represent regions with equal numbers of species. A similar pattern is seen in the Southern Hemisphere.

55.5 What Determines Species Richness in Ecological Communities?

The number of species living in a community is its **species richness**. Biologists have been aware of a geographic pattern in species richness for many years: more species are found in low-latitude than in high-latitude regions. **Figure 55.20** shows this latitudinal gradient in species richness for mammals in North and Central America. Similar patterns exist for birds, frogs, trees, and many groups of marine organisms. The figure also shows that more species are found in mountainous regions than in relatively flat areas because more vegetation types and climates exist within these topographically complex areas. Species richness on islands and peninsulas is always less than that in an equivalent area on the nearest mainland. Patterns of species richness on islands can be explained in large part by differences in immigration and extinction rates (see Section 52.5). What other processes determine species richness in mainland ecological communities?

Species richness is influenced by productivity

The species richness of ecological communities is correlated with ecosystem productivity, but the relationship between these two factors is complex. Ecologists first observed that species richness often increases with productivity up to a point, but then decreases (**Figure 55.21**). The increase occurs because the number of individuals an area can support increases with productivity, and with larger population sizes, species extinction rates are lower. But why should species richness decrease when productivity is still higher?

One hypothesis postulates that interspecific competition becomes more intense when productivity is very high, resulting in competitive exclusion of some species. This hypothesis is supported by the results of a long-term experiment at the Rothamstead Experiment Station in England begun in 1855. Fertilizer has been added regularly to selected plots of land to increase their productivity, and fertilized and unfertilized plots have been monitored continuously. Over this period, the number of plant species in the unfertilized plots has remained roughly constant, whereas species richness has declined in the fertilized plots.

Species richness and productivity influence ecosystem stability

We have seen that, up to a point, higher ecosystem productivity favors increased species richness. How might species richness in turn influence ecosystem productivity? Ecologists hypothesized that species richness might enhance productivity because no two

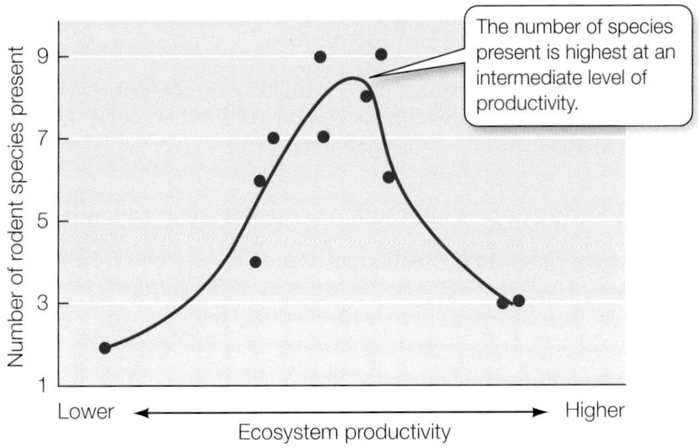

55.21 Species Richness Peaks at Intermediate Productivity The number of species of rodents that live on gravel and rocky plains in the Gobi Desert peaks at intermediate levels of productivity.

species in a community have the same relationship with the environment. Therefore, a richer mixture of species should result in a more complete use of the available resources. In addition, if the environment changes, a species-rich ecosystem is more likely to contain species that are already adapted to the new conditions than is a species-poor ecosystem. If that is the case, then a species-rich ecosystem should also be more *stable*; that is, it should change less over time in both productivity and species composition than a species-poor ecosystem.

To test this hypothesis, David Tilman and his colleagues at the University of Minnesota cleared several outdoor plots. In them they planted grasses in mixtures that ranged from a few to 25 species. At the end of each growing season, they measured total grass biomass and the population densities of all the grasses in each plot. Over a period of 11 years, which included a serious drought, the plots with more species were more productive, and their productivity varied less from one year to another, supporting the first part of the hypothesis (**Figure 55.22**). However, the population densities of individual species in the plots were not stable over the years (regardless of a plot's species richness), because different species performed better during drought years and wet years.

55.5 RECAP

More species are found in low-latitude than in high-latitude regions. Species richness correlates positively with ecosystem productivity up to a point, then declines at very high productivity levels.

- Why does species richness decline at very high levels of productivity? See p. 1200 and Figure 55.21
- Why are communities with many species more productive and more stable than communities with fewer species? See p. 1201 and Figure 55.22

EXPERIMENT

HYPOTHESIS: Communities with many species should have higher productivity and stability than communities with few species.

METHOD

Clear and plant plots with different numbers and mixtures of grass species. Measure productivity and species composition of the plots over 11 years.

RESULTS

(A) Productivity increases with species richness

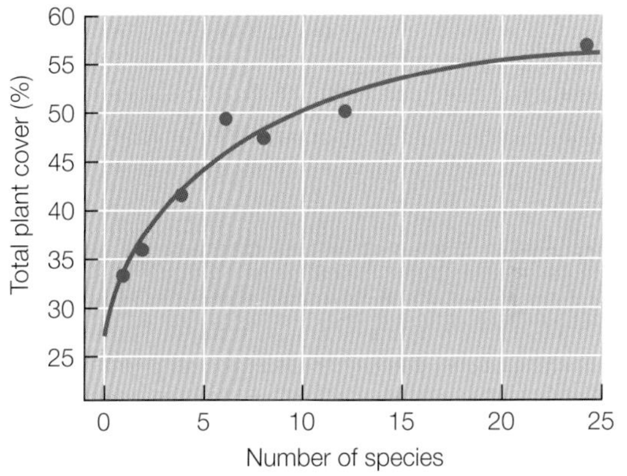

(B) Variation in productivity decreases with species richness

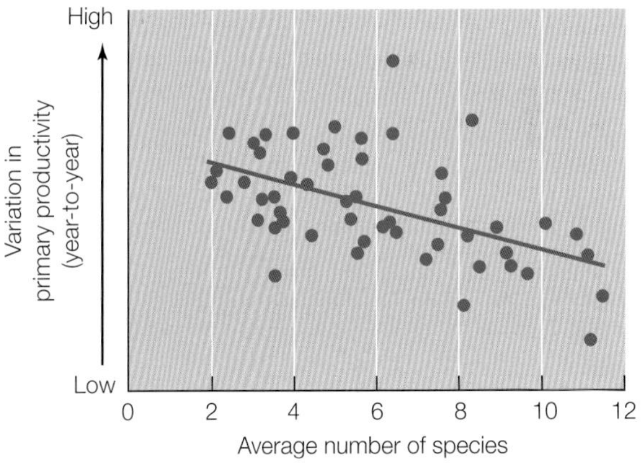

CONCLUSION: Plots with more species were more productive and varied less in productivity.

55.22 Species Richness Enhances Community Productivity Net primary productivity was greater, and variation in net primary productivity from one year to the next was less, in species-rich than in species-poor plots of grasses.

CHAPTER SUMMARY

55.1 What are ecological communities?

The species that live and interact in an area constitute an ecological **community**.

Ecological communities are loose assemblages of organisms. Review Figure 55.1

The organisms in a community can be divided into **trophic levels** based on the source of their energy. **Primary producers** get their energy from sunlight. Herbivores that get their energy by eating primary producers are **primary consumers**; organisms that get their energy by eating herbivores are **secondary consumers**; and so on. Review Table 55.1, Web/CD Activity 55.1

A **food chain** diagrams who eats whom. A **food web** shows how food chains are interconnected in an ecological community. Review Figure 55.2

Other kinds of diagrams show how energy decreases as it flows from lower to higher trophic levels and the **biomass** of organisms present at each trophic level. Review Figure 55.3

Most of the energy ingested by organisms that is converted to biomass is eventually consumed by **decomposers**.

55.2 What processes influence community structure?

Species interactions fall into five categories: **predation** or parasitism benefits the predator while harming the prey; **competition** is harmful to all participants; **mutualisms** are associations that benefit all participants; **commensal** interactions benefit one party while neither harming nor benefiting the other; **amensal** interactions have no effect on one party but harm another. Review Table 55.2, Web/CD Activity 55.2

Predator–prey interactions typically undergo oscillations. As predator populations grow, they may eat most of their prey; the predator population then crashes. Review Figure 55.4, Web/CD Tutorial 55.1

Predators may restrict the ways in which prey species make use of a habitat and may restrict the geographic distributions of their prey. Review Figure 55.6

Mimicry is an adaptation by prey to predation. In **Batesian mimicry**, a palatable species mimics an unpalatable species. In **Müllerian mimicry**, two or more unpalatable species converge to resemble one another. Review Figure 55.7

Microparasite populations can persist only if, on average, each infected host individual transmits the infection to one other individual.

Competition may restrict the abundances and ranges of species. **Interference competition** makes foraging more difficult; **exploitation competition** depresses prey populations. In **competitive exclusion**, one species prevents all members of another species from using a habitat. Review Figures 55.9 and 55.10

55.3 How do species interactions cause trophic cascades?

A predator, by reducing populations of its prey, may cause a **trophic cascade** of indirect effects across successively lower trophic levels. Trophic cascades may project across habitats. Review Figures 55.13 and 55.14

Ecosystem engineers are organisms that build structures that create environments for other species.

A **keystone species** affects the entire community out of proportion to its abundance.

55.4 How do disturbances affect ecological communities?

A **disturbance** is an event that changes the survival rate of one or more species in a community.

Ecological **succession** is a change in community composition following a disturbance.

Primary succession begins on sites that lack living organisms. **Secondary succession** begins on sites where some organisms have survived the most recent disturbance. Review Figures 55.17 and 55.18, Web/CD Tutorial 55.2

The **intermediate disturbance hypothesis** explains why communities with intermediate levels of disturbance often have more species than communities with very low or high levels of disturbance. Review Figure 55.19

Species that have already become established may facilitate or inhibit colonization by other species.

55.5 What determines species richness in ecological communities?

The number of species living in a community is its **species richness**. More species of most clades are found in low-latitude than in high-latitude regions. Review Figure 55.20

Species richness often increases with productivity, but only up to a point. Review Figure 55.21

Species-rich ecosystems tend to vary less in both productivity and species composition than species-poor ecosystems. Review Figure 55.22

SELF-QUIZ

1. An ecological community is
   a. all the species of organisms that live and interact with one another in an area.
   b. all the species that live and interact with one another in an area together with the abiotic environment.
   c. all the species in an area that belong to a particular trophic level.
   d. all the species that are members of a local food web.
   e. all of the above
2. A trophic level consists of the organisms
   a. whose energy source has passed through the same number of steps to reach them.
   b. that use similar foraging methods to obtain food.
   c. that are eaten by a similar set of predators.
   d. that eat both plants and other animals.
   e. that compete with one another for food.
3. Two organisms that use the same resources when those resources are in short supply are said to be
   a. predators.
   b. competitors.
   c. mutualists.
   d. commensalists.
   e. amensalists.

4. Damage caused to shrubs by branches falling from overhead trees is an example of
   a. interference competition.
   b. partial predation.
   c. amensalism.
   d. commensalism.
   e. diffuse coevolution.
5. The diagrams of energy and biomass distribution for forests and grasslands differ because
   a. forests are more productive than grasslands.
   b. forests are less productive than grasslands.
   c. large mammals avoid living in forests.
   d. trees store much of their energy in difficult-to-digest wood, whereas grassland plants produce few difficult-to-digest tissues.
   e. grasses grow faster than trees.
6. Keystone species
   a. influence the communities in which they live more than expected on the basis of their abundance.
   b. may influence the species richness of communities.
   c. may influence the flow of energy and nutrients through ecosystems.
   d. are not necessarily predators.
   e. all of the above
7. What is the general relationship between species richness and disturbance?
   a. Species richness peaks at low levels of disturbance.
   b. Species richness peaks at high levels of disturbance.
   c. Species richness peaks at intermediate levels of disturbance.
   d. Species richness is less at intermediate levels of disturbance.
   e. There is no general relationship between species richness and level of disturbance.
8. Ecological succession is
   a. the changes in species over time.
   b. the changes in community composition after a disturbance.
   c. the changes in a forest as the trees grow larger.
   d. the process by which a species becomes abundant.
   e. the buildup of soil nutrients.
9. Primary succession begins
   a. soon after a disturbance ends.
   b. at varying times after a disturbance ends.
   c. at sites where some organisms survived the disturbance.
   d. at sites were no organisms survived the disturbance.
   e. at sites where only primary producers survived the disturbance.
10. A latitudinal gradient in species richness
    a. is found in North America but not in South America.
    b. exists for birds, frogs, and mammals but not for plants.
    c. exists because tropical regions are more mountainous than high-latitude regions.
    d. exists because there are fewer peninsulas in the tropics.
    e. exists on land, but not in the oceans.

FOR DISCUSSION

1. Some evidence suggests that interspecific competition is responsible for the decrease in species richness at high levels of productivity. What other hypotheses might explain this puzzling relationship? How would you test them?
2. The increased productivity and stability of species-rich communities could be explained by ecological differences among the species or by the fact that the more species in a community, the greater the chance that it will contain an unusually productive species. How could you distinguish between these competing hypotheses?
3. If species-rich communities are more productive than species-poor communities, how can modern agriculture, which is based almost entirely on cultivating a single species on a plot, be so productive?
4. Figures 55.17 and 55.18 illustrated succession with two examples from forests. How might ecological succession differ in grasslands? In deserts? In the rocky intertidal zone?
5. Many conservationists believe that our greatest efforts should be expended to save undisturbed ecosystems. Many users of natural resources, on the other hand, argue that disturbing ecosystems to extract resources will actually improve species richness. Is the latter view an appropriate invocation of the intermediate disturbance hypothesis?

FOR INVESTIGATION

The experiments illustrated in Figure 55.5, even though they were done on enclosed plots 1 km^2 in size, were small in scale compared with the scale over which the natural hare–lynx cycle is synchronized in the boreal forest. How could scientists investigate the roles of food supply, predation, and other factors in synchronizing the cycle over large areas?

CHAPTER 56 Ecosystems and Global Ecology

Ecologists swing in the canopy to measure carbon's fate

Photosynthesis by terrestrial plants transforms (*fixes*) the carbon atoms in atmospheric carbon dioxide into the carbohydrate sugars that fuel the fungal and animal worlds. Concentrations of CO_2 in the atmosphere have been rising steadily over the last century, largely as a result of human activities, and some scientists have hypothesized that we can expect rates of photosynthesis—and thus CO_2 fixation and carbon storage in ecosystems—to increase. But ecologists are discovering that this "silver lining" scenario is probably not going to be the case.

Forest studies are important for determining the responses of vegetation to atmospheric carbon enrichment. Most photosynthesis in any forest takes place high up in tree canopies where sunlight is most intense. Until recently, making measurements in forest canopies required climbing or use of balloons. Then, in 1992, Alan Smith at the Smithsonian Tropical Research Institute in Panama realized that he could use a construction crane to gain access to the forest canopy. He recognized that by swinging the crane's boom in a circle, shuttling the gondola along its length, and lowering the cage to different heights, researchers could reach and measure biological processes over a large volume of the forest canopy. Smith's idea caught on quickly, and today ecologists on four continents use cranes to study forest canopies.

When they monitored growth, reproduction, and carbon storage in large trees, Swiss ecologists found that, as expected, canopy trees increased their growth rates in a carbon-enriched atmosphere; but unlike young, fast-growing saplings, they did not store large amounts of carbon in their trunks and branches. Instead, they transported more carbon first to the soil, and then to the atmosphere (via their roots). The leaf litter they produced was richer in starch and sugar, but poorer in lignin; the litter decomposed faster as soil microbes rapidly respired the carbon back to the atmosphere. These and other experiments suggest that young forests are likely to respond to greater atmospheric CO_2 concentrations by growing faster. However, mature forests, along with other types of vegetation such as grasslands, shrublands, and tundra, are not likely to store significantly more carbon in response to CO_2 enrichment.

Studying the Canopy A research crane 42 meters high allows scientists to study the forest canopy at the Smithsonian Tropical Research Institute in Parque Metropolitana, Panama. One aspect of these studies involves taking measurements that reveal the amount of carbon being stored in forest trees.

Enriching the Canopy Researchers from a team of Swiss ecologists install a mechanism of fine tubes that administers controlled amounts of CO_2 to enrich the trees' carbon supply. These experiments mimic what will happen as atmospheric CO_2 levels continue to increase.

As Section 52.3 described, the global ecosystem is composed of many different types of vegetation (biomes), many of which have few woody plants. To understand how ecosystems function at large scales, scientists must make observations and conduct experiments on many vegetation types. Moreover, because the responses of natural ecosystems to carbon enrichment will depend on many other factors—such as temperature, water, wind, and soil fertility—ecologists must study the interactions among these factors to understand how they influence ecosystem processes.

IN THIS CHAPTER we will begin describing the compartments of the global ecosystem. We then trace the flow of energy through the ecosystem, following key chemical elements such as carbon as they cycle through its compartments. We will highlight the many ways in which human activities have altered these cycles, and conclude by considering the many goods and services that ecosystems provide to human society.

CHAPTER OUTLINE

56.1 What Are the Compartments of the Global Ecosystem?

Earth is an essentially closed system with respect to atomic matter, but it is an open system with respect to energy. The sun delivers a nearly constant amount of energy to Earth every day and has done so for billions of years. Energy from the sun, combined with energy from the radioactive decay that melts the magma in Earth's interior, drives the processes that move materials around the planet. Many of these processes are cyclic (**Figure 56.1**). Almost all of the rocks that compose the continents have been processed at least once through a complex chemical and physical cycle involving weathering, formation of sediments, and movement into Earth's interior by means of the processes that result in continental drift (see Figure 21.2). Deep within the Earth, these rocks are subjected to great heat and pressure to form new rocks. The water in the oceans has evaporated, condensed in the atmosphere, precipitated, and returned to the oceans via rivers and groundwater flow millions of times. The carbon (C), nitrogen (N), oxygen (O), and sulfur (S) in atmospheric gases have been cycled repeatedly through living organisms.

We take for granted many features of Earth, but Earth is actually a very unusual planet. Its unusual properties include the presence of life, oceans, a moderate surface temperature, continental drift, and a large moon. The moon has had a profound effect on Earth's history. It stabilizes the tilt of Earth on its axis. The degree of tilt strongly influences climate. If Earth's tilt were 50° or more, for example, equatorial regions would receive less solar energy over the year than would polar regions! The moon also plays a major role in producing ocean tides, and it slows Earth's rotation.

The atmospheres of planets without life are in chemical equilibrium, but Earth's atmosphere is not. Oxygen gas (O_2), nitrogen gas (N_2), and water vapor (H_2O) are the main molecules of Earth's atmosphere, but without living organisms these gases would combine to produce nitric acid (HNO_3), which would dissolve in the oceans and remain there. Living organisms, by continually generating O_2, prevent this from happening.

All organisms depend on inputs of energy (in the form of sunlight or high-energy molecules), water, and nutrients for their metabolism and growth. As Section 55.1 explained, energy flows through ecosystems from producers to consumers.

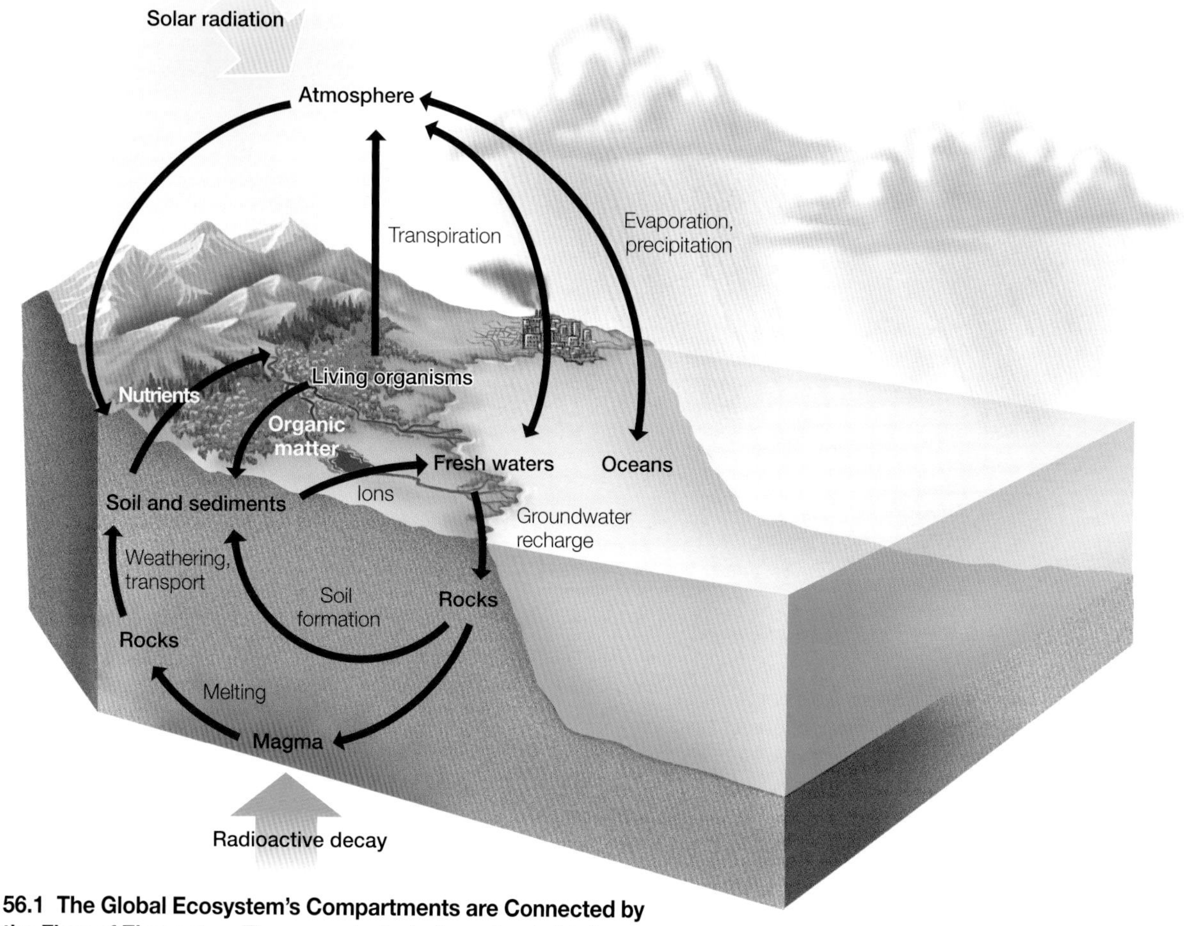

56.1 The Global Ecosystem's Compartments are Connected by the Flow of Elements The arrows indicate flow of materials between the compartments. The atmosphere exchanges some components rapidly. Exchange of materials between rocks, soils, and the oceans is generally much slower.

At each transformation, much of this energy is used to power metabolism and is dissipated as heat. Other organisms cannot use that heat to power their metabolism. Chemical elements, on the other hand, are not altered when they are transferred among organisms. The availability of the chemical elements of which living organisms are composed—carbon, nitrogen, phosphorus, calcium, sodium, sulfur, hydrogen, oxygen, and a few others—is strongly influenced by how organisms get them, how long they retain them, and what they do with them while they have them.

To understand the movement of energy and materials on Earth, it is convenient to divide the physical environment into four interacting compartments: oceans, fresh waters, atmosphere, and land. These four compartments, and the types of organisms living in them, are very different. Therefore, the quantities of different elements in each compartment, what happens to them, and the rates at which they enter and leave the compartments differ strikingly. After we have described the four compartments, we will discuss how energy flows and elements cycle among them.

Oceans receive materials from the land and atmosphere

Over time scales of hundreds to thousands of years, most materials that cycle through the four compartments end up in the oceans. The oceans are enormous, but they exchange materials with the atmosphere only at their surface, so they respond very slowly to inputs from other compartments. They receive materials from land primarily in runoff from rivers.

Except on continental shelves—the shallow ocean waters surrounding large land masses—ocean waters mix slowly. Most materials that enter the oceans from other compartments gradually sink to the seafloor. They may remain there for millions of years until the seafloor sediments are uplifted above sea level by mountain building. Therefore, concentrations of mineral nutrients are very low in most ocean waters. Some elements are brought back to the surface near the coasts of continents, where offshore winds push surface waters away from shore, causing cold bottom water to rise to the surface (**Figure 56.2**). The nutrient-rich waters in these **upwelling zones** support high rates of photosynthesis and dense animal populations. Most of the world's great fisheries are concentrated there.

Water moves rapidly through lakes and rivers

The freshwater compartment of the global ecosystem is comprised of rivers, lakes, and **groundwater** (water in soil and rocks). Only a small fraction of Earth's water resides at any one time in lakes and rivers, but water moves rapidly through them. Some mineral nutrients enter the freshwater compartment in rainfall, but most are released by the weathering of rocks. They are carried to lakes and rivers by surface flow or by movement of groundwater.

After entering rivers, mineral nutrients are usually carried rapidly to lakes or to the oceans. In lakes, they are taken up by organisms and incorporated into their cells. Those organisms eventually

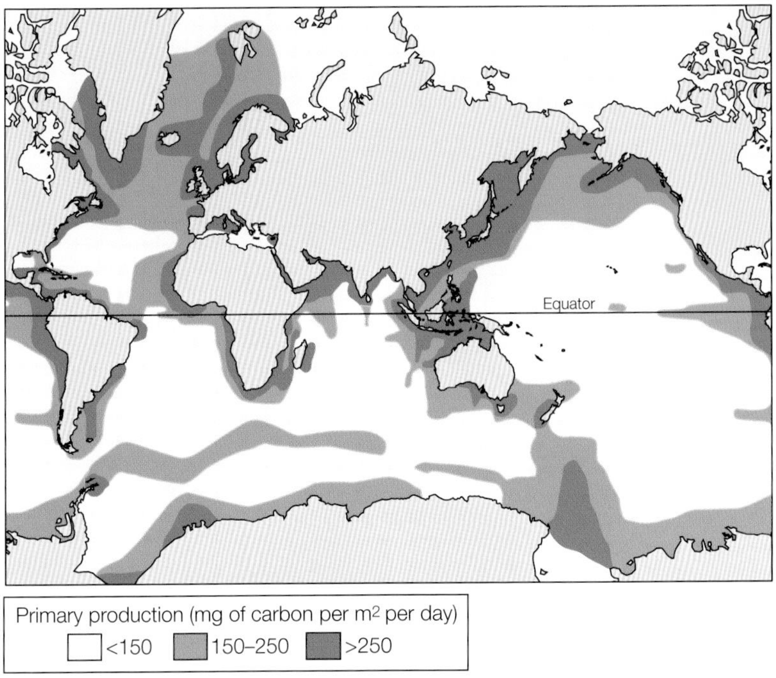

56.2 Zones of Upwelling are Productive Biological productivity is greatest near continents where surface waters, driven by prevailing winds, move offshore and are replaced by cool, nutrient-rich water upwelling from below.

die and sink to the bottom, taking the nutrients with them. Decomposition of their tissues by detritivores consumes the O_2 in the bottom water. The surface waters of lakes thus quickly become depleted of nutrients, while deeper waters become depleted of O_2. This process is countered, however, by vertical movements of water called **turnover**, which bring nutrients and dissolved CO_2 to the surface and O_2 to deeper water.

Wind is an important agent of turnover in shallow lakes, but in deeper lakes it usually mixes only surface waters. Deep lakes in temperate climates have an annual turnover cycle that is driven by temperature (**Figure 56.3**). These lakes turn over because water is most dense at 4°C; above and below that temperature, it expands. Thus in winter, the coldest water in a lake floats at its surface, often covered by a layer of ice. The waters below remain at 4°C. When the spring sun warms the surface of the lake, it brings the

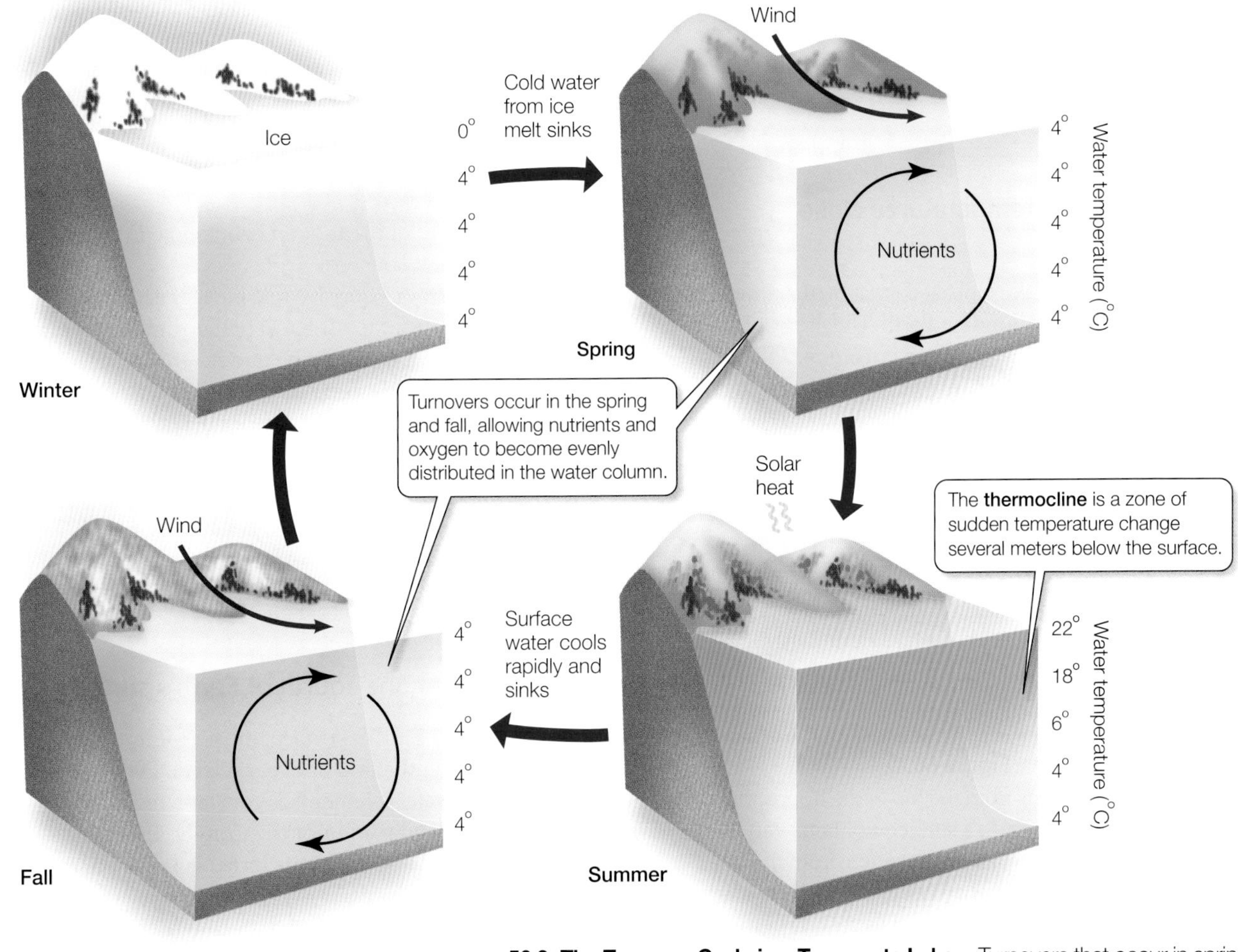

56.3 The Turnover Cycle in a Temperate Lake Turnovers that occur in spring and fall allow nutrients and oxygen to become evenly distributed in the water column. The vertical temperature profiles shown here are typical of temperate-zone lakes that freeze in the winter.

temperature of the surface water up to 4°C. At this point, the density of the water is uniform throughout the lake, and even modest winds readily mix the entire water column.

As spring and summer progress, the surface water becomes warmer, and the depth of the warm layer gradually increases. There is still a well-defined *thermocline*, however, where the temperature drops abruptly. Only if the lake is shallow enough to warm right to the bottom does the temperature of the deepest water rise above 4°C. Another turnover occurs in autumn as the surface of the lake cools, and the cooler surface water—which is now denser than the warmer water below it—sinks and is replaced by warmer water from below.

Arctic lakes turn over only once each year. Deep tropical and subtropical lakes may have permanent thermoclines because they never become cool enough to have uniformly dense water. Their bottom waters lack oxygen because decomposition quickly depletes any oxygen that reaches them. However, many tropical lakes are turned over at least periodically by strong winds, so that their deeper waters are occasionally oxygenated. If turnover does not happen, lake productivity can be drastically affected.

Lake Tanganyika is an extremely deep tropical lake in East Africa. Fishermen there have seen their catches decline by as much as 50 percent over the past 30 years, in part because warmer water temperatures and a decade of slack wind conditions have prevented the lake water from turning over.

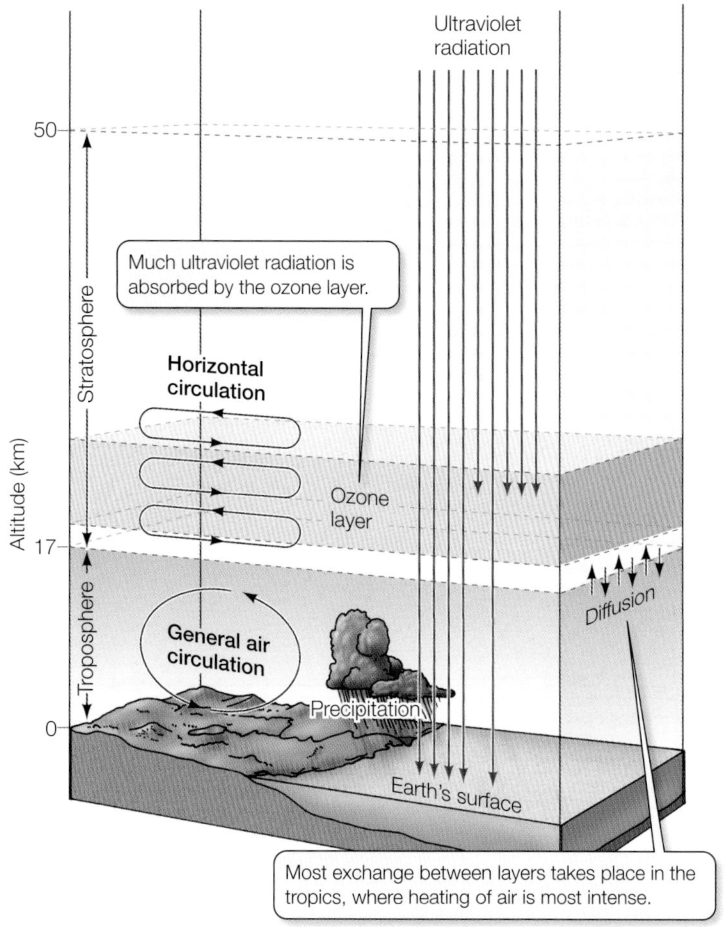

56.4 The Two Layers of Earth's Atmosphere The troposphere and the stratosphere differ in their circulation patterns, the amount of water vapor they contain, and the amount of ultraviolet radiation they receive.

The atmosphere regulates temperatures close to Earth's surface

The third major compartment of the global ecosystem, the atmosphere, is a thin layer of gases surrounding Earth. The atmosphere is 78.08 percent N_2, 20.95 percent O_2, 0.93 percent argon, and 0.03 percent carbon dioxide (CO_2). It also contains traces of hydrogen gas, neon, helium, krypton, xenon, ozone, and methane. The atmosphere contains Earth's biggest pool of nitrogen as well as a large proportion of its O_2. Although CO_2 constitutes a very small fraction of the atmosphere, it is the source of the carbon used by terrestrial photosynthetic organisms and of the dissolved carbonate in water used by marine primary producers.

About 80 percent of the mass of the atmosphere lies in its lowest layer, the **troposphere**, which extends upward from Earth's surface about 17 kilometers in the tropics and subtropics, but only about 10 kilometers at higher latitudes. Most global air circulation takes place within the troposphere, and virtually all of the atmospheric water vapor is found there (**Figure 56.4**).

The **stratosphere**, which extends from the top of the troposphere up to about 50 kilometers above Earth's surface, contains very little water vapor. Most materials enter the stratosphere from the troposphere in the intertropical convergence zone, where air heated by the sun rises to high altitudes. These materials tend to remain in the stratosphere for a relatively long time. A layer of ozone (O_3) in the stratosphere absorbs most of the biologically damaging short-wavelength ultraviolet radiation that enters the atmosphere, which is why the development of an ozone hole in the Southern Hemisphere is of great concern.

The atmosphere plays a decisive role in regulating temperatures at and close to Earth's surface. If Earth had no atmosphere, its average surface temperature would be about –18°C, rather than its actual +17°C. Earth has this warm temperature because the atmosphere is relatively transparent to visible light, but traps a large part of the heat that Earth radiates back toward space. Water vapor, carbon dioxide, and certain other gases, known as **greenhouse gases**, are especially important trappers of heat. When it is present in the troposphere, ozone also acts as a greenhouse gas. We will see how human-caused increases in greenhouse gases are contributing to global climate warming in Section 56.3.

Land covers about one-fourth of Earth's surface

About one-fourth of Earth's surface, most of it in the Northern Hemisphere, is currently land above sea level. Even though the global supply of chemical elements is constant, regional and local deficiencies of particular elements strongly affect ecosystem processes on land, because elements move slowly there and usually over only short distances.

The terrestrial compartment is connected to the atmospheric compartment by organisms that take chemical elements from and, some time later, release them back to the air. Chemical elements in soils are carried in solution into groundwater and eventually into the oceans, where they are unavailable to terrestrial organisms unless an episode of uplifting raises marine sediments and a new cycle of weathering begins. The type of soil that exists in an area depends on the underlying rock from which it forms, as well as on climate, topography, the organisms living there, and the length of time that soil-forming processes have been acting. Very old soils are much less fertile than most young soils because nutrients leach out of them over time.

Although land covers only a small proportion of Earth's surface, human life depends intimately on soil fertility and the productivity of terrestrial and freshwater ecosystems. The land and fresh waters are the global ecosystem compartments most strongly affected by human activities.

56.1 RECAP

Earth is a closed system with respect to atomic matter, but an open system with respect to energy. Earth's physical environment can be divided into four compartments—oceans, fresh waters, atmosphere, and land—through which energy flows and elements cycle.

- What keeps Earth's atmosphere from reaching chemical equilibrium? See pp. 1205–1206
- Can you describe the process of turnover in a temperate-zone lake in autumn? See p. 1207 and Figure 56.3
- How does the atmosphere keep temperatures at and close to Earth's surface warmer than they would be in its absence? See p. 1208 and Figure 56.4

Elements cycle through the global ecosystem, but energy flows through it unidirectionally. In the next section we introduce some concepts that help us track global energy flows.

56.2 How Does Energy Flow Through the Global Ecosystem?

We begin our discussion by explaining how solar energy drives ecosystem processes and describing the geographical distributions of the amount of energy assimilated by primary producers. We then show how human activities are modifying energy flow in the global ecosystem.

Solar energy drives ecosystem processes

With the exception of a few ecosystems in which solar energy is not the main energy source (some caves, deep-sea hydrothermal vent ecosystems), all energy utilized by organisms comes (or once came) from the sun. Even the fossil fuels—coal, oil, and natural gas—on which the economy of modern human civilization is based are reserves of captured solar energy locked up in the remains of organisms that lived millions of years ago.

Solar energy enters ecosystems by way of plants and other photosynthetic organisms. Only about 5 percent of the solar energy that arrives on Earth is captured by photosynthesis. The remaining energy is either radiated back into the atmosphere as heat or taken up by the evaporation of water from plants and other surfaces. **Gross primary productivity** (**GPP**) is the rate at which energy is incorporated into the bodies of photosynthetic organisms. The accumulated energy is called **gross primary production**. In other words, productivity is a rate; production is its product. Primary producers use some of this accumulated energy for their own metabolism; the rest is stored in their bodies or used for their growth and reproduction. The energy available to organisms that eat primary producers, called **net primary production** (**NPP**), is gross primary production minus the energy expended by the primary producers during their metabolism. Only the energy of an organism's net production—its growth plus reproduction—is potentially available to other organisms that consume it (**Figure 56.5**).

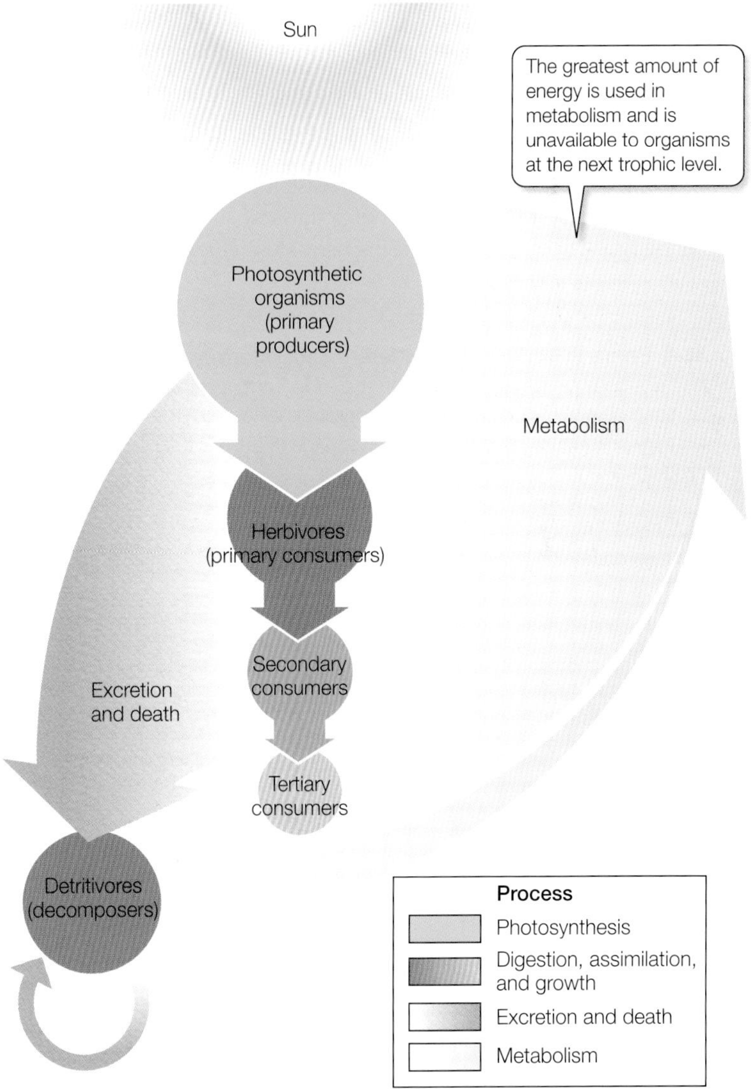

56.5 Energy Flow through an Ecosystem Only the energy of an organism's net production—its growth plus reproduction—is potentially available to other organisms that consume it. In this diagram, the width of each arrow is roughly proportional to the amount of energy flowing through that channel. Arrows indicate directions of energy flow.

The geographic distribution of the energy assimilated by primary producers reflects the distribution of land masses, temperature, and moisture on Earth (**Figure 56.6**). Close to the equator at sea level, temperatures are high throughout the year, and the water supply typically is adequate for plant growth much of the time. In these climates, productive forests thrive. In lower-latitude and mid-latitude deserts, where plant growth is limited by lack of moisture, primary production is low. At still higher latitudes, even though moisture is generally available, primary production is also low because it is cold much of the year (**Figure 56.7**). Production in aquatic ecosystems is limited by light, which decreases rapidly with depth; by nutrients, which sink and must be replaced by upwelling of water; and by temperature.

Human activities modify flows of energy

Human activities modify the ways in which energy flows between the global ecosystem compartments and how it is distributed among them. The effects of human activities on the flow of energy have accelerated remarkably in the last century and particularly in the last five decades. Some human activities decrease net global primary productivity (such as converting forests to grasslands and urban developments) and some increase it (such as intensifying agriculture). We also influence energy flows by supplementing solar energy with our ever-increasing use of fossil fuels. Satellite and climate data indicate that humans are appropriating about 20 percent of the average annual net primary production on Earth. The percentage varies strikingly among regions: urban areas consume as much as 300 times the NPP they generate, but people in sparsely inhabited parts of the Amazon Basin appropriate almost nothing.

56.2 RECAP

Nearly all energy utilized by living organisms comes from the sun. Human activities influence energy flows and appropriate a great deal of Earth's net terrestrial primary productivity.

- What is the difference between gross primary production and net primary production? See p. 1209
- What percentage of Earth's average annual net primary production is appropriated by humans, and how does that percentage vary regionally? See p. 1210

Understanding how human appropriation of net primary production affects quantities and patterns of energy flow requires knowing how the compounds of carbon, in which most energy is stored, are formed and move among Earth's physical compartments. In the next section we will consider how carbon, as well as the other chemical elements living organisms need, cycle through the compartments of the global ecosystem.

56.6 Primary Production in Different Ecosystem Types The contributions of different ecosystem types to Earth's primary production can be measured by (A) their geographic extent; (B) their net primary production; and (C) their percentage of Earth's total primary production.

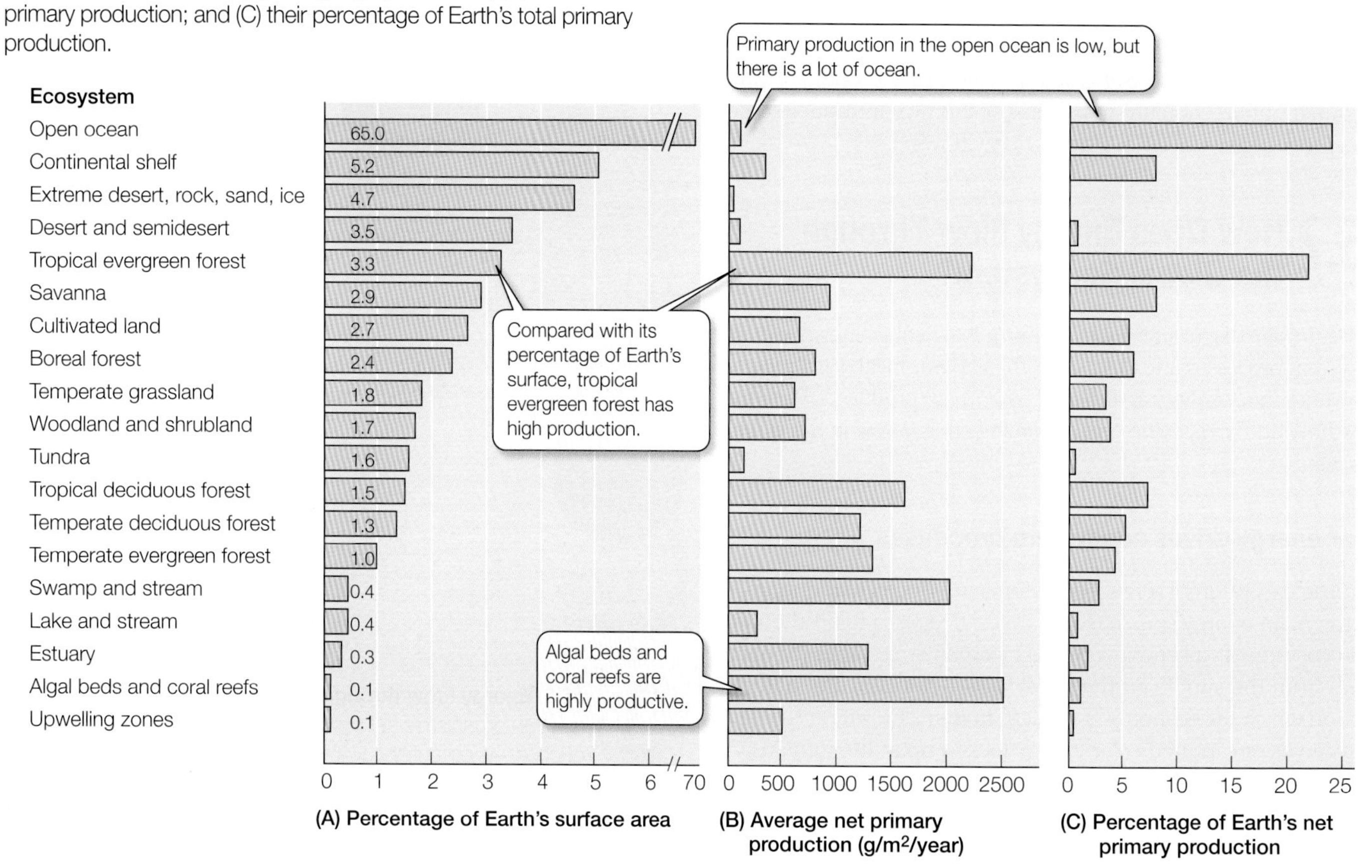

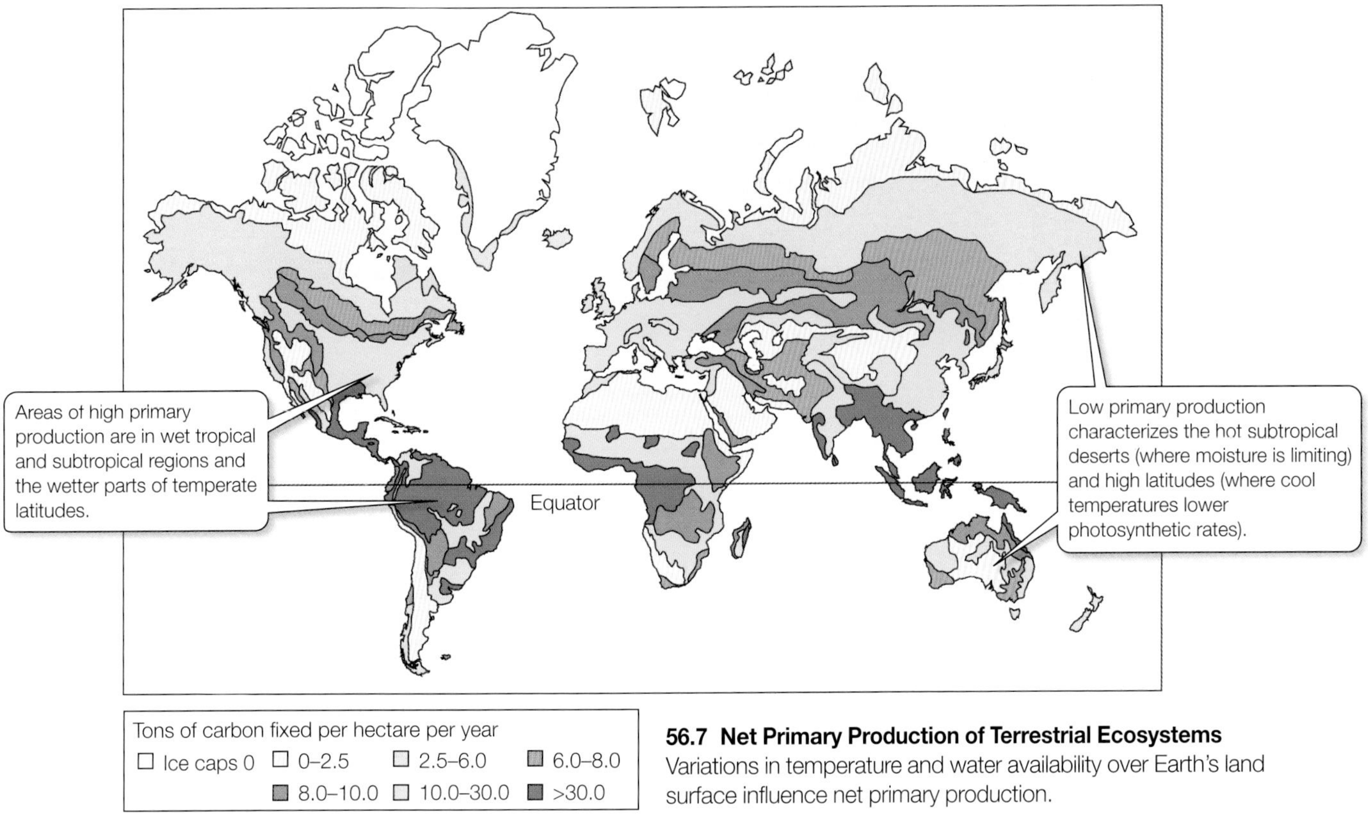

56.7 Net Primary Production of Terrestrial Ecosystems Variations in temperature and water availability over Earth's land surface influence net primary production.

56.3 How Do Materials Cycle Through the Global Ecosystem?

> An atom of phosphorus "X" had marked time in the limestone ledge since the Paleozoic seas covered the land. Time, to an atom locked in a rock, does not pass. The break came when a bur-oak root nosed down a crack and began prying and sucking. In the flash of a century the rock decayed, and X was pulled out and up into the world of living things. He helped build a flower, which became an acorn, which fattened a deer, which fed an Indian all in a single year.
>
> Aldo Leopold, *A Sand County Almanac*

The chemical elements organisms need in large quantities—carbon, hydrogen, oxygen, nitrogen, phosphorus, and sulfur—cycle through organisms to the physical environment and back again. The pattern of movement of a chemical element through organisms and the other compartments of the global ecosystem is called its **biogeochemical cycle**.

Each element has a distinctive biogeochemical cycle whose properties depend on the physical and chemical nature of the element and the ways in which organisms use it. All chemical elements cycle quickly through organisms because no individual, even of the longest-lived species, lives very long in geological terms.

Before we describe the biogeochemical cycles of individual elements, however, we must discuss the roles of water and fire.

- The movement of *water* transfers many chemical elements between the atmosphere, land, fresh waters, and oceans.
- *Fire* is a powerful agent that speeds the cycling of chemical elements.

Water transfers materials from one compartment to another

The cycling of water through the oceans, atmosphere, fresh waters, and land is known as the **hydrological cycle**. The hydrological cycle operates because more water is evaporated from the surface of the oceans than is returned to them as *precipitation* (rain or snow). The excess evaporated water is carried by winds over the land, where it falls as precipitation. Water also evaporates from soils, from lakes and rivers, and from the leaves of plants (transpiration), but the total amount evaporated from those surfaces is less than the amount that falls on them as precipitation. Excess terrestrial precipitation eventually returns to the oceans via rivers, coastal runoff, and groundwater flows (**Figure 56.8**).Earth's 16 largest rivers account for more than one-third of total water runoff from the land, more than half of which comes from the three largest rivers (the Amazon, Congo, and Yangtze). Despite their relatively small volume, rivers play a disproportionate role in the hydrological cycle because the average residence time of a water molecule in lakes and rivers is only 4.3 years, compared with 2,640 years in the oceans. The average residence time of water in living things is particularly short—about 5.6 days.

By building dams, canals, and reservoirs and by diverting huge quantities of water to irrigated fields, humans have had major effects on the temporal and spatial distribution of fresh water on Earth. The most important consequence of human activities is that

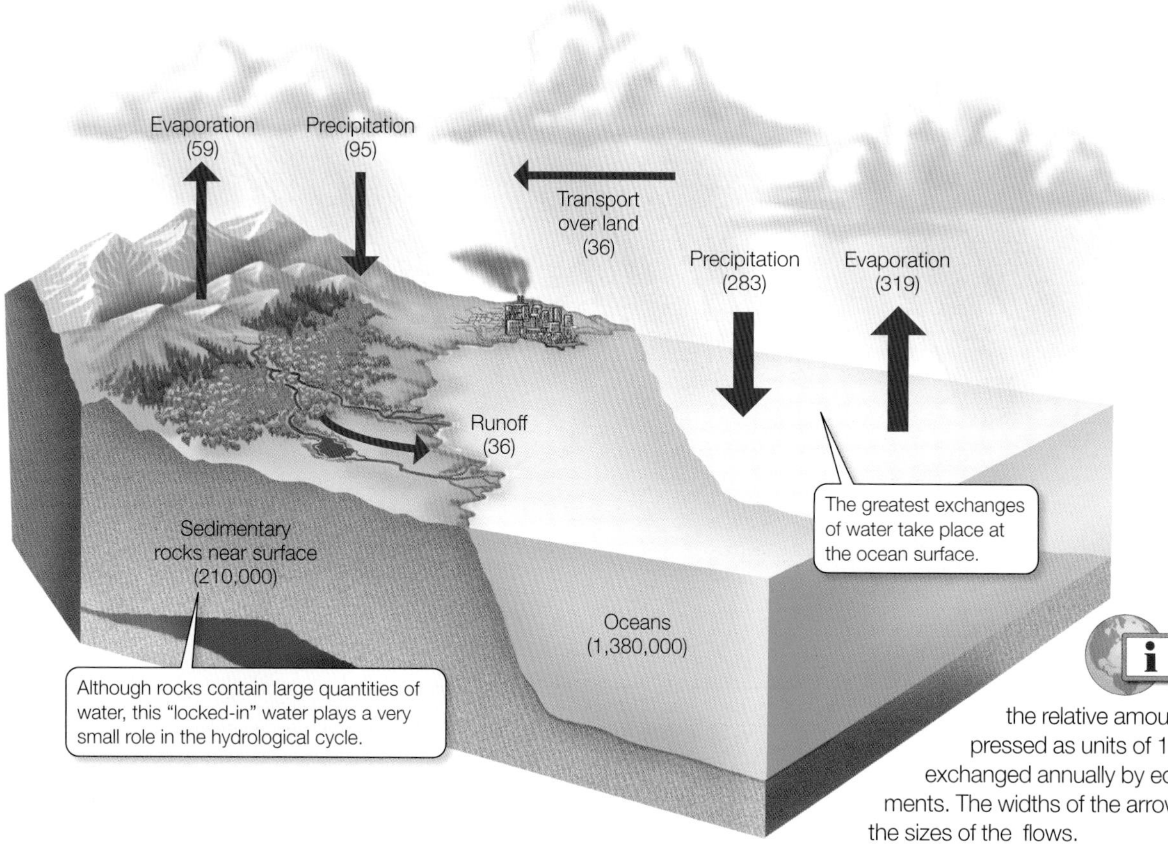

56.8 The Global Hydrological Cycle The numbers show the relative amounts of water (expressed as units of 10^{18} grams) held in or exchanged annually by ecosystem compartments. The widths of the arrows are proportional to the sizes of the flows.

more water now evaporates from land than before the Industrial Revolution. In addition, freshwater flow patterns are being seriously altered. For example, dams on the Columbia River in Washington State are managed to reduce the variation in water flow rates during the year (**Figure 56.9**). An unintended result of this practice is reduced autumn and winter ocean surface salinity from the mouth of the river north along the coast to the Aleutian Islands, with negative consequences for the growth and survival of salmon in that part of the Pacific Ocean.

Although large amounts of groundwater are present in underground pools called **aquifers**, this water has a long residence time underground and plays only a small role in the hydrological cycle. In some places, however, groundwater is being seriously depleted because humans are using it more rapidly than it can be replaced, primarily by pumping it for irrigation. In some areas of the High Plains of the central United States, more than half of the groundwater has been removed. On the North China Plain, depletion of shallow aquifers is forcing people to sink wells more than 1,000 meters deep to reach water.

Much of the groundwater used today in the Northern Hemisphere was deposited during the last Ice Age, when regional precipitation was much greater than it is now. Using this groundwater for irrigation and other purposes has increased flows of water to the oceans and has contributed to the sea level rise of the past century.

If current water consumption patterns continue, by 2025 at least 3.5 billion people (48 percent of the world's population) will live in areas with inadequate water supplies. The population in Asia will probably be most severely affected. Fortunately, current water consumption patterns need not continue. Per capita water consumption in the United States and Europe is dropping because of increasing use of water-efficient home appliances, increasing prices for water, and development of regulations that restrict water use. If such programs to improve water use efficiency were implemented vigorously, global water use in 2025 could be less than it is today, despite continued population growth.

Fire is a major mover of elements

Every year, 200–400 million hectares* of savannas, 5–15 million hectares of boreal forests, and lesser amounts of other biomes burn. Lightning ignites some fires, but humans start most fires to manage vegetation (as when they cut down and burn forests to clear land for growing crops). Fires consume the energy of, and release chemical elements from, the vegetation they burn. Some nutrients, such as nitrogen, sulfur, and selenium, are easily vaporized by fire. They are discharged to the atmosphere in smoke or carried into groundwater by rain falling on burned ground.

Fires also release large amounts of carbon into the atmosphere. The global annual flow of carbon to the atmosphere from savanna and forest fires is estimated to be in the range of 1.7 to 4.1 petagrams.* Biomass burning is responsible for about 40 percent of Earth's annual production of CO_2. It is also a significant contributor to the production of other greenhouse gases, such as carbon monoxide (CO), methane (CH_4), and tropospheric ozone (O_3).

*1 hectare = 10^4 square meters, or 2.5 acres; 1 petagram = 10^{15} grams

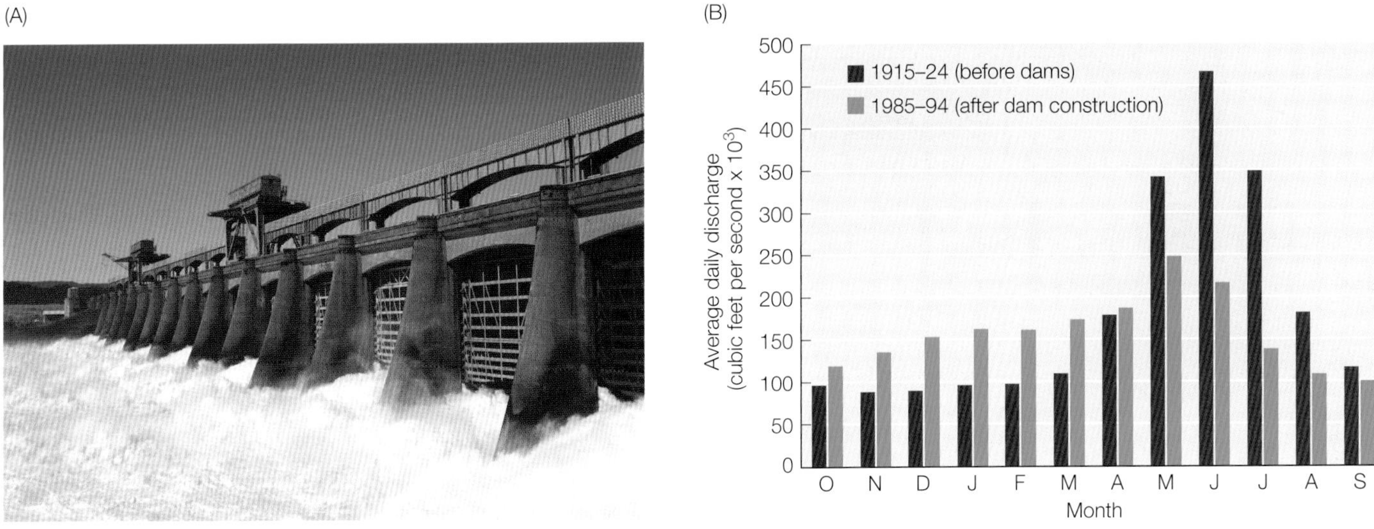

56.9 Columbia River Flows Have Been Massively Altered
(A) The Bonneville Dam, near Portland, Oregon, is just one of dozens of dams that have been built along the Columbia River and its tributaries to regulate the river's flow and generate hydroelectric power. (B) Before the construction of dams, rates of freshwater flow from the Columbia River to the Pacific Ocean varied greatly with the seasons. Human intervention has substantially reduced this variation.

Today's oceans absorb 20–25 million tons of CO_2 every day—a faster rate than at any time during the past 20 million years. As a result, water near the ocean surface is becoming more acidic. How this increasing acidity will affect marine organisms is unknown.

Because humans deliberately start most fires, we can take steps to reduce their undesirable consequences. Periodic burning of fire-prone vegetation can prevent the buildup of large fuel loads and reduce the frequency of high-intensity canopy fires, which discharge great quantities of materials to the atmosphere. Improvements in the productivity of tropical agriculture can reduce the rate at which forests are cleared and burned to create more agricultural land. Expanded use of simple solar cookers can reduce the need to harvest firewood for cooking meals.

Our discussion of water and fire shows that the biogeochemical cycles of different chemical elements are connected to one another. However, it is easier to understand the interactions of these cycles if we first discuss each one separately. Let's begin by tracing the biogeochemical cycle of carbon.

The carbon cycle has been altered by industrial activities

As we saw in Part Two of this book, all of the important macromolecules that compose living things contain carbon, and much of the energy that organisms need to fuel their metabolic activities comes from carbon-containing (organic) compounds. Carbon is incorporated into organic molecules by photosynthesis in the cells of autotrophs. All heterotrophic organisms get their carbon by consuming autotrophs or other heterotrophs, their remains, or their waste products.

Most of Earth's carbon is stored in rocks and marine sediments, and dissolved in ocean waters. In the terrestrial compartment, most carbon that is available to organisms is stored in soils. Biological processes move carbon between the atmospheric and terrestrial compartments, removing it from the atmosphere during photosynthesis and returning it to the atmosphere during metabolism (**Figure 56.10**).

At times in the remote past, great quantities of carbon were removed from the global carbon cycle when organisms died in large numbers and were buried in sediments lacking O_2. In such anaerobic environments, detritivores do not reduce organic carbon to CO_2. Instead, organic molecules accumulate and eventually are transformed into deposits of oil, natural gas, coal, or peat. Humans have discovered and used these deposits, known as **fossil fuels**, at ever-increasing rates during the past 150 years. As a result, CO_2, one of the final products of the burning of fossil fuels, is being released into the atmosphere faster today than it is dissolving in the surface waters of the oceans or being incorporated into terrestrial biomass. Based on a variety of calculations, atmospheric scientists believe that before the Industrial Revolution, the concentration of atmospheric CO_2 was probably about 265 parts per million. Today, it is 380 parts per million (**Figure 56.11**), representing a rate of increase more than ten times faster than at any other time for millions of years.

WHERE HAS ALL THE CARBON GONE? Less than half of the vast amount of CO_2 released to the atmosphere by human activities remains in the atmosphere. Where is the rest? Much of the remainder is dissolved in the oceans. Over decades to centuries, the oceans, which contain 50 times the amount of dissolved inorganic carbon as the atmosphere, determine atmospheric CO_2 concentrations. The rate at which CO_2 moves from the atmosphere to the oceans depends, in part, on photosynthesis by phytoplankton in

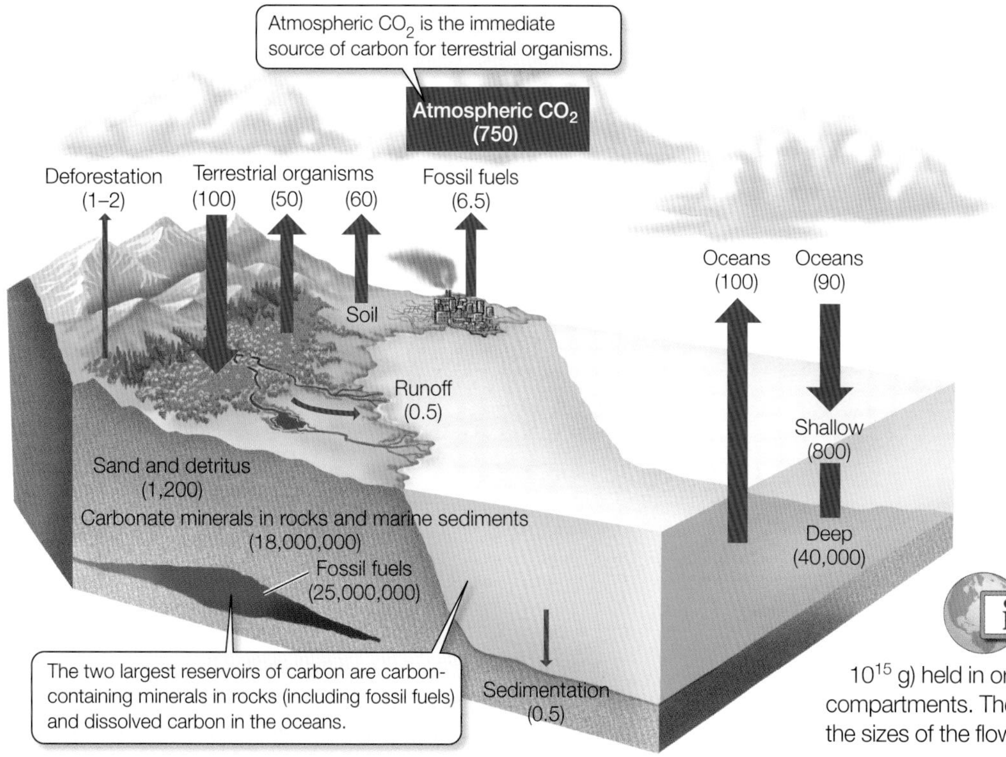

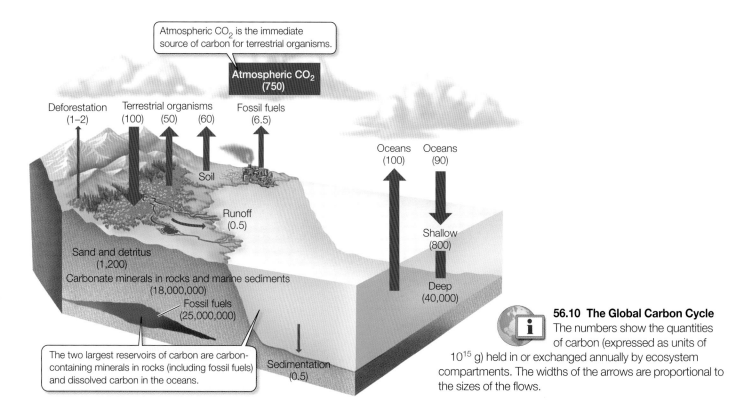

56.10 The Global Carbon Cycle The numbers show the quantities of carbon (expressed as units of 10^{15} g) held in or exchanged annually by ecosystem compartments. The widths of the arrows are proportional to the sizes of the flows.

the surface waters. These organisms remove carbon from the water, thereby increasing its absorption of carbon from the atmosphere. In addition, many marine organisms form calcium carbonate ($CaCO_3$) shells, which eventually sink to the ocean floor. Thus, by removing carbon from surface waters, marine organisms increase absorption of carbon from the atmosphere.

Between 1800 and 1994, the oceans absorbed a total of about 118 petagrams of carbon, accounting for nearly half of the carbon in the total CO_2 emissions from human activities, which would otherwise have remained in the atmosphere. This human-generated carbon is not evenly distributed in the oceans, however. The North Atlantic Ocean covers only 15 percent of the global ocean area, but it stores 23 percent of the human-generated carbon. By contrast, the much larger Southern Ocean, south of 50°S latitude, contains only 9 percent of the total carbon. The North Atlantic is an especially important carbon sink (a *sink* is a location in which an element is taken out of circulation) because of a circulation pattern called the *ocean conveyor belt*, which is driven by the sinking of dense, saline water between Greenland and Europe. By removing carbon-rich waters from the ocean surface, the ocean conveyor belt increases the transfer rate of carbon from the atmosphere to the oceans.

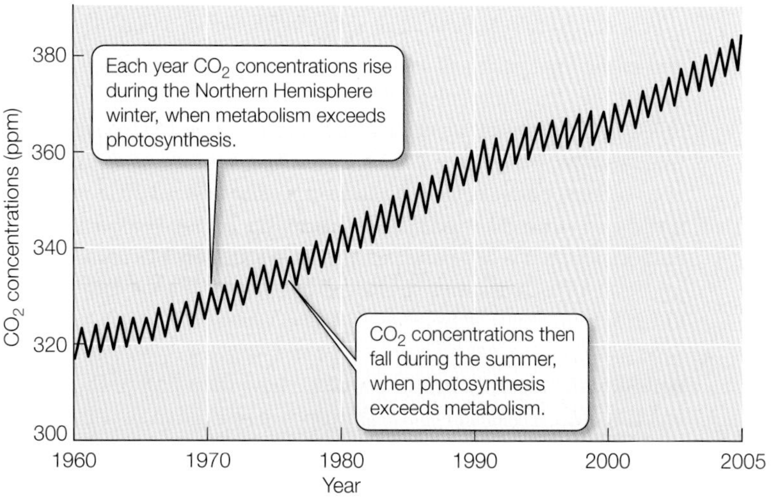

56.11 Atmospheric Carbon Dioxide Concentrations Are Increasing These carbon dioxide concentrations, expressed as parts per million by volume of dry air, were recorded on top of Mauna Loa, Hawaii, far from most sources of human-generated CO_2 emissions.

Some human-generated carbon is stored on land. Photosynthesis by terrestrial vegetation, principally in forests and savannas, typically absorbs about the same amount of carbon that is released by metabolism, about half by plants and half by microbes decomposing organic matter produced by the plants. Currently, however, the photosynthetic consumption of CO_2 exceeds the metabolic production of CO_2. Thus Earth's terrestrial vegetation is storing carbon that would otherwise be increasing atmospheric CO_2 concentrations.

Can terrestrial ecosystems help absorb the additional CO_2 that human activities are releasing into the atmosphere? As we saw at the opening of this chapter, recent investigations of forest trees suggest that we cannot count on terrestrial vegetation to store much more carbon for us. Christian Körner and his colleagues at the University of Basel in Switzerland monitored tree growth, reproduction, and carbon storage at these high CO_2 concentrations over a 4-year period. They found that the trees, which had reached only about two-thirds of their maximum size at the time of the experiment, did increase their growth rates, but that various species responded differently. For example, beech trees kept their leaves longer, but oaks held their leaves for less time.

The results of these and other experiments suggest that although *young* forests are likely to store more carbon in a carbon-enriched world, terrestrial vegetation, taken overall, may actually *lose* carbon rather than storing it under these conditions. For example, a recent intensive survey, based on almost 6,000 sites representing all types of land uses in England and Wales, showed that about 13 million tons of carbon had been lost from the soils of those countries during the past 25 years. Soil organic carbon was lost from all ecosystem types, including natural forests, but losses were greater from agricultural soils. Why the losses occurred is not clear, but climate warming (which increases plant metabolism) is the most likely factor causing losses of carbon from natural forests.

ATMOSPHERIC CO_2 AND GLOBAL WARMING How will global climates and ecosystems change in response to the rapid carbon dioxide enrichment caused by human activities? The buildup of atmospheric CO_2 that has resulted from the burning of fossil fuels is warming Earth. The concentration of CO_2 in air trapped in the Antarctic and Greenland ice caps during the last Ice Age—between 15,000 and 30,000 years ago—was as low as 200 parts per million. During a warm interval 5,000 years ago, atmospheric CO_2 may have been slightly higher than it is today. This long-term record shows that Earth has been warmer when CO_2 levels were higher and cooler when they were lower (**Figure 56.12**). The atmosphere would have warmed even more than it has if the oceans had not absorbed vast amounts of heat. The upper 100 to 200 meters of all the oceans have warmed dramatically, and warming has penetrated as deep as 700 meters in both the North Atlantic and South Atlantic oceans (**Figure 56.13**).

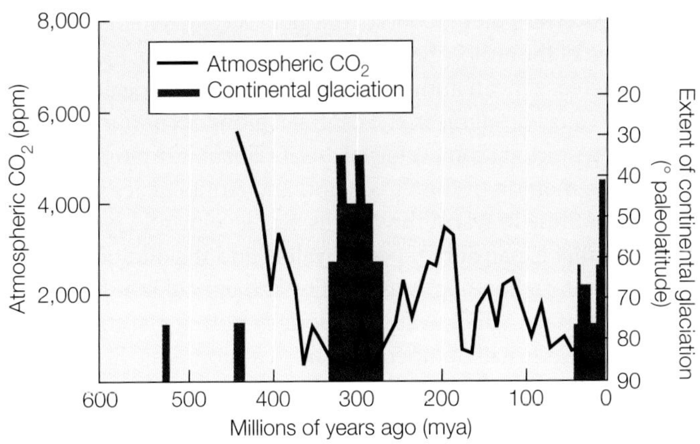

56.12 Higher Atmospheric CO_2 Concentrations Correlate with Warmer Temperatures The geological record for variations in carbon dioxide over the last 500 million years (black trace) shows that CO_2 levels have been lowest during times of extensive glaciation (red bars), and thus presumably lower temperatures.

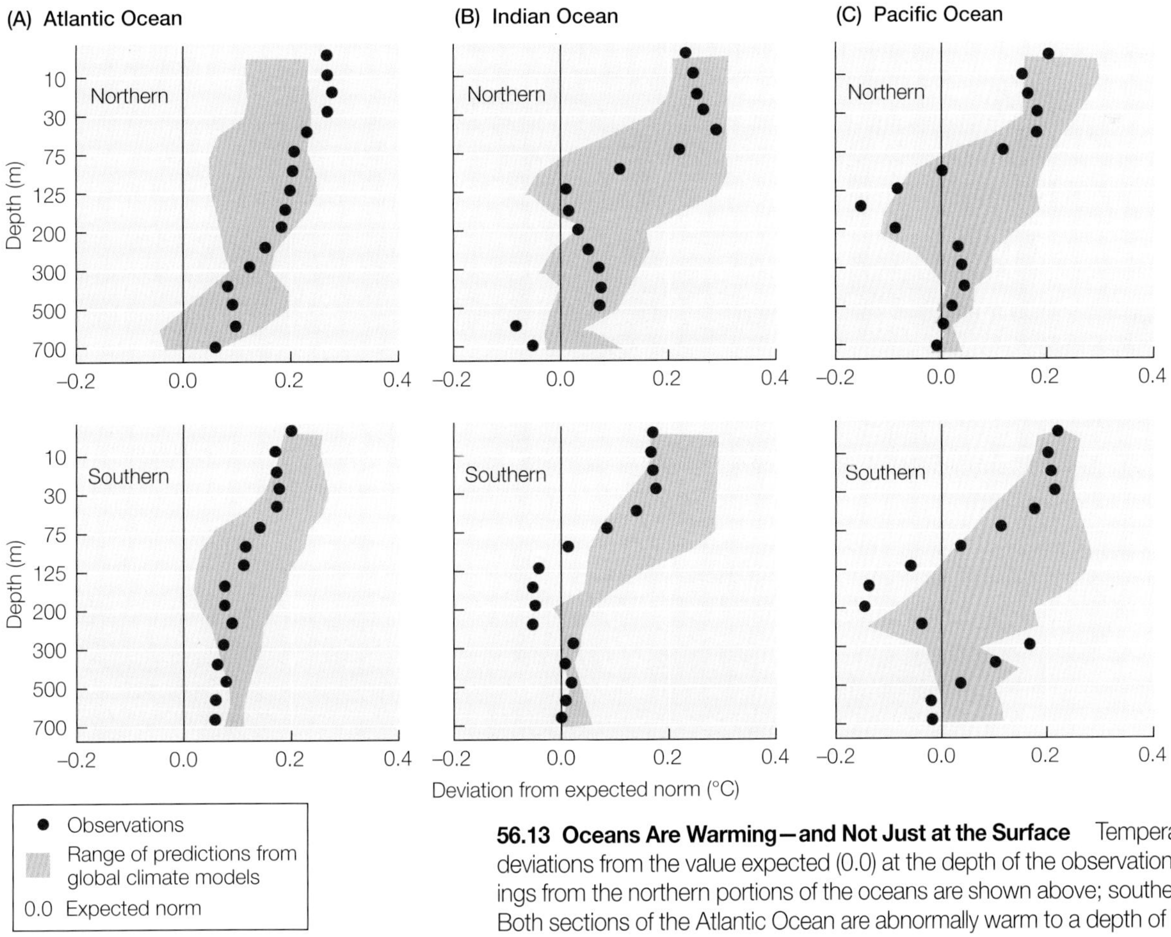

56.13 Oceans Are Warming—and Not Just at the Surface Temperatures are plotted as deviations from the value expected (0.0) at the depth of the observations. Temperature readings from the northern portions of the oceans are shown above; southern readings are below. Both sections of the Atlantic Ocean are abnormally warm to a depth of 700 meters.

Complex computer models of the global ecosystem indicate that a doubling of the atmospheric CO_2 concentration from today's level would increase mean annual temperatures worldwide and would probably cause droughts in the central regions of continents while increasing precipitation in coastal areas. Global warming could result in the melting of the Greenland and Antarctic ice caps, and would continue to warm the oceans. Sea level would rise (due to both thermal expansion of ocean waters and the addition of glacial meltwater), flooding coastal cities and agricultural lands. Because the rate at which water evaporates from the surface of the oceans increases with temperature, tropical storms are likely to become more intense. Indeed, during the past two decades, the number of hurricanes generated per year worldwide has not increased, but the number of severe (category 4 and 5) storms has.

Another result of global warming is increasing outbreaks of many diseases. Winter cold typically kills many pathogens, sometimes eliminating as much as 99 percent of a pathogen population. If climate warming reduces this population bottleneck, some diseases will become more common in temperate regions. For example, dengue fever and its mosquito vector are now spreading to higher latitudes where they were formerly absent. Several plant diseases are more severe after milder winters or during periods of warmer temperatures. For example, during a 39-year study in Maine, beech bark canker, caused by a fungus (*Nectria* spp.), was found to be more severe after mild winters or dry autumns. These weather conditions favor the survival and spread of the beech scale insect, which weakens the trees and predisposes them to the fungal infection. The spread of this pathogen poses a serious problem for the timber industry.

The 1990 Assessment Report of the Intergovernmental Panel on Climate Change (IPCC) paid little attention to risks to human health. In contrast, IPCC's 1996 Assessment Report gave detailed consideration to the potential effects of climate change on human health. Our increasingly interconnected global society enables disease organisms to travel rapidly around the world. Infected people carried the SARS virus, for example, from China to Canada within a few days. The combination of human mobility and climate warming poses serious challenges to human health throughout the world.

Recent perturbations of the nitrogen cycle have had adverse effects on ecosystems

Nitrogen gas (N_2) makes up 78 percent of Earth's atmosphere, but most organisms cannot use nitrogen in its gaseous form. Only a few species of microorganisms can convert atmospheric N_2 into forms that are usable by plants, a process called *nitrogen fixation*. Other microorganisms carry out *denitrification*, the principal process that removes nitrogen from the biosphere and returns it to the atmosphere (see Section 36.4 and Figure 36.10). These movements of nitrogen account for about 95 percent of all natural nitrogen flows on Earth (**Figure 56.14**).

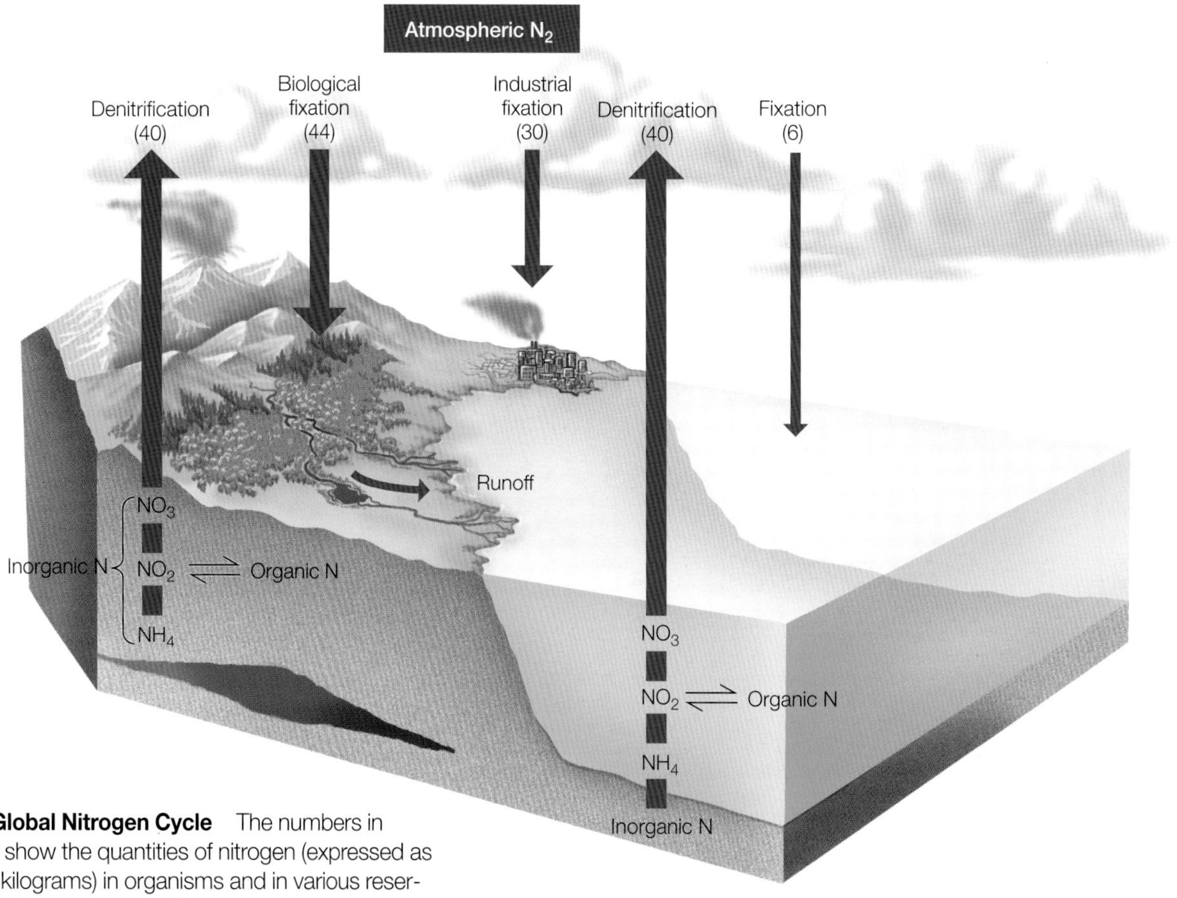

56.14 The Global Nitrogen Cycle The numbers in parentheses show the quantities of nitrogen (expressed as units of 10^9 kilograms) in organisms and in various reservoirs, and the amounts that move annually between ecosystem compartments. Arrow widths are proportional to the sizes of the flows.

Biologically usable nitrogen is often in short supply in ecosystems. That is why nearly all fertilizers contain nitrogen. Populations of nitrogen-fixing organisms rarely increase to such an extent that nitrogen is no longer limiting because nitrogen tends to be lost rapidly from ecosystems by leaching, vaporization of ammonia, and denitrification. In addition, fixing nitrogen requires large amounts of energy. Nitrogen-fixing organisms often lose out in competition with non-fixers when nitrogen becomes more readily available.

As a result of the extensive use of fertilizers on agricultural crops and the burning of fossil fuels (which generates nitric oxide), the total nitrogen fixation by humans today is about equal to global natural nitrogen fixation (**Figure 56.15**). This human-generated nitrogen flow is expected to continue to increase during the coming decades. A variety of adverse effects are associated with these large perturbations of the nitrogen cycle. They include the contamination of groundwater by nitrate (NO_3^-) from agricultural runoff, increases in atmospheric nitrous oxide (N_2O) and tropospheric O_3 (both greenhouse gases), and smog production.

When more nitrogen is applied to croplands than is taken up by plants, the excess nitrogen moves downward into groundwater or as runoff, into rivers, lakes, and oceans. The addition of nutrients to these bodies of water, known as **eutrophication**, can have a number of negative effects on aquatic ecosystems, as we'll see in our discussion of the phosphorus cycle. The "dead zone" in the Gulf of Mexico that has formed near the mouth of the Mississippi River is a result of flows of nitrogen-enriched water from agricultural fields in the Upper Midwest (**Figure 56.16**). Outbreaks of the toxic dinoflagellate *Pfiesteria* in estuaries on the Atlantic coast of North America are also triggered by nitrogen enrichment.

Some of the nitrogen that enters the atmosphere falls back to land in precipitation or as dry particles. This deposition of nitrogen affects the composition of terrestrial vegetation by favoring species that are better adapted to high nutrient levels. Atmospheric deposition of nitrogen has increased dramatically during recent decades due to human activity. Spatial variation in deposition rates allowed ecologists to determine that plant species richness is inversely correlated with the rate of nitrogen deposition in grasslands in the United Kingdom. Rates of nitrogen deposition are high enough over much of Europe and eastern North America to cause substantial reductions in species richness in grasslands on both continents.

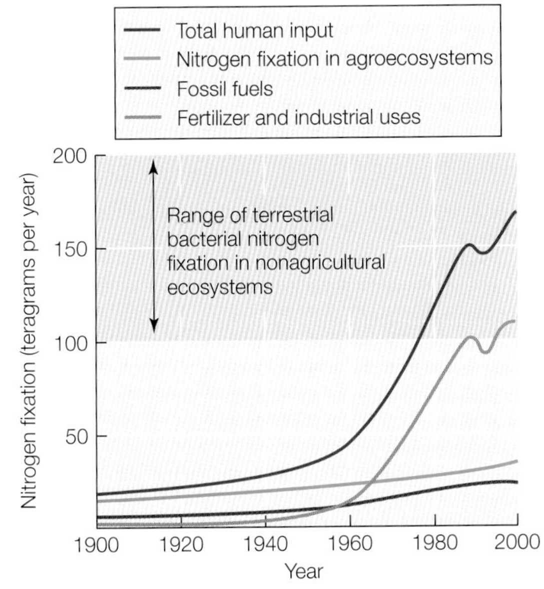

56.15 Human Activities Have Increased Nitrogen Fixation Most of the biologically reactive nitrogen produced by humans is manufactured for use in agriculture (synthetic fertilizers) and industry. Some nitrogen is also a by-product of burning fossil fuels.

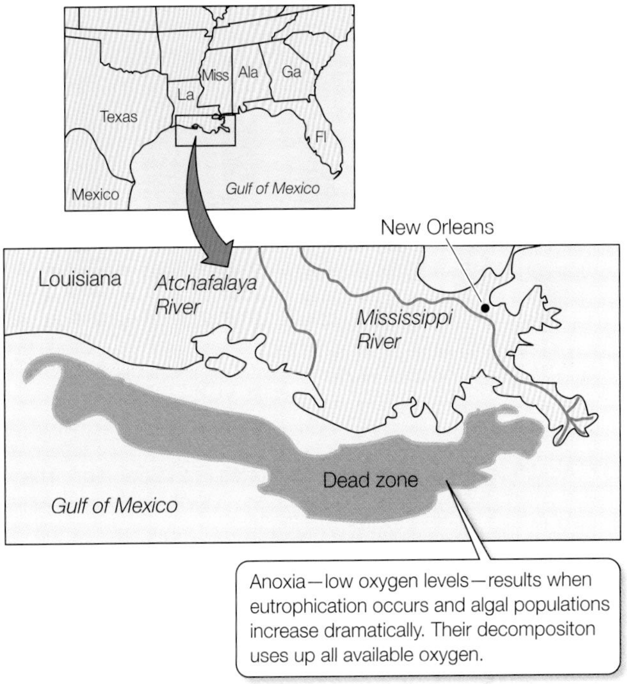

56.16 A "Dead Zone" Near the Mouth of the Mississippi River Rising amounts of nitrogen and phosphorus in the runoff from agricultural lands in the midwestern United States are carried by the Mississippi River to the Gulf of Mexico. This nutrient enrichment causes excessive algal growth, depleting the oxygen in the water and creating a deoxygenated "dead zone" in which most aquatic organisms cannot survive.

The burning of fossil fuels affects the sulfur cycle

Emissions of the gases sulfur dioxide (SO_2) and hydrogen sulfide (H_2S) from volcanoes and fumaroles (vents for hot gases) are the only significant natural nonbiological flows of sulfur. These sources release, on average, between 10 and 20 percent of the total natural flow of sulfur to the atmosphere, but they vary greatly in time and space. Large volcanic eruptions spread great quantities of sulfur over broad areas, but they are rare events. Terrestrial and marine organisms also emit compounds containing sulfur. Certain marine algae produce large amounts of dimethyl sulfide (CH_3SCH_3), which accounts for about half of the biotic component of the sulfur cycle (hence the smell of rotting seaweeds). Sulfur is apparently always abundant enough to meet the needs of living organisms.

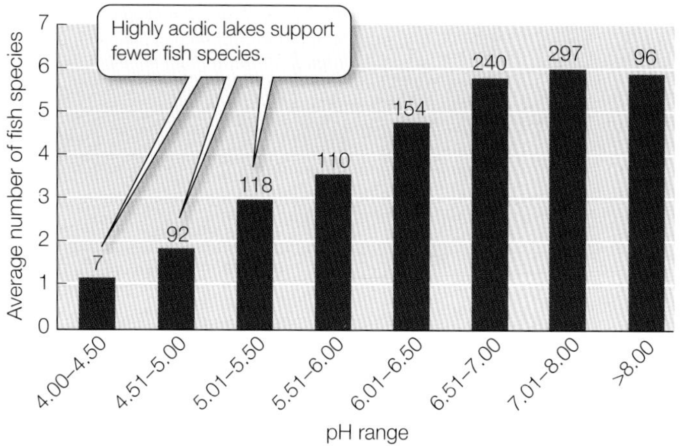

56.17 Acidification of Lakes Exterminates Fish Species The number of fish species found in lakes in the Adirondack region of New York State is directly correlated with pH. The numbers above the bars indicate the number of lakes in each pH range.

Sulfur plays an important role in global climate. Even if air is moist, clouds do not form readily unless there are small particles in the atmosphere around which water can condense. Dimethyl sulfide is the major component of such particles. Therefore, increases or decreases in atmospheric sulfur levels can change cloud cover and hence climate.

Humans have altered the sulfur cycle, as well as the nitrogen cycle, by the burning of fossil fuels. An important regional effect of these alterations is **acid precipitation**: rain or snow whose pH is lowered by the presence of sulfuric acid (H_2SO_4) and nitric acid (HNO_3). These acids may travel hundreds of kilometers in the atmosphere before they settle to Earth in precipitation or as dry particles.

Acid precipitation now falls in all major industrialized countries and is particularly widespread in eastern North America and Europe. The normal pH of precipitation in New England was about 5.6, but precipitation there now averages about pH 4.4, and there have been occasional rain- or snowfalls with a pH as low as 3.0. Precipitation with a pH of about 3.5 or lower causes direct damage to the leaves of plants and reduces rates of photosynthesis. Acidification of lakes in the Adirondack region of New York State has reduced fish species richness by causing the extinction of acid-sensitive species (**Figure 56.17**). Fortunately, as a result of the establishment of a flexible regulatory system under the 1990 Clean Air Act amendments, precipitation in much of the eastern United States is less acidic today than it was 20 years ago, primarily because of reduced sulfur emissions (**Figure 56.18**).

David Schindler at the University of Alberta studied the effects of acid precipitation on small Canadian lakes by adding enough H_2SO_4 to two lakes to reduce their pH from about 6.6 to 5.2. In both lakes, nitrifying bacteria failed to survive under these moderately acidic conditions. As a result, the nitrogen cycle was blocked, and ammonium ions accumulated in the water. When Schindler stopped adding acid to one of the lakes, its pH increased to 5.4, and nitrification resumed. After about a year, the pH of the lake returned to its original value. These experiments show that lakes are very sensitive to acidification, but that pH can return rapidly to normal values because water in lakes is exchanged rapidly.

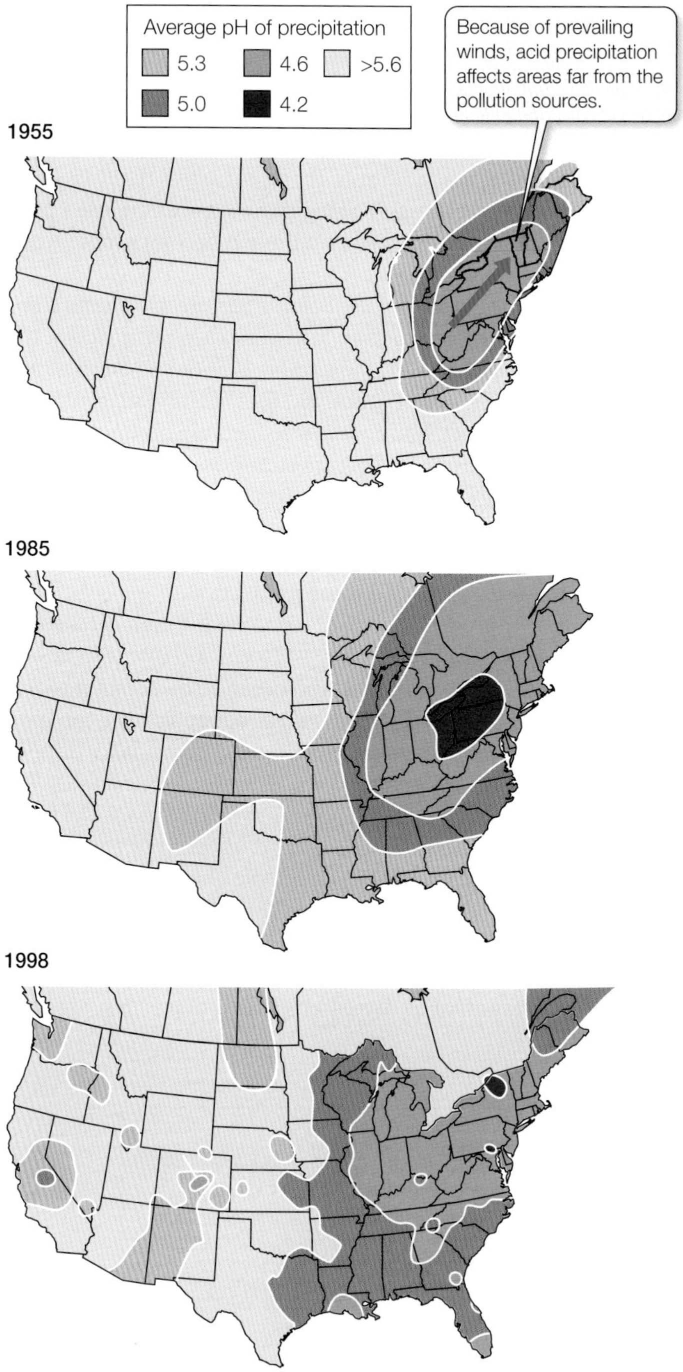

56.18 Acid Precipitation Is Decreasing in the Eastern United States Thanks to controls on sulfur emissions, precipitation in many parts of the eastern United States is less acidic than it was a decade ago, but acidity is increasing in parts of the West.

The global phosphorus cycle lacks an atmospheric component

Phosphorus accounts for only about 0.1 percent of Earth's crust, but it is an essential nutrient for all life forms. It is a key component of DNA, RNA, and ATP. Unlike the other elements discussed in this section, phosphorus lacks a gaseous phase. Some phospho-

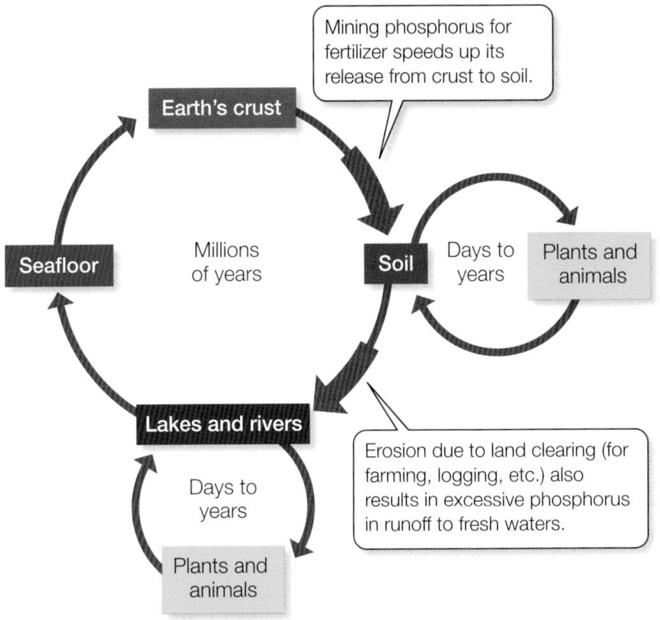

56.19 The Phosphorus Cycle The widths of the arrows are proportional to the flow rates. Two great increases in phosphorus flows (wider arrows) are due to human activities.

rus is transported on dust particles, but very little of the phosphorus cycle takes place in the atmospheric compartment. The global phosphorus cycle takes millions of years to complete because the processes of sedimentary rock formation on the ocean bottom, subsequent uplifting, and the weathering of rock into soil all act slowly (**Figure 56.19**). Phosphorus often cycles rapidly among organisms, however.

Human activity has radically accelerated some parts of the phosphorus cycle. About 90 percent of the phosphorus that is mined is used to produce fertilizers and animal feeds. One consequence of the massive use of phosphorus for fertilizer is that between 10.5 and 15.5 teragrams (1 teragram = 10^{12} grams) of phosphorus are accumulating in soils each year, primarily on agricultural lands. Increasing concentrations of phosphorus in agricultural soils are certain to lead to increased runoff of phosphorus into streams and lakes. Soil erosion due to the clearing of land for purposes such as agriculture and logging also increases the amount of phosphorus and other nutrients in runoff.

Phosphorus is often a limiting nutrient in soils and fresh waters, which is why adding phosphorus to agricultural soils and lakes increases rates of photosynthesis and biological productivity. Most extra phosphorus enters lakes as phosphates derived from fertilizers and household detergents. The resulting eutrophication allows algae and bacteria to multiply, forming blooms that turn the water green. When these organisms die, their decomposition consumes all of the O_2 in the lake, and anaerobic organisms come to dominate the bottom sediments. These anaerobic organisms do not break down carbon compounds all the way to CO_2. Their metabolic end products, many of which have unpleasant odors, build up.

Two hundred years ago, Lake Erie, one of the Great Lakes on the border between the United States and Canada, had only moderate rates of photosynthesis and clear, oxygenated water. Today, however, the more than 15 million people that live in the Lake Erie drainage basin pour more than 250 billion liters of sewage and industrial wastes into the lake annually. The entire basin is intensely farmed and heavily fertilized. In the early part of the twentieth century, when industrialization began, nutrient concentrations in the lake increased greatly, and algae proliferated. At the water filtration plant in Cleveland, Ohio, algae increased from 81 per milliliter in 1929 to 2,423 per milliliter in 1962. Populations of bacteria also increased: the numbers of the bacterium *Escherichia coli* increased enough to cause the closing of many of the lake's beaches as health hazards.

Since 1972, the United States and Canada have invested more than U.S. $9 billion to improve municipal waste treatment facilities and reduce discharges of phosphorus into Lake Erie. As a result, the amount of phosphate added to Lake Erie has decreased more than 80 percent from its highest level, and phosphorus concentrations in the lake have declined substantially. The deeper waters of Lake Erie still become poor in O_2 during the summer months, but the rate of O_2 depletion is declining.

Fortunately, the potential to recover and recycle phosphorus is high. The phosphorus contained in sewage and animal wastes could supply much of the needs of the detergent and fertilizer industries. Careful application of fertilizers on agricultural lands could reduce the rate of phosphorus accumulation in soils without reducing crop yields. However, reduction of phosphorus in soils takes many decades after remedial actions are initiated. During that time, increased eutrophication of lakes and streams is certain to happen.

Other biogeochemical cycles are also important

In addition to the common elements involved in biogeochemical cycles that we have just discussed, some rare elements are also very important ecologically because, in addition to macronutrients, there is an array of essential nutrients that organisms need only in very small amounts.

One important such nutrient is *iron* (Fe), which is required for photosynthesis. Iron is readily available on land, from which it is transported to coastal waters by rivers and to the open ocean by atmospheric dust. Because iron is insoluble in oxygenated water, it rapidly sinks to the ocean floor. Therefore, over much of the ocean, the rate of photosynthesis is limited by iron.

Three micronutrients that occur at particularly low concentrations, iodine (I), cobalt (Co), and selenium (Se), are of particular importance. Land plants require these micronutrients in minimal concentrations, but animals, particularly endothermic vertebrates, require I, Co, and Se in concentrations that exceed the supply in many environments. Large herbivores may need to obtain I, Co, and Se by drinking water at mineralized springs or by eating enriched earth.

- *Iodine* is an essential component of the hormone thyroxine, which governs many metabolic processes. The ocean is the main reservoir of iodine; fresh water is poor in iodine. Iodine, which is supplied to land primarily by marine algae that release it to the air, is therefore often deficient in continental interiors.

- *Cobalt* is an essential component of vitamin B_{12} and is necessary for the synthesis of proteins. Bacteria that fix atmospheric nitrogen require cobalamin and cobamide coenzymes. Herbivores cannot synthesize cobalamin, so microbes in their guts are essential not only for fixation of atmospheric nitrogen and digestion of plant fibers, but also for synthesis of cobalamin.
- *Selenium* is a component of several important enzymes, such as the antioxidant glutathione peroxidase, which protects tissues from oxidative damage.

Iodine, together with Co and Se, regulates the production and functioning of many metabolic agents, including hormones, vitamins, and enzymes that contain zinc and copper. Although I, Co, and Se are needed by animals in very small amounts, they are potentially deficient in their plant food, which has even lower concentrations than those found in soils and rocks. Deficiencies of I, Co, and Se are not frequent everywhere, but Australia, which is characterized by nutrient-poor soils, contains very few places where herbivores can obtained the necessary supplies of I, Co, and Se. Cobalt deficiency is prevalent among domestic ruminants in Western Australia.

Biogeochemical cycles interact

Because biogeochemical cycles interact in significant ways, alterations in any one cycle affect the others. We have already seen how human alterations of several biogeochemical cycles are increasing the concentrations of greenhouse gases in the atmosphere and warming Earth's climate. As we saw earlier, acid precipitation is a result of the combined influences of sulfuric and nitric acids released into the atmosphere by human activities. Nitrate released by humans is a powerful oxidant; by oxidizing iron, it increases the movement of both iron and arsenic in lakes. Arsenic is a toxic element of considerable importance in the United States and south central Asia. Iron interacts with silicon to influence phytoplankton productivity in some areas of the oceans. Every year, scientists are discovering new interactions of which they were previously unaware, and experiments exploring biogeochemical interactions are increasing in number.

The interaction between elevated concentrations of atmospheric CO_2 and nitrogen fixation was studied by a group of ecologists in scrub oak vegetation on sandy, acidic soils in central Florida. They artificially increased CO_2 concentrations surrounding a nitrogen-fixing vine, *Galactia elliottii*, and measured its response. Higher CO_2 levels increased nitrogen fixation during the first year of the experiment, but, surprisingly, the positive effect disappeared by the third year. Elevated CO_2 levels actually depressed nitrogen fixation during the fifth, sixth, and seventh years of the experiment (**Figure 56.20**). The experimenters suspected that shortages of other micronutrients, such as iron and molybdenum, caused the depression of nitrogen fixation, so they measured concentrations of those nutrients in the leaves of *G. elliottii*. Concentrations of molybdenum (Mo) were particularly low in the leaves because accumulation of soil organic matter and lowering of soil pH—both of which strengthen Mo attachment to soil particles—reduced the availability of Mo to the plants.

EXPERIMENT

HYPOTHESIS: Higher CO_2 levels enhance nitrogen fixation.

METHOD

Artificially increase concentration of atmospheric CO_2 around plants. Measure rates of nitrogen fixation over 7 years. Determine concentration of iron and molybdenum in leaves of experimental plants. Maintain normal CO_2 levels around control plants.

RESULTS

Although higher CO_2 levels initially increased the rate of nitrogen fixation, the effect was quickly reversed. By year 4, nitrogen fixation rates were lower than those in the control plants.

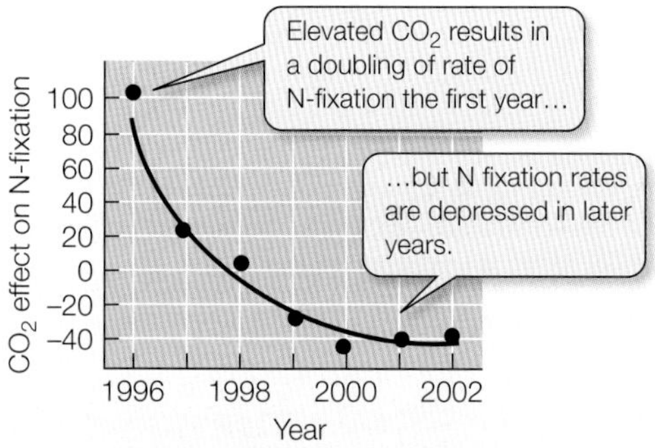

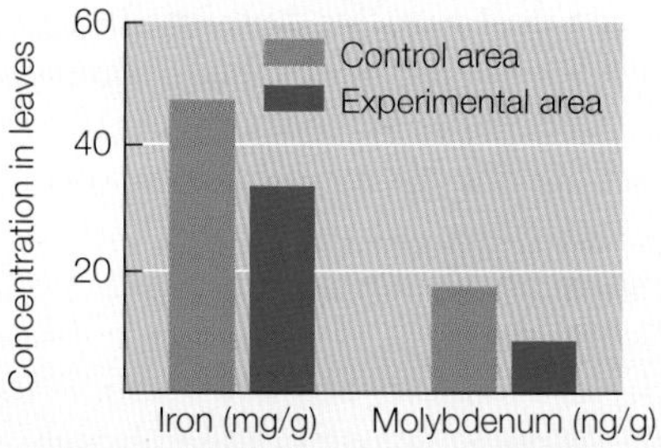

CONCLUSION: To predict the likely effects of CO_2 enrichment we need to investigate the interactions among many factors that might alter the processes we are studying.

56.20 High Atmospheric CO_2 Concentrations Negatively Affect Nitrogen Fixation Although the investigators' original expectation was that CO_2 enrichment would enhance nitrogen fixation, changes in soil pH made molybdenum (a key constituent of nitrogenase) less available to plants, thereby depressing nitrogen fixation.

56.3 RECAP

The pattern of movement of a chemical element through organisms and the physical environment is its biogeochemical cycle. Human activities have affected many biogeochemical cycles, especially those of water, carbon, nitrogen, sulfur, and phosphorus. Increasing concentrations of carbon dioxide and other greenhouse gases in the atmosphere are implicated in global warming.

- What drives the hydrological cycle? See pp. 1211–1212 and Figure 56.8
- Can you describe how biological processes move carbon from the atmospheric compartment to the terrestrial compartment and then return it to the atmosphere? See pp. 1213–1215 and Figure 56.10
- What are the results of human-induced alterations of the sulfur and nitrogen cycles? See pp. 1216–1218

The biogeochemical cycles of chemical elements are intimately connected with the functioning of ecosystems. Just as human alterations of these cycles are having many effects on ecosystems worldwide, the resulting changes in those ecosystems are having profound effects on human lives.

56.4 What Services Do Ecosystems Provide?

Ecosystems provide people with a variety of goods and services: food, clean water, clean air, flood control, soil stabilization, pollination, climate regulation, spiritual fulfillment, and aesthetic enjoyment, to name just a few. Most of these benefits either are irreplaceable or the technology necessary to replace them is prohibitively expensive. For example, potable fresh water can be provided by desalinating seawater, but only at great cost.

The rapidly expanding human population has greatly modified Earth's ecosystems to increase their ability to provide some of the goods and services it needs, particularly food, fresh water, timber, fiber, and fuel. These modifications have contributed substantially to human well-being and economic development. However, the benefits have not been equally distributed, and some people have actually been harmed by these changes. Moreover, short-term increases in some ecosystem goods and services have come at the cost of the long-term degradation of others. For example, efforts to increase the production of food and fiber have decreased the ability of some ecosystems to provide clean water, regulate flooding, and support biodiversity.

Increasing awareness of the complex linkages and trade-offs among the goods and services provided by ecosystems led to the establishment in 2001 of the Millennium Ecosystem Assessment (MA), a project that involved more than a thousand scientists and managers worldwide (**Figure 56.21**). The goals of the MA were to provide a global assessment of Earth's ecosystems, determine trends in the services they provide, and assess their importance for

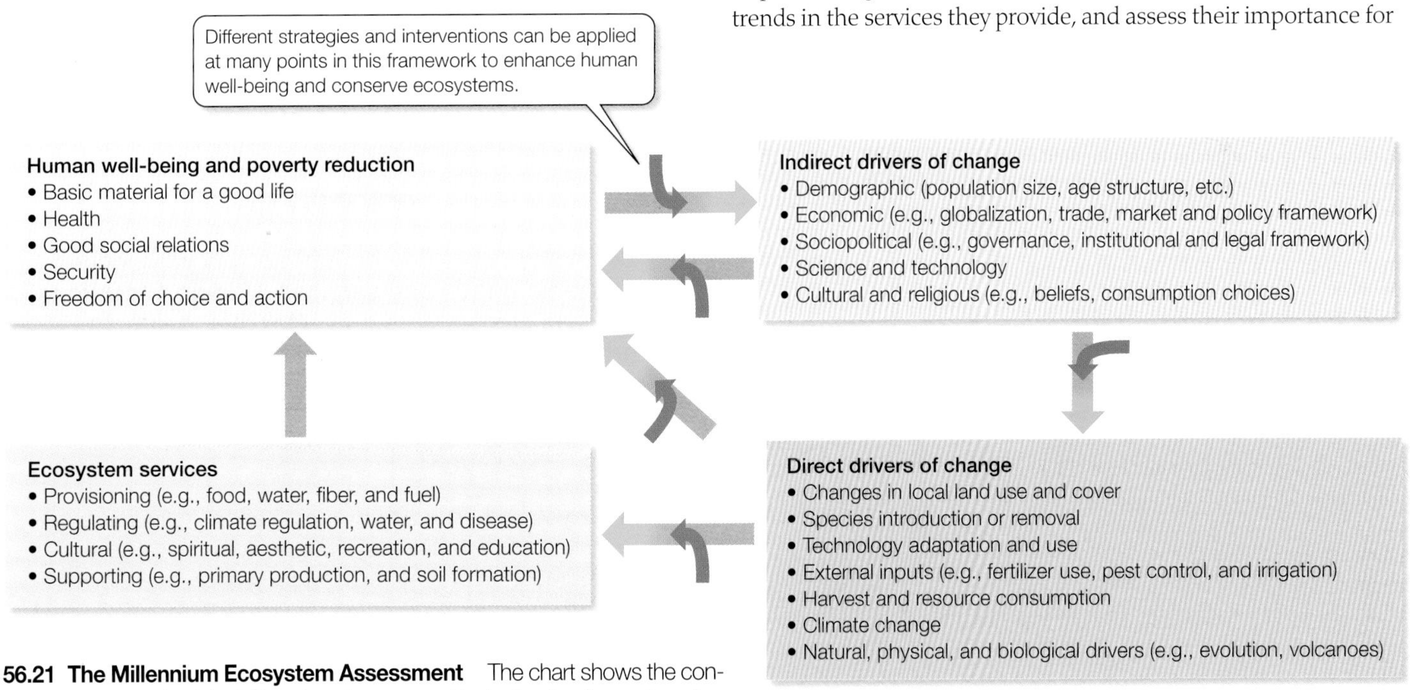
Different strategies and interventions can be applied at many points in this framework to enhance human well-being and conserve ecosystems.
Human well-being and poverty reduction
• Basic material for a good life
• Health
• Good social relations
• Security
• Freedom of choice and action
Indirect drivers of change
• Demographic (population size, age structure, etc.)
• Economic (e.g., globalization, trade, market and policy framework)
• Sociopolitical (e.g., governance, institutional and legal framework)
• Science and technology
• Cultural and religious (e.g., beliefs, consumption choices)
Ecosystem services
• Provisioning (e.g., food, water, fiber, and fuel)
• Regulating (e.g., climate regulation, water, and disease)
• Cultural (e.g., spiritual, aesthetic, recreation, and education)
• Supporting (e.g., primary production, and soil formation)
Direct drivers of change
• Changes in local land use and cover
• Species introduction or removal
• Technology adaptation and use
• External inputs (e.g., fertilizer use, pest control, and irrigation)
• Harvest and resource consumption
• Climate change
• Natural, physical, and biological drivers (e.g., evolution, volcanoes)

56.21 The Millennium Ecosystem Assessment The chart shows the conceptual framework of the Millennium Assessment including the four categories of ecosystem services, the connections of these services with human well-being, the drivers of changes in ecosystems that can affect their ability to provide these services, and several points where interventions (dark green, curved arrows) could prevent or mitigate undesirable changes.

human well-being. To achieve these goals, the MA divided ecosystem services into four categories: provisioning services, regulating services, cultural services, and supporting services. The MA determined the drivers of change in the ecosystems that provide these services, and it identified strong links among these services and components of human well-being. It also sought to identify options for mitigating undesirable changes in the availability of ecosystem services.

The most important driver of alterations in ecosystems and the services they provide has been changes in land use as natural ecosystems have been converted to other, more intensive uses. More land was converted to cropland between 1950 and 1980 than in the 150 years between 1700 and 1850. Conversions recently have been particularly rapid in tropical and subtropical terrestrial ecosystem types (**Figure 56.22**). Aquatic ecosystems have suffered losses as well: between 1970 and 2000 about 20 percent of the world's coral reefs were lost; an additional 20 percent were degraded. So much water is impounded behind dams that about six times as much water is held in reservoirs today as in natural rivers. The amount of water withdrawn from rivers has doubled since 1960, mostly for agriculture. More than half of all the synthetic nitrogen fertilizer ever used on Earth has been used since 1985. Changes in these components of the global ecosystem continue rapidly today.

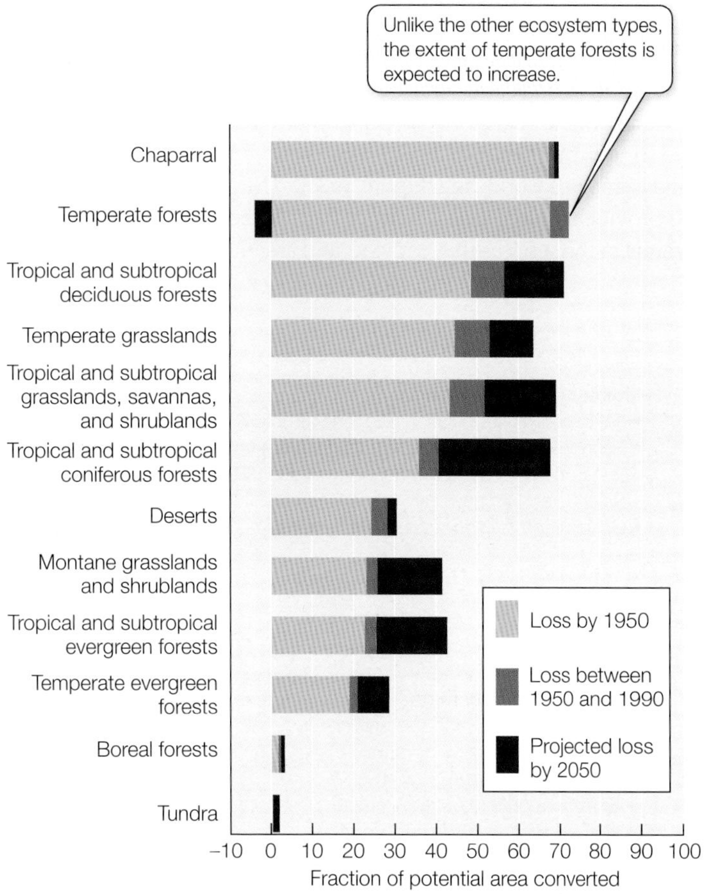

56.22 Some Ecosystem Types Have Suffered Extensive Losses Changes in land use have been the most important driver of changes in ecosystems, but some terrestrial ecosystem types have been converted to other uses more rapidly and extensively than others.

These altered ecosystems are delivering services that have contributed to many components of a good life—health, good social relations, and security. Food production contributes by far the most to economic activity and employment. The global market value of food production was $981 billion in 2000. About 22 percent of the world's population is employed in agriculture. More than 40 percent of people live in agriculturally based households, but the percentage is much lower in industrialized countries. Timber, marine fisheries, marine aquaculture, and recreational hunting and fishing are activities that use ecosystem services to make important contributions to human well-being.

Unfortunately, the modification of ecosystems to benefit human beings in one way has often resulted in the degradation of other services. For example, the spread of agriculture into marginal lands has increased soil degradation and reduced the ability of many ecosystems to provide clean water. Extensive use of pesticides controls pest insects, but also greatly reduces populations of pollinators and the services they provide to both crops and native plants. As described at the start of Chapter 52, the loss of wetlands and other natural buffers has reduced the ability of ecosystems to regulate natural hazards. The damage from the tsunami that hit Indonesia and other southeast Asian countries in December of 2004 was greater in many places than it would have been if the mangrove forests that protect the coast had not been cut down and converted to agriculture. Hurricane Katrina, which struck the U.S. Gulf Coast less than a year later, would not have caused as much flooding in New Orleans if the wetlands surrounding the city had been left intact (see p. 1112).

56.4 RECAP

Ecosystems provide human society with indispensable goods and services. Human alterations in these ecosystems may alter their ability to provide these goods and services.

- Can you describe some of the essential goods and services that ecosystems provide to humans? See pp. 1221–1222
- Do you understand how human efforts to increase the provision of some ecosystem services can cause degradation of other ecosystem services?

Human society faces the challenge of determining how we can obtain the goods and services we desire from ecosystems without compromising their ability to provide those goods and services over the long term. Fortunately, many options exist for such sustainable management of ecosystems.

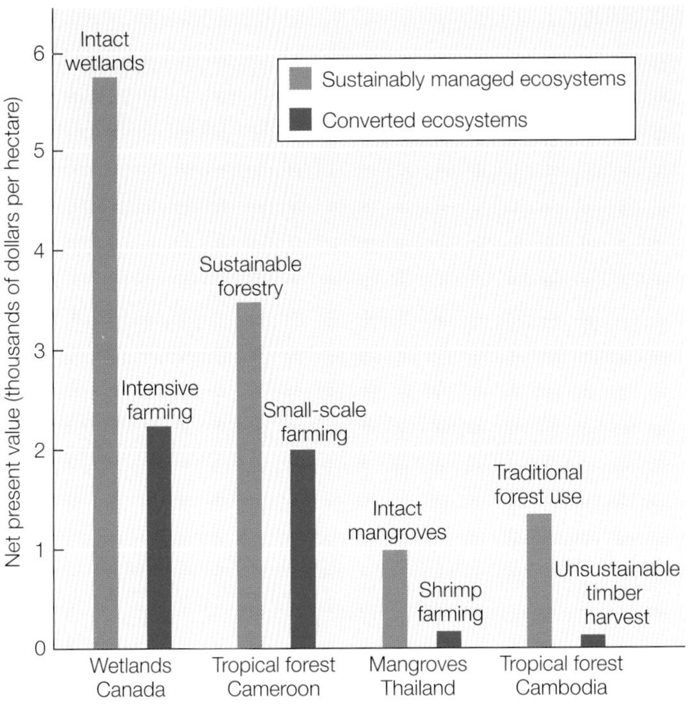

56.23 Economic Values of Sustainably Managed and Converted Ecosystems The Millenium Assessment estimates that sustainably managed ecosystems often provide more goods and services than those that have been converted to intensive uses.

56.5 What Options Exist to Manage Ecosystems Sustainably?

Despite the degradation of some important ecosystem services, there are many ways to conserve or enhance specific ecosystem services without compromising others. Often the total economic value of a sustainably managed ecosystem is higher than that of an intensively exploited ecosystem, such as one subjected to clear-cut logging or intensive agriculture (**Figure 56.23**). The long-term economic benefits of preventing overexploitation of marine fish populations, for example, are enormous; the collapse of the Newfoundland cod fishery in the 1990s due to overfishing resulted in the loss of tens of thousands of jobs and cost at least $2 billion in income support and retraining. As we saw in Section 54.5, it can take a long time for such fisheries to recover.

A major barrier to achieving these greater long-term benefits is that many ecosystem services are considered "public goods" that have no market value. Nobody has an incentive to pay for those services, whereas the individuals that own converted ecosystems can capture the great private economic benefits yielded by the conversions. It may take government action to create incentives encouraging sustainable ecosystem management. Elimination of subsidies that promote excessive exploitation of ecosystems is one such action governments might take. For example, between 2001 and 2003, governments of developed countries subsidized domestic agriculture each year by more than $324 billion. These subsidies led to greater food production in those countries than the global market warranted. They also promoted excessive use of fertilizers and reduced the profitability of agriculture in developing countries.

More sustainable use of fresh water can be achieved by charging users the full cost of providing the water, by developing methods to use water more efficiently in agriculture (the activity that consumes much more water than any other), and by altering the allocation of water rights so that the incentives favor conservation.

Sustainable use of marine fisheries can be achieved by establishing more marine protected reserves and flexible no-take zones where fish can readily grow to reproductive age. For some fisheries, establishing and implementing a limit to the number of fishing permits can increase the total economic value of the fishery and provide better fish to consumers. Such improvements followed the establishment of permits that could be bought and sold for the North Pacific halibut fishery.

Although many options exist to manage ecosystems in ways that will increase the total value of ecosystem goods and services and achieve a more equitable distribution of their benefits among people, implementing them will not be easy. Most people are not aware of how much human activities affect the functioning of ecosystems or of the great long-term value of ecosystem services, so active public education programs are needed. Maintaining and enhancing ecosystem services that have no established market value will be especially difficult. Probably the most difficult service to maintain in the face of increasingly intensive human use of global ecosystems is biological diversity. We devote the final chapter of this book to this important topic.

CHAPTER SUMMARY

56.1 What are the compartments of the global ecosystem?

Earth is a closed system with respect to atomic matter, but an open system with respect to energy.

Earth's physical environment can be divided into four compartments: oceans, fresh waters, atmosphere, and land. Review Figure 56.1

Most materials that cycle through the four compartments end up in the oceans. Cold, nutrient-rich water rises to the ocean surface in coastal **upwelling zones**.

Only a small fraction of Earth's water resides in the freshwater compartment, but water moves rapidly through it. Nutrients reach lakes and rivers by surface flow or by the movement of **groundwater**.

Vertical **turnover** of the water brings nutrients and dissolved CO_2 to the surface and O_2 to the deeper waters of freshwater lakes. Turnover occurs in temperate-zone lakes in spring and in autumn. Review Figure 56.3

The atmosphere plays a decisive role in regulating temperatures at and close to Earth's surface. Most global air circulation takes place in its lowest layer, the **troposphere**. A layer of ozone in the **stratosphere** absorbs ultraviolet radiation. Review Figure 56.4

Water vapor, carbon dioxide, and other **greenhouse gases** in the atmosphere are transparent to sunlight, but trap heat, warming Earth's surface.

Land covers only one-fourth of Earth's surface, but human life strongly depends on the productivity of that compartment.

56.2 How does energy flow through the global ecosystem?

See Web/CD Activity 56.1

Gross primary productivity is the rate at which energy is incorporated into the bodies of primary producers (which obtain their energy directly from the sun). The accumulated energy is **gross primary production** (**GPP**).

The energy available to organisms that eat primary producers, or **net primary production** (**NPP**), is gross primary production minus the energy expended by primary producers on their own metabolism.

The geographic distribution of primary production reflects the distribution of land masses, temperature, and moisture. See Figure 56.7

Humans appropriate about 20 percent of Earth's average annual net primary production, although the percentage varies regionally.

56.3 How do materials cycle through the global ecosystem?

See Web/CD Tutorial 56.1

The pattern of movement of a chemical element through organisms and the compartments of the global ecosystem is its **biogeochemical cycle**.

The **hydrological cycle** is driven by evaporation of water from the surface of the oceans. The excess terrestrial precipitation that falls on land eventually returns to the oceans, primarily through rivers. Groundwater in **aquifers** plays only a small role in the hydrological cycle, but is being seriously depleted by human activities. Review Figure 56.8, Web/CD Tutorial 56.2

Fires release large amounts of carbon and other elements into the atmosphere.

The oceans act as a carbon sink. The rate at which carbon moves from the atmosphere to the oceans depends in part on photosynthesis by phytoplankton. Carbon cycles between the atmospheric and terrestrial compartments as it is removed from the atmosphere by photosynthesis and returned to the atmosphere by metabolism. Review Figure 56.10, Web/CD Tutorial 56.3

The concentration of CO_2 in the atmosphere has greatly increased in the last 150 years, largely due to the burning of **fossil fuels**. This buildup of CO_2 is warming the global climate. Review Figures 56.11 and 56.12

Human activities have greatly altered the nitrogen cycle. As a result of agricultural use of fertilizers and the burning of fossil fuels, the total nitrogen fixation by humans is about equal to the amount of natural nitrogen fixation. Review Figures 56.14 and 56.15, Web/CD Tutorial 56.4

Human alterations of the nitrogen cycle have resulted in excess nitrogen compounds and other nutrients in bodies of water, leading to **eutrophication**. Review Figure 56.16

The burning of fossil fuels increases the amount of nitrogen and sulfur in the atmosphere, leading to **acid precipitation**.

Human activities such as agricultural use of fertilizers and clearing of land have dramatically increased the input of phosphorus into soils and fresh waters. Review Figure 56.19

56.4 What services do ecosystems provide?

Ecosystem services include food, clean water, soil stabilization by plants, pollination, climate regulation, spiritual fulfillment, and aesthetic enjoyment. Most ecosystem services either are irreplaceable or the technology necessary to replace them is prohibitively expensive.

Efforts to enhance the capacity of an ecosystem to provide some goods and services often come at the cost of its ability to provide others.

56.5 What options exist to manage ecosystems sustainably?

The economic value of a sustainably managed ecosystem is often higher than that of a converted or intensively exploited ecosystem. Review Figure 56.23

Recognition of the value of a sustainably managed ecosystem for "public goods" may induce organizations and government entities to protect them. Public education is needed to make people aware of how much human activities affect ecosystem sustainability.

SELF-QUIZ

1. Earth is not in chemical equilibrium because
   a. Earth has a moon.
   b. organisms dissipate energy as heat.
   c. most continents are in the Northern Hemisphere.
   d. Earth has living organisms that generate O_2.
   e. Earth is tilted on its axis.

2. Marine upwelling zones are important because
   a. they help scientists measure the chemistry of deep ocean water.
   b. they bring to the surface organisms that are difficult to observe elsewhere.
   c. ships can sail faster in these zones.

d. they increase marine productivity by bringing nutrients back to surface ocean waters.
e. they bring oxygenated water to the surface.

3. Which of the following is *not* true of the troposphere?
   a. It contains nearly all of the atmospheric water vapor.
   b. Materials enter it primarily at the intertropical convergence zone.
   c. It is about 17 kilometers thick in the tropics and subtropics.
   d. Most global air circulation takes place there.
   e. It contains about 80 percent of the mass of the atmosphere.
4. The hydrological cycle is driven by
   a. the flow of water into the oceans via rivers.
   b. evaporation (transpiration) of water from the leaves of plants.
   c. evaporation of water from the surface of the oceans.
   d. precipitation falling on the land.
   e. the fact that more water falls on the ocean as precipitation than evaporates from its surface.
5. Carbon dioxide is called a greenhouse gas because
   a. it is used in greenhouses to increase plant growth.
   b. it is transparent to heat, but traps sunlight.
   c. it is transparent to sunlight, but traps heat.
   d. it is transparent to both sunlight and heat.
   e. it traps both sunlight and heat.
6. Micronutrients whose cycles are particularly important for animals include
   a. iodine, cobalt, and molybdenum.
   b. selenium, iodine, and cobalt.
   c. molybdenum, iodine, and iron.
   d. iodine, zinc, and selenium.
   e. cobalt, selenium, and molybdenum .
7. The cycle of phosphorus differs from the cycles of carbon and nitrogen in that
   a. phosphorus lacks an atmospheric component.
   b. phosphorus lacks a liquid phase.
   c. only phosphorus is cycled through marine organisms.
   d. living organisms do not need phosphorus.
   e. The phosphorus cycle does not differ importantly from the carbon and nitrogen cycles.
8. The sulfur cycle influences the global climate because
   a. sulfur compounds are important greenhouse gases.
   b. sulfur compounds help transfer carbon from the atmosphere to the oceans.
   c. sulfur compounds in the atmosphere are components of particles around which water condenses to form clouds.
   d. sulfur compounds contribute to acid precipitation.
   e. The sulfur cycle does not influence the global climate.
9. Acid precipitation results from human modifications of
   a. the carbon and nitrogen cycles.
   b. the carbon and sulfur cycles.
   c. the carbon and phosphorus cycles.
   d. the nitrogen and sulfur cycles.
   e. the nitrogen and phosphorus cycles.
10. Maintaining the capacity of ecosystems to provide goods and services to humanity is important because
    a. most ecosystem services cannot be replicated by any other means.
    b. most ecosystem services can be replaced by technology, but only at great cost.
    c. technological substitutes take up valuable land.
    d. governments cannot function without taxing ecosystem services.
    e. It is not important; humans could survive quite well even if ecosystem services declined greatly.

FOR DISCUSSION

1. A string of powerful hurricanes struck the east coast of the United States over the course of a single year's hurricane season. Some people claim that this disaster was due to warming of the oceans caused by greenhouse gases in the atmosphere. Others assert that global warming is not responsible because hurricanes have occurred for many centuries. How would you evaluate these conflicting claims?
2. The waters of Lake Washington, adjacent to the city of Seattle, rapidly returned to their pre-industrial condition when sewage was diverted from the lake to Puget Sound, an arm of the Pacific Ocean. Would all lakes being polluted with sewage clean themselves up as quickly as Lake Washington if pollutant inputs were stopped? What characteristics of a lake are most important to its rate of recovery following reduction of pollutant inputs?
3. Tropical forests are being cut at a very rapid rate. Does this necessarily mean that deforestation is a major source of carbon dioxide inputs to the atmosphere? If not, why not?
4. What types of experiments would you conduct to assess the likely consequences of fertilization of the oceans with iron to increase rates of photosynthesis? At what spatial and temporal scales should they be conducted?
5. A government official authorizes the construction of a large coal-burning power plant in a former wilderness area. Its smokestacks discharge great quantities of combustion wastes. List and describe all likely effects on ecosystems at local, regional, and global levels. If the wastes were thoroughly scrubbed from the stack gases, which of the effects you have just outlined would still happen?
6. Many nations have signed the Kyoto Accord, which commits them to reducing their emissions of carbon dioxide to the atmosphere. The United States has so far refused to sign the treaty, claiming that it is against the country's interests. The United States has been severely criticized for this action. Is such criticism warranted? Justify your answer.

FOR INVESTIGATION

The experiment described in Figure 56.20 showed that interactions between elevated atmospheric carbon dioxide and nitrogen fixation can reduce the availability of other essential nutrients. The investigators suggested an expansion in the suite of elements that should be studied to determine whether elevated levels of carbon dioxide are likely to result in a large carbon sink in terrestrial ecosystems. What other elements should be considered, and how might their influence be investigated?

CHAPTER 57 Conservation Biology

Ultralight and Ultrahelpful

On October 10, 2004, a flock of young whooping cranes took to the air above Neceda, Wisconsin along with three ultralight aircraft operated by pilots from an organization called Operation Migration. The captive-reared birds had been trained from a young age to follow a slow-moving ultralight aircraft on the ground. Now the birds were ready to fly with the ultralights, which would guide them across seven states to their winter home near the Gulf of Mexico. Two months and 1,200 miles later, the ultralights and their avian entourage landed safely in a national wildlife refuge in Florida. Remarkably, the following spring, most of the cranes accomplished the return trip to Wisconsin on their own—they had learned the migration route during their southward journey.

When the first Europeans arrived in North America, large whooping crane populations bred across the Gulf Coast, in the Upper Midwest, and in Canada. By the middle of the twentieth century, hunting and habitat loss had reduced them to a tiny group of fewer than 20 individuals. These few remaining birds spent the winter in the Aransas National Wildlife Refuge on the Texas coast and bred in Wood Buffalo National Park in Canada's Northwest Territories.

The recovery of whooping cranes involves cooperation among the U.S. Fish and Wildlife Service, Operation Migration, the Whooping Crane Eastern Partnership, and the International Crane Foundation (ICF). The ICF raises young cranes in captivity in ways that ensure that the birds never see a human being. They are fed by people wearing crane costumes. These methods are vital to the recovery effort because young whooping cranes learn to know who they are by *imprinting* on their caretakers (see Section 53.2).

These were not the first efforts to save the endangered crane species. The organizations' early efforts to establish a whooping crane breeding population in Montana failed. Whooping crane eggs were placed in the nests of (non-endangered) sandhill cranes. The sandhill "foster parents" raised whooping crane chicks successfully, but there was no way the young whooping cranes could learn the specific courtship and mating behaviors of their own species—they simply behaved like sandhills, and thus were unable to mate successfully. A few of these adult whooping cranes still migrate with the sandhills to the Bosque del Apache National Wildlife Refuge in New

Operation Migration An ultralight aircraft piloted by a human leads young whooping cranes on their first long migration to their overwintering grounds. After being guided once in this fashion, the birds can find their own way along the migration route.

Saving a Species Young cranes imprint on the aircraft and on the human in crane costume who feeds them. They thus grow up primed to follow the aircraft as their "leader."

Mexico, but they are a genetic dead end that will leave no offspring.

Today there is a nonmigratory population of 50 whooping cranes in Florida, 64 birds in the migratory flock that summers in Wisconsin, and 214 birds that winter in Texas. Because so much of its habitat has been lost, the whooping crane will never be as common as it was when Europeans arrived in North America, but the species has been moved back from the brink of extinction. Its future is now secure, as long as we continue to protect its breeding habitat.

IN THIS CHAPTER we will describe the rapidly expanding field of conservation biology, which is devoted to preserving the diversity of life. We will show how conservation biologists predict species extinctions, and we will see how some human activities are causing those extinctions. Finally, we will describe the strategies conservation biologists use to reduce extinction rates, prevent species from becoming threatened with extinction, and help populations of endangered species recover.

CHAPTER OUTLINE

57.1 What Is Conservation Biology?

Conservation biology is an applied scientific discipline devoted to preserving the diversity of life on Earth. People have been practicing conservation biology for centuries, but conservation biology today differs from past efforts in two important ways. First, modern conservation biology is supported by, and integrated with, other scientific disciplines. Conservation biologists draw heavily on concepts and knowledge from population genetics, evolution, ecology, biogeography, wildlife management, economics, and sociology. In turn, the needs and goals of conservation biology are stimulating new research in those disciplines.

Second, until recently, humans directed conservation efforts primarily toward species from which they gained direct economic benefits. Management actions were designed to maximize yields of useful products from populations of wild plants and animals. Today, conservation biologists study the full array of goods and services that humans derive from species and ecosystems, including aesthetic and spiritual benefits. Understanding the global ecosystem and the effects of human activities on that system is now understood to be essential to our long-term well-being.

Nearly every natural ecosystem on the planet has been altered by the activities of the 6.5 billion humans that live on Earth today. Many habitats have disappeared completely, and many others have been greatly modified. Even Earth's climate and its great biogeochemical cycles have been altered. One consequence of these changes is a rapid increase in the rate at which species are becoming extinct.

Conservation biology is a normative scientific discipline

Conservation biology is a **normative** discipline—that is, it embraces certain values and applies scientific methods to the goal of achieving these values. Conservation biologists are not neutral with respect to the preservation of biodiversity. They are motivated by the belief that preservation of biodiversity is good and that its loss is bad. Some people have criticized the discipline on the grounds that science is "supposed" to be neutral, but in fact most applied sciences are normative. Medical science, for example, is motivated by a desire to improve human health—health is good, illness is bad. Good science is as easily carried out in normative as in value-free scientific disciplines, provided the practitioners carry out their investigations using the standards and practices of established scientific methods.

Conservation biology is guided by three basic principles:

- *Evolution is the process that unites all of biology.* To be effective in preserving biodiversity, we need to know how evolutionary processes generate and maintain that diversity.
- *The ecological world is dynamic.* There is no static "balance of nature" that can serve as a goal of conservation activities.
- *Humans are a part of ecosystems.* Human activities and needs must be incorporated into conservation goals and practices.

Conservation biology aims to prevent species extinctions

Why should we worry about species extinctions today? After all, many natural events have led to species extinctions—some of them massive, as we saw in Section 21.2. Organisms have always altered Earth's ecosystems. The very first organisms probably reduced the supply of energetically and structurally useful compounds (at least in their immediate vicinity), replacing these compounds with their waste products. Early photosynthetic prokaryotes and eukaryotes generated oxygen, making Earth's atmosphere unsuitable for anaerobic organisms. When plants colonized the land, they accelerated the weathering of rocks, thereby gaining access to rock-bound nutrients. The weathering of phosphorus increased global productivity, further contributing to the rise of atmospheric oxygen concentrations. The rise of vascular plants further increased atmospheric oxygen concentrations while decreasing carbon dioxide concentrations. All of these changes, and many others we have not mentioned, created conditions that favored some existing organisms but negatively affected others.

Human beings have likewise caused extinctions of other species for thousands of years. When people first arrived in North America about 20,000 years ago, they encountered a rich fauna of large mammals, including bison, camels, horses, mammoths, and giant ground sloths. Most of those species were exterminated—probably by overhunting—within a few thousand years. A similar extermination of large animals followed the human colonization of Australia, about 40,000 years ago. At that time, Australia had 13 genera of marsupials larger than 50 kilograms (**Figure 57.1**), a genus of gigantic lizards, and a genus of heavy, flightless birds. All of the species in 13 of those 15 genera had gone extinct by 18,000 years ago. When Polynesian people settled in Hawaii about 2,000 years ago, they exterminated, probably by overhunting, at least 39 **endemic** species of land birds—species that are found nowhere else.

If organisms have continually changed the nature of Earth's ecosystems, and if humans have been exterminating species for thousands of years, why should we worry about the changes humans are causing today? Life survived those changes. Indeed, both the productivity and richness of Earth's biota has increased during the long course of life's evolution. But, although a long-term historical perceptive might lead us to be complacent about today's human-caused changes, the current situation is unique. For the first time in history, all the major environmental changes are being caused by a single species. Unlike all previous ecosystem engineers, we are aware of what we are doing. If we deliberately or inadvertently exterminate a species, we have irreversibly destroyed a potentially valuable resource.

57.1 Extinct Australian Megafauna A number of large marsupial mammals, including *Diprotodon* (background), *Palorchestes* (center) and *Zygomaturus* (foreground), were found in Australia during the Pleistocene. These and other large-bodied species became extinct after the arrival of humans on the continent.

People are concerned about maintaining Earth's biodiversity because they recognize its value. People value biodiversity for many reasons:

- Humans depend on other species for food, fiber, and medicine. More than half of all medical prescriptions written in the United States contain or are based on a natural plant or animal product.
- Species are necessary for the functioning of ecosystems and the many benefits and services those ecosystems provide to humanity (such as clean water and purification of the air; see Section 56.4).
- Humans derive enormous aesthetic pleasure from interacting with other organisms. Many people would consider a world with far fewer species to be a less desirable place in which to live.
- Extinctions deprive us of opportunities to study and understand ecological relationships among organisms. The more species that are lost, the more difficult it will be to understand the structure and functioning of ecological communities and ecosystems.

- Living in ways that cause the extinction of other species raises serious ethical issues that increasingly concern philosophers, ethicists, and religious leaders because those species are judged to have intrinsic value.

Later in this chapter we will see how conservation biologists incorporate these values into strategies for preserving biodiversity.

57.1 RECAP

Conservation biology is an applied, normative scientific discipline devoted to preserving biodiversity.

- What is a normative discipline, and how does this influence its practices and its goals? See pp. 1227–1228
- If ecosystem change and species extinctions have always been part of Earth's evolutionary history, why are biologists today so concerned about preserving biodiversity? See p. 1228

To preserve biodiversity, conservation biologists must understand the biodiversity that exists today and how it is changing. We next consider how conservation biologists predict rates of extinctions and infer their causes.

57.2 How Do Biologists Predict Changes in Biodiversity?

As described Chapter 23, the rich array of species on Earth has been generated by speciation processes that have functioned for several billion years. Even if humans did not exist, many species would become extinct during the next 100 years. But human activities are causing more extinctions than would occur in our absence. To preserve Earth's biodiversity, we need to both maintain the processes that generate new species and provide conditions that will keep extinction rates no higher than typical levels.

For four reasons, scientists cannot accurately predict the number of extinctions that will occur during the coming century:

- *We do not know how many species live on Earth*. Many of the species that may become extinct during the next 50 years have not even been named and described!
- *We do not know where species live*. The ranges of most described species are poorly known.
- *It is difficult to determine when a species actually becomes extinct.*
- *We do not know what will happen in the future*. The number of extinctions that actually occur will depend both on what humans do and on unexpected natural events.

We live on a poorly explored planet, and even in the case of conspicuous species in biologically well-studied areas, our knowledge of biodiversity is incomplete. For example, North America's largest woodpecker, the ivory-billed woodpecker (**Figure 57.2A**), was believed to have been glimpsed in Arkansas late in 2004, after 60 years without any confirmed sightings. Most ornithologists had assumed that the species was extinct; a field guide to the birds of North America published in 2000 did not picture it. Although there have now been several reputably documented sightings of the woodpecker (sightings that require birdwatchers to sit motionless for long hours in swampy terrain; **Figure 57.2B**), a clear photograph of the living bird has so far eluded the watchers.

When the Laotian rock rat (*Laonastes aenigmamus*) was discovered and described in 2005, it was so distinct from all other living rodents that it was placed in a lineage all its own. In 2006, further research indicated that this animal actually belongs among the diatomyids, a rodent lineage known from the fossil record but believed to have been extinct for 11 million years.

(A) *Campephilus principalis*

(B)

57.2 Back from Extinction? (A) The ivory-billed woodpecker, shown here in a nineteenth century Audubon print, was presumed to be extinct until 2004, when sightings were reported in a wetland forest. (B) Ornithologists and amateur birdwatchers alike have spent long, silent hours searching for signs of the majestic bird in the Cache River National Wildlife Refuge. This naturalist camouflaged himself and his canoe for the wait.

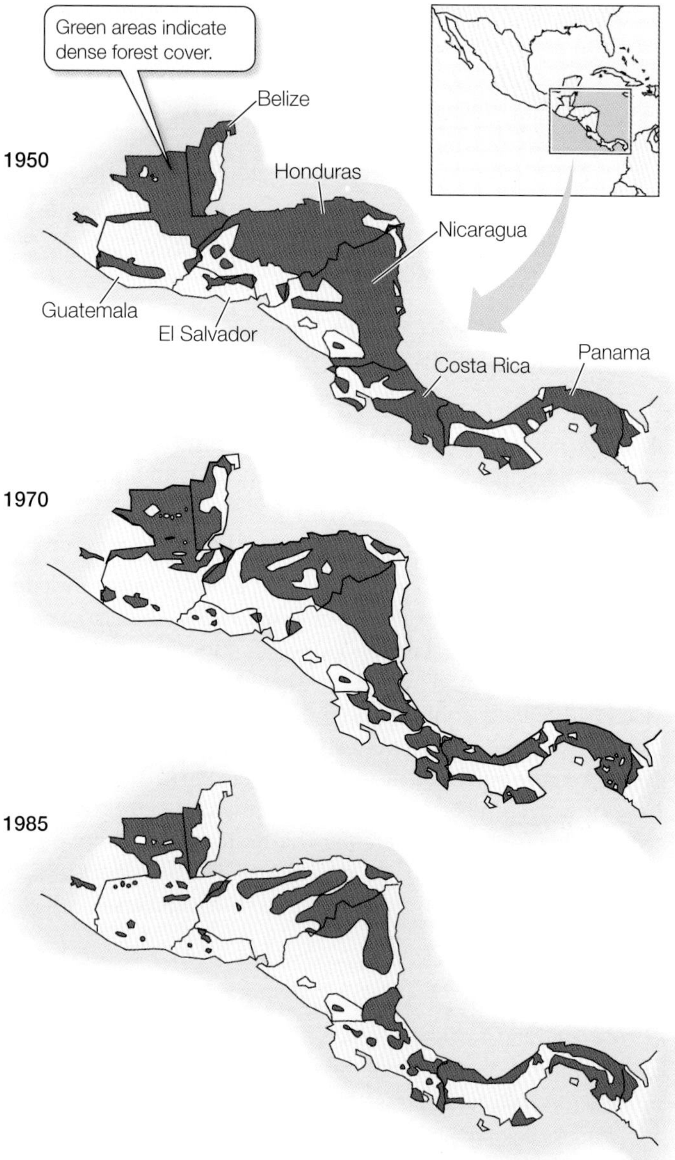

57.3 Deforestation Rates are High in Tropical Forests Less than half the tropical forest that existed in Central America in 1950 remained as of 1985. Much of the remaining forest is in small patches.

Despite these difficulties, methods exist for estimating probable rates of extinction resulting from human actions such as habitat destruction. One tool often used in making such estimations is the **species–area relationship**, a well-established mathematical relationship between the size of an area and the number of species that area contains. Conservation biologists have measured the rate at which species richness decreases with decreasing habitat patch size. Their findings suggest that, on average, a 90 percent loss of habitat will result in the loss of half of the species that live in and depend on that habitat.

Similar calculations can be made for the total area of a habitat type left on Earth. The current rate of loss of tropical evergreen forests—Earth's most species-rich biome—is about 2 percent of the remaining forest each year, due to the increasing demand for forest resources by a rapidly expanding human population (**Figure 57.3**). If this rate of loss continues, at least 1 million species that live in tropical evergreen forests could become extinct during this century.

To estimate the risk that a population will become extinct, conservation biologists develop statistical models that incorporate information about a population's size, its genetic variation, and the morphology, physiology, and behavior of its members.

Species in imminent danger of extinction in all or a significant part of their range are labeled *endangered species*. *Threatened species* are those that are likely to become endangered in the near future. Rarity in and of itself is not always a cause for concern, because some species that live in highly specialized habitats have probably always been rare and are well adapted to those conditions. However, species whose populations suddenly are shrinking rapidly—the "newly rare"—usually are at high risk. Species with special habitat or dietary requirements are more likely to become extinct than species with more generalized requirements.

Populations with only a few individuals confined to a small range can easily be eliminated by local disturbances such as fires, unusual weather, climate change, disease, habitat destruction, or predators. For example, the golden toad (*Bufo periglenes*), which lived only in the Monteverde Cloud Forest Preserve in Costa Rica, disappeared during the 1980s (see Figure 33.16B). At that time, the altitude at which clouds formed moved higher up the mountains, most likely as a result of climate warming. The forests became too dry for the toad, whose range was already near the summit of the mountains. Similarly, a group of scientists carrying out a rapid inventory of species in the western Andes of Ecuador found 90 endemic plant species on a single ridge. Shortly after they completed their survey, the entire ridge was cleared for agriculture. The populations of all 90 species were destroyed. Future surveys may reveal that some of the species exist elsewhere, but many of them are probably extinct.

57.2 RECAP

The species–area relationship can be used to predict rates of extinction in areas that are subject to habitat loss. Models that take into account variables such as population size, dietary and habitat requirements, and genetic variation help predict a species' risk of extinction.

- Why are species with only a few individuals confined to a small range especially vulnerable to extinction? See p. 1230
- Do you understand how a species can be rare but not necessarily in danger of extinction? See p 1230

Now that we've looked at some of the factors that place species at risk of extinction, let's see how human activities contribute to those risks.

57.3 What Factors Threaten Species Survival?

The human activities that threaten the survival of species include habitat destruction and modification, the introduction of exotic species, overexploitation, and climate change. Conservation biologists determine how these activities are affecting species and use that information to devise strategies to preserve species that are endangered or threatened.

Species are endangered by habitat loss, degradation, and fragmentation

Habitat loss is the most important cause of species endangerment in the United States, especially for species that live in fresh waters, the habitat that has been most extensively degraded (**Figure 57.4**). Obviously, when habitats are destroyed or made uninhabitable, the organisms living there are lost. But habitat loss affects even those remaining habitats that are not destroyed. As habitats are progressively lost to human activities, the remaining habitat patches become smaller and more isolated. In other words, the habitat becomes increasingly **fragmented**.

Small habitat patches are qualitatively different from larger patches of the same habitat in ways that affect the survival of species. Small patches cannot maintain populations of species that require large areas, and they can support only small populations of many of the species that can survive in small patches. In addition, the fraction of a patch that is influenced by factors originating outside it increases rapidly as patch size decreases (**Figure 57.5**). Close to the edges of forest patches, for example, winds are stronger, temperatures are higher, humidity is lower, and light levels are higher than they are farther inside the forest. Species from surrounding habitats often colonize the edges of patches to compete with or prey on the species living there. The effects of such factors are known as **edge effects**.

Often we do not know which organisms lived in an area before the habitat became fragmented. Timely surveys can provide some of this information. For example, a major research project in a tropical evergreen forest near Manaus, Brazil, was launched before logging took place. Landowners agreed to preserve forest patches of certain sizes and configurations (**Figure 57.6**). Biologists counted the species in those future "patches" while they were still part of the continuous forest. Soon after the surrounding forest was cut and converted to pasture, species began to disappear from isolated patches. The first species to be eliminated were monkeys that travel over large areas. Army ants and the birds that follow army ant swarms also soon disappeared.

Species that are lost from small habitat fragments are unlikely to become reestablished there because dispersing individuals are unlikely to find isolated fragments. As Section 54.4 points out, however, a species may persist in a small patch if it is connected to other patches by *corridors* of habitat through which individuals can disperse. Some of the pastures that surrounded the experimental forest fragments in Brazil have been

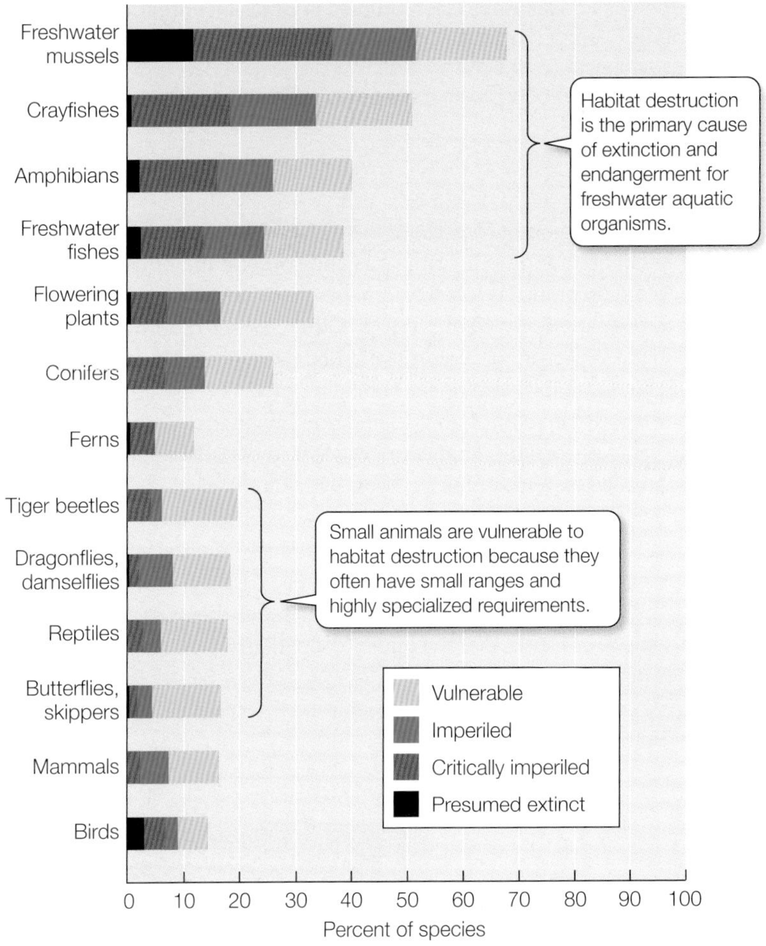

57.4 Proportions of U.S. Species Extinct or Threatened The groups of species that are most threatened with extinction—mussels, crayfishes, amphibians, and fishes—live in freshwater habitats, which have been extensively destroyed, degraded, and polluted.

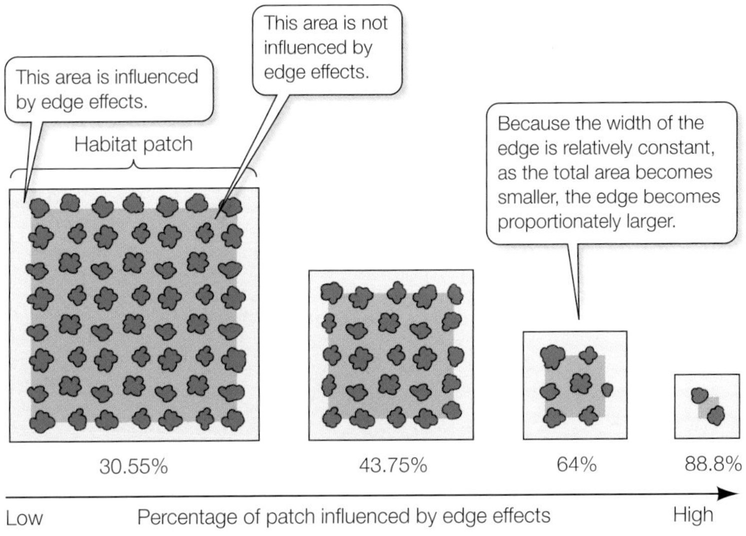

57.5 Edge Effects The smaller a patch of habitat, the greater the proportion of that patch that is influenced by conditions in the surrounding environment.

57.6 Species Losses Have Been Studied in Brazilian Forest Fragments Isolated patches lost species much more quickly than patches connected to the main forest. Even larger patches, such as the one in the foreground, were too small to maintain populations of some species.

abandoned, and young forests are now growing in them. Within 7–9 years of abandonment, army ants and some of the birds that follow them recolonized forest fragments that were connected to larger forest patches by young forests. Other species of birds that forage in the forest canopy also reestablished themselves. The young forest is not a suitable permanent habitat for most of these species, but they can disperse through it to find good habitat.

The value of corridors that connect patches of suitable habitat can be studied experimentally. A team of ecologists led by Douglas Levey and colleagues at the University of Florida and North Carolina State University established eight experimental sites in open pine forests to test whether bluebirds use corridors. By monitoring the movements of both birds and the seeds they dispersed, the investigators found that the bluebirds regularly used corridors to move among patches (**Figure 57.7**).

Overexploitation has driven many species to extinction

Humans have caused extinctions for thousands of years, but until recently, they did so primarily by overhunting. Some species are still threatened by overexploitation today. Elephants and rhinoceroses are threatened in much of Africa and Asia because poachers kill

EXPERIMENT

HYPOTHESIS: Bluebirds use corridors to move between open patches within a forest.

METHOD

1. Create suitable patches of bluebird habitat in pine forests by cutting down trees to create the open conditions favored by bluebirds. Connect two patches by habitat corridors.
2. In one of the connected patches, plant wax myrtle bushes with fluorescently tagged fruit. Because birds that eat the fruit carry seeds in their guts and later defecate them, fluorescently tagged seeds serve as a tracking device.
3. Observe birds as they fly from wax myrtle bushes until they fly out of sight. Record the direction and distance each bird travels. Check surrounding patches for fluorescent seeds defecated by the birds. Develop a computer model that predicts the movement of seeds based on observations of the birds.

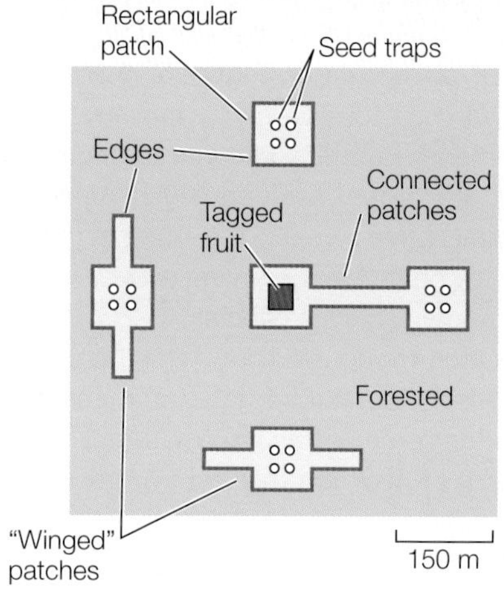

RESULTS

Bluebirds usually remained within a habitat patch when they encountered an edge, but they did move through corridors. Many more fluorescent seeds were moved between patches connected by corridors than between unconnected patches. The computer model correctly predicted the number of seeds that were moved.

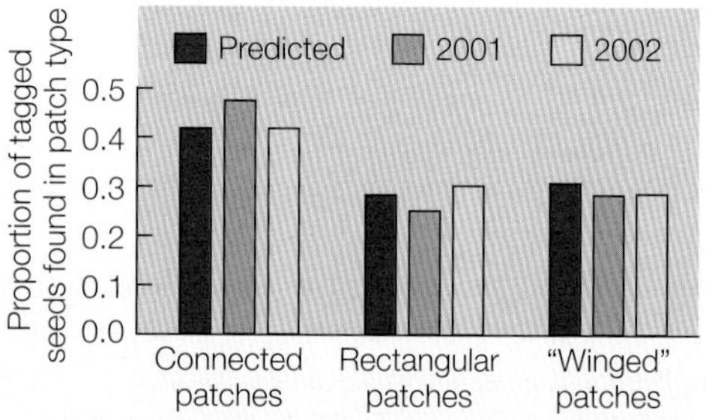

CONCLUSION: Bluebirds use corridors to move between patches of suitable habitat.

57.7 Habitat Corridors Facilitate Movement Bluebirds moved more frequently between habitat patches connected by corridors than between unconnected patches.

(A) *Dicerorhinus sumatrensis*

(B) *Saiga tatarica*

57.8 Endangered by Medical Practices (A) Endangered rhinoceros are killed for their horns, which are widely used in Oriental folk medicine. (B) An attempt to encourage the use of Mongolion saiga antelope horns in lieu of rhino horns succeeded too well—the antelope is now a threatened species.

them for their tusks and horns, which are used for ornaments and knife handles; in addition, some men believe that powdered rhinoceros horn enhances their sexual potency, and rhinoceros horn is used extensively in Chinese traditional medicine (**Figure 57.8**). The use of animal parts in traditional medical practices is a threat to some species. Massive international trade in pets, ornamental plants, and tropical forest hardwoods has decimated many species of tropical fishes, corals, parrots, reptiles, orchids, and trees.

Invasive predators, competitors, and pathogens threaten many species

People have moved many species to regions outside their original range, deliberately or accidentally. Some of these exotic (foreign) species become **invasive**—that is, they spread widely and become unduly abundant, often at a cost to the native species of the region. Weed seeds have been carried around the world accidentally in sacks of crop seeds, and marine organisms have been spread throughout the oceans by ballast water from ships. Europeans deliberately introduced rabbits and foxes to Australia for sport hunting. Nearly half of the small to medium-sized marsupials and rodents of Australia have been exterminated during the last 100 years by a combination of competition with introduced rabbits and predation by introduced cats, dogs, and foxes.

The brown tree snake, *Boiga irregularis* (**Figure 57.9**), arrived in air cargo on Guam sometime in the 1940s. Until then, the only snake on Guam was a tiny insectivorous species. For unknown reasons, brown tree snakes remained rare until the 1960s, when they began to appear in large numbers. Today they can be found at densities as high as 5,000 individuals per square kilometer. The snake has exterminated 15 species of birds, 3 of which were found only on Guam.

Invasive plants can have many negative effects on ecosystems. Most introduced species are imported without their natural enemies, while native plants must devote considerable energy to defending themselves against the native herbivores. The invasive plants typically have high rates of growth and reproduction, in part because they invest less energy in producing defensive compounds. The invaders thus tend to take advantage of soil nutrients in ways that the native plants cannot.

Disease-causing organisms may also proliferate quickly following their introduction to new continents. Introduced pathogens have destroyed whole populations of several eastern North American forest trees. The chestnut blight (*Cryphonectria parasitica*), caused by an introduced fungus, reduced the American chestnut (*Castanea dentata*), formerly an abundant tree in Appalachian Mountain forests, to an understory shrub. Dutch elm disease, caused by the fungus *Ophiostoma ulmi*, introduced in North America in 1930, has killed nearly all American elms (*Ulmus americana*) over large areas of the East and Midwest. Ecologists suspect that intercontinental movement of disease organisms caused extinctions in the past, but disease outbreaks usually leave no traces in the fossil record.

In the Hawaiian Islands, nearly all endemic bird species living below 1,500 meters elevation have been eliminated by avian malaria, which was introduced to the islands with exotic birds. The native birds, never having been exposed to malaria, were highly susceptible to the disease. Species that live above the current range

Boiga irregularis

57.9 Agent of Extinction Since its accidental introduction on Guam, the brown tree snake has eaten 15 species of land birds to extinction.

of the mosquitoes that transmit the disease have fared better, but the insects' range may be expanding upward as the climate warms.

Rapid climate change can cause species extinctions

Scientists predict that, as a result of human activites, average temperatures in North America will increase 2°C–5°C by the end of the twenty-first century. If the climate warms by only 1°C, the average temperature currently found at any particular location in North America today will be found 150 kilometers to the north. If the climate warms 2°C–5°C, some species will need to shift their ranges by as much as 500 to 800 kilometers within a single century. Some habitats, such as alpine tundra, could be eliminated as forests expand up mountain slopes.

Conservation biologists cannot alter rates of global warming, but their research can help us predict how the resulting climate changes will affect organisms and find ways to mitigate those effects. Their research activities include analyses of past climatic events and studies of sites currently undergoing rapid climate change. It is helpful to know, for example, how rapidly species ranges shifted during the last 10,000 years of postglacial warming. Which species were and were not able to keep pace with climate change by shifting their ranges? How much and in what ways did past ecological communities differ from those of today as a result of differences in the rates at which species changed their ranges?

Organisms that are able to disperse easily, such as most birds, may be able to shift their ranges as rapidly as the climate changes, provided that appropriate habitats exist in the new areas. However, the ranges of species with sedentary habits are likely to shift slowly. As the glaciers retreated in North America about 8,000 years ago, for example, the ranges of some coniferous trees expanded northward, so that today they grow as far north as the current climate permits (see Figure 21.22). On the other hand, some earthworm species spread only very slowly into the areas that had been covered by ice.

If Earth's surface warms as predicted, entirely new climates will develop, and some existing climates will disappear. New climates are certain to develop at low elevations in the tropics because a warming of even 2°C would result in climates near sea level that are warmer than those found anywhere in the humid tropics today. Adaptation to those climates may prove difficult even for many tropical organisms. Although there has been little recent climate warming in tropical regions, nights are now slightly warmer than they were only a few decades ago. Since the mid-1980s, the average minimum nightly temperature at the La Selva Biological Station, in the Caribbean lowlands of Costa Rica, has increased from about 20°C to 22°C. During these warmer nights, trees use more of their energy reserves. The result has been a reduction of about 20 percent in the average growth rates of six different tree species.

Staghorn coral, *Acropora cervicornis*, was recently discovered near Fort Lauderdale, Florida. Elkhorn coral, *A. palmata*, was discovered in the northern Gulf of Mexico. Both sites are 50 kilometers north of the previous range limits of these species, providing the first evidence of range expansions of Caribbean corals in response to climate warming.

In 1998, the highest sea surface temperatures ever recorded caused corals to lose their endosymbiotic dinoflagellates (a phenomenon called *bleaching*) and increased their mortality worldwide (**Figure 57.10**). If warming of the oceans continues as predicted, about 40 percent of coral reefs worldwide are likely to be killed by 2010. To identify possible ways to help preserve coral reefs, conservation biologists are measuring conditions in places where corals have escaped bleaching. They have found that reefs adjacent to cool, upwelling waters and reefs in cloudy waters, both of which have relatively low temperatures, are generally healthy. These reefs are receiving special protection because corals are likely to continue to survive well there. Corals from those reefs could be used as colonists for reestablishing bleached reefs if cooler ocean temperatures return in the future.

(A)

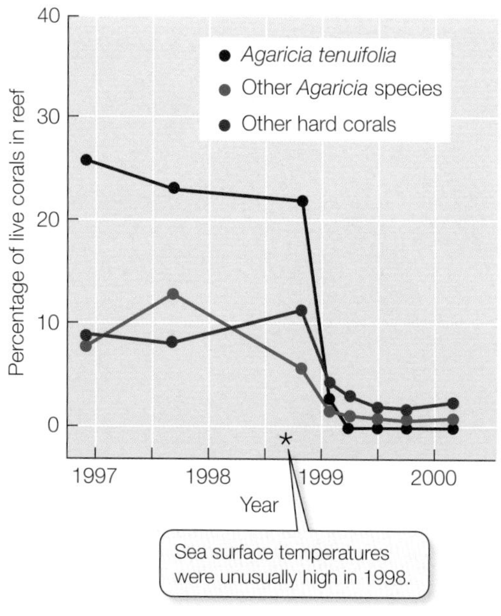

(B) *Lobophyllia* sp.

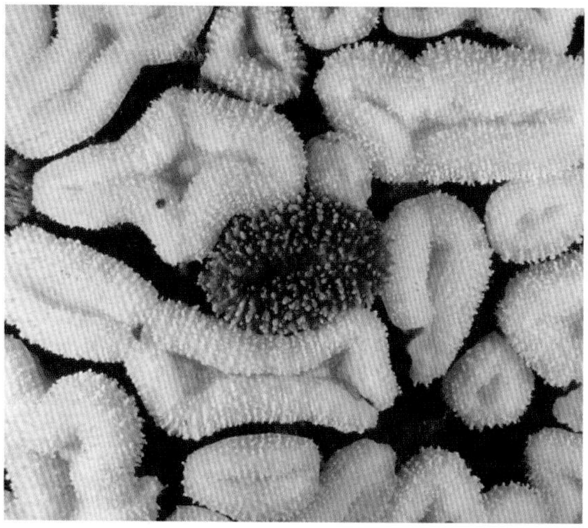

57.10 Global Warming Threatens Corals
(A) Unusually high sea surface temperatures in 1998 caused massive bleaching and death of corals on a reef in Belize. (B) A single photosynthesizing polyp (center) remains alive among the bleached remnants of this Indonesian coral colony.

57.3 RECAP

A number of human activities threaten the survival of species, including habitat destruction and fragmentation, the introduction of invasive species, overexploitation, and rapid climate change.

- Can you explain why extinction rates are high in small habitat patches? See pp. 1231–1232 and Figure 57.5
- Do you understand the particular worries conservation biologists have about climate change? See p. 1234

Acquiring data to demonstrate that species or communities are endangered is an empty exercise if we cannot implement a plan of action to save them. In the next section we consider some of the positive steps that can be taken to preserve biodiversity.

57.4 What Strategies Do Conservation Biologists Use?

Conservation biologists use data, concepts, and tools from a variety of disciplines to help preserve endangered and threatened species and communities. They determine what factors, including human activities, are affecting species health and numbers, and they use that information to devise a plan of action. What information is relevant depends on which of the factors that threaten biodiversity are most important in particular cases. Reducing threats to species often requires changes in national legislation or international rules and treaties. Therefore, biologists regularly join sociologists, political scientists, and environmental activists to influence legislation. Let's look at some of the actions that conservation biologists take to maintain biological diversity.

Protected areas preserve habitat and prevent overexploitation

Establishing *protected areas* is an important component of efforts to preserve biological diversity. Protected areas that preserve habitat while preventing the human exploitation of the species living there may serve as nurseries from which individuals disperse into exploited areas, replenishing populations that might otherwise become extinct. The importance of protected areas was highlighted in the United Nations Convention on Biological Diversity, a document generated by the Rio de Janeiro Earth Summit in 1992: "The fundamental requirement for the conservation of biological diversity is the *in situ* conservation of ecosystems and natural habitats and the maintenance and recovery of viable populations of species in their natural surroundings."

But how should we select the areas to be protected? Two obvious criteria are the number of species living in an area—its *species richness*—and the number of endemic species. Using these criteria, Norman Myers identified a number of biodiversity "hotspots" of unusual richness and endemism (**Figure 57.11**). These hotspots occupy only 15.7 percent of Earth's land surface, but they are home to 77 percent of Earth's terrestrial vertebrate species. Most of these hotspots are also regions of high human population density where habitat destruction is a major problem. Consequently, much effort is being devoted to saving particular habitats in those regions.

The hotspot concept successfully directs attention to places harboring unusual species richness, but hotspots do not represent all of Earth's biodiversity. Many areas with lower species richness and less endemism are nonetheless very important biologically. By using taxonomic uniqueness, unusual ecological or evolutionary phenomena, and global rarity, as well as species richness and endemism, scientists at the World Wildlife Fund identified 200 ecoregions of great conservation importance (**Figure 57.12**). Some of these "Global 200" ecoregions are marine areas. The list also includes tundra, boreal forests, and deserts, ecosystems that are missed by the hotspot approach.

Identifying focal areas for preservation is only the first step in a conservation program, however. Developing a conservation strategy for an ecoregion requires both a detailed analysis of the distributions of species and the locations of special resources, such as caves, springs, and migratory stopover areas for birds, and an

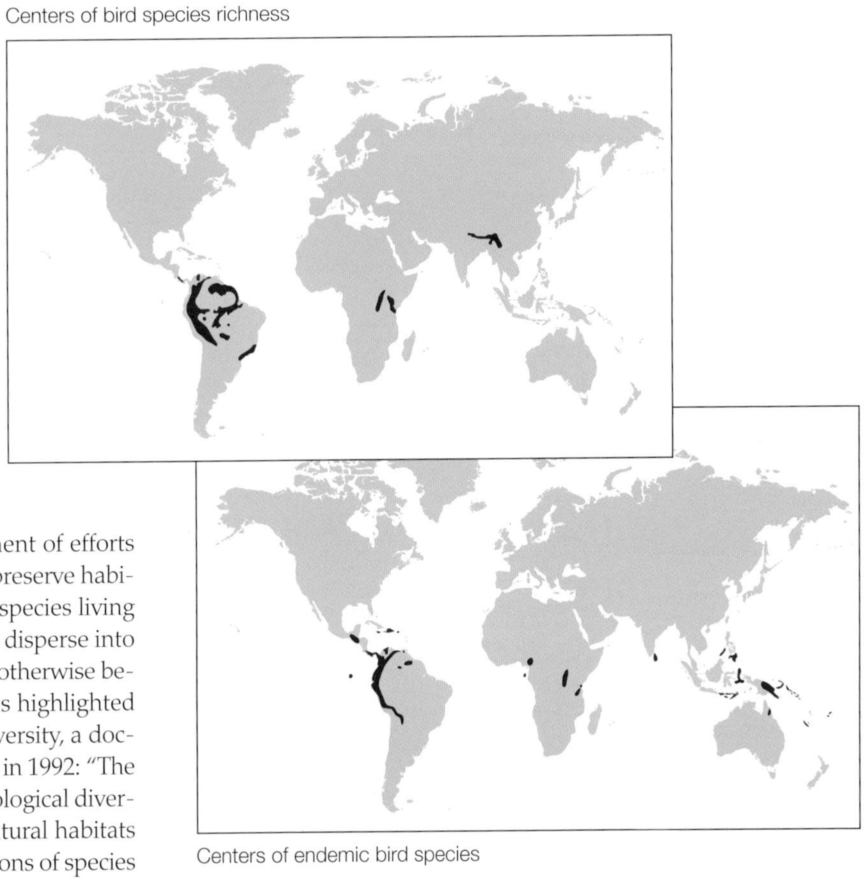

57.11 Hotspots of Avian Biodiversity The regions marked in red contain rich avian biodiversity in terms of either total number of species or number of endemic species (species found nowhere else in the world).

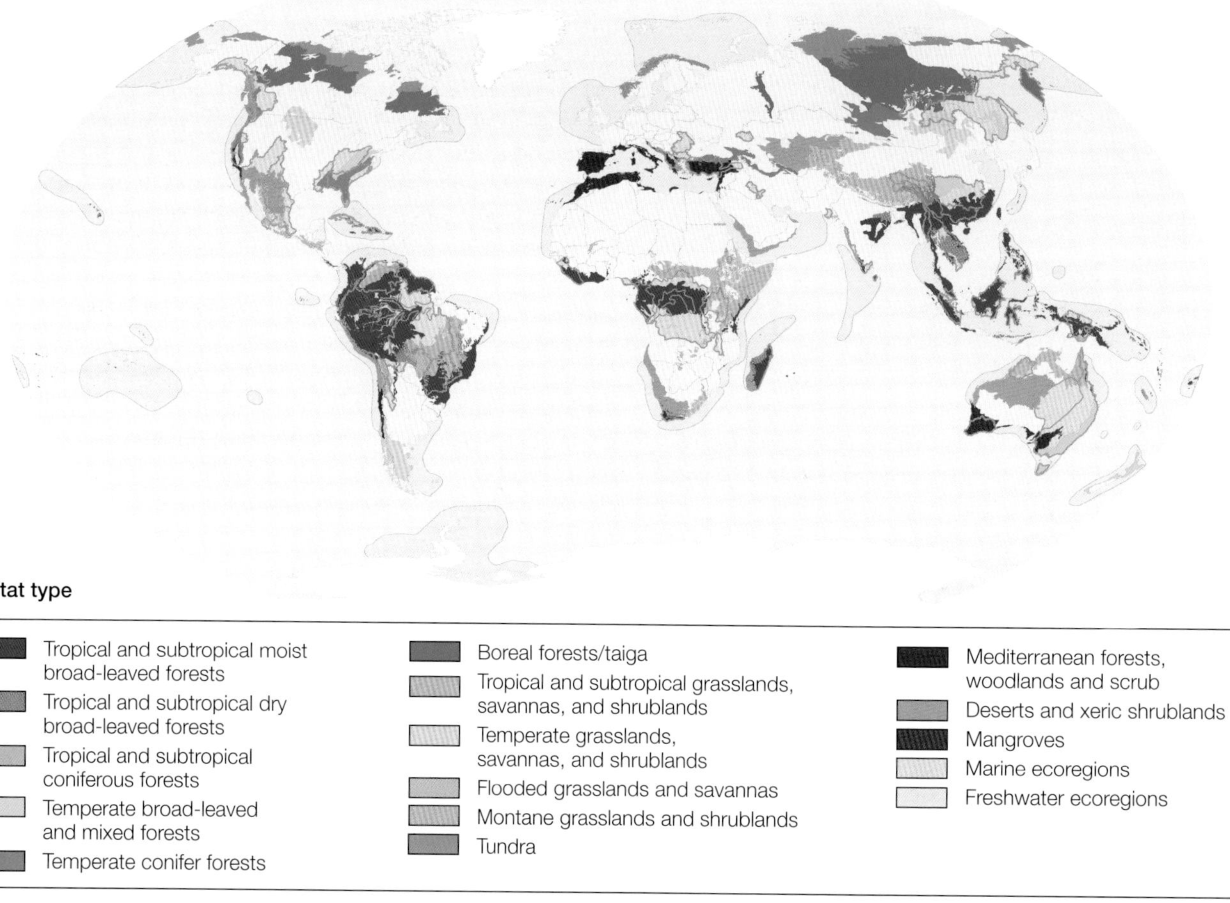

57.12 The "Global 200" Ecoregions The World Wildlife Fund has designated 200 areas worldwide (shown in color) as particularly important for the preservation of biodiversity.

analysis of the processes that both threaten and support biodiversity in the region. Then conservation biologists, together with other experts and local people, can develop an action plan to preserve the ecoregion.

In the effort to identify sites with threatened species that are found nowhere else, conservation biologists have analyzed distributions of mammals, birds, reptiles, amphibians, and conifers and have identified 595 "centers of imminent extinction." These centers are concentrated in tropical forests, on islands, and in mountainous regions (**Figure 57.13**). The sites harbor 794 species judged to be at serious risk of extinction—more than three times the number of species in those groups known to have become extinct since 1500. Only one-third of the sites are legally protected. Most of them are surrounded by rapid human development. Urgent action is needed in these areas if species extinctions are to be avoided.

Degraded ecosystems can be restored

If the cause of a species' endangerment is modification, rather than loss, of its habitat, preserving the species may require that the habitat be restored to its natural state. Practitioners of **restoration ecology** are developing methods that attempt to restore natural habitats. Such interventions are often needed because many degraded ecosystems will not recover, or will do so only very slowly, without human assistance.

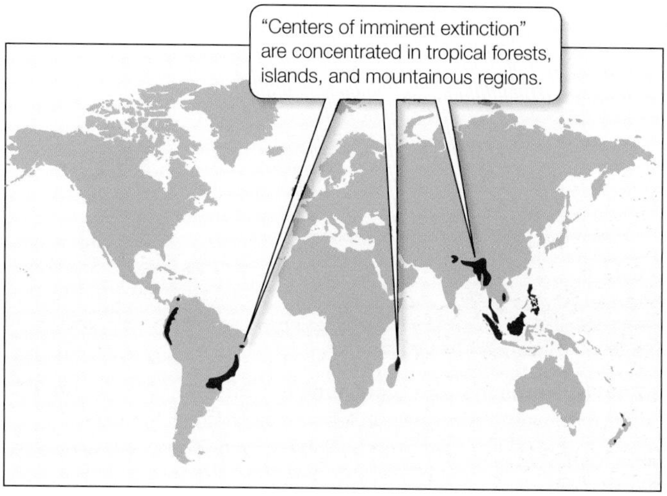

57.13 Centers of Imminent Extinction The areas shown in red include 595 "centers of imminent extinction." These regions are home to 794 species that are at serious risk of extinction.

Many grassland habitats grow on rich soils and humans have eagerly converted them to agricultural use. By the middle of the twentieth century, for example, most North American prairies had been converted to cropland or were heavily grazed by domesticated livestock. The populations of large mammals that roamed the prairies when Europeans arrived on the continent are now reduced to tiny remnants confined to small areas. Most of these remaining populations are too small to maintain their genetic diversity or their original ecological roles.

The species *have* survived, however, so opportunities exist to reintroduce them if the prairie habitat can be restored. A major prairie restoration project is under way in northeastern Montana. When Lewis and Clark mapped this region 200 years ago, it supported large herds of bison, elk, deer, and pronghorn, as well as their predators. The goal of the restoration project, run by the World Wildlife Fund and the American Prairie Foundation in cooperation with public land managers, is to restore the native prairie and its fauna in a 15,000-square-kilometer area near the Missouri River (**Figure 57.14A**).

This ambitious, multi-decade project is feasible for three reasons. First, the private land in the area is owned by a small number of ranchers. Each ranch owns extensive grazing leases on nearby public lands administered by the Bureau of Land Management, the Fish and Wildlife Service, or the state of Montana. Second, most of the land has never been plowed, so the native vegetation is likely to recover rapidly when grazing pressures are reduced. Third, the area is steadily losing its human population. In 1920 there were 9,300 people in Phillips County, which encompasses a large portion of the project area. Only 4,200 remained there in 2000. The ranchers are aging. Their children are choosing the excitement of cities and a regular paycheck rather than the hard labor and uncertain profits of ranching. The ranchers need to sell their land to fund their retirement; their grazing leases transfer automatically to the new owner.

The American Prairie Foundation is buying ranches from willing sellers with the objective of reintroducing bison, greatly expanding populations of prairie dogs, and restoring a viable population of black-footed ferrets, the most endangered mammal in North America. The prairie dogs, by digging extensive burrows and manipulating vegetation, in turn support dozens of species of birds, mammals, reptiles, and invertebrates (**Figure 57.14B**). In November 2005, the first 16 bison, obtained from Wind Cave National Park, were reintroduced to a ranch purchased by the Foundation (**Figure 57.14C**). Once a free-ranging herd of several thousand bison and large numbers of elk and their predators (wolves) have been established, nature-minded tourists should flock to the area to view the restored wildlife spectacle. The ecotourists will, in turn, spend money in the region, providing Phillips County with a new economic base. Over the long term, the restored ecosystem should deliver major economic benefits to the region.

Disturbance patterns sometimes need to be restored

Many species depend on particular patterns of disturbance on the landscape, such as fires, windstorms, and grazing. Conservation biologists work to assess whether reestablishment of historic disturbance patterns can help preserve biodiversity. Humans often try to reduce the frequency and intensity of such disturbances. For example, although many plant species require periodic fires for successful establishment and survival, for many years the official policy in the United States, symbolized by Smokey Bear, was to suppress all forest fires. Today, however, controlled burning is a common forest management tool, particularly in western North America. But to determine how to do this, we need to know the historical pattern of fires in an area.

Scars in the annual growth rings of trees preserve evidence of past fires that did not kill them. Tree ring researchers can determine when fires occurred, how severe they were, and when fire patterns changed. Annual growth rings on ponderosa pines show that low-intensity ground fires were common near Los Alamos, New Mexico, until about 1900 (**Figure 57.15A**). After that time, cattle and sheep grazing in pine forests and fire suppression greatly reduced the frequency of low-intensity fires. Without these fires,

(A)

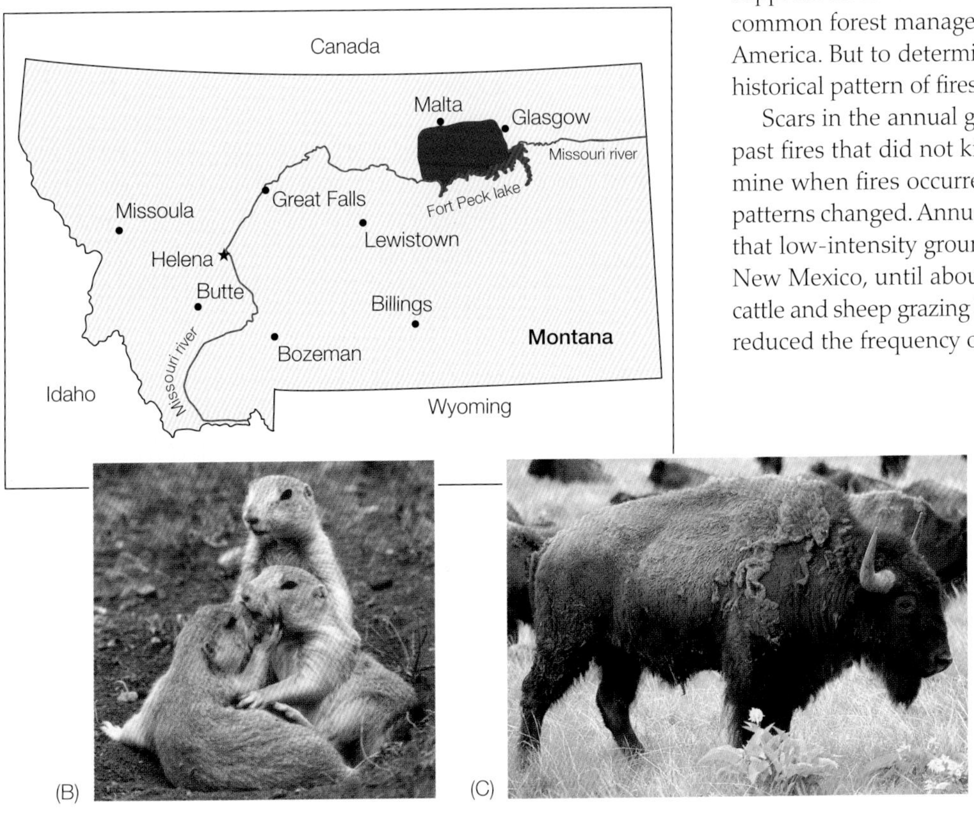

57.14 An American Prairie Is Being Restored (A) A major prairie restoration project is underway north of the Missouri River in the state of Montana. (B) Burrowing prairie dogs manipulate vegetation and are crucial in sculpting the natural ecosystem. (C) Bison have been reintroduced.

(A)

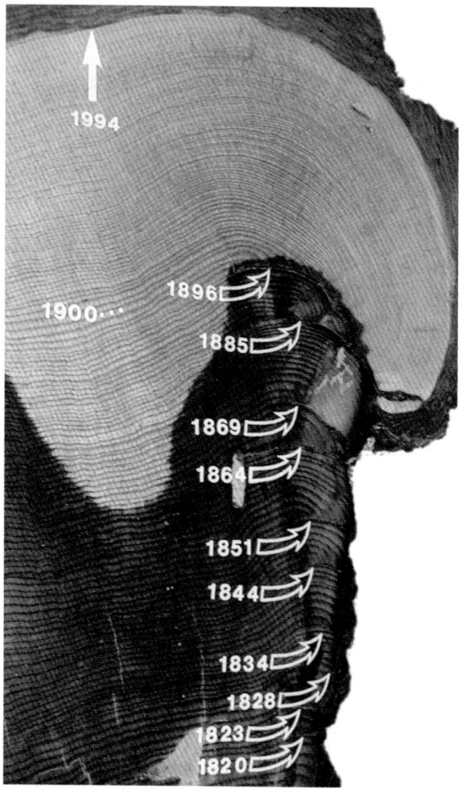

(B)

57.15 The Frequency and Intensity of Fires Affect Ecosystems (A) As revealed by scars (arrows) in the growth rings of this Ponderosa Pine tree, low-intensity ground fires were frequent in the pine forests of the southwestern United States prior to fire suppression. (B) Fire suppression results in the buildup of large quantities of fuel, so that subsequent fires are likely to spread to the canopy and kill most trees.

dead branches and needles accumulated in the forest. When fires inevitably did occur, the buildup of fuel made them more likely to become intense, tree-killing canopy fires (**Figure 57.15B**). Today, ground fires are deliberately started in many areas to reduce fuel loads and prevent destructive fires.

New habitats can be created

In the United States, the belief that humans know how to create functioning ecosystems has resulted in policies that make it easy to get permits for developments that destroy habitats. Developers need only state that they will create new habitats to substitute for the ones they are destroying. Of special concern is destruction of wetlands, which is permitted because regulators believe that alternative wetlands can be created. Creating new wetlands that can support all the species that live in those being destroyed is difficult, however, and requires detailed ecological knowledge.

In southern California, where 90 percent of the coastal wetlands have been destroyed, wetland restoration is a high priority. Because species have been lost from degraded coastal wetlands, restoration requires species introductions, but deciding which species should be introduced is not easy. Early attempts at restoration, in which one or two common, easily grown wetland species were planted, did not succeed; other wetland-associated species failed to re-colonize the "rehabilitated" wetlands. To understand why, conservation biologists established a large field experiment at the Tijuana Estuary to examine the effects of plant species richness on the success of wetland restoration. They found that experimental plots planted with species-rich mixtures developed a complex vegetation structure, which is important to insects and birds. The species-rich plots also accumulated nitrogen faster than species-poor plots (**Figure 57.16**).

We use markets to influence exploitation of species

Many people would like to consume only natural products that have been harvested in ways that protect biodiversity and ecosystem productivity. To enable consumers of forest products to exercise that choice, a consortium of environmental organizations and members of the forest products industry launched the Forest Stewardship Council (FSC) in 1993. FSC establishes criteria that a forest products company must meet for its products to be certified. Special certification companies determine whether a forestry operation meets the criteria and ensure that there is a chain of custody that tracks certified products on their way to market. By November 2005, about 34 million hectares of managed forests worldwide had been certified by FSC in 66 countries on five continents. FSC initially had its most significant impact on forest management in the temperate zone, but certification of tropical forests is growing rapidly. In October 2005, for example, more than 2 million hectares of Bolivian forests were certified by FSC. More than 400 companies in 18 countries have committed to purchasing certified wood products.

To serve the same function for marine products, the Marine Stewardship Council was formed through an alliance between the World Wildlife Fund and Unilever, one of the largest marketers of frozen seafood. Its first certified product, Australian rock lobster, came to market in 2000. Alaskan salmon has also been certified; other major fisheries are in the process of becoming certified. This action, combined with the elimination of government subsidies, can help reduce the current overexploitation of many marine fish stocks.

Ending trade is crucial to saving some species

Most endangered species cannot cope with any further reductions in their breeding populations. The legal mechanism for prohibiting

EXPERIMENT

HYPOTHESIS: Wetland plots to be restored that are planted with mixtures of species will approach their original condition more rapidly than single-species plots.

METHOD

Plant some plots with only one of each of the 8 plant species typical of wetlands in the region. Plant other plots with randomly chosen assemblages of 3 and 6 species. Plant the same density of seedlings in all plots. Replant and weed as necessary to compensate for early mortality of seedlings.

RESULTS

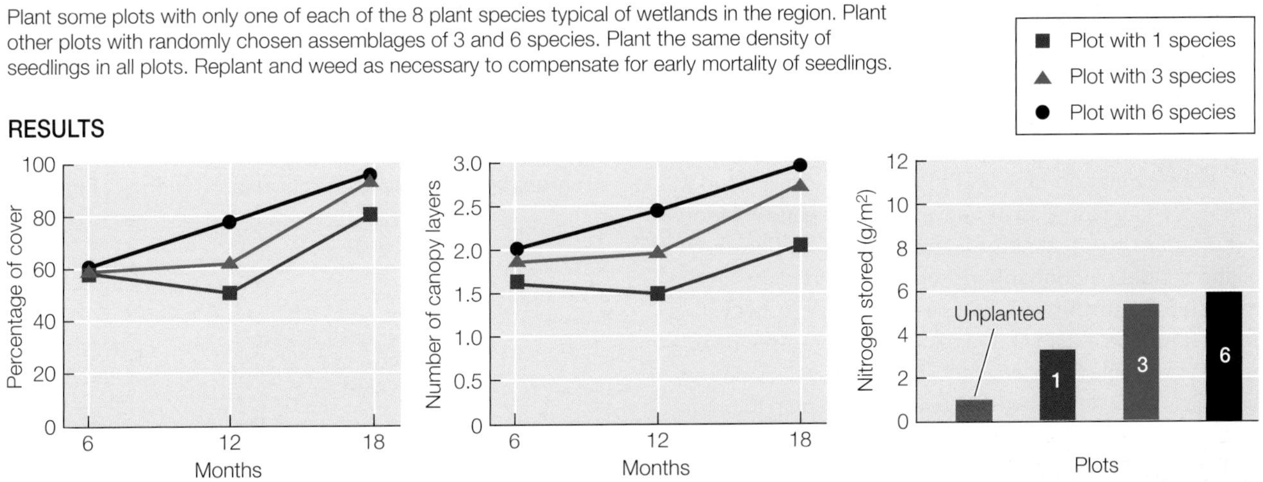

CONCLUSION: Vegetation cover, canopy complexity, and nitrogen accumulation are enhanced by species richness. In future wetland restoration attempts, a rich mixture of species should be planted.

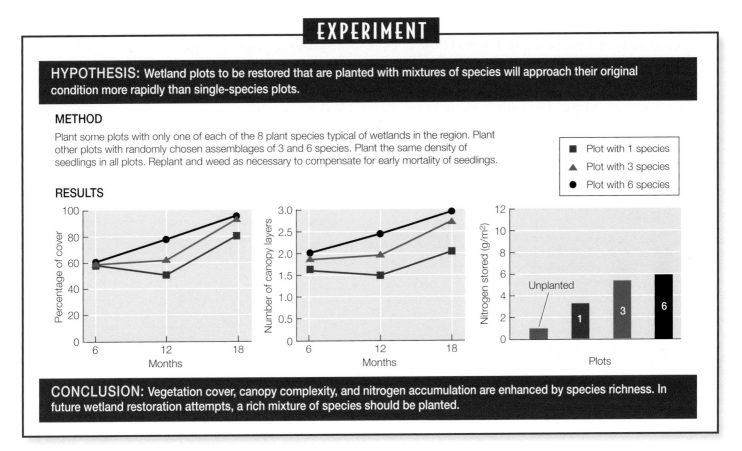

57.16 Species Richness Enhances Wetlands Restoration Both vegetation complexity and nitrogen accumulation are greater in species-rich than in species-poor experimental plots that had been degraded and lacked wetland plants.

exploitation of these species is an international agreement called the Convention on International Trade in Endangered Species (CITES). Most nations of the world are members of CITES. National representatives meet every two years to review the status of species on its protected species lists, to determine which species may no longer need protection, and to add new species. CITES rules currently prohibit international trade in items such as whale meat, rhinoceros horn, and many species of parrots and orchids.

CITES instituted a ban on international trade in elephant ivory in 1989, but demand for ivory remains strong, especially in Japan and China. As a result, poaching of elephants continues in the forests of central Africa and in East Africa. Southern African countries (Botswana, Namibia, South Africa, Zambia, and Zimbabwe), however, have so many elephants that government officials must kill many of them to control populations in the limited areas where they are allowed to roam. These countries would like to sell the ivory from the culled elephants to fund their conservation efforts. Other countries are worried that if trade restrictions are relaxed, poaching will escalate over the entire continent. Control and regulation of ivory trade might be possible if scientists could determine where the ivory comes from.

Samuel Wasser and his colleagues at the University of Washington have identified 16 microsatellite DNA markers that can be extracted from elephant feces. Park rangers in Malawi and Zambia were able to sample the elephant populations in their countries in just two weeks by collecting fresh scat while they were on routine patrols. The source of an elephant tusk can now be determined by matching the DNA extracted from the ivory to the geographically based frequencies of the 16 microsatellites. In June 2002, 6.5 tons of illegal ivory were seized in Singapore. DNA analyses showed that the elephants were killed in Zambia, providing an early warning system for directing anti-poaching efforts.

Controlling invasions of exotic species is important

Because some species are endangered by invasive exotic species, controlling these invasives is an important component of conservation biology. The best way to reduce the damage caused by invasive species, of course, is to prevent their introduction in the first place. Given the immense amount of global traffic between continents, it might seem impossible to curtail the spread of exotic species. Some promising options, however, do exist. For example, transoceanic transport of invasive species in ballast water could be largely eliminated by the simple procedure of deoxygenating ballast water before it is pumped out. This practice kills most organisms in the water. It also extends the life of ballast tanks, providing an economic benefit to shippers.

Regulating the importation and sale of exotic species can reduce deliberate introductions. In 2003, the Connecticut General Assembly established a penalty of $100 per plant for the sale of any of 81 plant species judged to be invasive in the state. In 2002, some members of the American horticultural industry crafted a voluntary code of conduct for their profession. The code states that the invasive potential of a plant should be assessed prior to introducing and marketing it. Horticulturists work with conservation biologists to determine which species are currently invasive, or likely to become so, and to identify suitable alternative species. Stocks of invasive species will be phased out, and gardeners will be encouraged to use noninvasive plants.

How do scientists assess the likelihood that a species will become invasive? One way is to compare the traits of species that have become invasive when introduced to a new area with those of species that have not. Such comparisons show that a plant species is more likely to become invasive if it has a high rate of growth, a short generation time and small seeds, is dispersed by vertebrates, has a large range in its native region, depends on non-specific mutualists (root symbionts, pollinators, and seed dispersers), and is not evolutionarily closely related to plants in the area to which it is introduced. The best predictor, however, is whether the species is already known to be invasive elsewhere.

Using the traits that characterize most invasive species, conservation biologists have developed a decision tree to help them determine whether an exotic species should be allowed into North America (**Figure 57.17**). Although following the protocols stipulated by this decision tree cannot eliminate the introduction of all potentially invasive species, if it is used conscientiously, its application can greatly reduce the risk.

Biodiversity can be profitable

Much of the value of ecosystems to humans depends on their biodiversity. But it has been difficult to assess the value of biodiversity in monetary terms. When an ecosystem is perceived to have economic value, industries and government agencies are given a greater incentive to preserve it than if no monetary values are evident. Fortunately, our ability to assess the potential economic value of ecosystem services is improving rapidly.

Enough is known about the value of biological diversity and its role in the functioning of natural ecosystems to establish markets for ecosystem services. In 2005, Ecosystem Marketplace, the first global clearinghouse for information on this emerging trade, was launched. The clearinghouse is sponsored not only by environmental organizations, such as The Nature Conservancy, but also by large corporations, such as Citigroup and the reinsurance company Swiss Re. Its Web site contains information on the basic work of forests, including water filtration, soil quality maintenance, habitats, and carbon sequestration. The Web site encourages trade by providing details of transactions to potential buyers and sellers who are considering entering into what may seem to be a risky market. It includes news features from around the world, and a "Market Watch" page that tracks money flows into ecosystem services. Ecosystem services may soon become big business!

There are many instances demonstrating the profit value of sustaining biodiversity. Let's take a closer look at three such cases.

FYNBOS OF SOUTH AFRICA Studies by a group of economists, ecologists, and land managers have attempted to calculate the value of the economic benefits provided by the biologically diverse native vegetation of the Western Cape Province, South Africa. The native vegetation of the highlands of this area is a species-rich community of shrubs known as *fynbos* (pronounced "fainbos"; **Figure 57.18A**).

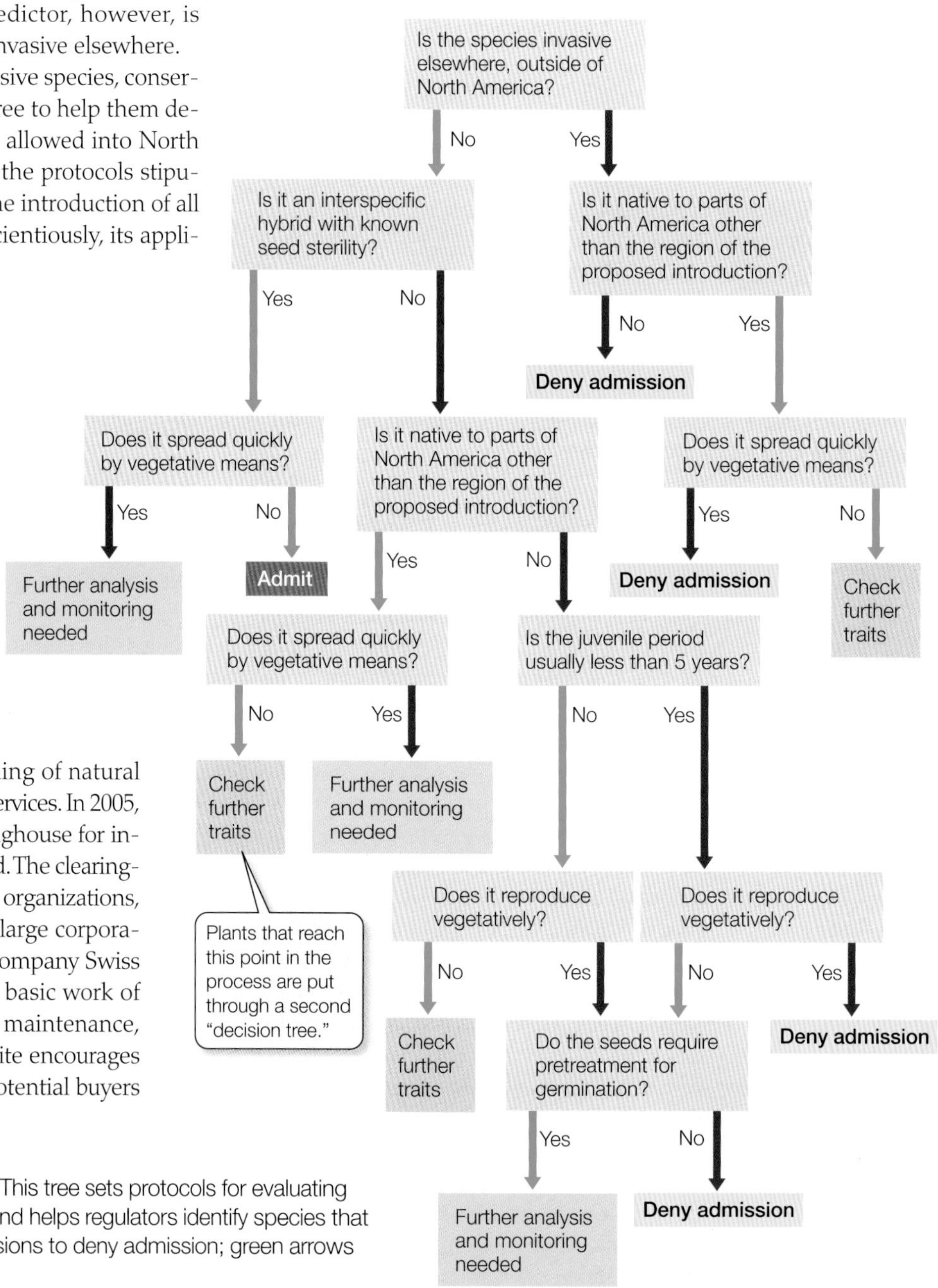

57.17 A "Decision Tree" for Exotic Species This tree sets protocols for evaluating the proposed introduction of an exotic species and helps regulators identify species that may become invasive. Red arrows indicate decisions to deny admission; green arrows can result in a decision to admit the species.

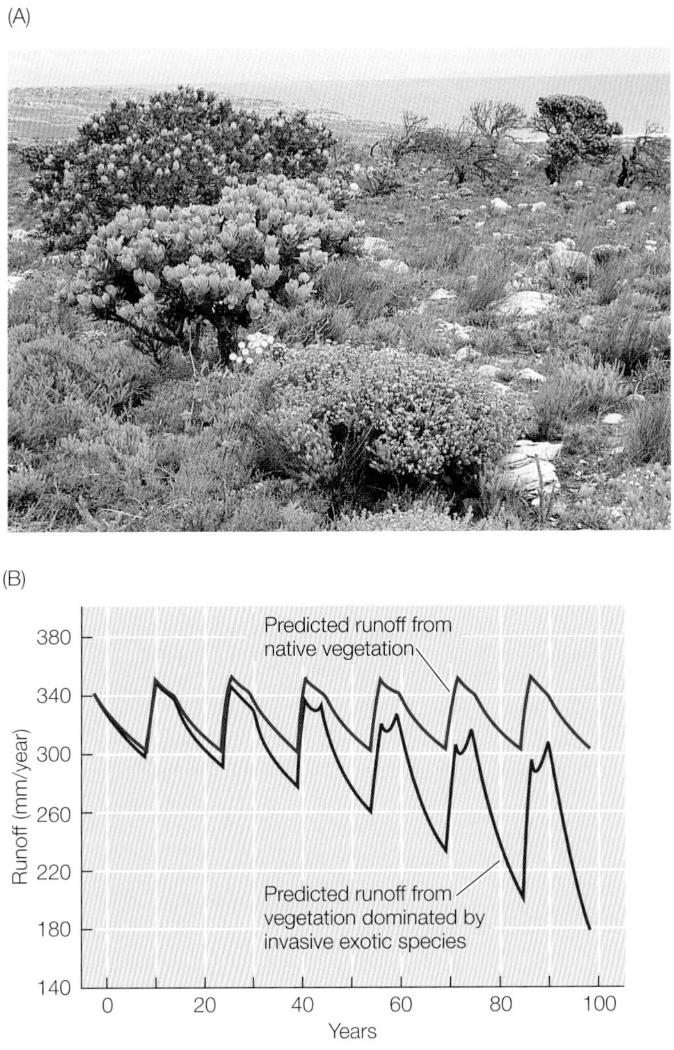

57.18 Invasive Species Disrupt Ecosystem Functioning (A) The unique fynbos ecosystem of the Western Cape Province of South Africa provides much of the area's water. (B) A computer simulation of stream flows from watersheds that have and have not been invaded by exotic trees.

These shrubs survive regular summer droughts, nutrient-poor soils, and the fires that periodically sweep through the highlands.

The fynbos-clad highlands provide a crucial economic service—about two-thirds of the Western Cape's water supply comes from them. In addition, some of the endemic plants are harvested for cut and dried flowers and thatching grass. The combined value of these harvests in 1993 was about $19 million. Some of the income from tourism in the region comes from people who want to see the fynbos. About 400,000 people visit the Cape of Good Hope Nature Reserve each year, primarily to see the endemic plants.

During recent decades, a number of plants introduced into South Africa from other continents have invaded the fynbos. Because they are taller and grow faster than the native plants, these exotics increase the intensity and severity of fires. By transpiring larger quantities of water, they decrease stream flows to less than half the amount flowing from mountains covered with native plants, reducing the water supply (**Figure 57.18B**). Removing the exotic plants by felling and digging out invasive trees and shrubs and managing fire costs between $140 and $830 per hectare, depending on the densities of invasive plants. Annual follow-up operations cost about $8 per hectare.

The services provided by fynbos vegetation could be replaced, but only at a much higher cost. A sewage purification plant that would deliver the same volume of water to the Western Cape Province as a well-managed watershed of 10,000 hectares would cost $135 million to build and $2.6 million per year to operate. Desalination of seawater would cost four times as much. Thus the available alternatives would deliver water at a cost between 1.8 and 6.7 times more than the cost of maintaining natural vegetation in the watershed. The technologically sophisticated methods that could substitute for the services provided by the biodiversity of the fynbos are more expensive than labor-intensive, employment-generating methods of maintaining those services.

WILD DOGS AND ECOTOURISM *Ecotourism* exists because of biodiversity and is a major source of income for many countries. For example, because diseases have caused populations of African wild dogs (*Lycaon pictus*) to plummet throughout Africa, tourists are now increasingly interested in seeing wild dogs when they sign up for a safari. South Africa has about 400 of Africa's remaining 5,700 dogs, most in Kruger National Park. A survey of South African tourists found that nearly three-fourths of them were willing to pay an extra U.S. $12 to see the dogs. In other words, a pack of 10 dogs would generate an added annual income of about U.S. $90,000. The investigators are now working with lodge owners and ranchers elsewhere in South Africa and in Kenya to engage them in efforts to reestablish wild dogs in areas from which they have disappeared.

COFFEE AND THE BEE Taylor Ricketts and colleagues at Stanford University assessed the economic value of the pollination services provided by the bees that live in, and depend on, tropical forest patches adjacent to a coffee plantation in Costa Rica. In a landscape where coffee plants are intermixed with forest patches, they found that coffee production was highest at the sites that were closest to forest patches (**Figure 57.19**). They also hand-pollinated some coffee plants to show that the difference in production was a result of pollination services rather than some other environmental condition. The investigators calculated that the value of pollination services to the plantation on which the experiments were carried out was about $60,000 per year, more than the current conservation incentive payments offered to landowners to preserve forest patches.

Living lightly helps preserve biodiversity

Protected areas, as we have seen, are an essential component of efforts to maintain biodiversity. More of them need to be established, but by themselves, protected areas cannot do the job. The extensive landscapes in which people live and extract resources must also play important roles in biodiversity conservation. The good news is that, carefully used, these lands can contribute much more to conservation than they currently do. The practice of using exploited lands in ways that sustain biodiversity is becoming known as **reconciliation ecology**.

57.19 Establishing the Economic Value of Forest Patches Coffee plants in plantations located close to forest patches, where they can benefit from the services of native pollinators (bees) living in the forests, produce more coffee beans (the seeds of coffee plants) than plantations farther from forest patches.

EXPERIMENT

HYPOTHESIS: Native pollinators (bees) of coffee flowers that live in forests increase both the quantity and quality of coffee seeds produced on adjacent plantations.

METHOD

Establish 12 sites that are near (50 m), intermediate (800 m), and far (1600 m) from forest patches. Measure the quality of seeds produced per plant and the frequency of small, misshapen seeds. To test whether the observed differences are due to the frequency of pollinator visits, hand-pollinate flowers.

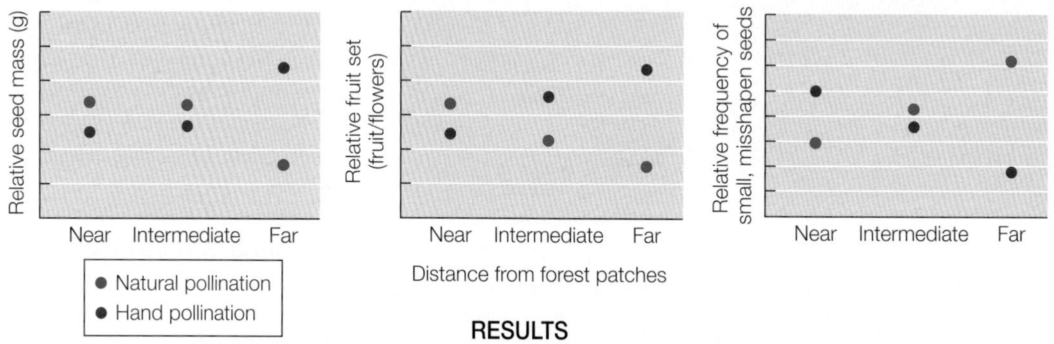

RESULTS

Forest-dwelling bees increased coffee yields by 20% at sites within 1 kilometer of a forest patch. There were sufficient visits from bees to support full production at the near and intermediate sites. At those sites, hand pollination did not increase seed production. However, at far sites, hand pollination increased the mass of harvested seeds and fruit set. Hand pollination also decreased the proportion of small, misshapen seeds.

CONCLUSION: Coffee plants close to forest patches produced more seeds and had fewer misshapen seeds than coffee plants farther from the forest patches.

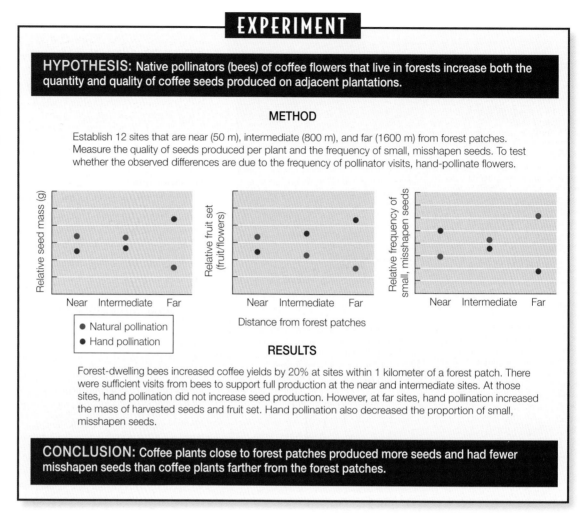

Most ecosystem services are provided locally. Fortunately, it is easier to motivate people to work to protect their local interests than it is to stimulate them to work on national or global issues. The National Wildlife Federation has established a very successful program in which people petition to have their backyards certified as wildlife-friendly. Criteria for certification include planting shrubs that provide food for birds and not using pesticides on lawns. The city of Tucson, Arizona, has launched a major project to make the city into an important habitat for many species of birds, not just the typical urban birds that live in most cities in North America.

Sometimes the alterations that make a big difference in biodiversity are remarkably simple. During the warmer part of the year, about 1.5 million Mexican free-tailed bats roost under the Congress Avenue Bridge in Austin, Texas. This is the largest population of bats in North America, much larger than the famous colony that lives in Carlsbad Caverns National Park. The nightly spectacle of their departure from the roost to feed is a major tourist attraction. Twenty years ago, however, only a few thousand bats roosted under the bridge. When the bridge was reconstructed a few years ago, engineers, with no intention of helping bats, added expansion joints that are about an inch wide and 16 inches deep. These crevices are perfect roosting sites for the bats, which quickly moved in by the tens of thousands.

The Turkey Point power plant in southern Florida consists of two fossil fuel generating units and two units powered by nuclear fuel. These four units generate lots of hot water. To cool the discharged water, the Florida Power & Light Company dug a system of 38 canals that covers 6,000 acres. The cooling canals are separated by low-lying berms that support a variety of native and exotic plants. Red mangroves grow along the edges of the canals. Today, the canals support a thriving population of American crocodiles, a highly endangered species. Crocodiles living in the canals yield about 10 percent of all young crocodiles in the United States. Having discovered the biodiversity value of its cooling system, the company employs biologists to monitor the crocodiles and does what it can to ensure their continued reproductive success.

The Elephant Pepper Development Trust of Africa promotes using chili peppers as a buffer crop around fields of maize, sorghum, and millet to deter elephants, buffalos, and other large mammals—who are repelled by the taste of the hot peppers—from invading the fields and destroying crops. Alternatively, farmers spray a mixture of capsaicin (the active ingredient in chilies) and axle grease on the string fences surrounding their fields.

Captive breeding programs can maintain a few species

A few of the world's many endangered species can be maintained in captivity while the external threats to their existence are reduced or removed. However, captive propagation is only a temporary measure that buys time to deal with those threats. Existing zoos, aquariums, and botanical gardens do not have enough space to maintain adequate populations of more than a small fraction of Earth's rare and endangered species. Nonetheless, captive propagation can play an important role by maintaining species during critical periods and by providing a source of individuals for reintroduction into the wild. Captive propagation projects in zoos also raise public awareness of threatened and endangered species.

As we saw at the beginning of this chapter, captive propagation helped save the whooping crane. The California condor, North America's largest bird, survives today only because of captive propagation (**Figure 57.20**). Two hundred years ago, condors ranged from southern British Columbia to northern Mexico, but by 1978, the wild population was plunging toward extinction—only 25 to 30 birds remained in southern California. Many birds died from ingesting carcasses containing lead shot.

To save the condor from certain extinction, biologists initiated a captive breeding program in 1983. The first chick conceived in captivity hatched in 1988. By 1993, nine captive pairs were producing chicks, and the captive population had increased to more than 60 birds. Six captive-bred birds were released in the mountains north of Los Angeles in 1992. These birds are provided with lead-free food in remote areas. They are using the same roosting sites, bathing pools, and mountain ridges as their predecessors did. Captive-reared birds have also been released in northern Arizona and Baja California. As of October 2005, there were 121 wild condors in California, Arizona, and Baja California. Three wild-born chicks are now flying free for the first time in 20 years in Arizona and California. Lead poisoning is still a problem, but an effort to encourage hunters to use non-lead ammunition is under way.

Gymnogyps californianus

57.20 A California Condor Soars California condors raised in captivity have successfully survived after being released into the wild. Numbered wing tags allow conservation biologists to identify the individual bird and track its movements. The survival of North America's largest bird species depends on this captive propagation project.

The legacy of Samuel Plimsoll

During the nineteenth century, many British merchant ships sailed Earth's oceans. At that time, there were no undersea telegraph cables or shipboard radios. Once a ship left a harbor, it was out of contact with the rest of the world; in the case of a shipwreck, rescue was impossible. Owners could maximize their profits by overloading their ships, even though this caused some of them to be unseaworthy and sink. Samuel Plimsoll, a member of England's Parliament, became concerned about the rate of loss of British vessels and sailors. He convinced Parliament to require that a "load line" be painted on the hull of every large oceangoing vessel. The position of the line was calculated using factors such as the structural strength of the vessel and the shape of its hull. If the load line was under water, the ship was not permitted to leave the harbor. The "Plimsoll line," as it has come to be known, dramatically reduced the rate of loss of British ships and sailors at sea.

The increasing loss of Earth's species suggests that the load of human activities has pushed the hull of Noah's Ark below the Plimsoll line. But where and how should society draw that line? The decision should be based on scientific information, but just as in Samuel Plimsoll's time, science cannot determine an "acceptable rate of loss." Ethical considerations will figure prominently in the decisions societies make concerning how much to change the way we use ecosystems so that other species can survive on Earth with us.

57.4 RECAP

To preserve biodiversity, we must be willing to set aside protected areas, restore habitat, develop programs to increase populations of endangered species, enforce laws that restrict transport of invasive species, and otherwise recognize the benefits of maintaining well-functioning ecosystems.

- Can you enumerate the priorities conservation biologists consider when establishing protected areas?
- Why is restoration of degraded ecosystems often necessary? See pp. 1236–1237
- Explain why controlling the importation of exotic species is an important component of conservation biology. See pp. 1239–1240 and Figure 57.17

CHAPTER SUMMARY

57.1 What is conservation biology?

Conservation biology is an applied scientific discipline devoted to preserving biodiversity.

Conservation biology is a normative discipline, but its practitioners adhere to scientific standards and practices.

Conservation biologists recognize that Earth's ecosystems are dynamic and that people are integral components of ecosystems.

There are many compelling reasons for preserving biodiversity, including maintaining well-functioning ecosystems that provide humans with many goods and services.

57.2 How do biologists predict changes in biodiversity?

The **species–area relationship** estimates the number of species that can be supported by an area of a given size.

To estimate probable rates of extinction, statistical models take into account data on population sizes, genetic variation, physiology, morphology, and behavior.

Rarity is not always a cause for concern, but species whose populations are shrinking rapidly are usually at risk.

Populations with only a few individuals confined to a small range may be eliminated by local disturbances such as fires, unusual weather, disease, habitat destruction, and predators.

57.3 What factors threaten species survival?

Habitat loss is the most important cause of endangerment of species worldwide. As habitats become increasingly **fragmented**, more species are lost from those habitats. Small habitat patches can support only small populations and are adversely influenced by **edge effects.** Review Figure 57.5, Web/CD Tutorial 57.1

Species introduced to regions outside their original range often become **invasive**, causing extinctions of native species by competing with them, eating them, or transmitting diseases to them.

Overexploitation has historically been the most important cause of species extinctions, and overexploitation continues today.

Climate change is likely to become an increasingly important cause of extinctions for those species that cannot shift their ranges as rapidly as the climate warms.

57.4 What strategies do conservation biologists use?

Establishing protected areas is crucial to preserving biodiversity. Protected areas are selected by taking into account species richness, **endemism**, imminence of threats, and the need to protect representative ecosystems.

Restoration ecology is an important conservation strategy because many degraded ecosystems will not recover, or will do so only very slowly, without human assistance. Review Figure 57.16

International trade in endangered species is controlled by regulations that most countries endorse.

Determining which species are likely to become invasive and preventing their introduction to new areas is an important component of conservation biology.

Calculating the economic value of ecosystem services is helping establish incentives to preserve them.

Even within the extensive landscapes where people live and extract resources, steps may be taken to preserve biodiversity. This approach is becoming known as **reconciliation ecology**.

Captive breeding programs can maintain selected endangered species while threats to their existence are reduced or removed.

See Web/CD Activity 57.1 for a concept review of this chapter.

SELF-QUIZ

1. Which of the following is *not* currently a major cause of species extinctions?
   a. Habitat destruction
   b. Rising sea levels
   c. Overexploitation
   d. Introduction of predators
   e. Introduction of diseases
2. The most important cause of endangerment of species in the United States currently is
   a. pollution.
   b. invasive species.
   c. overexploitation.
   d. habitat destruction.
   e. loss of mutualists.
3. People care about species extinctions because
   a. more than half of the medical prescriptions written in the United States contain a natural plant or animal product.
   b. people derive aesthetic pleasure from interacting with other organisms.
   c. causing species extinctions raises serious ethical issues.
   d. biodiversity helps maintain valuable ecosystem services.
   e. all of the above
4. As a habitat patch gets smaller, it
   a. cannot support populations of species that require large areas.
   b. supports only small populations of many species.
   c. is influenced to an increasing degree by edge effects.
   d. is invaded by species from surrounding habitats.
   e. all of the above
5. A plant species is most likely to become invasive when introduced to a new area if it
   a. grows tall.
   b. has become invasive in other places where it has been introduced.
   c. is closely related to species living in the area where it has been introduced.
   d. has specialized disseminators of its seeds.
   e. has a long life span.
6. Conservation biologists are concerned about global warming because
   a. the rate of change in climate is projected to be faster than the rate at which many species can shift their ranges.
   b. it is already too hot in the tropics.
   c. climates have been so stable for thousands of years that many species lack the ability to tolerate variable temperatures.
   d. climate change will be especially harmful to rare species.
   e. none of the above

7. Scientists can determine the historical frequency of fires in an area by
   a. examining charcoal in sites of ancient villages.
   b. measuring carbon in soils.
   c. radioactively dating fallen tree trunks.
   d. examining fire scars in growth rings of living trees.
   e. determining the age structure of forests.
8. Captive propagation is a useful conservation tool, provided that
   a. there is space in zoos, aquariums, and botanical gardens for breeding a few individuals.
   b. the areas of origin of all individuals are known.
   c. the threats that endangered the species are being alleviated so that captive-reared individuals can later be released back into the wild.
   d. there are sufficient caretakers.
   e. none of the above. Captive propagation should never be used because it directs attention away from the need to protect the species in their natural habitats.
9. Restoration ecology is an important field because
   a. many areas have been highly degraded.
   b. many areas are vulnerable to global climate change.
   c. many species suffer from demographic stochasticity.
   d. many species are genetically impoverished.
   e. fire is a threat to many areas.
10. The new discipline of reconciliation ecology has developed because
   a. all other methods of preserving biodiversity have failed.
   b. protected areas should be able to maintain biodiversity.
   c. protected areas alone are not sufficient to maintain biodiversity.
   d. scientists are unable to control diseases today.
   e. we are not reconciled with other species.

FOR DISCUSSION

1. Most species driven to extinction by humans in the past were large vertebrates. Do you expect this pattern to persist into the future? If not, why not?
2. Conservation biologists have debated extensively which is better: many small protected areas—which may contain more species—or a few large protected areas—which may be the only ones that can support populations of species that require large areas. What ecological processes should be evaluated in making judgments about the sizes and locations of protected areas?
3. During World War I, doctors adopted a "triage" system for dealing with wounded soldiers. The wounded were divided into three categories: those almost certain to die no matter what was done to help them, those likely to recover even if not assisted, and those whose probability of survival was greatly increased if they were given immediate medical attention. Limited medical resources were directed primarily at the third category. What are some implications of adopting a similar attitude toward species preservation?
4. Utilitarian arguments dominate discussions about the importance of preserving the biological richness of the planet. In your opinion, what role should ethical and moral arguments play?
5. The desert bighorn sheep of the southwestern United States is endangered. Its major predator, the puma, is also threatened in the region. Under what conditions, if any, would it be appropriate to suppress the population of one rare species to assist another rare species?

FOR INVESTIGATION

Forest-dwelling bees increased seed production in coffee plants growing within a kilometer of a forest patch in Costa Rica (see Figure 57.19). What forest organisms are likely to provide benefits, such as control of insect pests, to surrounding agricultural crops in addition to pollination? Over what distances are those effects likely to be felt? What experiments could be designed to test the importance of those effects and how they vary with distance from forest patches?

Appendix A: The Tree of Life

Phylogeny is the organizing principle of modern biological taxonomy, and a guiding principle of modern phylogeny is *monophyly*: a monophyletic group is considered to be one that contains an ancestral lineage and *all* of its descendants. Any such a group can be extracted from a phylogenetic tree with a single cut. The tree shown here provides a guide to the relationships among the major groups of the extant (living) organisms in the Tree of Life as we have presented them throughout this book. We do include three groups that are not believed to be monophyletic; these are designated with quotation marks.

The position of the branching "splits" indicates the relative branching order of the lineages of life, but the timing of splits in different groups is not drawn on a comparable time scale. In addition, the groups appearing at the branch tips do not necessarily carry equal phylogenetic "weight." For example, the ginkgo [55] is indeed at the apex of its lineage; this gymnosperm group consists of a single living species. In contrast, a phylogeny of the angiosperms [52] would continue on from this point to fill many more trees the size of this one.

The glossary entries that follow are informal descriptions of some major features of the organisms described in Part Six of this book. Each entry gives the group's common name, followed by the formal scientific name of the group (in parentheses). Numbers in square brackets reference the location of the respective groups on the tree.

It is sometimes convenient to use an informal name to refer to a collection of organisms that are not monophyletic but nonetheless all share (or all lack) some common attribute. We call these "convenience terms"; such groups are indicated in these entries by quotation marks, and we do not give them formal scientific names. Examples include "prokaryotes," "protists," and "algae." Note that these groups cannot be removed with a single cut; they represent a collection of distantly related groups that appear in different parts of the tree.

– A –

acorn worms (*Enteropneusta*) Benthic marine hemichordates [109] with an acorn-shaped proboscis, a short collar (neck), and a long trunk.

"algae" A convenience term encompassing various distantly related groups of aquatic, photosynthetic chromalveolates [5] and certain members of the Plantae [8].

alveolates (*Alveolata*) [7] Unicellular eukaryotes with a layer of flattened vesicles (alveoli) supporting the plasma membrane. Major alveolate groups include the dinoflagellates [49], apicomplexans [50], and ciliates [51].

ambulacrarians (*Ambulacraria*) [27] The echinoderms [108] and hemichordates [109].

amniotes (*Amniota*) [34] Mammals, reptiles, and their extinct close relatives. Characterized by many adaptations to terrestrial life, including an amniotic egg (with a unique set of membranes—the amnion, chorion, and allantois), a water-repellant epidermis (with epidermal scales, hair, or feathers), and, in males, a penis that allows internal fertilization.

amoebozoans (*Amoebozoa*) [76] A group of eukaryotes [4] that use lobe-shaped pseudopods for locomotion and to engulf food. Major amoebozoan groups include the loboseans, plasmodial slime molds, and cellular slime molds.

amphibians (*Amphibia*) [118] Tetrapods [33] with glandular skin that lacks epidermal scales, feathers, or hair. Many amphibian species undergo a complete metamorphosis from an aquatic larval form to a terrestrial adult form, although direct development is also common. Major amphibian groups include frogs and toads (anurans), salamanders, and caecilians.

amphipods (*Amphipoda*) Small crustaceans [106] that are abundant in many marine and freshwater habitats. They are important herbivores, scavengers, and micropredators, and are an important food source for many aquatic organisms.

angiosperms (*Anthophyta* or *Magnoliophyta*) [52] The flowering plants. Major angiosperm groups include the monocots, eudicots, and magnoliids.

animals (*Animalia* or *Metazoa*) [19] Multicellular heterotrophic eukaryotes. The majority of animals are bilaterians [21]. Other major groups are the cnidarians [87], ctenophores [86], calcareous sponges [85], demosponges [84], and glass sponges [83]. The closest living relatives of the animals are the choanoflagellates [82].

annelids (*Annelida*) [94] Segmented worms, including earthworms, leeches, and polychaetes. One of the major groups of lophotrochozoans [23].

anthozoans (*Anthozoa*) One of the major groups of cnidarians [87]. Includes the sea anemones, sea pens, and corals.

anurans (*Anura*) Comprising the frogs and toads, this is the largest group of living amphibians [118]. They are tailless, with a shortened vertebral column and elongate hind legs modified for jumping. Many species have an aquatic larval form known as a tadpole.

apicomplexans (*Apicomplexa*) [50] Parasitic alveolates [7] characterized by the possession of an apical complex at some stage in the life cycle.

arachnids (*Arachnida*) Chelicerates [104] with a body divided into two parts: a cephalothorax that bears six pairs of appendages (four pairs of which are usually used as legs) and an abdomen that bears the genital opening. Familiar arachnids include spiders, scorpions, mites and ticks, and harvestmen.

archaeans (*Archaea*) [3] Unicellular organisms lacking a nucleus and lacking peptidoglycan in the cell wall. Once grouped with the bacteria, archaeans possess distinctive membrane lipids.

archosaurs (*Archosauria*) [36] A group of reptiles [35] that includes dinosaurs and crocodilians [123]. Most dinosaur groups became extinct at the end of the Cretaceous; birds [122] are the only surviving dinosaurs.

arrow worms (*Chaetognatha*) [96] Small planktonic or benthic predatory marine worms with fins and a pair of hooked, prey-grasping spines on each side of the head.

arthropods (*Arthropoda*) The largest group of ecdysozoans [24]. Arthropods are characterized by a stiff exoskeleton, segmented bodies, and jointed appendages. Includes the chelicerates [104], myriapods [105], crustaceans [106], and hexapods (insects and their relatives) [107].

ascidians (*Ascidiacea*) "Sea squirts"; the largest group of urochordates [110]. Also known as tunicates, they are sessile (as adults), marine, sac-like filter feeders.

ascomycetes (*Ascomycota*) [78] Fungi that bear the products of meiosis within sacs (asci) if the organism is multicellular. Some are unicellular.

– B –

bacteria (*Eubacteria*) [2] Unicellular organisms lacking a nucleus, possessing distinctive ribosomes and initiator tRNA, and generally containing peptidoglycan in the cell wall. Different bacterial groups are distinguished primarily on nucleotide sequence data.

barnacles (*Cirripedia*) Crustaceans [106] that undergo two metamorphoses—first from a feeding planktonic larva to a nonfeeding swimming larva, and then to a sessile adult that forms a "shell" composed of four to eight plates cemented to a hard substrate.

basidiomycetes (*Basidiomycota*) [77] Fungi [17] that, if multicellular, bear the products of meiosis on club-shaped basidia and possess a long-lasting dikaryotic stage. Some are unicellular.

bilaterians (*Bilateria*) [21] Those animal groups characterized by bilateral symmetry and three distinct tissue types (endoderm, ectoderm, and mesoderm). Includes the protostomes [22] and deuterostomes [26].

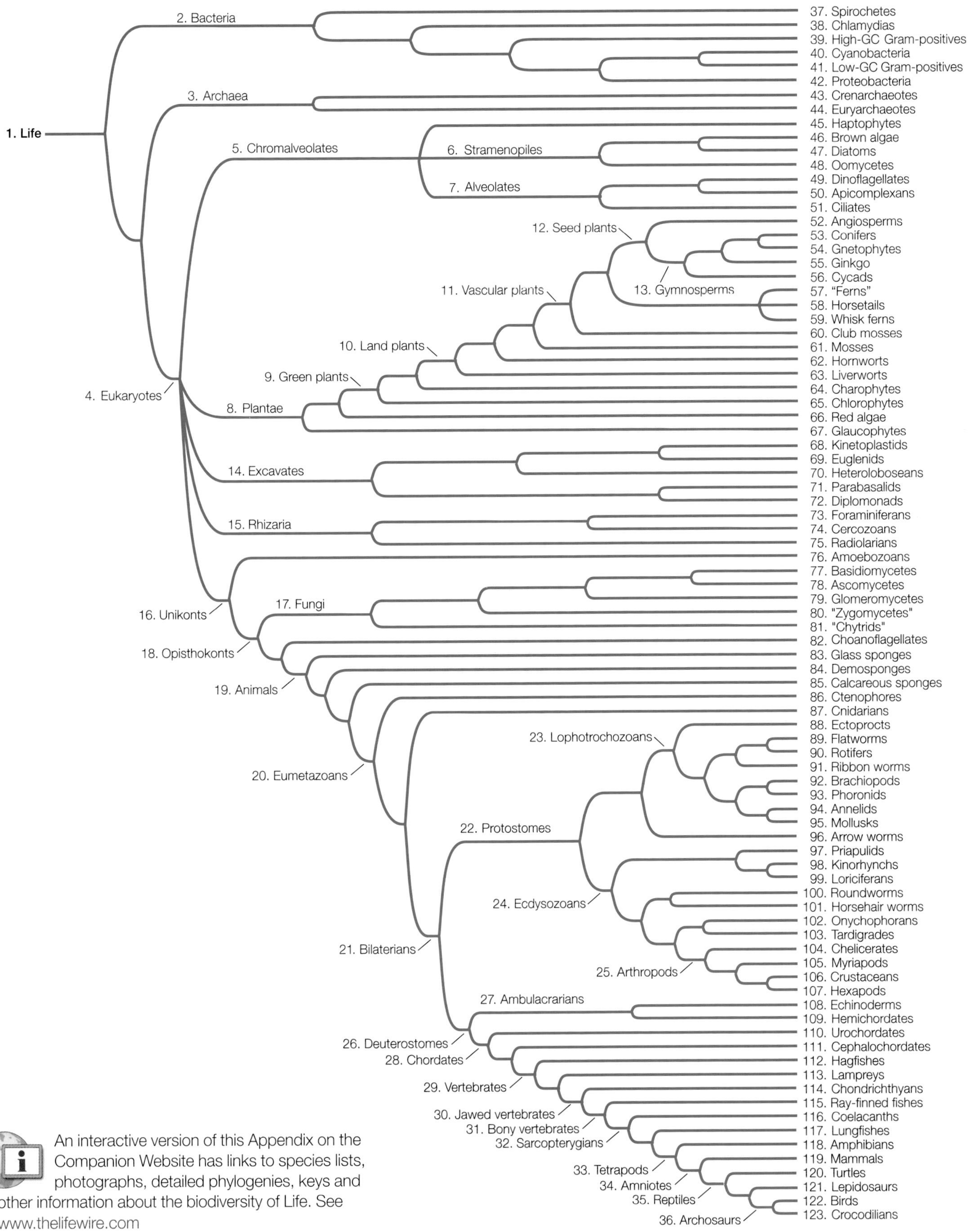

An interactive version of this Appendix on the Companion Website has links to species lists, photographs, detailed phylogenies, keys and other information about the biodiversity of Life. See www.thelifewire.com

birds (*Aves*) [122] Feathered, flying (or secondarily flightless) tetrapods [33].

bivalves (*Bivalvia*) Major mollusk [95]group; clams and mussels. Bivalves typically have two similar hinged shells that are each asymmetrical across the midline.

bony vertebrates (*Osteichthyes*) [31] Vertebrates [29] in which the skeleton is usually ossified to form bone. Includes the ray-finned fishes [115], coelacanths [116], lungfishes [117], and tetrapods [33].

brachiopods (*Brachiopoda*) [92] Lophotrochozoans [23] with two similar hinged shells that are each symmetrical across the midline. Superficially resemble bivalve mollusks, except for the shell symmetry.

brittle stars (*Ophiuroidea*) Echinoderms [108] with five long, whip-like arms radiating from a distinct central disk that contains the reproductive and digestive organs.

brown algae (*Phaeophyta*) [46] Multicellular, almost exclusively marine stramenopiles [6] generally containing the pigment fucoxanthin as well as chlorophylls *a* and *c* in their chloroplasts.

– C –

caecilians (*Gymnophiona*) A group of burrowing or aquatic amphibians [118]. They are elongate, legless, with a short tail (or none at all), reduced eyes covered with skin or bone, and a pair of sensory tentacles on the head.

calcareous sponges (*Calcarea*) [85] Filter-feeding marine sponges with spicules composed of calcium carbonate.

cellular slime molds (*Dictyostelida*) Amoebozoans [76] in which individual amoebas aggregate under stress to form a multicellular pseudoplasmodium.

cephalochordates (*Cephalochordata*) [111] A group of weakly swimming, eel-like benthic marine chordates [28]; also called lancelets. Thought to be the closest living relatives of the vertebrates [29].

cephalopods (*Cephalopoda*) Active, predatory mollusks [95] in which the molluscan foot has been modified into muscular hydrostatic arms or tentacles. Includes octopuses, squids, and nautiluses.

cercozoans (*Cercozoa*) [74] Unicellular eukaryotes [4] that feed by means of threadlike pseudopods. Together with foraminiferans [73] and radiolarians [75], the cercozoans comprise the group *Rhizaria* [15].

charophytes (*Charales*) [64] Multicellular green algae with branching, apical growth and plasmodesmata between adjacent cells. The closest living relatives of the land plants [10], they retain the egg in the parent organism.

chelicerates (*Chelicerata*) [104] A major group of arthropods [25] with pointed appendages (chelicerae) used to grasp food (as opposed to the chewing mandibles of most other arthropods). Includes the arachnids, horseshoe crabs, pycnogonids, and extinct sea scorpions.

chimaeras (*Holocephali*) A group of bottom-dwelling, marine, scaleless chondrichthyan fishes [114] with large, permanent, grinding tooth plates (rather than the replaceable teeth found in other chondrichthyans).

chitons (*Polyplacophora*) Flattened, slow-moving mollusks [95] with a dorsal protective calcareous covering made up of eight articulating plates.

chlamydias (*Chlamydiae*) [38] A group of very small Gram-negative bacteria; they live as intracellular parasites of other organisms.

chlorophytes (*Chlorophyta*) [65] The most abundant and diverse group of green algae, including freshwater, marine, and terrestrial forms; some are unicellular, others colonial, and still others multicellular. Chlorophytes use chlorophylls *a* and *c* in their photosynthesis.

choanoflagellates (*Choanozoa*) [82] Unicellular eukaryotes [4] with a single flagellum surrounded by a collar. Most are sessile, some are colonial. The closest living relatives of the animals [19]

chondrichthyans (*Chondrichthyes*) [114] One of the two main groups of jawed vertebrates [30]; includes sharks, rays, and chimaeras. They have cartilaginous skeletons and paired fins.

chordates (*Chordata*) [28] One of the two major groups of deuterostomes [26], characterized by the presence (at some point in development) of a notochord, a hollow dorsal nerve cord, and a post-anal tail. Includes the urochordates [110], cephalochordates [111], and vertebrates [29].

chromalveolates (*Chromalveolata*) [5] A contested group, said to have arisen from a common ancestor with chloroplasts derived from a red alga and supported by molecular evidence. Major chromalveolate groups include the alveolates [7] and stramenopiles [6].

"chytrids" [81] A convenience term used for a paraphyletic group of mostly aquatic, microscopic fungi [17] with flagellated gametes. Some exhibit alternation of generations.

ciliates (*Ciliophora*) [51] Alveolates [7] with numerous cilia and two types of nuclei (micronuclei and macronuclei).

clitellates (*Clitellata*) Annelids [94] with gonads contained in a swelling (called a clitellum) toward the head of the animal. Includes earthworms (oligochaetes) and leeches.

club mosses (*Lycophyta*) [60] Vascular plants [11] characterized by microphylls.

cnidarians (*Cnidaria*) [87] Aquatic, mostly marine eumetazoans [20] with specialized stinging organelles (nematocysts) used for prey capture and defense, and a blind gastrovascular cavity. The closest living relatives of the bilaterians [21].

coelacanths (*Actinista*) [116] A group of marine sarcopterygians [32] that was diverse from the Middle Devonian to the Cretaceous, but is now known from just two living species. The pectoral and anal fins are on fleshy stalks supported by skeletal elements, so they are also called lobe-finned fishes.

conifers (*Pinophyta* or *Coniferophyta*) [53] Cone-bearing, woody seed plants.

copepods (*Copepoda*) Small, abundant crustaceans [106] found in marine, freshwater, or wet terrestrial habitats. They have a single eye, long antennae, and a body shaped like a teardrop.

craniates (*Craniata*) Some biologist exclude the hagfishes [112] from the vertebrates [29], and use the term craniates to refer to the two groups combined.

crenarchaeotes (*Crenarchaeota*) [43] A major and diverse group of archaeans [3], defined on the basis of rRNA base sequences. Many are extremophiles (inhabit extreme environments), but the group may also be the most abundant archaeans in the marine environment.

crinoids (*Crinoidea*) Echinoderms [108] with a mouth surrounded by feeding arms, and a U-shaped gut with the mouth next to the anus. They attach to the substratum by a stalk or are free-swimming. Crinoids were abundant in the middle and late Paleozoic, but only a few hundred species have survived to the present. Includes the sea lilies and feather stars.

crocodilians (*Crocodylia*) [123] A group of large, predatory, aquatic archosaurs [36]. The closest living relatives of birds [122]. Includes alligators, caimans, crocodiles, and gharials.

crustaceans (*Crustacea*) [106] Major group of marine, freshwater, and terrestrial arthropods [25] with a head, thorax, and abdomen (although the head and thorax may be fused), covered with a thick exoskeleton, and with two-part appendages. Crustaceans undergo metamorphosis from a nauplius larva. Includes decapods, isopods, krill, barnacles, amphipods, copepods, and ostracods.

ctenophores (*Ctenophora*) [86] Radially symmetrical, diploblastic marine animals [19], with a complete gut and eight rows of fused plates of cilia (called ctenes).

cyanobacteria (*Cyanobacteria*) [40] A group of unicellular, colonial, or filamentous bacteria that conduct photosynthesis using chlorophyll *a*.

cycads (*Cycadophyta*) [56] Palmlike gymnosperms with large, compound leaves.

cyclostomes (*Cyclostomata*) This term refers to the possibly monophyletic group of lampreys [113] and hagfishes [112]. Molecular data support this group, but morphological data suggest that lampreys are more closely related to jawed vertebrates [30] than to hagfishes.

– D –

decapods (*Decapoda*) A group of marine, freshwater, and semiterrestrial crustaceans [106] in which five of the eight pairs of thoracic appendages function as legs (the other three pairs, called maxillipeds, function as mouthparts). Includes crabs, lobsters, crayfishes, and shrimps.

demosponges (*Demospongiae*) [84] The largest of the three groups of sponges, accounting for 90 percent of all sponge species. Demosponges have spicules made of silica, spongin fiber (a protein), or both.

deuterostomes (*Deuterostomia*) [26] One of the two major groups of bilaterians [21], in which the mouth forms at the opposite end of the embryo from the blastopore in early development (contrast with protostomes). Includes the ambulacrarians [27] and chordates [28].

diatoms (*Bacillariophyta*) [47] Unicellular, photosynthetic stramenopiles [6] with glassy cell walls in two parts.

dinoflagellates (*Dinoflagellata*) [49] A group of alveolates [7] usually possessing two flagella, one in an equatorial groove and the other in a longitudinal groove; many are photosynthetic.

diplomonads (*Diplomonadida*) [72] A group of eukaryotes [4] lacking mitochondria; most have two nuclei, each with four associated flagella.

– E –

ecdysozoans (*Ecdysozoa*) [24] One of the two major groups of protostomes [22], characterized by periodic molting of their exoskeletons. Roundworms [100] and arthropods [25] are the largest ecdysozoan groups.

echinoderms (*Echinodermata*) [108] A major group of marine deuterostomes [26] with fivefold radial symmetry (at some stage of life) and an endoskeleton made of calcified plates and spines. Includes sea stars, crinoids, sea urchins, sea cucumbers, and brittle stars.

ectoprocts (*Ectoprocta*) [88] A group of marine and freshwater lophotrochozoans [23] that live in colonies attached to substrata. Also known as bryozoans or moss animals.

elasmobranchs (*Elasmobranchii*) The largest group of chondrichthyan fishes [114]. Includes sharks, skates, and rays. In contrast to the other group of living chondrichthyans (the chimaeras), they have replaceable teeth.

eudicots (*Eudicotyledones*) A group of angiosperms [52] with pollen grains possessing three openings. Typically with two cotyledons, net-veined leaves, taproots, and floral organs typically in multiples of four or five.

euglenids (*Euglenida*) [69] Flagellate excavates characterized by a pellicle composed of spiraling strips of protein under the plasma membrane; the mitochondria have disk-shaped cristae. Some are photosynthetic.

eukaryotes (*Eukarya*) [4] Organisms made up of one or more complex cells in which the genetic material is contained in nuclei. Contrast with archaeans [3] and bacteria [2].

eumetazoans (*Eumetazoa*) [20] Those animals [19] characterized by body symmetry, a gut, a nervous system, specialized types of cell junctions, and well-organized tissues in distinct cell layers (although there have been secondary losses of some of these characteristics in some eumetazoans).

euphyllophytes (*Euphyllophyta*) This clade is sister to the club mosses [60] and includes all plants with megaphylls.

euryarchaeotes (*Euryachaeota*) [44] A major group of archaeans [3], diagnosed on the basis of rRNA sequences. Includes many methanogens, extreme halophiles, and thermophiles.

eutherians (*Eutheria*) A group of viviparous mammals [119], eutherians are well developed at birth (contrast to prototherians and marsupials, the other two groups of mammals). Most familiar mammals outside the Australian and South American regions are eutherians (see Table 33.1).

excavates (*Excavata*) [14] Diverse group of unicellular, flagellate eukaryotes, many of which possess a feeding groove; some lack mitochondria.

– F –

"ferns" [57] Vascular plants [11] usually possessing large, frond-like leaves that unfold from a "fiddlehead." Not a monophyletic group, although most fern species are encompassed in a monophyletic clade, the leptosporangiate ferns.

flatworms (*Platyhelminthes*) [89] A group of dorsoventrally flattened and generally elongate soft-bodied lophotrochozoans [23]. May be free-living or parasitic, found in marine, freshwater, or damp terrestrial environments. Major flatworm groups include the tapeworms, flukes, monogeneans, and turbellarians.

flowering plants *See* angiosperms.

flukes (*Trematoda*) A group of wormlike parasitic flatworms [89] with complex life cycles that involve several different host species. May be paraphyletic with respect to tapeworms.

foraminiferans (*Foraminifera*) [73] Amoeboid organisms with fine, branched pseudopods that form a food-trapping net. Most produce external shells of calcium carbonate.

fungi (*Fungi*) [17] Eukaryotic heterotrophs with absorptive nutrition based on extracellular digestion; cell walls contain chitin. Major fungal groups include the "chytrids" [81], "zygomycetes" [80], glomeromycetes [79], ascomycetes [78], and basidiomycetes [77].

– G –

gastropods (*Gastropoda*) The largest group of mollusks [95]. Gastropods possess a well-defined head with two or four sensory tentacles (often terminating in eyes) and a ventral foot. Most species have a single coiled or spiraled shell. Common in marine, freshwater, and terrestrial environments.

ginkgo (*Ginkgophyta*) [55] A gymnosperm [13] group with only one living species. The ginkgo seed is surrounded by a fleshy tissue not derived from an ovary wall and hence not a fruit.

glass sponges (*Hexactinellida*) [83] Sponges with a skeleton composed of four- and/or six-pointed spicules made of silica.

glaucophytes (*Glaucophyta*) [67] Unicellular freshwater algae with chloroplasts containing traces of peptidoglycan, the characteristic cell wall material of bacteria.

glomeromycetes (*Glomeromycota*) [79] A group of fungi [17] that form arbuscular mycorrhizae.

gnathostomes (*Gnathostomata*) *See* jawed vertebrates.

gnetophytes (*Gnetophyta*) [54] A gymnosperm [13] group with three very different lineages; all have wood with vessels, unlike other gymnosperms.

green plants (*Viridiplantae*) [9] Organisms with chlorophylls *a* and *b*, cellulose-containing cell walls, starch as a carbohydrate storage product, and chloroplasts surrounded by two membranes.

gymnosperms (*Gymnospermae*) [13] Seed plants [12] with seeds "naked" (i.e., not enclosed in carpels). Probably monophyletic, but status still in doubt. Includes the conifers [53], gnetophytes [54], ginkgo [55], and cycads [56].

– H –

hagfishes (*Myxini*) [112] Elongate, slimy-skinned vertebrates [29] with three small accessory hearts, a partial cranium, and no stomach or paired fins. *See also* craniata; cyclostomes.

haptophytes (*Haptophyta*) [45] Unicellular, photosynthetic stramenopiles [6] with two slightly unequal, smooth flagella. Abundant as phytoplankton, some form marine algal blooms.

hemichordates (*Hemichordata*) [109] One of the two primary groups of ambulacrarians [27]; marine wormlike organisms with a three-part body plan.

heteroloboseans (*Heterolobosea*) [70] Colorless excavates [14] that can transform among amoeboid, flagellate, and encysted stages.

hexapods (*Hexapoda*) [107] Major group of arthropods [25] characterized by a reduction (from the ancestral arthropod condition) to six walking appendages, and the consolidation of three body segments to form a thorax. Includes insects and their relatives (see Table 32.2).

high-GC Gram-positives (*Actinobacteria*) [39] Gram-positive bacteria with a relatively high G+C/A+T ratio of their DNA, with a filamentous growth habit.

hornworts (*Anthocerophyta*) [62] Nonvascular plants with sporophytes that grow from the base. Cells contain a single large, platelike chloroplast.

horsehair worms (*Nematomorpha*) [101] A group of very thin, elongate, wormlike freshwater ecdysozoans [24]. Largely nonfeeding as adults, they are parasites of insects and crayfish as larvae.

horseshoe crabs (*Xiphosura*) Marine chelicerates [104] with a large outer shell in three parts: a carapace, an abdomen, and a tail-like telson. Only five living species remain, but many additional species are known from fossils.

horsetails (*Sphenophyta* or *Equisetophyta*) [58] Vascular plants [11] with reduced megaphylls in whorls.

hydrozoans (*Hydrozoa*) A group of cnidarians [87]. Most species go through both polyp and mesuda stages, although one stage or the other is eliminated in some species.

– I –

insects (*Insecta*) The largest group within the hexapods [107]. Insects are characterized by exposed mouthparts and one pair of antennae containing a sensory receptor called a Johnston's organ. Most have two pairs of wings as adults. There are more described species of insects than all other groups of life [1] combined, and many species remain to be discovered. The major insect groups are described in Table 32.2.

isopods (*Isopoda*) Crustaceans [106] characterized by a compact head, unstalked compound eyes, and mouthparts consisting of four pairs of appendages. Isopods are abundant and widespread in salt, fresh, and brackish water, although some species (the sow bugs) are terrestrial.

– J –

jawed vertebrates (*Gnathostomata*) [30] A major group of vertebrates [29] with jawed mouths. Includes chondrichthyans [114], ray-finned fishes [115], and sarcopterygians [32].

– K –

kinetoplastids (*Kinetoplastida*) [68] Unicellular, flagellate organisms characterized by the presence in their single mitochondrion of a kinetoplast (a structure containing multiple, circular DNA molecules).

kinorhynchs (*Kinorhyncha*) [98] Small (< 1 mm) marine ecdysozoans [24] with bodies in 13 segments and a retractable proboscis.

korarchaeotes (*Korarchaeota*) A group of archaeans [3] known only by evidence from nucleic acids derived from hot springs. Its phylogenetic relationships within the Archaea are unknown.

krill (*Euphausiacea*) A group of shrimplike marine crustaceans [106] that are important components of the zooplankton.

– L –

lampreys (*Petromyzontiformes*) [113] Elongate, eel-like vertebrates [29] that often have rasping and sucking disks for mouths.

lancelets (*Cephalochordata*) *See* cephalochordates.

land plants (*Embryophyta*) [10] Plants with embryos that develop within protective structures; sporophytes and gametophytes are multicellular. Land plants possess a cuticle. Major groups are the liverworts [63], hornworts [62], mosses [61], and vascular plants [11].

larvaceans (*Larvacea*) Solitary, planktonic urochordates [110] that retain both notochords and nerve cords throughout their lives.

lepidosaurs (*Lepidosauria*) [121] Reptiles [35] with overlapping scales. Includes tuataras and squamates (lizards, snakes, and amphisbaenians).

life (*Life*) [1] The monophyletic group that includes all known living organisms. Characterized by a nucleic-acid based genetic system (DNA or RNA), metabolism, and cellular structure. Some parasitic forms, such as viruses, have secondarily lost some of these features and rely on the cellular environment of their host.

liverworts (*Hepatophyta*) [63] Nonvascular plants lacking stomata; stalk of sporophyte elongates along its entire length.

loboseans (*Lobosea*) A group of unicellular amoebozoans [76]; includes the most familiar amoebas (e.g., *Amoeba proteus*).

"lophophorates" Not a monophyletic group. A convenience term used to describe several groups of lophotrochozoans [23] that have a feeding structure called a lophophore (a circular or U-shaped ridge around the mouth that bears one or two rows of ciliated, hollow tentacles).

lophotrochozoans (*Lophotrochozoa*) [23] One of the two main groups of protostomes [22]. This group is morphologically diverse, and is supported primarily on information from gene sequences. Includes ectoprocts [88], flatworms [89], rotifers [90], ribbon worms [91], brachiopods [92], phoronids [93], annelids [94], and mollusks [95].

loriciferans (*Loricifera*) [99] Small (< 1 mm) ecdysozoans [24] with bodies in four parts, covered with six plates.

low-GC Gram-positives (*Firmicutes*) [41] A diverse group of bacteria [2] with a relatively low G+C/A+T ratio of their DNA, often but not always Gram-positive, some producing endospores.

lungfishes (*Dipnoi*) [117] A group of aquatic sarcopterygians [32] that are the closest living relatives of the tetrapods [33]. They have a modified swim bladder used to absorb oxygen from air, so some species can survive the temporary drying of their habitat.

– M –

magnoliids Major group of angiosperms [52] possessing two cotyledons and pollen grains with a single opening. The group is defined primarily by nucleotide sequence data; it is more closely related to the eudicots and monocots than to three other small angiosperm groups.

mammals (*Mammalia*) [119] A group of tetrapods [33] with hair covering all or part of their skin; females produce milk to feed their developing young. Includes the prototherians, marsupials, and eutherians.

marsupials (*Marsupialia*) Mammals [119] in which the female typically has a marsupium (a pouch for rearing young, which are born at an extremely early stage in development). Includes such familiar mammals as opossums, koalas, and kangaroos.

metazoans (*Metazoa*) *See* animals.

mollusks (*Mollusca*) [95] One of the major groups of lophotrochozoans [23], mollusks have bodies composed of a foot, a mantle (which often secretes a hard, calcareous shell), and a visceral mass. Includes monoplacophorans, chitons, bivalves, gastropods, and cephalopods.

monocots (*Monocotyledones*) Angiosperms [52] characterized by possession of a single cotyledon, usually parallel leaf veins, a fibrous root system, pollen grains with a single opening, and floral organs usually in multiples of three.

monogeneans (*Monogenea*) A group of ectoparasitic flatworms [89].

monoplacophorans (*Monoplacophora*) Mollusks [95] with segmented body parts and a single, thin, flat, rounded, bilateral shell.

mosses (*Bryophyta*) [61] Nonvascular plants with true stomata and erect,"leafy" gametophytes; sporophytes elongate by apical cell division.

multicellular eukaryotes *See* "protists."

myriapods (*Myriapoda*) [105] Arthropods [25] characterized by an elongate, segmented trunk with many legs. Includes centipedes and millipedes.

– N –

nanoarchaeotes (*Nanoarchaeota*) A hypothetical group of extremely small, thermophilic archaeans [3] with a much-reduced genome. The only described example can survive only when attached to a host organism.

nematodes (*Nematoda*) [100] A very large group of elongate, unsegmented ecdysozoans [24] with thick, multilayer cuticles. They are among the most abundant and diverse animals, although most species have not yet been described. Include free-living predators and scavengers, as well as parasites of most species of land plants [10] and animals [19].

neognaths (*Neognathae*) The main group of birds [122], including all living species except the ratites (ostrich, emu, rheas, kiwis, cassowaries) and tinamous (*see* palaeognaths).

– O –

oligochaetes (*Oligochaeta*) An annelid [94] group whose members lack parapodia, eyes, and anterior tentacles, and have few setae. Earthworms are the most familiar oligochaetes.

onychophorans (*Onychophora*) [102] Elongate, segmented ecdysozoans [24] with many pairs of soft, unjointed, claw-bearing legs. Also known as velvet worms.

oomycetes (*Oomycota*) [48] Water molds and relatives; absorptive heterotrophs with nutrient-absorbing, filamentous hyphae.

opisthokonts (*Opisthokonta*) [18] A group of unikonts [16] in which the flagellum on motile cells, if present, is posterior. The opisthokonts include the fungi [17], animals [19], and choanoflagellates [82].

ostracods (*Ostracoda*) Marine and freshwater crustaceans [106] that are laterally compressed and protected by two clam-like calcareous or chitinous shells.

– P –

palaeognaths (*Palaeognathae*) A group of secondarily flightless or weakly flying birds [122]. Includes the flightless ratites (ostrich, emu, rheas, kiwis, cassowaries) and the weakly flying tinamous.

parabasalids (*Parabasalia*) [71] A group of unicellular eukaryotes [4] that lack mitochondria; they possess flagella in clusters near the anterior of the cell.

phoronids (*Phoronida*) [93] A small group of sessile, wormlike marine lophotrochozoans [23] that secrete chitinous tubes and feed using a lophophore.

placoderms (*Placodermi*) An extinct group of jawed vertebrates [30] that lacked teeth. Placoderms were the dominant predators in Devonian oceans.

Plantae [8] The most broadly defined plant group. In most parts of this book, we use the word "plant" as synonymous with "land plant" [10], a more restrictive definition.

plasmodial slime molds (*Myxogastrida*) Amoebozoans [76] that in their feeding stage consist of a coenocyte called a plasmodium.

pogonophorans (*Pogonophora*) Deep-sea annelids [94] that lack a mouth or digestive tract; they feed by taking up dissolved organic matter, facilitated by endosymbiotic bacteria in a specialized organ (the trophosome).

polychaetes (*Polychaeta*) A group of mostly marine annelids [94] with one or more pairs of eyes and one or more pairs of feeding tentacles; parapodia and setae extend from most body segments. May be paraphyletic with respect to the clitellates.

priapulids (*Priapulida*) [97] A small group of cylindrical, unsegmented, wormlike marine ecdysozoans [24] that takes its name from its phallic appearance.

"progymnosperms" Paraphyletic group of extinct vascular plants [11] that flourished from the Devonian through the Mississippian periods. The first truly woody plants, and the first with vascular cambium that produced both secondary xylem and secondary phloem, they reproduced by spores rather than by seeds.

"prokaryotes" Not a monophyletic group; as commonly used, includes the bacteria [2] and archaeans [3]. A term of convenience encompassing all cellular organisms that are not eukaryotes.

proteobacteria (*Proteobacteria*) [42] A large and extremely diverse group of Gram-negative bacteria that includes many pathogens, nitrogen fixers, and photosynthesizers. Includes the alpha, beta, gamma, delta, and epsilon proteobacteria.

"protists" This term of convenience does not describe a monophyletic group but is used to encompass a large number of distinct and distantly related groups of eukaryotes, many but far from all of which are microbial and unicellular. Essentially a "catch-all" term for any eukaryote group not contained within the land plants [10], fungi [17], or animals [19].

protostomes (*Protostomia*) [22] One of the two major groups of bilaterians [21]. In protostomes, the mouth typically forms from the blastopore (if present) in early development (contrast with deuterostomes). The major protostome groups are the lophotrochozoans [23] and ecdysozoans [24].

prototherians (*Prototheria*) A mostly extinct group of mammals [119], common during the Cretaceous and early Cenozoic. The three living species—the echidnas and the duck-billed platypus—are the only extant egg-laying mammals.

pterobranchs (*Pterobranchia*) A small group of sedentary marine hemichordates [109] that live in tubes secreted by the proboscis. They have one to nine pairs of arms, each bearing long tentacles that capture prey and function in gas exchange.

pteridophytes (*Pteridophyta*) A group of vascular plants [11], sister to the seed plants [12], characterized by overtopping and possession of megaphylls. The pteridophytes include the horsetails [58], whisk ferns [59], and "ferns" [57].

pycnogonids (*Pycnogonida*) Treated in this book as a group of chelicerates [104], but sometimes considered an independent group of arthropods [25]. Pycnogonids have reduced bodies and very long, slender legs. Also called sea spiders.

– R –

radiolarians (*Radiolaria*) [75] Amoeboid organisms with needle-like pseudopods supported by microtubules. Most have glassy internal skeletons.

ray-finned fishes (*Actinopterygii*) [115] A highly diverse group of freshwater and marine bony vertebrates [31]. They have reduced swim bladders that often function as hydrostatic organs and fins supported by soft rays (lepidotrichia). Includes most familiar fishes.

red algae (*Rhodophyta*) [66] Mostly multicellular, marine algae characterized by the presence of phycoerythrin in their chloroplasts.

reptiles (*Reptilia*) [35] One of the two major groups of extant amniotes [34], supported on the basis of similar skull structure and gene sequences. The term "reptiles" traditionally excluded the birds [122], but the resulting group is then clearly paraphyletic. As used in this book, the reptiles include turtles [120], lepidosaurs [121], birds [122], and crocodilians [123].

rhizaria (*Rhizaria*) [15] Mostly amoeboid unicellular eukaryotes with pseudopods, many with external or internal shells. Includes the foraminiferans [73], cercozoans [74], and radiolarians [75].

rhyniophytes (*Rhyniophyta*) A group of early vascular plants [11] that appeared in the Silurian and became extinct in the Middle Devonian. Possessed dichotomously branching stems with terminal sporangia but no true leaves or roots.

ribbon worms (*Nemertea*) [91] A group of unsegmented lophotrochozoans [23] with an eversible proboscis used to capture prey. Mostly marine, but some species live in fresh water or on land.

rotifers (*Rotifera*) [90] Tiny (< 0.5 mm) lophotrochozoans [23] with a pseudocoelomic body cavity that functions as a hydrostatic organ and a ciliated feeding organ called the corona that surrounds the head. They live in freshwater and wet terrestrial habitats.

roundworms (*Nematoda*) [100] *See* nematodes.

– S –

salamanders (*Caudata*) A group of amphibians [118] with distinct tails in both larvae and adults and limbs set at right angles to the body.

salps *See* thaliaceans

sarcopterygians (*Sarcopterygii*) [32] One of the two major groups of bony vertebrates [31], characterized by jointed appendages (paired fins or limbs).

scyphozoans (*Scyphozoa*) Marine cnidarians [87] in which the medusa stage dominates the life cycle. Commonly known as jellyfish.

sea cucumbers (*Holothuroidea*) Echinoderms [108] with an elongate, cucumber-shaped body and leathery skin. They are scavengers on the ocean floor.

sea spiders *See* pycnogonids.

sea squirts *See* ascidians.

sea stars (*Asteroidea*) Echinoderms [108] with five (or more) fleshy "arms" radiating from an indistinct central disk. Also called starfishes.

sea urchins (*Echinoidea*) Echinoderms [108] with a test (shell) that is covered in spines. Most are globular in shape, although some groups (such as the sand dollars) are flattened.

"seed ferns" A paraphyletic group of loosely related, extinct seed plants that flourished in the Devonian and Carboniferous. Characterized by large, frond-like leaves that bore seeds.

seed plants (*Spermatophyta*) [12] Heterosporous vascular plants [11] that produce seeds; most produce wood; branching is axillary (not dichotomous). The major seed plant groups are gymnosperms [13] and angiosperms [52].

sow bugs *See* isopods.

spirochetes (*Spirochaetes*) [37] Motile, Gram-negative bacteria with a helically coiled structure and characterized by axial filaments.

"sponges" A term of convenience used for a paraphyletic group of relatively asymmetric, filter-feeding animals that lack a gut or nervous system and generally lack differentiated tissues. (*See* glass sponges [83], demosponges [84], and calcareous sponges [85].)

springtails (*Collembola*) Wingless hexapods [107] with springing structures on the third and fourth segments of their bodies. Springtails are extremely abundant in some environments (especially in soil, leaf litter, and vegetation).

squamates (*Squamata*) The major group of lepidosaurs [121], characterized by the possession of movable quadrate bones (which allow the upper jaw to move independently of the rest of the skull) and hemipenes (a paired set of eversible penises, or penes) in males. Includes the lizards (a paraphyletic group), snakes, and amphisbaenians.

starfish (*Asteroidea*) *See* sea stars

stramenopiles (*Heterokonta* or *Stramenopila*) [6] Organisms having, at some stage in their life cycle, two unequal flagella, the longer possessing rows of tubular hairs. Chloroplasts, when present, surrounded by four membranes. Major stramenopile groups include the brown algae [46], diatoms [47], and oomycetes [48].

– T –

tapeworms (*Cestoda*) Parasitic flatworms [89] that live in the digestive tracts of vertebrates as adults, and usually in various other species of animals as juveniles.

tardigrades (*Tardigrada*) [103] Small (< 0.5 mm) ecdysozoans [24] with fleshy, unjointed legs and no circulatory or gas exchange organs. They live in marine sands, in temporary freshwater pools, and on the water films of plants. Also called water bears.

tetrapods (*Tetrapoda*) [33] The major group of sarcopterygians [32]; includes the amphibians [118] and the amniotes [34]. Named for the presence of four jointed limbs (although limbs have been secondarily reduced or lost completely in several tetrapod groups).

thaliaceans (*Thaliacea*) A group of solitary or colonial planktonic marine urochordates [110]. Also called salps.

therians (*Theria*) Mammals [119] characterized by viviparity (live birth). Includes eutherians and marsupials.

theropods (*Theropoda*) Archosaurs [36] with bipedal stance, hollow bones, a furcula ("wishbone"), elongated metatarsals with three-fingered feet, and a pelvis that points backwards. Includes many well-known extinct dinosaurs (such as *Tyrannosaurus rex*), as well as the living birds [122].

trilobites (*Trilobita*) An extinct group of arthropods [25] related to the chelicerates [104]. Trilobites flourished from the Cambrian through the Permian.

tuataras (*Rhyncocephalia*) A group of lepidosaurs [121] known mostly from fossils; there are just two living tuatara species. The quadrate bone of the upper jaw is fixed firmly to the skull. Sister group of the squamates.

tunicates *See* ascidians.

turbellarians (*Turbellaria*) A group of free-living, generally carnivorous flatworms [89]. Their monophyly is questionable.

turtles (*Testudines*) [120] A group of reptiles [35] with a bony carapace (upper shell) and plastron (lower shell) that encase the body.

– U –

unikonts (*Unikonta*) [16] A group of eukaryotes [4] whose motile cells possess a single flagellum. Major unikont groups include the amoebozoans [76], fungi [17], and animals [19].

urochordates (*Urochordata*) [110] A group of chordates [28] that are mostly saclike filter feeders as adults, with motile larvae stages that resemble a tadpole.

– V –

vascular plants (*Tracheophyta*) [11] Plants with xylem and phloem. Major groups include the club mosses [60] and euphyllophytes.

vertebrates (*Vertebrata*) [29] The largest group of chordates [28], characterized by a rigid endoskeleton supported by the vertebral column and an anterior skull encasing a brain. Includes hagfishes [112], lampreys [113], and the jawed vertebrates [30], although some biologists exclude the hagfishes from this group (see craniates).

– W –

water bears *See* tardigrades.

whisk ferns (*Psilotophyta*) [59] Vascular plants [11] lacking leaves and roots.

– Y –

"yeasts" A convenience term for several distantly related groups of unicellular fungi [17].

– Z –

"zygomycetes" [80] A convenience term for a paraphyletic group of fungi [17] in which hyphae of differing mating types conjugate to form a zygosporangium.

Appendix B: Some Measurements Used in Biology

MEASURES OF	UNIT	EQUIVALENTS	METRIC → ENGLISH CONVERSION
Length	meter (m)	base unit	1 m = 39.37 inches = 3.28 feet
	kilometer (km)	1 km = 1000 (10^3) m	1 km = 0.62 miles
	centimeter (cm)	1 cm = 0.01 (10^{-2}) m	1 cm = 0.39 inches
	millimeter (mm)	1 mm = 0.1 cm = 10^{-3} m	1 mm = 0.039 inches
	micrometer (μm)	1 μm = 0.001 mm = 10^{-6} m	
	nanometer (nm)	1 nm = 0.001 μm = 10^{-9} m	
Area	square meter (m^2)	base unit	1 m^2 = 1.196 square yards
	hectare (ha)	1 ha = 10,000 m^2	1 ha = 2.47 acres
Volume	liter (L)	base unit	1 L = 1.06 quarts
	milliliter (mL)	1 mL = 0.001 L = 10^{-3} L	1 mL = 0.034 fluid ounces
	microliter (μL)	1 μL = 0.001 mL = 10^{-6} L	
Mass	gram (g)	base unit	1 g = 0.035 ounces
	kilogram (kg)	1 kg = 1000 g	1 kg = 2.20 pounds
	metric ton (mt)	1 mt = 1000 kg	1 mt = 2,200 pounds = 1.10 ton
	milligram (mg)	1 mg = 0.001 g = 10^{-3} g	
	microgram (μg)	1 μg = 0.001 mg = 10^{-6} g	
Temperature	degree Celsius (°C)	base unit	°C = (°F – 32)/1.8
			0°C = 32°F (water freezes)
			100°C = 212°F (water boils)
			20°C = 68°F ("room temperature")
			37°C = 98.6°F (human internal body temperature)
	Kelvin (K)*	°C – 273	0 K = –460°F
Energy	joule (J)		1 J ≈ 0.24 calorie = 0.00024 kilocalorie†

*0 K (–273°C) is "absolute zero," a temperature at which molecular oscillations approach 0—that is, the point at which motion all but stops.

†A *calorie* is the amount of heat necessary to raise the temperature of 1 gram of water 1°C. The *kilocalorie*, or nutritionist's calorie, is what we commonly think of as a calorie in terms of food.

Glossary

- A -

abdomen (ab′ duh mun) [L. *abdomin*: belly] In arthropods, the posterior segments of the body; in mammals, the part of the body containing the intestines and most other internal organs, posterior to the thorax.

abiotic (a′ bye ah tick) [Gk. *a*: not + *bios*: life] Nonliving. (Contrast with biotic.)

abscisic acid (ab sighs′ ik) A plant growth substance having growth-inhibiting action. Causes stomata to close.

abscission (ab sizh′ un) [L. *abscissio*: break off] The process by which leaves, petals, and fruits separate from a plant.

absorption (1) Of light: complete retention, without reflection or transmission. (2) Of liquids: soaking up (taking in through pores or cracks).

absorption spectrum A graph of light absorption versus wavelength of light; shows how much light is absorbed at each wavelength.

absorptive period When there is food in the gut and nutrients are being absorbed.

abyssal zone (uh biss′ ul) [Gk. *abyssos*: bottomless] The deep ocean, below the point that light can penetrate.

accessory pigments Pigments that absorb light and transfer energy to chlorophylls for photosynthesis.

acetylcholine A neurotransmitter substance that carries information across vertebrate neuromuscular junctions and some other synapses.

acetylcholinesterase An enzyme that breaks down acetylcholine.

acetyl coenzyme A (acetyl CoA) Compound that reacts with oxaloacetate to produce citrate at the beginning of the citric acid cycle; a key metabolic intermediate in the formation of many compounds.

acid [L. *acidus*: sharp, sour] A substance that can release a proton in solution. (Contrast with base.)

acid precipitation Precipitation that has a lower pH than normal as a result of acid-forming precursors introduced into the atmosphere by human activities.

acidic Having a pH of less than 7.0 (a hydrogen ion concentration greater than 10^{-7} molar).

acoelomate Lacking a coelom.

Acquired Immune Deficiency Syndrome *See* AIDS.

acrosome (a′ krow soam) [Gk. *akros*: highest + *soma*: body] The structure at the forward tip of an animal sperm which is the first to fuse with the egg membrane and enter the egg cell.

ACTH (adrenocorticotropin) A pituitary hormone that stimulates the adrenal cortex.

actin [Gk. *aktis*: ray] One of the two major proteins of muscle; it makes up the thin filaments. Forms the microfilaments found in most eukaryotic cells.

action potential An impulse in a neuron taking the form of a wave of depolarization or hyperpolarization imposed on a polarized cell surface.

action spectrum A graph of a biological process versus light wavelength; shows which wavelengths are involved in the process.

activating enzymes Enzymes that catalyze the addition of amino acids to their appropriate tRNAs. Also called aminoacyl-tRNA synthetases.

activation energy (E_a) The energy barrier that blocks the tendency for a set of chemical substances to react.

active site The region on the surface of an enzyme where the substrate binds, and where catalysis occurs.

active transport The energy-dependent transport of a substance across a biological membrane against a concentration gradient—that is, from a region of low concentration (of that substance) to one of high concentration. (See also primary active transport, secondary active transport; contrast with facilitated diffusion, passive transport.)

adaptation (a dap tay′ shun) (1) In evolutionary biology, a particular structure, physiological process, or behavior that makes an organism better able to survive and reproduce. Also, the evolutionary process that leads to the development or persistence of such a trait. (2) In sensory neurophysiology, a sensory cell's loss of sensitivity as a result of repeated stimulation.

adaptive radiation The proliferation of members of a single clade into a variety of different adaptive forms.

adenine (A) (a′ den een) A nitrogen-containing base found in nucleic acids, ATP, NAD, and other compounds.

adenosine triphosphate *See* ATP.

adenylate cyclase Enzyme catalyzing the formation of cyclic AMP (cAMP) from ATP.

adrenal (a dree′ nal) [L. *ad*: toward + *renes*: kidneys] An endocrine gland located near the kidneys of vertebrates, consisting of two glandular parts, the cortex and medulla.

adrenaline *See* epinephrine.

adrenocorticotropin *See* ACTH.

adsorption Binding of a gas or a solute to the surface of a solid.

aerobic (air oh′ bic) [Gk. *aer*: air + *bios*: life] In the presence of oxygen; requiring oxygen.

afferent (af′ ur unt) [L. *ad*: toward + *ferre*: to carry] Carrying to, as in a neuron that carries impulses to the central nervous system, or a blood vessel that carries blood to a structure. (Contrast with efferent.)

AIDS (acquired immune deficiency syndrome) Condition caused by a virus (HIV) in which the body's helper T lymphocytes are reduced, leaving the victim subject to opportunistic diseases.

alcoholic fermentation Breakdown of glucose in cells under anaerobic conditions to produce alcohol.

aldehyde (al′ duh hide) A compound with a —CHO functional group. Many sugars are aldehydes. (Contrast with ketone.)

aldosterone (al dohs′ ter own) A steroid hormone produced in the adrenal cortex of mammals. Promotes secretion of potassium and reabsorption of sodium in the kidney.

allantois (al lan′ to is) A sac-like extraembryonic membrane that contains nitrogen waste from embryo.

allele (a leel′) [Gk. *allos*: other] The alternate forms of a genetic character found at a given locus on a chromosome.

allele frequency The relative proportion of a particular allele in a specific population.

allergy [Ger. *allergie*: altered reaction] An overreaction to amounts of an antigen that do not affect most people; often involves IgE antibodies.

allometric growth A pattern of growth in which some parts of the body of an organism grow faster than others, resulting in a change in body proportions as the organism grows.

allopatric speciation (al′ lo pat′ rick) [Gk. *allos*: other + *patria*: homeland] The formation of two species from one when reproductive isolation occurs because of the interposition of (or crossing of) a physical geographic barrier such as a river. Also called geographic speciation. (Contrast with parapatric speciation, sympatric speciation.)

allopolyploidy The possession of more than two entire chromosomes sets that are derived from a single species.

allostery (al′ lo steer y) [Gk. *allos*: other + *stereos*: structure] Regulation of the activity of a protein by the binding of an effector molecule at a site other than the active site.

alpha (α) helix A prevalent type of secondary protein structure; a right-handed spiral.

alternation of generations The succession of multicellular haploid and diploid phases in some sexually reproducing organisms, notably plants.

alternative splicing A process for generating different mature mRNAs from a single gene by splicing together different sets of exons during RNA processing.

altruism Behavior that harms the individual who performs it but benefits other individuals.

alveolus (al ve′ o lus) (plural: alveoli) [L. *alveus*: cavity] A small, baglike cavity, especially the blind sacs of the lung.

amensalism (a men′ sul ism) Interaction in which one animal is harmed and the other is unaffected. (Contrast with commensalism, mutualism.)

amine An organic compound with an amino group. (Compare with amino acid.)

amino acid Organic compounds containing both NH_2 and COOH groups. Proteins are polymers of amino acids.

amino acid replacement A change in a protein sequence in which one amino acid is replaced by another.

ammonotelic (am moan′ o teel′ ic) [Gk. *telos*: end] Describes an organism in which the final product of breakdown of nitrogen-containing compounds (primarily proteins) is ammonia. (Contrast with ureotelic, uricotelic.)

amnion (am′ nee on) The fluid-filled sac in which the embryos of reptiles, birds, and mammals develop.

amniote egg A shelled egg surrounding four extraembryonic membranes and embryo-nourishing yolk. This adaptation allowed animals to colonize the terrestrial environment.

amphipathic (am′ fi path′ ic) [Gk. *amphi*: both + *pathos*: emotion] Of a molecule, having both hydrophilic and hydrophobic regions.

amylase (am′ ill ase) Any of a group of enzymes that digest starch.

anabolic reaction A single reaction that participates in anabolism.

anabolism (an ab′ uh liz′ em) [Gk. *ana*: upward + *ballein*: to throw] Synthetic reactions of metabolism, in which complex molecules are formed from simpler ones. (Contrast with catabolism.)

anaerobic (an ur row′ bic) [Gk. *an*: not + *aer*: air + *bios*: life] Occurring without the use of molecular oxygen, O_2.

anagenesis Evolutionary change in a single lineage over time.

analogy (a nal′ o jee) [Gk. *analogia*: resembling] A resemblance in function, and often appearance as well, between two structures that is due to convergent evolution rather than to common ancestry. (Contrast with homology.)

anaphase (an′ a phase) [Gk. *ana*: upward progress] The stage in nuclear division at which the first separation of sister chromatids (or, in the first meiotic division, of paired homologs) occurs.

anaphylactic shock A precipitous drop in blood pressure caused by loss of fluid from capillaries because of an increase in their permeability stimulated by an allergic reaction.

ancestral trait The trait originally present in the ancestor of a given group; may be retained or changed in the descendants of that ancestor.

androgens (an′ dro jens) The male sex steroids.

aneuploidy (an′ you ploy dee) A condition in which one or more chromosomes or pieces of chromosomes are either lacking or present in excess.

angiotensin (an′ jee oh ten′ sin) A peptide hormone that raises blood pressure by causing peripheral vessels to constrict. Also maintains glomerular filtration by constricting efferent vessels and stimulates thirst and the release of aldosterone.

animal hemisphere The metabolically active upper portion of some animal eggs, zygotes, and embryos; does not contain the dense nutrient yolk. (Contrast with vegetal hemisphere.)

anion (an′ eye on) [Gk. *ana*: upward progress] A negatively charged ion. (Contrast with cation.)

anisogamy (an eye sog′ a mee) [Gk. *aniso*: unequal + *gamos*: marriage] The existence of two dissimilar gametes (egg and sperm).

annual Referring to a plant whose life cycle is completed in one growing season. (Contrast with biennial, perennial.)

antenna system In photosynthesis, a group of different molecules that cooperate to absorb light energy and transfer it to a reaction center.

anterior pituitary The portion of the vertebrate pituitary gland that derives from gut epithelium and produces tropic hormones.

anther (an′ thur) [Gk. *anthos*: flower] A pollen-bearing portion of the stamen of a flower.

antheridium (an′ thur id′ ee um) [Gk. *antheros*: blooming] The multicellular structure that produces the sperm in nonvascular plants and ferns.

antibody One of the millions of proteins produced by the immune system that specifically binds to a foreign substance and initiates its removal from the body.

anticodon The three nucleotides in transfer RNA that pair with a complementary triplet (a codon) in messenger RNA.

antidiuretic hormone (ADH) A hormone that promotes water reabsorption by the kidney. ADH is produced by neurons in the hypothalamus and released from nerve terminals in the posterior pituitary. Also called vasopressin.

antigen (an′ ti jun) Any substance that stimulates the production of an antibody or antibodies in the body of a vertebrate.

antigenic determinant A specific region of an antigen, which is recognized by and binds to a specific antibody.

antiparallel Pertaining to molecular orientation in which a molecule or parts of a molecule have opposing directions.

antipodal cell At one end of the megagametophyte, one of the three cells which eventually degenerate.

antiport A membrane transport process that carries one substance in one direction and another in the opposite direction. (Contrast with symport.)

antisense nucleic acid A single-stranded RNA or DNA complementary to and thus targeted against the mRNA transcribed from a harmful gene such as an oncogene.

anus (a′ nus) Opening through which digestive wastes are expelled, located at the posterior end of the gut.

aorta (a or′ tah) [Gk. *aorte:* aorta] The main trunk of the arteries leading to the systemic (as opposed to the pulmonary) circulation.

aortic body A nodule of modified muscle tissue on the aorta that is sensitive to the oxygen supply provided by arterial blood.

apex (a′ pecks) The tip or highest point of a structure, as the apex of a growing stem or root.

apical (a′ pi kul) Pertaining to the apex, or tip, usually in reference to plants.

apical dominance Inhibition by the apical bud of the growth of axillary buds.

apical meristem The meristem at the tip of a shoot or root; responsible for the plant's primary growth.

apomixis (ap oh mix′ is) [Gk. *apo*: away from + *mixis*: sexual intercourse] The asexual production of seeds.

apoplast (ap′ oh plast) in plants, the continuous meshwork of cell walls and extracellular spaces through which material can pass without crossing a plasma membrane. (Contrast with symplast.)

apoptosis (ap uh toh′ sis) A series of genetically programmed events leading to cell death.

aquaporin A transport protein in plant and animal cells through which water passes in osmosis.

aquatic (a kwa′ tic) [L. *aqua*: water] Living in water. (Compare with marine, terrestrial.)

aqueous (a′ kwee us) Pertaining to water or a watery solution.

archegonium (ar′ ke go′ nee um) [Gk. *archegonos*: first, foremost] The multicellular structure that produces eggs in nonvascular plants, ferns, and gymnosperms.

archenteron (ark en′ ter on) [Gk. *archos*: first + *enteron*: bowel] The earliest primordial animal digestive tract.

area phylogenies Phylogenies in which the names of the taxa are replaced with the names of the places where those taxa live or lived.

arteriosclerosis *See* atherosclerosis.

artery A muscular blood vessel carrying oxygenated blood away from the heart to other parts of the body. (Contrast with vein.)

artficial selection The selection by plant and animal breeders of individuals with certain desirable traits.

ascus (ass′ cuss) [Gk. *askos*: bladder] In ascomycete fungi (sac fungi), the club-shaped sporangium within which spores (ascospores) are produced by meiosis.

asexual Without sex.

assortative mating A breeding system in which mates are selected on the basis of a particular trait or group of traits.

atherosclerosis (ath′ er oh sklair oh′ sis) [Gk. *athero*: gruel, porridge + *skleros*: hard] A disease of the lining of the arteries characterized by fatty, cholesterol-rich deposits in the walls of the arteries. When fibroblasts infiltrate these deposits and calcium precipitates in them, the disease become arteriosclerosis, or "hardening of the arteries."

atmosphere The gaseous mass surrounding our planet. Also a unit of pressure, equal to the normal pressure of air at sea level.

atom [Gk. *atomos*: indivisible] The smallest unit of a chemical element. Consists of a nucleus and one or more electrons.

atomic mass The average mass of an atom of an element; the average depends on the relative amounts of different isotopes of the element on Earth. Also called atomic weight.

atomic number The number of protons in the nucleus of an atom; also equals the number of electrons around the neutral atom. Determines the chemical properties of the atom.

ATP (adenosine triphosphate) An energy-storage compound containing adenine, ribose, and three phosphate groups. When it is formed from ADP, useful energy is stored; when it is broken down (to ADP or AMP), energy is released to drive endergonic reactions.

ATP synthase An integral membrane protein that couples the transport of proteins with the formation of ATP.

atrioventricular node A modified node of cardiac muscle that organizes the action potentials that control contraction of the ventricles.

atrium (a' tree um) [L. *atrium*: central hall] An internal chamber. In the hearts of vertebrates, the thin-walled chamber(s) entered by blood on its way to the ventricle(s). Also, the outer ear.

autocrine Referring to a cell signaling mechanism in which the signal binds to and affects the cell that makes it. An autocrine hormone, for example, is one that influences the cell that releases it. (Compare with endocrine gland;, paracrine)

autoimmune disease A disorder in which the immune system attacks the animal's own antigens.

autonomic nervous system The system that controls such involuntary functions as those of guts and glands. (Compare with central nervous system.)

autopolyploidy The possession of more than two chromosome sets that are derived from more than one species.

autosome Any chromosome (in a eukaryote) other than a sex chromosome.

autotroph (au' tow trow' fik) [Gk. *autos:* self + *trophe*: food] An organism that is capable of living exclusively on inorganic materials, water, and some energy source such as sunlight or chemically reduced matter. (Contrast with heterotroph.)

auxin (awk' sin) [Gk. *auxein*: to grow] In plants, a substance (the most common being indoleacetic acid) that regulates growth and various aspects of development.

auxotroph (awks' o trofe) [Gk. *auxein*: to grow + *trophe*: food] A mutant form of an organism that requires a nutrient or nutrients not usually required by the wild type. (Contrast with prototroph.)

Avogadro's number The number of atoms or molecules in a mole (weighed out in grams) of a substance, calculated to be 6.022×10^{23}.

***Avr* genes (avirulence genes)** Genes in a pathogen that may trigger defenses in plants. *See* gene-for-gene resistance.

axillary bud A bud occurring in the upper angle (axil) between a leaf and stem.

axon [Gk. *axon*: axle] The part of a neuron that conducts action potentials away from the cell body.

axon hillock The junction between an axon and its cell body, where action potentials are generated.

axon terminals The endings of an axon; they form synapses and release neurotransmitter.

axoneme (ax' oh neem) The complex of microtubules and their crossbridges that forms the motile apparatus of a cilium.

- B -

bacillus (bah sil' us) [L: little rod] Any of various rod-shaped bacteria.

bacterial artificial chromosome (BAC) A DNA cloning vector used in bacteria that can carry up to 150,000 base pairs of foreign DNA.

bacteriophage (bak teer' ee o fayj) [Gk. *bakterion*: little rod + *phagein*: to eat] One of a group of viruses that infect bacteria and ultimately cause their disintegration.

bacteroids Nitrogen-fixing organelles that develop from endosymbiotic bacteria.

balanced polymorphism [Gk. *polymorphos*: many forms] The maintenance of more than one form, or the maintenance at a given locus of more than one allele, at frequencies of greater than 1 percent in a population. Often results when heterozygotes are more fit than either homozygote.

bark All tissues outside the vascular cambium of a plant.

baroreceptor [Gk. *baros*: weight] A pressure-sensing cell or organ.

Barr body In female mammals, an inactivated X chromosome.

basal Pertaining to one end—the base—of an axis.

basal body Centriole found at the base of a eukaryotic flagellum or cilium.

basal metabolic rate (BMR) The minimum rate of energy turnover in an awake (but resting) bird or mammal that is not expending energy for thermoregulation.

base (1) A substance tha can accept a hydrogen ion in solution. (Contrast with acid.) (2) In nucleic acids, the purine or pyrimidine that is attached to each sugar in the backbone.

base pairing *See* complementary base pairing.

basic Having a pH greater than 7.0 (i.e., having a hydrogen ion concentration lower than 10^{-7} molar).

basidium (bass id' ee yum) In basidiomycete fungi, the characteristic sporangium in which four spores are formed by meiosis and then borne externally before being shed.

basophils One type of phagocytic white blood cell that releases histamine and may promote T cell development.

Batesian mimicry The convergence in appearance of an edible species (mimic) with an unpalatable species (model).

B cell A type of lymphocyte involved in the humoral immune response of vertebrates. Upon recognizing an antigenic determinant, a B cell develops into a plasma cell, which secretes an antibody. (Contrast with T cell.)

benefit An improvement in survival and reproductive success resulting from performing a behavior or having a trait. (Contrast with cost.)

benign (be nine') Referring to a tumor that grows to a certain size and then stops, uaually with a fibrous capsule surrounding the mass of cells. Benign tumors do not spread (metastasize) to other organs. (Contrast with malignant.)

benthic zone [Gk. *benthos*: bottom] The bottom of the ocean.

beta (β) pleated sheet Type of protein secondary structure; results from hydrogen bonding between polypeptide regions running antiparallel to each other.

biennial Referring to a plant whose life cycle includes vegetative growth in the first year and flowering and senescence in the second year. (Contrast with annual, perennial.)

bilateral symmetry The condition in which only the right and left sides of an organism, divided exactly down the back, are mirror images of each other. (Contrast with biradial symmetry.)

bilayer In membranes, a structure that is two lipid layers in thickness.

bile A secretion of the liver delivered to the small intestine via the common bile duct. In the intestine, bile emulsifies fats.

binary fission Reproduction by cell division of a single-celled organism.

binding domain The region of a receptor molecule where its ligand attaches.

binocular cells Neurons in the visual cortex that respond to input from both retinas; involved in depth perception.

binomial (bye nome' ee al) Consisting of two; for example, the binomial nomenclature of biology in which each species has two names (the genus name followed by the species name).

biodiversity crisis The current high rate of loss of species, caused primarily by human activities.

biofilm A community of microorganisms embedded in a polysaccharide matrix, forming a highly resistant coating on almost any moist surface.

biogeochemical cycle Movement of elements through living organisms and the physical environment.

biogeographic region One of several defined, continental-scale regions of Earth, each of which has a biota distinct from that of the others. (Contrast with biome.)

bioinformatics The use of computers and/or mathematics to analyze complex biological information, such as DNA sequences.

bioluminescence The production of light by biochemical processes in an organism.

biomass The total weight of all the living organisms, or some designated group of living organisms, in a given area.

biome (bye' ome) A major division of the ecological communities of Earth, characterized primarily by distinctive vegetation. A given biogeographic region contains many different biomes.

biosphere (bye' oh sphere) All regions of Earth (terrestrial and aquatic) and Earth's atmosphere in which organisms can live.

biota (bye oh' tah) All of the organisms—animals, plants, fungi, and microorganisms—found in a given area. (Contrast with flora, fauna.)

biotic (bye ah' tick) [Gk. *bios*: life] Alive. (Contrast with abiotic.)

biradial symmetry Radial symmetry modified so that only two planes can divide the animal into similar halves.

blastocoel (blass' toe seal) [Gk. *blastos*: sprout + *koilos*: hollow] The central, hollow cavity of a blastula.

blastocyst (blass' toe cist) An early embryo formed by the first divisions of the fertilized egg (zygote). In mammals, a hollow ball of cells.

blastodisc (blass' toe disk) A disk of cells forming on the surface of a large yolk mass, comparable to a blastula, but occurring in animals such as birds and reptiles, in which the massive yolk restricts cleavage to one side of the egg only.

blastomere A cell produced by the division of a fertilized egg.

blastopore The opening from the archenteron to the exterior of a gastrula.

blastula (blass' chu luh) An early stage in animal embryology; in many species, a hollow sphere of cells surrounding a central cavity, the blastocoel. (Contrast with blastodisc.)

blood–brain barrier A property of the blood vessels of the brain that prevents most chemicals from diffusing from the blood into the brain.

blood plasma (plaz' muh) [Gk. *plassein*: to mold] The liquid portion of blood, in which blood cells and other particulates are suspended.

blue-light receptors Pigments in plants that absorbs blue light (400–500 nm). These pigments mediate many plant responses including phototropism, stomatal movements, and expression of some genes.

body cavity Membrane-lined, fluid-filled compartment that lies between the cell layers of many animals.

Bohr effect The fact that low pH decreases the affinity of hemoglobin for oxygen.

bond *See* chemical bond.

bottleneck A stressful period during which only a few individuals of a once large population survive.

Bowman's capsule An elaboration of kidney tubule cells that surrounds a knot of capillaries (the glomerulus). Blood is filtered across the walls of these capillaries and the filtrate is collected into Bowman's capsule.

brain stem The portion of the vertebrate brain between the spinal cord and the forebrain.

brassinosteroids Plant steroid hormones that mediate light effects promoting the elongation of stems and pollen tubes.

bronchioles The smallest airways in a vertebrate lung, branching off the bronchi.

bronchus (plural: bronchi) The major airway(s) branching off the trachea into the vertebrate lung.

brown fat Fat tissue in mammals that is specialized to produce heat. It has many mitochondria and capillaries, and a protein that uncouples oxidative phosphorylation.

browser An animal that feeds on the tissues of woody plants.

budding Asexual reproduction in which a more or less complete new organism simply grows from the body of the parent organism and eventually detaches itself.

buffer A substance that can transiently accept or release hydrogen ions and thereby resist changes in pH.

bundle of His Fibers of modified cardiac muscle that conduct action potentials from the atria to the ventricular muscle mass.

bundle sheath cell Part of a tissue that surrounds the veins of plants; contains chloroplasts in C_4 plants.

- C -

C_3 photosynthesis Form of photosynthesis in which 3-phosphoglycerate is the first stable product, and ribulose bisphosphate is the CO_2 receptor.

C_4 photosynthesis Form of photosynthesis in which oxaloacetate is the first stable product, and phosphoenolpyruvate is the CO_2 acceptor. C_4 plants also perform the reactions of C_3 photosynthesis.

calcitonin A hormone produced by the thyroid gland; it lowers blood calcium and promotes bone formation. (Compare with parathyroid hormone.)

calmodulin (cal mod' joo lin) A calcium-binding protein found in all animal and plant cells; mediates many calcium-regulated processes.

calorie [L. *calor*: heat] The amount of heat required to raise the temperature of one gram of water by one degree Celsius (1°C) from 14.5°C to 15.5°C. Calorie spelled with a capital C refers to the kilocalorie (1 kcal = 1,000 cal).

Calvin cycle The stage of photosynthesis in which CO_2 reacts with RuBP to form 3PG, 3PG is reduced to a sugar, and RuBP is regenerated, while other products are released to the rest of the plant. Also known as the Calvin–Benson cycle.

calyx (kay' licks) [Gk. *kalyx*: cup] All of the sepals of a flower, collectively.

CAM *See* crassulacean acid metabolism.

cambium (kam' bee um) [L. *cambiare*: to exchange] A meristem that gives rise to radial rows of cells in stem and root, increasing them in girth; commonly applied to the vascular cambium which produces wood and phloem, and the cork cambium, which produces bark.

Cambrian explosion The rapid diversification of life that took place during the Cambrian period.

cAMP (cyclic AMP) A compound formed from ATP that mediates the effects of numerous animal hormones.

canopy The leaf-bearing part of a tree. Collectively, the aggregate of the leaves and branches of the larger woody plants of an ecological community.

capillaries [L. *capillaris*: hair] Very small tubes, especially the smallest blood-carrying vessels of animals between the termination of the arteries and the beginnings of the veins. Capillaries are the site of exchange of materials between the blood and the interstitial fluid.

capsid The outer shell of a virus that encloses its nucleic acid.

carbohydrates Organic compounds containing carbon, hydrogen, and oxygen in the ratio 1:2:1 (i.e., with the general formula $C_nH_{2n}O_n$). Common examples are sugars, starch, and cellulose.

carboxylase An enzyme that catalyzes the addition of carboxyl groups to a substrate.

carboxylic acid (kar box sill' ik) An organic acid containing the carboxyl group, —COOH, which dissociates to the carboxylate ion, $—COO^-$.

carcinogen (car sin' oh jen) A substance that causes cancer.

cardiac (kar' dee ak) [Gk. *kardia*: heart] Pertaining to the heart and its functions.

cardiac muscle One of the three types of muscle tissue, it makes up the vertebrate heart. Characterized by branching cells with single nuclei and striated (striped) appearance. (Contrast with smooth muscle, striated muscle.)

carnivore [L. *carn*: flesh + *vovare*: to devour] An organism that eats animal tissues. (Contrast with detritivore, herbivore, omnivore.)

carotenoid (ka rah' tuh noid) A yellow, orange, or red lipid pigment commonly found as an accessory pigment in photosynthesis; also found in fungi.

carpel (kar' pel) [Gk. *karpos*: fruit] The organ of the flower that contains one or more ovules.

carrier (1) In facilitated diffusion, a membrane protein that binds a specific molecule and transports it through the membrane. (2) In respiratory and photosynthetic electron transport, a participating substance such as NAD that exists in both oxidized and reduced forms. (3) In genetics, a person heterozygous for a recessive trait.

carrier protein *See* Carrier (1).

cartilage In vertebrates, a tough connective tissue found in joints, the outer ear, and elsewhere. Forms the entire skeleton in some animal groups.

Casparian strip A band of cell wall containing suberin and lignin, found in the endodermis. Restricts the movement of water across the endodermis.

caspase A member of a group of proteases that catalyze cleavage of target proteins and are active in apoptosis.

catabolic reaction A single reaction that participates in catabolism.

catabolism [Gk. *kata*: to break down + *ballein*: to throw] Degradational reactions of metabolism, in which complex molecules are broken down. (Contrast with anabolism.)

catabolite repression In the presence of abundant glucose, the diminished synthesis of catabolic enzymes for other energy sources.

catalyst (cat' a list) [Gk. *kata*: to break down] A chemical substance that accelerates a reaction without itself being consumed in the overall course of the reaction. Catalysts lower the activation energy of a reaction. Enzymes are biological catalysts.

catalytic subunit The polypeptide in an enzyme protein with quaternary structure that contains the active site for the enzyme. (Contrast regulatory subunit.)

cation (cat' eye on) An ion with one or more positive charges. (Contrast with anion.)

caudal [L. *cauda*: tail] Pertaining to the tail, or to the posterior part of the body.

cDNA *See* complementary DNA.

cecum (see′ cum) [L. *caecus*: blind] A blind branch off the large intestine. In many nonruminant mammals, the cecum contains a colony of microorganisms that contribute to the digestion of food.

cell adhesion molecules Molecules on animal cell surfaces that affect the selective association of cells during development of the embryo.

cell cycle The stages through which a cell passes between one division and the next. Includes all stages of interphase and mitosis.

cell division The reproduction of a cell to produce two new cells. In eukaryotes, this process involves nuclear division (mitosis) and cytoplasmic division (cytokinesis).

cell junctions Specialized structures associated with the plasma membranes of epithelial cells. Some contribute to cell adhesion, others to intercellular communication.

cell plate A structure that forms at the equator of the spindle in dividing plant cells.

cell recognition Binding of cells to one another mediated by membrane proteins or carbohydrates.

cell wall A relatively rigid structure that encloses cells of plants, fungi, many protists, and most prokaryotes. Gives these cells their shape and limits their expansion in hypotonic media.

cellular immune response Action of the immune system based on the activities of T cells. Directed against parasites, fungi, intracellular viruses, and foreign tissues (grafts). (Contrast with humoral immune system.)

cellular respiration *See* respiration.

cellulose (sell′ you lowss) A straight-chain polymer of glucose molecules, used by plants as a structural supporting material.

centimorgan In genetic mapping, a recombinant frequency of 0.01; a map unit.

central dogma The statement that information flows from DNA to RNA to polypeptide (in retroviruses, there is also information flow from RNA to cDNA).

central nervous system That part of the nervous system which is condensed and centrally located, e.g., the brain and spinal cord of vertebrates; the chain of cerebral, thoracic and abdominal ganglia of arthropods. (Compare with autonomic nervous system.)

centrifuge [L. *centrum*: center + *fugere*: to flee] A laboratory device in which a sample is spun around a central axis at high speed. Used to separate suspended materials of different densities.

centriole (sen′ tree ole) A paired organelle that helps organize the microtubules in animal and protist cells during nuclear division.

centromere (sen′ tro meer) [Gk. *centron*: center + *meros*: part] The region where sister chromatids join.

centrosome (sen′ tro soam) The major microtubule organizing center of an animal cell.

cephalization (sef ah luh zay′ shun) [Gk. *kephale*: head] The evolutionary trend toward increasing concentration of brain and sensory organs at the anterior end of the animal.

cerebellum (sair uh bell′ um) [L.: diminutive of *cerebrum*, brain] The brain region that controls muscular coordination; located at the anterior end of the hindbrain.

cerebral cortex The thin layer of gray matter (neuronal cell bodies) that overlays the cerebrum.

cerebrum (su ree′ brum) [L. *cerebrum*: brain] The dorsal anterior portion of the forebrain, making up the largest part of the brain of mammals. In mammals, the chief coordination center of the nervous system; consists of two cerebral hemispheres.

cervix (sir′ vix) [L. *cervix*: neck] The opening of the uterus into the vagina.

cGMP (cyclic guanosine monophosphate) An intracellular messenger that is part of signal transmission pathways involving G proteins. (*See* G protein.)

channel protein A membrane protein that forms an aqueous passageway though which specific solutes may pass.

chemical bond An attractive force stably linking two atoms.

chemical equilibrium A state reached in a reversible chemical reaction when the forward and reverse reactions balance each other and there is no net change.

chemical evolution The theory that life originated through the chemical transformation of inanimate substances.

chemical reaction The change in the composition or distribution of atoms of a substance with consequent alterations in properties.

chemiosmosis The formation of ATP in mitochondria and chloroplasts, resulting from a pumping of protons across a membrane (against a gradient of electrical charge and of pH), followed by the return of the protons through a protein channel with ATPase activity.

chemoautotroph *See* chemolithotroph.

chemoheterotroph An organism that must obtain both carbon and energy from organic substances. (Contrast with chemolithotroph, photoautotroph, photoheterotroph.)

chemolithotroph [Gk. *lithos*: stone, rock] An organism that uses carbon dioxide as a carbon source and obtains energy by oxidizing inorganic substances from its environment. (Contrast with chemoheterotroph, photoautotroph, photoheterotroph.)

chemoreceptor A sensory receptor cell or tissue that senses specific substances in its environment.

chemosynthesis Synthesis of food substances, using the oxidation of reduced materials from the environment as a source of energy.

chiasma (kie az′ muh) (plural: chiasmata) [Gk. *chiasmata*: cross] An X-shaped connection between paired homologous chromosomes in prophase I of meiosis. A chiasma is the visible manifestation of crossing over between homologous chromosomes.

chitin (kye′ tin) [Gk. *kiton*: tunic] The characteristic tough but flexible organic component of the exoskeleton of arthropods, consisting of a complex, nitrogen-containing polysaccharide. Also found in cell walls of fungi.

chlorophyll (klor′ o fill) [Gk. *kloros*: green + *phyllon*: leaf] Any of a few green pigments associated with chloroplasts or with certain bacterial membranes; responsible for trapping light energy for photosynthesis.

chloroplast [Gk. *kloros*: green + *plast*: a particle] An organelle bounded by a double membrane containing the enzymes and pigments that perform photosynthesis. Chloroplasts occur only in eukaryotes.

choanocyte (ko′ an uh site) The collared, flagellated feeding cells of sponges.

cholecystokinin (ko′ luh sis tuh kai′ nin) A hormone produced and released by the lining of the duodenum when it is stimulated by undigested fats and proteins. It stimulates the gallbladder to release bile and slows stomach activity.

chorion (kor′ ee on) [Gk. *khorion*: afterbirth] The outermost of the membranes protecting mammal, bird, and reptile embryos; in mammals it forms part of the placenta.

chromatid (kro′ ma tid) Each of a pair of new sister chromosomes from the time at which the molecular duplication occurs until the time at which the centromeres separate at the anaphase of nuclear division.

chromatin The nucleic acid–protein complex found in eukaryotic chromosomes.

chromatin remodeling Changes in chromatin structure to allow transcription, translation, and chromosome condensation.

chromatophore (krow mat′ o for) [Gk. *kroma*: color + *phoreus*: carrier] A pigment-bearing cell that expands or contracts to change the color of the organism.

chromosomal mutation Loss of or changes in position/direction of a DNA segment on a chromosome.

chromosome (krome′ o sowm) [Gk. *kroma*: color + *soma*: body] In bacteria and viruses, the DNA molecule that contains most or all of the genetic information of the cell or virus. In eukaryotes, a structure composed of DNA and proteins that bears part of the genetic information of the cell.

chylomicron (ky low my′ cron) Particles of lipid coated with protein, produced in the gut from dietary fats and secreted into the extracellular fluids.

chyme (kime) [Gk. kymus, juice] Created in the stomach; a mixture of ingested food with the digestive juices secreted by the salivary glands and the stomach lining.

cilium (sil′ ee um) (plural: cilia) [L.: eyelash] Hairlike organelle used for locomotion by many unicellular organisms and for moving water and mucus by many multicellular organisms. Generally shorter than a flagellum.

circadian rhythm (sir kade′ ee an) [L. *circa*: approximately + *dies*: day] A rhythm in behavior, growth, or some other activity that recurs about every 24 hours under constant conditions.

circannual rhythm [L. *circa*: approximately + *annus*: year) A rhythm of behavior, growth, or some other activity that recurs on a yearly basis.

citric acid cycle A set of chemical reactions in cellular respiration, in which acetyl CoA is oxidized to carbon dioxide, and hydrogen atoms are stored as NADH and $FADH_2$. Also called the Krebs cycle.

clade [Gk. *klados*: branch] A monophyletic group made up of an ancestor and all of its descendants.

class I MHC molecules These cell surface proteins participate in the cellular immune response directed against virus-infected cells.

class II MHC molecules These cell surface proteins participate in the cell-cell interactions (of helper T cells, macrophages, and B cells) of the humoral immune response.

class switching The process whereby a plasma cell changes the class of immunoglobulin that it synthesizes by changing the DNA region coding for the C segment.

clathrin A fibrous protein on the inner surfaces of animal cell membranes that strengthens coated vesicles and thus participates in receptor-mediated endocytosis.

cleavage First divisions of the fertilized egg of an animal.

climate The average of the atmospheric conditions (temperature, precipitation, wind direction and velocity) found in a region over time.

cline A gradual change in the traits of a species over a geographical gradient.

cloaca (klo ay' kuh) [L. *cloaca*: sewer] In some invertebrates, the posterior part of the gut; in many vertebrates, a cavity receiving material from the digestive, reproductive, and excretory systems.

clonal anergy Prevention of the synthesis of antibodies against the body's own antigens. When a T cell binds to a self-antigen, it does not receive signals from an antigen-presenting cell; thus the T cell dies (becomes anergic) rather than yielding a clone of active cells.

clonal deletion The inactivation or destruction of lymphocyte clones that would produce immune reactions against the animal's own body.

clonal selection The mechanism by which exposure to antigen results in the activation of selected T- or B-cell clones, resulting in an immune response.

clone [Gk. *klon*: twig, shoot] Genetically identical cells or organisms produced from a common ancestor by asexual means.

cnidocytes (nye' duh sites) The feeding cells of cnidarians, within which nematocysts are housed.

coacervate (ko as' er vate) [L. *coacervare*: to heap up] An aggregate of colloidal particles in suspension.

coated vesicle Cytoplasmic vesicle containing distinctive proteins, including clathrin.

coccus (kock' us) [Gk. *kokkos*: berry, pit] Any of various spherical or spheroidal bacteria.

cochlea (kock' lee uh) [Gk. *kokhlos* snail] A spiral tube in the inner ear of vertebrates; it contains the sensory cells involved in hearing.

codominance A condition in which two alleles at a locus produce different phenotypic effects and both effects appear in heterozygotes.

codon Three nucleotides in messenger RNA that direct the placement of a particular amino acid into a polypeptide chain. (Contrast with anticodon.)

coelom (see' lum) [Gk. *koiloma*: cavity] The body cavity of certain animals; the coelom is lined with cells of mesodermal origin.

coelomate Having a coelom.

coenocyte (seen' a sight) [Gk. *koinos*: common + *kytos*: container] A "cell" enclosed by a single plasma membrane but containing many nuclei.

coenzyme A nonprotein organic molecule that plays a role in catalysis by an enzyme.

cofactor An inorganic ion that is weakly bound to an enzyme and required for its activity.

cohesin Proteins involved in binding chromatids together.

cohesion The tendency of molecules (or any substances) to stick together.

cohort (co' hort) [L. *cohors*: company of soldiers] A group of similar-aged organisms, considered as it passes through time.

coleoptile A sheath that surrounds and protects the shoot apical meristem and young primary leaves of a grass seedling as they move through the soil.

collagen [Gk. *kolla*: glue] A fibrous protein found extensively in bone and connective tissue.

collecting duct In vertebrates, a tubule that receives urine produced in the nephrons of the kidney and delivers that fluid to the ureter for excretion.

collenchyma (cull eng' kyma) [Gk. *kolla*: glue + *enchyma*: infusion] A type of plant cell, living at functional maturity, which lends flexible support by virtue of primary cell walls thickened at the corners. (Contrast with parenchyma, sclerenchyma.)

colon [Gk. *kolon*] The large intestine.

common bile duct A single duct that delivers bile from the gallbladder and secretions from the pancreas into the small intestine.

communication A signal from one organism (or cell) that alters the functioning or behavior of another organism (or cell).

community Any ecologically integrated group of species of microorganisms, plants, and animals inhabiting a given area.

companion cell A specialized cell found adjacent to a sieve tube element in flowering plants.

comparative genomics Computer-aided comparison of DNA sequences between different organisms to reveal genes with related functions.

comparative method An experimental design in which two samples or populations exposed to different conditions or treatments are compared to each other.

compensation point The light intensity at which the rates of photosynthesis and of cellular respiration are equal.

competition In ecology, use of the same resource by two or more species, when the resource is present in insufficient supply for the combined needs of the species.

competitive exclusion A result of competition between species for a limiting resource in which one species completely eliminates the other.

competitive inhibitor A nonsubstrate that binds to the active site of an enzyme and thereby inhibits binding of substrate and reaction from part of the environment.

complement system A group of eleven proteins that play a role in some reactions of the immune system. The complement proteins are not immunoglobulins.

complementary base pairing The AT (or AU), TA (or UA), CG, and GC pairing of bases in double-stranded DNA, in transcription, and between tRNA and mRNA.

complementary DNA (cDNA) DNA formed by reverse transcriptase acting with an RNA template; essential intermediate in the reproduction of retroviruses; used as a tool in recombinant DNA technology; lacks introns.

complete metamorphosis A change of state during the life cycle of an organism in which the body is almost completely rebuilt to produce an individual with a very different body form. Characteristic of insects such as butterflies, moths, beetles, ants, wasps, and flies.

compound (1) A substance made up of atoms of more than one element. (2) A structure made up of many units, as the compound eyes of arthropods.

concerted evolution The common evolution of a family of repeated genes, such that changes in one copy of the gene family are replicated in other copies of the gene family.

condensation reaction A reaction in which two molecules become connected by a covalent bond and a molecule of water is released. ($AH + BOH \rightarrow AB + H_2O$.)

condensing A protein complex involved in chromosome condensation during mitosis and meiosis.

conditional mutations Mutations that show characteristic phenotype only under certain environmental conditions such as temperature.

conduction The transfer of heat from one object to another through direct contact. In neurophysiology, the progression of an action potential along an axon.

cone cells (1) In the vertebrate retina, photoreceptor cells that are responsible for color vision. (2) In gymnosperms, reproductive structures consisting of spore-bearing scales inserted on a short axis; the scales are modified branches. (Contrast with strobilus.)

conformation The three-dimensional shape of a protein or other macromolecule.

conidium (ko nid' ee um) [Gk. *konis*: dust] An asexual fungus spore borne singly or in chains either apically or laterally on a hypha.

conifer (kahn' e fer) [Gr. *konos*: cone + *phero*: carry] One of the cone-bearing gymnosperms, mostly trees, such as pines and firs.

conjugation (kon ju gay' shun) [L. *conjugare*: yoke together] The close approximation of two cells during which they exchange genetic material, as in *Paramecium* and other ciliates, or during which DNA passes from one to the other, as in bacteria.

conjugation tube Cytoplasmic connection between two bacterial cells through which DNA passes during conjugation.

connective tissue An animal tissue that connects or surrounds other tissues; its cells are embedded in a collagen-containing matrix.

connexon In a gap junction, a protein channel linking adjacent animal cells.

consensus sequences Short stretches of DNA that appear, with little variation, in many different genes.

conservation biology An applied, normative science that carries out investigations designed to help preserve the diversity of life on Earth.

constant region For a particular class of immunoglobulin molecules, the region with identical amino acid composition.

constitutive enzyme An enzyme that is present in approximately constant amounts in a system, whether its substrates are present or absent. (Contrast with inducible enzyme.)

consumer An organism that eats the tissues of some other organism.

continental drift The gradual movements of the world's continents that have occurred over billions of years.

contraception The act of preventing union of sperm and egg.

controlled experiment An experimental design in which a sample or population is divided into two groups; one group is exposed to a manipulated variable while the other group serves as a nontreated control. The two groups are compared to see if there are changes in a "dependent" variable as a result of the experimental manipulation.

convection The transfer of heat to or from a surface via a moving stream of air or fluid.

convergent evolution The independent evolution of similar features from different ancestral traits.

copulation Reproductive behavior that results in a male depositing sperm in the reproductive tract of a female.

coprophagy Ingesting ones own feces.

corepressor A low-molecular-weight compound that unites with a protein (the repressor) to prevent transcription in a repressible operon.

cork A waterproofing tissue in plants, with suberin-containing cell walls. Produced by a cork cambium.

cornea The clear, transparent tissue that covers the eye and allows light to pass through to the retina.

corolla (ko role' lah) [L. *corolla*: a small crown] All of the petals of a flower, collectively.

coronary (kor' oh nair ee) [L. *corona*: crown] Referring to the blood vessels of the heart.

coronary thrombosis A fibrous clot that blocks a coronary vessel.

corpus luteum (kor' pus loo' tee um) [L.: yellow body] A structure formed from a follicle after ovulation; it produces hormones important to the maintenance of pregnancy.

cortex [L. *cortex*: covering, rind] (1) In plants, the tissue between the epidermis and the vascular tissue of a stem or root. (2) In animals, the outer tissue of certain organs, such as the adrenal cortex and cerebral cortex.

corticosteroids Steroid hormones produced and released by the cortex of the adrenal gland.

cortisol A corticosteroid that mediates stress responses.

cost–benefit analysis An approach for explaining why the traits of organisms evolve as they do; assumes that an organism has a limited amount of time and energy to devote to its activities, and that each activity has costs (e.g., risks, expenditure of energy) as well as benefits (e.g., obtaining food, mating and successful reproduction). (*See also* trade-off.)

cotyledon (kot' ul lee' dun) [Gk. *kotyledon*: hollow space] A "seed leaf." An embryonic organ that stores and digests reserve materials; may expand when seed germinates.

countercurrent exchange An adaptation that promotes maximum exchange of heat or any diffusible substance between two fluids by the fluids flow in opposite directions through parallel tubes close together

countercurrent multiplier The mechanism of increasing the concentration of the interstitial fluid in the mammalian kidney as a result of countercurrent exchange in the loops of Henle and selective permeability and active transport of ions by segments of the loops of Henle.

covalent bond A chemical bond that arises from the sharing of electrons between two atoms. Usually a strong bond.

crassulacean acid metabolism (CAM) A metabolic pathway enabling the plants that possess it to store carbon dioxide at night and then perform photosynthesis during the day with stomata closed.

crista (plural: cristae) A small, shelflike projection of the inner membrane of a mitochondrion; the site of oxidative phosphorylation.

critical night length In the photoperiodic flowering response of short-day plants, the length of night above which flowering occurs and below which the plant remains vegetative. (The reverse applies in the case of long-day plants.)

critical period The age during which some particular type of learning must take place or during which it occurs much more easily than at other times. Typical of song learning among birds.

cross section A section taken perpendicular to the longest axis of a structure. Also called a transverse section.

crossing over The mechanism by which linked markers undergo recombination. In general, the term refers to the reciprocal exchange of corresponding segments between two homologous chromatids.

cryptic [Gk. *kryptos*: hidden] The resemblance of an animal to some part of its environment, which helps it to escape detection by predators.

cryptochromes [Gk. *kryptos*: hidden + *kroma*: color] Photoreceptors mediating some blue-light effects in plants and animals.

culture (1) A laboratory association of organisms under controlled conditions. (2) The collection of knowledge, tools, values, and rules that characterize a human society.

cuticle A waxy layer on the outer surface of a plant or an insect, tending to retard water loss.

cyanobacteria (sigh an' o bacteria) [Gr. *kuanos*: blue] A group of photosynthetic bacteria, formerly referred to as blue-green algae; they use chlorophyll *a* in photosynthesis.

cyclic AMP *See* cAMP.

cyclic electron transport In photosynthetic light reactions, the flow of electrons that produces ATP but no NADPH or O_2.

cyclins Proteins that activate cyclin-dependent kinases, bringing about transitions in the cell cycle.

cyclin-dependent kinase (Cdk) A kinase whose target proteins are involved in transitions in the cell cycle and which is active only when complexed with additional protein subunits, called cyclins.

cytochromes (sy' toe chromes) [Gk. *kytos*: container + *chroma*: color] Iron-containing red proteins, components of the electron-transfer chains in photophosphorylation and respiration.

cytokine A regulatory protein made by immune system cells that affects other target cells in the immune system.

cytokinesis (sy' toe kine ee' sis) [Gk. *kytos*: container + *kinein*: to move] The division of the cytoplasm of a dividing cell. (Compare with mitosis.)

cytokinin (sy' toe kine' in) A member of a class of plant growth substances playing roles in senescence, cell division, and other phenomena.

cytoplasm The contents of the cell, excluding the nucleus.

cytoplasmic determinants In animal development, gene products whose spatial distribution may determine such things as embryonic axes.

cytoplasmic domain The portion of a membrane bound receptor molecule that projects into the cytoplasm.

cytoplasmic segregation The asymmetrical distribution of cytoplasmic determinants in a developing animal embryo.

cytosine (C) (site' oh seen) A nitrogen-containing base found in DNA and RNA.

cytoskeleton The network of microtubules and microfilaments that gives a eukaryotic cell its shape and its capacity to arrange its organelles and to move.

cytosol The fluid portion of the cytoplasm, excluding organelles and other solids.

cytotoxic T cells (T_C) Cells of the cellular immune system that recognize and directly eliminate virus-infected cells. (Compare with helper T cells.)

- D -

DAG *See* diacylglycerol.

daughter chromosomes During mitosis, the separated chromatids from the beginning of anaphase onward.

deciduous [L. *deciduus*: falling off] Refers to a woody plant that sheds it leaves but does not die.

decomposers Organisms that metabolize organic compounds in debris and dead organisms, releasing inorganic material; found among the bacteria, protists, and fungi. *See also* detritivore.

defensin A type of protein made by phagocytes that kills bacteria and enveloped viruses by insertion into their plasma membranes.

degeneracy The situation in which a single amino acid may be represented by any of two or more different codons in messenger RNA. Most of the amino acids can be represented by more than one codon.

delayed hypersensitivity An increased immune reaction against an antigen that does not appear for 1–2 days after exposure. (Contrast with immediate hypersensitivity.)

deletion A mutation resulting from the loss of a continuous segment of a gene or chromosome. Such mutations never revert to wild type. (Contrast with duplication, point mutation.)

deme (deem) [Gk. *demos:* the populace] Any local population of individuals belonging to the same species that interbreed with one another.

demographic processes Events (births, deaths, immigration, and emigration) that change the number of individuals and the age structure of a population.

demographic stochasticity Random variations in the factors influencing the size, density, and distribution of a population.

denaturation Loss of activity of an enzyme or nucleic acid molecule as a result of structural changes induced by heat or other means.

dendrite [Gk. *dendron*: tree] A fiber of a neuron which often cannot carry action potentials. Usually much branched and relatively short compared with the axon, and commonly carries information to the cell body of the neuron.

denitrification Metabolic activity by which nitrate and nitrite ions are reduced to form nitrogen gas; carried out by certain soil bacteria.

density dependence The state in which changes in the severity of action of agents affecting birth and death rates within populations are directly or inversely related to population density.

density independence The state in which the severity of action of agents affecting birth and death rates within a population does not change with the density of the population.

deoxyribonucleic acid *See* DNA.

deoxyribose A five-carbon sugar found in nucleotides and DNA.

depolarization A change in the electric potential across a membrane from a condition in which the inside of the cell is more negative than the outside to a condition in which the inside is less negative, or even positive, with reference to the outside of the cell. (Contrast with hyperpolarization.)

derived trait A trait that differs from the ancestral trait. (Compare with shared derived trait.)

dermal tissue system The outer covering of a plant, consisting of epidermis in the young plant and periderm in a plant with extensive secondary growth. (Contrast with ground tissue system and vascular tissue system.)

desmosome (dez' mo sowm) [Gk. *desmos*: bond + *soma*: body] An adhering junction between animal cells.

desmotubule A membrane extension connecting the endoplasmic retituclum of two plant cells that traverses the plasmodesma.

determination Process whereby an embryonic cell or group of cells becomes fixed into a predictable developmental pathway.

determined The state of a cell prior to its differentiation but after its developmental fate has been channeled.

detritivore (di try' ti vore) [L. *detritus*: worn away + *vorare*: to devour] An organism that obtains its energy from the dead bodies and/or waste products of other organisms.

development Progressive change, as in structure or metabolism; in most kinds of organisms, development continues throughout the life of the organism.

developmental plasticity The capacity of an organism to alter its pattern of development in response to environmental conditions.

diacylglycerol (DAG) In hormone action, the second messenger produced by hydrolytic removal of the head group of certain phospholipids.

diaphragm (dye' uh fram) [Gk. *diaphrassein*: barricade] (1) A sheet of muscle that separates the thoracic and abdominal cavities in mammals; responsible for breathing. (2) A method of birth control in which a sheet of rubber is fitted over the woman's cervix, blocking the entry of sperm.

diastole (dye ass' toll ee) [Gk. dilation] The portion of the cardiac cycle when the heart muscle relaxes. (Contrast with systole.)

diencephalons The portion of the vertebrate forebrain that becomes the thalamus and hypothalamus.

differential gene expression The hypothesis that, given that all cells contain all genes, what makes one cell type different from another is the difference in transcription and translation of those genes.

differentiation Process whereby originally similar cells follow different developmental pathways. The actual expression of determination.

diffusion Random movement of molecules or other particles, resulting in even distribution of the particles when no barriers are present.

digestion Enzyme-catalyzed process by which large, usually insoluble, molecules (foods) are hydrolyzed to form smaller molecules of soluble substances.

dihybrid cross A mating in which the parents differ with respect to the alleles of two loci of interest.

dikaryon (di care' ee ahn) [Gk. *di*: two + *karyon*: kernel] A cell or organism carrying two genetically distinguishable nuclei. Common in fungi.

dioecious (die eesh' us) [Gk.: *di*: two + *oikos*: house] Refers to organisms in which the two sexes are "housed" in two different individuals, so that eggs and sperm are not produced in the same individuals. Examples: humans, fruit flies, date palms. (Contrast with monoecious.)

diploblastic Having two cell layers. (Contrast with triploblastic.)

diploid (dip' loid) [Gk. *diplos*: double] Having a chromosome complement consisting of two copies (homologs) of each chromosome. Designated $2n$.

diplontic A type of life cycle in which gametes are the only haploid cells and mitosis occurs only in diploid cells. (Contrast with haplontic.)

direct transduction A cell signaling mechanism in which the receptor acts as the effector in the cellular response. (Contrast with indirect transduction.)

directional selection Selection in which phenotypes at one extreme of the population distribution are favored. (Contrast with disruptive selection, stabilizing selection.)

disaccharide A carbohydrate made up of two monosaccharides (simple sugars).

displacement activity Apparently irrelevant behavior performed by an animal under conflict situations, especially when tendencies to attack and escape are closely balanced.

display A behavior that has evolved to influence the actions of other individuals.

disruptive selection Selection in which phenotypes at both extremes of the population distribution are favored. (Contrast with directional selection; stabilizing selection.)

distal Away from the point of attachment or other reference point. (Contrast with proximal.)

disturbance A short-term event that disrupts populations, communities, or ecosystems by changing the environment.

disulfide bridge The covalent bond between twosulfur atoms (–S—S–) linking to molecules or remote parts of the same molecule.

diverticulum (di ver tik' u lum) [L. *divertere*: turn away] A small cavity or tube that connects to a major cavity or tube.

division A term used by some microbiologists and formerly by botanists, corresponding to the term phylum.

DNA (deoxyribonucleic acid) The fundamental hereditary material of all living organisms. In eukaryotes, stored primarily in the cell nucleus. A nucleic acid using deoxyribose rather than ribose.

DNA chip A small glass or plastic square onto which thousands of single-stranded DNA sequences are fixed. Hybridization of cell-derived RNA or DNA to the target sequences can be performed.

DNA fingerprint An individual's unique pattern of DNA fragments produced by action of restriction endonucleases and separated by electrophoresis.

DNA helicase An enzyme that functions during DNA replication to unwind the double helix.

DNA ligase Enzyme that unites Okazaki fragments of the lagging strand during DNA replication; also mends breaks in DNA strands. It connects pieces of a DNA strand and is used in recombinant DNA technology.

DNA methylation The addition of methyl groups to DNA. Methylation plays a role in regulation of gene expression. Among bacteria, it protects DNA against restriction endonucleases.

DNA polymerase Any of a group of enzymes that catalyze the formation of DNA strands from a DNA template.

DNA sequencing Determining the precise sequence of nucleotides, and thus the sequence of amino acids encoded in a segment DNA.

DNA topoisomerase Enzymes that introduce positive or negative supercoils into the double-stranded DNA of continuous (circular) chromosomes.

docking protein A receptor protein that binds (docks) a ribosome to the membrane of the endoplasmic reticulum by binding the signal sequence attached to a new protein being made at the ribosome.

domain (1) Independent structural elements within proteins that affect the protein's function. Encoded by recognizable nucleotide sequences, a domain often folds separately from the rest of the protein. Similar domains can appear in a variety of different proteins across phylogenetic groups (e.g.,"homeobox domain";"calcium-binding domain.") (2) In phylogenetics, the three monophyletic branches of Life. Members of the three domains (Bacteria, Archaea, and Eukarya) are believed to have been evolving independently of each other for at least a billion years.

dominance In genetics, the ability of one allelic form of a gene to determine the phenotype of a heterozygous individual in which the homologous chromosomes carry both it and a different (recessive) allele. (Contrast with recessive.)

dormancy A condition in which normal activity is suspended, as in some seeds and buds.

dorsal [L. *dorsum*: back] Pertaining to the back or upper surface. (Contrast with ventral.)

dorsal lip In amphibian embryos, the dorsal segment of the blastopore. Also called the"organizer,"this region directs the development of nearby embryonic regions.

double fertilization Virtually unique to angiosperms, a process in which the nuclei of two sperm fertilize one egg. One sperm's nucleus combines with the egg nucleus to produce a zygote, while the other combines with the same egg's two polar nuclei to produce the first cell of the triploid endosperm (the tissue that will nourish the growing plant embryo).

double helix In DNA, the natural, right-handed coil configuration of two complementary, antiparallel strands.

duodenum (do' uh dee' num) The beginning portion of the vertebrate small intestine. (Compare with ileum, jejunum.)

duplication A mutation in which a segment of a chromosome is duplicated, often by the attachment of a segment lost from its homolog. (Contrast with deletion.)

dynein [Gk. *dynamis*: power] A protein that plays a part in the movement of eukaryotic flagella and cilia by means of conformational changes.

- E -

ecdysone (eck die' sone) [Gk. *ek*: out of + *dyo*: to clothe] In insects, a hormone that induces molting.

ecological community The species living together at a particular site.

ecological niche (nitch) [L. *nidus*: nest] The functioning of a species in relation to other species and its physical environment.

ecological succession The sequential replacement of one assemblage of populations by another in a habitat following some disturbance.

ecology [Gk. *oikos*: house + *logos*: study] The scientific study of the interaction of organisms with their living (biotic) and nonliving (abiotic) environment.

ecosystem (eek' oh sis tum) The organisms of a particular habitat, such as a pond or forest, together with the physical environment in which they live.

ectoderm [Gk. *ektos*: outside + *derma*: skin] The outermost of the three embryonic tissue layers first delineated during gastrulation. Gives rise to the skin, sense organs, nervous system, etc.

ectotherm [Gk. *ektos*: outside + *thermos*: heat] An animal unable to control its body temperature. (Contrast with endotherm.)

edema (i dee' mah) [Gk. *oidema*: swelling] Tissue swelling caused by the accumulation of fluid.

edge effect The changes in ecological processes in a community caused by physical and biological factors originating in an adjacent community.

effector Any organ, cell, or organelle that moves the organism through the environment or else alters the environment; for example, muscle, exocrine glands, chromatophores.

effector cell A cell responsible for the effector phase of the immune response.

effector phase Stage of the immune response, when cytotoxic T cells attack virus-infected cells, and helper T cells assist B cells to differentiate into plasma cells.

effector protein In cell signaling, a protein responsible for the cellular reponse to a signal transduction pathway.

efferent [L. *ex*: out + *ferre*: to bear] In physiology, conducting outward or away from an organ or structure. (Contrast with afferent.)

egg In all sexually reproducing organisms, the female gamete; in birds, reptiles, and some other vertebrates, a structure within which early embryonic development occurs. (Compare with amniote egg.)

elasticity The property of returning quickly to a former state after a disturbance.

electrocardiogram (ECG or EKG) A graphic recording of electrical potentials from the heart.

electroencephalogram (EEG) A graphic recording of electrical potentials from the brain.

electromagnetic radiation A self-propagating wave that travels though space and has both electrical and magnetic properties.

electromyogram (EMG) A graphic recording of electrical potentials from muscle.

electron A subatomic particle outside the nucleus carrying a negative charge and very little mass.

electron shell The region surrounding the atomic nucleus at a fixed energy level in which electrons orbit.

electron transport The passage of electrons through a series of proteins with a release of energy which may be captured in a concentration gradient or chemical form such as NADH or ATP.

electronegativity The tendency of an atom to attract electrons when it occurs as part of a compound.

electrostatic Pertaining to the attraction and repulsion of negative and positive charges on atoms due to the number and distribution of electrons.

electrophoresis (e lek' tro fo ree' sis) [L. *electrum*: amber + Gk. *phorein*: to bear] A separation technique in which substances are separated from one another on the basis of their electric charges and molecular weights.

element A substance that cannot be converted to simpler substances by ordinary chemical means.

elongation (1) In molecular biology, the addition of monomers to make a longer RNA or protein during synthesis. (2) Growth of a plant axis or cell primarily in the longitudinal direction.

embolus (em' buh lus) [Gk. *embolos*: inserted object; stopper] A circulating blood clot. Blockage of a blood vessel by an embolus or by a bubble of gas is referred to as an embolism. (Contrast with thrombus.)

embryo [Gk. *en*: within + *bryein*: to grow] A young animal, or young plant sporophyte, while it is still contained within a protective structure such as a seed, egg, or uterus.

embryophyte A photosynthetic organism that develops from a protected embryo; synonymous with"land plant."In this book,"plant,"when unmodified, is synonymous with"embryophyte."

embryo sac In angiosperms, the female gametophyte. Found within the ovule, it consists of eight or fewer cells, membrane bounded, but without cellulose walls between them.

emergent property A property of a complex system that is not exhibited by its individual component parts.

emigration The deliberate and usually oriented departure of an organism from the habitat in which it has been living.

3′ end (3 prime) The end of a DNA or RNA strand that has a free hydroxyl group at the 3′ carbon of the sugar (deoxyribose or ribose).

5′ end (5 prime) The end of a DNA or RNA strand that has a free phosphate group at the 5′ carbon of the sugar (deoxyribose or ribose).

embolism A blockage of a coronary vessel resulting from a circulating blood clot.

endemic (en dem' ik) [Gk. *endemos*: native, dwelling in] Confined to a particular region, thus often having a comparatively restricted distribution.

endergonic reaction A chemical reaction that requires the input of energy in order to proceed. (Contrast with exergonic reaction.)

endocrine gland (en' doh krin) [Gk. *endo*: within + *krinein*: to separate] Any gland, such as the adrenal or pituitary gland of vertebrates, that secretes certain substances, especially hormones, into the body through the blood. The *endocrine system* consists of all cells and glands in the body that produce and release hormones.

endocytosis A process by which liquids or solid particles are taken up by a cell through invagination of the plasma membrane. (Contrast with exocytosis.)

endoderm [Gk. *endo*: within + *derma*: skin] The innermost of the three embryonic tissue layers delineated during gastrulation. Gives rise to the digestive and respiratory tracts and structures associated with them.

endodermis In plants, a specialized cell layer marking the inside of the cortex in roots and some stems. Frequently a barrier to free diffusion of solutes.

endomembrane system Endoplasmic reticulum plus Golgi apparatus; also lysosomes, when present. A system of membranes that exchange material with one another.

endoplasmic reticulum (ER) [Gk. *endo*: within + L. *plasma*: form + L. *reticulum*: net] A system of membranous tubes and flattened sacs found in the cytoplasm of eukaryotes. Exists in two forms: rough ER, studded with ribosomes; and smooth ER, lacking ribosomes.

endorphins, enkephalins Naturally occurring, opiate-like substances in the mammalian brain.

endoskeleton [Gk. *endo*: within + *skleros*: hard] An internal skeleton covered by other, soft body tissues. (Contrast with exoskeleton.)

endosperm [Gk. *endo*: within + *sperma*: seed] A specialized triploid seed tissue found only in angiosperms; contains stored nutrients for the developing embryo.

endosymbiosis [Gk. *endo*: within + *sym*: together + *bios*: life] Two species living together, with one living inside the body (or even the cells) of the other.

endosymbiotic theory The theory that the eukaryotic cell evolved via the engulfing of one prokaryotic cell by another.

endotherm [Gk. *endo*: within + *thermos*: heat] An animal that can control its body temperature by the expenditure of its own metabolic energy. (Contrast with ectotherm.)

end product inhibition A control capacity of some metabolic pathways in which the final product produced inhibits an early enzyme in the pathway.

energetic cost The difference between the energy an animal expends in performing a behavior and the energy it would have expended had it rested.

energy The capacity to do work or move matter against an opposing force. The capacity to accomplish change.

energy budget A quantitative description of all paths of energy exchange between an animal and its environment.

enhancer In eukaryotes, a DNA sequence, lying on either side of the gene it regulates, that stimulates a specific promoter.

enterocoelous development A pattern of development in which the coelum is formed by an outpocketing of the embryonic gut (enteron).

enterokinase (ent uh row kine′ ase) An enzyme secreted by the mucosa of the duodenum. It activates the zymogen trypsinogen to create the active digestive enzyme trypsin.

enthalpy The sum of the internal energy of a system; the product of the volume multiplied by the pressure.

entrainment With respect to circadian rhythms, the process whereby the period is adjusted to match the 24-hour environmental cycle.

entropy (en′ tro pee) [Gk. *tropein*: to change] A measure of the degree of disorder in any system. Spontaneous reactions in a closed system are always accompanied by an increase in disorder and entropy.

environment Whatever surrounds and interacts with a population, organism, or cell. May be external or internal.

environmental carrying capacity (*K*) The number of individuals in a population that the resources of a habitat can support.

enzyme (en′ zime) [Gk. *zyme*: to leaven (as in yeast bread)] A protein, on the surface of which are chemical groups so arranged as to make the enzyme a catalyst for a chemical reaction.

enzyme–substrate complex The complex that forms when an enzyme binds to its substrate(s).

eosinophils Phagocytic white blood cells that attack multicellular parasites once they have been coated with antibodies.

epi- [Gk.: upon, over] A prefix used to designate a structure located on top of another; for example: epidermis, epiphyte.

epiblast The upper or overlying portion of the avian blastula which is joined to the hypoblast at the margins of the blastodisc.

epiboly The movement of cells over the surface of the blastula toward the forming blastopore.

epicotyl (epp′ i kot′ il) [Gk. *epi*: over + *kotyle*: something hollow] That part of a plant embryo or seedling that is above the cotyledons.

epidermis [Gk. *epi*: over + *derma*: skin] In plants and animals, the outermost cell layers. (Only one cell layer thick in plants.)

epididymis (epuh did′ uh mus) [Gk. *epi*: over + *didymos*: testicle] Coiled tubules in the testes that store sperm and conduct sperm from the seminiferous tubules to the vas deferens.

epinephrine (ep i nef′ rin) [Gk. *epi*: over + *nephros*: kidney] The "fight or flight" hormone produced by the medulla of the adrenal gland; it also functions as a neurotransmitter. (Also known as adrenaline.)

epiphyte (ep′ e fyte) [Gk. *epi*: over + *phyton*: plant] A specialized plant that grows on the surface of other plants but does not parasitize them.

epistasis Interaction between genes in which the presence of a particular allele of one gene determines whether another gene will be expressed.

epithelium Sheets of cells that line or cover organs, make up tubules, and cover the surface of the body.

equatorial plate In a cell undergoing mitosis, the region in the middle of a cell where the centromeres will align during metaphase.

equilibrium Any state of balanced opposing forces and no net change.

ER *See* endoplasmic reticulum.

erythrocyte (ur rith′ row site) [Gk. *erythros*: red + *kytos*: container] A red blood cell.

erythropoietin A hormone produced by the kidney in response to lack of oxygen. It stimulates the production of red blood cells.

esophagus (i soff′ i gus) [Gk. *oisophagos*: gullet] That part of the gut between the pharynx and the stomach.

essential element A mineral nutrient element required in order for a seed to develop and complete the plant's life cycle, producing viable new seeds.

essential acids Amino acids or fatty acids that an animal cannot synthesize for itself.

ester linkage A condensation (water-releasing) reaction in which the carboxyl group of a fatty acid reacts with the hydroxyl group of an alcohol. Lipids are formed in this way.

estivation (ess tuh vay′ shun) [L. *aestivalis*: summer] A state of dormancy and hypometabolism that occurs during the summer; usually a means of surviving drought and/or intense heat. (Contrast with hibernation.)

estrogen Any of several steroid sex hormones; produced chiefly by the ovaries in mammals.

estrus (es′ truss) [L. *oestrus*: frenzy] The period of heat, or maximum sexual receptivity, in some female mammals. Ordinarily, the estrus is also the time of release of eggs in the female.

ethology An approach to the study of animal behavior developed in Europe. Emphasizes the causes of the evolution of behavior.

ethylene One of the plant growth hormones, the gas $H_2C{=}CH_2$. Involved in fruit ripening and other growth and developmental responses.

euchromatin Chromatin that is diffuse and non-staining during interphase; may be transcribed. (Contrast with heterochromatin.)

eukaryotes (yew car′ ree oats) [Gk. *eu*: true + *karyon*: kernel or nucleus] Organisms whose cells contain their genetic material inside a nucleus. Includes all life other than the viruses, archaea, and bacteria.

eutrophication (yoo trofe′ ik ay′ shun) [Gk. *eu*: truly + *trephein*: to flourish] The addition of nutrient materials to a body of water, resulting in changes in ecological processes and species composition therein.

evaporation The transition of water from the liquid to the gaseous phase.

evolution Any gradual change. Organic or Darwinian evolution, often referred to as evolution, is any genetic and resulting phenotypic change in organisms from generation to generation. (*See* macroevolution, microevolution; compare with speciation.)

evolutionary agent Any factor that influences the direction and rate of evolutionary change.

evolutionarily conserved Refers to traits that have evolved very slowly and are similar or even identical in individuals of highly divergent groups.

evolutionary radiation The proliferation of species within a single evolutionary lineage.

evolutionary reversal The reappearance of an ancestral trait in a group that had previously acquired a derived trait.

excision repair The removal and damaged DNA and its replacement by the appropriate nucleotides.

excited state The state of an atom or molecule when, after absorbing energy, it has more energy than in its normal, ground state. (Compare with ground state.)

excretion Release of metabolic wastes by an organism.

exergonic reaction A reaction in which free energy is released. (Contrast with endergonic reaction.)

exocrine gland (eks' oh krin) [Gk. *exo*: outside + *krinein*: to separate] Any gland, such as a salivary gland, that secretes to the outside of the body or into the gut. (Contrast with endocrine gland.)

exocytosis A process by which a vesicle within a cell fuses with the plasma membrane and releases its contents to the outside. (Contrast with endocytosis.)

exon A portion of a DNA molecule, in eukaryotes, that codes for part of a polypeptide. (Contrast with intron.)

exoskeleton (eks' oh skel' e ton) [Gk. *exos*: outside + *skleros*: hard] A hard covering on the outside of the body to which muscles are attached. (Contrast with endoskeleton.)

exotoxins Highly toxic proteins released by living, multiplying bacteria.

expanding triplet repeat A three-base-pair sequence in a human gene that is unstable and can be repeated a few to hundreds of times. Often, the more the repeats, the less the activity of the gene involved. Expanding triplet repeats occur in some human diseases such as Huntington's disease and fragile-X syndrome.

experiment A testing process to support or disprove hypotheses and to answer questions. The basis of the scientific method. *See* comparative experiment, controlled experiment.

expiratory reserve volume The amount of air that can be forcefully exhaled beyond the normal tidal expiration. (Compares with inspiratory reserve volume, tidal volume, vital capacity.)

exploitation competition Competition in which individuals reduce the quantities of their shared resources. (Compare with interference competition.)

exponential growth Growth, especially in the number of organisms in a population, which is a geometric function of the size of the growing entity: the larger the entity, the faster it grows. (Contrast with logistic growth.)

expression vector A DNA vector, such as a plasmid, that carries a DNA sequence that includes the adjacent sequences for its expression into mRNA and protein in a host cell.

expressivity The degree to which a genotype is expressed in the phenotype; may be affected by the environment.

extensor A muscle the extends an appendage.

extinction The termination of a lineage of organisms.

extrinsic protein A membrane protein found only on the surface of the membrane. (Contrast with intrinsic protein.)

extracellular matrix In animal tissues, a material of heterogeneous composition surrounding cells and performing many functions including adhesion of cells.

extraembryonic membranes Four membranes that support but are not part of the developing embryos of reptiles, birds, and mammals, defining these groups phylogenetically as amniotes. (*See* amnion, allantois, chorion, and yolk sac.)

- F -

F_1 (first filial generation) The immediate progeny of a parental (P) mating.

F_2 (second filial generation) The immediate progeny of a mating between members of the F_1 generation.

facilitated diffusion Passive movement through a membrane involving a specific carrier protein; does not proceed against a concentration gradient. (Contrast with active transport, diffusion.)

facultative anaerobes (alternatively, facultative aerobes) Prokaryotes that can shift their metabolism between anaerobic and aerobic operations depending on the presence or absence of O_2.

FAD *See* flavin adenine dinucleotide.

fast-twitch fibers Skeletal muscle fibers that can generate high tension rapidly, but fatigue rapidly ("sprinter" fibers). Characterized by an abundance of enzymes of glycolysis.

fat A triglyceride that is solid at room temperature. (Contrast with oil.)

fate In an embryo, the type of cell that a particular cell will become in the adult.

fate map A diagram of the blastula showing which cells (blastomeres) are "fated" to contribute to specific tissues and organs in the mature body.

fatty acid A molecule with a long hydrocarbon tail and a carboxyl group at the other end. Found in many lipids.

fauna (faw' nah) All of the animals found in a given area. (Contrast with flora.)

feces [L. *faeces*: dregs] Waste excreted from the digestive system.

feedback information Information relevant to the rate of a process that can be used by a control system to regulate that process at a particular level.

feedforward information Information that can be used to alter the setpoint of a regulatory process.

fermentation (fur men tay' shun) [L. *fermentum*: yeast] The anaerobic degradation of a substance such as glucose to smaller molecules with the extraction of energy.

ferredoxin A protein containing iron that mediates the transfer of electrons in a number of pathways, including the light reactions of photosynthesis.

fertility factor (F factor) A plasmid that confers on a bacterium the ability to act as a DNA donor during conjugation.

fertilization Union of gametes. Also known as syngamy.

fertilization membrane A membrane surrounding an animal egg which becomes rapidly raised above the egg surface within seconds after fertilization, serving to prevent entry of a second sperm.

fetus Medical and legal term for the latter stages of a developing human embryo, from about the eighth week of pregnancy (the point at which all major organ systems have formed) to the moment of birth.

fiber An elongated, tapering cell of flowering plants, usually with a thick cell wall. Serves a support function.

fibrin A protein that polymerizes to form long threads that provide structure to a blood clot.

fibrinogen A circulating protein that can be stimulated to fall out of solution and provide the structure for a blood clot.

Fick's law of diffusion An equation that describes the factors that determine the rate of diffusion of a molecule from an area of higher concentration to an area of lower concentration.

filter feeder An organism that feeds upon much smaller organisms, that are suspended in water or air, by means of a straining device.

filtration In the excretory physiology of some animals, the process by which the initial urine is formed; water and most solutes are transferred into the excretory tract, while proteins are retained in the blood or hemolymph.

first law of thermodynamics Energy can be neither created nor destroyed.

fission Reproduction of a prokaryote by division of a cell into two comparable progeny cells.

fitness The contribution of a genotype or phenotype to the genetic composition of subsequent generations, relative to the contribution of other genotypes or phenotypes. (*See* also inclusive fitness.)

flagellum (fla jell' um) (plural: flagella) [L. *flagellum*: whip] Long, whiplike appendage that propels cells. Prokaryotic flagella differ sharply from those found in eukaryotes.

flavin adenine dinucleotide (FAD) A coenzyme involved in redox reactions and containing the vitamin riboflavin (B_2).

flexor A muscle that flexes an appendage.

flora (flore' ah) All of the plants found in a given area. (Contrast with fauna.)

floral meristem Meristem that forms the sexual parts of flowering plants (sepals, petals, stamens, and carpels).

florigen A plant hormone involved in the conversion of a vegetative shoot apex to a flower.

flower The total reproductive structure of an angiosperm; its basic parts include the calyx, corolla, stamens, and carpels.

fluid mosaic model A molecular model for the structure of biological membranes consisting of a fluid phospholipid bilayer in which suspended proteins are free to move in the plane of the bilayer.

fluorescence The emission of a photon of visible light by an excited atom or molecule.

follicle [L. *folliculus*: little bag] In female mammals, an immature egg surrounded by nutritive cells.

follicle-stimulating hormone A gonadotropic hormone produced by the anterior pituitary.

food chain A portion of a food web, most commonly a simple sequence of prey species and the predators that consume them.

food vacuole Membrane enclosed structure formed by phagocytosis in which engulfed food particles are digested by the action of lysosomal enzymes.

food web The complete set of food links between species in a community; a diagram indicating which ones are the eaters and which are eaten.

forb Any broad-leaved herbaceous plant. Especially applied to such plants growing in grasslands.

fossil Any recognizable structure originating from an organism, or any impression from such a structure, that has been preserved over geological time.

founder effect Random changes in allele frequencies resulting from establishment of a population by a very small number of individuals.

fovea [L. *fovea*; a small pit] In the vertebrate retina, the area of most distinct vision.

frame-shift mutation A mutation resulting from the addition or deletion of one or two consecutive base pairs in the DNA sequence of a gene, resulting in misreading mRNA during translation and production of a nonfunctional protein. (Compare with missense substitution, nonsense substitution, synonymous substitution.)

Frank–Starling law The stroke volume of the heart increases with increased return of blood to the heart.

free energy That energy which is available for doing useful work, after allowance has been made for the increase or decrease of disorder.

freeze-fracturing Method of tissue preparation for transmission and scanning electron microscopy in which a tissue is frozen and a knife is then used to crack open the tissue; the fracture often occurs in the path of least resistance, within a membrane.

frequency In population genetics, the proportion of all alleles or genotypes in a population composed of a particular allele or genotype.

frequency-dependent selection Selection that changes in intensity with the proportion of individuals in a population having the trait.

fruit In angiosperms, a ripened and mature ovary (or group of ovaries) containing the seeds. Sometimes applied to reproductive structures of other groups of plants.

fruiting body Any structure that bears spores.

functional genomics The assignment of functional roles to genes first identified by sequencing entire genomes.

functional group A characteristic combination of atoms that contribute specific properties when attached to larger molecules.

functional mRNA Eukaryotic mRNA that has been modified after transcription by the removal of introns and the addition of a 5′ cap and a 3′ poly(A) tail.

- G -

G cap A chemically modified GTP added to the 5′ end of mRNA; facilitates binding of mRNA to ribosome and prevents mRNA breakdown.

G_1 phase In the cell cycle, the gap between the end of mitosis and the onset of the S phase.

G_2 phase In the cell cycle, the gap between the S (synthesis) phase and the onset of mitosis.

G protein A membrane protein involved in signal transduction; characterized by binding GDP or GTP.

gametangium (gam uh tan′ gee um) [Gk. *gamos*: marriage + *angeion*: vessel] Any plant or fungal structure within which a gamete is formed.

gamete (gam′ eet) [Gk. *gamete/gametes*: wife, husband] The mature sexual reproductive cell: the egg or the sperm.

gametogenesis (ga meet′ oh jen′ e sis) The specialized series of cellular divisions that leads to the production of sex cells (gametes). (Contrast with oogenesis and spermatogenesis.)

gametophyte (ga meet′ oh fyte) In plants and photosynthetic protists with alternation of generations, the multicellular haploid phase that produces the gametes. (Contrast with sporophyte.)

ganglion (gang′ glee un) [Gk.: tumor] A cluster of neurons that have similar characteristics or function.

gap genes During insect development, the first step of segmentation genes to act organizing the anterior-posterior axis.

gap junction A 2.7-nanometer gap between plasma membranes of two animal cells, spanned by protein channels. Gap junctions allow chemical substances or electrical signals to pass from cell to cell.

gas exchange In animals, the process of taking up oxygen from the environment and releasing carbon dioxide to the environment.

gastrovascular cavity Serving for both digestion (gastro) and circulation (vascular); in particular, the central cavity of the body of jellyfish and other cnidarians.

gastrula (gas′ true luh) [Gk. *gaster*: stomach] An embryo forming the characteristic three cell layers (ectoderm, endoderm, and mesoderm) that give rise to all of the major tissue systems of the adult animal.

gastrulation Development of a blastula into a gastrula.

gated channel A membrane protein that opens and closes in response to binding of specific molecules or to changes in membrane potential. When open, it allows specific ions to move across the membrane.

gene [Gk. *genes*: to produce] A unit of heredity. Used here as the unit of genetic function which carries the information for a single polypeptide or RNA.

gene amplification Creation of multiple copies of a particular gene, allowing the production of large amounts of the RNA transcript (as in rRNA synthesis in oocytes).

gene cloning Formation of a clone of bacteria or yeast cells containing a particular foreign gene.

gene family A set of identical (or once-identical) genes derived from a single parent gene; need not be on the same chromosomes. The vertebrate globin genes constitute a classic example of a gene family.

gene flow Exchange of genes between different species (an extreme case referred to as hybridization) or between different populations of the same species caused by migration followed by breeding.

gene-for-gene resistance A mechanism for resistance to pathogens, in which resistance is triggered by the specific interaction of the products of pathogens′ *Avr* genes and plants′ *R* genes.

gene frequency *See* allele frequency.

gene library All of the cloned DNA fragments generated by action of a restriction endonuclease on a genome or chromosome.

gene pool All of the different alleles of all of the genes existing in all individual of a population.

gene therapy Treatment of a genetic disease by providing patients with cells containing functioning alleles of the genes that are nonfunctional in their bodies.

generative cell In a pollen tube, a haploid nucleus that undergoes mitosis to produce the two sperm nuclei that participate in double fertilization. (Contrast with tube cell.)

genetic code The set of instructions, in the form of nucleotide triplets, that translate a linear sequence of nucleotides in mRNA into a linear sequence of amino acids in a protein.

genetic drift Changes in gene frequencies from generation to generation as a result of random (chance) processes.

genetic map The positions of genes along a chromosome as revealed by recombination frequencies.

genetic screening The application of medical tests to determine whether an individual carries a specific allele.

genetic stochasticity Random variation in the frequencies of alleles and genotypes in a population over time. (Compare with demographic stochasticity.)

genetics The study of the structure, functioning, and inheritance of genes, the units of hereditary information.

genome (jee′ nome) All the genes in a complete haploid set of chromosomes. (Compare with proteome.)

genomics The study of entire sets of genes and their interactions.

genomic imprinting When a given gene′s phenotype is determined by whether that gene is inherited from the male or the female parent.

genotype (jean′ oh type) [Gk. *gen*: to produce + *typos*: impression] An exact description of the genetic constitution of an individual, either with respect to a single trait or with respect to a larger set of traits. (Contrast with phenotype.)

genus (jean′ us) (plural: genera) [Gk. *genos*: stock, kind] A group of related, similar species recognized by taxonomists with a distinct name used in binomial nomenclature.

geotropism *See* gravitropism.

germ cell [L. *germen*: to beget] A reproductive cell or gamete of a multicellular organism. (Contrast with somatic cell.)

germ line mutation A change in the genetic material in a germ cell; such mutations are heritable. (Contrast with somatic mutation.)

germ layers The three embryonic tissue layers formed during gastrulation (ectoderm, mesoderm, endoderm).

germination Sprouting of a seed or spore.

gestation (jes tay′ shun) [L. *gestare*: to bear] The period during which the embryo of a mammal develops within the uterus. Also known as pregnancy.

gibberellin (jib er el′ lin) A class of plant growth substances playing roles in stem elongation, seed germination, flowering of certain plants, etc. Named for the fungus *Gibberella*.

gill An organ for gas exchange in aquatic organisms.

gill arch A skeletal structure that supports gill filaments and the blood vessels that supply them.

gizzard (giz′ erd) [L. *gigeria*: cooked chicken parts] A muscular port of the stomach of birds that grinds up food, sometimes with the aid of fragments of stone.

gland An organ or group of cells that produces and secretes one or more substances.

glans penis Sexually sensitive tissue at the tip of the penis.

glia (glee′ uh) [Gk. *glia*: glue] Cells, found only in the nervous system, that do not conduct action potentials.

glomerular filtration rate (GFR) The rate at which the blood is filtered in the glomeruli of the kidney.

glomerulus (glo mare′ yew lus) [L. *glomus*: ball] Sites in the kidney where blood filtration takes place. Each glomerulus consists of a knot of capillaries served by afferent and efferent arterioles.

glucagon Hormone produced by alpha cells of the pancreatic islets of Langerhans. Glucagon stimulates the liver to break down glycogen and release glucose into the circulation.

glucocorticoids Steroid hormones produced by the adrenal cortex. Secreted in response to ACTH, they inhibit glucose uptake by many tissues in addition to mediating other stress responses.

gluconeogenesis The biochemical synthesis of glucose from other substances, such as amino acids, lactate, and glycerol.

glucose[Gk. *gleukos*: sugar, sweet] The most common monosaccharide; the monomer of the polysaccharides starch, glycogen, and cellulose.

glycerol (gliss′ er ole) A three-carbon alcohol with three hydroxyl groups; a component of phospholipids and triglycerides.

glycogen (gly′ ko jen) An energy storage polysaccharide found in animals and fungi; a branched-chain polymer of glucose, similar to starch.

glycolipid A lipid to which sugars are attached.

glycolysis (gly kol′ li sis) [Gk. *gleukos*: sugar + *lysis*: break apart] The enzymatic breakdown of glucose to pyruvic acid. One of the evolutionarily oldest of the cellular energy-yielding mechanisms.

glycoprotein A protein to which sugars are attached.

glycosidic linkage Bond between carbohydrate (sugar) molecules through an intervening oxygen atom (–O–).

glycosylation The addition of carbohydrates to another type of molecule, such as a protein.

glyoxysome (gly ox′ ee soam) An organelle found in plants, in which stored lipids are converted to carbohydrates.

Golgi apparatus (goal′ jee) A system of concentrically folded membranes found in the cytoplasm of eukaryotic cells; functions in secretion from cell by exocytosis.

gonad (go′ nad) [Gk. *gone*: seed] An organ that produces sex cells in animals: either an ovary (female gonad) or testis (male gonad).

gonadotropin A hormone that stimulates the gonads.

gonadotropin-releasing hormone (GnRH) Hypothalamic hormone that stimulates the anterior pituitary to secrete growth hormone.

Gondwana The large southern land mass that existed from the Cambrian (540 mya) to the Jurassic (138 mya). Present-day remnants are South America, Africa, India, Australia, and Antarctica.

graft Tissue artificially and viably transplanted from one organism to another. In agriculture, refers to the transfer of bud or stem segment from one plant onto another plant as a form of asexual reproduction.

Gram stain A differential purple stain useful in characterizing bacteria. The peptidoglycan-rich cell walls of Gram-positive bacteria stain purple; cell walls of Gram-negative bacteria generally stain orange.

granum (plural: grana) Within a chloroplast, a stack of thylakoids.

gravitropism A directed plant growth response to gravity.

grazer An animal that eats the vegetative tissues of herbaceous plants.

green gland An excretory organ of crustaceans.

greenhouse effect The heating of Earth's atmosphere by gases such as water vapor, carbon dioxide, and methane; such greenhouse gases are transparent to sunlight but opaque to heat; thus sunlight-engendered heat builds up at Earth's surface and cannot be dissipated into the atmosphere.

gross primary production The total energy captured by plants growing in a particular area.

gross primary productivity (GPP) The rate of assimilation of energy by plants growing in a particular area.

ground meristem That part of an apical meristem that gives rise to the ground tissue system of the primary plant body.

ground state The lowest energy state of an atom of molecule. (Compare with excited state.)

ground tissue system Those parts of the plant body not included in the dermal or vascular tissue systems. Ground tissues function in storage, photosynthesis, and support.

group transfer The exchange of atoms between molecules.

growth Irreversible increase in volume (an accurate definition, but at best a dangerous oversimplification).

growth hormone A peptide hormone of the anterior pituitary that stimulates many anabolic processes.

guanine (G) (gwan′ een) A nitrogen-containing base found in DNA, RNA, and GTP.

guard cells In plants, specialized, paired epidermal cells that surround and control the opening of a stoma (pore). *See* stoma.

gut An animal's digestive tract.

guttation The extrusion of liquid water through openings in leaves, caused by root pressure.

gymnosperm (jim′ no sperm) [Gk. *gymnos*: naked + *sperma*: seed] A plant, such as a pine or other conifer, whose seeds do not develop within an ovary (hence, the seeds are "naked").

gyrus The raised or ridged portion of the convoluted surface of the brain. (Contrast with sulcus.)

- H -

habitat The environment in which an organism lives.

habituation (ha bich′ oo ay shun) The simplest form of learning, in which an animal presented with a stimulus without reward or punishment eventually ceases to respond.

hair cell A type of mechanoreceptor in animals. Detects sound waves and other forms of motion in air or water.

half-life The time required for half of a sample of a radioactive isotope to decay to its stable, nonradioactive form, or for a drug or other substance to reach half its initial dosage.

halophyte (hal′ oh fyte) [Gk. *halos*: salt + *phyton*: plant] A plant that grows in a saline (salty) environment.

haploid (hap′ loid) [Gk. *haploeides*: single] Having a chromosome complement consisting of just one copy of each chromosome; designated $1n$ or n. (Contrast with diploid.)

haplontic A type of life cycle in which the zygote is the only diploid cell and mitosis occurs only in haploid cells. (Contrast with diplontic.)

Hardy–Weinberg equilibrium The allele frequency at a given locus in a sexually reproducing population that is not being acted on by agents of evolution; the conditions that would result in no evolution in a population.

haustorium (haw stor′ ee um) [L. *haustus*: draw up] A specialized hypha or other structure by which fungi and some parasitic plants draw nutrients from a host plant.

Haversian systems Units of organization in compact bone that reflect the action of intercommunicating osteoblasts.

heat of vaporization The energy that must be supplied to convert a molecule from a liquid to a gas at its boiling point.

heat-shock proteins Chaperone proteins expressed in cells exposed to high or low temperatures or other forms of environmental stress.

helical Shaped like a screw or spring; this shape occurs in DNA and proteins.

helper T cells (T_H) T cells that participate in the activation of B cells and of other T cells; targets of the HIV-I virus, the agent of AIDS. (Contrast with cytotoxic T cells.)

hematocrit (heme at′ o krit) [Gk. *heaema*: blood + *krites*: judge] The proportion of 100 cc of blood that consists of red blood cells.

hemizygous (hem′ ee zie′ gus) [Gk. *hemi*: half + *zygotos*: joined] In a diploid organism, having only one allele for a given trait, typically the case for X-linked genes in male mammals and Z-linked genes in female birds. (Contrast with homozygous, heterozygous.)

hemoglobin (hee′ mo glow bin) [Gk. *heaema*: blood + L. *globus*: globe] Oxygen-transporting protein found in the red blood cells of vertebrates (and found in some invertebrates).

Hensen's node In avian embryos, a structure at the anterior end of the primitive groove; determines the fates of cells passing over it during gastrulation.

hepatic (heh pat′ ik) [Gk. *hepar*: liver] Pertaining to the liver.

hepatic duct Conveys bile from the liver to the gallbladder.

herbivore (ur′ bi vore) [L. *herba*: plant + *vorare*: to devour] An animal that eats plant tissues. (Contrast with carnivore, detritivore, omnivore.)

heritable Able to be inherited; in biology, refers to genetically influenced traits.

hermaphroditism (her maf′ row dite ism) [Gk. Hermes (messenger god) + Aphrodite (goddess of love)] The coexistence of both female and male sex organs in the same organism.

hertz (abbreviated Hz) Cycles per second.

hetero- [Gk.: *heteros*: other, different] A prefix specifying that two or more different conditions are involved; for example, heterotroph, heterozygous.

heterochromatin Chromatin that retains its coiling during interphase; generally not transcribed. (Contrast with euchromatin.)

heterochrony Comparing different species, an alternation in the timing of developmental events, leading to different results in the adult.

heterocyst A large, thick-walled cell in the filaments of certain cyanobacteria; performs nitrogen fixation.

heterogeneous nuclear RNA (hnRNA) The product of transcription of a eukaryotic gene, including transcripts of introns.

heterokaryon In fungi, hypha containing two genetically different nuclei.

heteromorphic (het′ er oh more′ fik) [Gk. *heteros*: different + *morphe*: form] Having a different form or appearance, as two heteromorphic life stages of a plant. (Contrast with isomorphic.)

heterosporous (het′ er os′ por us) Producing two types of spores, one of which gives rise to a female megaspore and the other to a male microspore. (Contrast with homosporous.)

heterosis Situation in which heterozygous genotypes are superior to homozygous genotypes with respect to growth, survival, or fertility. Also called hybrid vigor.

heterotherm An animal that regulates its body temperature at a constant level at some times but not others, such as a hibernator.

heterotroph (het′ er oh trof) [Gk. *heteros*: different + *trophe*: food] An organism that requires preformed organic molecules as food. (Contrast with autotroph.)

heterotypic Referring to adhesion of cells of different types. (Contrast with homotypic.)

heterozygous (het′ er oh zie′ gus) [Gk. *heteros*: different + *zygotos*: joined] Of a diploid organism having different alleles of a given gene on the pair of homologs carrying that gene. (Contrast with homozygous.)

hexose [Gk. *hex*: six] A sugar containing six carbon atoms.

hibernation [L. *hibernum*: winter] The state of inactivity of some animals during winter; marked by a drop in body temperature and metabolic rate.

hierarchical sequencing An approach to DNA sequencing in which markers are mapped and DNA sequences are aligned by matching overlapping sites of known sequence.

highly repetitive DNA Short DNA sequences present in millions of copies in the genome, next to each other (in tandem). In reassociation experiments, denatured highly repetitive DNA reanneals very quickly.

hindbrain The region of the developing vertebrate brain that gives rise to the medulla, pons, and cerebellum.

hippocampus [Gr.: sea horse] A part of the forebrain that takes part in long-term memory formation.

histamine (hiss′ tah meen) A substance released by damaged tissue, or by mast cells in response to allergens. Histamine increases vascular permeability, leading to edema (swelling).

histology [Gk. *histos*: weaving] The study of tissues.

histone Any one of a group of basic proteins forming the core of a nucleosome, the structural unit of a eukaryotic chromosome. (Compare with nucleosome.)

hnRNA *See* heterogeneous nuclear RNA.

homeobox A 180-base-pair segment of DNA found in certain homeotic genes; regulates the expression of other genes and thus controls large-scale developmental processes.

homeostasis (home′ ee o sta′ sis) [Gk. *homos*: same + *stasis*: position] The maintenance of a steady state, such as a constant temperature or a stable social structure, by means of physiological or behavioral feedback responses.

homeotherm (home′ ee o therm) [Gk. *homos*: same + *thermos*: heat] An animal that maintains a constant body temperature by its own internal heating and cooling mechanisms. (Contrast with heterotherm, poikilotherm.)

homeotic genes (home ee ot′ ic) Genes that determine the developmental fate of entire segments of an animal.

homeotic mutations Mutations in homeotic genes that drastically alter the characteristics of a particular body segment, giving it the characteristics of other segments (as when wings grow from a *Drosophila* thoracic segment that should have produced legs).

homing The ability to return over long distances to a specific site.

homo- [Gk. *homos*: same] Prefix indicating two or more similar conditions, structures, or processes. (Contrast with hetero-.)

homolog (home′ o log′) [Gk. *homos*: same + *logos*: word] In cytogenetics, one of a pair (or larger set) of chromosomes having the same overall genetic composition and sequence. In diploid organisms, each chromosome inherited from one parent is matched by an identical (except for mutational changes) chromosome—its homolog—from the other parent.

homology (ho mol′ o jee) [Gk. *homologia*: of one mind; agreement] A similarity between two or more structures that is due to inheritance from a common ancestor. The structures are said to be homologous, and each is a homolog of the others. (Contrast with analogy.)

homoplasy (home′ uh play zee) [Gk. *homos*: same + *plastikos*: shape, mold] The presence in multiple groups of a trait that is not inherited from the common ancestor of those groups. Can result from convergent evolution, evolutionary reversal, or parallel evolution.

homosporous Producing a single type of spore that gives rise to a single type of gametophyte, bearing both female and male reproductive organs. (Contrast with heterosporous.)

homotypic Referring to adhesion of cells of the same type. (Contrast with heterotypic.)

homozygous (home′ oh zie′ gus) [Gk. *homos*: same + *zygotos*: joined] In a diploid organism, having identical alleles of a given gene on both homologous chromosomes. An individual may be a homozygote with respect to one gene and a heterozygote with respect to another. (Contrast with heterozygous.)

hormone (hore′ mone) [Gk. *hormon*: to excite, stimulate] A substance produced in minute amount at one site in a multicellular organism and transported to another site where it acts on target cells.

host An organism that harbors a parasite or symbiont and provides it with nourishment.

Hox genes Conserved homeotic genes found in vertebrates, *Drosophila*, and other animal groups. Hox genes contain the homeobox domain and specify pattern and axis formation in these animals.

human chorionic gonadotropin (hCG) A hormone secreted by the placenta which sustains the corpus luteum and helps maintain pregnancy.

Human Genome Project An effort to determine the DNA sequence of the entire human genome, understand the structures and functions of the genes, make comparisons with other organisms, and understand the social implications of this information.

humoral immune response The part of the immune system mediated by B cells that produce circulating antibodies active against extracellular bacterial and viral infections.

humus (hew′ muss) The partly decomposed remains of plants and animals on the surface of a soil.

hyaluronidase (high′ uh loo ron′ uh dase) An enzyme that digests proteoglycans. In sperm cells, it digests the coatings surrounding an egg so the sperm can enter.

hybrid (high′ brid) [L. *hybrida*: mongrel] (1) The offspring of genetically dissimilar parents. (2) In molecular biology, a double helix formed of nucleic acids from different sources.

hybridize To combine the genetic material of two distinct species or of two distinguishable populations within a species.

hybrid vigor *See* heterosis.

hybridoma A cell produced by the fusion of an antibody-producing cell with a myeloma cell; it produces monoclonal antibodies.

hybrid zone A narrow zone where two populations interbreed, producing hybrid individuals.

hydrocarbon A compound containing only carbon and hydrogen atoms.

hydrogen bond A weak electrostatic bond which arises from the attraction between the slight positive charge on a hydrogen atom and a slight negative charge on a nearby oxygen or nitrogen atom.

hydrological cycle The movement of water from the oceans to the atmosphere, to the soil, and back to the oceans.

hydrolysis (high drol' uh sis) [Gk. *hydro*: water + *lysis*: break apart] A chemical reaction that breaks a bond by inserting the components of water: $AB + H_2O \rightarrow AH + BOH$.

hydrophilic (high dro fill' ik) [Gk. *hydro*: water + *philia*: love] Having an affinity for water. (Contrast with hydrophobic.)

hydrophobic (high dro foe' bik) [Gk. *hydro*: water + *phobia*: fear] Having no affinity for water. Uncharged and nonpolar groups of atoms are hydrophobic; for example, fats and the side chain of the amino acid phenylalanine. (Contrast with hydrophilic.)

hydrostatic pressure Pressure generated by compression of liquid in a confined space. Generated in plants, fungi, and some protists with cell walls by the osmotic uptake of water. Generated in animals with closed circulatory systems by the beating of a heart.

hydrostatic skeleton The incompressible internal liquids of some animals that transfer forces from one part of the body to another when acted upon by the surrounding muscles.

hydroxyl group The —OH group found on alcohols and sugars.

hyper- [Gk. *hyper*: above, over] Prefix indicating above, higher, more.

hyperpolarization A change in the resting potential of a membrane so the inside of a cell becomes more electronegative. (Contrast with depolarization.)

hypersensitive response A defensive response of plants to microbial infection; it results in a "dead spot."

hypertension High blood pressure.

hypertonic Having a greater solute concentration. Said of one solution compared to another. (Contrast with hypotonic, isotonic.)

hypha (high' fuh) (plural: hyphae) [Gk. *hyphe*: web] In the fungi and oomycetes, any single filament.

hypo- [Gk. *hypo*: beneath, under] Prefix indicating underneath, below, less.

hypoblast The lower tissue portion of the avian blastula which is joined to the epiblast at the margins of the blastodisc.

hypocotyl [Gk. *hypo*: beneath + *kotyledon*: hollow space] That part of the embryonic or seedling plant shoot that is below the cotyledons.

hypothalamus The part of the brain lying below the thalamus; it coordinates water balance, reproduction, temperature regulation, and metabolism.

hypothesis A tentative answer to a question, from which testable predictions can be generated. (Contrast with theory.)

hypotonic Having a lesser solute concentration. Said of one solution in comparing it to another. (Contrast with hypertonic, isotonic.)

- I -

ileum The final segment of the small intestine.

imaginal disc [L. *imagos*: image, form] In insect larvae, groups of cells that develop into specific adult organs.

imbibition Water uptake by a seed; first step in germination.

immediate hypersensitivity A rapid, extensive immune reaction against an allergen involving IgE and histamine release. (Contrast with delayed hypersensitivity.)

immune system [L. *immunis*: exempt from] A system in vertebrates that recognizes and attempts to eliminate or neutralize foreign substances (e.g., bacteria, viruses, pollutants).

immunization The deliberate introduction of antigen to bring about an immune response.

immunoassay The use of labeled antibodies to measure the concentration of an antigen in a sample.

immunoglobulins A class of proteins, with a characteristic structure, active as receptors and effectors in the immune system.

immunological memory The capacity to more rapidly and massively respond to a second exposure to an antigen than occurred on first exposure.

immunological tolerance A mechanism by which an animal does not mount an immune response to the antigenic determinants of its own macromolecules.

implantation The process by which the early mammalian embryo becomes attached to and embedded in the lining of the uterus.

imprinting (1) In genetics, the differential modification of a gene depending on whether it is present in a male or a female. (2) In animal behavior, a rapid form of learning in which an animal comes to make a particular response, which is maintained for life, to some object or other organism.

inbreeding Breeding among close relatives.

inclusive fitness The sum of an individual's genetic contribution to subsequent generations both via production of its own offspring and via its influence on the survival of relatives who are not direct descendants.

incomplete dominance Condition in which the heterozygous phenotype is intermediate between the two homozygous phenotypes.

incomplete metamorphosis Insect development in which changes between instars are gradual.

incus (in' kus) [L. *incus*: anvil] The middle of the three bones that conduct movements of the eardrum to the oval window of the inner ear. (*See* malleus, stapes.)

independent assortment During meiosis, the random separation of genes carried on nonhomologous chromosomes. Articulated by Mendel as his second law.

indirect transduction A cell signaling mechanism in which a second messenger mediates the interaction between receptor binding and cellular response. (Contrast with direct transduction.)

individual fitness That component of inclusive fitness resulting from an organism producing its own offspring. (Contrast with kin selection.)

indoleacetic acid *See* auxin.

induced fit A change in enzyme conformation upon binding to substrate with an increase in the rate of catalysis.

induced mutation A mutation resulting from treatment with a chemical or other agent.

inducer (1) In enzyme systems, a small molecule which, when added to a growth medium, causes a large increase in the level of some enzyme. (2) In embryology, a substance that causes a group of target cells to differentiate in a particular way.

inducible enzyme An enzyme that is present in much larger amounts when a particular compound (the inducer) has been added to the system. (Contrast with constitutive enzyme.)

inflammation A nonspecific defense against pathogens; characterized by redness, swelling, pain, and increased temperature.

inflorescence A structure composed of several to many flowers.

inflorescence meristem A meristem that produces floral meristems as well as other small leafy structures (bracts).

inhibitor A substance that binds to the surface of an enzyme and interferes with its action on its substrates.

initial cells In plant meristems, undifferentiated cells that retain the capacity to divide producing both undifferentiated cells (initials) and cells committed to differentiation. (Compare with stem cells.)

initiation In molecular biology, the beginning of transcription or translation.

initiation complex Combination of a ribosomal light subunit, an mRNA molecule, and the tRNA charged with the first amino acid coded for by the mRNA; formed at the onset of translation.

initiation factors Proteins that assist in forming the translation initiation complex at the ribosome.

inner cell mass Derived from the mammalian blastula (bastocyst), the inner cell mass will give rise to the yolk sac (via hypoblast) and embryo (via epiblast).

inositol trisphosphate (IP_3) An intracellular second messenger derived from membrane phospholipids.

inspiratory reserve volume The amount of air that can be inhaled above the normal tidal inspiration. (Compares with expiratory reserve volume, tidal volume, vital capacity.)

instar (in' star) An immature stage of an insect between molts.

insulin (in' su lin) [L. *insula*: island] A hormone synthesized in islet cells of the pancreas that promotes the conversion of glucose into the storage material, glycogen.

integral membrane protein A membrane protein embedded in the bilayer of the membrane. (Contrast with peripheral membrane protein.)

integrase An enzyme that integrates retroviral cDNA into the genome of the host cell.

integrated pest management Control of pests by the use of natural predators and parasites in conjunction with sparing use of chemicals; an attempt to limit environmental damage.

integument [L. *integumentum*: covering] A protective surface structure. In gymnosperms and angiosperms, a layer of tissue around the ovule which will become the seed coat.

intercalary meristem A meristematic region in plants which occurs not apically, but between two regions of mature tissue. Intercalary meristems occur in the nodes of grass stems, for example.

intercostal muscles Muscles between the ribs that can augment breathing movements by elevating and suppressing the rib cage.

interference competition Competition in which individuals actively interfere with one another's access to resources. (Compare with exploitation competition.)

interference RNA (RNAi) A mechanism for reducing mRNA translation whereby a double-stranded RNA, made by the cell or synthetically, is processed to a small, single-stranded RNA, and binding of this RNA to a target mRNA results in the latter's breakdown.

interferon A glycoprotein produced by virus-infected animal cells; increases the resistance of neighboring cells to the virus.

interkinesis The period between meiosis I and meiosis II.

interleukins Regulatory proteins, produced by macrophages and lymphocytes, that act upon other lymphocytes and direct their development.

intermediate disturbance hypothesis The hypothesis that explains why species richness is lower in areas with both high and low rates of disturbance than in areas with intermediate rates of disturbance.

intermediate filaments Cytoskeletal component with diameters between the larger microtubules and smaller microfilaments.

internal environment The physical and chemical characteristics of the extracellular fluids of the body.

interneuron A neuron that communicates information between two other neurons.

ionotropic receptors A receptor that that directly alters membrane permeability to a type of ion when it combines with its ligand.

internode The region between two nodes of a plant stem.

interphase The period between successive nuclear divisions during which the chromosomes are diffuse and the nuclear envelope is intact. It is during this period that the cell is most active in transcribing and translating genetic information.

interspecific competition Competition between members of two or more species. (Contrast with intraspecific competition.)

intertropical convergence zone The tropical region where the air rises most strongly; moves north and south with the passage of the sun overhead.

intraspecific competition Competition among members of the same species. (Contrast with interspecific competition.)

intrinsic protein A membrane protein that is embedded in the phospholipid bilayer of the membrane. (Contrast with extrinsic protein.)

intrinsic rate of increase The rate at which a population can grow when its density is low and environmental conditions are highly favorable.

intron A portion of a DNA molecule that, because of RNA splicing, is not involved in coding for part of a polypeptide molecule. (Contrast with exon.)

invagination An infolding of cells during animal embryonic development.

inversion A rare 180° reversal of the order of genes within a segment of a chromosome.

invertebrate A "convenience term" that encompasses any animal that is not a vertebrate—that is, whose nerve cord is not enclosed in a backbone of bony segments.

in vitro [L.: in glass] In a test tube, rather than in a living organism. (Contrast with in vivo.)

in vitro evolution Evolution in the laboratory, as used to produce compounds for industrial and pharmaceutical purposes.

in vivo [L.: in the living state] In a living organism. Many processes that occur in vivo can be reproduced in vitro with the right selection of cellular components. (Contrast with in vitro.)

ion (eye' on) [Gk.: *ion*: wanderer] An atom or group of atoms with electrons added or removed, giving it a negative or positive electrical charge.

ion channel A membrane protein that can let ions diffuse across the membrane. The channel can be ion-selective, and it can be voltage-gated or ligand-gated.

ionic bond An electrostatic attraction between positively and negatively charged ions. Usually a strong bond.

iris (eye' ris) [Gk. *iris*: rainbow] The round, pigmented membrane that surrounds the pupil of the eye and adjusts its aperture to regulate the amount of light entering the eye.

irruption A rapid increase in the density of a population. Often followed by massive emigration.

islets of Langerhans Clusters of hormone-producing cells in the pancreas.

iso- [Gk. *iso*: equal] Prefix used two separate entities that share some element of identity.

isogamous Describes male and female gametes that are morphologically identical.

isolating mechanism Geographical, physiological, ecological, or behavioral mechanisms that lead to a reduction in the frequency of successful matings between individuals in separate populations of a species. Can lead to the eventual evolution of separate species.

isomers Molecules consisting of the same numbers and kinds of atoms, but differing in the bonding patterns by which the atoms are held together.

isomorphic (eye so more' fik) [Gk. *isos*: equal + *morphe*: form] Having the same form or appearance, as when the haploid and diploid life stages of an organism appear identical. (Contrast with heteromorphic.)

isotonic Having the same solute concentration; said of two solutions. (Contrast with hypertonic, hypotonic.)

isotope (eye' so tope) [Gk. *isos*: equal + *topos*: place] Isotopes of a given chemical element have the same number of protons in their nuclei (and thus are in the same position on the periodic table), but differ in the number of neutrons

isozymes Enzymes that have somewhat different amino acid sequences but catalyze the same reaction.

- J -

jasmonates Plant hormones that trigger defenses against pathogens and herbivores.

jejunum (jih jew' num) The middle division of the small intestine, where most absorption of nutrients occurs. (*See* duodenum, ileum.)

joule (jool, or jowl) A unit of energy, equal to 0.239 calories.

juvenile hormone In insects, a hormone maintaining larval growth and preventing maturation or pupation.

- K -

karyogamy The fusion of nuclei of two cells. (Contrast with plasmogamy.)

karyotype The number, forms, and types of chromosomes in a cell.

keratin (ker' a tin) [Gk. keras: horn] A protein which contains sulfur and is part of such hard tissues as horn, nail, and the outermost cells of the skin.

ketone (key' tone) A compound with a C=O group attached to two other groups, neither of which is an H atom. Many sugars are ketones. (Contrast with aldehyde.)

keystone species Species that have a dominant influence on the composition of a community.

kidneys A pair of excretory organs in vertebrates.

kin selection That component of inclusive fitness resulting from helping the survival of relatives containing the same alleles by descent from a common ancestor. (Contrast with individual fitness.)

kinase (kye' nase) An enzyme that transfers a phosphate group from ATP to another molecule. Protein kinases transfer phosphate from ATP to specific proteins, playing important roles in cell regulation.

kinesin Motor protein having the capacity to attach to organelles or vesicles and move them along microtubules of the cytoskeleton.

kinetic energy The energy associated with movement.

kinetochore (kin net' oh core) [Gk. *kinetos*: moving] Specialized structure on a centromere to which microtubules attach.

knockout A molecular genetic method in which a single gene of an organism is permanently inactivated.

Koch's posulates A set of rules for establishing that a particular microorganism causes a particular disease.

Krebs cycle *See* citric acid cycle.

- L -

lactic acid fermentation Fermentation whose end product is lactic acid (lactate).

lagging strand In DNA replication, the daughter strand that is synthesized in discontinuous stretches. (*See* Okazaki fragments.)

lamella (la mell' ah) [L. *lamina*: thin sheet] Layer.

larva (plural: larvae) [L. *lares*: guiding spirits] An immature stage of any invertebrate animal that differs dramatically in appearance from the adult.

larynx (lar' inks) [Gk. *larynx*: voice box] A structure between the pharynx and the trachea that includes the vocal cords.

lateral Pertaining to the side.

lateral gene transfer The transfer of genes from one species to another, common among bacteria and archaea.

lateral meristems The vascular cambium and cork cambium, which give rise to secondary tissue in plants.

laticifers (luh tiss' uh furs) In some plants, elongated cells containing secondary plant products such as latex.

leader sequence A sequence of amino acids at the amino-terminal end of a newly synthesized protein; determines where the protein will be placed in the cell.

leading strand In DNA replication, the daughter strand that is synthesized continuously. (Contrast with lagging strand.)

leaf primordium An outgrowth on the side of the shoot apical meristem that will eventually develop into a leaf.

lenticel (len' ti sill) Spongy region in a plant's periderm, allowing gas exchange.

leukocyte (loo' ko sight) [Gk. *leukos*: clear + *kytos*: container] A white blood cell.

lichen (lie' kun) An organism resulting from the symbiotic association of a true fungus and either a cyanobacterium or a unicellular alga.

life cycle The entire span of the life of an organism from the moment of fertilization (or asexual generation) to the time it reproduces in turn.

life history The stages an individual goes through during its life.

life table A table showing, for a group of equal-aged individuals, the proportion still alive at different times in the future and the number of offspring they produce during each time interval.

ligament A band of connective tissue linking two bones in a joint.

ligand (lig' and) Any molecule that binds to a receptor site of another (usually larger) molecule.

light reactions The initial phase of photosynthesis, in which light energy is converted into chemical energy.

light-independent reactions The phase of photosynthesis in which chemical energy captured in the light reactions is used to drive the reduction of CO_2 to form carbohydrates.

lignin The principal noncarbohydrate component of wood, a polymer that binds together cellulose fibrils in some plant cell walls.

limbic system A group of primitive vertebrate forebrain nuclei that form a network and are involved in emotions, drives, instinctive behaviors, learning, and memory.

limiting resource The required resource whose supply most strongly influences the size of a population.

linkage Association between genetic markers on the same chromosome such that they do not show random assortment and seldom recombine; the closer the markers, the lower the frequency of recombination.

lipase (lip' ase; lye' pase) An enzyme that digests fats.

lipids (lip' ids) [Gk. *lipos*: fat] Substances in a cell which are easily extracted by organic solvents; fats, oils, waxes, steroids, and other large organic molecules, including those which, with proteins, make up the cell membranes. (Compare with phospholipids.)

littoral zone The coastal zone from the upper limits of tidal action down to the depths where the water is thoroughly stirred by wave action.

liver A large digestive gland. In vertebrates, it secretes bile and is involved in the formation of blood.

lobes Regions of the human cerebral hemispheres; includes the temporal, frontal, parietal, and occipital lobes.

locus In genetics, a specific location on a chromosome. May be considered to be synonymous with *gene*.

logistic growth Growth, especially in the size of an organism or in the number of organisms in a population, that slows steadily as the entity approaches its maximum size. (Contrast with exponential growth.)

long-day plant (LDP) A plant that requires long days (actually, short nights) in order to flower.

long-term potentiation (LTP) A long-lasting increase in the sensitivity of a neuron resulting from a period of intense stimulation.

loop of Henle (hen' lee) Long, hairpin loop of the mammalian renal tubule that runs from the cortex down into the medulla, and back to the cortex. Creates a concentration gradient in the interstitial fluids in the medulla.

lophophore A U-shaped fold of the body wall with hollow, ciliated tentacles that encircles the mouth of animals in several different groups. Used for filtering prey from the surrounding water.

lumen (loo' men) [L. *lumen*: light] The open cavity inside any tubular organ or structure, such as the gut or a kidney tubule.

luteinizing hormone A gonadotropin produced by the anterior pituitary. It stimulates the gonads to produce sex hormones.

lymph [L. *lympha*: liquid] A clear, watery fluid that is formed as a filtrate of blood; it contains white blood cells; it collects in a series of special vessels and is returned to the bloodstream.

lymph nodes Specialized tissue regions that act as filters for cells, bacteria and foreign matter.

lymphocyte A major class of white blood cells. Includes T cells, B cells, and other cell types important in the immune response.

lymphoid tissue Tissues of the immune defense system dispersed throughout the body and consisting of: thymus, spleen, bone marrow, lymph nodes, blood, and lymph.

lysis (lie' sis) [Gk. *lysis*: break apart] Bursting of a cell.

lysogenic bacteria Bacteria that harbor a viral chromosome capable of the lysogenic cycle.

lysogenic cycle A form of viral replication in which the virus becomes incorporated into the bacterial chromosome and the host cell is not killed. (Contrast with lytic cycle.)

lysosome (lie' so soam) [Gk. *lysis*: break away + *soma*: body] A membrane-enclosed organelle found in eukaryotic cells (other than plants). Lysosomes contain a mixture of enzymes that can digest most of the macromolecules found in the rest of the cell.

lysozyme (lie' so zyme) An enzyme in saliva, tears, and nasal secretions that attacks bacterial cell walls, as one of the body's nonspecific defense mechanisms.

lytic cycle A form of viral reproduction that lyses the host bacterium releasing the new viruses. (Contrast with lysogenic cycle.)

- M -

M phase The portion of the cell cycle in which mitosis takes place.

macroevolution [Gk. *makros*: large, long] Evolutionary changes occurring over long time spans and usually involving changes in many traits. (Contrast with microevolution.)

macromolecule A giant polymeric molecule. The macromolecules are proteins, polysaccharides, and nucleic acids.

macronutrient A mineral element required by plant tissues in concentrations of at least 1 milligram per gram of their dry matter.

macrophage (mac' roh faj) A type of white blood cell that endocytoses bacteria and other cells.

MADS box A DNA-binding domain in many plant transcription factors that is active in development.

major histocompatibility complex (MHC) A complex of linked genes, with multiple alleles, that control a number of cell surface antigens that identify self and can lead to graft rejection.

malignant Referring to a tumor that can grow indefinitely and/or spread from the original site of growth to other locations in the body. (Contrast with benign.)

malleus (mal' ee us) [L. *malleus*: hammer] The first of the three bones that conduct movements of the eardrum to the oval window of the inner ear. (*See* incus, stapes.)

Malpighian tubule (mal pee' gy un) A type of protonephridium found in insects.

mantle A sheet of specialized tissues that covers most of the viscera of mollusks; provides protection to internal organs and secretes the shell.

mapping In genetics, determining the order of genes on a chromosome and the distances between them.

map unit The distance between two genes, a recombinant frequency of 0.01.

marine [L. *mare*: sea, ocean] Pertaining to or living in the ocean. (Contrast with aquatic, terrestrial.)

marker A gene of identifiable phenotype that indicates the presence on another gene, DNA segment, or chromosome fragment.

mass extinctions Periods of evolutionary history during which rates of extinction were much higher than during intervening times.

mass number The sum of the number of protons and neutrons in an atom's nucleus.

mast cells Typically found in connective tissue, mast cells can be provoked by antigens or inflammation to release histamine.

maternal effect genes These genes code for morphogens that determine the polarity of the egg and larva in the fruit fly *Drosophila melanogaster*.

maternal inheritance Inheritance in which the mother's phenotype is exclusively expressed. Mitochondria and chloroplasts are maternally inherited via egg cytoplasm. Also known as cytoplasmic inheritance.

mating type A particular strain of a species that is incapable of sexual reproduction with another member of the same strain but capable of sexual reproduction with members of other strains of the same species.

maximum likelihood A statistical method of determining which of two or more hypotheses (such as phylogenetic trees) best fit the observed data, given an explicit model of how the data were generated.

mechanoreceptor A cell that is sensitive to physical movement and generates action potentials in response.

medulla (meh dull' luh) (1) The inner, core region of an organ, as in the adrenal medulla (adrenal gland) or the renal medulla (kidneys). (2) The portion of the brain stem that connects to the spinal cord.

megaphyll The generally large leaf of a fern, horsetail, or seed plant, with several to many veins. (Contrast with microphyll.)

megaspore [Gk. *megas*: large + *spora*: to sow] In plants, a haploid spore that produces a female gametophyte.

meiosis (my oh' sis) [Gk. *meiosis*: diminution] Division of a diploid nucleus to produce four haploid daughter cells. The process consists of two successive nuclear divisions with only one cycle of chromosome replication. In *meiosis I*, homologous chromosomes separate but retain their chromatids. The second division *meiosis II*, is similar to mitosis, in which chromatids separate.

melatonin A hormone released by the pineal gland that is involved in photoperiodism and circadian rhythms.

membrane potential The difference in electrical charge between the inside and the outside of a cell, caused by a difference in the distribution of ions.

memory cells Long-lived lymphocytes produced by exposure to antigen. They persist in the body and are able to mount a rapid response to subsequent exposures to the antigen.

Mendelian population A local population of individuals belonging to the same species and exchanging genes with one another.

Mendel's first law *See* segregation.

Mendel's second law *See* independent assortment.

menstrual cycle The monthly sloughing off of the uterine lining if fertilization does not occur in the female. Occurs between puberty and menopause.

meristem [Gk. *meristos*: divided] Plant tissue made up of undifferentiated actively dividing cells.

mesenchyme (mez' en kyme) [Gk. *mesos*: middle + *enchyma*: infusion] Embryonic or unspecialized cells derived from the mesoderm.

mesoderm [Gk. *mesos*: middle + *derma*: skin] The middle of the three embryonic tissue layers first delineated during gastrulation. Gives rise to skeleton, circulatory system, muscles, excretory system, and most of the reproductive system.

mesophyll (mez' uh fill) [Gk. *mesos*: middle + *phyllon*: leaf] Chloroplast-containing, photosynthetic cells in the interior of leaves.

mesosome (mez' uh soam') [Gk. *mesos*: middle + *soma*: body] A localized infolding of the plasma membrane of a bacterium.

messenger RNA (mRNA) A transcript of one of the strands of DNA; carries information (as a sequence of codons) for the synthesis of one or more proteins.

meta- [Gk.: between, along with, beyond] A prefix used in biology to denote a change or a shift to a new form or level; for example, as used in metamorphosis.

metabolic compensation Changes in metabolic properties of an organism that render it less sensitive to temperature changes.

metabolic factor In bacteria, a plasmid that carries genes determining unusual metabolic functions, such as the breakdown of hydrocarbons in oil.

metabolic pathway A series of enzyme-catalyzed reactions so arranged that the product of one reaction is the substrate of the next.

metabolism (meh tab' a lizm) [Gk. *metabole*: to change] The sum total of the chemical reactions that occur in an organism, or some subset of that total (as in respiratory metabolism).

metabotropic receptor A receptor that that indirectly alters membrane permeability to a type of ion when it combines with its ligand.

metamorphosis (met' a mor' fo sis) [Gk. *meta*: between + *morphe*: form, shape] A change occurring between one developmental stage and another, as for example from a tadpole to a frog. (*See* complete metamorphosis, incomplete metamorphosis.)

metaphase (met' a phase) The stage in nuclear division at which the centromeres of the highly supercoiled chromosomes are all lying on a plane (the metaphase plane or plate) perpendicular to a line connecting the division poles.

metapopulation A population divided into subpopulations, among which there are occasional exchanges of individuals.

metastasis (meh tass' tuh sis) The spread of cancer cells from their original site to other parts of the body.

methylation The addition of a methyl group (—CH_3) to a molecule. Extensive methylation of cytosine in DNA is correlated with reduced transcription.

MHC *See* major histocompatibility complex.

micelle a particle of lipid covered with bile salts that is produced in the duodenum and facilitates digestion and absorption of lipids.

microbiology [Gk. *mikros*: small + *bios*: life + *logos*: discourse] The scientific study of microscopic organisms, particularly bacteria, protists, and viruses.

microevolution The small evolutionary changes typically occurring over short time spans; generally involving a small number of traits and minor genetic changes. (Contrast with macroevolution.)

microfilament Minute fibrous structure generally composed of actin found in the cytoplasm of eukaryotic cells. They play a role in the motion of cells.

micronutrient A mineral element required by plant tissues in concentrations of less than 100 micrograms per gram of their dry matter.

microphyll A small leaf with a single vein, found in club mosses and their relatives. (Contrast with megaphyll.)

micropyle (mike' roh pile) [Gk. *mikros*: small + *pylon*: gate] Opening in the integument(s) of a seed plant ovule through which pollen grows to reach the female gametophyte within.

micro RNA A small RNA, typically about 21 bases long, that binds to mRNA to reduce its translation.

microspore [Gk. *mikros*: small + *spora*: to sow] In plants, a haploid spore that produces a male gametophyte.

microtubules Minute tubular structures found in centrioles, spindle apparatus, cilia, flagella, and cytoskeleton of eukaryotic cells. These tubules play roles in the motion and maintenance of shape of eukaryotic cells.

microvilli (singular: microvillus) The projections of epithelial cells, such as the cells lining the small intestine, that increase their surface area.

middle lamella A layer of polysaccharides that separates plant cells; a shared middle lamella lies outside the primary walls of the two cells.

migration The regular, seasonal movements of animals.

mineral An inorganic substance other than water.

mineral nutrients Inorganic ions required by organisms for normal growth and reproduction.

mismatch repair When a single base in DNA is changed into a different base, or the wrong base inserted during DNA replication, there is a mismatch in base pairing with the base on the opposite strand. A repair system removes the incorrect base and inserts the proper one for pairing with the opposite strand.

missense substitution A change in a gene from one nucleotide to another that also results in a change in the amino acid specified by the corresponding codon. (Compare with frame-shift mutation, nonsense substitution, synonymous substitution.)

mitochondrial matrix The fluid interior of the mitochondrion, enclosed by the inner mitochondrial membrane.

mitochondrion (my' toe kon' dree un) [Gk. *mitos*: thread + *chondros*: grain] An organelle in eukaryotic cells that contains the enzymes of the citric acid cycle, the respiratory chain, and oxidative phosphorylation.

mitosis (my toe' sis) [Gk. *mitos*: thread] Nuclear division in eukaryotes leading to the formation of two daughter nuclei, each with a chromosome complement identical to that of the original nucleus.

mitotic center Cellular region that organizes the microtubules for mitosis. In animals a centrosome serves as the mitotic center.

moderately repetitive DNA DNA sequences that appear hundreds to thousands of times in the genome. They include the DNA sequences coding for rRNAs and tRNAs, as well as the DNA at telomeres.

modular organism An organism which grows by producing additional units of body construction (modules) that are very similar to the units of which it is already composed.

mole A quantity of a compound whose weight in grams is numerically equal to its molecular weight expressed in atomic mass units. Avogadro's number of molecules: 6.023×10^{23} molecules.

molecular clock The theory that macromolecules diverge from one another over evolutionary time at a constant rate; this rate may provide insight into the phylogenetic relationships among organisms.

molecular tool kit A set of developmental genes and proteins that is common to most animals and is hypothesized to be responsible for the evoltuion of their differing developmental pathways.

molecular weight The sum of the atomic weights of the atoms in a molecule.

molecule A particle made up of two or more atoms joined by covalent bonds or ionic attractions.

molting The process of shedding part or all of an outer covering, as the shedding of feathers by birds or of the entire exoskeleton by arthropods.

monoclonal antibody Antibody produced in the laboratory from a clone of hybridoma cells, each of which produces the same specific antibody.

monocytes White blood cells that produce macrophages.

monoecious (mo nee' shus) [Gk. *mono*: one + *oikos*: house] Describes organisms in which both sexes are "housed" in a single individual that produces both eggs and sperm. (In some plants, these are found in different flowers within the same plant.) Examples include corn, peas, earthworms, hydras. (Contrast with dioecious.)

monohybrid cross A mating in which the parents differ with respect to the alleles of only one locus of interest.

monomer [Gk. *mono*: one + *meros:* unit] A small molecule, two or more of which can be combined to form oligomers (consisting of a few monomers) or polymers (consisting of many monomers).

monophyletic (mon' oh fih leht' ik) [Gk. *mono*: one + *phylon*: tribe] Referring to a group that consists of an ancestor and all of its descendants. (Compare with paraphyletic, polyphyletic.)

monosaccharide A simple sugar. Oligosaccharides and polysaccharides are made up of monosaccharides.

monosomic Referring to an organism with one less than the normal diploid number of chromosomes.

monosynaptic reflex A neural reflex that begins in a sensory neuron and makes a single synapse before activating a motor neuron.

morphogen A diffusible substances whose concentration gradients determine patterns of development in animals and plants.

morphogenesis (more' fo jen' e sis) [Gk. *morphe*: form + *genesis*: origin] The development of form; the overall consequence of determination, differentiation, and growth.

morphology (more fol' o jee) [Gk. *morphe*: form + *logos*: study, discourse] The scientific study of organic form, including both its development and function.

mosaic development Pattern of animal embryonic development in which each blastomere contributes a specific part of the adult body. (Contrast with regulative development.)

motor end plate The modified area on a muscle cell membrane where a synapse is formed with a motor neuron.

motor neuron A neuron carrying information from the central nervous system to an effector such as a muscle fiber.

motor proteins Specialized proteins that use energy to change shape and move cells or structures within cells. *See* dynein, kinesin.

motor unit A motor neuron and the set of muscle fibers it controls.

mRNA *See* messenger RNA.

mucosa (mew koh' sah) An epithelial membrane containing cells that secrete mucus. The inner cell layers of the digestive and respiratory tracts.

Müllerian mimicry The convergence in appearance of two or more unpalatable species.

multifactorial In medicine, referring to a disease with many interacting causes, both genetic and environmental.

muscle fiber A single muscle cell. In the case of skeletal (striated) muscle, a syncitial, multinucleate cell.

muscle tissue Excitable tissue that can contract due to interactions of actin and myosin. Three types are striated, smooth, and cardiac.

muscle tone The degree of contraction of a muscle.

muscle spindle Modified muscle fibers encased in a connective sheat and functioning as stretch receptors.

mutagen (mute' ah jen) [L. *mutare*: change + Gk. *genesis*: source] Any agent (e.g., chemicals, radiation) that increases the mutation rate.

mutation A detectable, heritable change in the genetic material not caused by recombination.

mutation pressure Evolution (change in gene proportions) by different mutation rates alone (i.e., without the influence of natural selection).

mutualism The type of symbiosis, such as that exhibited by fungi and algae or cyanobacteria in forming lichens, in which both species profit from the association.

mycelium (my seel' ee yum) [Gk. *mykes*: fungus] In the fungi, a mass of hyphae.

mycorrhiza (my' ko rye' za) [Gk. *mykes*: fungus + *rhiza*: root] An association of the root of a plant with the mycelium of a fungus.

myelin (my' a lin) A material forming a sheath around some axons. Formed by Schwann cells that wrap themselves about the axon, myelin insulates the axon electrically and increases the rate of transmission of a nervous impulse.

myocardial infarction A blockage of an artery that carries blood to the heart muscle.

myofibril (my' oh fy' bril) [Gk. *mys*: muscle + L. *fibrilla*: small fiber] A polymeric unit of actin or myosin in a muscle.

myogenic (my oh jen' ik) [Gk. *mys*: muscle + *genesis*: source] Originating in muscle.

myoglobin (my' oh globe' in) [Gk. *mys*: muscle + L. *globus*: sphere] An oxygen-binding molecule found in muscle. Consists of a heme unit and a single globiin chain, and carrys less oxygen than hemoglobin.

myosin One of the two major proteins of muscle, it makes up the thick filaments. (*See* actin.)

- N -

NAD (nicotinamide adenine dinucleotide) A compound found in all living cells, existing in two interconvertible forms: the oxidizing agent NAD^+ and the reducing agent $NADH + H^+$.

NADP (nicotinamide adenine dinucleotide phosphate) A compound similar to NAD, but possessing another phosphate group; plays similar roles but is used by different enzymes.

natural killer cells A nonspecific defensive cell (lymphocyte) that attacks tumor cells and virus infected cells.

natural selection The differential contribution of offspring to the next generation by various genetic types belonging to the same population. The mechanism of evolution proposed by Charles Darwin.

necrosis (nec roh' sis) [Gk. *nekros*: death] Tissue damage resulting from cell death.

negative control The situation in which a regulatory macromolecule (generally a repressor) functions to turn off transcription. In the absence of a regulatory macromolecule, the structural genes are turned on.

negative feedback Information relevant to the rate of a process that can be used by a control system to return the outcome of that process to an optimal level.

nematocyst (ne mat' o sist) [Gk. *nema*: thread + *kystis*: cell] An elaborate, threadlike structure produced by cells of jellyfish and other cnidarians, used chiefly to paralyze and capture prey.

nephridium (nef rid' ee um) [Gk. *nephros*: kidney] An organ which is involved in excretion, and often in water balance, involving a tube that opens to the exterior at one end.

nephron (nef' ron) [Gk. *nephros*: kidney] The functional unit of the kidney, consisting of a structure for receiving a filtrate of blood, and a tubule that absorbs selected parts of the filtrate back into the bloodstream.

nephrostome (nef' ro stome) [Gk. *nephros*: kidney + *stoma*: opening] An opening in a nephridium through which body fluids can enter.

Nernst equation A mathematical statement; calculates the potential across a membrane permeable to a single type of ion that differs in concentration on the two sides of the membrane.

nerve A structure consisting of many neuronal axons and connective tissue.

net primary production Total photosynthesis minus respiration by plants.

neural plate A thickened strip of ectoderm along the dorsal side of the early vertebrate embryo; gives rise to the central nervous system.

neural tube An early stage in the development of the vertebrate nervous system consisting of a hollow tube created by two opposing folds of the dorsal ectoderm along the anterior–posterior body axis.

neurohormone A chemical signal produced and released by neurons; the signal then acts as a hormone.

neuromuscular junction The region where a motor neuron contacts a muscle fiber, creating a synapse.

neuron (noor' on) [Gk. *neuron*: nerve] A nervous system cell that can generate and conduct action potentials along an axon to a synapse with another cell.

neurotransmitter A substance produced in and released by one a neuron (the presynaptic cell) that diffuses across a synapse and excites or inhibits another cell (the postsynaptic cell).

neurula (nure' you la) Embryonic stage during the dorsal nerve cord forms from two ectodermal ridges.

neurulation A stage in vertebrate development during which the nervous system begins to form.

neutral allele An allele that does not alter the functioning of the proteins for which it codes.

neutral theory A view of molecular evolution that postulates that most mutations do not affect the amino acid being coded for, and that such mutations accumulate in a population at rates driven by genetic drift and mutation rates.

neutron (new' tron) One of the three most fundamental particles of matter, with mass approximately 1 amu and no electrical charge.

neutrophils Abundant, short-lived phagocytic leukocytes that attack antibody-coated antigens.

nitrate reduction The process by which nitrate (NO_3^-) is reduced to ammonia (NH_3).

nitric oxide (NO) An unstable molecule (a gas) that serves as a second messenger causing smooth muscle to relax. In the nervous system it operates as a neurotransmitter.

nitrification The oxidation of ammonia to nitrite and nitrate ions, performed by certain soil bacteria.

nitrogenase In nitrogen-fixing organisms, an enzyme complex that mediates the stepwise reduction of atmospheric N_2 to ammonia.

nitrogen fixation Conversion of nitrogen gas to ammonia, which makes nitrogen available to living things. Carried out by certain prokaryotes, some of them free-living and others living within plant roots.

node [L. *nodus*: knob, knot] In plants, a (sometimes enlarged) point on a stem where a leaf is or was attached.

node of Ranvier A gap in the myelin sheath covering an axon; the point where the axonal membrane can fire action potentials.

noncompetitive inhibitor An inhibitor that binds the enzyme at a site other than the active site. (Contrast with competitive inhibitor.)

noncyclic electron transport In photosynthesis, the flow of electrons that forms ATP, NADPH, and O_2.

nondisjunction Failure of sister chromatids to separate in meiosis II or mitosis, or failure of homologous chromosomes to separate in meiosis I. Results in aneuploidy.

nonpolar molecule A molecule whose electric charge is evenly balanced from one end of the molecule to the other.

nonrandom mating The selection by individuals of other individuals of particular genotypes as mates.

non-REM sleep A state of sleep characterized by low muscle tone, but not atonia, quiescence,

nonsense substitution A change in a gene from one nucleotide to another that prematurely terminates a polypeptide. Termination occurs when a codon that specifies an amino acid is changed to one of the codons (UAG, UAA, or UGA) that signal termination of translation. (Compare with frame-shift mutation, missense substitution, synonymous substitution.)

nonspecific defenses Immunologic responses directed against any invading agent without reacting to apecific antigens.

nonsynonymous substitution A change in a gene from one nucleotide to another that changes the amino acid specified by the corresponding codon (i.e., AGC → AGA, or serine → arginine). (Contrast with synonymous substitution.)

nonvascular plants Those plants lacking well-developed vascular tissue; the liverworts, hornworts, and mosses. (Contrast with vascular plants.)

norepinephrine A neurotransmitter found in the central nervous system and also at the postganglionic nerve endings of the sympathetic nervous system. Also called noradrenaline.

notochord (no' tow kord) [Gk. *notos*: back + *chorde*: string] A flexible rod of gelatinous material serving as a support in the embryos of all chordates and in the adults of tunicates and lancelets.

nuclear envelope The surface, consisting of two layers of membrane, that encloses the nucleus of eukaryotic cells.

nuclear lamina A meshwork of fibers on the inner surface of the nuclear envelope.

nuclear pore complex A protein structure situated in nuclear pores through which RNA and proteins enter and leave the nucleus.

nucleic acid (new klay' ik) A long-chain alternating polymer of deoxyribose or ribose and phosphate groups, with nitrogenous bases—adenine, thymine, uracil, guanine, or cytosine (A, T, U, G, or C)—as side chains. DNA and RNA are nucleic acids.

nucleic acid hybridization A technique in which a single-stranded nucleic acid probe is made that is complementary to, and binds to, a target sequence, either DNA or RNA. The resulting double-stranded molecule is a hybrid.

nucleoid (new' klee oid) The region that harbors the chromosomes of a prokaryotic cell. Unlike the eukaryotic nucleus, it is not bounded by a membrane.

nucleolar organizer (new klee' o lar) A region on a chromosome that is associated with the formation of a new nucleolus following nuclear division. The site of the genes that code for ribosomal RNA.

nucleolus (new klee' oh lus) A small, generally spherical body found within the nucleus of eukaryotic cells. The site of synthesis of ribosomal RNA.

nucleoplasm (new' klee o plazm) The fluid material within the nuclear envelope of a cell, as opposed to the chromosomes, nucleoli, and other particulate constituents.

nucleoside A nucleotide without the phosphate group.

nucleosome A portion of a eukaryotic chromosome, consisting of part of the DNA molecule wrapped around a group of histone molecules, and held together by another type of histone molecule. The chromosome is made up of many nucleosomes.

nucleotide The basic chemical unit in a nucleic acid. A nucleotide in RNA consists of one of four nitrogenous bases linked to ribose, which in turn is linked to phosphate. In DNA, deoxyribose is present instead of ribose.

nucleotide substitution A change of one base pair to another in a DNA sequence.

nucleus (new' klee us) [L. *nux*: kernel or nut] (1) In cells, the centrally located compartment of eukaryotic cells that is bounded by a double membrane and contains the chromosomes. (2) In the brain, an identifiable group of neurons that share common characteristics or functions.

null hypothesis The assertion that an effect proposed by its companion hypothesis does not in fact exist.

nutrient A food substance; or, in the case of mineral nutrients, an inorganic element required for completion of the life cycle of an organism.

- O -

obligate anaerobe An anaerobic prokaryote that cannot survive exposure to O_2.

odorant A molecule that can bind to an olfactory receptor.

oil A triglyceride that is liquid at room temperature. (Contrast with fat.)

Okazaki fragments Newly formed DNA making up the lagging strand in DNA replication. DNA ligase links Okazaki fragments together to give a continuous strand.

olfactory [L. *olfacere*: to smell] Having to do with the sense of smell.

oligomer [Gk.: *oligo*: a few + *meros*: units] A compound molecule of intermediate size, made up of two to a few monomers. (Contrast with monomer, polymer.)

oligosaccharide A polymer containing a small number of monosaccharides.

oligosaccharins Plant hormones, derived from the plant cell wall, that trigger defenses against pathogens.

ommatidium [Gk. *omma*: eye] One of the units which, collected into groups of up to 20,000, make up the compound eye of arthropods.

omnivore [L. *omnis*: everything + *vorare*: to devour] An organism that eats both animal and plant material. (Contrast with carnivore, detritivore, herbivore.)

oncogene [Gk. *onkos*: mass, tumor + *genes*: born] Genes that greatly stimulate cell division, giving rise to tumors.

one-gene, one-polypeptide The principle that each gene codes for a single polypeptide.

oocyte (oh' eh site) [Gk. *oon*: egg + *kytos*: container] The cell that gives rise to eggs in animals.

oogenesis (oh' eh jen e sis) [Gk. *oon*: egg + *genesis*: source] Female gametogenesis, leading to production of the egg.

oogonium (oh' eh go' nee um) In some algae and fungi, a cell in which an egg is produced.

operator The region of an operon that acts as the binding site for the repressor.

operon A genetic unit of transcription, typically consisting of several structural genes that are transcribed together; the operon contains at least two control regions: the promoter and the operator.

opportunity cost The sum of the benefits an animal forfeits by not being able to perform some other behavior during the time when it is performing a given behavior.

opsin (op' sin) [Gk. *opsis*: sight] The protein portion of the visual pigment rhodopsin. (*See* rhodopsin.)

optic chiasm [Gk. *chiasma*: cross] Structure on the lower surface of the vertebrate brain where the two optic nerves come together.

optical isomers Two isomers that are mirror images of each other.

orbital A region in space surrounding the atomic nucleus in which an electron is most likely to be found.

organ [Gk. *organon*: tool] A body part, such as the heart, liver, brain, root, or leaf. Organs are composed of different tissues integrated to perform a distinct function. Organs are in turn often integrated into systems, such as the digestive or reproductive system.

organ identity genes Plant genes that specify the various parts of the flower. *See* homeotic genes.

organ of Corti Structure in the inner ear that transforms mechanical forces produced from pressure waves ("sound waves") into action potentials that are sensed as sound.

organ system An interrelated and integrated group of tissues and organs that work together in a physiological function.

organelles (or gan els') Organized structures found in or on eukaryotic cells. Examples include ribosomes, nuclei, mitochrondria, chloroplasts, cilia, and contractile vacuoles.

organic Pertaining to any aspect of living matter, e.g., to its evolution, structure, or chemistry. The term is also applied to any chemical compound that contains carbon.

organism Any living entity.

organizer Region of an early embryo that directs the development of nearby regions. In amphibian early gastrulas, the dorsal lip of the blastopore is the organizer.

organogenesis The formation of organs and organ systems during development.

origin of replication DNA sequence at which helicase unwinds the DNA double helix and DNA polymerase binds to initiate DNA replication.

orthology (or thol' o jee) A type of homology applied to genes in which the divergence of homologous genes can be traced to speciation events. The genes are said to be *orthologous*, and each is an *ortholog* of the others. (Compare with paralogy)

osmoconformer An aquatic animal that maintains an osmotic concentration of its extracellular fluid that is the same as that of the external environment.

osmolarity The concentration of osmotically active particles in a solution.

osmoregulation Regulation of the chemical composition of the body fluids of an organism.

osmoreceptor Neuron that converts changes in the solute potential of interstial fluids into action potentials.

osmosis (oz mo' sis) [Gk. *osmos*: to push] The movement of water across a differentially permeable membrane, from one region to another region where the water potential is more negative.

ossicle (oss' ick ul) [L. *os*: bone] The calcified construction unit of echinoderm skeletons.

osteoblasts (oss' tee oh blast) [Gk. *osteon*: bone + *blastos*: sprout] Cells that lay down the protein matrix of bone.

osteoclasts (oss' tee oh clast) [Gk. *osteon*: bone + *klastos*: broken] Cells that dissolve bone.

otolith (oh' tuh lith) [Gk. *otikos*: ear + *lithos*: stone[Structures in the vertebrate vestibular apparatus that mechanically stimulate hair cells when the head moves or changes position.

oval window The flexible membrane that, when moved by the bones of the middle ear, produces pressure waves in the inner ear

ovary (oh' var ee) [L. *ovum*: egg] Any female organ, in plants or animals, that produces an egg.

oviduct [L. *ovum*: egg + *ducere*: to lead] In mammals, the tube serving to transport eggs to the uterus or to outside of the body.

oviparity Reproduction in which eggs are released by the female and development is external to the mother's body. (Contrast with viviparous.)

ovoviviparity Reproduction in which fertilized eggs develop and hatch within the body of the mother but are not attached to the mother by means of a placenta.

ovulation The release of an egg from an ovary.

ovule (oh' vule) In plants, a structure that contains a gametophyte and, within the gametophyte, an egg; when it matures, an ovule becomes a seed.

ovum (oh' vum) [L. *ovum*: egg] The egg; the female sex cell.

oxidation (ox i day' shun) Relative loss of electrons in a chemical reaction; either outright removal to form an ion, or the sharing of electrons with substances having a greater affinity for them, such as oxygen. Most oxidation, including biological ones, are associated with the liberation of energy. (Contrast with reduction.)

oxidative phosphorylation ATP formation in the mitochondrion, associated with flow of electrons through the respiratory chain.

oxidizing agent A substance that can accept electrons from another. The oxidizing agent becomes reduced; its partner becomes oxidized.

oxygenase An enzyme that catalyzes the addition of oxygen to a substrate from O_2.

-P-

P generation Parental generation. The individuals that mate in a genetic cross. Their immediate offspring are the F_1 generation.

pacemaker That part of the heart which undergoes most rapid spontaneous contraction, thus setting the pace for the beat of the entire heart. In mammals, the sinoatrial (SA) node. Also, an artificial device, implanted in the heart, that initiates rhythmic contraction of the organ.

Pacinian corpuscle A modified nerve ending that senses touch and vibration.

pair rule genes Segmentation genes that divide the *Drosophila* larva into two segments each.

pancreas (pan' cree us) A gland located near the stomach of vertebrates that secretes digestive enzymes into the small intestine and releases insulin into the bloodstream.

Pangaea (pan jee' uh) [Gk. *pan*: all, every] The single land mass formed when all the continents came together in the Permian period.

para- [Gk. *para*: akin to, beside] Prefix indicating association in being along side or accessory to.

parabronchi Passages in the lungs of birds through which air flows.

paracrine A substance, such as a hormone, that acts locally, near the site of its secretion. (Compare with autocrine, endocrine gland.)

parallel evolution Repeated evolutionary patterns of change that occur independently in multiple lineages.

paralogy (par al' o jee) A type of homology applied to genes in which the divergence of homologous genes can be traced to gene duplication events. The genes are said to be paralogous, and each is an paralog of the others. (Compare with orthology.)

parapatric speciation [Gk. *para*: along side + *patria*: homeland] Reproductive isolation between subpopulations arising from some non-geographic but physical condition, such as soil nutrient content. (Contrast with allopatric speciation, sympatric speciation.)

paraphyletic (par' a fih leht' ik) [Gk. *para*: beside + *phylon*: tribe] Referring to a group that consists of an ancestor and some (but not all) of its descendants. (Compare with monophyletic, polyphyletic.)

parasite An organism that attacks and consumes parts of an organism much larger than itself. Parasites sometimes, but not always, kill their host.

parasympathetic nervous system A portion of the autonomic (involuntary) nervous system. (Contrast with sympathetic nervous system.)

parathyroids Four glands on the posterior surface of the thyroid that produce and release parathormone.

parathyroid hormone Hormone secreted by the parathyroid glands. Stimulates osteoclast activity and raises blood calcium levels.

parenchyma (pair eng′ kyma) A plant tissue composed of relatively unspecialized cells without secondary walls.

parsimony The principle of preferring the simplest among a set of plausible explanations of any phenomenon.

parthenocarpy Formation of fruit from a flower without fertilization.

parthenogenesis (par′ then oh jen′ e sis) [Gk. parthenos: virgin + genesis: source] The production of an organism from an unfertilized egg.

partial pressure The portion of the barometric pressure of a mixture of gases that is due to one component of that mixture. For example, the partial pressure of oxygen at sea level is 20.9% of barometric pressure.

particulate theory In genetics, the theory that genes are physical entities that retain their identities after fertilization.

passive transport Diffusion across a membrane; may or may not require a channel or carrier protein. (Contrast with active transport.)

patch clamping A technique for isolating a tiny patch of membrane to allow the study of ion movement through a particular channel.

pathogen (path′ o jen) [Gk. *pathos*: suffering + *genesis:* source] An organism that causes disease.

pattern formation In animal embryonic development, the organization of differentiated tissues into specific structures such as wings.

pedigree The pattern of transmission of a genetic trait within a family.

penetrance Of a genotype, the proportion of individuals with that genotype who show the expected phenotype.

pentose [Gk. *penta:* five] A sugar containing five carbon atoms.

PEP carboxylase The enzyme that combines carbon dioxide with PEP to form a 4-carbon dicarboxylic acid at the start of C_4 photosynthesis or of crassulacean acid metabolism (CAM).

pepsin [Gk. *pepsis*: digestion] An enzyme in gastric juice that digests protein.

pepsinogen Inactive secretory product that is converted into pepsin by low pH or by enzymatic action.

peptide linkage The bond between amino acids in a protein. Formed between a carboxyl group and amino group (CO—NH^-) with the loss of water molecules.

peptidoglycan The cell wall material of many bacteria, consisting of a single enormous molecule that surrounds the entire cell.

perennial (per ren′ ee al) [L. *per*: throughout + *annus*: year] Refers to a plant that survives from year to year. (Contrast with annual, biennial.)

perfect flower A flower with both stamens and carpels, therefore hermaphroditic.

pericycle [Gk. *peri*: around + *kyklos*: ring or circle] In plant roots, tissue just within the endodermis, but outside of the root vascular tissue. Meristematic activity of pericycle cells produces lateral root primordia.

periderm The outer tissue of the secondary plant body, consisting primarily of cork.

period (1) A category in the geological time scale. (2) The duration of a single cycle in a cyclical event, such as a circadian rhythm.

peripheral membrane protein Membrane protein not embedded in the bilayer. (Contrast with integral membrane protein.)

peripheral nervous system Neurons that transmit information to and from the central nervous system and whose cell bodies reside outside the brain or spinal cord.

peristalsis (pair′ i stall′ sis) [Gk. *peri*: around + *stellein*: place] Wavelike muscular contractions proceeding along a tubular organ, propelling the contents along the tube.

peritoneum The mesodermal lining of the body cavity among coelomate animals.

permease A membrane protein that specifically transports a compound or family of compounds across the membrane.

peroxisome An organelle that houses reactions in which toxic peroxides are formed. The peroxisome isolates these peroxides from the rest of the cell.

petal [Gk. *petalon*: spread out] In an angiosperm flower, a sterile modified leaf, nonphotosynthetic, frequently brightly colored, and often serving to attract pollinating insects.

petiole (pet′ ee ole) [L. *petiolus*: small foot] The stalk of a leaf.

pH The negative logarithm of the hydrogen ion concentration; a measure of the acidity of a solution. A solution with pH = 7 is said to be neutral; pH values higher than 7 characterize basic solutions, while acidic solutions have pH values less than 7.

phage (fayj) Short for bacteriophage. A virus that infects bacteria.

phagocyte [Gk. *phagein*: to eat + *kystos*: sac] A white blood cell that ingests microorganisms by endocytosis.

phagocytosis Endocytosis by a cell of another cell or large particle.

pharmacogenomics The relaionship between an individual's genetic makeup and response to drugs.

pharming The use of genetically modified animals to produce medically useful products in their milk.

pharynx [Gk. *pharynx*: throat] The part of the gut between the mouth and the esophagus.

phenotype (fee′ no type) [Gk. *phanein*: to show] The observable properties of an individual resulting from both genetic and environmental factors. (Contrast with genotype.)

phenotypic plasticity Refers to the fact that the phenotype of a developing organism is determined by a complex series of processes that are affected by both its genotype and its environment.

pheromone (feer′ o mone) [Gk. *pheros*: carry + *hormon*: excite, arouse] A chemical substance used in communication between organisms of the same species.

phloem (flo′ um) [Gk. *phloos*: bark] In vascular plants, the tissue that transports sugars and other solutes from sources to sinks. It consists of sieve cells or sieve tubes, fibers, and other specialized cells.

phosphate group The functional group —OPO_3H_2. The transfer of energy from one compound to another is often accomplished by the transfer of a phosphate group.

phosphodiester linkage The connection in a nucleic acid strand, formed by linking two nucleotides.

phospholipids Lipids containing a phosphate group; important constituents of cellular membranes. (*See* lipids.)

phosphorylation The addition of a phosphate group.

photoautotroph An organism that obtains energy from light and carbon from carbon dioxide. (Contrast with chemolithotroph, chemoheterotroph, photoheterotroph.)

photoheterotroph An organism that obtains energy from light but must obtain its carbon from organic compounds. (Contrast with chemolithotroph, chemoheterotroph, photoautotroph.)

photon (foe′ ton) [Gk. *photos*: light] A quantum of visible radiation; a "packet" of light energy.

photoperiod (foe′ tow peer′ ee ud) The duration of a period of light, such as the length of time in a 24-hour cycle in which daylight is present.

photoperiodicity A condition in which physiological and behavioral changes are induced by changes in day length.

photoreceptor (1) A pigment that triggers a physiological response when it absorbs a photon. (2) A sensory receptor cell that senses and responds to light energy.

photorespiration Light-driven uptake of oxygen and release of carbon dioxide, the carbon being derived from the early reactions of photosynthesis.

photosynthesis (foe tow sin′ the sis) [literally, "synthesis from light"] Metabolic processes, carried out by green plants, by which visible light is trapped and the energy used to synthesize compounds such as ATP and glucose.

photosystem [Gk. *phos:* light + *systema:* assembly] A light-harvesting complex in the chloroplast thylakoid composed of pigments and proteins.

photosystem I In photosynthesis, the reactions that absorb light at 700 nm, passing electrons to ferrodoxin and thence to NADPH. Rich in chlorophyll *a*.

photosystem II In photosynthesis, the reactions that absorb light at 660 nm, passing electrons to the electron transport chain in the chloroplast. Rich I chlorphyll *b*.

phototropins A class of blue light receptors that mediate phototropism and other plant responses.

phototropism [Gk. *photos*: light + *trope*: turning] A directed plant growth response to light.

phycobilin Photosynthetic pigment that absorbs red, yellow, orange, and green light and is found in cyanobacteria and some red algae.

phylogenetic tree A graphic representation of lines of descent among organisms or their genes.

phylogeny (fy loj' e nee) [Gk. *phylon*: tribe, race + *genesis*: source] The evolutionary history of a particular group of organisms or their genes.

physiology (fiz' ee ol' o jee) [Gk. *physis*: natural form + *logos*: discourse, study] The scientific study of the functions of living organisms and the individual organs, tissues, and cells of which they are composed.

phytoalexins Substances toxic to pathogens, produced by plants in response to fungal or bacterial infection.

phytochrome (fy' tow krome) [Gk. *phyton*: plant + *chroma*: color] A plant pigment regulating a large number of developmental and other phenomena in plants.

pigment A substance that absorbs visible light.

pineal gland A gland located between the cerebral hemispheres that secretes melatonin.

pinocytosis Endocytosis by a cell of liquid containing dissolved substances.

pistil [L. *pistillum*: pestle] The structure of an angiosperm flower within which the ovules are borne. May consist of a single carpel, or of several carpels fused into a single structure. Usually differentiated into ovary, style, and stigma.

pith In plants, relatively unspecialized tissue found within a cylinder of vascular tissue.

pituitary A small gland attached to the base of the brain in vertebrates. Its hormones control the activities of other glands. Also known as the hypophysis.

pits Recessed cavities in the cell walls of a plant vascular element where only the primary wall is present. facilitating the movement of sap between cells.

placenta (pla sen' ta) [Gk. *plax*: flat surface] The organ found in most mammals that provides for the nourishment of the fetus and elimination of the fetal waste products.

placental (pla sen' tal) Pertaining to mammals of the subclass Eutheria, a group characterized by the presence of a placenta; contains the majority of living species of mammals.

plankton Free-floating small organisms inhabiting the surface waters of lakes and oceans. Photosynthetic members of the plankton are referred to as phytoplankton.

plant *See* embryophyte.

planula (plan' yew la) [L. *planum*: flat] The free-swimming, ciliated larva of the cnidarians.

plaque (plack) [Fr.: a metal plate or coin] (1) A circular clearing in a layer (lawn) of bacteria growing on the surface of a nutrient agar gel. (2) An accumulation of prokaryotic organisms on tooth enamel. Acids produced by these microorganisms can cause tooth decay. (3) A region of arterial wall invaded by fibroblasts and fatty deposits (*see* atherosclerosis).

plasma *See* blood plasma.

plasma cell An antibody-secreting cell that developed from a B cell. The effector cell of the humoral immune system.

plasma membrane The membrane that surrounds the cell, regulating the entry and exit of molecules and ions. Every cell has a plasma membrane.

plasmid A DNA molecule distinct from the chromosome(s); that is, an extrachromosomal element. May replicate independently of the chromosome.

plasmodesma (plural: plasmodesmata) [Gk. *plassein*: to mold + *desmos*: band] A cytoplasmic strand connecting two adjacent plant cells.

plasmogamy The fusion of the cytoplasm of two cells. (Contrast with karyogamy.)

plasmolysis (plaz mol' i sis) Shrinking of the cytoplasm and plasma membrane away from the cell wall, resulting from the osmotic outflow of water. Occurs only in cells with rigid cell walls.

plastid Organelle in plants that serves for food manufacture (by photosynthesis) or food storage; bounded by a double membrane.

plastoquinone A mobile electron carrier within the thylakoid membrane of the chloroplast linking photosystems I and II of photosynthesis.

platelet A membrane-bounded body without a nucleus, arising as a fragment of a cell in the bone marrow of mammals. Important to blood-clotting action.

pleiotropy (plee' a tro pee) [Gk. *pleion*: more] The determination of more than one character by a single gene.

pleural membrane [Gk. *pleuras*: rib, side] The membrane lining the outside of the lungs and the walls of the thoracic cavity. Inflammation of these membranes is a condition known as pleurisy.

pluripotent Of a stem cell, having the ability to differentiate into any of a limted number of cell types. (Compare with totipotent.)

podocytes Cells of Bowman's capsule of the nephron that cover the capillaries of the glomerulus, forming filtration slits.

poikilotherm (poy' kill o therm) [Gk. *poikilos*: varied + *thermos*: heat] An animal whose body temperature tends to vary with the surrounding environment. (Contrast with homeotherm, heterotherm.)

point mutation A mutation that results from a small, localized alteration in the chemical structure of a gene; can revert to wild type. (Contrast with deletion.)

polar body A nonfunctional nucleus produced by meiosis, accompanied by very little cytoplasm. The meiosis which produces the mammalian egg produces in addition three polar bodies.

polar molecule A molecule in which the electric charge is not distributed evenly in the covalent bonds.

polar nuclei In flowering plants, the two nuclei in the central cell of the megagametophyte; following fertilization they give rise to the endosperm.

polarity In development, the difference between one end and the other. In chemistry, the property that makes a polar molecule.

pollen [L. *pollin*: fine, powdery flour] In seed plants, the microscopic grains containing the male gametophyte (microgametophyte) and gamete (microspore).

pollination The process of transferring pollen from an anther to the stigma of a pistil in an angiosperm or from a strobilus to an ovule in a gymnosperm.

poly- [Gk. *poly*: many] A prefix denoting multiple entities.

poly(A) tail A long sequence of adenine nucleotides (50–250) added after transcription to the 3′ end of most eukaryotic mRNAs.

polygenes Multiple loci whose alleles increase or decrease a continuously variable phenotypic trait.

polymer [Gk. *poly*: many + *meros*: unit] A large molecule made up of similar or identical subunits called monomers. (Contrast with monomer, oligomer.)

polymerase chain reaction (PCR) An enzymatic technique for the rapid production of millions of copies of a particular stretch of DNA.

polymerization reactions Chemical reactions that generate polymers by linking monomers.

polymorphic Referring to a gene whose most frequent allele in a population is present less than 99% of the time.

polymorphism (pol' lee mor' fiz um) [Gk. *poly*: many + *morphe*: form, shape] In genetics, the coexistence in the same population of two distinct hereditary types based on different alleles.

polyp The sessile asexual stage in the life cycle of most cnidarians.

polypeptide A large molecule made up of many amino acids joined by peptide linkages. Large polypeptides are called proteins.

polyphyletic (pol' lee fih leht' ik) [Gk. *poly*: many + *phylon*: tribe] Referring to a group that consists of multiple distantly related organisms, and does not include the common ancestor of the group. (Compare with monophyletic, paraphyletic.)

polyploidy (pol' lee ploid ee) The possession of more than two entire sets of chromosomes.

polysaccharide A macromolecule composed of many monosaccharides (simple sugars). Common examples are cellulose and starch.

polysome (polyribosome) A complex consisting of a threadlike molecule of messenger RNA and several (or many) ribosomes. The ribosomes move along the mRNA, synthesizing polypeptide chains as they proceed.

polytene (pol' lee teen) [Gk. *poly*: many + *taenia*: ribbon] An adjective describing giant interphase chromosomes, such as those found in the salivary glands of fly larvae. The characteristic pattern of bands and bulges seen on these chromosomes provided a method for preparing detailed chromosome maps of several organisms.

pons [L. *pons*: bridge] Region of the brain stem anterior to the medulla.

population Any group of organisms coexisting at the same time and in the same place and capable of interbreeding with one another.

population bottleneck *See* bottleneck.

population density The number of individuals (or modules) of a population in a unit of area or volume.

population genetics The study of genetic variation and its causes within populations.

population structure The proportions of individuals in a population belonging to different age classes (age structure). Also, the distribution of the population in space.

portal blood vessels Blood vessels that begin and end in capillary beds.

positional cloning A technique for isolating a gene associated with a disease on the basis of its approximate chromosomal location.

positional information Signals by which genes regulate cell functions to locate cells in a tissue during development.

positive control The situation in which a regulatory macromolecule is needed to turn transcription of structural genes on. In its absence, transcription will not occur.

positive cooperativity Occurs when a molecule can bind several ligands and each one that binds alters the conformation of the molecule so that it can bind the next ligand more easily. The binding of four molecules of O_2 by hemoglobin is an example of positive cooperativity.

post [L. *postere*: behind, following after] Prefix denoting something that comes after.

postabsorptive period When there is no food in the gut and no nutrients are being absorbed.

posterior pituitary The portion of the pituitary gland that is derived from neural tissue.

postsynaptic cell The cell whose membranes receive neurotransmitter after its release by another cell (the presynaptic cell) at a synapse.

postzygotic reproductive barriers Barriers to the reproductive process that occur after the union of the nuclei of two gametes. (Contrast with prezygotic reproductive barriers.)

potential energy "Stored" energy not doing work, such as the energy in chemical bonds.

precapillary sphincter A cuff of smooth muscle that can shut off the blood flow to a capillary bed.

pre-mRNA (precursor mRNA) Initial gene transcript before it is modified to produce functional mRNA. Also known as the primary transcript.

predator An organism that kills and eats other organisms.

pressure flow model An effective model for phloem transport in angiosperms. It holds that sieve element transport is driven by an osmotically driven pressure gradient between source and sink.

pressure potential The hydrostatic pressure of an enclosed solution in excess of the surrounding atmospheric pressure. (Contrast with solute potential, water potential.)

presynaptic excitation/inhibition Occurs when a neuron modifies activity at a synapse by releasing a neurotransmitter onto the presynaptic nerve terminal.

prey [L. *praeda*: booty] An organism consumed as an energy source.

prezygotic reproductive barriers Barriers to the reproductive process that occur before the union of the nuclei of two gametes (Contrast with postzygotic reproductive barriers.)

primary active transport Form of active transport in which ATP is hydrolyzed, yielding the energy required to transport ions against their concentration gradients. (Contrast with secondary active transport.)

primary consumer An herbivore; an organism that eats plant tissues.

primary embryonic organizer *See* organizer.

primary growth In plants, growth produced by the apical meristems. (Contrast with secondary growth.)

primary immune response The first response of the immune system to an antigen, involving recognition by lymphocytes and the production of effector cells and memory cells. (Contrast with secondary immune response.)

primary motor cortex The region of the cerebral cortex that contains motor neurons that directly stimulate specific muscle fibers to contract.

primary producer A photosynthetic or chemosynthetic organism that synthesizes complex organic molecules from simple inorganic ones.

primary sex determination Genetic determination of gametic sex, male or female. (Contrast with secondary sex determination.)

primary somatosensory cortex The region of the cerebral cortex that receives input from mechanosensors distributed throughout the body.

primary succession Succession that begins in an area initially devoid of life, such as on recently exposed glacial till or lava flows. (Contrast with secondary succession.)

primary structure The specific sequence of amino acids in a protein.

primary wall Cellulose-rich cell wall layers laid down by a growing plant cell.

primase An enzyme that catalyzes the synthesis of a primer for DNA replication.

primer A short, single-stranded segment of DNA that is the necessary starting material for the synthesis of a new DNA strand, which is synthesized from the 3′ end of the primer.

primitive streak A line running axially along the blastodisc, the site of inward cell migration during formation of the three-layered embryo. Formed in the embryos of birds and fish.

primordium [L. *primordium*: origin] The most rudimentary stage of an organ or other part.

prion An infectious protein that can proliferate by converting other proteins.

pro- [L.: first, before, favoring] A prefix often used in biology to denote a developmental stage that comes first or an evolutionary form that appeared earlier than another. For example, prokaryote, prophase.

probe A segment of single stranded nucleic acid used to identify DNA molecules containing the complementary sequence.

procambium Primary meristem that produces the vascular tissue.

processive Referring to an enzyme that catalyzes many reactions each time it binds to a substrate, as DNA polymerase does during DNA replication.

progesterone [L. *pro*: favoring + *gestare*: to bear] A vertebrate female sex hormone that maintains pregnancy.

prokaryotes (pro kar′ ry otes) [L. *pro*: before + Gk. *karyon*: kernel, nucleus] Organisms whose genetic material is not contained within a nucleus: the bacteria and archaea. Considered an earlier stage in the evolution of life than the eukaryotes.

prometaphase The phase of nuclear division that begins with the disintegration of the nuclear envelope.

promoter The region of an operon that acts as the initial binding site for RNA polymerase.

proofreading The correction of an error in DNA replication just after an incorrectly paired base is added to the growing polynucleotide chain.

prophage (pro′ fayj) The noninfectious units that are linked with the chromosomes of the host bacteria and multiply with them but do not cause dissolution of the cell. Prophage can later enter into the lytic phase to complete the virus life cycle.

prophase (pro′ phase) The first stage of nuclear division, during which chromosomes condense from diffuse, threadlike material to discrete, compact bodies.

prostaglandin Any one of a group of specialized lipids with hormone-like functions. It is not clear that they act at any considerable distance from the site of their production.

prostate gland Glandular tissue that surrounds the male urethra at its junction with the vas deferens; contributes an alkaline fluid to the semen.

prosthetic group Any nonprotein portion of an enzyme.

proteasome In the eukaryotic cytoplasm, a huge protein structure that binds to and digests cellular proteins that have been tagged by ubiquitin.

protein (pro′ teen) [Gk. *protos*: first] One of the most fundamental building substances of living organisms. A long-chain polymer of amino acids with twenty different common side chains. Occurs with its polymer chain extended in fibrous proteins, or coiled into a compact macromolecule in enzymes and other globular proteins.

protein domain *See* domain (1)

protein kinase An enzyme that catalyzes the addition of a phosphate group from ATP to a target protein.

protein kinase cascade aA series of reactions in response to a molecular signal, in which a series of protein kinases activates one another in sequence, amplifying the signal at each step.

proteoglycan A glycoprotein containing a protein core with attached long, linear carbohydrate chains.

proteolysis [protein + Gk. *lysis*: break apart] An enzymatic digestion of a protein or polypeptide.

proteome The total of the different proteins that can be made by an organism. Because of alternate splicing of pre-mRNA, the number of proteins that can be made is usually much larger than the number of protein-coding genes present in the organism's genome.

protobiont [Gk. *protos*: first, before + *bios*: life] Aggregates of abiotically produced molecules that cannot reproduce but do maintain internal chemical environments that differ from their surroundings.

protoderm Primary meristem that gives rise to the plant epidermis.

proton (pro' ton) [Gk. *protos*: first, before] (1) A subatomic particle with a single positive charge. The number of protons in the nucleus of an atom determine its element. (2) A hydrogen ion, H^+.

proton pump An active transport system that uses ATP energy to move hydrogen ions across a membrane generating an electric potential (voltage).

proton motive force A force generated across a membrane expressed in millivolts having two components: a chemical potential (difference in proton concentration) plus an electrical potential due to the electrostatic charge on the proton.

proto-oncogenes The normal alleles of genes possessing oncogenes (cancer-causing genes) as mutant alleles. Proto-oncogenes encode growth factors and receptor proteins.

prototroph (pro' tow trofe') [Gk. *protos*: first + *trophein*: to nourish] The nutritional wild type, or reference form, of an organism. Any deviant form that requires growth nutrients not required by the prototrophic form is said to be a nutritional mutant, or auxotroph.

proximal Near the point of attachment or other reference point. (Contrast with distal.)

pseudocoelom [Gk. *pseudes*: false] A body cavity not surrounded by a peritoneum. Characteristic of nematodes and rotifers.

pseudogene [Gk. *pseudes*: false] A DNA segment that is homologous to a functional gene but is not expressed because of changes to its sequence or changes to its location in the genome.

pseudopod (soo' do pod) [Gk. *pseudes*: false + *podos*: foot] A temporary, soft extension of the cell body that is used in location, attachment to surfaces, or engulfing particles.

pulmonary [L. *pulmo*: lung] Pertaining to the lungs.

punctuated equilibrium An evolutionary pattern in which periods of rapid change are separated by longer periods of little or no change.

Punnett square A method of predicting the results of a genetic cross by arranging the gametes of each parent at the edges of a square.

pupa (pew' pa) [L. *pupa*: doll, puppet] In certain insects (the Holometabola), the encased developmental stage between the larva and the adult.

pupil The opening in the vertebrate eye through which light passes.

purine (pure' een) One of the types of nitrogenous bases. The purines adenine and guanine are found in nucleic acids. (Contrast with pyrimidine.)

Purkinje fibers Specialized heart muscle cells that conduct excitation throughout the ventricular muscle.

pyrimidine (per im' a deen) A type of nitrogenous base. The pyrimidines cytosine, thymine, and uracil are found in nucleic acids.

pyruvate A three-carbon acid; the end product of glycolysis and the raw material for the citric acid cycle.

pyruvate oxidation Conversion of pyruvate to acetyl CoA and CO_2 that occurs in the mitochondrial matrix in the presence of O_2.

- Q -

Q_{10} A value that compares the rate of a biochemical process or reaction over a 10°C range of temperature. A process that is not temperature-sensitive has a Q_{10} of 1; values of 2 or 3 mean the reaction speeds up as temperature increases.

quantum (kwon' tum) [L. *quantus*: how great] An indivisible unit of energy.

quaternary structure The specific three dimensional arrangement of protein subunits.

quiescent center In root meristem, central region where cells do not divide or divide very slowly.

- R -

R factor (resistance factor) A plasmid that contains one or more genes that encode resistance to antibiotics.

***R* genes** Resistance genes that function in plant defenses against bacteria, fungi, and nematodes. *See* gene-for-gene resistance.

R group The distinguishing group of atoms of a particular amino acid.

radial symmetry The condition in which two halves of a body are mirror images of each other regardless of the angle of the cut, providing the cut is made along the center line. Thus, a cylinder cut lengthwise down its center displays this form of symmetry. (Contrast with biradial symmetry.)

radioisotope A radioactive isotope of an element. Examples are carbon-14 (^{14}C) and hydrogen-3, or tritium (^{3}H).

radiometry The use of the regular, known rates of decay of radioisotopes of elements to determine dates of events in the distant past.

rain shadow The relatively dry area on the leeward side of a mountain range.

rapid-eye-movement *See* REM sleep

reactant A chemical substance that enters into a chemical reaction with another substance.

reaction A chemical change in which changes take place in the kind, number, or position of atoms making up a substance.

reaction center A group of electron transfer proteins that receive energy from light-absorbing pigments and convert it to chemical energy by redox reactions.

receptacle The end of a plant stem to which the parts of the flower are attached.

receptive field The area of visual space that activates a particular cell in the visual system.

receptor A site or protein on the outer surface of the plasma membrane or in the cytoplasm to which a specific ligand from another cell binds.

receptor-mediated endocytosis Endocytosis initiated by macromolecular binding to a specific membrane receptor.

receptor potential The change in the resting potential of a sensory cell when it is stimulated.

recessive In genetics, an allele that does not determine phenotype in the presence of a dominant allele. (Contrast with dominance.)

reciprocal crosses A pair of matings in one of which a female of genotype A mates with a male of genotype B and in the other of which a female of genotype B mates with a male of genotype A.

recognition site *See* restriction site.

recombinant An individual, meiotic product, or single chromosome in which genetic materials originally present in two individuals end up in the same haploid complement of genes. The reshuffling of genes can be either by independent segregation, or by crossing over between homologous chromosomes.

recombinant DNA DNA generated in vitro, from more than one source.

recombinant DNA technology The application of restriction endonucleases, plasmids, and transformation to alter and assemble recombinant DNA, with the goal of producing specific proteins.

recombinant frequency The proportion of offspring of a genetic cross that have phenotypes different from the parental phenotypes due to crossing over between linked genes during gamete formation.

reconciliation ecology The practice of making exploited lands more biodiversity-friendly.

rectum The terminal portion of the gut, ending at the anus.

redox reaction A chemical reaction in which one reactant becomes oxidized and the other becomes reduced.

reducing agent A substance that can donate electrons to another substance. The reducing agent becomes oxidized, and its partner becomes reduced.

reduction Gain of electrons by a chemical reactant; any reduction is accompanied by an oxidation. (Contrast with oxidation.)

reflex An automatic action, involving only a few neurons (in vertebrates, often in the spinal cord), in which a motor response swiftly follows a sensory stimulus.

refractory period Of a neuron, the time interval after an action potential, during which another action potential cannot be elicited.

regulative development A pattern of animal embryonic development in which the fates of the first blastomeres are not absolutely fixed. (Contrast with mosaic development.)

regulator sequence A DNA sequence to which the protein product of a regulatory gene binds.

regulatory gene A gene that codes for a protein that controls the transcription of another gene(s).

regulatory subunit The polypeptide in an enzyme protein with quaternary structure that does not contain the active site, but instead binds non-substrate molecules and changes its structure, in turn changing the structure and function of the active site. (Contrast catalytic subunit.)

regulatory system A system that uses feedback information to maintain a physiological function or parameter at an optimal level.

reinforcement The evolution of enhanced reproductive isolation between populations due to natural selection for greater isolation.

releaser A sensory stimulus that triggers the performance of a stereotyped behavior pattern.

releasing hormone One of several hypothalamic hormones that stimulates the secretion of anterior pituitary hormone.

REM sleep A sleep state characterized by vivid dreams, skeletal muscle relaxation, and rapid eye movements.

renal [L. *renes*: kidneys] Relating to the kidneys.

replication Pertaining to the duplication of genetic material.

replication complex The close association of several proteins operating in the replication of DNA.

replication fork A point at which a DNA molecule is replicating. The fork forms by the unwinding of the parent molecule.

replicon A region of DNA controlled by a single origin of replication.

reporter gene A marker gene included in recombinant DNA to indicate the presence of the recombinant DNA in a host cell.

repressible enzyme An enzyme whose synthesis can be decreased or prevented by the presence of a particular compound. A repressible operon often controls the synthesis of such an enzyme.

repressor A protein coded by the regulatory gene. The repressor can bind to a specific operator and prevent transcription of the operon.

reproductive isolating mechanism Any trait that prevents individuals from two different populations from producing fertile hybrids.

reproductive isolation The condition in which a population is not exchanging genes with other populations of the same species.

rescue effect The process by which a few individuals moving among declining subpopulations of a species and reproducing may prevent their extinction.

resolution Of an optical device such as a microscope, the smallest distance between two lines that allows the lines to be seen as separate from one another.

resource Something in the environment required by an organism for its maintenance and growth that is consumed in the process of being used.

respiration (res pi ra' shun) [L. *spirare*: to breathe] (1) Cellular respiration; the catabolic pathways by which electrons are removed from various molecules and passed through intermediate electron carriers to O_2, generating H_2O and releasing energy. (2) Breathing.

respiratory chain The terminal reactions of cellular respiration, in which electrons are passed from NAD or FAD, through a series of intermediate carriers, to molecular oxygen, with the concomitant production of ATP.

resting potential The membrane potential of a living cell at rest. In cells at rest, the interior is negative to the exterior. (Contrast with action potential, electrotonic potential.)

restoration ecology The science and practice of restoring damaged or degraded ecosystems.

restriction digestion Use of restriction enzymes to cleave DNA into fragments in a test tube.

restriction endonuclease *See* restriction enzyme.

restriction enzyme Any one of several enzymes, produced by bacteria, that break foreign DNA molecules at specific sites. Some produce "sticky ends." Extensively used in recombinant DNA technology.

restriction fragment length polymorphism *See* RFLP.

restriction point The specific time during G1 of the cell cycle at which the cell becomes committed to undergo the rest of the cell cycle.

restriction site A specific DNA base sequence recognized and acted on by a restriction endonuclease cutting the DNA.

reticular system A central region of the vertebrate brain stem that includes complex fiber tracts conveying neural signals between the forebrain and the spinal cord, with collateral fibers to a variety of nuclei that are involved in autonomic functions, including arousal from sleep.

retina (rett' in uh) [L. *rete*: net] The light-sensitive layer of cells in the vertebrate or cephalopod eye.

retinal The light-absorbing portion of visual pigment molecules. Derived from β-carotene.

retinoblastoma protein A protein that inhibits an animal cell from passing through the restriction point; inactivation of this protein is necessary for the cell cycle to proceed.

retrovirus An RNA virus that contains reverse transcriptase. Its RNA serves as a template for cDNA production, and the cDNA is integrated into a chromosome of the mammalian host cell.

reversal *See* evolutionary reversal.

reverse transcriptase An enzyme that catalyzes the production of DNA (cDNA), using RNA as a template; essential to the reproduction of retroviruses.

reversible reaction A chemical change that can occur in both the forward and reverse directions.

RFLP (restriction fragment length polymorphism) Coexistence of two or more patterns of restriction fragments (patterns produced by restriction enzymes), as revealed by a probe. The polymorphism reflects a difference in DNA sequence on homologous chromosomes.

rhizoids (rye' zoids) [Gk. *rhiza*: root] Hairlike extensions of cells in mosses, liverworts, and a few vascular plants that serve the same function as roots and root hairs in vascular plants. The term is also applied to branched, rootlike extensions of some fungi and algae.

rhizome (rye' zome) A special underground stem (as opposed to root) that runs horizontally beneath the ground.

rhodopsin A photopigment used in the visual process of transducing photons of light into changes in the membrane potential of photoreceptor cells.

ribonucleic acid *See* RNA.

ribose A five-carbon sugar in nucleotides and RNA.

ribosomal RNA (rRNA) Several species of RNA that are incorporated into the ribosome. Involved in peptide bond formation.

ribosome A small organelle that is the site of protein synthesis.

ribozyme An RNA molecule with catalytic activity.

risk cost The increased chance of being injured or killed as a result of performing a behavior, compared to resting.

RNA (ribonucleic acid) An often single stranded nucleic acid whose nucleotides use ribose rather than deoxyribose and in which the base uracil replaces thymine found in DNA. Serves as genome from some viruses. (*See* rRNA, tRNA, mRMA, and ribozyme.)

RNA editing The alteration of bases on mRNA prior to its translation.

RNA polymerase An enzyme that catalyzes the formation of RNA from a DNA template.

RNA primase A replication complex enzyme that makes the primer strand of DNA needed to initiate DNA replication.

RNA splicing The last stage of RNA processing in eukaryotes, in which the transcripts of introns are excised through the action of small nuclear ribonucleoprotein particles (snRNP).

rod cells Light-sensitive cell in the vertebrate retina; these sensory receptor cells are sensitive in extremely dim light and are responsible for dim light, black and white vision.

root The organ responsible for anchoring the plant in the soil, absorbing water and minerals, and producing certain hormones. Some roots are storage organs.

root cap A thimble-shaped mass of cells, produced by the root apical meristem, that protects the meristem; the organ that perceives the gravitational stimulus in root gravitropism.

root hair A long, thin process from a root epidermal cell that absorbs water and minerals from the soil solution.

rough ER That portion of the endoplasmic reticulum whose outer surface has attached ribosomes. (Contrast with smooth ER.)

rRNA *See* ribosomal RNA.

RT-PCR A technique in which RNA is first converted to cDNA by the use of the enzyme reverse transcriptase, then the cDNA is amplified by the polymerase chain reaction.

rubisco (ribulose bisphosphate carboxylase/ oxygenase) Acronym for the enzyme that combines carbon dioxide or oxygen with ribulose bisphosphate to catalyze the first step of the Calvin-Benson cycle.

rumen (rew' mun) The first division of the ruminant stomach. It stores and initiates bacterial fermentation of food. Food is regurgitated from the rumen for further chewing.

ruminant Herbivorous, cud-chewing mammals such as cows or sheep. The ruminant stomach consists of four compartments.

- S -

S phase In the cell cycle, the stage of interphase during which DNA is replicated. (Contrast with G_1 phase, G_2 phase, M phase.)

saprobe [Gk. *sapros*: rotten] An organism (usually a bacterium or fungus) that obtains its carbon and energy directly from dead organic matter.

sarcomere (sark′ o meer) [Gk. *sark*: flesh + *meros*: unit] The contractile unit of a skeletal muscle.

sarcoplasm The cytoplasm of a muscle cell.

sarcoplasmic reticulum The endoplasmic reticulum of a muscle cell.

saturated fatty acid A fatty acid usually containing from 12 to 18 carbon atoms and no double bonds.

scientific method A means of gaining knowledge about the natural world by making observations, posing hypotheses, and conducting experiments to test those hypotheses.

schizocoelous development [Gk. *schizo*: split + *koiloma*: cavity] Formation of a coelom during embryological development by a splitting of mesodermal masses.

Schwann cell A glial cell that wraps around part of the axon of a peripheral neuron, creating a myelin sheath.

sclereid [Gk. *skleros*: hard] A type of sclerenchyma cell, commonly found in nutshells, that is not elongated.

sclerenchyma (skler eng′ kyma) [Gk. *skleros*: hard + *kymus*: juice] A plant tissue composed of cells with heavily thickened cell walls, dead at functional maturity. The principal types of sclerenchyma cells are fibers and sclereids.

scrotum A sac of skin that encloses the testicles in male mammals.

second law of thermodynamics States that in any real (irreversible) process, there is a decrease in free energy and an increase in entropy.

second messenger A compound, such as cAMP, that is released within a target cell after a hormone (first messenger) has bound to a surface receptor on a cell; the second messenger triggers further reactions within the cell.

secondary active transport Form of active transport which does not use ATP as an energy source; rather, transport is coupled to ion diffusion down a concentration gradient established by primary active transport.

secondary consumer An organism that eat primary consumers.

secondary growth In plants, growth produced by vascular and cork cambia, contributing to an increase in girth. (Contrast with primary growth.)

secondary immune response A rapid and intense response to a second or subsequent exposure to an antigen, initiated by memory cells. (Contrast with primary immune response.)

secondary metabolite A compound synthesized by a plant that is not needed for basic cellular metabolism. Typically has an antiherbivore or antiparasite function.

secondary sex determination Formation of nongametic features of sex, such as external organs and body hair. (Contrast with primary sex determination.)

secondary structure Of a protein, localized regularities of structure, such as the α helix and the β pleated sheet.

secondary succession Ecological succession after a disturbance that did not eliminate all the organisms originally living on the site. (Contrast with primary succession.)

secondary wall Wall layers laid down by a plant cell that has ceased growing; often impregnated with lignin or suberin.

secretin (si kreet′ in) A peptide hormone secreted by the upper region of the small intestine when acidic chyme is present. Stimulates the pancreatic duct to secrete bicarbonate ions.

section A thin slice, usually for microscopy, as a tangential section or a transverse section.

seed A fertilized, ripened ovule of a gymnosperm or angiosperm. Consists of the embryo, nutritive tissue, and a seed coat.

seed plant Plants in which the embryo is protected and nourished within a seed; the gymnosperms and angiosperms.

seedling A young plant that has grown from a seed (rather than by grafting or by other means.)

segmentation genes In insect larvae, genes that determine the number and polarity of larval segments.

segment polarity genes Genes that determine the boundaries and front-to-back organization of the segments in the *Drosophila* larva.

segregation In genetics, the separation of alleles, or of homologous chromosomes, from each other during meiosis so that each of the haploid daughter nuclei produced by meiosis contains one or the other member of the pair found in the diploid parent cell, but never both. This principle was articulated by Mendel as his first law.

selective permeability Allowing certain substances to pass through while other substances are excluded; a characteristic of membranes.

self incompatability In plants, the rejection of their own pollen; promotes genetic variation and limits inbreeding.

selfish act A behavioral act that benefits its performer but harms the recipients.

semen (see′ men) [L. *semin*: seed] The thick, whitish liquid produced by the male reproductive organ in mammals, containing the sperm.

semiconservative replication The way in which DNA is synthesized. Each of the two partner strands in a double helix acts as a template for a new partner strand. Hence, after replication, each double helix consists of one old and one new strand.

seminiferous tubules The tubules within the testes within which sperm production occurs.

senescence [L. *senescere*: to grow old] Aging; deteriorative changes with aging; the increased probability of dying with increasing age.

sensory receptor cells Cells that are responsive to a particular type of physical or chemical stimulation.

sensory transduction The transformation of environmental stimuli or information into neural signals.

sepal (see′ pul) [L. *sepalum*: covering] One of the outermost structures of the flower, usually protective in function and enclosing the rest of the flower in the bud stage.

septum [L. *saeptum*: wall, fence] (1) A partition or cross-wall appearing in the hyphae of some fungi. (2) The bony structure dividing the nasal passages.

Sertoli cells Cells in the seminiferous tubules that nuture the developing sperm.

sessile (sess′ ul) [L. *sedere*: to sit] Permanently attached; not moving.

set point In a regulatory system, the threshold sensitivity to the feedback stimulus.

sex chromosome In organisms with a chromosomal mechanism of sex determination, one of the chromosomes involved in sex determination.

sex linkage The pattern of inheritance characteristic of genes located on the sex chromosomes of organisms having a chromosomal mechanism for sex determination.

sex pilus (pill′ us) [L. *pilus*: hair] A structure on the cell wall that allows one bacterium to adhere to another prior to conjugation.

sexual reproduction Reproduction involving union of gametes.

sexual selection Selection by one sex of characteristics in individuals of the opposite sex. Also, the favoring of characteristics in one sex as a result of competition among individuals of that sex for mates.

shared derived trait A trait that arose in the ancestor of a phylogenetic group and is present (sometimes in modified form) in all of its members, thus helping define that group. Also called a synapomorphy.

shoot system The aerial parts of a vascular plant, consisting of the leaves, stem(s), and flowers.

short-day plant (SDP) A plant that requires short days (or long nights) in order to flower.

short tandem repeat (STR) An inherited, short (2–5 base pairs), moderately repetitive sequence of DNA.

shotgun sequencing A relatively rapid method of analyzing DNA sequences in which a large DNA molecule is broken up into overlapping fragments, each fragment is sequenced, and computers are used to analyze and realign the fragments.

sieve tube A column of specialized cells found in the phloem, specialized to conduct organic matter from sources (such as photosynthesizing leaves) to sinks (such as roots). Found principally in flowering plants.

sieve tube element A single cell of a sieve tube, containing cytoplasm but relatively few organelles, with highly specialized perforated end walls leading to elements above and below.

signal A chemical (neurotransmitter or hormone) or light message emitted from a cell or cells or organism(s) and received by others to cause some change in function or behavior.

signal recognition particle (SRP) A complex of RNA and protein that recognizes both the signal sequence on a growing polypeptide and receptor protein on the surface of the ER.

signal sequence The sequence of a protein that directs the protein through a particular cellular membrane.

signal transduction pathway The series of biochemical steps whereby a stimulus to a cell (such as a hormone or neurotransmitter binding to a receptor) is translated into a response of the cell.

silencer sequence A sequence of eukaryotic DNA that binds proteins that inhibit the transcription of an associated gene.

silent substitution A change in gene sequence that, due to the redundancy of the genetic code, has no effect on the amino acid produced, and thus no effect on the protein phenotype. Also called a synonymous substitution.

similarity matrix A matrix used to compare the degree of divergence among pairs of objects. For molecular sequences, constructed by summing the number or percentage of nucleotidies or amino acids that are identical in each pair of sequences.

single nucleotide polymorphisms (SNPs) Inherited variations in a single nucleotide base in DNA.

single-strand binding protein In DNA replication, a protein that binds to single strands of DNA after they have been separated from each other, keeping the two strands separate for replication.

sinoatrial node (sigh' no ay' tree al) [L. *sinus*: curve + *atrium*: hall, chamber] The pacemaker of the mammalian heart.

sink In plants, any organ that imports the products of photosynthesis, such as roots, developing fruits, immature leaves. (Contrast with source.)

sinus (sigh' nus) [L. *sinus*: curve, hollow] A cavity in a bone, a tissue space, or an enlargement in a blood vessel.

sister chromatid In the eukaryotic cell, a chromatid resulting from chromosome replication during interphase.

sister groups Two phylogenetic groups that are each other's closest relatives.

skeletal muscle *See* striated muscle.

sliding DNA clamp A protein complex that keeps DNA polymerase bound to DNA during replication.

sliding filament theory A proposed mechanism of muscle contraction based on formation and breaking of crossbridges between actin and myosin filaments, causing them to slide together.

slow-twitch fibers Skeletal muscle fibers that generate tension slowly, but are resistant to fatigue ("marathon" fibers). They have abundant mitochondria, enzymes of aerobic metabolism, myoglobin, and blood supply.

slow-wave sleep A state of mammalian and avian sleep characterized by high amplitude slow waves in the EEG.

small intestine The portion of the gut between the stomach and the colon; consists of the duodenum, the jejunum, and the ileum.

small nuclear ribonucleoprotein particle (snRNP) A complex of an enzyme and a small nuclear RNA molecule, functioning in RNA splicing.

smooth muscle One of three types of muscle tissue. Usually consists of sheets of mononucleated cells innervated by the autonomic nervous system. (Compare with cardiac muscle, striated muscle.)

sodium–potassium pump (Na–K pump) The complex protein in plasma membranes that is responsible for primary active transport; it pumps sodium ions out of the cell and potassium ions into the cell, both against their concentration gradients.

solute A substance that is dissolved in a liquid (solvent) to form a solution.

solute potential A property of any solution, resulting from its solute contents; it may be zero or have a negative value. The more negative the solute potential, the greater the tendency of the solution to take up water through a differentially permeable membrane. (Contrast with pressure potential, water potential.)

solution A liquid (the solvent) and its dissolved solutes.

somatic [Gk. *soma*: body] Pertaining to the body. Somatic cells are cells of the body (as opposed to germ cells).

somatic mutation Permament genetic change in a somatic cell. These mutations affect the individual only; they are not passed on to offspring. (Contrast with germ line mutation)

somite (so' might) One of the segments into which an embryo becomes divided longitudinally, leading to the eventual segmentation of the animal as illustrated by the spinal column, ribs, and associated muscles.

source In plants, an organ exporting photosynthetic products in excess of its own needs. For example, a mature leaf or storage organ. (Contrast with sink.)

spatial summation In the production or inhibition of action potentials in a postsynaptic neuron, the interaction of depolarizations and hyperpolarizations produced by several terminal boutons.

spawning The direct release of sex cells into the water.

speciation (spee' shee ay' shun) The process of splitting one population into two populations that are reproductively isolated from one another.

species (spee' shees) [L. *species*: kind] The basic lower unit of classification, consisting of an ancestor–descendant lineage of populations of closely related and similar organisms. The more narrowly defined "biological species" consists of individuals capable of interbreeding freely with each other but not with members of other species.

species–area relationship The relationship between the sizes of areas and the numbers of species they support.

species diversity A weighted representation of the species of organisms living in a region; large and common species are given greater weight than are small and rare ones. (Contrast with species richness.)

species richness The total number of species living in a region. (Contrast with species diversity.)

specific defenses Defensive reactions of the immune system that are based on antibody reaction with a specific antigen.

specific heat The amount of energy that must be absorbed by a gram of a substance to raise its temperature by one degree centigrade. By convention, water is assigned a specific heat of one.

sperm [Gk. *sperma*: seed] A male gamete (reproductive cell).

spermatogenesis (spur mat' oh jen' e sis) [Gk. *sperma*: seed + *genesis*: source] Male gametogenesis, leading to the production of sperm.

sphincter (sfink' ter) [Gk. *sphinkter*: something that binds tightly] A ring of muscle that can close an orifice, for example at the anus.

spindle apparatus An array of microtubules stretching from pole to pole of a dividing nucleus and playing a role in the movement of chromosomes at nuclear division. Named for its shape.

spiracle (spy' rih kel) [L. *spirare*: to breathe] An opening of the treacheal respiratory system of terrestrial arthorpods.

spleen An organ that serves as a reservoir for venous blood and eliminates old, damaged red blood cells from the circulation.

spliceosome An RNA–protein complex that splices out introns from eukaryotic pre-mRNAs.

splicing Removal of introns and connecting of exons in eukaryotic pre-mRNAs.

spontaneous mutation A genetic change caused by internal cellular mechanisms, such as an error in DNA replication. (Contrast with induced mutation.)

spontaneous reaction A chemical reaction that will proceed on its own, without any outside influence. A spontaneous reaction need not be rapid.

sporangium (spor an' gee um) [Gk. *spora*: seed + *angeion*: vessel or reservoir] In plants and fungi, any specialized stucture within which one or more spores are formed.

spore [Gk. *spora*: seed] Any asexual reproductive cell capable of developing into an adult organism without gametic fusion. In plants, haploid spores develop into gametophytes, diploid spores into sporophytes. In prokaryotes, a resistant cell capable of surviving unfavorable periods.

sporocyte Specialized cells of the diploid sporophyte that will divide by meiosis to produce four haploid spores. Germination of these spores produces the haploid gametophyte.

sporophyte (spor' o fyte) [Gk. *spora*: seed + *phyton*: plant] In plants and protists with alternation of generations, the diploid phase that produces the spores. (Contrast with gametophyte.)

stabilizing selection Selection against the extreme phenotypes in a population, so that the intermediate types are favored. (Contrast with disruptive selection.)

stamen (stay' men) [L. *stamen*: thread] A male (pollen-producing) unit of a flower, usually composed of an anther, which bears the pollen, and a filament, which is a stalk supporting the anther.

starch [O.E. *stearc*: stiff] A polymer of glucose; used by plants to store energy.

start codon The mRNA triplet (AUG) that acts as a signal for the beginning of translation at the ribosome. (Contrast with stop codon.)

stasis [Gk. *stasis*: to stop, stand still] Period during which little or no evolutionary change takes place within a lineage or groups of lineages.

statocyst (stat' oh sist) [Gk. *statos*: stationary + *kystos*: cell] An organ of equilibrium in some invertebrates.

statolith (stat' oh lith) [Gk. *statos*: stationary + *lithos*: stone] A solid object that responds to gravity or movement and stimulates the mechanoreceptors of a statocyst.

stele (steel) [Gk. *stylos*: pillar] The central cylinder of vascular tissue in a plant stem.

stem Plant structure that holds leaves and/or flowers; it is the site for transporting and distributing material throughout the plant.

stem cells In animals, undifferentiated cells that are capable of extensive proliferation. A stem cell generates more stem cells and a large clone of differentiated progeny cells.

steroid Any of numerous lipids based on a 17-carbon atom ring system.

sticky ends On a piece of two-stranded DNA, short, complementary, one-stranded regions produced by the action of a restriction endonuclease. Sticky ends allow the joining of segments of DNA from different sources.

stigma [L. *stigma*: mark, brand] The part of the pistil at the apex of the style that is receptive to pollen, and on which pollen germinates.

stimulus [L. *stimulare*: to goad] Something causing a response; something in the environment detected by a receptor.

stolon [L. *stolon*: branch, sucker] A horizontal stem that forms roots at intervals.

stoma (plural: stomata) [Gk. *stoma*: mouth, opening] Small opening in the plant epidermis that permits gas exchange; bounded by a pair of guard cells whose osmotic status regulates the size of the opening.

stop codon Any of the thre mRNA codons that signal the end of protein translation at the ribosome: UAG, UGA, UAA.

stratosphere The upper part of Earth's atmosphere, above the troposphere; extends from approximately 18 kilometers upward to approximately 50 kilometers above the surface.

stratum (plural strata) [L. *stratos*: layer] A layer of sedimentary rock laid down at a particular time in the past.

striated muscle Contractile tissue characterized by multinucleated cells containing highly ordered arrangements of actin and myosin microfilaments. Also known as skeletal muscle. (Compare with cardiac muscle, smooth muscle.)

strobilus A conelike structure consisting of spore-bearing scales (modified leaves) inserted on an axis. (Contrast with cone.)

stroma The fluid contents of an organelle, such as a chloroplast.

stromatolites Composite, flat-to-domed structures composed of successive mineral layers produced by the action of cyanobacteria in water; ancient ones provide evidence for early life on the earth.

structural gene A gene that encodes the primary structure of a protein.

structural isomers Molecules made up of the same kinds and numbers of atoms, in which the atoms are bonded differently.

style [Gk. *stylos*: pillar or column] In flowering plants, a column of tissue extending from the tip of the ovary, and bearing the stigma or receptive surface for pollen at its apex.

sub- [L. *sub*: under] A prefix often used to designate a structure that lies beneath another or is less than another. For example, subcutaneous (beneath the skin); subspecies.

suberin A waxlike lipid that acts as a barrier to water and solute movement across the Casparian strip of the endodermis. Suberin is the waterproofing element in the cell walls of cork.

submucosa (sub mew koe' sah) The tissue layer just under the epithelial lining of the lumen of the digestive tract.

substrate (sub' strayte) The molecule or molecules on which an enzyme exerts catalytic action.

substrate-level phosphorylation Reaction in which ATP is formed from ADP by the addition of a phosphate group directly from a reactant.

substratum The base material on which a sessile organism lives.

succession In ecology, the gradual, sequential series of changes in species composition of a community following a disturbance.

sulcus [L. *sulcare*: to plow] The valleys or creases between the raised portions of the convoluted surface of the brain. (Contrast with gyrus.)

summation The ability of a neuron to fire action potentials in response to numerous subthreshold postsynaptic potentials arriving simultaneously at differentiated places on the cell, or arriving at the same site in rapid succession.

surface area-to-volume ratio For any cell, organism, or geometrical solid, the ratio of surface area to volume; this is an important factor in setting an upper limit on the size a cell or organism can attain.

surface tension The attractive intermolecular forces at the surface of liquid; especially important in water.

surfactant A substance that decreases the surface tension of a liquid. Lung surfactant, secreted by cells of the alveoli, is mostly phospholipid and decreases the amount of work necessary to inflate the lungs.

survivorship The proportion of individuals in a cohort that is alive at some time in the future.

suspensor In the embryos of seed plants, the stalk of cells that pushes the embryo into the endosperm and is a source of nutrient transport to the embryo.

symbiosis (sim' bee oh' sis) [Gk. *sym*: together + *bios*: living] The living together of two or more species in a prolonged and intimate ecological relationship. (Compare with parasitism, mutualism.)

symmetry Describes an attribute of an animal body in which at least one plane can divide the body into similar, mirror-image halves. (*See* bilateral symmetry, biradial symmetry, radial symmetry.)

sympathetic nervous system A division of the autonomic (involuntary) nervous system. (Contrast with parasympathetic nervous system.)

sympatric speciation (sim pat' rik) [Gk. *sym*: same + *patria*: homeland] Speciation due to reproductive isolation without any physical separation of the subpopulation. (Contrast with allopatric speciation, parapatric speciation.)

symplast The continuous meshwork of the interiors of living cells in the plant body, resulting from the presence of plasmodesmata. (Contrast with apoplast.)

symport A membrane transport process that carries two substances in the same direction across the membrane. (Contrast with antiport.)

synapomorphy *See* shared derived trait.

synapse (sin' aps) [Gk. *syn*: together + *haptein*: to fasten] The narrow gap between the terminal bouton of one neutron and the dendrite or cell body of another.

synapsis (sin ap' sis) The highly specific parallel alignment (pairing) of homologous chromosomes during the first division of meiosis.

synaptic vesicle A membrane-bounded vesicle containing neurotransmitter; the neurotransmitter is produced in and discharged by a presynaptic neuron.

synergids [Gk. *syn*: together + *ergos*: performing work] In flowering plants, the two cells accompanying the egg cell at one end of the megmagametophyte.

syngamy (sing' guh mee) [Gk. *syn*: together + *gamos*: marriage] Union of gametes. Also known as fertilization.

synonymous substitution A change of one nucleotide in a sequence to another when that change does not affect the amino acid specified (i.e., UUA → UUG, both specifying leucine). (Compare with nonsynonymous substitution, missense substitution, nonsense substitution.)

systematics The scientific study of the diversity of organisms, and of their relationships.

systemic circulation The part of the circulatory system serving those parts of the body other than the lungs or gills (which are served by the pulmonary circulation.)

systemic acquired resistance A general resistance to many plant pathogens following infection by a single agent.

systemin The only polypeptide plant hormone; participates in response to tissue damage.

systems biology The study of an organism as an integrated and interacting system of genes, proteins, and biochemical reactions.

systole (sis' tuh lee) [Gk. *systole*: contraction] Contraction of a chamber of the heart, driving blood forward in the circulatory system.

- T -

T cell A type of lymphocyte, involved in the cellular immune response. The final stages of its development occur in the thymus gland. (Contrast with B cell; see also cytotoxic T cell, helper T cell, suppressor T cell.)

T cell receptor A protein on the surface of a T cell that recognizes the antigenic determinant for which the cell is specific.

target cell A cell with the appropriate receptors to bind and respond to a particular hormone or other chemical mediator.

taste bud A structure in the epithelium of the tongue that includes a cluster of chemoreceptors innervated by sensory neurons.

TATA box An eight-base-pair sequence, found about 25 base pairs before the starting point for transcription in many eukaryotic promoters, that binds a transcription factor and thus helps initiate transcription.

taxis (tak' sis) [Gk. *taxis*: arrange, put in order] The movement of an organism or its part directly toward or away from the stimulus. For example, positive phototaxis is movement toward a light source, negative geotaxis is movement away from gravity).

taxon A biological group (typically a species or a clade) that is given a name.

telencephalon The frontmost division of the vertebrate brain; becomes the cerebrum.

telomerase An enzyme that catalyzes the addition of telomeric sequences lost from chromosomes during DNA replication.

telomeres (tee′ lo merz) [Gk. *telos*: end + *meros*: units, segments] Repeated DNA sequences at the ends of eukaryotic chromosomes.

telophase (tee′ lo phase) [Gk. *telos*: end] The final phase of mitosis or meiosis during which chromosomes became diffuse, nuclear envelopes reform, and nucleoli begin to reappear in the daughter nuclei.

template (1) In biochemistry, a molecule or surface upon which another molecule is synthesized in complementary fashion, as in the replication of DNA. (2) In the brain, a pattern that responds to a normal input but not to incorrect inputs.

template strand In double-stranded DNA, the strand that is transcribed to create an RNA transcript that will be processed into a protein. Also refers to a strand of RNA that is used to create a complementary RNA.

temporal summation [L. *tempus*: time; *summus*: highest amount] In the production or inhibition of action potentials in a postsynaptic neuron, the interaction of depolarizations or hyperpolarizations produced by rapidly repeated stimulation of a single point.

tendon A collagen-containing band of tissue that connects a muscle with a bone.

termination The end of protein synthesis triggered by a stop codon which binds release factor that causes the polypeptide to release from the ribosome.

terminator A sequence at the 3′ end of mRNA that causes the RNA strand to be released from the transcription complex.

terrestrial (ter res′ tree al) [L. *terra*: earth] Pertaining to the land. (Contrast with aquatic, marine.)

territory A fixed area from which an animal or group of animals excludes other members of the same (and sometimes other) species by aggressive behavior or displays.

tertiary structure In reference to a protein, the relative locations in three-dimensional space of all the atoms in the molecule. The overall shape of a protein. (Contrast with primary, secondary, and quaternary structures.)

test cross Mating of a dominant-phenotype individual (who may be either heterozygous or homozygous) with a homozygous-recessive individual.

testis (tes′ tis) (plural: testes) [L. *testis*: witness] The male gonad; the organ that produces the male sex cells.

testosterone (tes toss′ tuhr own) A male sex steroid hormone.

tetanus [Gk. *tetanos*: stretched] (1) A state of sustained maximal muscular contraction caused by rapidly repeated stimulation. (2) In medicine, an often fatal disease ("lockjaw") caused by the bacterium *Clostridium tetani*.

tetrad [Gk. *tettares*: four] During prophase I of meiosis, the association of a pair of homologous chromosomes or four chromatids.

thalamus [Gk. *thalamos*: chamber] A region of the vertebrate forebrain; involved in integration of sensory input.

thallus (thal′ us) [Gk. *thallos*: sprout] Any algal body which is not differentiated into root, stem, and leaf.

therapeutic cloning The use of cloning by nuclear transfer to produce an embryo that will provide embryonic stem cells to be used in therapy.

theory [Gk. *theoria*: analysis of facts] A far-reaching explanation of observed facts that is supported by such a wide body of evidence, with no significant contradictory evidence, that it is scientifically accepted as a factual framework. Examples are Newton's theory of gravity and Darwin's theory of evolution. (Contrast with hypothesis.)

thermoneutral zone [Gk. *thermos*: temperature] The range of temperatures over which an endotherm does not have to expend extra energy to thermoregulate.

thermoreceptor A cell or structure that responds to changes in temperature.

thoracic cavity [Gk. *thorax*: breastplate] The portion of the mammalian body cavity bounded by the ribs, shoulders, and diaphragm. Contains the heart and the lungs.

thoracic duct The connection between the lymphatic system and the circulatory system.

thorax [Gk. *thorax*: breastplate] In an insect, the middle region of the body, between the head and abdomen. In mammals, the part of the body between the neck and the diaphragm.

thrombin An enzyme that converts fibrinogen to fibrin, thus triggering the formation of blood clots.

thrombus (throm′ bus) [Gk. *thrombos*: clot] A blood clot that forms within a blood vessel and remains attached to the wall of the vessel. (Contrast with embolus.)

thylakoid (thigh la koid) [Gk. *thylakos*: sack or pouch] A flattened sac within a chloroplast. Thylakoid membranes contain all of the chlorophyll in a plant, in addition to the electron carriers of photophosphorylation. Thylakoids stack to form grana.

thymine (T) A nitrogen-containing base found in DNA.

thymus [Gk. *thymos*: warty] A ductless, glandular portion of the lymphoid system, involved in development of the immune system of vertebrates. In humans, the thymus degenerates during puberty.

thyroid [Gk. *thyreos*: door-shaped] A two-lobed gland in vertebrates. Produces the hormone thyroxin.

thyrotropin (thyroid-stimulating hormone, TSH) A hormone of the anterior pituitary that stimulates the thyroid gland to produce and release thyroxin.

thyrotropin-releasing hormone (TRH) A hypothalamic hormone that stimulates anterior pituitary cells to release TSH.

thyroxine The hormone produced by the thyroid gland that controls many metabolic processes.

tidal volume The amount of air that is exchanged during each breath when a person is at rest. (Compare with vital capacity.)

tight junction A junction between epithelial cells, in which there is no gap whatever between the adjacent cells. Materials may pass through a tight junction only by entering the epithelial cells themselves.

tissue A group of similar cells organized into a functional unit; usually integrated with other tissues to form part of an organ.

toll A member of a receptor family that responds to the binding of a molecule from a pathogen by initiating a protein kinase cascade, resulting in the synthesis of defensive proteins.

tonus (toe′ nuss) [L. *tonus*: tension] A low level of muscular tension that is maintained even when the body is at rest.

topsoil The uppermost soil layer; contains most of the organic matter of soil, but may be depleted of most mineral nutrients.

totipotent [L. *toto*: whole, entire + *potens*: powerful] Of a cell, possessing all the genetic information and other capacities necessary to form an entire individual. (Compare with pluripotent.)

toxic [L. *toxicum*: poison] Injurious to the tissues of the host organism.

trachea (tray′ kee ah) [Gk. *trakhoia*: tube] A tube that carries air to the bronchi of the lungs of vertebrates. When plural (*tracheae*), refers to the major airways of insects.

tracheary element Refers to either or both types of conductive xylem cells: tracheids and vessel elements.

tracheid (tray′ kee id) A distinctive conducting and supporting cell found in the xylem of nearly all vascular plants, characterized by tapering ends and walls that are pitted but not perforated. (Contrast with vessel element.)

tracheophytes [Gk. *trakhoia*: tube + *phyton*: plant] *See* vascular plants.

trade-off The relationship between the costs of performing a behavior or other trait and the benefits the individual gains from the trait or behavior. (*See also* cost–benefit analysis.)

trait In genetics, one form of a character: eye color is a character; brown eyes and blue eyes are traits. (Compare with character.)

transcription The synthesis of RNA using one strand of DNA as the template.

transcription factors Proteins that assemble on a eukaryotic chromosome, allowing RNA polymerase II to perform transcription.

transduction (1) Transfer of genes from one bacterium to another, with a bacterial virus acting as the carrier of the genes. (2) In sensory cells, the transformation of a stimulus (e.g., light energy, sound pressure waves, chemical or electrical stimulants) into action potentials.

transfection Uptake, incorporation, and expression of recombinant DNA.

transfer cell A modified parenchyma cell that transports solutes from its cytoplasm into its cell wall, thus moving the solutes from the symplast into the apoplast.

transfer RNA (tRNA) A family of double-stranded RNA molecules. Each tRNA carries a specific amino acid and anticodon that will pair with the complementary codon in mRNA during translation.

transformation Mechanism for transfer of genetic information in bacteria in which pure DNA extracted from bacteria of one genotype is taken in through the cell surface of bacteria of a different genotype and incorporated into the chromosome of the recipient cell.

transforming principle An early term for the as yet unidentified chemical substance responsible for bacterial tranformation.

transgenic organism An organism containing recombinant DNA incorporated into its genetic material.

transition-state species A short-lived, unstable intermediate with high potential energy in a chemical reaction.

translation The synthesis of a protein (polypeptide). Takes place on ribosomes, using the information encoded in messenger RNA.

transmembrane domain The portion of a protein that lies inside the membrane bilayer.

transmembrane protein An integral membrane protein that spans the lipid bilayer.

translocation (1) In genetics, a rare mutational event that moves a portion of a chromosome to a new location, generally on a nonhomologous chromosome. (2) In vascular plants, movement of solutes in the phloem.

transpiration [L. *spirare*: to breathe] The evaporation of water from plant leaves and stem, driven by heat from the sun, and providing the motive force to raise water (plus mineral nutrients) from the roots.

transposable element A segment of DNA that can move to, or give rise to copies at, another locus on the same or a different chromosome.

transposon Mobile DNA segment that can insert into a chromosome and cause genetic change.

triglyceride A simple lipid in which three fatty acids are combined with one molecule of glycerol.

triplet *See* codon.

triplet repeat The occurrence of repeated triplet of bases in a gene, often leading to genetic disease, as does excessive repetition of CGG in the gene responsible for fragile-X syndrome.

triploblastic Having three cell layers. (Contrast with diploblastic.)

trisomic Containing three rather than two members of a chromosome pair.

tRNA *See* transfer RNA.

trophic cascade The progression over successively lower trophic levels of the indirect effects of a predator.

trophic level [Gk *trophes*: nourishment] A group of organisms united by obtaining their energy from the same part of the food web of a biological community.

trophoblast [Gk *trophes*: nourishment + *blastos*: sprout] At the 32-cell stage of mammalian development, the outer group of cells that will become part of the placenta and thus nourish the growing embryo. (Comtrast with inner cell mass.)

trochophore (troke' o fore) [Gk. *trochos*: wheel + *phoreus*: bearer] The free-swimming larva of some annelids and mollusks, distinguished by a wheel-like band of cilia around the middle.

tropic hormones Hormones of the anterior pituitary that control the secretion of hormones by other endocrine glands.

tropism [Gk. *tropos*: to turn] In plants, growth toward or away from a stimulus such as light (phototropism) or gravity (gravitropism).

tropomyosin [troe poe my' oh sin] A protein that, along with actin, constitutes the thin filaments of myofibrils. It controls the interactions of actin and myosin necessary for muscle contraction.

troposphere The lowest atmospheric zone, reaching upward from the Earth's surface approximately 17 km in the tropics and subtropics but only to about 10 km at higher latitudes. The zone in which virtually all the water vapor in the atmosphere is located.

true-breeding A genetic cross in which the same result occurs every time with respect to the trait(s) under consideration, due to homozygous parents.

trypsin A protein-digesting enzyme. Secreted by the pancreas in its inactive form (trypsinogen), it becomes active in the duodenum of the small intestine.

T tubules A system of tubules that runs throughout the cytoplasm of muscle fibers, through which action potentials spread.

tube cell The larger of the two cells in a pollen grain; responsible for growth of the pollen tube. *See* generative cell.

tubulin A protein that polymerizes to form microtubules.

tumor [L. *tumor*: a swollen mass] A disorganized mass of cells, often growing out of control. Malignant tumors spread to other parts of the body.

tumor suppressor genes Genes which, when homozygous mutant, result in cancer. Such genes code for protein products that inhibit cell proliferation.

turgor pressure [L. *turgidus*: swollen] *See* pressure potential.

tympanic membrane [Gk. *tympanum*: drum] The eardrum.

- U -

ubiquinone (yoo bic' kwi known) [L. *ubique*: everywhere] A mobile electron carrier of the mitochondrial respiratory chain. Similar to plastoquinone found in chloroplasts.

ubiquitin A small protein that is covalently linked to other cellular proteins identified for breakdown by the proteosome.

umbilical cord Tissue made up of embryonic membranes and blood vessels that connects the embryo to the placenta in eutherian mammals.

understory The aggregate of smaller plants growing beneath the canopy of dominant plants in a forest.

unicellular (yoon' e sell' yer ler) [L. *unus*: one + *cella*: chamber] Consisting of a single cell, as in a unicellular organism. (Contrast with multicellular.)

uniport [L. *unus*: one + portal: doorway] A membrane transport process that carries a single substance. (Contrast with antiport, symport.)

unsaturated hydrocarbon A compound containing only carbon and hydrogen atoms, with one or more pairs of carbon atoms that are connected by double bonds.

upwelling The surfaceward movement of nutrient-rich, cooler water from deeper layers of the ocean.

uracil (U) A pyrimidine base found in nucleotides of RNA.

urea A compound serving as the main excreted form of nitrogen by many animals, including mammals.

ureotelic Describes an organism in which the final product of the breakdown of nitrogen-containing compounds (primarily proteins) is urea. (Contrast with ammonotelic, uricotelic.)

ureter (your' uh tur) A long duct leading from the vertebrate kidney to the urinary bladder or the cloaca.

urethra (you ree' thra) In most mammals, the canal through which urine is discharged from the bladder and which serves as the genital duct in males.

uric acid A compound that serves as the main excreted form of nitrogen in some animals, particularly those which must conserve water, such as birds, insects, and reptiles.

uricotelic Describes an organism in which the final product of the breakdown of nitrogen-containing compounds (primarily proteins) is uric acid. (Contrast with ammonotelic, ureotelic.)

urine (you' rin) In vertebrates, the fluid waste product containing the toxic nitrogenous by-products of protein and amino acid metabolism.

uterus (yoo' ter us) [L. *utero*: womb] The uterus or womb is a specialized portion of the female reproductive tract in certain mammals. It receives the fertilized egg and nurtures the embryo in its early development.

- V -

vaccination Injection of virus or bacteria or their proteins into the body, to induce immunization. The injected material is usually attenuated (weakened) before injection.

vacuole (vac' yew ole) [Fr.: small vacuum] A liquid-filled, membrane-enclosed compartment in cytoplasm; may function as digestive chambers, storage chambers, waste bins.

vagina (vuh jine' uh) [L.: sheath] In female mammals, the passage leading from the external genital orifice to the uterus; receives the copulatory organ of the male in mating.

van der Waals forces Weak attractions between atoms resulting from the interaction of the electrons of one atom with the nucleus of another. This type of attraction is about one-fourth as strong as a hydrogen bond.

variable regions The part of an immunoglobulin molecule or T-cell receptor that includes the antigen-binding site.

vasa recta Blood vessels that parallel the loops of Henle and the collecting ducts in the renal medulla of the kidney.

vascular (vas' kew lar) [L. *vasculum*: a small vessel] Pertaining to organs and tissues that conduct fluid, such as blood vessels in animals and phloem and xylem in plants.

vascular bundle In vascular plants, a strand of vascular tissue, including conducting cells of xylem and phloem as well as thick-walled fibers.

vascular cambium A lateral meristem giving rise to secondary xylem and phloem.

vascular rays In vascular plants, radially oriented sheets of cells produced by the vascular cambium, carrying materials laterally between the wood and the phloem.

vascular system The conductive system of the plant, consisting primarily of xylem and phloem.

vas deferens The duct that transfers sperm from the epididymis to the urethra.

vasopressin *See* antidiuretic hormone.

vector (1) An agent, such as an insect, that carries a pathogen affecting another species. (2) A plasmid or virus that carries an inserted piece of DNA into a bacterium for cloning purposes in recombinant DNA technology.

vegetal hemisphere The lower portion of some animal eggs, zygotes, and embryos, in which the dense nutrient yolk settles. The vegetal pole refers to the very bottom of the egg or embyro. (Contrast with animal hemisphere.)

vegetative Nonreproductive, nonflowering, or asexual.

vegetative reproduction Asexual reproduction.

vein [L. *vena*: channel] A blood vessel that returns blood to the heart. (Contrast with artery.)

ventral [L. *venter*: belly, womb] Toward or pertaining to the belly or lower side. (Contrast with dorsal.)

ventricle A muscular heart chamber that pumps blood through the lungs or through the body.

vernalization [L. *vernalis*: spring] Events occurring during a required chilling period, leading eventually to flowering.

vertebral column [L. *vertere*: to turn] The jointed, dorsal column that is the primary support structure of vertebrates.

vesicle A membrane enclosed compartment within the cytoplasm.

vessel element In plants, a nonliving water-conducting cell with perforated end walls. (Contrast with tracheid.)

vestibular apparatus (ves tib' yew lar) [L. *vestibulum*: an enclosed passage] Structures associated with the vertebrate ear; these structures sense changes in position or momentum of the head, affecting balance and motor skills.

vestigial (ves tij' ee al) [L. *vestigium*: footprint, track] The remains of body structures that are no longer of adaptive value to the organism and therefore are not maintained by selection.

vicariant distribution A population distribution resulting from the disruption of a formerly continuous range by a vicariant event.

vicariant event (vye care' ee unce) [L. *vicus*: change] The splitting of the range of a taxon by the imposition of some barrier to interchange among its members.

villus (vil' lus) (plural: villi) [L. *villus*: shaggy hair or beard] A hairlike projection from a membrane; for example, from many gut walls.

virion (veer' e on) The virus particle, the minimum unit capable of infecting a cell.

viroid (vye' roid) An infectious agent consisting of a single-stranded RNA molecule with no protein coat; produces diseases in plants.

virulent [L. *virus*: poison, slimy liquid] Causing or capable of causing disease and death.

virus Any of a group of ultramicroscopic particles constructed of nucleic acid and protein (and, sometimes, lipid) that require living cells in order to reproduce. Viruses probably evolved from eukaryotic cells, secondarily losing some cellular attributes.

vital capacity The sum total of the tidal volume and the inspiratory and expiratory reserve volumes.

vitamins [L. *vita*: life] Organic compounds that an organism cannot synthesize, but nevertheless requires in small quantity for normal growth and metabolism.

viviparous (vye vip' uh rus) [L. *vivus*: alive] Reproduction in which fertilization of the egg and development of the embryo occur inside the mother's body. (Contrast with oviparous.)

VNTRs (variable number of tandem repeats) In the human genome, short DNA sequences that are repeated a characteristic number of times in related individuals. Can be used to make a DNA fingerprint.

voltage-gated channels Ion channels in membranes that change conformation and therefore ion conductance when a certain potential difference exists across the membrane in which they are inserted.

- W -

waggle dance The running movement of a working honey bee on the hive, during which the worker traces out a repeated figure eight. The dance contains elements that transmit to other bees the location of the food.

water potential In osmosis, the tendency for a system (a cell or solution) to take up water from pure water through a differentially permeable membrane. Water flows toward the system with a more negative water potential. (Contrast with solute potential, pressure potential.)

wavelength The distance between successive peaks of a wave train, such as electromagnetic radiation.

wild-type Geneticists' term for standard or reference type. Deviants from this standard, even if the deviants are found in the wild, are usually referred to as mutant. (Note that this terminology is not usually applied to human genes.)

wood Secondary xylem tissue.

- X -

xanthophyll (zan' tho fill) [Gk. *xanthos*: yellowish-brown + *phyllon*: leaf] A yellow or orange pigment commonly found as an accessory pigment in photosynthesis, but found elsewhere as well. An oxygen-containing carotenoid.

X-linked A character that is coded for by a gene on the X chromosome; a sex-linked trait.

xerophyte (zee' row fyte) [Gk. *xerox*: dry + *phyton*: plant] A plant adapted to an environment with a limited water supply.

xylem (zy' lum) [Gk. *xylon*: wood] In vascular plants, the tissue that conducts water and minerals; xylem consists, in various plants, of tracheids, vessel elements, fibers, and other highly specialized cells.

- Y -

yolk [M.E. *yolke*: yellow] The stored food material in animal eggs, usually rich in protein and lipid.

yolk sac In reptiles, birds, and mammals, the extraembryonic membrane that forms from the endoderm of the hypoblast; it encloses and digests the yolk.

- Z -

Z-DNA A form of DNA in which the molecule spirals to the left rather than to the right.

zeaxanthin A blue-light receptor involved in the opening of plant stomata.

zona pellucida A jellylike substance that surrounds the mammalian ovum when it is released from the ovary.

zoospore (zoe' o spore) [Gk. *zoon*: animal + *spora*: seed] In algae and fungi, any swimming spore. May be diploid or haploid.

zygote (zye' gote) [Gk. *zygotos*: yoked] The cell created by the union of two gametes, in which the gamete nuclei are also fused. The earliest stage of the diploid generation.

zymogen An inactive precursor of a digestive enzyme secreted into the lumen of the gut, where a protease cleaves it to form the active enzyme.

Answers to Self-Quizzes

Chapter 2

1.	b	6.	a
2.	d	7.	d
3.	c	8.	a
4.	c	9.	c
5.	d	10.	b

Chapter 3

1.	e	6.	a
2.	e	7.	c
3.	c	8.	e
4.	d	9.	a
5.	b	10.	d

Chapter 4

1.	b	6.	e
2.	d	7.	a
3.	c	8.	d
4.	e	9.	b
5.	a	10.	d

Chapter 5

1.	e	6.	c
2.	c	7.	c
3.	a	8.	b
4.	d	9.	e
5.	c	10.	c

Chapter 6

1.	c	6.	e
2.	e	7.	d
3.	b	8.	b
4.	c	9.	d
5.	c	10.	e

Chapter 7

1.	d	6.	d
2.	d	7.	a
3.	e	8.	b
4.	e	9.	a
5.	c	10.	e

Chapter 8

1.	c	6.	c
2.	b	7.	c
3.	d	8.	d
4.	b	9.	d
5.	e	10.	b

Chapter 9

1.	d	6.	d
2.	c	7.	e
3.	b	8.	d
4.	d	9.	c
5.	c	10.	c

Chapter 10*

1.	e	6.	d
2.	a	7.	b
3.	d	8.	b
4.	d	9.	b
5.	d	10.	b

Chapter 11

1.	c	6.	b
2.	a	7.	d
3.	c	8.	d
4.	b	9.	c
5.	e	10.	c

Chapter 12

1.	c	6.	d
2.	d	7.	b
3.	e	8.	d
4.	b	9.	d
5.	a	10.	a

Chapter 13

1.	b	6.	d
2.	e	7.	d
3.	a	8.	c
4.	c	9.	b
5.	c	10.	d

Chapter 14

1.	c	6.	c
2.	c	7.	c
3.	a	8.	b
4.	a	9.	e
5.	c	10.	d

Chapter 15

1.	d	6.	a
2.	d	7.	e
3.	c	8.	b
4.	c	9.	c
5.	d	10.	a

Chapter 16

1.	b	6.	b
2.	a	7.	c
3.	a	8.	a
4.	c	9.	e
5.	e	10.	e

Chapter 17

1.	a	6.	b
2.	c	7.	e
3.	b	8.	d
4.	b	9.	c
5.	d	10.	b

Chapter 18

1.	a	6.	a
2.	b	7.	d
3.	e	8.	d
4.	e	9.	a
5.	c	10.	d

Chapter 19

1.	c	6.	c
2.	a	7.	d
3.	a	8.	b
4.	b	9.	a
5.	b	10.	b

Chapter 20

1.	d	6.	b
2.	e	7.	a
3.	c	8.	c
4.	b	9.	b
5.	c	10.	b

Chapter 21

1.	d	6.	a
2.	b	7.	c
3.	e	8.	b
4.	c	9.	c
5.	a	10.	e

*Answers to Chapter 10 Genetics Problems

1. Each of the eight boxes in the Punnett squares should contain the genotype *Tt*, regardless of which parent was tall and which dwarf.

2. See Figure 10.4, page 212.

3. The trait is autosomal. Mother *dp dp*, father *Dp dp*. If the trait were sex-linked, all daughters would be wild-type and sons would be *dumpy*.

4. Yellow parent = s^Ys^b; offspring 3 yellow (s^Y–): 1 black (s^bs^b).
Black parent = s^bs^b; offspring all black (s^bs^b). Orange parent = s^Os^b; offspring 3 orange (s^O–): 1 black (s^bs^b). Both s^O and s^Y are dominant to s^b.

5. All females wild-type; all males spotted.

6. F_1 all wild-type, *PpSwsw*; F_2 9:3:3:1 in phenotypes. See Figure 10.7, page 214, for analogous genotypes.

7*a*. Ratio of phenotypes in F_2 is 3:1 (double dominant to double recessive).

7*b*. The F_1 are $Pby\ pB^Y$; they produce just two kinds of gametes (*Pby* and *pBy*). Combine them carefully and see the 1:2:1 phenotypic ratio fall out in the F_2.

7*c*. Pink-blistery.

7*d*. See Figures 9.16 and 9.18 (pages 196–198). Crossing over took place in the F_1 generation.

8. The genotypes are:

PpSwsw
Ppswsw
ppSwsw
ppswsw

Ratio: 1:1:1:1

The phenotypes are:

wild eye, long wing	pink eye, long wing
wild eye, short wing	pink eye, short wing

Ratio: 1:1:1:1

9*a*. 1 black:2 blue:1 splashed white

9*b*. Always cross black with splashed white.

10*a*. $w^+ > w^e > w$

10*b*. Parents w^ew and w^+Y. Progeny $w{+}w^e$, $w{+}w$, w^eY, and wY.

11. All will have normal vision because they inherit dad's wild-type X chromosome, but half of them will be carriers.

12. Agouti parent *AaBb*. Albino offspring *aaBb* and *aabb*; black offspring *Aabb*; agouti offspring *AaBb*.

13. Because the gene is carried on mitochondrial DNA, it is passed through the mother only. Thus if the woman does not have the disease but her husband does, their child will not be affected. On the other hand, if the woman has the disease but her husband does not, their child *will* have the disease.

Chapter 22

1.	d	6.	d
2.	e	7.	b
3.	d	8.	e
4.	c	9.	b
5.	d	10.	c

Chapter 23

1.	c	6.	d
2.	e	7.	a
3.	d	8.	a
4.	c	9.	c
5.	a	10.	e

Chapter 24

1.	a	6.	e
2.	a	7.	e
3.	d	8.	e
4.	a	9.	b
5.	b	10.	e

Chapter 25

1.	b	6.	b
2.	e	7.	e
3.	a	8.	a
4.	b	9.	e
5.	e	10.	d

Chapter 26

1.	e	6.	b
2.	e	7.	d
3.	b	8.	d
4.	c	9.	c
5.	e	10.	d

Chapter 27

1.	a	6.	b
2.	e	7.	d
3.	c	8.	b
4.	d	9.	a
5.	c	10.	d

Chapter 28

1.	d	6.	e
2.	c	7.	c
3.	e	8.	b
4.	b	9.	b
5.	b	10.	d

Chapter 29

1.	d	6.	c
2.	c	7.	a
3.	d	8.	e
4.	a	9.	c
5.	d	10.	a

Chapter 30

1.	b	6.	a
2.	d	7.	e
3.	e	8.	a
4.	c	9.	c
5.	d	10.	c

Chapter 31

1.	c	6.	b
2.	d	7.	c
3.	c	8.	d
4.	d	9.	e
5.	c	10.	d

Chapter 32

1.	b	6.	e
2.	e	7.	b
3.	c	8.	d
4.	d	9.	d
5.	a	10.	e

Chapter 33

1.	b	6.	e
2.	a	7.	a
3.	c	8.	e
4.	c	9.	c
5.	d	10.	b

Chapter 34

1.	d	6.	b
2.	b	7.	b
3.	e	8.	c
4.	e	9.	a
5.	a	10.	d

Chapter 35

1.	c	6.	d
2.	d	7.	d
3.	b	8.	e
4.	b	9.	e
5.	b	10.	a

Chapter 36

1.	d	6.	c
2.	d	7.	e
3.	c	8.	a
4.	a	9.	d
5.	a	10.	e

Chapter 37

1.	a	6.	c
2.	e	7.	e
3.	c	8.	c
4.	d	9.	a
5.	b	10.	b

Chapter 38

1.	d	6.	e
2.	b	7.	a
3.	e	8.	b
4.	b	9.	c
5.	d	10.	d

Chapter 39

1.	e	6.	a
2.	b	7.	b
3.	c	8.	c
4.	c	9.	d
5.	d	10.	a

Chapter 40

1.	c	6.	b
2.	c	7.	e
3.	a	8.	a
4.	d	9.	e
5.	b	10.	b

Chapter 41

1.	b	6.	b
2.	a	7.	d
3.	b	8.	e
4.	e	9.	c
5.	e	10.	c

Chapter 42

1.	c	6.	d
2.	e	7.	d
3.	a	8.	c
4.	d	9.	d
5.	d	10.	a

Chapter 43

1.	a	6.	c
2.	c	7.	b
3.	e	8.	b
4.	a	9.	b
5.	d	10.	a

Chapter 44

1.	d	6.	e
2.	d	7.	c
3.	c	8.	c
4.	c	9.	d
5.	e	10.	d

Chapter 45

1.	d	6.	e
2.	d	7.	b
3.	a	8.	c
4.	b	9.	c
5.	e	10.	d

Chapter 46

1.	c	6.	c
2.	a	7.	a
3.	e	8.	c
4.	d	9.	a
5.	d	10.	c

Chapter 47

1.	e	6.	d
2.	a	7.	e
3.	b	8.	a
4.	c	9.	a
5.	b	10.	e

Chapter 48

1.	e	6.	b
2.	d	7.	c
3.	a	8.	c
4.	b	9.	a
5.	c	10.	d

Chapter 49

1.	d	6.	d
2.	a	7.	b
3.	c	8.	d
4.	d	9.	c
5.	c	10.	e

Chapter 50

1.	b	6.	d
2.	e	7.	a
3.	c	8.	b
4.	a	9.	d
5.	b	10.	d

Chapter 51

1.	d	6.	b
2.	a	7.	e
3.	d	8.	a
4.	a	9.	c
5.	d	10.	e

Chapter 52

1.	c	6.	c
2.	a	7.	d
3.	a	8.	c
4.	c	9.	d
5.	a	10.	a

Chapter 53

1.	c	6.	d
2.	e	7.	c
3.	c	8.	c
4.	a	9.	d
5.	d	10.	e

Chapter 54

1.	c	6.	b
2.	c	7.	d
3.	a	8.	c
4.	d	9.	a
5.	c	10.	d

Chapter 55

1.	a	6.	e
2.	a	7.	c
3.	b	8.	b
4.	c	9.	d
5.	d	10.	e

Chapter 56

1.	d	6.	b
2.	d	7.	a
3.	e	8.	c
4.	e	9.	d
5.	c	10.	b

Chapter 57

1.	b	6.	a
2.	d	7.	d
3.	e	8.	c
4.	e	9.	a
5.	b	10.	c

Illustration Credits

Cover *Amor de Madre* © Max Billder.

Frontispiece © Lynn M. Stone/Naturepl.com.

Part Openers and Back Cover:

Part 1 *Canopy researcher*: © Mark Moffett/ Minden Pictures.

Part 2 *Mitochondrion*: © Bill Loncore/SPL/Photo Researchers, Inc.

Part 3 *T4 bacteriophage*: © Dept. of Microbiology, Biozentrum/SPL/Photo Researchers, Inc.

Part 4 *Dividing cells*: © Steve Gschmeissner/SPL/ Photo Researchers, Inc.

Part 5 *Fossil mesosaur*: © John Cancalosi/AGE Fotostock.

Part 6 *Tiger moth*: © Piotr Naskrecki/Minden Pictures.

Part 7 *Protea flower*: © Gerry Ellis/Minden Pictures.

Part 8 *Great blue heron*: © Larry Jon Friesen.

Part 9 *Zebras and wildebeest*: © Gerry Ellis/ Minden Pictures.

Table of Contents:

p. xix *Leaf*: David McIntyre.

p. xxix *Passionflower*: © Larry Jon Friesen.

p. xxxv *Feather stars*: © André Seale/AGE Fotostock.

Chapter 1 *Frogs*: © Frans Lanting/Minden Pictures. *Pieter Johnson*: © Steven Holt/ stockpix.com. 1.1A: © Eye of Science/SPL/Photo Researchers, Inc. 1.1B: © Dennis Kunkel Microscopy, Inc. 1.1C: © Markus Geisen/NHMPL. 1.1D: © Frans Lanting/Minden Pictures. 1.1E: © Glen Threlfo/Auscape/Minden Pictures. 1.1F: © Piotr Naskrecki/Minden Pictures. 1.1G: © Tui De Roy/Minden Pictures. 1.2A: © Dr. Cecil H. Fox/ Photo Researchers, Inc. 1.2B: From R. Hooke, 1664. *Micrographia*. 1.2C: © Astrid & Hanns-Frieder Michler/Photo Researchers, Inc. 1.2D: © John Durham/SPL/Photo Researchers, Inc. 1.3 *Maple*: © Simon Colmer & Abby Rex/Alamy. 1.3 *Spruce*: David McIntyre. 1.3 *Lily pad*: © Pete Oxford/ Naturepl.com. 1.3 *Cucumber*: David McIntyre. 1.3 *Pitcher plant*: © Nick Garbutt/Naturepl.com. 1.5A: © Bob Bennett/Jupiter Images. 1.5B: © Frans Lanting/Minden Pictures. 1.6 *Organism*: © Sami Sarkis/Painet Inc. 1.6 *Population*: © LogicStock/ Painet Inc. 1.6 *Community*: © Georgie Holland/ AGE Fotostock. 1.6*Biosphere*: Courtesy of NASA. 1.7A: © Frans Lanting/Minden Pictures. 1.7B: © Marguerite Smits Van Oyen/Naturepl.com. 1.8: © Chris Howes/Wild Places Photography/Alamy. 1.10: © Dennis Kunkel Microscopy, Inc. 1.12: Courtesy of Wayne Whippen. 1.13 *Frog (inset)*: © Frans Lanting/Minden Pictures. 1.13, 1.14: After P. T. Johnson et al., 1999. *Science* 284: 802.

Chapter 2 *Enceladus*: Courtesy of the NASA Jet Propulsion Laboratory/Space Science Institute. *Rover*: Courtesy of the NASA Jet Propulsion Laboratory. 2.4: From N. D. Volkow et al., 2001. *Am. J. Psychiatry* 158: 377. 2.14: © Mitsuaki Iwago/ Minden Pictures. 2.15: David McIntyre.

Chapter 3 *Cookies*: David McIntyre. 3.8: Data from PDB 1IVM. T. Obita, T. Ueda, & T. Imoto, 2003. *Cell. Mol. Life Sci.* 60: 176. 3.9A: Data from PDB 2HHB. G. Fermi et al., 1984. *J. Mol. Biol.* 175: 159. 3.16C *left*: © Biophoto Associates/Photo Researchers, Inc. 3.16C *middle*: © Ken Wagner/ Visuals Unlimited. 3.16C *right*: © CNRI/SPL/Photo Researchers, Inc. 3.17 *Cartilage*: © Robert Brons/ Biological Photo Service. 3.17 *Beetle*: © Scott Bauer/ USDA. 3.26: Data from S. Arnott & D. W. Hukins, 1972. *Biochem. Biophys. Res. Commun.* 47(6): 1504. 3.27: Courtesy of NASA.

Chapter 4 *Fossil*: © Stanley M. Awramik/ Biological Photo Service. *Cyanobacterium*: © Dennis Kunkel Microscopy, Inc. *Protobionts*: © Sidney Fox/Science VU/Visuals Unlimited. 4.1: After N. Campbell, 1990. *Biology*, 2nd Ed., Benjamin Cummings. 4.3 *upper row, left*: Courtesy of the IST Cell Bank, Genoa. 4.3 *upper row, center and right, middle row, left*: © Michael W. Davidson, Florida State U. 4.3 *middle row, center*: © Dr. Gopal Murti/SPL/Photo Researchers, Inc. 4.3 *middle row, right*: © Richard J. Green/SPL/Photo Researchers, Inc. 4.3 *bottom row, left*: © Dr. Gopal Murti/Visuals Unlimited. 4.3 *bottom row, center*: © K. R. Porter/ SPL/Photo Researchers, Inc. 4.3 *bottom row, right*: © D. W. Fawcett/Photo Researchers, Inc. 4.4: © J. J. Cardamone Jr. & B. K. Pugashetti/Biological Photo Service. 4.5A: © Dennis Kunkel Microscopy, Inc. 4.5B: Courtesy of David DeRosier, Brandeis U. 4.7 *Mitochondrion*: © K. Porter, D. Fawcett/Visuals Unlimited. 4.7 *Cytoskeleton*: © Don Fawcett, John Heuser/Photo Researchers, Inc. 4.7 *Nucleolus*: © Richard Rodewald/Biological Photo Service. 4.7 *Peroxisome*: © E. H. Newcomb & S. E. Frederick/ Biological Photo Service. 4.7 *Cell wall*: © Biophoto Associates/Photo Researchers, Inc. 4.7 *Ribosome*: From M. Boublik et al., 1990. *The Ribosome*, p. 177. Courtesy of American Society for Microbiology. 4.7 *Centrioles*: © Barry F. King/Biological Photo Service. 4.7 *Plasma membrane*: Courtesy of J. David Robertson, Duke U. Medical Center. 4.7 *Rough ER*: © Don Fawcett/Science Source/Photo Researchers, Inc. 4.7 *Smooth ER*: © Don Fawcett, D. Friend/Science Source/Photo Researchers, Inc. 4.7 *Chloroplast*: © W. P. Wergin, E. H. Newcomb/ Biological Photo Service. 4.7 *Golgi apparatus*: Courtesy of L. Andrew Staehelin, U. Colorado. 4.8 *left*: From U. Aebi et al., 1986. *Nature* 323: 560. © Macmillan Publishers Ltd. 4.8 *upper right*: © D. W. Fawcett/Photo Researchers, Inc. 4.8 *lower right*: Courtesy of Dr. Ron Milligan, Scripps Research Institute. 4.9A: © Barry King, U. California, Davis/Biological Photo Service. 4.9B: © Biophoto Associates/Science Source/Photo Researchers, Inc. 4.10: © Don Fawcett/Visuals Unlimited. 4.11: © B. Bowers/Photo Researchers, Inc. 4.12: © Sanders/Biological Photo Service. 4.13: © K. Porter, D. Fawcett/Visuals Unlimited. 4.14: © W. P. Wergin, E. H. Newcomb/Biological Photo Service. 4.14 *inset*: © W. P. Wergin/Biological Photo Service. 4.15A: © John Durham/SPL/Photo Researchers, Inc. 4.15B: © Paul W. Johnson/ Biological Photo Service. 4.15C: © Gerald & Buff Corsi/Visuals Unlimited. 4.16A: David McIntyre. 4.16A *inset*: © Richard Green/Photo Researchers, Inc. 4.16B: David McIntyre. 4.16B *inset*: Courtesy of R. R. Dute. 4.17: © E. H. Newcomb & S. E. Frederick/Biological Photo Service. 4.18: Courtesy of M. C. Ledbetter, Brookhaven National Laboratory. 4.19: © Michael Abbey/Visuals Unlimited. 4.20: Courtesy of Vic Small, Austrian Academy of Sciences, Salzburg, Austria. 4.21: Courtesy of N. Hirokawa. 4.22A *top*: © SPL/Photo Researchers, Inc. 4.22A *bottom*, 4.22B, C: © W. L. Dentler/Biological Photo Service. 4.23: From N. Pollack et al., 1999. *J. Cell Biol.* 147: 493. Courtesy of R. D. Vale. 4.24: © Biophoto Associates/Photo Researchers, Inc. 4.25 *left*: Courtesy of David Sadava. 4.25 *upper right*: From J. A. Buckwalter & L. Rosenberg, 1983. *Coll. Rel. Res.* 3: 489. Courtesy of L. Rosenberg. 4.25 *lower right*: © J. Gross, Biozentrum/SPL/Photo Researchers, Inc.

Chapter 5 *Vibrio*: © Dennis Kunkel Microscopy, Inc. *Cholera treatment*: Courtesy of the WHO Global Consultation on Child and Adolescent Health and Development. 5.2: After L. Stryer, 1981. *Biochemistry*, 2nd Ed., W. H. Freeman. 5.3: Courtesy of L. Andrew Staehelin, U. Colorado. 5.7A: Courtesy of D. S. Friend, U. California, San Francisco. 5.7B: Courtesy of Darcy E. Kelly, U. Washington. 5.7C: Courtesy of C. Peracchia. 5.17: From M. M. Perry, 1979. *J. Cell Sci.* 39: 26.

Chapter 6 *Enzyme*: Data from PDB 1CW3. L. Ni et al., 1999. *Protein Sci.* 8: 2784. *Champagne*: © Purestock/Alamy. 6.1: Violet Bedell-McIntyre. 6.5B: © Darwin Dale/Photo Researchers, Inc. 6.11A: Data from PDB 1AL6. B. Schwartz et al., 1997. 6.11B: Data from PDB 1BB6. V. B. Vollan et al., 1999. *Acta Crystallogr. D. Biol. Crystallogr.* 55: 60. 6.11C: Data from PDB 1AB9. N. H. Yennawar, H. P. Yennawar, & G. K. Farber, 1994. *Biochemistry* 33: 7326. 6.13: Data from PDB 1A7K. H. Kim & W. G. Hol, 1998. *J Mol. Biol.* 278: 5.

Chapter 7 *Marathoners*: © Chuck Franklin/ Alamy. *Mouse*: © Royalty-Free/Corbis. 7.13: Courtesy of Ephraim Racker.

Chapter 8 *Rainforest*: © Jon Arnold Images/ photolibrary.com. *Meteor*: © Don Davis. 8.1: © Andrew Syred/SPL/Photo Researchers, Inc. 8.12: Courtesy of Lawrence Berkeley National Laboratory. 8.16, 8.18: © E. H. Newcomb & S. E. Frederick/Biological Photo Service. 8.20: © Aflo Foto Agency/Alamy.

Chapter 9 *Cells*: © Dr. Torsten Wittmann/Photo Researchers, Inc. *Henrietta Lacks*: © Obstetrics and Gynaecology/Photo Researchers, Inc. 9.1A: © David M. Phillips/Visuals Unlimited. 9.1B, C: © John D. Cunningham/Visuals Unlimited. 9.2B: © John J. Cardamone Jr./Biological Photo Service. 9.7: Courtesy of G. F. Bahr. 9.8 *inset*: © Biophoto Associates/Science Source/Photo Researchers, Inc. 9.9B: © Conly L. Rieder/Biological Photo Service. 9.10: © Nasser Rusan. 9.12A: © T. E. Schroeder/ Biological Photo Service. 9.12B: © B. A. Palevitz, E. H. Newcomb/Biological Photo Service. 9.13A: © Dr. John Cunningham/Visuals Unlimited. 9.13B:

© Garry T. Cole/Biological Photo Service. 9.14 *left*: © Andrew Syred/SPL/Photo Researchers, Inc. 9.14 *center*: David McIntyre. 9.14 *right*: © Gerry Ellis, DigitalVision/PictureQuest. 9.15: Courtesy of Dr. Thomas Ried and Dr. Evelin Schröck, NIH. 9.16: © C. A. Hasenkampf/Biological Photo Service. 9.17: Courtesy of J. Kezer. 9.21A: © Gopal Murti/Photo Researchers, Inc.

Chapter 10 *Bris*: © David H. Wells/CORBIS. *Eye test*: © Brand X Pictures/Alamy. 10.1: © the Mendelianum. 10.11: Courtesy the American Netherland Dwarf Rabbit Club. 10.14: Courtesy of Madison, Hannah, and Walnut. 10.15: Courtesy of Pioneer Hi-Bred International, Inc. 10.16: © Carolyn A. McKeone/Photo Researchers, Inc. 10.17: © Peter Morenus/U. of Connecticut. *Bay scallops*: © Barbara J. Miller/Biological Photo Service.

Chapter 11 *T. rex*: © The Natural History Museum, London. *Earrings*: David McIntyre/Model: Ashley Ying. 11.3: © Biozentrum, U. Basel/SPL/Photo Researchers, Inc. 11.6 *X-ray crystallograph*: Courtesy of Prof. M. H. F. Wilkins, Dept. of Biophysics, King's College, U. London. 11.6 *Franklin*: © CSHL Archives/Peter Arnold, Inc. 11.8A: © A. Barrington Brown/Photo Researchers, Inc. 11.8B: Data from S. Arnott & D. W. Hukins, 1972. *Biochem. Biophys. Res. Commun.* 47(6): 1504. 11.21C: © Dr. Peter Lansdorp/Visuals Unlimited.

Chapter 12 *Ricinus*: © Alan L. Detrick/Photo Researchers, Inc. *Ribosome*: Data from PDB 1GIX and 1G1Y. M. M. Yusupov et al., 2001. *Science* 292: 883. 12.4: Data from PDB 1I3Q. P. Cramer et al., 2001 *Science* 292: 1863. 12.8: Data from PDB 1EHZ. H. Shi & P. B. Moore, 2000. *RNA* 6: 1091. 12.9B: Data from PDB 1EUQ. L. D. Sherlin et al., 2000. *J. Mol. Biol.* 299: 431. 12.10: Data from PDB 1GIX and 1G1Y. M. M. Yusupov et al., 2001. *Science* 292: 883. 12.14: Courtesy of J. E. Edström and *EMBO J.* 12.18: © Stanley Flegler/Visuals Unlimited.

Chapter 13 *Market*: © Derek Brown/Alamy. *Scientist*: © Tek Image/SPL/Photo Researchers, Inc. 13.1A: © Dept. of Microbiology, Biozentrum/SPL/Photo Researchers, Inc. 13.1B: © Dennis Kunkel Microscopy, Inc. 13.2A: © Dr. Linda Stannard UCT/Photo Researchers, Inc. 13.2B: © E.O.S./Gelderblom/Photo Researchers, Inc. 13.2C: © BSIP Agency/Index Stock Imagery/Jupiter Images. 13.7: Courtesy of Roy French. 13.11: Courtesy of L. Caro and R. Curtiss. 13.22: Based on an illustration by Anthony R. Kerlavage, Institute for Genomic Research. *Science* 269: 449 (1995). *For Investigation*: Adapted from J. J. Perry, J. T. Staley, & S. Lory, 2002. *Microbial Life*. Sinauer Associates.

Chapter 14 *Cheetah*: © John Giustina/ImageState/Jupiter Images. *Panther*: © Lynn M. Stone/Naturepl.com. 14.7: From D. C. Tiemeier et al., 1978. *Cell* 14: 237. 14.18: Courtesy of Murray L. Barr, U. Western Ontario. 14.20: Courtesy of O. L. Miller, Jr.

Chapter 15 *Student*: © Ryan McVay, Photodisc Green/Getty Images. *Ethiopian*: © JTB Photo/photolibrary.com. 15.2: © Biophoto Associates/Photo Researchers, Inc. 15.3: From A. M. de Vos, M. Ultsch, & A. A. Kossiakoff, 1992. *Science* 255: 306. 15.14: © Stephen A. Stricker, courtesy of Molecular Probes, Inc.

Chapter 16 *Destruction*: © Suzanne Plunkett/AP Images. *Baby 81*: © Gemunu Amarasinghe/AP Images. 16.2: © Philippe Plailly/Photo Researchers, Inc. 16.5: Bettmann/CORBIS. 16.6: Courtesy of Keith Weller, De Wood, Chris Pooley, & Scott Bauer/USDA ARS. 16.19A: Courtesy of Ingo Potrykus, Swiss Federal Institute of Technology. 16.19B: Joan Gemme. 16.20: Courtesy of Eduardo Blumwald.

Chapter 17 *Champlain*: © North Wind Picture Archives/Alamy. *Family*: © David Sanger Photography/Alamy. 17.6: From C. Harrison et al., 1983. *J. Med. Genet.* 20: 280. 17.11: Courtesy of Harvey Levy and Cecelia Walraven, New England Newborn Screening Program. 17.14: © Dennis Kunkel Microscopy, Inc. 17.19B: © David M. Martin, M.D./SPL/Photo Researchers, Inc. 17.23: From P. H. O'Farrell, 1975. High resolution two-dimensional electrophoresis of proteins. *J. Biol. Chem.* 250: 4007-21. Courtesy of Patrick H. O'Farrell. 17.24: After N. M. Morel et al., 2004. *Mayo Clin. Proc.* 79: 651.

Chapter 18 *Vaccination*: Bettmann/CORBIS. *T cell*: © Dr. Andrejs Liepins/SPL/Photo Researchers, Inc. 18.3: © Dennis Kunkel Microscopy, Inc. 18.8: © Dr. Gopal Murti/SPL/Photo Researchers, Inc. 18.14: © David Phillips/Science Source/Photo Researchers, Inc.

Chapter 19 *Embryo*: © Dr. Yorgas Nikas/Photo Researchers, Inc. *Centrifuge*: Courtesy of Cytori Therapeutics. 19.4: © Roddy Field, the Roslin Institute. 19.5: Courtesy of T. Wakayama and R. Yanagimachi. 19.12: From J. E. Sulston & H. R. Horvitz, 1977. *Dev. Bio.* 56: 100. 19.15B, 19.16 *left*: Courtesy of J. Bowman. 19.16 *right*: Courtesy of Detlef Weigel. 19.17: Courtesy of W. Driever and C. Nüsslein-Vollhard. 19.18B: Courtesy of C. Rushlow and M. Levine. 19.18C: Courtesy of T. Karr. 19.18D: Courtesy of S. Carroll and S. Paddock. 19.20: Courtesy of F. R. Turner, Indiana U.

Chapter 20 *Fly head*: © David Scharf/Photo Researchers, Inc. *Mutant leg*: From G. Halder et al., 1995. *Science* 267: 1788. Courtesy of W. J. Gehring and G. Halder. 20.2A: Courtesy of E. B. Lewis. 20.2B: From H. Le Mouellic et al., 1992. *Cell* 69: 251. Courtesy of H. Le Mouellic, Y. Lallemand, and P. Brûlet. 20.5: Courtesy of J. Hurle and E. Laufer. 20.6: Courtesy of J. Hurle. 20.7 *Cladogram*: After R. Galant & S. Carroll, 2002. *Nature* 415: 910. 20.7 *Beetle*: © Stockbyte/PictureQuest. 20.7 *Centipede*: © Burke/Triolo/Brand X Pictures/PictureQuest. 20.8: Courtesy of S. Carroll and P. Brakefield. 20.9: © Erick Greene. 20.10: Photograph by C. Laforsch & R. Tollrian, courtesy of A. A. Agrawal. 20.11: © Nigel Cattlin, Holt Studios International/Photo Researchers, Inc. 20.13: © Simon D. Pollard/Photo Researchers, Inc. 20.14: Courtesy of Mike Shapiro and David Kingsley.

Chapter 21 *Capybara*: © Frans Lanting/Minden Pictures. *Grand Canyon*: © Robert Fried/Tom Stack & Assoc. 21.4A: David McIntyre. 21.4B: © Robin Smith/photolibrary.com. 21.7: © François Gohier/The National Audubon Society Collection/Photo Researchers, Inc. 21.8: © Jeff J. Daly/Visuals Unlimited. 21.9 *left*: © Ken Lucas/Visuals Unlimited. 21.9 *right*: © The Natural History Museum, London. 21.10: © Chip Clark. 21.11: © Hans Steur/Visuals Unlimited. 21.12: © Tom McHugh/Field Museum, Chicago/Photo Researchers, Inc. 21.13: Courtesy of Conrad C. Labandeira, Department of Paleobiology, National Museum of Natural History, Smithsonian Institution. 21.14: © Chase Studios, Cedarcreek, MO. 21.16: © The Natural History Museum, London. 21.18: © K. Simons and David Dilcher. 21.20: David McIntyre. 21.22: © John Worrall. 21.23: © Calvin Larsen/Photo Researchers, Inc.

Chapter 22 *Snake*: © Joseph T. Collins/Photo Researchers, Inc. *Newt*: © Robert Clay/Visuals Unlimited. *Pufferfish*: © Georgette Douwma/Naturepl.com. 22.1A, B: © SPL/Photo Researchers, Inc. 22.2: From W. Levi, 1965. *Encyclopedia of Pigeon Breeds*. T. F. H. Publications, Jersey City, NJ. (A, B: photos by R. L. Kienlen, courtesy of Ralston Purina Company; C, D: photos by Stauber). 22.9A: © S. Maslowski/Visuals Unlimited. 22.9B: © Anthony Cooper/SPL/Photo Researchers, Inc. 22.11, 22.17A: David McIntyre. 22.21A: © Marilyn Kazmers/Dembinsky Photo Assoc. 22.21B: © Paul Osmond/Painet Inc. 22.22 *upper*: © Jeff Foott/Naturepl.com. 22.22 *lower*: © franzfoto.com/Alamy.

Chapter 23 *Hummingbird*: © Tui De Roy/Minden Pictures. *Swift*: © Kim Taylor/Naturepl.com. 23.1A *left*: © Gary Meszaros/Dembinsky Photo Assoc. 23.1A *right*: © Lior Rubin/Peter Arnold, Inc. 23.1B: © Fi Rust/Painet Inc. 23.8: © Jan Vermeer/Foto Natura/Minden Pictures. 23.9A: © Virginia P. Weinland/Photo Researchers, Inc. 23.9B: © Pablo Galán Cela/AGE Fotostock. 23.10A: © J. S. Sira/photolibrary.com. 23.10B: © Daniel L. Geiger/SNAP/Alamy. 23.12: © Boris I. Timofeev/Pensoft. 23.14A: © Tony Tilford/photolibrary.com. 23.14B: © W. Peckover/VIREO. 23.15 *Madia*: © Peter K. Ziminsky/Visuals Unlimited. 23.15 *Argyroxiphium*: © Elizabeth N. Orians. 23.15 *Wilkesia*: © Gerald D. Carr. 23.15 *Dubautia*: © Noble Proctor/The National Audubon Society Collection/Photo Researchers, Inc.

Chapter 24 *Vaccination*: Courtesy of Chris Zahniser, B.S.N., R.N., M.P.H./CDC. *Virus*: © Dr. Tim Baker/Visuals Unlimited. 24.3 *Rice*: data from pdb 1CCR. *Tuna*: data from pdb 5CYT. 24.4: From P. B. Rainey & M. Travisano, 1998. *Nature* 394: 69. © Macmillan Publishers Ltd. 24.7A: © Barrie Britton/Naturepl.com. 24.7B: © M. Graybill/J. Hodder/Biological Photo Service.

Chapter 25 *HIV*: © James Cavallini/Photo Researchers, Inc. *Chimpanzees*: © John Cancalosi/Naturepl.com. 25.6A: © Mark Smith/Photo Researchers, Inc. 25.6B: After E. Verheyen et al., 2003. *Science* 300: 325. 25.7: © Larry Jon Friesen. 25.9: © Alexandra Basolo. 25.10: David M. Hillis and Matthew Brauer. 25.11A: © Helen Carr/Biological Photo Service. 25.11B: © Michael Giannechini/Photo Researchers, Inc. 25.11C: © Skip Moody/Dembinsky Photo Assoc.

Chapter 26 *River*: © Felipe Rodriguez/Alamy. *Salmonella* and *Methanospirillum*: © Kari Lounatmaa/Photo Researchers, Inc. 26.2A: © David Phillips/Photo Researchers, Inc. 26.2B: © R. Kessel & G. Shih/Visuals Unlimited. 26.2C: Courtesy of Janice Carr/NCID/CDC. 26.4: From F. Balagaddé et al., 2005. *Science* 309: 137. Courtesy of Frederick Balagaddé. 26.5A *left*: © David M. Phillips/Visuals Unlimited. 26.5A *right*: Courtesy of Peter Hirsch and Stuart Pankratz. 26.5B *left*: Courtesy of the CDC. 26.5B *right*: Courtesy of Peter Hirsch and Stuart Pankratz. 26.6A: © J. A. Breznak & H. S. Pankratz/Biological Photo Service. 26.6B: © J. Robert Waaland/Biological Photo Service. 26.7: © USDA/Visuals Unlimited. 26.8: © Steven Haddock and Steven Miller. 26.12: Courtesy of David Cox/CDC. 26.13: Courtesy of Randall C. Cutlip. 26.14: © David Phillips/Visuals Unlimited. 26.15A: © Paul W. Johnson/Biological Photo Service. 26.15B: © H. S. Pankratz/Biological Photo Service. 26.15C: © Bill Kamin/Visuals

Unlimited. 26.16: © Dr Kari Lounatmaa/Photo Researchers, Inc. 26.17: © Dr. Gary Gaugler/ Visuals Unlimited. 26.18: © Michael Gabridge/ Visuals Unlimited. 26.20: © Phil Gates/Biological Photo Service. 26.21: From K. Kashefi & D. R. Lovley, 2003. *Science* 301: 934. Courtesy of Kazem Kashefi. 26.23: © Krafft/Hoa-qui/Photo Researchers, Inc. 26.24: © David Sanger Photography/Alamy. 26.25: From H. Huber et al., 2002. *Nature* 417: 63. © Macmillan Publishers Ltd. Courtesy of Karl O. Stetter.

Chapter 27 *Leishmania*: © Dennis Kunkel Microscopy, Inc. *Macrocystis*: © Karen Gowlett-Holmes/photolibrary.com. 27.1: © Steve Gschmeissner/Photo Researchers, Inc. 27.2: © David Patterson, Linda Amaral Zettler, Mike Peglar, & Tom Nerad/micro*scope. 27.3: © London School of Hygiene/SPL/Photo Researchers, Inc. 27.4A: © Bill Bachman/Photo Researchers, Inc. 27.4B: © Markus Geisen/NHMPL. 27.5: © Astrid & Hanns-Frieder Michler/SPL/Photo Researchers, Inc. 27.9: © Mike Abbey/Visuals Unlimited. 27.12A: © Wim van Egmond/Visuals Unlimited. 27.12B: © Biophoto Associates/Photo Researchers, Inc. 27.18: © Dennis Kunkel Microscopy, Inc. 27.19A: © Mike Abbey/Visuals Unlimited. 27.19B: © Dennis Kunkel Microscopy, Inc. 27.19C: © Paul W. Johnson/Biological Photo Service. 27.19D: © M. Abbey/Photo Researchers, Inc. 27.21: © Manfred Kage/Peter Arnold, Inc. 27.22: After A. Ianora et al., 2005. *Nature* 429: 403. 27.23A: © Duncan McEwan/Naturepl.com. 27.23B, C: © Larry Jon Friesen. 27.23D: © J. N. A. Lott/Biological Photo Service. 27.24: © James W. Richardson/Visuals Unlimited. 27.25A: © Larry Jon Friesen. 27.25B: © Milton Rand/Tom Stack & Assoc. 27.26A: © Carolina Biological/Visuals Unlimited. 27.26B: © Larry Jon Friesen. 27.27A: © J. Paulin/Visuals Unlimited. 27.27B: © Dr. David M. Phillips/Visuals Unlimited. 27.29: © Andrew Syred/SPL/Photo Researchers, Inc. 27.30A: © William Bourland/ micro*scope. 27.30B: © David Patterson & Aimlee Laderman/micro*scope. 27.31: © Larry Jon Friesen. 27.32: Courtesy of R. Blanton and M. Grimson.

Chapter 28 *Fossils*: © Sinclair Stammers/SPL/ Photo Researchers, Inc. *Rainforest*: © Photo Resource Hawaii/Alamy. 28.2A: © Ronald Dengler/ Visuals Unlimited. 28.2B: © Larry Mellichamp/ Visuals Unlimited. 28.3: © Brian Enting/Photo Researchers, Inc. 28.5: © J. Robert Waaland/ Biological Photo Service. 28.6: After L. E. Graham et al., 2004. *PNAS* 101: 11025. Micrograph courtesy of Patricia Gensel and Linda Graham. 28.8: © U. Michigan Exhibit Museum. 28.10: Courtesy of the Biology Department Greenhouses, U. Massachusetts, Amherst. 28.11: After C. P. Osborne et al., 2004. *PNAS* 101: 10360. 28.13A, B: David McIntyre. 28.13C: © Harold Taylor/ photolibrary.com. 28.14: © Daniel Vega/AGE Fotostock. 28.15: © Danilo Donadoni/AGE Fotostock. 28.16A: © Ed Reschke/Peter Arnold, Inc. 28.16B: © Carolina Biological/Visuals Unlimited. 28.17A: © J. N. A. Lott/Biological Photo Service. 28.17B: © David Sieren/Visuals Unlimited. 28.18: Courtesy of the Biology Department Greenhouses, U. Massachusetts, Amherst. 28.19 A: © Rod Planck/Dembinsky Photo Assoc. 28.19B: © Nuridsany et Perennou/Photo Researchers, Inc. 28.19C: Courtesy of the Talcott Greenhouse, Mount Holyoke College. 28.20 *inset*: David McIntyre.

Chapter 29 *Masada*: © Eddie Gerald/Alamy. *Coconut*: © Ben Osborne/Naturepl.com. 29.1 *Seed fern*: David McIntyre. 29.1 *Cycad*: © Patricio Robles Gil/Naturepl.com. 29.1 *Magnolia*: © Plantography/ Alamy. 29.4: © Natural Visions/Alamy. 29.5A: © Patricio Robles Gil/Naturepl.com. 29.5B: © Dave Watts/Naturepl.com. 29.5C: © M. Graybill/J. Hodder/Biological Photo Service. 29.5D: © Frans Lanting/Minden Pictures. 29.7A *left*: David McIntyre. 29.7A *right*: © Stan W. Elems/Visuals Unlimited. 29.7B *left*: David McIntyre. 29.7B *right*: © Dr. John D. Cunningham/Visuals Unlimited. 29.10A: David McIntyre. 29.10B: © Aflo Foto Agency/Naturepl.com. 29.10C: © Richard Shiell/ Dembinsky Photo Assoc. 29.11A: © Plantography/ Alamy. 29.11B: © Thomas Photography LLC/ Alamy. 29.15A: © Inga Spence/Tom Stack & Assoc. 29.15B: © Holt Studios/Photo Researchers, Inc. 29.15C: © Catherine M. Pringle/Biological Photo Service. 29.15D: © blickwinkel/Alamy. 29.17A: Courtesy of Stephen McCabe, U. California, Santa Cruz, and UCSC Arboretum. 29.17B: David McIntyre. 29.17C: © Rob & Ann Simpson/Visuals Unlimited. 29.17D: © R. C. Carpenter/Photo Researchers, Inc. 29.17E: © Geoff Bryant/Photo Researchers, Inc. 29.17F: © José Antonio Jiménez/AGE Fotostock. 29.18A: © Andrew E. Kalnik/Photo Researchers, Inc. 29.18B: © Ed Reschke/Peter Arnold, Inc. 29.18C: © Adam Jones/ Dembinsky Photo Assoc. 29.19A: © Willard Clay/ photolibrary.com. 29.19B: © Adam Jones/ Dembinsky Photo Assoc. 29.19C: © Alan & Linda Detrick/The National Audubon Society Collection/Photo Researchers, Inc. 29.20: © Diaphor La Phototheque/photolibrary.com.

Chapter 30 *Fusarium*: © Dr. Gary Gaugler/ Visuals Unlimited. *Ant*: © L. E. Gilbert/Biological Photo Service. 30.3: © David M. Phillips/Visuals Unlimited. 30.5A: © G. T. Cole/Biological Photo Service. 30.6: © N. Allin & G. L. Barron/Biological Photo Service. 30.7: © Richard Packwood/photo-library.com. 30.8A: © Amy Wynn/Naturepl.com. 30.8B: © Geoff Simpson/Naturepl.com. 30.8C: © Gary Meszaros/Dembinsky Photo Assoc. 30.10A: © R. L. Peterson/Biological Photo Service. 30.10B: © Ken Wagner/Visuals Unlimited. 30.13A: © J. Robert Waaland/Biological Photo Service. 30.13B: © M. F. Brown/Visuals Unlimited. 30.13C: © Dr. John D. Cunningham/Visuals Unlimited. 30.13D: © Biophoto Associates/Photo Researchers, Inc. 30.14 *left*: © William E. Schadel/Biological Photo Service. 30.14 *right*: © Dr. John D. Cunningham/Visuals Unlimited. 30.15: © John Taylor/Visuals Unlimited. 30.16: © G. L. Barron/ Biological Photo Service. 30.17A: © Richard Shiell/Dembinsky Photo Assoc. 30.17B: © Matt Meadows/Peter Arnold, Inc. 30.18: © Andrew Syred/SPL/Photo Researchers, Inc. 30.19A: © Manfred Danegger/Photo Researchers, Inc. 30.19B: © Botanica/photolibrary.com.

Chapter 31 *Fossil*: Photograph by Diane Scott. *Dinosaur*: © Joe Tucciarone/Photo Researchers, Inc. 31.2A: Courtesy of J. B. Morrill. 31.2B: From G. N. Cherr et al., 1992. *Microsc. Res. Tech.* 22: 11. Courtesy of J. B. Morrill. 31.5A: © Jurgen Freund/ Naturepl.com. 31.5B: © Henry W. Robison/Visuals Unlimited. 31.6A: © Brian Parker/Tom Stack & Assoc. 31.6B: © Tui De Roy/Minden Pictures. 31.8: © Colin M. Orians. 31.9A: © T. Kitchin & V. Hurst/Photo Researchers, Inc. 31.9B: © Stockbyte/ PictureQuest. 31.10: Adapted from F. M. Bayerand & H. B. Owre, 1968. *The Free-Living Lower Invertebrates*, Macmillan Publishing Co. 31.11A: © Scott Camazine/Alamy. 31.11B, C: David McIntyre. 31.13A: © Stephen Dalton/Minden Pictures. 31.13B: © Gerald & Buff Corsi/Visuals Unlimited. 31.14A: © Dave Watts/Naturepl.com. 31.14B: © Larry Jon Friesen. 31.16A: © Jurgen Freund/Naturepl.com. 31.16B: © Larry Jon Friesen. 31.16C: © David Wrobel/Visuals Unlimited. 31.17B: © Larry Jon Friesen. 31.18: Adapted from F. M. Bayerand & H. B. Owre, 1968. *The Free-Living Lower Invertebrates*, Macmillan Publishing Co. 31.19A: © Larry Jon Friesen. 31.19B: © Michael Patrick O'Neill/Alamy. 31.19C, D: © Larry Jon Friesen. 31.20A: From J. E. N. Veron, 2000. *Corals of the World*. © J. E. N. Veron. 31.20B: © Jurgen Freund/Naturepl.com. 31.21: Adapted from F. M. Bayerand & H. B. Owre, 1968. *The Free-Living Lower Invertebrates*, Macmillan Publishing Co.

Chapter 32 *Strepsipterans*: Courtesy of Dr. Hans Pohl. *Wasp*: © Paulo de Oliveira/OSF/Jupiter Images. 32.2: © Robert Brons/Biological Photo Service. 32.3A: From D. C. García-Bellido & D. H. Collins, 2004. *Nature* 429: 40. Courtesy of Diego García-Bellido Capdevila. 32.3B: © Piotr Naskrecki/ Minden Pictures. 32.6: © Alexis Rosenfeld/Photo Researchers, Inc. 32.7A: © Ed Robinson/ photolibrary.com. 32.8B: © Robert Brons/Biological Photo Service. 32.9B: © Larry Jon Friesen. 32.10A: © Stan Elems/Visuals Unlimited. 32.11: © David J. Wrobel/Visuals Unlimited. 32.13A: © Jurgen Freund/Naturepl.com. 32.13B: Courtesy of R. R. Hessler, Scripps Institute of Oceanography. 32.13C: © Roger K. Burnard/Biological Photo Service. 32.13D: © Larry Jon Friesen. 32.15A: © Larry Jon Friesen. 32.15B: © Dave Fleetham/ photolibrary.com. 32.15C, D, E: © Larry Jon Friesen. 32.15F: © Orion Press/Jupiter Images. 32.16A: Courtesy of Jen Grenier and Sean Carroll, U. Wisconsin. 32.16B: Courtesy of Graham Budd. 32.16C: Courtesy of Reinhardt Møbjerg Kristensen. 32.17: © R. Calentine/Visuals Unlimited. 32.18B, C: © James Solliday/Biological Photo Service. 32.20A: © Michael Fogden/ photolibrary.com. 32.20B: © Diane R. Nelson/ Visuals Unlimited. 32.21: © Ken Lucas/Visuals Unlimited. 32.22A: © Frans Lanting/Minden Pictures. 32.22B: © Larry Jon Friesen. 32.22C: © Tom Branch/Photo Researchers, Inc. 32.22D: © Norbert Wu/Minden Pictures. 32.25: © Mark Moffett/Minden Pictures. 32.26: © Oxford Scientific Films/Jupiter Images. 32.27A: © Larry Jon Friesen. 32.27B: © Larry Jon Friesen. 32.27C: © Meul/ARCO/Naturepl.com. 32.27D: © Piotr Naskrecki/Minden Pictures. 32.27E: © Colin Milkins/Oxford Scientific Films/photolibrary.com. 32.27F: © Peter J. Bryant/Biological Photo Service. 32.27G: © Larry Jon Friesen. 32.27H: David McIntyre. 32.29A: © John Mitchell/Jupiter Images. 32.29B: © John R. MacGregor/Peter Arnold, Inc. 32.30A: © Norbert Wu/Minden Pictures. 32.30B: © Frans Lanting/Minden Pictures. 32.31A: © Kelly Swift, www.swiftinverts.com. 32.31B: © Larry Jon Friesen. 32.31C: © W. M. Beatty/Visuals Unlimited. 32.31D: Photo by Eric Erbe; colorization by Chris Pooley/USDA ARS.

Chapter 33 *Homo floresiensis*: © Christian Darkin/Alamy. *Tunicates*: © Larry Jon Friesen. 33.2: From S. Bengtson, 2000. Teasing fossils out of shales with cameras and computers. *Palaeontologia Electronica* 3(1). 33.4A: © Larry Jon Friesen. 33.4B: © Hal Beral/Visuals Unlimited. 33.4C: © Randy Morse/Tom Stack & Assoc. 33.4D: © Mark J. Thomas/Dembinsky Photo Assoc. 33.4E: © Peter Scoones/Photo Researchers, Inc. 33.4F: © Larry Jon Friesen. 33.5A: © C. R. Wyttenbach/Biological Photo Service. 33.6: © Robert Brons/Biological Photo Service. 33.7A: © Jurgen Freund/ Naturepl.com. 33.7B: © David Wrobel/Visuals Unlimited. 33.9A: © Brian Parker/Tom Stack &

Assoc. 33.9B: © Reijo Juurinen/Naturbild/Naturepl.com. 33.11B: © Roger Klocek/Visuals Unlimited. 33.12A: © Dave Fleetham/Tom Stack & Assoc. 33.12B: © Kelvin Aitken/AGE Fotostock. 33.12C: © Dave Fleetham/Tom Stack & Assoc. 33.13A: © Tobias Bernhard/Oxford Scientific Films/photolibrary.com. 33.13B: © Larry Jon Friesen. 33.13C: © Dave Fleetham/Visuals Unlimited. 33.13D: © Larry Jon Friesen. 33.14A, B: © Tom McHugh, Steinhart Aquarium/The National Audubon Society Collection/Photo Researchers, Inc. 33.14C: © Ted Daeschler/Academy of Natural Sciences/VIREO. 33.16A: © Ken Lucas/Biological Photo Service. 33.16B: © Michael & Patricia Fogden/Minden Pictures. 33.16C: © Gary Meszaros/Dembinsky Photo Assoc. 33.16D: © Larry Jon Friesen. 33.19A: © Dave B. Fleetham/Tom Stack & Assoc. 33.19B: © C. Alan Morgan/Peter Arnold, Inc. 33.19C: © Dave Watts/Alamy. 33.19D: © Larry Jon Friesen. 33.20A: © Frans Lanting/Minden Pictures. 33.20B: © Gerry Ellis, DigitalVision/PictureQuest. 33.21A: From X. Xu et al., 2003. *Nature* 421: 335. © Macmillan Publishers Ltd. 33.21B: © Tom & Therisa Stack/Painet, Inc. 33.23A: © Roger Wilmhurst/Foto Natura/Minden Pictures. 33.23B: © Andrew D. Sinauer. 33.23C: © Tom Vezo/Minden Pictures. 33.23D: © Skip Moody/Dembinsky Photo Assoc. 33.24A: © Ed Kanze/Dembinsky Photo Assoc. 33.24B: © Dave Watts/Tom Stack & Assoc. 33.25A: © Ingo Arndt/Naturepl.com. 33.25B: © JTB Photo Communications/photolibrary.com. 33.25C: © Jany Sauvanet/Photo Researchers, Inc. 33.26A: © Tim Jackson/OSF/Jupiter Images. 33.26B: © Claude Steelman/OSF/Jupiter Images. 33.26C: © Michael S. Nolan/Tom Stack & Assoc. 33.26D: © Erwin & Peggy Bauer/Tom Stack & Assoc. 33.28: © Pete Oxford/Minden Pictures. 33.29A: © mike lane/Alamy. 33.29B: © Cyril Ruoso/JH Editorial/Minden Pictures. 33.30A: © Frans Lanting/Minden Pictures. 33.30B: © Anup Shah/Dembinsky Photo Assoc. 33.30C: © Stan Osolinsky/Dembinsky Photo Assoc. 33.30D: © Anup Shah/Dembinsky Photo Assoc.

Chapter 34 *Poster*: Courtesy of NJSP Museum. *Wood*: © Werner H. Muller/Peter Arnold, Inc. 34.3: David McIntyre. 34.4A: © Joyce Photographics/Photo Researchers, Inc. 34.4B: © C. K. Lorenz/The National Audubon Society Collection/Photo Researchers, Inc. 34.4C: © Renee Lynn/Photo Researchers, Inc. 34.8: © Biophoto Associates/Photo Researchers, Inc. 34.9A: © Biodisc/Visuals Unlimited. 34.9B: © P. Gates/Biological Photo Service. 34.9C: © Biophoto Associates/Photo Researchers, Inc. 34.9D: © Jack M. Bostrack/Visuals Unlimited. 34.9E: © John D. Cunningham/Visuals Unlimited. 34.9F: © J. Robert Waaland/Biological Photo Service. 34.9G: © Randy Moore/Visuals Unlimited. 34.10: © R. Kessel & G. Shih/Visuals Unlimited. 34.11 *upper*: © Larry Jon Friesen. 34.11 *lower*: © Biodisc/Visuals Unlimited. 34.13: © D. Cavagnaro/Visuals Unlimited. 34.14B: © Microfield Scientific LTD/Photo Researchers, Inc. 34.16A: © Larry Jon Friesen. 34.16B: © Ed Reschke/Peter Arnold, Inc. 34.16C: © Dr. James W. Richardson/Visuals Unlimited. 34.17A *left*: © Ed Reschke/Peter Arnold, Inc. 34.17A *right*: © Biodisc/Visuals Unlimited. 34.17B: © Ed Reschke/Peter Arnold, Inc. 34.19: © John N. A. Lott/Biological Photo Service. 34.20: © Jim Solliday/Biological Photo Service. 34.21: From C. Maton & B. L. Gartner, 2005. *Am. J. Botany* 92: 123. Courtesy of Barbara L. Gartner. 34.22: © Biodisc/Visuals Unlimited. 34.23B: Courtesy of Thomas Eisner, Cornell U. 34.23C: © Larry Jon Friesen.

Chapter 35 *Sunflower*: © Aflo/Naturepl.com. *Sequoia*: © Jeff Foott/Naturepl.com. 35.5: David McIntyre. 35.8: After M. A. Zwieniecki, P. J. Melcher, & N. M. Holbrook, 2001. *Science* 291: 1059. 35.9A: © David M. Phillips/Visuals Unlimited. 35.10: After G. D. Humble & K. Raschke, 1971. *Plant Physiology* 48: 447. 35.12: © M. H. Zimmermann.

Chapter 36 *Family*: Dorothea Lange/Library of Congress, Prints & Photographs Division, FSA/OWI Collection. *Plow*: © Liz Hahn. 36.1: David McIntyre. 36.5: © Kathleen Blanchard/Visuals Unlimited. 36.7: © Hugh Spencer/Photo Researchers, Inc. 36.9 *left*: © E. H. Newcomb & S. R. Tandon/Biological Photo Service. 36.9 *right*: © Dr. Jeremy Burgess/SPL/Photo Researchers, Inc. 36.11A: © J. H. Robinson/The National Audubon Society Collection/Photo Researchers, Inc. 36.11B: © Kim Taylor/Naturepl.com. 36.12: Courtesy of Susan and Edwin McGlew.

Chapter 37 *Rubber ball*: © Luiz Claudio Marigo/Naturepl.com. *Tapping tree*: © J. D. Dallet/AGE Fotostock. 37.2: © Tom J. Ulrich/Visuals Unlimited. 37.3: © Joe Sohm/Pan America/Jupiter Images. 37.5: Courtesy of J. A. D. Zeevaart, Michigan State U. 37.11: © Ed Reschke/Peter Arnold, Inc. 37.13: © Biophoto Associates/Photo Researchers, Inc. 37.17, 37.20: David McIntyre. 37.21: Courtesy of Dr. Eva Huala, Carnegie Institution of Washington.

Chapter 38 *Rafflesia*: © Ingo Arndt/Minden Pictures. *Amorphophallus*: © Deni Bown/OSF/Jupiter Images. 38.2: © RMF/Visuals Unlimited. 38.4: From J. Bowman (ed.), 1994. *Arabiopsis: An Atlas of Morphology and Development.* Springer-Verlag, New York. Photo by S. Craig & A. Chaudhury, Plate 6.2. 38.8: David McIntyre. 38.14: After A. N. Dodd et al., 2005. *Science* 309: 630. 38.15: After M. J. Yanovsky & S. A. Kay, 2002. *Nature* 419: 308. 38.17: © Gerry Ellis/Minden Pictures. 38.18A: © Nigel Cattlin, Holt Studios International/Photo Researchers, Inc. 38.18B: © Maurice Nimmo/SPL/Photo Researchers, Inc.

Chapter 39 *Flood*: © Abir Abdullah/Peter Arnold, Inc. *Salty soil*: © STRDEL/AFP/Getty Images. 39.2: © Holt Studios International Ltd/Alamy. 39.5: After A. Steppuhn et al., 2004. *PLoS Biology* 2: 1074. 39.6: After S. Rasmann et al., 2005. *Nature* 434: 732. 39.8: Courtesy of Thomas Eisner, Cornell U. 39.9: © Adam Jones/Dembinsky Photo Assoc. 39.10: © John N. A. Lott/Biological Photo Service. 39.11 *left*: David McIntyre. 39.11 *right*: © Frans Lanting/Minden Pictures. 39.12: © Simon Fraser/SPL/Photo Researchers, Inc. 39.13: © imagebroker/Alamy. 39.14: © John N. A. Lott/Biological Photo Service. 39.15: © Jurgen Freund/Naturepl.com. 39.17: © Budd Titlow/Visuals Unlimited.

Chapter 40 *Radcliffe*: © Nick Laham/Getty Images. *Soldier*: Courtesy of Major Bryan "Scott" Robison. 40.3A: © Gladden Willis/Visuals Unlimited. 40.3B: © Ed Reschke/Peter Arnold, Inc. 40.4: © Manfred Kage/Peter Arnold, Inc. 40.5: © Ed Reschke/Peter Arnold, Inc. 40.6A: © James Cavallini/Photo Researchers, Inc. 40.6B: © Innerspace Imaging/SPL/Photo Researchers, Inc. 40.10B: © Frans Lanting/Minden Pictures. 40.12: © Gerry Ellis/PictureQuest/Jupiter Images. 40.14: Courtesy of Anton Stabentheiner. 40.17: © Gladden Willis/Visuals Unlimited. 40.18A: © Robert Shantz/Alamy. 40.18B: © Jim Brandenburg/Minden Pictures.

Chapter 41 *Bicyclists*: © Peter Dejong/AP Images. *Body builder*: © Tiziana Fabi/AFP/Getty Images. 41.9A: © Ed Reschke/Peter Arnold, Inc. 41.9C: © Scott Camazine/Photo Researchers, Inc. 41.14: Courtesy of Gerhard Heldmaier, Philipps U.

Chapter 42 *Queen bee*: David McIntyre. *Swarm*: © Stephen Dalton/Minden Pictures. 42.1A: © Biophoto Associates/Photo Researchers, Inc. 42.1B: © Constantinos Petrinos/Naturepl.com. 42.2A: © Patricia J. Wynne. 42.6: © SIU/Peter Arnold, Inc. 42.7: © ullstein - Ibis Bildagentur / Peter Arnold, Inc. 42.8A: © Mitsuaki Iwago/Minden Pictures. 42.8B: © Dave Watts/Naturepl.com. 42.10B: © Ed Reschke/Peter Arnold, Inc. 42.13: © P. Bagavandoss/Photo Researchers, Inc. 42.16B: © S. I. U. School of Med./Photo Researchers, Inc. 42.18: Courtesy of The Institute for Reproductive Medicine and Science of Saint Barnabas, New Jersey.

Chapter 43 *Whale*: © François Gohier/Photo Researchers, Inc. *Egg and sperm*: © Dr. David M. Phillips/Visuals Unlimited. 43.1: Courtesy of Richard Elinson, U. Toronto. 43.4: From J. G. Mulnard, 1967. *Arch. Biol. (Liege)* 78: 107. Courtesy of J. G. Mulnard. 43.20A: © CNRI/SPL/Photo Researchers, Inc. 43.20B: © Dr. G. Moscoso/SPL/Photo Researchers, Inc. 43.20C: © Tissuepix/SPL/Photo Researchers, Inc. 43.20D: © Petit Format/Photo Researchers, Inc.

Chapter 44 *Movie still*: © New Line Cinema/Courtesy of the Everett Collection. *Brain scan*: Courtesy of Dr. Kevin LaBar, Duke U. 44.4B: © C. Raines/Visuals Unlimited. 44.7: From A. L. Hodgkin & R. D. Keynes, 1956. *J. Physiol.* 148: 127.

Chapter 45 *Snake*: © James Gerholdt/Peter Arnold, Inc. *Bat*: © Stephen Dalton/Photo Researchers, Inc. 45.3A: © Hans Pfletschinger/Peter Arnold, Inc. 45.3B: David McIntyre. 45.10A: © P. Motta/Photo Researchers, Inc. 45.16A: © Dennis Kunkel Microscopy, Inc. 45.19: © Omikron/Science Source/Photo Researchers, Inc.

Chapter 46 *Neuron*: From H. van Praag et al., 2002. *Nature* 415: 1030. © Macmillan Publishers Ltd. *Meadowlark*: © Werner Bollmann/AGE Fotostock. 46.8: From J. M. Harlow, 1869. *Recovery from the passage of an iron bar through the head.* Boston: David Clapp & Son. 46.14A: © David Joel Photography, Inc. 46.16: © Wellcome Dept. of Cognitive Neurology/SPL/Photo Researchers, Inc.

Chapter 47 *Joyner-Kersee*: © AFP/Getty Images. *Frog*: © Michael Durham/Minden Pictures. 47.1: © Frank A. Pepe/Biological Photo Service. 47.2: © Tom Deerinck/Visuals Unlimited. 47.4: © Kent Wood/Peter Arnold, Inc. 47.7A: © Manfred Kage/Peter Arnold, Inc. 47.7B: © Innerspace Imaging/SPL/Photo Researchers, Inc. 47.7C: © SPL/Photo Researchers, Inc. 47.10: Courtesy of Jesper L. Andersen. 47.14: From the collection of Andrew Sinauer/photo by David McIntyre. 47.18: © Robert Brons/Biological Photo Service. 47.22B *upper*: © Ken Lucas/Visuals Unlimited. 47.22B *lower*: © Fred McConnaughey/The National Audubon Society Collection/Photo Researchers, Inc.

Chapter 48 *Geese*: © Konrad Wothe/Minden Pictures (geese) and John Warden/PictureQuest (mountains). *Barracudas*: © Fred Bavendam/Minden Pictures. 48.1A: © Larry Jon Friesen. 48.1B: © Norbert Wu/Minden Pictures. 48.1C: © Tom McHugh/Photo Researchers, Inc. 48.4B: © Skip Moody/Dembinsky Photo Assoc. 48.4C:

Courtesy of Thomas Eisner, Cornell U. 48.10A: © SPL/Photo Researchers, Inc. 48.10C: © P. Motta/Photo Researchers, Inc. 48.13: © Berndt Fischer/Jupiter Images. 48.17: After C. R. Bainton, 1972. *J. Appl. Physiol.* 33: 775.

Chapter 49 *Lewis*: © NBAE/Getty Images. *Emergency room*: © BananaStock/Alamy. 49.8A: © Brand X Pictures/Alamy. 49.9: After N. Campbell, 1990. *Biology*, 2nd Ed., Benjamin Cummings. 49.10B: © CNRI/Photo Researchers, Inc. 49.12: © Ed Reschke/Peter Arnold, Inc. 49.15A: © Chuck Brown/Science Source/Photo Researchers, Inc. 49.15B: © Biophoto Associates/ Science Source/Photo Researchers, Inc. 49.19: © Doc White/Nature Picture Library.

Chapter 50 *Pima*: © Marilyn "Angel" Wynn/Nativestock.com. *Fast food*: © Matt Bowman/FoodPix/Jupiter Images. 50.1A: © Gerry Ellis, DigitalVision/PictureQuest. 50.1B: © Rinie Van Meurs/Foto Natura/Minden Pictures. 50.3: © AP/Wide World Photos. 50.6: © Ace Stock Limited/Alamy. 50.9: © Dennis Kunkel Microscopy, Inc. 50.14: © Eye of Science/SPL/ Photo Researchers, Inc. 50.20: © Science VU/ Jackson/Visuals Unlimited.

Chapter 51 *Bat*: © Michael & Patricia Fogden/ Minden Pictures. *Kangaroo rat*: © Mary McDonald/ Naturepl.com. 51.1 *inset*: © Kim Taylor/ Naturepl.com. 51.2B: © Rod Planck/Photo Researchers, Inc. 51.8: From R. G. Kessel & R. H. Kardon, 1979. *Tissues and Organs.* W. H. Freeman, San Francisco. 51.11: From L. Bankir & C. de Rouffignac, 1985. *Am. J. Physiol.* 249: R643-R666. Courtesy of Lise Bankir, INSERM Unit, Hôpital Necker, Paris. 51.13: © Hank Morgan/Photo Researchers, Inc.

Chapter 52 *Katrina*: Courtesy of the NOAA Satellite and Information Service. *Swamp*: © Tim Fitzharris/Minden Pictures. 52.1: © Tim Fitzharris/ Minden Pictures. *Tundra, upper*: © Tim Acker/ Auscape/Minden Pictures. *Tundra, lower*: © Elizabeth N. Orians. *Boreal, upper*: © Carr Clifton/Minden Pictures. *Boreal, lower*: © Robert Harding Picture Library Ltd/Alamy. *Temperate deciduous*: © Paul W. Johnson/Biological Photo Service. *Temperate grasslands, upper*: © Robert & Jean Pollock/Biological Photo Service. *Temperate grasslands, lower*: © Elizabeth N. Orians. *Cold desert, upper*: © Edward Ely/Biological Photo Service. *Cold desert, lower*: © Robert Harding Picture Library Ltd./photolibrary.com. *Hot desert, left*: © Terry Donnelly/Tom Stack & Assoc. *Hot desert, right*: © Dave Watts/Tom Stack & Assoc. *Chaparral, left*: © Elizabeth N. Orians. *Chaparral, right*: © Larry Jon Friesen. *Thorn forest*: © Frans Lanting/Minden Pictures. *Savanna*: © Nigel Dennis/AGE Fotostock. *Tropical deciduous*: Courtesy of Donald L. Stone. *Tropical evergreen*: © Elizabeth N. Orians. 52.6: © Elizabeth N. Orians. 52.9 *Chameleon*: © Pete Oxford/ Naturepl.com. 52.9 *Tenrec*: © Nigel J. Dennis/ Photo Researchers, Inc. 52.9 *Fossa*: © Frans Lanting/ Minden Pictures. 52.9 *Lemur*: © Wendy Dennis/ Dembinsky Photo Associates. 52.9 *Baobab*: © Pete Oxford/Naturepl.com. 52.9 *Pachypodium*: © Michael Leach/Oxford Scientific Films/ photolibrary.com. 52.12: Courtesy of E. O. Wilson. 52.13A *left*: © Bill Lea/Dembinsky Photo Associates. 52.13A *right*: © Stephen J. Krasemann/ Photo Researchers, Inc. 52.13B *left*: © Frans Lanting/Minden Pictures. 52.13B *center*: © Kenneth W. Fink/Photo Researchers, Inc. 52.13B *right*: © Chris Gomersall/Naturepl.com.

Chapter 53 *Macaques*: © Frans de Waal, Emory U. *Spider*: © Mark Gibson/Jupiter Images. 53.2A: From J. R. Brown et al., 1996. *Cell* 86: 297. Courtesy of Michael Greenberg. 53.4A: © Nina Leen/Time Life Pictures/Getty Images. 53.4B: © Frans Lanting/ Minden Pictures. 53.9A: © Anup Shah/ Naturepl.com. 53.9B: © Mitsuaki Iwago/Minden Pictures. 53.9C: © Konrad Wothe/Minden Pictures. 53.11: © Frans Lanting/Minden Pictures. 53.13: Courtesy of John Alcock, Arizona State U. 53.16: © Tui De Roy/Minden Pictures. 53.18: © Cyril Ruoso/JH Editorial/Minden Pictures. 53.21: © Nigel Dennis/AGE Fotostock. 53.22: © Piotr Naskrecki/Minden Pictures. 53.23: © Steve & Dave Maslowski/Photo Researchers, Inc. 53.24: Courtesy of John Alcock, Arizona State U. 53.25: © José Fuste Raga/AGE Fotostock.

Chapter 54 *Caterpillars*: Courtesy of John R. Hosking, NSW Agriculture, Australia. *Farmer*: © Thomas Shjarback/Alamy. 54.1A: © Flip Nicklin/ Minden Pictures. 54.1B: © PhotoStockFile/Alamy. 54.1C: © Robert McGouey/Alamy. 54.5A: © Frans Lanting/Minden Pictures. 54.5B: © David Nunuk/ SPL/Photo Researchers, Inc. 54.6A: © Michael Durham/Minden Pictures. 54.6B: Courtesy of Colin Chapman. 54.7: © Larry Jon Friesen. 54.11: After P. A. Marquet, 2000. *Science* 289: 1487. 54.12: © Ed Reschke/Peter Arnold, Inc. 54.13: © Adam Jones/Photo Researchers, Inc. 54.14: © T. W. Davies/California Academy of Sciences. 54.18: © Kathie Atkinson/OSF/Jupiter Images.

Chapter 55 *Ant on spine*: © Piotr Naskrecki/ Minden Pictures. *Ant with Beltian bodies*: © Mark Moffett/Minden Pictures. 55.1: After R. H. Whittaker, 1960. *Ecological Monographs* 30: 279. 55.4: © Alan & Sandy Carey/Photo Researchers, Inc. 55.6: © Dave Watts/Naturepl.com. 55.7: © Lawrence E. Gilbert/Biological Photo Service. 55.8: © Shehzad Noorani/Peter Arnold, Inc. 55.11: © Mitsuaki Iwago/Minden Pictures. 55.12A: © Kim Taylor/Naturepl.com. 55.12B: © Perennou Nuridsany/Photo Researchers, Inc. 55.13D: Courtesy of William W. Dunmire/National Park Service. 55.14A: David McIntyre. 55.15: © Jim Zipp/Photo Researchers, Inc. 55.16: Courtesy of Jim Peaco/National Park Service. 55.18: After M. Begon, J. Harper, & C. Townsend, 1986. *Ecology*. Blackwell Scientific Publications.

Chapter 56 *Crane*: © Mark Moffett/Minden Pictures. *Researchers*: Courtesy of Christian Koerner, U. Basel. 56.1: After M. C. Jacobson et al., 2000. *Introduction to Earth System Science.* Academic Press. 56.9A: © Corbis Images/PictureQuest.

Chapter 57 *Flight*: © Tom Hugh-Jones/ Naturepl.com. *Crane suit*: © Mark Payne-Gill/ Naturepl.com. 57.1: © Michael Long/NHMPL. 57.2B: © Mark Godfrey/The Nature Conservancy. 57.6: Richard Bierregaard, Courtesy of the Smithsonian Institution, Office of Environmental Awareness. 57.7: © Mr_Jamsey/iStockphoto.com. 57.8A: © Terry Whittaker/Alamy. 57.8B: © Paul Johnson/Naturepl.com. 57.9: © Michael & Patricia Fogden/Minden Pictures. 57.10A: After R. B. Aronson et al., 2000. *Nature* 405: 36. 57.10B: © Fred Bavendam/Minden Pictures. 57.12: Courtesy of WWF-US, GIS Map provided by J. Morrison. 57.14B: © Jim Brandenburg/Minden Pictures. 57.14C: Courtesy of Jesse Achtenberg/U.S. Fish and Wildlife Service. 57.15A: Courtesy of Christopher Baisan and the Laboratory of Tree-Ring Research, U. Arizona, Tucson. 57.15B: © Karen Wattenmaker/Painet, Inc. 57.17: After S. H. Reichard & C. W. Hamilton, 1997. *Conservation Biology* 11: 193. 57.18A: © Elizabeth N. Orians. 57.20: © Tom Vezo/ Nature Picture Library.

Index

Numbers in **boldface** indicate a definition of the term. Numbers in *italic* indicate the information will be found in a figure, caption, or table.

- A -

Abalone, 1147
A band, *1006,* 1007
Abdomen, of crustaceans, 707
Abiotic factors, **1113**
ABO blood group system, **219**
Abomasum, **1085**
Abortion, **915–916**
Abscisic acid, 343, 774, *798,* **811**
Abscission, **806**
Abscission zone, 806
Absorption, of light, **163**
Absorption spectra
  chlorophyll, *165*
  cone cells, *981*
  pigments, **163**
Absorptive heterotrophs, 600, 651
Absorptive nutrition, **651**
Absorptive period, **1086**
Acacia, 1125, 1184–1185, 1194
Acacia ants, *1184–1185,* 1194
Accessory fruits, 643
Accessory pigments, **165**
Accessory sex organs, 904
Acclimatization, 861
Accommodation (vision), 979–980
Aceria tosichella, 289
Acetabularia, 602
Acetaldehyde, *40,* 118, 119
Acetate, 145
Acetic acid, 33, *40*
Acetone, *40*
Acetylcholine (ACh)
  actions of, 958, 959
  blood pressure regulation and, 1062, *1063*
  blood vessel relaxation and, 344
  cholinergic neurons and, 992
  clearing from synapses, 960–961
  heartbeat in mammals and, 1052, *1053*
  in neuromuscular junctions, **955,** 1008
  receptor for, 337
  smooth muscle cells and, 1011–1012
  in synapses, 956
Acetylcholine receptors, 337, 956, *957,* 958, 992, 1021
Acetylcholinesterase, 131, 134, *957,* 961
Acetyl CoA
  in citrate formation, 128
  citric acid cycle control point, 157
  fatty acids and, 154, 157
  in pyruvate oxidation, **144,** 145, *146*
N-Acetylglucosamine, *53*
Acetyl group, carbon skeletons and, 1072
ACh. *See* Acetylcholine
Acid-base catalysis, 128
Acid-base reactions, 33
Acid precipitation, 849, **1218**
Acids
  in chemical weathering of rock, 786
  described, **33**
  water as, 33–34
Acinonyx jubatus, 306
Acoelomates, **674,** *675*
Acorn woodpeckers, *1161*
Acorn worms, 721
Acquired immunodeficiency syndrome (AIDS), 274, **421–423,** 552, 653
Acropora, 1234
Acrosomal reaction, 901, *902*
Acrosome, **901**
Actin
  contractile ring and, 192
  in microfilaments, 86
  in muscles, 1005–1010
Actin filaments, **86–87**
  muscle contraction and, 1005–1010, 1014
  polarity, 436
Actinobacteria, 572–573, 788–789
Actinomyces sp., *572*
Action potentials
  characteristics, 947
  conduction along axons, 953–954
  generation of, 946, **951–953**
  heartbeat in mammals and, 1052, 1053, 1054, 1055
  muscle contraction and, 956, 1008–1009
  muscle twitches and, 1012
  nerve impulses, **943**
  neurons and, 944, 945, 946
  in postsynaptic neurons, 957
  release of neurotransmitters and, 950, 955–956
  saltatory conduction, 954–955
  in sensory receptor cells, 966–967
Action spectrum, **163–164,** *165*
Activation energy (E_a), **126**
Activation gate, **952, 953**
Activators, 132
Active sites, of enzymes, **127,** 129
Active transport, 105, **111–113,** 767–768
Actuarial tables, 1169
"Adam's apple," 1032
Adansonia grandidieri, 1130
Adaptation (evolutionary)
  meanings of, **489,** 497
  natural selection and, 6, 497–498
  sexual selection and, 496, 498–500
  trade-offs and, 486
Adaptation (sensory), **967**
Adapter hypothesis, 260
Adapter molecule, 260, 266
Adaptive radiations, 520–521, 528–529, *530*
  See also Evolutionary radiations
Adenine, 59, **238,** 239, *240*
Adenosine, 332, *959*
Adenosine deaminase, 392
Adenosine diphosphate (ADP), 124, 130
Adenosine monophosphate (AMP), *124*
Adenosine triphosphate (ATP)
  amount hydrolyzed in humans, 151
  in biochemical energetics, **123–125**
  Calvin cycle and, 170, *171*
  in catalyzed reactions, *129,* 130
  citric acid cycle and, 145, *146*
  energy for cellular work and, 1069
  fermentation and, 148
  function, 59
  generated in mitochondria, 82
  glycolysis and, 142, *143,* 144
  muscle contraction and, 1007–1008, 1012
  as neurotransmitter, *959*
  release of energy from, 124
  sources used for muscle contraction, 1014–1015
  structure of, *124*
  synthesis by chemiosmosis, 150–153
  synthesis by glucose metabolism, 140
  synthesis by oxidative phosphorylation, 148–153
  synthesis during photosynthesis, 166, 167, 168, *169*
Adenovirus, *409*
Adenylate cyclase, 116
Adenylyl cyclase, 341, 345, 346
Adiantum, 620, 626
Adipose tissue, 859
  brown fat, 867
  as insulation, 735
  insulin and, 1087
  leptin and, *880,* 1088
  stem cells derived from, 426, *427*
ADP. *See* Adenosine diphosphate
Adrenal cortex, *880,* **887,** 888–889
Adrenal gland, **887–889,** 992
Adrenaline, 338–339, *880,* 887, 888
Adrenal medulla, *880,* **887–888**
Adrenergic receptors, 888, *892,* 893
Adrenocorticotropic hormone (ACTH), *880*
Adriamycin, *391*
Adult-onset diabetes, 887
Adult stem cells, fat-derived, 426, *427*
Adventitious roots, 746
Aecial primordium, 663
Aeciospores, *662,* 663
Aedes aegypti, 1181
Aegopodium podagraria, 639
Aequopora victoriana, 361
Aerenchyma, **847**
Aerobes, 568
Aerobic metabolism, 11
Aerobic respiration, **140**
  See also Cellular respiration
Aerobic work
  effect on muscles, 1014
  muscle fuel supply and, 1015
  slow-twitch muscle fibers and, 1013
Aerotolerant anaerobes, **568**
Afferent arterioles, renal, **1100,** 1102, 1107
Afferent nerves, **985**
Afferent neurons, **943**
Affinity chromatography, **892**
Aflatoxins, 666
African green monkeys, *553*
African lungfish, 1048
African wild dogs, 1241
Afrosoricida, *736*
After-hyperpolarization, 953
AGAMOUS protein, 441
Agar, 602
Agaricia tenuifolia, 1234
Agaricus bisporus, 667
Agassiz, Louis, 706
Agave americana, 1172
Agelaius phoeniceus, 510
Aggregate fruits, 643
Aging, telomeres and, 248–249
Aglaophyton major, 619
Agnosias, 990
Agre, Peter, 1104
Agricultural soils
  global phosphorus cycle and, 1219
  salinization, 836–837, 848
Agriculture
  biotechnology and, 369–371, 430, 843–844
  reproduction of significant plant crops, 819
  significance to global economy, 1222
  soil salinization and, 836–837, 848
  subsidization and, 1223
  subtropical ecosystems and, 1222
  use of fertilizers and lime in, 787
  vegetative plant reproduction and, 833–834
Agrobacterium tumefaciens, 360–361, 575
Agrodiaetus, 517
Agrostis, 849
A horizon, **786,** 787
AIDS. *See* Acquired immunodeficiency syndrome
Ailuropoda melanoleuca, 1070
Air capillaries
  in birds, 1030, *1031*
  in insects, 1028, *1029*
Air circulation patterns, 1114–1115
Air pollution, 655
Air sacs, **1030,** *1031*
Air temperature, 1114
Aix galericulata, 734
Alanine, *43*
Alaskan salmon, 1238
Albinism, *216*
Alchemilla, 770
Alcoholic fermentation
  overview of, 140, *141,* 147, **148**

in water-saturated roots, 846
Alcoholism, 132, 1075
Alcohol(s)
absorption in the stomach, 1082
fermentation. *See* Alcoholic fermentation
functional group, *40*
inhibition of antidiuretic hormone release, 1108
sensitivity to, 118–119
Aldehyde dehydrogenase, 118–119, 132
Aldehydes, *40*
Alder, 1198
Aldosterone, *880,* **888**
Aleurone layer, **801**
Algae, 102, *103,* 591, 598
See also Brown algae; Green algae; Red algae
Algal beds, primary production in, *1210*
Alginic acid, 600
ALH 84001 (meteorite), 61
Alien species. *See* Exotic species
Alkaloids, *839*
Allantoic membrane, **935**
Allantois, 935, 936
Allard, H. A., 826
Allele frequencies
calculating, 491, *492*
gene flow and, 494
genetic drift and, 494–495
genetic structure and, 492
Hardy-Weinberg equilibrium, 492–493
Alleles, **211**
codominance, 219
incomplete dominance, 218–219
linked and unlinked, 214
multiple, 218
mutations and, 217–218
neutral, **501**
pleiotropic, 219
population genetics and, 490, 491–493
segregation during meiosis, *212*
wild-type, 217, 218
Allele-specific cleavage, 385
Allele-specific oligonucleotide hybridization, 385, *386*
Allergens, 420
Allergic reactions, **420,** *421*
Alligators, 732, *945,* 1048–1049
Allis, David, 322
Allomyces, 664
Allopatric speciation, **511–513**
Allopolyploidy, **513,** 514
Allosteric enzymes, 132–133
Allosteric regulation
compared to transcriptional regulation, *297*
in metabolic pathways, 156–157
overview of, 133, *134*
Allosteric regulators, 341
Allostery, **132**
Alluaudia procera, 1130
Alnus, 1198
Alper, Tikva, 378
Alpha (α) helix, *45,* **46**
Al-Qaeda, 256
Alternation of generations
in angiosperms, *820,* 821
in fungal life cycles, 659, 662
overview of, **194, 593–594,** 614–615
Alternative splicing, **324–325**
Altitude, oxygen availability and, 1026–1027
Altricial birds, 681
Altruism, 1159
Alu element, 312
Aluminum, plants and, 849
Alveolates, *584,* 596, *597,* 598
Alveoli, 596, **1033,** 1034
Alzheimer's disease, 48
Amacrine cells, 981
Amanita muscaria, 667
Amber, 232, *472*
Amblyrhynchus cristatus, 864
Amborella, 645
A. trichopoda, 644
Ambulacrarians, 718
Ambystoma tigrinum, 1027
Amensalisms, **1188,** 1193–1194
American chestnut, 666, 1233
American crocodiles, 1242
American elm, 666, 1233
American Prairie Foundation, 1237
American redstarts, 1168, *1169*
American sycamore, 515
Amine hormones, 878
Amines, *40*
Amino acid replacements
effects of, **526**
in positive and stabilizing selection, 532–533
sequence alignment technique and, 526–527
Amino acids
chemical structure, 42–43
essential, 1072
molecular evolution and, 526
as neurotransmitters, *959*
origin of life and, 61, 62
peptide bonds in, 43–44
primary structure of proteins, 44–45
proteins and, 39
sequence alignment, 526–527
synthesis through anabolic pathways, *154,* 155
Amino-acyl-tRNA synthetase, 267
Amino group, *40*
in amino acids, 42, *43*
amino sugars, 53
chemical characteristics, 33, 134
Aminopeptidase, *1084*
Amino sugars, 53
Ammocoetes, 724
Ammonia
chemical properties, 33
global nitrogen cycle and, 790, 791
as nitrogenous waste, **1095,** 1096
oxidation by bacteria, 569
Ammonium, 33, 790, *791*
Ammonoids, *10*
Ammontelic animals, **1096**
Amniocentesis, 385, **936**
Amnion, *925,* 932, **935**
Amniote egg, **730, 905**
Amniotes, 475, *723,* **730–731**
Amniotic cavity, 935
Amniotic fluid, 936
Amoebas, 592, *584, 593,* 603
A. proteus, 591, 605–606
Amoeboid motion, 591
Amoebozoans, *584, 597,* **605–607**
Amorphophallus titanum, 818, *819*
AMP. *See* Adenosine monophosphate
AMPA, 959
AMPA receptors, 959, 960
Amphibians
adaptations for water conservation in, 1099
circulatory system, 1048
gastrulation in, 928
major groups, 729
modern population decline in, 729–730
overview of, **728–729**
reproduction and, 904–905
in vertebrate phylogeny, *723*
Amphignathodon, 545
Amphioxus, 537
Amphipods, 706
Amphiprion percula, 904
Amphisbaenians, 731
Ampicillin, 361, 566
Amplitude, of cycles, 828
Ampulla, 719
Amygdala, *989*
fear memory and, 1000
functions of, **988**
human emotions and, 942–943
Amylase, **1080,** *1084*
Amylopectin, 51
Anabena, 573
Anabolic pathways, regulation, 299
Anabolic reactions, **120,** 122
Anabolic steroids, 874–875
Anabolism, 120, 154–155
Anaerobes, 568
Anaerobic metabolism
in diving mammals, 1064
in evolutionary history, 11
Anaerobic respiration, **140**
See also Fermentation
Anaerobic work, 1014
Anal warts, 387
Anaphase
first meiotic division, *197,* 198, *200*
mitosis, 190–191, *200*
second meiotic division, *196*
Ancestral traits, **545**
reconstructing, 550
Anchor cell, 438
Andersson, Malte, 499
Androgens, **889**
Anemia, 317, 376, *1074,* 1075
Anemone fish, 904
Anencephaly, 932
Aneuploidy, **199,** *201,* 226
Angel insects, *709*
Angelman syndrome, 383
Angina, 344
Angiogenesis, 387
Angiosperms
alternation of generations, *820,* 821
asexual reproduction, 832–834
characteristics, *612,* 638, 745
coevolution with animals, 641
double fertilization, **638,** *820,* **822–823**
embryo development, 823–824
evolution, 477, 478, **631,** *632, 644,* 645–646
floral structure, 638–639, 820–821
fruits, 643
gametophytes, *820,* 821
life cycle, 642–643, *820,* 821
major clades, 745
mating strategies, 822
meaning of term, 638
photoperiodism and, 826–828
pollen grain germination, 822
pollination in, 821–822
seed dispersal, 824
sexual reproduction in, *820,* 821
tissue systems, 748
transition to flowering, 825–831
vegetative organs, 745–748
Angiotensin, 1062, **1107**
Angular gyrus, 1000, 1001
Animal behavior
adaptive benefits of genetic control, 1144–1145
cooperative, 9
cost-benefit approach to, 1148–1149
courtship, 702, 1152
distinguishing between genetic and environmental influences on, 1142–1144
effects on populations and communities, 1161–1162
environmental cycles and, 1152–1154
foraging and feeding, 1149–1152, 1161–1162
habitat choice, 1147–1148
imprinting, 1145
influence of hormones on, 1146–1147
interactions between inheritance and learning, 1145–1146
mate choice and fitness, 1152
maternal, 1143–1144
migration, 1154–1156
pair bonding, 881
piloting and homing, 1154
proximate and ultimate mechanisms, 1141
reconstructing phylogenies and, 548
social, 1158–1160
species-specific, 1141–1142
stereotypic, 1140–1141
study of, 1141
Animal cells
cytokinesis in, 192
extracellular matrix, 91
structure of, *76*
Animal communication, *1148,* 1156–1157
Animal pole, 436, *437,* **922,** *923*
Animal-pollination, 641
Animal reproduction
asexual, 897–898
sexual, 899–905 *See also* Sexual reproduction
trade-offs in, 680–681
See also Human reproduction
Animals
appendages and locomotion, 676
body cavity structure, 674–675
body plans, 674–676
cloning, 431–432
developmental patterns, 672–673
distinguishing traits, 671
evolutionary history, 473–479
food requirements, 1069–1075
life cycle differences, 679–682
major groups, 682–687
monophyly of, 671–672
phylogenetic tree, *672*
segmentation, 675–676
strategies for acquiring food, 676–679
symmetry in, 674
totipotency in embryo cells, 430
Animal viruses, 287–289
Anions, **28,** 787
Anisogamy, **594**

Annelids, *683, 692*
characteristics, **698**
circulatory system, *1046*
clitellates, 699–700
excretory system, 1097
polychaetes, 698–699
See also Earthworms; Leeches
Annotation, **301**
Annual rings, 759
Annuals, **798, 826**
Anogenital cancers, 387
Anopheles, 585–586
A. gambiae, 585–586
Anorexia nervosa, 1071
ANS. *See* Autonomic nervous system
Anser indicus, 1024
Antabuse, 132
Antarctic ice cap, 1215, 1216
Antarctic Ocean, 470
Antelope jackrabbit, *868*
Antennapedia mutant, 444
Antennarius commersonii, 727
Antenna systems, in photosynthesis, **165–166**
Anterior pituitary gland
endocrine functions, *880,* **881–882**
hypothalamic control of, 882–883
male sexual function and, *908,* 909
negative feedback loops and, 883
in ovarian and uterine cycles, 911
thyrotropin and, 884
Anterior plane, **674**
Anterior-posterior axis
determination in *Drosophila,* 442, *443*
early definition of, 922
homeotic genes and, 450
vertebral column, 453
Antheridia, *615,* **616,** 618, 621, *622*
Anthers, **638,** *639, 820,* 821
Anthocerophyta, *612,* **623**
Anthoceros, 623
Anthocyanins, 86
Anthophyte hypothesis, 646
Anthopleura elegantissima, 686
Anthozoans, **686–687**
Anthrax, *409,* 574
Anthropoids, 738
Antibacterial antibiotics, 269
Antibiotic resistance, 294–295, 361
Antibiotics
bacteria that produce, 302
early commercial production of, 367
prokaryote ribosomes and, 269
prokaryotic cell wall and, 566
Streptomyces and, 573
Antibodies
ABO blood group system and, 219
antinuclear, 421
B cells and, 409
bivalent, 412
classes of, 412–413
generation of diversity in, 418–420
humoral immune response and, 408
monoclonal, 413–414
overview of, **403**
plasma cells and, 411
structure of, 411–412
Anticodons, **266**
Antidiuretic hormone (ADH), *880,* **881,** 1062, **1107–1108**
Antifreeze proteins, **847**
Antigenic determinants, **407–408**
Antigen-presenting cells, 415, *416,* 417
Antigens
ABO blood group system and, 219
antigenic determinants and, **407–408**
clonal selection process and, 409
polyclonal, 413
Antihamnion, 602
Antihistamines, 420
Antiparallel strands, in DNA, **238,** 239–240
Antipodal cells, 821, *823*
Antiports, **111,** 112
Antisense RNA, **365**
Antitranspirants, **774**
Antiviral vaccines, 387
Ant lions, *709*
Ants, *709*
feeding behaviors, 1161–1162
fungal farms, 658
mutualisms with plants, 1184–1185, 1194
population densities, 1161–1162
Anurans, 729
Anus, 673, 927, **1077**
Anza Borrego Desert, *1123*
Aorta
in amphibians, 1048
blood pressure regulation and, 1108
chemoreceptors, 1063
in humans, 1051
in reptiles, 1049
stretch receptors, 1062, 1108
Aortic body, **1040**
Aortic valve, **1050**
Apes, 738
Aphasia, **1000,** 1001
Aphids, *709,* 775
Aphytis, 1193
Apical buds, **747,** *753*
Apical cell division, in mosses, **624**
Apical complex, 596
Apical dominance, **806**
Apical hook, **810**
Apical meristems, *753,* 754, 825, 833–834
Apicomplexans, *584,* 585, **596,** *597*
Apis mellifera, 711
See also Honeybees
Apomixis, 832–833
Apoplast, **768,** *769*
Apoptosis, **202**
cellular immune response and, 417
in development, 428
genes determining, 439–440
loss of webbing between toes, 452
mechanisms in, 202–203
oncogenes and, 388
Aposematic coloration, 702
Appendages
in arthropods, 694, 706
locomotion in animals, 676
See also Limbs
Appendicular skeleton, 1017
Appendix, *1079,* **1085**
Apple trees, 513
Aptenodytes forsteri, 1145
AQP-2, 1108
Aquaporins, **109, 767,** 1104, 1108
Aquatic environments
biomass and energy distribution, 1187, *1188*
external fertilization systems, 903
See also Freshwater environments; Oceans
Aquatic plants, 846–847
Aqueous solutions, 32–34
Aquifers, **1212**
Aquilegia, 519
A. formosa, 516
A. pubescens, 516
Arabidopsis thaliana
clock genes and CO protein, 829, *830*
det2 mutant and brassinosteroids, 811
ethylene signal transduction pathway, 810
etiolation, *814*
flower development and organ identity genes, 440–441
flowering hormone and, 831
genes related to cell walls, 749
genome, 310–311, *533*
as model system, 9, 797
noncoding DNA in, *534*
phytochromes and, 813
pollen germination, *822*
salt-tolerance and, 371
Arabinocytosine, *391*
Ara chloroptera, 1151
Arachnids, 713
Arbuscular mycorrhizae, 657, 658, 665
Arcelin, 842–843
Arceuthobium americanum, 792–793
Archaea
cell wall, 565
chemolithotrophs, 568
compared to Eukarya and Bacteria, 561–562
evolution of Eukarya and, 590
major groups, 575, 576–577, *578*
membrane lipids, 575–576
overview of, 575
prokaryotes and, **12,** 72
shapes, 563
Archaefructaceae, 645
Archaefructus, 477, 645
Archaeognatha, 709
Archaeopteryx, 733
Archean Age, 35
Archegonia, *615,* **616,** 618, 621, *622, 623*
Archenteron, 927
Archosaurs, 550, 731, *731, 733*
Arctic foxes, 1189
Arctic ground squirrels, *8,* 1154
Arctic hare, *868*
Arctic lemmings, 1189
Arctic tundra, 1118
Arctium sp., *824*
Ardipithecines, 738
Area phylogenies, **1131**
Arginine, *43,* 127, 344, 841–842
arg mutants, 258, *259*
Argon, 1208
Argyroxiphium sandwicense, 520
Aristolochia grandiflora, 644
Aristotle, 764
Armillariella, 664
Army ants, *1160*
Arnold, Joseph, 818
Aromatase, 889
Arothron mappa, 487
Arrhythmia, 1044, 1045
Arrow worms, *672, 683, 692,* **694**
Arsenic, 145, 1220
Artemia, 1094–1095
Arteries, **1047**
atherosclerosis, 376, 396, *397,* 887, 1060, *1061*
autonomic nervous system and, 1062
blood flow and, 1057, *1058*
Arterioles, **1047**
autonomic nervous system and, 1062
blood flow and, 1057, *1058*
capillaries and, 1057–1058
renal, 1100, 1102, 1107
Arthobotrys, 654
Arthritis, 706
See also Rheumatoid arthritis
Arthropods, *692,* **694**
ancestors, 705–706
animal viruses and, 287
appendages, 694, 706
body plan, *694*
body segmentation, 676
chelicerates, 712–713
chemoreception, 968
circulatory system, *1046*
compound eyes, 978
crustaceans. *See* Crustaceans
exoskeleton, 694, 1016
fossil record, 473
hemocoel, 691
insects, 708–711
jointed legs appear in, 706
key features, 705
major groups, *683*
meaning of name, 706
molting, 693
myriapods, 712
parthenogenesis, 898
phylogenetic relationships among, *705*
Ubx gene and leg number in, 453, *454*
Artificial immunity, 409
Artificial insemination, **916**
Artificial kidneys, 1105–1106
Artificial life, 302–303
Artificial selection, **489**
Artificial sweeteners, 38
ARTs, **916**
The Ascent of Everest (Hunt), 1024
Ascidians, 722, *723*
Asclepias syriaca, 824, 844
Ascocarp, *661,* **665**
Ascomycetes (Ascomycota)
characteristics, 651, *664,* 665–666
component of lichens, 655
life cycle, *661*
phylogeny of the fungi and, *652*
Ascorbic acid, 1074
Ascospores, *661*
Ascus (asci), *661,* **665**
Asexual reproduction
animals, 897–898
flowering plants, 832–834
microbial eukaryotes, 593
mitosis and, **193**
Ashkenazic Jews, 395
Asparagine, *43*
Asparagus, white, 813
Aspartame, 391
Aspartic acid, *43*
Aspens, 1195–1196
Aspergillus
A. oryzae, 666
A. tamarii, 666
Aspirin, 839, 869, 1082
Assassin bug, 582, *604*

Assisted reproductive technologies (ARTs), 430, **916**
Association cortex, **990,** 997
Associative learning, **999**
Asthma, 395
Astrocytes, **946**
Astrolithium, 584, 585
Asymmetrical carbon, 40, 42
Asymmetry, in animals, 674
Atacama Desert, 846
Ateles geoffroyi, 739
Atherosclerosis, 376, 396, *397,* **1060,** *1061*
Athletes
 aerobic and anaerobic training, 1014
 anabolic steroids and, 874–875
 "cooling" and, 854–855
 heart failure and, 1044–1045
Atmosphere
 Earth's surface temperature and, 1208
 effect of living organisms on, 1205
 gases in, 1208
 layers, 1208
 nitrogen in, 1216
 See also Atmospheric carbon dioxide; Atmospheric oxygen
Atmospheric carbon dioxide
 concentrations, 471, 1213, 1215
 evolution of leaves and, 620–621
 fires and, 1212
 global warming and, 471, 1215–1216
 interactions with nitrogen fixation, 1220
 movement into the oceans, 1213–1214
 percent of total atmospheric gases, 1208
 photosynthesis and, 175
 plant carbon storage and, 1204–1205
Atmospheric oxygen
 altitude and, 1026–1027
 cyanobacteria and, 578
 effects of life on, 468–470
 effects of photosynthesis on, 11
 Mesozoic levels, 476
 origin of life and, 61
 percent of total atmospheric gases, 1208
 Permian levels, 475
Atomic mass unit (amu), **21**
Atomic number, **22**
Atomic weight, **23**
Atoms
 covalent bonding, 25–27
 hydrogen bonding, 29
 ionic bonding, 27–28
 overview, **21,** 22
ATP. *See* Adenosine triphosphate
ATP synthase, **150,** *151,* 152–153, 168, *169*
Atrial natriuretic peptide (ANP), 1108
Atrioventricular node, **1053,** 1054
Atrioventricular valves, **1050,** 1051
Atriplex halimus, 848
Atrium, **1047**–1049
 human heart, 1050, 1051, 1053
Atropine, *648*
Attenuated viruses, polio vaccine, 524–525
Auditory canal, 972
Auditory communication, 1157
Auditory systems, **972–974**
Australia
 desert vegetation, 1128
 human-caused extinctions, 1228
 human impact on environment, 1116–1167, 1181
Australian rock lobster, 1238
Australopithecines, 738–739
Australopithecus
 A. afarensis, 739, 740
 A. harhi, 740
Autocatalysis, **1080,** *1081*
Autocrine signals, **334**
Autoimmune diseases, 421
Autoimmunity, **421**
Autonomic nervous system (ANS), **985, 992,** *993*
 control of heartbeat, 1052, *1053*
 regulation of blood pressure, 1062, *1063*
Autonomic reflexes, gastrointestinal system and, 1085–1086
Autophagy, 82
Autophosphorylation, 334, 337
Autopolyploidy, **513–514**
Autosomal dominance, 379
Autosomes, **225**
Autotrophs, 13, **1069**
 as food sources for heterotrophs, *1070*
 nutrition and, **781–782**
 parasitic plants, 792
 trophic level, 1186
Auxin anion efflux carriers, **804**
Auxin(s)
 actions on plant cell walls, 807–808
 carrier proteins, 804, 805
 effects on plant growth, 805–806
 as herbicides, 807
 lateral redistribution, 805
 plant tissue culture and, 809
 polar transport, 803–804, *805*
 receptor proteins, 808–809
 typical activity, *798*
Auxotrophs, 258
Avery, Oswald, 234
Avian malaria, 1233
Avirulence genes, 839–840
Avogadro's number, **33**
Avr genes, **839–840**
Axel, Richard, 968, 969
Axial filaments, **566,** 571–572
Axial skeleton, 1017
Axillary buds, **747**
Axon hillock, *945,* **957**
Axons
 conduction of action potentials, 953–954
 functions of, **943,** 945, 985
 myelination, 946
 nerve impulses and, 859
 potassium channels and membrane potentials in, 948–950
 resting potential, 947
 spinal cord, 987
Axon terminal, **945**
Azolla, 788

- B -

Baby boom, 1170–1171
Bacillus, 563
Bacillus, 574
 B. anthracis, 409, 574, 579
 B. subtilis, 181
 B. thuringiensis, 370, 844
 nitrogen metabolism, 569
Bacillus thuringiensis toxin, 370, 371, 372
Back substitutions, 527
Bacteria, **12**
 in animal guts, 1077
 β-carotene pathway and, 370
 chemolithotrophs, 782
 clones, 290
 colonies, 290–291
 compared to Eukarya and Archaea, 561–562
 conjugation, 291–292, *293,* 294
 culturing, 290–291
 denitrifiers, 791
 genetic transformation experiments and, 233–235
 global nitrogen cycle and, 790–791
 major groups, 571–575
 nitrogen fixers, 477–478, 569, 788–789
 nitrogen metabolism in, 569
 as pathogens, 579
 plasmids and, 293–295
 recombinant DNA technology and, 359
 restriction enzymes and, 353–354
 studies in molecular evolution and, 528–529, *530*
 transduction, 293
 transformation, 292–293
 See also Prokaryotes
Bacteria (domain), 72
Bacterial artificial chromosome (BAC), **393**
Bacterial infections, treating, 286–287
Bacterial lawn, 291
Bacteriochlorophyll, 568
Bacteriophage
 construction of gene libraries and, 362–363
 Hershey-Chase experiment, 235–237
 λ (lambda), 290, 360, 362–363
 lytic and lysogenic cycles, **284–287,** 290
 T2, 235–237
 T4, *284*
 T7, 354, 549–550
Bacteriorhodopsin, 577
Bacteroids, **789**
Bakanae, 801–802
Baker's yeast, 665
Balaenoptera musculus, 920
Balance, organs of equilibrium, 974–975
Balanus balanoides, 1192–1193
Bald cypress, *1113*
Bald eagle, *678*
Baleen whales, 676
Bali, 1128, *1129*
Ball-and-socket joint, *1019*
Balloon-winged katydid, *4*
Bamboo, 832
*Bam*HI enzyme, 361
Bangladesh, 836
Banksia, 4, 518
Banteng, 432
Banting, Frederick, 887
Barberry, 663
Barhead geese, 1024
Bark, **754,** 760
Barklice, *709*
Barley, "malting," 801
Barnacles, 706, *707,* 1192–1193
Barn owl, *734,* 1129
Barometers, 1025
Barometric pressure, **1025**
Baroreceptors, **1062**
Barr, Murray, 323
Barracuda, *727, 1025*
Barr bodies, 323
Barrel cactus, *747*
Barrier contraception, 914
Basal body, of cilia and flagella, 89
Basal lamina, 91
Basal metabolic rate (BMR), **866–867,** 884, 1069–1070
Basement membrane, 91
Baseodiscus punnetti, 697
Base pairing, complementary. *See* Complementary base pairing
Bases
 in DNA, 238, 239, *240,* 277
 in nucleotides, **57,** 277
 pairing. *See* Complementary base pairing
 in RNA, 260
Bases (acid-base), **33**
Basidia, *661,* **667**
Basidiocarps, *661,* **667**
Basidiomycetes, 1198, 1199
Basidiomycetes (Basidiomycota)
 characteristics, 651, *664,* **667**
 dikaryosis, 662
 life cycle, *661*
 phylogeny of the fungi and, *652*
Basidiospores, *662,* 663, 667
Basilar membrane, *972,* **973**
Basophils, *403,* **406,** *1056*
Batesian mimicry, **1190–1191**
Bats, *737*
 echolocation, 964–965
 largest population in North America, 1242
 plant-animal mutualisms and, *1194*
 wing evolution, *458,* 545
 See also Vampire bats
Bay checkerspot butterfly, 1178
Bazzania trilobata, 623
B cells
 in autoimmunity, 421
 binding to antigens, 407–408
 clonal deletion, 410
 clonal selection, *408,* 409
 cytokines and, 404
 development into plasma cells, 411
 generation of antibody diversity, 418–420
 humoral immune response and, 408, 417
 in hybridomas, 413
 overview of, 402, **403**
BCL-2 protein, 440
Bdelloid rotifers, 696
Beach grasses, 832
Beach strawberry, *747*
Beadle, George W., 257–258
Beagle (ship), 487, *488*
Beans, 1085
"Bean sprouts," 813
Bean weevils, 842–843
Bears, 1089
Beauveria bassiana, 650
Beavers, 1196
Beech bark cancer, 1216
Bee-eaters, *905,* 1159, *1160*
Beer drinking, 1108
Bees, *709, 1194*
 See also Honeybees
Beetles, 709, *709*

Begonias, 827
Behavior. *See* Animal behavior
Behavioral thermoregulation, 863
Beijerinck, Martinus, 284
Belladonna, *648*
Belly button, 912
Beltian bodies, *1184*
Beluga sturgeon, 356, 903
Benign tumors, **386**
Benson, Andrew, 169
Bent grass, 849
Benzer, Seymour, 267
Benzpyrene, 277
Bergamia, 480
Beriberi, 1074–1075
Bermuda grass, 175, 747
Best, Charles, 887
Beta blockers, 888, 893
β-Cells, **1087**
Beta thalassemia, 317
Betula pendula, 633
Beutler, Ernest, 323
B horizon, **786**
Biased gene conversion, 536
Bicarbonate ions
  blood pH and, 1105, *1106*
  blood transport of carbon dioxide and, 1038
  blood volume in capillaries and, 1059
  buffers and, 34
  gastric pH and, 1080, *1081*
  pancreas and, 1083
  properties as a base, 33
"Bicarbonate of soda," 34
Biceps, 1005
bicoid gene, 442
Bicoid protein, 442, 443
Bicoordinate navigation, 1154
Bicyclus anynana, 454–455
Biennials, **802–803, 826**
Bighorn rams, 505
Bilateral symmetry, **674**
Bilateria (Bilaterians), *672,* **682–683**
Bilayers, of phospholipids. *See* Phospholipid bilayers
Bile, 56, **1082,** *1083*
Bile duct. *See* Common bile duct
Bile salts, 1082, 1084
Binary fission, 181–182, *183,* **566,** 593
Bindin, 901
Binding domain, of hormone receptors, **879**
Binocular cells, 996
Binocular vision, **996,** *997*
Binomial nomenclature, 12, 509, **554–555**
Bioaccumulation, **1089**
Biochemical energetics
  ATP in, 123–125
  enzymes in, 125–135
  physical laws underlying, 119–123
Biodiesel technology, 55
Biodiversity
  captive breeding programs and, 1243
  controlling exotic species and, 1239–1240
  creation of new habitats and, 1238
  current crisis in, 1243
  economic value of, 1240–1241
  ecosystems restoration, 1236–1237, 1238, *1239*
  establishment of protected areas and, 1235–1236
  hotspot concept, 1235
  reconciliation ecology's approach to sustaining, 1241–1242
  reducing overexploitation of species and, 1238–1239
  restoration of disturbance patterns and, 1237–1238
  value of, 1228–1229
Biofilms, **563–565,** 579
Biogeochemical cycles, **1211**
  carbon, 1213–1216
  interactions among, 1220
  iron, 1219
  micronutrients, 1219–1220
  nitrogen, 1216–1217
  phosphorus, 1218–1219
  sulfur, 1217–1218
Biogeography
  area phylogenies, 1131
  biotic interchange, **1134**
  continental drift and, 1131
  determining biogeographic regions, 1128
  dispersal and vicariance, 1133–1134, 1135
  endemics, 1129, *1130*
  founding of, 1128
  freshwater environments and, 1136
  major biogeographic regions, *1129*
  oceanic regions, 1135–1136
  See also Island biogeography
Bioinformatics, **393**
Biological catalysts, 125, 126
  See also Enzymes; Ribozymes
Biological clocks, 829, *830*
  See also Circadian rhythms; Molecular clocks
Biological membranes, **97–101**
  activities and properties of, 114–115
  dynamic nature of, 100–101
  passive transport, 105–110
  phospholipid bilayers, 55, *56*
  See also Endomembrane system; Extraembryonic membranes; Membranes; Plasma membrane
Biological rhythms
  melatonin and, 890
  photoperiodism, 826–828
  See also Circadian rhythms
Biological species concept, **510**
Biology, **3**
  public policy and, 16–17
  tools and methods used in, 13–16
Bioluminescence, 124, **567, 1021**
Biomass, **1187,** *1188*
Biomes, **1116–1117**
  boreal forests, 1119
  chaparral, 1124
  cold desert, 1122
  factors affecting plant distribution, 1117, 1128
  hot deserts, 1123
  temperate evergreen forests, 1119
  temperate grasslands, 1121
  temperature deciduous forests, 1120
  thorn forests, 1125
  tropical deciduous forests, 1126
  tropical evergreen forests, 1127
  tropical savannas, 1125
  tundra, 1118
Bioprospecting, 538
"Bioreactors," natural, 369
Bioremediation, 294, 370, 849
Biosphere, **1113**
Biota, **472,** 1128
  present-day number of species, 472
  provincialization, **476**
  through evolutionary history, 472–479
Biotechnology
  applications in agriculture, 369–371, 430, 843–844
  defined, **367**
  expression vectors, 367–368
  medically useful proteins from, 368–369
  public concerns about, 371–372
Biotic factors, **1113**
Biotic interchange, **1134**
Biotin, *129, 1074*
Bipedalism
  in dinosaurs, 670
  in primates, 738
Bipolar cells, *945,* 980–981, 995
Bipolar disorder, 342–343
Bird flu, 282
Birds
  altricial and precocial young, 681
  amniote egg, 905
  in amniote phylogeny, *731*
  circulatory system, 1049
  energy storage in fat, 1071
  evolution of feathers and flight, 733, *734*
  feeding behavior, 734
  flocking behavior, 1158
  gastrulation in, 931–932
  hibernation and, 870
  high-altitude flight, 1024–1025
  hotspots of biodiversity, *1235*
  major groups, 732
  nasal salt glands, 1095
  newly formed neurons in adults, 984–985
  number of species, 734
  promiscuous mating in, 508
  rates of speciation, 519
  relationship to dinosaurs, 731
  respiratory gas exchange in, 1030–1031
  sex chromosomes in, 225–226
  sexual selection in, 499–500, 508–509, 519
  wing evolution, *458,* 545
  See also Songbirds
Birds of paradise, 519
Birdsong, 984–985
Birkhead, Tim, 500
Birth, in humans, 912–913
Birth control pills, 912, 914–915
Birth defects, thalidomide and, 936
Birth rates
  estimating, 1169–1171
  in exponential growth, 1173
  population densities influence, 1174–1175
Bison, *1085,* 1237
2,3-Bisphosphoglyceric acid, 1037–1038
Bispira brunnea, 699
bithorax mutant, 444, 450
Bivalents, 197–198
Bivalves, *700,* **702**
Black-bellied seedcrackers, 498, *499,* 513
Blackberry, 832
Black bread mold, 664–665
Black cherry, 1175, *1176*
Black-footed ferrets, 1237
Black grouse, *1150*
Black jumping spider, *713*
Blackpoll warbler, 155
Black rhinoceros, *1193*
Black rockfish, 1171–1172
Blacksmith fish, *727*
Black stem rust, 653, 662–663
Bladder, urinary, 340, 1101
Blade (leaf), **747**
Blastocoel, **922,** *924,* 925, 931
Blastocyst
  embryonic stem cells and, 433
  formation, **910,** *924,* 925
  implantation, 910, 925
  sex steroids and, 912
  in vitro fertilization and, 916
Blastodisc, **923,** 931
Blastomere, **922,** 923, 925–926
Blastopore, 673, 691, **927**
  See also Dorsal lip of the blastopore
Blastula, **922,** 927, 931
Blattodea, 709
Bleaching
  of coral, 1234
  of photosensitive pigments, 976
Blending theory, 207, 210, 218
"Blind eels," *724*
Blind spot, *979,* 980
Blocks to polyspermy, **902**
Blood, **1045**
  ABO system, 219
  cell types in, *403*
  clotting. *See* Blood clotting
  composition of, 1055, *1056*
  as connective tissue, 859, 1055
  defensive roles of, 402
  glucose levels, 1087
  hormonal control of calcium levels, 885–886, 887
  phosphate level and, 887
  plasma. *See* Blood plasma
  platelets, 186–187, *403,* 1056, *1057*
  regulation of osmolarity in, 1107–1108
  regulation of pH in, 1105, *1106*
  transport of respiratory gases, 1036–1038
  turnover time, 858
Blood-brain barrier, **946, 1059**
Blood carbon dioxide
  autoregulatory mechanisms of blood flow and, 1061
  blood pH and, 1105, *1106*
  blood volume in capillaries and, 1059
  chemoreceptors and, 1063
  transport, 1038
Blood cells
  differentiation in bone marrow, 402, *403,* 432, 1055
  types of, *403*
Blood clotting
  endotoxins and, 73
  hemophilia and, 377
  process of, **1056,** *1057*
  tissue plasminogen activator and, 368–369
Blood glucose, factors affecting, 1087
Blood groups, 51, 219
Blood oxygen
  autoregulatory mechanisms of blood flow and, 1061, *1062*
  binding/dissociation dynamics with hemoglobin, 1036–1038
  chemoreceptors and, 1063

  plasma carrying capacity, 1036
Blood pH
  effect on hemoglobin, 1037
  regulation of, 1105, *1106*
Blood plasma
  described, 106, 402, 1046, **1055**
  as extracellular matrix, 859
  oxygen carrying capacity, 1036
Blood pressure
  hypertension, 1060
  measuring, 1052
  regulation of, 1061–1063, *1107,* 1108
  Starling's forces and, 1058, *1059*
  vasopressin and, 881
Blood vessels, **1045**
  anatomy of, *1058*
  atherosclerosis, 376, 396, *397,* 887, 1060, *1061*
  characteristics, 1057
  connecting the hypothalamus and pituitary, 882
  countercurrent heat exchange in fish, 865
  dilation by adenosine, 332
  inflammation and, 406
  See also individual vessel types
"Blots," 354, *355*
Blowhole, 920
Blubber, 735
Bluebirds, 1232
Bluefin tuna, *13,* 17, 865
Bluegill sunfish, 1150–1151
Blue-green bacteria, 68, 573
  See also Cyanobacteria
Blue light, guard cell activity and, 774
Blue-light receptors, **812,** 814, 827, 829
Blue-throated hummingbirds, 1161
Blue tits, *681,* 1178–1179
Blue whales, 676, *920,* 1180
Blumeria graminis, 654
BMP2. *See* Bone morphogenetic protein 2
BMP4. *See* Bone morphogenetic protein 4
Body builders, 874, *875*
Body cavity, in animals, **674–675**
Body plans, **674–676,** 700
Body segmentation, in early development, 933–934
Body size, population density and, 1175–1176
Body temperature
  acclimatization and, 861
  "cooling" athletes, 854–855
  elevation in fish, 864–865
  fevers, 869–870
  heat production in endotherms, 861–862
  heat stroke, 854
  hibernation, 870
  hypothalamus and, 868–869
  hypothermia, 870
  temperature sensitivity, 860, *861*
  thermal classification system, 861
  See also Thermoregulation
Bohadschia argus, 720
Bohr effect, **1037**
Boiga irregularis, 1233
Bolinopsis chuni, 685
Bolitoglossa, 451, *452*
Bolting, **803**
Bombina, 517–518
Bombykol, 968
Bombyx mori, 968
Bone marrow, blood cells and, 402, *403,* 432, 1055
Bone marrow transplantation, 432–433
Bone morphogenetic protein 2 (BMP2), 441
Bone morphogenetic protein 4 (BMP4), 452
Bone(s)
  cell types, 1017
  extracellular matrix in, 91, 859
  features of, **1017–1018**
  joints and, 1019, *1020*
  parathyroid hormone and, 885–886
  of theropod dinosaurs and birds, 733
  types of, 1018, *1019*
  Vitamin D and, 886
Bonner, James, 796, 827
Bonneville Dam, *1213*
Booklice, *709*
Boomsma, Jacobus, *658*
Bordetella pertusis, 409
Boreal forests, **1119,** 1175, *1210*
Boron, *782*
Bosque del Apache National Wildlife Refuge, 1226
Bossiella orbigniana, 602
Bothus lunatus, 504
Botrytis fabae, 839
Bottle cells, 928
Bottlenecks, cheetahs and, 306
Botulinum toxin, 955
Botulism, 579
Bovine spongiform encephalopathy (BSE), 378
Bowman's capsule, **1100–1101,** *1102*
Boyer, Herbert, 358
Brachiopods, 473, *683, 692,* 697, **698**
Brachiosaurus, *671*
Brachyphylla, 1194
Bracket fungi, 667
Bracts, 825
Brain
  acetylcholine receptors, 337
  amygdala, 942–943, 988, *989,* 1000
  blood-brain barrier, 946
  cerebrum, 987, 989–991
  comparative complexity in, *945*
  consciousness, 1001
  development in vertebrates, 986–987
  diseases of, 378–379
  human-chimpanzee comparisons, 396
  human evolution and, 740
  hypothalamus. *See* Hypothalamus
  language abilities and, 1000–1001
  learning and, 999
  limbic system, 988–989, 1000
  long-term potentiation, 960, *961*
  memory and, 999–1000
  newly formed neurons in adults, 984–985
  number of synapses in, 944
  programmed cell death and, 202
  in protostomes, 691
  in simple animals, **943**
  sleep and dreaming, 997–999
  tagging with radioisotopes, *23*
  thalamus, 987, 988, 995, 998
  visual cortex, 991, 992, 994–996, *997*
"Brain coral," *687*
Brain hormone, 876
Brain stem
  components of, **987**
  control of breathing, 1039
  parasympathetic nerves and, 992
  reticular formation, 988
  sleep and, 998, 999
Branches and branching, in plants, 619, 753
Branchiostoma spp., *722*
Branta canadensis, 681
Brassica oleracea, 490
Brassinosteroids, *798,* **811–812**
Braxton-Hicks contractions, 912
BRCA1 gene, 388
Bread, 51
Bread mold, 257–258
Breast cancer, 366, 388
Breathing
  in birds, 1030–1031
  effect on blood flow, 1059
  in humans, 1034–1035
  regulation of, 1039–1040
  See also Respiratory gas exchange
Brenner, Sydney, 309
Brevipalpus phoenicis, 713
Brewer's yeast, 665
Brewing industry, 801
Briggs, Robert, 430
Bright-field microscopy, *71*
Brine shrimp, 1094–1095
Bristlecone pines, 635, 755
Brittle stars, *720,* 721
Broad fish tapeworm, *682*
Broca's area, 1000, 1001
Broccoli, *490*
Bromelain, *648*
Bronchi
  in birds, **1030,** *1031*
  in humans, **1033**
Bronchioles, **1033–1034**
Broomrape, 793
Brown algae, *584,* 594, *597,* **599–600,** *601*
Brown fat, **867**
Brown molds, 666
Brown tree snake, 1233
Browsing, eutherian mammals and, 736–737
Brussels sprouts, *490*
Bryophyta (Bryophytes), 612, **623–624**
Bryopsis, 602
Bryozoans, 695
BSE. *See* Bovine spongiform encephalopathy
Buchsbaum, Ralph, 704
Buck, Linda, 968, 969
Buckwheat, 849
Budding, *309,* **593**
  in animals, **897,** *898*
  defined, 593
  in unicellular fungi, *309,* 659
Bud scales, 754
Buds (plant)
  axillary and apical, 747
  in mosses, 624
  plant growth and, 754–755
  primordia, **757**
Buffers
  chemistry of, 34, *35*
  regulation of blood pH and, 1105, *1106*
Bufo
  B. marinus, 1181
  B. periglenes, 730, 1230
Bulbourethral glands, *906,* **908**
Bulbs, 832, *833*
Bulk flow, 776
Bullhorn acacia, *1184*
Bundle of His, **1053**
Bundle sheaths, *173,* **174,** 761
α-Bungarotoxin, 1021
Burdock, *824*
Burgess Shale, 472
Burrows, of eusocial animals, 1160, *1162*
Bush katydid, *710*
"Bush meat," 1191
Bushmonkey flower, 641
Butane, 40
Buttercups, *756*
Butterflies, *709*
  eyespot development and, 454–455
  larvae, 679
  life cycle, *680*
  metamorphosis in, 710, 877
  wing development in, 451

- C -

C_3 plants, 173, 175
C_4 photosynthesis, 162
C_4 plants, 173–175, 348, 761
Cabbage, *490*
Cache River National Wildlife Refuge, *1229*
Cacti, *193, 646,* 747, 845–846
Cactoblastis cactorum, 1166–1167, 1181
Cadastral genes, **825**
Cadaverine, 818
Caddisflies, *709*
Cadmium, 849
Caecillians, 729, *730*
Caenorhabditis elegans
  genes affecting apoptosis in, 439
  genome, 309, *533*
  as model organism, 704
  noncoding DNA in, *534*
  vulval differentiation in, 437–439
  See also Nematodes
Caffeine, 332–333, 345, 1082
Caiman, 732
Calcareous sponges, *672, 683,* 684
Calciferol, 886
Calcite-based rocks, 62
Calcitonin, *880,* 883, **885,** 1017
Calcium
  bone and, 1017
  calmodulin and, 344
  cytoplasmic concentration, 343
  egg activation and, 903
  electric current, 947
  ionotropic glutamate receptors and, 959–960
  as macronutrient, 1073
  muscle contraction and, 1008–1010, 1012
  in plant nutrition, *782, 783,* 784
  regulation of intracellular concentrations, 345
  signal transduction pathways and, 343–344
  slow block to polyspermy and, 902
Calcium carbonate
  foraminiferans and, 604
  global carbon and, 1214
  red algae and, 602
Calcium channels
  cardiac muscle contraction and, 1010
  conotoxin and, 1021
  heartbeat and, 1052, 1054

IP_3/DAG signal transduction pathway and, 342
release of neurotransmitters and, 955
Calcium-induced calcium release, 1010
Calcium nitrate, 784
Calcium phosphate, 1017
Calcium phosphate salts, 887
Calidris alba, 1168
California condor, 1243
California fan palm, 495
Callus, 430
Callyspongia plicifera, 1027
Calmodulin, 344, 1012
Caloplaca flavescens, 656
Calorie (Cal) (nutrionists calorie), **1069**
Calorie (cal), 121, **1069**
energy budgets, 1070–1071
metabolic rates and, 1069–1070
Calumma parsonii, 1130
Caluromys philander, 736
Calvin, Melvin, 169
Calvin cycle, 162, **169–172,** 175, *176*
Calyx, **639,** 821
CAM. *See* Crassulacean acid metabolism
Camarhynchus spp., *512*
Cambrian explosion, **472**
Cambrian period, *466–467,* 472–473, 479
Camembert cheese, 666
cAMP. *See* Cyclic AMP
Campanula rotundifolia, 554
Campephilus principalis, 1229
CAM plants, **175**
See also Crassulacean acid metabolism
cAMP receptor (CRP) protein, 300
CAMs. *See* Cell adhesion molecules
Canada geese, *681*
Canadian lynx, 1189
Canary grass, 803
Canavanine, 841–842, 844
Cancellous bone, **1018**
Cancer Genome Anatomy Project, 395
Cancer(s)
abnormal cell division in, 386
acetylation of histones and, 322
cell cycle disruptions and, 186
creation of malignant cells, 389, *390*
forms of, 387
genetic mutations and, 387–388
glycolipids and, 101
growth factors and, 187
HeLa cells, 180–181
metastasis, 386–387
oncogenes and, 324, 388, 389
somatic mutations and, 277
telomerase and, 248
therapies for, 391, 392
treating with colchicine or taxol, 88
tumor suppressor genes and, 388–389
unregulated cell division, 327
virally induced, 387
Candida albicans, 653
Canines, 1076
Cannabis sativus, 750
Cannel coal, 625
Cannibalism, 378
Capacitance vessels, 1059
Cape Cod roses, *646*
Cape ground squirrel, *737*
Capillaries
blood-brain barrier, 1059
characteristics, **1047,** 1057
inflammation and, 406
peritubular, 1100, 1101, 1102, 1103
permeability, 1058, 1059
pressure and flow in, 1057–1058
Capillary beds, **1047**
autoregulation of blood flow, 1061, *1062*
exchange of materials, 1057–1059
Capsaicin, 1242
Capsids, **284,** 293
Capsules, of bacteria, 73
Captive breeding programs, 1243
Capybara, 464
Carbohydrases, 1077
Carbohydrates
Calvin cycle synthesis of, 169–172
functional groups and chemical modification, 53
glycosidic linkages, 50–51
membrane-associated, 97, *98,* 101
monosaccharides, 49–51
muscle glycogen and, 1015
overview of, 39, *41,* **49**
polysaccharides, 51–53
storage in animals, 1071
"Carbo-loading," 1015
Carbon
asymmetrical, 40, 42
atomic weight, 23
covalent bonding capability, *27*
electronegativity, *27*
fixed, human consumption of, 160–161
global cycle, 1213–1216
natural isotopes, 23
as nutrient, 781
Carbon-14, 169, 170, *466*
Carbon cycle, 655, 1213–1216
Carbon dioxide
effect on breathing rate, 1039–1040
fossil fuels and, 1213
partial pressures, 1038, 1039, 1040
in photorespiration, 172, 173
in photosynthesis, 162, 169–170
in plant nutrition, 781
plant uptake of, 773
respiratory gas exchange and, 1025, 1027
See also Atmospheric carbon dioxide; Blood carbon dioxide
Carbon fixation
atmospheric CO_2 levels and, 1204
in C_4 plants, 173–174
Calvin cycle, 169–170
in CAM plants, 175
in photorespiration, 173
photosynthesis and, 160
Carbonic acid, 33, 34, 786
Carbonic anhydrase, 1038, 1080, *1081*
Carboniferous period, *466–467*
formation of coal, 625
overview of, **475**
plant evolution during, 618, 620, 631
saprobic fungi and, 655
Carbon monoxide
in brown algae flotation bladders, 600
fires and, 1212
hemoglobin affinity for, 1036
as neurotransmitter, 958
Carbon monoxide poisoning, 1036
Carbon ratio, 466
Carbon sinks, 1214
Carbon skeletons, **1072**
Carboxylase, **172**
Carboxyl group, 33, *40,* 42, *43,* 134
Carboxylic acids, *40*
Carboxypeptidases, 127, *1084*
Carcinomas, 387
Cardiac arrest, *see* Heart attack
Cardiac cycle, **1050–1052**
Cardiac muscle
contraction in, 1010
ECGs and, 1055
epinephrine and, 339
Frank-Starling law, 1059
function, **858,** 1005
heartbeat, 1052–1054
Cardinalis cardinalis, 734
Cardinals, *734*
Cardiomyopathy, 1044
Cardiovascular disease
cardiac arrest, 879
heart failure in athletes, 1044–1045
overview of, 1060–1061
See also Heart attack
Cardiovascular system, **1045**
See Circulatory systems
Caribou, *737*
Carnivora, *736*
Carnivores, 482, **1075,** *1076*
Carnivorous plants, **792**
β-Carotene, 56, 370, *371*
Carotenoids, 56, **165,** 500
Carotid arteries, 1062, 1063, 1108
Carotid body, **1040**
Carpels, **638,** 640, 820–821, 823
Carpodacus mexicanus, 481
Carpolestes, 737–738
Carrier proteins, **110**
Carriers, of recessive alleles, **227**
Carrots, 746, *747*
Carson, Rachel, 457
Cartier, Jacques, 374
Cartilage, 859, **1017**
Cartilage bone, **1018**
Cartilaginous fishes, 1017, 1099
Cartilaginous skeletons, 726, 728
Casparian strips, **768,** *769*
Caspases, **202–203,** 439–440
Cassini (spacecraft), 20
Cassowary, 732
Castanea dentata, 1233
Castor bean, 256
Castor oil, 256
Catabolic pathways, regulation, 299
Catabolic reactions, 122
Catabolism, 120, 154, 155
Catabolite repression, **300**
Catagonus wagneri, 1134
Catalase, 130
Catalysts, **125**
Catalytic RNAs, 62–63
Catalytic subunit, **132**
β-Catenin, 922, *923,* 929
Caterpillars, 679, 710, 842
Catfish, electric, 1021
Cations, **28**
mineral nutrients in soils and, 786, *787*
Cats
coat color/pattern determination, 219, 221
homozygosity in, 306–307
Cattle, 579
Cattle egrets, *1193,* 1194
Caucasians, cystic fibrosis, 377
Cauliflower, *490*
Caviar, 356, 903
CBOL (Consortium for the Barcode of Life), 357
CCR5 protein, 423
CD4 protein, 415, 423
CD8 protein, 415
CD28 protein, 411, 417
Cdk. *See* Cyclin-dependent kinase
cDNA. *See* Complementary DNA
cDNA libraries, 363
Cecum, **1085**
ced genes, 439–440
Celery, 750
Cell adhesion molecules (CAMs), 103
Cell–cell adhesion, 101, 102–104
Cell–cell communication, 348–349
See also Cell signaling
Cell cycle
growth factors and, 186–187
overview of, **184**
regulation of, 184–186
See also Cell division; Mitosis
Cell death, programmed. *See* Apoptosis
Cell differentiation, **427**
cloning animals and, 431–432
metabolic pathways and, 157
pluripotent stem cells and, 432–433
role of gene expression in, 435
studies of developmental genes and, 435
totipotency in animals, 430
totipotency in plants, 429–430
Cell division
cancer and, 386
control of, 184–187
disrupting with colchicine or taxol, 88
in eukaryotes, 182–183, 184–187
key events in, 181
oncogenes and, 388
polyploids and, 200–202
in prokaryotes, 181–182, *183*
tumor suppressor genes and, 388–389
See also Meiosis; Mitosis
Cell expansion, in growth, 428
Cell fate
cytoplasmic segregation and, 436–437
determination and, 427, **428–429**
induction and, 436, 437–439
Cell fractionation, 75
Cell fusion, 100, 185, 430, 431
Cell junctions, **102–104**
Cell layers. *See* Germ layers
Cell membrane. *See* Biological membranes; Plasma membrane
Cellobiose, 50–51
Cell plate, 192
Cell recognition, 102, *103*
Cells
discovery of, 4
nutrients and, 7
size limitations, 69–70
specialization and multicellularity, 12
Cell signaling
chemical signals, 334
defense systems and, 407
effect on enzyme activities, 346, *347*

effect on gene transcription, 347
effect on ion channels, 345–346
environmental signals, 333–334
gap junctions, 348
receptors, 336–339
signal transduction, 334–335, 339–345
Cell theory, **5, 69**
Cellular immune response, **408,** 414–417
Cellular respiration, **140**
ATP cycling and, 124–125
citric acid cycle, 145–147
energy yield from, 153
mitochondria and, 82
oxidative phosphorylation, 148–153
pyruvate oxidation, 144–145, *146*
Cellular slime molds, *584, 597,* **607**
Cellular specialization, **12**
Cellulases, 579, 1084
Cellulose
digestion in ruminants, 1084–1085
in plant cell walls, 90, 807
structure and function, 51, **52–53**
Cell walls
in archaeans, 575
eukaryotes and, 588
in prokaryotes, 73, 565–566
See also Plant cell walls
Celsus, Aulus Cornelius, 406
Cenozoic era, *466–467,* 477–479
Cenozoic explosion, 528
"Centers of imminent extinction," 1236
Centimorgan (cM), **224**
Centipedes, 712
Central dogma, **260–261**
Central nervous system (CNS)
components of, **944, 985**
development in vertebrates, 986–987
nervous system information flow and, *986*
neurohormones and, 985–986
reticular formation, 988
See also Brain; Spinal cord
Central sulcus, 990
Centrifuges, 236
Centrioles, *76,* 89, **188,** 921, 922
Centromeres, **187,** 190
Centrosomes, **188–189,** 921
Cephalization, **674**
Cephalochordates, *718,* **722,** 723
Cephalopods, *700,* **702,** 978, *979*
Cercomonas, 584
Cercozoans, *584, 597,* 604
Cerebellum, **986–987,** 988, 999
Cerebral cortex, **989–991,** 997–999
Cerebral hemispheres, **987,** 989–991, 1000–1001
Cerebrum, **987,** 989–991
Certhidea olivacea, 512
Cervical cancer, 180, 181, 327, 387
Cervical caps, 914
Cervical vertebrae, 453
Cervix, **909,** 912, 915
Cetaceans, 737
Cetartiodactyla, *736*
cGMP. *See* Cyclic guanosine monophosphate
Chagas' disease, 582, *604*
Chambered nautilus, 479, *701*
Chameleons, 1020
Champlain, Samuel de, 374
Channeled wrack seaweed, *601*
Channel proteins, in facilitated diffusion, **108–109**
Chaoborus, 456
Chaparral, **1124**
Chaperonins, **48,** *49,* 273, 847
Chara, 584, 613
Characters, **208,** 221, 490
Charales, **602, 612,** 613
Charcharodon megalodon, 725
Charophytes, *584,* 589, *597,* **602**
Chase, Martha, 235–237
Cheetahs, 306, 495
Chelicerates, *683,* **712–713**
Chelonia mydas, 732, 905
Chemical bonds, **25**
covalent, 25–27, 38, 41, 45
hydrogen, 29. *See also* Hydrogen bonds
hydrophilic and hydrophobic interactions, 29–30
ionic, *26,* 27–28, 46, 48
partial charges, 29
Chemical communication, 1157
Chemical elements, 21–25
Chemical equilibrium, **122,** 123
Chemical evolution, **61–62**
Chemically gated channels, **950**
Chemical reactions
energy and, 30–31
rates, **125**
Chemical synapses, 950–951, **955**
Chemical weathering, 786
Chemiosmosis
described, 148, **150–153**
in photophosphorylation, 168, *169*
Chemiosmotic mechanism. *See* Chemiosmosis
Chemiosmotic pumping, 167
Chemoautotrophs, 568
Chemoheterotrophs, **568–569**
Chemolithotrophs, **568, 782**
Chemoreception, 967–970
Chemoreceptors
in blood pressure regulation, **1063**
types of, *966,* **967–970**
Chemotherapy, 391
Cherries, *643*
Chestnut blight, 665–666, 1233
Chestnuts, 1233
Chiasmata, **198**
Chicken
development of body segmentation in, *933*
egg, 905
engrailed gene, *537*
gastrulation in, 931–932
Chicken pox, *409*
Chief cells, 1080, *1081*
Childbirth, 912–913
Chili peppers, 1242
Chilling injury, 847
Chimaeras (finned fishes), 726
Chimpanzees, *739*
brain protein comparisons with humans, 396
evolutionary relationships to humans, 59
HIV and, *543,* 551, *552*
in primate evolution, 738
Chinese traditional medicine, 1233
Chipmunks, 980
Chiroptera, *736*
Chitin, 53, 652, **694**
Chitons, *700, 701,* **702**
Chlamydias, *571,* **572,** *917*
Chlamydia psittaci, 572
Chlamydia trachomatis, 301–302
Chlamydomonas, 602
Chlamydosaurus kingii, 732
Chlorella, 9
Chloride, 28, *767,* 768, 947
Chloride ion channels, 950
Chloride transporter, 377
Chlorine, *27, 782, 1073*
Chlorophyll
absorption spectrum, *165*
in chloroplasts, 83
in cyanobacteria, 568
molecular structure, *165*
in photosynthesis, 166–167, 168
plant mineral deficiencies and, 783
Chlorophyll *a,* 164, 166, 167
Chlorophyll *b,* 164
Chlorophytes, *584, 603*
chloroplasts and endosymbiosis, 589, 590, 602
described, **602**
in eukaryote phylogeny, *597*
life cycles, 594–595
Chloroplast DNA (cpDNA), 548
Chloroplasts
in apicomplexans, 596
ATP synthase in, 168
in dinoflagellates, 596
endosymbiosis and, 11–12, 92, 589–590
genetic code and, 264
in glaucophytes, 601
in hornworts, 623
light signals formation of, 347
movement of polypeptides into, 273
phototropins and, 814
in plant cells, *77*
structure and function, 75, **83,** *84, 85*
Choanocytes, **677,** 684
Choanoeca, 584, 605
Choanoflagellates, *584, 597,* **605,** 672, 677
Chocolate, 332
"Chocolate spot" fungus, 839
Cholecalciferol, 1074
Cholecystokinin, **1086**
Cholera
actions and effects of, 96–97, 1084, 1191
an exotoxin disease, 579
pathogenic agent, 96, *409,* 575, 1191
plasma membranes and, 96–97
reducing rates of infection, 1191, *1192*
Cholera toxin, 116, 579
Cholesterol
corticosteroid hormones and, *888*
familial hypercholesterolemia, 114, 376–377, 379, 391
lipoproteins and, 1086, 1087
in membranes, 98
overview of, 56
structure, *57*
uptake by cells, 113–114
Vitamin D and, 886
Cholinergic neurons, 992
Cholla, *193*
Chondrichthyans, *723,* **726**
Chondrocytes, 859
Chondrus, 584
C. crispus, 601
Chordamesoderm, 932
Chordates, *672, 683,* 722–723
Chordin protein, 932
Chorion, **935**
Chorionic villi, *925*
Chorionic villus sampling, 385, **936**
C horizon, **786**
Choughs, 1024
"Christmas tree worm," *677*
Chromalveolates
alveolates, 596, 598
chloroplast endosymbiosis and, 590
haptophytes, 596
major groups, *586*
overview of, **596**
red tide and, 586, *587*
stramenopiles, 598–600
Chromatids
genetic recombination, *198,* 199, 223–224
during meiosis, *196–197,* 197–199
during mitosis, 189–191
sister, 183, 187
tetrads, 197–198
Chromatin, 79, **187,** *188,* **322,** 323
Chromatography, 169
Chromatophores, **1020,** *1021*
Chromis punctipinnis, 727
Chromium, 849, *1073*
Chromoplasts, 83, *85*
Chromosomal mutations, **275,** 276, *277*
Chromosome 15 (human), 383
Chromosome 21 (human), 199–200
Chromosome maps, 224
Chromosomes, **182**
abnormalities during meiosis, 199–200
artificial, 360
chromatin and, 79
daughter, 190, *191*
deletions, 381
in eukaryotes, 183
genetic recombination, 223–224
homologous pairs, 193. *See also* Homologous chromosomes
human diseases and, 379
independent segregation, 214, *215*
linkage groups, 223
mapping, 224
during meiosis, 195–202
movements during mitosis, 189–191
mutations and, 275, 276, *277*
origins of replication, 244, *245*
plasmids as, 293–294
in prokaryotes, 182
in somatic cells, 193
structure during mitosis, 187, *188*
telomeres, 248
during telophase, 191
Chronic wasting disease, 378
Chrysolaminarin, 599
Chrysomela knabi, 678
Chthamalus stellatus, 1192–1193
Chumbe Coral Park, *687*
Chylomicrons, *1083,* **1084, 1086**
Chyme, **1082**
Chymotrypsin, *128, 1084*
Chytridiomycetes (Chytridiomycota)
alternation of generations, 659
decline of amphibians and, 729–730
described, **663–664**
life cycle, *660*
zygomycetes and, 651
Chytriomyces hyalinus, 664

Cicadas, *709*
Cichlids, 551
Cigarette smoke, 277
cholesterol and, 1087
emphysema and, 1034
Cilia
ctenes, 684
in eukaryotes, 591
rotifer corona, 696
structure and function, **88–89**
Ciliary muscles, 979–980
Ciliates, *584,* 593, *597,* **598**
Cinchona, 648
Cingulata, *736*
cI protein, 290
Circadian rhythms
animal navigation and, 1154–1156
body temperature and, 869
effects on animal behavior, 1153
overview of, **1152–1153**
in plants, **828–829**
See also Biological clocks
Circannual rhythms, 1154
Circular chromosomes
plasmids, 293–295
in prokaryotes, 182
replication, 244, *245*
Circular muscles, 675
Circulating hormones, **875,** *876*
Circulatory systems
amphibian, 1048
autoregulatory mechanisms, 1061, *1062*
in birds and mammals, 1049
blood flow through veins, 1059–1060
blood pressure regulation, 1061–1063
blood vessel characteristics, 1057, *1058*
closed, 402, 1046–1047
components of, **1045**
countercurrent heat exchange in fish, 865
diving mammals, 1063–1064
evolution in vertebrates, 1047–1049
exchange of materials in capillary beds, 1057–1059
lymph and lymphatic vessels, 1060
open, 1046
reptiles, 1048–1049
vascular disease, 1060–1061
Circumcision, 206, 908
Cisternae, 80
Citigroup, 1240
Citrate, 128, 145, *146, 1072*
Citrate synthase, 128
Citric acid cycle
allosteric regulation, 156–157
in anabolic and catabolic pathways, 154, *154,* 155
in cellular respiration, **140,** *141*
described, 145–147
energy yield from, *153*
Citrus trees, 441, 832
Clades, 543–544, 555
Cladonia subtenuis, 656
Cladophora, 602
Clams, 702, 1016
Classification systems, 554–556
Class switching, **419–420**
Clathrin, 113, *114*
Clathrin-coated vesicles, 113, *114*
Claviceps purpurea, 666
Clay, 62, 786, **786**
Clean Air Act, 1218
Cleavage
in mammals, 923, *924,* 925
overview of, **922**
types of, **672–673,** 923, *924*
Clements, Frederick, 1185
Climate, **1114**
changes through Earth's history, 470–471
impact of humans on, 1116
plant adaptations to extremities in, 845–847
solar energy and, 1114–1115
species extinctions and, 1234
warming. *See* Global warming
Clitellates, 699–700
Clitoris, 909
Clobert, Jean, 1148
Clonal anergy, **410–411**
Clonal deletion, **410,** 421
Clonal selection, **408–409**
Clones
asexual reproduction and, 193
plant, 429–430
in prokaryotes, 290
Cloning
of animals, 431–432
of plants, 429–430
positional, 381–382
recombinant DNA technology and, **359**
therapeutic, **434**
Closed circulatory systems, **1046–1047**
Clostridium, 574
C. botulinum, 574, 579
C. difficile, 574
C. tetani, 409, 579
Clotting factors, **1056,** *1057*
Clover, 828
Clown fish, 904
Club fungi, *664*
Club mosses, 474, *612,* 618, 620, **624–625**
Cnemidophorus uniparens, 898
Cnemidopyge, 480
Cnidarians, *672,* 673, *683*
described, **685–687**
gastrovascular cavity, 1076–1077
nematocysts, 678, *679*
nerve net, 943
radial symmetry, 674
Cnidocytes, *679*
CNS. *See* Central nervous system
Coal, 475, 618, 625, 655
Coastal redwoods, 634
Coast redwoods, *765,* 769
Coat color/pattern, genetic determination, 218, 219–220, 221
Coated pits, 113
Coated vesicles, 113, *114*
Cobalamin, *1074,* 1075, 1220
Cobalt, *1073,* 1219, 1220
Cobalt-60, 23
Cobamide, 1220
Cocaine, 650
Coca plant, 650
Coccolithophores, *4,* 476, 586, *587*
Coccus (cocci), 563
Cochlea, *972,* **973**
Cocklebur, 827, 829–830
Cockroaches, *709, 1077*
Coconut fruit, *631,* 646
Coconut milk, 809
Coconut palms, *646,* 824
Cocos nucifera, 646
Cod, 1180
Codominance, 219
Codons, **264**
Codosiga botrytis, 605
Coelacanths, 728
Coelom, **674–675**
in annelids, 1097
in protostomes, 691
Coelomates, **674–675**
Coenocytes, **600**
Coenocytic hyphae, **652,** *653*
Coenzyme A (CoA)
in catalyzed reactions, *129*
citric acid cycle, 145, *146*
pyruvate oxidation, 144, *146*
Coenzymes, *129,* **130**
Coevolution, herbivory in mammals and, 737
Cofactors, *129,* **130,** 341
Coffee, 332, *333*
Coffee plantations, 1241, *1242*
Cohen, Stanley, 358
Cohesins, **187,** 189, 190, *191,* 198, 199
Cohesion, of water, **32**
Cohorts, **1169**
Coincident substitutions, 527
Coitus interruptus, 908, **914**
Colchicine, 88
Cold
adaptations to, 867–868
perception of, 954
plants and, 847
Cold deserts, **1122**
Cold-hardening, **847**
Cold sensation, 954
Coleochaetales, *612,* 613
Coleochaete, 613
Coleoptera, 709
Coleoptiles, **803,** *804*
Coleoptology, 709
Coleus, 809
Colias, 502
C. eurydice, 711
Collagen
blood clotting and, 1056, *1057*
in connective tissue, **91,** 858, 1017
Collar, in hemichordates, 721
Collared flycatchers, 1148
Collar flagellates, 605
Collecting duct, **1102,** 1104–1105, 1108
Collenchyma, **750,** *751*
Colloidal osmotic pressure, 1058, *1059*
Colon, *1079,* **1084**
Colon cancer, 249, 389, *390*
Colonial birds, defense of nesting sites, 1149, *1150*
Colonies
bacterial, 290–291
eusocial animals, 1160, *1162*
Colony-stimulating factor, *368*
Color blindness, *207,* 227, *228*
Colossendeis megalonyx, 712
Columbia River, 1212, *1213*
Columbines, 516, 519
Coma, 988
Comb jellies, 684–685
Combustion, 139–140
Comets, 34–35, 61
Commensalisms, **1188,** 1193–1194
Commerson's frogfish, *727*
Common beans, 842–843
Common bile duct, 1082, *1083*
Communication, **1156–1157,** *1158*
Communities, 9, **1113, 1185**
effects of animal behavior on, 1161–1162
microbial, 563–565
See also Ecological communities
Community ecology
amensalisms, 1188, 1193–1194
biomass and energy distributions, 1187, *1188*
commensalisms, 1188, 1193–1194
competition, 1188, 1192–1193
disturbances, 1197
ecological community concept, 1185–1186
energy sources, 1186–1187
intermediate disturbance hypothesis, 1199
mutualisms, 1184–1185, 1188, 1194
parasitism, 1188, 1191, *1192*
predation, 1188, 1189–1190
succession, 1197–1200
trophic cascades, 1195–1197
types of ecological interactions, 1188–1194
Compact bone, **1018,** *1019*
Companion cells, **638,** *751,* **752**
Comparative experiments, **14,** 15
Comparative genomics, **301**
Compartmentalization, in cells, 69, 74–75
Competition, **1188,** 1192–1193
Competitive exclusion, **1192**
Competitive inhibitors, **131–132**
Complementary base pairing, **58, 239,** *240,* 260
Complementary diets, 1072
Complementary DNA (cDNA), 288, **363,** 366
Complement proteins, 405
Complement system, **405**
Complete cleavage, 673, **923,** *924*
Complete metamorphosis, **710**
Complex cells, visual cortex, 995
Complex ions, 28
Compound eyes, **978**
Compound leaves, 748
Compounds (chemical), **25**
Compound umbels, *639*
Concentration gradients, 105, 108
See also Ion concentration gradients; Proton concentration gradients
Concerted evolution, **535–536**
Condensation reactions, **41**
Condensins, **187**
Conditional mutants, **275**
Conditioned reflexes, **999**
Condoms, 914, 917
Condors, 1243
Conducting cells, 1010
Conduction, in heat transfer, **863,** 864
Conduction deafness, 974
Cone cells, **980,** *981*
Cones, of conifers, **636**
Cone snails, 679, 1021
Confocal microscopy, *71*
Conformation, 266
Congestive heart failure, 1105
Conidia, 659, 666
Coniferophyta, *612,* **634,** 636
Conifers, *612, 632,* **634,** 635
life cycle, 636, *637*
Coniosporium, 1198
Conjoined twins, 926
Conjugating fungi, *664*

Conjugation, **291–292,** *293,* 294, **593,** *594*
Conjugation tube, **291–292,** 294
Connective tissues, **858–859,** *860*
Connell, Joseph, 1193
Connexons, 104, **348,** 958
Conotoxin, 1021
Consciousness, 1001
Consensus sequences, **317**
Conservation biology
factors threatening species survival, 1231–1234
overview of, **1227–1228**
predicting species extinctions, 1229–1230
prevention of species extinctions, 1228–1229
strategies used to maintain biodiversity, 1235–1243
Conservative replication, DNA, 241, 242
Consortium for the Barcode of Life (CBOL), 357
CONSTANS gene, 829
Constellations, animal navigation and, 1156
Constipation, 1084
Constitutive proteins, **296**
Continental drift, **468,** 1131
Continental shelf, primary production in, *1210*
Contraception, **914–915**
Contractile ring, 192
Contractile vacuoles, 86, **591,** *592*
Contractions, during labor, 912
Contragestational pills, 915
Contralateral neglect syndrome, 991
Controlled burning, 1237–1238
Controlled experiments, **14–15**
Controlled systems, **856**
Convection, in heat transfer, **863,** 864
Convention on International Trade in Endangered Species (CITES), 1239
Convergent evolution, **545**
Cooksonia, 474
Cook Strait, 1135
"Cooling," exercise capacity and, 854–855
Cooperative behavior, 9
Copepods, 599, *600,* 706, *707,* 1187
Copper, *129, 782, 1073*
Copper mine tailings, 849
Coprophagy, **1085**
CO protein, 829, *830*
Copulation, **904,** 908
Coral reef fish, 727
Coral reefs
bleaching, 1234
decline in, 687, 1222
development of, 686
primary production in, *1210*
Corals, **686–687**
bleaching, 1234
dinoflagellates and, 585
dispersal, *1136*
mutualism with protists, 1194
red algae and, 602
Corepressors, **298,** 299
Coreus marginatus, 711
Cork, **760**
Cork cambium, *753,* **754,** 755, 760
Corms, 832
Corn
gibberellin and, 802
hybrid vigor, 220
root anatomy, *756*
Cornea, 437, **978,** *979*
Corn oil, 55
Cornus florida, 646
Corolla, **638–639,** 821
Corona, rotifers, 696
Coronary arteries, **1060**
Coronary thrombosis, **1060**
Coronavirus, 579
Corpora cardiaca, 876
"Corpse flower," 818
Corpus allatum, *878*
Corpus callosum, 1000
Corpus cavernosum, 344
Corpus luteum, 911, 912
Correns, Carl, 208
Corridors, 1231–1232
Cortex (plant), **756,** 757
Cortex (renal), **1101,** *1102*
Cortical cytoplasm, 922
Cortical granules, 902, 903
Corticosteroid hormones, **888–889**
Corticotropin, **887**
Corticotropin-releasing hormone, **888**
Cortisol, *57,* 339, *880,* **888–889**
Corynebacterium diphtheriae, 409, 579
Cost-benefit analyses
of animal social behavior, 1158
of feeding behavior, 1149–1151
overview of, **1148**
of territorial defense, 1148–1149, *1150*
Co-stimulatory signals, 411, 417
Co-transporter proteins, 1084
Cottonwood trees, 370
Cotyledons, **643,** 824
Cougars, *307*
Countercurrent flow, **1030**
Countercurrent heat exchange, **865,** 868
Countercurrent multiplier, **1103,** *1104*
Coupled transporters, 111
Courtship behavior
in cephalopods, 702
mate choice and fitness, 1152
Courtship displays
information content and, 1156
in mallard ducks, *1142–1143*
sexual selection and, 1143
Covalent bonds
characteristics, **25–27**
in condensation and hydrolysis reactions, 41
in macromolecules, 38
in protein primary structure, 45
Covalent catalysis, 128
Cowpox, 400
cpDNA. *See* Chloroplast DNA
Crabgrass, 175
Crabs, 1016, 1187
Craniates, 724
Crassulaceae, 175
Crassulacean acid metabolism (CAM), 162, **175,** 847, 848, *849*
Crayfish, *675, 710*
Creatine phosphate (CP), 1014, 1015
Crenarchaeota, *571,* **575,** 576–577
Cretaceous period, *466–467,* 471, 477, 482
Cretinism, 884
Crews, David, 898
Crichton, Michael, 232
Crick, Francis, 232, 238, *239,* 241, 260
Crick, Odile, 232
Crinoids, 475, 719
Cristae, 82, *83*
Critical day length, **826**
Critical night length, **827**
Crocodiles, 732, *733,* 1048–1049, 1242
Crocodilians, 732, 1048–1049
Crocodylus niloticus, 733
Crocuses, 832
Crohn's disease, 374
Cro-Magnons, 740
Crop (digestive organ), 533, **1077**
Cro protein, 290
Crops. *See* Agriculture
Crosses
dihybrid, 213–214, 216
monohybrid, 209–210, *211*
reciprocal, 207
test, 213
Crossing over, **198,** 286, 536
Cross-pollination, 822
Crowned lemur, *738*
Crown fungi, *652, 663,* 665
Crown gall, 575
CRP protein, 300
Crustaceans
chitin in, 53
insect wing evolution and, 710
nauplius larva, 680
number of living species, *683*
overview of, **706–707**
Crustose lichens, 656
Cryphonectria parasitica, 1233
Cryptochromes, **814,** 829
Cryptophytes, 590
Cryptoprocta ferox, 1130
Ctenes, **684**
Ctenophores, *672,* 673, 674, *683,* **684–685**
CTLA4 protein, 417, 421
Cud, 1085
"Cuddle hormone," 881
Cumulus, **901,** *903*
Cup fungi, 665, *666*
Cupulae, 974–975
Curare, *648*
Cuscuta, 793
Cuticle (animal), **693,** 1016
Cuticle (plant), 613, **748,** 845
Cutin, 748
Cuttings, 833
Cuttlefish, 1020, *1021*
CXCR4 protein, 423
Cyanide, 503, 844
Cyanobacteria, *571*
characteristics, **573**
chlorophyll *a,* 568
component of lichens, 655, 656
early photosynthesizers and, 11
endosymbiosis theory and, 589
fossil, 68
gas vesicles, *566*
hornworts and, 623
impact on atmospheric oxygen, 468–469, 578
motility, 566
nitrogen-fixation and, 788
Cyanophora, 584
Cycadophyta, *612,* **634**
Cycads, *612, 632,* **634,** *635*
Cyclic adenosine monophosphate (cAMP)
action on enzymes, 346
function, 59
regulation of, 345
regulation of promoters in bacteria, 300
as second messenger, 341
in signal transduction pathways, 346
Cyclic AMP. *See* Cyclic adenosine monophosphate
Cyclic electron transport, **168**
Cyclic GMP. *See* Cyclic guanosine monophosphate
Cyclic guanosine monophosphate (cGMP), 344, 908
Cyclin-dependent kinase (Cdk), **185–186**
Cyclins, **185–186,** 327
Cyclops sp., *707*
Cyclosporin, 417
Cyclostomes, 724
Cypresses, 846
Cysteine, 42–43
Cystic fibrosis, 377, 379, 380, 385, 1034
Cytochrome *c,* **149,** *150, 151, 528–529*
Cytochrome oxidase gene, 357
Cytochrome P450, 1089
Cytokines, 404, 406, 409, 419–420
Cytokinesis
in animal cells, 192
in cell division, **181**
in plant cells, 192
in prokaryotes, 182
Cytokinin oxidase, 810
Cytokinins, *798,* **809–810**
Cytoplasm
cytokinesis and, 192
in prokaryotes, **72–73**
See also Egg cytoplasm; Sarcoplasm
Cytoplasmic bridges, 899, *900*
Cytoplasmic determinants, **436–437**
Cytoplasmic domain, hormone receptors, **879**
Cytoplasmic dynein, 190
Cytoplasmic receptors, 336, **339**
Cytoplasmic segregation, **436–437**
Cytoplasmic streaming, 87, **606**
Cytosine
deamination, 326
in DNA, 59, **238,** 239, *240*
methylation, 323, 382
in RNA, 59
Cytoskeleton
in animal cells, *76*
attachment to membrane proteins, 100
eukaryotic evolution and, 588
functions of, **86**
and polarity in the egg, 436–437
in prokaryotes, 74
structural components, 86–88
Cytosol, **72**
Cytotoxic T cells (T_C), **414,** 415, *416,* 417*Dactylaria,* 654

- D -

Dactylella, 654
Daffodils, 370
Dahlias, 827
Daily torpor, **870**
Dali, Salvador, 232
Dalton (Da), 21
D-Amino acids, 42
Dams, 1212, *1213*
Damselflies, *709,* 710
Danaus plexippus, 680
Danchin, Etienne, 1148
Dance, honeybee, 1157, *1158*
Dandelions, 824, 832
Daphnia cuculata, 456

Dark reactions, 162
Darwin, Charles
ascidians, 722
community ecology, 1185
John Dawson and, 610
evolutionary theory, 5–6, 487–489
evolution of eyes, 448
origin of life, 68
origin of the angiosperms, 646
phototropism experiments, 803, *804*
plant breeding experiments, 208
"Sally Lightfoots," *707*
sexual selection, 498–500, 508
speciation, 509, 511
Darwin, Francis, 803, *804*
Darwin's finches, 511, *512*
Dasypus novemcinctus, 1134
Data, **16**
Dates, 630
Daughter chromosomes, **190,** *191*
Davidson, Eric, 435
Dawson, John William, 610
DAX1 gene, 226
Day-neutral plants, **826–827**
ddNTPs. *See* Dideoxyribonucleoside triphosphates
DDT, 457
Deafness, 974
Deamination, 277, *278*
Death rates
Darwin on, 489
estimating, 1169–1171
in exponential growth, 1173
population densities influence, 1174–1175
Decapods, 706, *707*
Decarboxylation, 174
Declarative memory, **1000**
Decomposers, **578,** 655, **1186**
Deductive logic, 14
Defense systems
blood and lymph tissue in, 402
nonspecific, 404–407
role of immune system proteins in, 403–404
self/nonself distinction and, 401
types of, 401
See also Specific immune system
Defensins, **405**
Deficiency diseases, **1074–1075**
Deficiency symptoms, in plants, 783
DEFICIENS protein, 441
Deforestation, 1230
Dehydration reactions, 41
Delay hypersensitivity, 420
Deletions, in chromosomes, **276,** *277, 380,* 381, 382
Delivery, 912
Demographic events, **1167**
Demography, 1167
Demosponges, *672, 683,* 684
Denaturation
of DNA, 250
of proteins, **48**
Dendrites, **945**
Dendritic cells, 402, *403,* 411, 422
Dendrotoxin, 1021
Dengue fever, 1181, 1216
Denitrification, **791,** 1216
Denitrifiers, **569,** 578, **791**
Density-dependent factors, **1174**
Density-independent factors, **1174**
Density labeling, 241–242
Dental plaque, 565
Dentine, **1076**
Deoxyribonucleic acid (DNA)
amplification through PCR, 250–251
analyzing, 353–357
artificial synthesis and mutagenesis, 363
central dogma and, 260, 261
chemical composition, 238
complementary, 288, 363, 366
complementary base pairing, 58
cutting and splicing, 358–359
differences from RNA, 260
double helix, **59–60,** 232, 233, 238–241
evolutionary relationships are revealed through, 60
identified as genetic material, 233–237
inserting into cells, 359–362
methylation, 249, 323, 354, 382
noncoding, 533–534
overview of, **7, 57,** *58, 59*
phylogenetic analyses and, 548
popularization of, 232–233
in prokaryotes, 181, 561–562
repair mechanisms, 249, *250*
repetitive sequences, 311–312
replication, **181,** 182, 241–249
sequencing, **251–253**
structure, 57–60, 238–241
transcription, 261–265
viruses and, 284
See also entries at DNA
Deoxyribonucleoside triphosphates (dNTPs), 251, 252
Deoxyribonucleotides, 59
Deoxyribose, 50
Dependent variables, 15
Depolarization, of membranes, **950,** 951
Depo-Provera, 915
Deprivation experiments, 1142
Derived traits, **545**
Dermal tissue system, **748**
Dermaptera, 709
Dermophis mexicanus, 730
Dermoptera, *736*
Descent of Man, and Selection in Relation to Sex, The (Darwin), 498
Desensitization, 420
Desert gerbil, 1105
Desert
Australian, 1128
cold, 1122
hot, 1123
plants of, 845–846
primary production in, *1210*
Desmazierella, 1198
Desmodus rotundus, 1092
Desmosomes, 88, **103–104**
Desmotubules, **348,** *349*
DET2, 811
det2 mutant, 811
Determinate growth, 754, 825
Determination, **427**
of blastomeres, **925**
cell fate and, 428–429
See also Sex determination
Detoxification, liver and, 1089
Detritivores, **1075,** 1186, **1187**
Detritus, 1187
Deuterium, *23*
Deuteromycetes, **663**
Deuterostomes, *672,* **683**
chordates. *See* Chordates
developmental patterns characterizing, **673,** 717
echinoderms, 718–721
hemichordates, 718, 721
major clades, *683,* 718
Development
activation by fertilization, 921–926
cell fate, 427, 428–429, 436–439
cleavage, 672–673, 922–925
differential gene expression and, 427–428
direct, 679
environmental modulation of, 454–457
extraembryonic membranes and the placenta, 934–936
gastrulation, 673, 926–932
in humans, 936–938
modularity and, 451–452
organogenesis, 932–934
origin of new species and, 453
overview of, 183, **427,** *428*
phylogenetic analyses and, 548
regulatory genes and, 449–450
See also Plant development
Developmental genes
constraints on evolution and, 457–459
eye development and, 448–449
genetic switches, 451
heterochrony, 451, *452*
modularity in embryos and, 451
morphological changes in species and, 453, *454*
overview of, 449–450
spatial patterns of expression, 452
Developmental plasticity, **454**
learning and, 456
in plants, 456, 457
in response to predators, 456
in response to seasonal changes, 454–455
Devonian period, *466–467*
finned fishes in, 726
overview of, **474**
plant evolution and, 618, 619, 620, 631
tetrapods and the colonization of land, 728
Devries, Hugo, 208
Dextral coils, 923
DHP receptor, 1009, 1010
Diabetes mellitus
among Native Americans, 1068
downregulation of insulin receptors, 892–893
type I, 421, 887
type II, 887, 892–893
Diacylglycerol (DAG), **342,** *343*
Diadophis punctatus, 732
Dialysis, 1105–1106
Diaphragm (in contraception), 914
Diaphragm (muscle), **1034–1035,** 1039
Diarrhea, 1084
Diastole, **1051,** *1053,* 1057
Diastolic pressure, 1052
Diatomaceous earth, 599
Diatom blooms, 599, *600*
Diatoms, *597*
amount of carbon fixed by, 584
cell surface, 592, *593*
described, **599**
diatom blooms, 599, *600*
Mesozoic era, 476
oil and, 586–587
Diatomyids, 1229
Dicerorhinus sumatrensis, 1233
Dicer protein, 325
Dichotomous branching, 619
Dictyostelium, 584
D. discoideum, 607
Dideoxyribonucleoside triphosphates (ddNTPs), 251, 252, 253
Diencephalon, **987,** 988
Diesel, Rudolf, 55
Differential gene expression
cell differentiation and, 435
development of organisms and, **427–428**
globin genes and, 315
overview of, 318–319
pattern formation and, 439–445
Differential gene transcription, 435
Differential interference-contrast microscopy, *71*
Differentiation, **427**
See also Cell differentiation
Diffusion
facilitated, 108–110
Fick's law of, **1026**
of gases, 1025–1026
overview of, **105–106**
Digestion
external, 1075
foregut fermentation, 532
in gastrovascular cavities, 1076–1077
liver and, 1082, *1083*
lysosomes and, 75, 81–82
pancreas and, 1083
small intestine and, 1082–1083
stomach and, 1080, *1081*
vacuoles and, 75, 82, 86
Digestive enzymes
amylase, 1080
bile and, 1082
in humans, *1084*
overview of, 1077
pancreas and, 1083
small intestine and, 1083, *1084*
Digestive systems. *See* Gastrointestinal systems; Guts
Digger wasps, *1144,* 1145
Digitalin, *648*
Digitalis, 1055
Digitalis purpurea, 1055
Dihybrid crosses, **213–214,** 216
Dihydropyridine (DHP) receptor, 1009, 1010
Diisopropylphosphorofluoridate (DIPF), 131, *132*
Dikaryon, 659, **662**
Dikaryosis, *661,* 662
Dimers, 88
Dimethyl sulfide, 1217, 1218
Dinoflagellates, *584, 597, 598*
coral and, 1234
described, **596**
endosymbiosis and, 585, 590
Mesozoic era, 476
red tide and, 586
Dinosaurs, 477, 670, *671,* 731–732
Dioecious organisms, 225, 639, 904
Diomedia chrysostoma, 1155
Dionaea muscipula, 792
Dipeptidase, *1084*
Dipeptides, 269
DIPF. *See* Diisopropylphosphorofluoridate
Diphtheria, *409*
Diphyllobothrium latum, 682

Diploblasts, *672,* **673**
Diploids, **193**
Diplomonads, *584, 597,* **603**
Diplontic life cycle, **194, 595**
Diploria labyrintheformes, 687
Dipodomys spectabilis, 1093
Diprotodon, 1228
Diptera, 709
Direct development, 679
Directional selection, 497, **498,** 1191
Direct transduction, **339,** *340*
Disaccharides, **49,** 50, 1083
Discoidal cleavage, **923**
Disease(s)
effects of global warming on, 1216
fungal, 653
Koch's postulates and, 579
sexually transmitted, 301–302, 423, 917
social behavior and, 1158
See also Genetic diseases
Disk flowers, *639*
Dispersal, **1115**
in animals, **680**
biogeographic patterns and, 1134, 1135
of freshwater organisms, 1136
of seeds, 824
Dispersive replication, of DNA, 241, 242
Displays, **1156**
See also Courtship displays
Disruptive selection, 497, **498,** *499*
Distal convoluted tubules, **1102,** 1104
distal-less (dll) gene, 453, 454, *454, 455*
Distance-and-direction navigation, 1154
Disturbance(s)
biodiversity and, 1237–1238
intermediate disturbance hypothesis, 1199
overview of, **1197**
Disulfide bonds, 172
Disulfide bridge, **43,** *44, 45,* 46
Diurnal animals, 1153
Diving animals
circulatory system in, 1063–1064
myoglobin and, 1037
Diving beetle, *710*
dll gene, 453, 454, *454, 455*
DNA barcode project, 356–357
DNA-binding proteins, 320, *321*
DNA chips, **365–366**
DNA fingerprinting, 352–353, **355–356,** *357*
DNA helicase, **244**
DNA ligase, **246,** *247,* 358, 359
DNA polymerase I, 246, *247*
DNA polymerase III, 246, *247*
DNA polymerase(s)
in DNA repair, 249, *250*
in DNA replication, **241,** 245–247
in DNA sequencing, 251, 252
errors in replication, 249
mutations and, 277
in PCR, 250, 251
DNA segregation
in cell division, **181**
during mitosis, 188–191
in prokaryotes, 182
DNA sequencing, **251–253**
DNA testing, **384–385,** *386*
DNA topoisomerase, **244**
DNA transposons, 312
DNA vaccines, 410
dNTPs. *See* Deoxyribonucleoside triphosphates
Dobsonflies, *709*
Dock bugs, *710*
Docking proteins, **273**
Dodd, Anthony, *829*
Dodders, 792, *793*
Dogface butterfly, *710*
Dogs
coat color determination, 219–220
conditioned reflex experiment, 999
estrus in, 911
olfaction in, 968
Doligez, Blandine, 1148
Dolly (cloned sheep), 431–432
Dolphins, 737
Domains (in classifications)
commonalties among, 561
differences between, 561–562
major groups, **12**
phylogenetic tree, *571*
Domains (molecular), 99, 315, 879
Dominance
of abnormal alleles in humans, *216,* 217
genetic, 213
in Mendelian traits, **210**
Dopamine, 958, *959*
Dormancy, seeds and, 457, 634, **798,** 799–800
Dorsal aorta, in fish, 865
Dorsal fin, *726*
Dorsal horn, spinal, 987
Dorsal lip of the blastopore, **928–929,** *930,* 931, 932
Dorsal plane, **674**
Dorsal root, spinal, 987
Dorsal-ventral axis, in early development, 922
Dose-response curves, 891, *892*
Double bonds, 27
Double fertilization, **638,** *820,* **822–823**
Double helix, of DNA, **59–60,** 232, 233, 238–241
Double-stranded RNA (dsRNA), 840
Doublets (microtubules), 88–89
Douching, 914
Douglas firs, 1176–1177
Down syndrome, 199
Downy mildews, 600
Dragonflies, *709,* 710, 1196
Dreaming, 998, 999
Dreissena polymorpha, 1176, *1177*
Driesch, Hans, 436
Drills, *739*
Drosera rotundifolia, 792
Drosophila, 584
allopatric speciation in, 511, *513*
cleavage in, 923, *924*
determination of body segmentation in, 442–444
developmental genes and, 450
gene duplications and, 534
genome size, *533*
homeotic mutations, 444
noncoding DNA in, *534*
wing, 451, *710*
Drosophila melanogaster
genetic studies with, 222, 223–224
genetic variation in, 490–491
genome, 309–310
sex determination in, 225
sex-linked inheritance in, 226–227
Drosophila subobscura, 495, *496*
Drought, gene expression in plants and, 321–322
Drummond, Thomas, 515
Dryas, 1198
Dry matter, 783n
Dryopithecus, 738
Dry weight, 783n
dsRNA. *See* Double-stranded RNA
Dubautia menziesii, 520
Duchenne muscular dystrophy, 377, 380, 381, 382, 385
Duck-billed platypus, 735, 969
Ducks, 452, *1142–1143,* 1143
Dugongs, 737
Dung beetles, 708
Duodenum, *1079,* **1082,** 1086
Duplications, in chromosomes, **276,** *277*
Dust Bowl, 780–781, 786
Dutch elm disease, 665, 666, 833, 1233
Dwarf mistletoe, 792–793
Dwarf plants, gibberellin and, 802
Dwarf rabbits, *218*
Dynein, 89, *90*
Dysentery, 294
Dystrophin, 377
Dystrophin gene, 382
Dytiscus marginalis, 711

- E -

E. coli. See Escherichia coli
Eagles, 1089
Eardrum, 972, 974
Early genes, 286
Earphones, hearing loss and, 974
Ears
equilibrium, 974, *975*
hearing, 972–974
pinnae, 972
"Ear stones," *975*
Earthworms, 904
anatomy of, *698*
body cavity, *675*
circulatory system, 1046–1047
digestive tract, *1077*
excretory system, 1097
feeding behavior, 699
hermaphroditism, *699*
hydrostatic skeleton, 1016
intestinal surface area, *1078*
nervous system, 943, *944*
Earwigs, *709*
Ecdysone, 876, 878
Ecdysozoans, *672, 683*
arthropods, 705–713
cleavage in, 673
exoskeletons, **693–694**
major groups, 702–705
phylogenetic tree, *692*
ECG. *See* Electrocardiogram
Echidnas, 735
Echiniscus springer, 706
Echinoderms, *672, 683*
cleavage in, 673
number of species, 718
overview of, 719–721
pentaradial symmetry, 718–719
regeneration in, 897–898
tube feet, 676
Echolocation, 964–965
Eciton hamatum, 1160
Ecological communities, **1185**
biomass and energy distributions, 1187, *1188*
disturbances and, 1197
food webs, 1187
intermediate disturbance hypothesis, 1199
as loose assemblages of species, 1185–1186
process affecting community structures, 1188–1194
succession, 1197–1200
trophic cascades, 1195–1197
trophic levels, 1186–1187
See also Communities; Community ecology
Ecology, 9, **1113–1114**
*Eco*RI enzyme, 354, 358, 381
Ecosystem engineers, **1196**
Ecosystem Marketplace, 1240
Ecosystems, 9, **1113**
atmospheric CO_2 and plant carbon storage, 1204–1205
goods and services examination of, 1221–1222
managing sustainability, 1223
productivity and species richness, 1200–1201
restoring, 1236–1237, 1238, *1239*
Ecosystem services, plants and, **646**
Ecotourism, 1237, 1241
Ectocarpus sp., *601*
Ectoderm, 673, *926,* **927**
Ectomycorrhizae, 657
Ectoparasites, 679
Ectopic pregnancy, 925
Ectoprocts, *683, 692, 693,* **695**
Ectotherms, **861,** 862–863, 864, 865, *866*
Edema, 1058, 1071
Edge effects, **1231**
Ediacarans, *473*
EEG. *See* Electroencephalogram
Effector cells, 409
Effector proteins, **338**
Effectors, **856,** 1004–1005, 1020–1021
See also Muscle contraction; Muscle(s); Skeletal systems
Efferent arterioles, renal, **1100,** 1102, 1107
Efferent nerves, **985**
Efferent neurons, **943**
"Efficiency genes," 1068
EGF growth factor, 439
Egg cytoplasm, fertilization and, 921–922
Egg(s)
amniote, 730, 905
blocks to polyspermy, 902
contribution to the zygote, 921–922
energy store/size trade-off, 680–681
enucleating, 430
external fertilization systems, 903
in human reproduction, 909, 910
internal fertilization systems, 903–904
meiosis and, 198
ovarian cycle, 910–911
oviparity, 905
polarity, 436–437
See also Ova
Egrets, *1193,* 1194
Ejaculation, **908,** 914
EKG. *See* Electrocardiogram
Elaphe guttata, 1076
Elasticity, 807
Elastin, 858, *1058*
Elderberry, *760*

Electrical synapses, **955,** 958
Electric catfish, 1021
Electric charge, effects on diffusion, 105
Electric currents, cell membranes and, 947, 951
Electric eels, 535, 1021
Electric organs, 992, 1021
Electric potential difference, 947
Electric signals/signaling, 535, 1157
Electrocardiogram (ECG or EKG), *1054,* **1055**
Electrochemical gradients
  component forces in, **948**
  described, 108–109
  plant uptake of mineral ions and, 767–768
Electrodes
  measuring membrane potential with, 947
  patch clamps, 950
Electroencephalogram (EEG), **998**
Electromagnetic radiation, **163**
Electromagnetic spectrum, 163, *164*
Electromyogram (EMG), 998
Electron carriers, 141–142
  See also Electron transport systems
Electron donor, in photosynthesis, 166
Electronegativity, **27**
Electron microscopy, *71,* 72
Electron probe microanalyzer, 774
Electrons
  characteristics, **21**
  in covalent bonds, 25–27
  orbitals and bonding, 23–25
Electron shells, **24–25**
Electron transport systems
  in cellular respiration, **140,** *141*
  cyclic, 168
  energy yield from, *153*
  noncyclic, 166–167
  in oxidative phosphorylation, 148, 149–150, *151*
  in photosynthesis, 166–168, *169*
Electrooculogram (EOG), 998
Electrophoresis, 253, 354, *355*
  See also Gel electrophoresis
Electroreceptors, 965, *966*
Elementary bodies, 572
Elements (chemical), **21–25**
Elephantiasis, 704
Elephant Pepper Development Trust, 1242
Elephants
  communication, 964
  ivory trade, 1239
  ovarian cycle, 910
  poaching, 1232–1233
  pygmy, 716
  teeth, 1076
  thermoregulation, *863*
Elephant seals, *9, 1063,* 1173, *1174*
Elephant's foot, *1130*
Elk, 841, 1195–1196
Elkhorn coral, 1234
Ellipsoid joint, *1019*
Elms, 1233
Elongation factors, 269
Embioptera, 709
Embolism, **1060**
Embolus, **1060**
Embryo
  human, *925,* 936
  insights into dinosaur evolution, 670
  modularity in, 451
  overview of, **427**
  in plants, 427, 613, *820,* 823–824
  totipotency in animals, 430
Embryonic induction
  lens differentiation in the vertebrate eye, 437
  vulval differentiation in the nematode, 437–439
Embryonic stem cells, 433–434, 926
Embryophytes, **611**
Embryo sac, in flowering plants, *820,* **821**
Emergent properties, 396
EMG. *See* Electromyogram
Emiliania, 584
  E. huxleyi, 586, *587*
Emission, **908**
Emlen, Stephen, 1156
Emotions
  amygdala and, 942–943
  limbic system and, 988
Emperor penguins, 1145
Emphysema, 1034
Emu, 732
Emulsifiers, 1082
Enamel, in teeth, **1076**
Enceldaus, 20
ENCODE project, 395
Encyclopedia of DNA Elements, 395
Endangered species, 1230, 1239
Endemic species, **1129**
  extinctions and, **1228**
  Hawaiian Islands, 521
  Madagascar, 1129, *1130*
  protected areas and, 1235
Endergonic reactions, **122,** 124–125
endo16 gene, 435
Endocrine cells, **875–876**
Endocrine glands, **876**
Endocrine system, **875**
  adrenal gland functions, 887–889
  gonads, 889
  of humans, *880*
  interactions with the nervous system, 880–883
  negative feedback loops and, 883
  pancreatic functions, 887
  pineal gland functions, 890
  regulation of blood calcium levels, 885–886, 887
  thyroid functions, 883–885
Endocytosis, *101,* **113–114**
Endoderm, 673, *926,* **927**
Endodermis, **756,** 768, *769*
Endomembrane system, **79–82**
Endometrium, *909,* **910,** 911, 912, 925
Endoparasites, 679
Endoplasmic reticulum (ER)
  movement of polypeptides into, 273–274
  in plant and animal cells, *76, 77*
  protein transport and, 80–81
  proteolysis and, 274
  structure and function, 75, **79–80**
  See also Sarcoplasmic reticulum
Endorphins, **882,** *959*
Endoskeletons, 605, **1017–1018**
Endosperm, *820*
  characteristic of angiosperms, **638**
  formation, 638, **822,** 823
  function, 642
  utilization by the embryo, **800–801**
Endospores, **573–574**
Endosymbiosis, **585**
  bacteria in animal guts, 1077
  chlorophytes and, 602
  chloroplasts and, 11–12, 92, 589–590
  digestion of cellulose in ruminants and, 1085
  eukaryotic evolution and, 588–590
  mitochondria and, 588
  transposons and, 312
Endosymbiosis theory, **92**
Endothelium, 113, 344
Endotherms, **861**
  adaptations to cold, 867–868
  basal metabolic rate, 866–867
  blood flow to the skin, 864
  heat production, 861–862
  thermoregulation, 863
Endotoxins, 73, **579**
End-product inhibition, **133**
Endymion non-scriptus, 554
Energetic costs, 1148
Energy, **30, 119**
  calorie measure, 1069
  chemical equilibrium and, 123
  chemical reactions and, 30–31
  distribution in ecosystems, 1187, *1188*
  forms, 119–120
  global flow in, 1209–1210, *1211*
  laws of thermodynamics, 120–122
  metabolic rates, 1069–1070
  storage in animals, 1071
  transformation by chloroplasts and mitochondria, 82–83, *84, 85*
  See also Biochemical energetics, Free energy
Energy barrier, 126, 127
Energy budgets, **863–864, 1070–1071**
Energy-coupling cycle, 124–125
Energy flow, global, 1209–1210, *1211*
Energy levels (electrons), 24–25
English elm, 833
engrailed gene family, 537–538
Enhancer sequences, **320**
Enkephalins, **882,** *959*
Enterococcus, 563
Enterokinase, 1083, *1084*
Enteromorpha, 1199
Enthalpy, **120–121**
Entrainment, **828,** 829, **1153**
Entropy, **120–121,** 122
Enucleated eggs, 430
Enveloped viruses, 287
Environment
  ecological meaning of, **1113**
  gene expression and, 220–221
  modulation of development in organisms, 454–457
  quantitative trait loci and, 221–222
  See also Internal environment
Environmental carrying capacity, **1173,** *1174*
Environmental toxicology, **1089**
EnvZ receptor protein, 334, 335
Enzymes, **118**
  active site, 127, 129
  as catalysts, **125**
  digestive, 1077, 1080, 1082, 1083, *1084*
  dysfunctional, 375–376
  effect of cell signaling on, 346, *347*
  nonprotein "partner" molecules, 129–130
  regulation, 131–135
  relationship to genes, 257–258, *259*
  restriction, 353–354, 358, 359, 381
  significance to metabolic pathways, 131
  specificity, 126–127
  structure-function relationship, 129
  substrate concentration and reaction rate, 130
  types of interactions with substrates, 128–129
Enzyme-substrate complex (ES), **127**
Eocene period, *478*
EOG. *See* Electrooculogram
Eosinophils, *403, 1056*
Ephedra, 648
Ephedrine, *648*
Ephemeroptera, 709
Epiblast, *925,* **931,** 932
Epiboly, **928**
Epidemiology, 96
Epidermal growth factor (EGF), 439
Epidermal hairs, xerophytes and, 845
Epidermis, **748, 756,** 761
Epididymis, *906,* **907**
Epiglottis, 1079, *1080*
Epilepsy, 999, 1000
Epinephrine, *880*
  activation of glycogen phosphorylase and, 340, 341
  blood pressure regulation and, 1062, *1063*
  characteristics, **887–888**
  fight-or-flight response, 333, 879, 888
  half-life, 891
  signal transduction and, 338–339, *892,* 893
  stimulation of glycogen metabolism, 346, *347*
Epiphyseal plates, *1018*
Epiphytes, 1127
Episodic reproduction, 1175, *1176*
Epistasis, **219–220,** 840
Epithelial cells, basement membrane and, 91
Epithelial tissues, 102–104, **857–858,** *860*
Epitopes, 407–408
Epstein-Barr virus, *387,* 422
Eptatretus stouti, 724
Equatorial plate, **189**
Equilibrium
  chemical, 122, 123
  Hardy-Weinberg, **492–493,** 494
  organs of, 974–975
  of solutions, 105
Equisetum, 625
ER. *See* Endoplasmic reticulum
era gene, 774
Erectile dysfunction, 908
Erection, penile, 344, **908**
Erethizon dorsatum, 1134
Ergot, 666
Ergotism, 666
Erinaceomorpha, *736*
Error signal, **856**
Erwin, Terry, 708–709
Erwinia, 370
Erysiphe sp., *666*
Erythrocytes, *403,* **1055**
  See also Red blood cells
Erythromycin, 302
Erythropoietin, 187, *368,* **1055**
Escherichia, 579
Escherichia coli (E. coli), 284, 563, 575
  cell division in, 181
  circular chromosome in, 182

comparative genomics, 301
conjugation, 291
contamination of Lake Erie, 1219
culturing, 290–291
DNA polymerases, 246
doubling time, 182, 290, 566
genome, 307, 309, *533*
lac operon, 297–298, 299–300
noncoding DNA in, *534*
recombinant DNA technology and, 358, 360, 361
regulation of lactose metabolism in, 296, 297–298
restriction enzymes and, 354
signal transduction pathway in, 334–335
speed of DNA replication in, 244, 247
strain O157:H7, 302
T7 phage and, 354
transposable elements and, 295
Esophagitis, 653
Esophagus
echinoderms, 719
vertebrates, **1079,** *1080*
Essay on the Principle of Population, An (Malthus), 488
Essential amino acids, **1072**
Essential elements, **782–785**
See also Macronutrients; Micronutrients
Essential fatty acids, **1072**
Ester linkages, *54,* **55**
Estivation, 1099
Estradiol, *888,* 889
Estrogen(s)
birth control pills and, 915
as cell signal, 336
in humans, *880,* 912
ovarian and uterine cycling, 911, 912
sex determination and, **889**
Estrus, **911**
Ethanol, *40*
Ether linkages, 576
Ethics, 16
Ethnobotanists, 648
Ethology, **1141**
Ethyl alcohol, 118–119, 147, 148
Ethylene, 797, *798,* **810,** *811*
Etiolation, **813,** *814*
Etoposide, *391*
Euascomycetes, **665**
Eucalyptus, 846
Eucalyptus regnans, 769
Eucera longicornis, 1194
Euchromatin, 323
Eudicots, *646*
in angiosperm phylogeny, *644,* **645**
early shoot development, *799*
key characteristics, **745,** *746*
leaf anatomy, *761*
root structure and anatomy, *756*
stem anatomy, *757*
Euglena, 584, 604
Euglenids, *584,* 589, 590, *597,* **603–604**
Eukarya, **12,** 72, 561–562
Eukaryotes, **72**
cell division in, 182–183, 184–187
compartmentalization in, 74–75
cytoskeleton, 86–90
endomembrane system and organelles, 75–86
endosymbiosis theory and, 92
evolution, 11–12, 92, 588–590
extracellular structures, 90–91
gene expression pathway, *308*
location of metabolic pathways in, *141*
location of phospholipid synthesis in, 100
major groups, *584,* 596–607
prokaryotes compared to, 561–562
ribosomes in, 268
structure and organization, *76–77*
Eukaryotic genome
compared to prokaryotic, 307–308
functions of, *310*
model organisms, 308–311
repetitive sequences, 311–312
Eulemur coronatus, 738
Eumetazoa, 673, **683**
Euphydryas editha bayensis, 1178
Euphylophytes, **619**
Euplectella aspergillum, 684
Euplotes sp., *598*
European sycamore, 515
Euryarchaeota, *571,* **575,** 577
Eurylepta californica, 1027
Eusocial insects, 1159–1160, 1161–1162
Eusociality, **1159–1160,** 1161–1162
Eustachian tube, *972,* 973
Eutherian mammals
cleavage in, 923, *924,* 925
overview of, 735–737
See also Placental mammals
Eutrophication, *573,* **1217,** 1219
Evans, Charles, 1024
Evans, Ron, 138
Evaporation, **32**
in heat transfer, **863,** 864
hydrological cycling and, 1211
in thermoregulation, 868
Evolution, **3**
adaptation, 489, 497–500
constraints on, 503–505
continental drift and, 468
convergent, 545
Darwinian theory, 5–6, 487–489
dating fossils, 464, 465–467
developmental genes constrain, 457–459
of eukaryotes, 11–12, 92, 588–590
extraterrestrial events and, 471
forces that maintain genetic variation, 501–503
genetic switches and, 453
global climate changes and, 470–471
human, 396, 479, 738–741
human effects on, 505
impact of photosynthesis on, 11
"living fossils," 479
major events in, 471–479
mechanisms of change, 494–497
of multicellularity, 12
mutation and, 277–278
origin of life, 10–11
parallel, 458–459
population genetics and, 489–493
radiations *See* Evolutionary radiations
rates of change, 479–482
Tree of Life concept, 12–13
in vitro, 538, *539*
volcanoes and, 471
Evolutionary developmental biology, 449
Evolutionary radiations
faunal, *478,* 479
of mammals, 734, 736
overview of, **520–521**
in primates, 737–738
See also Adaptive radiations
Evolutionary reversals, **545**
Evolutionary trends, 498
Excavates, *584, 597,* **602–604**
Excision repair, for DNA, **249,** *250*
Excitatory synapses, **957**
Excited state, **163,** *164*
Excretory organs, 1093, 1094, 1097–1098
See also Kidneys
Excretory systems, 1092–1093
adaptations to salt and water balance, 1094–1095
excretion of nitrogenous wastes, 1095–1096
of invertebrates, 1097–1098
principles of water movement, 1093–1094
of vertebrates, 1098–1101
See also Kidneys
Exercise
effect on bone, 1018
effect on breathing, 1035
effect on muscle, 1014
Exergonic reactions, **122,** 124–125, 126
Exocrine glands, **876**
Exocytosis, *101,* 113, **114**
Exons, **313**
Exoskeletons, **693–694, 1016**
Exosome, **325**
Exotic species
controlling, 1180–1181, 1239–1240
effects of, 1166–1167, 1233–1234, 1241
Exotoxins, **579**
Expanding triplet repeats, **382–383**
Expansins, 808
Experiments, types of, **14–15**
Expiratory reserve volume, **1032,** 1035
Exploitation competition, **1192**
Exponential growth, **1173,** *1174*
Expression vectors, **367–368**
Expressivity, **221**
Extensors, 988, **1019**
External fertilization, **903**
External gills, *718,* **1028**
Extinction. *See* Mass extinctions; Species extinctions
Extinction rates, 480, 482, 1132
Extracellular fluid, 1046
Extracellular matrix, **91,** 858, 859
Extraembryonic membranes
allantois, 936
amniote egg, 730
avian hypoblast and, 931
formation of, **934–935**
genetic screening and, 936
hormones from, 912
mammalian, 932
placenta, 935–396
Extrafloral nectar, 1184
Extraterrestrial life, 32, 61
Extreme halophiles, **577**
Eye-blink reflex, 999
Eye cups, **977**
eyeless gene, 448
Eyes
adaptations to diurnal *vs.* nocturnal habitats, 1153
anatomy of, 978–979
axon number and size in, 954
correction of misalignments, 938
development, 448–449
focusing, 979–980
lens differentiation in vertebrates, 437
vertebrate retina, 980–981

- F -

F_1 generation, 209
F_2 generation, 209
Face recognition, temporal lobe and, 990
Facilitated diffusion, 105, **108–109,** 110, *111,* 767–768
Factor VIII, *368*
Facultative anaerobes, **568**
FAD. *See* Flavin adenine dinucleotide
Fallopian tube, 909, 925
Familial hypercholesterolemia (FH), 114, 376–377, 379, 391
Family systems, 1159
Fan palms, 495
Far-red light, seed germination and, 812–813
Fas receptor, 417
Fast block to polyspermy, 902
Fast-twitch fibers, 138, **1013–1014,** 1015
Fate. *See* Cell fate
Fate maps, **925**
Fats
absorption in the small intestine, *1083,* 1084
digestion, 1082, *1083*
energy storage and, 155, 1071
liver and, 1086–1087
overview of, **55**
Fat tissue. *See* Adipose tissue
Fatty acids
essential, 1072
generation through anabolic pathways, 154–155
overview of, 55
synthesis, 157
Faunas, through evolutionary history, 472–479
Fear response, 942, 988, 1000
Feathers
evolution of, 733, *734*
as insulators, 868
phylogenetic analyses, 545
tracking birds and, 1168, *1169*
Feather stars, 475, 719
Feces, **1077,** 1084, 1085
Fecundity, 1169
Feedback
regulation of breathing and, 1039–1040
See also Negative feedback; Positive feedback
Feedback information, **856**
Feedback inhibition, **133,** *134*
Feedforward information, **857**
regulation of breathing and, 1040
Feeding behaviors, 1149–1152, 1161–1162
Feet. *See* Foot
Felis concolor coryi, 307
Female sex organs, human, 909–910
Fenestrations, in capillary walls, 1058
Fermentation
described, **140,** 147–148
energy yield from, 153
foregut, 532
in metabolic pathways, *141*
phosphofructokinase and, 156

in water-saturated roots, 846
Ferns
characteristics, *612,* **626**
evolution, 618, 619
fossil, *475*
life cycle, *194,* 626–627
Ferredoxin, **167,** 168, 172
Ferritin, 326
Ferroplasma, 576
Fertile Crescent, 371
Fertility factors, **294**
Fertilization
activation of development, 921–926
in alternation of generations, 594
in animals, **901–903**
blocks to polyspermy, 902
calcium ion channels and, 343
creation of zygotes, **193**
egg activation, 903
external systems, 903
in humans, 910
internal systems, 903–904
in plants, 615
specificity in sperm-egg interactions, 901–902
See also Double fertilization
Fertilization cone, 901, *902*
Fertilization envelope, 903
Fertilizers
global nitrogen cycle and, 1217
global phosphorus cycle and, 1219
industrial nitrogen fixation, 789–790
overview of, 787
to treat mineral deficiencies in plants, **783**
Fetal hemoglobin, 1037, 1039
Fetscher, Elizabeth, *641*
Fetus, **889**
development of, **936–938**
γ-globin and, 315, *316*
preimplantation screening, 385
sexual development and, 889
thalidomide deformities, 457
Fevers, 135, 406, 869–870
F factors, **294**
Fibers (muscle). *See* Muscle fibers
Fibers (plant), **638, 750,** *751*
Fibrin, **1056,** *1057*
Fibrinogen, 908, **1056,** *1057*
Fibrinolysin, 908
Fibrous root systems, 746, *747,* 846
Ficedula albicollis, 1148
Fick's law of diffusion, **1026**
"Fiddleheads," 626
Fight-or-flight response, **879,** 888, 891, 992
Figs, 810, 1197
Figure (wood), 760
Filamentous actinobacteria, 788–789
Filaments
associations of prokaryotes, **563**
axial, 566, 571–572
intermediate, 87–88
in skeletal muscle cells, *858*
of stamens, **638,** *639,* 640, *820,* 821
See also Actin filaments; Gill filaments; Myosin filaments
Filariasis, 704
Filopodia, 927
Filter feeders, **676–677, 1075**
Finches, 480, *481*
Finned fishes, 726
See also Ray-finned fishes
Fins, 725–726, 728, 904
Fiordland National Park, New Zealand, *1119*
Fire
biodiversity and, 1237–1238
cycling of chemical elements and, 1212–1213
humans and, 1212, 1213
seed germination and, 799, *800*
See also Forest fires
Fire-bellied toads, 517–518
Fireflies, 1021, 1157
Fireworms, *675*
Firmicutes, 573–574
First filial generation (F_1), **209**
First polar body, **900**
Fischer, Edmond, 340
Fischer, Emil, 129
Fish eggs, 903
Fisheries
managing sustainability, 1223
overharvesting, 1180
Fishes
acid precipitation and, 1218
body plan, *725*
cartilaginous, 1017, 1099
chondrichthyans and chimaeras, 726
circulatory system, 1047–1048
electric signaling, 1157
elevation of body temperature in, 864–865
evolutionary history, 474, 475, 476
fins, 725–726
gills, 1029–1030
gustation in, 969
jawless, *723,* 724
jointed-fins, 728
lateral line sensory system, 974–975
maintenance of salt and water balance in, 1099
numbers of ribosomes in, 324
plankton-eating, 1161
respiratory gas exchange and, 1025, *1027*
respiratory system, 1029–1030
in vertebrate phylogeny, *723*
See also Ray-finned fishes
Fishing industry. *See* Fisheries
Fission, in unicellular fungi, 659
Fitness, **497**
feeding behavior and, 1149–1152
habitat choice and, 1147–1148
individual and inclusive, 1159
navigational mechanisms and, 1154–1156
responding to environmental cycles, 1152–1154
territorial defense and, 1148–1149
Flagella
chytrid fungi and, 663
in eukaryotes, 591
in prokaryotes, **73–74, 566,** *567*
structure and function, 88–89
in unikonts, 605
Flagellin, 73, 566
Flame cells, 1097
Flamingos, 677
Flatworms, *675, 683, 692, 1027*
described, **695–696**
excretory system, 1097
eye cups, 977
Flavin, *129*
Flavin adenine dinucleotide (FAD)
in catalyzed reactions, *129*
citric acid cycle, 145, *146,* 147
electron carrier, **142**
electron transport chain and, 148, 149, *150*
Flavonoids, 789, *839*
Fleas, 679, *709,* 1004
Fleming, Alexander, 367
Flexors, 988, **1019**
Flies
chemoreceptors, 968
eye development genes and, 448
Flight
in birds, 733
insects and, 710
Flocking, 1158
Flora, **472**
through evolutionary history, 472–479
Floral meristems, **825**
Floral organ identity genes, **825–826**
Flores Island, 716
Florida panthers, 306–307
Florida scrub jays, 1159
Floridean starch, 602
Florigen, 798, **829–831**
Flotation bladders, 600
Flounder, 504, 1020
Flowering
apical meristems and, 825
circadian rhythms and, 828–829
florigen, 829–831
gene cascades affecting, 825–826
at night, 828
overview of, **798**
photoperiodic cues and, 826–828
vernalization, 831
Flowering dogwood, *646*
Flowering hormone, 798, 829–831
FLOWERING LOCUS T (FT) gene, 831
Flowering plants. *See* Angiosperms
Flowers
evolution in, 640–641
organ identity genes and, 440–441
pin and thrum, 496
structure of, **638–639,** 820–821
Flu, 282–283, *409,* 539
Fluid feeders, **1075**
Fluid mosaic model, **97–101**
Flukes, 696
Fluorescence, physics of, 165
Fluorescence microscopy, *71*
Fluorine, *1073*
5-Fluorouracil, *391*
Flu pandemics, 283
Flycatchers, 1148
Flying reptiles, 476–477
Flying squirrels, 980, 1153
FMR1 gene, 383
FMR1 protein, 383
Foliar sprays, 787
Folic acid, *1074*
Foliose lichens, 656
Folk medicines, exploitation of species and, 1233
Follicles
ovarian, **910,** 911
thyroid, 883, *884,* 885
Follicle-stimulating hormone (FSH)
anterior pituitary and, *880*
male sexual function and, *908,* 909
ovarian and uterine cycling, 911, 912
puberty and, **889,** 890
Fontanelle, 1018
Food
acquired carbon skeletons, 1072
animal strategies for acquiring, 676–679
calories, 1069–1070
energy budgets and, 1070–1071
energy storage in animals, 1071
foraging theory, 1149–1151
ingestion, 1075–1076
mineral elements, 1073
nonnutritive items, 1151–1152
nutrient deficiencies, 1074–1075
plants as primary sources, 646
regulating uptake of, 1087–1088
vitamins, 1073–1074
Food chains, **1187**
Food production, significance to global economy, 1222
Food tube. *See* Esophagus
Food vacuoles, 86, 113, **591,** *592*
Food webs, **1187**
"Foolish seedling" disease, 801–802
Foot
gastropods, 702
mollusks, 700
webbing, 451, 452, *453*
Foraging, 1149–1151, 1161
Foraging theory, 1149–1151
Foraminiferans, *584,* 587, 592, *597,* **604–605**
Forbs, 1121
Forebrain, **986,** 987, 988–989
Foregut fermentation, 532
Forelimbs, wing evolution and, *458*
Forensics, 356, *357*
Foreskin, 908
Forest fires, 1197, 1237–1238
Forest management, 1238
Forest patches, 1241, *1242*
Forests
atmospheric CO_2 and carbon storage, 1204–1205, 1214–1215
biomass and energy distribution, 1187, *1188*
boreal, 1119, 1175
mosaics following fires, 1197
primary production in, *1210*
stewardship management, 1238
synchronous and episodic reproduction of trees, 1175
temperate deciduous, 1120
temperate evergreen, 1119
thorn, 1125
tropical deciduous, 1126
tropical evergreen, 1127, 1230, 1231
Forest Stewardship Council (FSC), 1238
Formisano, Vittorio, 560
Forsythia, 832
fosB gene, 1143–1144
Fossa, *1130*
Fossil fuels
acid precipitation and, 1218
atmospheric carbon dioxide and, 471
formation of, **1213**
global nitrogen cycle and, 1217
global sulfur cycle and, 1218
origin of, 618
Fossils, *10,* **465**
dating, 464, 465–468
"living," 479
organisms not preserved, 472
phylogenetic analyses and, 548
rates of evolutionary change and, 479–480
Foster, Stephen, 220

Founder effect, **495,** *496,* 511, *513*
Fovea, *979,* **980**
Foxglove, *648*
Fragile-X syndrome, 379, 382–383
Fragmented habitats, **1231**
Frame-shift mutations, **276**
Franklin, Rosalind, 238
Frank-Starling law, **1059**
Free energy, **120–121**
 in anabolic and catabolic reactions, 122
 ATP hydrolysis and, 124
 changes during glycolysis, *144*
 chemical equilibrium and, 123
 citric acid cycle, *145*
 enzymatic reactions and, 127
 glucose metabolism and, 139–140
Free radicals, 277
Free-running circadian clock, 1153
Free-tailed bats, 1242
Freeze-fracture microscopy, *71,* **99,** *100*
Freezing, plant adaptations to, 847
French Canadians, 374
"French fry fuel,"55
Frequency, in population genetics, **491**
Frequency-dependent selection, **501–502**
Freshwater environments
 dispersal of organisms, 1136
 eutrophication, 1217
 mineral nutrient flow and turnover in, 1206–1208
 sustainable water use, 1223
Frigate bird, *734*
Frilled lizard, *732*
Frisch, Karl von, 1157
Frogs
 acquisition of poisons, 1151
 cell fate, 428–429
 cleavage in, 923, *924*
 estivation, 1099
 evolutionary reversal in, 545
 gastrulation in, 928
 jumping, 1004, *1005*
 life cycle, *729*
 limb abnormalities, 2, 14, 15
 neurulation in, *933*
 numbers of ribosomes in, 324
Fromia, 898
Frontal lobe, **990–991**
Fructose
 production in plants, 171
 in semen, 908
 structure of, *50*
 in sucrose, *51*
Fructose 1,6-bisphosphate, 53, 142, *143*
Fruit fly, 309–310
 See also Drosophila melanogaster
Fruiting bodies, 606, 652
Fruits
 angiosperms and, **638**
 auxin and, 806
 effect of ethylene on ripening, 810
 seed dispersal and, **824**
 types of, 643
Fruticose lichens, 656
Fucoxanthin, 599
Fuel metabolism. *See* Glucose metabolism
Fugu, 538
Fugu rubripes, 309
Fulcrums, 1019, *1020*
Fulton, M., 516
Fumaroles, 1217
Functional genomics, **301**
Functional groups, 39, *40,* 41, 53
Fungi, *584, 597*
 associations with animals, 658
 association with plants, 613, 614
 body of, 651–652
 characteristics of major groups, 663–667
 chitin in, 53
 decomposers, 655
 endophytic, 658
 environmental conditions and, 652
 evolutionary context, 651, *652*
 heterotrophs, 13
 "imperfect," 663
 lichens and, 655–656
 life cycles, *194,* 662–663
 mutualistic relationships, 651, 655
 mycorrhizae and, *664,* 789
 nutrient sources, 653–654
 opisthokont clade and, 677
 overview of, 651
 phylogenetic relationships, 651, *652*
 in secondary succession, 1198–1199
 sexual and asexual reproduction in, 659–662
 sizes attained by, 653
 unicellular, 651–652
 used against pests, 650
Funk, Casimir, 1075
Fur, 735, 864, 868
Furchgott, Robert, 344
Furcula, 732
Fusarium, 655
 F. moniliforme, 663
 F. oxysporum, 650, 666
fushi tarazu gene, 444
Fusicoccum, 1198
Fynbos, 1240–1241

- G -

G1 phase, **184,** 185, 186, 388–389
G2 phase, **184,** 185, 186
G3P. *See* Glyceraldehyde 3-phosphate
GABA, 958, *959*
Gage, Phineas, 991
Galactia elliottii, 1220
Galactosamine, 53
Galactose, *50*
β-Galactosidase, 296
β-Galactoside permease, 296
β-Galactoside transferase, 296
Galápagos Archipelago, 511, 1170
Galápagos hawk, *4*
Gallbladder, *1079,* **1082,** *1083*
Gametangia, 613, **616,** 618, *660,* 664
Gametes, **183**
 in alternation of generations life cycles, 594, *595*
 gametogenesis, 899–901
Gametocytes, *586*
Gametogenesis, **899–901**
Gametophyte generation, 615, 616
Gametophytes
 alternation of generations life cycles, 194, **594,** *595,* **615**
 flowering plants, *820,* 821
 seed plants, 631–632, **633**
γ-Aminobutyrate (GABA), 958, *959*
Gamma radiation, 23
Ganges River, 836
Ganglia (nerve cluster), 698, **943,** *944,* 992
Ganglion cells (retina), **980,** 981
Ganoderma applanatum, 667
Gap genes, **443**
Gap junctions, **104, 348,** 958, 1010
Gap phases. *See* G1 phase, G2 phase
Garden warblers, 1178–1179
Garner, W. W., 826
Garter snakes, 486, 498
Gases
 greenhouse, 577, 1208
 as neurotransmitters, 958, *959*
 respiratory, 1025
 See also specific gas
Gas exchange systems, **1028**
 See also Respiratory gas exchange; Respiratory systems
Gasterosteus aculeatus, 459
Gastric juice, 405
Gastric pits, **1080,** *1081*
Gastrin, **1086**
Gastritis, 1075
Gastrointestinal disease, 575
Gastrointestinal systems
 absorption of nutrients, 1083–1084
 absorption of water and ions, 1084
 absorptive and postabsorptive periods, 1086
 autonomic nervous system and, 992
 cellulose digestion, 1084–1085
 chemical digestion, 1080, *1081,* 1082–1083
 emptying of the stomach, 1082
 endocrine cells, 875–876
 epinephrine and, 339
 gastrovascular cavities, 1076–1077
 hormones and, 1086
 human, *1078*
 intrinsic nervous system, 1086
 large intestine, 1084
 liver, 1082, *1083,* 1084, 1086–1087
 peristalsis, 1079–1080
 prokaryotic fermentation and, 148
 rumen, 148, 579, 1084–1085
 small intestine, 1082–1084
 smooth muscle contraction, 1010–1012
 stomach ulcers, 1081–1082
 swallowing, 1079, *1080*
 tissue layers, 1078–1079
 tubular guts, 1077, *1078*
 turnover time of epithelium, 858
Gastropods, *700,* **702**
Gastrovascular cavities, **685, 1076–1077**
Gastrulation, **673**
 in birds, 931–932
 dorsal lip of the blastopore and, 928–929, *930*
 in frogs, 928
 overview of, **926–927**
 in placental mammals, 932
 primary embryonic organizer, 929–931
 vegetal pole invagination, 927–928
Gas vesicles, in cyanobacterium, *566*
Gated ion channels
 cardiac muscle contraction and, 1010
 effect on membrane potential, 950–951
 overview of, 108–109
 as signal receptors, 337
 See also Voltage-gated ion channels
G cap, **316,** 326
GDP. *See* Guanosine diphosphate
Geese
 high-altitude flight, 1024
 imprinting, 1145
Gehring, Walter, 448
Gel electrophoresis, **354,** *355*
 two-dimensional, 395, *396*
Gemmae, 622
Gemmae cups, 622
Gene amplification, **324,** 362
Gene duplication, **534–535**
Gene expression
 amplification of genes and, 324
 cell differentiation and, 435
 central dogma and, 260–261
 coordinating several genes, 320–322
 eukaryotic model of, *308*
 human evolution and, 396
 points for regulating, *318*
 posttranscriptional regulation, 324–326
 regulating after translation, 326–327
 regulating by changes in chromatin structure, 322–323
 regulating during translation, 326
 regulation in prokaryotes, 296–300
 regulation in viruses, 289–290
 See also Differential gene expression
"Gene expression signature," 366
Gene families
 concerted evolution and, 535–536
 engrailed genes, 537–538
 overview of, **315,** *316*
 See also Globin gene family
Gene flow
 allele frequencies and, 494
 allopatric speciation and, 511–513
 sympatric speciation and, 513–515
Gene-for-gene resistance, **839–840**
Gene knockout experiments, 364–365, 1143–1144
Gene libraries, **362–363**
Gene pool, **490**
Generative cell, 821, 822, *823*
Genes, 7, **212**
 amplification, **324**
 artificial life and, 302–303
 dominant and recessive, 213
 duplication, 534–535
 environmental effects and, 220–221
 epistasis and, 219–220
 in eukaryotes, 312–315
 evolution in (*see* Molecular evolution)
 genomes and, 525
 hybrid vigor, 220
 identified as DNA, 233–237
 inserting into cells, 359–362
 knockout experiments, 364–365, 1143–1144
 linkage groups, *222,* 223
 linked and unlinked, 214
 locus, **212**
 markers, 237
 Mendelian unit of inheritance, **210**
 multiple alleles, 218
 noncoding sequences, 313–315
 one-gene, one-polypeptide relationship, 258
 in organelles, 228–229
 orthologs and paralogs, 537–538

phenotype and, 212–213
relationship of proteins to, 257–258
strategies for isolating, 380–382
"thrifty," 1068
transcription (*see* Transcription)
transfer, 294–295
See also Alleles
Gene silencing, 365
Gene therapy, **391–392**
Genetically modified crops, 369–371
Genetic code, **263–264,** 265
"second," 267
Genetic diseases
of dysfunctional proteins, 375–377
genetic screening for, 384–385, *386*
genomic imprinting and, 383
identifying genes affecting, 380–382
multifactorial, **379**
mutations and, 382
patterns of inheritance in humans, 379
of protein conformation, 378–379
studies of French Canadians and, 374
treating, 391–392
Genetic diversity. *See* Genetic variation
Genetic drift, **494–495,** 531
Genetic engineering. *See* Biotechnology; Recombinant DNA technology
Genetic maps, **224**
Genetic markers, 359, 361–362, 381–382
Genetic mutations. *See* Mutation(s)
Genetic recombination, **223–224**
imprecise, 418–419
in prokaryotes, 290–295
in virulent viruses, 286
Genetics
alleles, 211, 217–219
animal behavior and, 1142–1147
early assumptions about inheritance, 207
epistasis, 219–220
gene-environment interactions, 220–221
genetic recombination, 223–224
human pedigrees, 216–217
hybrid vigor, 220
law of independent assortment, 213–214, *215*
law of segregation, 211–213
linkage groups, 223
Mendel's experiments, 207–210, 213
organelle genomes, 228–229
probability calculations, 214–216
sex-linked inheritance, 225–228
sex-linked traits, 206, *207*
Genetic screening, **384–385,** *386,* 395, 430, 936
Genetic structure, **492–493**
Genetic switches, 451, 453
Genetic variation
germ line mutations and, 277
mechanisms that maintain, 501–503
meiosis and, 199
mutation and, 494
in populations, 490–491
sexual reproduction and, 899
Gene transcription. *See* Transcription
Gene trees, 535, **537–538**
Genital herpes, *917*
Genitalia, **904**
Genital warts, 387, *917*
Genome, **7, 210, 525**
evolution in, 525–529, 533–534
model organisms, 308–311
of organelles, 228–229
prokaryote, 300–303, 307, *308*
"prospecting," 374, 382, 395
See also Eukaryotic genome; Human genome
Genome sequencing, 393, *394*
Genomic equivalence, **430,** 432
Genomic imprinting, **383**
Genotype, **211**
population evolution and, 490
population genetics and, 491–492
sexual recombination and, 501
Genotype frequencies
calculating, 491–492
nonrandom mating and, 495–496
Genus, 12, **554**
Geographic speciation, 511–513
Geology
continental drift, 468
major historical divisions, *466–467*
George III, 378
Georges Bank, 1180
Geospiza spp., *512*
Gerbils, 1105
German measles, *409*
Germ cells, **899,** *900*
Germination
of pollen grains, 822
of seeds, 630, **798,** 799, 800, *801,* 812–813
Germ layers, **927**
Germ line mutations, **275,** 277
Gestation, **936**
Gey, George and Margaret, 180
Gharials, 732
Ghrelin, **1088**
Giant baobob tree, *1130*
Giant clam, *701*
Giant kelp, *583,* 599
Giant panda, *1070*
Giant petrel, *1095*
Giant sequoias, *635,* 1176–1177
Giant tortoise, *4*
Giardia, 584, 603
Gibberella fujikuroi, 801–802
Gibberellins, 663, *798,* 801–803, 808–809
Gibbons, 738, *739*
Gigantism, 882
Gigartina canaliculata, 1199
Gill arches, *725,* 1048
Gill filaments, 1030
Gills
body temperature of fish and, 864–865
fish circulatory system and, 1047, 1048
gas exchange and, 1029–1030
of mollusks, 700
of ray-finned fishes, 726
surface area and, 1028
Ginkgo, *612, 632,* **634,** *635*
Ginkgo biloba, 479, *480, 584,* 634, *635*
Ginkgophyta, *612,* **634**
Girdling, 775
Gizzard, **1077**
Glacial moraines, 1197–1199
Glacier Bay, Alaska, 1197–1199
Glaciers, 470, 473, 478, 480, *481*
Gladiators, *709*
Gladioli, 832
Glands, **1021**
endocrine and exocrine, 876
See also specific glands
Glans penis, **908**
Glass sponges, *672, 683,* 684
Glaucophytes, *584,* 589, *597,* **600–601**
Glaxo-Smith/Kline, 139
Gleason, Henry, 1185, 1186
Glia, 859, **943,** 946, 954
Glial cells. *See* Glia
"Global 2000" ecoregions, 1235, *1236*
Global ecosystem, **1205–1206**
biogeochemical cycles, 1211
energy flow through, 1209–1210, *1211*
major compartments, 1206–1209
Global Surveyor (satellite), 32
Global warming, **1234**
carbon dioxide and, 471, 1215–1216
E. huxleyi and, 586
effect on diseases, 1216
See also Greenhouse effect; Greenhouse gases
Globigerina, 584
α-Globin, 1037
β-Globin, 275, 380–381, 385, 1037
γ-Globin, 315, *316,* 1037, 1039
β-Globin gene, *313,* 314–315, 317, 320
Globin gene family, 315, *316,* 535
Glomeromycetes (Glomeromycota), 614, 651, *652, 664,* 665
Glomerular filtration rate (GFR), **1107**
Glomeruli
olfactory, 969
renal, **1099,** 1100–1101, *1106*
Glomus, 664
Glucagon, *880,* **887, 1087**
Glucocorticoids, 888–889
Gluconeogenesis, **154,** 155, 175, 1086, 1087
Glucosamine, 53
Glucose
forms of, **49,** *50*
as fuel, 139
gluconeogenesis, **154,** 155, 175, 1086, 1087
in glycogen, 51–52
glycogen metabolism and, 346, *347*
in glycolysis, 142, *143*
glycosidic linkages, *51*
production in plants, 171
transport across membranes, 110
See also Blood glucose
Glucose metabolism
citric acid cycle, 145–147
energy yields from, 153
fermentation pathways, 147–148
free energy and, 139–140
glycolysis, 142–144
overview of, 140
oxidative phosphorylation, 148–153
pyruvate oxidation, 140, *141,* 144–145, *146, 153*
redox reactions, 140–141
Glucose transporter, 110
Glucuronic acid, 53
Glue lines, 1018, *1019*
Glutamate, 958, 995
Glutamate receptors, 959–960
Glutamic acid, *43,* 376
Glutamine, *43,* 125
Glutathione peroxidase, 1220
Glyceraldehyde, *50*
Glyceraldehyde 3-phosphate (G3P), 170, 171, 175
Glycerol, *54,* **55**
Glycine
chemical structure, 42–43
as neurotransmitter, 958, *959*
in photorespiration, 173
Glycogen
energy storage and, 155, 1071, 1086
fuel for muscles, 1015
insulin and, 1087
structure and function, **51–52**
synthesis, 1084
Glycogen metabolism, 346, *347*
Glycogen phosphorylase, 340–341, 346, *347*
Glycogen synthase, 346, *347*
Glycogen synthase kinase-3 (GSK-3), 922, *923*
Glycolate, 172, 173
Glycolipids, 101
Glycolysis
allosteric regulation, 156
energy yield from, 153
fermentation and, *147, 148*
in metabolic pathways, *141,* 154, 175
overview of, **140,** 142–144
production of ATP for muscular contraction, 1014, 1015
Glycolytic muscle, 1013
Glycoproteins, **101**
in cell adhesion and recognition, 102, *103*
formation, 79, 274
interferons, 405
MHC proteins, 415
T cell receptors, 414
Glycosides, *839*
Glycosidic linkages, **50–51**
Glycosylation, **274**
Glyoxysomes, 84
Glyphosate, 370
Gnathostomes, *723,* 724
Gnepine hypothesis, 635, *636*
Gnetifer hypothesis, 635, *636,* 646
Gnetophyta (Gnetophytes), *612, 632,* **634,** 635, *636,* 646
Gobi Desert, *1201*
Goiter, *884,* 885, 1075
Golden algae, 598
Golden feather sea star, *720*
Goldenrod, *793*
Golden toads, *730,* 1230
Goldman equation, 950
Golgi, Camillo, 80
Golgi apparatus
dynamic continuity of membranes and, 100–101
protein synthesis and, 273, 274
protein transport and, 80–81
structure and function, 75, **80–81**
Golgi tendon organs, *971,* 972
Gonadotropin-releasing hormone (GnRH)
discovery of, **883**
male sexual function and, *908,* 909
ovarian and uterine cycling, 911, 912
puberty and, 889–890
Gonadotropins
birth control pills and, 915

in human pregnancy, 912
ovarian and uterine cycles, 911
puberty and, **889–890**
Gonads, *880,* **889, 899**
Gondwana, 472, 473, 474, 475, 476, 477
Gongylidia, 658
Gonorrhea, *917*
Gonyaulax, 584
G. tamarensis, 587
goosecoid gene, 930
Goosecoid transcription factor, 930, 931
Gooseneck barnacles, *707*
Gorilla gorilla, 739
Gorillas, *739*
Goshawks, 1158
Gout, 1096
gp41 protein, 423
gp120 protein, 423
GPP. *See* Gross primary productivity
G protein-linked receptors, 337–339, 342
G proteins, **338,** 346, 958
Grafting, 833
Grains, complementary diets and, 1072
Gram-negative bacteria, *565,* **566,** 590
Gram-positive bacteria, **565–566**
Gram staining, **565–566**
Grana, 83, *84*
Grand Canyon, *465*
Granular cells, 402, *403*
Grapes, seedless, 802
Grapsus grapsus, 707
Grasses
endophytic fungi and, 658
fossils, 645
stolons, 832
Grasshoppers, *709*
Grasslands
atmospheric nitrogen deposition and, 1217
biomass and energy distribution, 1187, *1188*
temperate, 1121
Gravitropism, 805
Gray crescent, **922,** *923, 926,* 928
Gray-headed albatrosses, 1154, *1155*
Gray kangaroos, *736*
Graylag geese, *1145*
Gray matter, **987,** 989
Grazing
eutherian mammals and, 736–737
plants and, **840–841**
Great Barrier Reef, 686
Great Depression, 780
Great egret, 1129
Greater prairie chickens, 495
Great Lakes, PCBs and, 1089
Great white sharks, 865
Green algae, *603*
alternation of generations in, 594
component of lichens, 655, 656
endosymbiosis theory and, 589
major groups, 602
relationship to land plants, 611, 612–613
Green bottle fly, *710*
Green fluorescent protein, 361–362
Greenhouse effect, 577, 1208
See also Global warming
Greenhouse gases, 577, **1208**
Greenland ice cap, 1215, 1216
Green molds, 666
Green plants, **611,** *612*
Green sea turtles, *732, 905*
gremlin gene, 452, *453*
Gremlin protein, 452, *453*
Grevy's zebra, *1131*
Griffith, Frederick, 233–234
Gross primary production, **1209**
Gross primary productivity (GPP), **1209**
Ground meristem, **754,** 756
Ground squirrels, *870,* 1153
Ground state, **163,** *164*
Ground tissue system, **748**
Groundwater, 1212, 1217
Grouse, 1149
Growth, **427–428**
Growth factors, **186–187,** 325
Growth hormone (GH), *368, 880,* **882**
Grylloblattodea, 709
GTP. *See* Guanosine triphosphate
Guam, 1233
Guanine, 59, **238,** 239, *240*
Guanosine diphosphate (GDP), 338
Guanosine triphosphate (GTP)
citric acid cycle, 145, *146*
function, 59
G caps and, 316, *317*
G proteins and, 338
Guanylyl cyclase, 344
Guard cells
abscisic acid and, 811
function, 761
opening and closing of stomata, **773–774**
plant epidermis and, 748
zeaxanthin and, 814
Guayule, 796
Guillemin, Roger, 882–883
Gulf of Mexico, "dead zone," 1217
Gunnison National Forest, Colorado, *1119*
Guppies, 1172
Gustation, **969–970**
Guthrie, Robert, 384
Guts
mechanisms of smooth muscle contraction, 1010–1012
tissue layers in vertebrates, 1078–1079
tubular, 1077, *1078*
See also Gastrointestinal systems
Guttation, 770
GW501516 (drug), 139
Gymnogyps californianus, 1243
Gymnosperms
common name and characteristics, *612*
conifer life cycle, 636, *637*
conifer phylogeny, 635, *636*
evolutionary history, 474, **631,** *632*
major groups, 634
Gypsy moth caterpillars, 708
Gyri, 989
Gyrinophilus prophyriticus, 730
Gyrus, 989

- H -

H-2 proteins, 415
H5N1 flu virus, 282
HAART, **423**
Haber process, 790
Habitat fragmentation, **1231–1232**
"Habitat island," 1132
Habitat loss, species extinctions and, 1230, 1231–1232
Habitat patches
coffee plantations and, 1241, *1242*
decreasing sizes of, 1230
overview of, 1231–1232
subpopulations and, **1178**
Habitat, **1147**
creating new, 1238, *1239*
factors affecting animal choice in, 1147–1148
restoration, 1236–1237, 1238, *1239*
Habitat selection
factors affecting, 1147–1148
possible effects of, 1161
Haddock, 1180
Hadean Age, 34
Hadrobunus maculosus, 713
Haemophilus influenzae, 301, *409, 533*
See also Influenza virus
Hagfishes, *723,* 724, 904
Hair, 46, 735
Hair cells
inner ear, *972,* 973
in organs of equilibrium, **974–975**
Haldane, J. B. S., 709
Hales, Stephen, 764–765, 770
Half-life, **466**
Haliaeetus leucocephalus, 678
Halibut fishery, 1223
Hall, Donald, 1150
Halophiles, 577
Halophytes, 848
Hamilton, W. D., 1160
Hamner, Karl, 827
Hangingflies, 1152
Hangovers, 1108
Hantaviruses, 539
Hanuman langur, *532*
Haploids, **193**
Haplontic life cycle, **194, 595,** 659
Haptophytes, *584,* 590, **596,** *597*
Hardy-Weinberg equation, **492–493**
Hardy-Weinberg equilibrium, **492–493,** 494
Harpagochromis sp., *550*
Hartwell, Leland, 185
Harvestmen, *713*
Hashimoto's thyroiditis, 421
Hauptmann, Bruno Richard, 744–745
Haustoria, **653,** *654,* 792
Haversian bone, 1018, *1019*
Haversian systems, 1018, *1019*
Hawaiian crickets, 534
Hawaiian Islands
adaptive radiations, 520–521
allopatric speciation in, 511, *513*
avian malaria and, 1233
human-caused extinctions, 1228
speciation rates, 518
Hawkmoths, 516
Hawks, vision in, 980
Hawthorn, 513
hCG. *See* Human chorionic gonadotropin
HDLs. *See* High-density lipoproteins
Head, of crustaceans, 707
Headaches, 332
Hearing, 972–974
Heart
of amphibians, 1048
in circulatory systems, **1045**
of fish, 1047–1048
four-chambered, 1049
of lepidosaurs, 731
of lungfish, 1048
three-chambered, 1048
two-chambered, 1047–1048
Heart (mammalian), 735
anatomy of, 1050
athletes and, 1044–1045
atrial natriuretic peptide production, 1108
blood flow through, 1050–1052
blood pressure regulation and, 1062, *1063*
diving reflex and, 1064
electrocardiograms, *1054,* 1055
endocrine functions, *880*
murmurs, 1051
origin and control of heartbeat, 1052–1054
output when at rest, 1044
pacemaker region, 992, 1010, 1052
regulation of breathing and, 1040
stem cell therapy and, 433
Heart attack, 1060
athletes and, 1044–1045
epinephrine and, 879
familial hypercholesterolemia and, 376
tissue plasminogen activator and, 368–369
See also Cardiovascular disease
Heartbeat, in mammals, 1052–1054
Heart murmurs, 1051
Heart muscle. *See* Cardiac muscle
Heart-stage embryo, *823,* 824
Heartwood, 760
Heat
plants and, 847
sensing, 954
See also Body temperature
"Heat" (estrus), 911
Heat extractor, 854, *855*
Heat of vaporization, **32**
Heat shock proteins, **847**
Heat stroke, 854
Heavy metals, plants and, 849
hedgehog gene, 934
Hedrick, Marc, 426
Height, quantitative variation in, 221–222
HeLa cells, 180–181
Helianthus annuus, 639
Helical molecules, **238**
See also Double helix
Helicobacter pylori, 1081, 1082
Helicoma, 1198
Helium atom, *22*
α Helix, *45,* **46**
Helix-loop-helix motif, 320, *321*
Helix-turn-helix motif, 320, *321*
Helminthoglypta walkeriana, 701
Helper T cells (T_H)
in cellular immune response, **414,** 415, *416,* 417
development of plasma cells and, 411
HIV and, 421, 422
Hemagglutinin, 553–554
Hematocrit, 1055, *1056*
Heme, *129,* 130, 164, 378, 1036
Hemiascomycetes, **665**
Hemicentetes semispinosus, 1130
Hemichordates, *672, 683,* 718, 721
Hemipenes, 905
Hemiptera, 709
Hemizygous individuals, **226**
Hemocoel, 691, 700
Hemoglobin
abnormal, 376
affinity for carbon monoxide, 1036
beta thalassemia disease and, 317

fetal, 1037, 1039
forms of, 1037
globin gene family and, 315
number of polypeptide chains in, 42
oxygen-binding/dissociation properties, 535, **1036–1038**
protein domains and, 315
quaternary structure, 46, *47*
regulating translation of, 326
Hemoglobin C disease, 376
Hemophilia
blood clotting and, 377, 1056
form A, 206
inheritance, 225, 379
transposons and, 312
treating, 391, 392
Hemorrhaging, endotoxins and, 73
Hemp, 750
Henbane, *826*
Henricia leviuscula, 720
Hensen's node, *931,* **932**
Henslow, John, 487
Hepatic duct, 1082, *1083*
Hepatic portal vein, 1084
Hepatitis A, *409*
Hepatitis B, 387, *409, 917*
Hepatitis B virus, 387
Hepatophyta, *612,* **622**
Herbicides, 370, 807
Herbivores, **1075**
bioengineered resistance to insects, 842–844
conservation of sodium, 1095
effect of grazing on plants, 840–841
overview of, **677,** *678*
plant chemical defenses, 840–842
plant signaling pathways and, 842, *843*
as predators of plants, 1189
teeth, *1076*
trophic level, 1186
Herbivory. *See* Grazing
Heritable traits, **208, 490**
Mendelian inheritance, 208–214
Hermaphroditic species, 699, **904**
Hermodice carunculata, 675
Herpes virus, *285*
Hershey, Alfred, 235–237
Hershey–Chase experiment, 235–237, 285
Heterochromatin, 323
Heterochrony, **451,** *452*
Heterocysts, **573**
Heterokaryons, 662
Heteroloboseans, *584, 597,* 603
Heteromorphic life cycle, **594**
Heteropods, 702
Heterosis, **220**
Heterospory, **621,** *622,* 632
Heterotherms, **861**
Heterotrophs, **13, 1069**
classification of, 1075
food requirements, 1069–1075
ingestion of food, 1075–1076
nutrition and, 781, 782
parasitic plants, 792
trophic level, 1186
Heterotypic binding, **102,** *103*
Heterozygosis, 220
Heterozygotes, **211**
Hevea brasiliensis, 796, *797*
Hexaploids, 202
Hexapods, 453, *683*
See also Insects
Hexokinase, 129
Hexosephosphates, 175
Hexoses, 49–50
Hibernation, **870,** 1071
Hibernators, overnourishment and, 1071
Hierarchical sequencing, **393,** *394*
High blood pressure. *See* Hypertension
High-density lipoproteins (HDLs), **1087**
High-GC Gram-positive bacteria, *571,* **572–573**
Highly active antiretroviral therapy (HAART), **423**
Highly repetitive sequences, 311, 394
Himalayan bumblebees, 828
Hindbrain, **986,** 987
Hindgut, **1077**
Hinge joint, *1019*
Hippocampus
functions of, **988–989**
in learning and memory, 999–1000
newly formed neurons in adults, 984–985
Hippopotamus, 737
Hirudo, 1077
H. medicinalis, 699, 700
Histamine
edema and, 1058
inflammation response and, **406**
neurotransmitter, *959*
paracrine hormone, 875
Histidine, *43*
Histone acetyltransferases, 322
"Histone code," 322
Histone deacetylase inhibitor, 322
Histone deacetylases, 322
Histones, **187,** 322
HIV. *See* Human immunodeficiency virus
HIV protease, 274
H.M.S. *Beagle,* 487, *488*
Hoatzin, *532,* 533
Hodges, S., 516
Hodgkin, A. L., 948
Holdfasts, 600, *601*
Holley, Robert, 266
Holocene epoch, 478
Holometabola, *709,* 710
Homalozoans, 718
Homeobox, **445,** 933
Homeodomain, 445
Homeostasis, 131, **856–857**
Homeotherms, 861
Homeotic genes, anterior-posterior axis determination and, 450
Homeotic mutations, 444
Homing, **1154**
Hominids, 738–740
Homo
evolution of, 739–740
H. erectus, 716, *740*
H. ergaster, 740
H. floresiensis, 716–717, *740*
H. habilis, 740
H. heidelbergensis, 740
H. neanderthalensis, 740
H. sapiens, 12, 479, 554, 740
See also Humans
Homologous chromosomes
first meiotic division, 197–200
genetic recombination, 223–224
overview of, 193
Homologous pair, **193**
See also Homologous chromosomes
Homologous traits, **544**
Homologs (chromosomes), **193**
Homologs (genes), 537
Homology, 526
Homoplasies, **545**
Homospory, **621,** *622*
Homotypic binding, **102,** *103*
Homozygosity, in wild cats, 306–307
Homozygotes, **211**
Honeybees, *711*
dancing behavior, 1157, *1158*
heat production and, 865, *866*
reproduction in, 896–897, 898
swarming, 1148
Hooke, Robert, 4, 5
Horizons (soil), **786**
Horizontal cells, 981, 995
Horizontal transmission, of plant viruses, 289
Hormone receptors, 878–879, 892–893
Hormones, **797**
adrenal gland, 887–889
affecting the gastrointestinal system, 1086
anabolic steroids, 874–875
calcitonin, 885
can have many actions, 879
as cell signals, 334
circulating and paracrine, 875
controlling ovarian and uterine cycles, 911–912
control of molting in insects, 876–877
effects on animal behavior, 1146–1147
endocrine system and, **875**
half-life, 891
of humans, *880*
insect juvenile hormone, 877–878
location of receptors, 878–879
male sexual function and, *908,* 909
melatonin, 890
pancreatic, 887
parathyroid hormone, 885–886
pituitary, 881–882
receptors, 878–879, 892–893
regulation of blood calcium levels, 885–886, 887
sex steroids, 888, 889–890
signal transduction pathways and, *892,* 893
stress response and, 888–889
techniques for studying, 890–893
types of, 878
Vitamin D, 886
See also Plant hormones; *individual hormones*
Hornworts, 612, 614, 616, 623
Horsehair worms, *683, 692,* **703,** *704*
Horses, 220, *1131*
Horseshoe crabs, 479, 712–713
Horsetails, 474, 475, *612,* 618, 625, **625**
Hot deserts, **1123**
Hotspot concept, 1235
House finches, 480, *481*
Hoxc-6 gene, 453
Hoxc-8 gene, *450*
Hox genes
anterior-posterior axis and, **933–934**
body segmentation and, 442
in *Drosophila* development, **444,** 450
in ecdysozoans, 693
genetic switches and, 451
homeobox, 445
leg number in arthropods and, 453, *454*
suppression of limb formation in snakes, 458
HTLV-I. *See* Human T cell leukemia virus
Hubel, David, 995
Human chorionic gonadotropin (hCG), **912,** 937
Human genome, **393–395**
alternative splicing of mRNA and, 325
DNA content, 307
ethical concerns, 395
largest gene in, 315
sequencing, 393, *394*
size of, 533
uses of, 395
Human Genome Project, **393**
Human immunodeficiency virus (HIV)
criminal case concerning, 542–543
infection and replication, 422–423
infection "set point," 422
molecular evolution and, 539
overview of, **421–422**
phylogenetic analysis, 551-*553*
proteases and, 274
reproductive cycle, 288–289
RNA virus, 261
sexual transmission, *917*
treating, 423
Human leukocyte antigens (HLA), 415
Human papillomavirus, 327
Human reproduction
abortion, 915–916
childbirth, 912–913
contraception methods, 914–916
female reproductive tract, *909*
fertilization, 910
hormones and, *908,* 909, 911–912
infertility and reproductive technologies, 916
male reproductive tract, *906*
ovarian cycle, 910–911
production and delivery of semen, 906–908
sex organs, *889,* 906–910
sexually transmitted diseases, 917
sexual responses, 913–914
uterine cycle, 911
Humans *(Homo sapiens)*
ABO blood group system, 219
birth weight, 497, *498*
brain cell death, 202
brain size, 740
cell death genes, 439–440
chromosome abnormalities, 199–200, 226, 383
development of culture and language, 741
developmental stages, 936–938
digestive enzymes, *1084*
digestive system, *1078*
endocrine system, *880*
essential amino acids, 1072
evolution, 396, 479, 738–741
evolutionary relationship to chimpanzees, 60
fire and, 1212, 1213

genetic diseases, 375–379
globin genes and, 315, *316*
impact on animal extinctions, 482
impact on Earth's climate, 1116
impact on evolution, 505
impact on global energy flows, 1210-1212
karyotype, *195*
lactose-intolerance, 505
metabolic rates, 1070
neural tube defects, 932–933
noncoding DNA in, *534*
overnutrition in, 1071
pedigrees, 216–217
percent water, 856
phylogenetic tree, *740*
population density and, 1162
population management and, 1181
quantitative variation in height, 221–222
reproduction. *See* Human reproduction
respiratory system, 1032–1035
sex-linked characters, 227–228
sex organs, *889*, 906–910
skeleton, 1017–1018
skin color, 220, 1074
social behavior and, 1158
species name, 12
vitamin requirements, 1074
water consumption, 1212
Human T cell leukemia virus (HTLV-I), *387*
Humata tyermanii, 194
Humble, G. D., *774*
Hummingbirds, 508, 641, 870, 1161
Humoral immune response, **408,** 411–414
Humpback whales, 1157
Humus, **787–788**
hunchback gene, 442, 444
Hunchback protein, 443
Hunt, Sir John, 1024
Huntington's disease, *216*, 383
Hurricane Katrina, 1112, 1222
Hurricanes, 1197
Husa husa, 903
Huxley, A. F., 948
Huxley, Andrew, 1007
Huxley, Hugh, 1007
Hyalophora cecropia, 877–878
Hybridization, **514,** 1143
Hybridomas, **413**
Hybrids (nucleic acid hybridization), 314
Hybrids (organisms), 209
Hybrid vigor, 220
Hybrid zones, **517–518**
Hydra, 898
Hydride ion, 141
Hydrocarbons, 30, 294
Hydrochaeris hydrochaeris, 464
Hydrochloric acid, 33
gastric, 1078, 1080, *1081*
Hydrogen, *27*
isotopes, *23*
in plant nutrition, 781
Hydrogen bonds, *26*, **29**
in DNA, 239, 240
in nucleic acids, 59
protein function and, 48
in protein secondary and tertiary structure, *45*, 46
in water, *31*
Hydrogen ions, loosening of plant cell walls and, 807, 808
Hydrogen peroxide, 83, 903
Hydrogen sulfide
chemolithotrophs and, 782
intestinal bacteria and, 1085
nonbiological sources, 1217
photosynthetic bacteria and, 568
Hydrological cycle, **1211–1212**
Hydrolysis
breakdown of polymers, **41–42**
in chemical weathering of rock, 786
digestive enzymes and, 1077
Hydronium ion, 33–34
Hydrophilic molecules, **29**
Hydrophobic interactions, *26*, 30, 48
Hydrophobic molecules, **30**
Hydroponics, **784**
Hydrostatic skeletons, **675, 1016**
Hydrothermal vents
chemical evolution and, 62
detritivorous crabs and, 1187
pogonophorans and, 699
prokaryotes and, *576*, 577, *578*
Hydroxide ion, 33, *34*
Hydroxyl group, 39, *40*
Hydrozoans, 687
Hyla regilla, 2, 14, *1005*
Hylobates mulleri, 739
Hymen, 909
Hymenoptera, 709
Hymenopterans, *709*, 1159–1160
Hyperemia, **1061**
Hyperpolarization, **950,** *951*
Hypersensitive response (plants), **838–839**
Hypersensitivity (immune system), 420, 421
Hypertension, 1060
Hyperthermophiles, 576–577
Hyperthyroidism, **885**
Hypertonic solutions, **107**
Hyphae, **652,** 653–654, 662
Hypoblast, *925*, **931,** 932, 935
Hypometabolism, **1064**
Hypothalamus
birth control pills and, 912
control of the anterior pituitary, 882–883
endocrine system and, *880*
in the forebrain, **987**
functions of, 987, 988
male sexual function and, *908*, *909*
in ovarian and uterine cycles, 911, *912*
pituitary gland and, **880,** 881
regulation of food intake and, 1088
somatostatin and, 887
thyrotropin-releasing hormone and, 884–885
vertebrate thermostat and, **868–869**
Hypothermia, **870**
Hypotheses, 2, **14**
Hypothesis–prediction (H–P) method, 13–14
Hypothyroidism, *884*, **885,** 1075
Hypotonic solutions, **107**
Hypoxia, **1055**
Hyracoidea, *736*
H zone, *1006*, 1007

- I -

IAA. *See* Indoleacetic acid
I band, *1006*, 1007
Ice, 31–32
Ice ages, 478
Ice caps, 1215, 1216
Ice-crawlers, *709*
Ice plants, 521
ICSI. *See* Intracytoplasmic sperm injection
IgA antibodies, *412*
IgD antibodies, *412*
IgE antibodies, *412*, 420, *421*
i (inducibility) gene, 298
IgG antibodies, 412–413, 420
IgM antibodies, *412*
Igneous rocks, 467
Ignicoccus, 577, *578*
Ileum, *1079*, **1082**
Illicium floridanum, 644
Image-forming eyes, **978–980**
Imbibition, **800**
Imhoff, Mark, 160
Immediate hypersensitivity, **420,** *421*
Immediate memory, **999**
Immune system
adenosine deaminase and, 392
cortisol and, 888
epinephrine and, 879
lymph nodes and, 1060
See also Defense systems; Specific immune system
Immunity, natural and artificial, 409
Immunization, **409–410,** 414
Immunoassays, 414, **891**
Immunodeficiency diseases, 552, *553*
See also Acquired immunodeficiency syndrome
Immunodominance, 408
Immunoglobulins
classes of, 412–413
generation of diversity in, 418–420
structure of, **411–412**
Immunological memory, **408,** 409–410
Immunological tolerance, **410–411**
Immunotherapy, 414
Imperfect flowers, **639**
"Imperfect fungi," 663
Impermeability, 105
Implantation, **910,** 915, **925**
Impotence, 908
Imprinting, **1145,** 1226
Inactivated polio vaccine (IPV), 524–525
Inactivation gate, **952–953**
Inbreeding, **220,** 1160
Incisors, 1076
Inclusive fitness, **1159**
Incomplete cleavage, 673, **923,** *924*
Incomplete dominance, **218–219**
Incomplete metamorphosis, **710**
Incurrent siphon, 702
Incus, *972*, 973
Independent assortment, **199,** 213–214, *215*
Independent variables, 15
Indeterminate growth, 754, 825
Indian pipe, 792
Indirect transduction, **339,** *340*
Individual fitness, **1159**
Indoleacetic acid (IAA), 803
Induced fit, **129**
Induced mutations, 277, **277,** *278*
Inducers, **296,** 298, 299, **437,** 438
Inducible promoters, 367
Inducible proteins, **296**
Inducible systems, 298, 299
Induction, **436,** 437–439
Inductive logic, 14
Inductive treatments, 829
Infants, genetic screening, 384
Inferior vena cava, **1050**
Infertility, 916
Inflammations, **406,** 420
Inflorescence meristems, **825**
Inflorescences, **639,** 825
Influenza virus, *409*
hemagglutinin and, 553–554
molecular evolution, 539
predicting virulent strains of, 554
reproductive cycle, *287*, 288
typical size of, *284*
See also Haemophilus influenzae
Infrared radiation, rattlesnakes and, 964
Inheritance
early assumptions about, 207
of genetic diseases in humans, 379
interactions with learning, 1145–1146
Mendelian, 207–217
See also Sex-linked inheritance
Inherited behavior. *See* Stereotypic behavior
Inhibin, *908*, 909
Inhibition
experiments using, 86
influence on succession, 1199–1200
in retinal photoreceptors, 995
reversible and irreversible, 131–132
Inhibitors, 86, **131–132**
Inhibitory synapses, **957**
Initials, 753
Initiation complex, **268,** *269*
Initiation site, in transcription, **262**
Innate defenses, 401
See also Nonspecific defenses
Inner cell mass, **925,** 926
Inner ear
equilibrium and, 974, *975*
hearing and, 972, 973
Inoculation, 400–401, 409
Inorganic fertilizers, 787
Inositol trisphosphate (IP_3), **342,** *343*
Insect larvae, canavanine and, 842
Insects
chitin in, 53
compound eyes, 978
evolutionary history, 475
excretory system, 1098
flight and, 710
fossil record, 472
genitalia, 904
hormonal control of molting in, 876–877
juvenile hormone and, 877–878
major groups, 709–711
numbers of living species, 708–709
regulation of heat production, 865, *866*
structure of, 708
tracheal gas exchange system, 1028, *1029*
as vectors for plant viruses, 289
wing evolution, 710
Inspiratory reserve volume, **1032,** 1035
Instars, **710,** 876

Instincts, limbic system and, 988
Insulin, *368, 880*
actions of, **887,** 1087
as cell signal, 336, 337, *338*
Pima Indians and, 1068
Insulin-dependent diabetes mellitus, 421, 887
Insulin-like growth factors (IGFs), 882
Insulin receptors, 892–893
Insulin response substrates, 337, *338*
Integral membrane proteins, 99, *100*
Integral proteins, *98*
Integrase, HIV and, 423
Integuments, **632,** 633, 823
Interbreeding, 1143
Intercostal muscles, **1035**
Interference competition, **1192**
Interference RNA (RNAi), **323, 365**
Interferons, **405**
Intergovernmental Panel on Climate Change (IPCC), 1216
Interkinesis, **199**
Interleukins, 187
Intermediate disturbance hypothesis, **1199**
Intermediate filaments, **87–88**
Internal environment
multicellular animals and, 7–8, 855–856
physiological systems and, 855–856, 857–859
regulation of, 856–857
Internal fertilization, **903–904**
Internal gills, **1028**
International Crane Foundation (ICF), 1226
Interneurons, **943, 988**
Internodes, **745**
Interphase, mitosis, **184,** 187, 189, *190*
Intersexual selection, 498
Interspecific competition, 1192, 1200
Interstitial cells, 907
Interstitial fluid, 1046, 1060
Intertidal zonation, 1192–1193
Intertropical convergence zone, **1114**
Intestines
bacteria and digestion, 1085
endocrine functions, *880*
in guts, **1077**
large, 405, *1079,* 1084
surface area, *1078*
See also Small intestines
Intracytoplasmic sperm injection (ICSI), 916
Intrasexual selection, 498
Intraspecific competition, 1192
Intrauterine device (IUD), *914,* 915
Intrinsic factor, 1075
Intrinsic rate of increase, **1173**
Intron-exon boundary, 317
Introns, **313–315,** 394
Inuit peoples, 1074
Invasive species. *See* Exotic species
Inversions, in chromosomes, **276,** *277*
Invertase, 1184
Invertebrates
excretory systems, 1097–1098
nervous systems, 944
visual systems, 977–978
In vitro evolution, **538,** *539*
In vitro fertilization (IVF), 434, 916
Involuntary nerve pathways, 985
Involution, **928**
Iodide, 885
Iodine
deficiency, *884,* 885, 1075
micronutrient, *1073,* 1219
thyroxine and, 883, 1219
Ion, **28**
electric current and, 947
Ion channels
activation during the sense of smell, 345–346
cyclic AMP and, 341
cystic fibrosis and, 377
endotherms and, 862
gated, 950–951
hair cell stereocilia and, 974
heartbeat in mammals and, 1052, 1054
ionotropic receptors, 958
membrane potential and, 948–950, 950–591
overview of, **108–109**
patch clamp studies, 950
sensory receptor proteins and, 965–966
as signal receptors, **337**
types and functions of, 948
See also Voltage-gated ion channels
Ion concentration gradients, endotherms and, 862
Ion exchange, **786,** *787*
Ionic bonds, *26,* **27–28,** 46, 48
Ionic conformers, 1095
Ionic regulators, 1095
Ionizing radiation, 277
Ionotropic receptors, 958, 959–960, 965, *966*
Ion pumps
membrane potential and, 947–948
See also Proton pumps; Sodium-potassium pump
IP_3/DAG signal transduction pathway, 342–343, 344
Ipomopsis aggregata, 841
IPV. *See* Inactivated polio vaccine
Iridium, 471
Iris, **978–979**
Iron
in catalyzed reactions, *129*
deficiency, 783, 1073, 1075
global cycling, 1219
metabolism in archaeans, *576*
photosynthesis and, 1219
in plant nutrition, 781, *782,* 783, 784
Iron pyrite, 560
Island biogeography, **1132–1133**
Islets of Langerhans, *880,* **887**
Isobutane, 40
Isocitrate dehydrogenase, 156
Isogamy, **594**
Isoleucine, *43, 134*
Isomers
of amino acids, 42
overview of, **40,** *41*
Isometopids, *519*
Isomorphic life cycle, **594,** *595*
Isopentenyl adenine, 809
Isopods, 706, *707*
Isoptera, 709
Isotonic solutions, **107**
Isotopes, **23**
Isozymes, **134**
Ivanovsky, Dimitri, 283–284
IVF. *See* In vitro fertilization
Ivory, 1239
Ivory-billed woodpecker, 1229

- J -

Jacob, François, 297, 457
Jaegers, 1189
Janssen, Zaccharias and Hans, 4
Japan, whaling and, 1180
Japanese macaques, 1140, 1141
Japanese mint, *648*
Jasmonates, **842,** *843*
Jawless fishes, *723,* 724
Jaw(s)
evolution of, *725*
human evolution and, 740
as a lever system, 1019, *1020*
in vertebrates, 724–725
Jejunum, *1079,* **1082**
Jelly coat, **901**
Jellyfish, *685,* 686
See also Cnidarians
Jenner, Edward, 400
Jeyarajah, Abilass, 352–353
Jianfangia, 473
Jogging, 1014
Johnson, Pieter, 2, *3,* 14, 15
Jointed appendages, in arthropods, 706
Jointed-fins, 728
Jointed limbs, 676
Joint probability, 215
Joints, **1019,** *1020*
Jones, Steven, 301
Joppeicids, *519*
Joules (J), 121
Joyner-Kersee, Jackie, 1004
Judean dates, 630
Jumping, 1004–1005
Jumping bristletails, *709*
Jurassic Park (film), 232
Jurassic period, *466–467,* **476–477**
Juvenile hormone, **877–878**
Juvenile-onset diabetes, 887, 892–893

- K -

Kac, Eduardo, 232
Kalanchoe, 826
Kalanchoe, 832, *833*
Kale, *490*
Kamen, Martin, 161
Kangaroo rats, *1093,* 1144
Kangaroos, *8,* 735, *736,* 1004–1005
Kaposi's sarcoma, *387,* 422
Karenia brevis, 590
Karyogamy, *660, 661,* **662**
Karyotypes, 195
Kashefi, Kazem, *576*
Katz, Lawrence, 969
Kelp, *583,* 599
Kentucky bluegrass, 174
Keratins, 46, 87, 103–104
α-Ketobutyrate, *134*
Keto group, 39, *40*
Ketones, *40*
Kevlar, 46
Keystone species, **1196–1197**
Kidneys
amniotes and, 731
anatomy of, 1101–1102
antidiuretic hormone and, 881, 1062
endocrine functions, *880*
epithelial cells, *858*
erythropoietin and, 1055
filtering of water-soluble toxins, 1089
nephrons and, **1098,** 1101–1102
parathyroid hormone and, 887
regulation of blood pH and, 1105, *1106*
regulatory mechanisms, 1107–1108
renal failure and dialysis, 1105–1106
urine formation in, 1099–1105
Vitamin D and, 886
Kidney stones, 887
Kidston, Robert, 610
Kilocalories (kcal), **1069**
Kimura, Motoo, 531
Kinases, **144**
Kinesins, 89, *90,* 192, 881
Kinetic energy, **119–120**
Kinetin, 809
Kinetochore microtubules, 189, 190
Kinetochores, **189,** 190, 198
Kinetoplastids, *584,* 586, *597,* **603–604**
Kinetoplasts, 604
King, Thomas, 430
Kingdoms, **554**
Kinorhynchs, *683, 692,* 702, **703**
Kirschvink, Joseph, 61
Kiwi, 732
Klinefelter syndrome, 226
Knee-jerk reflex, 987–988, 1019
Knee joint, 1019, *1020*
Knife fish, 1021
Knockout experiments, **364–365,** 1143–1144
Knots (wood), 760
Knudson, Alfred, 388
Koch, Robert, 96, 579
Koch's postulates, **579,** 1081, *1082*
Kodiak brown bear, *678*
Koehler, Arthur, 744–745
Kohlrabi, *490*
Kölreuter, Josef Gottlieb, 207
Komodo dragon, 731
Korarchaeota, **575,** 577
Kornberg, Arthur, 241, 246
Körner, Christian, 1214
Krakatau, 1132–1133
Krebs, Charles, 1189
Krebs, Edwin, 340
Krebs cycle, 140
Kuffler, Stephen, 995
Kurosawa, Eiichi, 801–802
Kuru, 378
Kwashiorkor, 1071

- L -

Labia, 909
Labidomera clivicollis, 844
Labor contractions, 912
Labrador retrievers, 219–220
Lacewings, *709*
Lacks, Henrietta, 180–181
lac operon, 297–298, 299–300, 361
Lactase, 1083, *1084*
Lactic acid, 140, *141,* 147, 148
Lactic acid fermentation, 140, *141,* 147–148
Lactoglobulin, 369
Lactose, 50
Lactose intolerance, 505, 1083
Lactose metabolism, in *E. coli,* 296, 297–298
Lacunae, 1017
Lady's mantle, *770*
Laetiporus sulphureus, 667
Lagging strand, in DNA replication, **246,** *247*
Lagomorpha, *736*

Lake Erie, 1219
Lake Kivu, 551
Lakes
  acidification, 1218
  eutrophication, 1217, 1219
  mineral nutrient flow and turnover in, 1206–1208
  thermoclines, 1208
Lake Tanganyika, 1208
Lake Victoria, 551
Lamar Valley, 1196
Lamellae, of gills, 1030
L-amino acids, 42
Lamins, 79
Lampreys, *537, 723,* 724, 904
Lancelets, *722,* 723
Land bridges, 1134
Land plants
  adaptations to life on land, 613
  classification, *612*
  defining, **611**
  life cycles, 614–615
  major groups, 611–612
  nonvascular, 612, 614–616
  relationship to green algae, 611, 612–613
  seedless, 622–627
  soil formation and, 613
  vascular, *612,* 616–621, 624
Land snails, *701,* 702, 1016
Landsteiner, Karl, 219
Lang, William, 610
Lange, Dorothea, *780*
Language, 741, 1000–1001
Langurs, 533
Laonastes aenigmamus, 1229
Laotian rock rat, 1229
Large intestine, 405, *1079,* 1084
Larson, Gary, 984
Larvaceans, 722–723
Larvae
  in animal life cycles, **679**
  of butterflies, 877
  canavanine and, 842
Larynx, 1032, *1033,* 1079, *1080*
La Selva Biological Station, 1234
Lasix®, 1105
Late genes, 286
Lateral gene transfer, **570,** 590
Lateral hypothalamus, 1088
Lateralization, of language functions, 1000
Lateral line, **974**–975
Lateral meristems, *753*
  derivatives of, 754–755
Lateral roots, *755*
Latex, 796, *797,* 844
Latex condoms, 914
Laticifers, **844**
Latimeria
  L. chalumnae, 728
  L. menadoensis, 728
Latitudinal gradients, 1200
Laundry detergents, 127
Laupala, 534
Laurasia, 474, 476, 477
Law of independent assortment, **213–214,** *215*
Law of mass action, 34, 336
Law of segregation, **211–213**
LDLs. *See* Low-density lipoproteins
Leaching, **786,** *787*
  seed germination and, 799, *800*
  in soils
Lead, 849, 1089
Leading strand, in DNA replication, **246,** *247*
Leaf-cutting ants, 658
Leafhoppers, *709*
Leaflets, 748
leafy gene, 441
Leafy sea dragon, *727*
Learning
  acquisition of new neurons and, 938
  critical period in, **1145**
  glutamate receptors and, 959–960
  interactions with inheritance, 1145–1146
  in Japanese macaques, 1140, 1141
  neuronal basis of, 999
  response to environmental change, 456
  song-learning in birds, 1145–1146
  in wasps, *1144,* 1145
Leaves, **620**
  abscisic acid and, 811
  abscission, 806
  adaptation, *6*
  anatomical support for photosynthesis, 760–761
  anatomy of C_3 and C_4 plants compared, *173*
  blade and petiole, 747
  of carnivorous plants, *6,* 792
  evolution, 479, *480,* 620–621
  flowering hormone and, 829–831
  function, **745**
  guttation, 770
  as modular units, 753
  parasitic fungi and, *654*
  plant nutrition and, 782
  primordia, *753,* **757**
  salt glands, 848
  simple and compound, 748
  vegetative reproduction and, 832, *833*
  of xerophytes, 845, 846
  See also Stomata
Lecithin, *56*
Lederberg, Joshua, 291
Leeches, 699–700, 1077
Leghemoglobin, **789**
Legs. *See* Limbs
Legumes
  complementary diets and, 1072
  nitrogen-fixation and, 788, 789
Leishmania major, 582, *604*
Leishmaniasis, 582, *604*
LeMond, Greg, 229
Lemur catta, 1130
Lemurs, 738
Lens cells, gap junctions and, 348
Lens (eye)
  differentiation in vertebrates, 437
  focusing, **979–980**
Lens placode, 437
Lenticels, **760**
Leopold, Aldo, 1211
Lepas pectinata, 707
Lepidodendron, 618, 625
Lepidoptera, 709
Lepidosaurs, 731
Lepomis macrochirus, 1150
Leptin, **1088**
Leptoids, 624
Leptospira interrogans, 563
Leptosporangiate ferns, 625, 626
Lepus, 868, 1189
Leucilla nuttingi, 684
Leucine, *43,* 264
Leucine zipper motif, 320, *321*
Leucoplasts, 83, *85,* 750
Leukemias, 387
Leukocytes, 402
Leverage and lever systems, 1004, 1019, *1020*
Levey, Douglas, 1232
Levin, Donald, 517
Lewis, Edward, 444
Lewis, Reggie, 1044
Leydig cells, 907, *908,* 909
L'Hoest monkeys, *553*
Libellula
  L. luctuosa, 711
  L. pulchella, 1196
Lice, *709*
Lichens, **655–656,** *657*
Life
  artificial, 302–303
  autotrophs and heterotrophs, 13
  earliest evidence of, 68–69
  laws of thermodynamics and, 122
  major domains, 12
  mechanistic view of, 21
  origin, 10–11, 61–64
  phylogeny of, *562*
  water and, 34–35
Life cycles
  diplontic, 194, 595
  dispersal stage, 680
  haplontic, 194, 595, 659
  heteromorphic, 594
  isomorphic, 594, *595*
  of microbial eukaryotes, 593–595
  overview of, **679**
  of parasites, 681–682
  sexual, 193, 194
  trade-offs, 680–681
  See also Alternation of generations
Life histories, **1171–1172**
Life insurance, 1169
Life tables, **1169–1170,** 1171
Ligaments, *971,* 972, **1019**
Ligand-gated channels, 108
Ligands, **46, 336**
Light
  absorption by antenna systems, 165–166
  animal migration and, 1156
  developmental plasticity in plants and, 456
  physical characteristics, 163
  pigments and, 164–165
  plant seed dormancy and, 799
  regulation of plant development and, 812–814
  stimulation of the Calvin cycle, 172
  visible spectrum, 163, *164*
Light-independent reactions, **162,** *163*
Light microscopes, 71–72
Light reactions, **162,** *163*
Ligia occidentalis, 707
Lignin, 617, 749, 838
Lilies, *534,* 832
Lilium sp., *645*
Limb bud, 441
Limb development
  thalidomide deformities, 457
  in vertebrates, 441
Limbic system, **988–989,** 1000
Limbs
  abnormalities in frogs, 2, 14, 15
  evolution of wings and, 457, *458*
  jointed legs in arthropods, 706
  locomotion and, 676
Lime (calcium oxide), 787
Limes (fruit), 1074
Limestone, 587
"Limeys," 1074
Liming, **787**
Limulus polyphemus, 712
LIN-3 growth factor, 439
Linanthus, 551–552
Lind, James, 1074
Lindbergh, Charles, 744
Lindbergh kidnapping case, 744–745
LINEs, 312
Linkage groups, *222,* **223**
Linked alleles, 214
Linnaeus, Carolus, 509, 554, 555, 1131
Linoleic acid, 1072
Lions, prides, 1159
Lipases, 1077, **1082,** *1084,* 1086
Lipid monolayers, 576
Lipid rafts, 100
Lipids
  in archaean membranes, 575–576
  biological membranes and, 97–99
  as energy source, 154
  overview of, 39, *41,* **54–57**
  second messengers derived from, 341–343
Lipomas, 386
Lipoproteins, 377, 1084, **1086–1087**
Lithium, 342–343
Lithosphere, 468
Liu, Zhi-He, 1151
Liver, *1079*
  cholesterol and, 377
  in digestion, 1082, *1083*
  energy storage and, 1071
  glycogen phosphorylase and, 340–341
  hepatitis B and, 387
  insulin and, 1087
  interconversion of fuel molecules, 1086
  LDL recycling and cholesterol, 114–115
  lipoprotein production, 1086–1087
  metabolism of toxins and, 1089
  processing of absorbed nutrients, 1084
  turnover time of cells, 858
  undernourishment and, 1071
Liver cancer, 387
Liver diseases, *409,* 1058
Liverworts, 612, 614, 616, 622, *623,* 657
Live-virus vaccines, for polio, 524–525
Lizards
  characteristics, 731
  hemipenes, 905
  territorial defense behavior, 1148–1149
  thermoregulation in, 862–863
Llama guanaco, 1037
Llamas, 1037
Load arm, 1019, *1020*
Loams, **786**
Lobe-finned fishes, *723*
Loboseans, *584, 597,* 603, **605–606**
Lobster, 1238
Localization sequences, 273
Locus, of genes, **212**
Lodgepole pines, *481*
Lodging, 810
Lofenelac, 391

Logic, inductive and deductive, 14
Logistic growth, **1173–1174**
Logophyllia sp., *1234*
Lombok, 1128, *1129*
Long bones, 1018, *1019*
Long-day plants (LDPs), **826,** 827, 831
Long interspersed elements (LINEs), 312
Longitudinal muscles, 675
Long jumping, 1004
Long-short-day plants, 826
Long-tailed widowbird, 499–500
Long-term depression (LTD), **999**
Long terminal repeats (LTRs), 534
Long-term memory, **999**
Long-term potentiation (LTP), 960, *961,* **999**
Loop of Henle, **1102,** 1103, 1104, 1105
Lophodermium, 1198
Lophophorates, 692
Lophophore, **692,** 695, 697
Lophopus crystallinus, 693
Lophotrochozoans, *672, 683*
  characteristics, **692–693**
  major groups, 695–702
  phylogenetic tree, *692*
  spiral cleavage, 673
Lorenz, Konrad, 1143, 1145
Loriciferans, *683, 692,* 702, **703**
Lorises, 738
Lotus, 613
"Lotus effect," 613
Lovley, Derek, *576*
Low-density lipoproteins (LDLs), 377, **1087**
Lower esophageal sphincter, 1080, 1081
Low-GC Gram-positive bacteria, *571,* **573–574**
Loxodonta africana, 194
LTD. *See* Long-term depression
LTP. *See* Long-term potentiation
Luciferase, 1021
Luciferin, 124, 1021
"Lucy" (australopithecine), 739, *740*
Luehea seemannii, 708
Lumbricus terrestris, 699
Lumen
  of endoplasmic reticulum, 79
  of the gut, **1078,** *1079*
Lung cancer, *387,* 392
Lungfish, *534, 723,* **728,** 1048
Lungs
  in birds, 1030–1031
  in humans, 1032–1035
  in lungfish, 1048
  surface area and, **1028**
  tidal ventilation, 1031–1032
Luteinizing hormone (LH)
  anterior pituitary and, *880*
  male sexual function and, *908,* 909
  ovarian and uterine cycling, 911, 912
  puberty and, **889,** 890
Lycaon pictus, 1241
Lycoperdon sp., *655*
Lycophyta (Lycophytes), *612,* 618, **619,** 624–625
Lycopodium, 620
  L. obscurum, 625
Lygodium microphyllum, 626
Lymph, 402, **1060**
Lymphatic system, 402, **1060,** 1084
Lymph nodes, **402, 1060**
Lymphocytes, *1056*
  clonal anergy, 410–411
  clonal deletion, 410
  clonal selection, 408–409
  effector cells and memory cells, 409
  specific immune response and, 407–408
  types of, 402–403
Lymphoid tissue, 402
Lymphomas, 387, 422
Lynx canadensis, 1189
Lyon, Mary, 323
Lyperobius huttoni, 1135
Lyrurus tetrix, 1150
Lysenko, Trofim, 831
Lysine, *43*
Lysis, cellular immune response and, 417
Lysogenic bacteria, **286**
Lysogenic cycle, **285,** 286, 290
Lysosomal storage diseases, 82
Lysosomes, *101*
  glycoproteins and, 274
  in phagocytosis, 113
  in receptor-mediated endocytosis, 113
  structure and function, 75, **81–82**
Lysozyme, 46, *47,* 128, 130, **405,** 532–533
Lytic bacteriophage, 286–287
Lytic cycle, **285,** 286, 287, 290

- M -

Macaca fuscata, 1140
Macaques, learned behavior in, 1140, 1141
MacArthur, Robert, 1132
Macaws, *1151*
MacKinnon, Roderick, 109
MacLeod, Colin, 235
Macrocystis, 584, 599
  M. pyrifera, 583
Macromolecules, **38–42**
Macronectes giganteus, 1095
Macronucleus, 593
Macronutrients
  in food, **1073**
  identifying by experiment, 784
  plants and, *782,* **783**
  See also Mineral nutrients
Macrophages, 401
  cellular immune response and, 411, *415, 416*
  cytokines and, 404
  functions, **402,** *403*
Macropus giganteus, 736
Macroscelidea, *736*
Macula, *975*
Macular degeneration, 365
Madagascar, endemic species, 1129, *1130*
Madagascar ocotillo, *1130*
"Mad cow disease," 378
Madia sativa, 520
Madreporite, 719
MADS box, **441**
Magma, 468
Magnesium
  in chlorophyll, 164
  deficiency in plants, *783*
  as macronutrient, *1073*
  NMDA receptors and, 960
  in plant nutrition, 781, *782,* 784
Magnesium sulfate, 784
Magnetic poles, 333
Magnetism, 333
  animal migration and, 1156
Magnetite, 61, *576*
Magnolia watsonii, 640
Magnoliids, *644*
Maidenhair tree, 634, *635*
Maimonides, 206
Maize. *See* Corn
Major histocompatibility complex II (MHC II) proteins, 415, *416,* 417, 421
Major histocompatibility complex I (MHC I) proteins, 415, *416,* 417
Major histocompatibility complex (MHC) proteins, 403, **415,** 417
Malaria, 585–586, 648
Malathion, 131, 961
Malay Archipelago, 1128, *1129*
Male sex organs, human, 906–908
Malignant tumors, **386**
Mallards, courtship display, *1142–1143*
Maller, James, 185
Malleus, *972,* 973
Malnutrition, **1074–1075**
Malpighi, Marcello, 775
Malpighian tubules, **1098**
Maltase, 1083, *1084*
Malthus, Thomas, 488
"Malting," 801
Maltose, 50, *51*
Mammals
  allantois and umbilical cord, 936
  in amniote phylogeny, *731*
  circulatory system, 1049
  cleavage in, 923, *924,* 925
  evolutionary history, 477, 478
  extinctions through human predation, 482
  gastrulation in, 932
  latitudinal gradient of species richness, 1200
  major groups, 735
  overview of, 735
  placenta, 935
  radiations, **734,** 736
  sex chromosomes, 225
  size range in, 734–735
  teeth, 1076
  therians, 735–737
Mammary glands, 735
Mammoths, 479
Manatees, 737
Mandarin duck, *734*
Mandibles, **1077**
Mandrills, *553*
Mandrillus leucophaeus, 739
Manganese, *782, 783, 1073*
Mangold, Hilde, 929
Mangroves, 846, 848, 1242
Manic-depressive disorder, 342–343
Mannose, *50*
Man-of-war, *679*
Mantids, *709*
Mantle
  of cephalopods, 702
  of mollusks, 700
Mantle cavity, 700
Mantodea, 709
Mantophasmatodea, 709
Mantophasmatodeans, 710
Mantophasma zephyra, 711
Manucodes, 519
Manucodia comrii, 519
MAP kinases, 340, 347, 439
Map pufferfish, *487*
Map units, **224**
Marathon Mouse, 138–139
Marathon runners, 138
  muscle fibers and, 1013–1014, 1015
Marchantia, 622, 623
Marine iguana, 864
Marine Stewardship Council, 1238
Marker genes, **237**
Markers, **237**
Marking, of individual organisms, 1168, *1169*
Markov, Georgi, 256
Marler, Catherine, 1148
Mars
  formation of, 34
  life on, 560
  meteorites from, 61
  polar ice, 32
  question of life on, 560
  water on, 20, *21*
Marshall, Barry, 1081
Marsilea, 626
Marsileaceae, 627
Marsupials, 735, *736*
Masada, 630
Mass action, law of, 336
Mass extinctions
  changes in sea level and, 468, *469*
  Cretaceous, 471, 477, 482
  Devonian period, 474
  extraterrestrial events and, **470,** 471
  Permian, 476
  Triassic, 476
Mass number, **23**
Massospondylus carinatus, 670
Mass spectrometry, 396
Mastax, 696
Mast cells, *403,* **406**
Maternal behavior
  fosB gene and, 1143–1144
  pair bonding, 881
Maternal effect genes, **442**
Mating behavior, fitness and, 1152
Mating strategies, **903–904**
Mating types, fungal, **659**
Matthaei, J. H., 265
Maturation promoting factor, 185
Mawsonites, 473
Maximum likelihood analyses, **549**
Mayflies, *709,* 710
Mayr, Ernst, 510
M band, *1006,* 1007
McCarty, Maclyn, 235
McCulloch, Ernest, 432
MCMI protein, 441
Measles, *409*
Measles virus, *409*
Mechanically gated ion channels, **950**
Mechanical weathering, 786
Mechanistic view of life, 21
Mechanoreception, 970–975
Mechanoreceptors, 965, *966,* **970–975**
Mecoptera, 709
Medicinal leech, *699,* 700
Medicinal plants, 647–648
Medicine
  bone marrow transplantation, 432–433
  embryonic stem cells and, 433–434
  molecular evolution studies and, 538–539
  patient-derived stem cells and, 426
  See also Molecular medicine

Medulla (adrenal). *See* Adrenal medulla
Medulla (brain)
  control of heart rate and blood pressure, 1062, 1063
  location, **986**
  regulation of breathing and, 1039, 1040
Medulla (renal). *See* Renal medulla
Medusa, **685,** 686, *687*
Meerkats, *9*
Megagametophytes, **621,** *622,* 632, *820,* 821
Megapascals (MPa), 767
Megaphylls, **620**
Megapodes, 1190
Megasporangia, **621,** 632, 633, *820,* 821
Megaspores, **621,** *622, 820,* 821
Megasporocytes, *820,* 821
Meiosis, *196–197,* **197–199**
  compared to mitosis, *200–201*
  errors affecting chromosomes, 199–200
  first meiotic division, *196–197,* 197–199
  in gametogenesis, 899, *900*
  generation of mutations and, 277
  genetic diversity and, 193–195, 199
  genetic recombination, *223*
  overview of, **183,** 195
  polyploids and, 200–202
  second meiotic division, *196–197,* 199
  segregation of alleles and, *212*
Meissner's corpuscles, 970–971
Melanerpes formicivorus, 1161
Melanin, phenylketonuria and, 376
Melanocyte-stimulating hormone, 882
Melarsoprol, 582
Melatonin, *880,* **890**
Melospiza melodia, 1174–1175
Membrane potential, **109, 946**
  ion channels and, **108–109,** 948–951
  measuring, 947
  Nernst equation, 109, 948, *949*
  of plant cells, 768
  resting, 946, 947, 948–950, 952
  in sensory receptor cells, 966–967
Membrane proteins
  altered, 376–377
  integral, 99, *100*
  peripheral, *98,* 99, 100
Membranes
  cold-hardening and, 847
  organelles and, 74, 75
  origin of cells and, 11
  in prokaryotes, 73
  selective permeability, 105, 765
  See also Biological membranes; Endomembrane system; Extraembryonic membranes; Plasma membrane
Membrane transport
  active, 111–113
  carrier proteins, 110
  diffusion, 105–106, *111*
  ion channels and channel proteins, 108–109
  osmosis, 106–107
  passive, 105–110
Membranous bone, **1018**
Memory
  glutamate receptors and, 959–960
  limbic system and, 988–989
  methamphetamine abuse and, *23*
  neuronal basis of, 999–1000
Memory cells, 409
Menadione, *1074*
Mendel, Gregor, 207–210, 213, 490
Mendelian inheritance, 207–217
Mendelian populations, **491–492**
Meningitis, *409*
Menopause, **910**
Menstruation, **911,** 912
Mental retardation
  Down syndrome, 199
  fragile-X syndrome, 379, 382–383
  phenylketonuria and, 375–376
Menthol, *648*
MerA gene, 370
Mercaptan, 1021
Mercaptoethanol, *40*
Mercury (element), 370, 849
Meristem identity genes, **825**
Meristems, 432, **753**
Merkel's discs, 970, *971*
Merops
  M. apiaster, 905
  M. bullockoides, 1160
Merozoites, *586,* 596
Mertensia virginica, 554
Meselson, Matthew, 241, 242
Meselson–Stahl experiment, 241–242, *243*
Mesenchyme, 674, *675,* 927
Mesocyclops, 1181
Mesoderm, 673, *926,* **927**
Mesoglea, **684,** 686
Mesophyll, **173,** 174, **761**
Mesozoic era, *466–467,* **476–477,** 634, 733
Mesquite trees, 846
Messenger hypothesis, 260
Messenger RNA (mRNA)
  alternative splicing, **324–325**
  central dogma and, **260,** 261
  creation of cDNA and, 363
  differences in stability of, 325
  editing, **326**
  G cap and, 326
  genetic code and, 264
  translation. *See* Translation
Metabolic factors, **294**
Metabolic homeostasis, 155
Metabolic inhibitors, 391
Metabolic pathways
  anabolic and catabolic interconversions, 154–155
  commitment step, **133**
  feedback inhibition, 133, *134*
  overview of, 140, *141*
  principles of, 139
  regulation, 155–157
  role of enzymes in, 131
  systems approach to, 396, *397*
Metabolic pool, 155
Metabolic rates
  endotherms and, 861–862
  overview of, 1069–1070
  thyroxine and, 884
  See also Basal metabolic rate
Metabolism, **120**
  in the evolution of life, **11**
  liver and, 1086–1087
  thyroxine and, 884
  types of reactions in, 120
  See also Anabolism; Catabolism
Metabolites
  primary, 841
  secondary, *839,* 841, 844
Metabotropic receptors, 958, 965–966
Metal ion catalysis, 129
Metamorphosis, 679, **710,** 877–878
Metanephridia, **1097**
Metaphase
  first meiotic division, *197,* 198, *200*
  mitosis, *188,* **190,** *191, 200*
  second meiotic division, *196, 201*
Metaphase chromosomes, 195
Metaphase plate, 189
Metapopulation, **1178**
Metarhizium anisopliae, 650
Metastasis, **386–387**
Meteorites, 61, 477
Meteoroids, 471, 474
Methamphetamine, *23*
Methane
  bacteria and, 302, 1085
  covalent bonding in, 26
  Euryarchaeota and, 577
  fires and, 1212
  global warming and, 302, 577
  Martian atmosphere and, 560
  nonbiological reservoir of, 578
Methanococcus, 302
Methanogens, **577**
Methanopyrus, 577
Methanospirillum hungatii, 561
Methionine, *43,* 264, 268, 269, 270
"Methuselah" pine, 635
Methylamine, *40*
Methylases, 354
Methylation, of DNA, 249, 323, 354, 382
5-Methylcytosine, 382
N-Methyl-D-aspartate, 959
Methylococcus, 302
Methyl salicylate, 839
Mexico, opuntia and, 1166
MHC. *See* Major histocompatibility complex
Mice, 464
  atherosclerosis and, 396, *397*
  cloned, *432*
  engrailed gene, *537*
  eye development genes and, 448
  genome size, *533*
  knockout experiments, 364–365, 1143–1144
  newly formed neurons in adults, *984*
  olfactory receptor genes, 968
  ovarian cycle, 910
Micelles, **1082,** *1083*
Microbes, 563
Microbial communities, 563–565
Microbial eukaryotes, **583**
  cell surfaces of, 592, *593*
  endosymbiosis and, 585
  habitats, 591
  limestone deposits and, 587
  locomotion, 591
  major groups, *584,* 596–607
  nutrient sources, 591
  pathogenic, 582–583, 585–586
  petroleum deposits and, 587
  phytoplankton, 583, 584–585
  reproduction in, 593–595
  vacuoles in, 591, *592*
  See also Protists
Microchemostats, *564,* 565
Microfibrils, in plant cell walls, 807
Microfilaments
  polarity, 436
  structure and function, **86–87**
Microfluidics, *564,* 565
Microfossils, 616
Microgametophytes, **621,** *622, 820,* 821
Micronuclei, 593
Micronutrients
  foliar applications of, 787
  in food, **1073**
  global cycling, 1219–1220
  identifying by experiment, 784
  required by plants, *782,* **783**
  See also Mineral nutrients
Microparasites, 1189, 1191, *1192*
Microphylls, **620**
Micropyles, **636**
Microraptor gui, 733
Micro RNAs, **325,** 365
Microsatellites, 311
Microscopes, 70–72
  discovery of cells and, 4, *5*
Microsorum, 627
Microsporangia, **621,** *625,* 632, *820,* 821
Microspores, **621,** *622,* 632, 821
Microtubule organizing center, 88
Microtubules
  chromosomal movement and, 189
  in cilia and flagella, 88–89
  mitotic spindles and, 189
  polarity, 436
  structure and function, *87,* **88**
Microvilli, 87, **1077,** 1078, *1078,* 1083
Midbrain, **986,** 987
Middle ear, 972–973, 974
Middle lamella, **749**
Midgut, **1077**
Migration, **1115–1116,** 1154–1156
Milk, transgenic animals and, 369
Milkweed, 824, 844
Millennium Ecosystem Assessment project, 1221–1222
Miller, Stanley, 61
Miller-Urey experiments, 61–62
Millipedes, 712
Mimicry, 1190–1191
Mimulus aurantiacus, 641
Mineralcorticoids, 888–889
Mineral nutrients
  availability in soils, 786–787
  deficiencies, *783*
  elements included in, **781**
  foliar applications of, 787
  in food, 1073
  in plant nutrition, 782–785
Mine tailings, 849
Minimal nutrient media, 290
Minisatellites, 311
Miocene period, *478*
Mirids, *519*
Mirounga angustirostris, 1063, 1174
Miscarriages, 200, 915–916
Mismatch repair, in DNA, **249,** *250*
Missense mutations, **275**
Mistletoes, 792
Mites, 289, 713, 904
Mitochondria
  in animal cells, *76*
  ATP synthase and, 168
  in endosymbiosis theory, 92
  excavates and, 602–603
  genetic code and, 264
  genome and gene mutations, 228, 229
  in kinetoplasts, 604
  location of energy pathways in, *141, 146, 149, 151*

membrane-associated proteins and, 99
movement of polypeptides into, 273
structure and function, 75, **82–83**
Mitochondrial DNA (mtDNA), 548
Mitochondrial matrix, 83
Mitogen-activated protein (MAP) kinases, 340, 347, 439
Mitosis, **183**
cell division plane, 188–189
chromosome movements during, 189–191
compared to meiosis, 199, *200–201*
maintenance of genetic constancy, 193
phases of, 189–192
structure of chromosomes during, 187, *188*
Mitotic centers, 189
Mitotic spindles, 189, 190–191, 923
Mitter, C., 518
Mockingbirds, 59
Model systems/organisms, **9**, 257, 283, *284*, *797*
Moderately repetitive sequences, 311, *312*
Modules, **451–452**
Molarity, **34**
Molars, 1076
Molds, 665
Molecular clocks, 531, **551**
Molecular evolution
applications of, 536–539
concerted evolution in gene families, 535–536
determining evolutionary divergence, 527, *528–529*
experimental studies, 527–529
gene duplication, 534–535
in genome size and organization, 533–534
neutral theory of, 531, 532
overview of, **525–526**
positive and stabilizing selection, 532–533
sequence alignment technique, 526–527
Molecular markers, 1168, *1169*
Molecular medicine
genetic screening, 384–385, *386*
"genome prospecting," 374, 382
treating genetic diseases, 391–392
Molecular mimicry, autoimmune diseases and, 421
Molecular tool kit, **449–450**
Molecular weight, **25–26**
Molecules, **25**
isomers, 40, *41*
polar and nonpolar, 29–30
Mole (measurement), **33**
Moles (animals), 735
Mollusks, *683*, *692*, 698
body plan, **700**
circulatory system, *1046*
cleavage in, 923
coelom in, 691
image-forming eye in cephalopods, 978, *979*
major groups, 701–702
monoplacophorans, 700–701
Molting
arthropods and, 693, 1016
hormonal control in insects, 876–877
tracking birds and, 1168, *1169*
Molybdenum, *782*, *1073*, 1220
Monarch butterfly, *680*
Monoamines, 958, *959*
Monoclonal antibodies, **413–414**
Monocots, **745**, *746*
early shoot development, *799*
phylogeny, *644*, **645**
root structure and anatomy, *756*
stem anatomy, *757*
stem diameter, 759
Monocytes, *403*, 406, *1056*
Monod, Jacques, 297
Monoecious species, **639**, **904**
Monoecy, 225
Monogamy, 504, 508
Monogeneans, 696
Monohybrid crosses, **209–210**, *211*
Mono Lake, California, *1122*
Monomers, **39**
Monomorphic populations, 491
Mononucleosis, 422
Monophyletic groups, **555**
Monoplacophorans, **700–701**
Monosaccharides, **49–51**
Monosodium glutamate, 970
Monosomy, **200**, *201*
Monosynaptic reflex, **987–988**
Monoterpenes, *839*
Monotremes, 905
Monotropa uniflora, 792
Monounsaturated fatty acids, 55
Monozygotic twins, 926
Montverde Cloud Forest Preserve, 1230
Moore, Michael, 1148
Moraines, succession and, 1197–1199
Morchella esculenta, *666*
Morels, 665, *666*
Morgan, Thomas Hunt, 222, 223, 226
"Morning-after pills," 915
Morphine, *648*
Morphogenesis, **427**, 428
Morphogens, **441–442**, 804
Morphological species concept, **509**
Morphology, **465**
phylogenetic analyses and, 547–548
Morula, 922
Morus serrator, *1150*
Mosaic development, **925**
Mosquitoes, 587–588, 1181
"Moss animals," 695
Mosses, 612, 614, *615*, 616, **623–624**
Moths, 455, *709*, 968
Motifs, 320, *321*
Motile animals, **676**
Motor end plate, **955**
Motor neurons, 955, 1008
Motor proteins
actin filaments and, 87
microtubules and, 88
separation of chromatids and, 190–191
types and functions of, 89, *90*
Motor unit, **1008**
Mt. Kenya, 1118
Mountain, rain shadows and, 1115
Mountain zebra, *1131*
Mount Everest, 1024, 1026, 1027
Mouse. *See* Mice
Mouth, 673, **1077**, 1079, 1080
Movement proteins, 348, 777
M phase, **184**, 185, 186
mRNA. *See* Messenger RNA
mRNA codons, 264
*Mst*II enzyme, 385
mtDNA. *See* Mitochondrial DNA
Mucosa, 1078, *1079*
Mucosal epithelium, **1078**, *1079*, 1083
Mucous membranes, 405
Mucus
gastric, 1080, *1081*
mammalian lungs and, 1034
a nonspecific defense, 405
in semen, 908
in the vertebrate gut, 1078
Mucus escalator, 1034
Mudpuppy, *730*
Mule deer, 841
Muller, Hermann J., 501
Müllerian mimicry, **1190**, 1191
Muller's ratchet, 501
Mullis, Kerry, 251
Multicellular organisms, 4, 102
C. elegans genes essential to, 310
evolution of, 12
internal environment and, 7–8, 855–856
Multifactorial diseases, **379**
Multiple fission, 593
Multiple fruits, 643
Mumps, *409*
Mumps virus, *409*
Mung beans, 813
Muscarinic receptors, 958–959
Muscle cells
action potentials and, 956
energy storage and, 1071
Muscle contraction
calcium-mediated, 344
effect on blood flow, 1059, *1060*
mechanisms of, 1005–1012
sources of ATP for, 1014–1015
Muscle fibers
control of differentiation in, 435
mechanisms of contraction in, 1005–1010
structure, **1005**, *1006*
types, 138
See also Fast-twitch fibers; Slow-twitch fibers
Muscle(s)
anabolic steroids and, 874–875
antagonistic sets, 988
cardiac. *See* Cardiac muscle
contraction. *See* Muscle contraction
factors affecting strength and endurance in, 1013–1015
fibers. *See* Muscle fibers
joints and, 1019, *1020*
lactic acid buildup, 147–148
mechanoreceptors, 971–972
metabolic activity and, 157
types of, 1005, *1010*
Muscle spindles, *971*, 972
Muscle tissues
in organs, *860*
types of, **858**
Muscle tone, **1012**
Muscular dystrophy, 312
Duchenne, 377, 380, 381, 382, 385
Musculoskeletal system. *See* Muscle(s); Skeletal systems
Mushrooms, 667
Mutagens, 258, **277**, 936
Mutant alleles, 217–218
Mutants, conditional, 275
Mutation experiments, 86
Mutation rates, 494
immunoglobulin genes and, 419
Mutation(s), **7**, **530**
asexual reproduction and, 193
cancer and, 387–388
chromosomal, 275, 276, *277*
creation of new alleles, **217–218**
evolution and, 7, 277–278, 525
frequency of, 277–278
genetic diseases and, 375–377, 382
genetic variation and, 494
homeotic, 444
molecular evolution and, 526
neutral, 278, 501
nucleotide substitutions, 526, 527–528, 530–531
in organelle genes, 228–229
prokaryotes and, 571
synthetic DNA and, 363
transposons and, 312
types of, 274–277, *278*
Mutualisms, **1188**
ant-plant, 1184–1185, 1194
fungi and, 651, 655
mycorrhizae. *See* Mycorrhizae
nitrogen-fixation and, 788–789. *See also* Nitrogen-fixation
Mycelia, 652
Mycobacterium tuberculosis, 302, *409*, 573
Mycoplasma
M. gallisepticum, *574*
M. genitalium, 301, 303, 533
Mycoplasmas, *284*, **574**
Mycorrhizae, **655**, 657–658, 756, **789**, 1194
Mycorrhizal fungi, *664*
Myelin, 99, **946**, 987
Myelination, 955–955
Myers, Norman, 1235
Myliobatis australis, *726*
Myoblasts, 435, 1005
Myocardial infarction, **1060**
MyoD gene, 435
Myofibrils, **1006–1007**
Myogenic heartbeat, 1010
Myoglobin, 315, 535, 1013, **1014**, **1037**
Myosin
actin filaments and, 87
contractile ring and, 192
muscle contraction and, 344, 1005–1010, 1014
structure of filaments, 1007
Myosin phosphatase, 1012
Myotis sp., *737*
Myotonic dystrophy, 383
Myriapods, *683*, *705*, **712**
Mytilus californianus, 1197
Myxamoebas, 607

- N -

NAD. *See* Nicotinamide adenine dinucleotide
NADH-Q reductase, 149, *150*, *151*
NADP$^+$. *See* Nicotinamide adenine dinucleotide phosphate
Naegleria, *584*, 603
Nakano, Hiroyuki, 1151
Naked mole-rats, 1160
Nanaloricus mysticus, *703*
Nanoarchaeota, **575**, 577, *578*
Nanoarchaeum equitans, *578*
Nanos, 442
Nasal cavity, 968, *969*
Nasal epithelium, 984
Nasal salt glands, 1095
Nasopharyngeal cancer, *387*
National Wildlife Federation, 1242

Native Americans, 1068–1069, 1074
Natural gas, 585, 618
Natural immunity, 409
Natural killer cells, *403,* **405–406**
Natural selection, **7, 489**
  antibiotic resistance, 295
  Darwin's theory of, 5–6, 488–489
  ways of creating quantitative variation, 497–498
The Nature Conservancy, 1240
Nauplius larva, **680,** 707
Nautilus, 479, *701,* 702
Navel oranges, 819
Navigation, 1154–1156
Neanderthals, 740
Nebela collaris, 593
Necrosis, **202**
Nectar, 641, 1184, 1194
Nectar guides, 641
Nectria, 1216
Necturus sp., *730*
Negative control, **300**
Negative feedback
  control of hormone secretion, 883
  in physiological systems, **856–857**
  regulation of metabolic pathways, 156, 157
Neiserria, 73
Nematocysts, **678,** *679,* 1020
Nematodes, *683, 692*
  fungal predators, 654
  genes essential to multicellularity, *310*
  genome, 309
  overview of, **704**
  parasitic, 704–705
  recruitment in plant defenses, 842, *843*
  See also Caenorhabditis elegans
Nematus sp., *678*
Nemerteans, 697
Nemoria arizonaria, 455
Neognaths, 732
Neopterans, *709,* 710
Nephridiopore, 1097
Nephrons
  organization in the kidney, 1101–1102
  structure and function, **1098**
  urine formation in, 1099–1101
  water permeability and, 1104
Nephrostome, 1097
Nernst equation, 109, **948,** *949*
Nerve cells, 859, 943
  See also Glia; Neurons
Nerve deafness, 974
Nerve gases, 131, 961
Nerve impulses, 859
Nerve nets, 686, 943, *944*
Nervous systems
  blood glucose levels and, 1087
  brain regions, 989–991. *See also* Brain
  cell types in, 859, 943
  development of, 932, 933, 986–987
  evolutionary complexity in, 943–944
  functional organization, 985–986
  of the gastrointestinal tract, 1086
  interactions with the endocrine system, 880–883
  limbic system, 988–989
  production of new neurons in adults, 938
  in protostomes, 691
  reticular system, 988
  spinal cord. *See* Spinal cord)
  See also Autonomic nervous system; Central nervous system; Peripheral nervous system
Nervous tissues
  cell types in, **859,** 943
  in organs, *860*
Nest systems, of eusocial animals, 1160, *1162*
Net primary production (NPP), **1209,** 1210, *1211*
Net reproductive rate, 1173
Neural crest cells, 933
Neural ectoderm, *926*
Neural plate, 932
Neural tube, **932, 986–987**
Neural tube defects, 932–933
Neurogenesis, in adults, 984–985
Neurohormones, **881,** 882–883, **985–986**
Neuromuscular junctions, **955–956,** 1008–1009
Neuronal networks, 943–944
  See also Nervous systems
Neurons
  action potentials, 944, 945, 946
  autonomic nervous system, 992
  communication with other cells, 955–961
  excitatory and inhibitory synapses and, 957
  features of, 859, **943,** 985
  generation and conduction of signals, 946–955
  glucose requirement, 1087
  neuromuscular junctions, 955
  newly formed in adults, 984–985
  presynaptic and postsynaptic, 944
  production of stimulated by learning, 938
  summation in, 957
  structure and function, 944–946
Neuropterida, 709
Neurospora, 664
  N. crassa, 257–258
Neurotoxins, 486, 498
Neurotransmitters, 950–951
  clearing from synapses, 960–961
  metabotropic receptors and, 958
  at neuromuscular junctions, 955
  poisons and, 955, 1021
  receptors and, 958–959
  released by action potentials, 955–956
  response of postsynaptic membranes to, 956, *957*
  sensory receptor cells and, 967
  synaptic transmission and, **945–946**
  types of, 958, *959*
Neurulation, **932,** 933
Neutral alleles, **501**
Neutral mutations, 278, 501
Neutral theory of molecular evolution, 531, 532
Neutrons, **21,** 23
Neutrophils, *403,* 406, *1056*
Newborns, genetic screening, 384, 385
New Orleans, 1112
New World monkeys, 738, *739*
NF-kB transcription factor, 407
Niacin, *1074*
Niches, radiation of eutherian mammals and, 736
Nicholas II, 356, *357*
Nickel, *782, 784, 785*
Nicotinamide adenine dinucleotide (NAD)
  an electron carrier, **141–142**
  in catalyzed reactions, *129*
  cellular respiration and, 153
  in the citric acid cycle, 145, *146*
  electron transport chain and, 148, 149, *150*
  in fermentation, 147, *148*
  in glycolysis, *143,* 144
  in pyruvate oxidation, 145, *146*
Nicotinamide adenine dinucleotide phosphate ($NADP^{+}$)
  Calvin cycle and, 170, *171*
  in photosynthetic electron transport, 162, **166,** 167, 168, *169*
Nicotine, 337, 841, *842*
Nicotinic receptors, 958, 959
Night blindness, *1074*
Night length, plant growth regulation and, 812
Nightmares, 998
Night vision, 980
Nile crocodile, *733*
Nileids, 480
Nirenberg, Marshall W., 265
Nitrate reduction, **791**
Nitrate(s)
  bacteria and, 569
  global nitrogen cycle and, *790,* 791
  groundwater contamination and, 1217
Nitric acid, 1205, 1218
Nitric oxide
  fossil fuels and, 1217
  neurotransmitter, 958, *959*
  penile erection and, 908
  second messenger, 344, 345
Nitrification, **791**
Nitrifiers, **569,** 578, **790**
Nitrogen
  atmospheric, 477, 1208, 1216
  covalent bonding capability, *27*
  deficiency in plants, 783
  electronegativity, *27*
  global cycle, 790–791, 1216–1217
  in plant nutrition, 781, *782*
Nitrogenase, 789
Nitrogen cycle, 790–791, 1216–1217
Nitrogen deposition, 1217
Nitrogen-fixation
  effect of atmospheric CO_2 on, 1220
  global nitrogen cycle and, 790–791, 1216, 1217
  industrial, 789–790
  nitrogenase and, 789
  plant-bacteria symbiosis, 477–478, **788–789**
Nitrogen fixers, 477–478, **569,** 578, 788–789
Nitrogen metabolism
  excretion of wastes from, 1095–1096
  prokaryotes and, 569
Nitroglycerin, 344
Nitrosococcus, 569
Nitrosomonas, 569
Nitrous acid, 277
Nitrous oxide, 1217
Nitrum, 164
NMDA receptors, 959, 960
Nobiliasaphus, 480
Nocturnal animals, 1153
Nodes
  in phylogenetic trees, 543
  of plant stems, **745**
Nodes of Ranvier, *946,* **955**
"Nod" factors, 319
nod genes, 789
Nodulation (Nod) factors, 789
Noggin protein, 932
Noller, Harry, 269
Nominus novus, 720
Noncoding DNA, 307, 533–534
Noncompetitive inhibitors, 131–132
Noncyclic electron transport, **166–167**
Nondisjunction, **199,** *201,* 226, 277
Non-native species. *See* Exotic species
Nonpolar covalent bonds, 27
Nonpolar molecules, 29–30
Nonrandom mating, **495–496**
Non-REM sleep, **998,** 999
Nonsense mutations, **275–276**
Nonspecific defenses, 401, 404–407
Nonsynonymous substitutions, **530,** *531,* 532–533
Non-template strand, 261
Nonvascular plants, **612,** 614–616
Noradrenaline, *880,* 887, 888, 992
Noradrenergic neurons, 992
Norepinephrine
  adrenal gland and, *880,* **887**
  blood pressure regulation and, 1062, *1063*
  characteristics, 888, *959*
  heartbeat in mammals and, 1052, *1053*
  as a monoamine neurotransmitter, 958, *959*
  noradrenergic neurons and, 992
  signal transduction pathways and, *892,* 893
  smooth muscles and, 1012
Normal flora, **404**
Normative disciplines, **1227–1228**
Norplant, 915
Northern elephant seals, *1063*
Norway, whaling and, 1180
Norway rats, 464
Nose, of whales, 920
NO synthase, 344
Notechis, 455
Nothofagus, 1119
Notochord, 548, **722, 932**
Nottebohm, Fernando, 984
NPP. *See* Net primary production
Nuclear basket, *78*
Nuclear envelopes, 75, *78,* 191
Nuclear lamina, *78,* 79
Nuclear localization signal, 78
Nuclear matrix, 79
Nuclear pores, 75, 78
Nucleases, 1077, *1084*
Nucleic acid hybridization, **313–315**
Nucleic acids, 39, *41,* **57–60**
Nucleoid, **72,** *73,* 92
Nucleolus, **75,** *76, 77,* 191
Nucleoplasm, 79
Nucleosides, 57, 241
Nucleosomes, **187,** *188,* 322
Nucleotides, **7,** 57–59, 60, 238, 242–243
Nucleotide sequences, phylogenetic analyses and, 548
Nucleotide substitution rates, 528, 530–531
Nucleotide substitutions, **526,** 527–528, 530–531

Nucleus (atomic), 21
Nucleus (cells)
  cytoplasmic receptors and, 339
  eukaryotes and, **72**
  eukaryotic evolution and, **11,** 588
  movement of polypeptides into, 273
  structure and function, 75, *76,* 78–79
  telophase and, 191–192
Nucleus (neuron group), **988**
Nudibranchs, *701,* 702
Null hypothesis, **16**
Nutrient media, 290
Nutrients
  absorption in the small intestine, 1083–1084
  global cycling. *See* Biogeochemical cycles
  major elements, **781**
  significance of, 7
  See also Macronutrients; Micronutrients; Mineral nutrients
Nymphaea sp., *644*

- O -

Obesity, 1068–1069, 1087, 1088
Obligate aerobes, **568**
Obligate anaerobes, **568**
Obligate intercellular parasites, 284
Occam's razor, 547
Occipital lobe, **990,** 991, 995
Ocean conveyor belt, 1214
Oceans
  absorption of carbon dioxide, 1213–1214
  acidification of surface water, 1213
  biogeographic regions, 1135–1136
  biomass and energy distribution, *1188*
  currents, 1115, *1116,* 1135–1136
  global warming and, 1215, 1234
  iodine and, 1219
  primary production in, *1210*
  upwelling zones, 1206, *1207*
Ocorhynus, 1172
Ocotillo, 845
Octet rule, 25
Octopus, *701,* 702, *979*
Octopus bimaculoides, 701
Odocoileus virginianus, 1168
Odonata, 709
Odontochile rugosa, 706
Odorants, 345–346, **968**
Odor signal receptors, 345
Oedogonium, 602
Ogyginus, 480
Ogygiocarella, 480
Oil of wintergreen, 839
Oil (petroleum), 586–587
Oils, **55**
Oilseed rape, 1179
Oil spills, bioremediation, 294
Okazaki, Reiji, 246
Okazaki fragments, **246,** *247*
Old World monkeys, 738, *739*
Oleic acid, 55
Olfaction, **968–969**
  See also Smell
Olfactory bulb, 968, *969,* 984
Oligocene period, *478*
Oligochaetes, 699
Oligodendrocytes, **946**
Oligo dT, 363
Oligonucleotides, 385
Oligosaccharides, **49,** 51, 319, 838
Omasum, **1085**
Ommatidia, **978**
Omnivores, **679, 1075,** *1076,* 1186
ompC gene, 335
OmpC protein, 335
OmpR protein, 334, *335*
Onager, *1131*
Oncogenes, 324, **388,** 389
One-gene, one-enzyme hypothesis, **258,** *259*
One-gene, one-polypeptide relationship, **258**
Onions, 832
Onychophorans, *683, 692,* **705–706**
Oogenesis, **899–901**
Oogonia, **899,** *900*
Oomycetes, *584, 597,* **600,** *601*
Ootid, *900,* **901**
Open circulatory systems, **1046**
Open complex, 262
Open reading frames, 301
Operation Migration, 1226
Operator-repressor systems, 297–299, 300
Operators, 290, **297**
Opercular flaps, 1029
Operculum, 726
Operons, 297–300
Ophiostoma ulmi, 1233
Ophiothrix spiculata, 720
Ophrys scolopax, 1194
Opisthokonts, *584, 597,* **605,** 651, *652,* 677
Opium poppy, *648*
Opossums, 735, *736*
Opportunistic infections, 422
Opportunity costs, 1148
Opportunity (robotic vehicle), *21*
Opsins, 550, **976,** 980
Optical isomers, **40,** *41*
Optic chiasm, **996,** *997*
Optic cup, 437
Optic nerves, 995, 996
Optic vesicles, 437
Opuntia, 646
  O. stricta, 1166–1167
OPV. *See* Oral polio vaccine
Oral contraceptives, 914–915
Oral polio vaccine (OPV), 524–525
Orange sea pen, *686*
Orange trees, 819
Orangutans, 738, *739*
Orbitals, **23–25**
Orcas, *1168*
Orchids, 658, 1194
Orcinus orca, 1168
Orconectes williamsii, 675
Ordovician period, *466–467,* **473**
Organelles
  cytokinesis in plants and, 192
  endomembrane system, 79–82
  endosymbiosis and, 588–590
  eukaryote evolution and, **11**
  genomes, 228–229
  glyoxysomes, 84
  overview of, **74,** 75
  peroxisomes, 83–84, 85
  red blood cells and, 1055
  that process information, 75, 78–79
  that transform energy, 82–83
  vacuoles, 84–86
Organic chemicals, 39
Organic compounds, 781
Organic fertilizers, 787
Organic phosphates, *40*
Organ identity genes, **440–441**
Organisms
  cells and cell theory, 4–5
  common characteristics, 3
  control of internal environment, 7–8
  evolution and, 3, 5–6
  genomes and biological information, 7
  hierarchical perspective of, 8
  interactions between, 9
Organizer, **929–931**
Organ of Corti, *972,* **973,** 974
Organogenesis, **932–934**
Organophosphates, 370
Organs, 8, **859,** *860*
Organ systems, 8, 857 **859**
Organ transplant surgery, 417
Orgasm, **908,** 913, 914
Origin of life, 10–11, 61–64
Origin of replication *(ori),* 182, **243,** 244, *245,* 293
Origin of Species, The (Darwin), 68, 488, 489, 511, 1185
Origin of the Continents and Oceans, The (Wegener), 468
ori region. *See* Origin of replication
Ornithischia, *731*
Ornithorhynchus anatinus, 735
Orobanche, 793
Orthologs, **537–538**
Orthoptera, 709
Oryza sativa, 311, 646
  See also Rice
Osmoconformers, **1094**
Osmolarity, **1094**
Osmoreceptors, regulation of blood osmolarity and, 1107–1108
Osmoregulators, **1094–1095**
Osmosis, 106–107, 765–767, 1093–1094
Osmotic potential, 765–766
Osprey, 1129
Ossicles, *972,* **973,** 974
Ossification, 1017–1018
Osteoblasts, 885, 886, **1017,** *1018*
Osteoclasts, 885, 886, **1017,** *1018*
Osteocytes, **1017,** *1018*
Osteoporosis, 1018
Ostia, 1046
Ostracods, 706
Ostrich, 732, *733*
Otoliths, *975*
Otters, 737
Outcrossing, 551
Ova
  activation, 903
  blocks to polyspermy, 902
  a gamete, **899**
  oogenesis, 899–901
  specificity in interactions with sperm, 901–902
  See also Egg(s)
Oval window, *972,* **973**
Ovarian cycle
  hormonal control of, 911–192
  in humans, **910–911**
Ovaries (animal), *880,* **899,** 909
Ovary (plant), **638,** *639, 820,* 821
Overbeek, Johannes van, 809
Overdominance, 220
Overnourishment, **1071**
Overtopping growth, **619**
Oviducts, **909,** 910, 915, 925
Oviparity, **905**
Ovis canadensis, 505
Ovoviviparity, **905**
Ovulation
  in humans, **910**
  preventing, 914–915
Ovules, **632,** *820,* 821
Oxalic acid, 849
Oxaloacetate, 128, 173, 174
Oxidation, **140**
  in chemical weathering of rock, 786
Oxidation-reduction (redox) reactions, **140–141**
Oxidative muscle, 1013
Oxidative phosphorylation, 148–153
Oxidizing agents, 140
Oxpeckers, *1193,* 1194
Oxycomanthus bennetti, 720
Oxygen
  content in water and air, 1026
  covalent bonding capability, *27*
  diffusion in water, 1026
  electronegativity, *27*
  partial pressures, **1025,** 1036, 1037, 1039
  in photorespiration, 172, 173
  photosynthesis and, 11, 161–162
  in plant nutrition, 781
  as respiratory gas, 1025
  See also Atmospheric oxygen; Blood oxygen
Oxygenase, **172**
Oxytocin, *880,* **881,** 912, 913
Oysters, 903
Ozone, 1208, 1212
Ozone hole, 1208
Ozone layer, 11

- P -

p_1 promoter, 298
p21 protein, 186
p53 gene, 389
p53 protein, 186, 327
P_{680}, 167
P_{700}, 167, 168
Pääbo, Svante, 396
Pace, Norman, 570
Pacemaker, 992, 1010, **1052**
Pachypodium rosulatum, 1130
Pacific chorus frog, 2, 14, *1005*
Pacific yew, 647, *648*
Pacinian corpuscles, 971
Paedomorphosis, 729
Pain, 988
Paine, Robert, 1197
Pair bonding, 881
Pair rule genes, **443**
Palaeognaths, 732
Paleocene period, *478*
Paleomagnetic dating, 467
Paleontologists, 466
Paleontology, 548
Paleozoic era, *466–467,* **472,** 479, 617
Palindromes, 354
Palisade mesophyll, 173, 761
Palm fruit, 96
Palmitic acid, *1072*
Palms, 759
Palorchestes, 1228
Palo Verde National Park, Costa Rica, *1126*
Pan. See Chimpanzees
Pancreas, *880, 1079*
  diabetes and, **887**
  digestion and, **1083,** *1084*
  glucagon and, 887
  insulin and, 1087
  secretin and, 1086
  somatostatin and, 887

Pancreatic amylase, *1084*
Pancreatic duct, 1083
Pandas, 432
Pandemics, 283
Pangaea, **471,** 475, 476, 619
Panthera tigris, 1150
Panting, 868
Pantothenic acid, *1074*
Pan troglodytes, 739
Paper chromatography, 169
Paper wasps, *691*
Paphia lirata, 1016
Paphiopedilum maudiae, 640
Papilio sp., *678*
Papillomaviruses, 387
Pap test, 180
Parabasalids, *584, 597,* **603**
Parabronchi, **1030,** *1031*
Paracineta sp., *598*
Paracrine hormones, **875,** *876*
Paracrine signals, **334**
Paradisaea minor, 519
Paragordius sp., *704*
Parallel phenotypic evolution, **458–459**
Parallel substitutions, 527
Paralogs, **537–538**
Paralysis, 988
  nerve gases and, 961
Paramecium, 584
  anatomy of, 598, *599*
  cell surface, 592
  conjugation in, 593, *594*
  food vacuole, 591, *592*
  locomotion, 598
Paramecium bursaria, 598
Paranthropus, 739, *740*
Paraphyletic groups, **555**
Parapodia, 699
Parasites
  apicomplexans, 596
  fungi, **651,** 653, *654,* 662–663
  life cycles, 681–682
  obligate intercellular, 284
  overview of, **679,** 1189
  protostomes and, 691
  strepsipterans, 690–691
Parasitic plants, **792–793**
Parasitism
  microparasite–host interactions, 1191, *1192*
  overview of, **1188,** 1189
Parasitoids, 691
Parasitoid wasps, 1193
Parasympathetic nervous system
  features of, **992,** *993*
  heartbeat and, 1052, *1053*
  regulation of blood pressure and, 1062, *1063*
Parathormone, *880*
  See also Parathyroid hormone
Parathyroid glands, *880*
Parathyroid hormone (PTH), **885–886,** 887, 1017
Parenchyma, **749–750,** *751,* 758
Parental care, evolution of social systems and, 1159
Parental generation (P), **209**
Parent rock, **786**
Parietal cells, 1080, *1081*
Parietal lobe, **990,** 991
Parsimony principle, **546–547,** 1135
Parson's chameleon, *1130*
Parthenium argentatum, 796
Parthenocarpy, **806**
Parthenogenesis, **898**
Partial charges, 29
Partial pressure(s)
  carbon dioxide, 1038, 1039, 1040
  gradients, 1026, 1027, 1028
  oxygen, **1025,** 1036, 1037, 1039
Particulate theory, **210**
Parus caeruleus, 681
Passive immunization, 414
Passive transport, 105–110
Pasteur, Louis, 5, 63
Pasteurization, 5
Patch clamping, **950**
Pathogenesis-related (PR) proteins, **838,** 839
Pathogens, **401**
  bacterial, **579**
  fungi, 653
  plant defenses against, 837–840
  See also Plant viruses; Viruses
Pattern formation, **435**
  Drosophila body segmentation, 442–444
  genes determining apoptosis, 439–440
  morphogens and, 441–442
  plant organ identity genes, 440–441
Pauling, Linus, 238, 551
Pavlov, Ivan, 999
Pax3 gene, 934
Pax6 gene, 448
Pax genes, 934
pBR322 plasmid, *326,* 361
PC. *See* Plastocyanin
PCBs. *See* Polychlorinated biphenyls
PCR. *See* Polymerase chain reaction
pdm gene, *710*
Peanut allergies, 420
Peanut oil, 55
Peas, Mendel's experiments with, 208–210, 213
Peat, 624, 655
Pectoral fin, *726*
Pedigrees, **216–217**
Pegea socia, 723
Pelagia panopyra, 686
Pellagra, *1074*
Pellicle, 598
Pelvetia canaliculata, 601
Pelvic fin, *726*
Pelvis, renal, 1102
Penetrance, **221**
Penguins, *681*
Penicillin, 367, 566, 666
Penicillium, 367, *584,* 659, 666, *1198*
Penile erections, 344
Penis(es), **904,** 905, *906,* 908, 914
Pennisetum setaceum, 639
Pentaradial symmetry, **718**
Pentoses, **50**
PEP. *See* Phosphoenolpyruvate
PEP carboxylase, **174,** 175
Pepsin, **1080,** *1081, 1084*
Pepsinogen, **1080,** *1081*
Peptidases, 1077, 1083
Peptide bonds, **43–44,** *45,* 63
Peptide linkage, 43–44
Peptides
  as hormones, 878
  as neurotransmitters, 958, *959*
Peptidoglycan, 73, **565,** 566, 601
Peptidyl transferase activity, 269
Per capita growth rate, 1173
Peregrine falcon, 1129
Perennials, **798, 826**
Perfect flowers, **639**
Perforin, 417
Perfusion, 1028
Pericycle, **756**
Periderm, 748, **755,** 760
Perilla, 831
Period, of cycles, 828
Periodic table, 21, *22*
Peripatus sp., *706*
Peripheral membrane proteins, *98,* 99, 100
Peripheral nerves, 933
Peripheral nervous system (PNS), **944,** 946, **985,** *986*
Periplasmic space, *565,* 566
Perissodactyla, *736*
Perissodus microlepis, 502
Peristalsis, **1079–1080**
Peritoneum, 674–675, **1079**
Peritubular capillaries, **1100,** 1101, 1102, 1103
Periwinkle, *648*
Permafrost, 1118
Permeability, 105
Permian period, *466–467*
  described, **475–476**
  plant evolution and, 619, 634
  saprobic fungi and, 655
Pernicious anemia, *1074,* 1075
Peroxides, 83
Peroxisomes
  in animal cells, *76*
  in photorespiration, 172–173
  in plant cells, *77*
  structure and function, **83–84,** *85*
Persea sp., *644*
Pesticides, bioaccumulation, 1089
PET. *See* Positron emission tomography
Petals, **638–639,** *820,* 821
Petioles, **747,** 806
Petroleum, 586–587, 618
Petromyzon marinus, 724
Pfiesteria, 1217
  P. piscicida, 596
P generation, 209
pH, **34, 35**
  chemistry of, 34
  effect on enzyme activity, 133–134
  effect on hemoglobin, 1037
  effect on proteins, 48
  regulation in blood, 1105, *1106*
  of the stomach, 1080
  See also Soil pH
Phadnis, Shahikant, 38
Phaenicia sericata, 711
Phage, **284–287,** 360
  See also Bacteriophage
Phagocytes
  defensins and, 405
  inflammation and, 406
  in lymph nodes, 1060
  overview of, **402,** 405
Phagocytosis, 81, *82,* 101, **113,** 588
Phagosomes, 81, *82,* 113, 415
Phalaris canariensis, 803
Pharmacogenomics, **395**
Pharming, **369**
Pharyngeal basket, *722,* 723
Pharyngeal slits, 721, 722
Pharynx
  chordate, 722
  ecdysozoan, 693
  hemichordate, 721
  human, 1032, *1033,* 1079, *1080*
  nematode, 704
Phase-contrast microscopy, *71*
Phaseolus vulgaris, 842–843
Phase shifts, **828,** 829
Phasmida, 709
Phelloderm, **760**
Phenolics, in plant defenses, 838, *839*
Phenol-water, 169
Phenotype, **211,** 257, 489–490
  codominant alleles and, 219
  conditional mutants, 275
  genes and, 212–213
  incomplete dominance and, 218–219
  modifying to treat genetic disease, 391
  natural selection and, 497
  pleiotropic alleles and, 219
  quantitative variation, 221–222
  recombinant, 214
Phenylalanine, *43,* 375, 384, 391
Phenylalanine hydroxylase, 375, 384, 391
Phenylketonuria (PKU), 375–376, 379, 384, 391
Phenylobacterium immobile, 294
Phenylpyruvic acid, 375
Pheromones
  chytrid fungi and, 664
  communication and, **1157**
  detection of, 969
  mating in arthropods and, **968**
  overview of, 1021
Phidiana hiltoni, 701
Phillips, David, 129
Philodina roseola, 696
Phinney, Bernard O., 802
Phipps, James, 400
Phloem, *751,* 752
  in angiosperms, 638
  companion cells, 638
  functions of, **748**
  in roots, 757
  secondary growth and, 758
  translocation of substances, 775–777
  vascular cambium and, 754
  in vascular plants, **617**
Phloem sap, 752, 776–777
Phlox, 515, *517*
Phoberomys pattersoni, 464
Phoenicopterus ruber, 677
Phoenix dactyifera, 645
Pholidota, *736*
Phoronids, *683, 692,* **697**
Phoronopsis viridis, 697
Phosphate group, *40,* 53
Phosphate(s)
  ATP hydrolysis and, 124
  blood levels, 887
  eutrophication and, 1219
Phosphatidylcholine, *56*
Phosphatidyl inositol-bisphosphate (PIP2), **342,** *343,* 344
Phosphodiesterase (PDE), 345, 977, *978*
Phosphodiester linkages, **58**
Phosphoenolpyruvate (PEP), **174**
Phosphofructokinase, 156
Phosphoglucose isomerase (PGI), 502
3-Phosphoglycerate (3PG), *40,* 170, *171*
Phosphoglycolate, 172
Phospholipase C, 342, 343
Phospholipases, 341, 342, 343
Phospholipid bilayers, 55, *56,* 97–99, 106
Phospholipids, **55,** *56,* 97–99, 100
Phosphorus

covalent bonding capability, *27*
deficiency in plants, *783*
electronegativity, *27*
global cycle, 1218–1219
Hershey–Chase experiment, 235–236, 237
as macronutrient, *1073*
mycorrhizal associations and, 657
in plant nutrition, 781, *782,* 784
Phosphorus-32, 235–236, 237, *466*
Phosphorus cycle, 1218–1219
Phosphorylase kinase, 346, *347*
Phosphorylation
ATP and, 123
enzymes and, 784
of proteins, 185, **274**
substrate-level, 144
Photoautotrophs, **568**
Photoheterotrophs, **568**
Photomorphogenesis, 813
Photoperiod, animal behavior and, 1153–1154
Photoperiodism, **826–828,** 890
Photophosphorylation, **168**
Photoreceptor cells, **976**
See also Rod cells
Photoreceptors
biological clocks and, 829, *830*
plant growth and development, **797–798,** 812–814
in retinal receptive fields, 995
sensory, *966*
Photorespiration, **172–173**
Photosensitivity, **976–977**
Photosynthesis, **13, 161-162**
antenna systems of pigments, 165–166
atmospheric carbon dioxide and, 175
atmospheric oxygen and, 11
C_4 plants, 173–175
Calvin cycle, 169–172
CAM. *See* Crassulacean acid metabolism
chloroplasts, 83, *84*
connection to other metabolic pathways, 175–176
diatoms and, 584
electron transport systems in, 166–168, *169*
evolutionary impact of, **11**
iron and, 1219
leaf anatomy and, 760–761
pathways in, 162, *163*
percent of solar energy captured by, 1209
photophosphorylation, 168
photorespiration, 172–173
photosystems I &II, 167, 169
pigments used in, 164–165
production of oxygen during, 161–162
prokaryotes, 73
xerophytes and, 846
Photosynthesizers
autotrophs as, 781
trophic level, 1186
Photosynthetic bacteria, *284*
See also Cyanobacteria
Photosynthetic lamellae, 573
Photosystem I, 167, *169*
Photosystem II, 167, *169*
Photosystems, **167,** *169*
Phototransduction, 977, *978*
Phototropins, **814**
Phototropism, **803,** *804,* 814
Phthiraptera, 709
Phycobilins, **165**
Phycodurus eques, 727
Phycoerythrin, 601, 602
Phyla, **554**
Phylogenetic analyses
data sources, 547–548
forensic application, 542–543
mathematical modeling and, 548–549
parsimony principle, 546–547
reconstructing ancestral states, 550
testing accuracy of, 549–550
use of traits in, 544–545
Phylogenetic trees, **543–544**
classification systems and, 554–556
constructing, 545–551
forensic application, 542–543
Tree of Life, 544, Appendix A
uses of, 532, 551–554
Phylogeny, **543,** 544, 554–556
Phyrnus parvulus, 693
Physarum, 584, 606
Physiological regulation, 856–857
Physiological systems
components of, 857–859
internal environment and, 855–857
Phytoalexins, **838**
Phytochromes
biological clocks and, 829
forms of, 813
photoperiodism in plants and, 827–828
plant development and, 813–814
red light activation, 347, **812–813**
signal transduction pathways and, 347
Phytohemagglutinin (PHA), 500
Phytoplankton, **583,** 584–585
global carbon cycling and, 1213–1214
Mesozoic era, 476
Pibernat, Ricardo Amils, 560
Pigeons, *489,* 1155–1156, 1158
Pigments, **163**
absorption spectrum, 163, *165*
accessory, 165
antenna systems, 165–166
chromatophores and, 1020
land plants and, 613
in photosynthesis, 164–165
plant photoreceptors, 797–798
See also specific pigments
Pigs, brain, *945*
Pili, 74
Pilobolus sp., *664*
Pilosa, *736*
Piloting, **1154**
Pima Indians, 395, 1068–1069
Pimm, Stuart, 1161
Pineal gland, *880,* **890**
Pineapples, *643, 648*
Pines, life cycle, 636, *637*
Pin flowers, 496
Pinnae, 972
Pinocytosis, **113**
Pinus
P. longaeva, 755
P. ponderosa, 503
P. resinosa, 636
P. sylvestris, 1198
Piper nigrum, 644
Pisaster ochraceus, 1196–1197
Pisolithus tinctorius, 657
Pistil, **638,** *639,* 821
Pisum sativum. See Peas
Pitcher plants, 495, 792
Pith, **757,** 809
Pith rays, 757
Pit organs, 964
Pits, 750, *751*
Pituitary dwarfism, 882
Pituitary gland, **880**
birth control pills and, 912
endocrine functions, *880,* 881–882
ghrelin and, 1088
posterior, *880,* 881
See also Anterior pituitary gland
Pitx1 gene, 459
Pivot joint, *1019*
Pivots, 1019, *1020*
PKC. *See* Protein kinase C
Placenta, **905,** 932, **935**
Placental mammals, 735
cleavage in, 923, *924,* 925
gastrulation in, 932
See also Eutherian mammals
Placoderms, *725*
Plagiodera versicolora, 678
Plague, 575, 579
Plaice, 504
Plains zebra, *1131*
Planaria, 1097
Plane joint, *1019*
Planetesimals, 34
Plane trees, 515
Plankton, 472, **583,** 1161
Plankton blooms, 599, *600*
Plantae, *584, 597,* **600–601**
Plant cells
calcium signaling and, 343
cell wall characteristics. *See* Plant cell walls
chloroplasts. *See* Chloroplasts
cytokinesis in, 192
metabolic interactions in, 175–176
recombinant DNA technology and, 359
structure of, *77*
totipotency, 359
turgor pressure, 108
vacuoles, *77,* 85–86
Plant cell walls, *77*
defenses against pathogens, 837–838
effect of auxin on, 807–808
formation of, 749
microtubules and, 88
plasmodesmata. *See* Plasmodesmata
primary, 749, 750, *751*
secondary, 749, *751*
structure and function, 90, *91*
Plant development
embryonic growth, 800–801, 823–824
factors regulating, 797–798
generalized schema, *428*
light and, 456
major steps in, 798–799
morphogenesis in, 428
photoreceptors and, 812–814
plasticity in, 456, 457
signal transduction pathways and, 798
Plant diseases
crown gall, 575
effect of global warming on, 1216
Plant hormones, **797**
abscisic acid, 811
auxins, 803–809
brassinosteroids, 811–812
cytokinins, 809–810
ethylene, 810, *811*
florigen, 798, 829–831
gibberellins, 801–803
growth hormones, 797, *798*
salicylic acid, 839
systemin, 842
tissue culture experiments, 809
Plant nutrition
acquisition of nutrients, 781–782
carnivorous plants, 792
mineral elements, 781, 782–785
nitrogen fixation, 788–791
parasitic plants, 792–793
soils and, 785–788
Plant-pollinator interactions, 641
Plants
albino mutants, 792
apoplast and symplast, 768, *769*
atmospheric CO_2 and carbon storage, 1204–1205, 1214–1215
biotechnology and, 369–371
carnivorous, 792
circadian rhythms and, 828–829
climate extremes and, 845–847
cloning, 429–430
competition and, 1192
defenses against pathogens, 837–840
deficiency symptoms, 783
determinate and indeterminate growth, 754
development. *See* Plant development
drought stress and, 321–322
effect on soil fertility and pH, 787–788
embryonic growth, 800–801, 823–824
embryos, 427, 613, *820,* 823–824
endothermic, 862
environmental signals and, 333
evolution, 473–474, 475, 476, 477–478, *617*
fossil discoveries, 610–611
fungi and, 613, 614, 653, *654,* 658
global nitrogen cycle and, 790–791
gravitropism, 805
heavy metals and, 849
height, 769
herbivores and, 840–844
hormones. *See* Plant hormones
metabolic interactions in, 175–176
modular growth in, 753
mycorrhizae, 655, 657–658, 756, 789, 1194
nitrate reduction and, 791
nod factors and symbiosis, 319
nutrition. *See* Plant nutrition
organ identity genes, 440–441
parasitic, 792–793
phloem translocation, 775–777
plasmid vectors, 360–361
pluripotency of meristems, 432
polyploidy and, 514–515
primary and secondary body, 753
primary growth in, 754, 755–757
saline habitats and, 848
secondary growth, 754–755, 758–760
secondary metabolites and, *839,* 841, 844
seedless, 610–628

seed production and seed size in, 457
sun-following ability, 109
symbiosis with nitrogen-fixing bacteria, 477–478, 788–789
totipotency, 429–430
water and solute uptake, 765–768
water loss regulation, 773–774
ways of defining, 611, *612*
xylem transport of minerals and water, 769–772, *776*
See also specific plant groups
Plant tissue culture, 833–834
Plant viruses
movement through plasmodesmata, 348
plant immunity to, 840
transmission through vectors, 289
Planula, **685**
Plaque
atherosclerotic, **1060,** *1061*
dental, 565
Plaques
bacterial, 363
of desmosomes, 103–104
Plasma, 106, **1055**
See also Blood plasma
Plasma cells, **409,** 411, 417
Plasma membrane
in animal cells, *76*
in archaeans, 575–576
in cell adhesion and recognition, 102–104
cholera and, 96–97
endocytosis, 113–114
eukaryotic evolution and, 588
exocytosis, 113, 114
fast block to polyspermy in eggs, 902
flow of electric current and, 951
hyperpolarization and depolarization, 950, *951*
ion pumps and ion channels, 947–950
membrane potential. *See* Membrane potential
movement of water across, 1093–1094
receptors, 336–339
resting potential, 946, 947, 948–950, 952
roles of, **72**
Plasmids
genes carried by, 294
overview of, **293–294**
resistance factors, 294–295
vectors in DNA technology, 360–361
Plasmin, 368
Plasminogen, 368
Plasmodesmata
movement of plant viruses and, 289
plant defenses and, 837
pressure flow model of translocation and, 777
in sieve tube elements, 752
structure and function, **90,** 348–349, **749,** *750*
symplast and, 768, *769*
Plasmodial slime molds, *584, 597,* **606–607**
Plasmodium, 606
Plasmodium, 584, 585, 596
P. falciparum, 585–586
Plasmogamy, *660, 661,* **662**
Plastic food wrap, 413
Plasticity, plant cell walls and, **807**
Plastids, 83, *84, 85,* 228
Plastocyanin (PC), **168**
Plastoquinone (PQ), **168**
Platanus
P. hispanica, 515
P. occidentalis, 515
Plateau phase, 913
Platelet-derived growth factor, 187, *368*
Platelets, 186–187, *403,* **1056,** *1057*
Plate tectonics, 468
Platycalymene, 480
Platyfishes, 552–553
Platyspiza crassirostris, 512
Pleasure, limbic system and, 988
β Pleated sheets, *45,* **46**
Plecoptera, 709
Pleiotropy, **219**
Pleistocene epoch, 478, 479
Pleural membrane, **1034**
Pleural space, **1034,** *1035*
Plimsoll, Samuel, 1243
Plimsoll line, 1243
Pliocene period, *478*
Ploidy levels, 201
Pluripotency, **432**
Pluripotent stem cells, 402, *403,* 432–433
Pneumatophores, **846**
Pneumococcus, 234–235
Pneumocystis carinii, 422, 653
Pneumonia, *409,* 422, 653
PNS. *See* Peripheral nervous system
Podocytes, *1101,* **1101**
Poecilia reticulata, 1172
Poecilotheria metallica, 713
Pogonophorans, 699
Poikilotherms, 861
Point mutations, **275–276,** 494
"Point restriction" coat patterns, 221
Poison glands, 1021
Poison ivy, 420
Poisons
acquired by frogs, 1151
affecting neurotransmitters and ion channels, 955, 1021
inhibition of acetylcholinesterase, 961
See also Toxins
Polar bears, 868, *1070*
Polar bodies, in oogenesis, 900, 901
Polar covalent bonds, **27,** *28*
Polaris, 1156
Polarity, **436–437**
Polarized light, animal migration and, 1156
Polar microtubules, 189
Polar molecules, 29
Polar nuclei, **821,** 822, *823*
Polar substance, 48
Poles. *See* Mitotic centers
Polio, *409,* 524–525
Poliomyelitis, *409,* 524–525
Poliovirus, *284, 409*
Polistes, 691
Pollen grains
germination, 822
microgamethophytes, *820,* **821**
overview of, **632–633**
plant-animal mutualisms and, 1194
Pollen tubes, **633,** *820,* 822
Pollination, **633,** *820,* **821–822**
Pollination services, 1241
Poly A tail, **316,** *317,* 363
Polychaetes, 698–699
Polychlorinated biphenyls (PCBs), 1089
Polyclonal antigens, 413
Polycyclic aromatic hydrocarbons, 61
Polygynous species, 504
Polygyny, 504
Polymerase chain reaction (PCR), **250–251,** 356, 366
Polymerization, 62
Polymers, **39,** 62
Polymorphism(s), **218**
DNA fingerprinting and, 355–356
frequency-dependent selection and, 501–502
in populations, 491
Polyorchis penicillatus, 686
Polypeptide chains, 42
Polypeptides
as hormones, 878
peptide bonds, 43–44
posttranslational modifications, 274
protein primary structure, 44–45
proteolysis, 274
relationship to genes, 258
signal sequences and movement to cellular destinations, 272–274
translation of mRNA and, 265–271
Polyphyletic groups, **555**
Polyploidy, **513**
gene duplication and evolution, 534
meiosis and, **200–202**
speciation and, 514–515
types of, 513–514
Polyproteins, 274
Polyps, of cnidarians, **685,** 686, 687
Polyribosome, **270,** *271*
Polysaccharides, **49**
as energy source, 154
energy storage and, 155
in plant defenses against pathogens, 837–838
structures and functions, 51–53
Polysomes, **270,** *271*
Polyterpenes, *839*
Polyubiquitin complex, 327
Polyunsaturated fatty acids, 55
POMC, 882
Ponderosa pines, 503, 1237, *1238*
Pongo pygmaeus, 739
Pons, **986,** 988
Poor-wills, 870
Population bottlenecks, *494,* **495**
Population density, **1167**
body size and, 1175–1176
distant events may influence, 1178–1179
estimating, 1168–1169
estimating birth and death rates from, 1169–1171
factors influencing, 1173–1177
predation and, 1189
social animals and, 1161–1162
Population dynamics, **1167**
distant events may influence, 1178–1179
effects of foraging on, 1161
metapopulations, 1178
Population ecology, **1167**
effects of exotic species, 1166–1167
estimating birth and death rates, 1169–1171
estimating population density, 1168–1169
factors affecting population densities, 1173–1177
life histories, 1171–1172
population dynamics, 1161, 1167, 1178–1179
population management, 1180–1181
tracking individuals, 1167–1168, *1169*
Population genetics, **489–493**
Population growth
exponential, 1173, *1174*
logistic, 1173–1174
Population(s), **9, 489, 1167**
age distribution, 1170–1171
environmental carrying capacity and, 1173, *1174*
episodic reproduction in, 1175, *1176*
estimating birth and death rates, 1169–1171
evolution and, 490
exponential growth, 1173, *1174*
factors affecting species abundance, 1175–1177
gene flow, 494
genetic drift, 494–495
genetic structure, 492–493
logistic growth, 1173–1174
managing, 1180–1181
nonrandom mating, 495–496
predation and oscillations, 1189
rescue effect, 1178
resource fluctuations and, 1175
structure, 1167
tracking individuals in, 1167–1168, *1169*
p orbitals, 24
Porphyria, 378
Portal blood vessels, **882**
Portuguese man-of-war, *679*
Positional cloning, **381–382**
Positional information, **441**
Positive control, **300**
Positive cooperativity, **1036**
Positive feedback
in physiological systems, **857**
regulation of metabolic pathways, 156, 157
Positive selection, 532–533
Positron emission tomography (PET), of brain functioning, *1001*
Postabsorptive period, **1086,** 1087
Postelsia palmaeformis, 600, *601*
Posterior pituitary gland, *880,* **881**
Posterior plane, **674**
Postganglionic neurons, 992
Postsynaptic cells, 957, 958
Postsynaptic membrane, 945, 956, *957,* 958
Postsynaptic neurons, **944**
Posttranscriptional gene silencing, 840
Postzygotic reproductive barriers, **516–517**
Potassium
action potentials and, 952, 953
biological concentration gradients and, 108
deficiency in plants, *783*
electric current and, 947
electronegativity, *27*
equilibrium potential, 948
guard cell activity and, *773,* 774

ion channel specificity and, 109
as macronutrient, *1073*
in plant nutrition, *782*
plant transpiration rate and, 772
plant uptake of, 767–768
See also Sodium-potassium pump
Potassium-40, *466*
Potassium channels
axon membrane potentials and, 948–950
generation of action potentials and, 952, 953
heartbeat in mammals and, 1052
membrane resting potential and, 950
specificity, 109
Potassium equilibrium potential, **948**
Potassium phosphate, 784
Potatoes, 747, 832
Potential energy, **119–120**
Potentiation, 960, *961*
Pottos, 738
Poulsen, Michael, *658*
Powdery mildews, 666
Power arm, 1019, *1020*
Poxviruses, *284*
PPARd protein, 138, 139
PQ. *See* Plastoquinone
P_r, **813**
Prader-Willi syndrome, 383
Prairie dogs, 1237
Prairies, restoration projects, 1237
Precambrian age, *466–467*
Precapillary sphincters, **1061,** *1062*
Precipitation
acid, 849, 1218
hydrological cycling and, 1211
influence on plant distribution, 1117
Precocial birds, 681
Predation, **1188**
effect on life histories, 1172
impact on species' ranges, 1190
mimicry and, 1190–1191
population oscillations and, 1189
Predator–prey interactions, 1189
Predators
developmental responses to, 456
fungi, 653–654
overview of, **678**
predatory-prey interactions, 1189
trophic cascades and, 1195–1196
types of, **1075**
Preganglionic neurons, 992
Pregnancy
early test for, 937
in humans, 912, 936
methods of preventing, 914–915
spontaneous termination, 200
"water breaking," 936
Pregnancy testing, 912
Prehensile tails, 738
Preimplantation genetic diagnosis (PGD), 385, 916
Preimplantation screening, 385, 916
Premolars, 1076
Pre-mRNA, **313,** 316, *317*
Premutated individuals, 383
Prenatal screening, 384
Presbytis entellus, 532
Pressure chambers, **771–772**
Pressure flow model, **776–777**
Pressure gradients, pressure flow model of translocation and, 776
Pressure potential
bulk flow and, 767
in plant cells, **766**
transpiration–cohesion–tension mechanism and, 770, 771
Presynaptic membrane, 945, 958
Presynaptic neurons, **944**
Prey, **678**
predator-prey interactions, 1189
Prezygotic reproductive barriers, **515–516**
Priapulids, *683, 692,* 702, **703**
Priapulus caudatus, 703
Priapus, 703
Prickly-pear, 1166–1167
Prides, 1159
Primary active transport, **111,** *112*
Primary cell walls, **749,** 750, *751*
Primary consumers, **1186,** *1187*
Primary embryonic organizer, **929–931**
Primary endosymbiosis, **589,** *590*
Primary growth, **754,** 755–757
Primary immune response, **409**
Primary inducer, 438
Primary lysosomes, 81, *82*
Primary meristems, 754, 755
Primary mesenchyme, 927
Primary metabolites, 841
Primary motor cortex, **990**
Primary oocytes, **899–900,** 910
Primary plant body, **753**
Primary producers
global energy flow and, 1209–1210
phytoplankton, 584–585
plants and, **646**
trophic level, **1186,** *1187*
Primary production, global energy flow and, 1209–1210, *1211*
Primary sex determination, **226**
Primary sex organs, 904
Primary somatosensory cortex, **991**
Primary spermatocytes, **899,** *900*
Primary succession, **1197–1198**
Primase, **245,** *247*
Primates
evolutionary radiation in, **737–738**
human evolution, 738–741
major clades, 738
number of living species, *736*
Vitamin C and, 1074
Primer strand, in DNA replication, **245–246**
Primitive groove, 32
Primitive streak, 32, *931*
"Primordial soup" hypothesis, 62
Primrose, 496
Primula, 496
Prions, **378–379**
Probability, principles of, 214–216
Probes
in gel electrophoresis, 354, *355*
in nucleic acid hybridization, **314**
Proboscidea, *736*
Proboscis, 697, *703,* 721
Procambium, **754,** 756
Procedural memory, **1000**
Processive enzymes, **247**
Productivity
gross primary productivity, 1209
species richness and, 1200–1201
upwelling zones, 1206, *1207*
Products (in chemical reactions), **30**
Profibrinolysin, 908
Progesterones
birth control pills and, 915
contragestational pills and, 915
corpus luteum and, 911
in human pregnancy, 912
ovarian cycle and, 911
ovaries and, *880*
as sex steroids, **889**
Progestins, 915
Programmed cell death. *See* Apoptosis
Progymnosperms, 631
Prokaryotes
cell division in, 181–182, *183*
cell walls, 565–566
characteristics, **72–74,** 560–561
clones, 290
communication between, 566–567
compared to eukaryotes, 561–562
determining phylogenies, 569–571
DNA in, 181, 561–562
effect on Earth's atmosphere, 468–469
endosymbiosis. *See* Endosymbiosis
environmental roles of, 578
evolution of, **11**
evolution of eukaryotes from, 11–12, 92
fermentation, 147, 148
flagella, 566, *567*
gene expression in, *261*
generation times, 566
genetic recombination in, 290–295
genome, 300–303, 307, *308*
Gram staining, 565–566
lateral gene transfer, 570
major groups, 571–578
metabolic pathways, *141,* 567–569
microbial communities, 563–565
motility, 566
mutation and, 571
as pathogens, 579
positive relationships with eukaryotes, 579
regulation of gene expression in, 296–300
reproduction in, 566
ribosomes in, 79, 268, 269
shapes of, 563
transcription in, 296–300, 319
translation in, 262
Prolactin, *880,* **882**
Proline, 42–43, 848
Prometaphase
first meiotic division, *196,* 198
mitosis, 189–190
Promiscuous mating, in birds, 508
Promoters
control of protein synthesis and, 299–300
eukaryotic, 313, 319–320
inducible and tissue-specific, 367
prokaryotic, 319
transcription and, **262,** 296–297
Pronuclei, 383
Proofreading, in DNA repair, **249,** *250*
Pro-opiomelanocortin (POMC), 882
Propane, 30
Prophage, **286**
Prophase
first meiotic division, *196,* 198, *200*
mitosis, **189,** *190, 191, 200*
second meiotic division, *196*
Prosimians, 738
Prosopis, 846
Prostaglandins, 915
Prostate gland, *906,* **908**
Prosthetic groups, *129,* **130**
Proteads, 518
Proteases
in digestion, 1077
HIV and, 274, 423
in laundry detergents, 127
pancreas and, 1083
in proteasomes, 327
in proteolysis, 274
Proteasomes, **327**
Proteinaceous infective particles. *See* Prions
Protein deficiency, 1071
Protein kinase cascades, **339–341,** 346, *347*
Protein kinase C (PKC), 342
Protein kinases
action on enzymes, 346
cyclic AMP and, 341
EnvZ, 334
phosphorylation and, 274, **337**
Proteins
annotation, 301
as biologically significant macromolecules, 39
biological specificity and, 47–48
chaperonins and, 48, *49*
chemical structure of amino acids in, 42–43
conformation, 378
denaturation, 48
DNA-binding, 320, *321*
as energy source, 154
evolution. *See* Molecular evolution
genetic diseases and, 375–377
genetic screening and, 384
intracellular transport, 80–81
membrane-associated. *See* Membrane proteins
motor. *See* Motor proteins
nuclear localization signal, 78
overview of, **7, 42**
peptide bonds, 43–44
phenotype defined as, 257
phosphorylation, 185, 274
prions, 378–379
relationship between structure, shape, and function, 46–48
relationship of genes to, 257–258
relative proportion among all living things, *41*
structure. *See* Protein structure
studying through gene evolution, 538
synthesis. *See* Protein synthesis
undernourishment and, 1071
Protein starvation, edema and, 1058, 1071
Protein structure
functional qualities of proteins and, 46–48
primary, **44–45**
quaternary, *45,* **46,** *47*
secondary, **45–46**
sensitivity to environmental conditions, 48
tertiary, *45,* **46**
Protein synthesis
central dogma and, 260–261
control of promoter efficiency and, 299–300
posttranscriptional regulation, 324–326
posttranslational events, 272–274
transcriptional regulation in prokaryotes, 296–300
transcription of DNA, 261–265
translation of RNA, 265–272
Protenoids, *69*
Proteobacteria, *571,* **574–575**

Proteoglycans, **91**
Proteolysis, 274
Proteolytic enzymes, gastric, 1080
Proteome, **395–396**
Proteomics, 395–396
Prothoracic gland, 876
Prothoracicotropic hormone (PTTH), 876, *878*, 881
Prothrombin, **1056**, *1057*
Protists, **583**
  evolution of eukaryotes and, 12
  genetic code and, 264
  in mutualisms, 1194
  typical size of, *284*
  See also Microbial eukaryotes
Protoderm, **754, 756**
Protohominids, 738
Proton concentration gradients
  chemiosmosis and, 147, 150–152
  electron transport chain and, 147, 148–149
Protonema, *615*, 623
Protonephridium, **1097**
Proton-motive force, **150**
Proton pumps
  electron transport chain and, 150
  loading of sieve tube elements and, 777
  plant uptake of mineral ions and, **767–768**
Protons, **21**, 22–23
Protostomes, *672*, **683**
  arthropods, 705–713
  early development in, **673**
  ecdysozoans, 702–713
  lophotrochozoans, 692–693, 695–702
  overview of, 691, *692*
Prototherians, 735, 905
Prototrophs, 258
Protozoans, 591
Provincialization, **476**
Provirus, **288**
Proximal convoluted tubules, **1101**, 1103, 1104
Prozac, 961
PrP^c protein, 378
PR proteins, **838**, 839
Prunus serotina, 1176
Prusiner, Stanley, 378
Przewalski's horse, *1131*
Pseudacris regilla. See Hyla regilla
Pseudalopex griseus, 1134
Pseudoceros ferrugineus, 696
Pseudocoel, 674, *675*
Pseudocoelomates, **674**, *675*
Pseudogenes, **315**, 531, 534
Pseudomonas, 569
  P. aeruginosa, 73
  P. fluorescens, 528–529, *530*
Pseudomyrmex, 1184
  P. ferrugineus, 1184
  P. flavicornis, 1184
Pseudopeptidoglycan, 565
Pseudoplasmodium, 607
Pseudopods (pseudopodia), 87, **591**, 604, 605
Pseudouroctonus minimus, 713
Psilophyton, 610
Psilotum, 625
  P. flaccidum, 625, *626*
Psocoptera, 709
Psoriasis, 374
Pteridophyta (Pteridophytes), *612*, 618, **619**, 625–627
Pterobranchs, 721
Pterosaurs, *458*, 476–477, *731*
Pterygotes, 709–711
PTH. *See* Parathyroid hormone
Ptilosarcus gurneyi, 686
Ptychromis sp., *550*
Puberty, **889–890**, 909
Public policy, biological studies and, 16–17
Puccinia graminis, 653, 662–663
Puffballs, *655*, 667
Pufferfish, 309, *533*, 538
Pufferfish sushi, 538
Pullularia, 1198
Pulmonary artery, **1050**
Pulmonary circuit, **1047**, 1048–1049
Pulmonary valve, **1050**
Pulmonary veins, **1050**
Pulp cavity, **1076**
Pulse, 1051–1052
Punnett, Reginald C., 212, 492
Punnett square, **212**, 214
Pupa, **710**, 877
Pupfish, 1177
Pupil, **979**
Purines
  generation through anabolic pathways, *154*, 155
  as neurotransmitters, *959*
  origin of life and, 61
  structure, **57**, *58*
Purkinje cell, *945*
Purkinje fibers, **1054**
Purple bacteria, 574–575
Purple foxglove, 1055
Purple nonsulfur bacteria, 568
Purple sea urchin, *720*
Purple sulfur bacteria, *568*
Pus, 406
Putrescine, 818
Pycnogonids, 712
Pycnophyes kielensis, 703
Pygmy elephants, 716
Pygoscelis antarctica, 681
Pyloric sphincter, 1080, 1082
Pyramidal cell, *945*
Pyrenestes ostrinus, 498, *499*
Pyridoxine, *1074*
Pyrimidines
  generation through anabolic pathways, *154*, 155
  origin of life and, 61
  structure, **57**, *58*
Pyrogens, **869**
Pyruvate
  in fermentation, 147, 148
  in glycolysis, 140, **142**, *143*
  in metabolic pathways, *141*
  oxidation, 140, *141*, 144–145, *146*, *153*
  in plant metabolic pathways, 175
Pyruvate dehydrogenase complex, 145
Pyruvate oxidation, **140**, *141*, 144–145, *146*, *153*
Pyruvic acid. *See* Pyruvate

- Q -

Q_{10}, **860**, *861*
Quadriceps, 1019
Quadrupedalism, in dinosaurs, 670, *671*
Qualitative analysis, 32
Qualitative characters, 221
Quantitative analysis, 32–33
Quantitative trait loci, **221–222**
Quantitative variation, in populations, **221–222**
Quarantines, 1158
Quaternary period, *466–467*, 470, **478–479**
Quebec, 374
Queen bees, 896–897, 898
Queens, 1160
Quiescent center, 755
Quillworts, 624
Quinine, 648
Quinones, *839*
Quiring, Rebecca, 448
Quorum sensing, **567**

- R -

Rabbits
  coat color/pattern determination, 218, 221
  digestive tract, *1077*, 1085
  ovarian cycle, 910
Rabies, *409*
Rabies virus, *409*
Radcliffe, Paula, 854
Radial cleavage, 673, **923**
Radial symmetry, **674**
Radiation
  electromagnetic, 163
  gamma, 23
  in heat transfer, **863**, 864
  as mutagen, 277
  ultraviolet, 11, 277
Radiations. *See* Adaptive radiations; Evolutionary radiations
Radicle (embryonic root), **798**
Radioactive decay, 23
Radioimmunoassay, 891
Radioisotope labeling
  Calvin cycle and, 169–170
  Hershey–Chase experiment, 235–237
Radioisotopes, **23**, 466–467
Radiolarians, *584*, 585, *597*, **605**
Radula, 700, **1077**
Raffles, Sir Thomas Stamford, 818
Rafflesia arnoldii, 818, 819
Rainbow trout, 134
Rainey, Paul, 528–529
Rainfall. *See* Precipitation
Rain shadows, **1115**
Rananculus, 756
Rana temporaria, 681
Rangifer tarandus, 737
Rape (plant), 811
Rapid-eye-movement (REM) sleep, **998**, 999
Raschke, Klaus, *774*
Raspberries, *643*
Ras protein, 340, 439
Ras signaling pathway, 347
Ratfish, *726*
Rats
  controlling populations of, 1181
  ovarian cycle, 910
  regulation of food intake in, 1088
Rattlesnakes, 964, 1144
Ray-finned fishes
  body plan, *725*
  eggs, 726–727
  evolutionary history, 475, 476
  overview of, 726
  in vertebrate phylogeny, *723*
Ray flowers, *639*
Rays (marine animal), 504, 726, 904, 1099
Rb gene, 388–389
Reactants, **30**
Reaction center, **166**
Reactive atoms, *24*, 25
Receptacle, **639**, 821
Receptive fields
  retinal, *994*, **995**
  visual cortex, 995–996
Receptor-mediated endocytosis, 113–114
Receptor potential, **966**
Receptor proteins, **113**
Receptors, 333
  classified by location, 336–339
  ligand binding sites, 336
  nuclear, 78
  in signal transduction pathways, **334**, *335*
Recessive alleles, in human pedigrees, *216*, 217
Recessive genes, 213
Recessive traits, **210**
Reciprocal crosses, **207**
Reciprocal translocations, 276, *277*
Recognition sequence, **319, 353**
Recombinant chromatids, *198*, **199**, 223–224
Recombinant DNA, **358–359**
Recombinant DNA technology
  artificial DNA synthesis and mutagenesis, 363
  biotechnology and, 367–372
  cDNA, 363
  cutting and splicing DNA, 358–359
  gene libraries, 362–363
  inserting new genes into cells, 359–362
Recombinant frequencies, **223–224**
Recombinant phenotypes, **214**
Recombination, genetic. *See* Genetic recombination
Reconciliation ecology, **1241–1242**
Rectum, **1077**, 1084
Red abalone, 1147
Red algae, *584*, *597*, 611, *612*
  endosymbiosis theory and, 589
  overview of, **601–602**
Red blood cells
  ABO system and, 219
  in blood tissue, 402
  features of, 1055
  function, *403*
  life span, 1056
  malaria and, 585, *586*
  osmosis and, 106–107
  production of, 1055
  sickle-cell disease. *See* Sickle-cell disease
Red deer, 1156
Red-green color blindness, *207*, 227, *228*
Redi, Francesco, 63
Red light, seed germination and, 812–813
Red mangroves, 1242
Red muscle, 1013
Redox reactions
  coenzyme NAD and, 141–142
  described, 140–141
Redstarts, 1168, *1169*
Red tide, 586, *587*
Reducing agents, 141
  in photosynthesis, 166
Reduction, **140–141**
Reductionism, 396
Red-winged blackbirds, *510*

Reefs. *See* Coral reefs
Reflexes, 987–988
  See also Autonomic reflexes
Refractory period
  of male sexual response, 914
  of sodium ion channels, **952–953**
Regeneration, **897–898**
Regulative development, **925–926**
Regulator proteins, 320
Regulator sequences, **320**
Regulatory genes, **298**
  development and, 449–450
Regulatory subunits, **133**
Regulatory systems, **856**
Reindeer "moss," 656
Reinforcement, **517**
Reissner's membrane, *972,* **973**
Release factor, 270, *271*
Release-inhibiting hormones, 883
Releasers, **1144**
Releasing hormones, 883
Religion, 16
REM sleep, **998,** 999
Renaissance, 63
Renal artery, **1101,** *1102*
Renal dialysis machines, 1105–1106
Renal failure, 1105–1106
Renal medulla, **1101,** 1102, 1103, 1105
Renal pyramids, **1101,** *1102,* 1105
Renal tubules, **1099,** 1100–1101, 1102, 1104
Renal vein, **1101,** *1102*
Renin, **1107**
Repenomamus giganteus, 477
Replication complex, **243–244**
Replication forks, **243,** 244, *245*
Replication unit, 359
Replicons, **359**
Reporter genes, **359,** 361–362
Repressible proteins, **298**
Repressible systems, 298–299
Repressor proteins, 320
Repressors, **297,** 298, 299
  See also Operator-repressor systems
Reproductive incompatibility, 515–518
Reproductive isolation, **510**
Reproductive signals, 181–182
Reproductive technologies, 916
Reptiles, *732*
  adaptations for water conservation in, 1099
  in amniote phylogeny, *731*
  circulatory system, 1048–1049
  evolutionary history, 475, 476
  overview of, 731
  reproduction, 905
RER. *See* Rough endoplasmic reticulum
Rescue effect, **1178**
Residual volume, 1032
Resistance factors, **294–295**
Resistance genes, 839–840
Resistance vessels, 1057
Resolution (microscopes), **70–71**
Resolution phase (human sexual response), 913
Respiration, 140
  See also Cellular respiration
Respiratory chain, 140
Respiratory diseases, infectious agents, *409*
Respiratory distress syndrome, 1034
Respiratory gases, **1025**
  diffusion, 1025–1026
  transport in blood, 1036–1038
Respiratory gas exchange
  adaptations maximizing, 1028–1032
  control of breathing, 1039–1040
  in extreme conditions, 1024–1025
  in humans, 1032–1035
  physical factors governing, 1025–1027
Respiratory systems
  in birds, 1030–1031
  control of breathing, 1039–1040
  in fish, 1029–1030
  in humans, 1032–1035
  in insects, 1028, *1029*
  maximization of partial pressure gradients, 1028
  surface area of respiratory organs, 1028
  tidal ventilation, 1031–1032
Responders, in signal transduction pathways, **334–335**
Resting potential, **946,** 947, 948–950, 952
Restoration ecology, **1236–1237,** 1238, *1239*
Restriction digestion, **353**
Restriction endonucleases, 353
Restriction enzymes, **353–354,** 358, 359, 381
Restriction fragment length polymorphisms (RFLPs), **381,** 382
Restriction point, **186**
Restriction site, **353**
Reticular activating system, 988
Reticular formation, 998
Reticular system, **988**
Reticulate bodies, 572
Reticulum, **1084–1085**
Retina, **979**
  adaptations to diurnal *vs.* nocturnal habitats, 1153
  information flow in, 980–981
  layering of cells in, 980, *981*
  photoreceptor cells in, 980
  receptive fields, *994,* 995
Retinal, *129,* 577, **976**
Retinoblastoma, 186
Retinoblastoma protein (RB), **186**
Retinol, *1074*
  See also Vitamin A
Retrotransposons, 312, 534
Retroviruses, 63, **261,** 288–289
Reverse transcriptase
  action of, 63
  cDNA and, 363
  DNA chip technology and, 366
  HIV and, **288,** 423
Reverse transcription, **261**
Reversible reactions, **33**
Reznick, David, 1172
R factors, 294–295
RFLPs, **381,** 382
R genes, **839–840**
R group, **42**
Rhagoletis pomonella, 513
Rhea, 732
Rheumatoid arthritis, 421
Rhinoceroses, 1193, 1194, 1232–1233
Rhinovirus, *285*
Rhizaria, *584, 597,* **604–605**
Rhizobium
  gene transfer efforts, 790
  in mutualisms, 1194
  nitrogen-fixation with legumes, **788,** 789
  proteobacteria, 575
  R. japonicum, 790
Rhizoids, *615,* **619,** 624, 652, *664*
Rhizomes, 619, **619,** 832
Rhizopus
  R. microsporus, 665
  R. oligosporus, 194
  R. stolonifer, 660, 664–665
Rhizoxin, 665
Rhodinus, 876, 877
Rhodopsins, 56, **976–977,** 980
Rhopalaea spp., *723*
Rhynchocoel, 697
Rhynia, 610, *619*
  R. gwynne-vaughani, 610
Rhyniophytes, **619,** 625
"Rhythm method," 914
Ribbon worms, *683, 692,* 695, **697**
Ribeiroia, 14, 15
Riboflavin, *1074*
Ribonucleases, 316
Ribonucleic acid (RNA), **57**
  central dogma and, 260–261
  complementary base pairing, 58
  differences from DNA, **260**
  nucleotide structure, 57–59
  in origin of life theories, 62–63
  primer strand in DNA replication, 245–246
  synthesis of, 261–265
  viruses and, 284
Ribonucleotides, 59
Ribose, 50
Ribosomal RNA genes, concerted evolution in, 535–536
Ribosomal RNA (rRNA), 79
  gene amplification and, 324
  moderately repetitive DNA sequences and, 311, *312*
  peptidyl transferase activity and, 269
  subunits, 268, 311
  synthesis of, 261
  in translation, **268**
  using in evolutionary studies, 569–570
Ribosomes
  in eukaryotes, 79
  membrane proteins and, 100
  numbers in fish and frogs, 324
  in plant and animal cells, *76, 77*
  polysomes, 270, *271*
  in prokaryotes, **73,** 79, 268, 269
  ricin and, 257
  rough endoplasmic reticulum and, *79, 80*
  sex-linked proteins contribute to, 228
  in translation, 269, 270
Ribozymes
  as catalysts, **63,** 125
  specificity, 126
  in vitro evolution, 538, *539*
Ribulose 1,5-bisphosphate (RuBP), **170,** 171
Ribulose bisphosphate carboxylase/oxygenase (Rubisco), **170,** *171,* 172–173, 174
Ribulose monophosphate (RuMP), 170
Rice, 646
  beriberi and, 1074–1075
  cytokinins and, 810
  "foolish seedling" disease, 801–802
  genome, 311
  nitrogen-fixation and, 788
  quantitative trait loci, 222
  Rhizopus microsporus and, 665
  transgenic, 370, *371*
Ricin, 256–257
Ricinus communis, 256
Rickets, *1074*
Ricketts, Taylor, 1241
Rickettsia prowazekii, 302, 533
Riftia sp., *699*
Rift valleys, 468
Rigor mortis, 1008
Ring canal, 719
Ringneck snake, *732*
Ring-tailed lemur, *1130*
Rio de Janeiro Earth Summit, 1235
Río Tinto, 560
Risk costs, 1148
Rivers, hydrological cycling and, 1211
RNA-dependent RNA polymerase, 288
RNA editing, **326**
RNA interference (RNAi), 840
RNA polymerase(s), **262, 319**
  promoter sequence recognition, 313, 319
  protein synthesis in prokaryotes and, 299–300
  in transcription, 261, 262, *263*
  virulent viruses and, 286
RNA splicing, **316–318**
RNA viruses
  characteristics of, 261
  nucleotide substitution rates, 528
  plant immunity to, 840
  reproductive cycles, 288–289
"RNA world," 63
Robins, 1144
Robotic vehicles, 20, *21*
Rock, soil formation and, 786
Rockfish, 1171–1172
Rock lobster, 1238
"Rock louse," *707*
Rod cells, **977,** *978,* 980, *981*
Rodentia, *736*
Rodents
  diversity in size, 464
  Laotian rock rat, 1229
  number of species, 735
  species richness and ecosystem productivity, *1201*
Root apex, *823,* 824
Root apical meristem, *753,* 754, 755–756
Root cap, *753,* **755**
Root hairs, *755,* **756**
  mycorrhizae and, 657–658. *See also* Mycorrhizae
Root nodules, **788,** 789, *790*
Root pressure, **770**
Roots, **755–757**
  adaptations to dry environments, 846
  auxin-induced initiation, 805
  effect of auxin on elongation, 806
  effect on soil pH, 788
  embryonic, 798
  eudicot *vs.* monocot, *756*
  evolution of, 619–620
  gravitropism in, 805
  mycorrhizae, 655, 657–658, 756, 789, 1194
  plant nutrition and, 782
  types of, *746*
  uptake of water and mineral ions, 768, *769*
  water-saturated soils and, 846–847
Root systems
  absorptive surface of, 756
  functions of, **745**

roots as modules, 753
Roquefort cheese, 666
Rosa rugosa, 646
Rosenzweig, Michael, 1161
Rotational cleavage, 923, *924*
Rothamstead Experiment Station, 1200
Rotifers, *683, 692,* 695, **696**
Rough endoplasmic reticulum (RER)
in animal cells, *76*
insertion of membrane-associated proteins, 100, *101*
structure and function, **79,** *80*
Rough-skinned newt, 486
Round dance, 1157
Roundup, 370
Roundworm, 309, *675*
Rous, Peyton, 387
R plasmids, 294–295
rRNA. *See* Ribosomal RNA
RT-PCR, **366**
RU-486, *914,* 915
Rubber, 796–797, 810
Rubber tires, 796
Rubber trees, 796–797
Rubella virus, *409*
Ruben, Samuel, 161
Rubisco. *See* Ribulose bisphosphate carboxylase/oxygenase
RuBP. *See* Ribulose 1,5-bisphosphate
Ruffini endings, 971
Rumen, 148, 579, **1084–1085**
Ruminants, 532, **1084–1085**
RuMP. *See* Ribulose monophosphate
Runcaria, 474
Runners, of plants, 747, 832
RUNX1 transcription factor, 421
Rushes, 832
Russell, Liane, 323
Rust fungi, 667
Ryanodine receptors, 1009, 1010
Rye, 831

-S-

Sabin, Albert, 524
Saccharides, 39
Saccharomyces, 359, *652*
cerevisiae, 309, 665
See also Yeasts
Saccoglossus kowalevskii, 721
Saccule, *975*
Sac fungi, 655, *664*
Saddle joint, *1019*
Safflower, 650
Saiga antelope, *1233*
Saiga tatarica, 1233
Saint-John's-wort, 1196
Sake, 666
Salamanders, *730*
heterochrony and foot morphology in, 451, *452*
key features of, 729
larval gills, *1027*
noncoding DNA in, *534*
Salicylic acid, 839
Saline habitats, plants and, 848
Salinization, of soils, 836–837, 848
Salivary amylase, *1084*
Salivary glands, *1079,* 1080
Salivation, 1085
Salix, 839
S. sericea, 678
Salk, Jonas, 524
"Sally Lightfoots," *707*
Salmon, 727, 903, *1172,* 1238
Salmonella, 73, 302, *567,* 579
S. gallinarum, 287
S. typhimurium, 561, 575
S. typhi, 409
Salps, 722, *723*
Salsifies, 514–515
Saltatory conduction, **954–955**
Salt glands, **848,** 1095
Salt licks, 1095
Salts, 28
soil salinization, 836–837, 848
Salt-tolerant plants, 371
Salviniaceae, 627
Sambucus, 760
Sand, in soils, 786
Sand County Almanac, A (Leopold), 1211
Sanderlings, *1168*
Sand flies, 582, *604*
Sandhill cranes, 1226
Sand verbena, 832
Sap. *See* Phloem sap; Xylem sap
Saprobes, **600, 651,** 655, **1075**
Saprolegnia, 584, 600, *601*
Sapwood, 760
Sarcolemma, *1008*
Sarcomas, 387
Sarcomeres, **1006–1007,** 1014
Sarcophilus harrisii, 736
Sarcoplasm, **1009,** 1010, 1012
Sarcoplasmic reticulum, **1009**
Sarcopterygians, **728**
Sarcoscypha coccinea, 666
Sargasso Sea, 302, 600
Sargassum, 302, 600
Sarin, 131
Sarracenia, 792
S. purpurea, 495
SARS, 301, 539, 1216
Satellite cells, 1014
Saturated fatty acids, **55**
Saturation
of catalyzed reactions, 130
of membrane-transport systems, 110
Saturn, 20
Saurischia, *731*
Saussurea, 828
Savanna, primary production in, *1210*
Scale-eating fish, 502
Scales, of ray-finned fishes, 726
Scandentia, *736*
Scanning electron microscopy, *71*
Scarab beetle, 865
Scarlet banksia, *4*
Scarlet gilia, 841
Schally, Andrew, 882–883
Schindler, David, 1218
Schleiden, Matthias, 5
Scholander, Per, 771, 772
Schools, of fishes, 726
Schopf, J. William, 68
Schwann, Theodor, 5
Schwann cells, **946**
Schwarzkopf, H. Norman, 744
Scientific method, 2–3, **13–14**
Scientific names, 12, 554, 555–556
Scion, 833
Sclera, 978, *979*
Sclereids, **750,** *751*
Sclerenchyma, **750,** *751*
Scleria
S. goossensii, 498
S. verrucosa, 498
Sclerotium, 606
Scolopendra gigantea, 712
Scorpionflies, *709*
Scorpions, *693, 713*
Scots pine, 1198
Scouring rushes, 625
Scrapie, 378
Scrotum, **906**
Scudderia mexicana, 711
Scurvy, 1074
Scyphozoans, *685,* 686
Sea anemones, 83, *84,* 686, *944*
Seabirds, defense of nesting sites, 1149, *1150*
Sea butterflies, 702
Sea cucumbers, 719, *720,* 721
Sea daisies, 720–721
Sea lamprey, 724
Sea lettuce, 594, *595,* 602, *603*
Sea levels, changes in, 468, *469*
Sea lilies, 475, 719, *720,* 721
Sea lions, 737
Seals, 737, 1037, 1064
Sea palms, 600, *601*
Sea pens, 686
Sea slugs, *701,* 702
Sea spiders, 712
Sea squirts, *533,* 548, *717,* 722, *723*
Sea stars, *717*
description of, 719–720
feeding behavior, 721
keystone species, 1196–1197
regeneration in, 897–898
tube feet, 676
Sea turtles, 731, *732*
Sea urchin
cleavage in, 923, *924*
description of, 719
egg, 185, 901, 902
engrailed gene, *537*
feeding behavior, 721
fertilization in, 901, 902
gastrulation in, 927–928
polarity in early development, 436
purple, *720*
studies of developmental genes in, 435
tube feet, 676
Sebastes melanops, 1171–1172
Secondary active transport, **111–112**
Secondary cell walls, **749,** *751*
Secondary consumers, **1186,** *1187*
Secondary endosymbiosis, **589–590,** 602
Secondary growth, **631, 754–755,** 758–760
Secondary immune response, **409**
Secondary inducer, 438
Secondary lysosomes, 81, *82*
Secondary mesenchyme, 927
Secondary metabolites, *839,* 841, 844
Secondary oocytes, **900,** 901, 910
Secondary phloem, 754, 758
Secondary plant body, **753**
Secondary sex determination, **226**
Secondary sexual characteristics, 909
Secondary spermatocytes, **899,** *900*
Secondary succession, **1197,** 1198–1199
Secondary xylem, 754, 758
Second filial generation (F_2), **209**
Second genetic code, 267
Second messengers, **339**
derived from lipids, 341–343
nitric oxide, 344
phototransduction, 977, *978*
stimulation of protein kinase cascades, 340–341
Second polar body, *900,* **901**
Secretin, **1086**
Securin, 190, *191*
Sedimentary rocks, 465, 467
Seed coat, 633, 634, 823
Seed ferns, 631
Seedless grapes, 802
Seedless plants, 622–627
Seedlings, **798,** *799*
etiolation, 813, *814*
Seed plants
common name and characteristics, *612*
evolutionary history, 631, *632*
as food sources, 647
life cycle, 631–633
major living groups, *632*
medicinal value of, 647–648
overview of angiosperms, 638–643
overview of gymnosperms, 634–637
Seeds
abscisic acid and, 811
aleurone layer, 801
arcelin and resistance to weevils, 842–843
characteristics, **633–634**
constancy of size in, 457
development, 823–824
dispersal, 824
dormancy, 457, 634, 798, 799–800
environmental effects on production of, 457
germination, 630, 798, 799, 800, *801,* 812–813
sensitivity to red and far-red light, 812–813
Segmentation, in animals, **675–676**
Segmentation genes, **442–444,** 445
Segment polarity genes, **443**
Segregation, law of, 211–213
Seizures, 999
Selective breeding, 1142
Selective herbicides, 807
Selectively permeable membranes, 765
Selective permeability, **105**
Selenium, *1073,* 1219, 1220
Self-compatibility, 551
Self-fertilization, 496, 821
Self-incompatibility, 551–552, **822**
Selfing, 496, 551
Self-pollination, 208
Self-tolerance, cellular immune response and, 417
Semen, **906,** 908
Semicircular canals, **974,** *975*
Semiconservative replication, **241–242**
Semideserts, primary production in, *1210*
Seminal fluid, 908
Seminal vesicles, *906,* **908**
Seminiferous tubules, **907**
Senescence, **798–799**
Senescence hormone, 810
Sensation, cellular basis of, 967
Sensory cells. *See* Sensory receptor cells
Sensory exploitation hypothesis, 552–553
Sensory neurons, **943,** 967
Sensory organs, 967
Sensory receptor cells, **965–967**
Sensory receptor proteins, 965–966
Sensory systems, **967**
actions of sensory cells in, 965–967
adaptation, 967
chemoreception, 967–970

detection of light, 976–978
mechanoreception, 970–975
Sensory transduction, **965–967**
Sepals, **638–639,** *820,* 821
Separase, 190, *191*
Sepia latimanus, 1021
Sepsis, 406
Septa, of hyphae, **652,** *653*
Septate hyphae, **652,** *653*
Sequential hermaphrodites, 904
Sequential secondary endosymbiosis, 590
Sequoiadendron giganteum, 1177
Sequoia sempervirens, 765, 769
SER. *See* Smooth endoplasmic reticulum
Serine, *43, 132*
Serotonin, 958, *959*
reuptake inhibitor, 961
Sertella septentrionalis, 695
Sertoli cells, **907,** *908,* 909
Serum, 219
Serum immunity, 524
Sesquiterpenes, *839*
Sessile animals, **676**
competition and, 1192–1193
dispersal, 680
Sessile organisms, 782
Setae, 699
Setophaga ruticilla, 1168, *1169*
Set points, **856,** 869
Seven-transmembrane-spanning G protein-linked receptors, 337–339
Severe Adult Respiratory Syndrome (SARS), 301
Sex, hypotheses for the existence of, 501
Sex chromosomes
abnormalities, 226
cytoplasmic bridges between spermatocytes and, 899, *900*
sex determination, **225–226,** 1160
sex-linked inheritance, 206, *207,* 226–228
Sex determination
chromosomal, **225–226**
in hymenopterans, 1160
Sex hormones, 984
See also Sex steroids
Sex-linked inheritance
hemophilia A, 206
in humans, 227–228
overview of, **226–227**
red-green color blindness, *207,* 227, *228*
Sex organs, human, *889,* 906–910
Sex pilus, **291,** *292,* 294
Sex steroids, 888, 889–890
See also individual hormones
Sexual behavior
asexual reproduction in animals and, 898
external fertilization systems and, 903
in humans, 913–914
positive feedback and, 857
Sexual dimorphisms, 504, 519
Sexual excitement, in humans, 913–914
Sexual life cycles, 193, 194
Sexually transmitted diseases (STDs), 301–302, 423, **917**
Sexual reproduction
evolutionary adaptations to life on land, 904–905
fertilization, 901–903
gametogenesis, 899–901
genetic diversity and, **193–195,** 501, 899
hermaphroditism, 904
mating systems, 903–904
meiosis, 195–202
oviparity, 905
steps in, 899
viviparity, 905
See also Human reproduction
Sexual selection
in birds, 499–500, 508–509, 519
courtship displays and, 1143
effects of, 496
reproductive success and, **498–500**
S gene, 822
Sharks
brain, *945*
characteristics of, 726
clasper fins, 904
intestinal surface area, *1078*
jaws and teeth, *725*
maintenance of salt and water balance in, 1099
viviparity, 905
Sheep, cloned, 431–432
Shelled egg. *See* Amniote egg
Shells, of mollusks, 1016
Shigella, 73, 294, 302
Shine–Dalgarno sequence, 268
Shivering, 867
Shoots
apical meristems, 754, 757
early development, *799*
embryonic apex, *823,* 824
gravitropism in, 805
See also Stems
Shoot systems, components of, **745**
Short-day plants (SDPs), **826,** 827, 831
Short interspersed elements (SINEs), 312
Short-long-day plants, 826
Short tandem repeats (STRs), 356
Short-term memory, **999**
Shotgun sequencing, **393,** *394*
Shrews, 735
Shrubland, primary production in, *1210*
Shull, G. H., 220
Siamese cats, 219, 221
siamois gene, 930
Siamois transcription factor, 930
Sickle-cell disease
abnormal hemoglobin and, 376
autosomal recessive mutant allele, 379
DNA testing for, 385, *386*
identifying genetic basis of, 380–381
missense mutation, 275, *276*
Side chains, of amino acids, **42–43**
Sieve plates, **752,** 776, 777
Sieve tube elements, *751,* **752,** 776, 777
Sieve tubes, 752
Sigmoria trimaculata, 712
Signal recognition particles, **273**
Signals, **1156**
environmental, 333–334
types of, 1157
Signal sequences, **272–274,** 367–368
Signal transduction
calcium ions and, 343–344
direct and indirect, 339, *340*
protein kinase cascades, 339–340, *341*
regulation, 345
Signal transduction pathways
affecting glycogen metabolism, 346, *347*
auxin and, 808–809
calcium ions and, 343–344
defense systems and, 407
E. coli model for, **334–335**
effects on gene transcription, 347
ethylene and, 810, *811*
hormones and, *892,* 893
plant defensive secondary metabolites and, 842, *843*
plant development and, 798
receptors, 336–339
for smell, 345–346
Silencer sequences, **320**
Silent mutations, **275**
Silent Spring (Carson), 457
Silent substitutions, **530–531**
Silicates, 62
Silicon, 25
Silk, 878
Silkworm moths, 877–878, 968
Silky willow, *678*
Silurian period, *466–467,* **473–474,** 618, 619
Silverfish, 709
Silver salts, 810
Silverswords, *520,* 521
Simberloff, Daniel, 1133
Similarity matrix, *526,* **527**
Simple cells, visual cortex, 995
Simple diffusion, 105, **106,** *111*
Simple fruits, 643
Simple leaves, 748
Simple lipids, 55
Simple sugars. *See* Monosaccharides
Simpson Desert, Australia, *1123*
Simultaneous hermaphrodites, 904
SINEs, 312
Single bonds, 27
Single nucleotide polymorphisms (SNPs), 356, 374, 381–382, 394
Single-strand binding proteins, **244**
Sinistral coils, 923
Sinks, **775,** 1214
carbon, 1214
pressure flow model of translocation and, 776–777
Sinoatrial node, **1052,** 1053
Sinusoids, 1084
Siphonaptera, 709
Sirenia, *736*
Sirius Passet, 472
siRNAs. *See* Small interfering RNAs
Sister chromatids, **183, 187**
Sister clades, **544**
Sister species, **544**
SIVagm, *553*
SIVcpz virus, *553*
SIVhoest, *553*
SIVmdn, *553*
SIVsm, *553*
SIVsun, *553*
SIVsyk, *553*
Skates, 504, 726
Skeletal muscles
functions, **858,** 1005
length/tension relationship, 1014
mechanisms of contraction in, 1005–1010
neuromuscular junction and, 955
slow-twitch and fast-twitch, 138, 157, 1013–1014, 1015
stretch receptors, *971,* 972
summation of twitches in, 1012
Skeletal systems, **1016–1019**
Skeletons
cartilaginous, 726, 728
of echinoderms, 719
hydrostatic, 675
overview of, 1016–1019
of ray-finned fishes, 726
See also Exoskeletons
Skin
as an organ, 15
endocrine functions, *880*
epithelial cells, *858*
nonspecific defense against pathogens, 404
pigmentation, 220, 1074
tactile mechanoreceptors in, 970–971
in thermoregulation, 864
Skin cancers, 249
Skin color, 220, 1074
Skunk cabbage, 818, 862
Skunks, 1021
SLE. *See* Systemic lupus erythematosis
Sleep, 997–999
Sleeping sickness, 582, *604*
"Sleep movements," in plants, 828
Sliding DNA clamp, **247,** *248*
Sliding filament theory, **1007–1008**
Slime, 73
Slime molds, 584, 591, *597,* 606–607
Slime nets, 598–599
Slow block to polyspermy, 902
Slow-twitch fibers, 138, 157, **1013–1014**
Slow-wave sleep, **998**
Slug, of cellular slime molds, 607
Slugs (mollusk), 702
Small interfering RNAs (siRNAs), 365, 840
Small intestines
absorption of nutrients in, 1083–1084
digestion in, **1082–1083**
digestive enzymes, *1084*
in humans, *1079*
nonspecific defense against pathogens, 405
Smallmouth bass, 1177
Small nuclear ribonucleoprotein particles (snRNPs), **317**
Smallpox, 400, *409,* 410
Small RNAs, 325
Smell, 968–969
ion channel activity and, 345–346
Sminthurides aquaticus, 709
Smith, Alan, 1204
Smith, Hamilton, 300
Smith, William, 465
Smithsonian Tropical Research Institute, 1204
Smog, 1217
Smoker's cough, 1034
Smoking. *See* Cigarette smoking
Smooth endoplasmic reticulum (SER)
in animal cells, *76*
in plant cells, *77*
structure and function, **80**
synthesis of phospholipids and, 100, *101*
Smooth muscles
in blood vessels, 1057, *1058*

capillary blood flow and, 1061, *1062*
epinephrine and, 339
functions, **858,** 1005
mechanisms of contraction in, 1010–1012
in the vertebrate gut, 1079
Smut fungi, 667
Snails, 679, 923, 1016
Snakes
adaptations for feeding, 1075, *1076*
characteristics of, 731
evolutionary history, 477
hemipenes, 905
suppression of limb formation in, 458
venoms, 1021
vomeronasal organ in, 969
Sneezing, 405
Snow, John, 96
Snowshoe hares, 1189
Snowy owls, 1189
SNPs. *See* Single nucleotide polymorphisms
snRNPs. *See* Small nuclear ribonucleoprotein particles
Soap, 98
Social animals, population density and, 1161–1162
Social behavior, 729, 1158–1160
"Social feathers," *699*
Sodium
action potentials and, 951–953
conservation in animals, 1095
electric current and, 947
electronegativity, *27*
fast block to polyspermy and, 902
halophytes and, 848
ion channel specificity and, 109
ionotropic glutamate receptors and, 959, 960
as macronutrient, *1073*
nutrient absorption in the small intestine and, 1084
unit of charge, 28
See also Sodium-potassium pump
Sodium bicarbonate, 164
Sodium channel genes, 538
Sodium channels
acetylcholine receptor, 337
generation of action potentials and, 951–953, 954
heartbeat and, 1052, 1054
in phototransduction, 977, *978*
specificity, 109
Sodium chloride, 28, *29*
Sodium hydroxide, 33
Sodium-potassium ATPase, **947–948**
Sodium-potassium pump, **111,** *112,* **947–948**
Soft drinks, 124
Soft palate, 1079, *1080*
Soft-shelled crabs, 1016
Soil erosion, global phosphorus cycle and, 1219
Soil fertility, 786
Soil pH
effect of plants on, 788
effect on mineral nutrients, 787
effect on nitrogen uptake, 791
Soil profiles, 785–786
Soils
availability of mineral nutrients in, 786–787
Dust Bowl, 780–781
effect of plants on fertility and pH, 787–788
formation, 613, 786
heavy metals and, 849
organic carbon losses, 1215
permafrost, 1118
primary succession and, 1198
salinization, 836–837, 848
structure and complexity of, 785–786
water-saturated, 846–847
Soil solution, mineral nutrients and, **781,** 782
Solar compass, time-compensated, 1156
Solar energy
global climates and, 1114–1115
global ecosystem processes and, 1209–1210
utilization and loss, *176*
Solar system, age and formation of, 34
Soldiers, 1159–1160
Sole, 504, 1020
Solidago, 793
Solowey, Elaine, 630
Solute potential, **765–766**
Solutes, **32**
diffusion of, 105, *106*
Solutions, *29,* **32–33**
Solvent drag, 1084
Solvents, water as, **32–33**
Somatic cells
animal cloning and, 431–432
chromosomes in, **193**
mutations and, 275
Somatic mutations, **275**
Somatomedins, 882
Somatostatin, *880,* **887**
Somites, **933,** 934
Songbirds
effect of testosterone on singing behavior, 1146–1147
song learning in, 1145–1146
survivorship curves, *1170*
Songs
of whales, 1157
See also Birdsong
Song sparrows, 1174–1175
sonic hedgehog gene, 934
Sooty mangabeys, 551, *552*
s Orbitals, 24
Soredia, 656, *657*
Sorghum, 844
Sori, 627
Soricomorpha, *736*
Source, **775**
pressure flow model of translocation and, 776–777
Sousa, Wayne, 1199
South Africa, fynbos vegetation, 1240–1241
Southern beeches, 1119
Southern blots, 354, *355*
Sow bugs, 706
Soybean, root nodules, *790*
Soy sauce, 666
Spadix, 818
"Spanish flu," 283
Sparrows, song-learning in, 1145–1146
Spastic muscle paralysis, 961
Spathe, 818
Spatial summation, in neurons, **957**
Speciation, 12, **510**
allopatric, 511–513
early density of species, 1177
genetic switches and, 453
as a gradual process, 510–511
hybrid zones, 517–518
prezygotic reproductive barriers and, 515–516
rates of, 518–519
reproductive incompatibility, 515–518
sympatric, 513–515
Species, **5–6,** 509
classification by appearance, 509
endemic, 1129, *1130*
extinctions. *See* Species extinctions
factors threatening survival of, 1231–1234
formation, 510–515. *See also* Speciation
invasive, 1233–1234
present-day number of, 472
reproductive incompatibility, 515–518
richness. *See* Species richness
sister, 544
Species-area relationship, **1230**
Species extinctions
"centers of imminent extinction," 1236
conservation biology's goal to prevent, 1228–1229
extinction rates, 480, 482, 1132
factors contributing to, 1231–1234
predicting, 1229–1230
species affected in the U.S., *1231*
strategies used to limit, 1235–1243
Species name, 12
Species pool, **1132**
Species richness, **1117, 1200**
ecosystem productivity and, 1200–1201
establishment of protected areas and, 1235
factors affecting, 518–519, 1200–1201
intermediate disturbance hypothesis, 1199
island biogeographic theory and, 1132
keystone species and, 1196–1197
latitudinal gradients, 1200
Species-specific behavior, 1141–1142
Specific defenses, 401, 418–420
Specific heat, **32**
Specific immune system
AIDS and, 421–423
allergic reactions, 420, *421*
autoimmune diseases, 421
cellular immune response, 408, 414–417
clonal selection, 408–409
humoral immune response, 408, 411–414
immune deficiency disorders, 421
immunological memory, 408, 409–410
immunological tolerance, 410–411
key traits, 407–408
primary and secondary immune response, **409**
Speech, brain areas participating in, 1001–1001
Spemann, Hans, 928–929
Spergula vernalis, 1170
Sperm
acrosome and acrosomal reaction, 901, *902*
contribution to the zygote, 921–922
external fertilization systems, 903
fertilization in humans and, 910
gametogenesis and, **899**
in humans, 906, 907, 908
internal fertilization systems, 903–904
in plants, 616, *820,* **822,** *823*
specificity in interactions with the egg, 901–902
spermatogenesis. *See* Spermatogenesis
spermatophores, 729, 904
vasectomies and, 915
viability, 914
Spermatia, 663
Spermatids, **899,** *900*
Spermatogenesis
hormones and, 909
overview of, **899,** *900*
problems with suppressing, 916
seminiferous tubules and, 907
temperature and, 906
Spermatogonia, **899,** *900*
Spermatophores, 729, 904
Spermicides, 914
Sperry, Roger, 1000
Sphagnum, 624
S phase, **184,** 185, 186
Sphenodon punctatus, 732
Spherical symmetry, **674**
Sphincter muscles, 1080, 1101
Sphinx moth, *1029*
Sphygmomanometer, 1052
Sphyraena
S. barracuda, 727
S. genie, 1025
Spices, antimicrobial qualities, 1151–1152
Spicules, **684**
Spider monkeys, *739*
Spiders
external digestion of prey, 1075
fossil record, 472
nervous system, 944
overview of, 713
spermatophores, 904
system of movement, 1017
web spinning behavior, 1140–1141
Spider silk, 46
Spike mosses, 624
Spikes (inflorescence), *639*
Spina bifida, 932
Spinach, 826
Spinal cord
autonomic nervous system and, 992, *993*
development, 932, 934, 986
structure and function, 987–988
in vertebrates, **944**
Spinal nerves, development in vertebrates, 986
Spinal reflex, **987–988**
Spindle checkpoint, 190
Spindles (mitotic), **189,** 190–191, 923
Spindles (muscle), *971,* 972
Spines, of xerophytes, 846
Spinifex, 1128
Spinner dolphins, *737*
Spiracles, 708, 1028, *1029*
Spiral bacteria, *4*
Spiral cleavage, 673, 692, **923**
Spiralians, 673, 692
Spirobranchus sp., *677*
Spirochetes, *284,* 566, *571,* **571–572**
Spirometer, 1031–1032
Spleen, **1056**
"Splenda," 38

Spliceosome, **317**
Splicing, of pre-mRNA, 316–318
Spohr, H. A., 792
Sponge industry, 684
Sponges, *672*
asymmetry in, 674
cell recognition and adhesion in, 102, *103*
described, **683–684**
filter feeding, 677
gas exchange and, *1027*
Spongy mesophyll, 761
Spontaneous generation, 5, 63–64
Spontaneous mutations, 277, **277,** *278*
Sporangia, *664*
of fungi, 659, *660*
in homospory, *622*
of land plants, **614–615**
of plasmodial slime molds, 606
Sporangiophores, 625, *660,* **664**
Spore mother cells, in ferns, 626
Spores
of cyanobacteria, **573**
of fungi, 654, *655,* 659
haplontic life cycle and, **194**
microbial eukaryotic reproduction and, 593
of plants, 613, 624, 632
Sporocytes, **594**
Sporophylls, 624
Sporophytes
in alternation of generations life cycles, 194, **594,** *595*
in land plant life cycle, **614–615**
in plant evolution, *633*
in vascular plants, 617
Sporopollenin, 613, 633
Sporozoites, *586,* 596
Spriggina flounderosi, 473
Spring salamander, *730*
Spring wheat, 831
Spruces, 1198
Squamates, 731
Squamous cells, *858*
Squids
chromatophores, 1020
giant axon, 948, *949*
nervous system, *944*
spermatophores, 904
Squirrels, 464
Srf protein, 441
SRY gene, 226, 228
SRY protein, 226
Stabilizing selection, **497,** *498,* 532–533
Stable atoms, *24,* 25
Staghorn coral, 1234
Stahl, Franklin, 241, 242
Stained bright-field microscopy, *71*
Stamens, **638,** *639,* 640, 820–821
Stanley, Wendell, 284
Stapes, *972,* 973
Staphylococci, 574
Staphylococcus, 574
S. aureus, 574
Star anise, *644*
Starch
production in plants, 171
in red algae, 602
structure and function, **51,** *52*
Starling, E. H., 1058
Starlings, 59
Starling's forces, 1058, *1059*
Stars, navigational systems of animals and, 1156
Start codon, **264**
Starvation, 155, 1071
Statin, 391
Statistical methods, 15–16
estimating population density and, 1169
STDs. *See* Sexually transmitted diseases
Stele
plant uptake of ions and, 768
of roots, **756**
Stem cells, **364, 432**
fat-derived, 426, *427*
in knockout experiments, 364, 365
pluripotent, 402, *403,* 432–433
See also Embryonic stem cells
Stems
abscisic acid and, 811
auxin-induced elongation, 806
effect of ethylene on growth, 810
eudicot *vs.* monocot, *757*
functions of, **745, 746–747**
girdling, 775
meristematic growth in, 757
See also Shoots
Stenella longirostris, 737
Steno, Nicolaus, 465
Stephanoceros fimbriatus, 696
Stephen Hales Prize, 765
Steppuhn, Anke, *842*
Stereocilia, 973, 974–975
Stereotypic behavior, 1140–1142
Sterilization, to control fertility, 915
Steroid drugs. *See* Anabolic steroids
Steroid hormones, 878
See also Sex steroids
Steroids
structure and function, 56, *57*
used in plant defenses, *839*
Stick insects, *709*
Sticklebacks, 459
Sticky ends, **358–359**
Stigmas, **638,** *639, 820,* 821, 822
Stingrays, 726
Stinkhorn mushrooms, *4*
Stock, 833
Stolons, 832
Stomach
digestion and, 1080, *1081*
emptying of, 1082
endocrine functions, *880*
food storage and, 1080
function, **1077**
ghrelin and, 1088
in humans, *1079*
nonspecific defense against pathogens, 405
pepsin, *1084*
pernicious anemia and, 1075
tissues in, 859, *860*
ulcers, 1081–1082
Stomach acid, 34, 1080, 1081
Stomata
abscisic acid and, 811
cycling in CAM plants, 848, *849*
evolution, 620
function, 620, 761
guard cells, 748, 761
of hornworts, 623
mechanisms of opening and closing, **773–774,** 814
photosynthesis and, 161
xerophytes and, 845
Stomatal crypts, **845**
Stomatal cycles, 848, *849*
Stone cells, 750
Stoneflies, *709*
Stoneworts, 613
Stop codons, **264,** 270, 313
Strabismus, 938
Stramenopiles, *584, 597,* **598–600,** 611
Strasburger, Eduard, 770
Strata, **465**
Stratigraphic dating, 465–466
Stratosphere, **1208**
Strawberries (fruit), *643*
Strawberries (plant), 747, 819, 832
Streams, primary production in, *1210*
Strength, muscle and, 1014
Strepsiptera, 709
Strepsipterans, 690–691, *709*
Streptococcus pneumoniae, 234–235, *409*
Streptokinase, 368
Streptomyces, 573
S. coelicolor, 302
Streptomycin, 302, 573
Streptophytes, **611,** *612*
Stress response, cortisol and, 888–889
Stress response element (SRE), 321
Stretch receptors
action potentials and, 966–967
regulation of blood pressure and, 1062, *1063*
in skeletal muscles, *971,* **972**
Striated muscle, 1005
See also Skeletal muscles
Striga, 650, 658, 793
Strip mining, *849*
Strobili, 624, *625,* **636**
Stroke
familial hypercholesterolemia and, 376
mechanism of, **1060**
tissue plasminogen activator and, 368–369
Stroma, 83, *84*
Stromatolites, 468–469, 564
Strongylocentrotus purpuratus, 720
STRs. *See* Short tandem repeats
Structural genes, **296–297**
Structural isomers, **40**
Structural proteins, altered, 377
Struthio camelus, 733
Sturgeon, 356, 903
Sturtevant, Alfred, 224
Style, **638,** *639,* 640, *820,* 821
Stylets, 775
Suberin, 749, 755, 756
Submucosa, **1078–1079**
Subpopulations, 503, 1178
Subsoil, **786**
Substance P, *959*
Substrate-level phosphorylation, **144**
Substrates
effects of concentration on reaction rates, 130, 133
enzyme-catalyzed reactions, **126–127**
enzyme interactions with, 128–129
Subtropical ecosystems, conversion to cropland, 1222
Succession, **1197–1200**
Succulence, **845,** 848
Succulent Karoo, 521
Succulents, 175, 521, 1128
Suckers, 832
Sucralose, 38
Sucrase, 1083, *1084*
Sucrose
a disaccharide, 50
in plant metabolic pathways, 175
plant nectars and, 1184
pressure flow model of translocation, 776–777
production in plants, 171
structure, *51*
Sucrose-proton symport, 777
Sudden Acute Respiratory Syndrome (SARS), 539
Sugar phosphates, 53
Sugars
amino, 53
chemical properties that affect people, 38
monosaccharides, 49–51
plant nectars and, 1184
See also Fructose; Glucose; Sucrose
Sulcus (sulci), 989
Sulfhydryl group, *40*
Sulfolobus, 4, 576
Sulfur
adding to soils, 787
covalent bonding capability, *27*
deficiency in plants, *783*
global cycle, 1217–1218
Hershey–Chase experiment, 235–236, 237
as macronutrient, *1073*
metabolism in prokaryotes, 568, 569
in plant nutrition, 781, *782*
Sulfur-35, 235–236, 237
Sulfur cycle, 1217–1218
Sulfur dioxide, natural sources of, 1217
Sulfuric acid, 33, 1218
Sulfur springs, 576
Sulston, Sir John, 232
Sumatra, 818
Sun, navigational systems of animals and, 1155–1156
Sunbirds, 1070–1071
Sundews, 792
Superficial cleavage, **923,** *924*
Superior vena cava, **1050**
Surface area-to-volume ratio, 69–70
Surface tension, 32, **1034**
Surface winds, 1114–1115
Surfactants, **1034**
Survivorship, **1169,** 1170–1171
Sushi, 538
Suspensor, **823–824**
Sutherland, Earl, 340
Svedberg unit, 311
Swallowing, 1079, *1080*
Swamps, primary production in, *1210*
Swarm cells, 607
Swarming, in honeybees, 896, *897,* 1148
Sweat glands, 735
Sweating, 32, 868
Swifts, 508, *509*
Swim bladders, 726
Swiss Re, 1240
Swordtails, 552–553
Sykes' monkeys, *553*
Symbiosis
with chloroplasts, 83, *84*
fungi and, **655**
lichens and, 656
nitrogen-fixation and, 477–478, 788–789
plant-microbe, 319
Symmetry, in animal body plans, **674**
Sympathetic nervous system
features of, **992,** *993*
heartbeat and, 1052, *1053*

regulation of blood pressure and, 1062, *1063*
Sympatric speciation, **513–515**
Symplast, 348, **768,** *769*
Sympodiella, 1198
Symports, **111,** 112
Synapomorphies, **545,** 611
Synapses
actions of neurotransmitters at, 950–951
chemical, 945–946
electrical, 958
excitatory and inhibitory, 957
fast and slow, 958
in humans, **944**
neuromuscular junctions, 955–956, 1008–1009
neurotransmitters and the postsynaptic membrane, 956
potentiation, 960, *961*
terminating neurotransmitter action, 960–961
types of, 955
Synapsis, **197,** 199
Synaptic cleft, **955**
Synaptonemal complex, 197
Synchronous reproduction, 1175
Syncytium, 442, 923
Synergids, 821, 822, *823*
Syngamy, 594, 615
Synonymous substitutions, **530–531,** 532–533
Syphilis, *917*
Systematics, **544**
Systemic circuit, **1047,** 1048–1049
Systemic lupus erythematosis (SLE), 421
Systemin, **842,** *843*
Systems biology, **131, 396,** *397*
Systole, **1051,** *1053,* 1057
Systolic pressure, 1052

- T -

Tachyglossus aculeata, 735
Tactile receptors, 970–971
Tactile signals, 1157
Tadpoles, 1096
Taeniopygia guttata, 500
Taeniura lymma, 504
Tailings, 849
Tails, prehensile, 738
Tamarisk, 848
Tannins, *839*
Tapeworms, *682,* 696, 904
TAP protein, 318
Taproot system, 746, *747*
Tardigrades, *683, 692,* **706**
Target cells, **875,** *876*
Target proteins, 337
Taricha granulosa, 486
Tarsiers, 738
Tarweeds, *520,* 521
Tasmanian devil, *736*
Taste, 969–970
Taste buds, **969–970**
TATA box, **319**
Tatum, Edward L., 257–258, 291
Tautomers, 277, *278*
Taxodium distichum, 1113
Taxol, 88, *391,* 647, *648*
Taxon, **543,** 554
Taxoplasma, 596
Taxus brevifolia, 647
T cell leukemia, *387*
T cell receptors, 403, 408, 414–415
T cells, 401
in autoimmunity, 421
binding to antigens, 407–408
cellular immune response and, 408
clonal anergy, 410–411
clonal deletion, 410
clonal selection, 409
cytokines and, 404
effectors, 409
overview of, 402, **403**
T cell receptors and, 414–415
"testing" of, 417
types of, 414
Tcf-3 transcription factor, 930
Tea, 332
Tectorial membrane, *972,* 973
Teeth
evolutionary reversal in frogs, 545
herbivory and, 737
in mammals, 735, 737
in vertebrates, 724–725, 1076
Teleki Valley, Kenya, *1118*
Telencephalization, 987
Telencephalon
consciousness and control of behavior, 989–991
limbic system, 988–989
location, *986,* **987**
Teliospores, 662–663
Telomerase, **248–249,** 386
Telomeres, 247–249
Telophase
first meiotic division, *197,* 198, *201*
mitosis, **191–192**
second meiotic division, *197*
Temperate deciduous forest, **1120,** *1210*
Temperate evergreen forests, **1119,** *1210*
Temperate grasslands, **1121,** *1210*
Temperate viruses, **286**
Temperature
basal metabolic rate and, 866–867
butterfly eyespot, 454–455
diffusion and, 105
Earth's atmosphere and, 1208
enzyme activity and, 134–135
flowering and, 831
effect on plant distribution, 847, 1117
effect on proteins, 48
gene expression and, 221
Q_{10}, 860, *861*
respiratory gas exchange in water and, 1026
See also Body temperature
Template strand, **241, 261**
Temporal lobe, **990**
Temporal summation, in neurons, **957**
Tendons, **1019**
golgi tendon organs and, *971,* 972
jumping and, 1004–1005
Tendrils, 748
Tension, muscles and, 1012, 1014
See also Transpiration–cohesion–tension mechanism
Tentacles, 702
Tepals, **639**
Terminal transferase, 419
Terminator, **313**
Termites, *709,* 1161, *1162,* 1194
Terpenes, 801, 811, 838, *839*
ter region, 182
Terrestrial ecosystem, 1208–1209
Territoriality, 9, *1150*
cost-benefit analyses, 1148–1149
effect on community structure, 1161
Tertiary consumers, 1186
Tertiary endosymbiosis, 590
Tertiary period, *466–467,* **478,** 726
Test crosses, **213**
Testes
endocrine functions, *880*
gametogenesis in, **899**
meiosis in, 198
sperm production in, 906–908
Testospongia testudinaria, 684
Testosterone
abuse of, 874–875
blood levels, 891
a circulating hormone, 875
effect on singing behavior in birds, 1146, 1147
effect on territorial defense behavior in lizards, 1149
in humans, *880*
male puberty and, 890
male sexual function and, *908,* 909
present in males and females, 889
a steroid, 56, *57*
structure, *888*
"Test-tube babies," 916
Tetanus (disease), *409,* 579
Tetanus (muscular state), **1012**
Tetanus toxin, 955
Tetracycline, 302, 361
Tetrads, **197–198**
Tetraploids, 514
Tetrapods, **728**
Tetrastigma tuberculatum, 819
Tetrodotoxin (TTX), 486, 498, 538, 1021
resistance to, 486, 498
TFIID protein, 319
Thalamus, **987,** 988, 995, 998
Thalassiosira, 584, 585
Thale cress. *See Arabidopsis thaliana*
Thaliaceans, 722
Thalidomide, 457, 936
Thalloid gametophytes, 622
Thallus, 656
Thamnophis sirtalis, 486, 498
Therapeutic cloning, **434**
Therians, 735–737
Thermoclines, 1208
Thermodynamics, 120–122, 163
Thermogenin, 867
Thermoneutral zone, **866–867**
Thermophiles, **571,** 576–577, *578*
Thermoplasma, 577
Thermoreceptors, 965, *966*
Thermoregulation
adaptations to cold, 867–868
behavioral, 863
blood flow to the skin, 864
ectotherms and, 862–863, 865, *866*
endotherms and, 861–862, 863, 867–868
energy budgets and, 863–864
"hot" fish, 864–865
hypothalamus and, 868–869
panting and sweating, 868
See also Body temperature
Thermostat, vertebrate, 868–869
Thermotoga maritima, 570
Thermus aquaticus, 251
Theropods, 732, 733
Thiamin, *1074,* 1075
Thick filaments, 1005
See also Myosin filaments
Thin filaments, 1005
See also Actin filaments
Thiols, *40*
Thioredoxin, 172
Thompson, Richard, 999
Thoracic cavity, **1034–1035**
Thoracic ducts, 402, **1060**
Thoracic vertebrae, 453
Thorax, of crustaceans, 707
Thorn forests, **1125**
Threatened species, 1230
Three-spined sticklebacks, 459
Threonine, *43, 134*
Threshold potential, **952**
"Thrifty genes," 1068
Thrips, *709*
Thrombin, **1056,** *1057*
Thrombus, **1060,** *1061*
Thrum flowers, 496
Thylakoids, 83, *84,* 573
Thymine, 59, 260, 382
in DNA, **238,** 239, *240*
Thymosin, *880*
Thymus gland, *402, 880*
Thyroglobulin, 883, *884,* **885**
Thyroid gland, 421, *880,* **883,** 884
Thyroid-stimulating hormone (TSH), *880,* **884,** 885
Thyrotropin, **884,** 885, 887
Thyrotropin-releasing hormone (TRH), **883,** 884–885
Thyroxine, *880,* **883–885,** 1219
Thysanoptera, 709
Thysanura, 709
Ticks, 679
Tidal ventilation, **1031–1032**
Tidal volume, **1032**
Tigers, defense of territory, 1149, *1150*
Tiger snakes, 455
Tiger swallowtail butterfly, *678*
Tight junctions, **103,** *104*
Till, James, 432
Tilman, David, 1201
Time-compensated solar compass, 1156
Tinamous, 732
Tinbergen, Niko, 1141, 1145
Tingids, *519*
Ti plasmid, *326,* 361
Tip layers, 832
Tires, rubber, 796
Tissue culture, of plants, 809, 833–834
Tissue plasminogen activator (TPA), 368–369
Tissue rejection, 434
Tissues, *8,* **748**
diffusion in, 105
types of, **857**
Tissue-specific promoters, 367
Tissue systems, of plants, **748**
Titan arum, 818
Titanosaurs, 645
Titin, 315, **1006**
Tmesipteris, 625
Toads
Bufo marinus in Australia, 1181
hybrid zones and, 517–518
Tobacco hornworm, 841
Tobacco mosaic disease, 283–284
Tobacco mosaic virus (TMV), *285,* 777, 839
Tobacco plants
Maryland Mammoth, 826
nicotine as defense, 841, *842*
plasmodesmata in, 777
Tocopherol, *1074*

- Toll protein, **407**
- Tomato plants, *371,* 847, 848
- Tongue, 1079, *1080*
  - in humans, 969–970
  - in snakes, *964*
- *Tonicella* sp., *701*
- Tonoplast, 752
- Topsoils, 786
- Torpedo rays, 992, 1021
- Torpedo-stage embryo, *823,* 824
- Total lung capacity, **1032**
- Totipotency, **429**
  - in animal embryos, 430–432
  - plant cells, 359, 429–430, 833
- Touch
  - as communication, 1157
  - sensations of heat and cold, 954
  - tactile receptors, 970–971
- Tour de France, 874
- Toxigenicity, **579**
- "Toxin Alert" food wrap, 413
- Toxin genes, biotechnology and, 370, 371, 372
- Toxins
  - bacterial, 579
  - bioaccumulation, 1089
  - metabolism of, 1089
  - neurotransmitters and, 955
  - as weapon of predators, 678, *679*
  - *See also* Poisons
- TPA. *See* Tissue plasminogen activator
- Tracheae (of insects), 708, **1028,** *1029*
- Trachea (of animals)
  - in birds, **1030,** *1031*
  - in humans, 1032, *1033,* 1079, *1080*
- Tracheary elements, **750,** *751,* 752
- Tracheids
  - evolutionary significance of, 617
  - in gymnosperms, 634
  - overview of, **750,** *751*
  - rhyniophytes and, 619
  - vascular plants and, **611**
- Tracheoles, 1028, *1029*
- Tracheophytes, 611
- Trade-off, **680**
  - adaptations and, 486
  - in animal reproduction, 680–681
  - constraints on evolution, 504
- Trade winds, 1115
- Traditional medicines, exploitation of species and, 1233
- *Tragopogon,* 514–515
- Traits, **208,** 490
  - dominant and recessive, 210
  - Mendelian inheritance and, 208–214
  - quantitative variation in, 221–222
  - true-breeding, 209
  - used in phylogenetic analyses, 544–545
- Transcription, **260**
  - amplification of genes and, 324
  - central dogma and, 260, 261
  - coordinating several genes, 320–322
  - differential, 435
  - effects of signaling pathways on, 347
  - in eukaryotes, *308, 313,* 316–318, 319
  - genetic code, 263–265
  - overview of, 261, 262, *263*
  - in prokaryotes, 296–300, 319
  - regulating by changes in chromatin structure, 322–323
  - RNA polymerases and, 261, 262, *263*
  - selective, 318–322
  - start and stop signals, *270*
  - *See also* Gene expression
- Transcription complex, 319
- Transcription factors
  - cytoplasmic receptors and, 339
  - homeodomain, 445
  - overview of, **319–320**
  - regulation of cell differentiation and, 435
- Transcription initiation complex, 441
- Transducin, 977, *978*
- Transduction, **293**
- Transfection, **237, 359**
- Transfer RNA (tRNA)
  - central dogma and, **260,** 261
  - functions of, 266
  - moderately repetitive DNA sequences and, 311
  - molecular structure, 266
  - synthesis of, 261
  - in translation of mRNA, 265–270
- Transformation
  - in bacteria, **292–293**
  - in eukaryotes, 237
  - experiments in, 233–235
- Transforming growth factor-β (TGF-β), 930
- Transforming principle, **234–235**
- Transgenes, 367
- Transgenics, **359**
  - in agriculture, 369–371
  - created by transfection, 237
  - pharming, 369
- Transitions, 527
- Transition-state species (chemistry), **126**
- Translation, **260**
  - central dogma and, 260, 261
  - key events, 265–266, 268–270
  - polysomes, 270, *271*
  - in prokaryotes, 262
  - regulating, 326
  - ribosomes and, 268
  - role of tRNA in, 265–270
  - start and stop signals, *270*
- Translocase, 572
- Translocations, **200, 276,** *277,* **775–777**
- Transmembrane domain, of hormone receptors, **879**
- Transmembrane proteins, **99–100**
- Transmissible spongiform encephalopathies (TSEs), 378
- Transmission electron microscopy, *71*
- Transpiration, **770,** 771, 774, 847
- Transpiration–cohesion–tension mechanism, 770–771
- Transplant surgery, tissue acceptance and, 417
- Transport proteins, nutrient absorption in the small intestine, 1084
- Transposable elements, 278, 295, 311–312, 534
- Transposition, transposons and, *312*
- Transposons, **295, 311–312**
- Transverse tubules, *1008,* **1009**
- Transversions, 527
- Travisano, Michael, 528–529
- Tree ferns, 474, 475
- Tree of Life, 544, 555, 1246–1251
- Trees
  - atmospheric CO_2 and carbon storage, 1214–1215
  - height, 769
  - synchronous and episodic reproduction, 1175
- Tree squirrels, 1142
- Tremblay, Pierre and Anne, 374
- *Treponema pallidum, 572*
- *Triaenodon obesus, 726*
- Triassic period, *466–467,* **476,** 479
- Tricarboxylic acid cycle, 140
- *Trichinella spiralis,* 704–705
- Trichinosis, 704–705
- Trichocysts, 598
- *Trichoderma, 1198*
- *Trichomonas, 584*
  - *T. vaginalis,* 603
- *Trichoptera, 709*
- *Tridacna gigas, 701*
- *Trifolium repens,* 503
- Triglycerides, *54,* **55,** 155, *1083,* 1084
- Triiodothyronine, 884
- Trilobites, 473, 480
- Trimesters, 936
- Trimethylamine oxide, 1099
- *Triodia, 1128*
- Triose phosphate, 171
- Triple bonds, 27
- Triploblasts, *672,* **673**
- Triploids, 202
- Trisomy, **199,** *201*
- *Triticum* sp., *645*
- Tritium, *23, 466*
- Trochophore, **680,** 692
- Trophic cascades, **1195–1197**
- Trophic levels, **1187–1188**
- Trophoblast, *924,* **925,** 935
- Trophosome, 699
- Trophy hunting, 505
- Tropical alpine tundra, 1118
- Tropical deciduous forests, **1126,** *1210*
- Tropical ecosystems, conversion to cropland, 1222
- Tropical evergreen forests
  - current rate of loss in, 1230
  - described, **1127**
  - habitat fragmentation and, 1231
  - primary production in, *1210*
- Tropical forests, certification, 1238
- Tropical savannas, **1125**
- Tropic hormones, *880,* **881–882**
- Tropomyosin, 325, 1007, 1008, **1009,** 1010
- Troponin, 1007, 1008, **1009,** 1010
- Troposphere, **1208**
- *trp* operon, 298–299, 300
- True-breeding traits, **209**
- True bugs, *709*
- True flies, *709*
- Truffles, 665
- Trumpet cells, 600
- Trunk, in hemichordates, 721
- *Trypanosoma, 584*
  - *T. brucei,* 326, 582, *604*
  - *T. cruzi,* 582, *604*
- Trypanosomes, 582–583, 595, *604*
- Trypsin, *132,* **1083,** *1084*
- Trypsinogen, **1083**
- Tryptophan
  - genetic code and, 264
  - indoleacetic acid and, 803
  - structure, *43*
  - *trp* operon, 298–299, 300
- Tschermak, Erich von, 208
- TSEs, 378
- Tsetse flies, 582, *604*
- Tsien, Joe, 960
- *Tsix* gene, 323
- Tsunami of 2004, *352–353,* 1222
- T tubules, *1008,* **1009,** 1010, 1012
- Tetrodotoxin (TTX),
- TTX (tetrodotoxin), 486, 498, 538, 1021
- TTX-resistance, 486, 498
- Tuataras, 731, *732*
- Tubal ligation, *914,* **915**
- Tubal pregnancy, 925
- Tube cell, 821, 822, *823*
- Tube feet, 676, **719–720,** 721
- Tuberculosis, 302, *409,* 573
- Tubers, 747, 832
- Tubocurarine, *648*
- Tubular guts, 1077, *1078*
- Tubulidentata, *736*
- Tubulin, 88, 189
- Tulip bulbs, 832
- Tumor cells
  - cellular immune response and, 417
  - cytokines and, 404
  - HeLa cells, 180–181
  - in hybridomas, 413
- Tumors, **386**
- Tumor suppressor genes, **388–389**
- Tumor suppressors, 186
- Tumor viruses, 387
- Tuna, *13*
- Tundra, **1118,** *1210*
- Tunicates, 722, *723*
- Turbellarians, 696
- Turgidity, 766
- Turgor pressure, **107,** 108, 766
- Turner syndrome, 226
- Turnover, **1207–1208**
- Turtles, 731
- Twigs, growth in, 754–755
- Twinning, 926
- Twitches, **1012**
- Two-dimensional gel electrophoresis, 395, *396*
- 2,4-Dichlorophenoxyacetic acid (2,4-D), 807
- "Two-hit" hypothesis, 388, *389*
- Tympanic membrane, **972,** 974
- *Tympanuchus cupido, 495*
- Type I diabetes, 421, 887
- Type II diabetes, 887, 892–893
- Typhlosole, *1078*
- Typhoid fever, *409*
- *Tyrannosaurus rex,* 232, 732
- Tyrosine, *43,* 376
- *Tyto alba, 734*

- U -

- Ubiquinone (Q), **149,** *150, 151*
- Ubiquitin, **327**
- *Ubx* gene, 450, 451, 453, *454*
- Ubx protein, 453
- Udx1 enzyme, 903
- Ulcers, stomach, 1081–1082
- *Ulmus americana,* 1233
- *Ulmus procera,* 833
- *Ulothrix,* 595
- Ultrabithorax protein, *451*
- *Ultrabithorax (Ubx)* gene, 451
- Ultraviolet radiation, 11, 277
- *Ulva, 568, 584,* 1199
  - *U. lactuca,* 594, *595,* 602, *603*
- Umami, **970**
- Umbels, *639*
- Umbilical cord, 912, 936
- Umbilicus, 913
- Unconscious reflexes, gastrointestinal system and, 1085–1086
- Undernourishment, **1071**
- Undernutrition, 155
- Unequal crossing over, 536

Unicellular organisms, 4, 11, 102
Unikonts, *584, 597,* **605–607**
Unilever, 1238
Uniports, **111**
United Nations Conference on Sustainability, 161
United Nations Convention on Biological diversity, 1235
United States
  baby boom, 1170–1171
  survivorship curves, *1170*
U.S. Fish and Wildlife Service, 1226
Universal genes, 302–303
Unsaturated fatty acids, **55,** 847
Upregulation, of hormone receptors, **892**
Upwelling zones, **1206,** *1207, 1210*
Uracil, 59, **260,** 382
Uranium-238, *466*
Urchin dual oxidase (Udx1), 903
Urea, 48, **1096,** 1105
Urease, 1082
Uredospores, 662, 663
Ureotelic animals, **1096**
Ureter, **1101,** *1102*
Urethra, **907–908, 1101**
Urey, Harold, 61
Uric acid, **1096**
Uricotelic animals, **1096**
Urinary bladder, **1101**
Urine
  amniotes and, 731
  concentrated in mammals, 1101–1105
  excretory physiology of vampire bats and, 1092–1093
  formation in kidneys, 1099–1101
  in invertebrates, 1097
  output of excretory organs, **1094**
  phenylketonuria and, 375
  storage in the bladder, 1101
Urochordates, *718,* **722–723**
Ursus arctos, 678
Ursus maritimus, 1070
Uterine cycle, **911–912**
Uterus, 735, **905,** 909, 912
Utricle, *975*

- V -

Vaccinations
  immunological memory and, 408, **409–410**
  origin of, 400–401
Vaccines
  antiviral, 387
  immunological memory and, 409–410
  for polio, 524–525
  produced through biotechnology, *368*
Vaccinia virus, *409*
Vacuoles
  digestive ability of, 75, 82, 86
  in eukaryotic cells, 591
  in plant cells, *77,* 85–86, 844
  storage of toxic materials, 844
  structure and function, 75, 84–86
  types, 86
Vagina
  accessory sex organ, **904**
  douching, 914
  in humans, **909,** 912, 913
Vaginal ring, 915
Valence shell, 25
Valine, *43,* 376
Vampire bats, 735
  excretory physiology, 1092–1093
van der Waals forces, *26,* **30,** 46
van Leeuwenhoek, Antony, 4, 5
Van'T Veer, Laura, 366
Van Valkenburgh, Blaire, 482
Varicella-zoster virus, *409*
Varicose veins, 1059
Vasa recta, **1102,** 1103
Vascular bundles, **757**
Vascular cambium, *753,* **754,** 758–759
Vascular disease, 1060–1061
  See also Heart attack
Vascular plants, *612,* 616–617
  characteristics of rhyniophytes, 619
  evolution, 618–621
  heterospory, 621
  life cycles, 617
  members of, **611**
  overtopping growth, 619
  sporophyte generation, 617
  vascular tissue, 617
  without seeds, 624
Vascular rays, **758–759**
Vascular system, plant, **617,** 619, **748**
Vas deferens, *906,* **908,** 915
Vasectomy, *914,* **915**
Vasopressin, 881, 1062, 1107–1108
Vectors
  for animal viruses, **287**
  inserting new DNA into cells, **359–361**
  transmission of plant viruses, 289
  See also Expression vectors
Vegetal pole
  cytoplasmic segregation and, 436, *437,* **922,** *923*
  invagination, 927–928
Vegetarian diet, 1072
Vegetative cells, of cyanobacteria, **573**
Vegetative reproduction
  adventitious rooting and, 746
  disadvantages of, 833
  in flowering plants, 832–833
  importance to agriculture, 833–834
  mitosis and, 193
Veins (animal), **1047**
  anatomy, *1058*
  blood flow through, 1059–1060
  defined, **1047**
  varicose, 1059
Veins (leaves), 761
Velvet worms, 705–706
Venoms, 1021
Venter, Craig, 300, 303
Ventilation, 1028
  tidal, 1031–1032
Ventilators, 765
Ventral horn, spinal, 987
Ventral plane, **674**
Ventral root, spinal, 987
Ventricles, **1047**
  heartbeat and, 1053, 1054
  human heart, 1050, 1051
  vertebrate hearts, 1047, 1048, 1049
Ventromedial hypothalamus, 1088
Venules, **1047,** 1059
Venus flytraps, 792
Vernalization, **831**
Vertebrae, Vertebral column, 453, **723**
Vertebrates, **722**
  adaptations to life on land, 728–737
  amniotes, 730–731
  body plan, 724, *725*
  circulatory system evolution, 1047–1049
  in deuterostome phylogeny, *718*
  gene duplications and, 534
  Hox genes and, 450, 453
  jaws and teeth, 724–725
  nervous system, 944
  notochord and, 722
  number of species, 722
  phylogeny, *723*
  skeletal system, 1017–1018
  tidal ventilation, 1031–1032
  vertebral column and, **723**
Vertebrate thermostat, 868–869
Vertical transmission, 289
Very-low-density lipoproteins (VLDLs), **1087**
Vesicles
  dynamic continuity of cellular membranes and, 101
  in endocytosis, 113
  in evolution of cells, 11
  in exocytosis, 114
Vessel elements, **638, 750,** *751,* 752
Vestibular apparatus, **974,** *975*
Vetulicosystids, 718
Viagra, 344, 345
Vibrio, 567
  V. cholerae, 96, *409,* 575, 579, 1191, *1192*
Vicariant events, **1133–1134,** 1135
Vicugna vicugna, 1134
Vicuñas, 1037
Villi, **1077,** *1078*
Vincristine, *391, 648*
Vinograd, Jerome, 242
Virginia opossum, 735
Virions, **284**
Viruses, 3, **283–284**
  autoimmune diseases and, 421
  genome size, 307
  lytic and lysogenic life cycles, 284–287
  as model organisms, 283, *284*
  regulation of gene expression, 289–290
  reproductive cycles, 287–289
  RNA-type, 261
  that cause cancer, 387
  vectors for, 287
  as vectors in DNA technology, 360
  virulent, 286
  See also Plant viruses
Visceral mass, of mollusks, 700
Visible spectrum, 163, *164*
Vision
  binocular, 996, *997*
  correction of eye misalignments and, 938
  occipital lobe and, 991
  retinal receptive fields, *994,* 995
  systems, 976–978
  visual cortex information processing, 995–996
Visual communication, 1157
Visual cortex, 991, 992, 994–996, *997*
Visual systems, 976–978
Vital capacity, **1032**
Vital force, 21
Vitamin A, 56, 57, *1074*
Vitamin B_1, *1074,* 1075
Vitamin B_2, *1074*
Vitamin B_6, *1074*
Vitamin B_{12}, 579, *1074,* 1075, 1220
Vitamin C, 1074
Vitamin D, 57, **886,** 1074
Vitamin D_2, *57*
Vitamin E, 57, *1074*
Vitamin K, 57, 579, *1074*
Vitamins, **57, 886, 1073**–*1074*
  coenzymes and, 130
Vitelline envelope, **901,** 902, 903
Vitreous humor, *979*
Viviparity, 729, 811, **905**
VLDLs. *See* Very-low-density lipoproteins
VNO, **969**
Voice box, 1079, *1080*
Volcanoes, 62, 475, 1217
Voltage, 947
Voltage-gated ion channels, **950**
  characteristics, 108–109
  generation of action potentials and, 951–953, 954
  release of neurotransmitters and, 955
Voluntary nerve pathways, 985
Volvox, 602, *603*
Vomeronasal organ (VNO), **969**
von Frisch, Karl, 1157
von Tschermak, Erich, 208
"Voodoo lily," 818
Vorticella sp., *598*
vp mutants, 811
Vulva, differentiation in nematodes, 437–439

- W -

Waggle dance, honeybee, 1157, *1158*
Wallace, Alfred Russel, 488, 1128
"Wallace's line," 1128, *1129*
Walruses, 737
Warren, Robin, 1081
Warts, 387
Washington, George, 400
Washingtonia filifera, 495
Wasps, *709*
  competition and, 1193
  learning in, *1144,* 1145
  strepsipteran parasites, 690, *691*
Wasser, Samuel, 1239
Water
  absorption in the intestines, 1084
  aqueous solutions, 32–34
  desert plants and, 845–846
  excretory physiology of animals and, 1092–1093
  extraterrestrial, 20
  in the global ecosystem, 1206–1208
  human consumption, 1212
  hydrogen bonds in, 29, *31*
  hydrological cycle, 1211–1212
  life and, 20–21, 34–35
  movement across plasma membranes, 1093–1094
  osmosis and, 106–107, 765–767
  polar covalent bonding, 27, *28*
  properties of, 31–32
  as respiratory medium, 1026, 1027
  structure of, 31
  uptake by plants, 765–767, 768
Water bears, 706
Water contamination, cholera and, 1191, *1192*
Water fleas, 1150–1151
Water lilies, *644*
Water molds, 600, *601*
Water potential, **766,** 848
Water-saturated soils, 846–847
Water vascular system, echinoderm, **719**

Watson, James, 232, 238, *239,* 241, 260
Wavelength, light, **163**
Waxes, 57
W chromosome, 225–226
Weather, **1114**
Weathering, rock, 786
Webbing
duck feet, 452
gremlin gene and, 452, *453*
salamander feet, 451, *452*
Webs and web spinning, 713, 1140–1141
Webspinners, *709*
Weevils, 1135
Wegener, Alfred, 468, 1131
Weight-bearing exercise, 1018
Weight lifting, 1014
Weiss, Benjamin, 61
Welwitschia, 634, *635*
Went, Frits W., 803
Werner, Earl, 1150
Wernicke's area, 1000, 1001
Westerly winds, 1115
Wetlands, 1238, *1239*
coastal, 1112–1113
Whales
blowhole, 920
evolution, 737
filter feeding, 676
population management, 1180
songs, 1157
Whaling industry, 1180
Wheat
black stem rust, 653, 662–663
flowering controlled by temperature, 831
ploidy level, 202
Wheat streak mosaic virus, 289
Whip scorpion, *693*
Whiptail lizard, 898
Whisk ferns, *612,* **625,** *626*
White asparagus, 813
White blood cells
in blood and lymph tissues, 402
defensive roles of, **402–403**
gene therapy and, 392
types of, *403*
White clover, 503, 826
White-crowned sparrows, 1145–1146
White fat, *867*
Whitefish, 1175, *1176*
White-fronted bee eaters, 1159, *1160*
White matter, **987,** 989
White muscle, 1013
White-tailed deer, *1168*
Whitetip reef shark, *726*
Whitman, Charles, 942
Whittaker, Robert, 1186
Whittardolithus, 480
Whooping cough, *409*
Whooping cranes, 1226–1227
Wiesel, Torsten, 995
Wigglesworth, Sir Vincent, 876, 877
Wild dogs, 1241
Wild type alleles, **217,** 218
Wilkesia hobdyi, 520
Wilkins, Maurice, 238
Willows, *756,* 839, 1195–1196
Willow sawfly, *678*
Wilmut, Ian, 431, 432
Wilson, Edward O., 1132, 1133
Wind, turnover in lakes and, 1207
Windpipe, 1079, *1080*
See also Trachea
Wind pollination, 821
Winds, 1114–1115
Wings
in *Drosophila,* 451
evolution of, 457, *458,* 545
in insects, 710
as modified limbs, 457, *458,* 676
Wintergreen, oil of, 839
Winter wheat, 831
"Wishbone," 732
Witchcraft trials, 666
Witchweed, 650, 793
Withdrawal reflex, 988
Wobble, tRNA and, 266
Wolf, Larry, 1070
Wolpert, Lewis, 673
Wolves, 1195, 1196
Womb, 905
Wood
annual rings, 759
heartwood/sapwood, 760
plant evolution and, 631, **634**
secondary growth and, 758, 759–760
secondary xylem, **754**
Woodlands
primary production in, *1210*
See also Forests
Woodpeckers, *1161,* 1229
Wood pigeons, 1178–1179
Woolly mammoths, 479
Work, **119**
Worker bees, 896, 898
World Wildlife Fund, 1235, 1238

- X -

Xanthoria sp., *656*
X chromosomes
cytoplasmic bridges between spermatocytes and, 899, *900*
deletions, *380,* 381, 382
evolution of, 323
fragile-X syndrome, 379, 382–383
heterochromatin, 323
in sex determination, 225, 226
sex-linked color blindness, *207,* 227, *228*
sex-linked inheritance, 226–227
X-linked recessive alleles, 379, 381
Xeroderma pigmentosum, 249
Xerophytes, **845–846,** 848
Xerox, 845
Xerus inauris, 737
Xiphophorus, 552–553
Xist gene, 323
X-linked recessives, 379, 381
X-ray crystallography, 238
X rays, 258, 277
Xylem
in angiosperms, 638
functions of, **748**
in gymnosperms, 634
in roots, 757
secondary growth and, 631, 758
transport of water and minerals in, 769–772, *776*
vascular cambium and, 754
in vascular plants, **617**
vessel elements, 638
Xylem sap, 768–772, *776*

- Y -

YACs, **360**
Yalow, Rosalyn, 891
Yarrow's spiny lizards, 1148–1149
Y chromosomes
evolution of, 323
in sex determination, 225, 226
Yeast artificial chromosomes (YACs), **360**
Yeasts, **651,** 652
ascomycetes, 665
cell cycle experiments in, 185
gene duplications and, 534
genome, 309, *533*
noncoding DNA in, *534*
recombinant DNA technology and, 359
Yellow-bellied toads, 517–518
Yellowing, in plants, 783
Yellow pigments, 814
Yellowstone National Park
forest fires and, 1197
trophic cascades and, 1195–1196
Yellow-streaked tenrec, *1130*
Yersinia pestis, 575, 579
Yolk
amniote egg and, 730
chicken egg and, 905
chicken yolk sac and, 935
cleavage patterns and, 672–673
polarity in eggs and, 436
in reptilian and avian eggs, 931
Yolk sac, **934–935**
Young, J. Z., 948
Yuccas, 641
Yunnanozoans, 718
Yunnanozoon lividum, 718

- Z -

Zambia, 1239
Z chromosome, 225–226
Zea mays. See Corn
Zeatin, 809
Zeaxanthin, **814**
Zebra finches, 500
Zebrafish, *engrailed* gene, *537*
Zebra mussel, 1176, *1177*
Zebras, *1131*
Zeevaart, Jan A. D., 831
Zenith (hot air balloon), 1024
Zinc, *129, 782, 783, 1073*
Zinc finger motif, 320, *321*
Z lines, 1006, 1007, 1014
Zona pellucida, **901,** 902, *903*
Zone of cell division, **755**
Zone of cell elongation, **755–756**
Zone of maturation, **756**
Zone of polarizing activity, 441
Zoonotic diseases, 552
Zoospores, 594, 664
Zoraptera, 709
Z scheme, **167**
Zuckerkandl, Emile, 551
Zwieniecki, Maciej, 772
Zygomaturus, 1228
Zygomycetes, 651, *652, 660,* **664–665**
Zygosporangia, *660,* **665**
Zygospores, 665
Zygotes
chromosome abnormalities in humans, 200
cleavage, 672–673, 922–925
contribution of egg and sperm to, 921–922
cytoplasmic rearrangements, 922
in flowering plants, *820*
formation, **193, 901**
polarity, 436
totipotency, 429
Zymogens, **1077,** 1083